TAXONOMY AND ECOLOGY OF WOODY PLANTS IN NORTH AMERICAN FORESTS

Forest regions of North American: (1a) Northern Conifer-Hardwood or Boreal (Spruce-Fir) Forest, closed; (1b) Northern Conifer-Hardwood Forest, open savanna and barrens; (1c) Northern Conifer-Hardwood Forest, Aspen Parkland; (2a) Northern Hardwood-Conifer Forest (Sugar Maple-Yellow Birch-Eastern Hemlock-White Pine Forest), Great Lakes section; (2b) Northern Hardwood-Conifer Forest, New England section; (3a) Central Hardwood Forest (Oak-Hickory Beech-Maple, and Western Mesophytic), plateaus section; (3b) Central Hardwood Forest, Appalachian Mountain section; (4a) Southeastern Pine-Hardwood Forest, Upper Coastal Plain; (4b) Southeastern Pine-Hardwood Forest, Lower Coastal Plain; (4c) Southeastern Pine-Hardwood Forest, Piedmont Plateau; (5a) Central and Southern Rocky Mountain Mixed Conifer Forest; (5b) Southwestern Juniper-Pinyon Savanna; (6) Northern Rocky Mountain Conifer Forest; (7a) Northern Coastal Conifer Forest; (7b) Sierra Nevada/Southern Coastal Conifer Forest. Other zones include grassland (G), desert (D), and tundra (T), as well as the Mississippi River valley (MRV), which has largely been cleared for farming. Base map adapted from Little (1971).

TAXONOMY AND ECOLOGY OF WOODY PLANTS IN NORTH AMERICAN FORESTS

(Excluding Mexico and Subtropical Florida)

JAMES S. FRALISH, Ph.D.
Southern Illinois University at Carbondale

SCOTT B. FRANKLIN, Ph.D.
University of Memphis

Illustrations by Hyosun Park

John Wiley & Sons, Inc.

This book is printed on acid-free paper. ♾

Published simultaneously in Canada.

Library of Congress Cataloging-in-Publication Data

Fralish, James S. (James Steven)

Taxonomy and ecology of woody plants in North American forests (excluding Mexico and subtropical Florida) / James S. Fralish and Scott B. Franklin ; illustrations by Hyosun Park.

p. cm.

Includes bibliographical references (p.).

ISBN 0-471-16158-6 (cloth : alk. paper)

1. Woody plants—United States—Classification. 2. Woody plants—Canada—Classification. 3. Woody plants—Ecology—United States. 4. Woody plants—Ecology—Canada. I. Franklin, Scott B. II. Title.

QK115 .F73 2002

582.16′0973—dc21 2001024031

Printed in the United States of America

10 9 8 7 6 5 4 3 2 1

Contents

Preface

During the past 30 years, forestry, wildlife management, and the other ecological sciences have become progressively more advanced and complicated. In part, this change is due to voluminous amounts of new scientific information and, in part, to the development of conflicting ecological and social values—for example, conservation of endangered species versus timber production versus recreation. However, the identification of major tree species has been and remains a common component across scientific disciplines. Various new tree identification books for regions and states and new editions of older books covering species for the United States and/or Canada continue to be published in moderate numbers. The majority of these books concentrate primarily on timber-producing tree species.

However, ecological research has shown that shrubs, vines, and midcanopy trees (stems with genetically limited growth potential) are important for erosion control, maintaining high water quality, retaining nutrients on the site, increasing soil organic matter, and providing wildlife food and cover while maintaining biodiversity and aesthetics. In addition, there is strong interest that rare and endangered plant and animal species be conserved as gene pools. Society now often values these benefits of a sustainable ecosystem as more important than timber production. Concurrent with these values is a strong commitment to preserve and restore presettlement-type (ca.1800) plant and animal communities in an effort to maintain biological diversity across the landscape.

Within this new ecology context, *Taxonomy and Ecology of Woody Plants in North American Forests* has developed out of undergraduate courses the senior author taught for 27 years at Southern Illinois University-Carbondale (Fralish 1981). In developing this book, we have emphasized and integrated several important scientific areas of study so it will be strongly relevant to and match contemporary professional needs. First, in addition to considering commercially important trees, we have included information on and illustrations of a large number of noncommercial tree, shrub, and vine species in recognition that they are part of the natural diversity within the forest community. Field ecologists, wildlife biologists, naturalists, and similar field-oriented professional workers should be familiar with these species, as these workers are increasingly being asked to sustain biological diversity (species richness) through management decisions. Approximately 90 percent of both commercial and noncommercial woody species found in North American forest communities are considered in this book. The included species are derived, in part, from lists used in taxonomy/dendrology courses taught in resource management, forestry, botany, and environmental studies departments.

Second, the coverage of species provided here is ultimately compartmentalized into communities within natural vegetation regions or zones of North America. Within each region, we have used cover types (Society of American Foresters 1980), forest alliances of the National Vegetation Classification System (The Nature Convervancy), and ecological/land classes or habitat types to group species into communities and to tie communities to site environment. Much new ecological information regarding woody shrubs, vines, and midcanopy (arborescent) trees has centered on the identification of forest site environment and their associated forest community. Moreover, small noncommercial tree species and shrubs are the basis of every habitat typing and ecological/land classification system developed in North America. This system uses arborescent and/or shrub species in combination with herbaceous and tree species to identify land types of different productivity and stability levels, and thus the identification of these species has become a necessity.

Ecological/land classification is important because forest production, species preservation, and community restoration management require that the site environment be evaluated. Regardless of how it is viewed, maintaining high forest productivity is a necessity; to this end, resource managers must recognize that species have ecological requirements that must be matched with the site environment. Most small-statured woody plants (shrubs and arborescents) are relatively site-specific as compared with the major timber-producing trees that may be found in both compositionally stable (climax) communities on xeric sites and compositionally unstable (successional) communities located on mesic sites. In addition, common herbaceous plants, as well as rare and endangered species, are usually isolated on sites having specific environments. Thus, shrubs and herbs may be used to characterize growing conditions. Using common shrub and herb species in combination with tree species, an ecological/land system has been completed for nearly all national forests and much of the Cannadian forest. In Canada, ecosystem classification and habitat typing was begun years before it became popular in the United States. Many of the systems for national forests have been developed under contract with university researchers, and others with The Nature Conservancy. Recently, Secretary of Interior Bruce Babbit endorsed a standardized National Vegetation Classification System developed by The Nature Conservancy.

Two additional areas of science have been considered in this book. One thrust is in the area of basic plant taxonomy. Although forestry students are required to take at least one course in dendrology or woody plant identification, the approach is usually oriented toward general information on species in lectures and simple identification of trees and some shrubs during laboratory periods (Fralish 1981). The taxonomic principles and non-commercial plant identification important to the field sciences are not generally included in these courses, even though the subject matter is basic to effective communication. Usually, family or genus characteristics of the woody species do not appear in basic tree identification books and field manuals. Currently, basic plant taxonomic principles, as they apply to woody plant identification, continue to be relegated to textbooks in advanced botany and plant biology courses. Unfortunately, these courses usually do not appear in most college programs on forestry or other natural resources.

In selecting the species to include here, the term *forest* was liberally interpreted. Although most of the illustrated species are components of forest communities, some are found in the arctic, some in savanna, and others in desert-shrub situations. Some species are found on the forest edge or in disturbed places like old fields or along fence rows. From a purely taxonomic standpoint, a few species were included so the genus or family would be complete. Conversely, several genera containing woody species are sufficiently large that it was possible to include only the most important, or the only species, that adequately represents the range of physical characteristics within the genus or family.

Finally, we have included a number of introduced species. Most of these have been and are being planted in order to rebuild the "urban forest." Urban forestry, or arboriculture, is important in maintaining a high-quality city environment through planting of trees and shrubs; moreover, a number of introduced species have become economically important in a positive sense (e.g., commercial plantations) or, in a negative sense, where they have escaped cultivation to become naturalized and create a potentially and actively serious ecological problem (noxious invaders and competitors).

We believe that *Taxonomy and Ecology of Woody Plants in North American Forests* is a new generation dendrology book that will guide the student beyond plant taxonomy and identification into community ecology and site-environment relationships. To accomplish this task, the book is divided into four primary parts. The initial introductory chapters orienting the student to plant taxonomy and ecology are followed by short chapters in Part I that outline basic taxonomic concepts relating to classification, nomenclature, herbarium specimens, and construction of plant identification keys. The chapters of Part II on gymnosperms (conifers) include an introductory chapter on general taxonomic characteristics, regional distribution of coniferous biomes, and general ecological relationships, followed by chapters on leaf/twig and cone characteristics. The final chapters in Part II describe the general features of the family, genus (genera), and major characteristics for each species. Both major commercial trees and noncommercial midcanopy trees and shrubs are illustrated. Part III, on angiosperm (hardwood or deciduous) species, follows a format similar to that of Part II. The organization of families follows Cronquist's (1981) treatise on taxonomic classification (Chapter 13).

Part IV groups trees, shrubs, and vines from various taxonomic families together into community (cover) types found in each of several large geographic regions whose boundaries are based on well-defined forest vegetation relationships. These communities are described by ecoregions (Bailey 1998) and in terms of cover types (Society of American Foresters 1980) and the new National Vegetation Classification System (Grossman 1998).

The authors are deeply indebted to Hyosun Park, a graduate of the Southern Illinois University School of Art, who illustrated nearly all of the book, and to the assistance of Karen Grady, who provided a few early illustrations. The illustrations developed by Ms. Park were made from fresh material or specimens stored at herbaria located at Southern Illinois University, Carbondale (Michael Mibb, Curator), and the Missouri Botanical Garden, St. Louis (Jim Solomon, Curator). Additional information was obtained from herbaria and curators at The University of Florida-Gainesville, Tulane University-New Orleans, The University of Utah-Salt Lake City, and The University of Wisconsin-Madison. Additional fruit and cone specimens were obtained from the Tree Identification Laboratory of the Department of Forestry at Southern Illinois University. The contribution of these herbaria and their associated curators is greatly appreciated. Figure 2.1 was developed by Steve Mueller of the IMAGE unit at Southern Illinois University. Our sincere thanks to Kathleen B. Fralish Ph.D. and Deanna Franklin J.D., who assisted with this work.

James S. Fralish
Scott B. Franklin

TAXONOMY AND ECOLOGY OF WOODY PLANTS IN NORTH AMERICAN FORESTS

CHAPTER 1

Introduction to Taxonomy

THE ***best management practices applied to a forest are based on forest ecology concepts, and forest ecology is first and foremost based on correct field identification of woody plant species and knowledge of their ecological characteristics.***

IMPORTANCE OF CORRECT PLANT IDENTIFICATION

The study of every science begins with the identification of basic units. In chemistry, these are the elements; in geology, minerals and rocks; in soil science, the profile and texture; in biology, cells and organs; and in wildlife management, the basic units are animals. In other sciences such as ecology, forestry, and many other land or resource management fields, study usually begins with basic identification of trees and shrubs and progresses toward a knowledgeable understanding of their ecological tolerances, how they interact and compete, where they are located in time and space, and what combinations make up community types.

Trees and shrubs are usually adapted to a range of light and soil moisture conditions. Within this range, individual woody plants are able to survive, grow, and develop (complete their life cycle). Outside this range, the mortality rate is high. Thus, before variation in forest composition as influenced by soil and topography can be studied, before silvicultural activities such as tree planting, thinning, or timber stand improvement can be applied, and before wildlife habitat can be manipulated, the resource manager must be able to identify the important plants of the community. The success of management, whether for economic return and sustainability or for ecosystem restoration, in terms of survival, growth, and development of trees and animals, often depends on correct plant identification.

The next step after identification is the development of a flora, a list of all plants within an area possibly as large as a national forest or park or as small as a watershed, natural area, or protected nature preserve. Actually, a list of species is just the beginning of a flora. A flora as presently perceived includes considerably more, such as plant ranges, descriptions, phenology (timing of seasonal and life cycle events such as bud break, leafing out, flowering, pollination, fertilization, seed-set, seed distribution, senescence, etc.), names, infraspecific variation, endemism, environmental conditions, and information on genetics (e.g., chromosome numbers, hybrids), cytology, and morphology (Morin et al. 1988).

Although this book deals with identification of woody plants only, they are an important component of nearly every flora. During the past several decades, there has been considerable discussion at the local, state, and federal levels regarding a national biological survey as part of a study of biological diversity (Kosztarab 1988). Unless the current situation in forest (plant and animal) communities is assessed, there is no reliable method for determining the loss of species or change over time. Paralleling this discussion is the development of a *Flora of North American*, initiated by the Missouri Botanical Garden in St. Louis. Several volumes of this *Flora* have already been published (Flora of North America Editorial Committee 1993a, 1993b, 1997).

Floras can be used in a number of ways by various professional workers. A flora can be used in environmental assessment and land management, in management of reserves and restoration, in determining the naturalness of a forest community following distribution and expansion of native and exotic species that are becoming naturalized, and in supporting protection status and identifing wetlands and ecologically sensitive sites. A flora can be used by botanists, foresters, silviculturalists, ecologists, zoologists, biologists, wildlife managers, ecological consultants, environmental engineers, construction (e.g., road) engineers, environmental lawyers, restoration biologists, researchers, and teachers, to name but a few professionals (Morin et al. 1988).

DEFINITION AND SCOPE OF TAXONOMY

The identification of woody plants falls within the oldest area of science dealing with plants, that is, systematic botany or, as it is commonly called, plant taxonomy or plant systematics. The term *taxonomy* comes from the Latin word *taxis*, which means an arrangement or order, and the ending *nomy*, which refers to knowledge. Thus, plant taxonomy is the study of the arrangement or classification of plants. However, as a discipline, taxonomy is much broader, for in order to classify plants, they must be described and assigned names at the same time they are arranged into the classification system.

To a large extent, description and identification are the primary concern of the beginning botanist, taxonomist, ecologist, or forester, and much information is drawn from the field of plant morphology, the study of form (Latin: *morpho* = form). The taxonomist concentrates on the areas of classification and nomenclature and generally deals with the placement of a larger plant group or taxon (plural, taxa), such as an order, family, or genus, within a classification system. Inasmuch as most plants have been described, identified, classified, and named, except for those in tropical regions, it might appear that there is a reduced need for taxonomic expertise.

The need for modern taxonomy has not changed, but modern studies have changed taxonomy. As a result, concepts and methodology have been incorporated from fields such as ecology, genetics, cytology, biochemistry, phytogeography, paleobotany, computer science, and geology. Ecology is the study of the interrelationships between plants and environment; it is common knowledge that differences in environment can cause changes in the appearance (form) of two individuals of the same species and can change the entire species population through the natural selection process. Genetics and cytology provide principles and information on hybridization, inheritance, reproduction, and chromosome number, which assist in determining how plants are related and classified.

Geology has furnished considerable information on past life, fossils, climate, and landforms (paleotaxonomy). For example, during times of glacial advance, ocean levels were greatly reduced and land bridges, such as that across the Bering Sea between North America and Asia (Beringia), permitted the migration of some plants. A notable example of a recent major change in taxonomic thinking occurred some 30 years ago with the development of the science of plate tectonics and the concept of continental drift. Until that time, taxonomists believed that similar plant species on different continents were the product of a parallel evolutionary sequence. The concept of a single land mass existing millions of years ago suggests that many similar species came from a common ancestor.

More recent developments (since the 1960s) in plant taxonomy have greatly assisted in understanding the relationships between taxa (groups). A number of books review such approaches in more detail (Stace 1980; Sivarajan 1991; Quicke 1993), but these methodologies are of limited importance here and are only briefly noted. Numerical taxonomy developed in the 1960s. This approach applied mathematical procedures to encode various qualitative states for a given characteristic. The idea here was to have a more objective evaluation of differences between taxa so they could be arranged into a hierarchy. However, the approach did not provide satisfactory information.

In the 1960s and 1970s the earlier work of Henning, a German entomologist, was first applied to plant taxonomy. His analysis was based on the concept that evolutionary relationships (a phylogenetic tree diagram) can be based only on shared derived characteristics. This approach became known as cladistics, and the diagrams that were developed came to be known as cladograms (Quicke 1993). This area of taxonomy is called evolutionary taxonomy.

More recently, biochemistry has provided a means to solve some taxonomic and evolutionary puzzles. This approach, called chemotaxonomy, uses various compounds produced within the plant to identify evolutionary relationships and improve classification; it is one of the more rapidly expanding areas of plant taxonomy (Stace 1980; Quicke 1993). With the development of advanced laboratory methods, information-carrying molecules such as DNA, RNA, and protein are being analyzed through amino acid sequence mapping to construct evolutionary relationships. This approach appears to hold considerable promise for providing information, but much interpretation will continue to remain with the researcher. Most taxonomists must still measure visible plant characteristics, but in evaluating the data, statistics and computer analysis are invaluable; this field is known as taxometrics.

All of these fields assist in working toward several major objectives of plant taxonomy: 1) the identification of all plants and their placement in a classification system, 2) the development of a classification system based on genetic and evolutionary relationships, and 3) the preservation of gene pools to maintain biodiversity within the ecosystem.

HISTORY OF CLASSIFICATION

The inherent curiosity of the human mind, the need for orderliness in knowledge, and the desire to communicate effectively has stirred interest in plant study and classification for centuries. According to Lawrence (1951), the earliest system of classification, proposed by the Greeks and herbalists, was based on the forms of plants: trees, shrubs, herbs, vines, and so on. This system prevailed from about 300 B.C. to the middle of the eighteenth century and became somewhat more elaborate as new information and concepts were incorporated.

From about 1500 to 1700, gross flower characters such as ovary position (superior vs. inferior) and petal structure (petalous vs. apetalous, polypetalous vs. sympetalous, and regular vs. irregular flowers) became important. During this period, Joseph Pitton de Tournefort (1656–1708) developed the modern "genus" concept, and many of his names, including *Salix, Populus, Fagus*, and *Acer* are still valid. Another botanist, John Ray (1628–1705), developed a system based on gross morphology of plant structures. He divided plants into woody and non-woody and recognized monocots and dicots within each division. Further subdivisions were based on fruit type—coniferous, berry or berry-like, nut-bearing, and so forth—and leaf and flower characters. This approach subsequently formed the basis for the more modern classification systems developed by Bernard de Jussieu and Carolus Linnaeus.

Carolus Linnaeus (1707–1778) is considered the father of modern plant and animal taxonomy. Prior to 1700, lengthy descriptive names were assigned to plants and it was difficult to relocate plants that had been previously described. Linnaeus was a prolific writer and an extraordinary taxonomist. In *Hortus uplandicus* (1730), he proposed what he called a sexual system of plant classification. This system was a major contribution because it was simple and based on plant taxonomic relationships. However, the system was artificial, because the emphasis was primarily on numerical relationships of flower parts; thus, similar plants often fell into widely separated classes. Later, in *Genera plantarum* (1737), he revised this system and included a list of natural plant orders and their genera. The publication of his famous *Species plantarum* in 1753 is considered the starting point of present-day nomenclature (naming of plants). In this treatise, Linnaeus proposed assigning two names, a Latin binomial, to each plant. Many of the names he assigned to plants during his lifetime remain valid and in use today, as can be seen by observing the species names found in the chapters of Parts II and III.

Near the end of the eighteenth century, systems based on form (morphological) relationships began to appear. These new systems were called natural systems, as plants having similar combinations of characters were placed together. They were not based on evolutionary relationships, but ordered plants along more natural lines while serving as an aid in identification. The taxonomists of this period included Antoine Laurent de Jussieu (1748–1836), three generations of de Candolles, the most important being Augustin Pyrame, and two botanists who collaborated on a system, George Bentham (1800–1884) and Sir Joseph Dalton Hooker (1817–1911). There was a marked similarity in the systems of these taxonomists. In particular, all were based on the concept of species immutability; that is, species were created and therefore they could not change. However, the systems still represented scientific progress.

The systems of de Jussieu and Bentham and Hooker divided the seed plants into three classes: Dicotyledoneae, Monocotyledoneae and Gymnospermae. The Dicotyledoneae class was further divided in the Polypetalae (corolla of separate petals), Gamopetalae (corolla of fused or partially fused petals, sympetalae), and Monochlamydeae (no petals, apetalous). These classes appeared relatively natural, and the divisions remained part of taxonomic classification until about 40 years ago (refer to Harlow et al. 1978).

The theories of Alfred Wallace and Charles Darwin around the middle of the nineteenth century began a new period in taxonomic investigation and classification. The new systems were based on phylogeny or on the evolutionary development and genetic relationships between plants. The Engler and Prantl system of classification was developed during this period, although it was not phylogenetic in a modern sense. This system divided the seed plants into the Gymnospermae and the Angiospermae, the latter being divided into the Monocotyledoneae and the Dicotyledoneae. The dicots were then divided into subclasses (Apetalae, Gamopetalae, and Polypetalae). Each subclass was divided into orders of presumably related families. There was some objection to the system, but it was widely accepted because Engler and Prantl applied it to the plants of the world, the result being a 20-volume work.

Charles Bessey (1845–1915) was the first American taxonomist to make a contribution to classification by developing a system that was truly phylogenetic (Lawrence 1951). In essence, he realigned the system developed by Bentham and Hooker according to evolutionary principles but also included some of the principles of Engler and Prantl. Other systems that were developed include one by John Hutchinson of England, which is closely aligned with that of Bentham and Hooker and that of Bessey, and one by Oswald Tippo of the University of Illinois (Fuller and Tippo 1954), which incorporated the latest developments in phylogeny.

The classification systems of Bentham and Hooker, Engler, Bessey, Hutchinson, and Tippo are in use throughout the world. The Engler system is widely used in herbaria in the United States, but the other systems are more accurate. No doubt these systems will continue to be revised as new scientific information becomes available.

The most recent classification system was proposed by Cronquist (1981). He considered only the division Magnoliophyta (angiosperms), which was split into two classes: Magnoliopsida (dicots; six subclasses) and Liliopsida (monocots; five subclasses). The species are clustered into these subclasses based on best available evolutionary evidence, so that some orders and families once considered relatively close have now been separated. Perhaps the most notable change is the breaking up of the old Amentiferae (catkin-bearing woody plants); the Fagaceae (oak, beech), Juglandaceae (hickory, walnut), and Betulaceae (birch) families are in subclass Hamamelidae, while the Salicaeae (aspen, cottonwood, poplar) has been placed in subclass Dilleniidae, because although the inflorescences are similar, the origins of the unisexual flowers are apparently different. The Cronquist system has been readily accepted and used; thus, we also have used it in this book.

Before leaving this short historical account of the development of taxonomy, the reader should contemplate for a moment the problems and frustrations encountered by early botanists. Prior to the modern systems, it must have been extremely difficult to describe, identify, and classify plants. It is likely that for a long period, every new plant added to the list of known species somehow changed the ideas and the concepts of the botanist, and the classification system changed accordingly. Consider the problem of keeping track of plants and their names before the standardized system of nomenclature was developed. Letters to distant colleagues must have been filled with long descriptions and probably many sketches. As compared with the effort required during the early 1700s, the identification of an unknown plant and relocation of specimens within a herbarium is now relatively simple.

CHAPTER 2

Introduction to Forest Ecology

THE IMPORTANCE OF FOREST ECOLOGY

EVERY forest plant, be it a tree, shrub, vine, or herb, is tolerant of (is able to grow and develop within) a specific range of environment conditions. Because both the reproductive and vegetative attributes (life history) and the ecological characteristics of a plant are largely determined by the genetic system, it is not surprising that just as two species have different physical characteristics, they are also ecologically different. Outside a specific environmental range, usually at the extremes, conditions prove lethal to a plant's existence. Plants may occur on different soil types, on sites of different moisture or nutrient regimes, respond differently to varying light levels or disturbances, have different insect and disease susceptibilities, and/or be found in different community types or in different regions. These characteristics are referred to as silvical characteristics; the root *silva* means "of the forest." Knowledge of silvical characteristics is extremely important in a variety of forest management activities, including regeneration, timber harvesting, thinning to improve growth rate, prescribed burning, increasing mast (nuts, acorns, fruits) for wildlife, increasing biodiversity (species number or richness), maintaining rare or endangered species or community types, and forest restoration.

In general, we can view forest ecology as the study of relationships between abiotic and biotic components, as well as biotic-biotic and abiotic-abiotic interactions, of tree-dominated communities. However, forest land managers have not always incorporated ecological knowledge into their decisions. Indeed, the development of best forestry practices usually follows a period of exploitation that created actual or potential timber shortages (Kimmins 1987); in this regard, the United States has been no different from other countries (Franklin 1998). Early colonists exploited a seemingly endless forest resource as they moved west across North America in the 1800s. Around the turn of the century state forestry commissions, forest reserves, and the U. S. Forest Service were created, but eastern lands continued to be denuded, first by logging the forest and then by wildfire. When attention was called to the dwindling resource, fire control laws were enacted in the 1920s and 1930s. In the past half century, United States forestry professionals have steadily integrated ecological principles as they have become available.

Forest ecology provides the essential ecological basis for successful forest management (Kimmins 1987). Foresters, now more than ever, are expected to protect as well as manage forest lands to ensure a renewable forest resource for the future. It is generally because of management interests that the classification of forests systems is undertaken. Of necessity, land stewards work with unit areas, and thus to manage each area properly, they need to know the present forest composition and the potential of the future forest.

DEFINITION AND SCOPE

Forest ecology is the science that deals with forest relationships and interrelationships. Study may concentrate on (1) relationships between individuals, species, populations, or communities (biotic factor vs. biotic factor), (2) relationships between forest plant and animal species and their physical environment (biotic factor vs. abiotic factor), or (3) the interaction between environmental factors (abiotic factor vs. abiotic factor). Living forest organisms include shrubs, vines, herbs, lichens, fungi, bacteria, and a variety of animals, as well as species that appear as seedlings, saplings, and trees. Similar organisms are clustered together in communities because of similar resource requirements. The primary nonliving environmental factors can be grouped under climate (e.g., precipitation, temperature, relative humidity, and wind), soil (water- and nutrient-holding capacity, depth, stoniness, and bulk density), and disturbance factors such as fire, tornadoes, frost, drought, and flooding. Insects and disease are part of the environment of other living species, so the plant environment is not wholly abiotic. The presence of another plant, living or dead, is also considered part of the environment, as it may physically injure, compete with, or chemically interfere with the growth and development of an adjacent plant.

Because relationships are often complex and intertwined, they are referred to as **interrelationships**. The concept of interrelationships may be the most important in the science of forest ecology and for developing a basis for management and environmental impact assessment. The idea here is that organisms, or organisms and environment, are closely tied together and influence each other in a reciprocal fashion. An alternative term, and one that suggests an even stronger relationship, is **interaction**. Interaction may encompass both positive and negative reciprocal actions of organisms on organisms (e.g., symbiosis, competition, mutualism, parasitism), organisms on environment (climate amelioration or soil modification), and environment on organisms (plant stress). At first glance, these relationships appear simple and easy to understand, but in application they are relatively complex.

SILVICAL FEATURES

The value of forest ecology concepts and supporting information lies in their application; applied forest ecology is called **silvics**. The formal definition of silvics is study of the life history and ecological characteristics of forest tree species with special reference to environmental conditions (Ford-Robertson 1983). Included are not only climate, soil and topography, and associated species (communities or cover types), but also factors such as geographic range, flowering and fruiting, seed production and dissemination, seedling development, vegetative reproduction, stem growth and yield, rooting habits, reaction to competition, disease, insects, and genetic variation (races and hybrids).

Knowledge of interrelationships forms the basis of an environmental impact statement (EIS), which must be developed by a public agency before management activities in a forest or other community can proceed. After the EIS is complete, knowledge of forest ecology is basic to the development of a set of ecosystem management objectives and the successful implementation of silvicultural techniques to attain those objectives. **Silviculture** is defined here as the application of silvics to the care and tending of a forest for continued production of goods and services and maintenance of biological diversity (Ford-Robertson 1983). Silvicultural objectives may center on any or all of the five resources of the forest (wood, water, wildlife, recreation, and forage) or concentrate on maintaining a functional, sustainable ecosystem.

RELATIONSHIP TO OTHER SCIENCES

Determining the nature of a relationship makes the study of ecology interesting and exciting and often leads to a study of other ecological disciplines. One soon finds that forest ecology is not a separate science but an integral part of **plant ecology**, which includes the study of the prairie, steppe, marsh, fen, glade, barren, savanna, chaparral, desert, heath, arctic, alpine, wetland, estuary, and other communities and environments. In turn, plant ecology must be viewed as a subset of the relatively broad field of **general ecology**, which encompasses animal (insects, birds, and mammals) as well as plant species and communities in both terrestrial and aquatic environments. **Terrestrial ecology** deals with species and communities associated with upland surfaces. **Aquatic ecology** deals with those associated with streams, lakes, and oceans.

An important difference between terrestrial and aquatic environments and their associated communities is that upland surfaces are erosional by nature; that is, they tend to lose soil, organic material, and nutrients, whereas aquatic systems are depositional, or a sink for these materials. Important wetlands associated with streams or water bodies are found at the interface of terrestrial and aquatic ecosystems and may have some characteristics of both. For example, wetlands can be periodically flooded, which may result in either erosion of soil material from, or the addition of such material to, the land surface. In general, however, wetlands are usually thought of as depositional and important in filtering and purifying water flowing from uplands before it reaches streams, lakes, or oceans. In northern United States and Canada, the most common forested wetlands referred to as lowlands, are swamps and bogs; marshes are dominated by grasses and sedges. In central and southern United States, forested wetlands are called bottomlands, which can include terraces, flats, swamps, and sloughs. Terrestrial and aquatic ecology, along with atmospheric science, combine to form the science of **global ecology**.

It is important to remember that the activities of the ecologist, soil scientist, or resource manager not only have consequences at the local level, but may also produce global consequences. These activities may affect air and water quality, wildlife habitat and population density, and possibly even atmospheric temperature (global warming).

LEVELS OF INTEGRATION

In 1961, Rowe, a forest ecologist, proposed a series of ecological levels, a hierarchy, each with its own body of knowledge beginning at the most elementary and progressing to the most complex level within living systems. He called this series "levels of integration," because basic information developed at one or more lower levels could be used to explain an observed response at a higher level, thereby providing a scientific understanding of how the higher level functions. The importance of these levels of complexity is that they provide a framework for understanding structure, function, process, or response at one or more of the higher, more complex levels where management occurs. It is interesting to note that answers to ecological questions and problems often come from scientists who may not be community or ecosystem ecologists.

Our understanding of the forest and its management begins with a knowledge of atoms, molecules, cells, tissues, and organs at the elementary (lower) levels of integration and progresses to the secondary levels of how organisms, communities, and ecosystems function. It is inappropriate to refer to atoms and molecules as elementary, because by themselves they are relatively complex and, like cells, tissues, and organs, they are components whose processes can explain events at a higher level. At the lower levels, physical and chemical phenomena are reasonably, often precisely, predictable. However, in progressing from the organism through the population, community, and ecosystem, response to a management activity is considerably more difficult to predict accurately. For example, when high light intensity saturates the chlorophyll molecules in the leaves of quaking aspen (*Populus tremuloides*), white oak (*Quercus alba*), shortleaf pine (*Pinus echinata*), or lodgepole pine (*Pinus contorta*), maximum photosynthetic activity occurs. Thus, a resource manager may cut a forest to increase light levels at the forest floor so as to enhance the survival, growth, and development of seedlings of these species. However, the operation may still be a failure for reasons not related to the physical requirements of the seeds and seedlings. Other environmental factors, such as soil water deficits or high evaporation rates, may kill seedlings. Weather may have affected flowering or the amount of seed produced, harvesting may have removed the seed source, or squirrels, mice, deer, or rabbits may have eaten the seed or seedlings. A disease or insect may have attacked and killed the stem. It is apparent that the resource manager must have a good understanding of various sciences in order to properly manage the organism, population, community, and ecosystem.

The most important levels for the forest resource manager are the individual, population, community, and ecosystem. An **individual** is a single living functioning unit, and a **species** is a group of interbreeding individuals that have similar morphological (appearance), ecological, and genetic characteristics and produce fertile offspring. Each species has its own specific set of ecological relationships: range; soil, drought, shade, and climate tolerances; growth rate; insect and disease problems; and associated species and community types. Thus, it should be apparent that **plant taxonomy** and species identification are prerequisite to applying ecological principles and information, as these vary with the species found in a forest. **Autecology** (Greek: *autos*, "self," interpreted to mean "one") is the study of individuals or species of plants (e.g., redwood, *Sequoia sempervirens*; Douglas-fir, *Pseudotsuga menziesii*) or animals (*Odocoileus virginiana*, white-tailed deer) but in the present context must be more broadly defined to include individual environmental factors (soil texture or depth, precipitation, wind, evapotranspiration, etc.) and how these factors affect living individuals. Much of this body of knowledge comes from the area of physiological ecology (atoms, molecules, cells, tissues). However, life history studies on individual species also are autecological in character and considerably more applicable to solving management problems. Such phenological studies provide information on time of flowering, pollination, fertilization, seed-set, seed production, seed dissemination, time of germination, seedling establishment, and early growth, senescence, and mortality. For example, Forest Service scientists are studying the physiology and life history of oak (*Quercus*) in the Central States in order to understand the reproductive process. Their goal is to increase oak regeneration, which has greatly declined in the past 50 years or so.

A **population** is a group of individuals of a single species within a single locale. A **community** is a group of interacting populations. **Synecology** (Greek: *syn*, meaning "together"), is the study of relationships between groups of species, populations or communities, and environmental factors. The idea is that individuals and populations within a given area interact with other individuals and populations within the same area. They do so by competing for or modifying available resources such as light, water, nutrients, and growing space. Individuals and organisms also respond to variations in environment not caused by the presence of other plants. A particular environment (xeric, mesic, hydric) is the product of different interacting factors such as soil available water capacity (AWC) and rate of evapotranspiration (ET). The former is determined by soil texture, depth, stoniness, and bulk density, the latter by solar radiation, temperature, heat, wind, and relative humidity. Here again, for estimates of AWC and ET, data on a number of factors must be integrated.

An **ecosystem** is a community and environment acting as a functional unit through which there is a flow and interchange of energy and nutrients (food web) and water. It is not surprising that resource managers have experienced the most difficulty when trying to manage at the ecosystem level. There are several reasons for this problem. First, data and information on management at the ecosystem level for specific forest types are somewhat limited as a result of the inherent high complexity at this level. Often, there is a 5- to 10-year delay before community response can be evaluated. This lack of information has led to the development of long-term ecological research studies (LTERS), which are used to monitor the effect of human activity on an ecosystem. The collection of data and analysis are time-consuming and require trained personnel, which costs money; thus, a modest number of large, intensive projects on ecosystems have been established. Second, there is often a breakdown in the transfer of information from the researcher to the resource manager. The responsibility of transferring concepts usually falls to state extension agents; all too often, the concepts fail to provide sufficient information for on-the-ground operations. Most journal articles reporting this information are sufficiently sophisticated that an understanding of special terminology and advanced concepts is usually required; many universities do not teach a course in forest ecosystems, and thus resource managers are not equipped to understand such journal articles, nor do they have time to seek out the articles. Perhaps the greatest problem for resource managers is the change in attitude that must accompany ecosystem management. The ecosystem has always been there, but too often the resource manager has been concerned with only one part of the system (timber, wildlife, water, insects, birds, etc.). Now all components must be considered simultaneously and with nearly equal weight given to each.

One level that did not appear in Rowe's original (1961) list and has not been addressed until recently is the **landscape**. It is safe to suggest that concepts regarding ecology and management at this level are still developing. Forman and Godron (1986) and Forman (1995) describe landscape structure as consisting of patches and corridors networked within a matrix, the most extensive and most connected landscape element type that plays a major role in landscape (ecosystem) function. Considerable attention is being directed to the landscape because of the ecological problems associated with forest fragmentation, particularly the loss of herbaceous plants and neotropical birds.

HISTORICAL DEVELOPMENT OF PLANT (FOREST) ECOLOGY

The history of forest ecology is an integral part of the development of plant ecology. As with the development of other sciences, simple concepts were identified early, and the most complex relationships only in the last few years. It appears that the first ecological term was used by Kerner in 1863 when he used the word *succession* in his study of the vegetation of the Danube Basin. The term *oecology* was introduced by Haeckel in 1865 (Oosting 1956) to describe this science of living organisms and their environment. The term *ecology* is derived from the Greek words *oikos*, which means "a place to live," and *logia*, or "the study of"; thus, literally interpreted, ecology is a study of organisms and their habitats or environment. During the later 1800s, progress on concepts was slow and concentrated primarily on succession in peat bogs, development of the community concept, and the idea of interaction. The formalization of ecology began with European geographers such as Warming (1909), whose book on ecological plant geography was based on Haekel's definition, and with Schimper (1889), whose book on plant geography had a physiological orientation, primarily an examination of the relationship between water and plant adaptations.

In the United States, Cowles published the first formal treatises (1899, 1901) on the development, succession, and classification of vegetation, including forest communities as social units on the Lake Michigan sand dunes. A major figure in the development of plant ecology concepts during the early part of the 1900s was Frederick E. Clements (1907, 1928, 1934, 1936). He had a relatively strong dogmatic approach to vegetation/community composition and classification. His concepts sharply contrasted with the individualistic concept (Gleason 1926, 1935) later integrated into the idea of gradient analysis.

After World War II, developments in plant ecology as a science were slow, but major progress began in the 1950s and has continued into the early twenty-first century. The continuum concept and gradient analysis, an extension of Gleason's individualistic concept, were developed at about the same time by Curtis and McIntosh (1951) and Whittaker (1951). Simply stated, the continuum implies that community composition slowly but consistently changes with small but directional change in environment from xeric to mesic to hydric conditions (a gradient). *The Vegetation of Wisconsin* (Curtis 1959) was the first and perhaps the best work describing and applying the approach. However, the continuum concept was not universally accepted, and for years scientists have hotly debated continuum versus classification approaches. Gradient analysis was developed in the eastern United States and has been applied primarily in this region. Concurrently, Daubenmire (Daubenmire and Daubenmire 1968), following the concepts of Clements, was developing ideas on forest classification that were accepted and applied primarily in the western mountain regions. Although the community concept of Clements was far too rigid to be widely applied, some of his terminology (e.g., climax, postclimax, edaphic climax, sere) can still be found in the current literature. Unfortunately, people on one side of the question often do not recognize the importance of the other approach. Classification as now used in ecological-land classification (refer to Chapter 25) is a valuable management tool, and gradient analysis is used to identify and understand species and community response to variation in environment. Both approaches continue to be widely applied.

From the 1950s until the present there have been major advances in concepts and the understanding of relationships every few years. Beyond Curtis and Whittaker, the time of the 1950s and early 1960s was a period of radioecological research in which plants and genetic systems were tested for their reactions to radioactive substances (Cs 239) to determine how organisms would respond to fallout from a nuclear war.

The 1960s ushered in the era of ecosystem research and large intensive projects, beginning with the International Biological Program (IBP), which advanced ecology from community research to ecosystem research. Under IBP, energy, water, and nutrient flow and storage were studied in a variety of forest regions (biomes), including several watersheds in the eastern hardwood forest and several in the western conifer forest, as well in prairie, arctic and alpine vegetation, and aquatic systems. When IBP ended in the early 1970s, research on ecosystem components continued at a smaller scale and at a variety of universities and public agencies. During the 1970s, and later in the 1980s, considerable ecosystem information was developed on aboveground biomass (standing crop), nutrient storage of leaves, twigs, stems and roots, rates of litter decomposition, root growth, and related components in a variety of forest communities. Rates of nutrient loss from small forest ecosystems due to timber harvesting and herbicide applications were studied at Hubbard Brook (New Hampshire) and the Coweeta Experimental Forest (North Carolina). In a new application of radioecology, rates of nutrient storage and movement were studied using radioactive elements that were similar to nutrients required by a tree (e.g., cesium and calcium). These studies also brought forth computer programmers and mathematicians to create software for handling and analyzing large volumes of data and to develop structural and mathematical models to simulate the natural world. Models and modeling are now an integral part of ecology.

THE PRESENT DOMAIN OF FOREST ECOLOGY

As indicated by the foregoing review of plant and forest ecology, the field has evolved into a number of sciences, each with its encompassing philosophy and approach to identifying interrelationships and solving environmental problems. There are literally hundreds of books on specific aspects of ecology. However, the focus of this book includes forest ecology, and it is appropriate to examine it both as a science and as a problem-solving approach. Not surprisingly, it overlaps with and draws from various other areas of ecology.

Forest Ecology as a Science

The magnitude and scope of forest ecology as a science can be addressed by answering a number of questions modeled after those suggested by Egler (1977). The first five of the following questions encompass much of the material in this book.

1. *What is it?* "It" could be an individual of a species, a forest community (stand), or an ecosystem as previously described. Keep in view that most questions to follow may apply equally well to any of these three entities.
2. *What is its composition?* Organisms of the forest include trees, shrubs, vines, herbaceous plants, fungi, bacteria, mammals, birds, and insects, many of which live not only aboveground but also in the soil system. However, the dominant plants in the forest are trees, and thus traditionally the community name is based on tree composition (white spruce-balsam fir, black spruce-tamarack, aspen-paper birch, jack pine-aspen). These community names represent cover types (Eyre 1980). Cover types may be subdivided on the basis of variations in site environment; these habitat types (ecological/land classes) are identified and named using the names of tree, shrub, and herbaceous plants (Kotar et al. 1988; Kotar and Burger 1996). For this reason, shrubs and herb species have been included with cover types in Chapters 26 to 30.

   A site is defined as an area of land having a specific environment (e.g., soil and microclimate) which restricts the types of tree species and communities that will survive within the environmental conditions. A site may be as small as 1 or 2 hectares. Some common sites are a ridgetop, stream terrace, and south and north slopes, with slope position (high, mid, low) further modifying the environment.

   At the other end of the scale are the major North American forest regions or biomes, also called ecosystems (Figure 2.1). These regions have been identified by a number of common species that form community types that are repeated throughout the geographic area. For example, *Picea glauca* (white spruce), *Abies balsamea* (balsam fir), *Pinus banksiana* (jack pine), *Picea mariana* (black spruce), *Larix laricina* (tamarack), *Populus tremuloides* (trembling aspen), and *Betula papyrifera* (paper birch) are the major species in the Northern Conifer-Hardwood Forest, also called the boreal forest or the spruce-fir forest, that stretches from Nova Scotia across Canada into central Alaska. Some of these species may form a pure community type (e.g., jack pine, aspen, and black spruce) or a mixed community type in combination with one or more additional species (e.g., jack pine-aspen, aspen-paper birch, black spruce-tamarack, white spruce-balsam fir).
3. *What is its environment?* Forest ecology includes the study of environment and environmental factors. The concept of "Environment" is usually viewed as the abiotic, nonliving or physical component of the ecosystem that influences the biotic or living

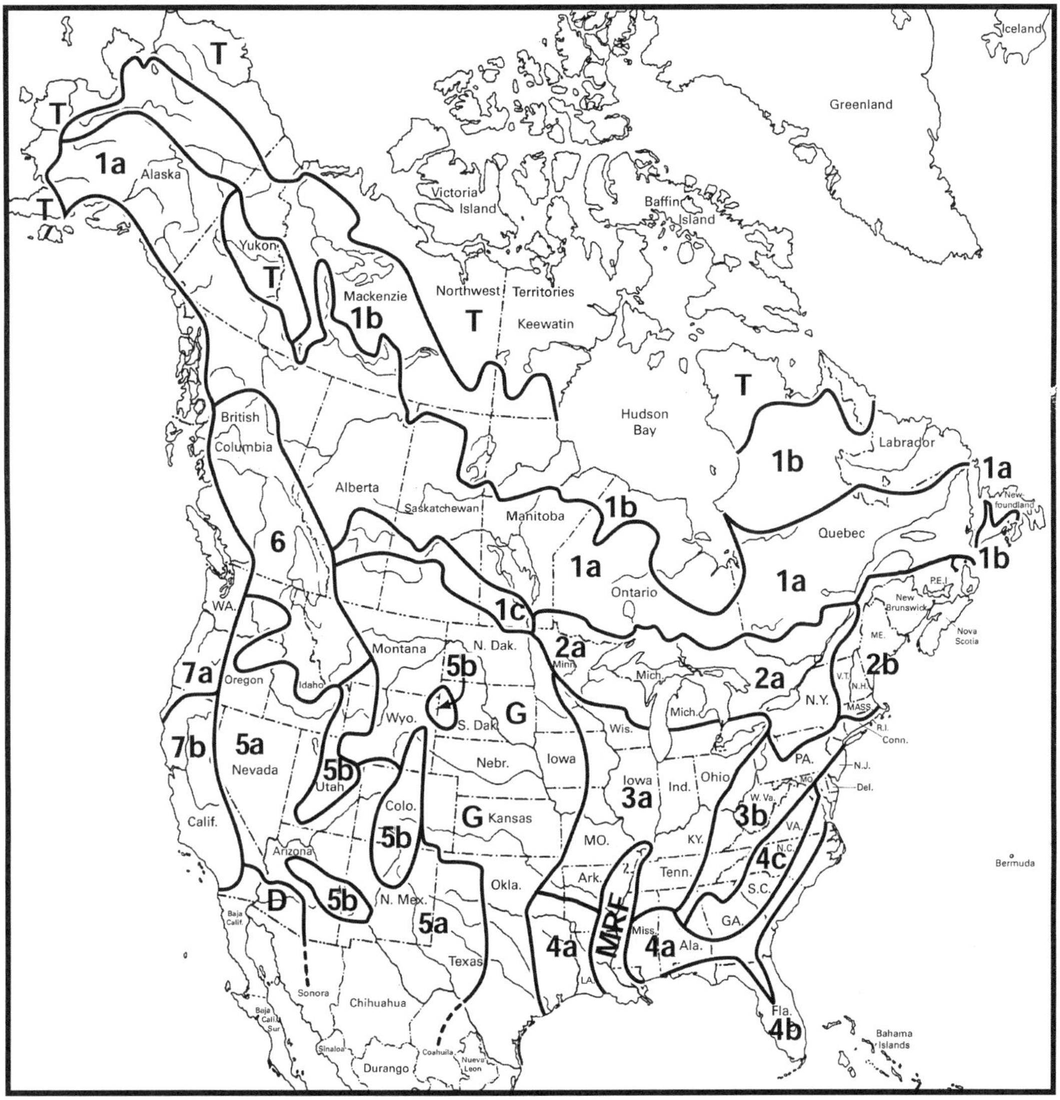

FIGURE 2.1 Forest regions of North American: (1a) Northern Conifer-Hardwood or Boreal (Spruce-Fir) Forest, closed; (1b) Northern Conifer-Hardwood Forest, open savanna and barrens; (1c) Northern Conifer-Hardwood Forest, Aspen Parkland; (2a) Northern Hardwood-Conifer Forest (Sugar Maple-Yellow Birch-Eastern Hemlock-White Pine Forest), Great Lakes section; (2b) Northern Hardwood-Conifer Forest, New England section; (3a) Central Hardwood Forest (Oak-Hickory Beech-Maple, and Western Mesophytic), Interior Plateaus section; (3b) Central Hardwood Forest, Appalachian Mountain section; (4a) Southeastern Pine-Hardwood Forest, Upper Coastal Plain; (4b) Southeastern Pine-Hardwood Forest, Lower Coastal Plain; (4c) Southeastern Pine-Hardwood Forest, Piedmont Plateau; (5a) Southwestern Juniper-Pinyon Savanna; (5b) Central and Southern Rocky Mountain Mixed Conifer Forest; (6) Northern Rocky Mountain Conifer Forest; (7a) Northern Coastal Conifer Forest; (7b) Sierra Nevada/Southern Coastal Conifer Forest. Other zones include grassland (G), desert (D), and tundra (T), as well as the Mississippi River floodplain (MRF), which has largely been cleared for farming. Base map adapted from Little (1971).

component. The major abiotic physical factors can be grouped into broad categories such as climate, soil, topography, and disturbance (e.g., fire, windstorms, floods). The biotic factors include the presence of other plants that create competition, insects, fungi, bacteria, and various animals such as deer, porcupine, and beaver.

Macroclimate or regional climate over extremely large geographic areas (i.e., forest regions) controls not only the composition of the forest community found in a region but also soil development. Macroclimatic factors include precipitation (amount and type of rain or snow), light (type and amount of solar radiation), temperature, heat (or lack of it), and wind. Macroclimate in the Northern Hemisphere is determined by latitude (closeness to the equator or the north pole), location relative to land, oceans or Great Lakes, mountain ranges, elevation, and direction of the prevailing winds.

On a macroscale, there is a near perfect correspondence between climate, soil, and regional vegetation (Table 2.1). For example, the Canadian Northern Conifer-Hardwood Forest has a cold climate, mostly conifer-dominated communities, and Spodosol (podzol, sandy, leached, ashy gray *E* horizon), Inceptisol (poorly developed horizons), or Histosol (organic) soils. The Central Hardwood Forest of the Midwest (Ohio to Missouri) has a moist continental climate (characterized by extreme diurnal and seasonal variations) and Alfisol soils (dark *A* and clayey *B* horizons). The Southeastern Pine-Hardwood Forest has a semitropical climate and Ultisol soils (highly weathered, due in part to the warm temperature and high rainfall). Some areas of United States are semiarid, have Aridisol soils (accumulation of calcium carbonate or sodium), and are dominated by drought-tolerant *Juniperus* (juniper) and *Pinus* (pinyon pine).

These climatic factors may be greatly modified by mountain ranges and large water bodies. On the windward side of mountains, temperatures decrease with elevation and the air becomes denser and more humid until precipitation begins (orographic precipitation). In North America, this precipitation is largely confined to the western slopes. As the dried air passes over to the lee, or eastern side, of a mountain range and descends, it warms quickly to create a relatively arid climate. These variations in air temperature and precipitation result in substantially different forest types on the windward and leeward sides. The Sierra Nevada and Rocky Mountains are the cause of the arid intermountain region and the prairie region, respectively, east of these ranges.

Large water bodies moderate the climate and reduce fluctuations in temperature and precipitation. Along the west coast from California to southeast Alaska there exist a fog belt, areas of extremely high precipitation, and community types not found elsewhere: *Sequoia sempervirens* (redwood of California and Oregon), temperate rain forest of *Pseudotsuga menziesii* (Douglas fir of Washington and British Columbia), and *Picea sitchensis* and *Chamaecyparis nootkatensis* (Sitka spruce and Alaska–cedar, respectively, of British Columbia and Alaska).

Climate and soil also are modified at the site level. Site environment is more commonly called microclimate. Although level terrain may largely reflect macroclimate, hilly or mountainous terrain can be substantially different. For example, south-, southwest-, or west-facing slopes may have higher temperatures, lower relative humidity, and higher evapotranspiration rates than the regional climates. *Evapotranspiration*, a factor not usually considered a climatic factor, is a term used to describe the amount of water evaporated from soil and transpired from plant leaves. Because of the westerly winds over the United States, these sites also have higher and more constant wind velocities, which increases evapotranspiration. Conversely, north, northeast, and east slopes are cooler, more humid, and have reduced wind and water loss from soil and leaves. In much of the eastern United States, tree and forest growth is slowest on the south and southwest slopes, and highest on the north and northeast slopes. Low sites such as stream terraces and drainages may be extremely cool and humid because of cold air drainage during the night resulting in low evapotranspiration. Ridgetop sites usually have the highest wind velocities.

Soil influences plant growth and community composition by providing available water and nutrients, growing space for roots, and anchorage for trees. Available soil water-holding capacity is determined by soil parent material (texture: sand, silt, clay), depth, structure, bulk density, organic matter,

Table 2.1. Climate types (Domains; Bailey 1998) and corresponding regional forest vegetation and soil orders for North America

The vegetation and soil of the regions are described later in Part IV. Prairie (grassland) is included here only to demonstrate the close correspondence.

Climatic Domain	Regional Vegetation	Soil Orders
Polar Subarctic	Northern Conifer-Hardwood Forest	Spodosols, Entisols, Histosols
Humid Temperate-Warm Continental	Northern Hardwood-Conifer Forest	Spodosols, Histosols
Humid Temperate-Hot Continental	Central Hardwood Forest	Alfisols
Humid Temperate-Hot Continental, Mountains	Appalachian Hardwood Forest	Inceptisols
Humid Temperate-Subtropical	Southern Conifer-Hardwood Forest	Ultisols
Subhumid Temperate	Prairie	Molisols
Dry, Temperate	Juniper-Pinyon Savanna and Desert	Aridisols
Dry, Temperate, Mountains	Central Rocky Mountain Conifer Forest	Alfisols, Entisols
Humid Temperate, Marine, Mountains	Northern West Coast Conifer Forest	Alfisols, Ultisols, Andisols
Humid Temperate-Mediterranean, Mountains	Southern West Coast Conifer Forest	Alfisols

and stoniness. Nutrient availability is determined by local climate (input of nutrients with precipitation) and soil parent material (initial nutrient level), texture, depth, and organic matter. Other factors that influence soil include topography and time.

Fire and various other types of disturbance are major factors influencing the composition and structure of vegetation. Fire has a considerable impact on a forest. Only in recent years has fire been viewed as beneficial and necessary to the maintenance of fire-dependent communities; thus, the insistence that "all fire is bad" has been diminished. Most of the destructive wildfires in western conifer forests result from the absence of small frequent fires over the past century that would have eliminated the large fuel volumes now on the ground.

A consideration of environment must also include the effect of one organism on another. This effect can be beneficial such as the symbiotic relationship that occurs in root nodules (trees and bacteria), which fix nitrogen, or in micorrhizal relationships (trees and fungi), which extend tree rooting volume and increase nutrient absorption. The effect can also be negative (competition), as in the case of one tree shading a smaller stem or dense grass or herbaceous cover absorbing water and nutrients so that tree seedlings cannot grow.

4. *When was it there?* There is more than one answer to this question. Forest communities are characterized by change, but the process of change may differ, depending on the time frame. Over geologic time, communities and forest regions have migrated considerable distances. The expansion of glaciers into eastern North America during the Holocene shifted species, communities, and entire forest regions southward, with spruce-fir forest eventually located south of Illinois and Indiana. The pollen record indicates that when the glaciers melted in Illinois and Indiana approximately 10,000 to 11,000 years ago, the sequence of communities began with *Populus tremuloides* and *Betula papyrifera* (white birch) and progressed through the boreal spruce-fir (*Picea-Abies*) forest to northern hardwoods (*Acer saccharum*, sugar maple; *Tilia americana*, basswood; *Fagus grandifolia*, beech; *Betula allegheniensis*, yellow birch) and conifers (*Tsuga canadensis*, eastern hemlock; *Pinus resinosa*, red pine; and *Pinus strobus*, white pine.

   Within a short time frame, about equal to the life of a tree (about 40 to 400 years), forests are dynamic. Some forest types may largely disappear if they are composed of early-successional species. Communities composed of these short-lived, shade-intolerant pioneer species may temporarily dominate a stand (or occupy a site) for less than 70 years. In the absence of disturbance, most forest communities dominated by pioneer species are replaced in 100 to 300 years. These successional communities develop after major disturbances such as fire, tornadoes, downbursts, or hurricanes, and insect and disease attack may kill an existing community and open the land to invasion and colonization by an opportunistic species. Timber harvesting may have much the same effect, particularly in combination with a natural disturbance.

Generally, succession is restricted to mesic sites—that is, where soil moisture is readily available. Pioneer species are usually replaced by a community composed of more shade-tolerant, moisture-demanding species, which have extremely dense, deep crowns. Because their seedlings and saplings grow well in shade, these shade-tolerant species generally form a stable community that may occupy the land until another major disturbance occurs. For example, on sites where there is an optimum amount of available soil water, a *Liriodendron tulipifera* (yellow-poplar)-dominated stand in Tennessee or an *Quercus-Carya* stand in Indiana would likely be replaced by *Acer saccharum* and *Fagus grandifolia. Liriodendron, Quercus,* and *Carya* would eventually disappear from the site, as seedlings of these species die from lack of sunlight within a dense forest. However, frequently when a community of shade-tolerant species is harvested for timber, disturbed by fire, or receives severe wind damage, seedlings can become established in the high light environment, and under such circumstances these shade-intolerant species are likely to be a substantial component of the new stand.

With the advent of fire prevention laws in the 1920s and 1930s and the effectiveness of the Smokey Bear campaign, disturbance of forest communities by fire has been nearly absent. The result has been major changes in the location and composition of North American forest communities. Fire-dependent communities include trembling aspen (*Populus tremuloides*) and jack pine (*Pinus banksiana*) in the Lake States and Canada, northern red oak (*Quercus rubra*) in the Lake States, white and black oak (*Quercus alba* and *Q. velutina*) in the central states, shortleaf pine (*Pinus echinata*), loblolly pine (*Pinus taeda*) and other pines species in the southeastern United States, and lodgepole pine (*Pinus contorta*) and giant sequoia (*Sequoiadendron giganteum*) in the west. Perhaps the species most affected in southeastern United States is longleaf pine (*Pinus palustris*), whose range along the Atlantic Coast has been reduced by tens of thousands of hectares since 1850.

5. *How can it be classified*? Several systems have been used to classify plant and forest communities. For practical purposes, the approaches have been used to describe the forest community and ecosystem variability. Forest ecologists have classified forest communities and regions according to their present tree species composition (Eyre 1980) or their anticipated future composition (climax status; Küchler 1964); infrequently, there is substantial disagreement on the latter classification (Braun 1950). Resource managers consider age, development (e.g., density, rate of growth), and condition (insect, disease, wind damage) in addition to tree species composition in delineating stands. In recent years, habitat typing and ecological land classification systems have been widely used. Tree, shrub, and herbaceous species are used in both systems, but there is considerable emphasis on geologic, topographic, and soil factors in ecological/land classification.

There are several other questions regarding forest ecology that can be asked, but within the scope of this book they are not as important as the preceding questions.

6. *What is its structure*? The general term *structure* can be interpreted in a variety of ways, in as much as structure is an inherent property of ecosystems, plants, and cells. At the macroscale (ecosystem, region, community), classification and mapping are based on the form and structure (physiognomy) of the units: evergreen versus nonevergreen, shrub- versus tree-dominated, savanna versus forest, forest versus prairie.

   To biologists and ecologists, structure may suggest the arrangement and function of the range of organisms within a trophic level or across several levels. Trees, shrubs, vines, and algae are autotrophs, as they produce food for themselves as well as for other organisms. Animals such as rabbits, mice, and deer, in addition to insects that graze on plant material, are herbivores or primary consumers. Secondary consumers such as the wolf, cougar, and eagle feed on herbivores. These three groups form the base of the food chain or web. Fungi and bacteria also are consumers, because they use plant and animals remains for food; these decomposers are responsible for the breakdown of organic matter into minerals and nutrients that are reused by plants or recycled through the ecosystem.

Ecologists may also view structure as the internal organization of a plant, community, or ecosystem, whereas to resource managers structure may mean the arrangement of woody stems of different sizes (trees, saplings, seedlings, and shrubs). At times structure has referred to tree crown shape, which results from the branching pattern or the angle of the branches on the stem of the tree. Physiologists and anatomists may regard structure as a reference to the internal organization of a organelle, cell, or plant.

7. *What is its function?* There are four primary functions that can be ascribed to a healthy forest community and ecosystem. For centuries, the primary function of the forest has been viewed as productivity, the production of wood or cellulose. In broad terms, productivity is the conversion of light energy and carbon dioxide into sugar, cellulose, and other compounds storing carbon dioxide. Healthy forest ecosystems also conserve nutrients by storing them on-site in plant tissue. More recently, ecosystem functions have been described in terms of conservation, protection, and maintenance. Since the 1930s "dust bowl," protection of the land surface and soil and maintenance of water quality, as well as conservation, have been recognized as important ecological issues. Even more recently, as forest communities have received increasingly heavy use, the preservation of species and biological and genetic diversity, including the common as well as rare and/or endangered species, has become a major concern.
8. *What are the processes that permit it to complete its function?* Forest ecology is concerned with individuals, communities, and ecosystems and their biology and functional processes. Any process or activity that occurs within a cell, tissue, organ, or organism may ultimately affect community function. These processes include photosynthesis, respiration, transpiration, mechanisms to tolerate water stress, translocation of organic compounds, accumulation of food and cellulose, mineral nutrition, growth and growth regulators, and genetic and reproductive processes. Other biological considerations are strategies for survival (e.g., shade tolerance, sprouting ability), flowering, pollination, seed-set, distribution, germination, and seedling establishment. The effect of these species factors interacting with environmental factors is the basis for the development of successional and stable community patterns. At the ecosystem level, the processes of energy, water, and nutrient storage and flow are of major interest.
9. *How can it be maintained?* This question has many answers because it moves us into a consideration of silviculture and management, which vary for regions of the country, forest cover types, and species. Although there is considerable pressure for "preservation" from environmental groups, there are a large number of forest community types that cannot be preserved because of the successional process; few are compositionally stable. Intact old-growth stands of trembling aspen, jack pine, upland black spruce, white and black oak, southern pine, lodgepole pine, and giant sequoia, seldom occur.

Forest Ecology as an Environmental Problem-Solving Approach

Forest ecology has been a science since about the 1930s, when uncontrolled timber harvesting, fire, and extensive land clearing for agriculture came to an end in eastern United States. Since the 1950s, ecology has increasingly become an approach or a viewpoint to solving actual or avoiding potentially serious environmental problems. At the present time, few activities can occur on federal land without consideration of their impact on environmental quality; most activities are addressed in an EIS. These activities range from the digging of a soil pit or collecting a plant to development of a hiking trail, planting trees, or timber harvesting. An ecological approach to resource management combined with an EIS has been successful in avoiding some potential environmental problems. Management planning, which involves the development of an EIS, also provides an opportunity for public input in the planning process. However, the need to avoid ecological problems may be difficult for some segments of the public to accept. For example, prohibiting horseback riding in ecologically sensitive natural areas and restricting off-road vehicles to a limited number of trails in the Shawnee National Forest (Illinois) has raised a furor over land management policies and law enforcement.

The problems presently facing the resource specialist are unparalleled in the history of renewable resource management. At times, resource management for production seems to pose a dilemma, in that the National Forests were mandated by Congress to produce wood for public consumption. Their reason for being is unlike that of the National Parks, which were set aside for scenic beauty or observing natural phenomena.

RECENT DEVELOPMENTS IN FOREST ECOLOGY

For the last 25 years or so, there has been growing concern about the effect of air pollution on forest health. The problem of forest decline was first identified in the heavily industrialized areas of Europe but also began to appear in the Ohio River Valley region, where numerous coal-burning electricity-generating power plants are located. Before pollution control laws were enacted, hundreds of thousands of tons of sulfur dioxide and other pollutants were released into the atmosphere, later to be washed from the sky in the form of acid precipitation. In the mid 1980s, in response to the apparent decline of forests, the United States Forest Service and the Environmental Protection Agency initiated the Forest Response Program to examine the effect of long-term air pollution on forest health. This program was replaced in the 1990s by the Forest Health Monitoring Program, which uses remeasurement data from permanent plots to examine long-term changes in forest health.

Interest in rare, threatened, and endangered (RTE) species has been on the increase during the past 15 or 20 years. A large number of these plants and animals are forest inhabitants. Developing concurrently with the concern for RTE species was a movement for a national biological survey to catalog the present species (a species list) and thereby have some base or framework for examining species loss rates. This interest has expanded into a relatively new field of science, **conservation biology** (Soule 1986; Wilson 1988). Because of the fragmentation of forest and other plant communities, the field of **landscape ecology** has been developing to examine the effect of the present landscape pattern on species survival and movement (Forman and Godron 1986; Ingram and Moss 1992; Naveh and Lieberman 1994).

The ideas for preserving species have been extended to include plant communities and their associated animals. To preserve plant communities, it was deemed necessary to classify and catalog the present natural communities, including rare communities and/or those thought to represent pre-European settlement conditions. This ambitious program has been spearheaded primarily by The Nature Conservancy. However, public land management agencies will also benefit from an examination of presettlement communities through the identification of RTE species and communities. In order to catalog these plant communities, a classification system had to be developed, and it is interesting that although classification has been throught of as part of the "old" ecology, it has been resurrected for use in modern research and management. The search for near-pristine communities resulted in the location of somewhat degraded communities that could be restored. Subsequent attempts to regenerate high-quality remnants have led to the development of **restoration ecology** (Jordan et al. 1987) and, now, restoration forestry (Pilarski 1994). Presently there is much activity on the restoration of wetlands, prairies, savannas, and barrens.

Superimposed on the major ecological projects and the development of new science areas over the past 30 years were a number of smaller investigations, of which some were basic research and some were applied research. In both cases, the research provided much valuable information for our understanding of forests and their management. Topics included community diversity and diversity indices, the importance of seed pools in forest redevelopment, forest fertilization, the use of the forest as a natural filter for sewage effluent, stream bank stabilization, plant mineral nutrition and uptake, water relationships and the effect of deficits, the effect of pollution on soil, roots, and mineral nutrition, microclimatology, stem architecture, plant physiology, soil carbon storage, fire ecology, population ecology, and the all-time perennial favorite, forest and old-field succession.

THE SUSTAINABLE ECOSYSTEM CONCEPT

As a result of the new ecology movement, terms such "new directions" and "new perspectives" reflect increased interest in the application of ecology to land management. To quote Swanson and Berg (1991): "In the broadest sense, we are moving from forest management as a series of discreet operations to viewing forests as ecological systems." In June 1992, the U.S. Forest Service initiated the "new perspectives" approach to management (Cohn 1992). Included in a consideration of the forest ecosystem are, among other factors, effects of disturbance, soil, standing and down dead wood, coarse woody debris in streams, water quality, ecological processes in early and late stages of succession, biodiversity, nongame birds, canopy invertebrates, and tree roots. Jerry Franklin, a forest ecologist at Oregon State University, emphasizes that ecosystem management is more than a modification in procedure, it is a major change in attitude and perspective, a new way of looking at the forest community (Ecosystem Management Conference, Stevens Point, Wisconsin, 1994.). According to Jack Ward Thomas (Statement Before the Subcommittee on National Parks, Forests and Public Lands, 1994), "This ethic accepts short-term constraints on human treatment of land so as to ensure long-term preservation of the integrity, stability and beauty of the biotic community. Unless we implement management strategies that truly conserve biodiversity and maintain aesthetic values while producing needed commodities, we have changed only labels, not actual management."

The new, broader concept now being promoted by the Ecological Society of America and similar professional groups, and accepted by forest land managers, is that of a sustainable forest ecosystem linked to ecosystem management (National Research Council 1990; Lubchenco et al. 1991; Gale and Corday 1991; Mathews 1991; Le Master and Sedjo 1992; Covington and DeBano 1994; Gerlach and Bengston 1994; Salwasser 1994), with some authors using the concept of sustainable forestry (Maser 1994). Within this concept, a new paradigm (model) is being developed where by humans are integrated into and viewed as part of the ecosystem as well as using the productivity from it (landscape ecology).

PART I

Taxonomic Concepts and Methodology

CHAPTER 3

Plant Classification

INTRODUCTION

The orderly classification of plants has been and continues to be one of the major challenges to taxonomists and morphologists, and, as noted in Chapter 1, a number of hierarchial taxonomic classification systems have been published. *Hierarchial* refers to a system or organization of things (in this case, plants) arranged into higher and lower ranks or classes, with the lower rank a part of the next higher rank. Upon close examination, classification systems usually reflect one of two approaches, based on the origin and evolutionary development of plants. These approaches are referred to as the monophyletic (Greek: *monos* = one; *phyletikos* = tribe, phylum) and polyphyletic (*polys* = many) systems. The former is based on the idea that all or most plants had a common ancestor or origin, while the polyphyletic school promotes the concept that plants evolved from two to several ancestors along independent evolutionary lines. The arguments and fossil evidence for and against each school and associated classification system are beyond the scope of this book. It will suffice here that the reader recognize the difference between the two systems.

HIGHER TAXA

The highest level in plant classification is the kingdom Plantae (plants), which includes all plants, both green (plants with chlorophyll, called autotrophs) and non-green (fungi and bacteria, called heterotrophs). The kingdom includes all trees, shrubs, vines, and herbs, which are vascular plants that reproduce by seed. Below the kingdom are several categories of "higher taxa." These categories are division, subdivision, class, and order. Below the order are "lower taxa" called family, genus, and species.

Several systems illustrate the monophyletic versus polyphletic debate in regard to the Division level. Tippo (1942) place all vascular plants in the single division Tracheophyta, Bold (1977) separates them into five divisions, and Cronquist (1981) into two divisions (Table 3.1). In the monophyletic system, there is only one division, indicating that plants in the classes below it come from a common ancestor. In the polyphyletic system, the classes found in the monophyletic system are moved up to divisions, thus indicating more than one common ancestor. Concurrently, taxa in the subclass level in the monophyletic system are elevated to classes in the polyphyletic system. Cronquist's (1981) major treatise and plant classification system was an important contribution; it has been widely adopted and is followed in this book. Takhtajan (1966) also developed a system that parallels that of Cronquist.

A general feature of most modern systems is the natural separation of seed plants into two taxa: angiosperms and gymnosperms. The angiosperms are listed as a class (Angiospermopsida) in the monophyletic system and as a division (Anthophyta or Magnoliophyta) in the polyphyletic system. The gymnosperm group is listed as classes (Cycadopsida,

Table 3.1. A comparison of the monophyletic and polyphyletic systems of classification

Taxon	Monophyletic System (After Tippo 1942)	Polyphyletic System (Bold 1977)	Polyphyletic System (Cronquist 1981)
Division	Spermatophyta or Tracheophyta	Cycadophyta Ginkgophyta Coniferophyta Anthophyta	Pinophyta* Magnoliophyta
Class	Gymnospermae Angiospermae	Monocotyledonopsida Dicotyledonopsida	Liliopsida (Monocots) Magnoliopsida (Dicots)
Subclass	Monocotyledonidae Dicotyledonidae		Magnolidae Hamamelidae Caryophyllidae Dilleniidae Rosidae Asteridae

*Not considered by Cronquist.

Ginkgopsida, Coniferopsida) or one division (Pinophyta) in the latest polyphyletic system. In yet another classification system, the angiosperms and gymnosperms are placed together into the division Spermatophyta (Foster and Gifford 1959). Regardless of the level at which angiosperms and gymnosperms are placed, they represent a valid concept. We use the informal names "angiosperms" and "gymnosperms" in later chapters. The characteristics of these two groups appear in Table 3.2.

The Magnoliophyta is split into two classes: the Liliopsida (Monocotyledonidae, monocots) and Magnoliopsida (Dicotyledonidae, dicots). They are distinct as groups, although single characters occasionally are not restricted to only one taxon; usually several characters will place a plant in one of the two groups (Table 3.3). Most woody plants are members of the Dicotyledonidae; only a few woody plants such as palms and some vines (*Smilax*) are found in the Monocotyledonidae. There are no analogous classes for the gymnosperms.

The general concept of an order is that its members will have some general characteristics in common and have definite affinities and evolutionary trends. An order can also be defined as a group of related families.

LOWER TAXA

Fortunately, the monophyletic-versus-polyphyletic problem in classification does not affect the order, family, genus, and species levels. These taxa do not change much, regardless of the approach taken in classification at the higher levels, although, for other reasons, specific plant groups or individual species may be shifted from one order or family to another from time to time.

Family

The next level in the hierarchy, in descending order, is the family. Within each family there are usually general combinations of morphological characters (especially flowers and fruits) that are common to all members. For instance, all members of the birch family, the Betulaceae, have catkins, small winged seeds, and alternate, simple leaves. Although composed of 550 genera, all members of the pea family, the Fabaceae, have a similar fruit (a legume or pod). Some families are relatively large; an exceptionally large family of many genera such as the Fabaceae may be divided into subfamilies or into several families. Other families such as the Platanaceae and Ginkgoaceae are quite small and

Table 3.2. Morphological characteristics of angiosperms and gymnosperms

Angiosperms	Gymnosperms
Trees, shrubs, and herbs	Trees and woody shrubs
Leaf blade expanded	Leaf needle-like, linear, or fanlike.
Leaves deciduous or persistent	Leaves persistent, with a few exceptions
Wood composed of vessels and fibers	Wood composed mostly of tracheids
Flowers composed of anthers and pistils (sporophyll)	Flowers absent; pollen and ovule bearing structures (mega- and microsporangia)
Fruit present; seeds enclosed by the ovary of the pistil	Fruit absent; seed naked, borne exposed, often on the scale of a cone

contain only one genus. Worldwide, more than 75 families contain woody plants. A simple definition of a family is a group of naturally related genera.

Genus

The concept of a genus is one of the oldest known, as it is present in the history and language of nearly all societies. The idea of an oak, pine, spruce, or birch is relatively easy to grasp because these are natural groupings, and a genus may be defined as a group of naturally related species. The species within a valid genus have morphological similarities and often resemble one another so much that one species may be mistaken for another. There is also enough genetic similarity that hybrids can occur. Like a family, a genus may be large or small. When only one species occurs in a genus, it is called monotypic (e.g., *Ginkgo*); infrequently, a large genus may be subdivided into subgenera, for example *Quercus* into *Lepidobalanus* (white oak group; also called *Leucobalanus*) and *Erythrobalanus* (red or black oak group). Subsections of a genus have also been considered for regions. Nixon (1997) breaks the North America section of *Quercus* into subsections: *Lobatae* (red oak group), *Protobalanus* (intermediate or golden oak group), and *Quercus* (white oak group)

Species

A species is a natural population of actually or potentially interbreeding individuals that agree in essential morphological and ecological characters and produce fertile offspring. Implied here is that a species is reproductively isolated from other species, and that two individuals of a species are genetically similar but not identical. No two individuals in a freely interbreeding population can be identical. On the other hand, when genetic and morphological variation becomes too great, the population may be separated into two or more species.

A species remains a species for a variety of reasons that are collectively known as reproductive isolation. Usually, but not always, reproductive isolation implies no exchange of genetic material through pollen transfer. Two species may be geographically separated. Stace (1980) provides the example of *Larix decidua* (European larch) in Europe and *Larix kaempferi* (Japanese larch) in Japan. Hybrids naturally occur when these species are grown together in cultivation. Temporal or seasonal isolation occurs when either pollen release or pistil receptiveness occurs at different times of the year (e.g., *Acer rubrum* (red maple) flowers appear in the spring several weeks before those of *A. saccharum*). Many species of woody plants are anemopholous (wind pollinated), but some are entomopholous (insect pollinated); it is unlikely that insects would carry pollen to the wind-pollinated individuals, assuming the flowers appeared at the same time. Mechanical isolation may occur between two closely related species; if, for example, the pollen grain from one species produces a tube that is too short for the style in the flower of the other species, the ovule can not be fertilized.

The are a number of other mechanisms that can be broadly classed as genetic (gene or chromosome) incompatibility. This results in the development of

Table 3.3. General morphological characteristics of magnoliopsida (dicots) and liliopsida (monocots)

Class	No. of Cotyledons	Leaf Venation	Flower Parts	Plant Life Forms
Magnoliopsida	2	Netted	4 or 5 meris	Herbs, vines, shrubs, trees
Liliopsida	1	Parallel	3 meris	Grasses, palms, herbs, vines

nonviable seed or infertile offspring (either F_1 or F_2 hybrid). However, even if offspring are fertile, the habitat for a hybrid between ecologically different species may not be available. As land in eastern Canada was cutover or cleared and intermediate habitats were opened for establishment of hybrid seedlings, hybrids of *Picea mariana* (lowland black spruce) and *Picea rubens* (upland red spruce) began to appear.

A species population may be subdivided into infraspecific taxa: subspecies (ssp.), variety (var.), and forma (f.), in descending order of magnitude.

Subspecies

The subspecies taxon is a rank between species and variety and has been defined as a group of related varieties (Benson 1957). This taxon is not easily distinguished or separated from a variety, and thus these taxa are used more or less interchangeably (Porter 1967). Although the subspecies taxon is recognized in nonwoody plant species, only a few subspecies have been identified for woody plants.

Variety

A variety is defined as a group of individuals that have some morphological differences but are considered ecologically similar to the general species population. Because individuals vary genetically, it is expected that some variation will occur. If the variation is not excessive, these individuals are still considered part of the species population. However, when obvious inheritable differences are present, then the taxonomist must decide whether these differences constitute a variety or a new species. Such decisions are relatively common occurrences, which has led to the thought that a species, subspecies, or a variety is based largely on the perceptions of a competent taxonomist.

Forma

Forma is the lowest recognized category that may be identified by a scientific name. A small variation in tree or leaf form may occur from one population to another as a result of differences in site environment. There may be some question about whether a variation in form is an inheritable characteristic; some variations may be, other may not.

Ecotype

Although taxonomists have included the concept of a variety and other ranks below that of species in classification, they have only recently considered the problem of an ecotype. An ecotype is composed of individuals that are morphologically similar to the general population but are found in (adapted to) substantially different habitats. Research has shown that although a species may appear uniform, two individuals of the same population may respond differently to the same habitat. The classic example is *Thuja occidentalis* (northern white cedar) in Wisconsin. Typically, it grows on low, relatively wet sites, but it can also be found on mesic (moist) upland slopes. Research by Habeck (1958) showed that the offspring from either population did not grow well in the habitat of the other. Other species that appear on both dry and wet sites include *Pinus banksiana* (jack pine), *Pinus strobus* (eastern white pine), and *Acer rubrum.*

Moreover, as early as 1922, Turesson pointed out that species found over large geographic areas may be composed of ecotypes (Turesson 1922, 1925). The idea of an ecotype is inherent to the policy of not moving nursery-grown seedlings more than 160 km (100 mi) latitudinally. If they are to be moved, they should be moved south of their origin. Seedlings moved to the north remain active longer because they have adapted to a longer growing season. Therefore, they are susceptible to frost damage in late spring or early fall. Moving seedlings east or west of their origin apparently is not a problem. For example, *Pinus strobus* from the Appalachian Mountains grows exceptionally well in southern Illinois and Indiana.

CHAPTER 4

Nomenclature

INTRODUCTION

The term *nomenclature* means "the act of calling by name." The purpose of nomenclature is to provide all populations of organisms with names. During the past two centuries, taxonomists have developed a standardized system for naming plants. After the publication of *Species plantarum* in 1753, in which Linnaeus proposed assigning a Latin binomial to each plant, the first organized effort toward standardizing rules on nomenclature was made at the First International Botanical Congress held in Paris (1867). Since that time, the International Rules of Botanical Nomenclature have undergone revision at each congress, the last ones being held in Germany in 1988 and at St. Louis, Missouri, in 1999. These rules are lengthy, and only the salient points as they relate to native woody plants are considered here.

One important rule is called the Principle of Priority. Under this rule, each taxon can bare only one valid name. This name usually is based on the earliest date of publication, although there are some exceptions; certain names are retained over older names because they have been in common use for a long period. However, such names become fewer as taxa are renamed with each meeting of the Botanical Congress.

THE LATIN BINOMIAL

Because the classification and naming of plants and higher taxa are based on a hierarchy (Chapter 3), it is easier to understand nomenclature by beginning with the two scientific names (the Latin binomial: *bi* meaning "two"; *nomial* meaning "names") assigned to a plant species. The binomial consists of the generic name (usually a noun) and a specific epithet (usually an adjective); a name combination is unique and assigned to all the plants of a given species. To ensure that the binomial remains with one species, the name is permanently associated with a single plant specimen (type specimen or lectotype) stored in a herbarium (a plant museum containing pressed plants attached to sheets of special paper; see Chapter 5). As shown later in this chapter, a knowledge of name meanings is important because, in part, a name describes the plants that belong to a species.

Most of the names come from the Latin language (hence the term *Latin binomial*), but some are derived from the Greek. Unfortunately, names from foreign languages are not in common usage nor their meanings generally known, and, as a result, binomials tend not to be used in everyday communications. On the other hand, the common names so familiar to us often cause confusion because there may be many names for a single species or several species may have the same common name. For example, a common tree of the spruce-fir forest, *Populus tremuloides* Michaux, has many common or regional names: aspen, quaking aspen, trembling aspen, popple, Vancouver aspen, golden aspen, and mountain aspen. These common names make the binomial indispensable. Note that because binomials are taken from a foreign language, if they are not written in italics, they must be underlined.

Genus Name

The first name of a binomial, a noun, places a species within a genus. All species in a genus are assigned this name, which has been developed in a number of ways. It may be taken from the aboriginal or Latin common name): *Acer*—maple; *Quercus*—oak; *Fagus*—beech; *Betula*—birch; *Pinus*—pine; *Picea*—spruce; *Tsuga*—hemlock. The generic name may be descriptive: *Liriodendron*—lily or tulip tree; *Pseudotsuga*—false hemlock. Or the name may be in honor of a person, real or imaginary: *Magnolia*—Magnol, an early botanist; *Menziesia*—Menzie, an early botanist; *Nyssa*—nurse of Bacchus, Greek god of wine; *Juglans*, an abbreviation of "Jupiter's glans."

Species Name

The specific epithet, an adjective or modifier of the generic name, is assigned to only one species in a genus and usually refers to a particular attribute of that species. In the examples given in the following lists, the genus (first) name is included to show how the species name is used as a modifier.

The specific epithet may refer to a physical characteristic. A large number of species names were assigned on this basis.

Amelanchier alnifolia—serviceberrry with an alder-like leaf
Betula papyrifera—birch with paper white bark; paper or white birch
Carya tomentosa—hickory with long (tomentose) hair on leaves; mockernut hickory
Populus deltoides—poplar with deltoid (triangular) leaves; cottonwood
Quercus macrocarpa—oak with large acorns; bur oak
Quercus rubra—oak with red leaves; northern red oak
Picea rubens—spruce with red leaves; red spruce
Acer rubrum—maple with red leaves; red maple

> *Name Endings.* In the last three examples listed, the specific epithet is based on the root word *rub*, which means "red." Because Latin is an inflective language, the ending of the adjective must correspond to the gender of the noun; that is, masculine, feminine, or neuter. The result is three different endings.

The name may describe the habitat.

Pinus palustris—pine of swamps; longleaf pine
Quercus palustris—oak of swamps; pin oak
Carya aquatica—hickory of wetlands; water hickory

The name may refer to the geographic location where the species was first described.

Tsuga canadensis—hemlock of Canada; eastern hemlock
Carya texana—hickory of Texas; Texas or black hickory
Pinus virginiana—pine of Virginia; Virginia pine
Juglans californica—walnut of California; California walnut
Castanea ozarkensis—chestnut of the Ozark plateau; Ozark chinkapin

The name may refer to a person, often an early explorer or botanist.

Abies fraseri—Fraser fir
Pinus jefferyi—Jeffrey pine
Picea brewerana—Brewer spruce
Quercus shumardii—Shumard oak

> *Usage.* As these examples indicate, the generic name is always capitalized and precedes the specific epithet. It is usually spelled out completely but may be abbreviated when the meaning is understood. For example, after the first time a complete binomial is used in a journal article, it may abbreviated (e.g., *Quercus alba* becomes *Q. alba; Acer rubrum* becomes *A. rubrum*). The specific epithet is not capitalized, although some writers capitalize a personal name to indicate respect (e.g., *Abies Fraseri*—Fraser fir).

Authority Name

In technical books and articles, binomial names are accompanied by the name of the taxonomist who identified the species. It is written in complete or abbreviated form after the binomial; it is not italicized or underlined. When the rank of a taxon is changed or when a name is changed, the name of the original authority is placed in parentheses and the name of the taxonomist who initiated the change is placed outside the parentheses, as shown in the last two species in the following list.

Celtis occidentalis L.—hackberry named by Linnaeus
Quercus imbricaria Michx.—shingle oak named by Michaux
Carya ovata (Mill.) K. Koch—shagbark hickory, originally named by Miller
Picea glauca (Moench) Voss—white spruce, originally named by Moench

Synonyms. Name changes sometimes cause confusion and there may be some reluctance to accept a new name, or there may be some uncertainty as to priority. To have priority, a name must be the first name to have been effectively published. As a result of oversight regarding publication, synonyms may develop. For example:

Northern red oak	*Quercus borealis* Michaux f.
	Quercus rubra Linnaeus
Yellow birch	*Betula alleghaniensis* Britton
	Betula lutea Michaux f.
Poison ivy	*Rhus radicans* Linnaeus
	Toxicodendron radicans (Linnaeus) Kuntze
	Rhus toxicodendron (American author)

Variety Designation and Names

Frequently there may be two or more recognized populations within a species. If they are sufficiently distinct in morphology, variety names may be applied to these populations. The variety name for the type species is usually dropped because the designation is understood.

The scientific name becomes a trinomial when varieties and subspecies appear in a species population. The term "var." indicates a variety or a part of the species population that has characteristics that vary from those typical of the species. Fortunately, among woody plants there are substantially fewer varieties than among herbaceous (nonwoody) plants such as wildflowers. However, some taxonomists subdivide a species into varieties to reflect small variations in the stems, leaves, and fruits. The division of a species into varieties results in the scientific name becoming a trinomial. For example:

Pinus contorta var. *contorta*—shore pine or coastal lodgepole pine
Pinus contorta var. *latifolia*—Rocky Mountain lodgepole pine
Pinus contorta var. *murrayana*—Sierra lodgepole pine
Acer saccharum var. *saccharum*—sugar maple
Acer saccharum var. *schneckii*—Schneck's sugar maple

Quercus falcata var. *pagodaefolia* (cherrybark oak) was once considered a variety of *Quercus falcata* (southern red oak) because of essentially identical acorns even though its leaf structure was somewhat different. However, the former is found primarily on relatively wet or periodically flooded sites along streams, and the latter only on xeric upland sites. Possibly because of this habitat difference, cherrybark oak is now considered a separate species known as *Quercus pagoda*. Because of the difficulty in identifying true ecotypes and separating them from the natural tolerances and variations within a species, names have not been used to designate ecotypes.

Some taxonomists have divided species and varieties based on form (forma, Chapter 3). In some cases, recognizing variation in form also results in a trinomial or, sometimes, a quadrinomial: *Acer nigrum* f. *nigrum; Acer saccharum* var. *saccharum* f. *glaucum* (Steyermark 1963).

As noted in Chapter 3, few populations within a woody species have been identified as subspecies, but the designation "ssp." does frequently appear in regard to herbaceous plant populations.

Hybrid Designation

Hybrids between closely related species are common. Such hybrids are designated by placing an "x" between the genus name and the specific epithet:

Quercus x *fernowii* Trel.—*Quercus alba* x *Q. stellata*
Quercus x *bushii* Sarg.—*Quercus velutina* x *Q. marilandica*

Hybrids are not uncommon in nature. About 15 oak hybrids are recognized in the state of Missouri alone (Settergren and McDermott 1969). One reason for the appearance of hybrids over the past 100 years is the clearing of intermediate habitats for hybrids of two closely related species that may have substantially different site (environmental) requirements. For example, in the northeastern United States, hybrids of *Picea rubens* (red spruce) of high uplands and mountains and *Picea mariana* (black spruce) of swamps now are relatively common.

FAMILY NAME

A family name is formed by adding -aceae (pronounced a-c-e) to the root name of a genus within a family; for example: *Fagus*, Fagaceae (beech); *Salix*, Salicaceae (willow); *Betula*, Betulaceae (birch); *Ulmus*, Ulmaceae (elm); *Pinus*, Pinaceae (pine, spruce, fir).

SUBFAMILY NAME

A subfamily name is designated by the ending -oideae. There are only a few subfamilies found among the woody species of North America. For example, the pea family, Fabaceae (legumes; in older books listed as Leguminosae) is a relatively large family that has been split into three subfamilies based on differences in flower structure: Mimosoideae, Caesalpinoideae, and Papilionoideae. Some taxonomists have broken the Fabaceae into three separate families: the Mimosoideaceae, Caesalpinoideaceae, and Papilionoideaceae.

HIGHER TAXA

Higher in the classification hierarchy, only the name of the order is based on the genus name. The order name is usually formed by adding -ales to the root of a name of a family within the order. For example, *Fagus*, Fagaceae, Fagales (beech and oak) and *Salix*, Salicaceae, Salicales (willow and poplar). The exception to this rule occurs with the conifers; the general term "conifer" is used as the root for the order Coniferales (cone-bearing plants) in an older classification system. Another classification system uses the generic name *Pinus* (Pinaceae) and calls the order Pinales. Suborders may also be designated with the ending -ineae.

As noted in Chapter 3, the higher taxa have specific endings for each level in the hierarchy: kingdom, -ae; division, -phyta; class, -opsida; subclass, -idae. The names at these levels generally are not based on genus names.

SUMMARY

The nomenclatural system (following the classification system developed by Cronquist 1981) for flowering plants may be summarized as follows:

- Kingdom—Plantae, plants (conifers do not have true flowers)
  - Division—Magnoliophyta, flowering plants, seeds enclosed in an ovary
    - Class—Magoliopsida, seedling with two seed leaves, dicots
      - Subclass—Dilleniidae, leaves simple and entire, flowers polypetalous
        - Order—Malvales, sepals valvate, stamens united into a tube
          - Family—Tiliaceae, linden or basswood family
            - Genus—*Tilia*, basswood
              - Species—*americana*, American basswood
                - Variety—var. *heterophylla*, white basswood

The Cronquist system presently is widely accepted and used, thus we have followed it in this book. The subclasses, orders, families, genera, and species of woody plants are listed beginning in Chapter 9 for Pinophyta, Chapter 17 for Magnoliopsida, and in Chapter 24 for Liliopsida.

Cronquist (1981) did not consider conifers (gymnosperms) in his system. The generally accepted naming and classification of gymnosperms, using *Pinus echinata* (shortleaf pine) as an example, is as follows:

- Kingdom—Plantae, plants
  - Division—Pinophyta
    - Class—Pinopsida
      - Order—Pinales, Coniferales
        - Family—Pinaceae
          - Genus—*Pinus*
            - Species—*echinata*

CHAPTER 5

Collecting and Preserving Specimens

INTRODUCTION

MANY amateur and professional biologists, botanists, ecologists, and foresters collect plants for identification, for study, and to associate common and scientific names with plant specimens. For general and short-term use, specimens collected from trees or shrubs are usually put into a plastic bag and, either immediately or after a few days, placed between sheets of newspapers with a few books piled on top and pressed until dry. Sometimes they may be placed in a refrigerator for storage until pressed, but these specimens may become moldy and dark after a week or two. This casual handling of specimens may be acceptable for general short-term use but not for long-term preservation and research, as leaves can quickly become brown, brittle, and easily broken and/or become moldy and thus lose their important physical characteristics.

To preserve their essential characteristics, most types of plant material (except large cones) that are to be retained for long-term study have to be immediately and properly pressed, dried, and mounted on special paper. These near permanent herbarium specimens are stored in a plant museum called a herbarium.

Herbarium specimens are important. They are used in instruction in plant taxonomy, particularly for introducing students to family and genus characteristics, and in simple identification of unknown plant specimens collected in the field. Although most living woody plants are best studied during the growing season when leaves and, often, fruits are available, herbarium specimens with leaves, fruits, and buds can be studied at any time. Herbarium specimens also are important in taxonomic, evolutionary, and genetic research. They are used to verify or modify plant descriptions. The size, shape, hairiness, and other characteristics of leaves and twigs reported in taxonomy and dendrology texts and field guides have largely been obtained from herbarium specimens. Herbarium specimens document the occurrence of a species at a given time and place and thus are used in the development of national (e.g., Little 1971), regional (e.g., Petrides 1992), and state (e.g., Viereck and Little 1975) range and distribution maps. Herbarium specimens are used in the development of identification keys and to examine hybridization between two similar species or several species within a genus (e.g., *Quercus*, oak). Moreover, information on environmental relationships can be obtained from herbarium specimens if the collector has recorded the location information in sufficient detail. Protocol now requires that in addition to the name of the plant, the name of the collector, date, specific location of the collection and information on the habitat be included on the label. A field notebook is essential.

The world's largest herbaria include the Royal Botanical Gardens in Kew, England (6 million mounted specimens), Komarov Institute of Botany, St. Petersberg, Russia (6 million), New York Botanical Garden (5.5 million), Harvard Herbarium (5 million), Missouri Botanical Garden, St. Louis (4.8 million), Smithsonian Institution, Washington, D. C. (4.7 million), and the Illinois Field Museum,

Chicago (2.5 million). Much of the data from these herbaria are now available on the Internet. For example, the Missouri Botanical Garden has placed nearly 900,000 scientific names, more than 1,300,000 specimen records, and 75,000 citations in Internet data bases. Many universities have smaller herbaria (several hundred thousand specimens). Some herbarium specimens stored at these herbaria date back to the eighteenth century. Unfortunately, because curator position vacancies have not been filled in recent years, a few university herbaria are no longer maintained, and some universities have given away their collections.

FIGURE 5.1 A plant press of side panels, straps, corrugated cardboard, blotters, and newspapers.

FIELD EQUIPMENT AND NOTES

Basic field equipment for collecting woody plant specimens need not be extensive. The simplest equipment includes a field guide, a notebook, pruning shears, and one or two large (15 to 30 gallon) plastic trash bags. If seedling specimens are to be collected, a garden trowel is also a necessary tool. Clipped specimens (twigs with attached leaves, flowers, and/or fruit) should be put in the bag with a small amount of water to reduce evaporation loss and deterioration. It is a good procedure to collect two or three twig/leaf specimens of the same species unless they are rare or endangered. If a flower or the inflorescence is present on the same twig as the leaves or on a separate twig, it should be collected as well. Bark specimens may also be collected if representative samples can be obtained. Bark and dry woody cones may be placed in a bag without water. The bag should be kept closed and fastened except when adding specimens. Some field biologists carry a plant press and immediately place specimens in it to prevent wilting and keep shape and color loss to a minimum. A plant press consists of a combination of newspapers, blotters, and corrugated cardboard sandwiched between either wood or metal endplates (22 cm × 42 cm); two straps around the entire assemblage are tightened to press the plants (Figure 5.1). However, using a plant press in the field usually takes time and is less efficient than putting the specimens in a bag and later pressing them in the laboratory. Bagged specimens may be put in a cooler overnight without loss of tissue or moisture; they can then be placed in a plant press the next day.

If a specimen cannot be identified in the field, it should be given a number, with the same number used both in the notebook and on a tag attached to the specimen; masking tape wrapped around the stem usually makes a adequate tag, or standard tags may be tied to the stems. For each specimen, the species' common and scientific names, if known, date of collection, name of collector, specific location of collection (country, state, county, road, distance from a landmark), unusual morphological characteristics, and environmental conditions must be recorded. Environmental conditions include type of community (forest, woodland, prairie, old field, roadside, associated species, etc.), soil (dry, moist, wet, rocky, periodically flooded), topography (elevation, slope, aspect, steep and hilly, gently rolling, etc.), and other information that may be important.

HERBARIUM PROCEDURES

Regardless of whether specimens are pressed in the field or in the laboratory, the procedures for pressing and the combination of newspaper, blotters, and corrugated cardboard in the press remain nearly the same. The twig, leaves, flowers, and/or fruit are placed between the pages of a folded newspaper sheet. At this time it may be necessary to trim the specimen, as an excessive amount of plant material on a single herbarium sheet usually does not add to its quality. Unlike herbaceous species, for which the entire plant (root, shoot, flowers, fruit) is typically collected and attached to a blank herbarium sheet, only a relatively small specimen of an entire woody plant can be fitted on the sheet.

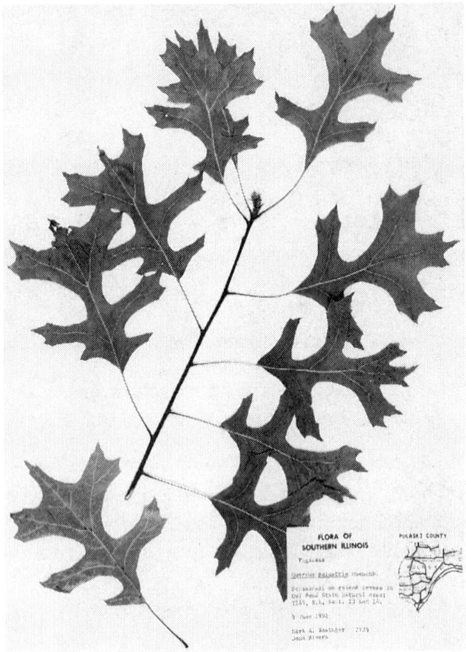

FIGURE 5.2 Herbarium sheet displaying a *Quercus palustris* (pin oak) specimen.

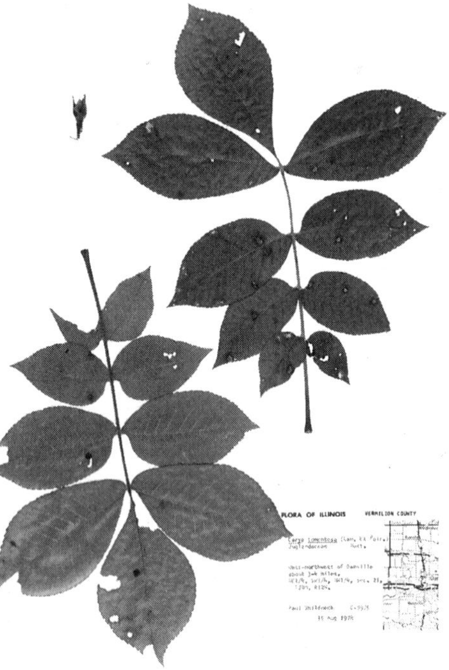

FIGURE 5.3 Herbarium sheet containing only leaves of *Carya tomentosa* (mockernut hickory).

A herbarium sheet will easily accommodate a maximum twig length of about 30 cm, depending somewhat on leaf size. Usually a 20 to 30 cm long section of twig with 8 to 20 average-size (5 to 15 cm long) leaves is an adequate specimen (Figure 5.2). Care should be taken to arrange the specimen so that the leaves are flat and spread out; most leaves should have their upper surfaces exposed with some leaves turned to show the lower surface. Leaves and flowers must be turned at this time because after they are dry, they cannot be turned without breaking the petiole, peduncle, or petals. If the leaves are extremely large, only one or two may be retained. The leaf of *Carya tomentose* (mockernut hickory) is usually 30 or 40 cm long and so large that only two will fit (Figure 5.3); in this case, one should be mounted with the upper surface visible and one with the lower surface visible. The bipinnately compound leaf of *Gymnocladus dioica* (Kentucky coffee tree) may be nearly 110 cm in length and must be carefully folded to fit the sheet.

If an inflorescence is large or on a separate twig, it may be mounted on a separate sheet. Flowers are often pressed as they are, but a small number may be turned or sectioned to expose the stamens and pistil. Most small fruits may be placed between newspaper sheets along with the leaves and twigs. Small (length <10 cm) green cones may be sectioned in half lengthwise for mounting on a herbarium sheet with a twig and leaves. Large fruits and cones that are already generally dry may be allowed to air dry or can be dried in a dryer. Because they are too bulky to press, they may be stored in a container or drawer. Bark may be permitted to air dry.

After the specimen is arranged in the newspaper, a blotter sheet or two is placed on each side of the newspaper, and a sheet of corrugated cardboard is placed outside each blotter. The newspaper is used to protect the blotter so it can be reused later and to protect the specimen after it is dried but before it is mounted on the blank herbarium sheet. The blotter is used to absorb moisture and plant sap. The space in corrugated cardboard permits the movement of dry air into and moisture out of the press so that the specimens can dry evenly and quickly. The press must be placed in a dryer. A commercial dryer consists of a cabinet heated at the bottom by several light bulbs (not more than 100 watts total). Air holes at the top and bottom permit convection flow to remove the moisture and prevent overheating. In an emergency, such as might occur in remote locations where a dryer is not available, a cloth sheet draped over the backs of chairs placed several feet apart will create a temporary drying chamber. A desk lamp on the floor under the sheet may be used to dry plant materials. After drying, the specimen is removed from the press and mounted on a 22 cm × 42 cm herbarium sheet of acid-free paper.

Specimens are attached to the herbarium sheet by one of several different methods. A common method is to utilize an adhesive or glue designed for long-term use, but the problem here is that the method is irreversible; once the glue is attached to plant tissue it cannot be separated from it. Small strips of a gummed linen tape (0.5 to 1.0 cm × 5 cm) moistened only at the ends is another preferred method. A third suitable method is to stitch specimens to sheets using linen thread or waxed dental floss, which is strong and easily tied in knots. Sectioned cones and bark samples are usually placed on the same sheet as the twig and leaves. More often than not, an envelope is attached to the herbarium sheet so that parts of the specimen that become loose over time may be retained.

Special care must be taken to preserve the twigs/leaves of *Picea* and *Tsuga* and cones of *Abies* and *Taxodium*. The leaves of *Picea* and *Tsuga* drop from the twig upon drying (*Picea*, immediately and completely; *Tsuga*, a little later and a few at a time). Therefore, the leaves must be collected and saved in an envelope when a dried specimen is removed from the plant press. The cones of *Abies* and *Taxodium* fracture (disintegrate) when mature and dry. *Abies* cones seldom reach the ground intact and must usually be collected from the twigs before reaching full maturity. To keep them intact, the cone can be longitudinally wrapped with narrow linen strips or fine wire, or placed in a glass jar containing a preservative. *Taxodium* cones will slowly

FLORA OF SOUTHERN ILLINOIS

Fagaceae

Quercus palustris Muenchh.

Occasional on raised levees in Owl Pond State Natural Area; T14S, R1E, Sect. 13 and 14.

9 June 1992

Mark A. Basinger 2725
John Rivera

PULASKI COUNTY

FIGURE 5.4 The information tag attached to a herbarium sheet.

fracture as they dry, particularly if they are used for instruction and receive a moderate amount of handling. These cones are small and difficult to maintain.

Along with the specimen, a 10 × 15 cm tag placed in the lower right-hand corner of the herbarium sheet should include the name of the specimen, date of collection, state and specific location where it was collected, a small map, information on site environment, and name and number of the collector (Figure 5.4).

Once mounted, herbarium specimens must be protected from insects that feed on plant tissue. Storage is usually in an airtight cabinet that has numerous compartments. A few mothballs (paradichlorobenzene) placed in each cabinet is generally sufficient protection. However, in large herbaria (*e.g.*, Missouri Botanical Garden; University of Florida, Gainesville), where adequate storage for large numbers of sheets may be a problem, compactors are used. These compactors are doorless cabinets approximately 5 to 10 m long, 3 to 4 m tall, set on rails so they can be moved to fit snugly together to eliminate aisle space. Here, chemicals cannot be used, because accessing the sheets would permit large amounts of insecticide to escape to various work areas of the building. To prevent infestations, new specimens and herbarium sheets and other dry specimens that are on loan are frozen before being returned to the compactor.

The arrangement of herbarium sheets depends on the purpose of the herbarium. In a university herbarium, which generally has a regional or state concentration, the sheets are grouped into a single family location, with the genus and species alphabetically arranged and specimens from the various regions or states placed in separate folders. In a large herbarium, any number of sheets from a perticular state (e.g., where the herbarium is located) may be found in a single folder and the remaining sheets from various states, regions (e.g., Central America), or countries (e.g., North America, Mexico) are clustered in one or more folders.

Duplicate sheets of a species may be exchanged for species from a herbarium in another region. Sheet exchange often permits a small herbarium to have a relatively complete collection from large sections of its own country or from other countries.

CHAPTER 6

Key Construction and Use

INTRODUCTION

A common method for quickly and accurately identifying an unknown plant is to use a key. A key is an analytical construct used to determine the name of an unknown plant. Of course, the unknown plant can be compared with herbarium specimens, but rather than comparing the unknown with a herbarium specimen, which may not be available and certainly cannot be carried into the field, identification may be made by using a key.

An "artificial" key consists of a series of statements systematically describing the physical characteristics of known plants and, in effect, summarizes the major physical identifying characteristics for plants included in it. It is constructed from a successive series of two opposing statements (a couplet). Such a key is called a dichotomous (two choices) or forked key and is based on the principle of elimination. Some keys with three or more alternatives have been published, but keys of this type usually are more difficult to use because the unknown plant must be compared with each statement. Another type of key, not discussed here, is the "natural" key by which the taxonomist attempts to outline evolutionary relationships between taxa.

FORMAT

Identification keys are constructed along one of two general forms. In the indented form (Table 6.1), couplets are progressively indented, with the each succeeding couplet listed directly below the alternative that fits the characteristics of the unknown plant. The advantage of this format is that it isolates groups of plants that have common characteristics and the characteristics of the plant are clustered within a few lines of text. Moreover, when keying out an unknown specimen, it is easier to backtrack in the indented key when an obvious error has been made. The disadvantage is that if the key is long, the couplets will shift progressively toward the right, creating a wider left margin of wasted space. And if a large number of species are included in the key, the two opposing statements of a couplet are often widely separated by other couplets.

The couplets used in Figure 6.1 have been rearranged into the bracketed key shown in Table 6.2. In the bracketed key the alternatives of the couplet are never split; that is, the two opposing statements remain next to each other for easy comparison. A number at the end of each alternative will direct the user to the next set of alternatives or couplet. The disadvantage to this key is that it is difficult to backtrack to correct an error. Unlike the arrangement of the indented key, only the even-numbered couplets are indented several spaces from the left margin, and thus in long keys, loss of space is not a problem. Recently, some authors have elected to start all couplets at the left margin (e.g., Mohlenbrock 1986).

CONSTRUCTION

Construction of a key can be a difficult and time-consuming task. When a large number of species are

Table 6.1. Indented key to seven *Acer* species of the eastern United States

1. Leaf trifoliate or pinnately compound with leaflets. *A. negundo*
1. Leaf simple
  2. Margin smooth except for large, irregularly spaced teeth.
    3. Petiole and adaxial surface heavily pubescent, lobes and large teeth reduced and rounded . . . . *A. nigrum*
    3. Petiole glabrous; adaxial surface glabrous or with scattered hair; lobes and teeth usually sharply pointed. . . . . *A. saccharum*
  2. Margin serrated (about 10 teeth per 2 cm length)
    4. Margin ciliated. . . . . *A. spicatum*
    4. Margin aciliate
      5. Sinus depth more than half the distance to midrib; sinus narrow, hairpin shaped; margin 1 to 2 teeth/cm . . . . *A. saccharinum*
      5. Sinus depth less than half the distance to midrib; sinuses open and broad or cleft
        6. Margin toothed and intermediately serrated, with 2 to 4 teeth/cm; lobes 3 to 5; apices acute. *A. rubrum*
        6. Margin finely serrated, with 6 to 8 teeth/cm; lobes 3; apices long acuminate. . . . . *A. pensylvanicum*

included, the key is frequently broken into several smaller keys or a series. For example, the first key may lead to the certain genus, with a subsequent key used to identify the species within the genus. However, regardless of the number of taxa to be included, the general approach is to divide all the species into two groups (corresponding to the two alternatives in a couplet), and each of these groups must be split into two groups, and so forth. This approach is illustrated in the dendrogram of Figure 6.1. The first division should be based on a major characteristic that creates two groups, with subsequent divisions generally relying on progressively less noticeable but no less important characteristics (Tables 6.1 and 6.2). However, there may be a situation in which one species is so distinct from the others in a genus that it can be directly identified in the first alternative of the first couplet, with all other species having characteristics that place them within the second alternative.

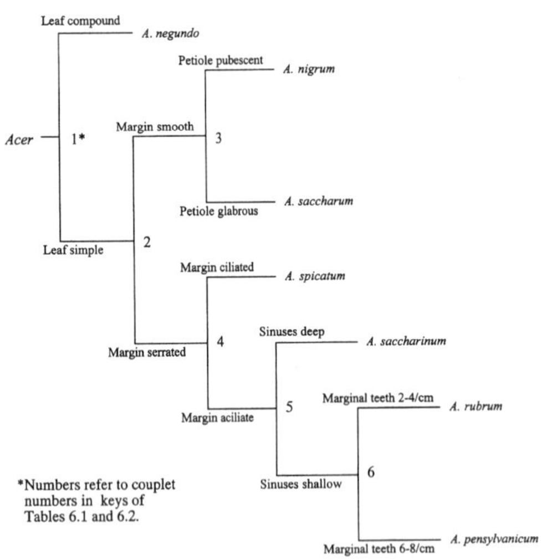

FIGURE 6.1 Dendrogram illustrating the dichotomous keys found in Tables 6.1 and 6.2.

There are various errors that may be committed in attempting to construct a key. The following list of suggestions should eliminate most errors and ambiguities.

- Develop a clear, concise title; e.g., "Leaf Key to Woody Angiosperms of North America."
- Provide a list of species included in the key.
- Include only two (2), not one (1) or three (3) or more, alternatives in a couplet. As noted earlier, the two groups of species created by the alternatives need not be of equal size. For example, in Figure 6.1, the use

Table 6.2. Bracketed key to seven *Acer* species of the eastern United States

1. Leaf trifoliate or pinnately compound with 5 leaflets.	*A. negundo*
1. Leaf simple.	2
2. Margin smooth except for large, irregularly spaced teeth.	3
2. Margin serrated; about 10 teeth/2 cm length.	4
3. Petiole glabrous; leaf adaxial surface glabrous or with scattered hair; lobes and teeth usually sharply pointed.	*A. saccharum*
3. Petiole and adaxial surface heavily pubescent; lobes and large, usually bluntly pointed teeth.	*A. nigrum*
4. Margin ciliated.	*A. spicatum*
4. Margin aciliate.	5
5. Sinus depth more than half the distance to midrib; sinus narrow, hairpin shaped; margin 1 to 2 teeth/cm.	*A. saccharinum*
5. Sinus depth less than half the distance to midrib; sinuses open or cleft.	6
6. Margin toothed and intermediately serrated with 2 to 4 teeth/cm; lobes 3 to 5; apex acute.	*A. rubrum*
6. Margin finely serrated with 6–8 teeth/cm; lobes 3; apices long acuminate.	*A. pensylvanicum*

of leaf complexity splits the seven *Acer* species into groups of 1 and 6, but use of another characteristic might have split *Acer* into groups of 2 and 5 or 3 and 4.

- Use only correct technical terminology.
- Use quantitative measurements where possible. "Leaf length long" vs. "leaf length short "conveys no information and is not usable for opposing alternatives. Quantitative measurements include length, width, length to width ratio, etc. Usually, a range is necessary (i.e., "leaf length 5 to 10 cm"). Use only the metric system. The next item indicates the type of information required.
- In each couplet, the alternative descriptions must be set in opposition. For example, the statements "Leaf margin serrate or dentate" vs. "Leaf margin smooth or wavy," or "Leaf length 3 to 6 cm" vs. "Leaf length 8 to 12 cm" are distinct alternatives as they do not overlap. However, the combination of "Leaf margin serrate or dentate" vs. "Leaf length 3 to 6 cm" creates an unworkable situation. Likewise, "Leaf length 3 to 8 cm" vs. "Leaf length 6 to 12 cm" breaks the key if leaf length of the unknown species is 7 cm.
- Avoid the use of negatives when stating qualitative characters, such as in "Leaf margin not serrate or dentate." Use a description that indicates the characteristic present: "Leaf margin smooth or crenate."
- Arrange each statement so that the structure name is placed before the modifier, as shown in the preceding couplet and keys. For example, attention is drawn to the leaf margin before the characteristic is provided.
- Check for continuity or "dead ends" when developing a bracketed key. Be certain that there is a number that leads to every couplet. Except for the first couplet (No. 1) where the key begins, every couplet number on the left side of a bracketed key should be preceded by one on the right-hand side. A dendrogram would help eliminate the problem.
- There should be approximately the same number of couplets in a key as there are numbers of species identified in the key. A much larger number of couplets usually means that the key may not function well, if at all, and that the key certainly contains some unnecessary material or "deadwood."
- Do not overload the statements of a couplet with numerous trivial characteristics. Within reason, a simple key is the best key. Occasionally, it may be necessary to add one or two characteristics (clinchers), particularly when two species are difficult to separate.
- If a leaf key is being developed, then use only leaf characteristics; however, if a leaf is present, a bud is usually also present and twig characteristics can be

mixed with leaf characteristics for summer keys. The most frustrating type of key to use (because most of the time it is not usable) is one, for example, that mixes leaf characteristics with fruit or flower characteristics. The problem is that for woody species and for many woodland herbaceous species, the flower and fruit are present for only an extremely short period of time. Reproductive structures are absent about 80 percent of the time during the summer months.

USING A KEY

The approach to using a key is the same, regardless of the format or number of alternatives. Assuming that the genus of an unknown specimen has been identified as *Acer*, its characteristics would be compared with the statements in the first couplet in Table 6.1 or Table 6.2. If the specimen has a simple leaf, the second alternative would be the correct one and lead to couplet No. 2. If the specimen characteristics match the first alternative of couplet No. 2, then couplet No. 3 should be consulted. If the specimen characteristics match the second alternative, then one should proceed to couplet No. 4. This procedure continues until a species name appears at the end of the statement describing the unknown.

When using a key it is necessary to:

- Know the technical terms used in the key. If a term is new to the reader, then a glossary should be consulted.
- Read completely all alternatives and all statements in a couplet. Some keys are poorly constructed and/or several species difficult to separate; thus although the first alternative appears to describe the unknown specimen, the second statement may contain a better description.
- If possible, use several examples of the unknown specimen to observe the range of variability.
- Use a ruler if measurements are given.
- Use a 10× to 20× magnifier to observe minute characteristics.

PART II

Gymnosperms (Division Pinophyta)

CHAPTER 7

The Gymnosperms : An Overview

INTRODUCTION

GYMNOSPERMS are woody seed plants that have neither true flowers nor fruits. Rather, the reproductive structure of most gymnosperms is a cone composed of fibrous or woody scales tightly arranged on a central stalk. The seeds are not enclosed and protected in an ovary as in angiosperms, and it is from this condition that the name "gymnosperm" is derived (Greek *gymnos* = naked; *sperma* = seed). Literally interpreted, the seed is born "naked" on a scale; hence, the name "conifer" or "cone-bearing."

As with the angiosperms, general names applied to gymnosperms often are misleading. The term *conifer* refers to species that bear true cones. Included here are members of the pine family, Pinaceae (e.g., *Pinus*, pine; *Abies*, fir; *Picea*, spruce, and *Larix*, larch), and members of the redwood, cedar, and cypress family, Cupressaceae (e.g., *Taxodium*, bald cypress; *Juniperus*, juniper; *Thuja*, white-cedar or arborvitae; *Cupressus*, cypress). However, not all gymnosperms have true cones. In *Ginkgo* (ginkgo) and *Taxus* (yew), each seed is individually attached to a peduncle and surrounded by a fleshy or gelatinous integument; no scales or cones are present.

Most species are evergreen, but *Larix* and *Taxodium*, as well as *Ginkgo*, are deciduous and lose their leaves during the autumn season. However, even in evergreen species, there is a limited annual leaf fall. Depending on the species, leaves that are three to eight years old terminate photosynthetic activity and drop from the tree. *Pinus strobus* (eastern white pine) is pronounced in this regard, as the oldest leaves simultaneously turn a distinct yellow in September or October. The event usually generates a number of inquiries from landowners who believe their trees are dying.

Gymnosperms are known to have needle-like leaves. However, only *Pinus*, *Picea*, and *Larix* have the long, round, or square leaves that could be called "needle-like." *Abies*, *Tsuga* (hemlock), *Pseudotsuga* (Douglas-fir), *Taxodium*, and *Taxus* have narrow but flattened linear leaves. Even more extreme is the fan-like leaf of *Ginkgo* and the small awl-shaped (juvenile) and scale-like (adult) leaves of *Juniperus*.

The gymnosperms also are known as "softwoods." The term probably developed from early loggers who found that pine, spruce, and fir trees were relatively easy to cut down. There are exceptions here also, as the wood of "hard pines" (e.g., *Pinus palustris*, longleaf pine; *Tsuga canadensis*, eastern hemlock) is harder and stronger than that of many angiosperms. However, the wood of many gymnosperms (e.g., *Pinus monticola*, western white pine; *Picea*; and *Abies*) is relatively soft; that is, as compared with the wood of most angiosperms, it is easier to cut, split, or carve.

Because these general names do not precisely fit all the species in the Pinophyta, Arnold (1948) and Sporne (1965) believe that the term *gymnosperm* itself has outlived any usefulness and should be abandoned. However, in *Flora of North America: Volume 2, Pteridophytes and Gymnosperms* (Flora of North America Editorial Committee 1993b), the term is retained. It is also used here as a broad term that encompasses the five North American families whose members have naked seeds.

EVOLUTIONARY HISTORY

As a group, the gymnosperms are evolutionarily old. The present flora is the remnant of a once large and diverse group that flourished during the Mesozoic Era. The first fossils of extinct species appeared during the Middle Devonian Period of the Paleozoic Era some 375 million years before present (B.P.). According to Sporne (1974), at least four orders of gymnosperms are extinct. Coniferales (Cupressaceae) and Taxales, which contain most of the modern North American genera and species, first appeared during the Jurassic Period of the Mesozoic Era about 200 million years B.P. (Mirov and Hasbrouck 1976). However, it was not until the Cretaceous Period (136 to 65 millions years B.P.) that pine fossils became common. Most taxonomists consider gymnosperms to be more primitive then angiosperms.

CLASSIFICATION

At present, there is a relatively small number of gymnosperm taxa as compared with the number of angiosperm taxa. Worldwide, there are about 44 genera and 370 species, nearly all of which are found in the Northern Hemisphere. Most are restricted to the temperate zone, but a few are subtropical in distribution. In the North American flora, there are 5 families, 19 genera, and 100 to 105 species. Several *Pinus* and *Larix* species have been introduced for timber production and for horicultural use. *Ginkgo biloba, Metasequoia glyptostroboides* (dawn redwood), and *Chamaecyparis pisifera* (sawara-cypress) are introduced ornamental trees.

The division of gymnosperms into orders, families, and genera is considerably less complicated than that of the angiosperms (Magnoliophyta). In North America, the division Pinophyta contains 4 orders, 5 families, and 18 genera. Still, there continues to be some debate over classification. The most recent modifications have been the combining of the Taxodiaceae (redwood family) with the Cupressaceae (cypress family), the reduction of a number of named *Zamia* to one North American species, and some splitting/combining of western *Pinus* and *Abies* species. The following taxa and their respective species are described and illustrated in subsequent chapters.

Division : Pinophyta
- Order: Pinales (Coniferales)
  - Family: Pinaceae (Pine)
    - Genera: *Pinus* (pine), *Larix* (larch), *Picea* (spruce), *Abies* (fir), *Tsuga* (hemlock), *Pseudotsuga* (Douglas-fir)
  - Family: Cupressaceae (cypress family)
    - Genera: *Cupressus* (cypress), *Juniperus* (juniper), *Calocedrus* (incense-cedar), *Thuja* (arborvitae), *Chamaecyparis* (white-cedar), *Sequoia* (redwood), *Sequoiadendron* (giant sequoia), *Taxodium* (bald-cypress), *Metasequoia* (dawn redwood)
- Order: Taxales
  - Family: Taxaceae (yew family)
    - Genus: *Taxus* (yew)
- Order: Ginkgoales
  - Family: Ginkgoaceae (ginkgo family)
    - Genus: *Ginkgo* (ginkgo)
- Order: Cycadales
  - Family: Zamiaceae (Cycadaceae; cycad family)
    - Genus: *Zamia*

Additional important gymnosperm taxa are found in other parts of the world, as well as in southwest North America. In the order Pinales, the Araucariaceae (araucaria) family of 2 genera and about 34 species are found in New Caledonia and the temperate zone of South America and Australia. The Podocarpaceae (podocarp) family of 13 genera and about 140 species is distributed through the temperate and tropical regions of the Southern Hemisphere. The order Gnetales includes the pantropical family Gnetaceae (1 genus, *Gnetum*; 40 species), the Welwitchiaceae family, whose members are found in the deserts of South Africa (1 genus, 1 species), and the Ephraceae family found in the deserts of North and South America (1 genus, Ephedra; 42 species) (Woodland 1991).

REPRODUCTIVE MORPHOLOGY

Nearly all gymnosperms are monoecious, with both male and female cones produced on the same individual. In many gymnosperms, most notably the Pinaceae, the male structure (Figure 7.1a) does not

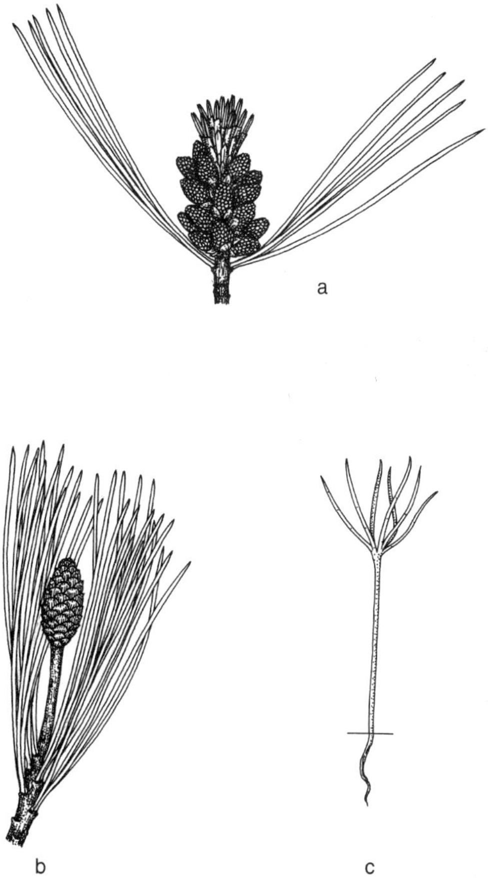

FIGURE 7.1 *Pinus strobus*: (a) twig apex with mature leaves, male strobili, and developing new leaves; (b) twig with mature leaves and immature seed cone; (c) seedling with numerous cotyledons.

resemble a cone, but when expanded, appears more like a catkin or ament found in several angiosperm families (e.g., Betulaceae, birch; Salicaceae, willow and poplar; Fagaceae, beech and oak). This 2 to 3 cm long structure, called a strobilis (plural strobili), is defined as a cluster of unisexual sporophylls (pollen-producing leaves) attached to a central axis (stalk). Each sporophyll has attached microsporangia that produce many thousands of pollen cells. Literally interpreted, a sporophyll is a spore-producing leaf or leaf-like structure.

The strobilis of megasporophylls (ovule-producing leaves) bearing megasporangia with accompanying bracts eventually develops into a typical woody cone. The megasporophylls become ovuliferous scales, usually on which are borne one or two seeds. The cone may be borne at or near the tip of the twig (Figure 7.1b) or along its axis and mature in 1 to 3 years. Fertilization is by nonmotile pollen grains, except in *Ginkgo*, which also is dioecious (male and female cones borne on different trees).

VEGETATIVE MORPHOLOGY AND LIFE FORM

Vegetative structures include the main stem, cotyledons, branches, twigs, buds, and leaves above ground and roots below ground. In contrast to woody angiosperms, which are subdivided into monocots (seedlings with 1 cotyledon) and dicots (seedlings with 2 cotyledons), some gymnosperm species may have as many as 20 cotyledons (Figure 7.1c). The adult leaf form may be acicular, linear, or scale-like, except in *Ginkgo*, which has a fan-shaped leaf, and *Zamia integrifolia* (Florida arrowroot), which has a fern-like compound leaf. Resin ducts are found in the wood and leaves of most species. At maturity, gymnosperms may have a number of life forms.

Subterranean

In subterranean gymosperms most of the vertical woody stem is below ground, but the apex is exposed and may extend up to 30 cm above the soil surface. Superficially, the leaf resembles a fern or that of a palm. In North America, only *Zamia* has this form (Chapter 12).

Trailing (Prostrate, Creeping)

In North America, only *Juniperus horizontalis* (creeping juniper) has a trailing life form. The main stem of this woody plant rests on and grows along the soil surface. Lateral twigs may extend upward as much as 20 cm (Figure 11.5). A true crown is absent.

Shrub

A shrub is a woody plant composed of several to many stems; maximum height is only about 2 m. Branches

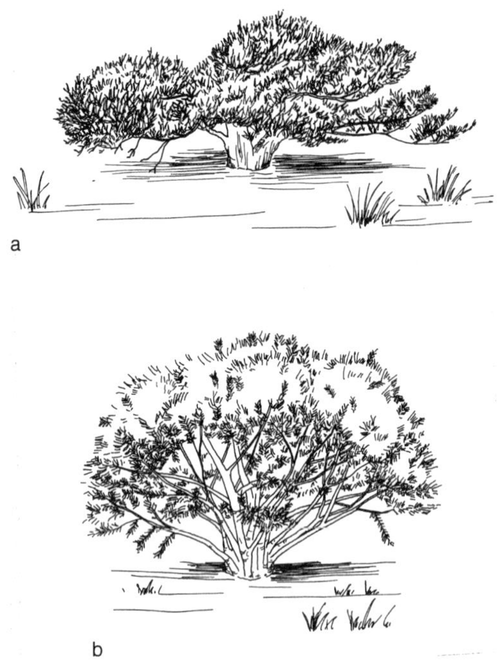

FIGURE 7.2 Shrub forms: (a) low-spreading *Juniperus communis* (oldfield juniper); (b) upright form of *Pinus mugo* (mugo pine).

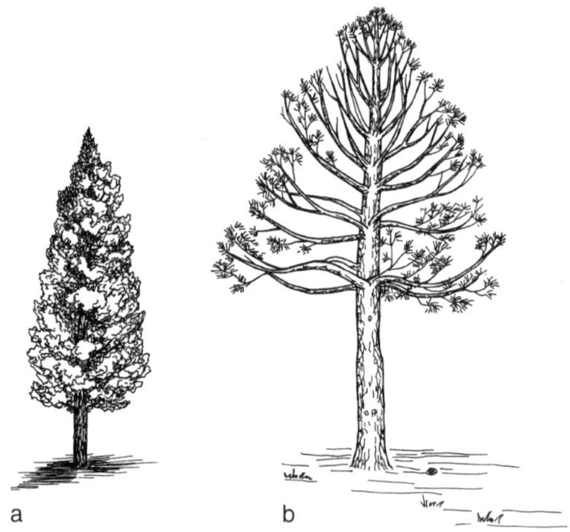

FIGURE 7.3 Tree forms: (a) arborescent form of *Juniperus virginiana* (eastern redcedar); (b) arboreous (tree) form of *Pinus resinosa* (red pine).

are more or less horizontal in the spreading form (Figure 7.2a) or somewhat horizontal in the upright form (Figure 7.2b). The crown may be poorly defined. Most shrubby species are found in the Cupressaceae (junipers) and Taxaceae (yew). In the Pinaceae (pine), only the introduced *Pinus mugo* (Swiss mountain pine) has this genetically programmed form.

Arborescent (Tree-like)

An arborescent gymnosperm is a woody plant composed of one to several short stems and usually a well-developed crown (Figure 7.3a). It may reach a height of 5 to 10 m at maturity. Species with this life form usually are members of the Cupressaceae.

Arboreous (Tree)

An arboreous gymosperm is a woody plant composed of a single main bole and a well-defined crown (Figure 7.3b). The height at maturity is not less than about 10 m and may be as great as 80–100 m in a few species. The main stem of most coniferous trees is typically a straight central bole that extends without forking to the apex of the tree; this condition is called excurrent. Deviations from the excurrent form occur when insects, disease, or frost kill the central leader and stop the flow of auxin to the closest side branches. These branches are released to grow into one or more new leaders. Unfortunately, several leaders may develop simultaneously, resulting in a multi-stemmed form and/or a crooked stem. Deformed trees are important for wildlife cover and erosion control but have lost much of their value for timber or wood production.

The majority of large gymnosperm trees have a crown that resembles an inverted cone. Typically, side branches extend out at right angles to the main stem and become progressively younger and thus shorter with height. The inverted cone structure permits the leaves of lower branches to receive nearly as much light as the leaves of upper branches.

Special Forms

Two unusual forms develop as a result of extreme environmental conditions. One form, called "krummholz" (meaning "dwarf forest") develops near "tree line," the upper limit of tree distribution in mountainous regions or at the northern (latitudinal) limit. Here, cold temperatures limit growth and ice crystals blown along the ground or snow surface shear off leaves, buds, and twigs on the windward side of stems, or at least remove the cutin layer and cause winter desiccation of leaves. Growth is severely restricted, except on the lee side. The result is a shrubby or near prostrate stem that is often less than 1 m in height and has a definite orientation with the direction of the wind ("wind-trained"). This form usually occurs in members of the genus *Abies.* At lower elevations where climatic conditions are less harsh, the same species may be represented by trees 20 to 30 m in height.

The second form is a deformed shrubby or arboreous form that develops, for example, on the low-nutrient sand soil of the pine plains within the New Jersey pine barrens area (Gibson et al. 1999). Here, tree species such as *Pinus rigida* (pitch pine), *Quercus marilandica* (blackjack oak), and *Quercus ilicifolia* (bear oak) seldom grow to a height of more than 3 m.

WOOD COMPOSITION

The softness and often low strength (as compared with hardwoods) of most conifer wood is due to cell structure. Gymnosperm wood (xylem) is relatively simple and composed primarily of tracheids, although some fiber-tracheids may occur, which have (1) large pits like angiosperm vessels to facilitate water and nutrient transport and (2) a thickened secondary wall of structure similar to fibers, which provide strength. The tracheid is a relatively long cell that handles the function of two specialized cell types in angiosperms. Conifers also have resin canals interspersed throughout the xylem; these canals are not cells but spaces formed after the development of the stem (Kramer and Kozlowski 1979). Fibers are absent in *Pinus* phloem but are consistently present in Cupressaceae and Taxaceae (Esau 1977). Because of the relatively great length of tracheids and fibers, conifers such as spruce and pine are the primary woods used to produce paper.

RANGE AND ECOLOGY

North American forests composed primarily of conifers tend to grow at nutrient, moisture, and temperature extremes. They are found on low-nutrient and/or droughty soils, in extremely wet habitats, in semiarid regions, and at high (cold) latitudes and altitudes. *Pinus* is the major component of upland forest on the droughty, sandy, infertile soil of the Gulf and Atlantic Coastal regions and of the Lake States and New England regions. *Picea* and *Pinus* dominate the sandy soils of the Canadian boreal (northern latitude) forest, as well as the higher altitude forests of the Appalachian, Rocky Mountain, and west coast (Sierra Nevada and Cascade) mountain regions. *Pinus* and *Juniperus* are the major species in the semiarid Southwest and Intermountain Plateau regions. *Taxodium distichum* is one of the major dominant species of southern swamps, and *Thuja, Abies,* and *Picea* grow in northern swamps and bogs. In contrast, the only naturally occurring gymnosperm over most of the central hardwood region is *Juniperus virginiana* (eastern redcedar), which has limited commercial importance. This species is generally associated with rock outcrops (barrens) and other extremely xeric site conditions (hill prairies) as well as abandoned fields (Baskin and Baskin 1999; Fralish et al. 1999; Heikens 1999).

There is no simple explanation for the dominance of confers (*Pinus, Picea, Abies, Larix,* and *Thuja*) in cold climates. An early idea was that because conifer leaves remain green and viable on the stem through the dormant season, they begin to photosynthesize as soon as soil and air temperatures rise above freezing. Deciduous trees expend considerable energy flushing (producing a new set of leaves) before photosynthesis and rapid stem growth can begin in the spring. This depletion of the carbohydrate (food) reserve is a serious problem in areas where the growing season is relatively short; thus, deciduous trees must not only produce sufficient food for growth but must replenish the reserve for flushing the following spring. In Canada the average growing season length ranges from 2 to 4 months (Woodward 1995), but freezing temperatures may occur more frequently, possibly monthly, in low areas such as conifer swamps. The result is that few angiosperms are found in cold climates. In early autumn, leaves of deciduous trees cease photosynthesis and avoid damage from the first frost, but conifer leaves can continue to fix carbon as long as air and soil temperatures are favorable; the evergreen habit effectively gives the conifer a longer growing season. The evergreen habit places deciduous angiosperms at a competitive disadvantage, but the explanation goes beyond this simple idea.

Conifer survival mechanisms are reviewed by Woodward (1995) and Pallardy et al. (1995). One of the major problems to survival in cold climates is the freezing of cells at extremely low temperatures, and conifers have been shown to endure temperatures below −70° C (Woodward 1987, 1995). At below-zero temperatures, the protection of overwintering buds and living xylem is important. Thus, in addition to the evergreen habit, survival in a northern climate is strongly tied to the presence of tracheids in conifer wood, recovery from cavitation after freeze-thaw episodes, and prevention of desiccation. In the stem, the smaller internal volume of tracheids (as compared with the volume in hardwood vessels) is important in resistance to cavitation (the formation of air bubbles). Freezing creates air bubbles in water. Once the bubbles form and enlarge, the liquid column is broken and water cannot be pulled up into the leaves except under great pressure. Conifer needles may suffer considerable frost damage, but substantial repair apparently occurs in the spring, and once cold hardened, needles are more resistant to damage. The deciduous *Betula* and *Populus* also survive these extremes, hence the name "Northern Conifer-Hardwood Forest" for the region. Chlorophyll in the bark of *Populus* may photosynthesize whenever temperatures permit. Conifer needles also have a thick cutin to protect against winter desiccation. When both the xylem and ground are frozen, replacement water is not available to the leaf. The cutin, and probably the tough epidermis, is protection against blowing wind and ice crystals.

Under a low soil nutrient regime as found in the old soils of the Southern Conifer-Hardwood Region, evergreeness is an effective adaptation for storing nutrients. Because nutrients remain in the live leaves and stem, smaller amounts must be obtained from soil when flushing occurs during the following spring season. In addition, as the 2- to 8-year-old leaves begin to deteriorate prior to drop in the fall, a large proportion of the nutrients in them is translocated back into branches, stems, and, probably, roots for use the following spring.

Conifers tend to dominate plant communities in areas of low precipitation such as the southwest and on droughty sites (thin soils, south slopes, ridgetops) elsewhere. The thick cutin layer on the epidermis of conifer leaves prevents excessive water loss. The retention of nutrients internally and the low transpiration rate make these species ideally suited for reforesting areas such as old fields where the soil has been degraded or eroded, and where the evapotranspiration rate is high because of the openness or windiness of the site.

Conifer leaves have a slow decay rate as a result of their low nutrient and high lignin content. Needles generally remain on the forest floor for many years and accumulate into a litter mat that may be 10 to 15 cm or more thick. From this mat develop organic (fulvic and humic) acids that leach the top layer of mineral soil. Iron, aluminum, organic matter, clay, and calcium and magnesium carbonates are mobilized and move from the uppermost mineral soil layer. This layer develops into the light ashy gray, strongly podzolized (A_2 or E) horizon of coarse-textured (sandy) northern soils. Subsequent deposition of the materials in a subsurface zone creates the reddish brown spodic (B_2) horizon. Such soil is usually classed as a Spodosol under the new soil classification system and as a Podzol under the old system.

Only a few conifer species are adapted to survive severe fire by sprouting (*Pinus rigida*, pitch pine; *Pinus serotina*, pond pine; *Pinus echinata*, shortleaf pine). A number of pine species that are adapted to fire release seed from serotinous cones. The term *serotinous* means "late opening." These cones are closed, sealed by resin, and remain on the tree for several years. When high temperatures caused by fire break the seal, the scales open (late or after the fire has passed) and the seed is released to produce seedlings that revegetate the area. This adaptation is found in *Pinus clausa* (sand pine of Florida), *Pinus contorta* (lodgepole pine of the western mountains), and *Pinus banksiana* and *Picea mariana* (jack pine and black spruce of northern United States and most of Canada). If a cone is near the ground, direct light and light reflected from soil may create a temperature sufficiently high to break the seal in the absence of fire. *Pinus* species that have evolved with fire also tend to retain their low branches ("ladder fuel"), although they may be dead and dry. The absence of such natural pruning is viewed as a mechanism to create a hot crown fire from a low-intensity surface fire. The increased heat from a crown fire, as compared with that of a surface fire, ensures the opening of cones and release of seed. Exemplifying another characteristic related to survival under a fire regime, *P. banksiana* and *P. contorta* seedlings and saplings produce cones and seeds as early as age 6 or 7 years. Thus, at least some seed is nearly always available to regenerate the forest, although the fire return time may be relatively short.

ECONOMIC IMPORTANCE

The wood of conifers is used primarily for lumber, pulp and paper, plywood, and chemicals such as naval stores, from which turpentine, paint thinners, and a large variety of industrial chemicals are produced. Various *Pinus* species are important in reforestation and stabilization of denuded areas and old fields of low nutrient status. Recently, a natural chemical (taxol) in *Taxus brevifolia* (Pacific yew) was found to be important in the treatment of ovarian cancer; thus far, efforts to produce the drug synthetically have been unsuccessful.

The forests of high mountain regions are nearly completely dominated by *Pinus*, *Abies*, *Larix*, and *Picea*. Here such species grow slowly and are relatively short at maturity because of the cold, wind, and eroding effect of moving snow and ice crystals, but they play an important economic and environmental role in protecting the watershed and maintaining high water quality. Many of the cities along the Rocky Mountain Front Range (including the major cities of Colorado Springs, Denver, Boulder, and Fort Collins) use middle-altitude reservoirs for their water supply. The land around these reservoirs is often owned by the cities, and city governments are extremely protective of the land and forest above them. Much of the city land is posted to keep out campers and campfires. The greatest negative effect on water quality comes from fire and logging.

Two members of the Coniferales are the largest and oldest living organisms in the world. *Sequoiadendron giganteum* (giant sequoia) trees have the largest girth. Diameters in excess of 8 m at a height of 5 m above the ground have been recorded for living trees in California. Trees of *Sequoia sempervirens* (coast redwood) have smaller diameters but are taller, with measured heights of more than 100 m; a few approach 110 m. Stems of *Pinus longaeva* (Intermountain bristlecone pine) living on the east slope of the Sierra Nevada (Inyo National Forest) are among the oldest living organisms; their ages range from 4300 to 5000 years. Cross sections of the ring width patterns in these stems have been matched against those of wood found in archeological sites in order to date the time of habitation and to determine the prevailing climatic conditions when the rings were formed. Ring patterns have been used to date archeological material back nearly 12,000 years. The use of tree rings in this manner is called dendrochronology (tree ring dating).

CHAPTER 8

Gymnosperm Vegetative and Reproductive Structures

INTRODUCTION

Among the gymnosperms, there is a relatively limited number of leaf characteristics that can be used to identify a species. This range is usually sufficient, however, to separate genera easily and to separate most, but often not all, species within a genus. For positive identification, it is occasionally necessary to examine other structures such as leaf scars, buds, and cones and branching characteristics. Fortunately, except for a small number of species, the combination of leaf and twig characteristics is present throughout the year. This chapter, reviews the range of vegetative and reproductive characteristics that may be used for identification.

VEGETATIVE CHARACTERISTICS

Leaf Clustering and Arrangement (Phylotaxy)

The clustering and arrangement of leaves on a twig often constitute an important identifying characteristic. *Pinus* leaves are grouped into fascicles of twos, threes, fours, or fives, depending on the species, with the fascicles spirally arranged on the twig (Figure 8.1a). One *Pinus* species (*P. monophylla*) bears single needles. The leaves in other gymnosperm genera are also individually attached. Some species, such as *Picea* (spruce) and some Abies (fir), have the leaves arranged spirally around the twig or mostly pointing upward like bristles on a hairbrush (Figure 8.1b). *Abies balsamea* (balsam fir) has two-ranked but alternate leaves (Figure 8.1c). *Abies amabilis* (Pacific silver fir) and *Tsuga canadensis* (eastern hemlock) essentially have three-ranked leaves (Figure 8.1d), although most identification books report them as two-ranked. In *Larix* and *Ginkgo*, leaves are clustered on spur (short) shoots (Figure 8.1e).

The *Cupressaceae, Thuja, Calocedrus,* and *Chamaecyparis* have scale-like leaves forming a flattened branchlet (Figure 8.2a). *Juniperus* and *Cupressus* have decussate scale-like (Figure 8.2b) or awl-shaped (Figures 8.2c) leaves. An exception is *Juniperus communis,* which has a ternate leaf arrangement (whorls of three) (Figure 8.2d). *Calocedrus decurrens* (incense-cedar) has leaves in whorls of four (Figure 8.2e), and *Taxodium distichum* (baldcypress) has two-ranked opposite or subopposite leaves (Figure 8.2f). The leaves of *Torreya* are opposite or subopposite and two-ranked. (Figure 8.2g).

Leaf Shape

With two exceptions (anomalous leaves), the gymnosperm leaf is a relatively simple structure typically consisting of a single narrow, undivided blade. Given the limited range in leaf characteristics for a large proportion of gymnosperms, it is often difficult to separate one species from another within some genera.

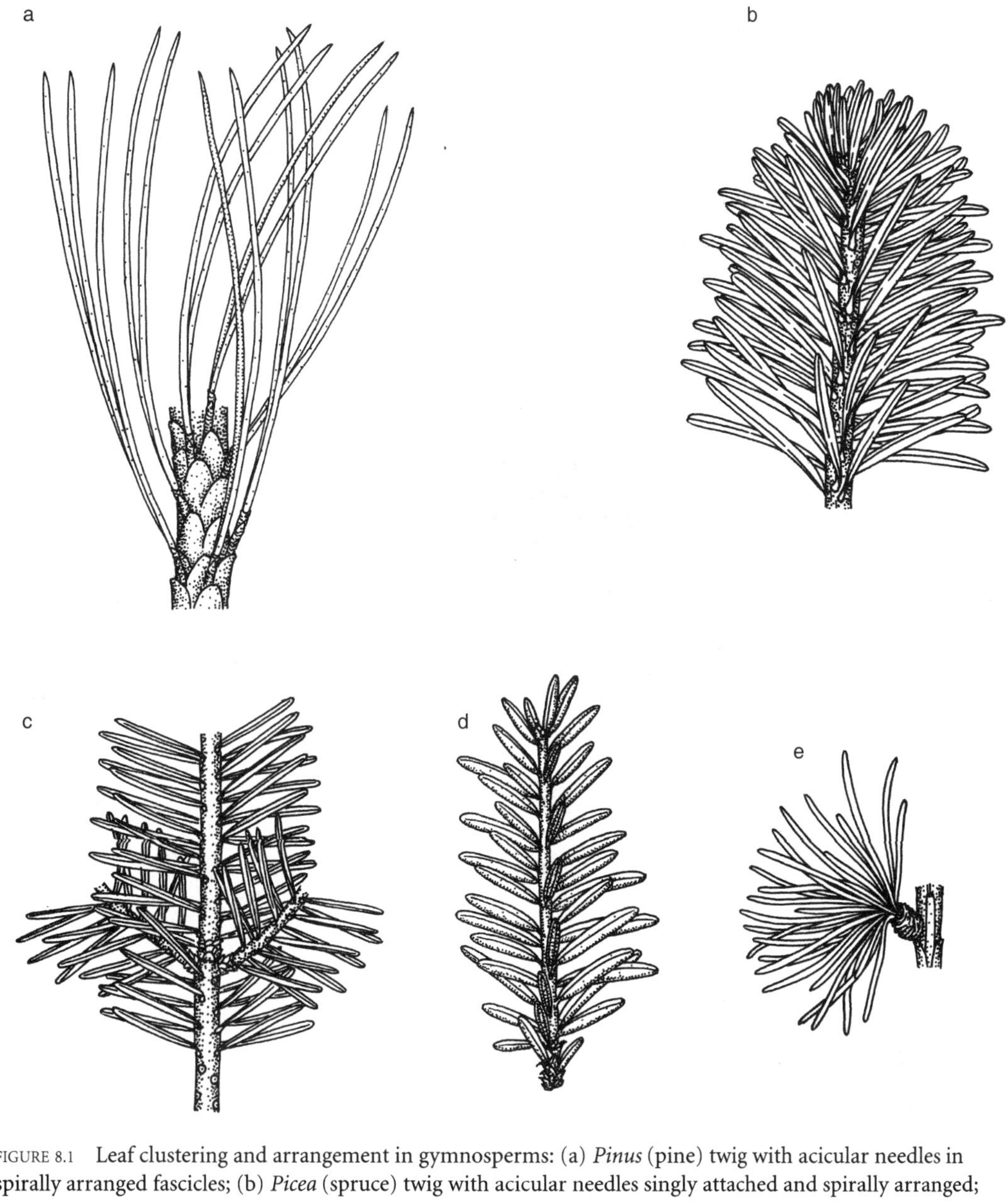

FIGURE 8.1 Leaf clustering and arrangement in gymnosperms: (a) *Pinus* (pine) twig with acicular needles in spirally arranged fascicles; (b) *Picea* (spruce) twig with acicular needles singly attached and spirally arranged; (c) *Abies balsamea* (balsam fir) twig with two-ranked, alternate linear leaves; (d) *Tsuga canadensis* (eastern hemlock) twig with three-ranked linear leaves; (e) *Larix* (larch) twig with acicular needles borne individually on spur (short) shoots.

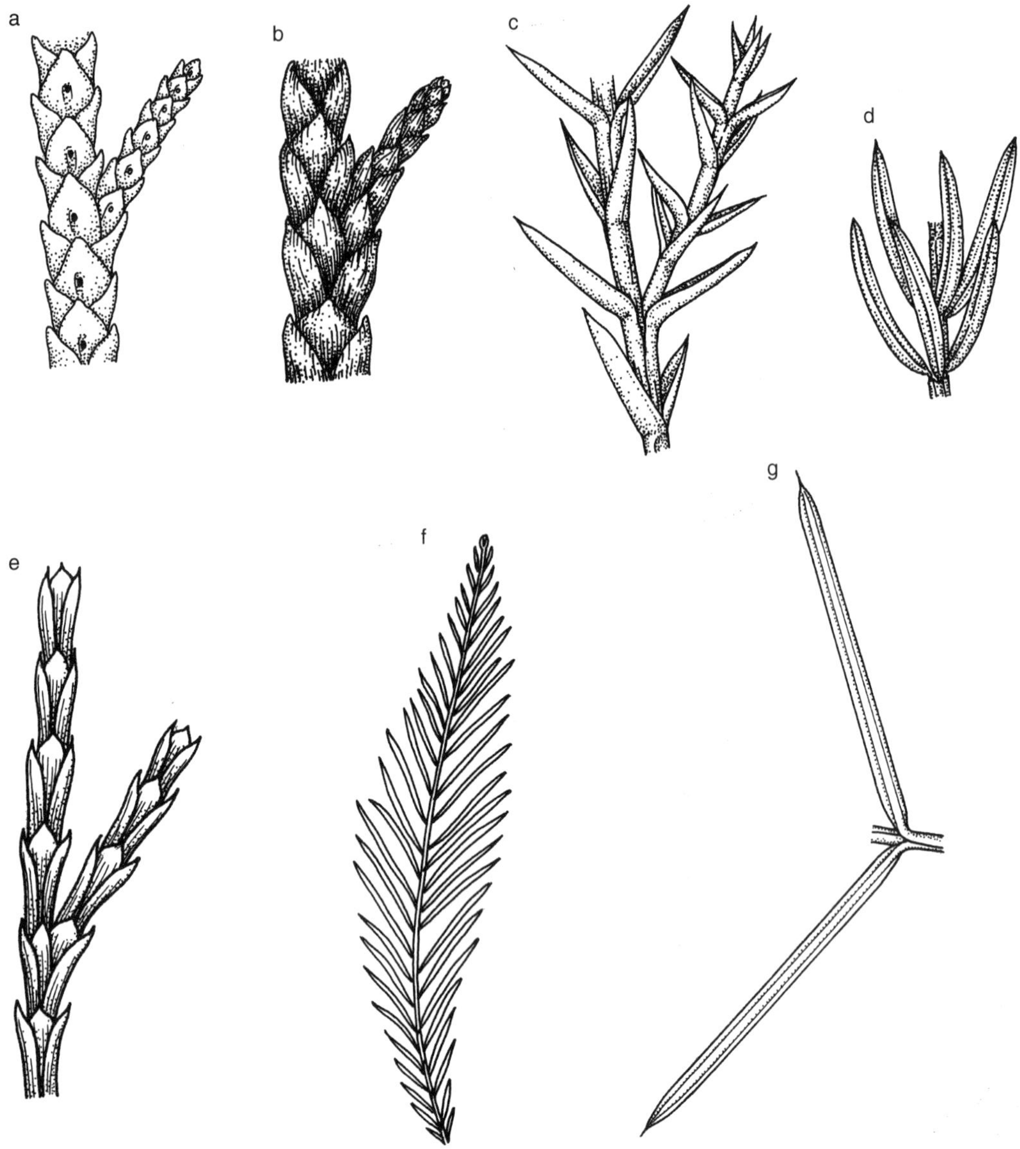

FIGURE 8.2 Leaf clustering and arrangement in gymnosperms: (a) *Thuja occidentalis* (northern white-cedar) twig with four-ranked, decussate scale-like leaves; (b) *Juniperus virginiana* (eastern redcedar) twig with four-ranked decussate scale-like leaves; (c) *Juniperus virginiana*, four-ranked decussate subulate (awl-shaped) leaves; (d) *Juniperus communis* (oldfield juniper) twig with ternate (three at a node) awl-shaped leaves; (e) *Calocedrus decurrens* (incense-cedar) twig with, whorled scale-like leaves; (f) *Taxodium districhum* (baldcypress) twig with two-ranked linear leaves; (g) *Torreya* (torreya) twig with subopposite linear leaves.

Although *Pinus*, *Picea*, *Larix*, and the other genera in the Pinaceae family are easily separated on the basis of leaf shape, leaf clustering, and twig characteristics, separating species based on leaf characteristics may be considerably more difficult. The problem is accentuated in the Cupressaceae family, in which *Juniperus*, *Cupressus*, and several *Chamaecyparis* species have nearly identical leaf form and structure. Six leaf shapes are found among gymnosperms.

Acicular

The beginning botanist, ecologist, or forester will probably think in terms of a "needle-like structure that describes the leaves of *Pinus*, *Picea*, and *Larix*" (Figure 8.1a; 8.1e), about 50 to 55 (excluding varieties) native and several exotic species commonly planted in North America. The general shape of the acicular leaf restricts variations to length, cross-sectional shape, and occasionally width and color. Length varies from as short as 2 cm in *Picea* and *Larix* to as long as 48 cm in *Pinus*. This long, narrow (0.7 to 2.0 mm), needle-like leaf may vary in cross section from semicircular in *Pinus* species, with two leaves per fascicle, to triangular in *Pinus* species, with three to five leaves per fascicle, to square in *Picea* and essentially round in *Larix*.

Linear

The linear leaf shape is found in *Abies* (fir), *Tsuga* (hemlock), *Pseudotsuga* (Douglas-fir), *Sequoia* (redwood), *Taxodium* (baldcypress), and *Taxus* (yew), a total of about 20 to 23 species. This leaf shape is flattened, with distinct upper (adaxial) and lower (abaxial) surfaces (Figure 8.1b, c, d; 8.2f, g). Maximum length is 6 to 8 cm, with a maximum width of 3 mm.

Awl-shaped

The awl-shaped leaf is one of two types that characterize *Juniperus* and *Sequoiadendron* of the Cupressaceae family (Figure 8.2c, d), about 31 species. The term "awl" is the name for a long, slender, sharp instrument used for making holes in leather. However, the leaf shape is essentially triangular, and the leaf is occasionally referred to as spear-shaped, lanceolate, or subulate. This leaf type is individually attached to but usually divergent from the twig. Length does not exceed 1.5 cm.

Scale-like

This is the second leaf type that characterizes a number of genera of the Cupressaceae family. The scale-like leaf is usually only 2 or 3 mm long and is typically appressed or adnate to (flattened against) the branchlet (Figure 8.2 a, b, e). It may be as narrow as 1 mm (*Juniperus; Cupressus*, cypress) or 2 to 3 mm broad and flattened (*Thuja and Chamaeparis*, cedar). In *Sequoiadendron*, the leaf is somewhat triangular in cross section. This leaf type is individually attached to but closely appressed to the branchlet. Scale-like leaves are typically paired, opposite in arrangement, and decussate; that is, each pair of leaves is rotated 90° from the next pair on the branchlet. This arrangement produces a four-ranked structure on *Juniperus* and *Cupressus* branchlets (Figure 8.2b), but in *Thuja* and *Chamaecyparis*, every other pair of leaves is folded to produce a flattened branchlet (Figure 8.2a). The scale-like leaves of *Calocedrus* (incense-cedar) are large (length up to 8 mm), appressed, and in whorls of four (Figure 8.2e).

As compared with awl-shaped and scale-like leaves, acicular and linear leaves are large and usually attached to stout woody twigs. The smaller awl-shaped, scale-like, and more delicate linear leaves are attached to relatively slender woody twigs, commonly called branchlets. In deciduous *Taxodium* trees, the branchlet and attached leaves die and drop as a unit in autumn (Figure 8.2f). In other species, the scale-like leaves are flattened against or adnate to the branchlets, completely cover the woody surface, and cannot easily be separated from it.

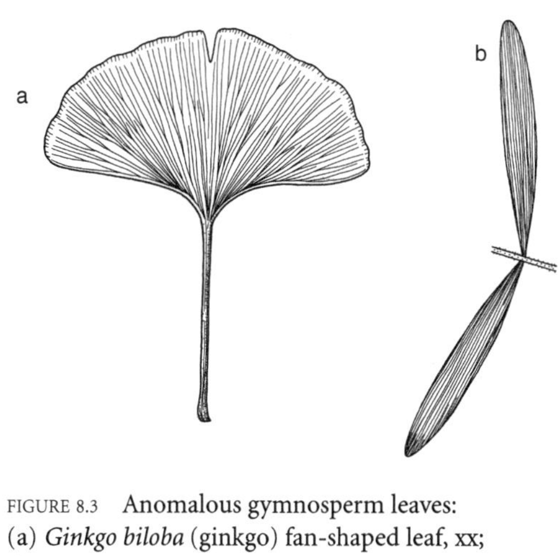

FIGURE 8.3 Anomalous gymnosperm leaves: (a) *Ginkgo biloba* (ginkgo) fan-shaped leaf, xx; (b) *Zamia*, (arrowroof) leaflets of a pinnately compound leaf, xx.

Two leaf types are found on the same tree in species such as *Juniperus virginiana* (eastern red-cedar) (Figure 8.2b, c), *Sequoia sempervirens* (redwood), and *Sequoiadendron giganteum* (giant sequoia) (refer to Figure 11.16). The trees of these species are called dimorphic. In *Juniperus*, juvenile leaves are awl-shaped or lanceolate and adult leaves are scale-like. In *Sequoia*, linear leaves occur on sterile shoots, and shorter awl-shaped (triangular) leaves on cone-producing shoots. *Sequoiadendron* has awl-shaped leaves on older twigs and more scale-like leaves on new twigs.

Anomolous Leaves
Given the number of gymnosperm species with long, thin, and/or small leaves, it is unusual to find two additional leaf types in this taxon. The leaf of *Ginkgo biloba* (ginkgo) has a fan-like blade similar to that of an angiosperm (Figure 8.3a), and the leaf of *Zamia* (a cycad) is composed of sclerophyllus leaflets (Figure 8.3b).

Leaf Stoutness

Pinus needles vary considerably in stoutness (diameter or thickness). The narrowest acicular needles with diameters of 0.3 to 0.6 mm occur in *Larix*. Most pines with four and five needles per fascicle (e.g., soft pines; *P. strobus*, *P. monticola*, *P. aristata*) (width about 0.8 to 1.0 mm) (Figure 8.1a) and a few hard pines (*P. echinata*, *P. virginiana*, *P. glabra*) also have relatively thin needles. Many hard pines (e.g., *P. resinosa*, *P. taeda*, *P. palustris*, *P. ponderosa*, *P. jeffreyi*) have needles of medium thickness (width 1.0 to 1.5 mm). Stout acicular needles (width 1.5 to 2.5 mm) often occur in *P. contorta*, *P. jeffreyi*, and *P. banksiana*. Linear leaves are flattened and are wider than the acicular needles of *Pinus*. The narrow linear leaves of *Pseudotsuga* and *Tsuga* (Figure 8.1d) have a width of 1.0 to 2.0 mm, with slightly wider leaves of 1.5 to 2.0 mm on average in *Taxus* and *Abies* (Figure 8.1b). The widest linear leaves (width 2 to 3 cm) occur in *Torreya* (refer to Figure 12.3); however, *Zamia* has the widest linear-type leaflet (width 6 to 8 mm) (refer to Figure 12.5).

Leaf Twisting and Curvature

The needles of a number *Pinus* species are characteristically twisted, although green and on the tree. Species whose needles are distinctly twisted include *P. sylvestris* (Figure 8.4a), *P. pungens*, *P. rigida*, *P. banksiana*, *P. contorta*, and *P. virginiana*. Additional species may have needles that occasionally are slightly twisted: *P. serotina*, *P. glabra*, *P. clausa*, *P. echinata*, *P. palustris*, *P. ponderosa*, and *P. jeffreyi*. These species are referred to as "hard pines," which have fascicles of two and/or three needles (refer to Chapter 9).

The long needles of many *Pinus* species curve or droop because they are extremely thin and flexible. Conversely, it could be expected that short (<3 cm) and stiff needles should be relatively straight. However, short, distinctly curved leaves occur in *Pinus longaeva* (Figure 8.4b), *Pinus edulis*, *Pinus cembroides*, *Pinus monophylla*, and *Pinus balfouriana* (refer to Chapter 9).

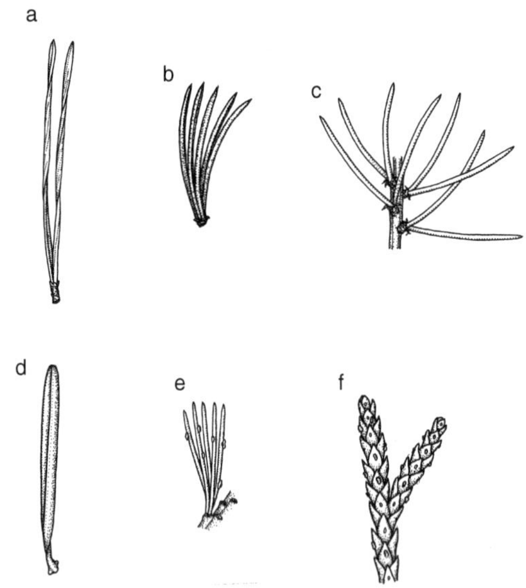

FIGURE 8.4 Variation in gymnosperm leaves: (a) *Pinus sylvestris*, twisted needles; (b) *Pinus longaeva*, curved needles; (c) *Pinus banksiana*, paired divergent needles in fascicles; (d) *Abies*, twisted petiole creating two-ranked leaf arrangement; (e) *Pinus aristata*, resin droplets; (f) *Juniperus*, leaf glands and resin droplets.

The two leaves in each fascicle of *Pinus banksiana* diverge from each other (Figure 8.4c). The leaves of *Abies concolor* (white fir), *Abies amabilis* (Pacific silver fir), and *Abies grandis* (grand fir) twist directly above the point of attachment to the twig (Figure 8.4d). This twist angles the leaves so that they are two-ranked on the twig (Figure 8.1c).

Resin Droplets

Resin is produced by all species in the Pinaceae and about half of the species in the Cupressaceae. In 1970, Bailey separated *Pinus longaeva* from *Pinus aristata*, and although difficult to otherwise separate, *P. aristata* leaves produce tiny resin droplets (Figure 8.4e). A number of closely related western *Juniperus* (Figure 8.4f), as well as several *Cupressus* species, in part, are separated by the presence or absence of resin droplets (exudate) on the leaves.

Leaf Color and Odor

Infrequently, leaf color may assist in identification. The linear needles of some *Tsuga* and *Abies* species are marked by two white bands of stomata on the lower surface, one on each side of the midrib. The acicular leaves of many "soft pine" species have as many as six to ten fine white lines of stomata (as seen under a microscope) running their entire length. Some *Picea* species have a glaucous bloom on the needles that may give them a distinct whitish or bluish appearance. *Chamaecyparis pisifera* has whitened areas on the lower leaf surfaces that appear to form a cross or an X. Odor is a poor characteristic to use in identification, as most gymnosperms leaves are distinctly aromatic.

Leaf Brittleness

Pinus resinosa (red pine) is the only pine whose needles are extremely brittle when green. If bent, they snap cleanly, creating two sections with square ends (Figure 9.16). The strong vascular tissue in the needles of other *Pinus* prevents them from breaking cleanly.

Branchlet/Twig Arrangement

For convenience, a range of characteristics are considered here, including the arrangement of branchlets, fan-like sprays, and pendant branches. In *Taxodium distichum* (baldcypress), twigs with attached needles (branchlets) fall as a unit (Figure 8.2f). The scars that remain after branchlets have dropped are scattered along the twig in a pattern similar to alternate leaf scars on angiosperm twigs. In *Metasequoia glyptostroboides* (dawn redwood), the small buds that remain after branchlets have dropped are paired and usually regularly spaced at 1 to 2 cm intervals along the twig (refer to Figure 11.18). *Thuja* and *Chamaecyparis* have a small cluster of twigs oriented (flattened) into a single vertical plane, thus creating a fan-like structure. Pendent or drooping secondary branches characterize *Picea abies* (Norway spruce) and create the graceful appearance so desirable in a shade tree. The primary and secondary branches of most other gymnosperms are horizontally oriented. A drooping or "buggy whip" lead is characteristric of *Tsuga* (refer to Figure 10.28).

Leaf Scar and Bud Characteristics

Two to ten years after development, gymnosperm leaves die and fall, leaving a scar that can be used for identification of some genera (Figure 8.5). *Abies* and *Pseudotsuga* twigs have a bull's-eye leaf scar that is flush with the epidermis; thus the twig surface remains relatively smooth. In *Picea*, each leaf is borne on a 1 to 2 mm long raised peg (sterigma), and these (sterigmata) collectively create an extremely rough surface after the leaves drop from older twigs. *Tsuga* leaves are borne on smaller pegs, but the density of *Tsuga* leaves

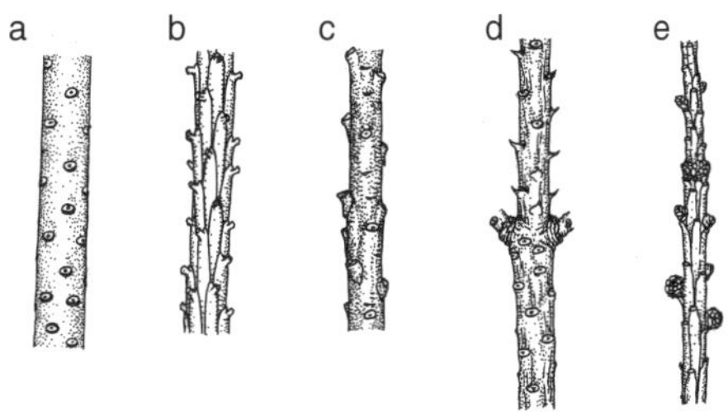

FIGURE 8.5 Leaf scars and twig surfaces of gymnosperm genera: (a) *Abies*, smooth twig with bull's-eye leaf scars; (b) *Picea*, rough twig due to sterigmata (pegs); (c) *Tsuga*, twig with raised leaf scars; (d) *Pinus*, twig with raised leaf scars; (e) *Larix*, first-year and second-year twig with spur shoots.

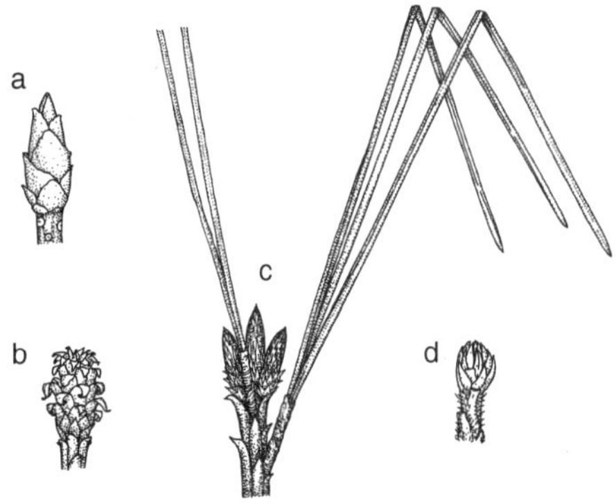

FIGURE 8.6 Common bud shapes and characteristics: (a) sharp-pointed bud of *Pseudotsuga menziesii*; (b) *Picea* bud with reflexed scales; (c) *Pinus* elliptically shaped bud; (d) *Tsuga* globose bud.

is substantially less than that of *Abies*, so twig roughness is reduced by an order of magnitude. In *Larix* and *Ginkgo*, a bud is located at the end of each spur shoot. *Pinus* twigs vary from extremely smooth in some soft pines to rough in the hard pines.

Bud Characteristics

The terminal bud may identify a gymnosperm as to genus, but seldom can be used to identify a species (Figure 8.6). The sharp, dark brown bud of *Pseudotsuga menziesii* (Figure 8.6a) is one exception.

Bark

The identification of any tree by means of bark characteristics is an art developed only with field experience. Bark characteristics vary with age and growth rate and generally defy description, with a few exceptions. The red bark of *Pinus resinosa* (red pine) and the orange bark of *Pinus sylvestris* or *Pinus ponderosa* identify these species. A smooth bark dotted with pitch blisters is characteristic of *Abies*. *Juniperus*, *Thuja*, and *Calocedrus* bark is broken into loose, narrow vertical strips.

REPRODUCTIVE STRUCTURES

Cone Size

Cones size varies from <1 cm to nearly 65 cm in length, and the number of scales from 2 to 3 pairs to about 200 (Figure 8.7). However, most cones range between 5 and 15 cm in length, with number of scales ranging from 50 to 100.

Cone Shape

Cones vary in shape (Figure 8.7) from cylindrical, in which the cone is about three to five times longer than wide, to oval and ovoid, in which it is about twice as long as wide, to orbicular and globose, in which length and width are essentially equal. An oval cone differs from an ovoid cone in that the base of the ovoid cone is flattened: a truncated oval. The width of a depressed cone is greater than its length.

Cone Scale Characteristics

The standard (long and wide but thin) scale of *Pinus* cones usually has an apophysis swollen to varying thickness (Figure 8.8). Depending on the species, the umbo (a protuberance) may be terminal (at the tip of the apophysis or scale), as found in most soft pine, or dorsal (on the back or outer surface of the scale), as found in most hard pine (refer to Chapter 9 for features of these two groups). In the dorsal position the umbo in centrally located, and sometimes the term "central" is preferred to "dorsal" (Flora of North America Committee 1993b). A dorsal umbo may be unarmed and blunt, pyramidal, or armed with a prickle or claw.

In cones of other genera of the Pinaceae (*Larix*, *Picea*, *Abies*, *Tsuga*, and *Pseudotsuga*) the apophyses are not swollen and lack an umbo (Figure 8.8 g, h). The scales usually are fibrous and have a smooth edge, but in most western *Picea*, the scale is thin and brittle with an erose (uneven) margin. The apophysis is differentiated by color.

In the Pinaceae, two seeds are borne on the scale. Although these seeds may have been released at the

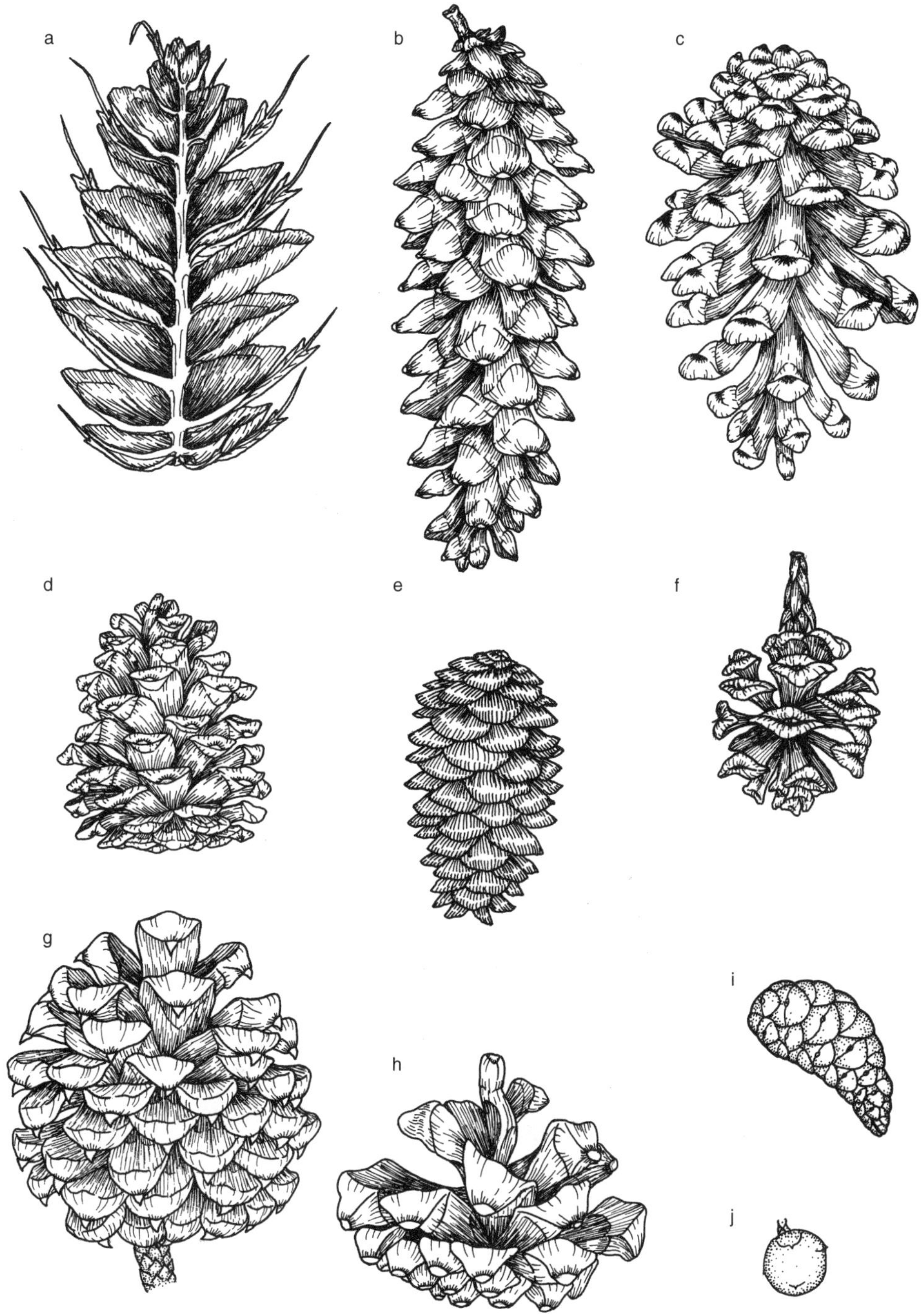

FIGURE 8.7 Gymnosperm cone shapes: (a) *Pseudotsuga menziesii*, cross section of cone showing the stalk, scales, and exserted bracts; (b) *Pinus strobus*, cyclindrical or long elliptical cone; (c) *Pinus balfouriana*, ovate cone; (d) *Pinus resinosa*, ovoid cone; (e) *Picea glauca*, ovate-elliptical cone; (f) *Sequoia sempervirens*, orbicular cone; (g) *Pinus attentuata*, orbicular cone; (h) *Pinus cembroides*, compressed cone; (i) *Pinus banksiana*, asymmetrical serotinous cone; (j) *Juniperus*, globose cone.

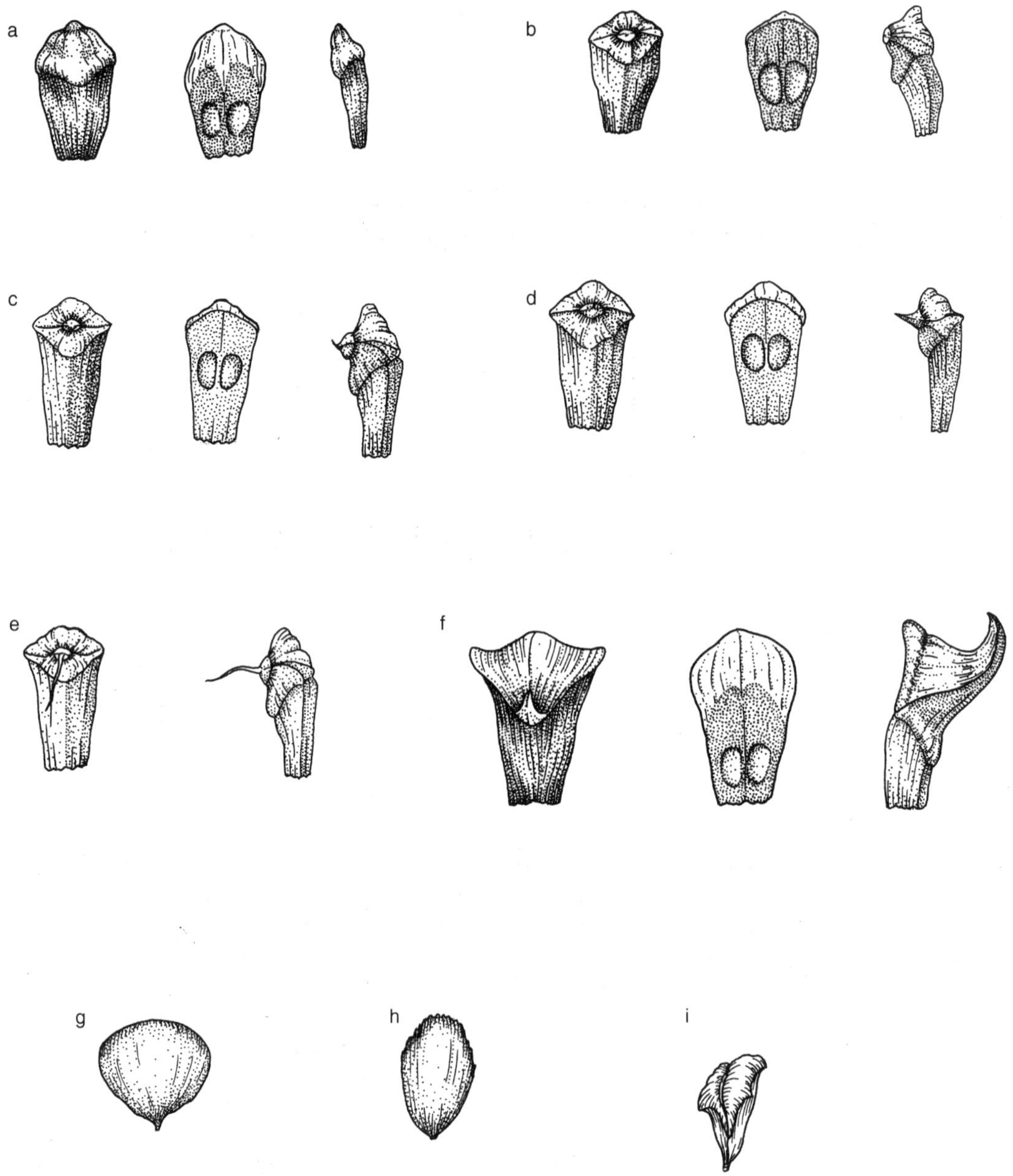

FIGURE 8.8 Variation in *Pinus* cone scales as seen from the dorsal (lower) surface, adaxial (upper) surface (showing indentations of seeds and attached wings), and side view: (a) *Pinus strobus* scale, apophysis slightly swollen, umbo terminal and unarmed; (b) *Pinus cembroides* scale, apophysis swollen, umbo dorsally positioned and pyramidal; (c) *Pinus virginiana* scale, apophysis swollen, umbo dorsally positioned and armed with a slender prickle; (d) *Pinus pungens* scale, apophysis swollen, umbo dorsally located and armed with a stout prickle; (e) *Pinus aristata* scale, apophysis swollen, umbo dorsally located and armed with a long prickle (arista); (f) *Pinus sabiniana* scale, apophysis swollen, umbo dorsally positioned and armed with a stout claw; (g) *Picea glauca*, apophysis not swollen, color differentiated, margin smooth; (h) *Picea sitchensis*, apophysis not swollen, color differentiated, margin erose; (i) *Sequoiadendron giganteum* peltate scale, apophysis terminal.

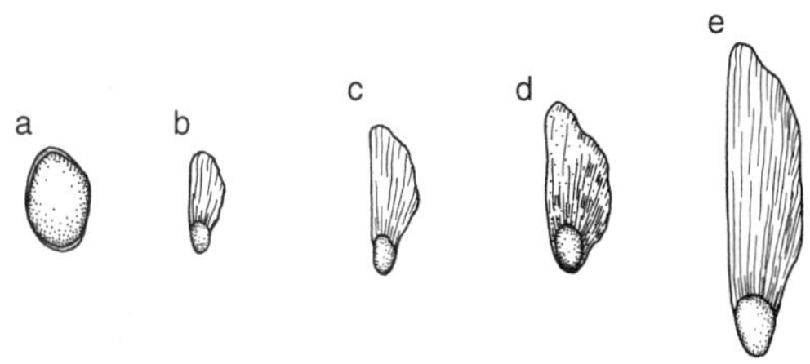

FIGURE 8.9 Variation in gymnosperm seed: (a) *Pinus edulis*, unwinged seed (pine nut); winged seeds of (b) *Pinus muricata*, (c) *Pinus monticola*, (d) *Pinus taeda*, (e) *Pinus palustris*.

time the cone opened, two distinct seed cavities or depressions indicate their location and that of the wings (Figure 8.8a–f). Seeds may be winged or unwinged and can vary greatly in size (Figure 8.9).

In the Cupressaceae family, cones may have standard or peltate (mushroom-shaped) scales (Figure 8.8i), depending on the species. The apophysis may be swollen or differentiated by color. The cone scales of *Sequoia*, *Sequoiadendron*, *Taxodium*, and *Cupressus* are peltate with the apophysis in a terminal position; the umbo is centrally located. The standard cone scale of *Biota orientalis* is armed with a claw. Unlike those of other gymnosperms, the cone scales of some species of *Chamaecyparis* and *Thuja* are paired and decussate. The semifleshy cones of *Juniperus* are small (diameter <2 cm) as compared with those of other gymnosperms. This cone type is composed of one to three pairs of semifleshy, tightly coalesced scales enclosing one or two seeds (Figure 8.7j).

Bract Characteristics

In gymnosperms, bracts are fibrous structures that accompany cone scales (Figure 8.10). They vary from vestigial (small and poorly formed) and inserted (shorter than the accompanying scale) to a structure that is relatively large and exserted (they are longer than the accompanying scale). Some bracts are about the same size and shape of the scale, some long and extremely narrow like a needle, and others are three-lobed. In *Pseudotsuga*, the bract is three-lobed and longer than the scale, and, in the Rocky Mountains, the bracts are reflexed over the scale.

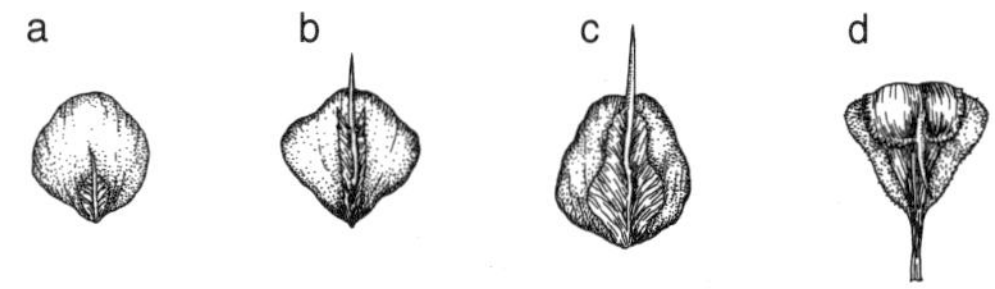

FIGURE 8.10 Gymnosperm bracts: (a) *Larix laricina*, inserted bract; (b) *Pseudotsuga menziesii*, exserted three-lobed bract; (c) *Larix occidentalis*, exserted straight bract; (d) *Abies procera*, exserted reflexed bract.

Aril

An aril is a nonreproductive structure that develops from swollen fleshy integuments to surround the seed of *Taxus* and *Ginkgo*. Each seed is borne on an individual stalk. *Taxus* seeds are partially enclosed by the aril (Figure 8.11a). The soft and fleshy aril of *Ginkgo biloba* completely surrounds the seed and has an outer skin similar to that of a drupe (Figure 8.11b), but the origin of this layer differs from those of a typical drupe.

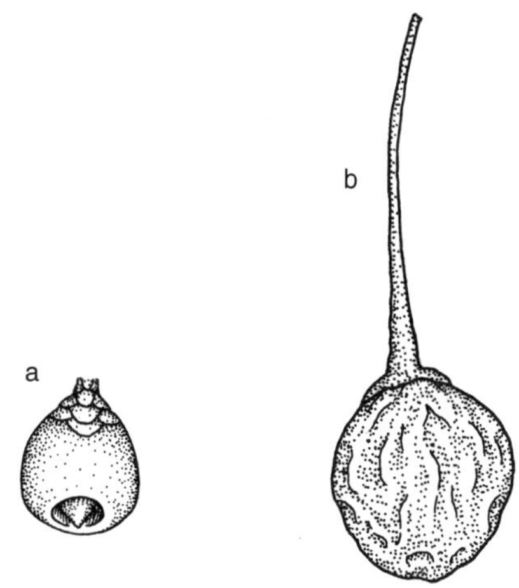

FIGURE 8.11 Anomalous seed and aril (enclosing fleshy structure): (a) *Taxus*, distally open aril and exposed seed; (b) *Ginkgo*, drupaceous structure of aril and seed.

CHAPTER 9

Order: Pinales I: Pinaceae: Pinus

INTRODUCTION

THE Pinaceae family consists of six genera, but because of the number of species, only *Pinus* (pine) species are considered in this chapter. *Pinus* is the largest genus in the Pinaceae. It contains about 100 species worldwide, primarily in the Northern Hemisphere, where it is circumpolar, but the distribution of species in the Southern Hemisphere has greatly expanded with plantation development in southern Africa, New Zealand, and Australia. About 40 species are native to North America. They are a major natural component of forests in Canada, the United States, Mexico, and Central America and important in the production of timber, pulp, paper, naval stores, construction lumber, paneling, and a variety of other products. As noted in Chapter 7, many species are extremely drought tolerant, can survive on the poorest sites, and grow rapidly when soil moisture is available. They have been widely planted, particularly in eastern North America where the land has been degraded and/or must be quickly revegetated, such as on old fields.

All members are resinous trees of medium to large size (heights up to 100 m) with a central upright stem (bole). The branches may be scattered along the bole or in whorls located at nodes. In the latter structure, one new whorl is produced each year, and the number of whorls on a tree less than about 25 years old can often be counted to provide an estimate of age. The distances between nodes (internodes) may be used as an estimate of growth rate. The introduced *Pinus mugo* (mugo pine) is one of the few species that does not have the typical pine form; it is generally a short, shrubby species with numerous branches spreading in all directions.

Pinus buds are large (length 1.0 to 2.5 cm), pointed, resinous, and located directly on the apex of the twig; normally, few auxiliary buds are formed. The leaves are persistent, acicular in shape, and vary in length from 2 to 45 cm. Unlike those of all other members of the Pinaceae, the needles are borne in clusters (fascicles) of two to five, except in one pine species (*Pinus monophylla*). The fascicles are spirally arranged on the twig. The leaves may live up to 10 or 12 years, but in most species they die in 3 to 6 years.

Pinus trees are monoecious. In spring, clusters of 5 to 25 male stobili (pollen cones) surround the twig at the base of the candle (the new growth) before the developing leaves have fully expanded (Figure 9.1a). Microsporangia often produce pollen in such density that yellow smoke may emanate from the tree when a branch is shaken by wind gusts. Pollen from trees growing near water may form a yellow film on the water's surface. After pollen is released, the male strobili become senescent and soon drop from the tree. Pollen may also cause sinus problems.

At the time of pollen release, as many as three or four female strobili appear near the apex of the candle, the new shoot bearing undeveloped leaves. After pollination, the developing woody cones remain closed and on the tree until maturity is reached in 2 years (3 years in *Pinus leiophylla*, *Pinus torreyana*, and *Pinus muricata*). The scales are spirally arranged around a central axis (stalk). At maturity, each fibrous or woody scale is

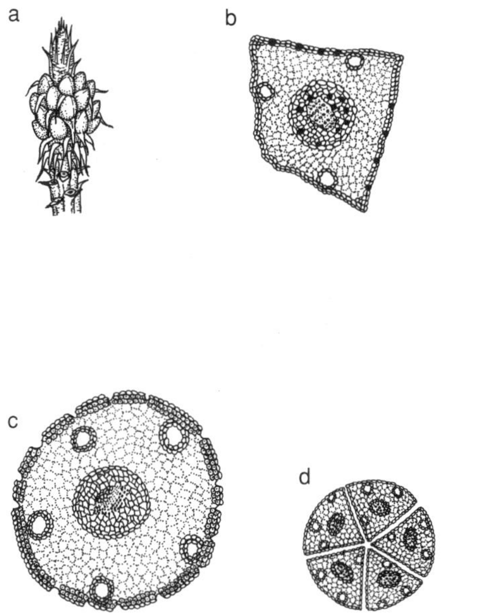

FIGURE 9.1 *Haploxylon* (Subgenus Strobus) (a) male strobili surrounding the terminal bud, (b) cross section of needle showing a single vascular bundle, (c) cross section of *Pinus monophylla* needle with a single vascular bundle, (d) cross section of needles in a fascicle of five. Illustrations b and c adapted from Farjon (1984).

subtended by a bract that is fused to the scale and not readily distinguishable. The seeds, borne in pairs on the cone scale, typically have a terminal wing. Depending on the species, the mature cones may open to release seed or may remain sealed for several years (serotinous) unless opened by heat from fire (e.g., *P. banksiana*, *P. clausa*, and *P. contorta*) or natural weathering. New seedlings may have as many as 15 to 20 cotyledons.

Members of *Pinus* are classed as shade intolerant to extremely shade intolerant, except for one or two pines that are listed as shade intermediate (e.g., *Pinus strobus*, eastern white pine). Shade tolerance, defined as the ability of a plant to grow and develop (complete a life cycle) while growing in shade, has a major influence on the regeneration of pine species and forest community stability. In a slightly different approach, it may be said that shade intolerant species are light demanding; that is, high light levels are required for individuals to survive and grow more or less unrestricted. Any amount of shade will reduce photosynthesis and growth to the point where death may occur. In forested regions, such as the Southern Pine-Hardwood Region, pine is naturally replaced because developing hardwoods create shade to the extent that pine seedlings are unable to survive. Therefore, it usually takes a relatively severe disturbance—such as a fire, windstorm, or timber harvest—to remove overstory shade so that pine seedlings may survive. In addition to being extremely shade intolerant, new seedlings also are highly susceptible to "damping off" disease, a fungus that attacks the root system.

The genus is considered to have two distinct subgenera, which have been recognized by various names that continue to appear in taxonomy texts. The recognition of two distinct sections is credited to Shaw (1914), and the raising of these sections to subgenera (*Haploxylon* and *Diploxylon*) is credited to both Rehder (in Baily 1923) and Pilger (in Engler and Prantl 1926; Little and Critchfield 1969). Critchfield and Little (1966) and Little and Critchfield (1969) subsequently revised and subdivided the two groups. The 'soft pine' group has typically been called the Haploxylon (subgenus *Strobus*). The 'hard pine' group is referred to as the Diploxylon (subgenus *Pinus*). Little and Critchfield (1969) provide historical background as well as the complete classification of pines of the world, which includes a third subgenus (*Ducampopinus*). Although there are obvious external morphological differences between the two major subgenera considered here, the names *Haploxylon* and *Diploxylon* are based on internal leaf differences. The original subgenus *Haploxylon* was so named because of the single (*haplo-*, single or simple) vascular bundle (*xylo*, wood or xylem) within the vascular strand of the leaf (Figure 9.1b,c). *Diploxylon* has two (*di* = two) vascular bundles.

Pinus twigs vary from relatively smooth in some soft pine to rough in the hard pines and thus are seldom used in identification.

SPECIES DESCRIPTIONS

Haploxylon (Subgenus *Strobus* Lemmon, Section *Strobus*), Soft Pines

In North America this subgenus is represented by four groups, generally referred to as white pines,

FIGURE 9.2 Classic outline of *Pinus strobus* extending above the northern Wisconsin hardwood forest. Photo by J. Fralish.

stone pines, pinyon pines, and foxtail pines; additional groups are found in Europe and Asia. This subgenus is referred to as the "soft pines" because the wood, particularly that of the white pine group is relatively soft. However, when the other groups are considered, there is sufficient similarity in wood hardness between the *Haploxylon* and *Diploxylon* that the term "soft pine" may be technically inappropriate. Yet there are various other differences. Within the growth ring or layer of wood developed during the growing season in Haploxylon species, there is a gradual transition, with no color difference, from the larger tracheids formed in the spring to the smaller tracheids developed in summer. Although generally distinguishable, the boundary between summer wood of one year and spring wood of the following growing season appears as an extremely thin dense line. These characteristics give the wood a uniform texture and color that make it valuable as 'millwork' (doors, door frames, baseboards, molding, etc.) in houses because it can be nailed easily, can be stained to a uniform color, and holds paint well. Its softness also makes it easy for cutting and fitting. The white pine group accounts for a large proportion of the timber volume cut within the *Haploxylon*.

The needles are nearly always borne in fascicles (clusters) of four or five (Figure 9.1d), except in one species (*P. monophylla*), where they are borne singly. The fascicle sheath that surrounds the developing needles early in the growing season is deciduous and drops soon after the leaves are developed. The twigs usually are relatively smooth.

Subsection Strobi Loud (White Pines)

Pinus strobus Linnaeus, eastern white pine (Figure 9.2; Figure 9.3); large tree, height to 50 m; leaves 5 per fascicle, length 7.6–12.5 cm, rows of stomates appearing as whitish lines; terminal bud length about 1 cm, covered with numerous reddish orange scales; cone length 10–20 cm, cylindrical, apophysis slightly thickened, umbo terminal, unarmed, scales spotted with white pitch; older twigs smooth; Regions 2 and 3, Lake States and New England south along the Appalachian

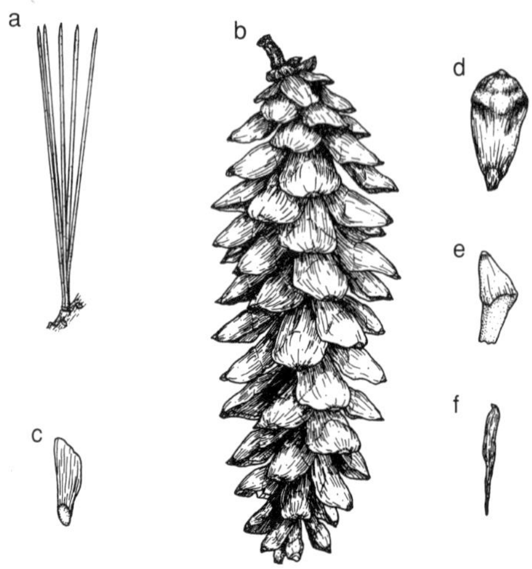

FIGURE 9.3 *Pinus strobus* (a) fascicle of five needles × 1/3, (b) cone × 1/3, (c) winged seed × 1/3, (d) front view of cone scale × 1/3, (e) side view of cone scale × 1/3, (f) cross section of cone scale × 1/3.

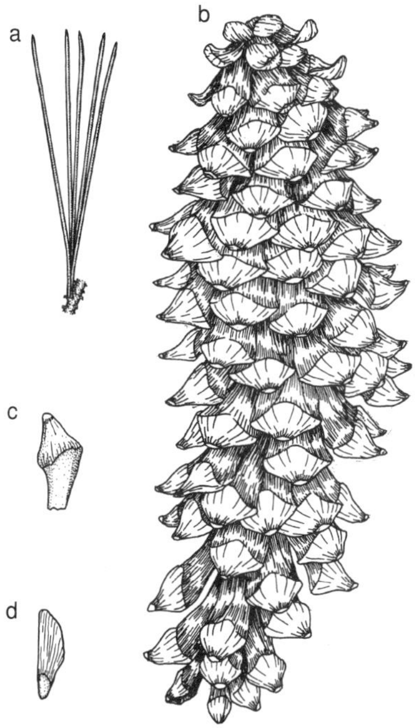

FIGURE 9.4 *Pinus monticola* (a) fascicle of five needles × 1/3, (b) cone × 1/2, (c) side view of cone scale × 1/2, (d) winged seed × 1/2.

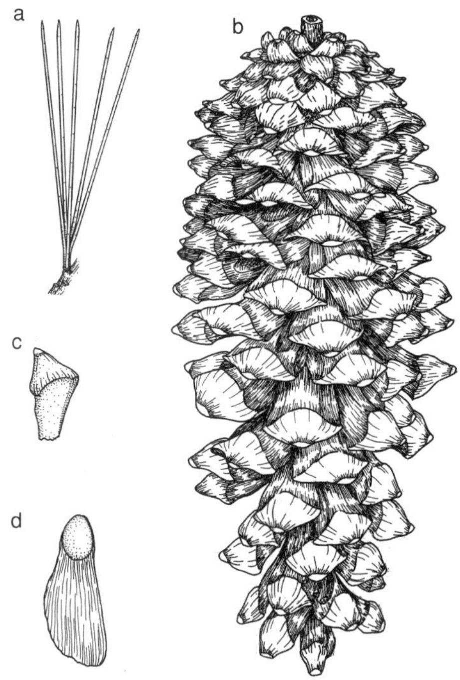

FIGURE 9.5 *Pinus lambertiana* (a) fascicle of five needles × 1/3, (b) cone × 1/3, (c) side view of cone scale × 1/3, (d) winged seed × 1/2.

Mountains to Georgia, disjunct populations in Iowa, southern Wisconsin and southern Michigan; common on dry to moist upland sites, in mountain coves infrequent in swamps. Cover type **26.19, 26.27, 26.28, 26.29,** 18.24, 26.8, 26.9, 26.10, 26.14, 26.17, 26.18, 26.21, 26.22, 26.23, 26.26, 27.15, 27.22.

Pinus monticola Douglas ex D. Don, western white pine (Figure 9.4); large tree, height to 60 m; leaves five per fascicle, length 6 to 12 cm, slender and flexible, glaucous with stomata rows, fascicle sheath deciduous; bud length 1 to 2 cm, covered with several imbricated scales; cone length 12 to 25 cm, open diameter 7 to 10 cm, cylindrical, apophysis slightly thicker than scale shaft, pyramidal, umbo terminal, unarmed; older twigs smooth; Regions 6 and 7, Sierra Nevada (elevation to 3000 m), Pacific Northwest (elevation to 1050 m), and Inland Empire to Montana and Idaho; moist sites, northern exposures, middle slopes. Cover type **29.32**, 29.36, 30.1, 30.3, 30.18, 30.19, 30.27, 30.32.

Pinus lambertiana Douglas, sugar pine (Figure 9.5); large tree, height to 75 m; leaves five per fascicle, length 7 to 10 cm, stomata rows on all sides, fascicle sheath deciduous; terminal bud covered with several reddish brown scales; cone length 15 to 50 cm, open diameter 12 to 20 cm, cylindrical, apophysis slightly thicker than scale shaft, umbo pyramidal, terminal, unarmed; older twigs smooth; bark smooth on saplings and immature trees; Region 7, central Oregon south into southern Sierra Nevada and Mexico (southern California); moist to dry exposed sites at middle elevations (up to 1200 m north) to high elevations (up to 3100 m south). Cover type **30.44**, 30.7, 30.18, 30.25, 30.26, 30.27, 30.28, 30.30, 30.32, 30.34, 30.36, 30.37, 30.39, 30.43.

Pinus flexilis E. James, limber pine (Figure 9.6); medium-sized tree, height to 20 m; leaves five per fascicle, length 4 to 8 cm, flexible, stomata rows on all sides, fascicle sheath deciduous; cone length 9 to 18 cm, cyclindrical; apophysis thicker than scale shaft, umbo terminal, unarmed; older twigs smooth; bark smooth on saplings and immature trees; Regions 5b,

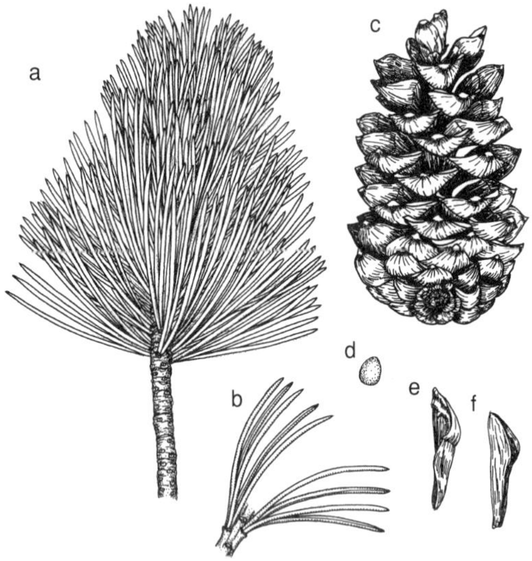

FIGURE 9.6 *Pinus flexilis* (a) leaves clustered at the twig tip × 1/3, (b) twig with fascicles of four and five needles × 1/3, (c) cone × 1/3, (d) seed × 1/3, (e) side view of cone scale × 1/3, (f) cross section of cone scale × 1/3.

6, and 7, Rocky Mountains from British Columbia to Idaho, Montana, South Dakota, Colorado, Arizona, New Mexico, Nevada, and California; exposed dry rocky slopes and ridges at middle elevations (up to 1000 m north) to high elevations (3050 m south). Cover type **29.31,** 29.3, 29.16, 29.17, 29.18, 29.26, 29.28, 29.29, 29.30, 29.35, 29.36.

Pinus strobiformis Engelmann, southwestern white pine (Figure 9.7), large tree, height to 30 m; leaves five per fascicle, length 4 to 9 cm, soft, slightly twisted, stomatal lines absent; leaves ascending on twig; cone length 15 to 25 cm, shape cylindrical; scale tips distinctly reflexed, apophysis thicker than scale shaft, pyramidal, umbo unarmed, terminal; older twigs smooth; Region 5, disjunct populations on mountain slopes of Arizona, New Mexico and Texas, primary range in Mexico; elevations 1910 to 3020 m. Cover type 29.15, 29.6.

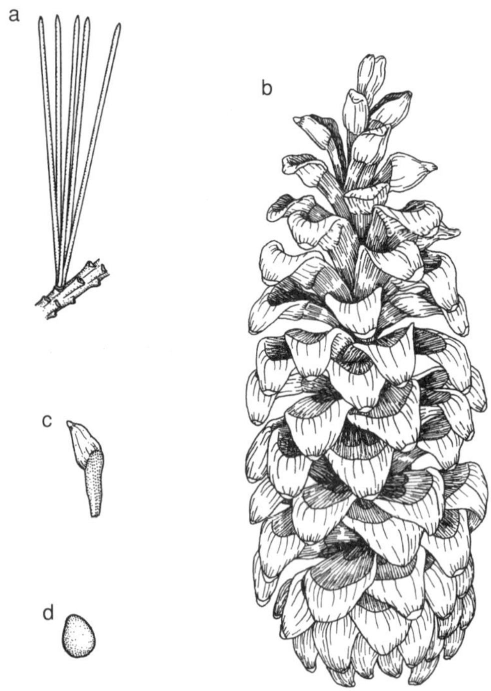

FIGURE 9.7 *Pinus strobiformis* (a) twig with fascicle of five needles × 1/2, (b) cone × 1/2, (c) side view of cone scale × 1/2, (d) seed × 1/2.

Subsection Cembrae Loud (Stone Pines)

Pinus albicaulis Engelmann, whitebark pine (Figure 9.8); medium-sized tree, height to 20 m but may be nearly prostrate on exposed sites; leaves five per fascicle, length 3 to 7 cm, stiff, stout, ascending on twig, fascicles clustered toward end of branch; cone length 4 to 8 cm, apophysis thicker than scale shaft, pyramidal, umbo terminal, armed; Regions 6 and 7, mountain regions from British Columbia and Alberta to Wyoming, Washington, California, and Nevada;

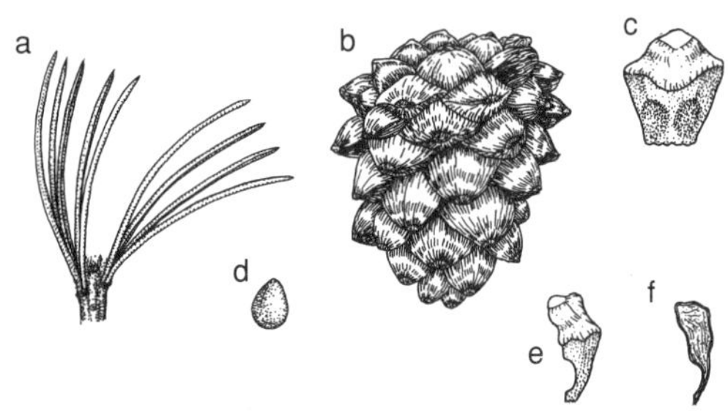

FIGURE 9.8 *Pinus albicaulis* (a) fascicles of five needles × 1/2, (b) cone × 1/2, (c) front view of cone scale × 1/2, (d) seed × 1/2, (e) side view of cone scale × 1/2, (f) cross section of cone scale × 1/2.

subalpine region near timberline, at elevations 1310 to 2300 m in British Columbia, 2400 to 3700 in California.

Subsection Cembroides Engelmann (Pinyon or Nut Pines)

A number of pinyon pine that occur only in Mexico are not described here; these include *Pinus discolor* Bailey and Hawksworth (border pinyon), *Pinus nelsonii* Shaw (Nelson pinyon), *Pinus pinceana* Gordon (Pince pinyon), *Pinus culminicola* Andresen and Beaman (Potosi pinyon), *Pinus maximartinezii* Rzedowski (Martinez pinyon), *Pinus johannis* Robert (Johannis pinyon), and *Pinus juarezensis* Lanner (Sierra Juarez pinyon) (Lanner 1981).

Pinus cembroides Zuccarini, Mexican pinyon (Figure 9.9); tree, medium-sized height to 20 m; leaves three per fascicle, infrequently two and four per fascicle, length 2 to 6 cm, stomata bands pale to white; cone length 1.5 to 3.5 cm, resinous, maturing in 2 years, shape a depressed ovoid after opening, apophysis considerably thicker than scale shaft, diamond shaped, raised, umbo raised, level or sunken, unarmed, scale number 30 to 40, seed length 7 to 20 mm; the edible seeds are harvested and sold in stores; Region 5a, Arizona, New Mexico, Texas, Mexico; xeric mountain slopes and plateaus, at 700 to 2300 m. The populations on the Edwards Plateau in Texas have been described as *Pinus remota* (Little) Bailey and Hawksworth (Texas pinyon) by Lanner (1981) and as *P. cembroides* var. *remota* Little; however, there may be too much overlap with parent population characteristics for them to be considered a species or a variety (Kral 1993). Cover type **29.4.**

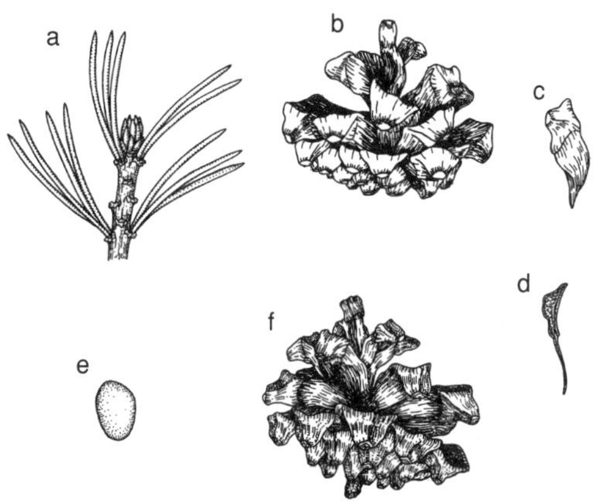

FIGURE 9.9 *Pinus cembroides* (a) twig with fascicles of two, three, and four needles × 1/2, (b) compressed cone × 4/5, (c) side view of cone scale × 4/5, (d) cross section of cone scale × 4/5, (e) seed × 1/2; (f) *Pinus quadrifolia* compressed cone × 2/3.

Pinus quadrifolia Parlatore ex Sudworth, Parry pinyon (Figure 9.9), small tree, height to 10 m; leaves most commonly four per fascicle but also two, three, and five per fascicle, length 3 to 5 cm; cone length 4 to 7 cm, shape a depressed oval-elliptical after opening, apophysis considerably thicker than scale shaft, diamond shaped, raised, umbo level or sunken, unarmed; Region 7, extreme southern California and Baja; at 1200 to 1750 m, rocky sites. Lanner (1981) demoted *Pinus quadrifolia* to hybrid status, as he believed that the three- and four-needle pines of this group resulted from the crossing of *P. monophylla* with *P. juarenzensis* (of Baja and extreme southern California); however, the Kral (1993) maintained it as a species.

Pinus edulis Engelmann, Colorado pinyon (Figure 9.10); medium-sized tree, height to 20 m; leaves two per fascicle, infrequently one and three per fascicle, length 3 to 4 cm, stomata bands pale; cone length 3.5 to 5.0 cm, resinous, maturing in 2 years, shape a depressed ovoid after opening, apophysis considerably thicker than scale shaft, diamond shaped, raised, umbo raised, level or sunken, unarmed, scale number 40 to 55, seed length 10 to 15 mm; seeds are harvested and eaten; Region 5a, Colorado, Utah, Arizona, New Mexico; xeric mountain slopes and plateaus, at 1500 to 2350 m. State tree of New Mexico. Cover type **29.5**, 29.1, 29.1, 29.2, 29.3.

Pinus monophylla Torrey & Fremont, singleleaf pinyon (Figure 9.11); small tree, height to about 12 m;

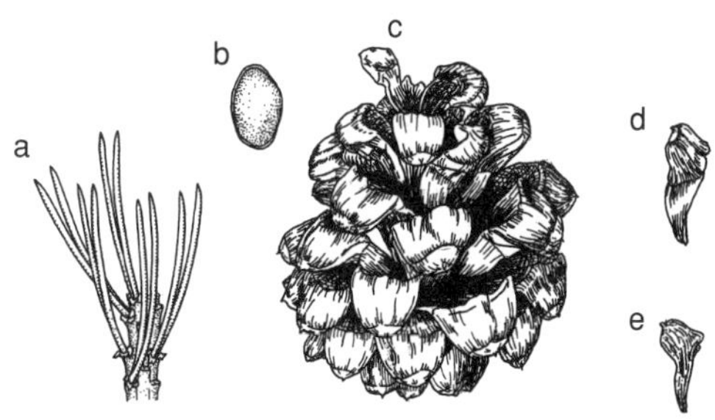

FIGURE 9.10 *Pinus edulis* (a) twig with fascicles of needles × 1/2, (b) seed with a vestigial wing × 1/2, (c) cone × 4/5, (d) side view of cone scale × 4/5, (e) cross section of cone scale × 4/5.

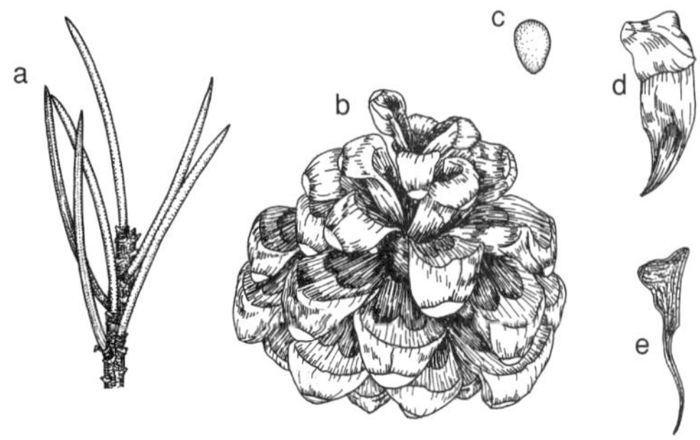

FIGURE 9.11 *Pinus monophylla* (a) twig with needles singly attached × 1/1, (b) depressed cone × 1/1, (c) seed × 1/3, (d) side view of cone scale × 1/1, (e) cross section of cone scale × 1/1.

leaves one, infrequently two, per fascicle, length 2 to 6 cm, curved, stomata lines on all sides; cone length 2.5 to 6 cm, resinous, maturing in 2 years, shape depressed ovoid after opening, apophysis considerably thicker than scale shaft, diamond shaped, raised, umbo raised, level or sunken, unarmed, scale number 40 to 55; Regions 5 and 7, southeast Idaho, Nevada, Utah, California Sierra Nevada, and Nevada into Mexico (Baja California); xeric shallow or rocky soil, at 1000 to 2350 m. Cover type 30.36.

Subsection Balfourianae Engelmann (Foxtail Pines)
Pinus balfouriana Greville & Balfour, foxtail pine (Figure 9.12); medium-sized tree, height to 20 m but short and shrub-like at tree line; leaves five per fascicle, length 1.5 to 4.0 cm, white stomate bands present; cone length 6 to 10 cm, shape elliptical, apophysis greatly thickened, umbo central, armed with a minute deciduous prickle, thus sometimes unarmed; Region 7, California; alpine regions, scattered patches at tree line in the Sierra Nevada.

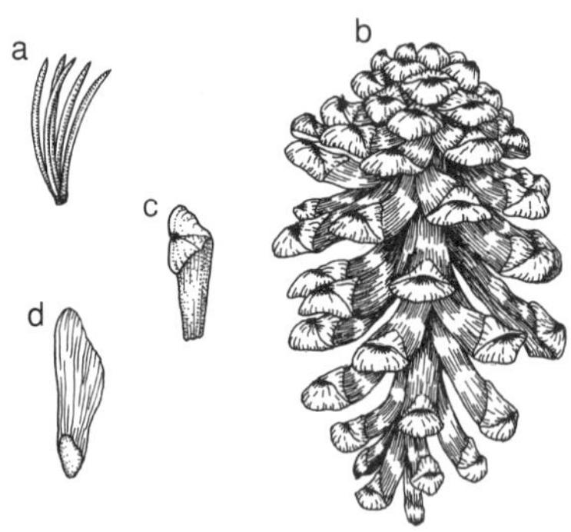

FIGURE 9.12 *Pinus balfouriana* (a) fascicle of five needles × 1/2, (b) cone × 1/2, (c) side view of cone scale × 1/2, (d) winged seed × 1/2.

Pinus aristata Engelmann, Rocky Mountain bristlecone pine (Figure 9.13); small tree, height to 13 m; leaves five per fascicle, length 2 to 4 cm, stomate lines conspicuous, often with resin droplets; cone length 6 to 11 cm, oval but flattened at base, apophysis considerably thicker than scale shaft, umbo central, armed with a slender prickle, prickle length 0.6 to 1.0 cm; Region 5, Colorado, New Mexico, and Arizona; subalpine, at elevations of 2300 to 3400 m. Individual trees have been aged at about 2000 years (Harlow et al. 1996). Cover type **29.17**.

Pinus longaeva D. K. Bailey [P. *aristata* var. *longaeva* (D. K. Bailey) Little], Intermountain bristlecone pine (Figure 9.13); medium-sized tree, height to 17 m; leaves five per fascicle, length 1.5 to 3.5 cm, white stomate bands present; cones narrowly ovoid, apophysis considerably thicker than scale shaft, umbo central, armed with a slender prickle, prickle length 1 to 6 mm; Regions 5 and 7, Utah, Nevada, and east slopes of California Inyo Mountains; subalpine, 1700 to 3300 m, extreme xeric sites. Individuals have been aged at nearly 5000 years, making them the oldest living organisms. These trees have been used in dendrochronological (tree ring dating) research to age archeological sites and examine climatic patterns over the past 10,000 to 12,000 years.

Cone characteristis of the four soft pine groups are listed in Table 9.1.

Diploxylon (Subgenus *Pinus*, Section *Pinus*), Hard Pines

Species of the subgenus *Diploxylon* have relatively hard wood, often harder than that of some angiosperm species, hence the name "hard pine"; but they are also referred to as "yellow pine" because of the resinous color of the wood. Unlike *Haploxylon* species, the transition from springwood to summerwood within the growth ring, and from summerwood to springwood between rings is quite abrupt and distinct, with

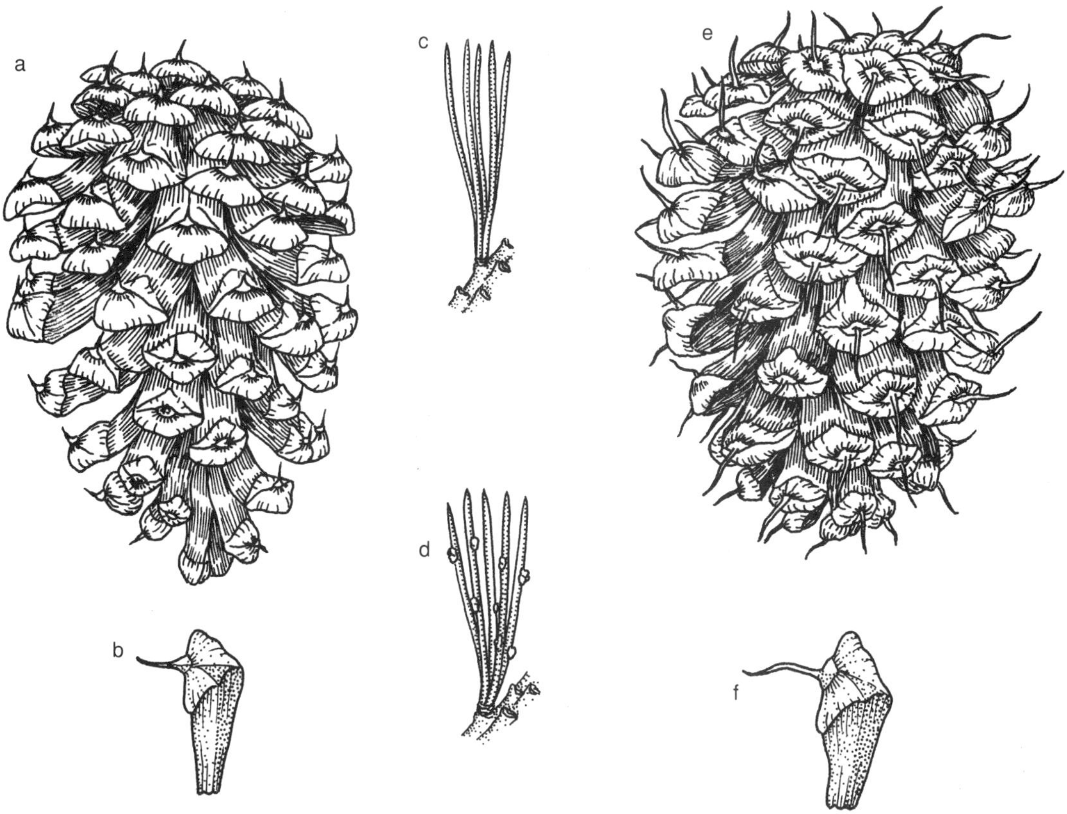

FIGURE 9.13 *Pinus longaeva* (a) cone × 1, (b) side view of cone scale × 1, (c) fascicle of five needles; *Pinus aristata* (d) fascicle of five needles showing resin droplets × 1, (e) cone × 1, (f) side view of cone scale × 1.

sharp color and density changes. Wood hardness, high strength, and color change make the wood of species in this taxon less desirable for fine work and more appropriate for medium to heavy construction, rough use, poles, posts, and kraft paper such as that used in grocery bags and corrugated cardboard.

Needles of the *Diploxylon* are borne in fascicles of two or three (Figure 9.14); the fascicle sheath is usually persistent. The internal structure of the needles differs from that of the *Haploxylon* in that there are two vascular strands within the bundle. In addition, there are differences in leaf diameter, cross-sectional shape, and flexibility between the two groups. These differences are due to the number of needles in each fascicle. With only two needles per fascicle, there is a larger cross-sectional area, with one long flat side and one curved surface per needle. With five needles per fascicle, the cross-sectional area is reduced and the shape is more triangular. The increased thickness, combined with two vascular strands, increases the stiffness of *Diploxylon* needles. *Diploxylon* cones generally have a thick apophysis, a dorsal umbo, and are usually armed. In contrast to those of the terminal white and stone pines of the *Haploxylon*, the cones are positioned perpendicular to the branch.

The *Diploxylon* of North America have been split into seven major groups (Critchfield and Little 1966). The species of these groups, which include several common introduced species, are briefly described.

Table 9.1. Cone and Seed Characteristics for White, Stone, Pinyon, and Foxtail Pine Groups Found in North America. Data from Schropmeyer (1974).
The low number of seeds/kg in the Pinyon pine group is indicative of their large size. These seeds or "nuts" were extensively gathered and used as food by Native Americans and early settlers; they are still widely used and sold in stores of the southeastern United States (New Mexico, Arizona, Utah, Nevada) where juniper-pinyon woodland covers millions of hectares.

	Pine Group			
Cone Characteristic	White	Stone	Pinyon	Foxtail
Cone Length (cm)	10–45	3–9	4–8	8–12
Cone Length:Width	4:1	2-3:1	1:1	2:1
Cone Stalk Length	Long	Short	Short	Short
Apophysis Swollen	No	Yes	Yes	Yes
Umbo Position	Terminal	Terminal	Dorsal	Dorsal
Umbo Armed	No	No	No	Yes
Wing Length	>seed	<seed or absent	<seed or absent	>seed
No. Seed/kg	57,200–59,400*	15,500–11,000	2,200–4,400	35,200–41,800

*Except *P. lambertiana*, which has an exceptionally large seed and averages only 2,100 seeds per pound.

Subsection Leiophyllae
Pinus leiophylla Schiede & Deppe var. *chihuahuana*, Chihuahua pine (Figure 9.15); tree, height to 25 m; leaves usually three per fascicle but may also have two or four, length 6 to 15 cm, fascicle sheath deciduous; cone length 4 to 7 cm, ovoid, symmetrical, apophysis considerably thicker than scale shaft, umbo central, armed, prickle often deciduous; Region 5, extreme southern Arizona and New Mexico into Mexico; xeric slopes, 1500 to 2000 m. The var. *leiophylla* is found only in Mexico. Cover type 29.2.

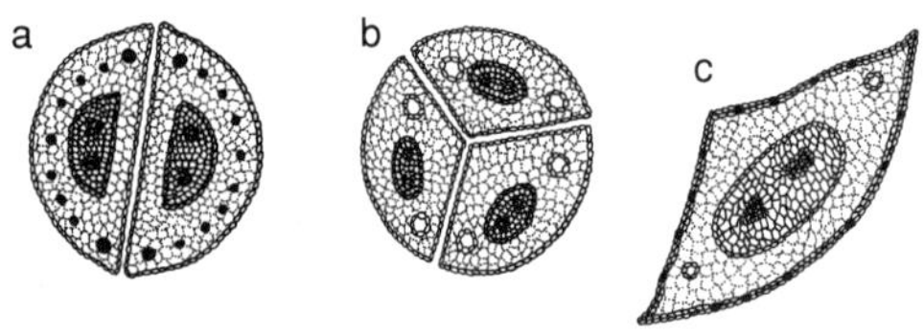

FIGURE 9.14 *Diploxylon* (Subgenus Pinus)
(a) cross section of needles in a fascicle of two,
(b) cross section of needles in a fascicle of three,
(c) cross section of single needle showing two vascular bundles. Adapted from Farjon (1984).

Subsection Sylvestres
Pinus resinosa Aiton, red pine (Figure 9.16); tree, height to 30 m, bark of mature trees flaking, bark and branches distinctly reddish; leaves two per fascicle, length 10 to 15 cm, breaking cleanly when bent, fascicle sheath persistent; cone length 3 to 6 cm, ovoid with a flat base, symmetrical, inner surface dark reddish purple, umbo central, unarmed; Region 2, occasionally disjunct in Region 3b; xeric, sandy upland sites. The name "Norway pine" is frequently misapplied to this species. Cover type **26.19, 26.26,** 26.14, 26.18.

Pinus sylvestris Linnaeus, Scotch pine, Scots pine (Figure 9.17); large tree, height to 25 m, North American trees usually with a deformed or crooked trunk, bark of upper stem orange; leaves two per fascicle, 3.8 to 7.6 cm, distinctly twisted at least 180° from base to apex, fascicle sheath persistent; cone length 4 to 6 cm, pointing toward base of branch, apophysis considerably thicker than scale shaft, umbo central, unarmed but raised and pyramidal, exotic; widely planted and harvested for Christmas trees.

Pinus nigra Arnold, Austrian pine, European black pine (Figure 9.17); large tree, height to 20 m, bark dark brown to black, often marked by horizontal rows of sapsucker holes; leaves two per fascicle, length 10 to 16 cm, stiff and sharp pointed to touch, fascicle sheath

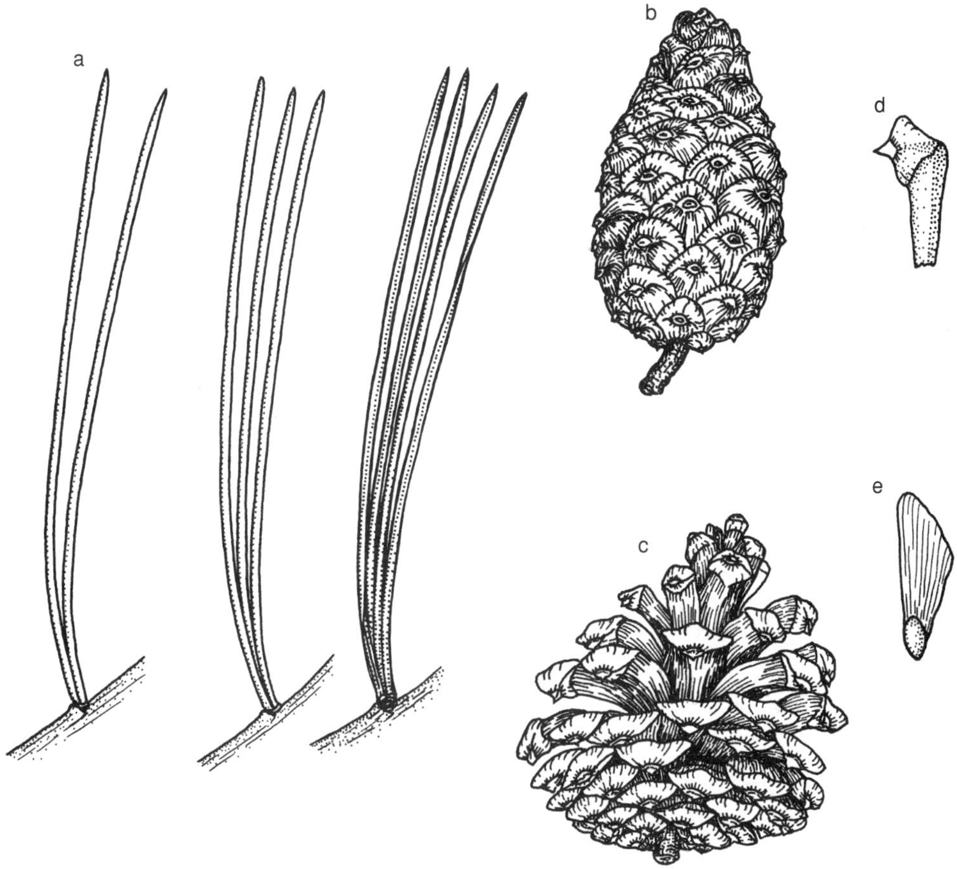

FIGURE 9.15 *Pinus leiophylla* (a) fascicles of two, three, and four needles × 1, (b) closed cone × 1, (c) open cone× 1, (d) side view of cone scale × 1, (e) winged seed × 1.

persistent; cone length 5 to 8 cm, ovoid, apophysis considerably thicker than scale shaft, umbo armed with a slender prickle; exotic, commonly planted as a lawn or park tree.

Pinus densiflora Siebold & Zuccarini, Japanese red pine (Figure 9.18); large tree, height to 25 m, bark of upper stem orangish red; leaves two per fascicle, length 7 to 12 cm, fascicle sheath persistent; cone length 3 to 10 cm, ovoid, apophysis considerably thicker than scale shaft, armed with a small prickle; exotic, infrequently planted as an ornamental.

Pinus mugo Turra, mugo pine, Swiss mountain pine (Figure 9.18); North American material shrubby, height up to 6 m, as wide as tall, only multiple-stemmed pine; leaves two per fascicle, rigid, curved, length 3 to 7 cm, fascicle sheath persistent; cone length 3 to 6 cm, oval, apophysis considerably thicker than scale shaft, umbo central, armed with a slender prickle; exotic, commonly planted as a lawn or park tree.

Subsection Australes (Southern Yellow Pine)

Pinus echinata Miller, shortleaf pine (Figure 9.19); large tree, height 35 m; leaves primarily two, but occasionally three, per fascicle, 4 to 13 cm long, fascicle sheath persistent; leaves often produced from dormant buds on the bole; cone length 4 to 6 cm, apophysis

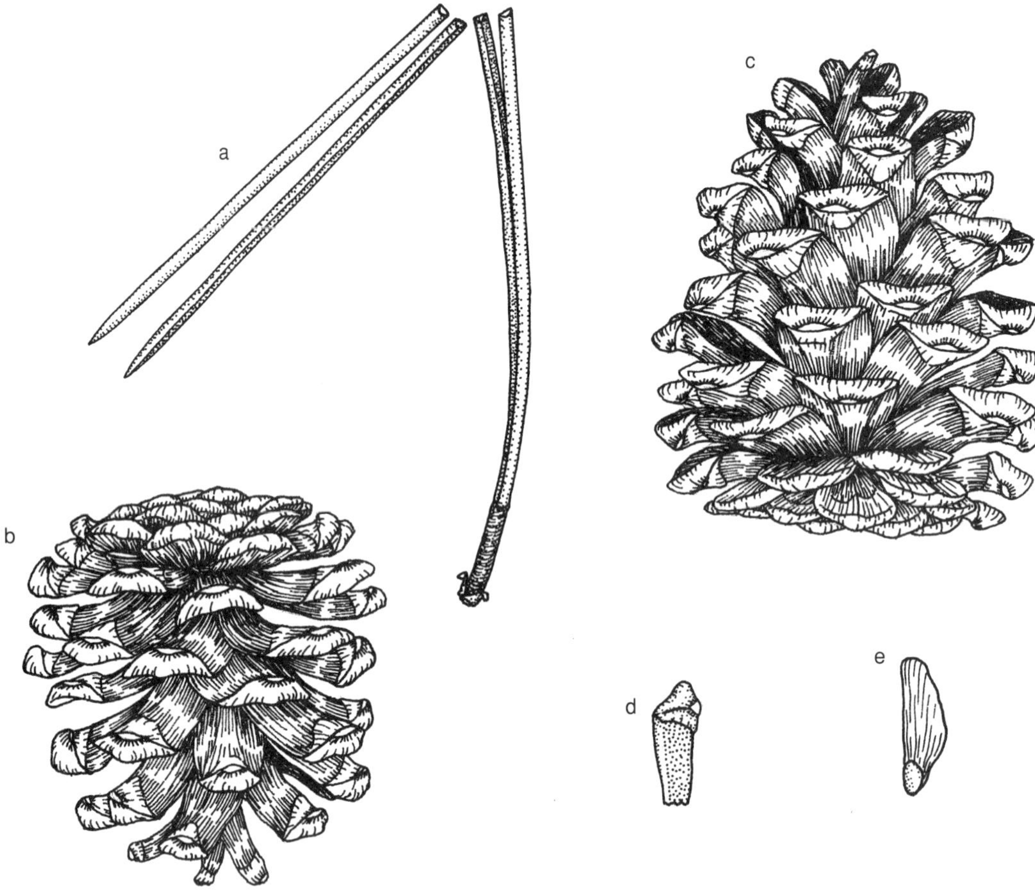

FIGURE 9.16 *Pinus resinosa* (a) fascicle of two needles showing clean breakage × 1, (b), (c) ovoid cone × 1, (d) side view of cone scale × 1, (e) winged seed × 1.

considerably thicker than scale shaft, umbo central, armed with a slender prickle; Region 4 and southern half of Region 3, central and southern states except Lower Coastal Plain; xeric sandy sites. Sprouts after bole is killed by fire. Cover type **27.13, 27.32, 28.1, 28.2, 28.13, 28.39,** 27.24, 27.26, 28.16.

Pinus taeda Linnaeus, loblolly pine (Figure 9.20); large tree, height to 50 m; leaves primarily three, but occasionally two, per fascicle, length 11–19 cm fascicle sheath persistent; cone length 6 to 13 cm, shape ovoid-cylindrical, apophysis considerably thicker than scale shaft, umbo central, armed with a stout prickle; Region 4a and b, Maryland, Delaware and Tennessee to Texas and central Florida, xeric to moist sites. This species is the leading commercial timber tree in southeastern United States. Cover type **28.2, 28.3, 28.13, 28.14, 28.29, 28.41, 28.42,** 28.1, 28.16, 28.26, 28.27, 28.33, 28.40, 28.45, 28.51, 28.53.

Pinus elliottii Engelmann var. *elliottii*, slash pine (Figure 9.21); large tree, height to 45 m; leaves two or three per fascicle, length 15 to 25 cm, fascicle sheath persistent; cone length 8 to 19 cm, stalk length 3 cm, shape ovoid-cylindrical, apophysis thickened, smooth and shiny, umbo central, armed with a stout prickle; Region 4a and b, Coastal Plain from Mississippi to South Carolina except southern Florida;

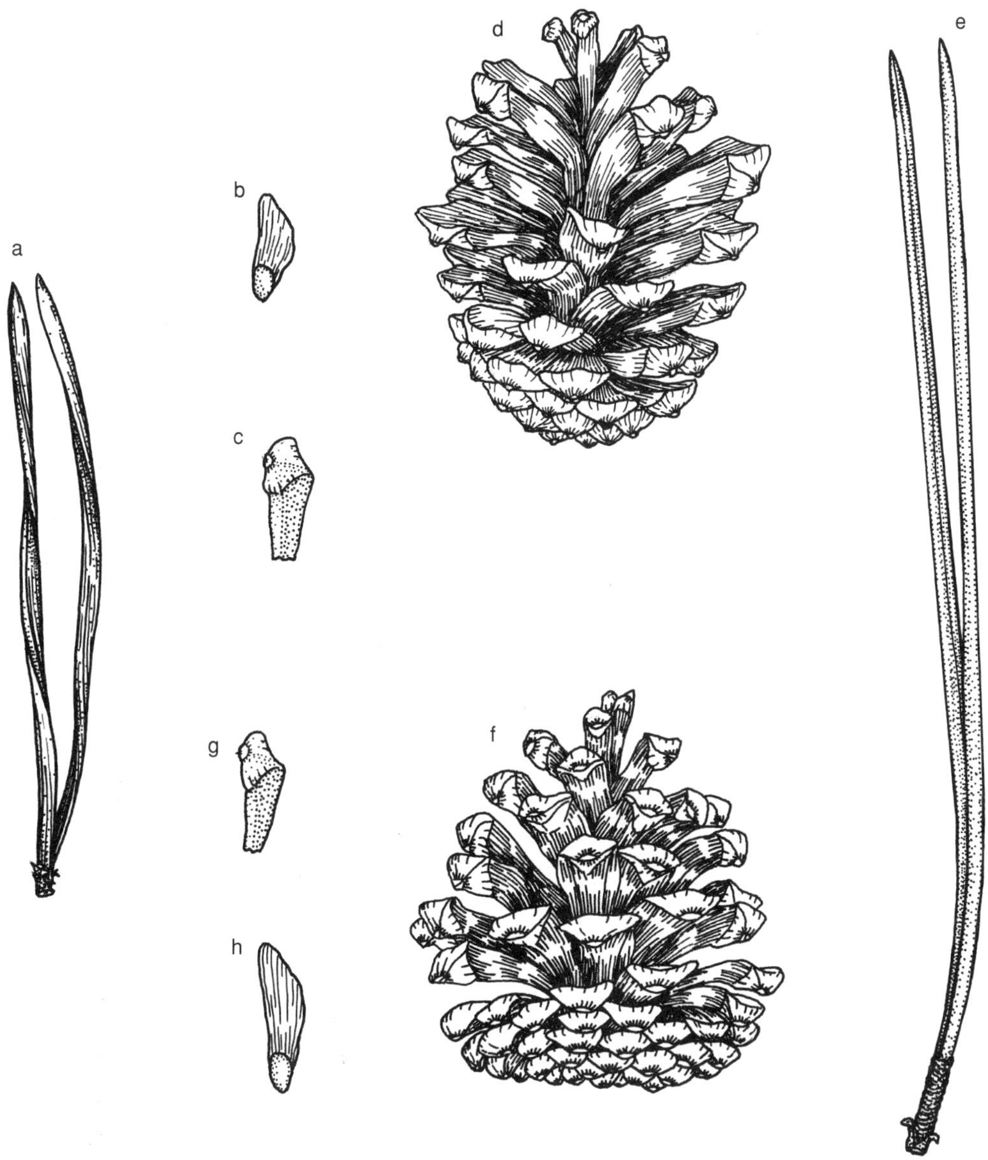

FIGURE 9.17 *Pinus sylvestris* (a) fascicle of two twisted needles × 1, (b) winged seed × 1, (c) side view of cone scale × 1, (d) cone; *Pinus nigra* (e) fascicle of two needles × 1, (f) cone × 1, (g) side view of cone scale × 1, (h) winged seed × 1.

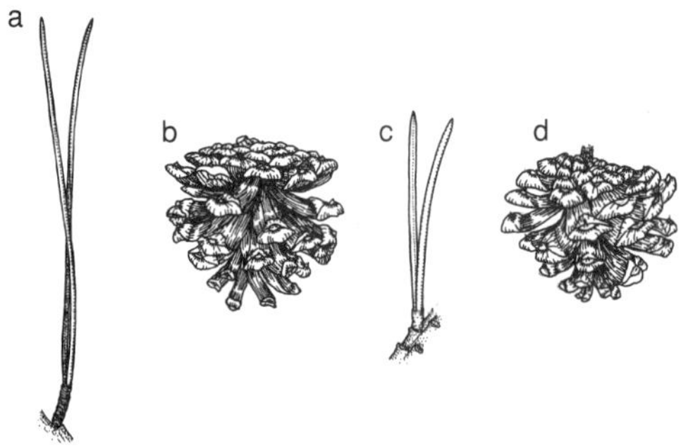

FIGURE 9.18 *Pinus densiflora* (a) fascicle of two needles × 1/2, (b) cone × 1/3, *Pinus mugo* (c) fascicle of two needles × 1/3, (d) cone × 1/4.

moist sites, swamps, and flatwoods but also upland sandy soils and old fields. This species is the fastest growing of the southern pines and is a major producer of naval stores, industrial chemicals, and wood. Cover type **28.35, 28.36, 28.39,** 28.14, 28.23, 28.24, 28.25, 28.28, 28.34, 28.40, 28.53.

Pinus elliottii Engelmann var. *densa* Little & Dorman, South Florida slash pine; similar to var. *elliottii* except (1) leaves two, rarely three, per fascicle, (2) seedlings proceed through a "grass stage" similar to those of *P. palustris*, and (3) the wood is denser and harder than var. *elliottii*; Region 4b, southern half of Florida; dry sites usually over limestone. This

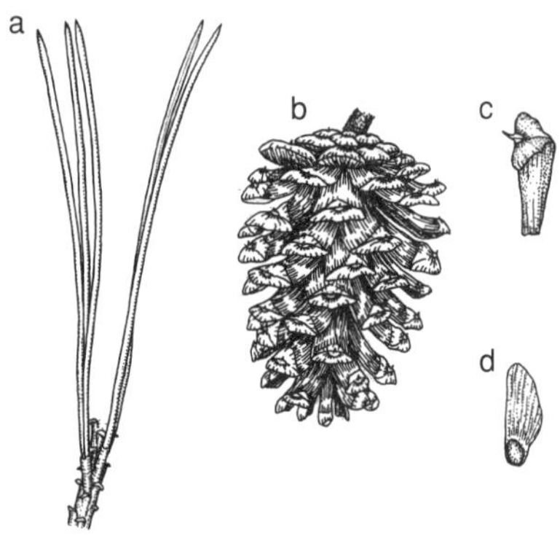

FIGURE 9.19 *Pinus echinata* (a) twig with fascicles of two and three needles × 1/2, (b) cone × 1/2, (c) side view of cone scale × 1/2, (d) winged seed × 2/3.

variety is not used for naval stores and is not commercially planted. The name *Pinus caribaea*, the name of a species found in the Bahaman Islands, Cuba, and Central America, was previously incorrectly applied to this species (Kral 1993).

Pinus palustris Miller, longleaf pine (Figure 9.22); large tree, height to 47 m; leaves three per fascicle, length 18 to 45 cm, fascicles concentrated at end of the twig; cone length 15 to 25 cm, ovoid-cylindrical, apophysis thicker than shaft, scales oriented perpendicular to stalk, umbo central, armed with a slender prickle; Region 4a and b, Coastal Plain, Virginia to Texas except southern Florida; xeric to moist sites and flatwoods. Cover type **28.13, 28.38, 28.39**, 28.2, 28.14, 28.34, 28.35, 28.40, 28.46.

Pinus rigida Miller, pitch pine (Figure 9.23); large tree, height to 30 m, generally considerably shorter; leaves three per fascicle, length 6 to 14 cm, slightly twisted, stiff, leaves often produced from dormant buds on the bole; cone length 4 to 8 cm, shape ovoid with a flattened base, frequently serotinous, apophysis thickened, umbo central, armed with a slender prickle; Region 3b and 2b, Georgia and South Carolina to New York and Maine; xeric, sandy to loamy soils: Sprouts after bole is killed by fire. Cover type **27.32, 27.33**, 27.22, 27.38, 28.4.

Pinus serotina Michaux, pond pine (Figure 9.24); large tree, maximum height 21 m; leaves three per fascicle, length 14 to 21 cm, flexible after twisted; cone length 5 to 7 cm, shape oval, serotinous, apophysis thickened, umbo dorsal, armed, weak prickles often deciduous; Region 4a and b, primarily Atlantic Coastal Plain from New Jersey to Alabama and central Florida, mesic to hydric sites; found on organic soil of pocosins, elevation to 200 m. Physically similar to *Pinus rigida* but has a more southern/coastal distribution. Sprouts after bole is killed by fire. Cover type **28.40**, 28.23, 28.24, 28.25, 28.28, 28.29, 28.30, 28.53.

Pinus pungens Lambert, table-mountain pine (Figure 9.25); medium-sized tree, height to about 13 m; leaves two per fascicle, length 3 to 6 cm., twisted, rigid; cone length 5 to 9 cm, ovoid, asymetrical, serotinous, apophysis thicker than shaft, umbo armed with a stout spine or claw; Region 3b, Appalachian Mountains from Pennsylvania to Georgia; xeric rocky sites, outcrops and ridges at high altitudes hence the name "table-mountain", elevation 500 to 1350 m. Cover type 27.38, 28.4.

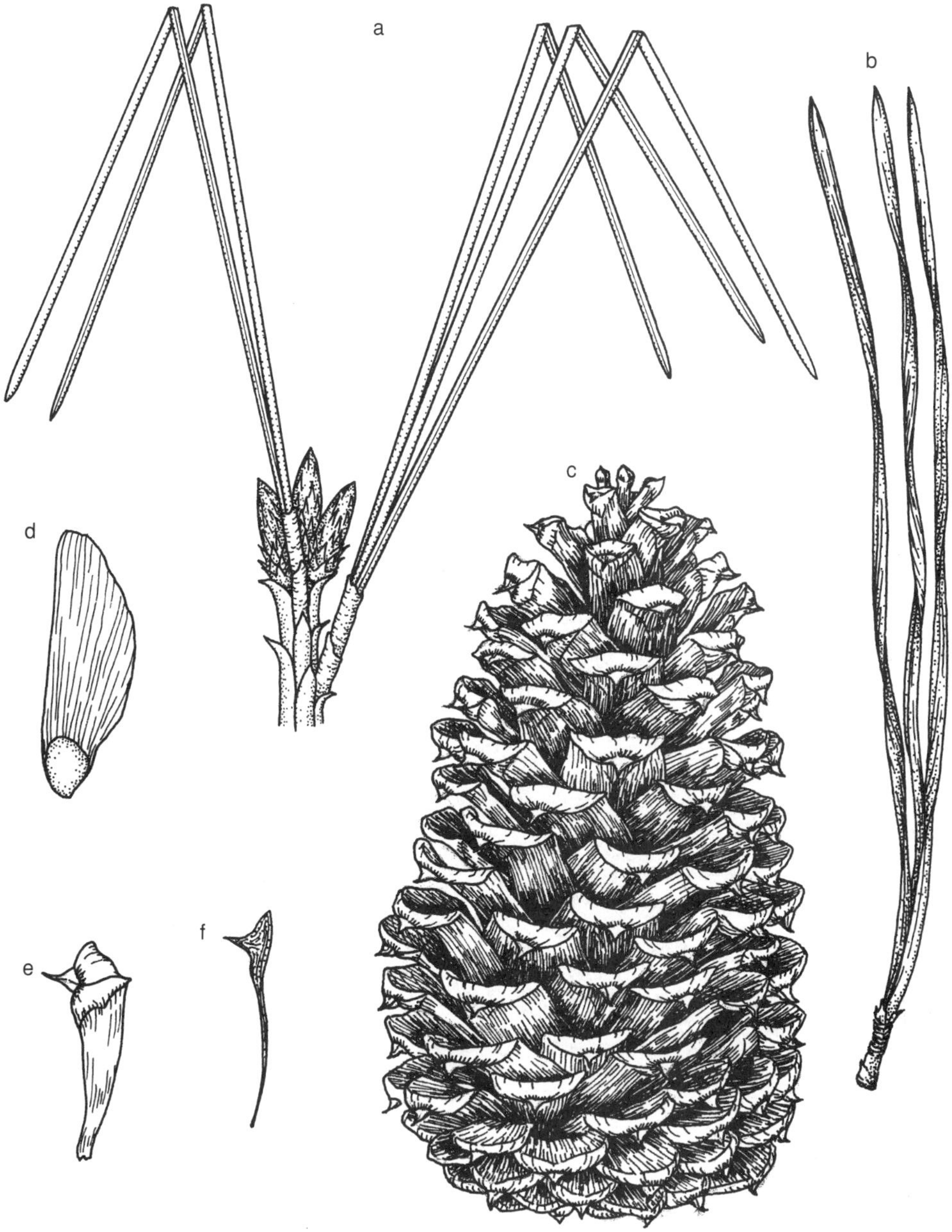

FIGURE 9.20 *Pinus taeda* (a) twig with fascicles of two and three green needles × 1, (b) fascicle of three dead needles × 1, (c) cone × 1, (d) winged seed × 1.5, (e) side view of cone scale × 1, (f) cross section of cone scale × 1.

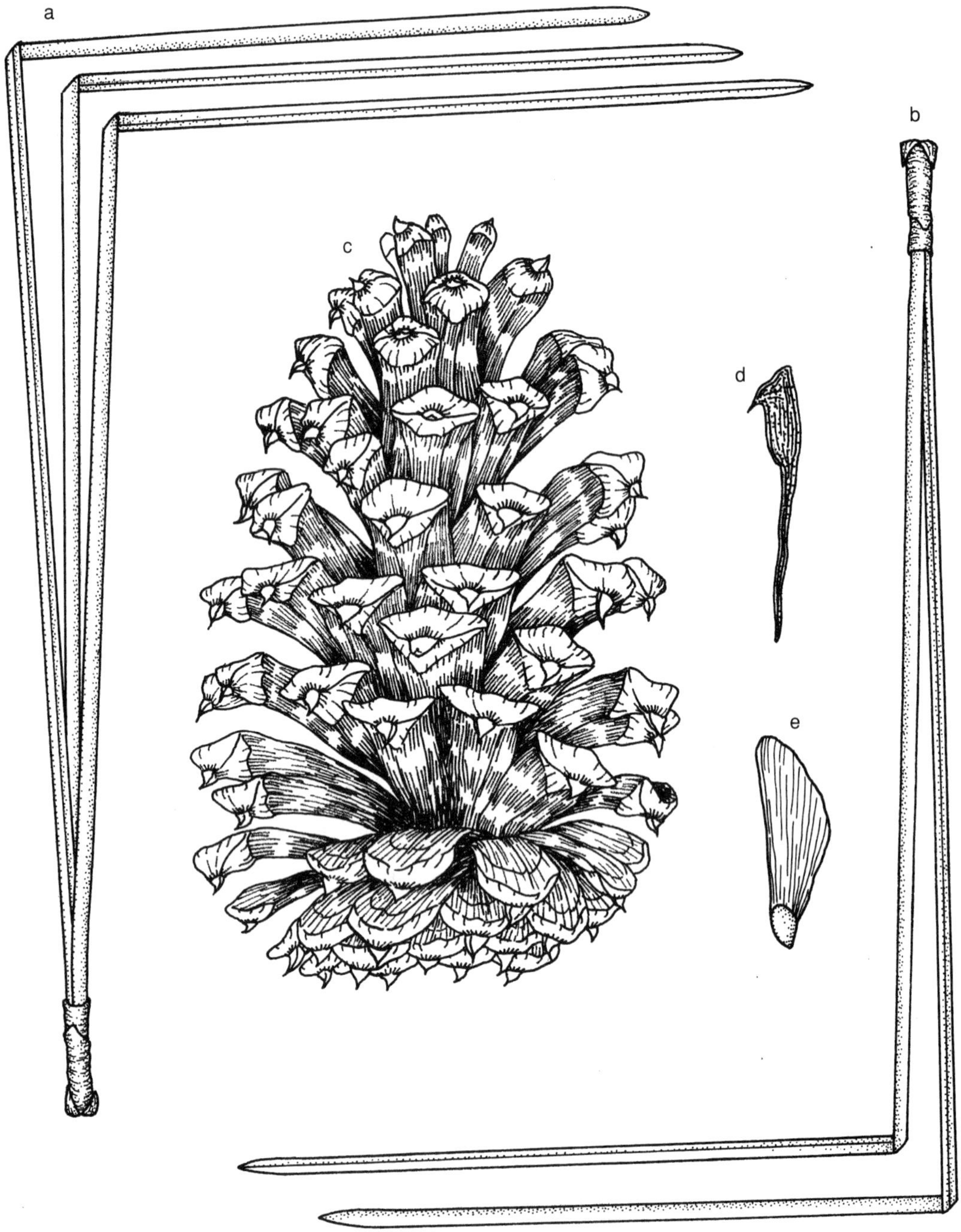

FIGURE 9.21 *Pinus elliottii* (a), (b) fascicles of two and three needles × 1, (c) cone × 1, (d) cross section of cone scale × 1, (e) winged seed × 1.

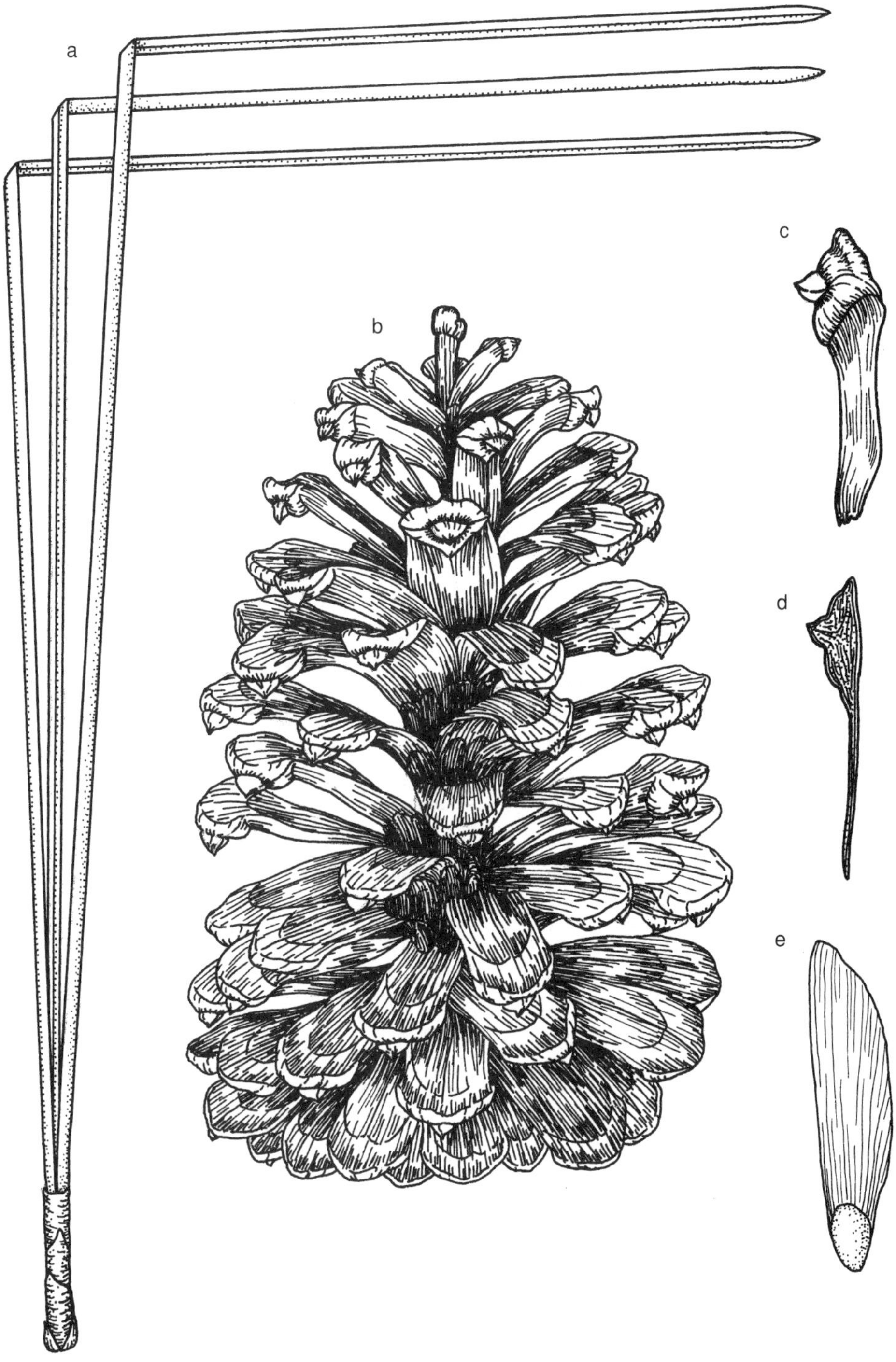

FIGURE 9.22 *Pinus palustris* (a) fascicle of three needles × 1, (b) cone × 2/3, (c) side view of cone scale × 1, (d) cross section of cone scale × 1, (e) winged seed × 1.

FIGURE 9.23 *Pinus rigida* (a) fascicles of two needles × 1/2, (b) cone × 2/3, (c) side view of cone scale × 2/3, (d) cross section of cone scale × 2/3, (e) winged seed × 1/2.

Pinus glabra Walter, spruce pine (Figure 9.26); large tree, height to 25 m; leaves two per fascicle, length 4 to 10 cm, twisted, flexible; cone (similar to *P. echinata* cone) length 3 to 7 cm, ovoid, apophysis slightly thicker than shaft, umbo armed with a slender prickle; Regions 4a and b, Coastal Plain from Louisiana to northern Florida and South Carolina; mesic-hydric and alluvial sites; elevation 0 to 150 m. Cover type **28.37**, 28.27.

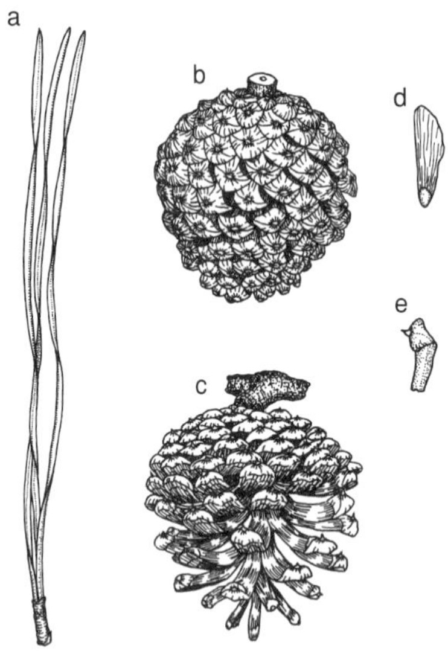

FIGURE 9.24 *Pinus serotina* (a) fascicles of three twisted needles × 1/2, (b) closed cone × 1/2, (c) open cone × 1/2, (d) winged seed × 1/2, (e) side view of cone scale × 1/2.

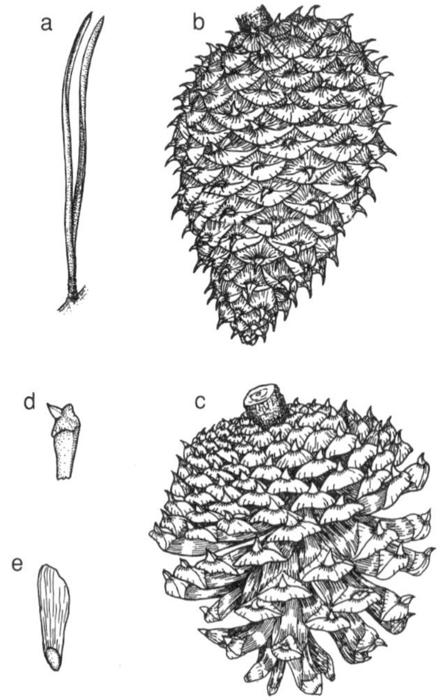

FIGURE 9.25 *Pinus pungens* (a) fascicle of two twisted needles × 3/4, (b) closed cone × 3/4, (c) open cone × 3/4, (d) side view of cone scale × 3/4, (e) winged seed × 3/4.

Subsection Contortae

The four species of subsection Contortae are relatively closely related and may have had a common ancestor. They are generally medium sized, have two leaves per fascicle and serotinous cones (adapted to fire), are extremely shade intolerant (light demanding), and are early successional. They are ecologically equivalent and have the same function in each region where they are found. They rapidly reforest an area after a major fire.

Pinus banksiana Lambert, jack pine (Figure 9.27; Figure 9.28); tree, height about 24 m; leaves two per fascicle, length 2 to 4 cm, hemispherical in cross section, stout (width 1.5 to 2.0 mm), the two needles

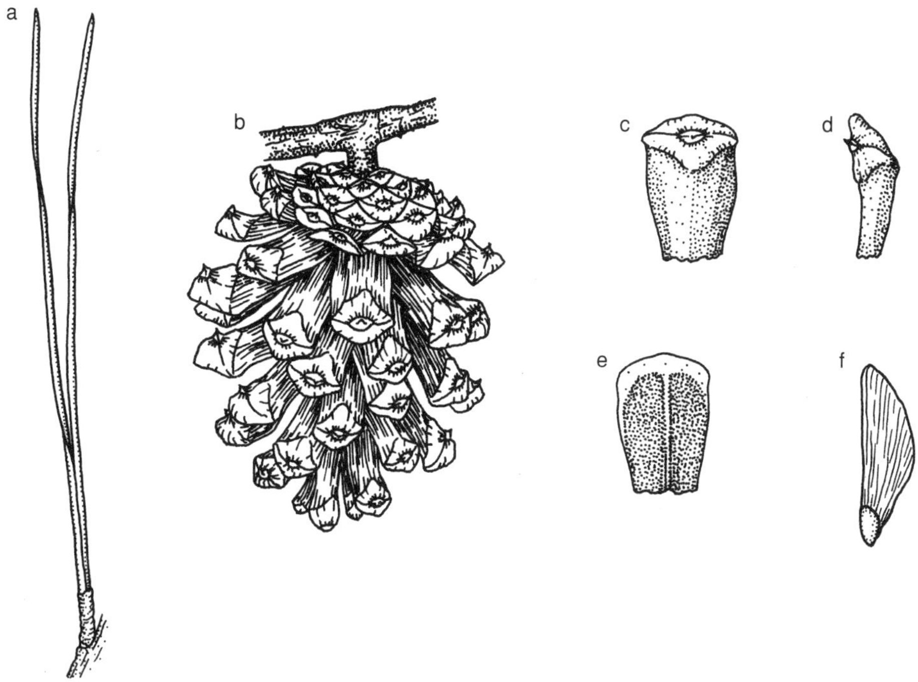

FIGURE 9.26 *Pinus glabra* (a) fascicle of two twisted needles × 1, (b) cone × 1, (c) front view of cone scale × 1, (d) side view of cone scale × 1, (e) top view of cone scale showing location of seed wings × 1, (f) winged seed × 1.

in each fascicle widely divergent; cone length 3 to 6 cm, serotinous, strongly asymmetrical, apophysis thickened, umbo unarmed, cone pointing toward end of branch; Regions 1 and 2, in Canada from Yukon-Northwest Territories border and Rocky Mountains in Alberta to Nova Scotia; xeric rocky or extremely sandy sites, individual trees found on mesic-hydric sites. Hybridizes with *P. contorta* in Alberta, British Columbia and Northwest Territories. Cover type **26.18, 26.11, 26.12,** 26.7, 26.14, 26.26.

Pinus virginiana Miller, Virginia pine (Figure 9.29); large tree, height to 20 m; leaves two per fascicle, length 3 to 7 cm, twisted; cone length 4 to 7 cm, apophysis thicker than shaft, armed with a slender prickle, cone perpendicular to branch; Regions 3b and 4c, Appalachian Mountains and Piedmont Plateau into Region 3a, central Tennessee and Kentucky; xeric rocky outcrops, mineral soils, or strip mine spoil banks. Cover **28.4, 28.5,** 27.24, 27.26, 27.32.

Pinus clausa (Chapman ex Engelmann) Sargent, sand pine (Figure 9.30); large tree, height to 20 m; needles two per fascicle, length 5 to 9 cm, flexible; cone length 4 to 8 cm, serotinous, asymmetrical, apophysis thicker than shaft, umbo central, armed with a stout spine; Region 4b, Lower Coastal Plain, central and northern Florida to Alabama; xeric sandy sites. Two varieties are recognized. *Pinus clausa* var. *clausa* of northern and central Florida usually has serotinous cones whereas var. *immuginata* D. B. Ward (Choctawhatchee sand pine) has open cones. Cover type **28.34.**

Pinus contorta Douglas. As a species, *Pinus contorta* is recognized as being composed of three (Farjon 1984; Kral 1993) or four varieties (Harlow et al. 1996). All varieties have several common characteristics (Figure 9.31): leaves two per fascicle, length 3 to 7 cm, frequently twisted; cone length 2 to 7 cm, strongly asymmetrical, apophysis on basal scales frequently pyramidal, umbo dorsal with a long prickle.

FIGURE 9.27 Open-grown *Pinus banksiana* trees in northern Wisconsin. Photo by J. Fralish.

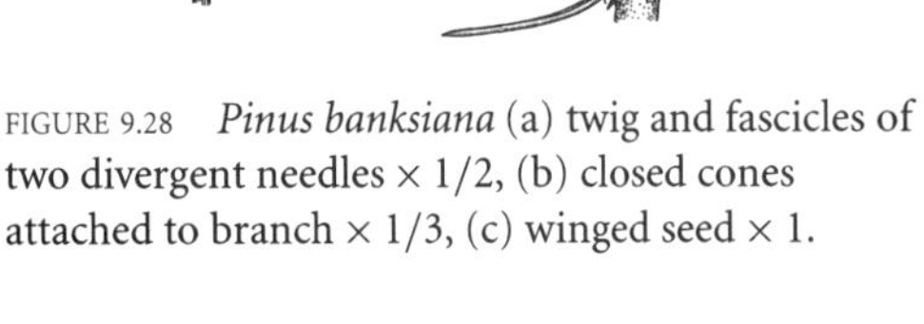

FIGURE 9.28 *Pinus banksiana* (a) twig and fascicles of two divergent needles × 1/2, (b) closed cones attached to branch × 1/3, (c) winged seed × 1.

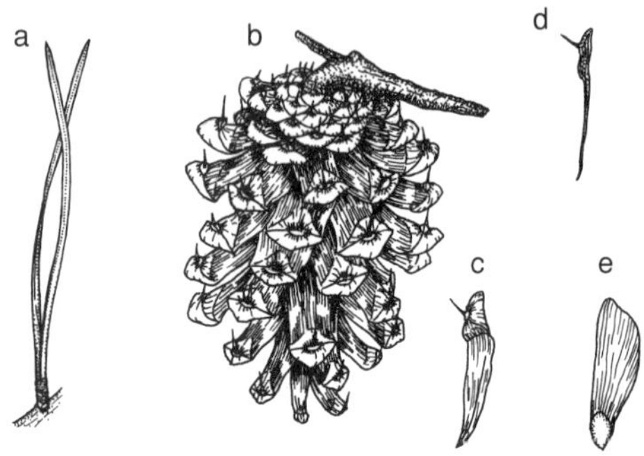

FIGURE 9.29 *Pinus virginiana* (a) fascicle of two twisted needles ×2/3, (b) cone ×2/3, (c) side view of cone scale ×2/3, (d) cross section of cone scale ×2/3, (e) winged seed ×3/4.

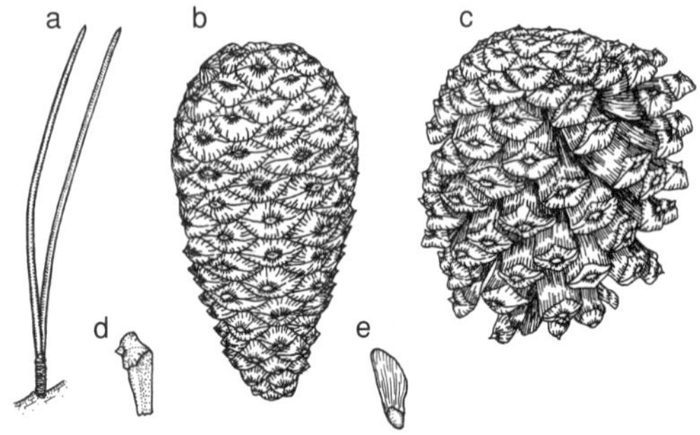

FIGURE 9.30 *Pinus clausa* (a) fascicle of two needles × 1/2, (b) closed cone × 1/2, (c) open cone × 1/2, (d) side view of cone scale × 1/2, (e) winged seed × 1/2.

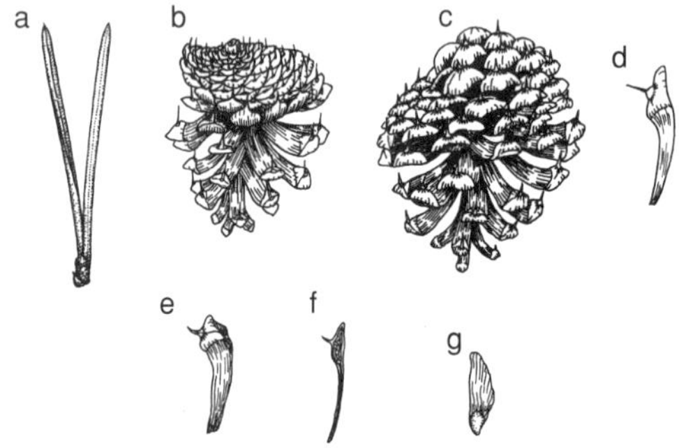

FIGURE 9.31 *Pinus contorta* var. *murrayana* (a) fascicle of two needles × 1/2, (b), (c) open cones × 1/2, (d), (e) side view of cone scales × 1/2, (f) cross section of cone scale × 1/2, (g) winged seed × 1/2.

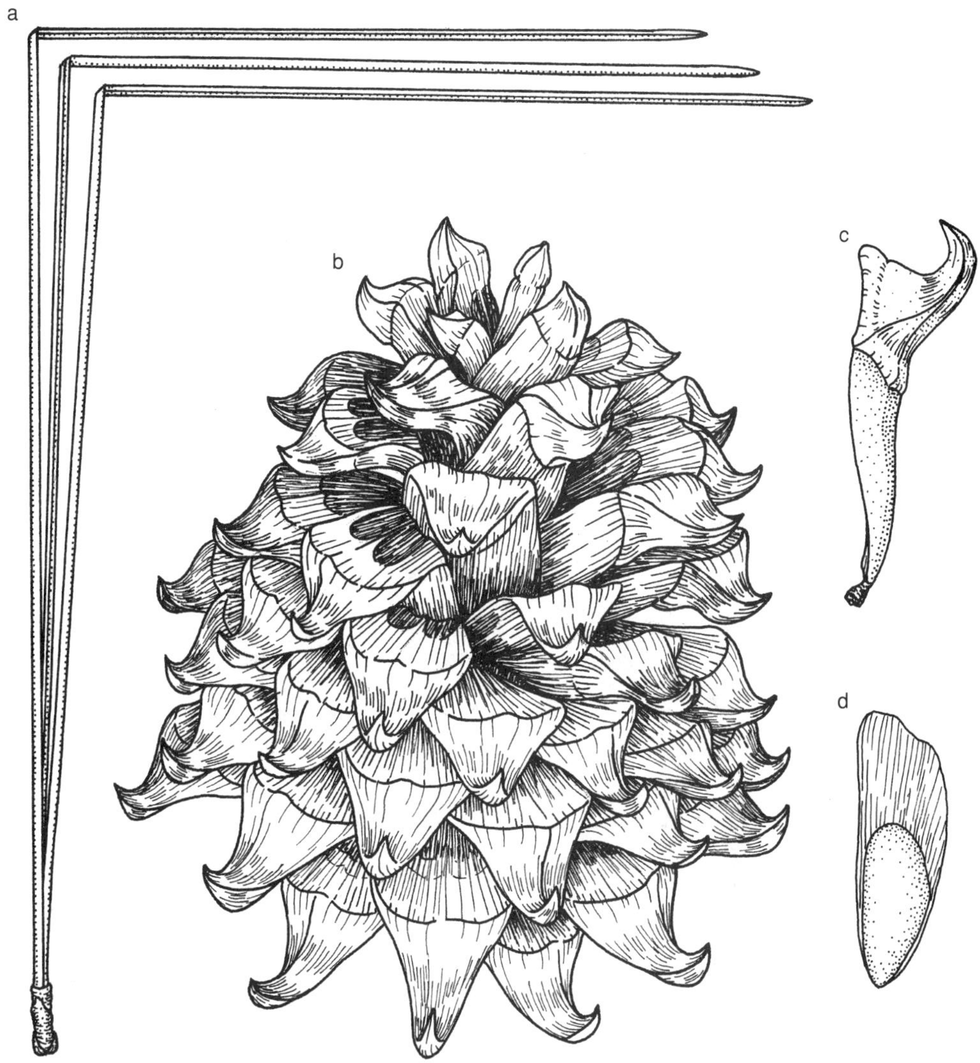

FIGURE 9.32 *Pinus sabiniana* (a) fascicle of three needles × 1, (b) cone × 1/2, (c) side view of cone scale × 1, (d) winged seed × 1.

Pinus contorta Douglas var. *contorta*, shore lodgepole pine (the original population described by Douglas): tree or shrubby, height to 10 m, often deformed and leaning inland, shrubiness (wind trained) due to salt spray and prevailing onshore wind; leaf length 2 to 7 cm; cones infrequently serotinous and in whorls, oriented at right angles to twig or pointing toward the base of the branch, strongly asymmetrical; Region 7a, coastal mountains from southeast Alaska and British Columbia to Oregon and northern California; maritime or coastal forest bogs, sand dunes, rocky ridges, elevation 0 to 600 m. Cover type 30.7 *Pinus*

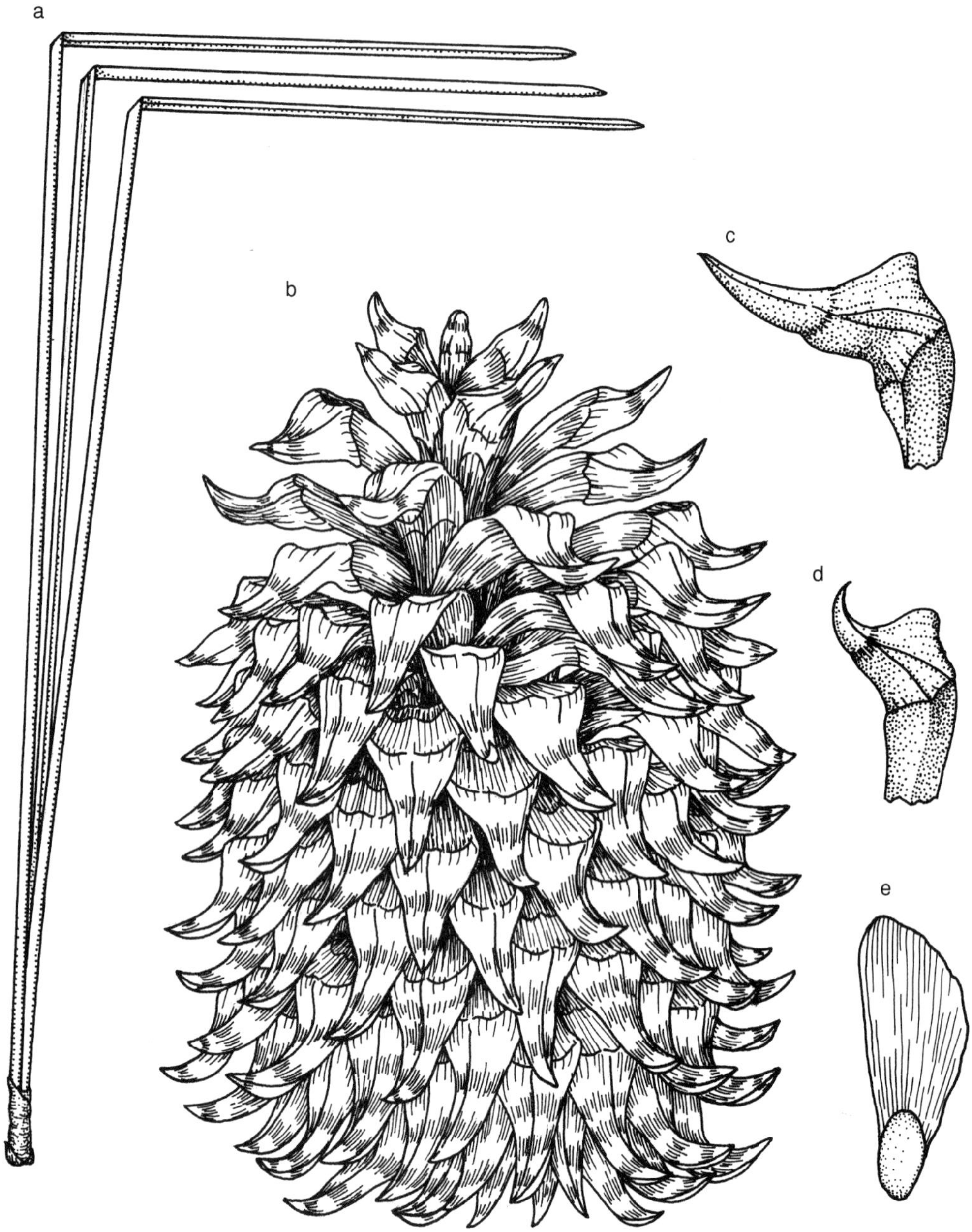

FIGURE 9.33 *Pinus coulteri* (a) fascicle of three needles × 1, (b) cone × 1, (c), (d) side view of cone scales × 1, (e) winged seed × 1.

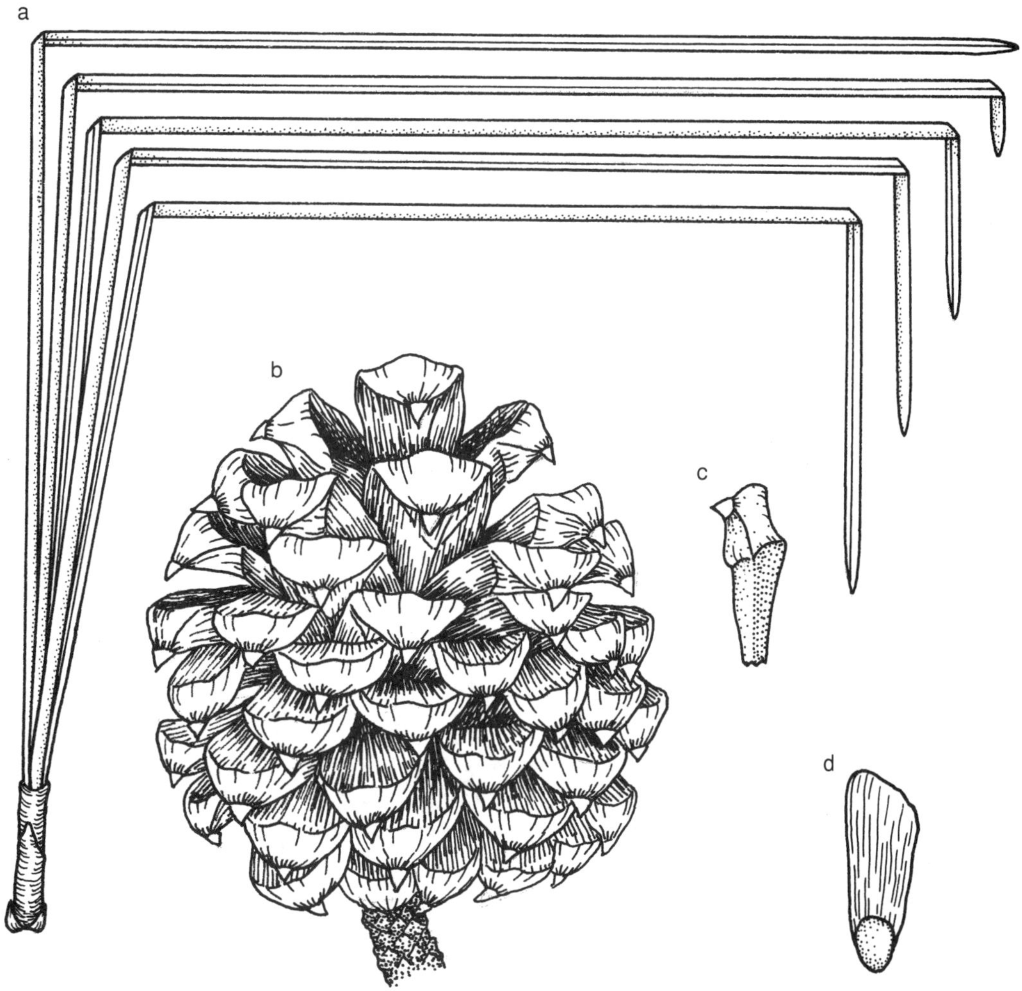

FIGURE 9.34 *Pinus torreyana* (a) fascicle of five needles × 1, (b) cone × 2/3, (c) side view of cone scale × 2/3, (d) winged seed × 1.

contorta var. *bolanderi* (Parlatore) Vasey is isolated in Mendocino County of northern coastal California and is considered to be part of the shore pine population; however, it has a heavier serotinous cone and lacks the characteristic leaf resin canals (Harlow et al. 1996).

Pinus contorta Douglas ex Loudon var. *latifolia* Engelmann, Rocky Mountain lodgepole pine; large tree, height to 45 m; leaf length 5 to 8 cm; cones serotinous, closed for several years, solitary or in twos, strongly asymmetrical oriented at right angles to twig or pointing toward the base of the branch; Regions 5 and 6, Rocky Mountains, southern Colorado to Montana, Oregon, Washington, Alberta, British Columbia, the Yukon, and Northwest Territories; montane forest, xeric to mesic sites, elevation up to 3500 m. The term "lodgepole" comes from the use of

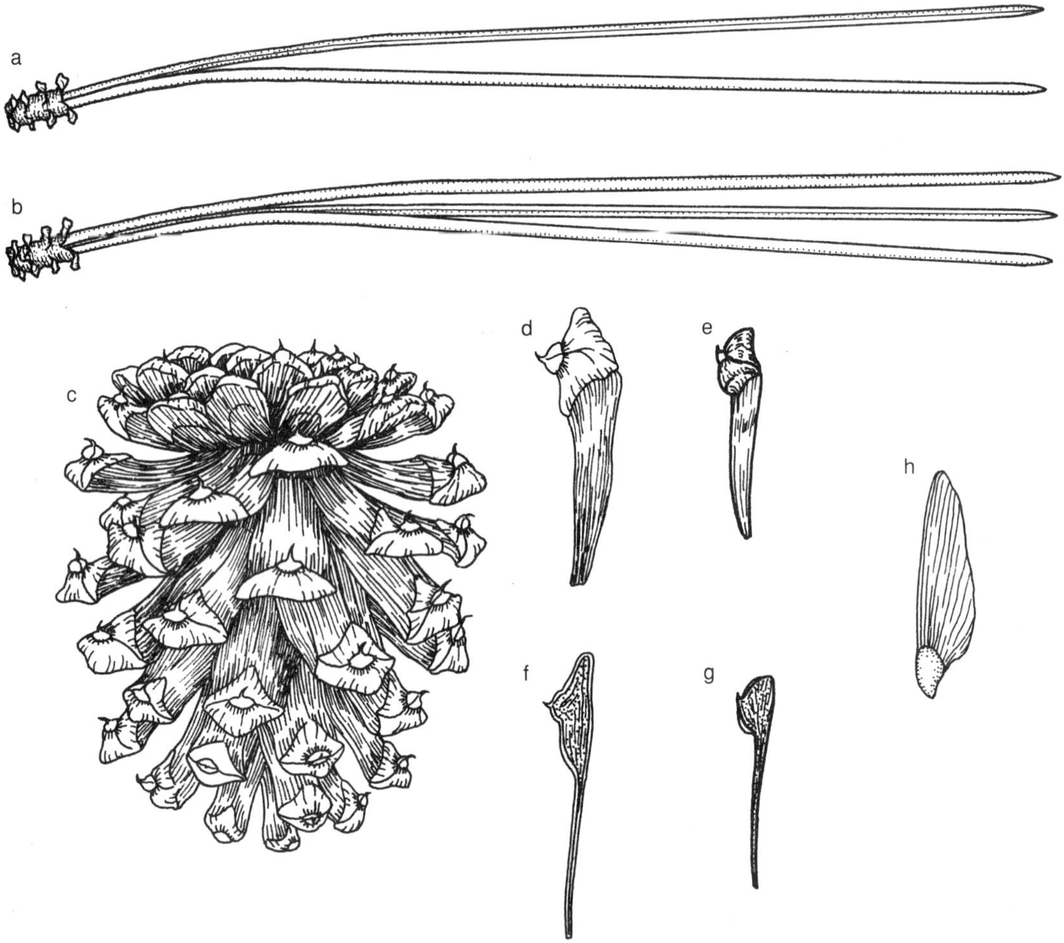

FIGURE 9.35 *Pinus ponderosa* var. PONDEROSA (a), (b) fascicles of two and three needles × 1, (c) cone × 2/3, (d), (e) side view of cone scales × 1, (f), (g) cross section of cone scales × 1, (h) winged seed × 1.

tall, thin trees for lodges by Native Americans living on the plains. Cover type **29.18**, **29.30**, 29.11, 29.15, 29.16, 29.26, 29.28, 29.29, 29.34, 29.35, 29.36.

Pinus contorta Douglas ex Loudon var. *murrayana* (Greville and Balfour) Engelmann, Sierra lodgepole pine; similar to *P. contorta* var. *latifolia*, except for seed cones opening and releasing seed at maturity; Region 7, Cascade Mountains of inland southern Washington to the Sierra Nevada of central California, disjunct populations in southern California and Baja California; mid and high elevation forests, from 400–3600 m; hydridizes with *P. banksiana* in Alberta, British Columbia, and Northwest Territories. Cover type **30.13**, **30.14**, 30.17, 30.27, 30.32, 30.43.

Subsection Sabinianae

Pinus sabiniana Douglas ex D. Don; gray (formerly called Digger) pine (Figure 9.32); large tree, height to 20 m; leaves three per fascicle, length 18 to 28 cm, slightly drooping, stomate lines present, fascicle sheath

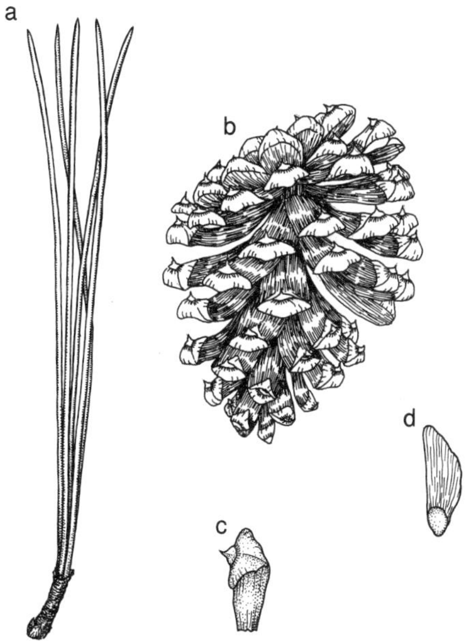

FIGURE 9.36 *Pinus arizonica* (a) fascicle of five needles × 3/4, (b) asymmetrical cone × 1/2, (c) side view of cone scale × 1/2, (d) winged seed × 3/4.

length ~2.5 cm; cone large and heavy, pendent, ovoid, length 18 to 25 cm, width 13–20 cm, apophysis dorsal, umbo substantially enlarged and extended to form a distinctly strongly reflexed claw, claw length 1.5 to 2.0 cm; Region 7, northern California (Mount Shasta area) south along west slope of the Sierra Nevada and intermittently in the Coastal Ranges to Tehachapi Mountains and coastal islands (e.g., Santa Catalina, San Clemente); xeric slopes up to 1500 m. Seeds were an important food source for the California "Digger Indians"; thus "digger" has become a derogatory ethnic term to be avoided. Cover type 30.29, 30.31, 30.38, 30.41.

Pinus coulteri D. Don, Coulter pine (Figure 9.33); large tree, height to 25 m; leaves three per fascicle, length 16 to 31 cm, stomate lines present, fascicle sheath length ~ 4 cm; cone large and heavy, pendent, length 20 to 35 cm, width 13 to 20 cm, weight to 2 kg (heaviest cone), ovoid, apophysis dorsal, umbo enlarged and extended to form a sharp claw, claw length 1.5 to 2.0 cm; Region 7, central and southern California Coastal Ranges and islands (e.g., Santa Catalina, San Clemente), Baja; elevation to 2300 m, xeric slopes and rocky soils. Cover type 30.25, 30.36, 30.39.

Pinus torreyana Parry ex Carriere, Torrey pine (Figure 9.34); small tree, but occasionally arboreous, height to 15 m; leaves five per fascicle, length 16 to 30 cm, stiff, stomate lines present, fascicle sheath length ~2 cm; cone length 10 to 15 cm, broadly ovoid, apophysis greatly thickened, five distinct convergent keels, umbo armed with an exceptionally stout prickle or beak, maturing in 3 years; Region 7, southern coastal California and islands; rare, smallest range of North American pines; xeric slopes, bluffs. Kral (1993) reports two subspecies.

Pinus torreyana Parry ex Carriere subspecies *torreyana*, mainland Torrey pine; mature tree height greater than crown width; cone length < 13.5 cm, umbo length < 6 cm, tip of spine curved outward.

Pinus torreyana Parry ex Carriere subspecies *insularis*, J. R. Haller; island Torrey pine; mature tree crown width greater than height; cone length > 13.5 cm, umbo length > 6 cm, complete umbo curved outward.

Subsection Ponderosae (Western Yellow Pine)

Pinus ponderosa Douglas ex Lawson, ponderosa pine (Figure 9.35). Three varieties have been recognized; however, Farjon et al. (1997) have raised one variety (*P. ponderosa* Douglas ex. Lawson var. *arizonica* (Engelmann) Shaw to species status (*P. arizonica* Engelmann) based on material from Mexico.

Pinus ponderosa Douglas ex Lawson var. *ponderosa*; ponderosa pine; tree, height to 70 m; leaves three, and sometimes two, per fascicle, length 12 to 28 cm, fascicle sheath length 1.5 to 3.0 cm; cone length 7 to 15 cm, symmetrical apophysis diamond shaped, umbo central, pyramidal, armed, prickle deciduous; Region 7, southern California and the Sierra Nevada to Washington, Oregon, and southern British Columbia; xeric to mesic sites, open forest and savanna, 0 to 2300 m. Individuals of this species are the largest of the hard (yellow) pines. Hybridizes with *P. jeffreyii*. Cover type **30.34, 30.44**, 30.3, 30.7, 30.15, 30.18, 30.20, 30.25, 30.26, 30.28, 30.31, 30.32, 30.35, 30.36, 30.37, 30.38, 30.43.

FIGURE 9.37 *Pinus jeffreyi* (a), (b) fascicles of three and two needles × 1, (c) cone × 1, (d) winged seed × 1, (e) side view of cone scale × 1.

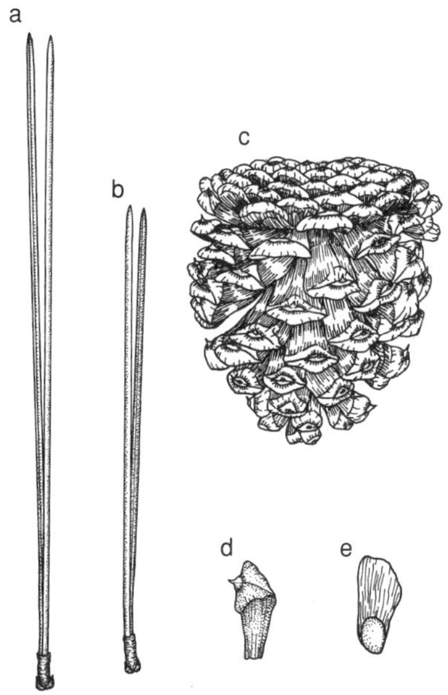

FIGURE 9.38 *Pinus washoensis* (a), (b) fascicles of two needles × 1/2, (c) cone × 1/2, (d) side view of cone scale × 1/2, (e) winged seed × 1/2.

Pinus ponderosa Douglas ex Lawson var. *scopulorum* Engelmann, Rocky Mountain ponderosa pine; large tree, height to 24 m; leaves two and three per fascicle, length 10 to 20 cm, fascicle sheath length 1.5 to 3.0 cm; cone length 5 to 10 cm, symmetrical, umbo central, pyramidal, armed, prickle deciduous; Regions 5 and 6, Texas, Arizona, and New Mexico north through the Rocky Mountains to South Dakota (Black Hills), Montana, Idaho, Oregan, Washington, and British Columbia; xeric to mesic sites, open forest and savanna, foothills to 3000 m. There is hybridization with var. *ponderosa* and *P. arizonica*. Cover type **29.6, 29.33**, 29.1, 29.3, 29.6, 29.10, 29.15, 29.24, 29.35, 29.36.

Pinus arizonica Engelmann, Arizona pine (Figure 9.36); large tree, height to 35 m; leaves three to five per fascicle, length 8 to 16 cm, slightly curved, stomata lines on all surfaces, fascicle sheath persistent; cones paired or in whorls of three, length 5 to 10 cm, broadly ovoid, asymmetrical, apophysis strongly raised, umbo central and pyramidal; twigs glaucous; Region 5, extreme southern Arizona and New Mexico, primary range in Mexico; xeric sites, 2000 to 3000 m. Farjon et al. (1997) recognize three varieties. Cover type 29.2.

Pinus jeffreyi Greville & Balfour in A. Murray, Jeffrey pine (Figure 9.37); large tree, height to 40 m, occasionally taller; leaves three, and occasionally two, per fascicle, length 10 to 23 cm, stomate lines on all surfaces, glaucous with a pineapple odor when crushed, fascicle sheath length 1.5 to 3.0 cm; cone length 12 to 30 cm, apophysis thickened, not keeled, umbo central, armed with a sharp spine; twigs purplish; Region 7, southwestern Oregon through the Sierra Nevada of California, western Nevada, into Baja, Mexico. Cover type **30.32**, 30.27, 30.28, 30.37, 30.43.

Pinus washoensis Mason & Stockwell, Washoe pine (Figure 9.38); large tree, height to 60 m; leaves two or three per fascicle, length 10 to 15 cm, stomate lines on all surfaces, fascicle sheath length 1 to 2 cm; cone length 7 to 10 cm, symmetrical, apophysis slightly thickened, umbo central, pyramidal, prickle reflexed; Region 7, isolated in northeast California and northwest Arizona; xeric montane forest and savanna; 2100 to 2500 m. Often considered a part of the *P. ponderosa* population.

Pinus engelmannii Carrierre, Apache pine (Figure 9.39); large tree, height to 35 m; leaves usually three per fascicle but also two, four, and five per fascicle, length 22 to 45 cm, often drooping, stomate lines on all surfaces, fascicle sheath length 3 to 4 cm, margin coarsely serrulate; cone length 11 to 14 cm, ovoid, apophysis rhombic, slightly thicker than shaft, keeled, umbo central, armed, raised, tapering to an outcurved claw; Region 5, extreme southeast Arizona and southwest New Mexico, primary range in Mexico; dry montane forest, 1500 to 2500 m. Similar in appearance to the eastern *P. palustris*; both have a "grass" stage and similar needles. Cover type 29.2.

Subsection Oocarpae

Pinus radiata D. Don, Monterey pine (Figure 9.40); large tree, height to 30 m; leaves three per fascicle, infrequently two per fascicle, length 9 to 19 cm, stomate lines on all surfaces, fascicle sheath length 1.5 to 2.0 cm; cone length 7 to 14 cm, extremely asymmetrical, serotinous, basal apophyses nipple-like, umbo armed with a prickle; Region 7, three coastal counties of central California. Although a poorly formed

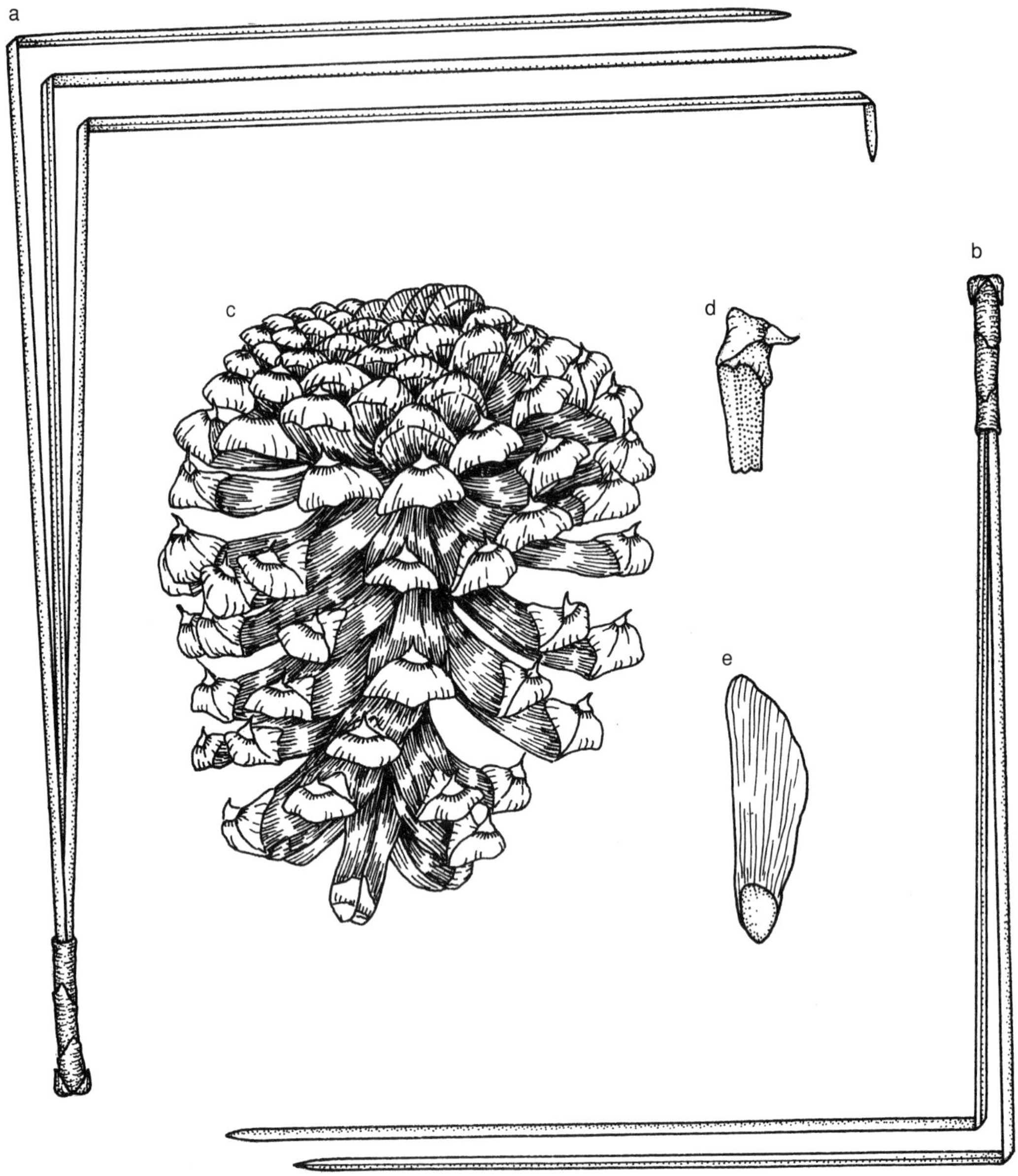

FIGURE 9.39 *Pinus engelmannii* (a), (b) fascicles of three and two needles ×3/4, (c) cone × 3/4, (d) side view of cone scale × 1, (e) winged seed × 1.5.

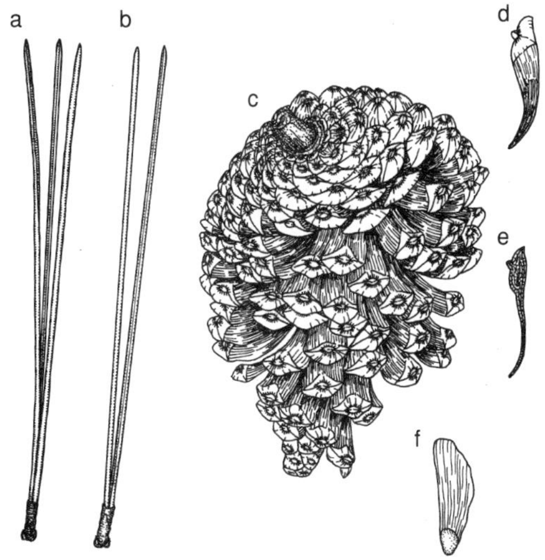

FIGURE 9.40 *Pinus radiata* (a), (b) fascicles of three and two needles × 1/2, (c) asymmetrical cone × 1/2, (d) side view of cone scale × 1/2, (e) cross section of cone scale × 1/2, (f) winged seed × 1/2.

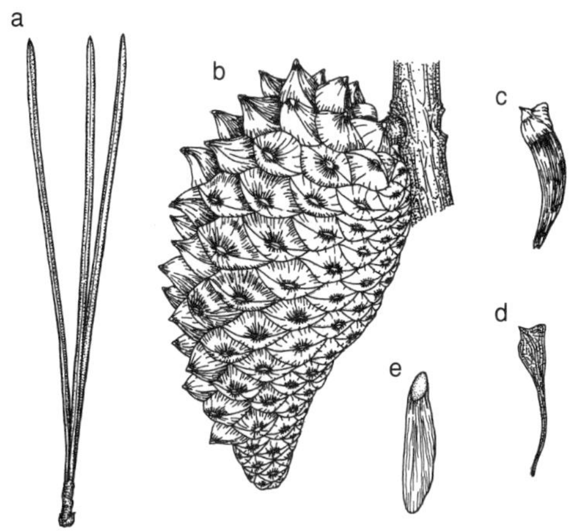

FIGURE 9.41 *Pinus attenuata* (a) fascicle of three needles × 1/3, (b) closed asymmetrical cone × 1/3, (c) side view of cone scale × 1/3, (d) cross section of cone scale × 1/3, (e) winged seed × 1/2.

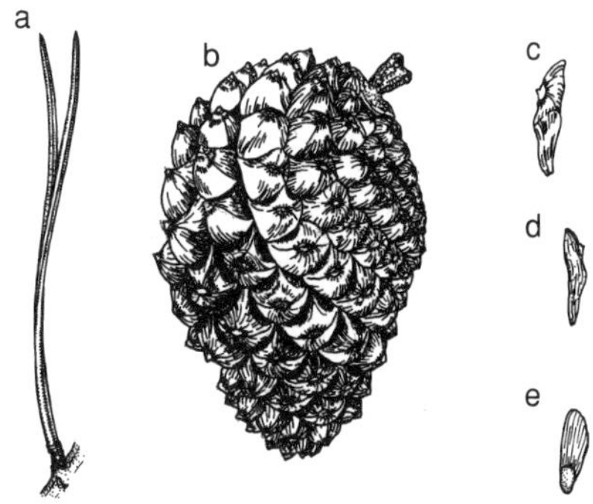

FIGURE 9.42 *Pinus muricata* (a) fascicle of two needles × 1/3, (b) closed asymmetrical cone × 1/2, (c) side view of cone scale × 1/2, (d) cross section of cone scale × 1/2, (e) winged seed × 1/2.

tree in California, it is taller, straighter, and an important timber-producing plantation tree in New Zealand, Australia, Africa, and Europe. Cover type **30.35**.

Pinus attenuata Lemmon, knobcone pine (Figure 9.41); large tree, height to 24 m; leaves three per fascicle, length 9 to 19 cm, stomate lines on all surfaces, fascicle sheath length 1.5 to 2.0 cm; cone length 8 to 15 cm, extremely asymmetrical, serotinous, hard and heavy, basal apophyses nipple-like, rhombic, umbo central, pyramidal, often with a short prickle; Region 7, southern California to southern Oregon; xeric slopes and foothills, primarily the Sierra Nevada, open woods and savannas, often on serpentine soil, 300 to 1200 m. Cover type **30.31**, 30.29.

Pinus muricata D. Don, bishop pine (Figure 9.42); large tree, height to 24 m; leaves two per fascicle, length 8 to 15 cm, margin strongly serrulate, stomate lines on all surfaces, fascicle sheath length to 1.5 cm; cone length 4 to 9 cm, extremely asymmetrical, maturing in 3 years, serotinous, hard and heavy, in whorls, apophysis considerably thicker than shaft, basal apophyses becoming dome-like, umbo central, becoming a stout claw; Region 7, California and the Baja; Coastal Ranges, xeric slopes to near bog conditions, elevation to 900 m. Cover type **30.33**, 30.35.

CHAPTER 10

Order: Pinales II: Pinaceae: Larix, Picea, Abies, Tsuga, Pseudotsuga

INTRODUCTION

THE leaf structure and arrangement and the cones of *Larix* (larch), *Picea* (spruce), *Abies* (fir), *Tsuga* (hemlock), and *Pseudotsuga* (Douglas-fir) are distinct from those of other gymnosperm families. As compared with *Pinus* species (Chapter 9), the leaves are shorter, usually have a flattened or square cross-sectional shape (except in *Larix*, which is similar to *Pinus*), and are individually attached to the twigs. Although many of the Cupressaceae genera (Chapter 11) have extremely short (scale-like) leaves appressed to the branchlet that sharply contrast with those of the genera of Pineaceae considered here, *Sequoia, Sequoiadendron, Taxodium*, and *Glyptostroboides* have linear leaves similar to those of *Abies, Tsuga*, and *Pseudotsuga*. In *Larix*, a large number of leaves are clustered on a spur shoot, whereas in most species of *Abies, Tsuga*, and *Pseudotsuga*, the leaves alternate (spiral) along the twig or may be two-ranked. The cones have only a color apophysis (no swelling) and no umbo. The bracts are often longer than the scales. Two seeds are borne on each scale, and although the seeds are released soon after the cone opens, the paired indentations or color differences are easily visible.

SPECIES DESCRIPTIONS

Genus: *Larix*, Larch

The circumpolar genus *Larix* (Figure 10.1) contains ten species worldwide, but only three are native to North America: *Larix laricina* (eastern larch or tamarack) of Canada and the northern United States, and *Larix occidentalis* (western larch) and *Larix lyallii* (subalpine larch) of the northern Rocky Mountains (Parker 1993). The ranges of *L. occidentalis* and *L. lyallii* overlap, but they do not overlap with *L. laricina*. *Larix kaempferi* (Japanese larch) and *Larix decidua* (European larch) are introduced.

Larix needles are soft, slender, short (usually less than 4 cm in length), and triangular to square (*L. lyallii*) in cross section. They are borne singly and alternately (spirally) on twigs formed during the current growing season, or on spur (short) shoots of older twigs. Unlike most other gymnosperms, *Larix* trees are deciduous and bare through the dormant season; however, the presence of numerous spur shoots is positive identification. On older twigs, leaf scars are confined to the sides of the spur shoot, giving it a rough appearance. A small globose bud is situated at the end of each spur shoot. The bark is rough, scaly, and loose.

The male and ovulate strobili appear before the leaves in the spring and mature during the same year. The mature cones are nearly round to oblong, less than 3 cm in length, and remain on the twig for several growing seasons. The cone scales are thin with a color apophysis. Each scale is subtended by a narrow (2 mm wide) bract that is longer (exserted) in *L. occidentalis* and *L. lyallii*) or shorter (inserted) in *L. laricina* than the scale. The seeds have long wings; the seedlings have six cotyledons. Larch is used for pulpwood, rough construction lumber, poles, and specialty items.

FIGURE 10.1 A *Larix laricina* tree growing in an *Alnus incana*-dominated swamp in northern Wisconsin. The fine-textured graceful crown is typical. *Photo by J. Fralish.*

Eastern Species

Larix laricina (DuRoi) K. Koch, eastern larch, tamarack (Figure 10.2); medium-sized tree, height to 20 m; deciduous; leaf length 1 to 2 cm, rounded on one side, leaves borne singly on new twigs or clustered

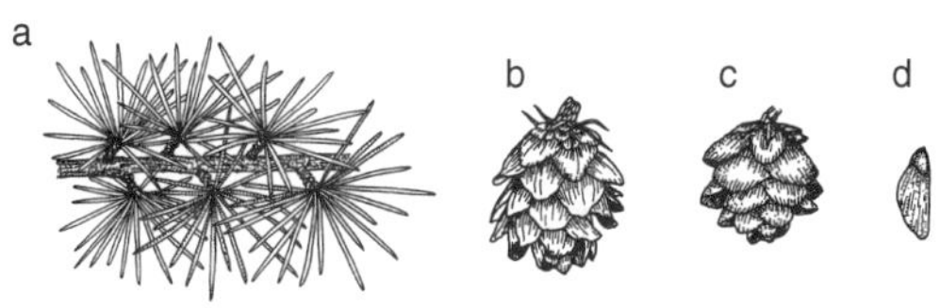

FIGURE 10.2 *Larix laricina* (a) 3- or 4-year-old twig with spur shoots and attached leaves × 1/2, (b), (c) cones × 1/2, (d) winged seed × 1/2.

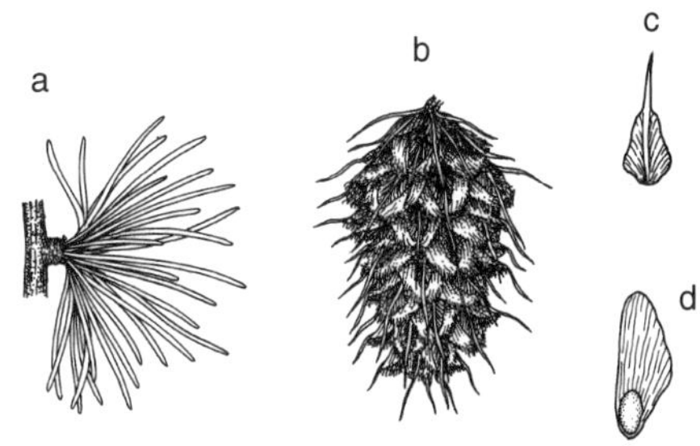

FIGURE 10.3 *Larix occidentalis* (a) 3- or 4-year-old twig with spur shoot and attached leaves × 1/2, (b) cone with exserted bracts × 1, (c) bract × 1, (d) winged seed × 1.

on spur shoots of second-, third-, and fourth-year twigs; cone brown, length 1 to 2 cm, stalk length 2 to 5 mm, color apophysis, margin smooth; bracts hidden by scales, terminate in a sharp point, twigs orange brown, glabrous, spur shoots present; Regions 1 and 2, Alaska to Nova Scotia, Lake States, and New England, extends to southern edge of permafrost; swamps and wet sites but also on upland sites. Important for pulpwood and locally for posts and fuel. Cover type **26.3, 26.6**, 26.1, 26.5, 26.16, 26.25.

Western Species

Larix occidentalis Nuttall, western larch (Figure 10.3); large tree, height to 50 m; deciduous; leaf length 2 to 5 cm, borne singly on new twigs or on spur shoots of second-, third-, and fourth-year twigs; cone purplish, length 2 to 3 cm, stalk length 2.5 to 4.5 mm, color apophysis, margin smooth, bracts longer than scales by 3 to 4 mm, terminate in a sharp 4 to 5 mm long point; twigs orange brown, new twigs pubescent becoming glabrous, spur shoots present; Regions 6 and 7, British Columbia, Washington, Oregon, and Montana; mountain slopes, 500 to 1600 m. Commercially important timber-producing tree. Cover type **29.27**, 29.11, 29.23, 29.28, 29.30, 29.32, 30.3, 30.21.

Larix lyallii Parlatore, subalpine larch (Figure 10.4); medium-sized tree, height to 25 m, sometimes stunted; deciduous; leaf length 2.0 to 3.5 cm, leaves borne singly on new twigs or on spur shoots of second-, third-, and fourth-year twigs; cone purplish length 2.5 to 5.0 cm, color apophysis, margin erose, bracts longer than scales by 6 mm, terminate in a sharp 4 to 5 mm long point; twigs white to yellow

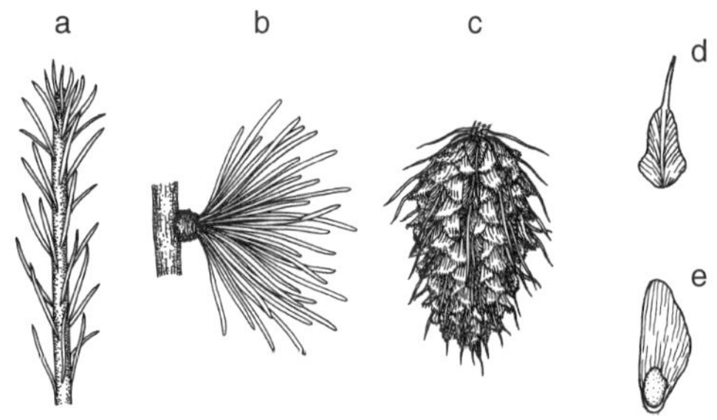

FIGURE 10.4 *Larix lyallii* (a) first-year twig and leaves × 1/3, (b) 3- or 4-year-old twig with spur shoot and attached leaves × 1/2, (c) cone with exserted bracts × 1/2, (d) bract × 1/2, (e) winged seed × 1.

tomentose for several years, spur shoots present; Region 6, British Columbia, Washington, Idaho, and Montana; high mountain slopes, 1800 to 2400 m. Important for protection of high-altitude watersheds. Cover type **29.25**, 29.11, 30.12.

Introduced Species
Larix decidua Miller, European larch (Figure 10.5); large tree, height to 30 m; leaf and twig characteristics similar to those of *L. laricina*; cone brown, length 2 to 4 cm, bracts shorter than scales, but long point usually visible on abaxial sides of scales; introduced, found in plantations planted as a lawn tree.

Larix kaempferi (Lambert) Carriere, Japanese larch (Figure 10.5); large tree, height to 30 m; leaf and twig characteristics similar to those of *L. laricina*; cone length 2 to 4 cm, bracts shorter than scale tips, distinctly reflexed; introduced, found in plantations or planted as a lawn tree.

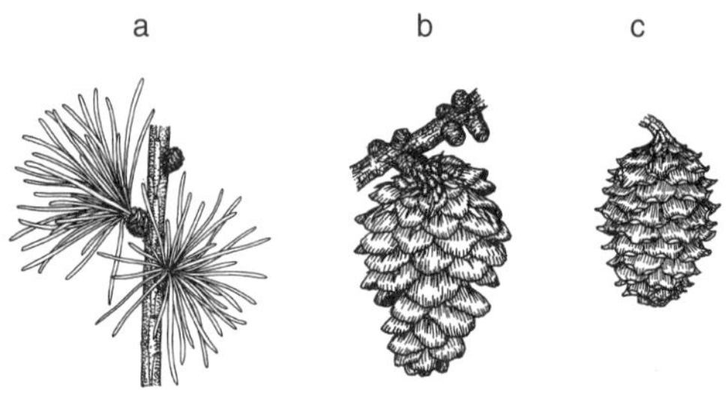

FIGURE 10.5 *Larix decidua* Miller (a) 3- or 4-year-old twig with spur shoot and attached leaves × 2/3, (b) cone × 1/2; (c) *L. kaempferi*, cone × 1/2; note reflexed scale tips.

Genus: *Picea*, Spruce

Picea (Figure 10.6) is the second largest genus in the Pinaceae, with a total of approximately 30 species; about half are native to China and only seven to North America. Northern/eastern species include *Picea glauca* (white spruce) and *Picea mariana* (black spruce), primarily found in Canada, and *Picea rubens* (red spruce) found in the northeast and at the highest elevations in the Appalachian Mountains. Western mountain species include *Picea pungens* (blue spruce), *Picea engelmannii* (Engelmann spruce), *Picea sitchensis* (Sitka spruce), and *Picea breweriana* (Brewer spruce). In general, individuals of western species are considerably larger than those of northern/eastern species. The most common introduced species is *Picea abies* (Norway spruce), which is regarded as an extremely desirable lawn tree/ornamental because of its graceful form developed from drooping branches with upturned tips.

Like *Larix*, the genus is circumpolar. The needles are typically four-angled in cross section (Figure 10.7),

FIGURE 10.6 *Picea mariana* community in northern Wisconsin. The characteristic club-top crowns are indicative of slow growth on wet sites and in swamps. *Photo by J. Fralish.*

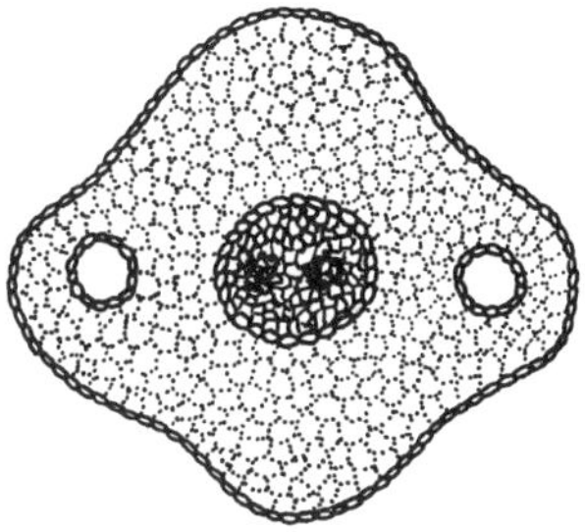

FIGURE 10.7 Cross section of a *Picea* needle. The vascular strand and two resin canals are aligned in the mesophyll. Note the relatively square shape.

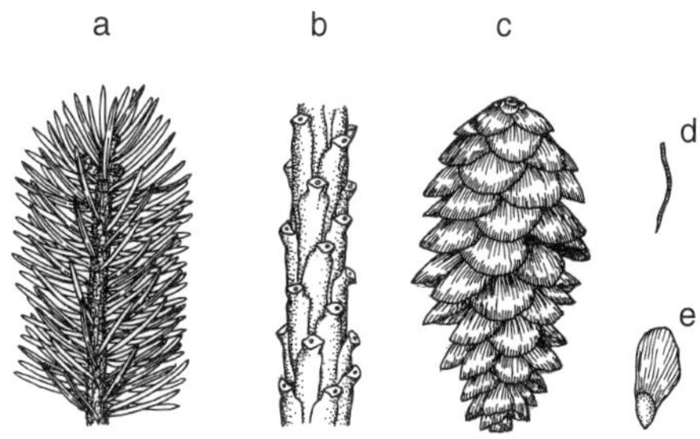

FIGURE 10.8 *Picea glauca* (a) twig with attached leaves × 1/2, (b) bare twig with sterigmata (pegs) × 3, (c) cone × 2/3, scale margins smooth, (d) cone scale cross section × 2/3, (e) winged seed × 1.

relatively stiff, and sharp pointed; they may remain on a tree for 8 to 10 years. The leaf is individually attached to the twig on a small peg called a sterigma (plural: sterigmata). The pegs remain on the twig when the oldest leaves finally die, thus the surface of an old twig is quite rough and easy to identify. The buds are not resinous and when dry, the tips of the scales may reflex (curl back), giving the bud an appearance somewhat like that of a rose.

The cones vary in shape from oval or ovate to cylindrical, and in length from 2 to 18 cm. Species may be divided into two groups, based on the cone scale margin. The northern/eastern *Picea* have more or less smooth, rounded margins and a slightly thickened leathery scale. The western species generally have a thin and extremely brittle scale when dry. The irregular, erose margins converge in a wedge-like manner to a truncated or blunt apex. The bark is rough and broken into thin, scaly plates. Crown shape usually resembles an inverted cone.

Picea wood, which is extremely versatile, is used for pulpwood, plywood, lumber, millwork, shipbuilding, and specialty products. During World War II, *P. sitchensis* wood was used for airplane construction because of its light weight and high strength.

Northern/Eastern Species

Picea glauca (Moench) Voss, white spruce (Figure 10.8); large tree, height to 30 m; evergreen; leaf shape acicular, length 1.3 to 2.0 cm, four-angled in cross section, stiff, sharp pointed, stomata lines on all sides, color blue green, glaucous; cone length 3 to 6 cm, scales fan-shaped, margin smooth, apophysis thin and differentiated by color; twig yellowish brown, glabrous, sterigmata present; Regions 1 and 2, transcontinental, Alaska to Newfoundland, Lake States, and New England; primarily moist upland sites such as stream terraces and loamy soils. At least two varieties of *P. glauca* are recognized: *P. glauca* var. *albertiana* (S. Brown) Sargent and *P. glauca* var. *porsildii.* Taylor (1993b) briefly describes these varieties and indicates that their characteristics are intermediate between those of *P. glauca* and *P. engelmannii.* Apparently hybridization and introgression occur between these two species not only where their ranges overlap, but at distances beyond their ranges. Cover type **26.4**, 26.1, 26.2, 26.3, 26.5, 26.8, 26.9, 26.13, 26.14, 26.23, 26.26.

Picea rubens Sargent, red spruce (Figure 10.9); large tree, height to 40 m; evergreen; leaf shape acicular, length 0.8 to 2.5 cm, four-angled in cross section, stiff, sharp pointed, stomata lines on all sides, color dark green, not glaucous; cone length 3 to 5 cm, scales fan-shaped, margin smooth, apophysis thin and differentiated by color; twigs generally pubescent, color yellowish brown, sterigmata present; Regions 2b and 3b, Nova Scotia, New Brunswick, and Quebec through New England into New York, disjunct communities appear along the high Appalachian mountain peaks south into North Carolina and Tennessee; sea level to 2000 m, forms nearly pure stands at high altitude. Cover type **26.8, 26.9, 26.10, 27.31**, 26.1, 26.2, 26.4, 26.15, 26.21, 26.23, 26.29, 27.29.

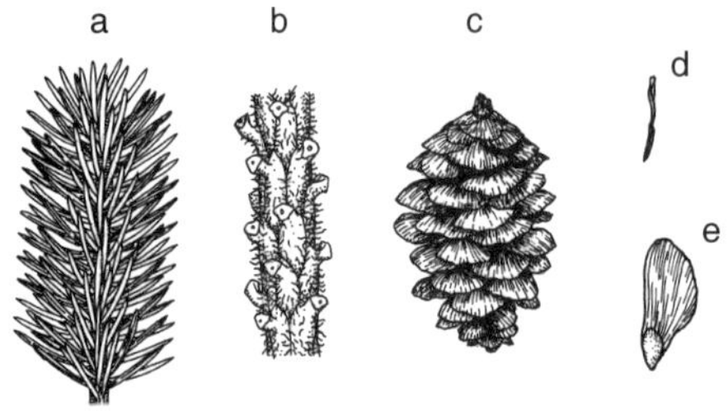

FIGURE 10.9 *Picea rubens* (a) twig with attached leaves × 1/2, (b) bare twig with sterigmata (pegs) × 3, lightly pubescent, (c) cone × 2/3, scale margins slightly erose, (d) cone scale cross section × 2/3, (e) winged seed × 1.

Picea mariana (Miller) Britton, Sterns & Poggenburg, black spruce (Figure 10.10); large tree, height to 25 m; evergreen; leaf shape acicular, length 0.6 to 1.5 cm, four-angled in cross section, stiff, sharp pointed, stomata lines on all sides; cone length 1.5 to 2.5 cm, scales fan-shaped, margin smooth, apophysis thin and differentiated by color; twigs pubescent, color yellowish brown, sterigmata present; Regions 1 and 2, transcontinental, Alaska to Newfoundland, Lake States, and New England; bogs and swamps, but also on upland sites. Cover type **26.5, 26.6, 26.7**, 26.1, 26.3, 26.4, 26.10, 26.13, 26.14, 26.16, 26.25.

Western Species

Picea pungens Engelmann, blue spruce, Colorado blue spruce (Figure 10.11); large tree, height to 50 m; evergreen; leaf shape acicular, length 1.8 to 3.0 cm, strongly four-angled in cross section, extremely stiff and sharp pointed, stomata lines on all sides, blue green, glaucous; cone length 6 to 12 cm, scale shape elliptic, widest near the middle, extremely brittle, margin erose, apophysis thin and differentiated by color; twigs stout, glabrous, color yellowish brown, sterigmata present; Region 5, Rocky Mountains, Idaho, Wyoming, Colorado, Utah, Arizona, and New Mexico; midelevation slopes, 1800 to 3000 m. Cover type **29.15, 29.16,** 29.6, 29.11, 29.18, 29.26, 29.30, 29.35.

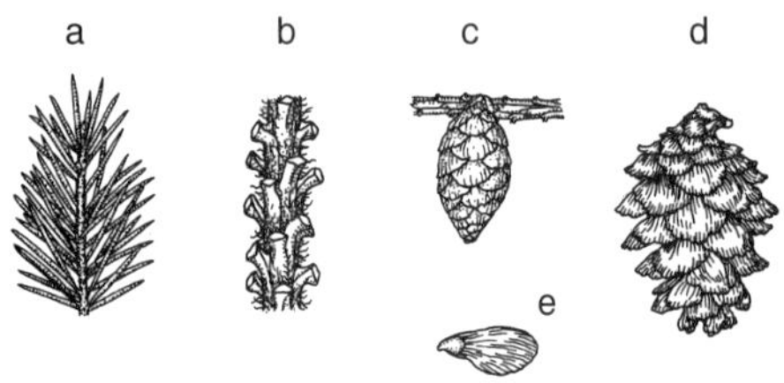

FIGURE 10.10 *Picea mariana* (a) twig with attached leaves × 3/4, (b) bare twig with sterigmata (pegs) × 2, lightly pubescent, (c) mature closed cone × 1, (d) mature open cone, scale margins erose × 1, (e) winged seed × 1.

Picea engelmannii Parry ex. Engelmann, Engelmann spruce; large tree (Figure 10.12), height to 60 m; evergreen; leaf shape acicular, length 1.6 to 3.2 cm, strongly four-angled in cross section, extremely stiff and sharp pointed, stomata lines on all sides, blue green; cone length 4 to 8 cm, scale shape elliptic, widest near the middle, extremely brittle, margin erose, apophysis thin and differentiated by color; twigs stout, usually pubescent, color yellowish brown, sterigmata present; Regions 5 and 6, Rocky Mountains from central British Columbia to New Mexico; middle elevation slopes, 600 to 1000 m in British Columbia (north), 1200 to 6000 m in New Mexico (south). Cover type **26.28**, 29.11, 29.15, 29.18, 29.23, 29.26, 29.27, 29.29, 29.30, 29.31, 29.32, 29.35, 29.36, 30.12, 30.24.

Picea sitchensis (Bongard) Carriere, Sitka spruce (Figure 10.13); large tree, height to 80 m; evergreen; leaf shape acicular, length 1.3 to 2.8 cm, somewhat flattened to triangular in cross section, extremely stiff and sharp pointed, blue green, abaxial surface darker with narrow or absent stomata lines; cone length 5 to 10 cm, scale shape elliptic, widest near the middle, extremely brittle, margin erose, apophysis thin and differentiated by color; twigs stout, glabrous, color yellowish brown, sterigmata present; Region 7, coastal Alaska to coastal northern California; middle elevation slopes, sea level to 600 m; fog belt. *Picea* × *lutzii* Little is a recognized hybrid between *P. sitchensis* and *P. glauca* (Little 1979a). Cover type **30.10, 30.11**, 30.6, 30.7, 30.18, 30.21, 30.22.

Picea breweriana S. Watson, Brewer spruce (Figure 10.14); large tree, height to 40 m; evergreen; leaf shape acicular, length 1.5 to 3.0 cm, somewhat flattened to triangular in cross section, extremely stiff and

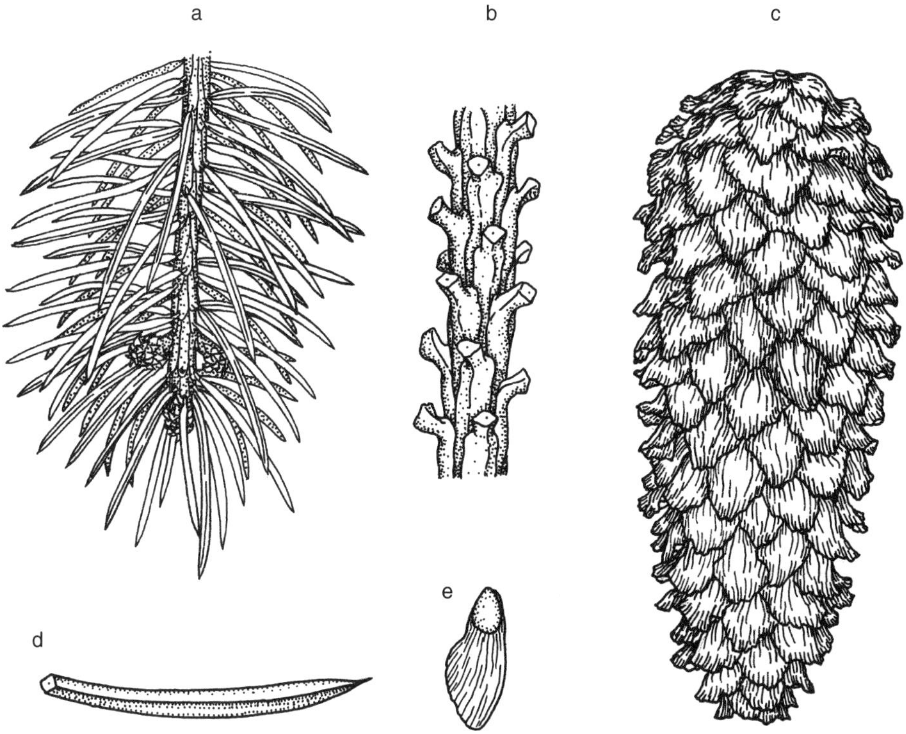

FIGURE 10.11 *Picea pungens* (a) twig with attached leaves × 1, (b) bare twig with sterigmata (pegs) × 5, (c) cone × 1, scale margins strongly erose, (d) leaf × 2, cross section square, tip sharp, (e) winged seed × 2.

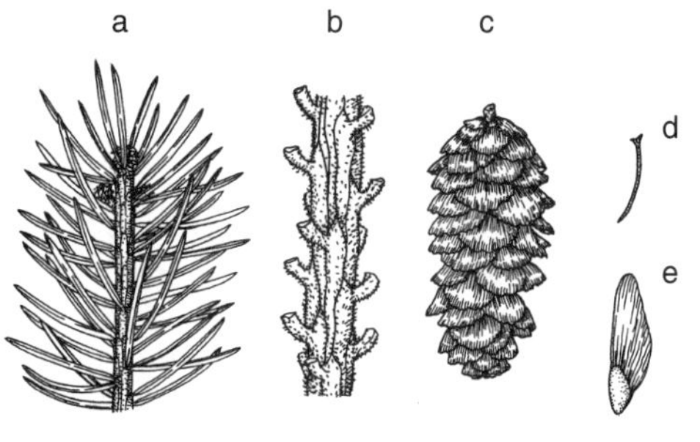

FIGURE 10.12 *Picea engelmannii* (a) twig with attached leaves × 1/2, (b) bare twig with sterigmata (pegs) × 2, lightly pubescent, (c) cone × 1/2, scale margins erose, (d) cone scale cross section × 1/2, (e) winged seed × 1.

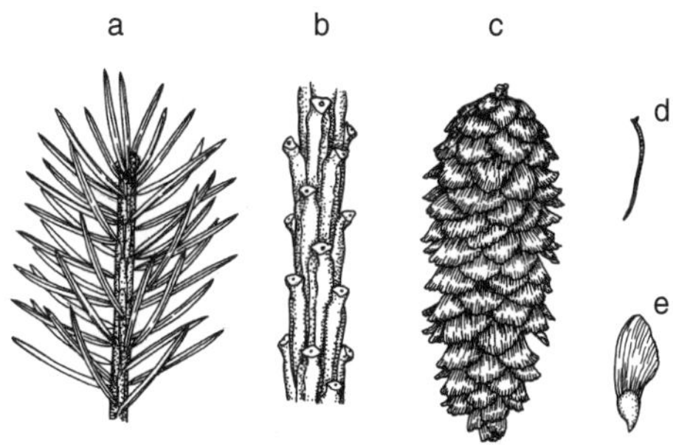

FIGURE 10.13 *Picea sitchensis* (a) twig with attached leaves × 1/2, (b) bare twig with sterigmata (pegs) × 2, (c) cone × 1/3, scale margins erose, (d) cone scale cross section × 1/2, (e) winged seed × 1.

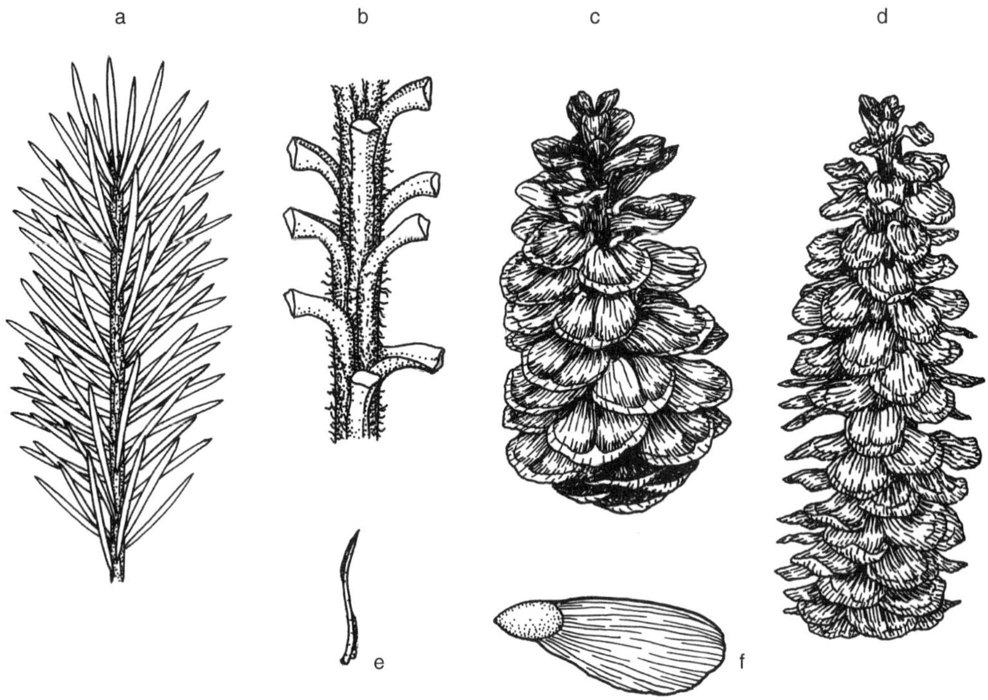

FIGURE 10.14 *Picea breweriana* (a) twig with attached leaves × 1, (b) bare twig with sterigmata (pegs) × 4, lightly pubescent, (c), (d) cone × 1, scale margins smooth, scales slightly retrorse, (e) cross section of cone scale × 1, (f) winged seed × 2.

sharp pointed, blue green, abaxial surface darker with stomata lines; cone length 5 to 12 cm, scale fan-shaped, margin generally smooth, apophysis thin and differentiated by color; twigs slender, pendent, pubescent, color gray brown, sterigmata present; Region 7; montane to subalpine forest, elevation 1000 to 2300 m. Cover type **30.30**.

Introduced Species
Picea Abies (Linnaeus) H. Karsten, Norway spruce (Figure 10.15); large tree, height to 30 m; evergreen; leaf length 1 to 2.5 cm, four-angled in cross section, stiff, dark green, stomata on all surfaces; cone length 10 to 16 cm, scales thin, margin erose or toothed; large primary branches drooping but with tips ascending, secondary twigs essentially pendent, glabrous, sterigmata present; introduced and widely planted for timber/pulpwood production in Regions 1 and 2; desirable lawn and shade tree.

Genus: *Abies*, Fir

There are about 48 species of *Abies* (true fir) worldwide. Eleven *Abies* are native to North America. Only *Abies balsamea* (Figure 10.16) and *Abies fraseri* are found in northern/eastern United States and Canada; the other nine species are found in the western mountains and coastal regions. A number of the North American species are closely related and sometimes difficult to separate (e.g., *A. balsamea* and *A. fraseri*; *Abies magnifica* and *Abies procera*) or have been recently split into two species (*Abies concolor* and *Abies lowiana*; *Abies lasiocarpa* and *Abies bifolia*) (Hunt 1993). The companion species in these three groups have ranges

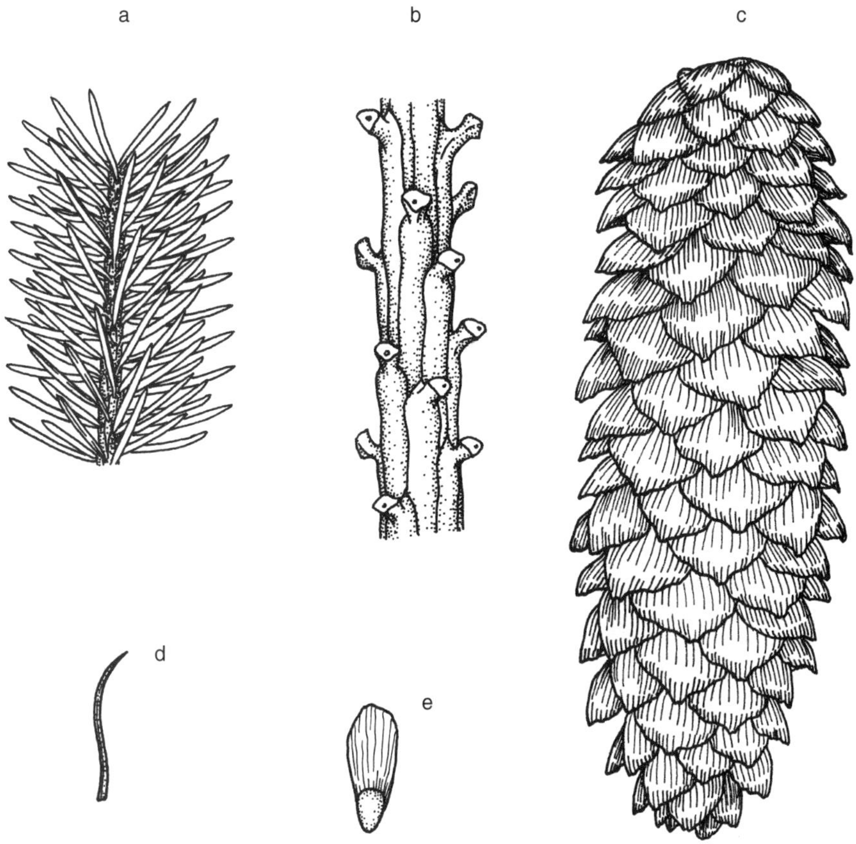

FIGURE 10.15 *Picea abies* (a) twig with attached leaves × 1, (b) bare twig with sterigmata (pegs) × 5, (c) cone × 1, scales pointed and slightly erose, (d) cross section of cone scale × 1, (e) winged seed × 2.

that are generally allopatric (do not overlap) and thus probably represent genetic drift within a common population. Hunt's (1993) coverage closely follows Little (1979).

Species of this the genus are restricted to the cold climates of the Northern Hemisphere. Their distribution is circumpolar, as they occur on all continents (North and Central America, Europe, and Asia) between 45° and 60° north latitude. They also are found in the cold climate of high altitudes; thus, *Abies* species grow in the mountains extending southward to Guatemala, northern Africa, the Himalayas, and Taiwan. The prostrate *Abies* stems known as krummholz ("dwarf forest") of alpine regions are the result of extremely cold weather and blowing ice crystals that shear off upright leaders.

Abies needles are linear (flattened in cross section; Figure 10.17a), and when lost after 5 to 10 years, leave a disclike scar resembling a bull's-eye: a circle with a dot in the center (Figure 10.17b). The leaves may be arranged in two ranks (rows), which form a horizontal plane, or in "hairbrush" fashion, whereby they all curve or point upward. Twig arrangement is usually opposite.

Each spring male and ovulate strobili develop on twigs of the previous year, but only on the uppermost branches. Four to five weeks elapse between pollination and fertilization. The cones mature late in the

FIGURE 10.16 A cluster of *Abies balsamea* trees in northern Wisconsin. *Photo by J. Fralish.*

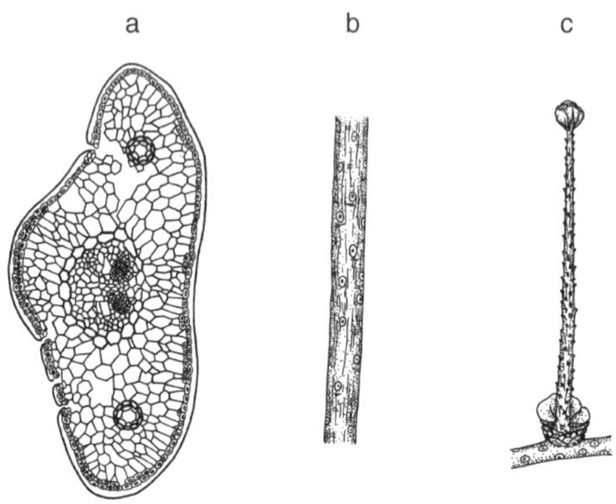

FIGURE 10.17 *Abies* (a) leaf cross section showing vascular strand and two resin canals, (b) twig with bull's-eye leaf scars, (c) upright cone stalk on twig after cone disintegrates.

season and remain on the tree only a short time. Typically, cones are composed of triangular, closely packed ovuliferous scales, which alternate with bracts. In many of the western species, the bracts are longer than the scales (exserted). *Abies* is unlike other taxa in the Pinaceae in that the immature green or purplish cones are upright on the branches and disintegrate at maturity, leaving only the central stalk on the twig over winter (Figure 10.17c). Seeds are usually paired on the scale. Cotyledon numbers range from five to seven.

Abies species are used for pulpwood, boxes, crates, general construction, millwork (siding, molding, sash, doors, trim), novelties, and Christmas trees. Low-stature, high-altitude forests are composed primarily of *Abies* and are important for watershed protection.

Northern/Eastern Species

Abies balsamea (Linnaeus) Miller, balsam fir (Figure 10.16; Figure 10.18); medium-sized tree, height to 23 m; evergreen; leaves usually two-ranked on twig, length 1.2 to 2.5 cm, width 1.5 to 2.0 mm, linear, upper surface dark green, lower surface with four to eight white parallel stomata rows, apex round to notched aromatic; branches at right angles to main stem; immature cones upright on twig, disintegrating at maturity, length 4 to 7 cm, scales rounded, bracts sharp pointed, usually shorter than scales; twigs smooth; leaf scar a bull's-eye; bark smooth, highly resinous, pitch blisters filled with clear resin; Regions 1 and 2, Canada from the eastern slope of the Rocky Mountains to Newfoundland, Lake States, and New England, disjunct areas in Pennsylvania, West Virginia and Virginia. Cover type **26.1, 26.4, 26.9**, 25.13, 26.2, 26.3, 26.5, 26.6, 26.10, 26.11, 26.14, 26.15, 26.16, 26.20, 26.21, 26.23, 26.25, 26.26, 26.29.

Abies fraseri (Pursh) Poiret in Lamark, Fraser fir (Figure 10.19), extremely similar to *A. Balsamea* except for minor leaf and cone differences; leaf lower surface with 0 to 3 rows of stomata, apex pointed; leaves arranged in hair brush fashion; immature cones purple, bracts greenish, exserted and reflexed; Region 3b, high peaks of the Appalachian Mountains, Virginia, Tennessee and North Carolina; elevation 1400 to 2000 m. Matzenko (1968), Hunt (1993), and Harlow et al. (1996) list *A. fraseri* as a separate species, whereas Thor and Barnett (1974) consider it a variety of *A. balsamea.* Based on electophoresis studies, Jacobs et al.

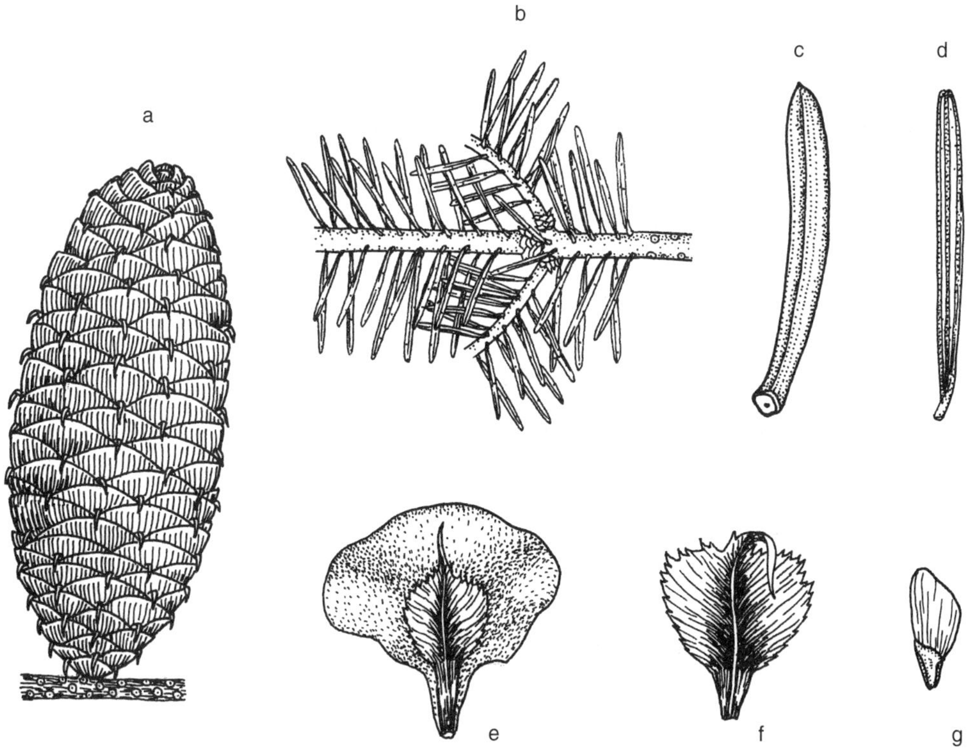

FIGURE 10.18 *Abies balsamea* (a) cone prior to disintegration × 1, (b) twig with divergent two-ranked leaf arrangement × 1, (c) leaf × 2, adaxial surface (d) abaxial surface of leaf with twisted petiole × 2, (e) cone scale with bract × 2, bract tip inserted, reflexed, (f) bract with reflexed aristate tip × 3, (g) winged seed × 1. Redrawn from Liu (1971).

(1984) indicate that *A. fraseri* may be only part of a cline in *A. balsamea. Abies fraseri* may be endangered as the result of attack by the balsam woolly adelgid and air pollution. Cover type 27.31.

Western Species

Abies concolor (Gordon & Glendinning) Hildebrand, Rocky Mountain white fir (Figure 10.20); large tree, height to 40 m; evergreen; leaves usually two-ranked on twig, length 1.5 to 6.0 cm, width 2 to 3 mm, linear, upper surface glaucous, lower surface with four to seven white parallel stomata rows each side of midrib, apex round to notched, strongly aromatic; branches at right angles to main stem; immature cones upright on twig, green, disintegrating at maturity, length 7 to 12 cm, scales rounded, bracts not visible; twigs smooth, leaf scar a bull's-eye; bark smooth, highly resinous; Region 5, Colorado, New Mexico, Arizona, Utah, Nevada, northern California, and southern Oregon; mountains, elevation 1700 to 3400 m. According to Hunt (1993), this species is composed of various populations that have green cones and glaucous upper leaf surfaces; excluding *A. lowiana*, Hunt briefly describes several morphologically distinct populations (Utah; Colorado and northern New Mexico; northern California, Baja); the southern California population is closely aligned with the southern New Mexico and Arizona population. Cover type **29.12**, 29.6, 29.11, 29.15, 29.16, 29.26, 29.30, 29.31, 29.34, 29.35.

Abies lowiana (Gordon) A. Murray, Sierra white fir (Figure 10.21); large tree, height to 60 m; evergreen; leaves usually two-ranked on twig, length 2 to 6 cm,

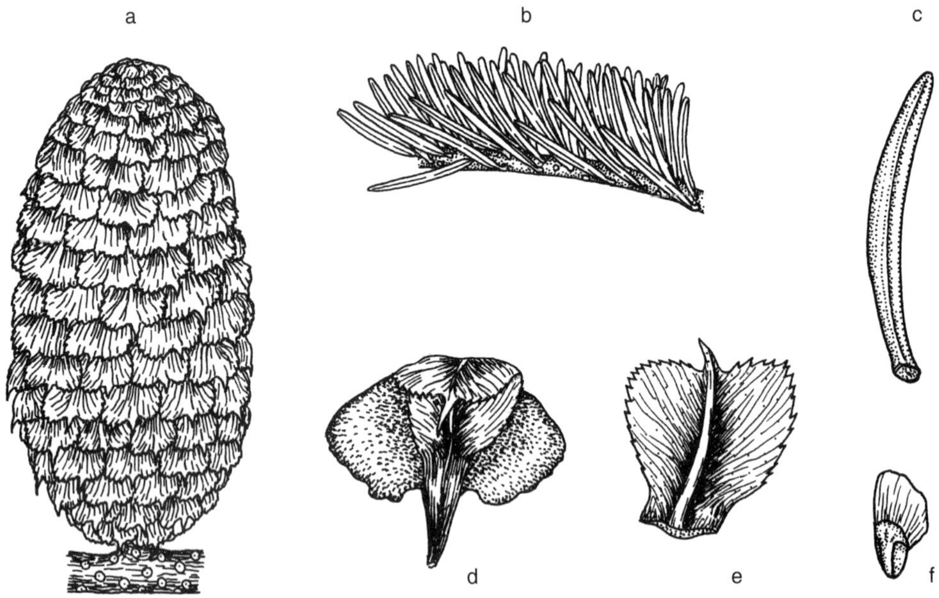

FIGURE 10.19 *Abies fraseri* (a) cone prior to disintegration × 1, (b) twig with compact cluster of upright leaves × 1, (c) leaf × 2, adaxial surface, (d) cone scale with exserted × 3, reflexed bract, (e) bract with aristate tip × 1, (f) winged seed. Redrawn from Liu (1971).

width 2 to 3 mm, linear, upper surface light green, not glaucous, lower surface with five to eight white parallel stomata rows each side of midrib, apex round to notched, strongly aromatic; branches at right angles to main stem; immature cones upright on twig, green, disintegrating at maturity, length 8 to 10 cm, scales rounded, bracts not visible; twigs smooth, leaf scar a bull's-eye; bark smooth, highly resinous; Region 7, the Sierra Nevada of central and northern California, elevation 900 to 2300 m. Cover type **30.2, 30.26, 30.44**, 30.7, 30.27, 30.28, 30.30, 30.32, 30.34, 30.36, 30.39, 30.43.

Abies lasiocarpa (Hooker) Nuttall, west coast subalpine fir (Figure 10.22); medium-sized tree, height to 20 m; evergreen; leaves usually spiral and turned skyward on twig (hairbrush-like), length 1.8 to 3.2 cm, width 1.5 to 2.0 mm, linear, upper surface strongly glaucous, lower surface with four to five parallel stomata rows each side of midrib, apex round to notched, strongly aromatic; immature cones upright on twig, dark purple, length 6 to 12 cm, scales rounded, densely pubescent, bracts inserted disintegrating at maturity; twigs smooth, leaf scar a bull's-eye, new leaf scars show red periderm; branches at right angles to main stem; bark smooth, highly resinous; Regions 1, 6 and 7, Yukon to southern Oregon and northern California; coastal mountains, subalpine, elevation 1100 to 2300 m. Throughout most of its range, *A. lasiocarpa* introgresses with *A. bifolia.* Cover type **29.11, 29.25, 29.26**, 29.15, 29.18, 29.23, 29.27, 29.28, 29.29, 29.30, 29.34, 29.35, 29.36, 30.8, 30.10, 30.12, 30.24.

Abies bifolia A. Murray, corkbark fir, Rocky Mountain subalpine fir; large tree, height to 30 m; evergreen; leaves usually spiral and turned skyward on twig (hairbrush-like), length 1.1 to 2.5 cm, width 1.2 to 1.5 mm, linear, upper surface glaucous, lower surface with three to five parallel stomata rows each side of midrib, apex round to notched, strongly aromatic; branches at right angles to main stem; immature cones upright on twig, dark purple, length 5 to 10 cm, scales rounded, bracts overlay scales, densely pubescent, disintegrating at maturity; twig smooth, leaf scar a

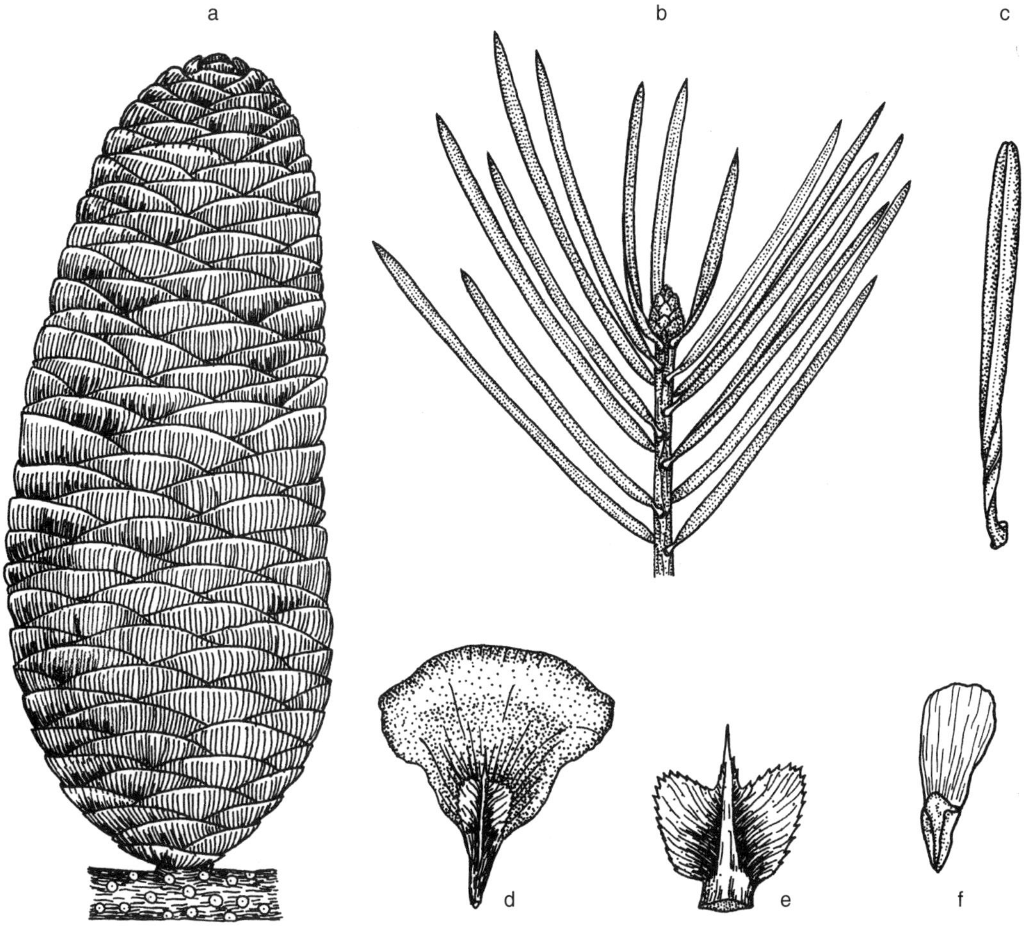

FIGURE 10.20 *Abies concolor* (a) cone prior to disintegration × 1, (b) twig with two-ranked leaf arrangement × 1, (c) leaf × 1; note twisted petiole, (d) cone scale with inserted straight bract × 1.5, (e) bract with aristate tip × 4, (f) winged seed × 1. Redrawn from Liu (1971).

bull's-eye, new leaf scars show light brown periderm; bark smooth, highly resinous; Regions 5 and 6, Yukon to Montana, and Idaho, and western Washington to Arizona and New Mexico; Rocky Mountains, subalpine, elevation 600 to 3600 m. Recent studies show that this species is distinct from *A. lasiocarpa* (Hunt 1993), although similar in morphology (Figure 10.22).

Abies amabilis Douglas ex. J. Forbes, Pacific silver fir, silver fir (Figure 10.23); large tree, height to 75 m; evergreen; leaves three-ranked, length 1.0 to 2.5 cm, width 1 to 3 mm, some leaves angled forward with longer leaves at right angles to twig, linear, lower surface with five to six parallel rows of stomata, apex distinctly notched, strongly aromatic; immature cones upright, length 8 to 12 cm, purple, pubescent, scales broadly rounded, bracts not visible, disintegrating at maturity; twigs generally opposite, leaf scar a bull's-eye; Region 7, northern California to southeast Alaska; coastal, moist sites, elevation 0 to 2000 m. Cover type **30.1, 30.2, 30.23**, 30.8, 30.10, 30.19, 30.24.

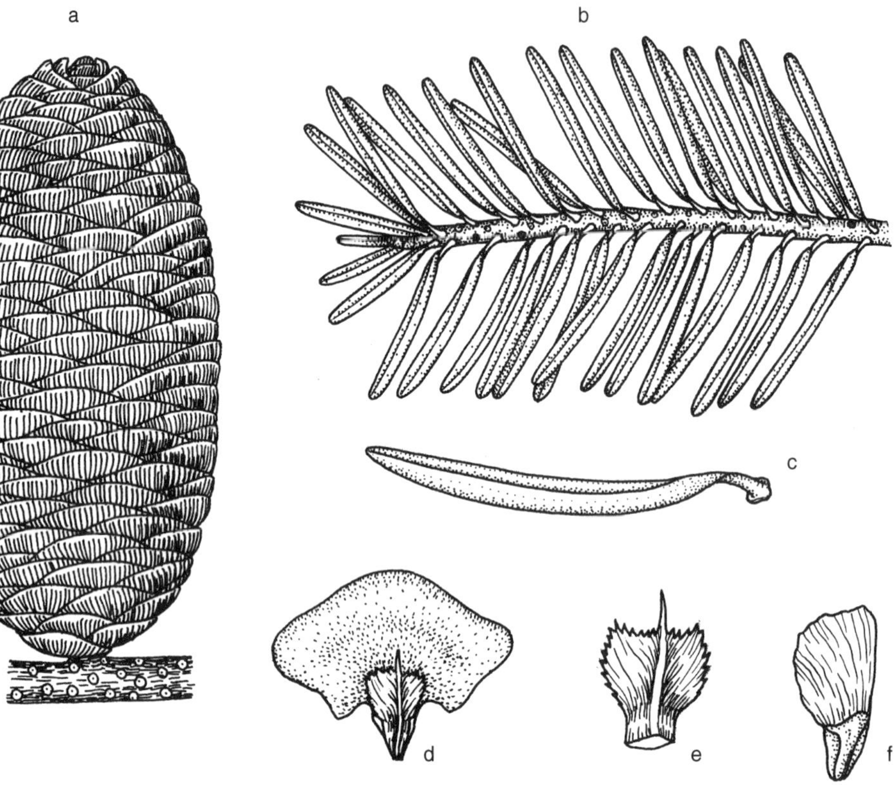

FIGURE 10.21 *Abies lowiana* (a) cone prior to disintegration × 1, (b) twig with two-ranked leaf arrangement × 1, (c) leaf × 1, (d) cone scale with inserted straight bract × 1, (e) bract with aristate tip × 2, (f) winged seed × 1. Redrawn from Liu (1971).

Abies grandis (Douglas ex. D. Don) Lindley, grand fir (Figure 10.24), lowland white fir; large tree, height to 75 m; evergreen; leaves on lower branches two-ranked, spiral on higher branches, length 2 to 6 cm, linear, width 1.5 to 2.5 mm, lower surface with five to seven parallel rows of stomata, upper surface green, apex distinctly notched, strongly aromatic; immature cones upright, length 6 to 10 cm, green, blue, or purple, scales broadly rounded, densely pubescent, bracts not visible; twigs generally opposite, leaf scar a bull's-eye; Region 6, Idaho, Montana, Washington, Oregon, California, British Columbia; moist mountain slopes, elevation 0 to 1500 m. Cover type **29.23, 29.24, 30.3**, 29.32, 30.6, 30.7, 30.18, 30.19, 30.21, 30.41.

Abies procera Rehder, noble fir (Figure 10.25); large tree, height to 80 m; evergreen; leaves angled upward (hairbrush-like), length 1 to 3 cm, linear but nearly four-sided on cone-bearing twigs, width 1.5 to 2.0 mm, lower surface with two to four parallel rows of stomata, glaucous, upper surface bluish green, apex rounded to notched, strongly aromatic; immature cones upright, length 10 to 15 cm, green, red, or purple, scales broadly rounded, densely pubescent, bracts exserted and reflexed, bract apex narrow acuminate; twigs generally opposite, leaf scar a bull's-eye; Regions 6 and 7, Washington, Oregon, California; moist, cool mountain slopes and tops, elevation 60 to 2700 m. Cover type 29.11, 30.1, 30.2, 30.18, 30.32.

Abies magnifica A. Murray, California or Shasta red fir (Figure 10.26); large tree, height to 55 m; evergreen; leaves angled upward (hairbrush-like), the basal 3 mm often appressed to twig, leaf length 2.0

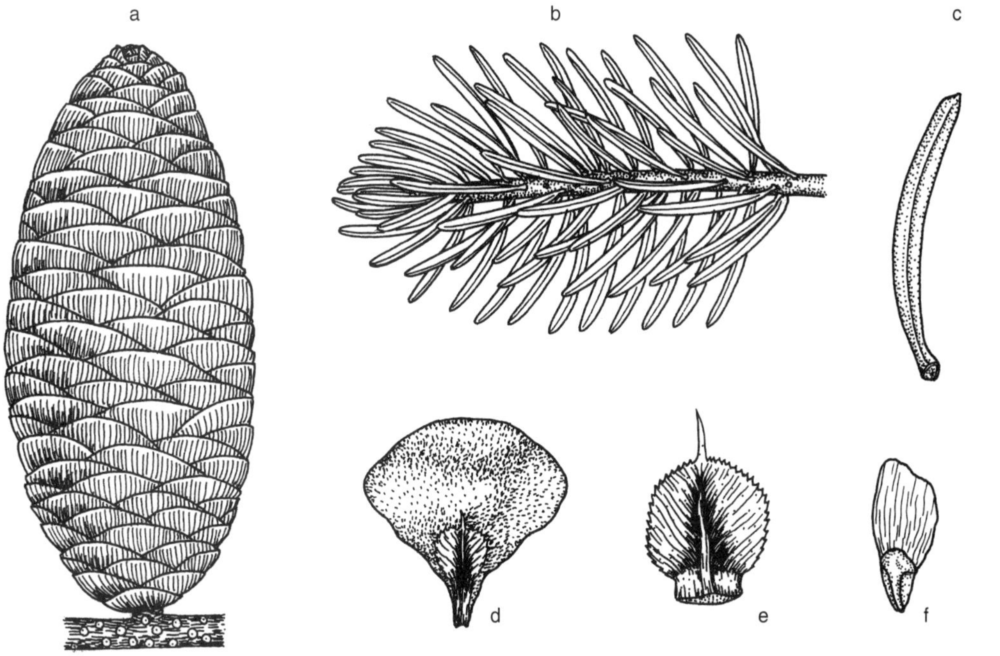

FIGURE 10.22 *Abies lasiocarpa* (a) cone prior to disintegration × 1, (b) twig with spreading upright leaf arrangement × 1, (c) leaf, adaxial surface × 1.5, (d) cone scale with inserted straight bract × 1, (e) bract with aristate tip × 3, and (f) winged seed × 1. Redrawn from Liu (1971).

to 3.5 cm, width about 2 mm, linear but three to four-sided on cone-bearing twigs, lower surface with two white bands, each band with four to five parallel rows of stomata expanding to eight to ten rows at mid-leaf, upper surface blue green to silvery blue, apex rounded to pointed, strongly aromatic; immature cones upright, length 15 to 22 cm, purplish to greenish, scales broadly rounded, densely pubescent, bracts not visible; twigs smooth, leaf scar a bull's-eye; Region 7, Oregon, California, Nevada; mountain slopes and ravines, elevation 1400 to 2700 m. A population of *A. magnifica* has cones with exserted bracts. Harlow et al. (1996) list this population as a variety (*A. magnifica* var. *shastensis* Lemmon). However, Hunt (1993) indicates that this population occurs where the ranges of *A. magnifica* and *A. procera* overlap in southern Oregon and northern California and that var. *shastensis* is likely to be a hybrid; Parker (1963) suggests that this population be named *A. magnifica* × *procera*. Refer to Hunt (1993) for an expanded discussion.

Abies bracteata D. Don ex. Poiteau, bristlecone fir (Figure 10.27); large tree, height to 25 m; evergreen; leaves two-ranked to spiral, length 2.5 to 6.0 cm, width about 3 mm, linear, upper surface green, lower surface with eight to ten parallel stomata rows each side of midrib, apex sharply pointed, strongly aromatic; immature cones upright on twig, color purple, length 7 to 10 cm, scales rounded, glabrous, bracts exserted but not reflexed, disintegrating at maturity; bark smooth, highly resinous; twigs smooth, leaf scar a bull's-eye; branches at right angles to main stem; Region 7, coastal Santa Lucia Mountains of central California. Rare. Cover type **30.25**.

Genus: *Tsuga*, Hemlock

Four *Tsuga* species are found in North America and six in Asia (Taylor 1993a). *Tsuga canadensis* and *Tsuga caroliniana* are found in eastern mesophytic forests,

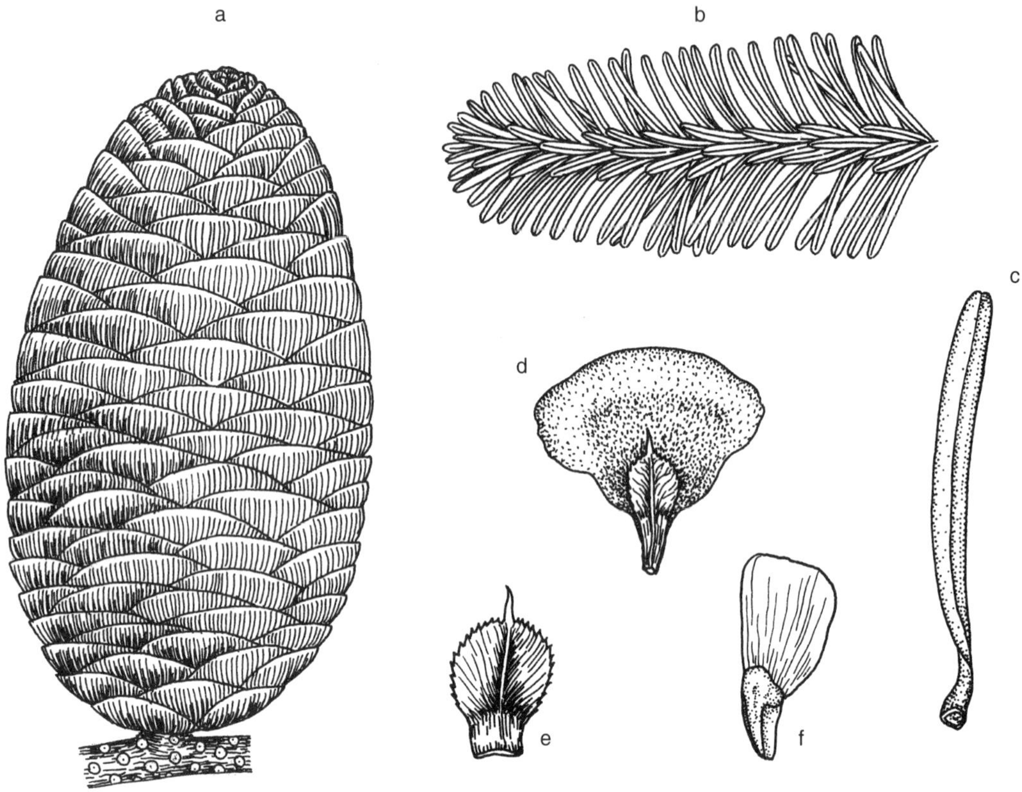

FIGURE 10.23 *Abies amabilis* (a) cone prior to disintegration × 1, (b) twig with three-ranked leaf arrangement × 1, (c) leaf × 2, adaxial surface, (d) cone scale with inserted straight bract × 1, (e) bract with aristate tip × 2, (f) winged seed × 1. Redrawn from Liu (1971).

and *Tsuga heterophylla* and *Tsuga mertensiana* grow in western mid-altitude mountain forests.

Hemlock growth and form are distinct in that the central terminal leader droops like a "buggy whip" (Figure 10.28); as the whip thickens and straightens, the tip grows at the same rate, so it always remains bent. Other taxa of the Pinaceae have a stiff, upright terminal leader. The needles have a linear shape, and there are frequently two white stomata bands on the lower surface, one on each side of the midrib. Leaf length varies from 0.5 to 2.0 cm. The needles have a petiole, a rounded or emarginate apex, and may be arranged on the twig in a spiral or two- or three-ranked fashion. On *Tsuga canadensis* twigs the larger needles are two-ranked, but a third row of smaller reflexed needles is located on the upper twig surface. Because of the reflexed position, the rows of stomata are visible without inverting the twig. The twigs are rough after the old leaves fall, but the remaining pegs (where petioles were attached) are smaller and fewer than on *Picea* twigs. In three species, cone length does not exceed 2.5 cm, but in *Tsuga mertensiana* it may be up to 8 cm. The cone scales have a rounded to somewhat irregular margin, a striated surface, and a leathery texture. The scale apex tends to reflex after the cone opens.

The wood is used for pulpwood, general construction, and other purposes for which thin boards are not

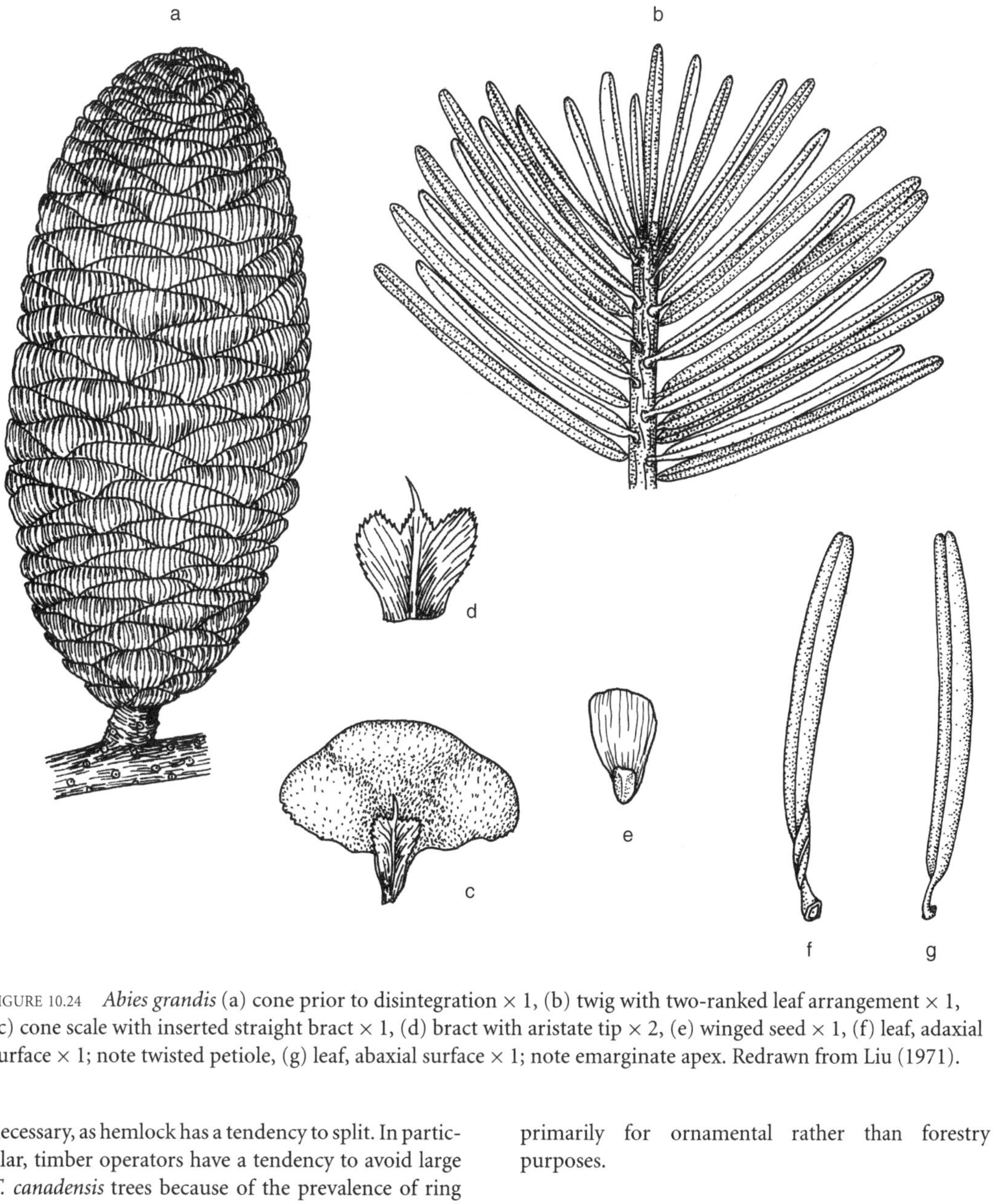

FIGURE 10.24 *Abies grandis* (a) cone prior to disintegration × 1, (b) twig with two-ranked leaf arrangement × 1, (c) cone scale with inserted straight bract × 1, (d) bract with aristate tip × 2, (e) winged seed × 1, (f) leaf, adaxial surface × 1; note twisted petiole, (g) leaf, abaxial surface × 1; note emarginate apex. Redrawn from Liu (1971).

necessary, as hemlock has a tendency to split. In particular, timber operators have a tendency to avoid large *T. canadensis* trees because of the prevalence of ring shake (splitting of wood along the growth ring). Hemlock bark has a high tannin level and was once a major commercial source of this chemical. Several Asian species have been brought into North America primarily for ornamental rather than forestry purposes.

Eastern Species

Tsuga canadensis (Linnaeus) Carrierre, eastern hemlock (Figure 10.29); large tree, height to 30 m; evergreen; leaf arrangement three-ranked; leaf length 0.8

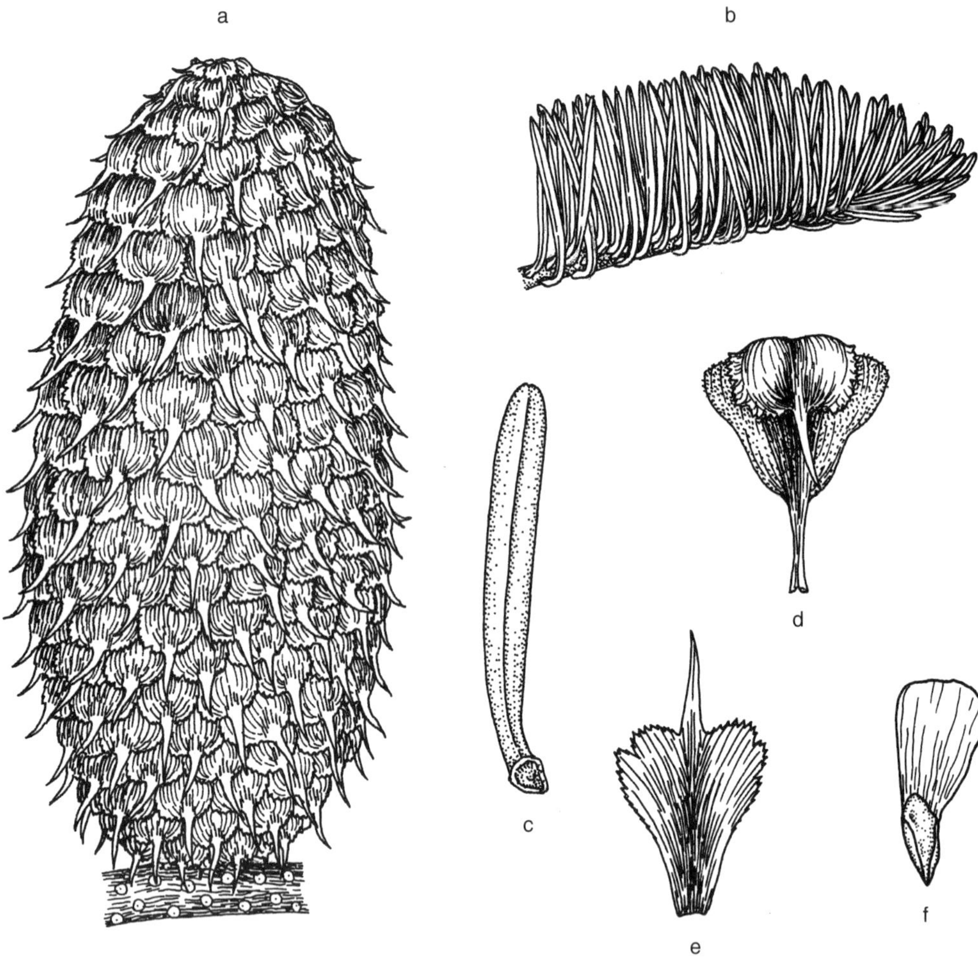

FIGURE 10.25 *Abies procera* (a) cone prior to disintegration × 1, (b) twig with compact cluster of upright leaves × 1, (c) leaf × 2, adaxial surface, (d) cone scale with exserted, reflexed bract × 1, (e) bract with aristate tip × 1.5, (f) winged seed × 1.5. Redrawn from Liu (1971).

to 1.5 cm, lower surface with two parallel white stomata bands, margin uneven, leaf shape linear, constricted at the base to form a short petiole, the twisted petioles creating a two-ranked arrangement and relatively flat sprays, short (0.3 to 0.5 cm) reflexed leaves somewhat appressed on the upper twig surface create a weak third rank; cone shape ovoid, length 1.5 to 2.5 cm, scales spatulate, margins slightly uneven, surface striated, angled forward; twigs rough; central leader drooping; Regions 2 and 3b, Lake States, New England, and the Appalachian Mountains south to Georgia; dense woods on slopes and in ravines; moist soil ranging from sandy to silt loam. The state tree of Pennsylvania. Cover type **26.21, 26.22, 26.29, 27.36, 27.37**, 26.2, 26.8, 26.9, 26.10, 26.17, 26.23, 26.27, 26.28, 27.12, 27.15, 27.31, 28.24.

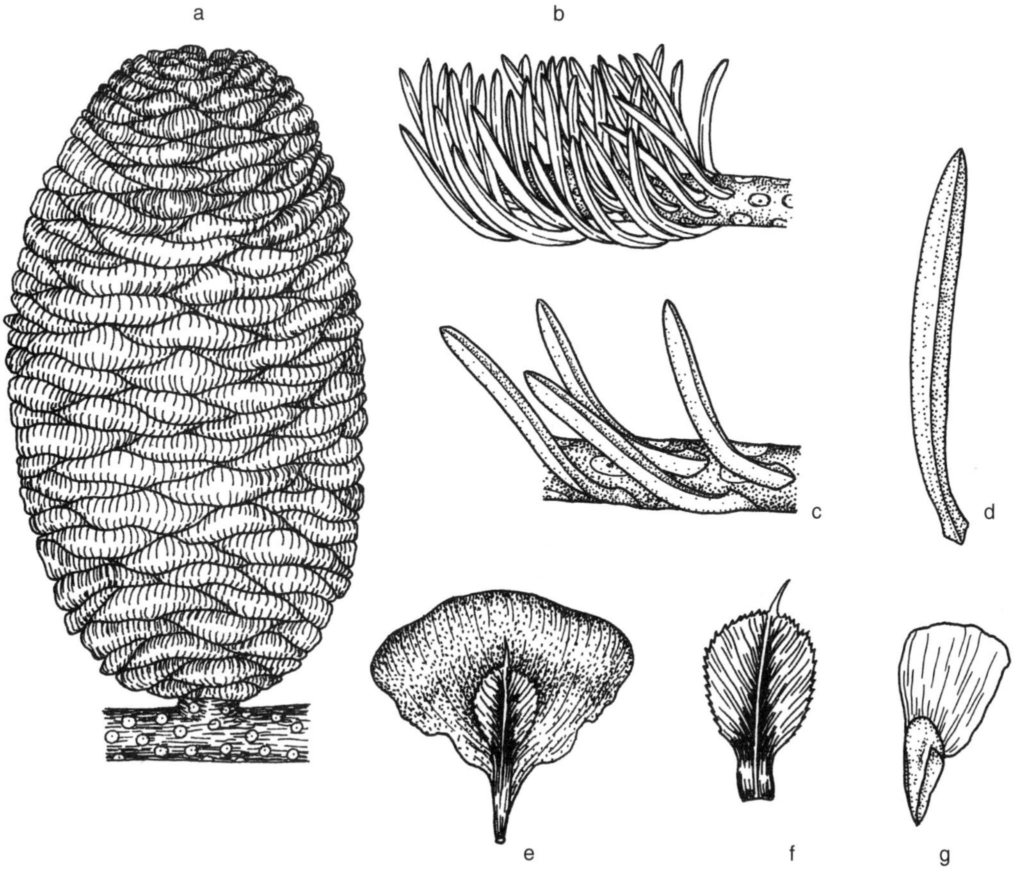

FIGURE 10.26 *Abies magnifica* (a) cone prior to disintegration × 1/2, (b) twig with compact cluster of upright leaves × 1, (c) twig with attached upright leaves × 1.5, note appressed leaf bases, (d) adaxial surface of leaf × 2, (e) cone scale with inserted straight bract × 1, (f) bract with aristate tip × 2, (g) winged seed × 1. Redrawn from Liu (1971).

Tsuga carolinana Engelmann, Carolina hemlock (Figure 10.29); large tree, height to 30 m; evergreen; leaves spreading in all directions (bottlebrush-like); leaf length 1.0 to 2.0 cm, linear but slightly revolute, lower surface with two parallel white stomata bands, margin smooth; cone shape ovoid to oblong, length 2.5 to 4.0 cm, scale shape oblong, scales angled forward; twigs rough; central leader drooping; Region 3b, Appalachian Mountains, Tennessee and Virginia to Georgia; dry slopes, elevation 700 to 1200 m. An Appalachian endemic. Cover type **27.38**.

Western Species

Tsuga heterophylla (Rafinesque) Sargent, western hemlock (Figure 10.30); large tree, height to 50 m; evergreen; leaf arrangement generally two-ranked, leaf length 0.7 to 2.2 cm, linear, lower surface with two parallel white stomata bands, margin uneven; cone shape ovoid, length 1.5 to 2.5 cm, scale shape ovate, scales angled forward; twigs rough; central leader drooping; Regions 6 and 7, Oregon, Washington, Idaho, and Montana to southeast Alaska; coastal and montane forest, elevation 0–1500 m. Cover type **30.19, 30.22,**

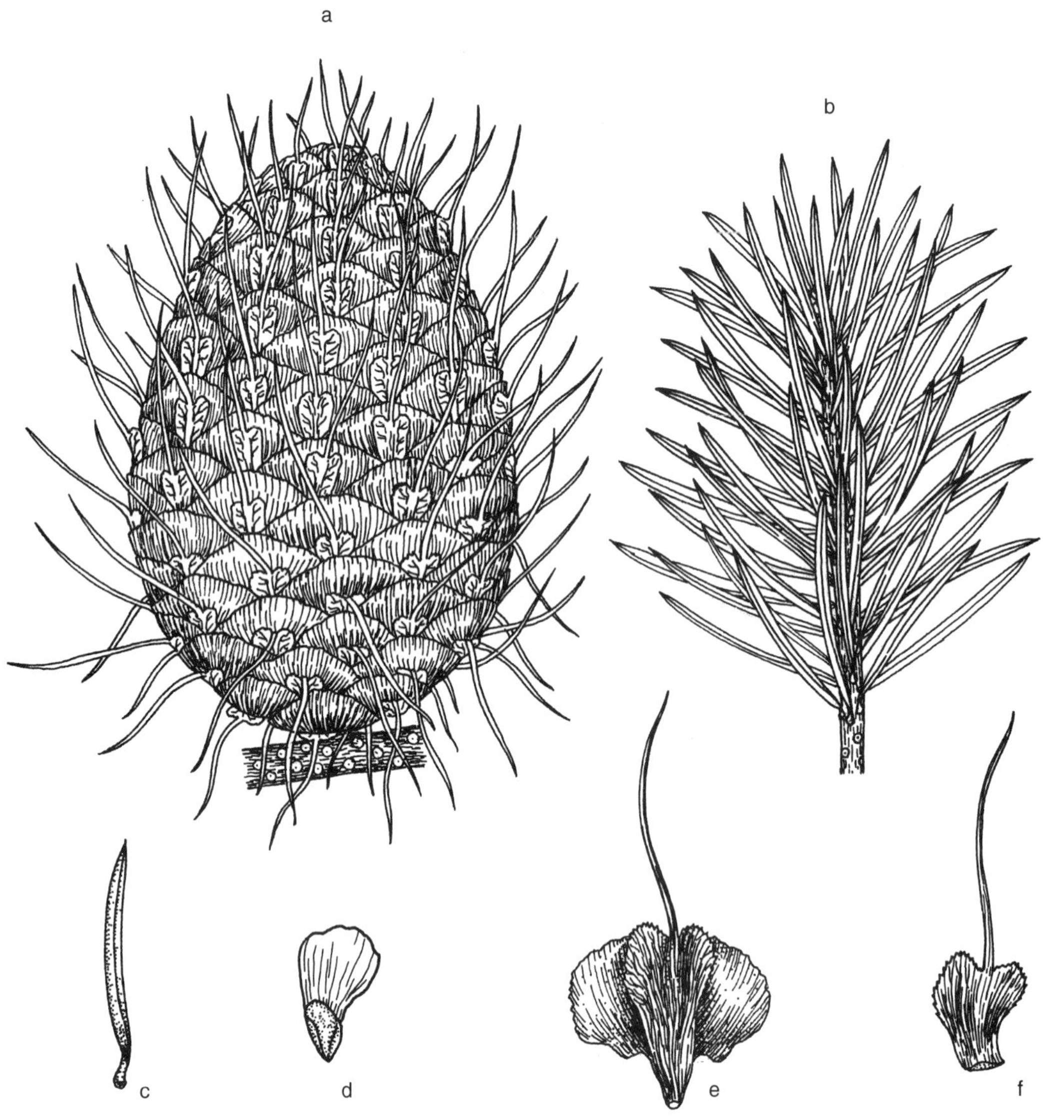

FIGURE 10.27 *Abies bracteata* (a) cone prior to disintegration × 1; note bracts exserted but not reflexed except at the base of the cone, (b) twig with spreading leaf arrangement × 1, (c) leaf, adaxial surface × 1, note characteristically sharp apex, (d) winged seed × 1, (e) cone scale with exserted straight bract × 1, (f) bract with long aristate tip × 1. Redrawn from Liu (1971).

FIGURE 10.28 Northern Wisconsin *Tsuga canadensis* trees with "buggy whip" leaders. *Photo by J. Fralish.*

29.28, 29.32, 30.1, 30.6, 30.7, 30.8, 30.10, 30.16, 30.18, 30.21, 30.23, 30.41.

Tsuga mertensiana (Bongard) Carriere, mountain hemlock (Figure 10.30); large tree, height to 40 m; evergreen; leaves spreading in all directions (bottlebrush-like) or three-ranked, curved toward twig apex, leaf length 0.8 to 2.5 cm, cross-sectional shape rounded, semicircular, or slightly four angled, lower surface with green stomata bands, margin smooth; cone shape oblong to cylindrical, length 3 to 8 cm, scales fan-shaped, perpendicular to stalk or slightly reflexed; twigs rough; central leader drooping; Region 7, California Sierra Nevada north into southeast Alaska; coastal and interior mountains into subalpine zone, elevation from sea level to 2400 m; part of krummholz. Cover type **30.23, 30.24**, 20.27, 29.29, 30.1, 30.10.

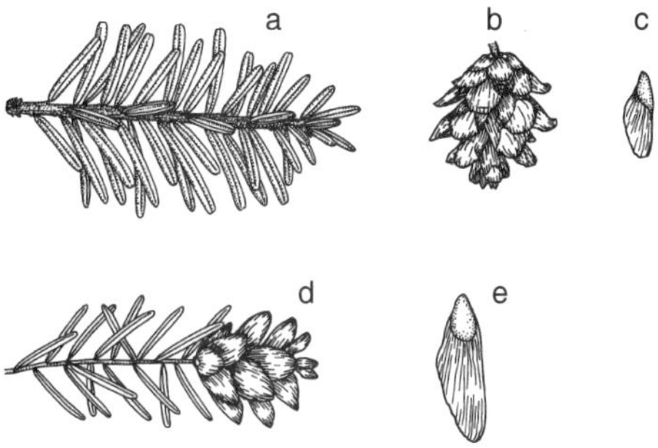

FIGURE 10.29 *Tsuga canadensis* (a) twig with three-ranked leaf arrangement × 1, (b) cone × 1, (c) winged seed × 1; *T. caroliniana* (d) twig with spreading leaves and terminal cone × 1/2, (e) winged seed × 1.

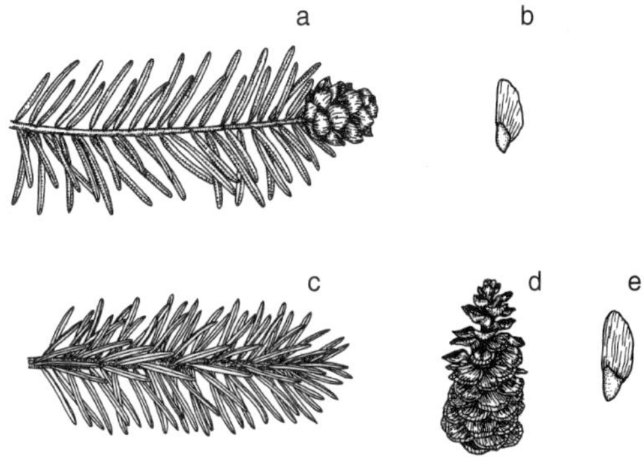

FIGURE 10.30 *Tsuga heterophylla* (a) twig with two-ranked leaf arrangement and terminal cone × 2/3; note alternating long and short leaves, (b) seed × 1; *T. mertensiana* (c) twig with generally three-ranked leaf arrangement × 1/2, (d) cone × 1/3, note slightly retrorse scales, (e) winged seed × 1.

Genus: *Pseudotsuga*, Douglas-Fir

This is a small genus of six species, two of which are located in the western United States and three in Asia (Lipscomb 1993). *Pseudotsuga menziesii* (Douglas-fir) is a widespread major species of western forests, but two varieties are recognized; *P. menziesii* var. *menziesii* (coast Douglas-fir) is the second largest tree in the west coast mountains, (*Sequoiadendron giganteum*, giant sequoia, is the largest), and var. *glauca* grows in the Rocky Mountains. *Pseudotsuga macrocarpa* (bigcone Douglas-fir) is found as scattered trees or in small groves of the low, southern California coastal mountains that support chaparral.

Literally interpreted, the name *Pseudotsuga* means "false hemlock," but the taxon was never considered close to *Tsuga*; rather, at various times it was classified with *Picea*, *Abies*, or *Pinus*. *Pseudotsuga* is one

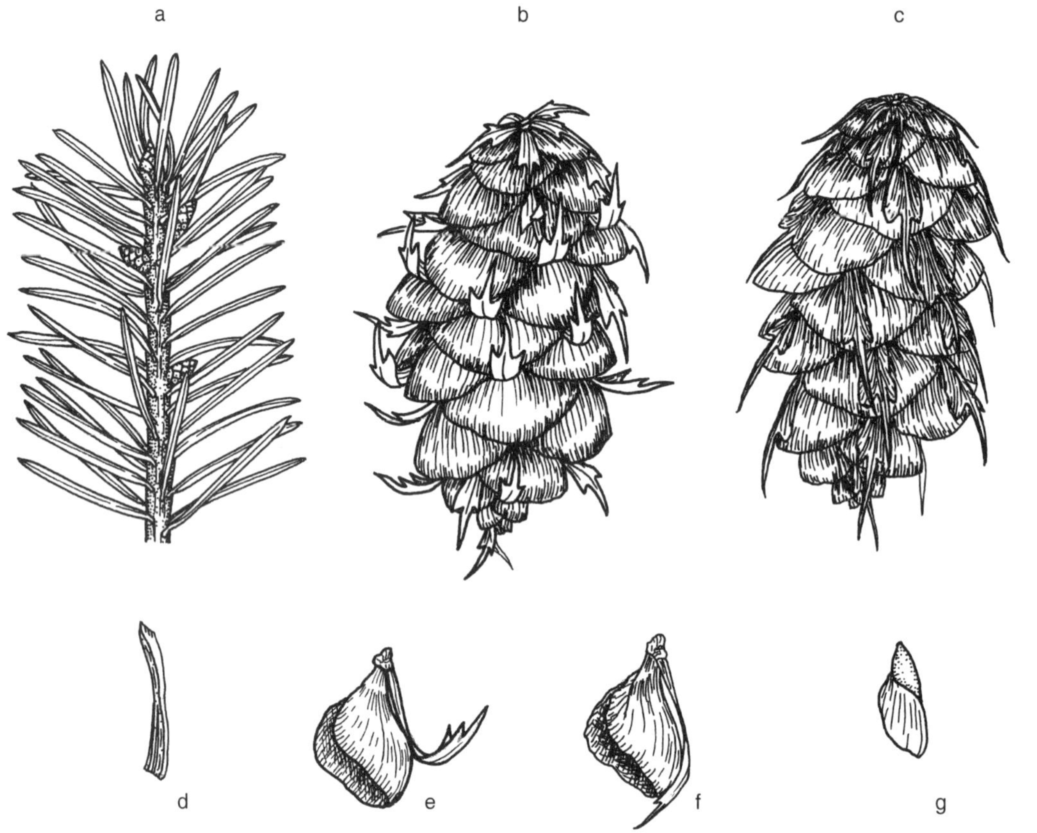

FIGURE 10.31 *Pseudotsuga menziesii* (a) twig with leaves and pointed buds × 1, (b) cone of var. *glauca* showing exserted reflexed bracts × 1, (c) cone of var. *menziesii* with exserted straight bracts × 1, (d) cross section of cone scale × 1, (e) cone scale and reflexed bract of var. *glauca* × 1, (f) cone scale and straight bract of var. *menziesii* × 1, (g) winged seed × 1.

of three genera of the Pinaceae to have flat (linear) petioled needles, the others being *Tsuga* and *Abies*. The leaves are 2 to 3 cm long, 2 to 3 mm wide, and somewhat aromatic. On the lower leaf surface there is a broad stomata band on each side of the midrib. The leaves remain functional for up to 8 years and leave a round bull's-eye-like scar that is similar to that of *Abies*. The terminal bud of each twig is reddish brown and sharp pointed. The cones are oval in shape with three-lobed exserted bracts. In the Rocky Mountain form (var. *glauca*) the bracts are reflexed but are straight and appressed in the coastal form (var. *menziesii*). The cones on both forms are pendent.

The wood is extensively used for plywood, construction and building lumber, timbers, piling, millwork, furniture, and a variety of other purposes. It is a popular wood because of its strength and the availability of quality material in large sizes.

Pseudotsuga menziesii (Mirbel) Franco var. *menziesii*, Coast Douglas-fir (Figure 10.31); large tree, height to 90 to 100 m; evergreen; leaves extending from all sides of twig (bottlebrush-like), shape

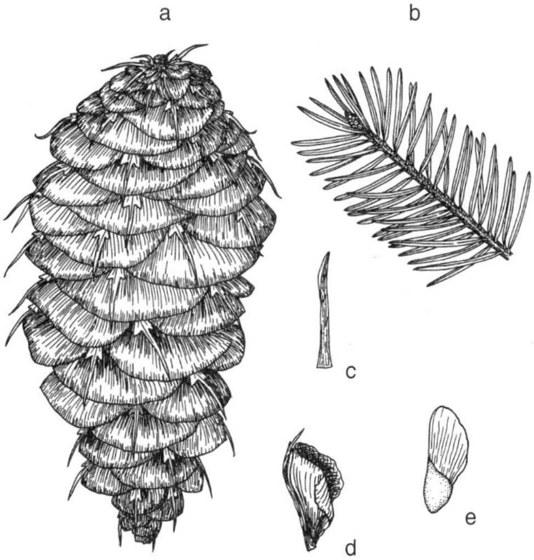

FIGURE 10.32 *Pseudotsuga macrocarpa* (a) cone with exserted straight bracts × 1/2, (b) twig with leaves and pointed bud × 1/2, (c) cross section of cone scale × 1/2, (d) cone scale and bract × 1/2, (e) winged seed × 1/2.

linear, length 1.5 to 3.5 cm, width 1.0 to 1.5 cm, color yellowish green; cone pendent, length 5 to 10 cm, scales rounded, margin smooth, bracts exserted, straight and appressed, three-lobed, middle lobe (apex) linear and pointed; new twigs pubescent; Region 7, central California and Nevada to British Columbia; moist sites, elevation 0 to 1800 m. Cover type **30.16, 30.17, 30.18, 30.19, 30.28, 30.37, 30.44,** 30.1, 30.2, 30.6, 30.10, 30.20, 30.21, 30.22, 30.23, 30.26, 30.30, 30.31, 30.34, 30.35, 30.38, 30.39, 30.41, 30.43.

Pseudotsuga menziesii (Mirbel) Franco var. *glauca* (Mayr) Franco, Rocky Mountain Douglas-fir (Figure 10.31); large tree, height to 40 to 60 m; evergreen; leaves extending from all sides of twig (bottlebrush-like), shape linear, length 2.5 to 4.5 cm, width 1.0 to 1.5 cm, color bluish green; cone pendent, length 5 to 10 cm, scales rounded, margin smooth, bracts exserted, moderately to strongly reflexed, three-lobed, middle lobe (apex) linear and pointed; new twigs pubescent, leaf scar a bull's-eye; Regions 5 and 6, Mexico, Texas, New Mexico, and Arizona to western Oregon, Washington, and British Columbia; Montane forest, elevation 600 to 3050 m. Cover type **29.35, 29.36,** 29.11, 29.15, 29.18, 29.23, 29.24, 29.26, 29.27, 29.28, 29.29, 29.30, 29.31.

Pseudotsuga macrocarpa (Vasey) Mayr, bigcone Douglas-fir (Figure 10.32); large tree, height to 44 m; evergreen; leaves extending from all sides of twig (bottlebrush-like), shape linear, length 1.5 to 3.5 cm, width 1.0 to 1.5 cm, color bluish green; cone pendent, length 10 to 20 cm, scales rounded, margin smooth, bracts exserted, spreading, and often reflexed, three-lobed, middle apex lobe linear and pointed; twigs smooth, leaf scar a bull's-eye; Region 7, San Bernadino Mountains and surrounding area of southern California; xeric slopes and canyons, often with chaparral, elevation 200 to 2400 m. Cover type **30.36.**

CHAPTER 11

Order: Pinales III: Cupressaceae

INTRODUCTION

In recent years, the Cupressaceae family (*Cupressus, Chamaecyparis, Thuja, Calocedrus, Juniperus*) has been combined with the Taxodiaceae (*Sequoia, Sequoiadendron,* and *Taxodium*), with the Taxodiaceae no longer being recognized as a family (Eckenwalder 1976; Watson 1993). Although the cones of the Taxodiaceae are generally larger than those of most species of the Cupressaceae, the *Taxodium* cone closely resembles that of *Cupressus,* so there is a cline in cone characteristics from one family to the other, or from *Sequoia* to *Cupressus* to *Juniperus.* Leaf shape follows a similar pattern. The juvenile awl-shaped *Juniperus* leaf resembles those of *Sequoiadendron* and the linear leaves of a variety of *Taxodium* (var. *imbricarium,* pond cypress), and the adult scale-like leaf is extremely similar to the *Cupressus* leaf. Combining these two families solved the problem of where to draw the boundary between them. Chemical and subcellular evidence supports the joining of these genera into one family (Brunsfeld et al. 1994; Price and Lowenstein 1989).

During evolutionary time, the members of this family were important constituents in forests of the Northern Hemisphere. Today the total number of genera and species is 25 to 30 and 110 to 130, respectively; a number of genera are monotypic. In North America, there are 9 genera and about 30 species (Watson and Eckenwalder 1993).

Species of the original cypress or cedar family (excluding the Taxodiaceae) are trees and shrubs with scale-like or awl-shaped leaves that are substantially smaller in size as compared with those of *Pinus.* The Taxodiaceae have linear as well as awl-shaped leaves. Leaf length seldom exceeds 1.5 cm, and most are less than 5 mm in length. Most species have persistent, opposite and decussate leaves; only *Calocedrus* (incense cedar) has a whorled leaf arrangement. Some species are dimorphic. Species of *Taxodium* and *Metasequoia* are deciduous.

The individuals of this family are monoecious except for those of *Juniperus,* which usually are dioecious. The male strobili are small and terminal or auxiliary, or a cluster of stamens born on a peltate microsporophyll. Female cones generally have fewer than 12 flattened or peltate, leathery or woody scales (megasporophylls). In *Juniperus,* the cones are semi-fleshy and the scales are joined (connate) and difficult to distinguish. Most species have two cotyledons.

SPECIES DESCRIPTIONS

Genus: *Cupressus,* Cypress

The cypress genus contains approximately 20 species. These are found in warm, dry summer, wet winter (Mediterranean-type) climates in the Northern Hemisphere (China, Himalayan mountain region, Mediterranean region). Eckenwalder (1993) lists 7 North American species, although the number of taxa with varieties could be as high as 15. One additional species is found in Central America (Little 1950; United States Department of Agriculture 1974). The ranges of these relic species populations are extremely small, and seldom do they form pure or extensive stands. All species are found from northern California to Texas and Mexico.

Although one or a combination of characteristics related to seeds, presence or absence of leaf glands, resin droplets, growth form, cones, and bark are sufficient to separate populations, the foliage generally resembles that of *Juniperus* (scale-like leaves in four ranks), whereas the cones are similar to those of *Taxodium* (globose with peltate scales) but have a distinct raised boss (umbo). Cypress is used mainly for landscaping, Christmas trees, erosion control, windbreaks, fuelwood, fence posts, and lumber (Little 1950; United States Department of Agriculture 1974). Sargent (1965) provides relatively specific locations for each of the following species.

Cupressus arizonica Greene, Arizona cypress (Figure 11.1); medium-sized tree, height to 20 m; evergreen; secondary side branchlets widely separated, length usually 1.0 to 2.5 cm, tertiary branchlets few or absent, width 1.0 to 1.3 mm; leaves scale-like, short (3 mm), appressed, decussate, four-ranked, often strongly glaucous, a single pitlike gland in the center of lower (abaxial) leaves, often with white resin droplets; cones globose or slightly oblong, diameter 1.5 to 2.5 cm, scales woody, peltate, number of scales six to eight, paired, resin blisters scattered or absent, umbo (boss) protruding to 4 mm from the middle of the apophysis of distal scales, sometimes glaucous; bark smooth but developing layers; Regions 5 and 7, California, Baja California, New Mexico, Arizona, Texas, and Mexico (largest range of the New World cypress species); juniper-pinyon woodland and chaparral. Cover type **29.2**, 29.1, 29.6.

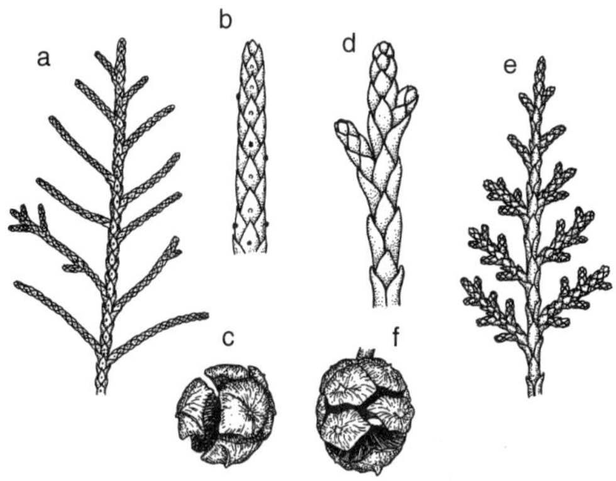

FIGURE 11.1 *Cupressus arizonica* (a) spray of undivided secondary branchlets ×3/4, (b) showing shape, centered glands, and decussate arrangement of scale-like leaves ×3, (c) globose seed cone of peltate scales ×1/2; *C. goveniana* (d) branchlet showing shape and decussate arrangement of scale-like leaves ×3, (e) spray of secondary and short tertiary branchlets ×1, (f) globose seed cone of peltate scales ×1/2.

Cupressus goveniana Gordon, Gowen or Mendocino cypress (Figure 11.1); small and shrubby to large-sized tree, height usually to 10 m but may range from 20 cm on thin soil to 50 m on moist soil; evergreen; short side branchlets dense, length 2 to 5 mm, diameter 1 mm; leaves short (3 mm), appressed, decussate, four-ranked, gland (pit) on the lower surface usually lacking or indistinct, resin lacking, not glaucous; cone globose, diameter 1.5 to 3.0 cm, scales woody, peltate, number of scales six to ten, paired, umbo slightly or not raised at maturity, not glaucous; Region 7, restricted to dry pine barrens (sterile or alkaline soil) of central coastal California; elevation 60 to 800 m.

Cupressus bakeri Jepson; Modoc cypress (Figure 11.2); large tree, height up to 30 m; evergreen; secondary side branchlets widely separated, length usually 1.0 to 2.5 cm, width 1.0 to 1.3 mm; tertiary branchlets few or absent, length 2 to 5 mm, diameter 1 mm; leaves short (3 mm), appressed, decussate, four-ranked, slightly glaucous, a distinct resin gland (pit) and white resin droplets on the lower surface; cone globose, diameter 1.0 to 2.0 cm, silvery but not glaucous, diameter 2.5 to 4.0 cm, scales woody, peltate, number of scales six to eight, paired, umbo raised at maturity, resin blisters common; Region 7, isolated groves in the Siskiyou Mountains of northern California and southern Oregon; montane evergreen forest; elevation 1100 to 2000 m. Rare and may be endangered. Included with *Cupressus macnabiana* by some researchers.

Cupressus macnabiana A. Murray, MacNab cypress (Figure 11.2); medium sized tree, often shrubby, height to 15 m; evergreen; short side branchlets dense, length 2 to 5 mm, diameter 1 mm; leaves short (3 mm), appressed, decussate, four-ranked, often glaucous with a distinct resin gland (pit) and white resin droplets on the lower surface; cone globose to slightly oblong,

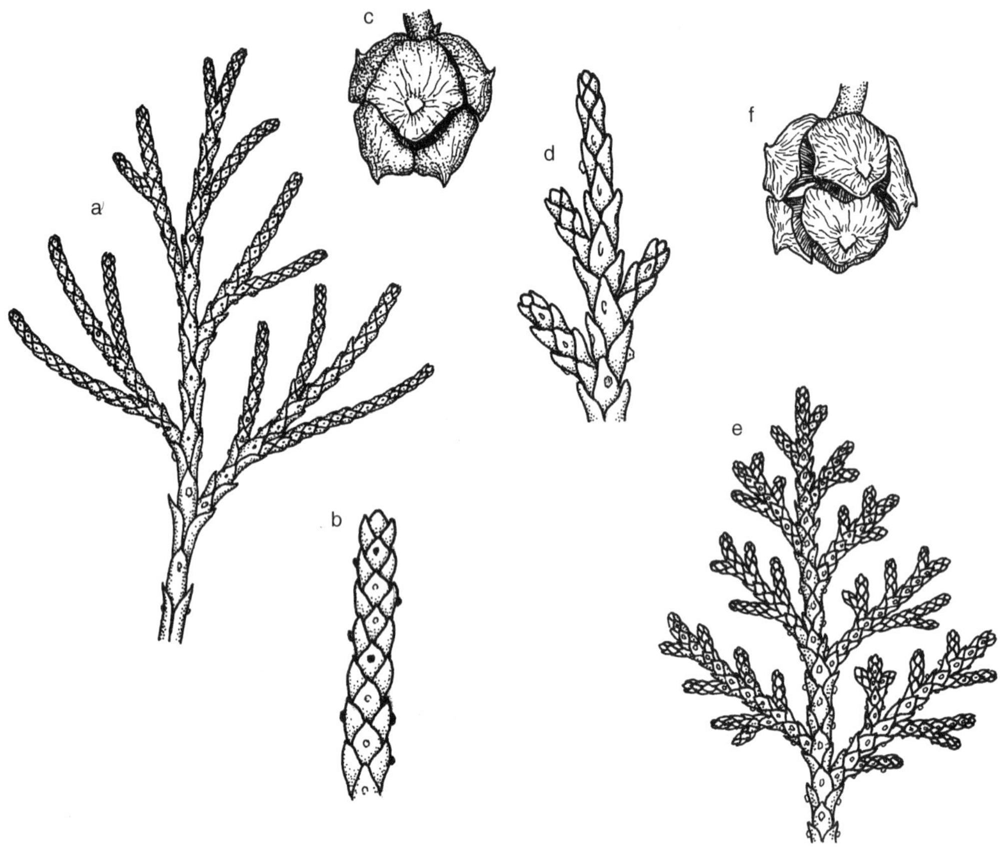

FIGURE 11.2 *Cupressus bakeri* (a) spray of forked secondary branchlets ×1, (b) branchlet showing shape, centered glands, resin droplets, and decussate arrangement of scale-like leaves ×4, (c) globose seed cone of peltate scales ×3/4; *C. macnabiana* (d) branchlet showing shape, centered glands, resin droplets, and decussate arrangement of scale-like leaves ×4, (e) spray of secondary and short tertiary branchlets ×1, (f) globose seed cone of peltate scales ×1.

length 1.5 to 3.0 cm, not glaucous, scales woody, peltate, number of scales six to eight, paired, umbo (boss) protruding to 4 mm from the middle of apophysis of distal scales; Region 7, restricted to mountains of northern California; chaparral and woodland, occasionally on serpentine soil, elevation 300 to 850 m. Cover type **30.29**.

Cupressus sargentii Jepson, Sargent cypress (Figure 11.3); medium-sized tree, often shrubby, height 10 to 25 m; evergreen; short side branchlets dense, length 2 to 5 mm, diameter 1.0 mm; leaves short (3 mm), appressed, decussate, four-ranked, gland (pit) on the lower surface indistinct, resin lacking, often glaucous; cone globose, diameter 2.0 to 2.6 cm, scales woody, peltate, number of scales six to eight, paired, umbo protruding at maturity, not glaucous, resin blisters scattered; Region 7, restricted to dry pine barrens and chaparral of central coastal California; found on serpentine soil, elevation 200 to 1150 m. Included with *Cupressus goveniana* by some researchers. Cover type 30.29.

Cupressus macrocarpa Hartweg, Monterey cypress (Figure 11.3); large tree, height to 25 m;

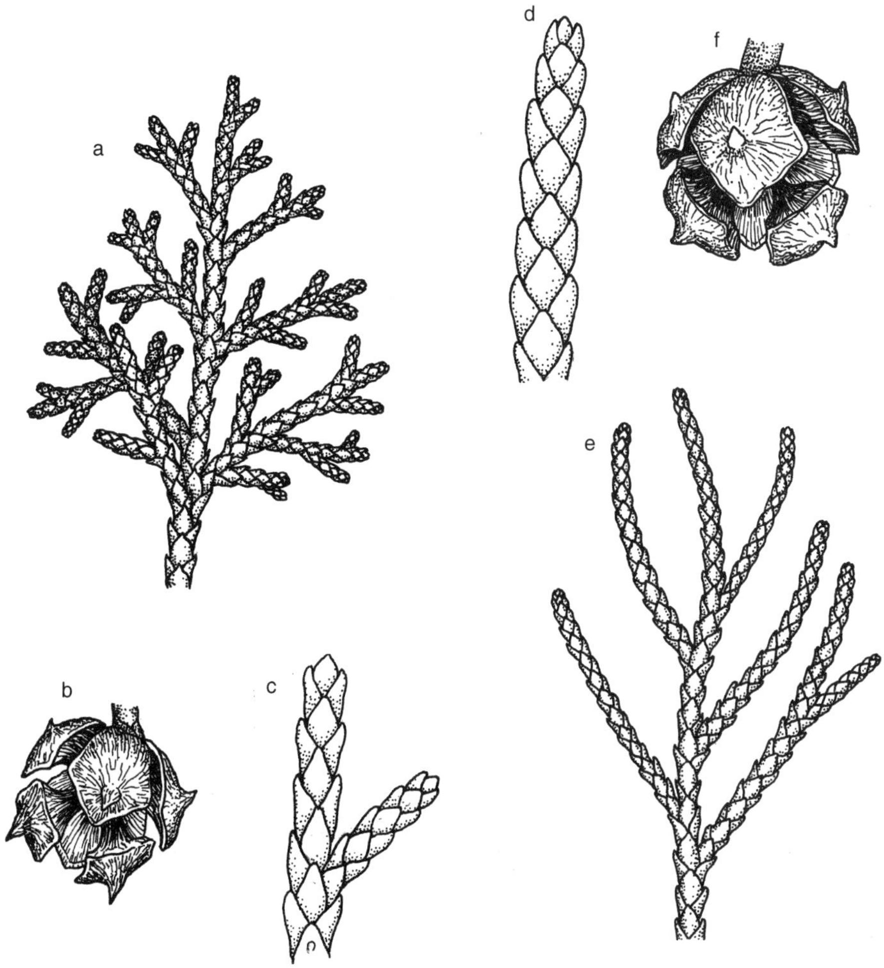

FIGURE 11.3 *Cupressus sargentii* (a) spray of secondary and short tertiary branchlets ×1.5, (b) globose seed cone of peltate cone scales ×1, (c) branchlet showing shape and decussate arrangement of scale-like leaves ×4; *C. macrocarpa* (d) branchlet showing shape and decussate arrangement of scale-like leaves ×5, (e) spray of undivided secondary branchlets ×2, (f) globose seed cone of peltate cone scales ×1.1.

evergreen; secondary side branchlets widely separated, length usually 1.0 to 2.5 cm, width 1.0 to 1.3 mm, tertiary branchlets few or absent; leaves, scale-like, short (3 mm), appressed, decussate, four-ranked, often strongly glaucous, gland (pit) on the lower surface lacking or indistinct, resin lacking, not glaucous; cone globose, diameter 2.5 to 4.0 cm, scales woody, peltate, number of scale pairs four to six, umbo protruding at maturity; Region 7, restricted to Monterey County of central coastal California; chaparral and woodland, elevation 300 to 850 m. Rare and may be endangered.

Cupressus guadalupensis S. Watson, Tecate cypress (not illustrated); small tree, shrubby, height to 10 m; tertiary branchlets few or absent, width 1.0 to 1.3 mm; leaves, scale-like, short (3 mm), appressed, decussate, four-ranked, nonresinous gland indistinct on the lower surface, not glaucous; cone globose, diameter 2.0 to 3.5 cm, scales woody, peltate, number of scales eight to ten, paired, resin blisters scattered or absent, umbo (boss) protruding to 4 mm from the middle of the apophysis of distal scales; Region 7, restricted to chaparral of southern and Baja California; coastal, elevation 5 to 30 m. Rare and may be endangered.

Genus: *Juniperus*, Juniper

This a large genus containing approximately 50 to 70 species of small to medium-sized trees and shrubs. Most are found in North and Central America, the West Indies, northern and eastern Africa south of the equator, and the Far East. Thirteen species are recognized in North America (Little 1953, Adams 1993).

Individual trees often are dimorphic. The juvenile leaves, which may be persistent for long periods under extremely poor growing conditions, are awl-shaped (subulate) and 0.5 to 1.0 cm in length. The adult leaves are scale-like, decussate, four-ranked, and appressed; length is only 1 to 3 mm. *Juniperus* twigs and branchlets do not form a fan-like spray as found in *Thuja*, *Chamaecyparis*, and *Biota*. In species that grow to tree size, the branches spread or ascend from the bole. The bark varies from ashy gray to brown and exfoliates in long vertical fibrous strips.

All species have dioecious trees but a few are also monoecious (*Juniperus occidentalis*, *J. osteosperma*, *J. californica*). The cone resembles a berry in that it is composed of three to eight semifleshy scales that are fused at maturity. In the immature state, they are green and may be covered with a whitish bloom; when mature, they vary in color from nearly black to dark blue to brownish and reddish brown. Seed number per cone is usually one to four. Cotyledons range from two to six. *Juniperus* wood is used for fence posts, chests, wardrobes, millwork, pencils, and novelties. The cones are used to flavor gin.

The northern/eastern species are associated with forest and special communities such as glades, and the western species are often found in near desert conditions or mixed with various pinyon pine on dry, rocky soil.

Northern/Eastern Species

Juniperus virginiana Linnaeus var. *virginiana*, eastern redcedar (Figure 11.4); medium-sized tree, height to 20 m; evergreen; leaves may be dimorphic, juvenile leaves awl-shaped, length 3 to 7 mm, sharp-pointed, divergent, mature leaves scale-like, length 2 to 3 mm, appressed to branchlet, both leaf types decussate; cone globose, diameter 4 to 6 mm, mature cones bluish to black, glaucous, semifleshy; outer bark breaking into loose vertical strips; Regions 2, 3, and 4, eastern United States except Lower Coastal Plain of Atlantic and Gulf coasts, Minnesota and South Dakota to Quebec and Maine to Georgia and Texas; found on a wide variety of sites, mostly xeric to mesic soil conditions, old fields, fencerows, sandstone and limestone glades. Cover type **27.11, 28.2**, 27.17, 27.19, 27.24, 28.6.

Juniperus virginiana Linnaeus var. *silicicola* (Small) E. Murray; southern redcedar (Figure 11.4); small tree, height to 10 m; evergreen; leaves and cones similar to those of var. *virginiana*, except cone diameter smaller, 3 to 4 mm; branches often drooping; Region 4, Lower Coastal Plain of Atlantic and Gulf coasts; coastal dunes and ridges.

Juniperus horizontalis Moench, creeping juniper; prostrate shrub (Figure 11.5); evergreen; leaves dimorphic, juvenile leaves awl-shaped, length 4 to 8 mm, sharp-pointed, divergent, mature leaves scale-like, length 1 to 2 mm, imbricate, appressed to branchlet, both leaf types decussate; cone globose, diameter 5 to 7 mm, mature cones bluish to black, glaucous, semi-fleshy; Regions 1 and 2, Alaska to Wyoming, Newfoundland to New York and Illinois, various disjunct areas in Region 3 and northern plains; sandy soil, sand dunes, rock outcrops.

Juniperus communis Linnaeus var. *depressa* Pursh, common or old-field juniper (Figure 11.5); low multistemmed shrub, height to 2 m; evergreen; leaves arranged in whorls of three, awl-shaped, sharp-pointed, length 1.0 to 1.5 cm, adaxial white stomata bands; cones globose, diameter 0.6 to 1.6 cm, color bluish black, glaucous; Regions 1, 2, 3, 5 and 6, throughout most of North America except Atlantic and Gulf Coastal plains, Alaska to New Mexico, Labrador and Nova Scotia to Georgia; dry rocky soil, sand dunes, limestone, serpentine. Two other varieties are recognized: var. *megistocarpa* Fernald & St. John, found in several isolated areas of Nova Scotia and New Brunswick; and var. *montana* Aiton, found in Washington, Oregon, and northern California and the Canadian Rocky Mountains of British Columbia (Adams 1993). According to Adams (1993), the major identifying character is the width of the white stomata

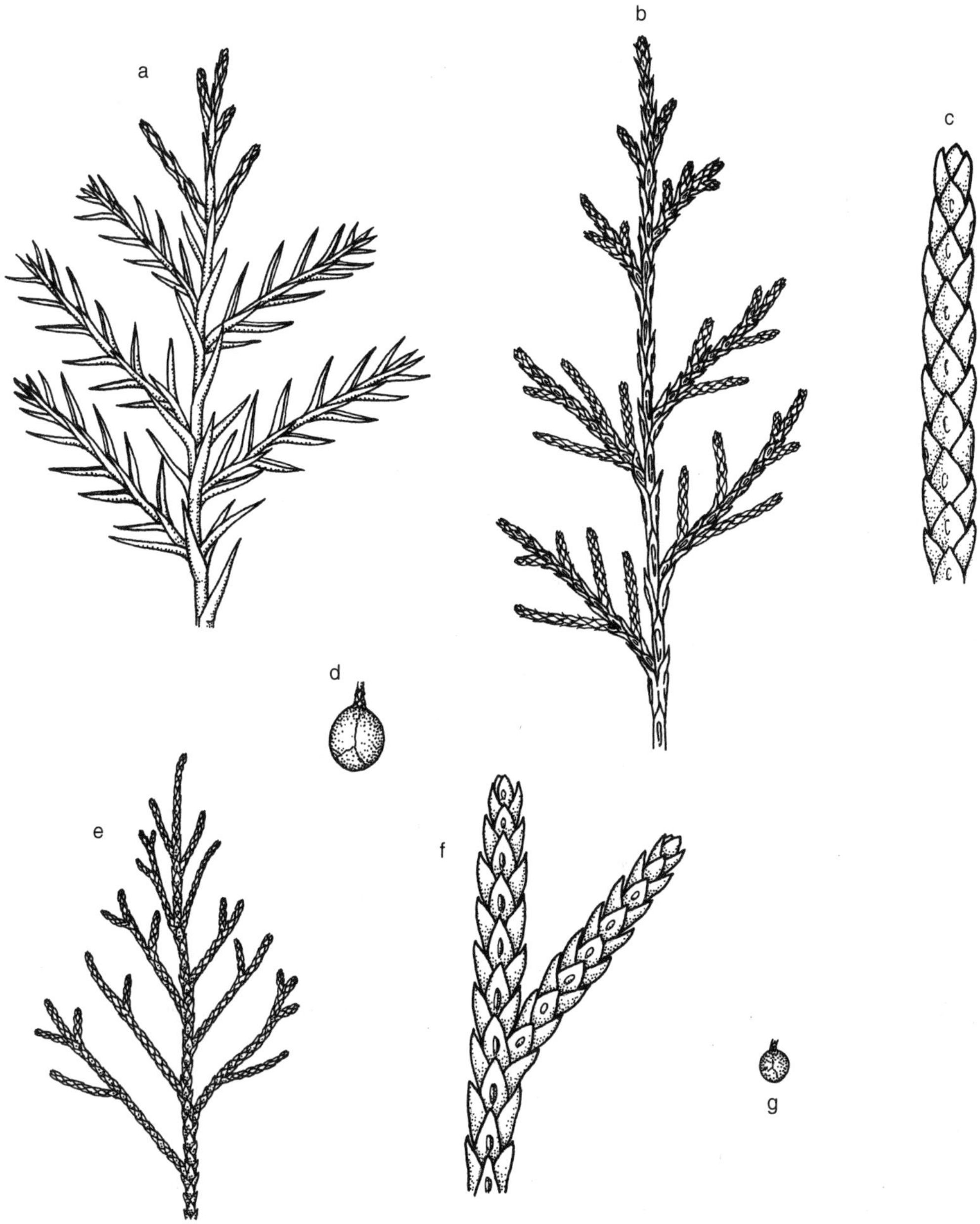

FIGURE 11.4 *Juniperus virginiana* (a) spray of branchlets with juvenile subulate leaves ×1, (b) spray of branchlets with mature scale-like leaves ×1, (c) branchlet showing decussate arrangement of scale-like, glandular leaves ×4, (d) semi-fleshy globose seed cone ×1; *J. virginiana* var. *silicicola* (e) slender branchlet of mature scale-like leaves ×1, (f) branchlet showing decussate arrangement of scale-like glandular leaves ×5, (g) semi-fleshy globose seed cone ×1.

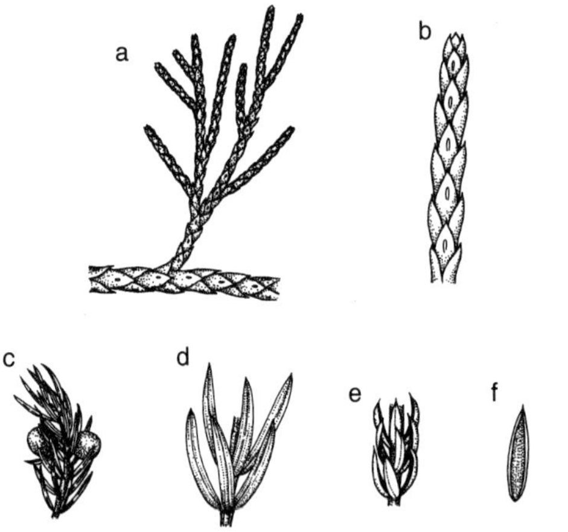

FIGURE 11.5 *Juniperus horizontalis* (a) spray of branchlets from trailing stem ×3/4, (b) branchlet showing shape, glands, and decussate arrangement of scale-like leaves ×4; *Juniperus communis* var. *depressa* (c) branchlet with ternate arrangement of subulate leaves and semi-fleshy globose seed cones ×1/2, (d) ternate arrangement of subulate leaves ×1; *Juniperus communis* var. *montana* (e) ternate arrangement of curved subulate leaves ×3/4, (f) subulate leaf shape, stippled area indicates a broad white band on the abaxial surface ×1.

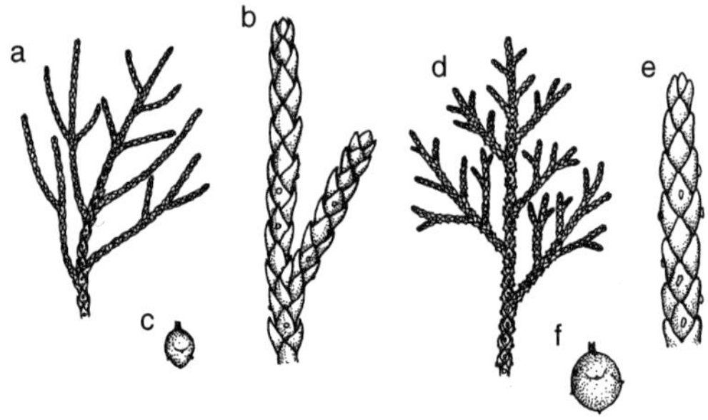

FIGURE 11.6 *Juniperus ashei* (a) spray of forked secondary branchlets ×1/2, (b) branchlet showing shape, glands, and decussate arrangement of scale-like leaves ×2, (c) semi-fleshy globose seed cone ×1/2; *J. deppeana* (d) spray of secondary and tertiary branchlets ×1/2, (e) branchlet showing leaf shape, glands, and decussate arrangement of scale-like leaves ×2, (f) semi-fleshy globose seed cone ×1/2.

band on the adaxial leaf surface. In var. *megistocarpa* the band is 1.5 times the width of the green margin, whereas in var. *montana* the band is 2 times the width. The latter variety has leaves that are shorter than var. *depressa* and tend to be curled (Figure 11.5). Cover type 29.11, 29.12, 29.15, 29.17, 29.18, 29.26, 29.28, 29.29, 29.31, 29.32, 29.35.

Western/Southwestern Species

Juniperus ashei J. Buchholz, Ashe juniper (Figure 11.6); medium-sized tree, height to 13 m; single-stemmed; evergreen; leaves dimorphic, four-ranked, juvenile leaves awl-shaped, length 3 to 6 mm, adult leaves scale-like, appressed, length 1 to 2 mm, glaucous, dark green, glands present, resin droplets absent; cone semifleshy, globose, slightly angled, diameter 6 to 8 mm, glaucous, dark blue; Regions 3 and 5, southern Missouri, Arkansas, Oklahoma, and Texas into Mexico; savanna and woodland on dry rocky soils, elevation 175 to 600 m. Cover type 29.9.

Juniperus deppeana Steudel, Alligator juniper (Figure 11.6); medium-sized tree, height to 20 m, single-stemmed; evergreen; leaves dimorphic, four-ranked, juvenile leaves awl-shaped, length 3 to 6 mm, not glaucous, adult leaves scale-like, appressed, length 1 to 2 mm, dark green, glands raised, resin droplets absent; branchlets stiff or flaccid and drooping; cone semifleshy, globose, slightly angled, diameter 0.9 to 1.3 cm, glaucous, reddish; Region 5, Arizona, New Mexico and Texas into Mexico; savanna and woodland on dry rocky soils in mountains, elevation 2000 to 3000 m. Cover type 29.1, 29.2, 29.4, 29.6, 29.22.

Juniperus scopulorum Sargent, Rocky Mountain juniper, Rocky Mountain redcedar, weeping juniper (Figure 11.7); medium-sized tree, height to 15 m, shorter on exposed sites, usually single-stemmed; evergreen; leaves dimorphic, four-ranked, appressed, decussate, juvenile leaves awl-shaped, length 3 to 6 mm, adult leaves scale-like, length up to 3 mm, glaucous, bluish, glands absent or indistinct, resin droplets absent; twigs often pendent; cone semifleshy, globose, slightly angled, diameter 0.6 to 1.0 cm, glaucous, light blue to dark blue; Regions 5 and 6, extending across the Great Plains, British Columbia to Oregon, Nevada, New Mexico, Colorado, and North and South Dakota; savanna and dry woodland, low mountain slopes, dry

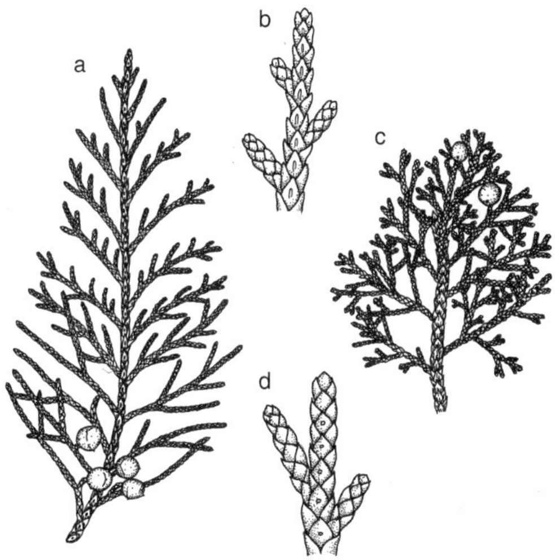

FIGURE 11.7 *Juniperus scopulorum* (a) spray of forked branchlets and semi-fleshy globose seed cones ×1/2, (b) branchlet showing shape, glands, and decussate arrangement of scale-like leaves ×2; *J. osteosperma* (c) spray of mature of forked branchlets and semi-fleshy globose seed cones ×1/2, (d) branchlet showing leaf shape, glands, and decussate arrangement of scale-like leaves ×2.

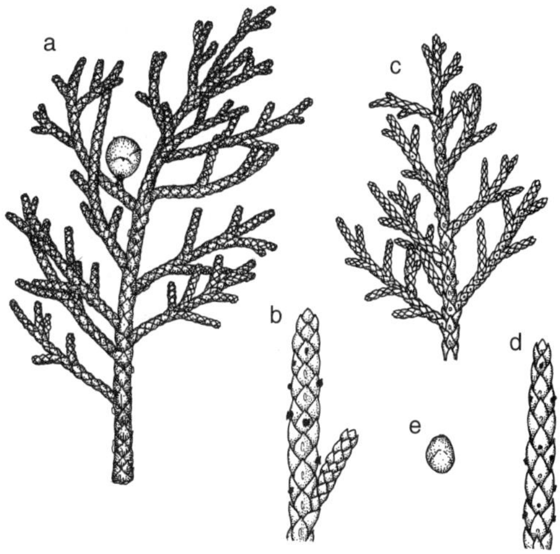

FIGURE 11.8 *Juniperus monosperma* (a) spray of forked branchlets and semi-fleshy globose seed cone ×2/3, (b) branchlet showing shape, glands, resin droplets, and decussate arrangement of scale-like leaves ×2; *J. pinchotii* (c) spray of mature scale-like leaves ×1/2, (d) branchlet showing shape, glands, resin droplets, and decussate arrangement of scale-like leaves ×2, (e) semi-fleshy globose seed cone ×1/2.

rocky soils, talus slopes, elevation 1300 to 2600 m. Cover type **29.3**, 29.5, 30.17.

Juniperus osteosperma (Torrey) Little, Utah juniper (Figure 11.7); small single- or multistemmed tree, height to 10 m; evergreen; leaves dimorphic, yellowish green, four- or six-ranked, appressed, decussate, juvenile leaves awl-shaped, length 3 to 5 mm, adult leaves scale-like, length up to 1 to 2 mm, glaucous, glands indistinct, resin droplets absent; from a distance foliage usually appears yellowish green because of mistletoe infestation; cone semifleshy, globose, slightly angled, diameter 0.6 to 1.0 cm, glaucous, reddish brown; Region 5, extending across the Great Basin, Idaho, Wyoming, Utah, Colorado, California, Nevada, Arizona, and New Mexico; savanna and woodland, dry rocky soils, elevation 1300 to 2500 m. Cover type 29.5.

Juniperus monosperma (Engelmann) Sargent, one-seed juniper, New Mexico cedar (Figure 11.8); medium-sized tree, height to 17 m; evergreen; leaves dimorphic, four-ranked, appressed, juvenile leaves awl-shaped, length 4–6 mm, adult leaves scale-like, length 1 to 3 mm, glaucous, dark green, glands present, resin droplets present; cone semifleshy, globose, slightly angled, diameter 5 to 7 mm, glaucous, reddish or rusty, sometimes blue; Region 5, Colorado, New Mexico, Arizona, and Texas into Mexico; savanna and woodland on dry rocky soils, elevation 1000 to 2400 m. Colorado, New Mexico, Arizona and Utah, extremely slow growing; trees with a bole diameter of about 15 cm and about 25 cm are reported to be nearly 200 and 375 years old, respectively (Peattie 1953). Cover type 29.7.

Juniperus pinchotii Sudworth, Pinchot juniper (Figure 11.8); small multistemmed tree, height to 6 m; evergreen; leaves dimorphic, four-ranked, yellow green, juvenile leaves awl-shaped, length 3 to 6 mm, adult leaves scale-like, appressed, length 1 to 2 mm, glaucous, glands present, resin droplets present; cone semifleshy, globose, slightly angled, diameter 6 to 8 mm, not glaucous, copper to red; Region 5, west Texas, Oklahoma, and New Mexico into Mexico; hills

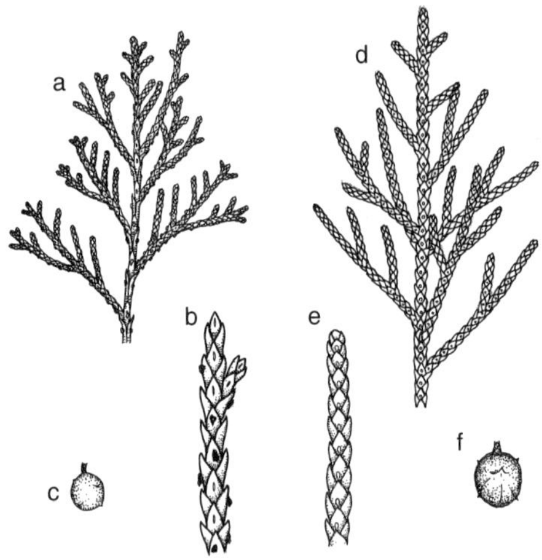

FIGURE 11.9 *Juniperus occidentalis* (a) spray of forked branchlets ×1/2, (b) branchlet showing shape, glands, resin droplets, and decussate arrangement of scale-like leaves ×2, (c) semi-fleshy globose seed cone ×2/3; *J. californica* (d) spray of forked branchlets ×1/2, (e) branchlet showing shape, glands, and decussate arrangement of scale-like leaves ×2, (f) semi-fleshy globose seed cone ×1.

and ravines, savanna and woodland on dry rocky soils, elevation 300 to 1200 m.

Juniperus occidentalis Hooker, western juniper, Sierra juniper (Figure 11.9); medium-sized to large tree, height to 25 m; evergreen; leaves dimorphic, four-ranked, appressed, juvenile leaves awl-shaped, length 3 to 6 mm, adult leaves scale-like, length 1 to 3 mm, glaucous, dark green, glands present, resin droplets present; cone semifleshy, globose, slightly angled, diameter 0.5 to 1.0 cm, glaucous, dark blue; Regions 5 and 7, two varieties are recognized: var. *occidentalis* centered in Oregon but extending a short distance into Washington, Idaho, Nevada, and northern California; var. *australis* (Vasek) Holmgren and Holmgren of the California Sierra Nevada; foothills and mountain slopes, savanna and woodland on dry rocky soils, elevation 1000 to 3000 m. Cover type 30.32.

Juniperus californica Carriere, California juniper (Figure 11.9); shrub or small multistemmed tree, height to 10 m; evergreen; leaves dimorphic, four-ranked, juvenile leaves awl-shaped, length 3 to 6 mm, not glaucous, adult leaves scale-like, appressed, length 1 to 2 mm, dark green, glands present, resin droplets absent; cone semifleshy, globose, slightly angled, diameter 6 to 8 mm, glaucous, blue-brown; Regions 5 and 7, Arizona and California into Baja Mexico; savanna and woodland on dry rocky soils, elevation 700 to 1600 m.

Juniperus coahuilensis (Martinex) Gaussen ex. R. P. Adams, roseberry juniper (not illustrated), small multi-stemmed tree, height to 8 m; evergreen; leaves dimorphic, four-ranked, green, juvenile leaves awl-shaped, length 3 to 6 mm, adult leaves scale-like, appressed, length 1 to 3 mm, glaucous, glands present, resin droplets present; cone semifleshy, globose, slightly angled, diameter 6 to 8 mm, not glaucous, rose, pink, orange, or red; Region 5, west Texas, Oklahoma, and New Mexico into Mexico; hills and flats, savanna and woodland on dry rocky soils, elevation 950 to 2000 m.

Juniperus flaccida Schlechtendal, Mexican drooping juniper (not illustrated); small tree, height to 12 m, multistemmed above 1 m; evergreen; leaves dimorphic, four-ranked, juvenile leaves awl-shaped, length 3 to 6 mm, not glaucous, adult leaves scale-like, appressed, length 1 to 2 mm, dark green, glands absent, resin droplets absent, cone semifleshy, globose, slightly angled, diameter 0.9 to 1.9 cm, glaucous, brownish to purple; Region 5, Texas (Big Bend National Park) into Mexico; savanna and woodland on dry rocky soils in mountains, elevation 900 to 3000 m. Cover type 29.4, 29.6.

Genus: *Calocedrus* (formerly *Libocedrus*), Incense-Cedar

There are three species in this genus (Fowell 1965); one is found in western North America, one in Taiwan, and one in China. *Calocedrus* wood is durable, splits with ease, is soft, uniform in texture, and has excellent painting characteristics. The North American species is used for fence posts, ties, pencils, venetian blinds, millwork, novelties, chests, and furniture; the United States is a major supplier of wood pencil stock.

Calocedrus decurrens (Torrey) Florin, incense-cedar (Figure 11.10); large tree, height to 55 m; leaves and branchlets flattened in a fan-like spray; leaves scale-like, broader distally with a mucronate tip, leaf

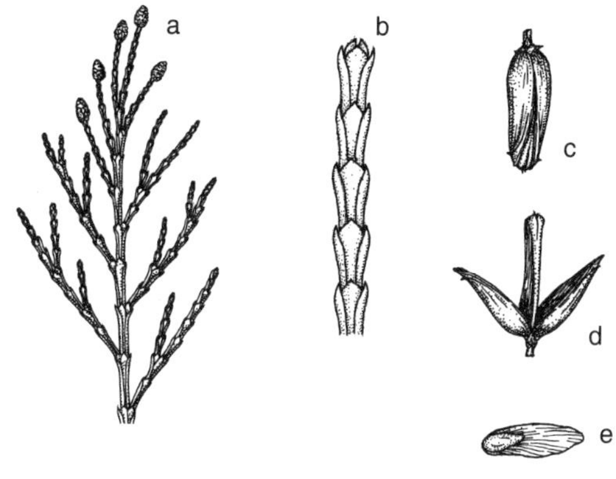

FIGURE 11.10 *Calocedrus decurrens* (a) spray of branchlets with terminal pollen cones ×1/2, (b) branchlet showing shape and whorled arrangement (four ranks) of leaves ×2, (c) closed seed cone ×1/2, (d) open seed cone showing septum and two scales ×1/2, (e) winged seed ×1/2.

length 0.3 cm on new branchlets to 1.2 cm on old twigs, in whorls of four but appressed to branchlet, appearing four-ranked; male and female strobili at the ends of different branches; cone length 1.5 to 3.0 cm, elliptical, distinguished by two large paired woody scales separated by a septum; Region 7, Oregon to Baja California; montane forest, elevation 500 to 2800 m. Cover type **30.28, 30.44,** 30.2, 30.7, 30.18, 30.20, 30.26, 30.32, 30.34, 30.36, 30.39, 30.43.

FIGURE 11.11 *Thuja occidentalis* growing in an *Alnus incana*-dominated swamp in northern Wisconsin. *Photo by J. Fralish.*

Genus: *Thuja*, White Cedar

Four species of this small genus are native to Asia and two to North America. The North American species have scale-like leaves that are 2 to 3 mm in length, decussate, and appressed. Male and ovulate strobili develop on the ends of different twigs. The mature cones, composed of 8 to 12 paired scales, remain on the tree through the dormant season. The scales are thin, fibrous, and may be tipped with a small (1 to 2 mm) spine at the apex. Up to eight seeds per cone are produced, these being borne on the middle scales.

The wood is extremely resistant to decay and relatively soft and weak but light in weight. It is used for poles, posts, cabins, paneling, furniture, shingles, canoes and boats, and novelties. Many cultivars have been developed for landscaping. In the Lake States and New England, communities of *Thuja occidentalis* provide excellent winter cover and food for wildlife, particularly deer (Figure 11.11). The western species, *T. plicata*, is an important wood-producing tree. *Biota orientalis*, oriental arborvitae, a common introduced species, is discussed here because it is a close relative of *Thuja* and at times has been included in the genus.

Thuja occidentalis Linnaeus, northern white-cedar, arborvitae (Figure 11.12); small to medium-sized tree, height to about 20 m; evergreen; branchlets collectively form a vertical fan-like spray; leaves scale-like, length 2 to 5 mm, decussate, appressed, a glandular protrusion appears in the center of the leaf; cone shape elliptical, length 0.9 to 1.4 cm, number of scales six to

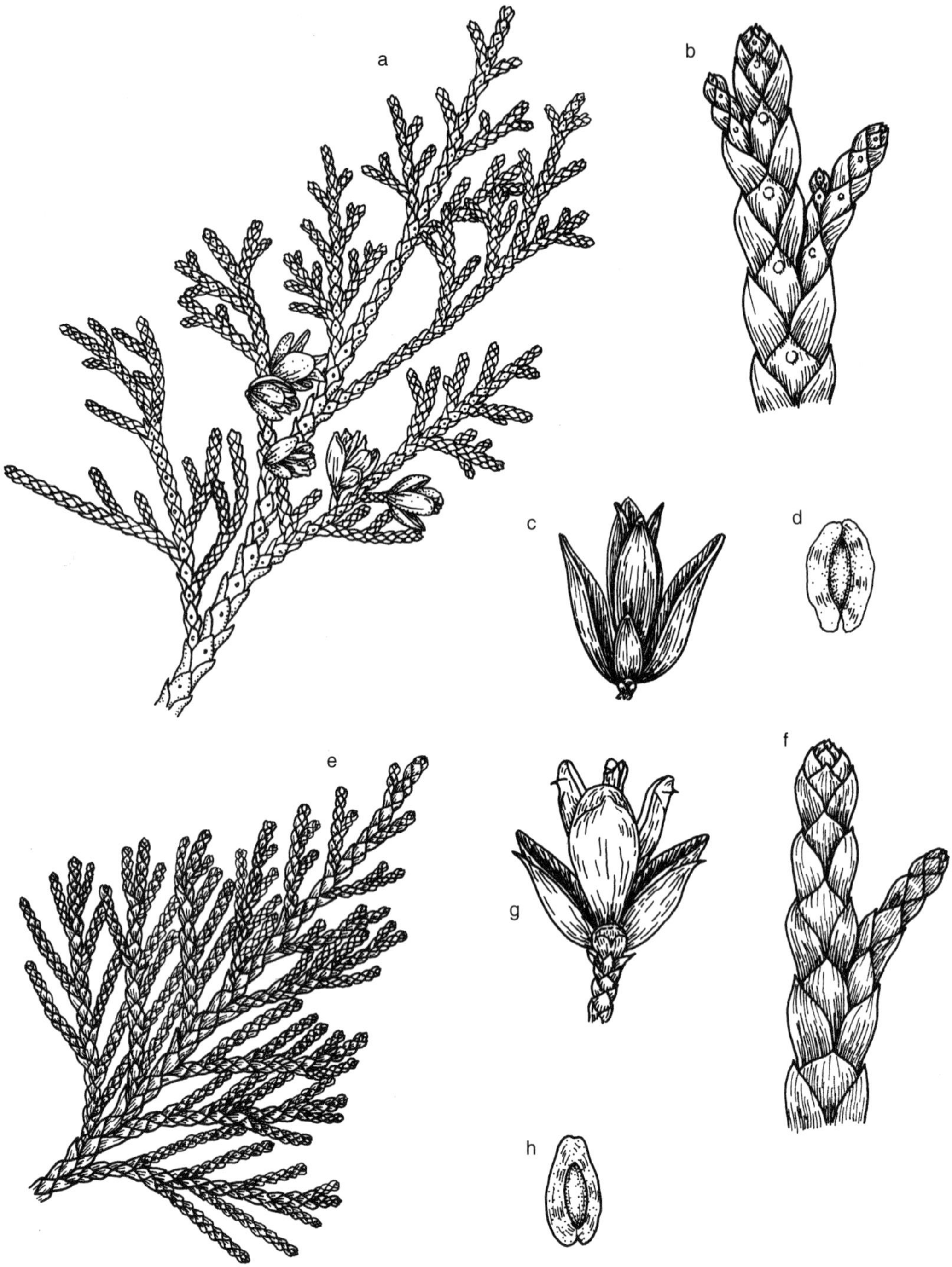

FIGURE 11.12 *Thuja occidentalis* (a) flat spray of branchlets and seed cones ×1, (b) branchlet showing shape, glands, and arrangement of scale-like leaves ×3, (c) seed cone of eight paired scales ×3, (d) winged seed ×3; *T. plicata* (e) flat spray of branchlets ×3/4, (f) branchlet showing shape and arrangement of scale-like leaves ×3, (g) open seed cone of eight paired scales ×3, (h) winged seed ×3.

eight, paired, only two pairs fertile, scales unarmed; seed winged; bark fibrous, splitting into vertical strips; Regions 1 and 2, Lake States, New England, and southern Ontario, Quebec to Prince Edward Island; lowland tree, typically found in mesic-hydric soil moisture conditions, ecotypes appear on calcarious uplands and rock outcrops. Cover type **26.16, 26.20**, 26.1, 26.6, 26.8, 26.9, 26.10, 26.14, 26.25.

Thuja plicata Donn ex. D. Don, western redcedar (Figure 11.12); large tree, height 50 to 70 m; evergreen; branchlets collectively form a vertical fanlike spray; leaves scale-like, length 2 to 6 mm, decussate, appressed, young leaves with white abaxial lines, glandular protrusion absent; cone shape elliptical, length 1.0 to 1.5 cm, number of scales six to eight, paired, each scale armed with a small dorsal projection; seed winged; bark fibrous, splitting into vertical strips; Regions 6 and 7, Montana, Idaho, Washington, Oregon, California, Alberta, British Columbia, and Alaska; montane forest, seldom forming a pure stand, elevation sea level to 200 m. Cover type **30.21,** 30.1, 30.6, 30.7, 30.8, 30.10, 30.18, 30.19, 30.22

Biota orientalis Endler (*Thuja orientalis* Linnaeus), oriental arborvitae (Figure 11.13); small tree, height to 10 m; evergreen; crown compact and cylindrical, branchlets form distinct vertical sprays; leaves scale-like, pointed, decussate, length 1 to 2 mm on branchlets tips to 5 mm on older branchlets, appressed,

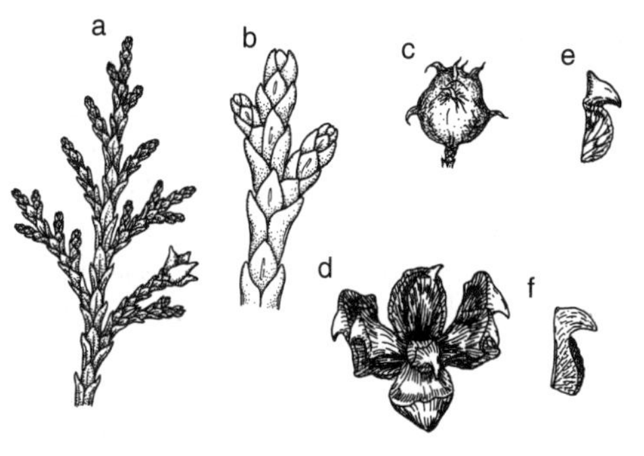

FIGURE 11.13 *Biota orientalis* (a) flat spray of branchlets and terminal pollen cone ×1/2, (b) branchlet showing shape, glands, and arrangement of scale-like leaves ×1.5, (c) immature closed seed cone ×3/4, (d) open seed cone of eight paired scales ×3/4, (e) side view of beaked cone scale ×3/4, (f) cross section of beaked cone scale ×3/4.

except tips raised on older leaves, glands absent; first-year cones closed, globose, diameter 1 to 2 cm, tipped with a slender projection (length 2 to 3 mm); second-year cones open, roughly globose, number of scales six or eight, scales thick, paired, outer surface bluish purple, distinctly beaked; seed unwinged; native to Asia but commonly planted for landscaping in North America. Distinguished from *Thuja* by nearly vertical sprays, large-beaked cone scales, and wingless seeds.

Genus: *Chamaecyparis*, Cedar

Three species of this genus are native to North America and three to eastern Asia. *Chamaecypress thyoides* (Atlantic white-cedar) is found in swamps along the Atlantic coast, and *Chamaecypress nootkatensis* (Alaska-cedar) is located along the humid coastal region of the west coast from central Oregon to southern Alaska. *Chamaecypress lawsoniana* (Port-Orford-cedar) is restricted to a small area of northern California and southern Oregon.

The needles are scale-like, 1.5 to 3.0 mm in length, decussate, glandular, and appressed. The leaves and branchlets are flattened, with the numerous small branches and branchlets forming vertical sprays. Leaves on seedlings are awl-shaped and approximately 1 cm in length.

The cones are round, 6 to 12 mm in diameter, and composed of 6 to 12 woody peltate scales; the cone remains intact at maturity. Two to five small seeds with small lateral wings are borne on each scale. The cones remain on the tree during the dormant season.

The North American species are large trees at maturity and an important source of wood. The wood is light, holds paint well, has moderate shrinkage, a uniform texture, and good working and machining qualities. It is used for a wide variety of products, including poles, piling, furniture, cabinetwork, millwork, small boat hulls, canoe paddles, sounding boards in musical instruments, caskets, and novelties. The wood has been used as storage battery separators because of its high electrical resistance to acids.

Chamaecyparis nootkatensis (D. Don) Sudworth, Alaska-cedar, yellow-cypress (Figure 11.14); large tree, height to 40 m; evergreen; leaves scale-like, length 2 to 3 mm, decussate; leaves arranged into four ranks, appressed, a glandular protrusion appears in the center

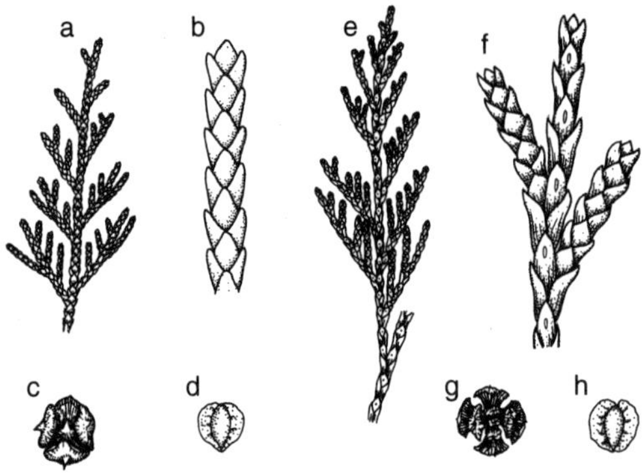

FIGURE 11.14 *Chamaecyparis nootkatensis* (a) flat spray of branchlets ×1/2, (b) branchlet showing shape and arrangement of scale-like leaves ×2, (c) open seed cone of four paired peltate scales ×1/2, (d) winged seed ×1; *C. lawsoniana* (e) flat spray of branchlets ×1/2, (f) branchlet showing shape and arrangement of scale-like leaves ×2, (g) open cone of eight paired peltate scales ×1/2, (h) winged seed ×1.

of the leaf; leaves on margins folded; leaves and branchlets flattened to form a vertical fanlike spray; cone shape ovoid, length 8 to 12 mm, reddish brown, glaucous, scales four to six, scale shape peltate with a boss; bark fibrous, splitting into vertical strips; Region 7, Coastal Ranges, Oregon to southeast Alaska, disjunct in northern California and eastern British Columbia; mesic or wetter sites, fog belt, mountain slopes between sea level and 1500 m. Cover type **30.8**, 30.1, 30.10, 30.23, 30.24.

Chamaecyparis lawsoniana (A. Murray), Port-Orford-cedar (Figure 11.14); large tree, height to 50 m; evergreen; leaves scale-like, length 2 to 3 mm, decussate; leaves arranged into four ranks, appressed, a glandular protrusion appears in the center of the leaf; leaves on margins folded; leaves and branchlets flattened to form a vertical fan-like spray; cone shape ovoid, length 8 to 12 mm, purplish to reddish brown, glaucous, scales five to nine, scale shape peltate with a boss; bark fibrous, splitting into vertical strips; Region 7, Siskiyou Mountains of southern Oregon to Mount Shasta in northern California; dry soils, fog belt, mountain slopes between sea level and 1500 m. Cover type **30.7**, 30.10, 30.18, 30.30, 30.31, 30.37.

Chamaecyparis thyoides (Linnaeus) Britton, Stearns & Poggenburg, Atlantic white-cedar (Figure 11.15);

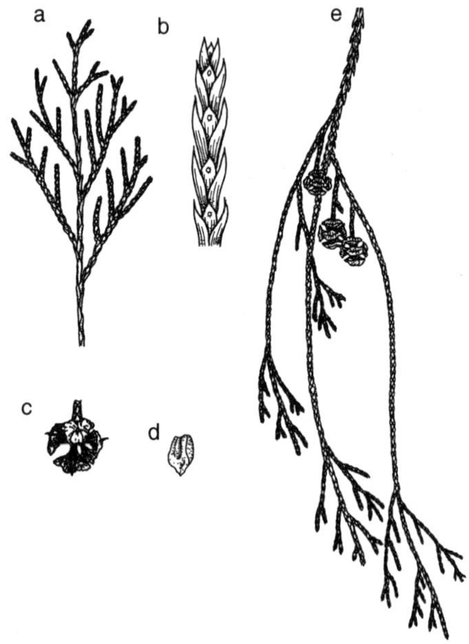

FIGURE 11.15 *Chamaecyparis thyoides* (a) spray of scale-like leaves ×1/2, (b) branchlet showing shape, glands, and arrangement of scale-like leaves ×2, (c) open seed cone of four peltate scales×1/2, (d) angular seed ×1; (e) *Chamaecyparis pisifera* pendent spray of scale-like leaves and seed cones ×1/2.

large tree, height to 25 m; evergreen; leaves scale-like, length 2 to 3 mm, decussate; leaves arranged into four ranks, appressed, a glandular protrusion appears in the center of the leaf, leaves on margins folded; leaves and branchlets form a vertical fanlike spray; cone shape ovoid, length 4 to 9 mm, glaucous, bluish purple to brownish, scales six to eight, shape peltate with a reflexed boss; bark fibrous, splitting into vertical strips; Regions 2b and 4a, b, New England and Coastal Plain, southern Maine and Massachusetts to Florida and Mississippi; lowland tree, typically found in bogs, pocosins, and swamps, elevation sea level to 500 m. Cover type **28.41**, 28.29, 28.33.

Chamaecyparis pisifera (Siebold & Zuccarini) Endlicher, Sawara-cypress (Figure 11.15), is native to Japan but commonly planted in North America. Its cones and foliage are similar to those of other *Chamaecyparis* species, except that the leaves have a

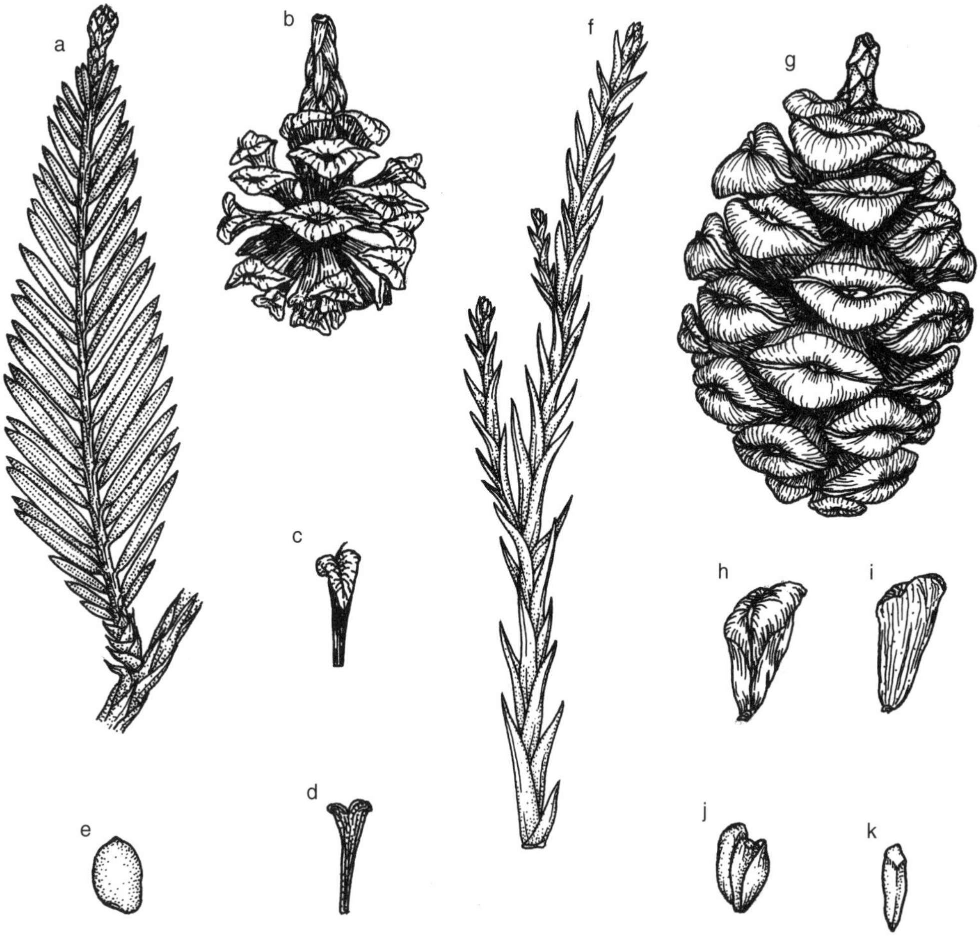

FIGURE 11.16 *Sequoia sempervirens* (a) branchlet of linear two-ranked leaves ×1, (b) orbicular seed cone of peltate scales ×1, (c) side view of peltate cone scale ×1, (d) cross section of peltate cone scale ×1, (e) seed ×2; *Sequoiadendron giganteum* (f) branchlets of subulate leaves ×1, (g) oval seed cone of peltate scales ×1, (h) side view of peltate cone scale ×1, (i) cross section of peltate cone scale ×1, (j) seed with lateral wings ×1, (k) seed without wings ×1.

white cross on the lower surface. The cone is small and orbicular (diameter 5 to 7 mm).

Genus: *Sequoia*, Redwood

Sequoia sempervirens is the only species in the genus (monotypic). For years, the wood of this species has been in high demand because of its strength and exceptional resistance to decay. It is easily planed and used for lumber, millwork, furniture, decorative fences, decks, and similar structures. Shrinkage is low and there is little twisting and warping. As long as the wood is not in direct contact with moist soil, exterior material can be expected to last as long as 25 years. However, *Sequoia sempervirens* has a small range, large mature trees are

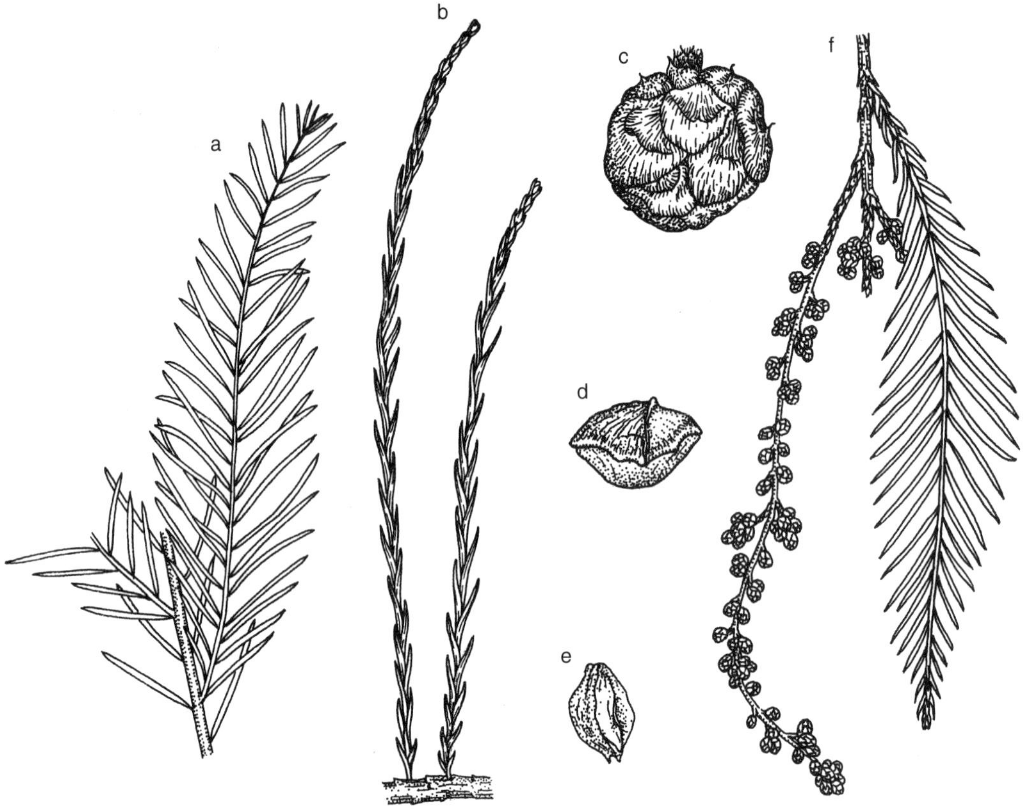

FIGURE 11.17 (a) *Taxodium distichum* var. *distichum* branchlet with alternately arranged two-ranked linear leaves ×1; (b) *Taxodium distichum* var. *imbricarium* upright branchlets with attached appressed subulate leaves ×1, (c) globose cone of peltate scales ×1, (d) peltate cone scale ×1, (e) seed ×1, (f) branchlet and pollen cone strobilus ×1.

in short supply, and stands have aesthetic properties; thus, where rot-resistant wood is required it has been largely replaced by treated pine lumber.

Considerable controversy has arisen as old-growth stands have been removed over extensive areas. Modest acreages were set aside as California state parks by the 1960s, and there is now a national park. Recently, under pressure from environmentalists, a timber company was persuaded to sell a large acreage to the federal government. The highly aesthetic atmosphere created by these dense, tall trees cannot be described; it must be experienced. Trees of more than 70 m in height and 3 to 4 m in diameter are often separated by only a few meters. The luxuriant ground flora adds to the atmosphere.

Sequoia trees are the tallest organisms in the world. Tree heights of 103 to 111 m (340 to 364 ft) with diameters of 6 to 7 m (20 to 22 ft) have been reported (Peattie 1953). Some individual trees may live a 1000 years; Peattie (1953) has reported a tree with 1728 rings.

Sequoia sempervirens (D. Don) Endlicher, redwood (Figure 11.16); large tree, height to 110 m; evergreen; branchlets alternate; leaves dimorphic, leaf shape on lateral branchlets linear, maximum length 2.5 cm, leaves two-ranked; leaf shape on strobili-

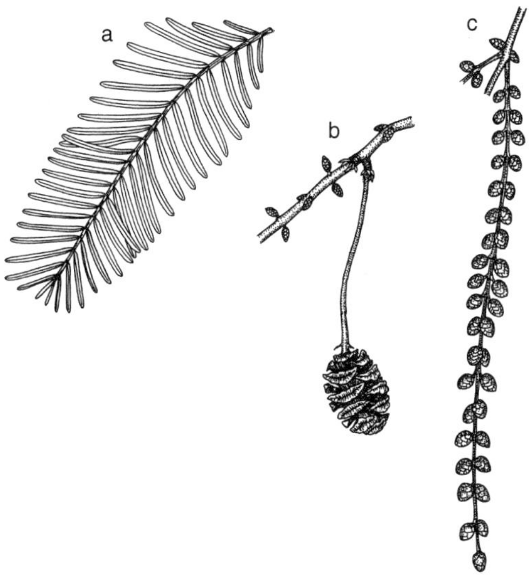

FIGURE 11.18 *Metasequoia glyptostroboides*
(a) branchlet with oppositely arranged two-ranked leaves ×1/2, (b) winter twig with opposite buds and oval seed cone of peltate scales ×1, (c) winter twig with pollen cones ×2.

bearing branchlets somewhat awl-shaped, length 1.0 cm, leaves spiral, somewhat appressed to the twig; cone shape globose, length 2.0 to 2.5 cm, scales woody and peltate; bark thickness up to 30 cm, dark brown, fibrous; Region 7, central California to southern Oregon; coastal fog belt, elevation usually below 300 m; alluvial terraces. State tree of California. Cover type **30.41, 30.42**, 30.3, 30.7, 30.18, 30.35, 30.45.

Genus: *Sequoiadendron*, Giant Sequoia

This genus is monotypic, the only member being *Sequoiadendron giganteum*, giant sequoia. In 1939, Bucholtz recommended that one of the two species then in the genus *Sequoia* be placed in the genus *Sequoiadendron.* Although there has been some debate as to the status of *S. giganteum*, taxonomists have generally accepted the new genus. The remaining trees of *S. giganteum* are protected in a few groves on the west slope of the Sierra Nevada of California. Although the giant trees of Sequoia National Park (e.g., General Sherman tree: height = 85 m, diameter = 9.8 m) are the most well known, several groves are also found in the adjacent Sequoia National Forest.

Sequoiadendron giganteum (Lindley) J. Buchholz, giant sequoia (Figure 11.16); large tree, height to 90 m; evergreen; branchlets alternate; leaves somewhat dimorphic, leaf length on lateral branchlets 0.7 cm, scale-like, and appressed to branchlet, leaf length on leaders 1.5 cm, lanceolate, and spreading; cone length 4 to 9 cm, scales woody and peltate; bark thickness up to 60 cm, cinnamon color, fibrous; Region 7, Sierra Nevada west slope; montane forest, middle elevation of 1500 to 2500 m. Cover type **30.43,** 30.26.

Several interesting comparisons between *Sequoia* and *Sequoiadendron* may be made for these two closely related species. The 2N chromosome numbers of *Sequoia* and *Sequoiadendron* are 66 and 22, respectively. Thus, it may be suggested that *Sequoiadendron* is a polyploid of *Sequoia.* Polyploids are generally more robust than the diploids, and this relationship appears to apply to these two species. *Sequoiadendron* attains nearly the same height but a much larger diameter than redwood, the cone is two to three times larger but otherwise very similar in appearance, and *Sequoia* has two (two to four) cotyledons, whereas *Sequoiadendron* has four (three to six).

Genus: *Taxodium*, Baldcypress

During the Tertiary period, *Taxodium* was widely distributed, but now only one or two species are known. One species with two varieties is native to southeastern United States. Both varieties grow on wet to continuously inundated sites, are deciduous, produce pneumatophores ("knees") from the root system, and have a buttressed trunk (flared at the base). Their cones are indistinguishable. The number of cotyledons ranges from four to nine. The status of a possible third taxon (*Taxodium mucronatum*; *T. distichum* var. *mexicanum*, Montezuma baldcypress) in Texas, Mexico, Central America, and Guatemala has been recognized by some scientists, but at present its status remains uncertain. Whereas *Taxodium distichum* var. *distichum* and var. *imbricarium* are deciduous and bare for several months during the growing season, *T. mucronatum* is seldom "bald" because new needles develop before the older ones drop.

The cones are irregularly globose, composed of peltate scales, mature in one season, and fracture at maturity in the early fall. The apophysis of the scale has four sides of equal length that create the diamond shapes on the cone surface. The apophysis is called actinomorphic; that is, it is as wide as it is long and thus may be divided along several different planes to yield equal halves, in contrast to the zygomorphic (parallelogram or oval) apophysis of *Sequoia* and *Sequoiadendron* cone scales.

The wood is extremely resistant to decay and is often used where wood is exposed to the weather: construction, posts, siding, timbers, coffins, and a variety of special products. Defective wood with holes harvested from standing or down trees is used to make "pecky cypress" paneling and may be sold for a higher price than sound wood.

Taxodium distichum (Linnaeus) Richard var. *distichum*, baldcypress (Figure 11.17); large tree, height to 50 m, flat-topped and distorted at maturity; branchlet arrangement alternate, oriented horizontally; branchlets and leaves deciduous, falling as a unit; leaves two-ranked; leaf shape linear, length 0.5 to 1.7 cm; cones globose, fracturing at maturity, diameter 2 to 4 cm, scales peltates Regions 3a and 4; Coastal Plain from east Texas to Delaware, Mississippi River floodplain north to southern Illinois and Indiana; blackwater swamps, ponds, stream margins. Cover type **28.31, 28.54, 28.55,** 27.10, 27.28, 28.21, 28.29, 28.30, 28.38, 28.40, 28.43, 28.47, 28.49, 28.50, 28.53.

Taxodium distichum (Linnaeus) Richard var. *imbricarium* (Nuttall) Croom, (*Taxodium distichum* var. *nutans*, *Taxodium ascendens*), pondcypress (Figure 11.17); large tree, height to 30 m, flat-topped and distorted at maturity; branchlets vertically oriented, arrangement alternate, branchlets and leaves deciduous, falling as a unit; leaves imbricate and appressed to twig; leaf more or less lanceolate or awl-shaped, length 0.5 to 1.7 cm; cones globose, fracturing at maturity, diameter 2 to 4 cm, scales peltate; Region 4; Coastal Plain from Louisiana to North Carolina; black water swamps, ponds, stream margins, often found forming a "dome" in circular depressions of fields; the major *Taxodium* of the Okefenokee Swamp. Cover type **28.36, 28.53,** 28.29, 28.30, 28.38, 28.40, 28.43, 28.54.

Genus: *Metasequoia*, Dawn Redwood

This genus is monotypic. *Metasequoia glyptostroboides* (dawn redwood) was introduced into the United States some years ago. Once thought to be extinct, it was rediscovered in China in 1948. Fossil evidence indicates that at least the genus, if not the species, was once widely distributed across North America, Greenland, and parts of the Arctic, as well as Asia (Sporne 1965). *Metasequoia* is now a commonly planted ornamental, somewhat resembling the native *Taxodium distichum.* A major distinguishing character is that the deciduous branchlets and leaves have an opposite arrangement, whereas *Taxodium* has alternate branchlets.

Metasequoia glyptostroboides Hu & W. C. Cheng, dawn-redwood (Figure 11.18); large tree, height to 35 m, crown shape inverted cone; branchlets opposite and horizontally oriented; branchlets and leaves deciduous, falling as a unit; leaves two-ranked, appearing opposite, shape linear, length 0.5 to 2 cm; cones globose, length 0.7 to 1.2 cm, scales paired, not fracturing at maturity. Introduced; planted as a lawn tree or ornamental.

CHAPTER 12

Order: Taxales, Ginkgoales, and Cycadales

ORDER: TAXALES

TAXALES is a small order of one family (Taxaceae), five genera, and approximately 20 species. Only members of the genera *Taxus* and *Torreya* are native to North America. Originally the family was included in the Coniferales, but there are sufficient differences in the reproductive structures to place it in a separate order.

The members are evergreen shrubs and small to medium-sized trees with linear leaves; the leaves are arranged spirally on the twig, although occasionally they are twisted at the base to form two ranks. Some genera have dimorphic (two forms) leaves. The ovules are solitary, and after fertilization the integument forms a fleshy aril that surrounds the seed. New seedlings have two cotyledons. The wood is non-resinous.

Genus: Taxus (Yew)

The *Taxus* genus of approximately nine species is widely distributed in the Northern Hemisphere; three species are native to North America, and several exotic species are commonly used as ornamentals. The leaves are linear, 1 to 3 cm in length, and sharp pointed. On some species there are two stomata bands on the lower leaf surface. The seed is approximately 5 mm long and enclosed, except at the distal end, by an aril that is a bright red to orange red, soft, fleshy inverted cup formed from the expanded integument. The mature seed and surrounding fleshy aril somewhat resemble a berry. The seed is more or less oval and is set in a circular disk. Most species are dioecious.

Of the North American species, *Taxus brevifolia* (Pacific yew) and *Taxus floridana* (Florida yew) reach tree size. The wood of *T. brevifolia* has traditionally been used for cabinets, carvings, bows, canoe paddles, and poles. However, the importance of this species has recently increased with the discovery that a chemical extracted from the bark can be effectively used to treat ovarian cancer. There is increasing concern that *Taxus brevifolia* may become rare or endangered because this extract (taxol) has not been synthesized in the laboratory and larger trees are killed to remove the taxol-containing bark. *Taxus floridana* is a rare endemic, listed as endangered by the Florida Department of Agriculture. Its population numbers are limited, and it should not be collected from its natural habitat. Even with continuous care in the urban setting, it is difficult to maintain. *Taxus canadensis* (Canadian yew) is a multistemmed shrub. As an important deer food, it has been nearly extirpated in the Lake States, where herds are excessively large.

Taxus brevifolia Nuttall, Pacific yew (Figure 12.1); small to medium-sized tree, height to 15 m; evergreen; leaf arrangement alternate to subopposite; leaves becoming two-ranked; leaf shape linear, length 1 to 3 cm, width 1.0 to 2.5 mm, flexible, pale yellow green, lower surface with two parallel greenish stomata bands, apex mucronate but not sharp to touch, petiole length 1.0 to 1.5 mm; aril ovoid, fleshy, bright red to orange red, length about 7 mm; seed oval, length about 5 mm; first-year twigs green, second- and third-year twigs

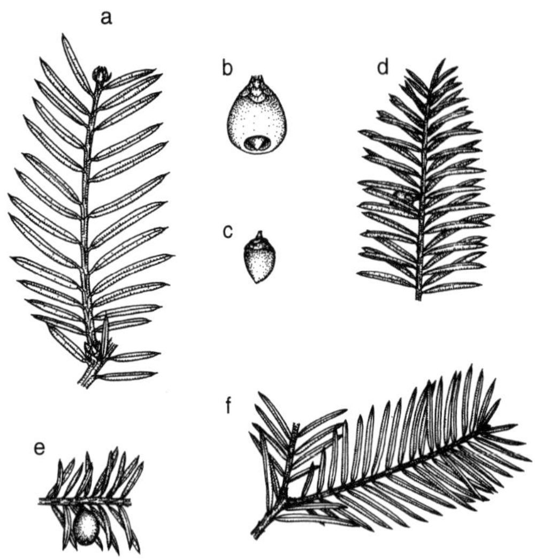

FIGURE 12.1 *Taxus brevifolia* (a) branchlet with leaves and bud ×1/2, (b) aril, seed visible at the distal end ×1, (c) seed ×1; (d), (e) *Taxus canadensis* branchlet with leaf and aril ×1/2; (f) *Taxus floridana* branchlet and leaf ×1/2.

reddish brown; Regions 6 and 7, northern California through Oregon, Washington to central British Columbia and southeastern Alaska, also a disjunct population in northern Idaho, western Montana and Alberta, eastern Washington and British Columbia; scattered in open to closed forest, moist slopes, shady canyons and drainages, elevation sea level to 2200 m. Cover type 29.24, 30.7, 30.17, 30.21, 30.37.

Taxus canadensis Marshall, Canadian yew, ground hemlock (Figure 12.1); straggling shrub developing into loosely formed patches, height to 2 m; evergreen; leaf arrangement alternate to subopposite; leaves two-ranked; leaf shape linear, length 1.0 to 2.5 cm, width 1.0 to 2.5 mm; flexible, dark green, lower surface with two parallel greenish stomata bands, apex mucronate but not sharp to touch, petiole length 1.0 to 1.5 mm; aril ovoid, fleshy, bright red to orange red, length about 7 mm; seed oval, length about 5 mm; leaf sheath present on second- and third-year twigs; twigs dark brown; Regions 1 and 2; moist soil, closed woods.

Taxus floridana Nuttall ex. Chapman, Florida yew (Figure 12.1); small tree, height to 9 m; evergreen; leaf arrangement alternate to subopposite; leaves two-ranked; leaf shape linear but sides frequently converging slightly toward apex, length 1 to 2 cm, width 1.0 to 2.5 mm; flexible, dark green, lower surface with two parallel greenish stomata bands, apex mucronate but not sharp to touch, petiole length 1.0 to 1.5 mm; aril ovoid, fleshy, bright red to orange red; seed oval, length 5 mm; first-year twigs light brown; leaf sheath present on second- and third-year twigs, twigs gray brown; Region 4b, Florida panhandle; moist drainages, bluffs and ravines along the Appalachicola River. Cover type 28.27. Rare endemic.

Two *Taxus* species have been introduced into North America for landscaping and ornamental purposes. There are about 15 known cultivars for each species.

Taxus cuspidata Siebold & Zuccarini, Japanese yew (Figure 12.2); shrub forming a dense patch, height to 3 m (10 m in some cultivars); leaf arrangement alternate to subopposite; leaves spreading or two-ranked; leaf shape linear, length 1.0 to 2.5 cm, width 1.0 to 2.5 mm, flexible, dark green, lower surface with two parallel greenish stomata bands, apex mucronate but not sharp, petiole length 1.0 to 1.5 mm; aril ovoid, fleshy, bright red to orange red; seed oval, length about 5 mm; plant widely planted but not naturalized.

Taxus baccata Linnaeus, English yew (Figure 12.2); tree, height to 20 m; evergreen; leaf arrangement alternate to subopposite; leaves spreading (hairbrush fashion); leaf shape linear, length 0.8 to 3 cm, width 3–6 mm, flexible, dark greenish black, lower surface

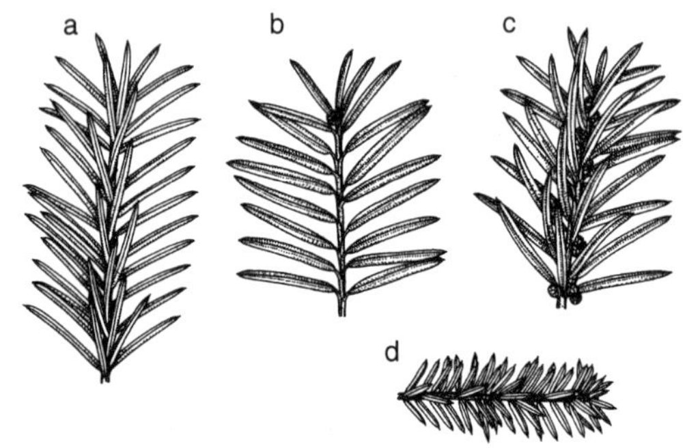

FIGURE 12.2 *Taxus cuspidata* (a) branchlet with spreading leaves ×1/2, (b) branchlet with two-ranked leaves ×1/2; *Taxus baccata,* (c) long leaf form ×1/2, (d) short leaf form ×1/2.

with poorly defined stomata bands, apex mucronate but not sharp, petiole length 1.0 to 1.5 mm; aril ovoid, length about 7 mm, fleshy, bright red to orange red; seed oval, length 5 mm; widely planted but not naturalized.

Genus: Torreya

Prior to glaciation and the accompanying colder climate, *Torreya* was circumpolar at high latitudes (Sporne 1965; Sudworth 1908). Six species have been described; four are found in eastern Asia and Japan. In North America, one relict species has been found in California and one in Florida. The wood is used locally for fence posts, cabinets, and novelties. Like *Taxus floridana, Torreya taxifolia* is a rare endemic whose populations are in decline. Nelson (1996) indicates that this species is in serious trouble and may be the most endangered species in Florida.

Torreya californica Torrey, California torreya or nutmeg (Figure 12.3); medium-sized tree, height to 20 m; evergreen; leaf arrangement decussate but alternate set of leaves bent to form two ranks; leaf length 3.0 to 5.5 cm, width 2.3 mm, linear, rigid and stiff, apex acute and sharp-spine tipped, lower surface with two parallel light green stomate bands, emitting a strong odor when crushed; basal sheath on new twigs extending 0.5 to 1.5 cm below point of attachment; sheath line on older twigs tan; new twigs light green, turning reddish brown; aril ovoid, fleshy, length 2.0 to 3.5 cm, diameter 1 to 2 cm, light green with purple streaks, completely enclosing seed, drupe-like; seed ovoid; Region 7; cool slopes, mountain stream edges, canyons, elevation 0 to 1900 m. Relatively rare.

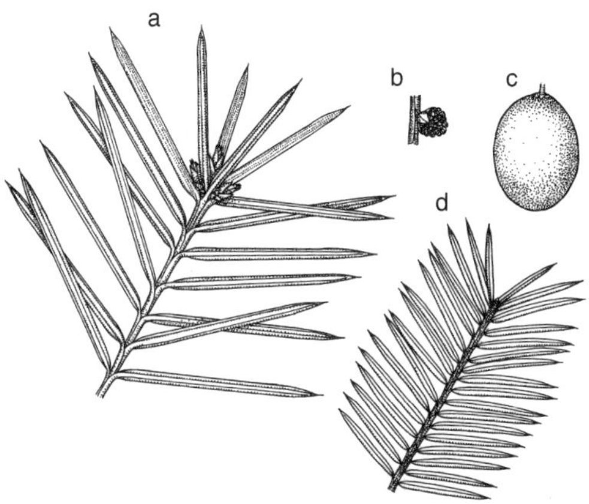

FIGURE 12.3 *Torreya californica* (a) branchlet with leaves and buds ×1/2, (b) male strobilis ×1, (c) aril ×1/2; (d) *Torreya taxifolia* branchlet with leaves ×1/3.

Torreya taxifolia Arnott, Florida or stinking cedar (Figure 12.3); medium-sized tree, small to height to 15 m; leaf arrangement decussate but alternate set of leaves bent to form two ranks; leaf length 2.0 to 3.5 cm, width 2 mm, linear, rigid and stiff, apex acute and spine tipped, lower surface with two parallel gray stomate bands, emitting a strong odor when crushed; new twigs dark green; basal sheath on new twigs extending 0.5 to 1.5 cm below point of attachment; sheath line on older twigs tan; new twigs light green, turning reddish brown; aril ovoid, fleshy, length about 2.5 cm, diameter 1.5 to 2.0 cm, completely enclosing seed, drupe-like, dark green to purple, sometimes streaked; Region 4b, Florida panhandle; moist drainages, bluffs, and ravines along the Appalachicola River and in extreme southwestern Georgia. Rare and endangered. Cover type 28.27.

ORDER: GINKGOALES

The Ginkgoales order contains one family (Ginkgoaceae) and one genus (*Ginkgo*) with a single living member, *Ginkgo biloba*, also known as the ginkgo or maidenhair tree. The species is the only surviving member (a relic) of a once widespread and old group of gymnosperms. Although probably eliminated from natural forest for centuries, *G. biloba* was saved from extinction because it was considered a sacred tree and cultivated in Chinese temple gardens (Lawrence 1951). The species was introduced from Asia into Europe at about 1730 and into North America at about 1784. Lawrence (1951) notes that the name "ginkgo" is a translation error and that the correct name should be "ginkyo." However, the present name is so firmly established that it is not likely to be changed.

The species is dioecious. Male trees have paired stamens in a catkin-like strobilus attached in the leaf axil; two anthers (microsporangia) are found on each stalk (sporophyll). Female trees bear two ovules per peduncle, but only one develops into a seed. As in the Cycadaceae, fertilization is by motile sperm, an unusual phenomenon in higher plants; fertilization may occur after the ovule has fallen from the tree. The fruit is about the size of a plum, including the outer fleshy or gelatinous layer (the old integument), which surrounds an inner hard, pit-like center. The seed inside the pit is considered a delicacy in Asia. However, when ripe, the outer layer has a strong odor similar to that of rancid butter. This fleshiness and odor make female trees undesirable for the lawn and street. Cuttings from known male trees are widely planted in cities because they do not produce fruit. Ginkgo is extremely tolerant of polluted city air. This species is deciduous, and the wood has resin canals but no true vessels.

Ginkgo biloba Linnaeus, ginkgo (Figure 12.4); large tree, height to 30 meters; deciduous; leaf fan-shaped, blade length 4 to 6 cm, width 8 to 11 cm, often notched along the margin by a small sinus suggesting two lobes (hence the name *biloba*), venation dichotomous, petiole length 3 to 6 cm; leaves borne on stout spur shoots tipped with a distinct bud; aril essentially globose, diameter 2 to 3 cm; seed elliptical to slightly obovate; length 1.5 to 2.5 cm, width 1.2 to 1.7 cm; widely planted as an ornamental.

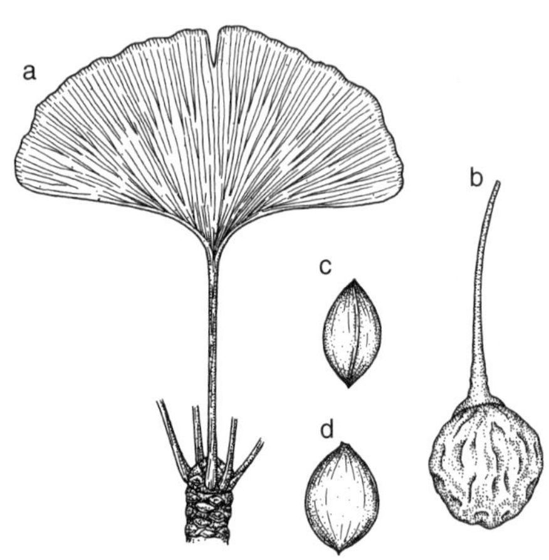

FIGURE 12.4 *Ginkgo biloba* (a) fan-shaped leaf ×1/2, (b) aril with seed enclosed ×1/2, (c) seed, side view ×1/2, (d) seed, front view ×1/2.

ORDER: CYCADALES

The Cycadaceae (cycads) is the only family of the Cycadales with living representatives. Species of this family have pinnately compound leaves, which form an upright cluster on a thick tuberous underground stem or on a short trunk such as found in the palms. The cycads differ from other gymnosperms because of their compound leaves, palmlike appearance, and their numerous scattered microsporangia. However, ovulate reproductive structures closely resemble cones. The species are dioecious; fertilization is by motile sperm. The cycads are represented by nine genera and approximately 100 species, primarily confined to tropical and subtropical regions. According to Nelson (1996), only one species of the genus *Zamia* is found in North America, although a number of scientific names have been applied to it. Native

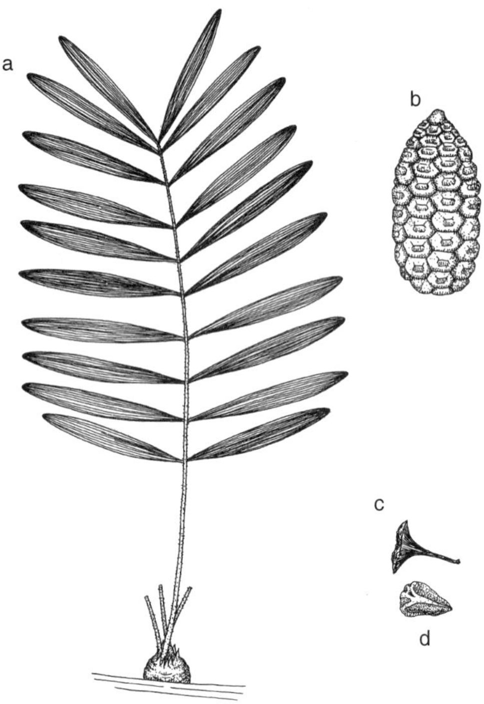

FIGURE 12.5 *Zamia integrifolia* (a) pinnately compound leaf ×1/10, (b) seed cone ×1/5, (c) peltate cone scale ×1/2, (d) angular seed ×1/2.

Americans of the Seminole tribe once used the root as a source of starch by crushing it into a powder, washing it to remove the toxins, and letting it dry. Eaten raw, the stem can be fatal. The underground stem is broad and more or less pointed, hence the name Florida arrowroot. *Zamia* is widely planted as an ornamental. The species identified by de Candolle in the 1800s and Small (1933) have now been combined into one species.

Zamia integrifolia Linnaeus (*Zamia floridana* A. de Candolle; *Zamia angustifolia* Small; *Zamia integrifolia* Small; *Zamia silvicola* Small; *Zamia umbrosa* Small), Florida arrowroot, Seminole bread, coontie (Figure 12.5); primary stem below ground; above-ground stem short, seldom reaching 1.0 m, leaves borne from the shoot apex, pinnately compound, two-ranked, appearing somewhat fern-like, length 30 to 150 cm, arching; leaflet number 18 to 60, length 6 to 15 cm, stiff, margin revolute, essentially opposite, female cones reddish brown, borne on the shoot apex, scales peltate; seeds angled; Region 4b, Florida and adjacent Alabama.

PART III

Woody Angiosperms (Division Magnoliophyta)

CHAPTER 13

The Woody Angiosperms: An Overview

INTRODUCTION

ANGIOSPERMS, or plants of the Magnoliophyta (Cronquist 1981), are known as flowering plants because they produce "true flowers." Literally, the term *angiosperm* means a sperm or seed enclosed in a "vessel," or protective structure, the ovary. Collectively, the mature ovary with the seeds and sometimes with other external structures is the fruit, a structure that does not appear in gymnosperms. The Magnoliophyta (Table 13.1) is split into two taxa, the Magnoliopsida (plants with two cotyledons), commonly called dicots (Chapters 17–23) and the Liliopsida (plants with one cotyledon), or monocots (Chapter 24). There are major differences in both stem and flower characteristics between these two groups. In this chapter, we are concerned only with woody dicotyledonous plants.

Several terms are applied to the woody dicots. Possibly the best term is "broadleaf," as it refers to the expanded blade of the leaf common to nearly all woody dicots. The term "deciduous" is also used, particularly in the temperate zone of eastern North America, but outside this region a substantial number of angiosperms have persistent leaves (an evergreen habit). The deciduous habit of most angiosperms has evolved as a mechanism to avoid winter frost injury in regions of severe winters. In Canada and the northern and central United States, angiosperms are deciduous; they develop leaves in the late spring and drop them in the early fall. In the Central States, where winters are of moderate severity, *Fagus* (beech) and many species of *Quercus* (oak) retain the dead leaves on the trees through the winter months (marcescent). Plants of other species retain green leaves well into the fall or winter months and may be considered semi-evergreen. Because of the mild winter of the Atlantic and Gulf coastal regions, several species of *Quercus*, *Magnolia* (magnolia), *Ilex* (holly), and a variety of shrubs maintain healthy leaves throughout the year (evergreen plants). The number of evergreen broadleaf trees and shrubs continues to increase in south (subtropical) Florida and Central America.

The term "hardwood" is applied strictly to trees, not to shrubs or vines, and is the term most commonly used in forestry and ecology to indicate a woody dicot. However, it may also be inappropriate, depending on the species. The wood of *Populus* (cottonwood and aspen) is softer than the wood of many gymnosperms such as *Tsuga* (hemlock) and some *Pinus* (e. g., *P. palustris*, longleaf pine).

MORPHOLOGY

The structures of the woody dicot are classed as vegetative (cotyledons, branches, twigs, buds, and leaves above ground; roots below ground) and reproductive structures (flowers and fruits). Dicots usually, but not always, have two cotyledons. The cotyledons, or the primary seed leaves, are temporary organs lasting only a few weeks, usually until new leaves are formed. Two types of cotyledons are found in dicots. In species that have large seeds and fruits (*Quercus*, *Fagus*, *Carya* (hickory), and *Aesculus*, buckeye), the cotyledons are primarily storage organs that provide food to the

Table 13.1. The classification of woody dicots

Woody dicots are classified here according the system developed by Cronquist (1981). The common name of each family follows the scientific name.

- Class: Magnoliopsida
  - Subclass I: Magnoliidae
    - Order: Magnoliales
      - Families: Magnoliaceae—magnolia
      - Annonaceae—custard apple
    - Order: Laurales
      - Family: Lauraceae—laurel
      - Calycanthaceae—strawberry shrub
    - Order: Aristolochiales
      - Family: Aristolochiaceae—birthwort
    - Order: Illiciales
      - Families: Illiciaeae—star anise
      - Schisandraceae—schrandra
    - Order: Ranunculales
      - Families: Ranunculaceae-crowfoot, buttercup
      - Berberidaceae—barberry
      - Menispermaceae—moonseed
  - Subclass II: Hamamelidae
    - Order: Hamamelidales
      - Family: Platanaceae—plane tree
      - Hamamelidaceae—witch hazel, sweetgum
    - Order: Urticales
      - Families: Ulmaceae—elm
      - Moraceae—mulberry
    - Order: Juglandales
      - Family: Juglandaceae—walnut, hickory
    - Order: Myricales
      - Family: Myricaceae—bayberry
    - Order: Casuarinales
      - Family: Casuarinaceae—she-oak
    - Order: Fagales
      - Families: Fagaceae—beech
      - Betulaceae—birch
  - Subclass III: Dilleniidae
    - Order: Theales
      - Family: Theaceae—tea
      - Clusiaceae—Saint-John's-wort
    - Order: Malvales
      - Family: Tiliaceae—linden, basswood
    - Order: Salicales
      - Family: Salicaceae—willow
    - Order: Ericales
      - Families: Cyrillaceae—cyrilla
      - Clethraceae—clethra
      - Ericaceae—heath
    - Order: Ebenales
      - Family: Sapotaceae—sapodilla
      - Ebenaceae—ebony
      - Styracaceae—storax
      - Symplocaceae—sweetleaf
  - Subclass IV: Rosidae
    - Order: Rosales
      - Family: Hydrangeaceae—hydrangea
      - Rosaceae—rose, blackberry
      - Grossulariaceae—current, gooseberry
      - Saxifragaceae—saxifrage
    - Order: Fabales
      - Family: Mimosaceae—mimosa
      - Caesalpiniaceae—caesalpinia
      - Fabaceae—pea or bean
    - Order: Proteales
      - Family: Elaeagnaceae—oleaster, autumn-olive
    - Order: Myrtales
      - Family: Thymelaeaceae—mezereum
      - Myrtaceae—myrtle
    - Order: Cornales
      - Families: Nyssaceae—black gum, tupelo
      - Cornaceae—dogwood
      - Garryaceae—silk tassel
    - Order: Santalales
      - Families: Loranthaceae—showy mistletoe
      - Viscaceae—Christmas mistletoe
    - Order: Celastrales
      - Families: Celastraceae—bittersweet
      - Aquifoliaceae—holly
    - Order: Rhamnales
      - Families: Rhamnaceae—buckthorn
      - Vitaceae—grape
    - Order: Apiales
      - Family: Araliaceae—ginseng
    - Order: Sapindales
      - Families: Staphyleaceae—bladdernut
      - Sapindaceae—soapberry
      - Hippocastanaceae—horse-chestnut
      - Aceraceae—maple
      - Anacardiaceae—sumac

(Continued)

Table 13.1. (*cont.*)

Simaroubaceae—quassia, tree of heaven	Class: Liliopsida
Rutaceae—rue	Subclass I: Arecidae
Subclass V: Asteridae	Order: Arecales
Order: Scrophulariales	Family: Arecaceae—palm
Families: Oleaceae—olive	Subclass II: Commelinidae
Scrophulariaceae—figwort	Order: Cyperales
Bignoniaceae—trumpet-creeper	Family: Poaceae—grass
Order: Rubiales	Subclass III: Liliidae
Family: Rubiaceae—madder	Order: Liliales
Order: Dipsacales	Family: Liliaceae—lily
Families: Caprifoliaceae—honeysuckle	Smilacaceae-greenbriar
Adoxaceae—elderberry, viburnum	

developing root and shoot. They remain in or on the soil as the seedling develops. In some small-seeded species (*Tilia* (basswood); *Liriodendron*, yellow-poplar), the cotyledons are green and photosynthetic, producing food for the growing seedling until the regular photosynthetic leaves are developed. Seed leaves can often be used in identification, but, unfortunately, the form is frequently different from that of mature leaves. A helpful guide to the identification of juvenile seedlings was written by Maisenhelder (1969).

The flowers and fruits of most trees do not develop until the individual is several years old; some species may not flower until 30 or more years of age. For this reason and because of the perennial nature of the woody plant and availability of other organs, the flower is not generally used in identification. It is, however, the primary structure used in plant classification. Flower parts are usually in sets of four or five; the flower parts of monocots occur as sets of three or six.

The woody dicot has three meristems. There is an apical meristem in each bud, which is responsible for height growth. The vascular cambium (lateral meristem) produces secondary (diameter) growth in both the stem and the root. Xylem cells for water transport from the roots to the leaves are produced on the inside of the cambium, and phloem cells for food transport from the leaves to the roots are produced outside the cambium. Eventually, a second cambial layer (cork cambium or phelogen) develops on the stem to produce the typical bark characteristics found in mature trees. In contrast, the woody monocot has only an apical meristem, thus palms and grasses grow in height but not in diameter.

Dicots may have a large taproot or two or more large roots that continue to subdivide and spread in all directions into a heartroot or only horizontally to form a plateroot. All roots eventually developing into fine roots with root hairs. Most have a typical leafblade with adaxial (upper) and abaxial (lower) surfaces. The blade is usually attached to the stem by a petiole. Stipules often subtend the leaf petiole.

Two types of wood are found in dicot trees: ring porous and diffuse porous. In both types, growth is first initiated in the early spring when there are ample water and nutrients and temperature is favorable. The developing leaves produce auxin, a growth substance that moves to the cambium layer to initiate cell development. In ring-porous wood species (i.e., *Quercus*, *Carya*, *Ulmus* (elm), *Fraxinus* (ash)), the major flush of leaves occurs in the spring. The result is the development of large cells, mostly vessels for rapid water transport to the crown when water is available. Through the rest of the growing season there is a much reduced but even production of leaves that produce smaller but stronger cells, such as fibers that create and support the woody stem. A cross-sectional (radial) view of the growth ring (also called an annual ring) in a stem of a ring-porous species shows an abrupt change from a zone of mostly large pore vessels (spring- or earlywood) to a zone of small pore fibers (summer- or latewood). Diffuse-porous wood species (i.e., *Acer* (maple), *Fagus*, *Tilia*, *Nyssa* (blackgun), *Betula* (birch), *Populus* (poplar)) have a more even development of leaves and auxin through the growing season and produce a wood of a relatively homogeneous texture.

DISTRIBUTION AND ECOLOGY

Woody angiosperms such as species of *Fagus, Quercus, Carya, Acer, Fraxinus, Populus, Tilia, Salix* (willow), *Nyssa* and *Ulmus* dominate the forest communities in the temperate zone of the eastern United States. In Canada, these species are restricted to the southern portions of Manitoba, Ontario, and Quebec. Several *Populus* and *Betula* are important pioneer species across Canada and Alaska. A small number of hardwood tree species are found in western mountains.

In contrast to conifers, hardwood trees generally inhabit the more productive sites of the temperate zone. Here, the long growing season permits the development of surplus food sufficient to sustain life through the winter and to allow for a "flush" of leaves the following spring. Several species of *Populus* are important components of forests in Canada and Alaska, probably because they have photosynthetic bark that begins to produce food before the leaves develop. In addition, soil moisture and nutrients in these regions are adequate for growth and development. The term "development" implies the completion of the life cycle: seed germination, seedling establishment, growth to maturity, flowering, pollination, and the setting and distribution of seed.

Hardwood forest communities are found on moderately deep (depth >40 cm) and deep (depth >1 m) upland soils and nearly all bottomland sites that are not permanently flooded. Poorly developed forests such as "scrub oak" may be found on soils of low water- or nutrient-holding capacity. Hardwood trees are important timber producers, providing material for construction lumber, paper, furniture, and veneer, as well as nuts and fruits. Both game and nongame wildlife species depend on *Fagus, Quercus, Carya,* and *Juglans* (walnut), which are known as the "mast producers," and many other trees, shrubs, and vines for nuts and fruit; buds and new twigs are an important wildlife food. A hardwood forest provides excellent watershed protection and soil stability.

LIFE FORM

The aboveground portion of an angiosperm may take several different forms; however, within a species form will usually be the same for most if not all mature stems, although stems of some species such as *Toxicodendron radicans* (poison-ivy) may have several forms. Including herbaceous plants, there are five general life forms found within the angiosperm group.

1. ***Herbaceous.*** A nonwoody plant usually not more than 30 cm in height (an herb). The leaves of several woody species such as *Toxicodendron radicans* and *Parthenocissus quinquefolia* (Virginia creeper) may project from the forest floor and give the appearance of an herb. Hundreds of herbs are forest understory plants.
2. ***Liana.*** A climbing or twinning plant having a woody main stem. Also called a vine. The stem may twine along a fence or the ground or climb a tree trunk. Occasionally, a vine such as *Vitis* (grape) may intertwine with the crown of a seedling and grow as the stem grows, to heights of 20 or 30 m. In this situation, the cylindrical woody stem can be seen dangling free of the tree. Climbing structures are of several forms; *Vitis* has tendrils or thin, corkscrew-like modified twigs that wrap around small twigs, stems, or other objects, whereas *Toxicodendron radicans* has numerous slender (aerial) rootlets. *Parthenocissus quinquefolia* also produces aerial rootlets; each may have a suction-cup-like structure at the tip.

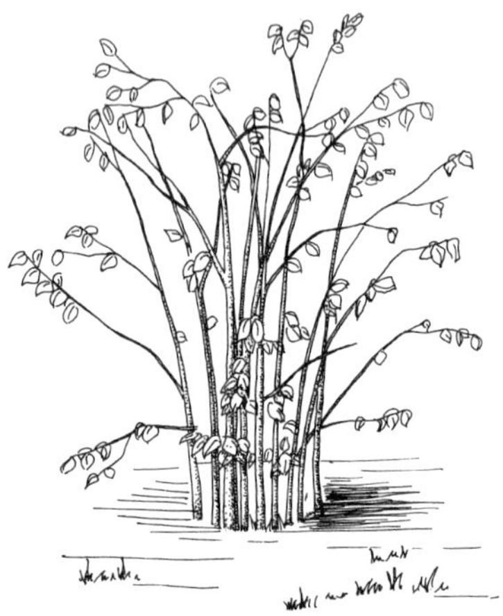

FIGURE 13.1 *Corylus americana*, American hazel, a multistemmed shrub.

FIGURE 13.2 *Cercis canadensis*, eastern redbud, an arborescent with low-branching form.

3. ***Fruiticose.*** Also called a shrub. A more or less woody plant, composed of one or two to many main stems, that generally has a poorly defined crown. At maturity, it is less than about 3 m in height (Figure 13.1).
4. ***Arborescent.*** The term refers to a tree-like structure. A woody plant composed of one to several short stems having a moderately well defined crown. At maturity, it may reach a height of 10 m. Form may vary somewhat (Figures 13.2 and 13.3).

FIGURE 13.3 *Malus pumila* apple an arborescent with a short distinct bole.

FIGURE 13.4 A *Quercus alba* lawn tree grown in high sunlight. Open-grown trees have full, rounded crowns with live branches relatively near the ground.

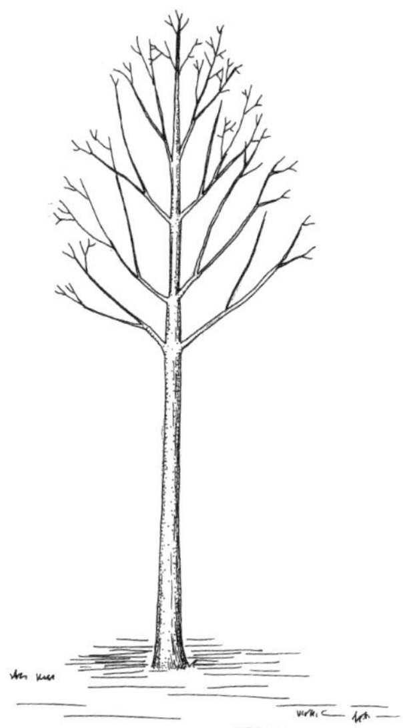

FIGURE 13.5 A *Liquidambar styraciflua* tree grown in dense forest where sunlight is reduced. Forest-grown trees tend to have smaller crowns, be taller, and have few low branches, as compared with open-grown trees of the same age.

5. ***Arboreous.*** A tree. A woody plant composed of a single large main bole and a well-defined crown. The height at maturity is not less than 10 m and may be as great as 45 m. Most hardwood trees are deliquescent; that is, they have a main stem that forks or divides within the main crown area and does not reach to the top of the tree as a single unit. However, form will vary considerably, depending on whether the individual was grown alone in full sunlight that permitted a multiforked main stem and many side branches close to the ground (Figure 13.4), or in the forest where it was shaded and crowded to form a single bole without low side branches (Figure 13.5).

IDENTIFICATION

The leaves, twigs, fruits, and bole are commonly used in field identification; flowers are seldom used. Leaves and twigs have numerous individual characters that permit a positive identification of most individuals. When present, fruit provides a means for unequivocal identification. The bole and twigs are the primary structures used for identification during the dormant season.

In field work, the form (e.g., tree, shrub) of the main stem (also called the bole or trunk) often provides the greatest amount of information for identification. However, unlike leaves, twigs, and fruit, bark characteristics are not discussed in a separate chapter of this book, as the structure and the differences between species are not easily described and bark varies somewhat from tree to tree within a species and with age of individual stems. Most trees do not have typical bark characteristics until they are near maturity. As a result, identifying tree species by bark is not easily learned by using a book; it is an art developed only with field experience.

CLASSIFICATION

Five subclasses of the Magnoliopsida contain woody plants; the sixth subclass, Caryophyllidae, includes only herbaceous species. The woody plants in the five subclasses are found in 32 orders and 67 families (Table 13.1).

CHAPTER 14

The Woody Angiosperm Leaf

INTRODUCTION

THE leaf is a most important organ, as it produces food for tree and shrub growth from oxygen, carbon dioxide, and water. It also is one of the most conspicuous structures of the woody angiosperm and the one most often used in identification. However, because of the deciduous habit of most woody species, its usefulness is restricted to the summer months, as only a comparatively small number of species remain green (evergreen species) or at least retain dead leaves through the winter months (marcescent species). An additional problem in the use of the leaf for identification is that some variability (e.g., size, number of lobes) is encountered not only between individuals of the same species but also in a single individual. Within a species, leaves on stems in the forest understory (shaded conditions) may be substantially different from those in the overstory (high light levels). On the same individual, leaves on the lower branches may be different from those in the upper portion of the crown; a number of *Quercus* species (*Q. velutina*, black oak; *Q. rubra*, northern red oak; *Q. coccinea*, scarlet oak) exhibit such variability. These changes are caused by environmental differences. Light is an important factor in modifying leaf structure (sun vs. shade leaves), but the internal water balance may control the rate and amount of blade expansion and, hence, the size and extent of sinus development.

In this chapter, the range of variability is examined for each leaf structure. However, be aware that when identifying a species, it will be necessary to look for unique character combinations. Within a range of variability, a character combination has high fidelity for a given species. Thus, using leaf characteristics, individuals of *Quercus alba* (white oak) can be identified as belonging to that species, just as individuals of *Ulmus americana* (American elm) can be so identified.

LEAF ARRANGEMENT (PHYLOTAXY)

Technically, leaf arrangement on the twig is not a leaf characteristic per se, but when examined with the pattern of leaves is often one of the most obvious characteristics to consider in woody plant identification. Because leaf arrangement does not fit neatly into any other chapter, it is discussed here. However, because the leaf also is identified by the position of the axillary bud, bud arrangement in a given species follows the same pattern as leaf arrangement. Bud types and arrangement are covered in Chapter 15 on twigs.

As the twig grows, the apical meristem produces leaves at various intervals. The place where the leaf is attached to the twig (or a twig to a branch or a branch to a tree trunk) is called the node; the distance between two nodes is an internode. Leaf patterns formed on the twig are of three types: alternate, opposite, and whorled. These types are easy to identify and they assist in plant identification, but there can be some confusion if the leaflet of a pinnately compound leaf is incorrectly identified as a simple leaf. Leaflets on a pinnately compound leaf are nearly always opposite (rarely

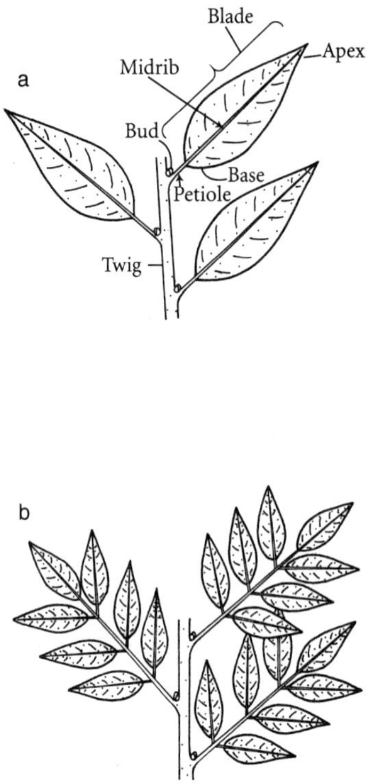

FIGURE 14.1 Alternate leaf arrangement: (a) simple leaves, (b) compound leaves. To separate these two conditions, the bud must be located on the twig at the base of the leaf petiole (the axil).

alternate), as described in the later section on blade complexity. Many species (e.g., *Carya*, hickory; *Rhus*, sumac; *Gleditsia*, honey locust) with compound leaves alternately arranged on the twig also have opposite leaflets. Moreover, the leaflets of a palmately compound leaf may be mistaken for a whorled arrangement. These problems are solved by locating the axillary bud, which is always at the base of the leaf petiole. The axil is the zone (not a structure) between the axis (twig or branchlet) and the base of the petiole at the point of attachment to the twig.

Alternate Arrangement

In an alternate arrangement, the leaves are positioned along the twig in an apparently alternate fashion; that

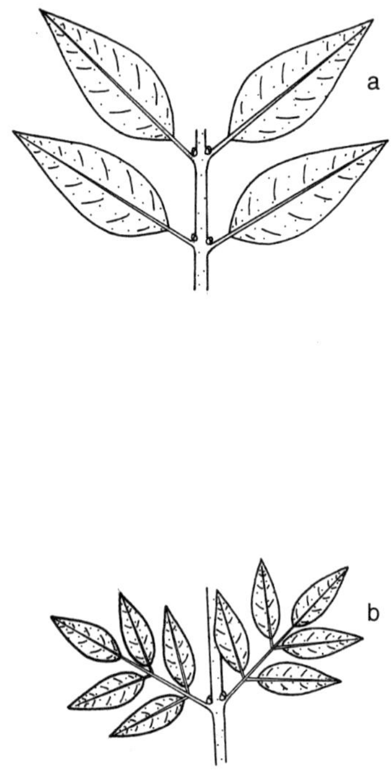

FIGURE 14.2 Opposite leaf arrangement: (a) simple leaves, (b) compound leaves. To separate these two conditions, the bud must be located on the twig at the base of the leaf petiole (the axil).

is, the placement along the twig is at regular intervals alternating from one side to the other (Figure 14.1). Actually, the pattern is more complex, as the apical meristem produces the leaves in a spiral pattern

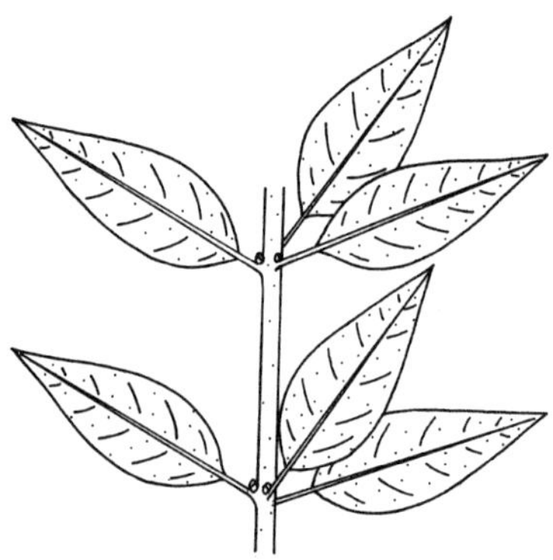

FIGURE 14.3 Whorled leaf arrangement.

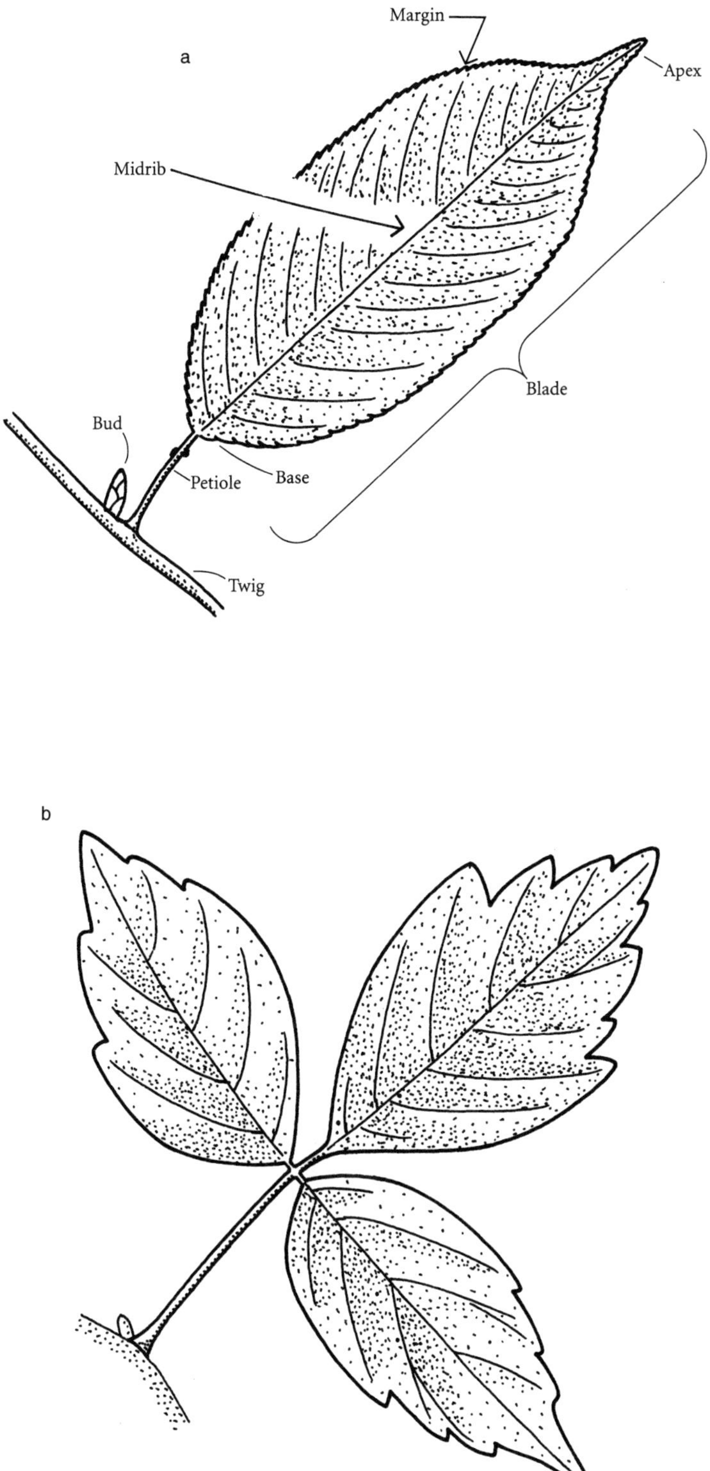

FIGURE 14.4 (a) Simple leaf, *Prunus virginiana*; (b) trifoliate (pinnately compound) leaf, *Acer negundo*.

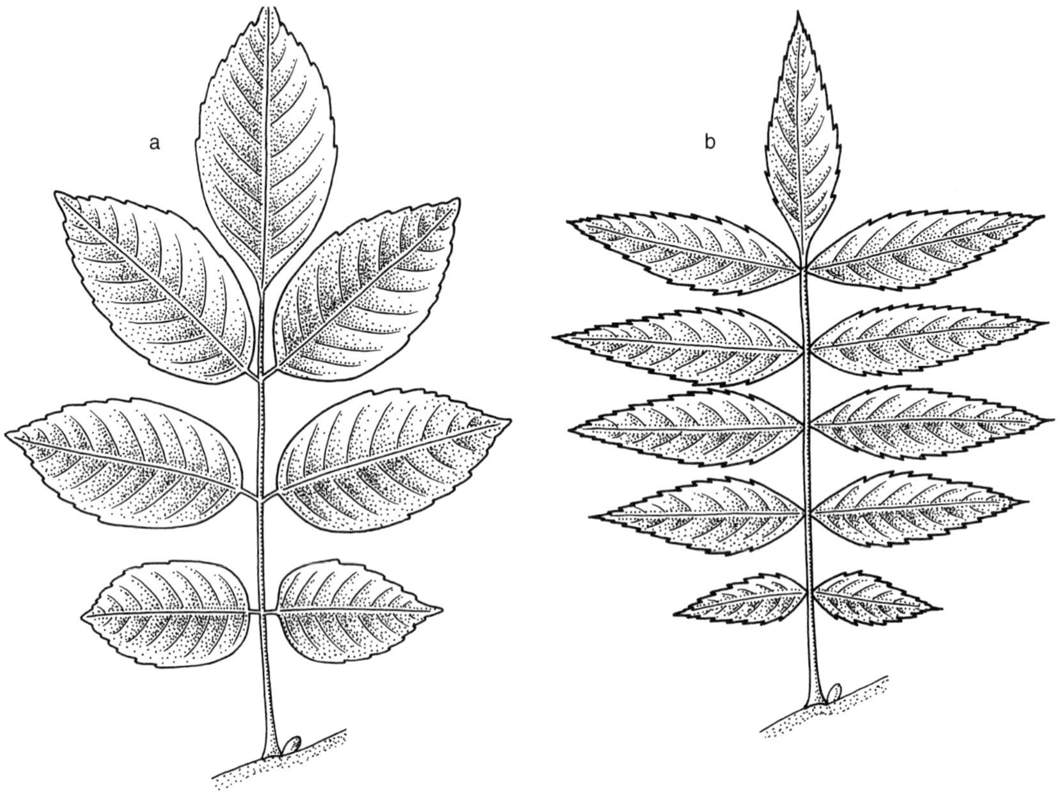

FIGURE 14.5 Odd-pinnately compound (a) opposite, petioluled, leaflets, *Fraxinus americana*; (b) opposite sessile leaflets, *Fraxinus nigra.*

with some degree of rotation before the next leaf is produced. In an alternate arrangement, only one leaf is produced at a node. A variation of this arrangement is the location of multiple leaves at the tip of a spur (short) shoot in several Rosaceae (cherry, apple, pear) species.

Opposite Arrangement

In an opposite arrangement, two leaves are positioned at a node but on opposite sides of the twig (Figure 14.2) Typically, leaves at one node are rotated 90 degree from leaves at adjacent nodes (those directly higher and lower on the twig).

Whorled (Verticilate) Arrangement

In a whorled arrangement, three leaves are positioned at a node (Figure 14.3).

The number of species groups with opposite or whorled leaves is relatively small, as compared with those with an alternate leaf arrangement. Therefore, a simple mnemonic device will assist in remembering a large proportion of the more important species with opposite and whorled arrangements: M.A.D. Buck but Ad. a Big. Cap. Staph. and Rubi. Picture an angry white-tail deer with a large cap on top of his head, a ruby stone on a ribbon hanging around his neck, and one hoof holding a walking stick. Now that the image is set, here are the species/genera/families it represents:

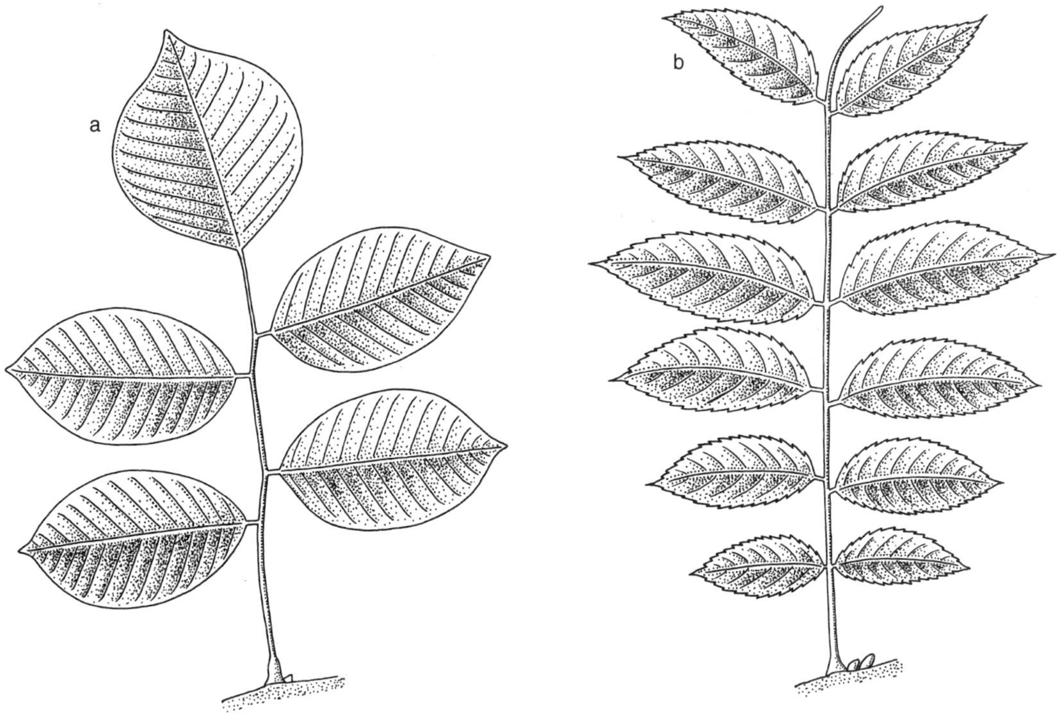

FIGURE 14.6 (a) Odd-pinnately compound leaf of alternate leaflets, *Cladrastis lutea*; (b) even-pinnately compound leaf, *Juglans nigra*; the terminal leaflet is usually greatly reduced in size or missing.

M, maple; A, ash; D, dogwood, Ad., Adoxaceae family, elderberry and viburnum; Big., Bignoniaceae family or catalpa; Cap., Caprifoliaceae family or honeysuckle; Staph., Staphyleaceae family or bladdernut; and Rubi., Rubiaceae family which includes buttonbush.

BLADE CHARACTERISTICS

The blade is the green, expanded, flat structure attached to the twig by a petiole. There are a variety of leaf and blade characteristics used for identification.

Complexity (Macrocharacteristics)

Complexity refers to the division of the leaf blade into leaflets or pinnules. Again note that the leaf is defined by the position of the axillary bud regardless of how the blade is divided. The leaf may be simple where there is a single undivided blade (Figure 14.4a) or trifoliate where the blade is divided into three leaflets (Figure 14.4b), or it may pinnately or palmately compound. In the odd-pinnately compound leaf, the blade is divided into five or more leaflets (also called pinna) arranged along two sides of a common axis (the rachis), which is attached to the petiole (Figure 14.5). A terminal leaflet is present, hence there is an odd number (five, seven, nine, eleven, etc.) of leaflets. Leaflets may be opposite and have a petiolule (leaflet stalk) (Figure 14.5a), sessile (no leaflet stalk or petiolule) and opposite (Figure 14.5b), or, rarely, petioluled and alternate (Figure 14.6a). An even-pinnately compound occurs when the terminal leaflet is absent (Figure 14.6b) and hence has an even number of leaflets, as found in *Juglans nigra*, black walnut. In the bipinnately

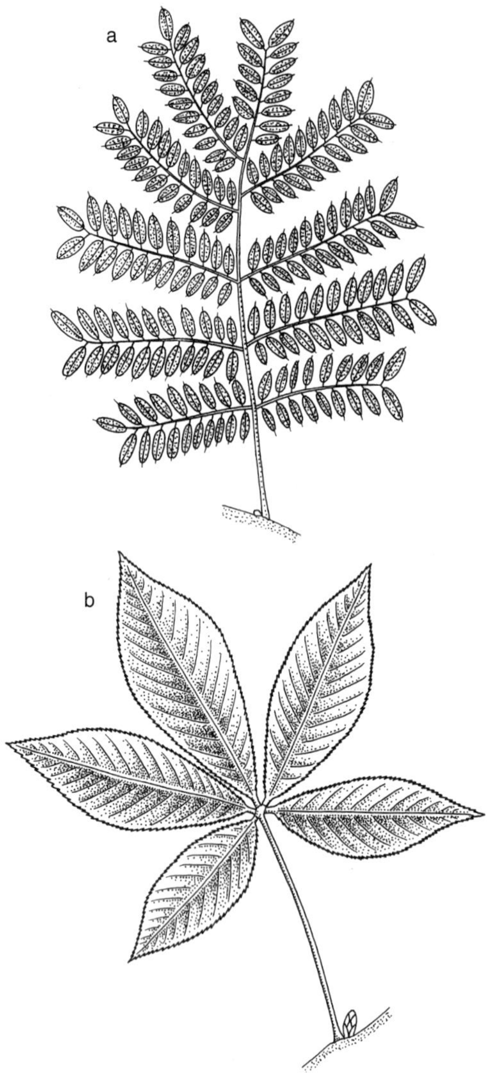

FIGURE 14.7 (a) Bipinnately compound leaf, *Gleditsia triancanthos*; (b) palmately compound leaf, *Aesculus* spp.

compound leaf (Figure 14.7a), the leaflets (pinna) of the pinnately compound leaf are again subdivided into smaller units called pinnules. Another term for this type of leaf is "twice pinnately compound." A few species, like *Aralia spinosa* (devil's walking stick), have a tripinnately compound leaf. In the palmately compound leaf (Figure 14.7b) the blade is divided into five, seven, or nine leaflets, which are attached to a central point at the end of the petiole. Other terms for this leaf type are "radiately divided" and "digitate."

Simple Blade Types

A simple blade may have considerable variation in form (Figure 14.8). The lobe is that part of the blade separated by an open area or a sinus or split.

Apex Types

The tip (distal section) of the leaf may take a variety of shapes (Figure 14.9).

Base Types

The blade near the attachment of the petiole (proximal section) may take a variety of shapes (Figure 14.10).

Blade Shape Types

Various combinations of apex and base shapes frequently determine leaf shape. The sequence of shapes shown in Figure 14.11 generally begins with the most accentuated base and ends with the most accentuated apex. The elliptical and orbicular types may represent the actual form or the general outline of the leaf.

Margin Mesocharacteristics

There are a large number of margin characteristics beyond lobes and sinuses that are common to trees, shrubs and vines, as shown in Figure 14.12. The margin characters are important because they are often among the first to be used in identification.

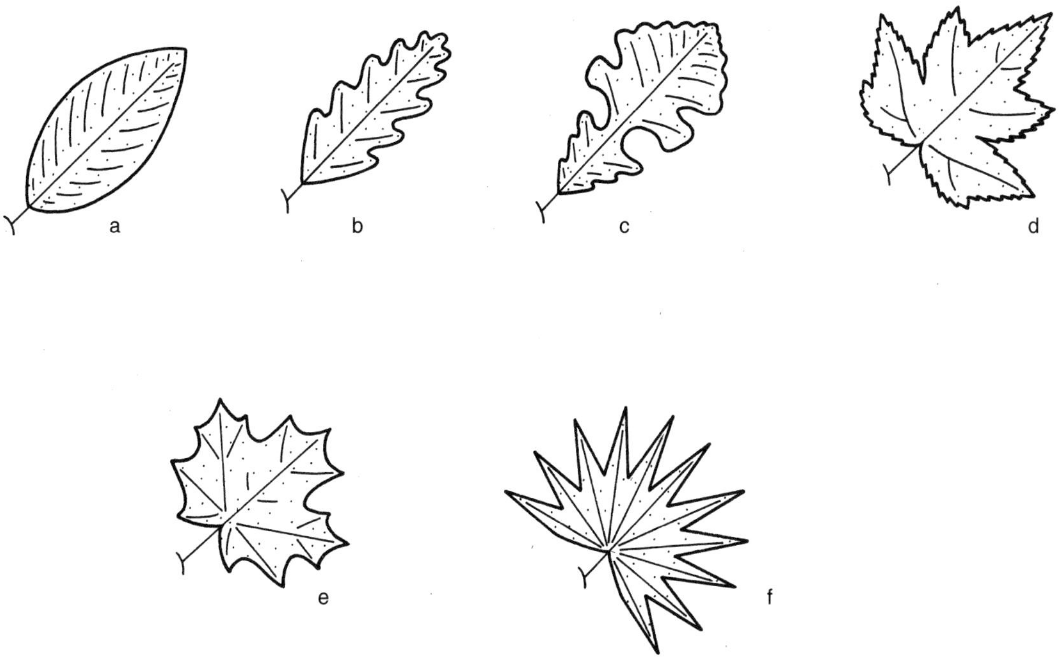

FIGURE 14.8 Leaf types (macrocharacteristics): (a) entire, *Diospyros virginiana*; (b) pinnately lobed, *Quercus alba*; (c) parted, *Quercus macrocarpa*; (d) cleft, *Acer rubrum*; (e) palmately lobed, *Acer saccharum*; (f) flabelliformis (fan-shaped), *Serenoa repens*, saw palmetto.

Margin Microcharacteristics

Occasionally the margin of a leaf has a row of fine, short hairs that can distinguish two closely related species (Figure 14.13). For example, the presence of cilia separates the leaves of *Carya ovata* (shagbark hickory) from *Carya ovalis* (red hickory) or *C. glabra* (pignut hickory), and the leaves of *Acer rubrum* (red maple) from those of *Acer spicatum* (mountain maple). The cilia may be seen by silhouetting the leaf against the sky and/or using a 10× hand lens.

Venation Types

There are three general types of vein arrangements in leaves of the dicots: palmate, pinnate, and a combination of palmate and pinnate (mixed) (Figure 14.14). Pinnate venation may be further subdivided depending on whether the veins are straight or curved. Parallel venation occurs only in the monocots.

Blade Surface Characteristics

A glabrous leaf surface is smooth, often shiny, and not hairy. A glaucous surface is covered with a white, waxy substance, a bloom usually on the abaxial (lower) surface or petiole. Often yellowish, rusty, or silvery scales appear on the leaf surface of some species; this surface may appear rough or scurfy. The term "pubescent" is used to refer to soft, short hair (Figure 14.15); however, the term is often used in a general sense to indicate any type of surface covered with hair. "Scabrous" refers to short, stiff hair or sharp epidermal projections; when such a leaf is rubbed, the sensation and sound indicate a sandpaper-like surface. Strigose is the condition where short stiff hairs are aligned in a given

FIGURE 14.9 Apex types: (a) emarginate, (b) truncate, (c) rounded, (d) obtuse, (e) acute, (f) mucronate, (g) cuspidate, (h) acuminate, (i) aristate.

direction and are some what appressed. Tomentose hair is intermediate-length soft, interwoven hair, and stellate hair is short, branched hair that somewhat resembles a star when examined with a microscope. A villous or pilose surface is a surface covered by long, soft hair. The rugose or wrinkled leaf upper surface is caused by sunken veins (Figure 14.15 f).

Blade Texture

Leaf texture or thickness is relative and depends on the amount of palisade tissue; thus determination of thickness comes only with field experience. If the leaf blade is "papery," it is thin, extremely smooth, and may "crackle" when bent or crushed (*Fagus grandifolia*, American beech). Most tree and shrub species have a standard leaf of medium thickness. Many evergreen species have a coriaceous leaf that is thick, leathery, hard and/or stiff (*Ilex opaca*, American holly; *Magnolia grandiflora*, southern magnolia).

PETIOLE CHARACTERISTICS

Length

The length of the petiole is distance from the blade base to the petiole base where it is attached to the twig. This characteristic may occasionally be important in separating two closely related species.

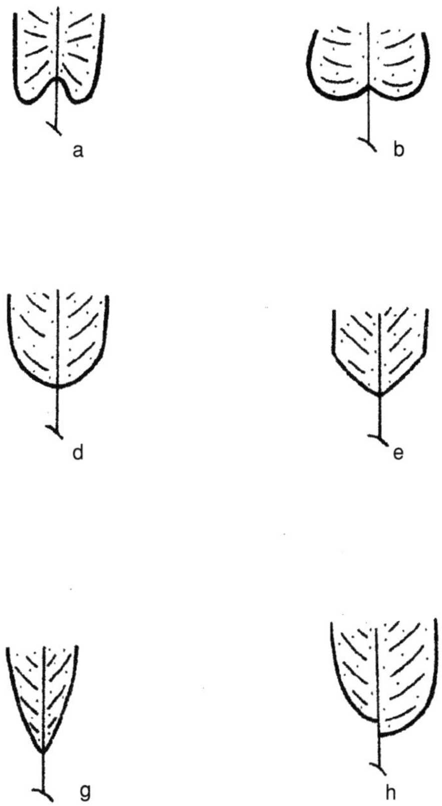

FIGURE 14.10 Base types: (a) auriculate, (b) cordate, (c) truncate, (d) rounded, (e) cuneate, (f) obtuse, (g) acute, (h) inequilateral (also called oblique and asymmetrical).

Shape

The cross-sectional shape of the petiole, as shown in Figure 14.16, is sometimes a useful characteristic.

Surface

The petiole surface may be glabrous, glaucous, pubescent, scabrous, tomentose, or hairy in a manner similar to the blade surface.

Base

The part of the petiole attached to the twig directly below the bud is usually solid (Figure 14.17a). The hollow petiole is essentially the only variation from the standard petiole base that can be used in identification. Here the petiole base caps the bud, effectively hiding it until the leaf falls (Figure 14.17b; *Platanus occidentalis*, American sycamore; *Rhus glabra*, smooth sumac).

MODIFIED STRUCTURES

Modified leaf blades or petioles are common plant structures. Modified or special structures may also be attached to the petiole or rachis or to the twig. Flower petals, sepals, and bracts are modified leaves, as are bud scales. The entire flower structure (Chapter 16) is

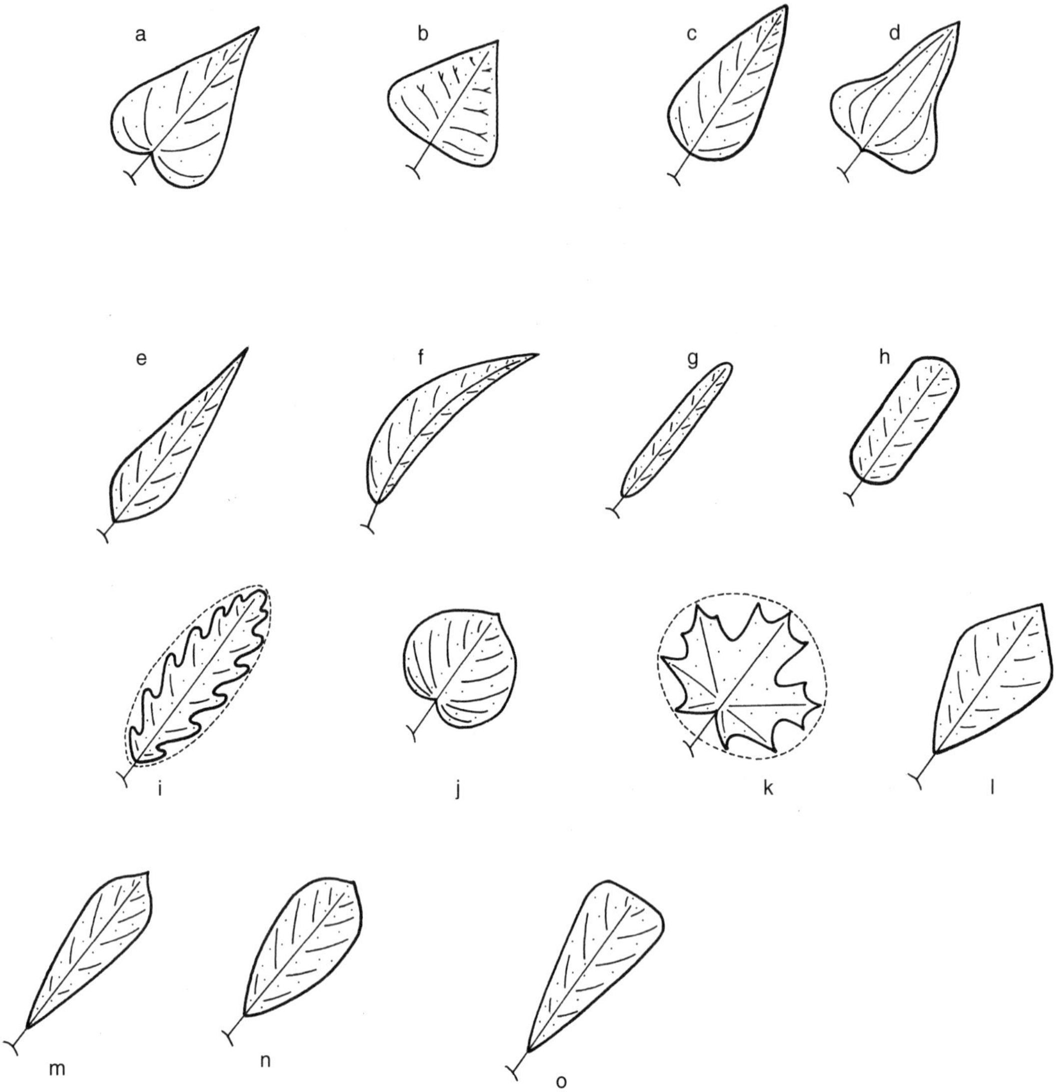

FIGURE 14.11 Leaf shapes: (a) cordate, (b) deltoid, (c) ovate, (d) panduriform (also called hastate or violin-shaped) (e) lanceolate, (f) falcate, (g) linear, (h) oblong, (i) elliptical, (j) reniform, (k) orbicular, (l) rhomboid, (m) oblanceolate, (n) obovate, (o) spatulate.

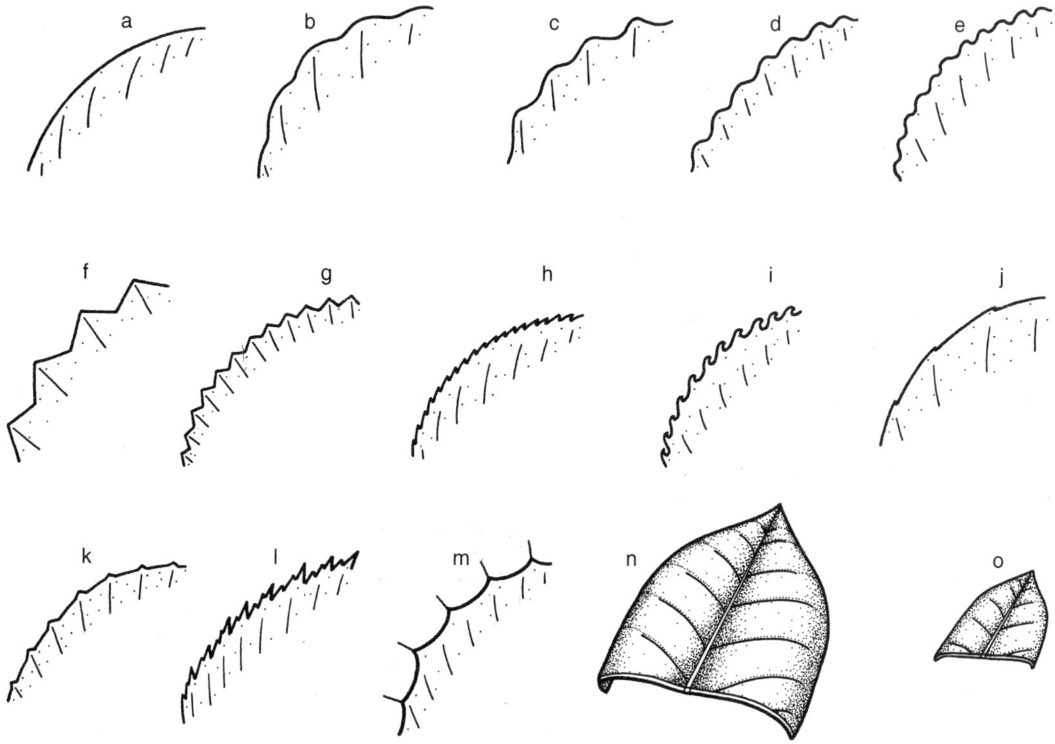

FIGURE 14.12 Margin types (mesocharacteristics): (a) smooth, (b) undulate, (c) sinuate, (d) crenate, (e) crenulate, (f) dentate, (g) denticulate, (h) serrate, (i) serrate with incurved teeth, (j) serrulate, (k) remotely serrate, (l) doubly serrate, (m) spiny, (n) revolute, (o) minutely revolute.

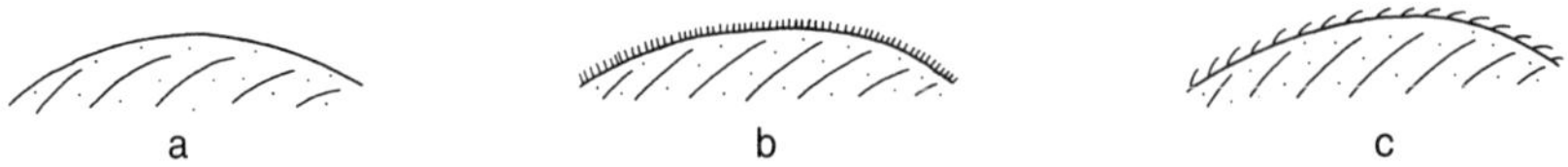

FIGURE 14.13 Margin microcharacteristics: (a) aciliate (no hair), (b) ciliated (lined with soft hair), (c) strigose or antrorse (lined with stiff, tapered hair directed forward).

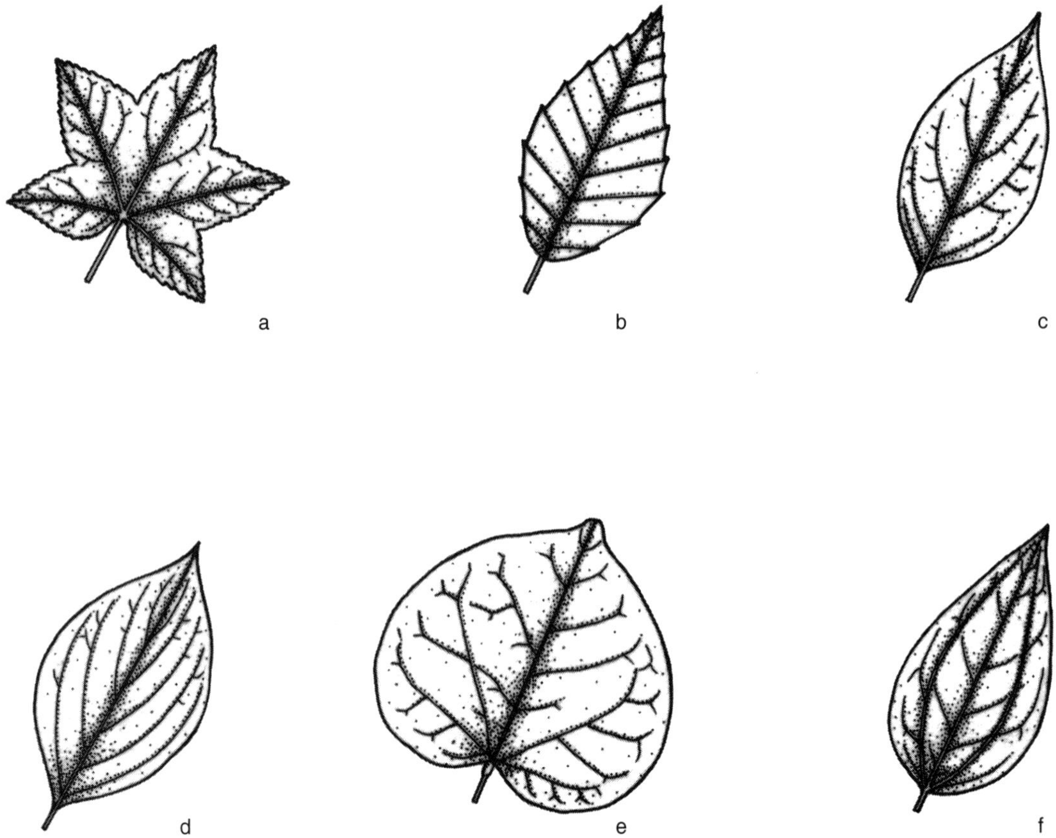

FIGURE 14.14 Venation types (stylized): (a) palmate, *Liquidambar styraciflua*; (b) pinnate-secondary veins straight, *Fagus grandifolia*; (c) pinnate-secondary veins curved, *Diospyros virginiana*; (d) arcuate, *Cornus florida*; (e) mixed pinnate-palmate, *Cercis canadensis*; (f) parallel (monocots), *Smilax glauca*.

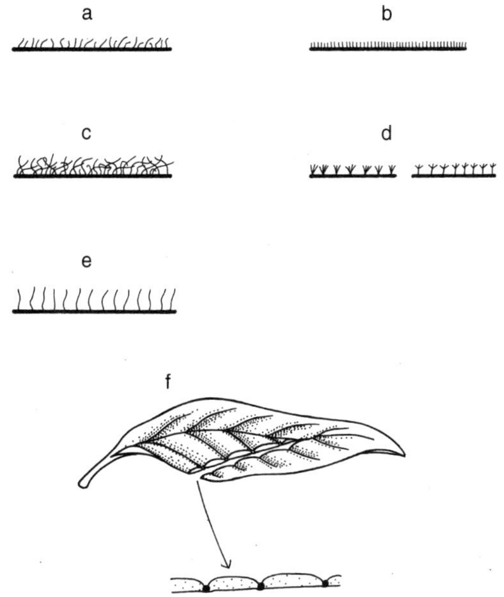

FIGURE 14.15 Blade surface characteristics: (a) pubescent (short, soft hair), (b) scabrous (short, stiff hair), (c) tomentose (long, interwoven hair), (d) stellate (branched hair), (e) villous or pilose (long, soft hair), (f) rugose (wrinkled surface, usually due to sunken veins).

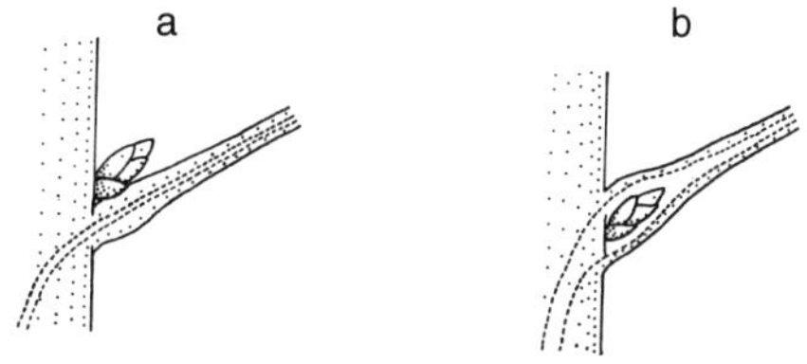

FIGURE 14.17 Petiole base: (a) solid, *Ulmus americana*; (b) hollow, *Platanus occidentalis*.

considered to be a modified branch with attached leaves. Wings are found on some leaves; these are usually leafy appendages attached to the rachis (Figure 14.18a). Glands may appear as small, round dark protrusions on the petiole (Figure 14.18b) or on the leaf surface, vein, or margin. A blade (leafy appendage) attached to the twig near the bud or petiole base is a stipule (Figure 14.18c). Various types of stipule scars may an aid in identification. Bud scales are modified leaves. Several layers of scales protect the dormant shoot from low temperatures or desiccation (drying out). See Chapter 15 for stipule scar and bud scale characteristics. The tendril is a corkscrew-like structure for climbing (Figure 14.18d). On *Ampelopsis cordata* (a vine similar to *Vitis*, grape), it is green and nonwoody, much like a petiole. On *Vitis*, it is woody and appears to be a modified twig.

ENVIRONMENTAL MODIFICATIONS

Leaves that develop in high light conditions (sun leaves) are usually thicker (more palisade tissue) smaller in size and have deeper sinuses and a thicker cutin layer than shade leaves. Occasionally, these modifications may cause problems in identification in certain species, particularly in *Quercus*. For example, the sun leaves of *Quercus velutina* (black oak) located high in the crown and in full sunlight are substantially different from the shade leaves located in the lower crown and on seedlings and saplings in the understory (Figure 14.19).

FIGURE 14.16 Petiole shape: (a) round, (b) grooved, (c) ridged, (d) flattened.

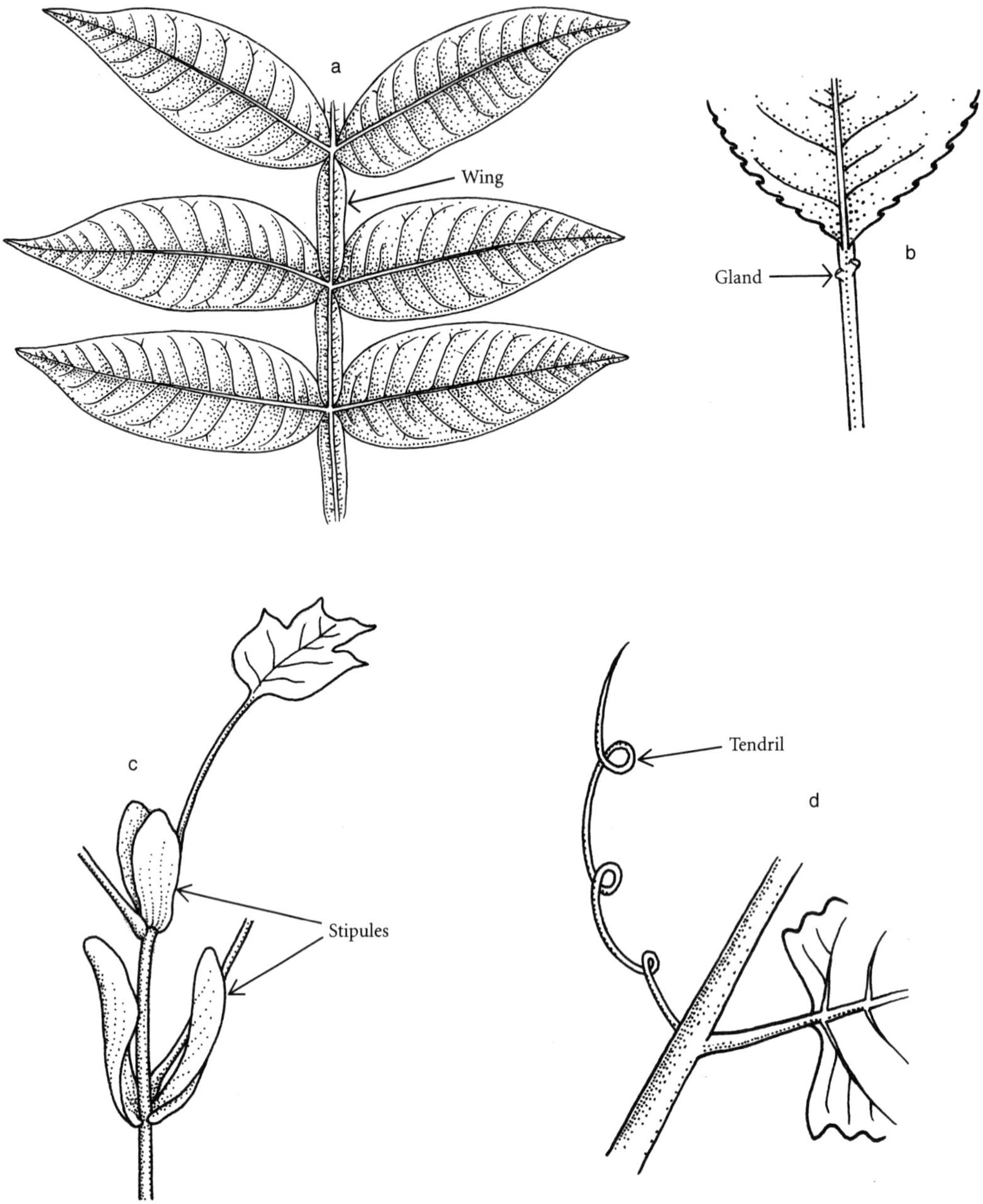

FIGURE 14.18 Modified structures: (a) rachis wings, *Rhus copallina*; (b) petiole glands, *Prunus serotina*; (c) stipules of *Liriodendron tulipifera*; (d) tendril of *Vitis*.

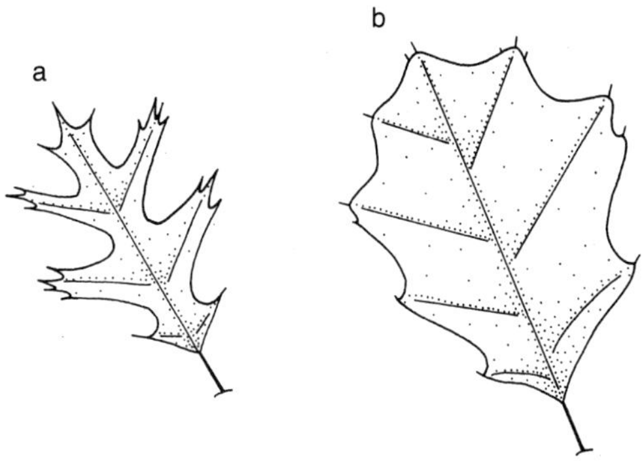

FIGURE 14.19 Effect of shade on leaf shape: *Quercus velutina* (a) leaf developed in full sunlight, (b) leaf developed in shade.

ODOR

Several species have a distinctive identifying odor. These include *Myrica cerifera* (bayberry), *Comptonia perigrina* (sweetfern), *Lindera benzoin* (spicebush), and *Asimina* (pawpaw). Although the twigs of *Sassafras albidum* (sassafras) and *Prunus serotina* (black cherry) are strongly scented, their leaves have little or no odor.

CHAPTER 15

The Woody Angiosperm Twig

INTRODUCTION

THE twig is that portion of the woody plant stem on which the bud, leaf, flower, fruit, and other structures such as the stipule, spine, thorn, and tendril are located. It is unfortunate that field manuals tend to concentrate on the leaf, for the twig has nearly as many physical characters that may be used in identification. The problem is that a hand lens or 10× magnifier is often needed to examine the minute characteristics of twig structure and, moreover, another set of terminology must be learned.

The importance of the twig in identification varies with the season of year. Extensive physical changes begin in the spring when the buds "break" to allow leaf, flower, and fruit growth and development. Concurrent with shoot elongation, the twigs of the previous year begin to increase in diameter. These stages of growth are not completed until midsummer or later in the growing season. As a result, the twig is not easily used for identification purposes during the period from early spring until midsummer or, in some species, until leaf fall.

However, once the twig is developed, it normally has less variability within an individual and between individual trees of the same species than leaf characteristics. This is fortunate, for in autumn after the leaves have fallen, it is the major identifying structure of small trees that have not developed typical mature bark characteristics; it is the primary structure for identifying shrubs, vines, and seedlings and saplings of tree species. The bud, leaf scar, stipule scar, twig surface, and pith are the major structures used in identification, but other characters such as odor and the presence of modified leaves and twigs can also be used. Only infrequently are lenticels of value.

The most important structure on the twig from the standpoint of both identification and growth is the bud (Figure 15.1), which contains the apical meristem that must be protected and remain healthy. This meristematic tissue, as well as the miniature shoot, leaves, and/or flowers, are protected from cold and desiccation during the dormant season because they are enclosed by several layers of modified leaves called bud scales. Unlike typical leaves of deciduous species that drop in autumn, bud scales are lost as growth begins in the spring. However, as with typical leaves when they fall, scars remain at the point of attachment. These are a series of short concentric ridges or marks collectively encircling the twig and marking the end of last year's growth. They also mark the beginning of new growth. Twigs that developed many years ago may often be aged by the number of bud scale scars, and the rate of stem elongation or growth may be estimated by measuring the distance between scars of successive years. However, one must be cautious, because auxin produced in the terminal leader of angiosperm trees inhibits the growth of lateral branches. The best estimate of tree height growth is indicated by the terminal leader or leaders.

There are two general growth (twig elongation) patterns found in angiosperm trees: determinant and indeterminant. In species of *Acer* (maple), *Fraxinus* (ash), *Quercus* (oak), and *Carya* (hickory), stem

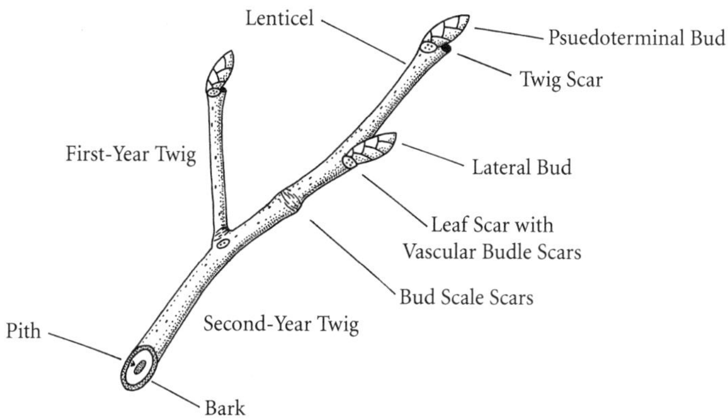

FIGURE 15.1 Structures of a woody angiosperm twig.

growth resulting from new cell formation may cease midway through the warm season because of the development of a bud at the end of the twig; this bud is a true terminal bud. Thereafter, twig elongation can take place only by cell enlargement. Under certain weather conditions, the newly formed bud may break to begin a second period of growth. This type of growth is called determinant.

Species of other genera such as *Betula* (birch), *Populus* (aspen and cottonwood), *Platanus* (sycamore), and *Ulmus* (elm) have indeterminant or continuous growth, with new cells being produced as long as environmental conditions are favorable. With this type of growth, the last lateral bud formed in late summer becomes the terminal bud and is appropriately called the false or pseudoterminal bud. It is usually accompanied by a portion of the dead twig developed after the bud was formed or by a twig scar. The original apical meristem is lost when the twig dies, but it is replaced by the meristem in the pseudoterminal bud. The common structures of an angiosperm twig, *Ulmus americana* (American elm), are shown in Figure 15.1.

BUD CHARACTERISTICS

Bud/Twig/Leaf Arrangement

The arrangement of buds and leaf scars follows the arrangement of leaves on the twig. By definition, a bud must be located in the axis formed by the leaf and the twig, thus buds and leaf scars, as well as next year's twigs, follow the arrangement of the leaves.

- ***Alternate.*** Leaf scars and buds appear to alternate from one side to the other along the twig axis. Actually, they are arranged in a spiral fashion, because that is the way they form from the apical meristem (Figure 15.2).
- ***Opposite.*** The buds appear in pairs, each bud positioned directly opposite the other on the twig (Figure 15.3a,b,c).
- ***Subopposite.*** The nodes are separated along the twig axis but not enough to be considered alternate

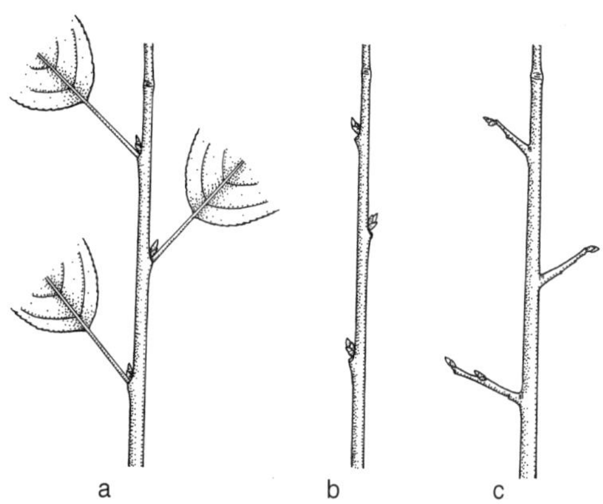

FIGURE 15.2 Alternate bud/twig arrangement in *Prunus serotina*: (a) last summer, (b) this winter, (c) next winter.

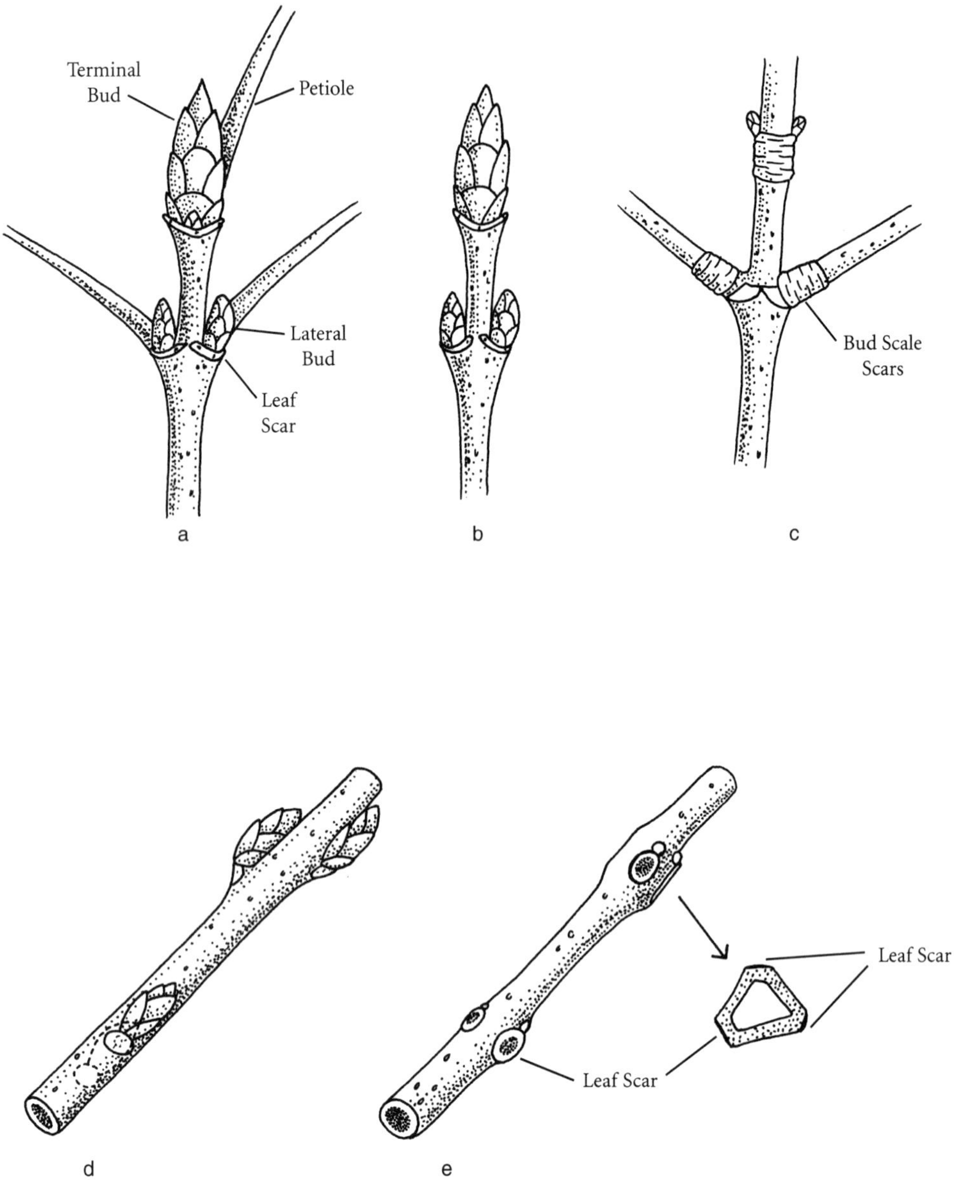

FIGURE 15.3 Opposite bud/twig arrangement in *Acer saccharum*: (a) last summer, (b) this winter, (c) next winter; (d) subopposite arrangement in *Rhamnus*; (e) whorled arrangement in *Catalpa*. Leaf scar locations are shown in twig cross section.

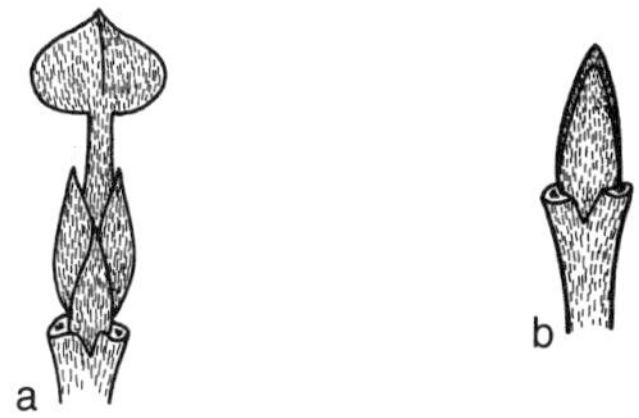

FIGURE 15.4 Bud types in *Cornus*: (a) flower bud, (b) foliage bud.

(Figure 15.3d). The buds are positioned on opposite sides of the twig but are separated by a distance of less than 1 cm.

- ***Whorled.*** Three leaf scars are present at a node (Figure 15.3e). In species that have this arrangement (e.g., *Catalpa*), one of the three leaf scars and accompanying buds may be reduced in size or absent, giving the remaining scars and buds an opposite or near opposite appearance.

Bud Function

Species may have two bud types (Figure 15.4). *Cornus florida* (flowering dogwood), *Cercis canadensis* (eastern redbud), and *Acer rubrum* (red maple) have flower buds that produce only flowers, which appear in the early spring before leaf flush; these buds are usually larger than foliage buds. Foliage buds, which contain the leaves, shoot, and meristem, open several weeks later. The sequence of flower and leaf development varies with the species. *Hamamelis virginiana* (witch-hazel) also has flower buds, but these open in the late autumn after the leaves have fallen. *Carya* (hickory) has a separate flower bud superimposed directly above the foliage bud.

Bud Position

Position is the term used to indicate the relative location of the bud on a twig (Figure 15.5).

- ***True terminal bud.*** This bud is formed at the extreme tip of the twig and indicates the end of growth; no twig scar is present.
- ***Lateral bud.*** A solitary bud located below the leaf scar on the side of the twig.
- ***Pseudoterminal bud.*** A pseudoterminal bud is formed when a lateral bud assumes the position and function of a terminal bud in species that do not form a true terminal. It may also be referred to as a "false terminal" bud. It is accompanied by a twig scar.
- ***Superimposed bud.*** In *Carya* (hickory) and *Juglans* (walnut), a second lateral bud is located directly above the normal lateral bud. In these genera, the superimposed bud produces flowers (catkins) and the lateral bud associated with a leaf scar expands to form twigs and leaves.
- ***Accessory bud.*** This bud type is located on either side of a lateral bud (*Acer rubrum*), or several may be clustered around a terminal bud as in *Quercus* (oak). Each bud is accompanied by a leaf scar, indicating a clustering of leaves around the terminal during the growing season. Accessory buds may contain flowers.

Bud Shape

The general outline (silhouette) of a bud may be helpful in identification (Figure 15.6). Infrequently, cross-sectional shape may identify a taxon to genus; for example, the terminal bud of *Quercus velutina* has a distinctive pentangle (five-sided) shape.

- ***Linear.*** The bud is many times longer than wide. The apex sides converge at a narrow (acute) angle.
- ***Elliptical.*** The bud is only two to three times longer than it is wide. The apex is described as obtuse.
- ***Globose.*** The bud is as long as it is wide. The apex may be rounded or obtuse.
- ***Onion-like.*** A globose bud with an acute apex and a slightly constricted base.
- ***Asymmetrical.*** The two sides of the bud have different shapes; inequilateral.
- ***Cone.*** Similar to the inverted shell of an ice cream cone. The sides taper evenly.
- ***Pentangle.*** Each of the five rows of scales of the *Quercus velutina* bud is slightly folded, creating a five-sided structure as viewed from the apex.

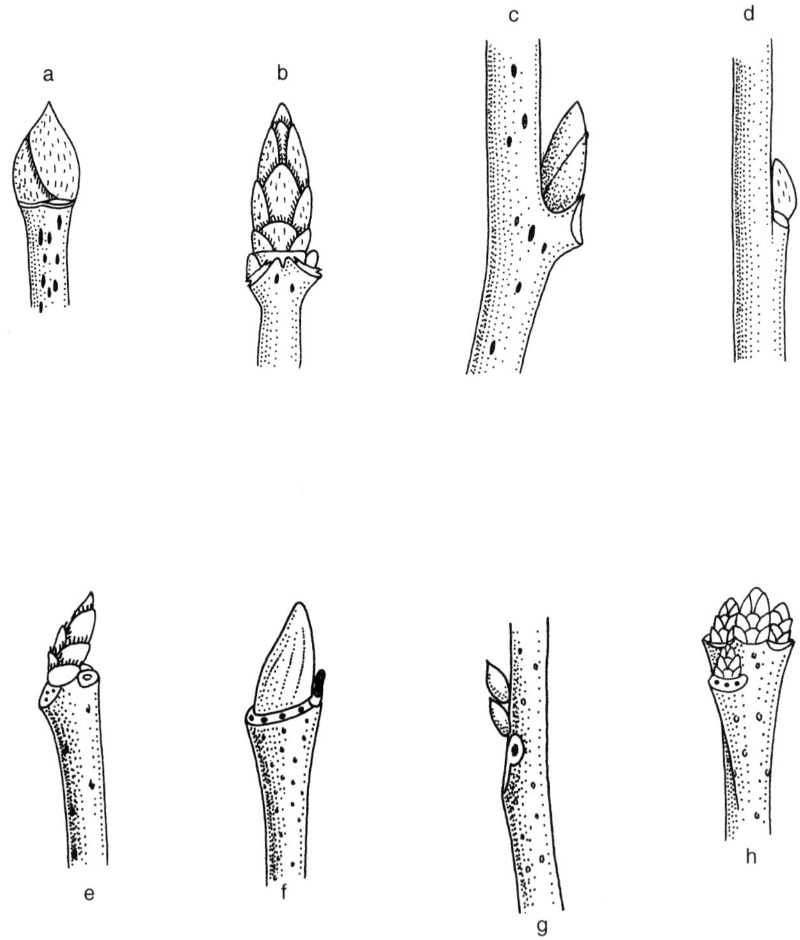

FIGURE 15.5 Bud position: (a) true terminal bud, *Carya glabra*; (b) true terminal bud, *Acer saccharum*; (c) standard lateral bud, *Diospyros virginiana*; (d) appressed lateral bud, *Salix nigra*; (e) pseudoterminal bud, *Ulmus americana*; (f) pseudoterminal bud, *Platanus occidentalis*; (g) superimposed (upper) bud, *Carya illinoensis*; (h) accessory buds around the terminal bud, *Quercus alba*.

Bud Scale Arrangement

Nearly all species have scales covering the miniature leaves and/or flowers of the next growing season. These scales, arranged in a systematic fashion that is consistent for a given species, are frequently an important aid in identification (Figure 15.7). Their arrangement usually but not always follows the arrangement of leaves on the twig—that is, spiral (alternate) or opposite. A few species have one or two vestigial scales or none at all.

- *Imbricate bud scales.* These scales overlap like shingles on the roof of a house (Figure 15.7). There are

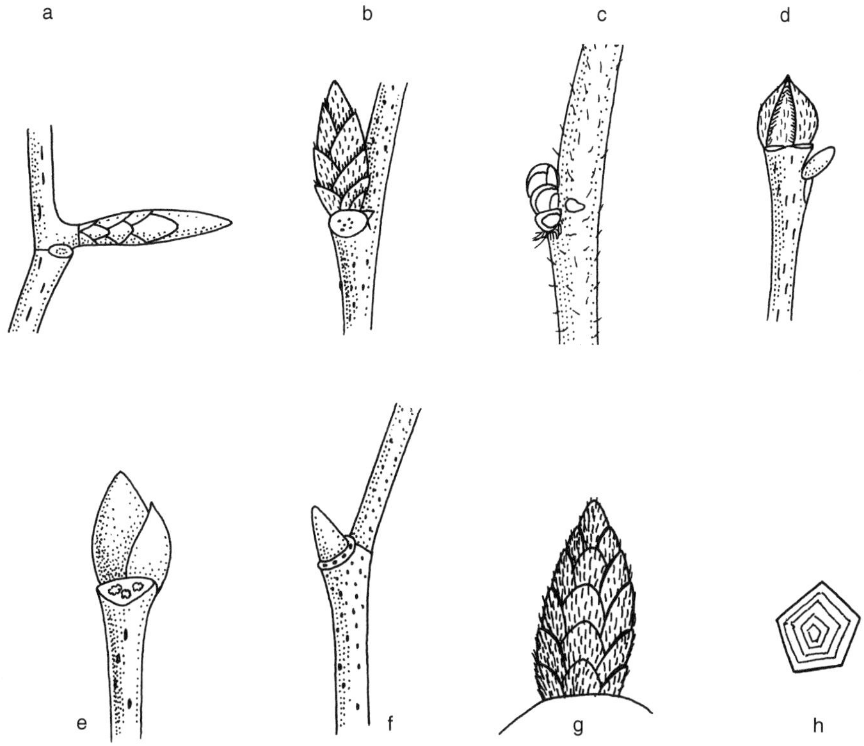

FIGURE 15.6 Bud shape: (a) linear, *Fagus grandifolia*; (b) elliptical, *Ulmus americana*; (c) globose, *Corylus americana*; (d) onion-shaped, *Carya ovalis*; (e) asymmetrical, *Tilia americana*; (f) cone, *Platanus occidentalis*; (g) pentangle, *Quercus velutina*, side view; (h) *Quercus velutina*, top view with inside pentagons representing rows of scales.

a number of variations, based on the number of visible scales and whether they were formed from the meristem in a spiral or decussate manner. Additional scales may be hidden from view.

- *Non-Imbricate bud scales.* In the valvate scale arrangement, the two outer scales do not overlap but come together along the margin of the scales (Figure 15.8). The capped bud has a single rolled scale that covers the miniature shoot and leaves.

Some buds have no scales (naked) or, if present, they are small and do not cover the miniature shoot and leaves. With this condition, the immature leaves are usually covered by extremely dense hair to protect the bud from frost damage and desiccation.

Bud Scale Margin and Surface

Occasionally, the surface characteristics of bud scales are important in distinguishing between two closely related species (Figure 15.9).

- *Glabrous* (not illustrated). No hair is present on the surface or margin.
- *Ciliated.* The bud scale has a distinct fringe of hair along the margin.
- *Pubescent.* This general term is applied to a wide variety of short to medium-length hair. If the hair is sufficiently dense, the surface will appear dull.
- *Tomentose.* Covered with extremely dense, long, often interwoven hair. The hair may be very dense in some species.

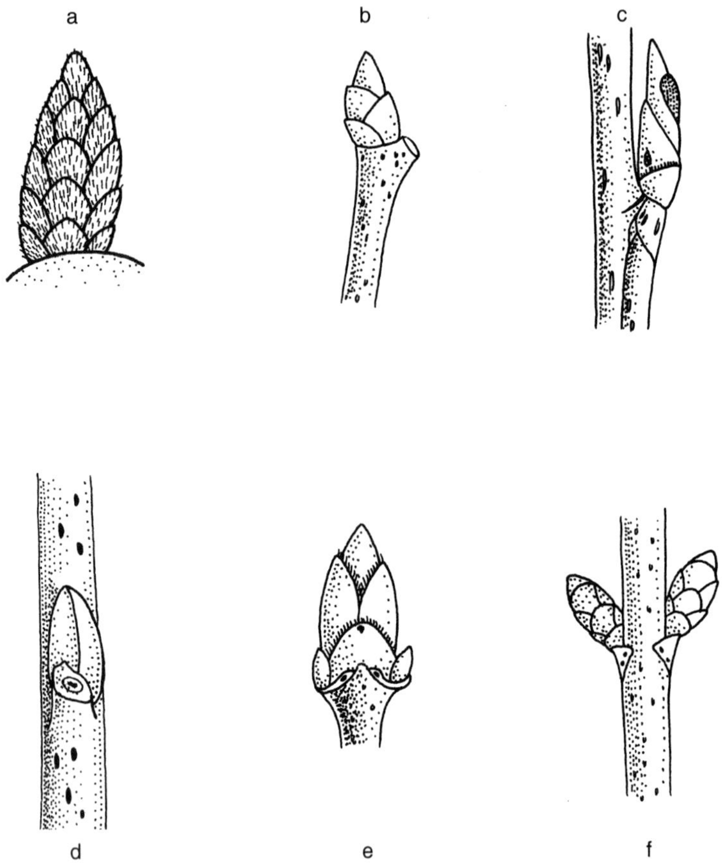

FIGURE 15.7 Imbricate bud scale arrangement: (a) spiral, with ten or more scales of the same shape, *Quercus velutina*; (b) spiral, with four to nine scales of the same shape, *Prunus serotina*; (c) spiral, with three or four scales of different shapes and sizes, *Populus deltoides*; (d) two scales of unequal size, *Diospyros virginiana*; (e) decussate, with two to four pairs of scales, *Acer saccharum*; (f) decussate, with four to six pairs of scales, *Aesculus glabra*.

- *Scurfy*. Covered with scale-like or bran-like particles.
- *Striated*. The striations appear as lines (actually ridges formed by veins) aligned in a more or less parallel fashion from the base to the tip of the scale.
- *Shellacked* (not illustrated). An extremely shiny surface that appears varnished.
- *Resinous* (not illustrated). Droplets of yellowish resin or sticky material may be present. The surface also may appear shellacked.

LEAF SCAR CHARACTERISTICS

Size

The length of a leaf scar may range from as little as 2 to 3 mm to nearly 2 cm (*Juglans nigra*, black walnut).

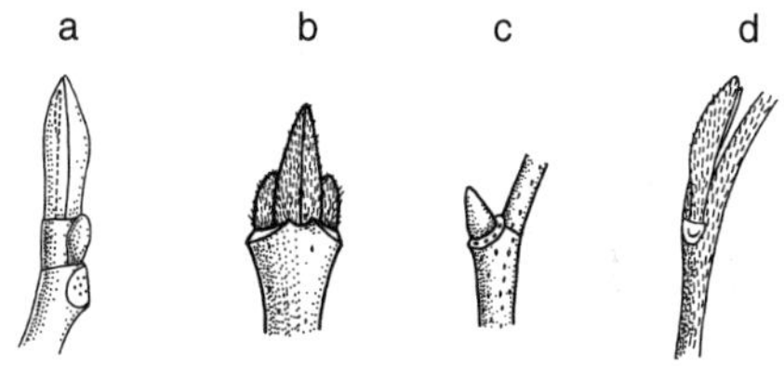

FIGURE 15.8 Non-imbricate bud scale arrangement: (a) valvate, *Liriodendron tulipifera*; (b) valvate, *Viburnum rufidulum*; (c) capped, *Platanus occidentalis*; (d) naked, *Hamamelis virginiana*.

Shape

Leaf scars are found in a variety of shapes; most can be grouped into a variety of types (Figure 15.10), but it should recognized that there are gradations between types.

FIGURE 15.9 Bud scale margin and surface: (a) ciliated margin, *Ulmus americana*; (b) pubescent surface, *Quercus stellata*; (c) tomentose surface, *Carya tomentosa*; (d) scurfy surface, *Carya cordiformis*; (e) striated surface, *Aesculus glabra*.

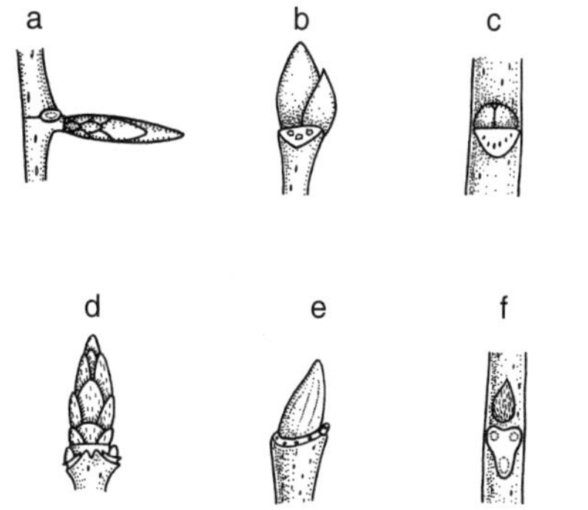

FIGURE 15.10 Leaf scar shape: (a) round, *Fagus grandifolia*; (b) elliptical, *Tilia americana*; (c) shield-shaped, *Fraxinus pennsylvanica*; (d) V-shaped, *Acer saccharum*; (e) broken circle, *Platanus occidentalis*; (f) skull-like (lobed), *Juglans nigra*.

Vascular Bundle Scars

Vascular bundle scars within the leaf scar are tissues that developed to seal the vascular system (xylem and phloem or leaf traces) when the leaf died (Figure 15.11).

STIPULE SCAR CHARACTERISTICS

Like leaf scars, stipule scars are formed when a small modified leaf dies (Figure 15.12). These leafy appendages or the scars are located at the intersection of the bud and leaf scar. If present, stipule scars may

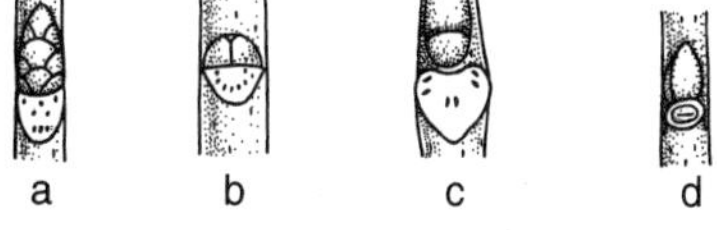

FIGURE 15.11 Vascular bundle scar pattern: (a) random, *Aesculus glabra*; (b) U-shaped, *Fraxinus pennsylvanica*; (c) clustered, *Populus deltoides*; (d) linear, *Diospyros virginiana*.

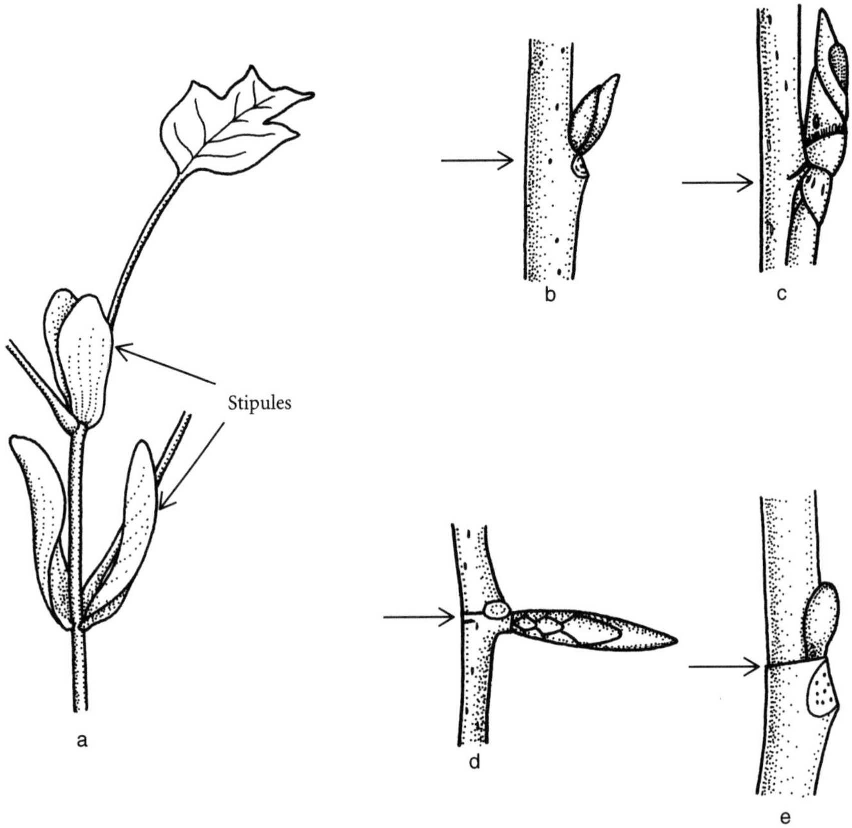

FIGURE 15.12 Stipules and stipule scars: (a) leafy stipules of *Liriodendron tulipifera*; (b) absent, *Diospyros virginiana*; (c) discontinuous, *Populus deltoides*; (d) overlapping, *Fagus grandifolia*; (e) continuous, *Liriodendron*, *Magnolia* spp.

be short, long and overlap without touching opposite the bud, or form a continuous scar depending on the species.

PITH CHARACTERISTICS

The innermost portion of first- and second-year twigs (the pith) is generally composed of a spongy material, septa (cross walls), or a combination of both, or may be empty. In older twigs, the pith generally has been crushed and is no longer present. The types of pith discussed in the following section are common to woody species.

Composition

The characteristics of the material in the pith zone are variable (Figure 15.13). Spongy pith has relatively large cells, so that it resembles a household sponge. Density varies from firm (*Fraxinus*) to soft (*Sambucus*). In diaphragmed pith, cross-walls that divide the pith into sections are more or less regularly spaced through the spongy material (*Nyssa sylvatica*, blackgum). When the spongy material is absent but the cross-walls remain to create small chambers, the pith is chambered (*Juglans nigra*, black walnut). Finally, in an excavated pith, there is an empty pith column that has neither cross-walls nor spongy material. (*Diospyros virginiana* persimmon; *Forsythia suspensa*, Forsythia).

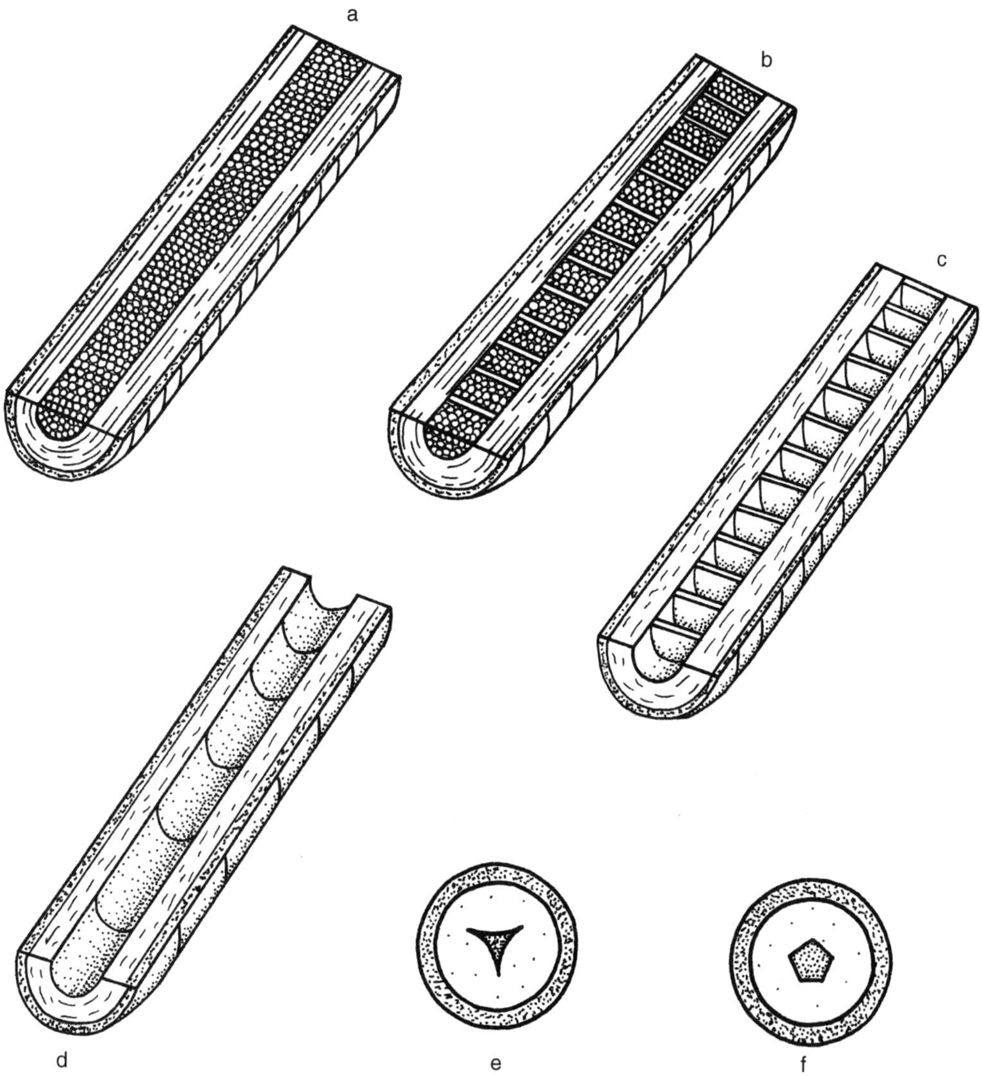

FIGURE 15.13 Pith characteristics: (a) spongy, *Sambucus canadensis*; (b) diaphragmed, *Nyssa sylvatica*; (c) chambered, *Juglans nigra*; (d) hollow, *Diospyros virginiana*; (e) triangular cross section, *Alnus* spp.; (f) pentangle cross section, *Quercus* spp. The outer stippled zone represents bark.

Cross-sectional Shape

The twigs of some taxa have a distinctive radial pattern of pith in cross section that can identify the genus (Figure 15.13e,f).

Pith Color

Color may be helpful in separating two species whose leaves are identical when flowers or fruits are absent. For example, *Sambucus canadensis* (American elderberry) has white pith and *S. pubens* (red elderberry) has pink pith.

MISCELLANEOUS TWIG/STEM CHARACTERISTICS

Corky Excrescenses

Cork may be formed as a continuous layer on twig surfaces of *Quercus macrocarpa* (bur oak), and on *Ulmus alata* (winged elm) and *Liquidambar styraciflua* (sweetgum) the cork forms rather sharp protrusions, referred to as wings (Figure 15.14a); however, cork is not a consistent feature and may be entirely absent in some individuals. Cork on the *Q. macrocarpa* twig forms a layer that substantially increases its diameter.

Prickles

Prickles or spines are epidermal outgrowths scattered along the stem. The cane (stem) of *Rubus* can provide a classic example of prickles (Figure 15.14b). Unlike thorns, they readily break off and can become embedded in clothes and skin.

Thorns

A thorn is a modified twig that is sharply pointed and bears no leaves (Figure 15.14c–f). It may be small and appear singly, or there may be paired stipular thorns at the leaf node. *Crataegus* (hawthorn) characteristically has unbranched thorns up to 8 cm long. *Gleditsia triacanthos* (honey locust) may have barbed thorns up to 15 cm long.

Tendril

A tendril is a modified twig, usually associated with climbing vines (Figure 15.15a, b). In *Vitis* (grape), *Ampelopsis*, and *Smilax*, the tendril resembles a corkscrew and provides support by wrapping around a twig or petiole of another stem. In *Parthenocissus* (Virginia creeper), the tendril terminates in a flat adhesive disk that attaches to a supportive surface.

Aerial Rootlets

Aerial rootlets provide a mechanism by which a stem may climb from the ground into the crown of a tree (Figure 15.15c). The roots of *Toxicodendron radicans* (poison-ivy) and English *helix* (Boston ivy) often emerge from the ground and grow upward along the surface of a tree trunk. This stem is attached to the bark by numerous aerial rootlets.

Spur Shoots

Spur shoots are short twigs (length <5 cm) on which leaves are borne (Figure 15.15d). The surface is a continuous series of leaf scars; a bud is found at the apex. Morphologists and physiologists now call them short shoots, as opposed to long shoots or typical twigs. *Betula* (birch), *Liquidambar* (sweetgum), and *Malus* (apple) have spur shoots.

Epidermal Surface

Twig surfaces of some species are covered with a whitish, waxy bloom; the whitish cast disappears when rubbed into the shiny waxy surface. Twigs of *Rhus glabra* (smooth sumac) and several species of *Rubus* (blackberry, raspberry) have this characteristic. Hair types found on twigs are similar to those on leaves (Figure 14.15) and buds (Figure 15.9). Some twigs are glabrous (no hair).

Color

Twig color may vary with season, between individuals, and from one side of the twig to the other; thus, for most species, color must be used for identification only with discretion. A distinctive twig is that of *Acer pensylvanicum* (striped maple); it is green with white stripes.

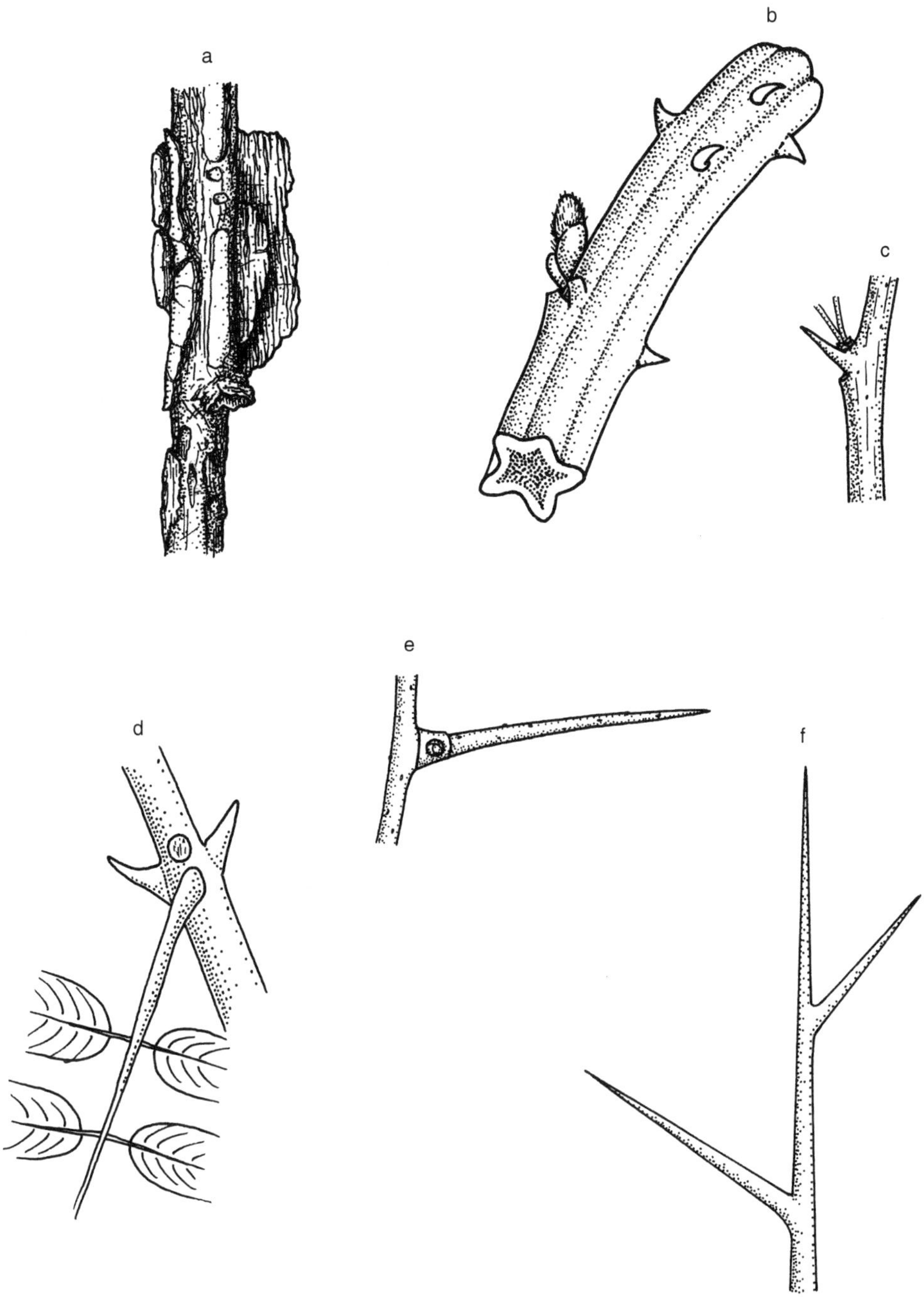

FIGURE 15.14 Modified twig structures: (a) corky excrescences, *Liquidambar stryaciflua*; (b) prickles, *Rubus allegheniensis*; (c) single stipular thorn, *Maclura pomifera*, *Ribes oxyacanthoides*; (d) double stipular thorn; *Robinea pseudoacacia*; (e) simple thorn from modified twig, *Crataegus* spp.; (f) branched thorn of modified twig, *Gleditsia triacanthos*.

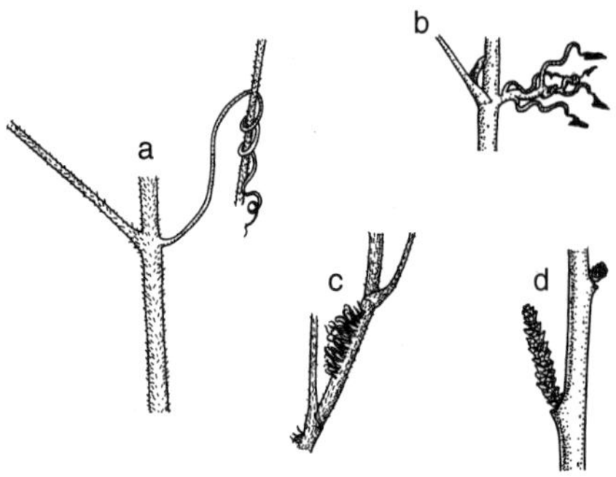

FIGURE 15.15 Modified twigs and roots: (a) slender-tipped tendril, *Vitis* spp.; (b) disk-tipped tendril, *Parthenocissus quinquefolia*; (c) aerial rootlets, *Hedera helix*; (d) spur shoot, *Malus* spp.

Lenticels

Lenticels are openings that permit O_2 and CO_2 exchange between the twig and the atmosphere. They are only infrequently used in species identification. Species of *Prunus* have long horizontal lenticels on the older twigs and bark. *Ulmus americana* has an extremely rough twig with small visible holes or openings in each lenticel.

Odor

Twigs of some species contain resins and oils that give them distinct odors. *Sassafras albidum* (sassafras) and *Lindera benzoin* (spicebush) have a pleasant spicy odor due to the presence of the oil of sassafras. *Prunus serotina* and *Prunus virginiana* (chokecherry) have a strong pungent odor caused by the presence of prussic acid (hydrocyanic acid), a derivative of cyanide that makes the twig taste as bad as it smells. *Ailanthus altissima* (tree-of-heaven) has a strong repulsive odor.

Sap

In most species, the watery exudate from a freshly broken twig is clear. *Maclura pomifera* (Osage-orange) and *Morus rubra* (red mulberry), members of the Moraceae family, have milky sap; some tropical species produce natural latex. *Acer platanoides* (Norway maple) and *Rhus glabra* (smooth sumac) also have milky sap.

Twig Cross Section

The cross section of most twigs is either round or ridged in some way; these characteristics are of limited use in identification. However, the cane (stem) of *Rubus alleghaniensis* (blackberry) is distinctly pentangle and provides an excellent character for identification during the winter months (Figure 15.14b).

Twig Thickness

The use of twig thickness in identification is applicable to only certain species or genera that have exceptionally thin or stout twigs. The foremost example of a species with exceptionally long, thin, flexible twigs is *Salix nigra* (black willow). Stout twigs (about 1.0 cm in diameter) are typical of *Fraxinus*, *Carya*, and *Juglans*, but the thickest twigs appear on *Rhus glabra*, *Ailanthus altissima*, *Catalpa speciosa*, and *Paulownia tomentosa* (princess-tree). Large trees of *Fraxinus* and *Juglans* are easy to identify, particularly during the dormant season, because of their stout twigs and branching habit.

Inflorescences

The inflorescence may be a modified petiole attached to the twig or may be a modified twig. In *Rhus* (sumac), the inflorescence is terminal (at the tip of the twig) and relatively open. A catkin (an ament; a spike of unisexual flowers) may be found on twigs of *Betula* (birch), *Alnus* (alder), *Carpinus* (bluebeech), *Ostrya* (ironwood), and *Corylus* (hazelnut) through the winter months (Chapter 16).

CHAPTER 16

Angiosperm Flower and Fruit Structures

INTRODUCTION

Sexual reproduction (the uniting of male and female gametes in the flower) is one of two primary reproductive systems in woody plants. Although all angiosperms have flowers, those of most tree and shrub species in North American forests are relatively inconspicuous. These flowers are often small, with greenish or yellowish green petals that blend in with the foliage; some species do not have petals. The few species that have conspicuous flowers with large, showy, or colorful petals are usually cultivated and are commonly used in landscaping. These include the well-known *Magnolia* (magnolia), *Cercis* (redbud), *Albizzia* (mimosa), *Catalpa* (catalpa), *Cornus* (dogwood), *Viburnum* (viburnum), *Rhododendron* (azalea), and *Hamamelis* (witch-hazel). The number of woody plants with showy flowers substantially increases from north to south (Foote and Jones 1989).

Sexual reproduction in plants involves a large number of distinct processes, and a breakdown in any one of them will influence the amount of seed produced and seedlings established. The processes begin with bud break (expansion), usually in the spring or early summer months, and is followed by pollen production and release, pollen transfer to stigma, pollination (deposition of pollen and elongation of pollen tubes), fertilization, seed development, dispersal, germination, and seedling establishment. Climate and weather may play an important role in reproduction not only during the current year when these processes are occurring, but also in the previous year. Weather may so exceptionally unfavorable the previous year that flowers in the bud may not develop. In addition, weather during flowering, pollination, and seed-set will affect these processes. Weather at and after the time of seed germination will affect establishment of seedlings. A disruption of the reproductive process may be the primary reason that a species remains a species and does not cross with another closely related species.

Distinct from sexual reproduction is parthenogenesis, an unusual type of reproduction also associated with flowers. Here the development of an embryo occurs without fertilization by male gametes. This condition led to the genus name for Virginia creeper, *Parthenocissus* (*parthenos*, virgin or unfertilized; *cissos*, ivy).

Asexual (vegetative) reproduction is also common in woody plants. This type of reproduction includes the development of stump or root collar sprouts, and root sprouts also referred to as suckers. Most angiosperm woody plants develop sprouts after the main stem is suddenly killed by cutting or fire. In contrast, only a few gymnosperms reproduce vegetatively.

FLOWER STRUCTURE

There are are a variety of important flower characteristics that are considered in classification. The detailed flower characteristics used for classification are

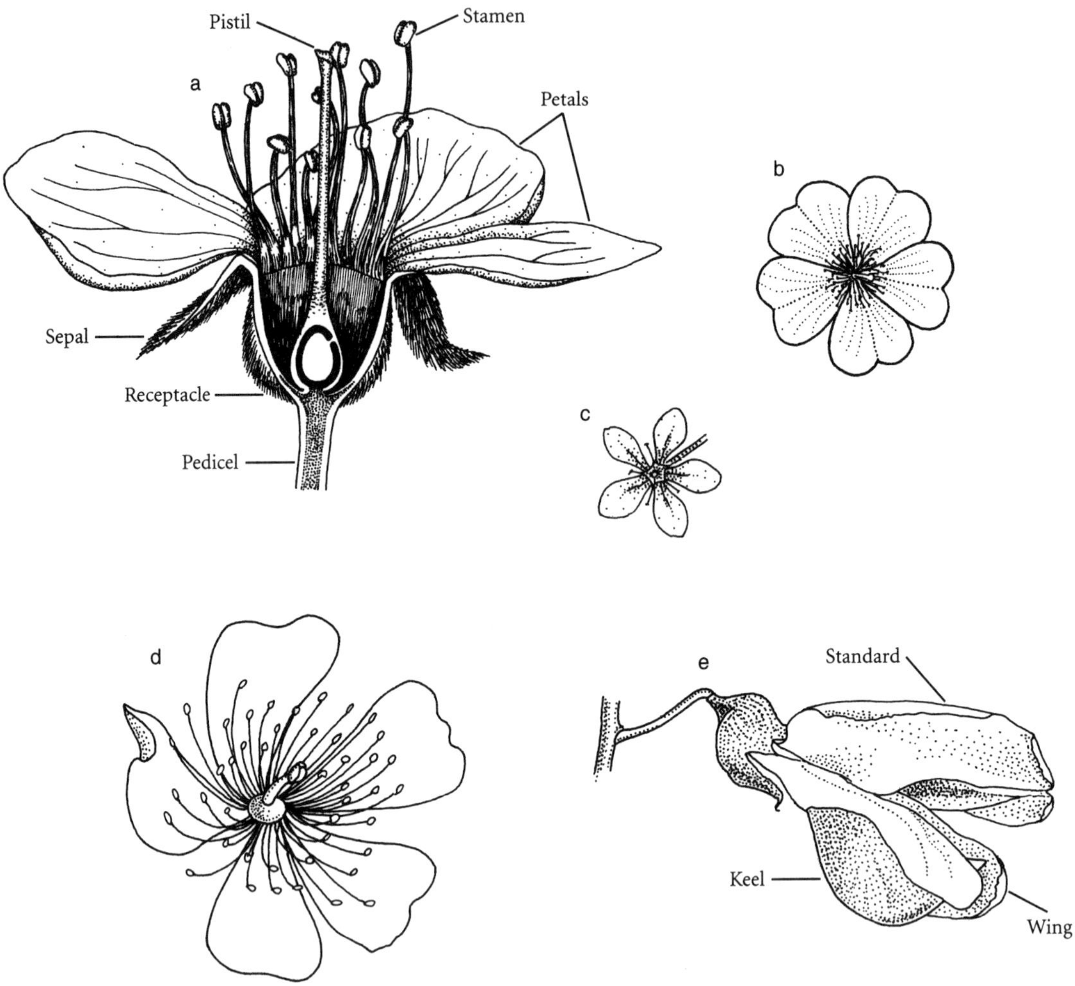

FIGURE 16.1 (a) Cross section of a *Prunus* (plum) flower showing its component parts (redrawn from Fuller and Tippo 1954); polypetalous, actinomorphic flowers of (b) *Rosa multiflora* (multiflora rose); (c) *Prunus serotina* (black cherry); (d) *Malus* (apple); (e) polypetalous, zygomorphic flower of *Cercis canadensis* (eastern redbud).

not considered here. The reader is referred to a standard college-level class in taxonomy for a detailed discussion; Judd et al. (1999) provide an excellent review. For woody plants, it will suffice here to consider only the easily visible structural characteristics, which include the presence or absence and condition of the various flower parts.

The flower consists of a receptacle (a modified twig) or a pedicel (a petiole) on which are borne the sepals (calyx), petals (corolla), pistil (stigma, style, and ovary), and stamens (anther and filament) (Figure 16.1a). The pistil is composed one or more carpels, which contain the ovules; the stamen produces pollen to fertilize the ovules. The sepals, petals, pistil, and stamens are modified leaves and exhibit variations in color, number, and/or character, such that have been and remain major structures used in plant classification.

Complete Flowers

A "complete" flower has sepals, petals, stamens, and a pistil (Figure 16.1a). If a flower lacks one or more of these parts, it is "incomplete." A flower is called "perfect" if it has both stamens and a pistil, the reproductive parts. A complete flower is also a perfect flower, but perfect flowers may or may not be complete. In classification the character of the corolla also is considered.

Polypetalous Corolla

Polypetalous flowers, flowers with many petals, may be categorized as actinomorphic and zygomorphic. A radial view of an actinomorphic flower illustrates its symmetrical (or regular) character (Figure 16.1a–d). Such flowers can be divided equally in any number of different planes. Woody plants such as *Prunus* (cherry, plum), *Malus* (apple, pear), *Rosa* (rose), and *Rubus* (blackberry) have complete, polypetalous actinomorphic flowers.

A zygomorphic (also called an asymmetrical or irregular) flower can be divided into two equal halves along only one plane. The petals are of different sizes and shapes as in the Fabaceae (legume) family (Figure 16.1e). The large petal on top is called the standard or banner, the side petals are called wings, and the bottom petals are fused to form the keel that encloses the stamens and pistil. These flowers are characteristic of the Fabaceae (bean or legume family), which includes *Cladrastis lutea* (yellowwood) and *Robinea pseudoacacia* (black locust), and a few members of the Caesalpiniaceae (Caesalpinia family, another legume family), which includes *Cercis canadensis* (eastern redbud).

Sympetalous Corolla

In some species, flower petals are fused together to form a floral tube. Woody plants with flowers of this structure are found in the Bignoniaceae (*Catalpa*, catalpa), Scrophulariaceae (*Paulownia tomentosa*, princess-tree), Caprifoliaceae (*Lonicera*, honeysuckle), Ericaceae (heath family) and Hippocastanaceae (buckeye family) (Figure 16.2). The flowers of these species also are zygomorphic.

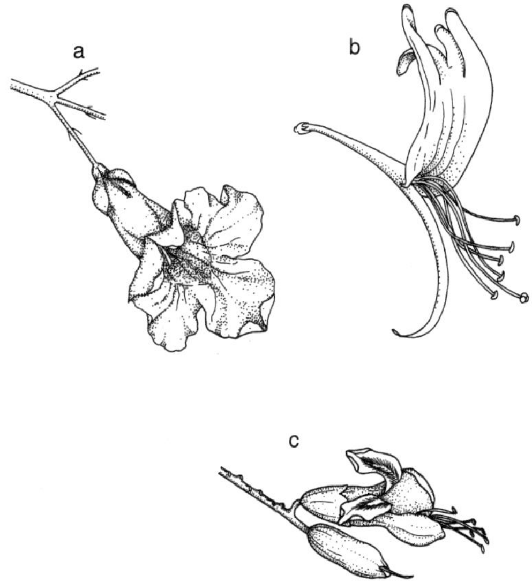

FIGURE 16.2 Sympetalous, zygomorphic flowers of (a) *Catalpa speciosa* (catalpa); (b) *Lonicera japonica* (Japanese honeysuckle); (c) *Aesculus* spp. (buckeye).

Incomplete Flowers

Incomplete flowers lack one of the primary flower parts: sepals, petals, stamens, or the pistil. Two common types of incomplete flowers are apetalous (no petals) and imperfect (lacking a pistil or stamens). Flowers that lack petals are found in several *Acer* species (*A. saccharum* (sugar maple); *A. saccharinum* (silver maple)) as well as in Betulaceae (birch, alder, hazel), Fagaceae (beech, oak), and Juglandaceae (hickory, walnut) (Figure 16.3). The last three species also have imperfect flowers. Imperfect flowers lack either a pistil or stamens and are thus unisexual flowers. This characteristic in particular may have considerable importance not only in plant classification but also in practical forestry, as noted later in the section

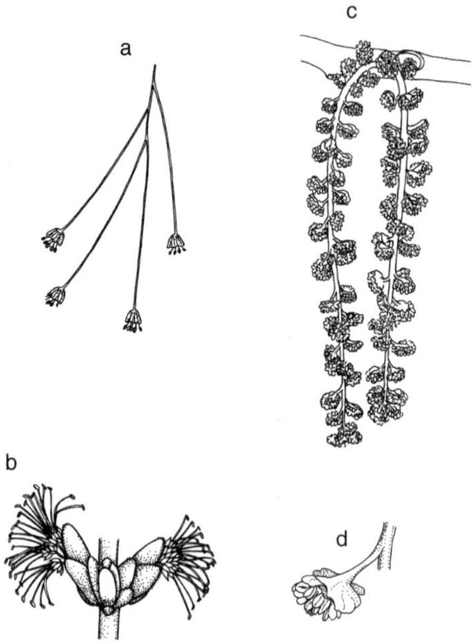

FIGURE 16.3 (a) Apetalous flower of *Acer saccharum* (sugar maple); (b) *Acer rubrum* (red maple) flower with petals reduced to sepal-like, minute appendages; (c) catkin of unisexual (staminate) flowers of *Juglans nigra* (black walnut); (d) staminate flower of *Juglans nigra.*

on tree sex. Unisexual flowers in catkins (aments) distinguish the Amentiferae (now a discarded name) such as *Populus* (aspen cottonwood), *Salix* (willow), *Quercus* (oak), *Juglans* (walnut), and *Carya* (hickory) from other woody plant groups, but unisexual flowers also occur in *Fraxinus* (ash).

TREE SEX

Individual woody plants that bear only one kind of flower are called homogamous. Some species have flowers that are hermaphroditic; that is, they produce complete flowers (pistil and stamens present) that produce seed. Other species produce only unisexual flowers (flowers with either stamens or pistils), thus differentiating individual trees into male and female. This condition is called dioecious ("two houses"), because two individual trees are needed to produce seed. *Fraxinus* and *Ilex opaca* (American holly) are dioecious species.

Heterogamous plants bear two or three types of flowers. When both male (staminate) and female (pistillate) (unisexual) flowers are found on one individual, the species is monoecious ("one house"), as that individual may produce seed (*Quercus; Betula,* birch). Dioecious species have unisexual flowers on separate trees thus two trees of different sex must be relatively close for pollination to occur. (*Salix, Populus, Fraxinus*). Polygamous species have complete and unisexual flowers on the same or different individuals. Species that have both complete flowers and both types of unisexual flowers on a single individual are polygamo-monoecious. Species that are polygamo-dioecious have complete flowers with male flowers on some individual trees, and perfect and female flowers an others trees.

INFLORESCENCE TYPES

The flowers on some woody plants are borne individually on a stalk (a pedicel) and are not associated with other flowers. These solitary flowers usually occur as axillary flowers attached to the twig directly above the leaf petiole. However, some flowers are borne in an aggregation or cluster called an inflorescence (Figures 16.4 and 16.5). The inflorescence may be axillary or it may be terminal, that is, borne at the tip of the twig. Among angiospermous plants, there are eight general inflorescence types, based on the flower arrangement pattern. In woody plants, there are six or seven inflorescence types. The most common are the catkin (ament), raceme, panicle, thyrse, cyme, head (capitulum), and corymb. The stalk of an inflorescence or of an individual flower in an inflorescence is called a peduncle.

POLLINATION VECTORS

Pollination is the transfer of pollen from the stamen to the pistil. The flowers of most woody plants in the temperate zone are anemophilous (pollinated by wind)

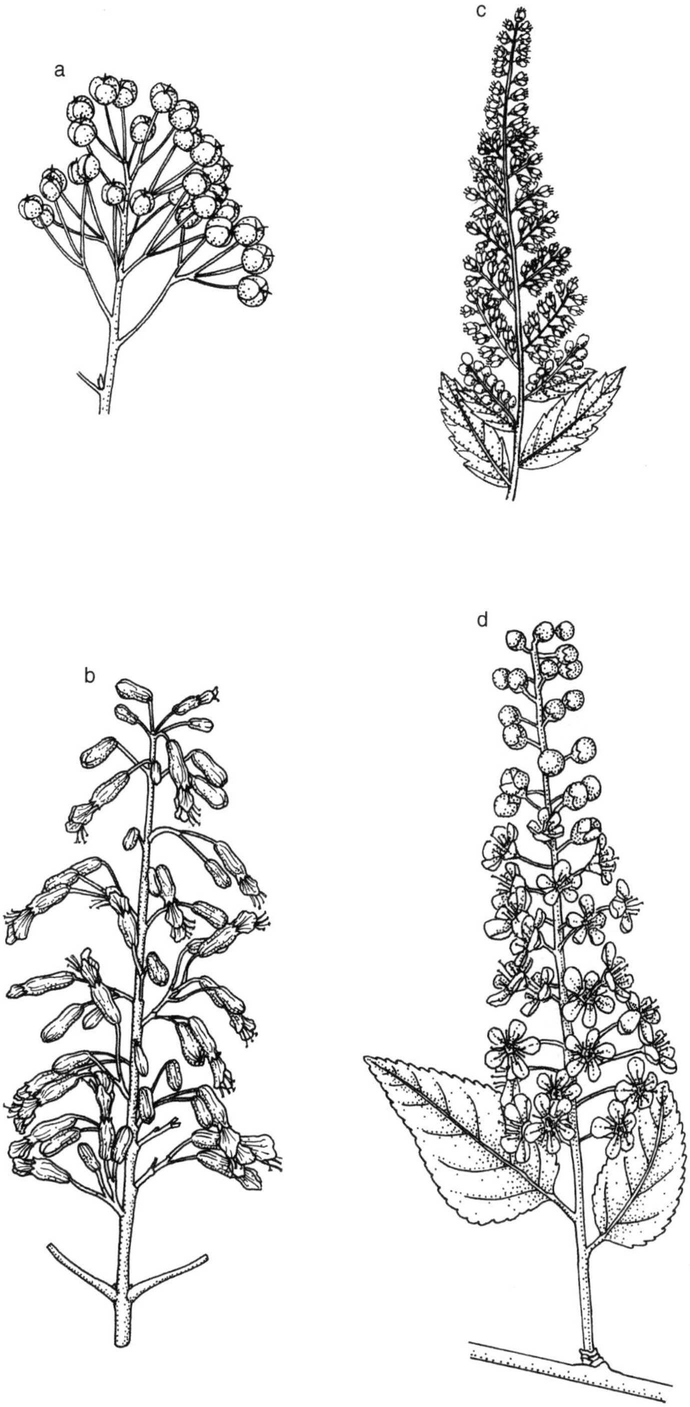

FIGURE 16.4 Inflorescences: (a) thryse of *Ceanothus* (New Jersey tea) with fruit (b) thyrse of *Aesculus* (buckeye); (b) panicle of *Spiraea* (spiraea); (d) panicle of *Prunus serotina* (block cherry).

FIGURE 16.5 Inflorescences: (a) catkin of *Corylus* (hazel) in winter; (b) catkin of *Juglans nigra* (black walnut) after expansion in spring; (c) compound raceme of *Rhus glabra* (sumac) with fruits; (d) corymb of *Spiraea* (spriraea); (e) head of *Cephalanthus occidentalis* (buttonbush).

or entomophilous (pollinated by insects). Flowers in the tropics are frequently pollination by birds or bats.

FRUIT TYPES

As an ovary ripens into a fruit, its outer layers or pericarp may thicken and become differentiated into three rather distinct tissue layers. The exocarp, or outermost layer, forms the "skin," which may consist of one to several layers of epidermal cells. The mesocarp, the middle layer of tissue, varies from a single layer of cells to a mass of soft or fibrous tissue several centimeters thick. The innermost layer is the endocarp, and like the mesocarp, it also varies in structure, texture, and thickness, depending on the plant species. In some species, the endocarp becomes a hard or bony structure (a "pit"), which encloses the seed(s). In some types of fruit the layers of pericarp are quite thin, not readily distinguishable, and appear to be a seed coat.

The many different kinds of fruits are classified primarily according to their structure and number of carpels and ovaries. The following brief classification and description of major fruit types in woody plants is adapted from Fuller and Tippo (1954).

Simple Fruit

A simple fruit consists of a single ripened ovary (carpel) containing seeds and in some species may include the old sepals and stamens. Most woody angiosperms have simple fruit. Simple fruits may be fleshy or dry.

Fleshy Fruit

In fleshy fruit, one or more tissue layers of the pericarp are soft or fleshy at maturity.

- **Berry**. A fruit with an outer skin that covers the fleshy mesocarp containing the seeds; the endocarp is undifferentted (*Vitis*, grape, and *Diospyros virginiana*, persimmon; also tomato and banana (Figure 16.6a).
- **Hesperidium**. A berry with a thick leathery rind (*Citrus sinensis*, orange; also grapefruit and lemon) (Figure 16.6b).
- **Drupe**. Similar to a berry, but the endocarp is differentiated into a hard stony pit (*Prunus*, cherry, plum, peach, apricot; *Celtis*, hackberry) (Figure 16.6c).

Dry, Dehiscent Fruit

In dry dehiscent fruit, the entire pericarp is dry, fibrous, and often brittle at maturity. The following dry fruit types are dehiscent; that is, the fruit splits open along one or two seams at maturity. The fruit may contain several to many seeds.

- **Legume**. A fruit consisting of one carpel, which splits open along two seams (*Cercis canadensis; Gleditsia triacanthos* (honeylocust); *Robinea pseudoacacia*, black locust; also pea and bean) (Figure 16.6d).
- **Follicle**. A fruit consisting of one carpel, which splits open along a single seam at maturity. Few woody plants produce a follicle; however, follicles are found in aggregate fruits.
- **Capsule**. A fruit consisting of two or more fused carpels, splitting open in various ways at maturity (*Populus deltoides*, cottonwood; *Catalpa; Hamamelis virginiana*, witch-hazel) (Figure 16.6e, f, and g).

Dry, Indehiscent Fruit

The following dry fruit types are indehiscent, in that they do not split open along definite seams at maturity. The fruit usually contains only one or two seeds (Figure 16.7).

- **Achene**. A fruit bearing only one seed, which is separable from the ovary wall. Few woody plants have an achene as a fruit; achenes do appear in aggregate fruits.
- **Nut**. A one- to three-seeded fruit with a thick, extremely hard pericarp, as in an acorn or hickory nut (*Quercus; Carya; Castanea*, chestnut; *Fagus*, beech; *Aesculus*, buckeye; *Tilia*, basswood; *Corylus*, hazel). The acorn cupule (cap) is an involucre (many whorls of bracts) (Figure 16.7a, b, c and d).
- **Samara**. A one- or two-seeded fruit, the pericarp of which bears a flattened winglike outgrowth (*Ulmus*, elm; *Fraxinus* (ash); *Ptelea trifoliata*, hop tree) (Figure 16.7e and f).

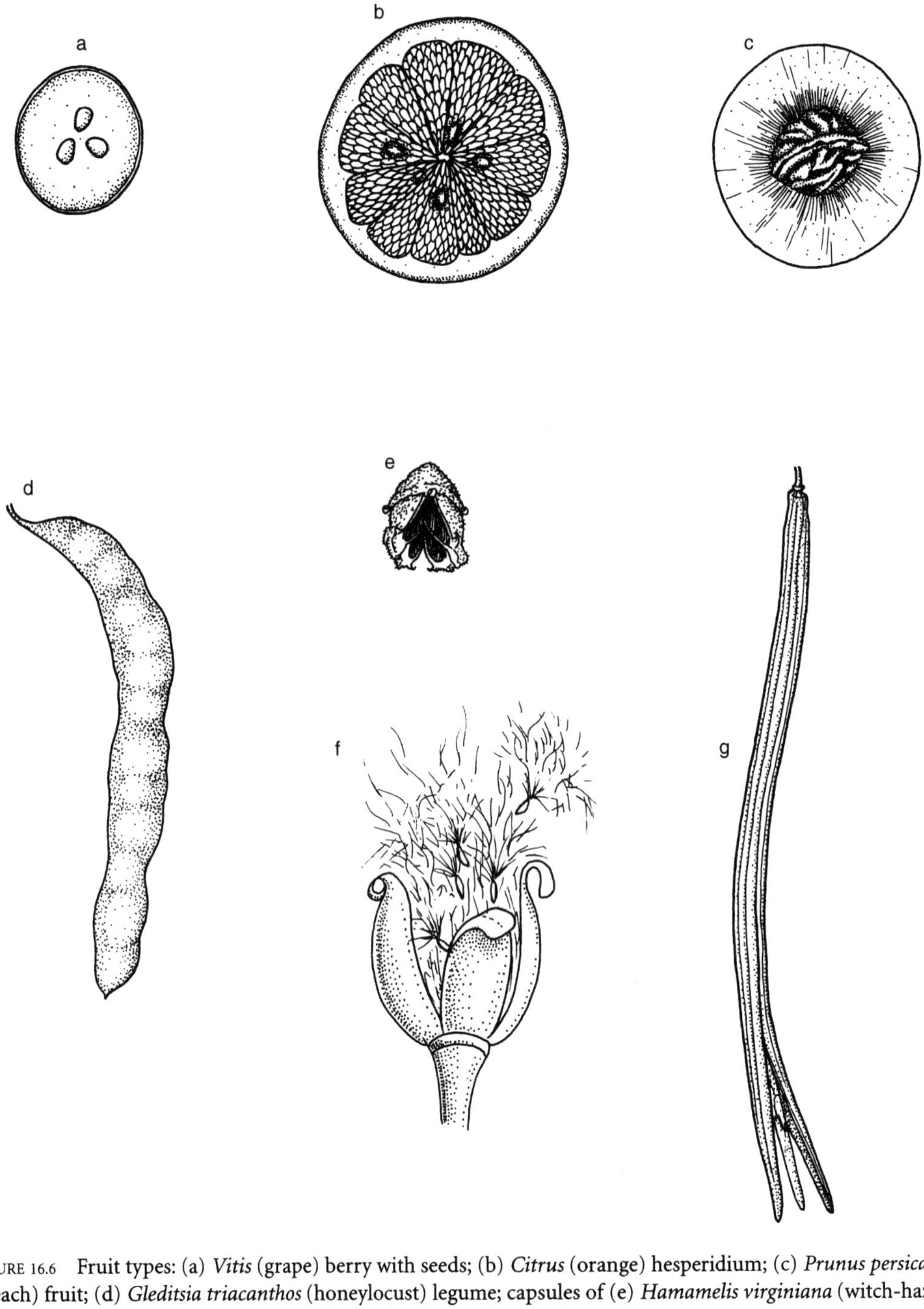

FIGURE 16.6 Fruit types: (a) *Vitis* (grape) berry with seeds; (b) *Citrus* (orange) hesperidium; (c) *Prunus persica* (peach) fruit; (d) *Gleditsia triacanthos* (honeylocust) legume; capsules of (e) *Hamamelis virginiana* (witch-hazel); (f) *Populus deltoides* (eastern cottonwood); (g) *Catalpa speciosa* catalpa.

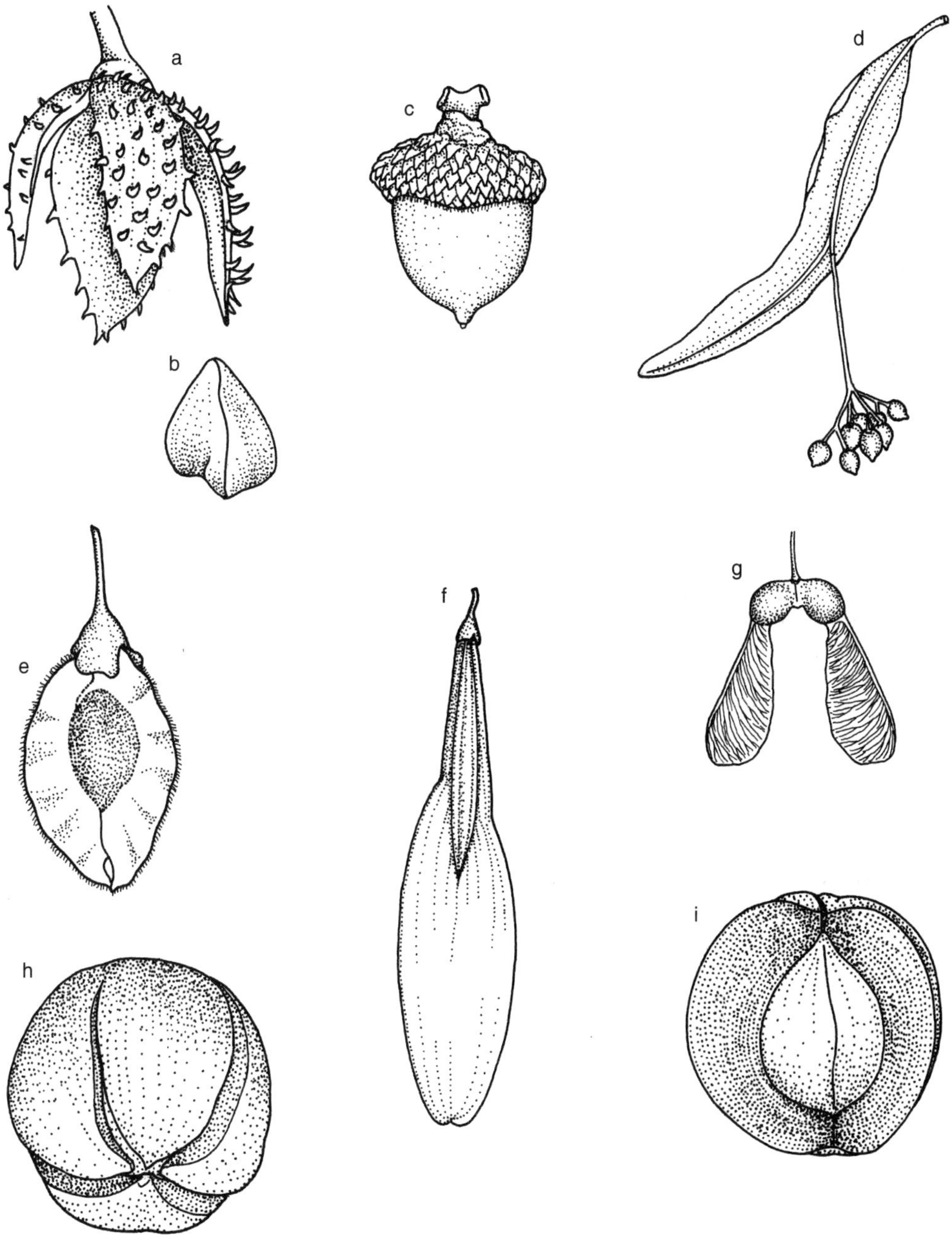

FIGURE 16.7 *Fagus grandifolia* (American beech) (a) cupule (bracts) and (b) nutlet; (c) *Quercus* cupule (cap) and acorn (nut); (d) *Tilia* (basswood) nutlets attached to a leafy bract; (e) *Ulmus americana* (American elm) samara with a lateral wing; (f) *Fraxinus pennsylvanica* (green ash) samara with a terminal wing; (g) *Acer saccharum* (sugar maple) schizocarp; *Carya ovata* (shagbark hickory) (h) nut enclosed by the husk, (i) nut with the husks removed.

- **Schizocarp**. A two-seeded fruit that splits into two 1-seeded mericarps, a double samara (Figure 16.7g).

Aggregate Fruit

An aggregate fruit consists of several to many ripened ovaries produced by a single flower on one receptacle (Figure 16.8a, b, c). The individual ripened ovaries may be drupes (*Rubus*, raspberry and blackberry), samaras (*Liriodendron tulipifera*, yellow-poplar), or follicles (*Magnolia*, magnolia). A head is a dense cluster of sessile flowers or fruit on an extremely short receptacle (Figure 16.5e).

Multiple (Compound) Fruit

A multiple fruit consists of a cluster of many ripened ovaries produced by several flowers in the same inflorescence (Figure 16.8d, e, f). The unit of a compound fruit in woody species may be a beaked capsule (*Liquidambar sytraciflua*, sweetgum), a nutlet or linear achene (*Platanus occidentalis*, American sycamore), or an achene (*Maclura pomifera*, Osage-orange; *Morus rubra*, red mulberry). Although the fruits of *Maclura* and *Morus* appear to be a multiple of drupes, they actually are a multiple of achenes buried in a greatly enlarged fleshy or berry-like calyx.

In accessory fruits, parts other than ovaries may adhere to or enclose the ovaries or true fruit. A well-known accessory fruit is the strawberry (*Fragaria*); here the individual units are achenes, borne upon a fleshy, red edible receptacle. The pome is an accessory fruit, exemplified by apple and pear, in which the ovaries in the core are surrounded by an enlarged edible receptacle and floral tissue. In pomes, the ripened ovaries constitute the core.

MISLEADING FRUIT NAMES

A study of the preceding classification and descriptions indicates that the common names of some fruits may not reflect their botanical nature. Blackberry and raspberry fruit are not berries, but aggregates of tiny drupes; mulberry fruit is not a berry, but a multiple of achenes surrounded by fleshy sepals; the strawberry is a much enlarged receptacle, bearing an aggregation of achenes. The true fruit of the apple is embedded in the succulent receptacle and floral tissue.

For many species, the popular term "nut" does not reflect the botanical nature of the fruit to which it is applied. A walnut is drupaceous; the exocarp and mesocarp form the fibrous husk and the endocarp becomes the walnut "shell"; the "meat" is the seed, composed of two large convoluted cotyledons, an epicotyl, a hypocotyl, and a thin, papery seed coat. A Brazil nut is a seed borne with several other seeds in a large thick-walled capsule. The coconut is a drupe, as is the almond; the almond we buy from a store is the "stone," or hardened endocarp, containing one or two seeds. The peanut is a legume, and the filbert is the seed of a nut. In western United States, pine (pinyon) nuts may be purchased in stores; they are actually seeds from a pine cone.

SEED TRANSPORT

Gravity has a major influence on seed distribution in forested upland areas; better soil on level ground is used primarily for agriculture, while the steeper slopes remain forested. Heavy seed from trees high on a slope may travel downward after they are released. The higher the elevation and steeper the slope, the farther seed is likely to travel. More or less round seed from *Quercus*, *Carya*, and *Juglans* move relatively easily downslope.

Animals also transport seed. *Quercus*, *Carya*, *Juglans*, and *Corylus* (hazel) seed may be transported moderate distances by squirrels. However, probably the most important transport vector for *Quercus* seed is the common bluejay, who may carry the seed for some distance. Birds tend to transport bright red seed or fleshy fruit that is less than one centimeter in diameter. These criteria would include seed from *Ilex* (holly), *Magnolia*, *Sassafras albidum* (sassafras), *Rhus* (sumac), and *Juniperus* (redcedar); the latter three species are generally found in old fields or along fence rows, as these are frequented by birds. The seed transported by birds usually germinates quickly, and the seedling has a good chance of establishment as the seed has passed through the alimentary tract (the gut) of the birds, which softens the seed coat. Ultimately, the seed is deposited with a lump of fertilizer, usually in an open area.

Wind is an important mode of transport for seed of many species. *Populus* (cottonwood) and *Salix*

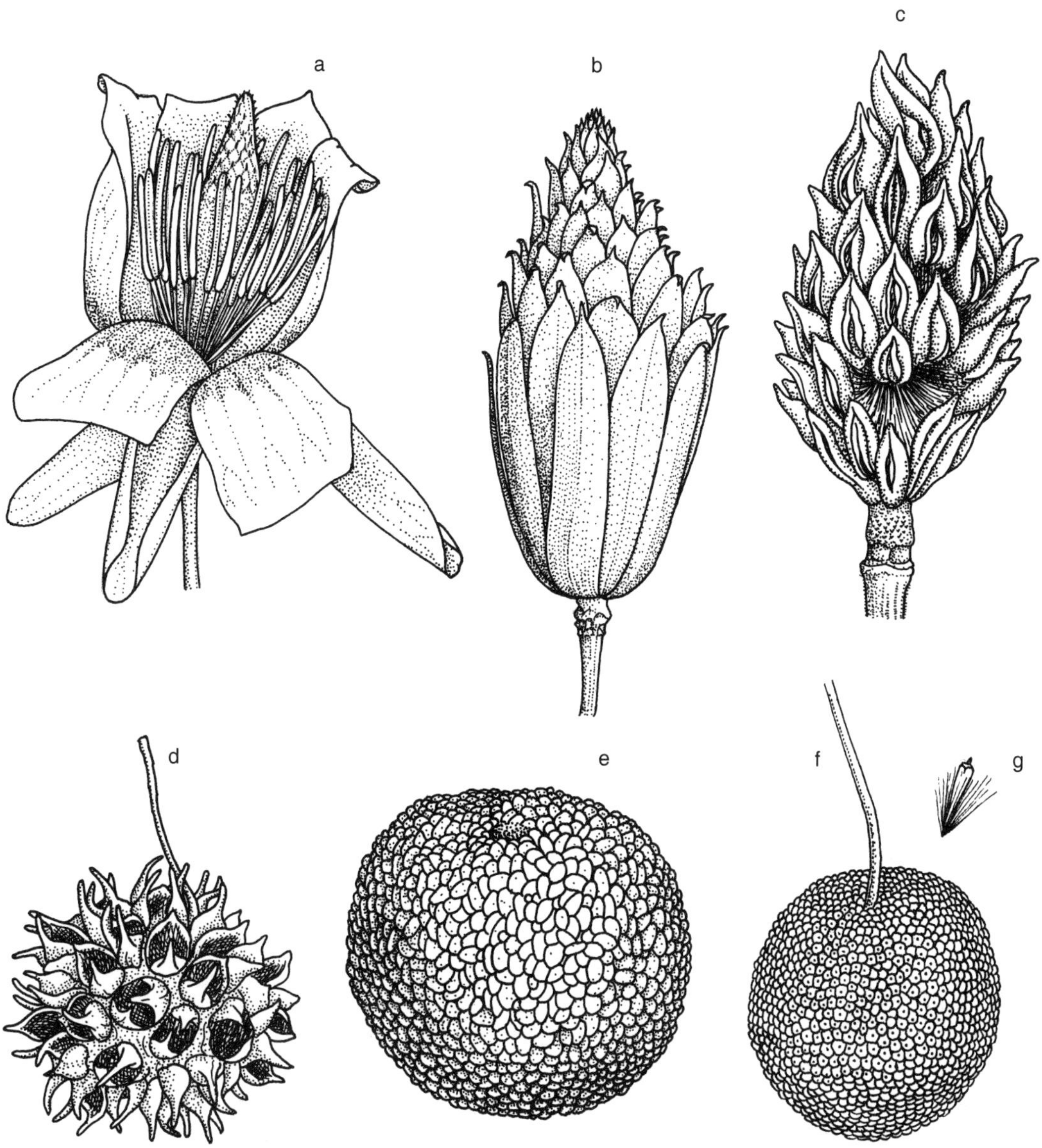

FIGURE 16.8 (a) *Liriodendron tulipifera* (yellow-poplar) flower showing a cluster of ripening ovaries in the center; (b) *Liriodendron* aggregate (cone) of samaras; (c) *Magnolia* (magnolia) aggregate of follicles; (d) *Liguidamber styraciflua* (sweetgum) multiple of capsules; (e) *Maclura pomifera* (Osage-orange) multiple of achenes; *Platanus occidentalis* (eastern sycamore) (f) multiple of achenes and (g) individual achene.

(willow) seed, with their light, cottony hair, may travel for hundreds of miles. It is believed that directly after the glacier receded, *Populus tremuloides* (quaking aspen) revegetated the Lake States (Minnesota, Wisconsin, and Michigan) with seed transported by wind from the Rocky Mountains. A variety of species have winged seed. *Acer, Fraxinus,* and *Liriodendron* have seed with a terminal wing extending to one side of the seed cavity, and *Ulmus* has a lateral wing surrounding the seed cavity. Although such seed is considerably heavier than that of *Populus* or *Salix,* it will travel some distance, perhaps several hundred meters in a strong wind.

In lowland and bottomland species, water transport can be important. In the spring rainy season, when flooding frequently occurs, seed can be transported long distances from and often to an elevation higher than the parent tree. Seed of *Nyssa aquatica* (water tupelo), *Carya illinoensis* (pecan), *Carya aquatica* (water hickory), *Quercus bicolor* (swamp white oak), and *Quercus michauxii* (swamp chestnut oak) are transported by water.

CHAPTER 17

Class: Magnoliopsida; Subclass: Magnoliidae

ORDER: MAGNOLIALES

Magnoliaceae (Magnolia Family)

THE Magnoliaceae is a well-defined family of trees and shrubs. Worldwide, there are 12 genera and about 220 species, located primarily in temperate and tropical southeast Asia, with a few in South America and southeastern United States. The larger genera are *Magnolia* (120 species), *Talauma* (50 species) and *Michelia* (40 species). *Magnolia* (8 to 10 species) and *Liriodendron* (1 species) are common in eastern United States forests. One other *Liriodendron* species occurs in southeast Asia.

The flowers are solitary, perfect, actinomorphic, and showy. The large (up to 30 cm diameter) flowers are composed of three to five whorls of undifferentiated white, greenish, or rose-colored petals and sepals (usually called tepals), with the outermost whorl considered the calyx. The flowers appear in late spring as the leaves are expanding. The fruit is a conelike aggregate of follicles in *Magnolia* and samaras in *Liriodendron tulipifera.* The fruit is formed at the tip of the twig. The seeds of *Magnolia* are bright red or rose colored and dangle from the individual follicles by a thin thread, 1 to 2 cm long. This color and display of seeds appear ideally designed for bird distribution, whereas the samaras of *Liriodendron* are winged and wind distributed. *Liriodendron* seeds persist for a time in the duff layer and readily germinate after a disturbance, such as a fire that kills the overstory trees. Clusters of *Liriodendron* trees or unusually high numbers in a forest community suggest a previous disturbance.

Magnolia leaves are typically arranged alternately on the twig, although they are sometimes clustered into false terminal whorls. The blade is simple, alternate, smooth margined, and unlobed; some species have persistent leaves (evergreen). Large leafy stipules are often persistent, encircling the twig from one end of the leaf scar around the back side of the twig to the other end of the leaf scar; the resulting encircling and continuous stipule scar is a characteristic of the family. Another identifying characteristic of *Magnolia* during the dormant season is the large (length 2 to 3 cm) flower buds, in some species covered with long, silvery tomentose hair. Individuals of most *Magnolia* species have strongly aromatic bark. The bud of *Liriodendron* is a 1 to 2 cm long glabrous valvate bud that somewhat resembles a duckbill.

In North America, *Magnolia* species occur in and are restricted to the southeastern United States, either the Appalachian Mountains or the Coastal Plain, whereas *Liriodendron* and *M. acuminata* are distributed across much of eastern United States. At maturity, individuals of six major species are large trees and individuals of the other three are arborescents or large shrubs.

Arboreous Species

Liriodendron tulipifera Linnaeus, yellow-poplar (Figure 17.1); large tree, height to 40 m; deciduous; leaf arrangement alternate; leaf simple, length 15 to

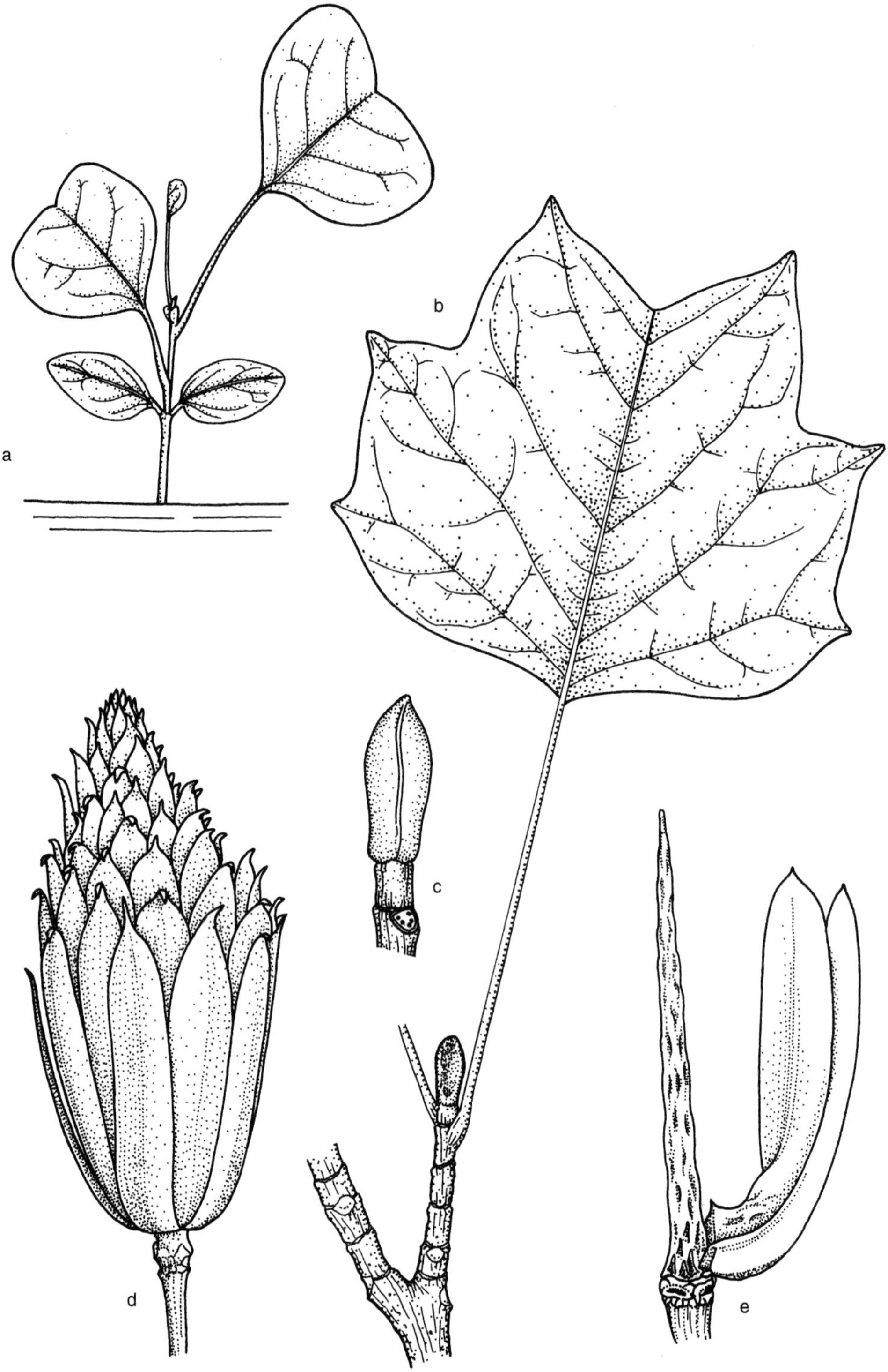

FIGURE 17.1 *Liriodendron tulipifera* (a) seedling ×1, (b) twig and simple leaf ×1/2, (c) valvate terminal bud ×2, (d) aggregate of samaras ×1, (e) flower stalk after dispersion of seed ×1.

30 cm, width 8 to 17 cm, shape tulip-like, mostly four- but sometimes six-lobed, sinuses shallow, margin smooth, apex emarginate, base truncate; bud valvate, length 1 to 2 cm, glabrous, visible scales two, terminal bud resembling a duckbill; twig and buds glabrous, stipule scar continuous; flower large, tepals 6 to 9, greenish yellow, length 4 to 6 cm; fruit a cone-like aggregate of samaras, length 5 to 10 cm, samara length 2.5 to 5.5 cm; Regions 2, 3, and 4, southern Michigan into Ontario and New England south through the central and southern states to Texas and Florida; cool, moist sites. Cover type **27.8, 27.12, 27.15, 27.30, 27.37, 28.12, 28.33**, 26.21, 26.24, 27.5, 27.6, 27.22, 27.34, 27.36, 28.2, 28.4, 28.6, 28.14, 28.23, 28.29, 28.36.

Magnolia acuminata Linnaeus, cucumber-tree (Figure 17.2); large tree, height to 35 m; deciduous; leaf arrangement alternate; leaf simple, length 15 to 25 cm, width 5 to 15 cm, broadly oval or elliptical, margin smooth, apex acuminate, base obtuse; petiole glabrous; twig glabrous, buds tomentose, terminal bud length 1.0 to 1.7 cm, stipule scar continuous; flower tepals 6, yellowish green, length 5 to 8 cm; fruit a cone-like aggregate of follicles, length 5 to 8 cm; seeds reddish, length 1 cm; Region 3b, concentrated from Lake Erie to northern Georgia, scattered in Regions 3a and 4a, c; cool, moist sites and drainages. Cover type 27.9, 27.12, 27.37, 28.11.

Magnolia virginiana Linnaeus, sweetbay (Figure 17.2); large tree, height to 25 m; evergreen; leaf arrangement alternate; leaf simple, length 8 to 15 cm, width 3.5 to 5.0 cm, elliptical to oval, coriaceous, margin smooth, apex mucronate, base obtuse to acute, abaxial surface glabrous and somewhat whitish, spicy when crushed; petiole pubescent; twig and buds pubescent, terminal bud length 1.0 to 1.5 cm, stipule scar continuous; flower tepals white, 9 to 12, length 3 to 5 cm; fruit a cone-like aggregate of follicles, length 3 to 5 cm; seeds red, length 5 mm; Region 4, scattered in 4c (Piedmont), Massachusetts and New York south to Florida and Texas; mesic-hydric woods, swamps, and stream banks. Cover type **28.29, 28.30, 28.36**, 27.33, 28.14, 28.23, 28.24, 28.28, 28.32, 28.36, 28.38, 28.40, 28.53, 28.55.

Magnolia grandiflora Linnaeus, southern magnolia (Figure 17.3); large tree, height to 25 m; evergreen; leaf arrangement alternate; leaf simple, length 15 to 20 cm, width 6 to 12 cm, elliptical, distinctly coriaceous, margins smooth, revolute, lower surface rusty tomentose, apex slightly mucronate, base obtuse; petiole pubescent; twig and buds pubescent, terminal bud length 1.0 to 1.7 cm, stipule scar continuous; flowers showy, tepals white, 6 to 12, length 7 to 13 cm; fruit a cone-like aggregate of follicles, length 7 to 10 cm; seeds red, length 1 to 1.5 cm; Region 4a and b, mostly Lower Coastal Plain from North Carolina and Florida to Louisiana and Texas but widely planted as a lawn tree north to Illinois, Indiana, and Ohio; cool, moist sites and drainages under natural conditions. Cover type 27.9, 28.11, 28.16, 28.29, 28.45.

Magnolia pyramidata Bartram, pyramid magnolia (Figure 17.4); medium-sized tree, height to 20 m; deciduous; leaf arrangement alternate; leaf simple, length 15 to 25 cm, width 8 to 14 cm, obovate, margin smooth, apex obtuse, blade tapering to a narrow auriculate base; petiole glabrous; twig and buds glabrous, terminal bud length 1.0 to 1.7 cm, stipule scar continuous; flowers tepals white, six, length 5 to 8 cm; fruit a cone-like aggregate of follicles, length 5 to 10 cm; seeds red, length 7 mm; Region 4, scattered and somewhat rare from Georgia and Florida to east Texas. Similar to *M. fraseri*, a mountain species that has a larger leaf and pale yellow petals. Cover type 28.27.

Magnolia fraseri Walter, Fraser magnolia (Figure 17.4); height to 15 m; deciduous; leaf arrangement alternate; leaf simple, length averaging 25 to 30 cm, maximum 55 cm, width 10 to 25 cm, obovate to rhomboid, margin smooth, apex slightly mucronate, base distinctly auriculate (long-eared); petiole glabrous; twig and buds glabrous, terminal bud length 1.0 to 1.7 cm, stipule scar continuous; flower tepals six to nine, pale yellow to creamy white, length 7 to 10 cm; fruit a cone-like aggregate of follicles, length 5 to 10 cm; seeds red, length 0.7 to 1.0 cm; Region 3b, mountains from West Virginia to Georgia and South Carolina; mountain forest; cool, moist sites and stream banks. Cover type 27.12, 27.30.

Arborescents

Magnolia macrophylla Michaux, bigleaf magnolia (Figure 17.5); height to 12 m; deciduous; leaf arrangement continuous; leaf simple, length 50 to 100 cm, width 15 to 30 cm, obovate, margin smooth, base narrowly cordate, abaxial surface white, tomentose; petiole tomentose; twig tomentose, bud tomentose, terminal bud length 1.0 to 1.7 cm, stipule scar continuous; flower tepals six, white, rose at the base, length 12 to 18 cm; fruit a cone-like aggregate of follicles, length 5 to 10 cm; seeds reddish orange, length 0.9 to 1.2 cm; concentrated in Regions 3 and 4 (Ohio and Kentucky to Mississippi and Alabama, Georgia, and

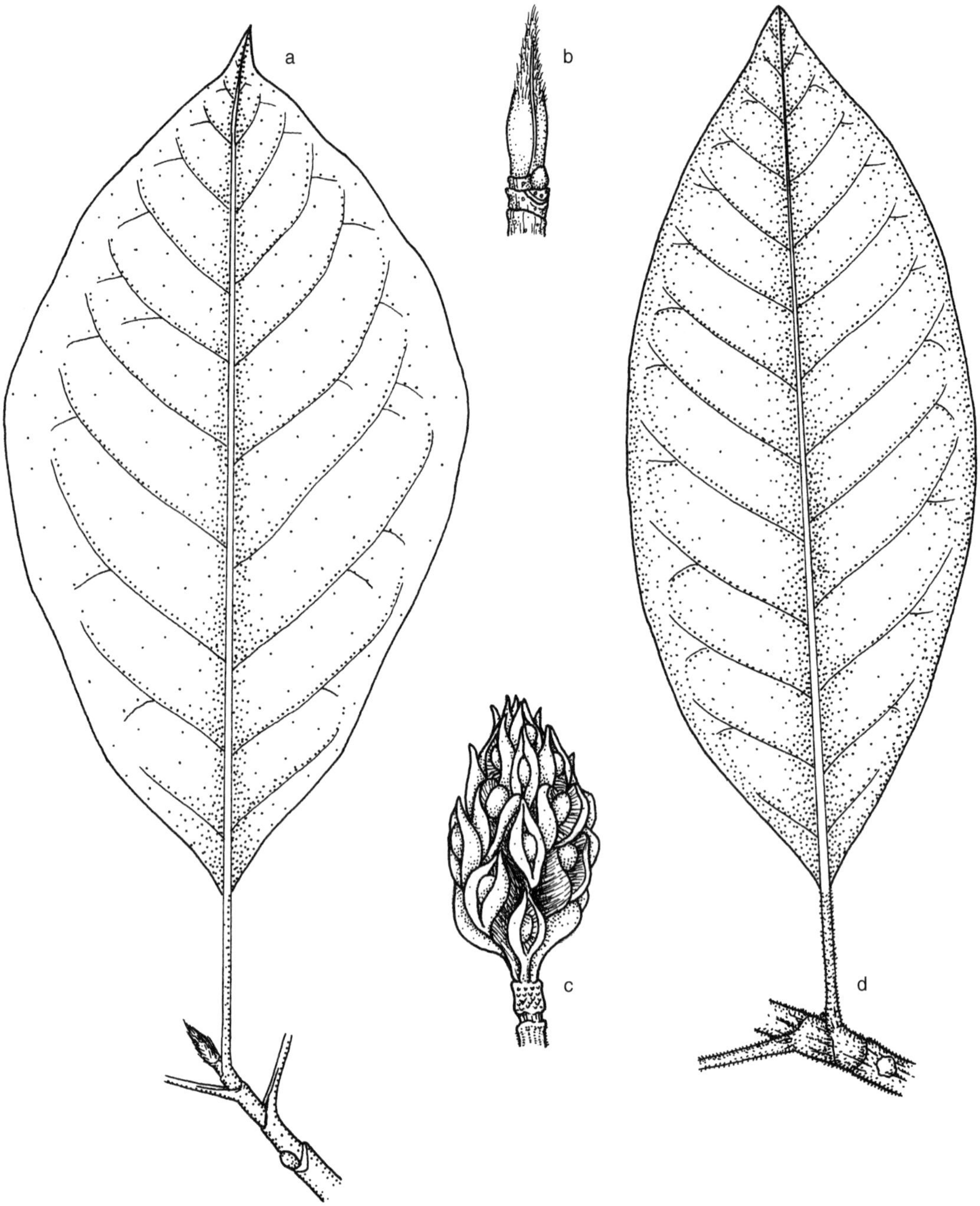

FIGURE 17.2 *Magnolia acuminata* (a) simple leaf ×1, (b) terminal bud ×2, (c) aggregate of follicles ×3/4; (d) *M.virginiana* twig and leaf ×1.

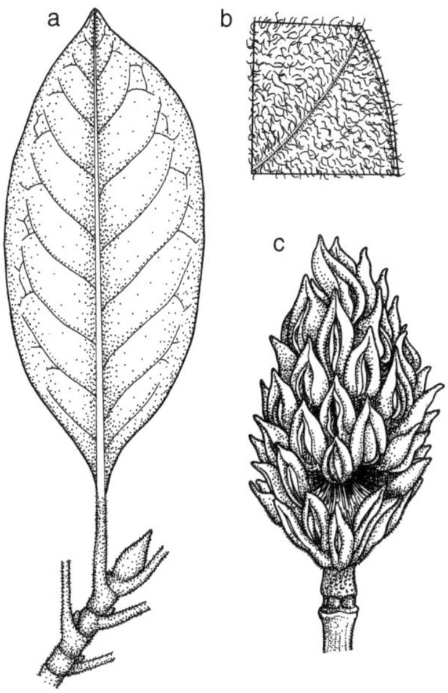

FIGURE 17.3 *Magnolia grandiflora* (a) simple leaf ×1/2; (b) tomentose abaxial surface ×1; aggregate of follicles ×1/2.

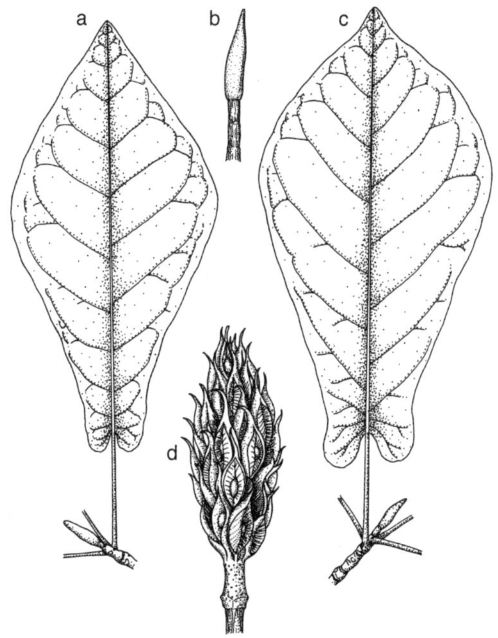

FIGURE 17.4 *Magnolia pyramidata* (a) twig and leaf of, (b) glabrous terminal bud ×1/3; *M. fraseri* (c) leaf, (d) aggregrate of follicles. Note auriculate leaf bases.

North Carolina), moist soil of valleys and ravines. Cover type 28.16, 28.27.

Magnolia macrophylla ssp. *ashei* (Wheatherby) Spongberg, (*M. ashei* Wheatherby), Ashe or sandhill magnolia; large, multistemmed shrub or small tree, height to 8 m; deciduous; leaf arrangement alternate; leaf simple, length 20 to 60 cm, width 15 to 30 cm, obovate, margin smooth, apex slightly mucronate, base auriculate, abaxial surface white; flower tepals six, white with purple marking, length 10 to 14 cm; fruit a cone-like aggregate of follicles, length 12 to 18 cm; seeds reddish orange, length 0.8 to 1.0 cm; Region 4c, restricted to the Florida panhandle; hardwood forest on mesic sites, bluffs, and ravines. Similar in appearance to *M. macrophylla.* Cover type 28.27.

Magnolia tripetala Linnaeus, umbrella magnolia (Figure 17.5); medium-sized tree, height to 13 m; deciduous; leaf arrangement alternate; leaf simple, length 25 to 70 cm, width 10 to 25 cm, elliptical to obovate, margin smooth, apex acuminate, base acute; petiole glabrous, twig glabrous, bud glabrous, terminal bud length 2.0 to 2.5 cm, leaf scar continuous; flower tepals six to nine, white, length 8 to 14 cm; fruit a cone-like aggregate of follicles, length 7 to 12 cm; seed pink to red; length 1 to 1.2 cm; Regions 3 and 4, primarily 3b (Appalachian Mountains) and 4a (Piedmont), disjunct in Arkansas and Oklahoma; low slopes of mountains, along streams, and in ravines.

Magnolia × *soulangiana* Soulange-Bodin, saucer magnolia (Figure 17.6); tall multistemmed shrub, height to 9 m; deciduous; leaf arrangement alternate; leaf simple, length 10 to 20 cm; width 7 to 12 cm, elliptical to obovate, margin smooth, apex and base acuminate; lower surface finely pubescent; buds silvery tomentose, length 1.0 to 1.5 cm, stipule scar continuous; flower tepals white to pink or purple, length 5 to 12 cm; fruit a cone-like aggregate of follicles, length 6 to 11 cm; introduced; a hybrid between two overseas species, but authors differ on the parent species (Phillips 1978; Coombes 1992); cultivated as a lawn tree.

Only *Liriodendron* is a major, often dominant, component of forest communities in the Central States and Appalachian Mountains. The other species are relatively minor components of forests within their

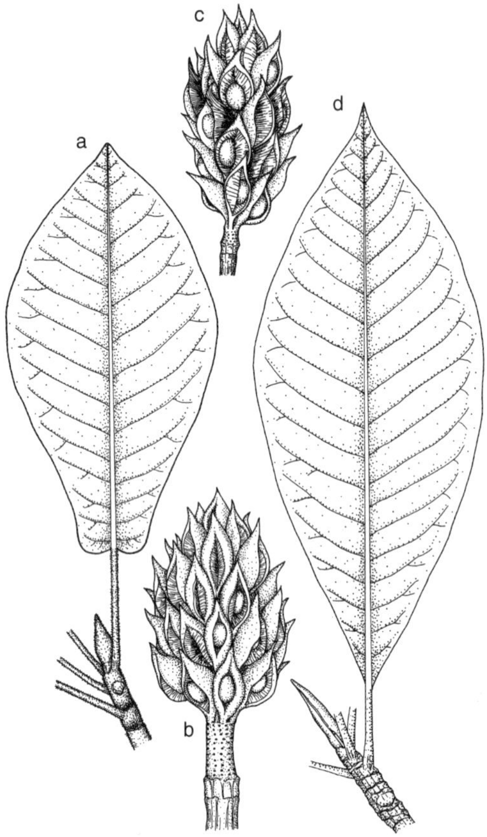

FIGURE 17.5 *Magnolia macrophylla* (a) twig and leaf of leaf, (b) aggregate of follicles; *M. tripetala* (c) aggregate of follicles, (d) twig and leaf.

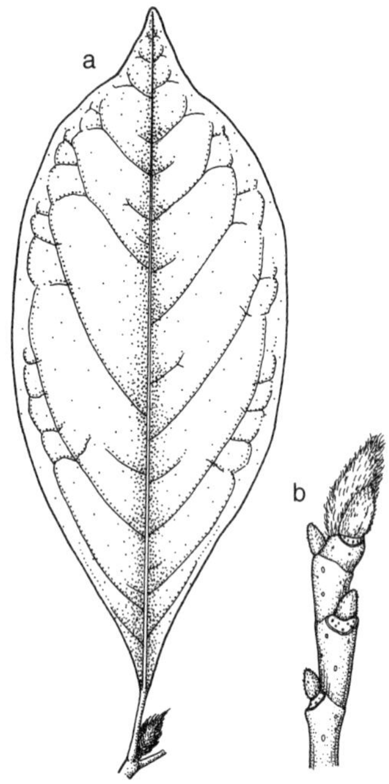

FIGURE 17.6 *Magnolia* × *soulangiana* (a) leaf and tomentose bud, (b) pubescent terminal bud.

botanical range. Both *Magnolia* and *Liriodendron* have been widely planted as street and lawn trees in the eastern United States. *Magnolia grandiflora* is one of the most commonly planted, but *Magnolia* × *soulangiana* is a multistemmed arborescent that is widely planted as a lawn tree and ornamental both in the eastern United States and Europe; there are 10 to 12 cultivars of this species based on the flower color. Both of these species, as well as the other *Magnolia* species, are adapted to a warm climate, thus they may sustain substantial winter frost damage. In the Central States (Illinois, Indiana, and Ohio; Region 3), *Magnolia grandiflora* is often damaged, but *M. soulangiana* is planted as far north as southern Ontario and Quebec.

Annonaceae (Custard-Apple Family)

The custard-apple family is a large pan-tropical family of about 120 genera and 2300 species of aromatic trees, shrubs, and vines. In North America the family is represented by 3 genera and about 15 species of small trees and shrubs, most of which are located in the southeastern United States. There are 8 *Asimina* species, about 5 *Annona* species, and 2 *Deeringothamnus* species.

Asimina leaves are deciduous, alternate, simple, entire, and have a smooth margin. The lateral veins do not intersect the leaf margin but curve to connect to the next vein. *Asimina* can be separated from other similar-appearing species by the presence of naked, rusty to reddish, silky pubescent buds. Stipules are absent.

The flower of *Asimina* is regular, solitary, perfect (plant synoecious), and borne in the axil of the leaf and twig. There are three sepals and a corolla of six petals aligned in two rows. The edible fruit is a somewhat peanut-shaped (up to 3 to 4 cm wide, 10 cm long in some species) pulpy berry that contains several flat seeds. Nelson (1996) separated *Asimina* into two groups, based on flower color and odor.

Maroon/Purple Flowers: Disagreeable Fetid Odor

Asimina triloba (Linnaeus) Dunal, common pawpaw, dog banana (Figure 17.7); medium-sized tree, height to 13 m; deciduous; leaf arrangement alternate; leaf simple, entire, length 15 to 30 cm, width to 5 to 8 cm, obovate or oblanceolate, margin smooth, apex acuminate, base acute; twig reddish brown, silky pubescent; buds naked, covered with rusty-colored pubescence, terminal bud length ~1 cm, lateral buds nearly globose; fruit a berry, length 7 to 13 cm; Regions 3 and 4, New York to southern Ontario, Michigan, and Illinois, south to Florida and Texas; dense woods on moist sites, primarily north slopes and stream terraces. Cover type 27.1, 27.3, 27.4, 27.6, 27.8, 27.9, 27.10, 27.12, 27.1, 27.15, 28.11, 28.12, 28.16, 28.18, 28.42.

Asimina tetramera Small, four-petal pawpaw, opossum pawpaw (Figure 17.7); shrub, height to 3 m; deciduous; leaf arrangement alternate; leaf simple, entire, length 5 to 12 cm, width 1 to 3 cm, oblong to somewhat oblanceolate, margin smooth, minutely revolute, apex and base obtuse; twig reddish brown, pubescent; buds naked, dark red, silky pubescent; fruit a large berry, greenish yellow, length 4 to 9 cm; Region 4c, Lower Coastal Plain of central Florida; sand scrub and xeric sites.

Asimina parviflora (Michaux) Dunal, small-flower pawpaw, dwarf pawpaw (Figure 17.8); arborescent, height to 6 m; deciduous; leaf arrangement alternate; leaf simple, entire, length 6 to 17 cm, width 3 to 6 cm, obovate or oblanceolate, lower surface somewhat reddish, hairy; twig reddish brown, pubescent; buds naked, dark reddish brown, silky pubescent; fruit a

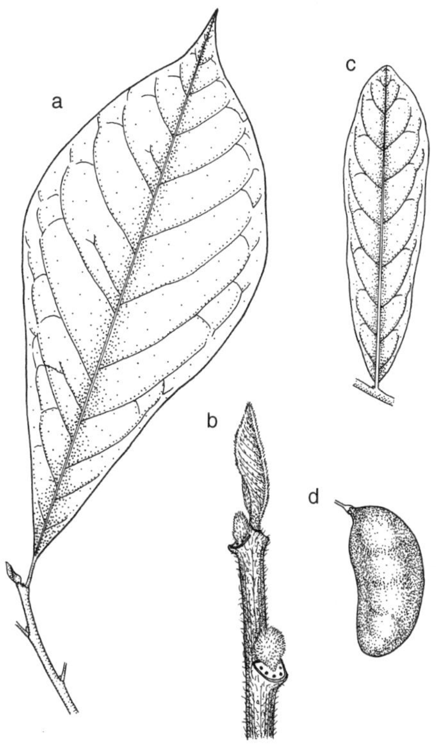

FIGURE 17.7 *Asimina triloba* (a) leaf ×1/3, (b) tomentose twig with tomentose naked terminal and lateral buds ×1.5; *A. tetramera* (c) leaf ×1/2, (d) fruit ×1/3, a fleshy aril enclosing seeds.

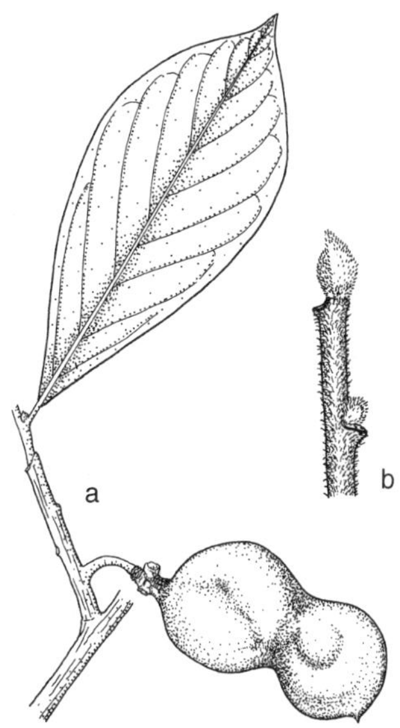

FIGURE 17.8 *Asimina parviflora* (a) twig with leaf and fruit (fleshy aril) ×1/3, (b) tomentose twig with tomentose naked terminal and lateral buds ×1.

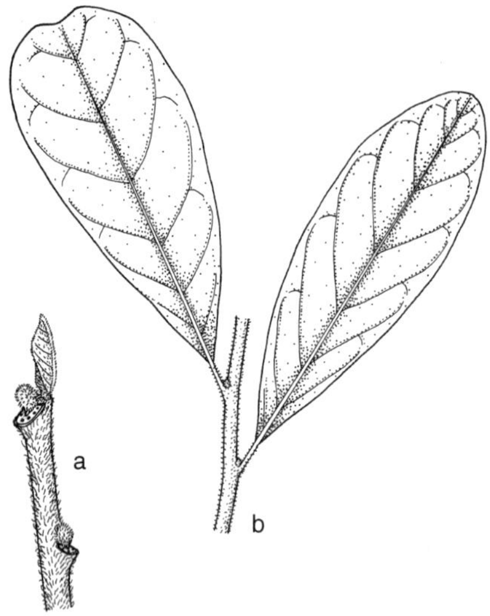

FIGURE 17.9 *Asimina incana* (a) leaves ×1/2, (b) tomentose twig with tomentose naked terminal and lateral buds ×1.0.

large berry, greenish yellow, length 3 to 8 cm; Region 4, from Virginia to Florida and Mississippi, north to Tennessee; pine and oak woods, moist sites and floodplains. This species is an important wildlife food (Foote and Jones 1989). Cover type 28.27.

White Flowers: Pleasant Odor

Asimina incana (Bartram) Exell (*A. incarna*), polecat bush (Figure 17.9); shrub, height to 1.5 m; semideciduous; leaf arrangement alternate; leaf simple, entire, length 4 to 10 cm, width 2 to 4 cm, elliptical, ovate or obovate, slightly coriaceous, margin smooth, slightly undulate, apex rounded or emarginate; twig and buds silky reddish pubescent, buds naked; fruit a large berry, yellowish green, length 5 to 9 cm; Region 4b, primarily the Lower Coastal Plain, southern Georgia and Florida; xeric sites, sand scrub oak and pine savanna.

Asimina longifolia Kral, small pawpaw (Figure 17.10); shrub, height to 2 m; deciduous; leaf arrangement alternate; leaf simple, entire, length 7 to 20 cm, width 1 to 3 cm, linear to narrowly obovate or spatulate, slightly coriaceous, apex obtuse, base acute; twig and buds silky reddish pubescent, buds naked; Region 4a and b, Alabana, Georgia, and Florida; xeric sites, sand scrub and pine savanna. Two varieties are

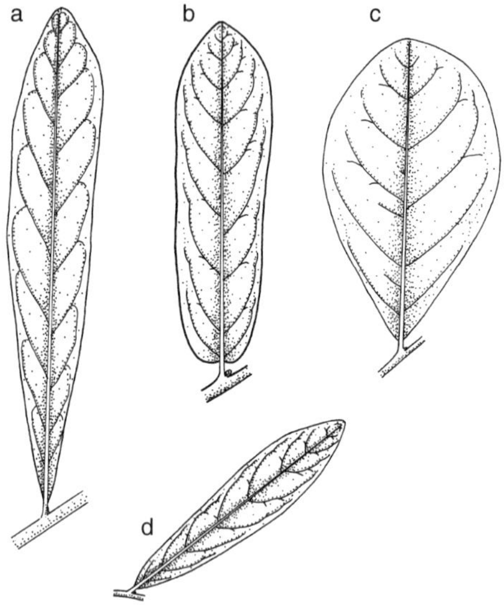

FIGURE 17.10 Leaves of (a) *Asimina longifolia* ×1/2; (b) *A. reticulata* ×1/2; (c) *A. obovata* ×1/2; (d) *A. pygmaea* ×1/2.

recognized by Kral (1977): individuals of var. *longifolia* have stiff, ascending, and somewhat tomentose stems, and var. *spatulata* has flexible, nearly decumbent, and glabrous twigs.

Asimina reticulata Shuttleworth ex. Chapman, pawpaw (Figure 17.10); shrub, height to 2 m; deciduous; leaf arrangement alternate; leaf length 5 to 10 cm, width 1 to 3 cm, oblong, slightly coriaceous, margin smooth, minutely revolute, abaxial surface slightly pubescent; twig and bud reddish brown, pubescent, buds naked; fruit a large berry, yellowish green, length 4 to 7 cm; Region 4b, Lower Coastal Plain, central and southern Florida; pine flatwoods and xeric scrub.

Asimina obovata (Willdenow) Nash, flag pawpaw (Figure 17.10); arborescent, height to 5 m; deciduous; leaf arrangement alternate; leaf simple, entire, length 5 to 12 cm, width 3 to 5 cm, broadly oval to slightly oblong, apex rounded, base obtuse; twig and buds reddish brown, silky pubescent, buds naked; fruit a large berry, greenish yellow, length 5 to 10 cm; Region 4b, Lower Coastal Plain, central Florida; xeric sites, sand scrub, dunes, and pine savanna.

Asimina pygmaea (Bartram) Dunal, dwarf pawpaw (Figure 17.10); shrub, height to 30 cm; deciduous;

leaf arrangement alternate; leaf simple, entire, length 5 to 10 cm, width 1.5 to 3.0 cm, oblong to oblanceolate, margin minutely revolute, apex acute to nearly obtuse, base acute; petiole twisted to hold leaves erect; twig and buds reddish brown, silky pubescent, buds naked; fruit a large berry, yellowish green, length 3 to 6 cm; Region 4b, Lower Coastal Plain; northern and central Florida; xeric sites, sand scrub, and pine savanna.

The various *Asimina* species are common in woodland and forest communities. In the central hardwood forest (Region 3), a dense thicket of *A. triloba* interferes with lesser plant growth by creating a lethal environment. The forest floor below a thicket usually is bare of other woody seedlings and herbs. The general absence of understory plants may be due to a reduced light level created by the combined effect of a dense midcanopy thicket and overstory canopy, or it may be due to the strong aromatic (possibly allelopathic) compounds that may leach from the leaves or the roots of *A. triloba* and in Region 4, possibly from the aromatic *A. parviflora* and *A. tetramera* as well. Woodland (1991) indicates that the aromatic compounds in the wood and leaves of all species come from ethereal oil cells in parenchyma tissue.

Related species such as *Annona glabra* Linnaeus (pond apple) and *A. squamosa* Linnaeus (sugar apple) are arborescents (height < 6 m) of the subtropical Florida keys. *Deeringothamnus pulchellus* Small (beautiful pawpaw) and *D. rugelii* (Robinson) Small (Rugels pawpaw) are rare species and, at least in Florida, are listed as endangered by the U. S. Fish and Wildlife Service and the Florida Department of Agriculture.

ORDER: LAURALES

Lauraceae (Laurel Family)

Estimates of the number of genera and species of the Lauraceae vary widely but often fall at about 49 (40 to 50) genera and 2000 (1000 to 2500) species. Most members are aromatic evergreen trees and shrubs distributed in tropical and subtropical regions but concentrated in Brazil and Asia. Seven genera and eleven species are native to the United States.

The leaves are alternate simple and have a smooth margin. Most species have deciduous leaves, but a number have leaves that persist for several seasons. The fruit is a blue or red drupe, usually formed singly at the end of a 2 to 7 cm long peduncle.

Sassafras albidum (Nuttall) Nees, sassafras (Figure 17.11); tree, height to 30 m, usually considerably shorter; deciduous; leaf arrangement alternate; leaf simple, length 10 to 15 cm, width 5 to 10 cm, elliptical, entire, two-lobed (mitten-shaped), or three-lobed, margin smooth, emitting a pleasant spicy odor when crushed; twigs light green, becoming darker and mottled green with age; fruit a drupe, dark blue, borne singly, nearly globose, diameter 4 to 7 mm; Regions 2b, 3, and 4, Illinois, Michigan, and New England south; undisturbed and cutover forest of xeric-mesic and mesic slopes and terraces. Most often found as a seedling or sapling in the forest understory. A component of old field vegetation and along fence rows. May invade prairie remnants and grassland. Cover type 27.17, 27.26, 28.13, 28.14.

Lindera benzoin Linnaeus, spicebush (Figure 17.12); multistemmed shrub, height to 4 m; deciduous; leaf arrangement alternate; leaf simple, entire, length 6 to 12 cm, width 3 to 5 cm, obovate to broadly oblanceolate, glabrous, margin smooth, apex cuspidate; twig and leaf emitting a pleasant spicy odor when crushed; twig dark green; fruit a drupe, red, borne singly, oval, length 0.7 to 1.0 cm; Regions 2, 3, and 4, Illinois, Michigan, and New England south; common understory component of forest on mesic, cool slopes and stream terraces. Cover type 27.3, 27.4, 27.6, 27.8, 27.23, 27.30, 28.12.

Lindera melissifolia (Walter) Blume, Jove's-fruit (Figure 17.12); multistemmed shrub, height to 2 m; deciduous; leaf arrangement alternate; leaf simple, entire, shape oval to oblanceolate, length 4 to 8 cm, width 2 to 4 cm, surfaces pilose to densely pubescent, margin smooth, apex cuspidate; twig and leaf emitting a pleasant, somewhat lemonlike odor when crushed; twig dark green, pubescent; fruit a drupe, red, oval, length 0.7 to 1.0 cm; Regions 4a, b, Coastal Plain from North Carolina south to southern Georgia; disjunct populations in the Mississippi River Valley; forest of stream terraces and swamps; somewhat rare and on the conservation watch list.

Lindera subcoriacea Wofford, bog spicebush; multistemmed shrub, height to 4 m; deciduous; similar to *Lindera melissifolia*, except the leaf is somewhat leathery, becoming odorless late in the growing season; Regions 4a and b, Coastal Plain from North Carolina

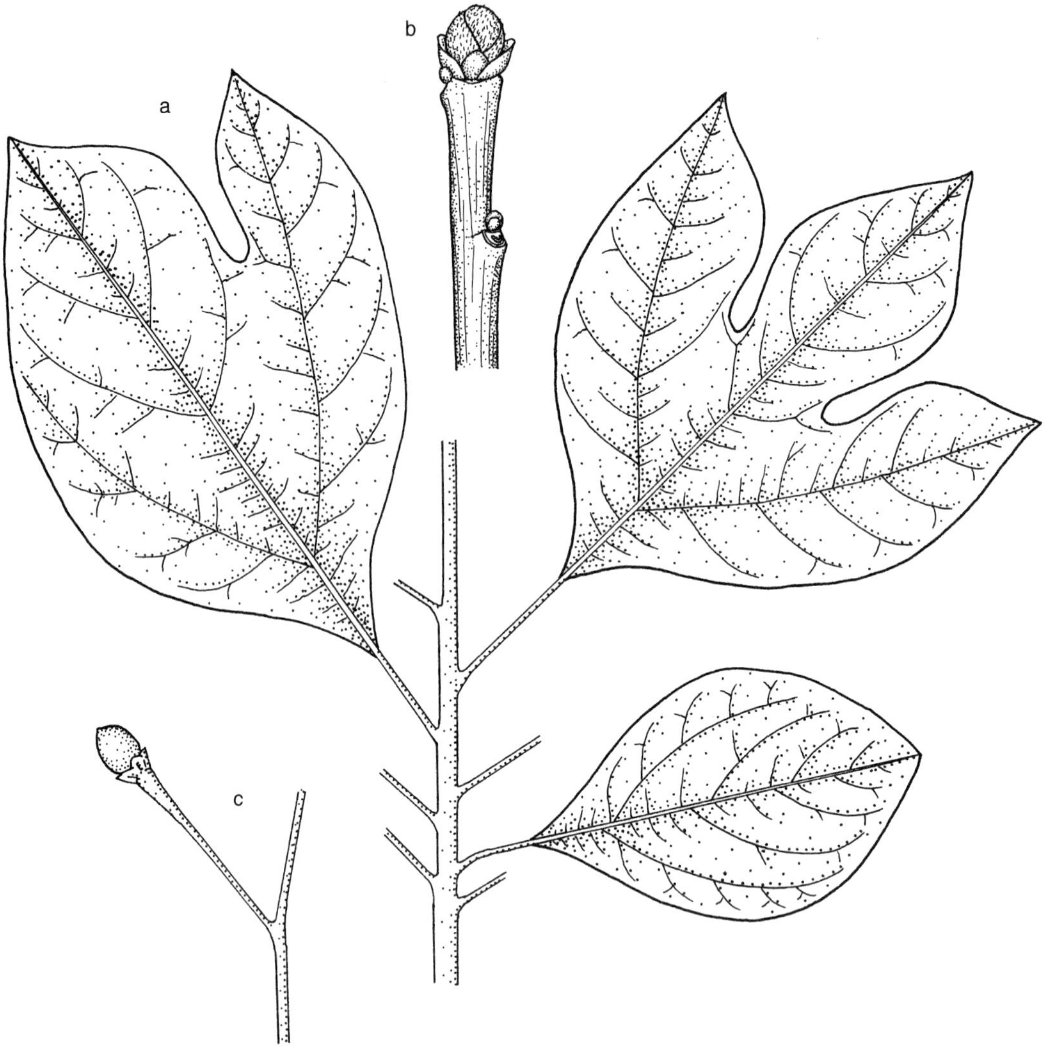

FIGURE 17.11 *Sassafras albidum* (a) leaves ×1/2–2/3, (b) terminal bud ×2, (c) individual (blue) drupe on a peduncle ×1.

south to southern Mississippi; acid swamps and bogs. This most recently described third *Lindera* is listed as rare and endangered (Foote and Jones 1989).

Persea borbonia (Linnaeus) Sprengel, redbay (Figure 17.13); medium-sized tree, height to 18 m; evergreen; leaf arrangement alternate; leaf simple, length 5 to 15 cm, width 2 to 5 cm, narrowly elliptical to lanceolate-oblong, coriaceous, apex acute but tip rounded, base acute, lower surface glabrous or sparsely covered with short appressed pubescence appearing as light brown or grayish flecks, aromatic when crushed; twigs light brown, buds naked; fruit a drupe, dark blue to nearly black, globose, diameter 1.0 to 1.5 cm, borne singly; Regions 4a, b, Coastal Plain, east Texas to

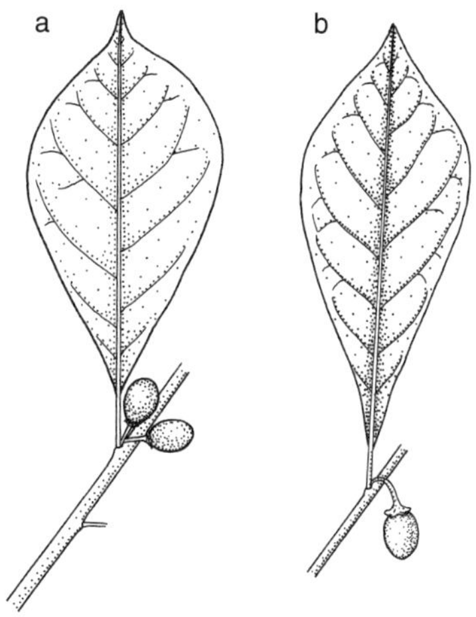

FIGURE 17.12 (a) *Lindera benzoin* leaf and (red) drupes ×1/2; (b) *L. melissifolia* leaf and (red) drupe ×1/2.

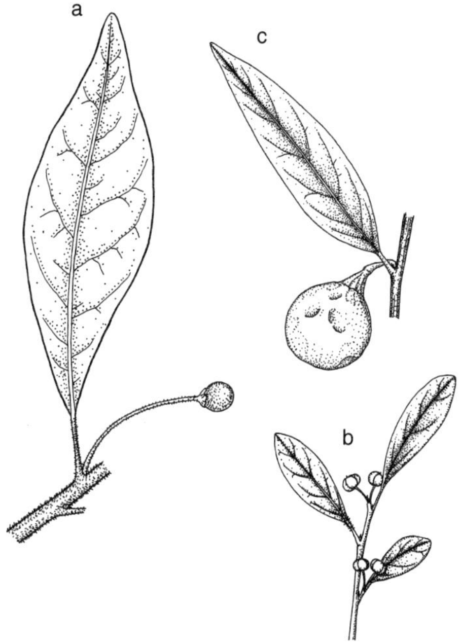

FIGURE 17.13 *Persea borbonia* (a) twig with leaf and (blue black) drupe ×1/2; (b) *Litsea aestivalis* leaves and flower buds ×1; (c) *Umbellularia californica* leaf and (purple) drupe ×1/2.

Delaware; mostly wet sites, swamps, and bays. Cover type 28.14, 28.24, 28.27, 28.28, 28.29, 28.40, 28.44, 28.53.

Persea palustris (Rafinesque) Sargent, swamp bay; tree, height to 14 m; evergreen; leaf arrangement alternate; leaf simple, entire, shape narrowly elliptical, length 7 to 20 cm, similar to *P. borbonia* except lower surface is brown tomentose, especially along the midrib, aromatic when crushed; buds naked; fruit a drupe, blue, oblong, borne singly, length about 1 cm; Region 4a, b; lowland pine forest, swamps, and wet areas. Cover type **28.30,** 28.23, 28.28, 28.33, 28.36.

Persea humulis Nash, silk bay; small tree, height to 10 m; evergreen; leaf arrangement alternate; leaf simple, entire, length 4 to 10 cm, width 1 to 3 cm narrowly elliptical, similar to *P. borbonia*, except lower surface densely pubescent, velvety, aromatic when crushed; bud naked; fruit a dark blue globose drupe, diameter 1.3 to 1.7 cm; branches and twigs black; Region 4b, restricted to the Florida peninsula; xeric sand scrub.

Litsea aestivalis (Linnaeus) Fernald, pond-spice (Figure 17.13); shrub, height to 5m; deciduous; leaf arrangement alternate; leaf entire, length 1 to 4 cm, width 0.5 to 1.0 cm, narrowly elliptical to lanceolate, coriaceous, aromatic when crushed; fruit a drupe, bright red, globose, diameter 1 cm; Regions 4a, b, Coastal Plain; wet sites and woods. Listed as an endangered species.

Umbellularia californica (Hooker & Arnott) Nuttall, California laurel; Oregon myrtle (Figure 17.13); large tree, height 25 to 35 m; evergreen; leaf arrangement alternate; leaf simple, entire, length 5 to 12 cm, width 1 to 3 cm, lanceolate to ovate-lanceolate, coriaceous; fruit a drupe, dark purple globose, diameter about 2 to 3 cm, borne singly; Region 7, southern Oregon though the Sierra Nevada to Southern California. *U. californica* is the only western member of the family. *Umbellularia* is considered a monotypic genus, although two varieties of *U. californica* are recognized (van der Werff 1997). Cover type **30.45,** 30.7, 30.16, 30.37, 30.38, 30.39, 30.41, 30.42.

Other members of the family include *Licaria triandra* (Swartz) Griesbach (Gulf licaria), which is extremely rare and restricted to Miami, Florida, and *Nectrandra coriacea* (Swartz) Grisebach (lancewood)—also listed as *Ocotea coriacea* (Swartz)

Britton—which is restricted to the subtropical region of southern Florida.

Aromatic oil in the wood is responsible for the fragrance of most species and the economic importance of several species. The most valuable members of the Laurel family are found in tropical and subtropical regions. Worldwide, the most important members are *Cinnamonum zeylandicum* (cinnamon from bark; Asia), *Cinnamonum camphora* (Linnaeus) J. Presley (camphor from leaves; Asia), and *Persea americana* Miller (avocado fruit; tropical America) (Woodland 1991). The North American species have limited economical importance. Spices are produced from *Umbellularia californica.* Until about 1960, the roots of *Sassafras albidum* were used to make tea, but the flavoring has been shown to be carcinogenic.

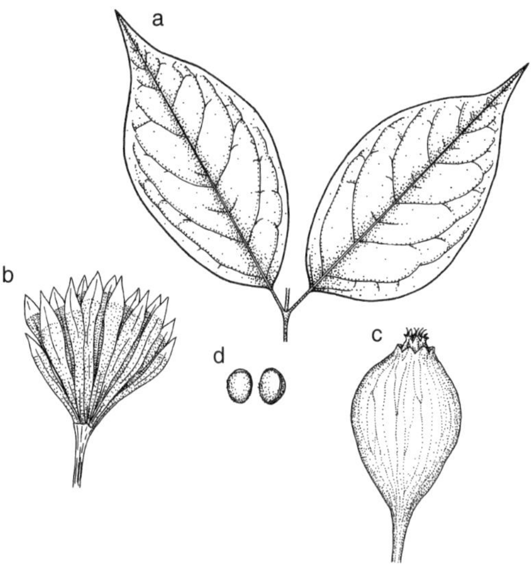

FIGURE 17.14 *Calycanthus floridus* (a) leaves × 1/2, (b) flower × 1/2, (c) hypanthium (floral cup) × 1/2, (d) seed × 1/2.

Calycanthaceae (Strawberry-Shrub Family)

The Calycanthaceae is a small family of three genera and about five species widely disjunct in China, Australia, and eastern and western North America. Members are deciduous shrubs with opposite, simple leaves. Unlike other members of the Magnoliales, they are estipulate. The flowers are perfect and regular; plants are synoecious.

Calycanthus floridus Linnaeus, sweet-shrub, Carolina allspice (Figure 17.14); shrub, height to 3 m; leaf arrangement opposite; leaf simple, entire, length 8 to 15 cm, width 2 to 6 cm, shape lanceolate to ovate-lanceolate, apex acuminate, base obtuse, aromatic when crushed; fruit a leathery floral cup containing one to 35 achenes seeds, length and width 2 to 3 cm, borne singly; Regions 3b and 4, Pennsylvania and Ohio south and east to Florida and Mississippi; near openings in upland hardwood and mixed pine. Cover type 27.5, 27.30.

Calycanthus occidentalis Hooker & Arnott; California spicebush, sweetshrub; similar to *C. floridus*; height to 4 m; leaf shape ovate-elliptical to lanceolate, length 5 to 14 cm, width 2 to 8 cm; Region 7, northern Coastal Range, southern Cascade Mountains, and western Sierra Nevada; streams and canyon slopes.

Aristolochiaceae (Birthwort Family)

The Aristolochiaceae is a family of shrubs, woody vines, and herbs. The five to eight genera and 400 to 600 species have a pantropical distribution but are not found in Australia. Members are aromatic because of oil from ethereal cells; they have a peppery-tasting yellow sap (Zomlefer 1994). There are several *Aristolochia* species in the eastern United States, but only two are woody vines. The species are well known for their flower, a yellowish green to dark purple three-lobed calyx tube that is constricted and twisted into a U shape. The shape has been compared to a smoker's pipe. Petals are lacking.

The leaves are simple and entire. They have three primary veins and a cordate to auriculate base. Leaf arrangement is alternate. The stems of the climbing, somewhat woody, perennial vines are ribbed.

Aristolochia macrophylla Lamarck (synonym *A. durior* Hill), Dutchman's pipe (Figure 17.15); high-climbing vine; deciduous; leaf blade simple, entire, 10 to 35 cm long and wide, blade width equal to length, broadly cordate to nearly round, margin smooth, base cordate, lower surface minutely hairy; petiole long, length 3 to 5 cm; flowers borne on axillary glabrous peduncles, deep purple; fruit a capsule, cylindrical, length 5 to 8 cm; Region 3b, Appalachian Mountains from Pennsylvania to Georgia; dense moist mountain forest; widely cultivated. Cover type 27.29.

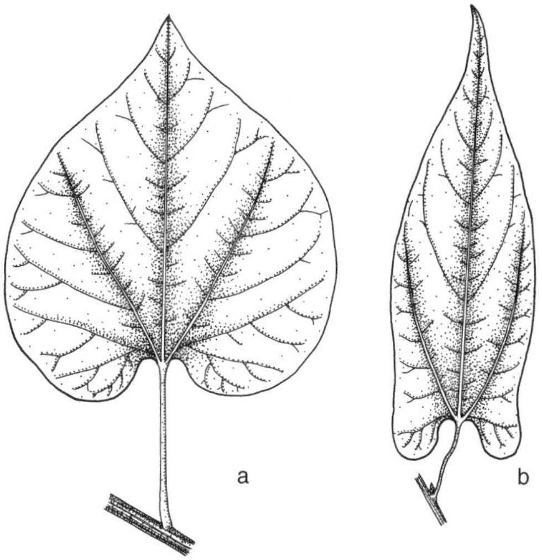

FIGURE 17.15 Leaves of (a) *Aristolochia macrophylla* ×1/5, (b) *A. serpentaria* ×.

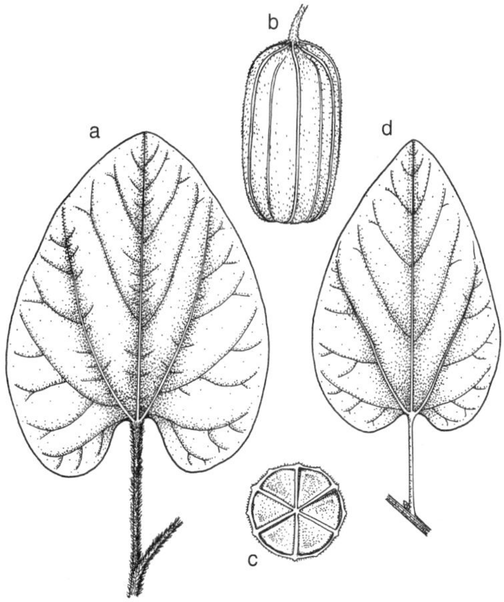

FIGURE 17.16 *Aristolochia tomentosa* (a) leaf ×1/4, (b) capsule ×1/2, (c) capsule cross section ×1/2; *A. californica* leaf ×1/2.

Aristolochia serpentaria Linnaeus, Virginia snakeroot (Figure 17.15); semiwoody, height to 60 cm; deciduous; leaf arrangement alternate; leaf blade length 6 to 12 cm, length three to four times width, lanceolate, margin smooth, apex long acuminate, base cordate to auriculate; petiole short, length 2 to 3 cm; fruit a globose capsule, length 1 to 2 cm; Regions 3 and 4, south and east from Illinois, Iowa, and Pennsylvania; forest on xeric-mesic and mesic sites.

Aristolochia tomentosa Sims, pipe-vine (Figure 17.16); climbing woody vine; similar to *A. macrophylla*; deciduous; leaf arrangement alternate; leaf blade simple, entire, length 10 to 20 cm, blade width less than length, ovoid, apex obtuse to rounded, base deeply cordate, abaxial surface tomentose; petiole tomentose, length 3 to 5 cm; flower densely tomentose, calyx long, length 1.0 to 1.5 cm, strongly curved, fruit a capsule, length 6 to 8 cm, cylindrical; Regions 3a and 4, Indiana to Kansas, Texas, and Florida; dense woods on alluvial sites.

Aristolochia californica Torrey, California snakeroot, Dutchman's pipe (Figure 17.16); climbing vine, length to 4 m; deciduous; leaf arrangement alternate; leaf blade simple, entire, length 5 to 12 cm, length two times width, ovate to somewhat cordate, margin smooth, apex obtuse, base slightly cordate; petiole short, length 2 to 3 cm; flower calyx greenish, veins purple, length 5 cm, lobed; fruit a capsule, length 4 to 6 cm; Region 7, California; in woods and chaparral, along streams.

The eastern pipe-vine swallowtail butterfly lays eggs on eastern *Aristolochia* species, and California species are used by the western swallowtail butterfly. The larvae feed on the leaves. Apparently, the butterfly uses the leaf image when looking for *Aristolochia* (Barringer 1997).

ORDER: ILLICALES

Illiciaceae (Star-Anise Family)

The Illiciaceae family includes only the genus *Illicium*, with about 35 to 40 species. Most of the species are found in the more tropical regions of southeast Asia, but some are found in Korea, Cuba, and Mexico. Members of the family are shrubs or small-stature trees and are aromatic. The flowers are small; sepal and petal components, which are not well differentiated, occur in several rows. Those in the outer row are shorter and bract- or sepal-like structures, those in the middle series are larger and petal-like, and the innermost

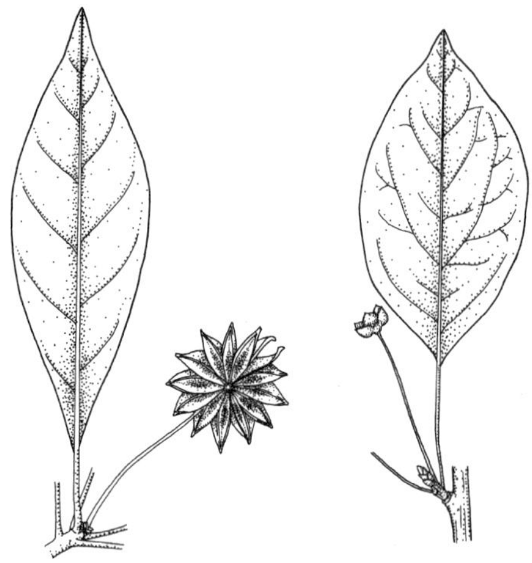

FIGURE 17.17 (a) Leaf, twig, and fruit of *Illicium floridanum*; ×1/2 (b) leaf, twig, and flower of *Schisandra glabra.*

structures are smaller. Japanese anise (*I. anisatum* Linnaeus) is the traditional source of anise oil used in perfumes, dentifrices, flavoring, industrial chemicals, and medicine (Cronquist 1981). Two *Illicium* species are restricted to southeastern United States; they are important as ornamentals.

Illicium floridanum Elliott, Florida anise (Figure 17.17); multiple-stemmed shrub, height to 3 m; evergreen; leaves persistent for several seasons, alternate but sometimes appearing whorled at end of twig; leaf simple, entire, narrowly elliptical, length 7 to 15 cm, margin smooth, apex acuminate, base acute, aromatic; flower red; fruit star-shaped, an aggregate of one-seeded follicles; Region 4a, b, Coastal Plain of Alabama, Mississippi, Louisiana, and Florida panhandle; ravines, swamps, and wet areas. Cover type 28.30, 28.49.

Illicium parviflorum Michaux, yellow anise; similar to *I. floridanum*, except that leaf apex is rounded and flowers are yellow; distribution restricted to central Florida and a disjunct population in southwest Georgia.

Schisandraceae (Star-Vine Family)

The Schisandraceae is a family of two genera and about 50 species of twining and climbing woody vines located primarily in eastern Asia. Like the Lauraceae and Illiciaceae, members of this family are aromatic. The flower is small and unisexual. Sepal and petals (5 to 24) are not well differentiated, with the outer series appearing somewhat sepal-like and the inner row somewhat petal-like. The only species found in North America, *Schisandra glabra*, is listed as threatened.

Schisandra glabra (Bunkell) Rehder, star-vine (Figure 17.17); trailing or climbing vine; deciduous; leaves widely alternate along twig or concentrated on short shoots; leaf simple, entire, length 4 to 13 cm, elliptical, margin smooth or remotely serrate, apex acute to acuminate; twig estipulate; flower small, diameter 1 cm, five petal-like structures, greenish white outside, reddish inside; fruit a berry, red, oval; Region 4, Piedmont and Coastal Plain; stream terraces, moist woods, ravines, and bluffs.

ORDER: RANUNCULALES

Ranunculaceae (Crowfoot Family)

The Ranunculaceae is a large family of about 50 to 60 genera and 2000 to 2500 species worldwide. In North America there about 22 genera and 275 species, mostly herbaceous species; many are components of woodland and forest. The genus *Clematis* in North America contains 30 to 32 species; all are herbaceous or semiwoody vines. About 28 species have a small range or are found in nonforested regions.

Clematis virginiana, the most common and widespread species, and *C. viorna* are found throughout most of eastern United States. Both are climbing semiwoody vines; the tendrils appear to be modified petioles. The leaves are deciduous, compound, and have an opposite arrangement. The flower petals are absent or extremely small.

Clematis virginiana Linnaeus, virgin's bower (Figure 17.18); trailing or climbing vine; deciduous; leaf arrangement opposite; leaf trifoliate, leaflets smooth or coarsely toothed with three to four large teeth per side, length 4 to 9 cm, palmately veined, number of veins three to five; petiole length 7 to 10 cm; flower showy, creamy white, in clusters; fruit an achene, small, diameter 2 to 4 mm, long style attached, length 3 to 5 cm, densely villous; Regions 2, 3, and 4; southern Manitoba to New England, Florida, and Texas; disturbed woodland and open places.

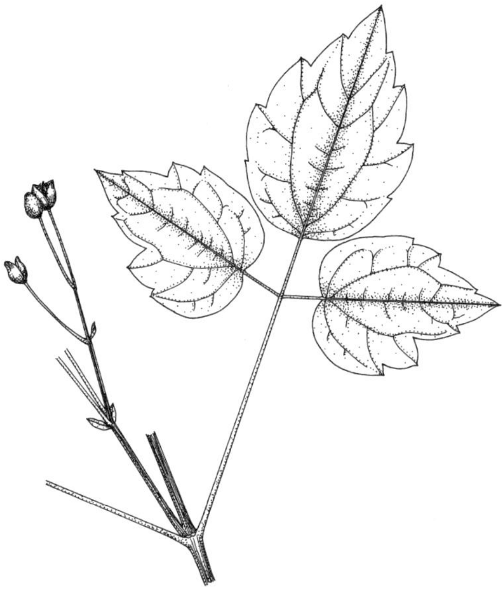

FIGURE 17.18 *Clematis virginiana* trifoliate leaf and inflorescence with achenes ×1/2.

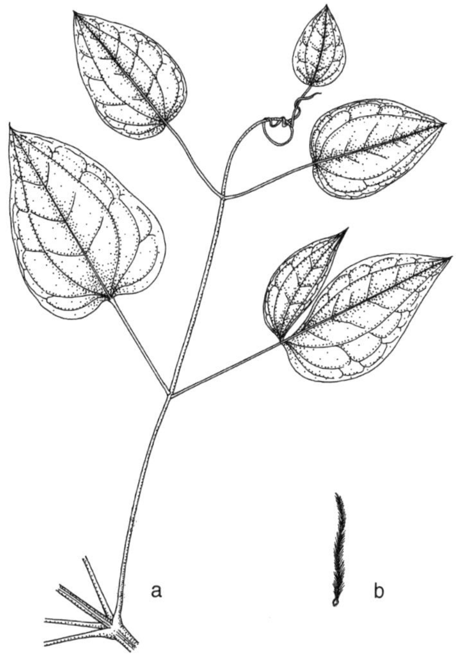

FIGURE 17.19 *Clematis viorna* (a) pinnately compound leaf, (b) plumose style of flower.

Clematis viorna Linnaeus, leather-flower (Figure 17.19); trailing or climbing vine; deciduous; leaf arrangement opposite; leaf simple and entire or compound with three to seven leaflets; leaflets palmately veined, length 4 to 12 cm, veins five to seven; achene small, ovate, diameter 2 to 4 mm, attached style villous or plumose, featherlike; Regions 3 and 4; east and south from Illinois; wooded cliffs and streams.

There are a number of other *Clematis* species. *Clematis reticulata* Walter, found in dry upland forest, and *C. crispa* Linnaeus, of swamps and bottomlands, are species generally restricted to Region 4, Coastal Plain. *Clematis occidentalis* (Hornemann) de Candolle var. *occidentalis* is found in woods and on cliffs and rock outcrops in Regions 2 and 3b, Minnesota to New England and south along the Appalachian Mountains. *Clematis occidentalis* (Hornemann) de Candolle var. *grosseserrata* (Rydberg) is the western counterpart found in the Rocky Mountains (Regions 5 and 6, Colorado to British Columbia). The most common and widespread virgin's bower of riparian deciduous woods in the western United States is *C. ligusticifolia* Nuttall (cover type 29.33); it ranges from New Mexico and California to British Columbia and Manitoba. (Pringle 1997).

Berberidaceae (Barberry Family)

The Berberidaceae family of shrubs and perennial herbs is one of few families that is pantemperate in the Northern Hemisphere; some members occur in the Southern Hemisphere. Shrubs occur in both the Northern and Southern Hemisphere, but herbaceous members occur only in the north temperate zone. The family contains about 15 to 16 genera and 650 species. The flowers are actinomorphic and have six sepals, six stamens, and six bright yellow petals; the flowers appear in dense clusters. The fruit is a blue berry about 1.0 cm in diameter; the surface is glabrous.

The major North American genera, which includes woody plants, is now included in the genus *Berberis* (Oregon-grape); previously a group of western species with pinnately compound leaves

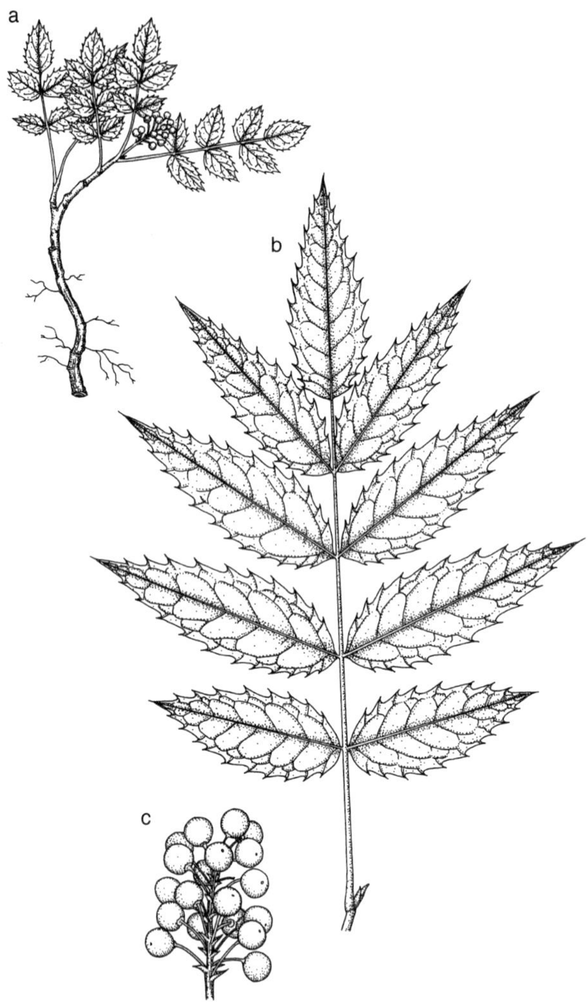

FIGURE 17.20 *Mahonia aquifolium* (a) shrub life form ×1/10, (b) pinnately compound leaf ×1/2, (c) inflorescence of berries ×1/2.

was listed as *Mahonia* (Whittemore 1997). However, because of the major difference in leaf structure, we have elected to retain the name *Mahonia* for species with pinnately compound leaves. In North America there are about 20 native species of *Berberis* and *Mahonia* and 2 to 3 introduced species of *Berberis*. *Berberis canadensis* is the common native species found in eastern United States, but it is not in Canada as its name would imply.

Mahonia aquifolium (Pursh) Nuttall (*Berberis aquifolium* Pursh), tall mahonia, mountain-grape (Figure 17.20), Oregon-grape; upright shrub, height to 1 m; evergreen; leaf arrangement alternate; leaf pinnately compound, length 15 to 25 cm; five to nine leaflets, length 5 to 9 cm, lanceolate, one-veined, margin with 5 to 21 teeth, adaxial and abaxial surfaces glossy; twigs unarmed; inflorescence a dense raceme; flowers yellow, clustered; fruit a berry, blue, oblong,

FIGURE 17.21 *Mahonia* (a) *nervosa* pinnately compound leaf × 1/2, (b) shrub life form × 1/6.

length 0.6 to 1.0 cm; Regions 6 and 7, Oregon to Montana and British Columbia; open woods and shrubland. This plant is the state flower of Oregon.

Mahonia nervosa (Pursh) Nuttall (*Berberis nervosa* Pursh), Cascades barberry, Oregon-grape (Figure 17.21); low shrub, height to 70 cm; evergreen; leaf arrangement alternate; leaf pinnately compound, length 18 to 25 cm; leaflet number 9 to 21, length 3 to 8 cm, margin with 6 to 13 spines, palmately veined, three to five veins, adaxial and abaxial surfaces dull or glaucous; twigs unarmed; inflorescence a raceme, elongated, flowers yellow; fruit a berry, blue oblong, length 7.0 to 11.0 cm; Regions 6 and 7, Central California to British Columbia; open and closed woods

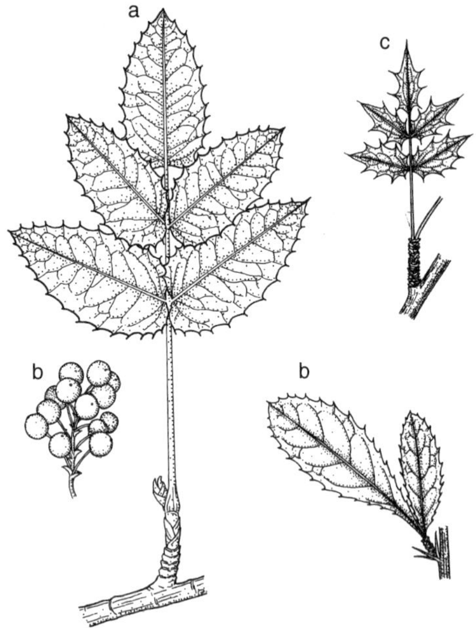

FIGURE 17.22 *Mahonia repens* (a) compound leaf ×1/3, (b) inflorescence of berries ×1/3; (c) *Mahonia fremontii* leaf on a spur shoot ×1/2; (d) *Berberis canadensis* simple leaves on a spur shoot.

and rocky sites. Cover type 30.2, 30.18, 30.21, 30.22, 30.28, 30.30.

Mahonia repens (Lindley) G. Don (*Berberis repens* Lindley), creeping mahonia (Figure 17.22), Oregon-grape; low or trailing shrub, height to 30 cm; evergreen; leaf arrangement alternate; leaf pinnately compound, length 10 to 20 cm, often on a spur shoot; leaflet number five to seven (infrequently three), length 4 to 7 cm, one-veined, 10 to 28 teeth per side, adaxial and abaxial surfaces dull, appearing similar to *Ilex opaca* leaves; twigs unarmed; inflorescence a dense raceme; flowers yellow; fruit a berry, blue, oblong, length 0.6–1.0 cm; Regions 5 and 6, Arizona and New Mexico north to Central British Columbia; montane, dry open aspen and spruce-fir forests. Cover type 29.11, 29.12, 29.15, 29.30, 29.31, 29.35.

Mahonia fremontii (Torrey) Fed (*Berberis fremontii* Torrey), Fremont barberry, desert barberry (Figure 17.22); shrub, height to 4 m; evergreen; leaf arrangement alternate; leaf pinnately compound, length 4 to 6 cm, often on a spur shoot; five to nine leaflets, length about 2 cm, elliptical, margin spiny, three to eight spines per side; twigs unarmed; inflorescence a raceme, number of flowers three to nine; fruit a berry, yellow, red, or brown, globose, diameter 1.0 to 1.8 cm; Region 5, New Mexico and Colorado to Nevada; juniper-pinyon woodland. Cover type 29.22.

Berberis canadensis Miller, American barberry (Figure 17.22); shrub, height to 2 m; deciduous; leaves clustered at nodes; leaf simple, length 3 to 6 cm, shape elliptical to oblanceolate, margin spiny-serrate; twigs brown and warty, armed with three-point spines subtending the leaves; fruit a red berry, oblong, diameter 0.5 to 1.0 cm; Regions 3b and 4a, West Virginia to Georgia, disjunct populations in Missouri, Illinois, Indiana, and Kentucky.

A number of other *Berberis* species are found in the west, primarily the southwest. They are components of shrub and grassland communities. The most widespread are *B. trifoliolata* Moricand, *B. dicyota* Jepson, *B. haematocarpa* Wooton, *B. piperiana* (Abrams) McMinn, and *B. pinnata* Lagasca. Several additional species are listed by Whittemore (1997).

A number of introduced *Berberis* and related species have escaped and become naturalized. *Berberis vulgaris* and *B. thunbergii* are probably the most common exotic species.

Berberis vulgaris Linnaeus, common barberry (Figure 17.23); shrub, height to 2 m; leaves clustered at nodes; leaf simple, entire, length 1 to 3 cm, obovate, margin bristly serrulate; twigs armed with 1 cm long simple or three-point spines subtending the leaves; fruit a berry, bright red; introduced from Europe, naturalized in both eastern and western United States.

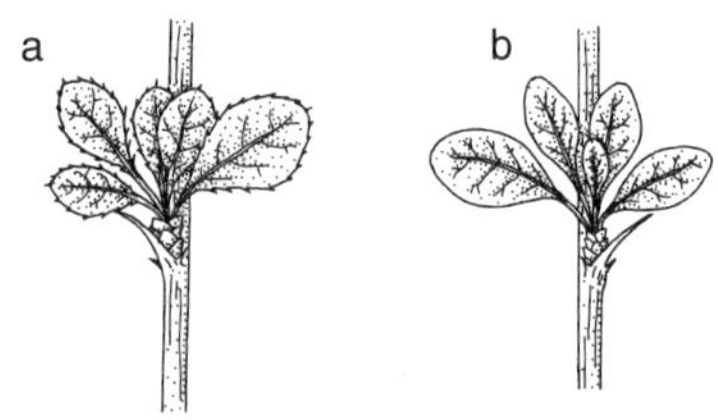

FIGURE 17.23 Simple leaves of (a) *Berberis vulgaris* ×1/2 and (b) *B. thunbergii* ×1/2. Leaves are clustered on a spur shoot.

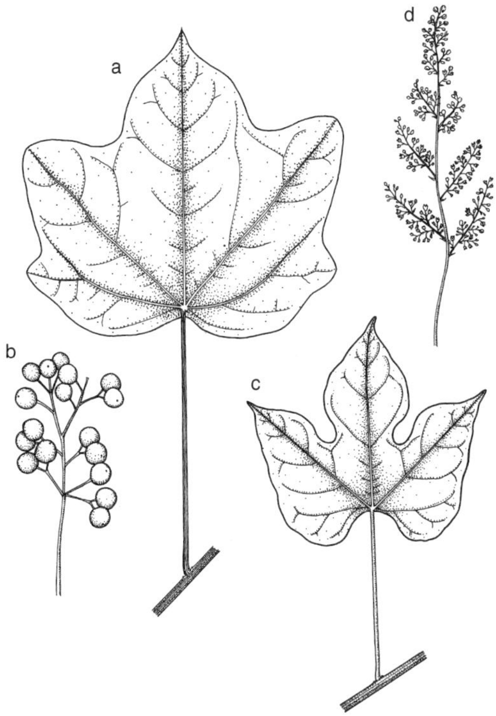

FIGURE 17.24 *Menispermum canadense* (a) leaf ×1/6, (b) inflorescence with drupes ×1/3, (c) *Calycocarpum lyonii* leaf ×1/10, (d) inflorescence with flowers ×1/10.

Berberis thunbergii DeCandolle, Japanese barberry (Figure 17.23); shrub, height to 1.5 m; leaves clustered at nodes; leaf simple, entire, length 1 to 3 cm, shape elliptical to obovate, margin smooth; twigs armed with 1 cm long simple spines subtending the leaves; fruit a berry, red, oval, length 8.0 to 1.0 cm; introduced from Japan; naturalized in eastern North America but moving west.

Other naturalized species include *Nandina domestica* Thunberg (nandina or heavenly bamboo), an evergreen shrub with pinnately compound, deltoid-shaped leaves and red berries. *Berberis bealei* Fortune [*Mahonia bealei* (Fortune) Carr.] has spiny evergreen leaves, racemes of bright yellow flowers, and dark blue berries. Both *N. domestica* and *B. bealei* are restricted to the far eastern United States. The fruits are distributed by birds, which assists in expanding their distribution.

Three species are of economic importance to American agriculture. The native *B. canadensis* and the introduced *B. vulgaris* and *B. thunbergii* are alternate hosts of wheat rust, a seriously destructive disease. There has been an attempt to systematically eliminate these species to prevent crop damage; until they are eliminated, new varieties of rust-resistant wheat must be continuously developed.

Native herbaceous plants in this family include common eastern woodland species such as *Podophyllum peltatum* Linnaeus (mayapple), *Caulophyllum thalictroides* (Linnaeus) Michaux (blue cohosh), *Diphylleia cymosa* Michaux (umbrella-leaf), and *Jeffersonia diphylla* (Linnaeus) Persoon (twinleaf). Some taxonomists include *Podophyllum* and *Jeffersonia* in the family Podophyllaceae. In the west, three *Vancouveria* species (inside-out-flower) and two *Achlys* species (vanilla leaf) are distributed from British Columbia to northern California.

Menispermaceae (Moonseed Family)

The 65 genera of the Menispermaceae family are confined primarily to Old World (Africa, Asia) tropical rain forest, but a few of the 350 species have their ranges in North America. Members are semi-woody or herbaceous vines with alternate, simple, entire, or sometimes palmately lobed leaves. The flowers are

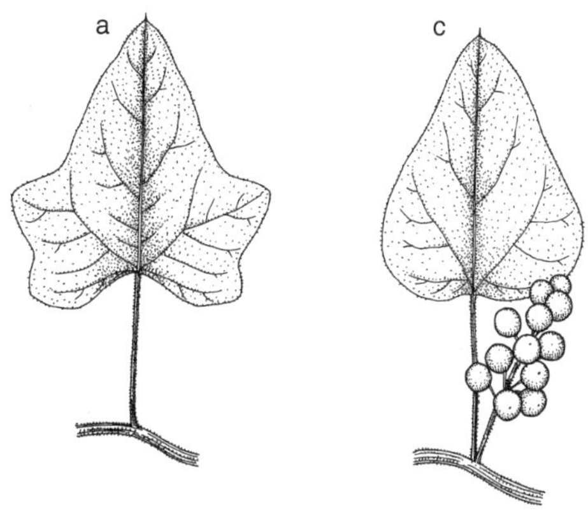

FIGURE 17.25 *Cocculus carolinus* leaf shapes and inflorescence with drupes ×1/3.

white to greenish, small, and inconspicuous. The fruit is a drupe. Three species are found extensively in the eastern United States. The only western species, *Cocculus diversifolia* de Candolle, is found in the prairie and desert shrubland of Arizona, Texas, and Mexico (Rhodes 1977).

Menispermum canadense Linnaeus; moonseed (Figure 17.24); twining vine, length to 7 m; deciduous; leaf arrangement alternate; leaf blade simple, orbicular, length 18 to 25 cm, palmately five- to seven-veined and lobed, margin smooth, apex short acuminate, petiole inserted inside blade margin (slightly peltate) length 18 to 25 cm; stem green to reddish, new growth tomentose, older growth with scattered hair, tendrils absent; inflorescence an open cyme; flower with sepals and petals, six to eight; fruit a drupe, blue black, glaucous, length 0.6 to 1.0 cm, occurring in grape-like clusters, poisonous; Regions 2, 3, and 4, southern Manitoba and Quebec south to Oklahoma and Florida and east to the Atlantic Ocean, not found directly along the Gulf Coast; deciduous forest along streams and on slopes; often grown as an ornamental but difficult to eradicate once established.

Calycocarpum lyonii (Pursh) A. Gray; cupseed (Figure 17.24); high-climbing vine; deciduous; leaf arrangement alternate; leaf blade simple, length 12 to 22 cm, palmately three- to five-veined and lobed, sinuses C-shaped, margin slightly undulate, apices acuminate, base cordate, petiole length 20 to 26 cm, inserted at edge of blade margin; stem green, slender, finely ridged, tendrils absent; inflorescence a cyme, flower small, diameter 5 mm, six sepals, petals absent; fruit a dark green to black nearly drupe, oval, length 1.0 to 2.5 cm, endocarp indented like a cup; Regions 3a and 4, Kentucky and Missouri to Oklahoma, Louisiana, and Florida; deciduous forest communities adjacent to rivers and streams.

Cocculus carolinus (Linnaeus) de Candolle; coralbeads, red-berried moonseed (Figure 17.25); twining vine, length to 7 m; deciduous; leaf arrangement alternate; leaf simple, length 15 to 22 cm, palmately three- to five-veined, entire or lobed, ovate to deltoid and panduriform, margin smooth, apex obtuse to rounded, base truncate or slightly cordate, petiole inserted at edge of blade margin; stem color green, slender, finely ridged, new growth tomentose, older growth with scattered hair, tendrils absent; inflorescence a cyme; flowers small, diameter 3 to 4 mm, six sepals, six petals, greenish; fruit a drupe, red, nearly globose, diameter 5 to 8 mm; Regions 3 and 4, Missouri and Texas to Florida and North Carolina; forest and stream edges and disturbed sites; fencerows.

CHAPTER 18

Class: Magnoliopsida; Subclass: Hamamelidae I

ORDER: HAMAMELIDALES

Platanaceae (Plane-Tree or Sycamore Family)

THE only genus in the small (monotypic) Platanaceae family of six to eight species is *Platanus*, but all species are large trees distributed in North America and the Himalaya Mountains of southern Asia. *Platanus occidentalis* (American sycamore), with its distribution throughout the eastern United States and the southern tip of Ontario, is the major species of North America. *Platanus racemosa* (California sycamore) is restricted to California and Baja California, and *P. wrightii* (Arizona sycamore) is found in Arizona, southwestern New Mexico, and northern Mexico. *Platanus acerifolia* (London plane tree) appears to be a cross between *P. occidentalis* and *P. orientalis* (Oriental plane tree). This cross closely resembles *P. occidentalis* in bark and leaf characteristics, and is widely planted in this country as a sidewalk and lawn tree.

All species are deciduous. The flowers are small, regular, unisexual, and wind pollinated. Pistilate flowers are packed into globose (diameter = 2 to 4 cm) heads (multiple fruit) of hairy achenes. The heads of *P. occidentalis* are borne singly on the peduncle, in twos in *P. acerifolia* and in threes or more in *P. orientalis.* At times the number of heads is helpful in separating these species. The plants are monoecious. In *P. occidentalis*, the exposed bark layer is generally cream colored to nearly white, whereas in *P. acerifolia* it is somewhat greenish. North American species generally are fast growing. The seedlings require relatively high light conditions to survive. The wood is hard and commercially important. Early presettlement pictures show that *P. occidentalis* can grow extremely large (diameter = 4 to 5 m).

Native Species

Platanus occidentalis Linnaeus, American sycamore, plane-tree (Figure 18.1); large tree, height to 45 m; deciduous; leaf simple, length 20 to 35 cm, blade width slightly greater than length, palmately lobed, lobes usually five, sometimes three, prominently three-veined, apices acuminate, sinuses shallow, margin coarsely toothed, lower surface tomentose, base strongly cordate, petiole base hollow and capping the bud; twigs zigzag, leaf scar present; bud covered by a single visible scale, bud smooth, leaf scar nearly surrounding bud; stipules leafy, large (length = 1 to 2 cm), toothed, stipule scars encircling twig; fruit a globose head of achenes, heads usually one per stalk, diameter 2 to 3 cm, each achene subtended by numerous brown hairs; outer bark brown, exfoliating to expose the nearly white to light green bark beneath; Regions 3 and 4, Kansas to New York and south to Texas and Florida; stream banks, stream terraces, or disturbed moist uplands where there is little competition. Two varieties are recognized. *Platanus occidentalis* var. *glabrata* (Fernald) Sargent has lobes with smooth margins and occurs from Iowa to Mexico. *Platanus occidentalis* var. *attenuata* has deep sinuses and a long cuneate base; it occurs over the general range. Cover type **27.6, 27.14, 27.15**, 27.3, 27.10, 27.28, 28.7, 28.12, 28.15, 28.21, 28.42.

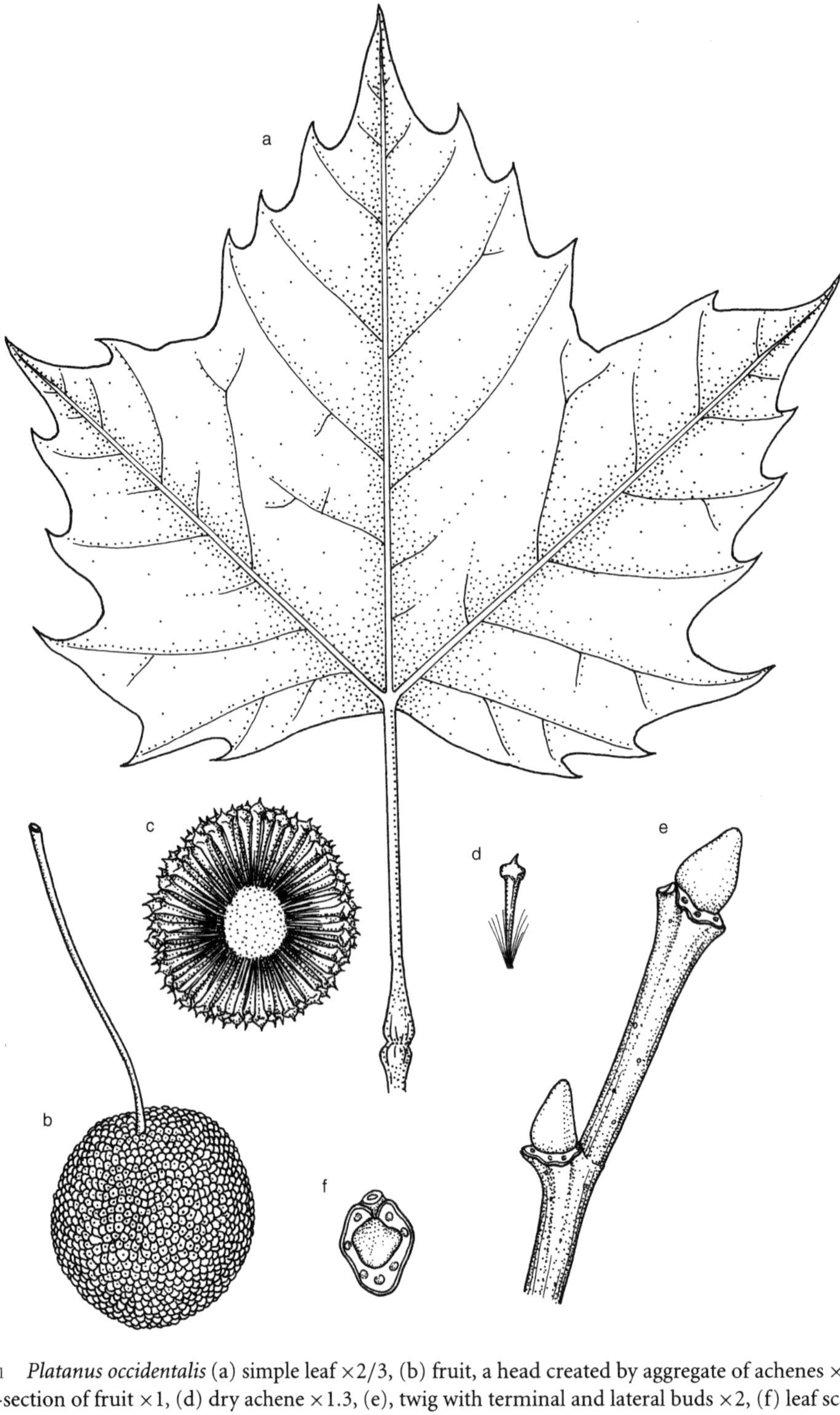

FIGURE 18.1 *Platanus occidentalis* (a) simple leaf ×2/3, (b) fruit, a head created by aggregate of achenes ×1, (c) cross-section of fruit ×1, (d) dry achene ×1.3, (e), twig with terminal and lateral buds ×2, (f) leaf scar with seven bundle scars, pseudoterminal bud, and solitary twig scar ×2.

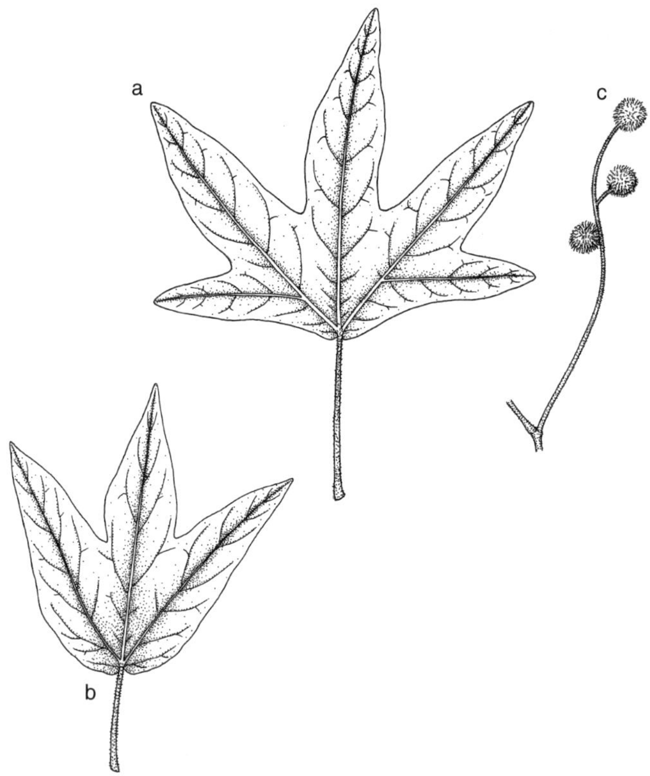

FIGURE 18.2 *Platanus racemosa* (a) five-lobed simple leaf × 1/3, (b) three-lobed leaf × 1/3, (c) a stalk of fruits, each head created by an aggregate of achenes × 1/4.

Platanus racemosa Nuttall, California sycamore (Figure 18.2); large tree, height to 30 m; deciduous leaf simple, length 15 to 25 cm, width slightly greater than length, palmately lobed, lobes usually five, sometimes three, prominently three- to five-veined, apices acuminate to slightly rounded, margins smooth to crenulate, without large teeth, sinuses extend about one half to one third the distance from the apices to the midrib, base essentially truncate, abaxial surface tomentose, petiole base hollow and capping the bud; twigs zigzag, terminal bud absent, twig scar present; bud covered by a single visible scale, scale smooth, leaf scar nearly surrounding bud; stipules, if present, leafy, large (length = 1 to 2 cm), toothed, stipule scars encircling twig; fruit a globose head of achenes, diameter 2 cm, heads two to six per stalk, each achene subtended by numerous brown hairs, outer bark brown, exfoliating to expose the cream to light green bark beneath; Region 7, California, stream bottoms and riverbanks.

Platanus wrightii S. Watson, Arizona sycamore (Figure 18.3); Region 5, Arizona, New Mexico, and Mexico. This species is similar to *P. racemosa*, but the sides of lobes are nearly straight and uniformly converging, and apices are long acuminate. It is disjunct from *P. racemosa*. However, when the two are grown together, hybrids occur (Benson 1943). Some taxonomists consider *Platanus wrightii* as *Platanus racemosa* var. *wrightii* (S. Watson) Benson. Cover type **29.19.**

Introduced Species

Platanus x *acerifolia* (Aiton) Willdenow, London planetree (Figure 18.4); large tree, height to 25 m; deciduous; leaf simple, length 15 to 25 cm, width slightly

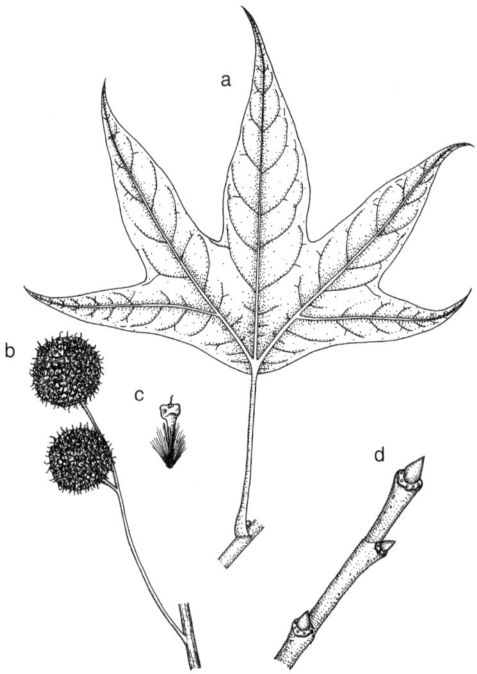

FIGURE 18.3 *Platanus wrightii* (a) five-lobed simple leaf ×1/3, (b) a stalk of two fruit, each head created by an aggregate of achenes ×1/2, (c) dry achene ×1, (d), twig with terminal and lateral buds ×1.

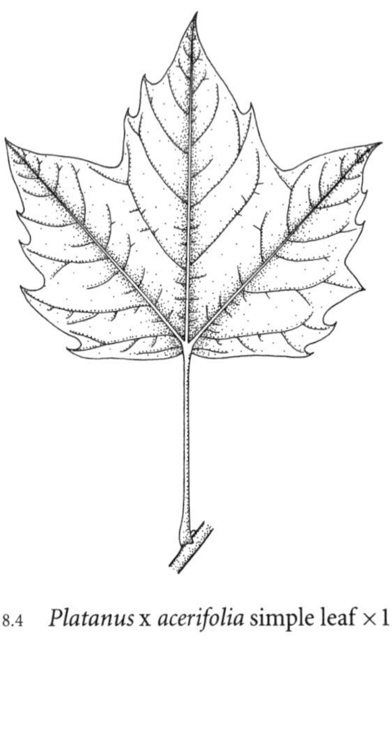

FIGURE 18.4 *Platanus* x *acerifolia* simple leaf ×1/3.

greater than length, palmately lobed, lobes usually five, sometimes three, apices long acuminate, prominently three-veined, sinuses shallow, margin coarsely toothed, lower surface tomentose, petiole base hollow and capping the bud; twigs zigzag, leaf scar present, bud covered by a single visible scale, scale smooth, leaf scar nearly surrounding bud; stipules leafy, large (length = 1 to 2 cm), toothed, stipule scars ring twig; fruit a globose head of achenes, diameter 2 to 3 cm, each achene subtended by numerous brown hairs, outer bark brown, exfoliating to expose the cream to medium green bark beneath. Introduced and widely planted; not naturalized. This species is a cross between *P. orientalis* Linnaeus and *P. occidentalis* but strongly resembles *Platanus orientalis*, which also is introduced (Dirr 1998).

Hamamelidaceae (Witch-Hazel Family)

The 25 to 26 genera and about 100 tree and shrub species of the Hamamelidaceae family are distributed worldwide and generally restricted to warm-temperate and subtropical regions (Asia, Africa, Australia). Only a total of five species in the genera *Hamamelis, Liquidambar,* and *Fothergilla* extend into the eastern United States.

All species of the Hamamelidaceae are planted as shade trees or border shrubs. The leaves are simple and alternate. The flowers are perfect, often polygamous and sometimes monoecious. Petals are absent in *Fothergilla* and *Liquidambar.* The slender (2 mm wide; 2 cm long) yellow petals of *Hamamelis virginiana* flowers appear in the late autumn, often after the leaves have fallen, and the deep red to orange flowers of *Hamamelis vernalis* appear from January to March. The capsules mature in 1 year at the same time that new flowers appear. The fruit of *Fothergilla* and *Hamamelis* is a hard, fibrous two-celled capsule 1 to 2 cm long. When the capsule is mature, it splits open suddenly and propels the seeds a distance of up to several m. The fruit of *Liquidambar styraciflua* is a head composed of capsules.

Hamamelis virginiana Linnaeus, witch-hazel (Figure 18.5); small tree, often multistemmed, height to about 6 m; leaf arrangement alternate; deciduous; leaf simple, length 4 to 16 cm, width 3 to 12 cm, shape broadly elliptical, margin repand to sinuate, base strongly oblique; apex; buds naked, terminal and

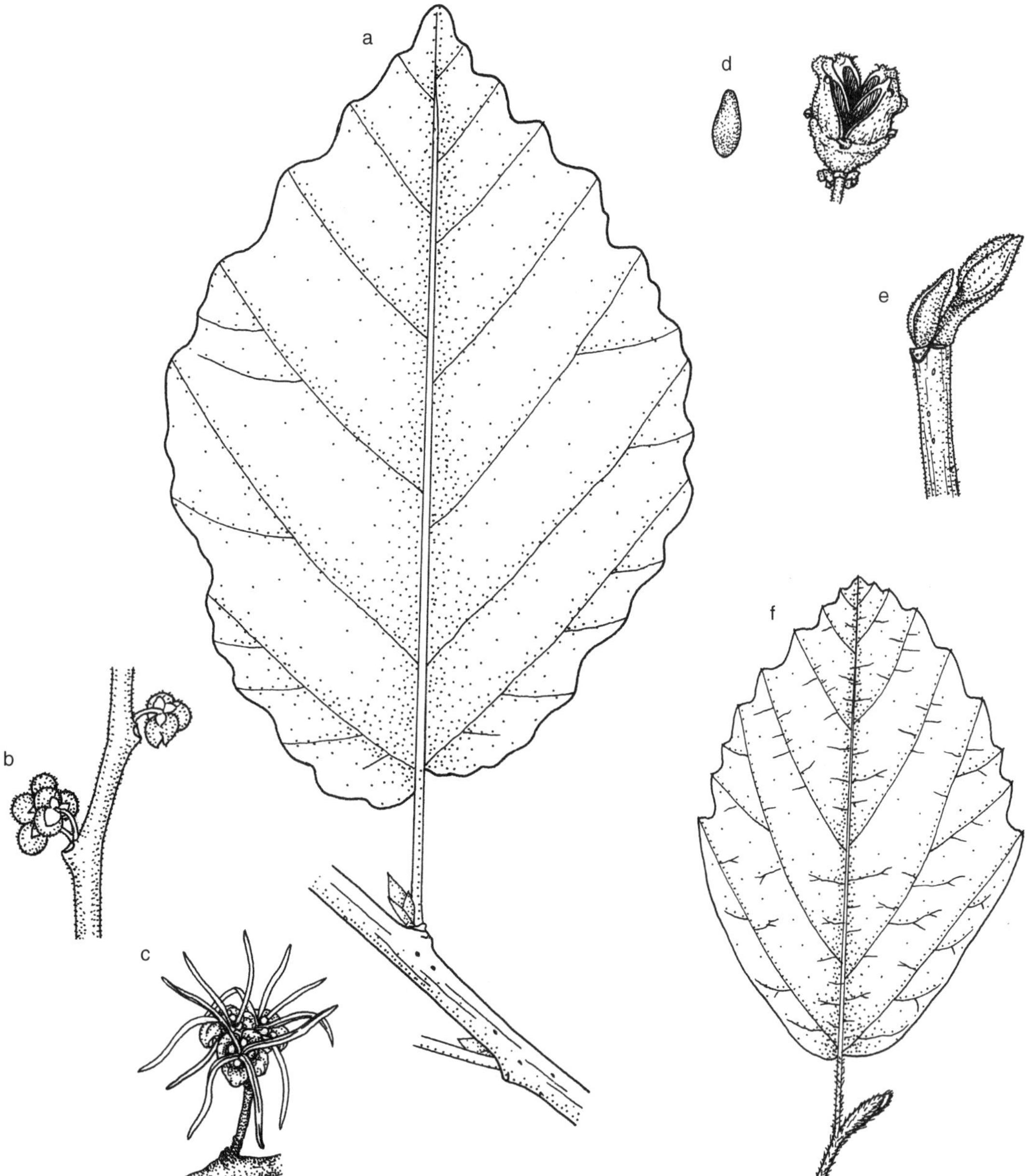

FIGURE 18.5 *Hamamelis virginiana* (a) simple leaf ×1, (b) closed flowers ×2, (c) open flowers ×1, (d) capsule and seed ×1.5, (e) stalked terminal bud ×1.5; (f) *H. vernalis* simple leaf ×3/4.

obtuse lateral bud at each node naked, length 1.5 cm, stalked; flowering in late fall, petals slender, yellow; fruit a capsule, solitary or paired, length 1.0 to 1.5 cm, explosive when dry, seeds propelled; Regions 2, 3, and 4, throughout the eastern United States and southern Ontario and Quebec; dry and moist upland forest. *Hamamelis virginiana* is the source of witch-hazel astringent and lotion; widely used by Native Americans. Witch-hazel lotion is still available in stores. The pioneers used witch-hazel branches for "water witching." Cover type 26.21, 26.23, 26.24, 26.28, 27.22, 27.23, 27.26, 27.34, 27.35, 28.6, 28.11, 28.12, 28.16, 28.27.

Hamamelis vernalis Sargent, springtime witch-hazel (Figure 18.5); multistemmed shrub, height to 4 m; leaf arrangement alternate; deciduous; leaf simple, length 8 to 13 cm, width 7 to 12 cm, shape obovate or rhomboid, margin near apex repand to sinuate, margin near base generally smooth, base somewhat rounded to cuneate; apex obtuse; buds naked, terminal bud and lateral bud at each node naked, stalked; flowering in winter and early spring, petals deep red to orange; fruit a capsule, solitary or paired, length 1.0 to 1.5 cm, explosive when dry, seeds propelled; Region 3, Ozark Hills of southern Missouri, Arkansas, and Oklahoma; often grows in close proximity to *H. virginiana*, but species distinctions are maintained (Meyer 1997).

Liquidambar styraciflua Linnaeus, sweetgum (Figure 18.6); large tree, height to 40 m; leaf arrangement opposite; deciduous; leaf simple, length 12 to 20 cm, width 10 to 14 cm, palmately five-veined and five- or seven-lobed, lobes uniformly tapered to a long acuminate apex, star-shaped, margin serrate, teeth rounded; twigs usually with corky ridges; bud length about 1.0 cm, surface shiny, scale margin dark reddish brown, central area greenish; fruit a head (multiple) of capsules, each capsule strongly beaked; diameter 3 to 5 cm; Region 4, Coastal Plain, and Region 3, southern portion of the Central States Region south of a line from southeast Missouri and southern Illinois to southern Ohio and West Virginia; restricted to mesic-hydric bottomlands and disturbed uplands, occasionally planted on or invading old fields. *Liquidambar styraciflua* is a major timber-producing species. Trees of *L. styraciflua* growing near a stream are often damaged or completely girdled by beaver, which like the taste of the sweet, fragrant resin (hence the name sweetgum). *Liquidambar styraciflua* is also widely planted as a lawn tree, and although it creates pleasant shade and has an outstanding deep red autumn color, it is not a good lawn or street tree (near sidewalks). It produces copious numbers of 3 to 5 cm diameter sweetgum balls (fruit) that make walking difficult or dangerous and thus must be removed every spring at the time of leaf expansion. Cover type **27.15, 28.12, 28.42**, 27.2, 27.3, 27.6, 27.8, 27.10, 27.21, 28.7, 28.11, 28.13, 28.14, 28.15, 28.16, 28.27, 28.29, 29.32, 28.33, 28.36, 28.37, 28.38, 28.40, 28.43, 28.49, 28.50, 28.55.

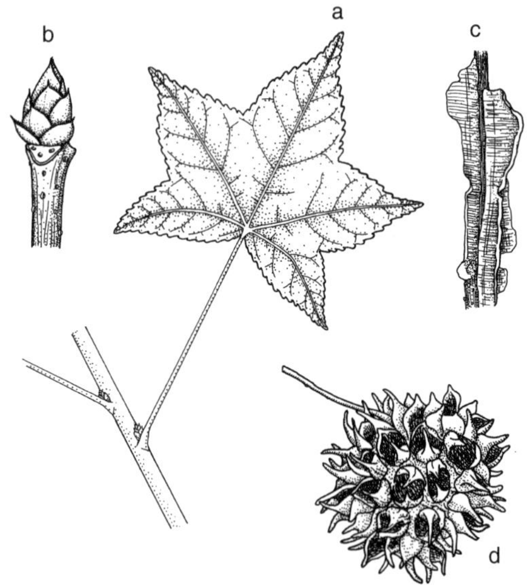

FIGURE 18.6 *Liquidambar styraciflua* (a) simple leaf ×1/3, (b) twig with terminal bud and leaf scar ×1.5, (c) branch with corky wings ×1/2, (d) multiple of capsules ×3/4.

Fothergilla major (Sims) Loddiges, witch-alder (Figure 18.7); multistemmed shrub, height to 2 m; branches stout; leaf arrangement alternate; deciduous; leaf simple, length 3 to 13 cm, width 4 to 12 cm, shape oblong to orbicular; margin wavy to undulate to smooth, blade surfaces stellate pubescent, base cordate, apex obtuse to rounded, venation pinnate, veins alternate; spring flowering, flower apetalous, stamens showy white; fruit a capsule, length 0.6 to 1.3 cm, pubescent; Region 3b, North and South Carolina, Tennessee, Georgia, Alabama, and, recently, Arkansas; xeric, rocky woods, bluffs, riverbanks, and balds. This species is considered rare and endangered.

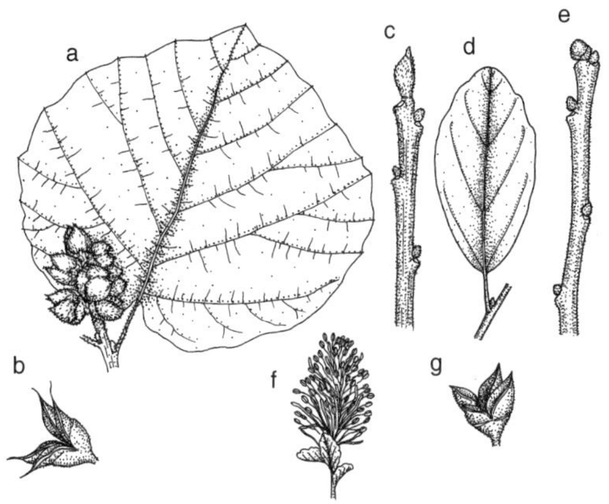

FIGURE 18.7 (a) *Fothergilla major* simple leaf and developing capsules ×1/2 (b) open capsule ×3/4, (c) twig with terminal and lateral buds ×1; *F. gardenii* (d) simple leaf ×1/2, (e) twig with terminal and lateral buds ×1, (f) inflorescence ×1/2, (g) open capsule ×3/4.

Fothergilla gardenii Linnaeus, dwarf witch-alder (Figure 18.7); multistemmed shrub, height to 1.5 cm; branches slender; leaf arrangement alternate; deciduous; leaf simple, length 2 to 6 cm, width 1.5 to 4.5 cm, shape elliptical to obovate; margin generally smooth but distally undulate to denticulate, blade surfaces stellate pubescent, base obtuse, apex acute to rounded; veins paired, pairs four to five; spring flowering, flower apetalous, stamens showy white; fruit a capsule, length 0.6 to 1.2 cm; Region 4, North Carolina to Alabama; found in xeric pine woods or savanna, pocosins (upland swamps), and bogs.

ORDER: URTICALES

Ulmaceae (Elm Family)

The Ulmaceae is a family composed of about 18 genera and 150 species of trees and shrubs, distributed mostly in the temperate, tropical, and subtropical regions of the Northern Hemisphere. Four genera (*Ulmus*, elm; *Celtis*, hackberry, *Planera*, water elm, and *Trema*, nettle tree) are found in North America. Based on leaf and fruit characteristics, the family has at times been broken into two subfamilies. The Ulmoideae includes the genus *Ulmus*, which has doubly serrated, pinnately veined leaves, and samaras for fruit. The Celtidoideae contains *Celtis, Planera*, and *Trema*, which have singly serrated, three-veined leaves and drupes (Zomlefer 1994).

Plants of *Ulmus* have perfect flowers and are monoecious, whereas the other genera are polygamous or polygamodioecious. The flowers are small, radially symmetrical, solitary, imperfect (no petals), and occur in axillary clusters. The calyx is short, usually a tube with four to eight lobes or separate sepals. The purplish or yellowish flowers of *Ulmus* generally appear in April or May, except for *Ulmus crassifolia* and *Ulmus serotina*, whose flowers appear in September or October as do the greenish *Celtis* flowers. The flowers are anemophilous. The one-seeded fruit of *Ulmus* is a 1 to 2 cm diameter samara with a lateral wing surrounding the seed cavity. *Celtis* produces a 1.0 cm diameter orange-red to purple drupe. The fruit of *Planera* is an ellipsoid nutlet.

The leaves alternate along the twig. The leaf is simple, usually less than 15 cm long, elliptical, obovate or lanceolate in shape, and has an asymmetrical base. Apex shape is acuminate on some species. The *Ulmus* leaf margin is doubly serrate, and the leaf margins of other genera are singly serrate. The leaves of *Celtis* are three-veined at the base. Buds are slender and divergent from the twig in *Ulmus*, appressed in *Celtis*. There are no stipules. *Ulmus alata, U. crassifolia, U. thomasii*, and *U. serotina* have corky wings on the smaller branches. The trunks of *Celtis* trees have corky warts and ridges. The bark of *Ulmus rubra* is uniformly reddish brown, and that of the other *Ulmus* species is layered reddish brown alternating with light tan.

All *Ulmus* species grow to a large tree size and were important for timber or as shade trees until the introduction of Dutch elm disease, which attacks all North American elm species. However, *Ulmus americana* and *U. rubra* appear most susceptible, and *U. alata* least susceptible. The disease has decimated most urban shade tree and natural populations as far west as the Rocky Mountains. Within existing forest communities, individuals of *Ulmus* are found mostly as seedlings and saplings, with occasional scattered trees. The wood of *Ulmus* has an interlocking grain that makes it exceedingly difficult to split but makes excellent bentwood furniture.

The family belongs to the order Urticales, which also contains the Maraca (fig), Cannabinaceae (hemp) and Urticaceae (nettle) families. The latter, which contains mostly herbaceous plants, includes *Urtica dioica* Linnaeus (stinging nettle), a lowland plant well known to North American field biologists.

Six of approximately 20 species of *Ulmus* are native to North America. All species occur naturally east of the Rocky Mountains. *Ulmus alata* and *U. crassifolia* grow on dry, often rocky ridge tops and south slopes. The other four species are found on cool, moist north slopes, stream terraces, and/or bottomlands. Because of Dutch elm disease, *Ulmus* is now found as a scattered tree in forest communities.

Native Ulmus Species

Ulmus americana Linnaeus, American or white elm (Figure 18.8); large tree, height to 35 m; deciduous; leaf arrangement alternate; leaf simple, blade shape ovate to elliptical, length 7 to 12 cm, width 4 to 7 cm, doubly serrate with curved teeth, base inequilateral, apex acuminate, upper surface only glabrous, pubescent, or scabrous (sandpapery), lower surface pubescent; petiole length 5 mm, glabrous or pubescent; twigs brown, surfaces glabrous to slightly pubescent, buds chocolate brown, length 4 to 6 mm, divergent, elliptical; fruit a samara with a lateral wing, oval, diameter about 1 cm, margin ciliated, a shallow notch distally; bark of larger trees with alternating reddish and tan layers; Regions 2, 3, and 4, eastern United State and Canada; scattered on north slopes and stream drainages, once a major component of bottomland forest. Cover type **27.10**, 26.17, 26.23, 26.25, 27.1, 27.2, 27.4, 27.8, 27.9, 27.14, 27.15, 27.19, 27.20, 27.23, 27.28, 28.7, 28.9, 28.10, 28.12, 28.14, 28.15, 28.21, 28.17, 28.32, 28.42, 28.55.

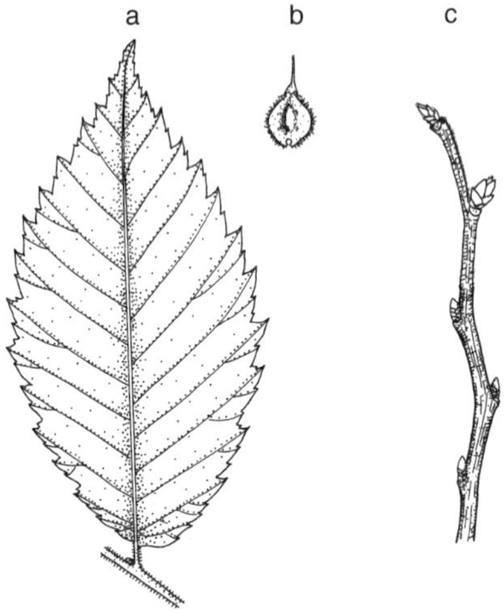

FIGURE 18.8 *Ulmus americana* (a) simple leaf ×1/2, (b) samara ×1/2, (c) twig with pseudoterminal and lateral buds ×1/2. The large lateral bud is a flower bud.

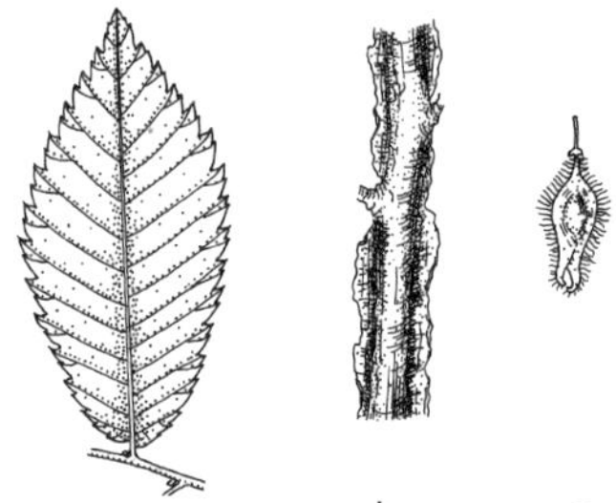

FIGURE 18.9 *Ulmus alata* (a) simple leaf ×1/2, (b) branch with corky wings ×1, (c) elongated samara ×3/4.

Ulmus alata Michaux, winged elm (Figure 18.9); large tree, height to 35 m; leaf arrangement alternate; leaf simple, blade shape ovate to elliptical, length 3 to 7 cm, width 0.7 to 3.0 cm, surfaces with scattered hair, doubly serrate with curved teeth, base inequilateral, rounded; petiole length 2.0 to 5 mm, pubescent twigs often with corky wings; buds divergent, length 3–5 mm, narrow elliptical, brownish; fruit a samara, with a lateral wing, elongated, length 1 to 2 cm, width 3 to 6 mm, margin ciliated, deeply notched distally; bark of larger trees with alternating reddish and tan layers; Regions 3 and 4, Missouri to Virginia, south to Texas and Florida and 4; most common on rock outcrops, xeric slopes, and old fields, scattered on north slopes and stream drainages. Cover type 27.17, 27.24, 28.4, 28.7, 28.27.

Ulmus rubra Muhlenberg, slippery or red elm (Figure 18.10); large tree, height to 35 m; deciduous; leaf arrangement alternate; leaf simple, blade shape ovate to obovate, length 7 to 14 cm, width 4 to 7 cm, doubly serrate with curved teeth, base inequilateral, apex long acuminate, upper and lower surfaces scabrous (sandpapery); petiole length 5 mm, pubescent; twigs gray, surfaces pubescent; buds chocolate brown, length 4 to 7 mm, divergent, elliptical; fruit a flat samara with a lateral wing, pubescent, oval to

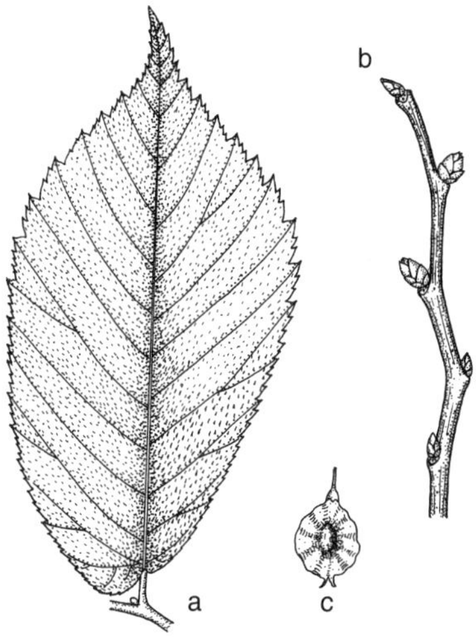

FIGURE 18.10 *Ulmus rubra* (a) simple leaf ×2/3, (b) twig with pseudoterminal and lateral buds ×3/4, (c) samara ×1/2.

FIGURE 18.11 *Ulmus serotina* (a), (b) simple leaves ×1/3, (c) branch with corky wings ×1, (d) samara ×1/3; *Ulmus crassifolia* (e) simple leaf ×1/2, (f) samara ×1/2.

round, diameter about 1.3 to 1.8 cm, margin aciliate, distal notch lacking; bark of larger trees uniform reddish layers; Regions 2, 3, and 4, eastern United States except Florida peninsula; north slopes, stream drainages, and terraces. The name "slippery" is related to the mucilaginous inner bark that was widely used by Native Americans for a variety of ailments. Cover type 26.24, 27.1, 27.4, 27.9, 28.33.

Ulmus crassifolia Nuttall, cedar elm (Figure 18.11); large tree, height to 27 m; leaf arrangement alternate; deciduous; leaf simple, blade shape ovate to elliptical, length 2 to 5 cm, width 1.2 to 2.0 cm, doubly serrate with curved teeth, base round or inequilateral, apex rounded, adaxial surface nearly smooth to scabrous (sandpapery), abaxial surface densely pubescent; petiole length 3 to 5 mm, pubescent; twigs reddish brown, pubescent, older branches often with opposite corky wings, terminal bud absent, twig scar prominent; buds reddish to chocolate brown, divergent, elliptical, pubescent, length 2 mm; fruit a flat samara, oval, lateral wing surrounding seed cavity, diameter about 0.5 to 1.0 cm, margin ciliated, distal notch lacking; Regions 3 and 4, Arkansas, Mississippi, Louisiana, Oklahoma, and Texas; slopes, stream drainages, terraces, glades, roadsides. Cover type 29.9, 28.9.

Ulmus serotina Sargent, September elm (Figure 18.11); medium-sized tree, height to 20 m; leaf arrangement alternate; deciduous; leaf simple, blade shape ovate to elliptical, length 7 to 11 cm, width 3 to 4 cm, doubly serrate with curved teeth, base inequilateral, apex acuminate, adaxial surface glabrous, abaxial surface hairy along the veins; petiole length 5 mm, glabrous or pubescent; twigs reddish brown to gray, older branches often with corky wings, buds reddish brown, divergent, elliptical, glabrous; fruit a samara with a lateral wing, oval, length 1 to 2 cm, margin densely ciliated, distally notched; bark of larger trees with alternating reddish and tan layers; Region 3a, Kentucky, Tennessee, Georgia, Alabama, Mississippi, and Arkansas; mesic woods and drainages.

Ulmus thomasii Sargent, rock elm (Figure 18.12); large tree, height to 30 m; leaf arrangement alternate; deciduous; leaf simple, blade shape ovate to elliptical, length 7 to 14 cm, width 3 to 7 cm, doubly serrate with curved teeth, base inequilateral, apex acuminate, upper surface glabrous, lower surface pubescent; petiole length 5 mm, pubescent; twigs reddish, surfaces glabrous to slightly pubescent, older branches often with corky wings, terminal bud absent, twig prominent; buds reddish brown, divergent, elliptical, pubescent; fruit a samara, lateral wing surrounding seed cavity, oval, length 2.0 to 2.5 cm, width 1.5 to

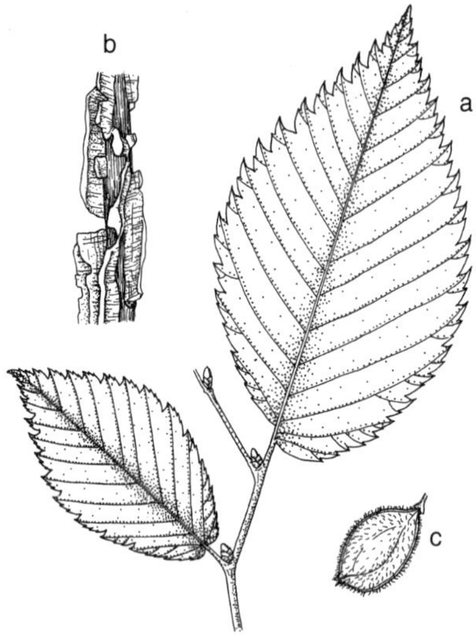

FIGURE 18.12 *Ulmus thomasii* (a) twig with simple leaves and lateral buds ×1/2, (b) branch with corky wings ×1, (c) samara ×1/2.

2.0 cm, margin densely ciliated, apex bearing a blunt mucro; bark of larger trees with alternating reddish and tan layers; Regions 2 and 3, more heavily concentrated in Minnesota, Wisconsin, Michigan, Indiana, Ohio, and New York; mesic woods, floodplains, drainages. Cover type 26.23.

Introduced Ulmus Species

Four species have been introduced as ornamentals from Europe and Asia.

Ulmus procera Salisbury (*Ulmus compestris* Linnaeus), English elm (Figure 18.13); height to 35 m; leaf arrangement alternate; deciduous; leaf simple, blade length 4 to 10 cm, width 4 to 7 cm, margin doubly serrate, base strongly inequilateral, not covering the petiole, apex obtuse to acuminate, adaxial surface glabrous to scabrous (sandpapery), abaxial surface villous; petiole length 3 to 10 mm; twigs reddish brown, densely pubescent; corky wings on branches; fruit a flat samara, length 0.9–1.5 cm wing lateral surrounding the seed cavity, glabrous, shape elliptical, ovate, or nearly round, margin ciliated near apex, apex rounded, notch shallow; native of Europe. This species is widely planted and has naturalized in a number of eastern and western states. Hillier Nurseries (1998) consider

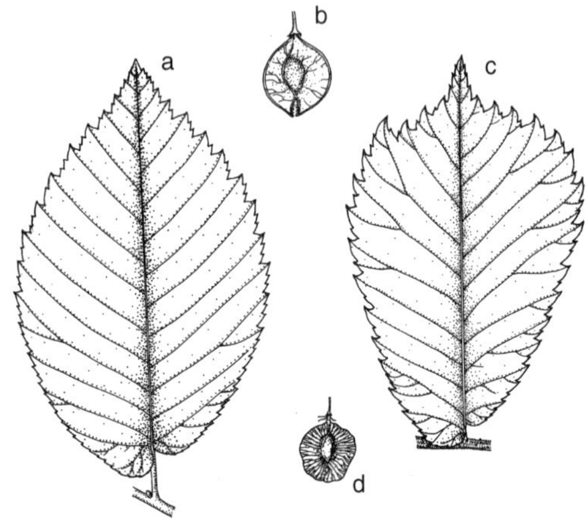

FIGURE 18.13 *Ulmus procera* (a) simple leaf ×1/2, (b) samara ×3/4; *U. glabra* (c) simple leaf ×1/2, (d) samara ×1/3.

U. procera as part of *U. minor* Miller, specifically var. *vulgaris* (Aiton) Richens.

Ulmus glabra Hudson, Scotch or wych elm, a "weeping" elm (Figure 18.13); large tree, height to 35 m; leaf arrangement alternate; deciduous; leaf simple, blade length 5 to 15 cm, width 4 to 7 cm, margin doubly serrate, base strongly inequilateral, largest basal lobe covering the petiole, primary apex long acuminate, frequently accompanied by one or two secondary apices located 2 to 3 cm from the primary, adaxial surface scabrous (sandpapery), abaxial surface villous; petiole length 3 to 6 mm, pubescent; twigs ashy gray or reddish brown, branches often pendent, corking wings absent; fruit a flat samara, length 1.5 to 2.5 cm, lateral wing surrounding the seed cavity, glabrous, shape elliptical, ovate, or round, margin aciliate, apex lacking a notch or small and obscured; native of Europe and Asia. This species is widely planted and has naturalized in a number of eastern and western states. The leaf is similar to that of *U. rubra*; however, both surfaces of the *U. rubra* leaf are scabrous. According to Sherman-Broyles (1997), the smooth bark and glabrous samaras of *U. glabra* also separate the two species. The common name *wych* is Gaelic and means "drooping," widely interpreted to mean "weeping."

Ulmus pumila Linnaeus, Siberian elm (Figure 18.14); large tree, height to 35 m; deciduous; leaf

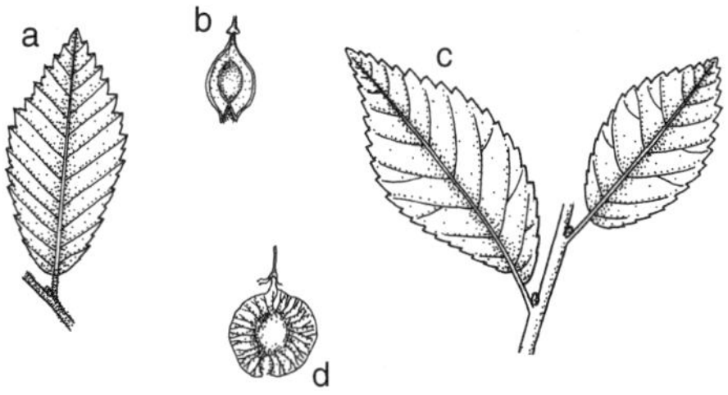

FIGURE 18.14 *Ulmus parvifolia* (a) simple leaf ×1/2, (b) samara ×1/2; *U. pumila* (c) simple leaf ×1/2, (d) samara ×1/2.

arrangement alternate; leaf simple, blade length 1 to 7 cm, width 1 to 4 cm; margin generally singly serrate, minutely revolute, lateral veins forking two to four times near margin each side of midrib, base rounded and generally symmetrical, apex acute, adaxial surface glabrous, abaxial surface with axillary hair tufts; petiole length 2 to 4 mm; branches lacking corky wings; twigs gray brown, pubescent; buds dark brown, glabrous, scale margin ciliated; flowers appear in early spring; fruit a flat samara, lateral wing surrounding the seed cavity, notched at the distal end, glabrous except ciliated in the notch, shape orbicular, diameter 1.0 to 1.5 cm; native of Asia. This species is widely planted and has locally naturalized in most states; disturbed areas, fencerows, roadsides.

Ulmus parvifolia Jacquin, Chinese elm (Figure 18.14); large tree, height to 25 m; deciduous; leaf arrangement alternate; leaf simple, blade length 2 to 6 cm, width 1 to 3 cm; margin singly serrate, occasionally weakly doubly serrate, base inequilateral, apex acute, adaxial surface glabrous, abaxial surface with scattered hair; petiole length 2 to 6 mm; twigs gray brown; buds reddish brown, pubescent; flowers appear in late summer; fruit a flat samara, lateral wing surrounding the seed cavity, notched at the distal end, glabrous except ciliated within the notch, shape elliptical to ovate, diameter about 1 cm; native of China and Japan. This species is widely planted and has locally naturalized in many states; disturbed areas.

Celtis Species

Four species of *Celtis* occur in the eastern and two in the western United States. Reports of *Celtis iguanaea* (Jacquin) Sargent, a Mexican species, in the southeastern United States are unconfirmed (Barker 1997). *Celtis occidentalis* and *Celtis laevigata* are larger trees used for timber production, but also are planted as ornamentals (street and lawn trees). The drupes of *Celtis* are used extensively by wildlife.

Celtis occidentalis Linnaeus, hackberry (Figure 18.15); large tree, height to 35 m; deciduous; leaf arrangement alternate; leaf simple, shape lanceolate to somewhat ovate, blade length 6 to 11 cm, width 3 to 6 cm, margin serrated except near base, apex acuminate, base strongly inequilateral, adaxial surface scabrous (sandpapery); petiole length 0.6 to 1.2 cm; bark gray, older trees with corky warts or ridges; fruit a drupe, dark purple to nearly black, globose, diameter 0.8 to 1.2 cm; Regions 2 and 3, Manitoba to New England, panhandle of Texas to Georgia and North Carolina; moist soil, cool north slopes, and stream terraces. Populations of *Celtis pumila* Pursh and *Celtis occidentalis* var. *pumila* (Pursh) are included with this species.

Celtis laevigata Willdenow, sugarberry (Figure 18.15); large tree, height to 35 m; deciduous; leaf arrangement alternate; leaf simple, shape lanceolate to somewhat ovate; blade length 6 to 11 cm, width 3 to 6 cm, margin mostly smooth, with widely scattered minute teeth, apex acute to long acuminate, base strongly inequilateral to rounded, surfaces glabrous; petiole length 0.6 to 1.2 cm; bark gray, older trees with corky warts or ridges; fruit a drupe, yellow to reddish, globose, diameter 6 to 8 mm; Regions 3 and 4, southern Missouri, Illinois, and Indiana south and east to Gulf and Atlantic coasts; cool low slopes, along stream terraces, and bottomlands. Cover type **27.10, 27.14, 28.9, 28.10,** 27.6, 27.9, 27.11, 27.14, 27.28, 28.7, 28.8, 28.12, 28.15, 28.17, 28.21, 28.22, 28.27, 28.42, 28.49, 28.50, 29.9.

Celtis tenuifolia Nuttall, dwarf hackberry (Figure 18.15); small tree, height to 8 m; deciduous; leaf arrangement alternate; leaf simple, shape lanceolate to somewhat ovate, blade length 4 to 8 cm, width 3 to 6 cm, margin mostly smooth, with widely scattered minute teeth near apex, apex acute or acuminate, base inequilateral to rounded, surfaces scabrous; petiole length 0.5 to 1.0 cm; bark gray, older trees with corky warts or ridges; fruit a drupe, orange to red, globose, diameter 6 to 8 mm; Regions 2, 3 and 4, Missouri, Illinois, and Michigan south to the Gulf and Atlantic coasts; cool low slopes, along streams. This species includes the populations listed as *Celtis georgiana* Small, *Celtis*

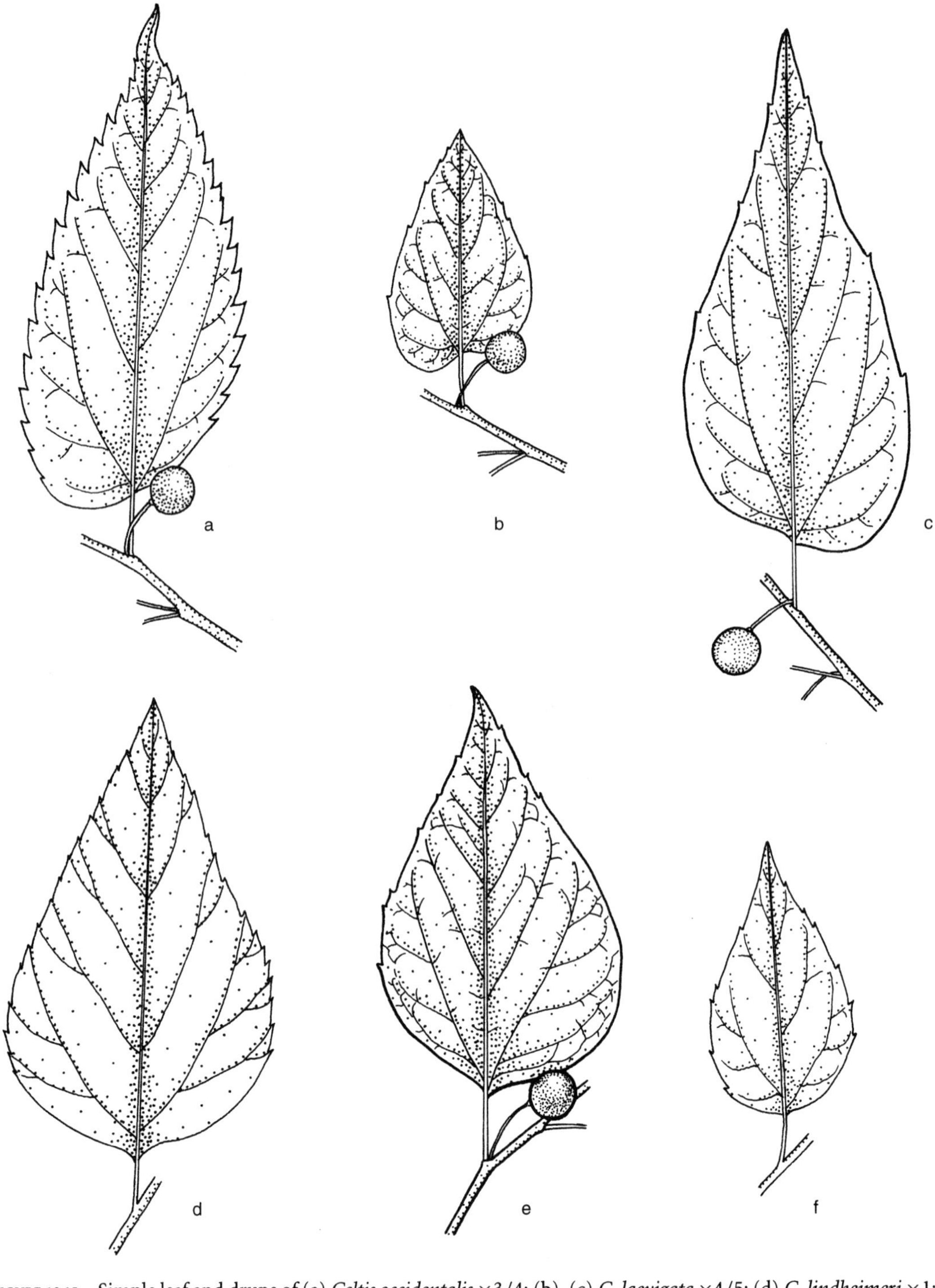

FIGURE 18.15 Simple leaf and drupe of (a) *Celtis occidentalis* ×3/4; (b), (c) *C. laevigata* ×4/5; (d) *C. lindheimeri* ×1; (e) *C. tenuifolia* ×1, (f) *C. tenuifolia* var. *georgiana* ×1.

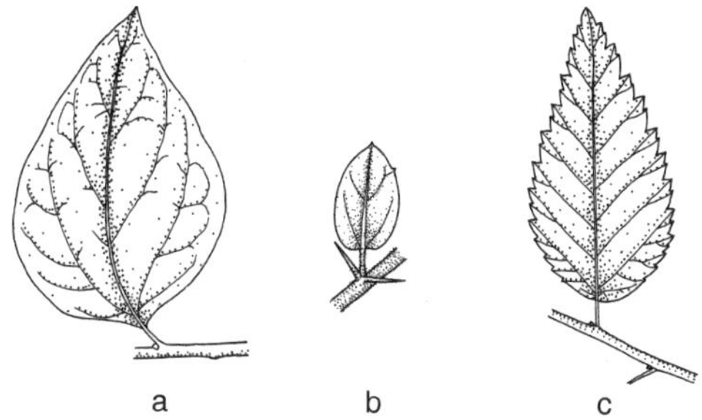

FIGURE 18.16 Simple leaf of (a) *Celtis reticulata*; (b) *C. pallida*; (c) *Planera aquatica.*

tennuifolia var. *georgiana* (Small) Fernald & Schubert, *Celtis occidentalis* Linnaeus var. *georgiana* (Small) Ahles (cover type **27.10**), and *Celtis pumila* Pursh var. *georgiana* (Small) Sargent. Cover type 27.11.

Celtis lindheimeri Engelmann ex. K. Koch, Lindheimer hackberry (Figure 18.15); small tree, height to 14 m; deciduous; leaf arrangement alternate; leaf simple, shape lanceolate to somewhat ovate, blade length 4 to 9 cm, width 2 to 5 cm, margin with irregularly scattered minute teeth, margin smooth near base, apex acute or acuminate, base inequilateral to rounded, adaxial surface scabrous; petiole length 0.4 to 1.0 cm; bark gray, older trees with corky warts or ridges; twigs pubescent; fruit a drupe, light brown, globose, diameter 6 to 9 mm; Region 4, Texas (Edwards Plateau) and Mexico; canyons and savanna.

Celtis reticulata Torrey, netleaf hackberry (Figure 18.16); small tree, height to about 10 m; deciduous; leaf arrangement alternate; leaf simple, shape lanceolate to somewhat ovate, blade length 3 to 6 cm, width 1 to 4 cm, margin mostly smooth, may have widely scattered minute teeth near apex, smooth near base, apex acute or acuminate, base inequilateral to somewhat cordate, adaxial surface scabrous; petiole length 0.3 to 9.0 cm; bark gray, older trees with corky warts or ridges; fruit a drupe, red to black, globose, diameter 6 to 9 mm; Regions 5 and 6, southwestern North America from Mexico, Texas, Oklahoma, and Kansas to Nevada and Utah, disjunct populations in California, Nevada, Idaho, Oregon, and Washington; xeric sites, canyons, rock outcrops. Cover type 29.19, 29.22, 29.2, 30.5.

Celtis pallida Torrey, desert hackberry (Figure 18.16); shrub, height to 3 m; deciduous; leaf arrangement alternate; leaf simple, shape ovate to oblong, blade length 2 to 4 cm, width 1 to 2 cm, margin mostly smooth, with widely scattered blunt teeth (crenulate) near apex, apex acute obtuse, base inequilateral to rounded, surfaces scabrous; petiole length 0.4 to 1.5 cm; bark gray, older trees with corky warts or ridges, twigs thorny; fruit a drupe, orange to red, globose, diameter 6 to 7 mm; Region 5, Texas, New Mexico, and Arizona into Mexico and Central and South America; canyons and washes, deserts and savanna.

Other Ulmaceae

Planera aquatica Walter ex J. F. Gmelin, water elm (Figure 18.16); large shrub to medium-sized tree, height to 15 m; deciduous; leaf arrangement opposite; leaf simple, blade length 4 to 8 cm, width 2 to 4 cm, shape ovate, base rounded or inequilateral, apex acute, margin serrate or doubly serrate; petiole length 3 to 6 mm; fruit a leathery drupe, globose, diameter 7 to 9 mm; Region 4, Coastal Plain, but extending northward up the Mississippi floodplain to southern Illinois; locally abundant in bottomlands, backwaters, and riverbanks of streams flowing to the Gulf of Mexico. The species has limited wildlife and economic value. The leaf structure and drupe of *Planera aquatica* closely resembles those of *Celtis.* Cover type **28.43,** 27.10, 27.28, 28.8, 28.17, 28.21, 28.31, 28.55.

Two species of *Trema* occur in the subtropical forests at the extreme southern tip of Florida. They are small-to-medium-sized trees that grow in or near disturbed areas. The drupes are eaten and distributed by birds.

Trema micrantha (Linnaeus) Blume, Florida trema; shrub or small tree, height to 8 m; deciduous; leaf arrangement alternate; leaf simple, length 5 to 8 cm, width 2 to 4 cm, shape narrowly ovate, margin singly serrated, base cordate or inequilateral, apex acute to long acuminate; twigs tomentose, with numerous warts; fruit a drupe, red-orange or yellow; southern Florida, hammocks, disturbed areas.

Trema lamarckiana (J. A. Schultes) Blume, West Indian trema; shrub, height to 3 m; deciduous; leaf arrangement alternate; leaf simple, length 2 to 5 cm, width 2 to 3 cm, shape narrowly ovate, margin crenate, base nearly truncate, apex acute; twigs strongly pubescent; fruit a drupe, pink; Florida keys; disturbed areas.

Moraceae (Fig Family)

Like several other families, the members of the Moraceae (approximately 50 genera and 1400 species) are distributed in tropical and subtropical regions. However, the family is closely aligned and at times included with the Urticaceae (nettle) family (Zomlefer 1994), which increases the number to more than 90 genera and nearly 2500 species. The largest and perhaps most economically important genus is *Ficus*, with more than 500 species. The common edible fig is produced by *Ficus carica* Linnaeus, and *Ficus elastica* Roxburgh (Indian rubber tree) and *Ficus lyrata* (fiddleleaf fig) are widely grown as house plants. *Ficus aurea* Nuttall (strangler fig) and *Ficus citrifolia* Miller (banyan tree) are native to southern Florida.

The Moraceae is composed primarily of trees, shrubs, and vines, with species of only two genera, *Morus* and *Maclura*, native to the temperate zone. Other genera and species have been introduced. All have simple, alternate leaves and milky sap. The flowers are minute, without petals, and unisexual. In *Morus*, the flowers are clustered on a stalk, which results in a multiple fruit of drupes and accessory tissue that somewhat resembles a blackberry. It is edible and an important food for birds, small animals, and other wildlife. The genus *Morus* is represented by a single native species in the eastern United States and one in the southwest.

Morus rubra Linnaeus, red mulberry (Figure 18.17); medium-sized, height to 15 m; deciduous; leaf arrangement alternate; leaf simple, broadly ovate, leaf of forest-grown tree moderately large, 10 to 25 cm, width 8 to 15 cm, leaf smaller on open-grown trees, lobing highly variable, entire to strongly multiple lobed, margin serrated, teeth rounded, distinctly palmately three-veined, upper surface dull and somewhat scabrous, apex acuminate, a second apexlike projection infrequently appears on one side; twigs light yellowish brown; fruit a multiple of fleshy drupes, red, purple, or black; throughout Regions 3 and 4, southern Wisconsin, Michigan and New York and south; upland forest, cutover forest, fencerows where birds drop the seeds with their natural fertilizer. For entire leaves lacking the extra projection, the rounded teeth distinguish the species from *Tilia*, which has sharply pointed teeth. The upper leaf surface is dull and slightly scabrous which distinguishes *Morus rubra* from the glossy, smooth, usually smaller leaf of *M. alba*. Cover type 27.1, 27.4, 27.9, 28.10.

Morus alba Linnaeus, white mulberry (Figure 18.18); medium-sized tree, height to 20 m; deciduous; leaf arrangement alternate; leaf simple, ovate, length 6 to 10 cm, width 3 to 6 cm, lobing highly variable, usually strongly multiple lobed, occasionally entire, margin serrate or crenate, teeth rounded, distinctly palmately three-veined, upper surface glabrous, shiny, base truncate to cordate, apex acute to acuminate; twigs light yellowish brown, buds ovate; fruit a multiple of fleshy drupes, purple or black; Introduced, naturalized throughout Regions 3 and 4; natural and cutover forest, fencerows where birds drop the seeds with their natural fertilizer. Introduced into the United States in an attempt to develop the silk industry. This species is the primary food of silkworms, and silk was needed during World War II for parachutes. Subsequently, it escaped cultivation and is presently nearly as common as *M. rubra*.

Morus microphylla Buckley, Texas mulberry; shrubby, height to 7 m; deciduous; leaf arrangement alternate; leaf simple, shape ovate, length 3 to 7 cm, width 1.5 to 4.0 cm, lobing highly variable, entire or three- to five-lobed, margin serrate or crenate, teeth rounded, distinctly palmately three-veined, upper surface glabrous, shiny, lower surface strongly scabrous, base truncate to cordate, apex acute to acuminate; twigs light yellowish brown, buds ovate; fruit a multiple of fleshy drupes, red, purple or black; Region 5, west Texas, Oklahoma, Nevada, New Mexico, and Mexico; along streams and in mountain canyons to 2100 m. Cover type 29.2, 29.22.

Maclura pomifera (Rafinesque) Schneider, Osage-orange, hedge-apple (Figure 18.19); medium-sized tree, height to 17 m, often multistemmed; deciduous; leaf arrangement alternate; leaf simple, length 5 to 15 cm, blade glabrous and shiny, dark green, shape lanceolate; new leaves accompanied by a single stipular thorn at the base of the petiole, thorn length 5 to 12 mm, three to six leaves are clustered on spur shoots of older twigs; fruit a multiple of achenes, globose, larger than an apple, diameter 8 to 12 cm, green, surface pebble-like, internally saturated with white sticky latex sap; Region 4, original range was restricted; moist bottomland soil in an area of southern and eastern Texas, Oklahoma, Arkansas, and Louisiana. Because it was widely planted as a "living fence" or hedge, its range now extends throughout Regions 3 and 4. Posts of this species sprout after planting and develop into umbrella-shaped trees. The sticky latex material in the fruit may irritate the skin; thus the fruit is not edible. The common names are based on the fruit size and outer appearance.

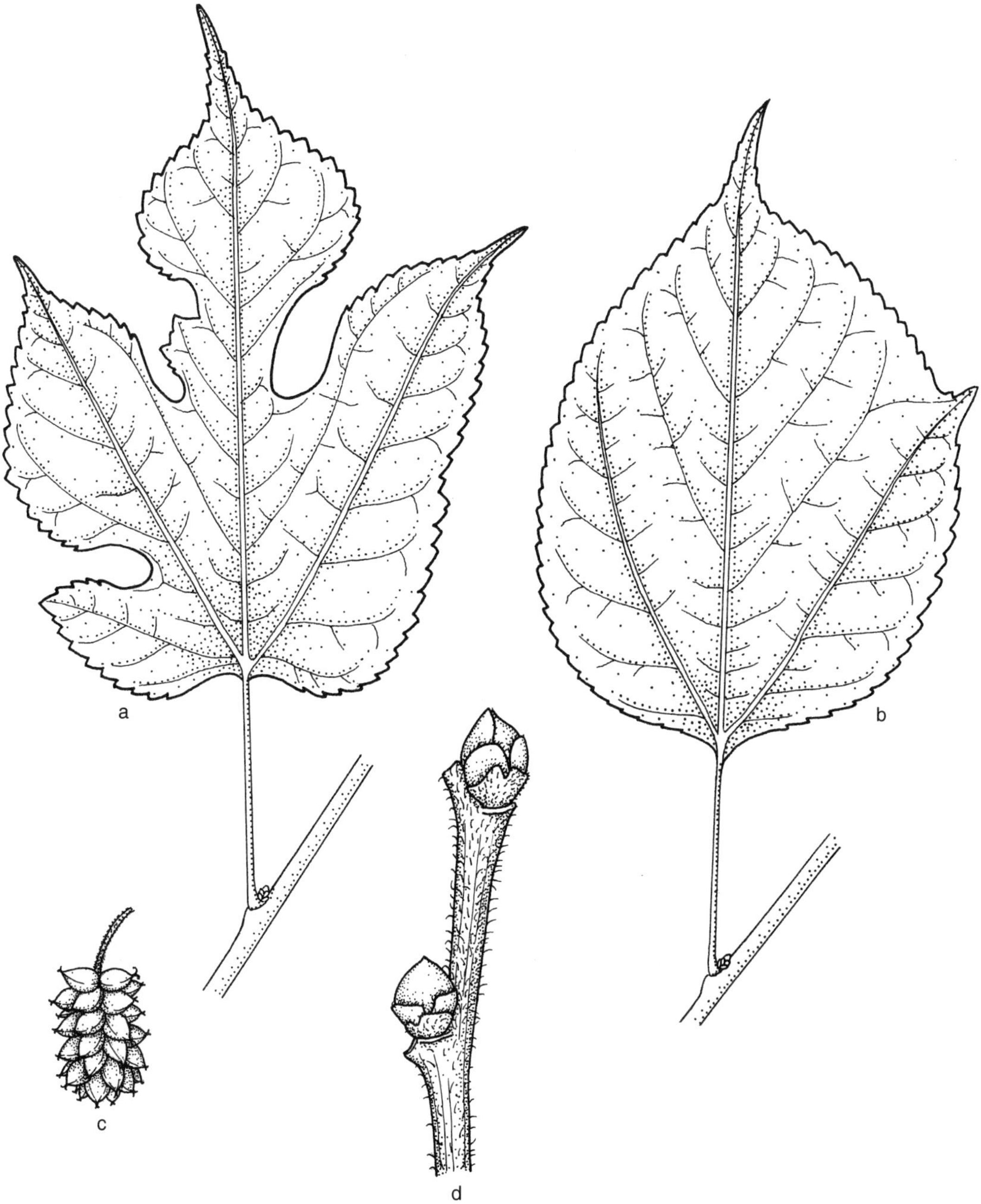

FIGURE 18.17 *Morus rubra* (a), (b) simple leaf ×1/2, (c) multiple of drupes ×1, (d) twig with pseudoterminal and lateral buds ×2.

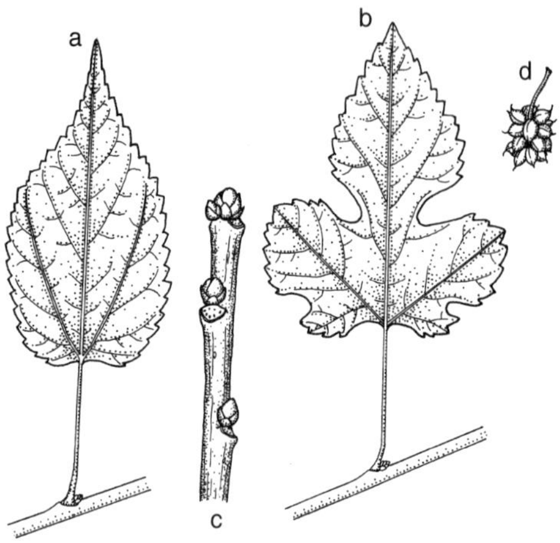

FIGURE 18.18 *Morus alba* (a), (b) simple leaves ×1/2, (c) fruit, a multiple of drupes ×1/2, (d) twig with terminal and lateral buds ×1.5.

Broussonetia papyrifera (Linnaeus) Ventenat, paper mulberry, elephant ear (Figure 18.20); small tree, height to 14 m; deciduous; leaf arrangement alternate; leaf large, length 10 to 30 cm, blade entire or variably lobed, prominently three-veined, all surfaces with long pubescent hair, somewhat scabrous, base truncate, rounded, or cordate, lobe apices long acuminate; fruit globose, multiple of drupes, diameter 2 to 3 cm. This *Morus* relative was introduced from Asia during the eighteenth century and has become naturalized. It is primarily found in disturbed areas—near homes, fences, and roads. Once a seedling is established, the roots grow aggressively just below the soil surface, produce sprouts to form a clone, and may completely occupy an area within 2 or 3 years.

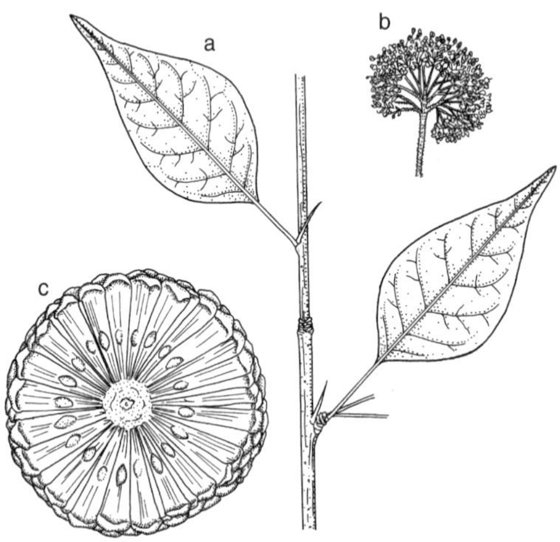

FIGURE 18.19 *Maclura pomifera* (a) twig with leaves and stipular thorns ×1/3, (b) inflorescence ×1/3, (c) cross section of fruit showing location of seeds ×1/3. See Figure 16.8e for fruit.

ORDER: JUGLANDALES

Juglandaceae (Walnut/Hickory Family)

This genus of trees and a few shrubs consists of only 8 genera and about 60 species primarily distributed in the north temperate zone, but with one species extending through Central America into Argentina. Two genera (*Juglans* and *Carya*) and 18 species are native to North America; all grow to tree size. Two *Juglans* (walnut) and all 12 *Carya* (hickory) species are restricted to the eastern United States and southern Canada; 4 *Juglans* species occur in the western United States.

In spring, unisexual flowers are borne in a catkin that may lengthen to 20 cm. The catkin comes from the upper bud of superimposed lateral buds, the lower bud producing leaves. The fruits develop in late summer. Although they are "nut"-producing, *Juglans* species produce a fruit that closely resembles a drupe with a continuous outer skin (epiderm), a spongy/fibrous inner layer (mesoderm), and an inner hard shell (endoderm or pit) similar to that of a peach or cherry. The *Carya* nut differs from a drupe and the *Juglans* fruit in that the epiderm and mesoderm are composed of nearly the same cell tissue, and at maturity the husks split along definite sutures. The husk does not develop from the ovary wall but from the involucre and calyx (Elias 1972). The leaves of both genera are pinnate compound, serrated, and aromatic when crushed; the distinctive odor has been compared to that of turpentine. Both genera have stout twigs. The leaf scars are lobed or skull-shaped, with prominent vascular bundle scars usually in clusters. *Juglans* twigs have a chambered pith; in *Carya* the pith is continuous.

Both *J. nigra* (black walnut) and *J. cincerea* (butternut) woods are noted for their attractive chocolate brown wood, which is widely used for quality furniture and specialty items. The trees are important as mast producers as the nuts are sought by wildlife. In recent

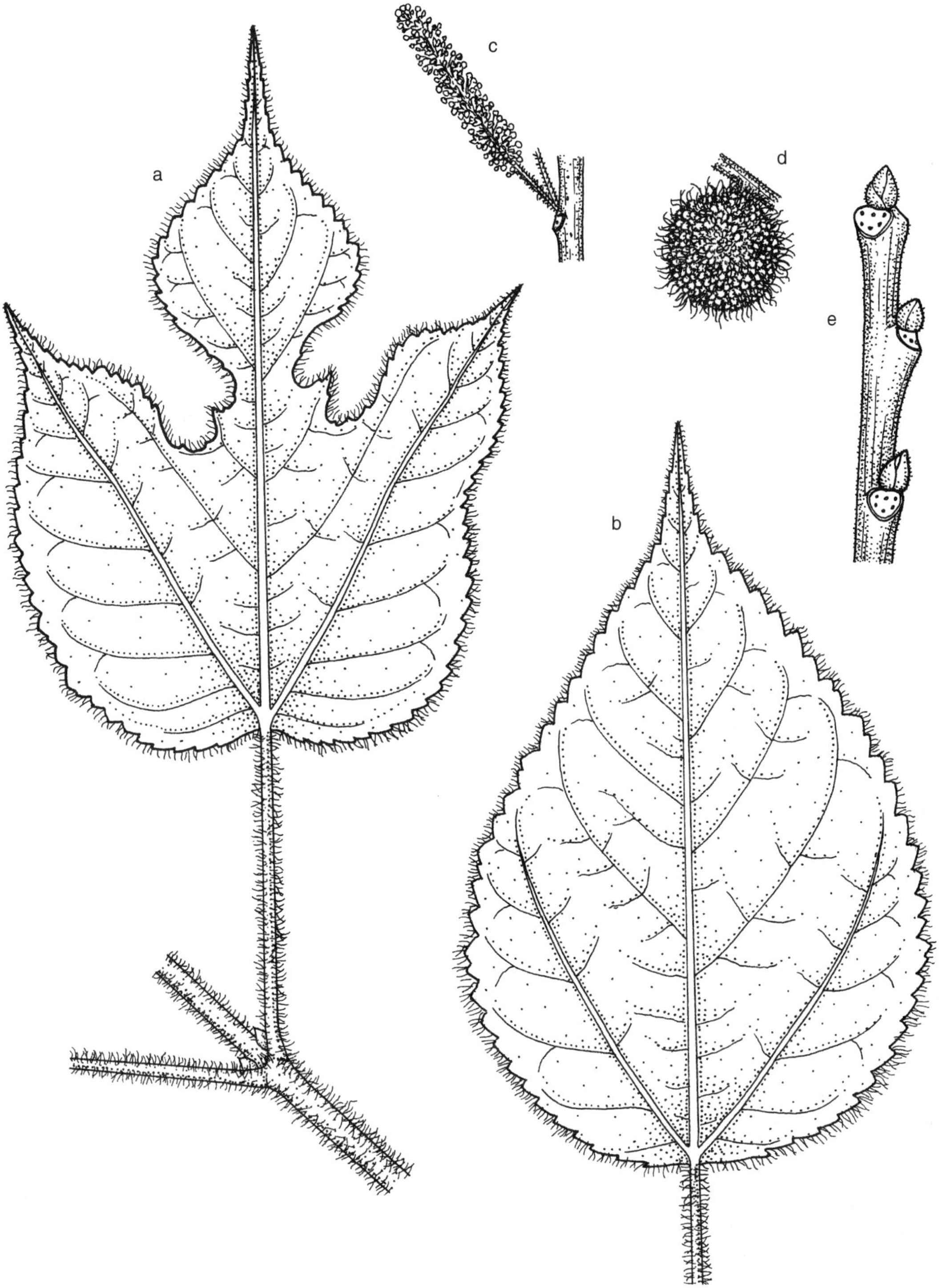

FIGURE 18.20 *Broussonetia papyrifera* (a), (b) simple leaves ×1, (c) inflorescence ×1, (d) fruit structure ×1, (e) twig with terminal and lateral buds ×2.

years the U.S. Forest Service has experimented with plantations to increase the supply of *J. nigra* wood and as a possible source of income for rural landowners. Because the chocolate brown wood is extremely valuable, some individual trees grown in yards and natural forest have sold for thousands of dollars. The result is that theft of both forest and lawn trees has been a problem.

The "meat" (seed) of *J. nigra* nuts is noted for its flavoring quality in cooking and is also eaten uncooked. An allelopathic chemical (juglone) produced by *J. nigra* has been isolated and is known to inhibit growth of other plants; therefore, planting walnut in gardens or using the husks as a mulch is definitely not recommended. *Juglans cinerea* has typically occurred as individual trees scattered through the forest. The species has become so rare that it is now listed as endangered or threatened in several states.

Juglans Species

Six species are recognized: two in the eastern United States (*Juglans nigra, Juglans cinerea*), two in the southwest (*Juglans major, Juglans microcarpa*), and two in California (*Juglans hindsii, Juglans californica*).

Juglans nigra Linnaeus, black walnut (Figure 18.21); large tree, height to 40 m; deciduous; leaf arrangement alternate; leaf pinnately compound, length 25 to 60 cm, petiole, rachis, and surfaces variously hairy; leaflet number 11 to 19, sometimes 9, 21, or 23, terminal leaflet much reduced, rudimentary, or represented by a vestigial petiole, leaflet shape slightly falcate, margin serrated, length 7 to 12 cm, width 3 to 6 cm; pith chambered, tan; terminal bud gray brown, pubescent, dull, length 0.8 to 1.5 cm, asymmetrical, scales imbricate; fruit strongly drupaceous, shape globose, length 4 to 7 cm, brown to dark reddish brown; husks not splitting, thickness to 2 cm, yielding a chocolate brown dye when mature; seed distinctly flavored; mature bark dark reddish brown, split into furrows and interlacing ridges; Regions 3 and 4, southern Minnesota to New York and south, extending west into South Dakota, Nebraska, Kansas, Colorado, Oklahoma, and central Texas; mesic slopes and stream terraces. Cover type 26.24, 28.9, 28.12, 28.27.

Juglans cinerea Linnaeus, butternut, white walnut (Figure 18.22); large tree, height to 30 m; deciduous; leaf arrangement alternate; leaf pinnately compound, length 25 to 60 cm, petiole, rachis, and surfaces variously hairy; leaflet number 11 to 17, sometimes 9, terminal leaflet present, only slightly smaller than lateral leaflets, leaflet shape slightly lanceolate to elliptical, margin serrated, length 6 to 12 cm, width 4 to 7 cm; pith chambered, chocolate brown; terminal bud gray brown, pubescent, dull, length 1.0 to 1.5 cm, asymmetrical, scales imbricate; fruit strongly drupaceous, shape oval; length 4 to 9 cm, brown to dark reddish brown; husks not splitting, thickness to 1.5 cm; mature bark gray, relatively smooth with a few fissures; Region 3, but extending into Regions 2, northern Wisconsin, southern Ontario, and Maine, and 4, North and South Carolina; mesic slopes and stream terraces.

Juglans major (Torrey) Heller, Arizona walnut (Figure 18.23); medium-sized tree, height to 18 m; deciduous; leaf arrangement alternate; leaf pinnately compound, length 20 to 35 cm, petiole, rachis, and surfaces variously hairy; leaflet number 9 to 15, terminal leaflet less than one half the size of the lateral leaflets, rudimentary, or represented by a vestigial petiole, leaflet shape lanceolate or slightly falcate, margin serrated, length 6 to 11 cm, width 2 to 4 cm; pith chambered, brown; terminal bud gray brown, pubescent, dull, length 4 to 8 mm, asymmetrical, scales imbricate; fruit strongly drupaceous, shape globose; length 2 to 4 cm, brown to dark reddish brown; husks not splitting, thickness 4 to 6 mm; seed distinctly flavored; mature bark relatively smooth with a few fissures; Region 5, Texas, New Mexico, and Arizona; near streams, in canyons, and on low mesic slopes. Cover type 29.2, 29.19, 29.22.

Juglans microcarpa Berlandier, little walnut (Figure 18.24); small tree, height to 10 m; deciduous leaf arrangement alternate; leaf pinnately compound, length 15 to 38 cm, petiole, rachis, and surfaces variously hairy; leaflet number 17 to 25, terminal leaflet about one half the size of the lateral leaflets, leaflet shape lanceolate or falcate, margin serrated to nearly smooth, length 6 to 10 cm, width 1 to 2 cm; pith chambered, brown; terminal bud gray brown, pubescent, dull, length 3 to 5 mm, asymmetrical, scales imbricate; fruit strongly drupaceous, shape globose, length 1.5 to 2.5 cm, brown to dark reddish brown; husks not splitting, thickness to 4 mm; seed distinctly flavored; mature bark relatively smooth with a few fissures; Region 5, south-central Kansas, Oklahoma, Texas, and New Mexico; near streams and rivers.

Juglans hindsii Jeps ex. R. E. Smith, northern California walnut, Hinds walnut (Figure 18.25); medium-sized tree, height to 23 m; deciduous; leaf

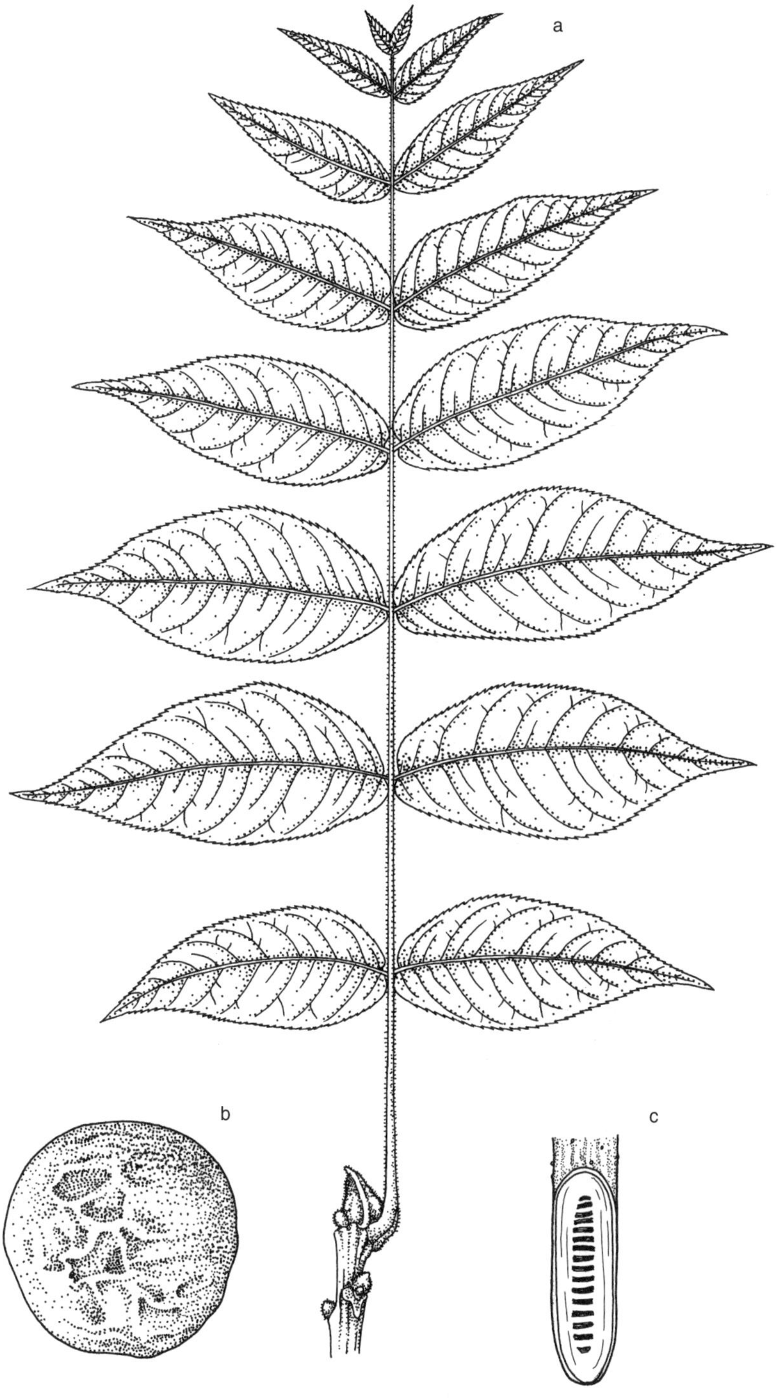

FIGURE 18.21 *Juglans nigra* (a) pinnately compound leaf $\times 1/2$, (b) globose nut enclosed by the husk $\times 1/2$, (c) twig showing chambered pith $\times 1.5$.

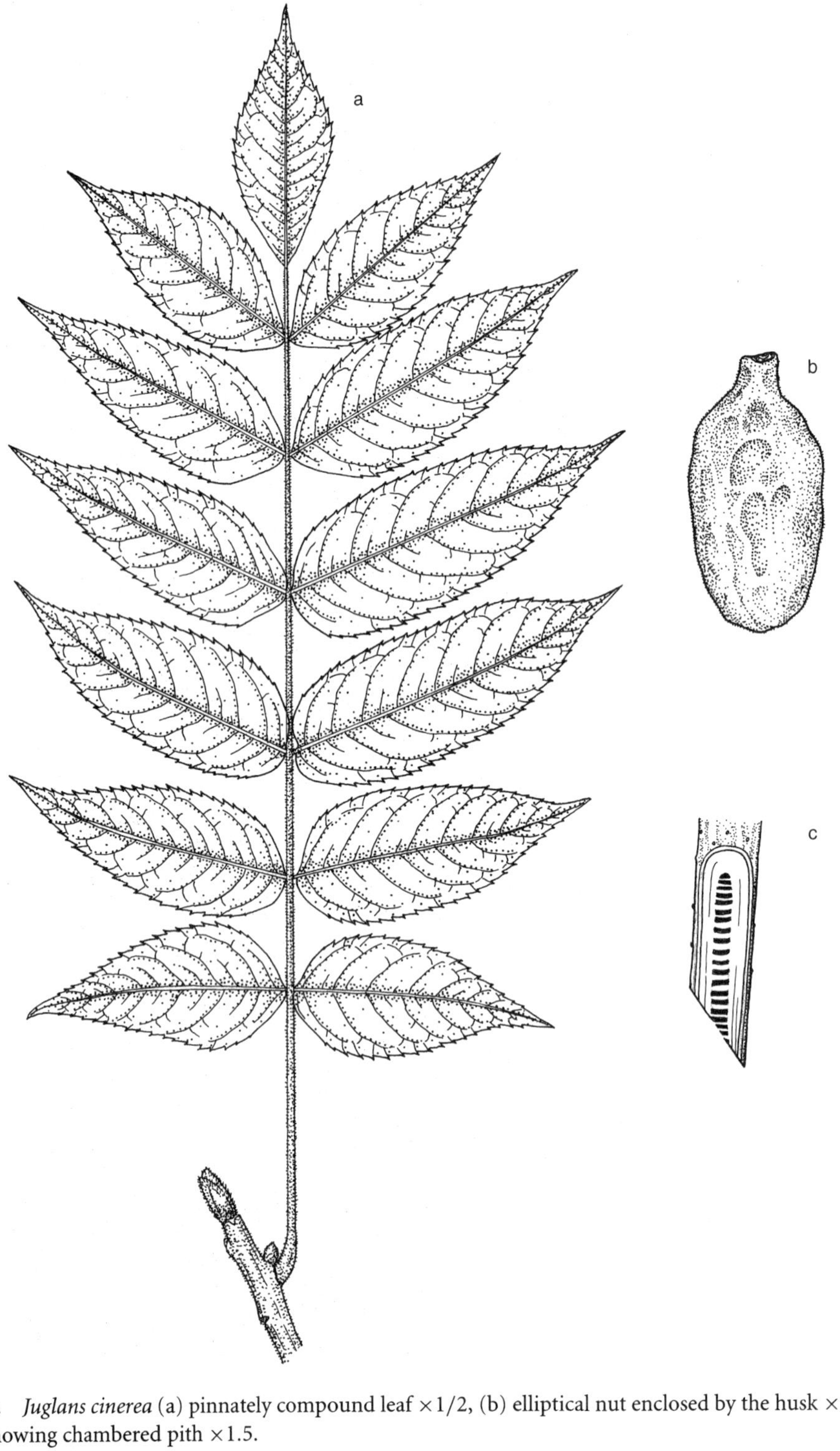

FIGURE 18.22 *Juglans cinerea* (a) pinnately compound leaf ×1/2, (b) elliptical nut enclosed by the husk ×1/2, (c) twig showing chambered pith ×1.5.

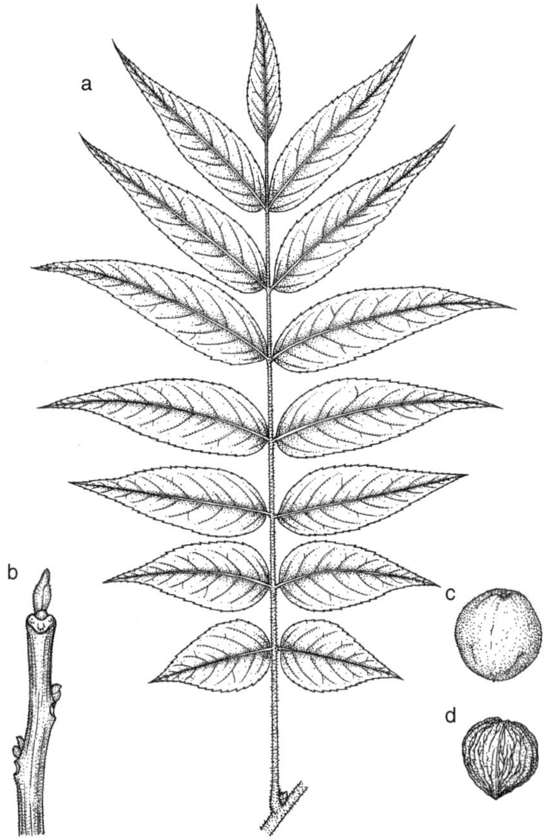

FIGURE 18.23 *Juglans major* (a) pinnately compound leaf × 1/3, (b) twig with terminal and lateral buds × 1, (c) nut enclosed by husk, (d) oval nut without the husk × 1/3.

arrangement alternate; leaf pinnately compound, length 25 to 40 cm, petiole, rachis, and surfaces variously hairy; leaflet number 13 to 21, terminal leaflet nearly the size of the lateral leaflets, leaflet shape narrowly lanceolate to slightly falcate, margin serrated, length 7 to 13 cm, width 2 to 3 cm; pith chambered, light brown; terminal bud gray brown, pubescent, dull, length 6 to 9 mm, asymmetrical, scales imbricate; fruit strongly drupaceous, shape globose, length 3.5 to 5.0 cm, brown to dark reddish brown; husks not splitting, thickness to 1.0 cm; seed distinctly flavored; mature bark light gray, relatively smooth with a few fissures; Region 7, restricted to north-central California; near streams and rivers.

Juglans californica Watson, southern California walnut (Figure 18.26); small tree, height to 9 m; deciduous leaf arrangement alternate; leaf pinnately compound, length 15 to 25 cm, petiole, rachis, and surfaces variously hairy; leaflet number 9 to 15, sometimes 17, terminal leaflet equal to the size of the lateral leaflets, leaflet shape oblong or elliptical, margin serrated, length 3 to 10 cm, width 1.5 to 2.8 cm; pith chambered, brown; terminal bud gray brown, pubescent, dull, length 5 to 7 mm, asymmetrical, scales imbricate; fruit strongly drupaceous, shape globose, length 2.0 to 3.5 cm, brown to dark reddish brown; husks not splitting, thickness to 5 mm; seed distinctly flavored; mature bark light gray, relatively

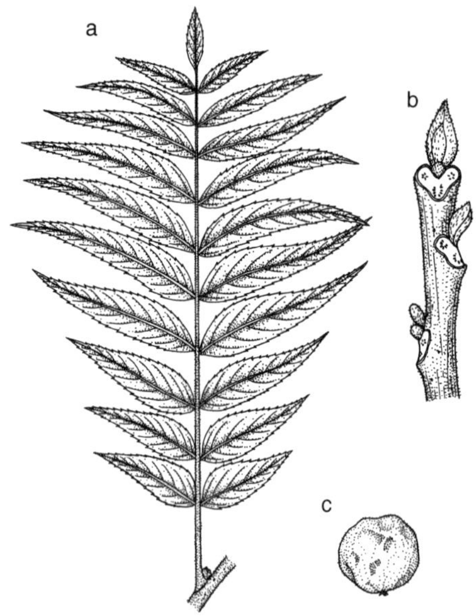

FIGURE 18.24 *Juglans microcarpa* (a) pinnately compound leaf ×1/4, (b) twig with terminal and lateral buds ×1.3, (c) globose nut enclosed by the husk ×1/2.

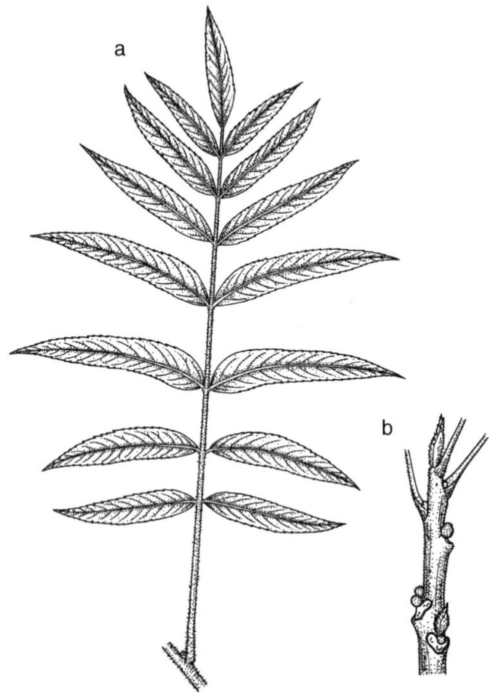

FIGURE 18.25 *Juglans hindsii* (a) pinnately compound leaf ×1/4, (b) twig with terminal and lateral buds ×1.

smooth with a few fissures; Region 7, southern California; slopes and canyons, elevation 25 to 875 m.

***Carya* Species**

Carya (hickory) species are economically and ecologically important. Although originally restricted to a small range in the central United States, plantations of *C. illinoensis* have been widely established throughout the south for pecan production. The nuts of *C. myristiciformis* and *C. ovata* are sweet, edible, and have been used locally for cooking. The common names for *C. aquatica* (bitter pecan), *C. cordiformis* (bitternut hickory), and *C. glabra* (pignut hickory) indicate that their nuts may contain tannins or other substances that make them extremely bitter and not readily edible. Although there are 11 or 12 species of *Carya*, they are not generally distinguished (except for *Carya illinoensis*) during harvesting operations because there is little difference in the wood, which is hard and strong. It is used for tool handles, furniture, fuel, and, perhaps more important, for professional baseball bats. Ecologically, *Carya* provides considerable mast for wildlife, and the endangered Indiana bat uses hickory trees for nesting and reproduction. Timber harvesting of *Quercus-Carya* (oak-hickory) forest in the Central States (Region 3) appears to have increased the proportion of *Carya* in the overstory as *Quercus* trees are selectively removed. *Carya* seedlings and saplings appear to be more shade tolerant and often outnumber those of *Quercus*.

Two subgenera (*Apocarya, Eucarya*) of *Carya* are recognized (sensu Harlow et al. 1996). The four *Apocarya* species have a larger number of leaflets (generally 7 to 17) per leaf, slightly to strongly falcate leaflets, pseudovalvate bud scales (except *C. mryisticiformis*), and wings or ridges along the fruit husk sutures. *Apocarya* species grow on sites that vary from mesic north slopes to wet, often flooded bottomland sites. The eight *Eucarya* species have 3 to 9 leaflets per leaf, elliptical, lanceolate, ovate, or obovate leaflets, imbricate bud scales, and no wings along

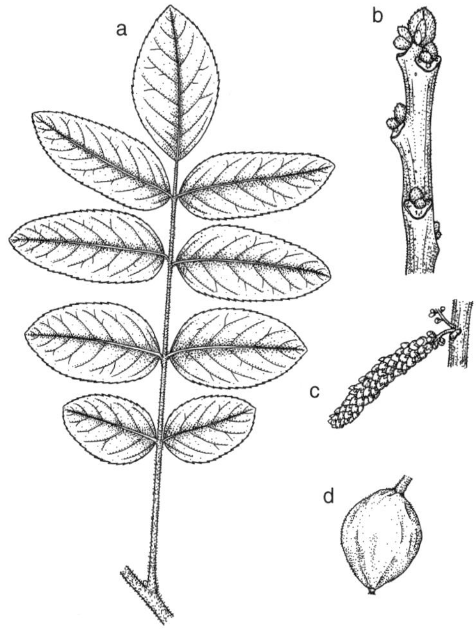

FIGURE 18.26 *Juglans californica* (a) pinnately compound leaf ×1/3, (b) twig with terminal and lateral buds ×1, (c) inflorescence ×1, (d) oval nut enclosed by the husk ×1/2.

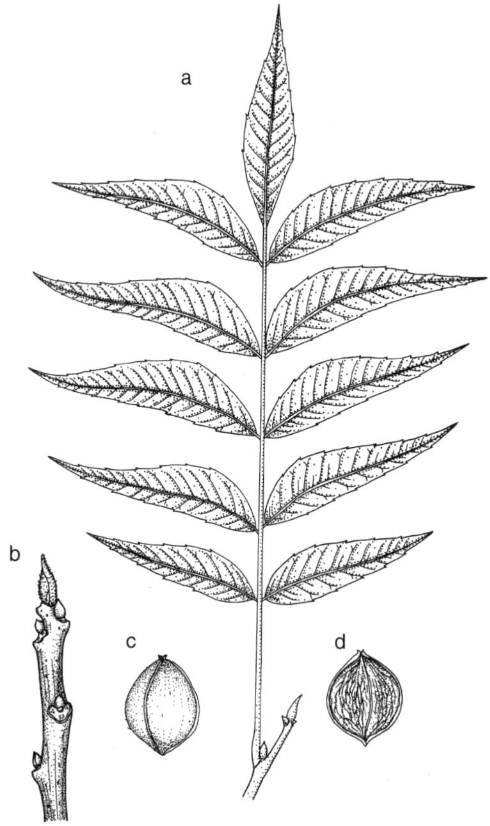

FIGURE 18.27 *Carya aquatica* (a) pinnately compound leaf ×1/3, (b) twig with terminal and lateral buds ×1/2, (c) nut enclosed by a winged husk ×1/2, (d) nut with one half of husk removed ×1/2.

the fruit husk sutures. Most *Eucarya* species grow on sites that range from dry south slopes to moist north and alluvial sites; one species (*Carya laciniosa*) is a species of bottomland communities. Hybridization between species is common. No *Carya* species naturally occur in western North America.

Subgenus: Apocarya *Carya aquatica* (Michaux f.) Nuttall, water hickory, bitter pecan (Figure 18.27); large tree, height to 45 m; deciduous; leaf arrangement alternate; leaf pinnately compound, length 15 to 40 cm; leaflet number 9 to 11, sometimes 5 or 13, shape slightly falcate, length 5 to 12 cm, width 2 to 4 cm, margin serrated; twigs glabrous; buds brown to nearly black, pubescent, valvate with scales of unequal size; fruit drupaceous, husk splitting along four winged sutures, shape essentially oval to elliptical, length 1.5 to 3.0 cm, husk thickness 1 to 2 mm; seed bitter; bark splitting in long, broad exfoliating plates; Region 4, Atlantic and Gulf Coastal Plains and north in the Mississippi floodplain to southern Illinois; mesic-hydric sites of bottomlands and swamps. Cover type **28.8, 28.17**, 17.10, 28.10, 28.32, 28.49, 28.50, 28.55.

Carya illinoensis (Wangenheim) K. Koch, pecan (Figure 18.28); large tree, height to 45 m; deciduous; leaf arrangement alternate; leaf pinnately compound, length 35 to 55 cm; leaflet number 9 to 13, sometimes 5, 15, or 17, shape strongly falcate, length 5 to 15 cm, width 2 to 6 cm, margin serrated; twigs hairy, scaly; buds yellowish brown, hirsute, strongly scaly, valvate with scales of unequal size; fruit drupaceous, shape

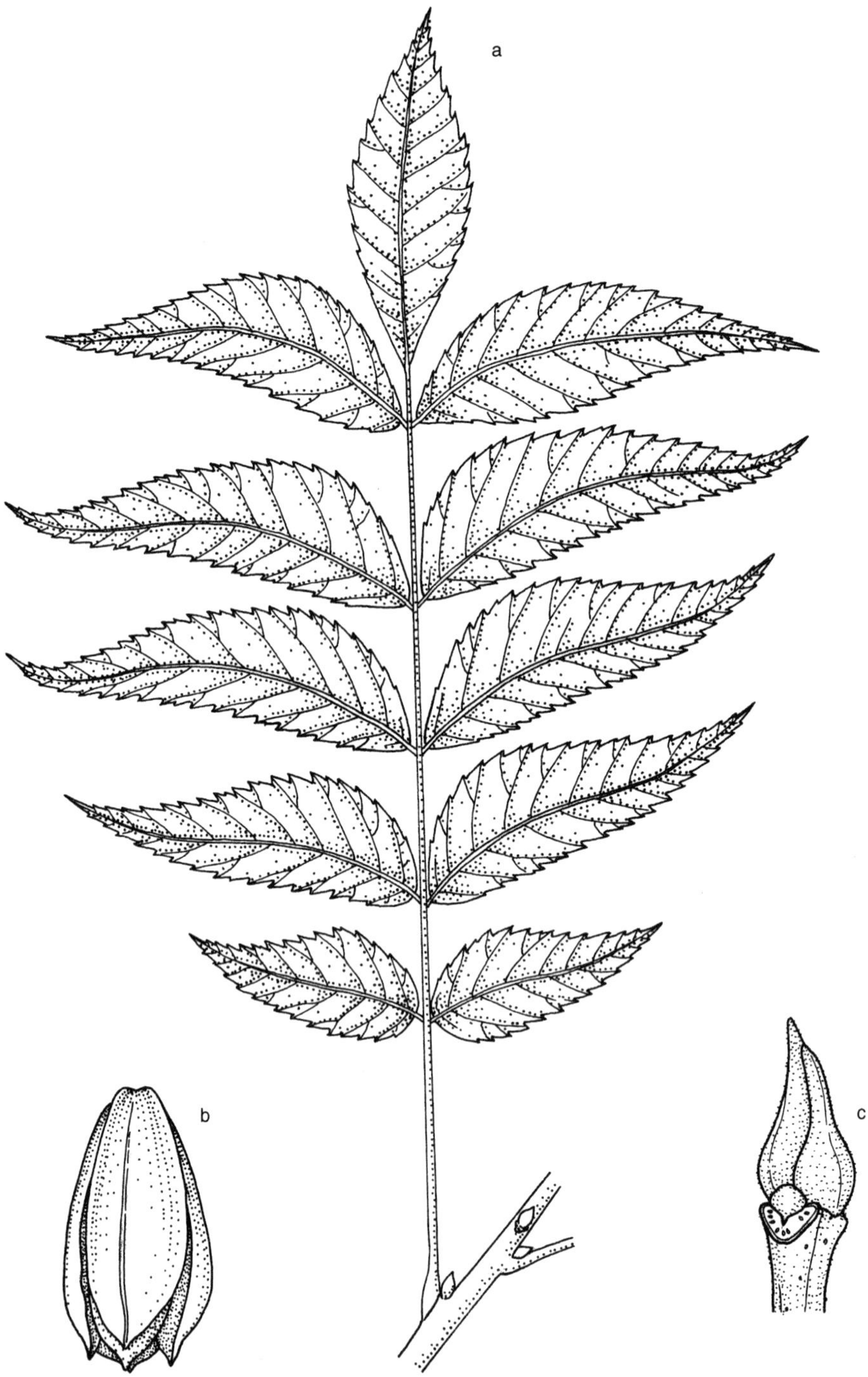

FIGURE 18.28 *Carya illinoensis* (a) pinnately compound leaf ×1/2, (b) nut enclosed by a winged husk ×1, (c) twig with terminal bud and leaf scar ×3.

oval to elliptical, length 2.5 to 6.0 cm, width 1.5 to 2.0 cm; husk splitting along four winged sutures, thickness 3 to 4 mm; bark tight, eventually breaking into narrow exfoliating plates; Region 4, Gulf Coastal Plain of Mississippi, Louisiana, and Texas, north along the floodplains of the Mississippi River and western tributaries, southern Indiana and Iowa and central Illinois, widely planted in the southeast; mesic-hydric sites of bottomlands and swamps. Cover type **28.9**, 27.14, 27.20, 27.28, 28.7, 28.21.

Carya cordiformis (Wangenheim) K. Koch, bitternut or yellowbud hickory (Figure 18.29); large tree, height to 48 m; deciduous; leaf arrangement alternate; leaf pinnately compound, length 20 to 35 cm; leaflet number 7 to 9, sometimes 5 or 11 to 13, shape slightly ovate to lanceolate, sometimes slightly falcate, length 6 to 15 cm, width 2 to 6 cm, margin serrated, terminal leaflet two to three times the length of proximal leaflets; twigs glabrous; buds sulfur yellow, scales of unequal size, rough scaly; fruit drupaceous, husk splitting along four winged sutures, shape essentially oval, length 2 to 3 cm; seed bitter; bark slightly rough, becoming smoother with age with numerous oval indentations, each bisected by vertical yellowish mark (length 2 to 3 cm); Regions 3 and 4, occasionally in southern half of Region 2; mesic sites, north slopes, and high stream terraces. Cover type **27.1, 27.4,** 27.30, 28.12, 28.18, 28.42.

Carya myristiciformis (Michaux f.) Nuttall, nutmeg hickory (Figure 18.30); large tree, height to 35 m; deciduous; leaf arrangement alternate; leaf pinnately compound, length 15 to 35 cm, petiole densely scaly; leaflet number 7 to 9, sometimes 5, shape elliptical to obovate, length 5 to 15 cm, width 2 to 7 cm, margin serrated; twig scaly; buds bronze, scales imbricate; fruit drupaceous, bronze; husk splitting along four winged sutures, shape essentially oval to elliptical, length 2 to 3.5 cm, husk thin, thickness 2 mm; seed sweet; bark splitting in long, broad, exfoliating plates; scattered in Region 4, disjunct populations in Texas, Oklahoma to Alabama, North and South Carolina, not reported in Georgia and Florida, mesic slopes, stream terraces, bottomlands. This species is the rarest of the hickories. Cover type 27.9, 28.19, 28.22.

Subgenus: Eucarya *Carya glabra* (Miller) Sweet, pignut hickory (Figure 18.31); large tree, height to 30 m; deciduous; leaf arrangement alternate; leaf pinnately compound, length 20 to 35 cm, rachis glabrous or nearly so; leaflet number five to seven, shape ovate, obovate, or elliptic, margin serrated, lower surface glabrous, length 4 to 15 cm, width 2 to 8 cm; buds and twigs reddish brown, glabrous; terminal bud length 0.5 to 1.2 cm, scales imbricate, glabrous, outer scales greenish, each completely deciduous at the base; fruit a nut enclosed by a husk, length 2.0 to 4.5 cm, width 2 to 3 cm, tan to reddish brown, shape distinctly obovate (teardrop-like); husks with four unwinged sutures, not splitting or splitting only one half the distance between the apex and base, thickness 2 to 4 mm; seed bitter; bark relatively smooth with small flat plates, becoming smoother with age, often distinguished by numerous small white spots; Regions 3 and 4, into southern Michigan and New York; xeric to mesic slopes and ridges, occasionally on high stream terraces. Cover type **27.7,** 27.16, 27.21, 27.22, 27.26, 28.3, 28.6, 28.18.

Carya ovalis (Wangenheim) Sargent, red hickory (Figure 18.31). The status of this species has been regarded with uncertainty. Most taxonomists include this population as part of *Carya glabra* without an intraspecific designation (Elias 1980; Stone 1997) or as a variety [*Carya ovalis* var. *odorata* (Marshall) Little] (Little 1979a; Harlow et al. 1996). Boiven (1966) considered it an interspecific hybrid between *Carya glabra* and *Carya ovata*, but this population has little to relate it to *Carya ovata.* Although it is similar to *Carya glabra* in most leaf and bud characteristics (but may have some hair), in the western Central Hardwood Forest region (Illinois, Missouri, west Kentucky), we maintain that this hickory population may be regarded as a distinct species although in ecological research we have treated the species as a *C. glabra-C. ovalis* complex. As a seedling, sapling, or young tree, its characteristics are nearly identical to those of *Carya glabra*, but when mature the fruit and bark are distinctly different. Specifically, the *Carya ovalis* fruit is oval and the husk splits along the entire suture, whereas the *Carya glabra* fruit has a teardrop (obovate) shape, a short persistent basal knob, and the husk splits only about one half the distance from the apex to the base. The nut of each species is covered by a thin husk (thickness about 2 to 4 mm). The bark of *C. glabra* tends to become smoother with age, much like that of *C. cordiformis*, but *C. ovalis* bark breaks into interlacing (metastasizing) 2 to 4 cm wide

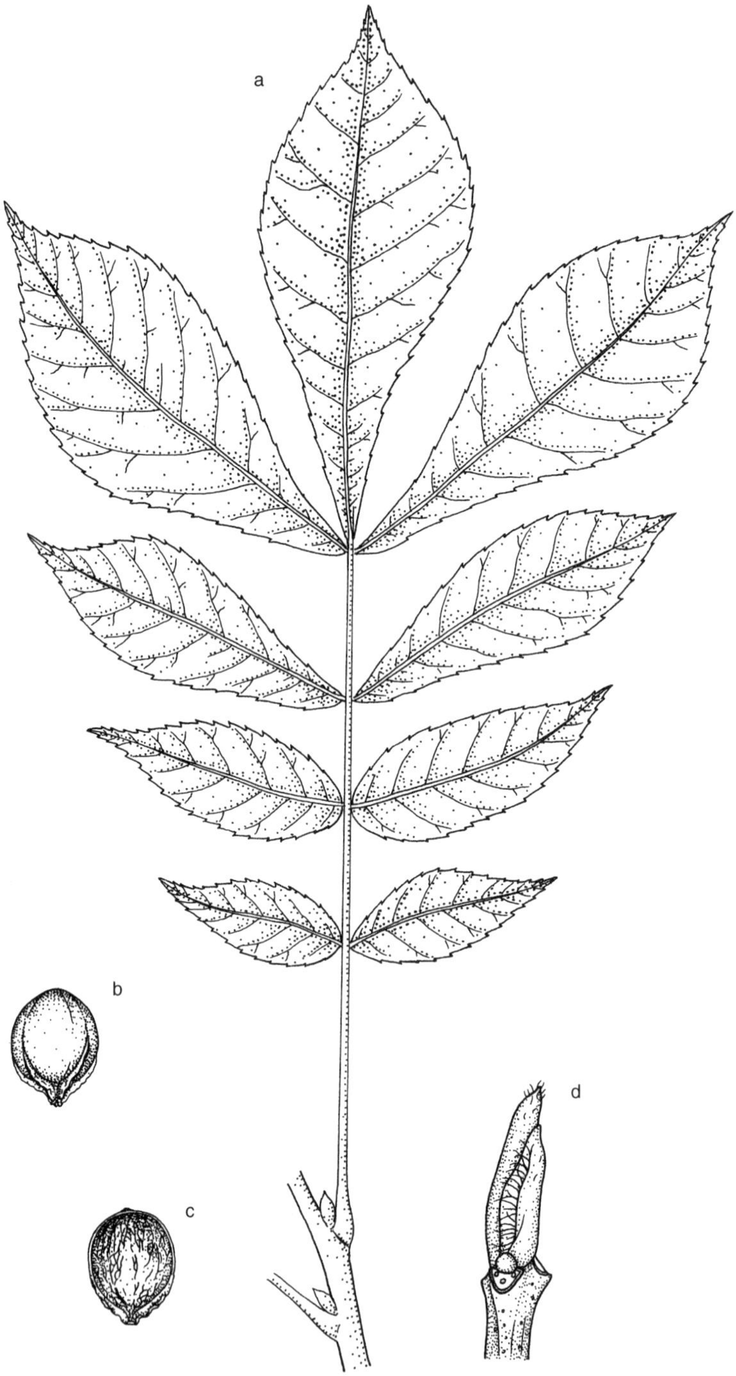

FIGURE 18.29 *Carya cordiformis* (a) pinnately compound leaf ×1/2, (b) nut enclosed by a distally winged husk ×2/3, (c) nut with one half of husk removed ×2/3, (d) twig with terminal bud and leaf scar ×2.

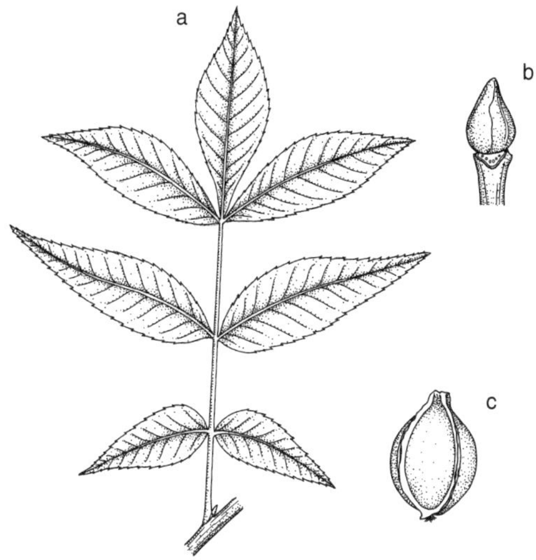

FIGURE 18.30 *Carya myristiciformis* (a) pinnately compound leaf ×1/5, (b) twig with terminal bud and leaf scar ×1.5, (c) nut with winged husk ×2/3.

flat-topped ridges, separated by moderately deep V-shaped fissures or furrows. Cover type 27.7, 28.6, 28.16.

Carya ovata (Miller) K. Koch var. *ovata*, shagbark hickory (Figure 18.32); large tree, height to 45 m; deciduous; leaf arrangement alternate; leaf pinnately compound, length 20 to 45 cm, rachis glabrous or slightly pubescent; leaflet number five with three at the twig apex, shape ovate, obovate, or elliptic, margin serrated and ciliated, lower surface hirsute, length 6 to 20 cm, width 2 to 12 cm; twig stout, buds reddish brown, glabrous or pubescent; terminal bud pubescent, length 1.0 to 1.6 cm, scales imbricate, brittle, each persistent at the base; fruit a nut enclosed by a husk, shape nearly spherical; length 2 to 5 cm, brown to reddish brown; husks with four unwinged sutures, splitting from the apex to the base, thickness to 1.2 cm; seed sweet; mature bark with large, wide, flat exfoliating plates; Regions 3 and 4, except absent on the Gulf Coastal Plain; xeric-mesic and mesic slopes and ridges, occasionally on stream terraces and bottomland sites. Cover type **27.7**, 27.9, 27.16, 29.17, 27.22, 27.26, 28.3, 28.6, 28.18.

Carya ovata (Miller) K. Koch var. *australis* (Ashe) Little [*Carya carolinae-septentrionalis* (Ashe) Engler & Graebner], Carolina shagbark hickory. This species is similar to *Carya ovata* var. *ovata*, but is a slightly smaller tree (height to 35 m), has a smaller leaf (length 20 to 30 cm), a glabrous lower leaf surface and terminal bud, and a more slender twig. Its range is sympatric with that of *Carya ovata* var. *ovata*. Regions 3b, 4c, southern Appalachians, North and South Carolina piedmont, also Tennessee, northern Georgia, Alabama, and Mississippi; xeric uplands, mesic uplands and bottomlands.

Carya floridana Sargent, Florida hickory; large tree, height to 25 m; deciduous; leaf arrangement alternate; leaf pinnately compound, length 15 to 30 cm, rachis with coarse hair in spring, glabrous later; leaflet number five to seven, sometimes three, length 4 to 10 cm, width 2 to 4 cm, shape ovate, obovate, or elliptic, margin serrated, lower surface rusty brown because of peltate scales; buds and twigs rusty brown, densely scaly; terminal bud length 3 to 9 mm, scales imbricate, pubescent; fruit a nut enclosed by a husk, length 2.0 to 4.5 cm, width 2 to 3 cm, tan to reddish brown, shape oval to oblong, husks with four slightly winged sutures, splitting to base, thickness 2 to 3 mm; seed sweet; bark relatively smooth; Region 4, restricted to central Florida; sandy sites of pine-scrub region. Cover type 28.46.

Carya tomentosa (Poiret) Nuttall, mockernut hickory (Figure 18.33); large tree, height to 35 m; deciduous; leaf arrangement alternate; leaf pinnately compound, length 25 to 40 cm, densely tomentose; leaflet number seven to nine, sometimes five, shape ovate, obovate, or elliptic, margin serrated and ciliated, lower surface pubescent, length 4 to 18 cm, width 2 to 8 cm; buds reddish brown, twig densely tomentose, terminal bud densely tomentose, length 1.5 to 2.2 cm, scales imbricate, brittle, each persistent at the base; fruit a nut enclosed by a husk, shape nearly spherical, length 3 to 7 cm, brown to reddish brown; husks with four unwinged sutures, splitting from the apex to the base, thickness 0.4 to 1.0 cm; seed sweet; mature bark with obscure ridges and furrows; Regions 3 and 4, except central and southern Florida and southern Wisconsin and Michigan; xeric and xeric-mesic slopes and ridges, occasionally on stream terraces and bottomland sites.

Carya laciniosa (Michaux f.) Loudon, shellbark or kingnut hickory (Figure 18.34); large tree, height to 40 m; deciduous leaf arrangement alternate; leaf pinnately compound, length 45 to 80 cm, rachis glabrous

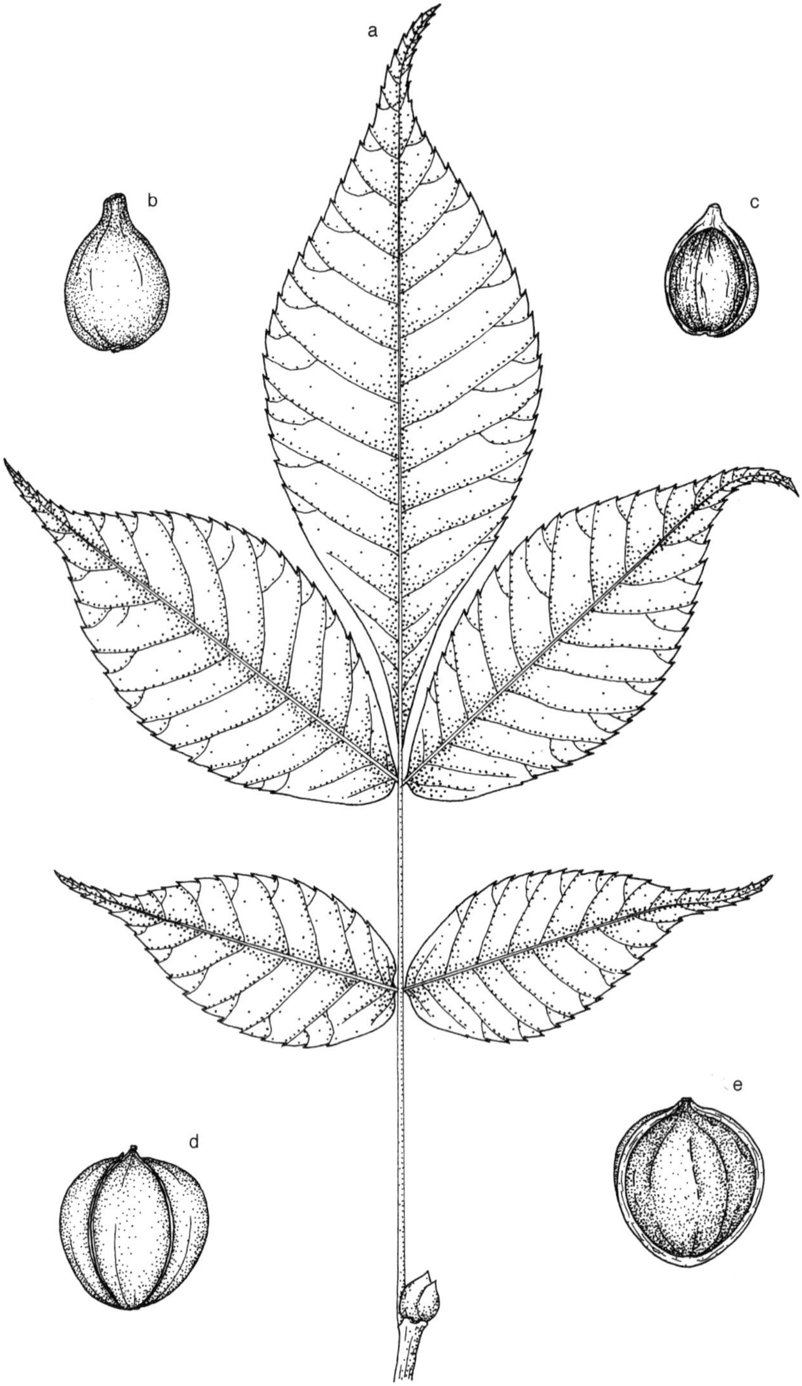

FIGURE 18.31 *Carya glabra* (a) twig with terminal bud and pinnately compound leaf ×1/2, (b) teardrop-shaped nut enclosed by thin husk ×2/3, (c) nut with one half of husk removed ×2/3; *C. ovalis* (d) oval nut enclosed by thin husk ×2/3, (e) nut with one half of husk removed ×2/3.

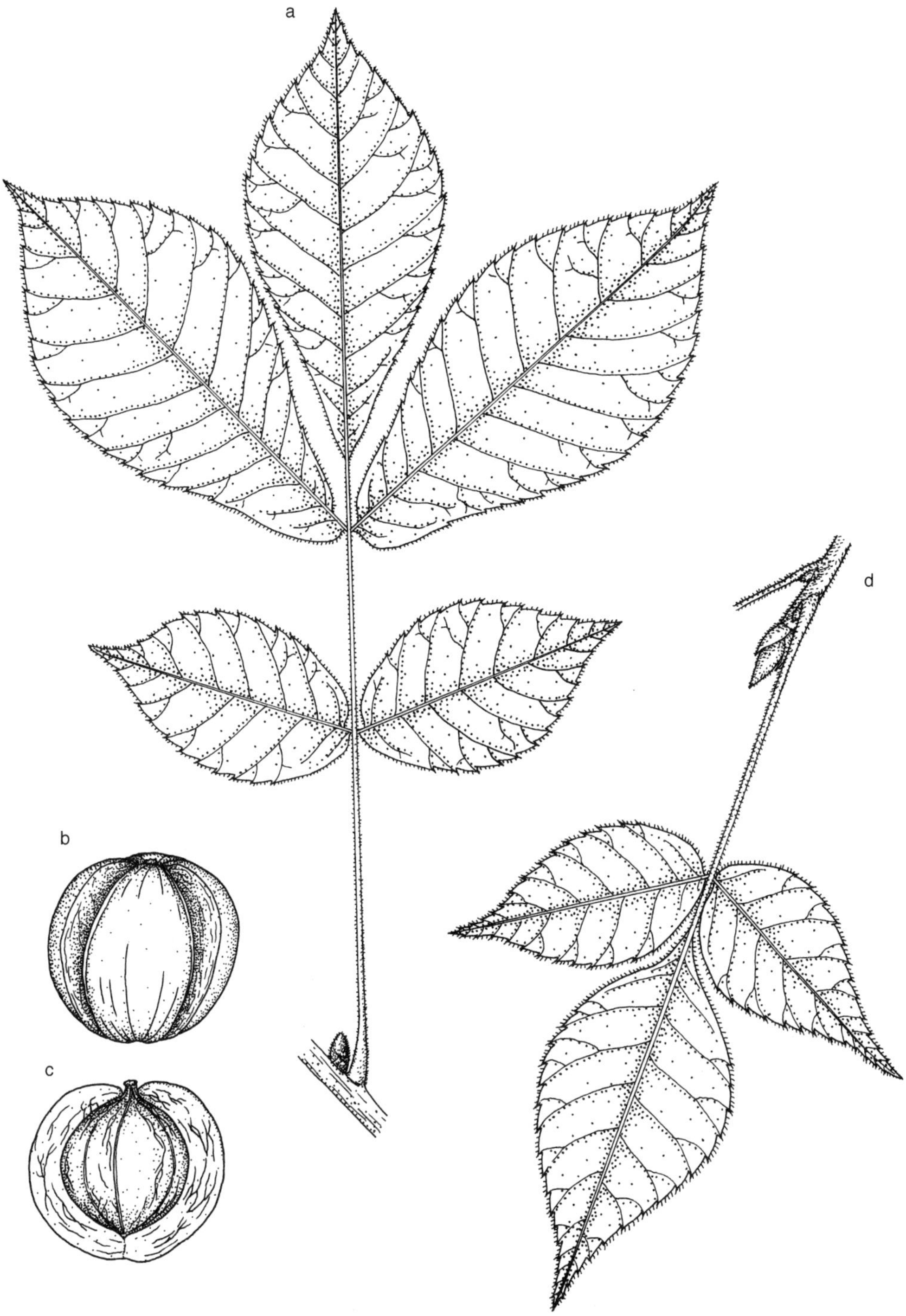

FIGURE 18.32 *Carya ovata* (a) twig with pinnately compound leaf and lateral bud $\times 1/2$, (b) twig with terminal bud and trifoliate leaf $\times 1/2$, (c) nut enclosed by thick husk $\times 2/3$, (d) nut with one half of husk removed $\times 2/3$.

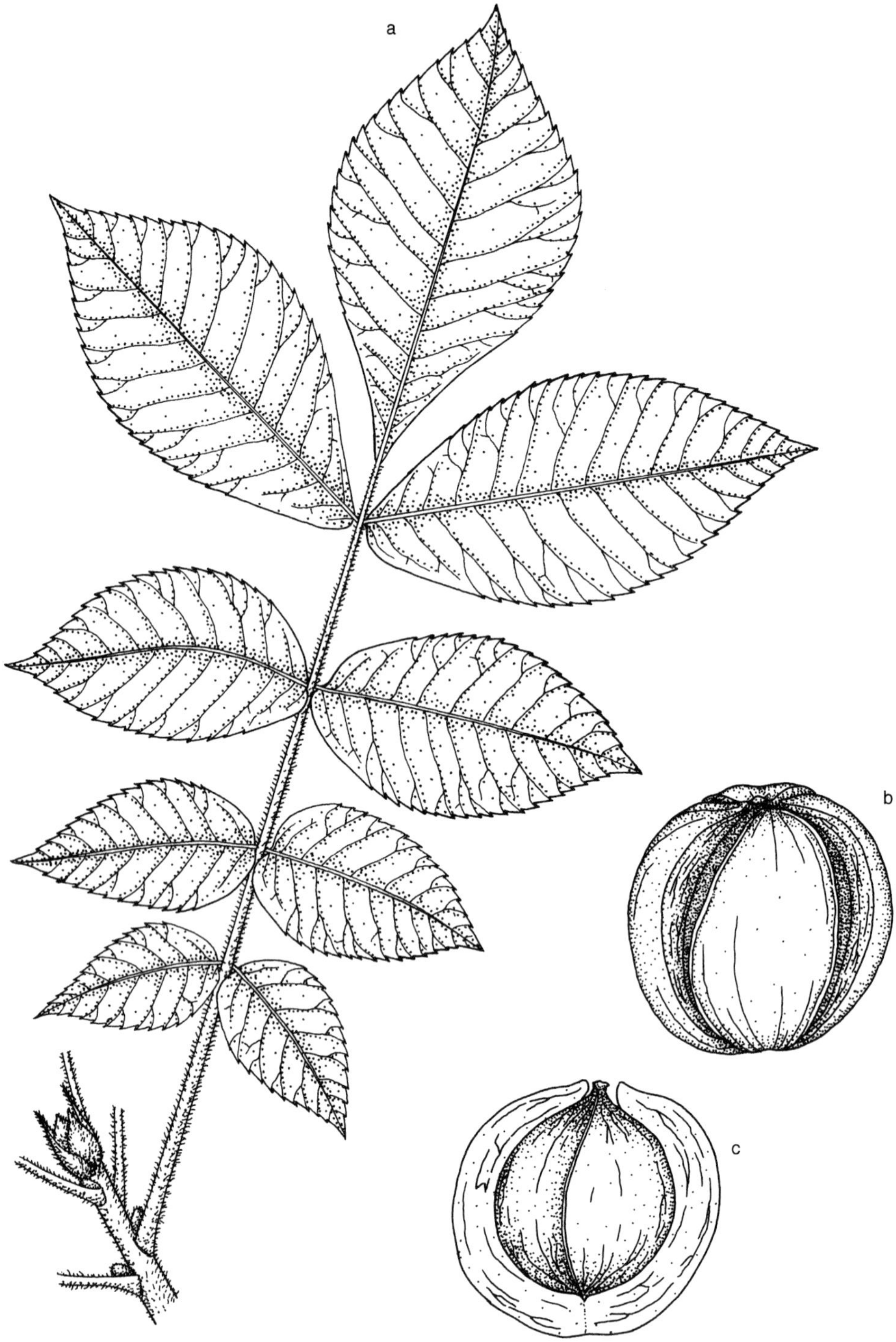

FIGURE 18.33 *Carya tomentosa* (a) twig with terminal bud and pinnately compound leaf ×1/2, (b) nut enclosed by thick husk ×3/4, (c) nut with one half of husk removed ×3/4.

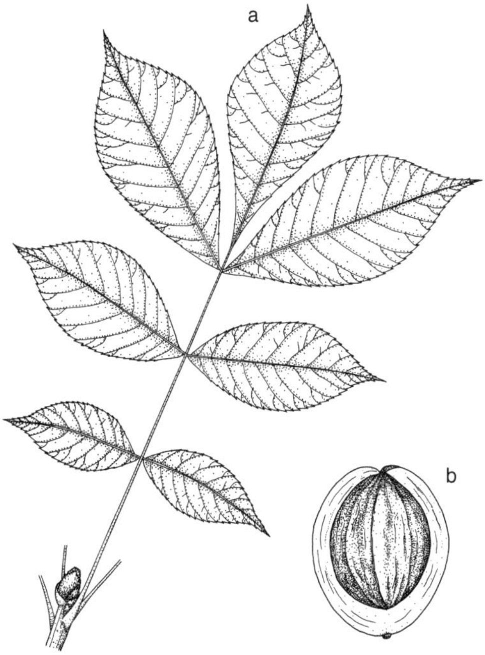

FIGURE 18.34 *Carya laciniosa* (a) twig with terminal bud and pinnately compound leaf ×1/8, (b) nut with one half of thick husk removed ×1/2.

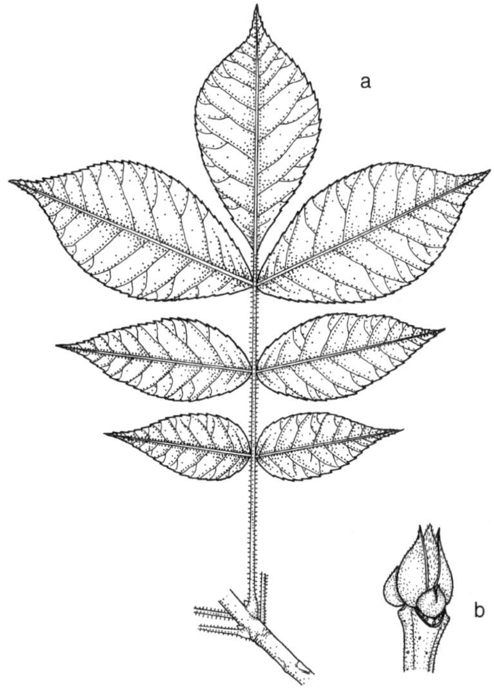

FIGURE 18.35 *Carya texana* (a) twig with pinnately compound leaf ×1/3, (b) twig with terminal and accessory buds ×1.3.

or slightly pubescent; leaflet number seven to nine, sometimes five or eleven, shape ovate, obovate, or elliptic, margin serrated and ciliated or strigose, lower surface hirsute, length 8 to 18 cm, width 3 to 12 cm; terminal bud reddish brown, glabrous or pubescent, length 1.5 to 2.3 cm, scales imbricate, brittle; fruit a nut enclosed by a husk, shape spherical to slightly oblong, length 4 to 6 cm, brown to reddish brown; husks with four unwinged sutures, splitting from the apex to the base, thickness to 1.3 cm; seed sweet; mature bark with large, wide, flat exfoliating plates; primarily Region 3, but extending into Region 4, absent on the Lower Coastal Plain; mesic to wet bottomland sites. Cover type **27.20,** 28.18, 28.22.

Carya texana Buckley (*C. buckleyi* Durand), black hickory (Figure 18.35); large tree, height to 40 m; deciduous; leaf arrangement alternate; leaf pinnately compound, length 20 to 35 cm, petioles with yellowish scales; leaflet number five to seven, sometimes nine, shape ovate, obovate, or elliptic, margin serrated, lower surface densely scaly developing a rusty color, length 4 to 14 cm, width 2 to 5 cm; buds and twigs rusty or yellowish stellate, scales imbricate, densely scaly, terminal bud length 5 to 9 mm; fruit a nut enclosed by a husk, length 3.0 to 4.5 cm, width 2.5 to 3.5 cm, bronze to brown, ovate to round; husks splitting along four unwinged sutures, thickness 2 to 4 mm; seed sweet; bark ridged and furrowed; Regions 3 and 4, central Illinois and central Missouri south to central Texas and Louisiana. This species appears in only a few locations east of the Mississippi River; xeric sites, rocky slopes and ridges. The population of *C. arkansana* Sargent is now included with this species (Stone 1997). Cover type 27.9, 27.11, 27.24.

Carya pallida (Ashe) Englmann & Graebner, sand hickory (Figure 18.36); large tree, height to 30 m; deciduous; leaf arrangement alternate; leaf pinnately

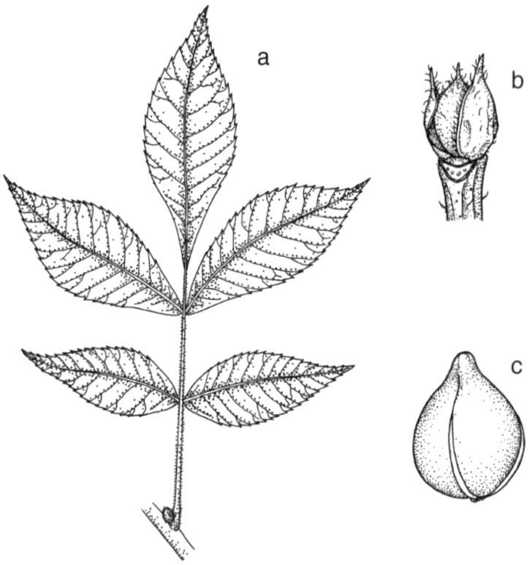

FIGURE 18.36 *Carya pallida* (a) pinnately compound leaf ×1/5, (b) twig with terminal bud and leaf scar ×1.3, (c) teardrop-shaped nut enclosed by distally winged husk ×2/3.

compound, length 20 to 40 cm, rachis slightly hirsute; leaflet number five to seven, sometimes nine, shape ovate, obovate, or elliptic, margin serrated, lower surface glabrous, scaly, silvery, length 4 to 14 cm, width 2 to 5 cm; buds and twigs reddish brown, scaly, scales imbricate, terminal bud length 5 to 9 mm; fruit a nut enclosed by a husk, length 3 to 4 cm, width 2 to 3 cm, tan to reddish brown, ovate to teardrop; husks splitting along four unwinged sutures, thickness 2 to 4 mm; seed sweet; bark ridged and furrowed; Region 3, western Tennessee and western Kentucky south to Louisiana and Florida panhandle; xeric, rocky ridges and slopes. This species extensively intergrades with *Carya texana*, which it strongly resembles, and also crosses with *Carya glabra* and *Carya tomentosa*. Cover type 27.22, 28.6, 28.16, 28.27, 28.43.

ORDER: MYRICALES

Myricaceae (Bayberry or Wax-Myrtle Family)

Myricaceae is a small family composed of three genera and about 50 species of north temperate to subtropical regions, including North and South America, Europe, Africa, and Asia. *Canacomyrica* is restricted to New Caledonia. The two North American genera are *Myrica* (bayberry) and *Comptonia* (sweet-fern).

Members are shrubs or small trees. The unisexual flowers are borne in short, stubby catkins. The flowers are anemophilous. In general, the flowers and catkins have allied the Myricaceae with other amentiferous families. The leaves are simple, alternate on the twig, often coriaceous, and have peltate yellow gland-scales or glandular hair (appearing as yellowish dots). These glands secrete an aromatic oil that produces the odor typical of family members. The peltate gland-scales are homologous with those of the Juglandaceae (Cronquist 1981), to which this family has been considered most closely related; moreover, some taxonomists have included the Myricaceae in the Juglandaceae. Stipules are absent.

There are eight *Myrica* species in North America. All have glands (white dots) on the leaf and, except for *Myrica inodora*, a strong odor of bayberry. *Comptonia* is also aromatic. The fruit of *Myrica* is a drupe, often with a thick waxy coating. The waxy covering of *M. cerifera* and *M. pensylvanica* fruit is removed to make bayberry candles and soap. The fruit is used as food by wildlife. Species of *Myrica* and *Comptonia* form root nodules with nitrogen-fixing bacteria (actinomycetes). This symbiotic relationship undoubtedly assists *C. peregrina* in surviving on the extreme xeric, low nutrient sites of sand plains and glacial outwash of the northern hardwood-conifer forest.

Myrica cerifera Linnaeus, southern bayberry, southern wax-myrtle (Figure 18.37); shrub to small tree, height to 14 m; forming dense colonies from rhizomes, semi-evergreen, leaves persist until the second year; leaf arrangement alternate; leaf simple, length 3 to 10 cm, shape narrowly elliptical to oblanceolate, margin serrated with two or three pairs of teeth near apex, apex acute, base cuneate to long cuneate, blade coriaceous, strongly aromatic when crushed, both surfaces are densely dotted with yellowish or orange glands; fruit drupaceous, globose, diameter 2 to 4 mm, covered with wax, drying blue white. Regions 4a, b, Coastal Plain from New Jersey to Florida and Texas; wet sandy soil, swamps, and pine barrens. *Myrica pusilla* Rafinesque and *M. pumila* Michaux, dwarf wax-myrtle, are included with *M. cerifera*. Cover type 28.2, 28.13, 28.32, 28.36, 28.39, 28.44, 28.51, 28.52, 28.53, 28.55.

FIGURE 18.37 Simple leaf and drupaceous fruit of (a) *Myrica cerifera* ×1; (b) *M. gale* ×1; (c) *M. pensylvanica* ×1; (d), (e) *M. heterophylla* ×1.

Myrica gale Linnaeus, sweet gale (Figure 18.37); multistemmed shrub, height to 2 m; deciduous; leaf arrangement alternate; leaf simple, length 3 to 7 cm, shape oblanceolate to somewhat spatulate, margin serrated with two or five pairs of teeth near apex, apex rounded, base cuneate to acute, blade coriaceous, strongly aromatic when crushed, both surfaces densely dotted with yellowish or orange dots; fruit drupaceous, globose, diameter 2 to 3 mm, wax covering absent; Regions 1 and 2, transcontinental in Canada into Alaska, Lake States, and New England, and 7a, Northern Coastal Conifer Forest, Alaska to Washington; shallow water, lake margins, and swamps.

Myrica heterophylla Rafinesque, evergreen bayberry (Figure 18.37); shrub to small tree, forming dense colonies from rhizomes, height to 3 m; semi-evergreen, leaves persist until the second year; leaf arrangement alternate; leaf simple, length 3 to 12 cm, width 2 to 5 cm; shape elliptical to oblanceolate, margin serrated with one or three pairs of teeth near apex, apex acute to obtuse, base acute to long cuneate, blade usually coriaceous, occasionally membranous, strongly aromatic when crushed, yellowish or orange glands lacking or sparse on blade surfaces; twigs nearly black, occasionally pilose; fruit drupaceous, globose, diameter 3 to 5 mm, covered with wax, drying white; Regions 4a, b, Coastal Plain, New Jersey to Louisiana. Cover type 28.33, 28.36.

Myrica pensylvanica Loiseleur-Deslongchamps, northern bayberry (Figure 18.37); shrub to small tree, forming dense colonies from rhizomes, height to 4 m; deciduous; leaf arrangement alternate; leaf simple, length 3 to 8 cm, width 1.5 to 3.0 cm, shape elliptical to oblanceolate, margin serrated with one or three pairs of teeth near apex, apex acute to obtuse, base acute to long cuneate, blade usually membranous, occasionally coriaceous, strongly aromatic when crushed, lower blade surface lacking or with scattered yellowish or orange glands, upper surface moderately glandular; young twigs reddish brown and glandular; older branches gray, pilose; fruit drupaceous, globose, diameter 3 to 6 mm, covered with wax, drying white; Region 2b, New Jersey to Nova Scotia, inland to Pennsylvania, New York, and the eastern shore of Lake Erie; bogs, swamps, pond edges, and other wet areas. Difficult to separate from *M. heterophylla*, which has larger persistent leaves and glands on both leaf surfaces.

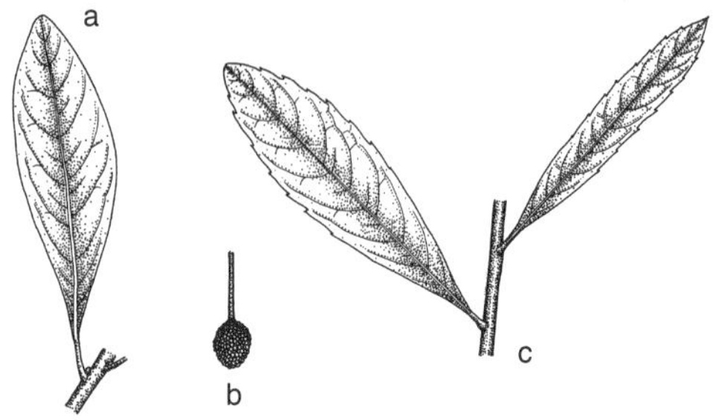

FIGURE 18.38 *Myrica inodora* (a) simple leaf × 1/2, (b) drupaceous fruit covered with wax × 1/2; (c) *M. californica* simple leaves × 1/2.

Myrica inodora Bartram, odorless bayberry (Figure 18.38); shrub to small tree, height to 7 m; semi-evergreen, leaves persist until the second year; leaf arrangement alternate; leaf simple, length 3.5 to 9.5 cm, width 2 to 5 cm; shape elliptical to oblong, margin smooth, rarely serrate, apex acute to rounded, base cuneate, blade usually coriaceous, not aromatic when crushed, blade surfaces with minute colorless glands; twigs reddish brown, glandular; fruit drupaceous, globose, diameter 4 to 7 mm, densely pubescent, covered with wax, drying white; Region 4, Coastal Plain, Florida panhandle to Louisiana; pond edges and bogs dominated by pine. Cover type 28.25.

Myrica californica Chamisso, Pacific bayberry (Figure 18.38); shrub to small tree, height to 10 m; semi-evergreen, leaves persist until the second year; leaf arrangement alternate; leaf simple, length 4 to 12 cm, width 0.7 to 3 cm; shape elliptical, narrowly elliptical, or oblanceolate, margin smooth or serrate with five to nine pairs of teeth, base smooth, apex acute, base cuneate, blade usually coriaceous, blade aromatic when crushed, both surfaces with black to colorless glands; twigs reddish brown, later black, villous, glandular; fruit drupaceous, globose, diameter 4 to 7 mm, villous, white wax covered, absent to thick; Region 7, Coastal Conifer Forest, California to Washington and Vancouver Island.

Myrica hartwegii S. Watson, Sierra sweetbay; shrub, height to 2 m; deciduous; leaf arrangement

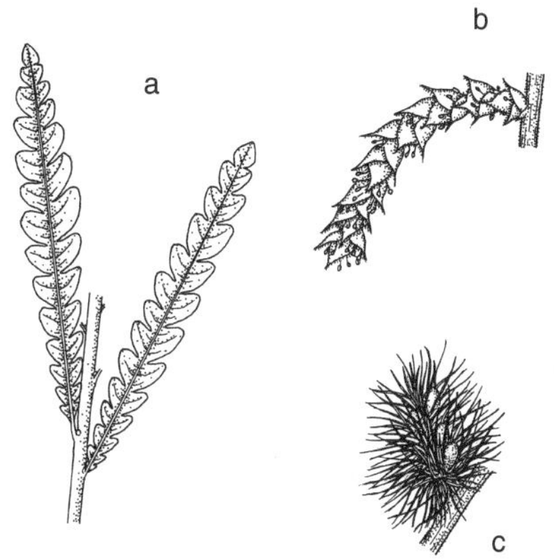

FIGURE 18.39 *Comptonia peregrina* (a) simple pinnatified leaves ×1/2, (b) catkinlike inflorescence of male (staminate) flowers ×1/2, (c) bur with enclosed nutlets ×1/2.

alternate; leaf simple, length 3 to 9 cm, width 1 to 3 cm, shape elliptical to oblanceolate, margin serrate with four to twelve pairs of teeth toward apex, apex acute to obtuse, base cuneate, blade not coriaceous, aromatic when crushed, blade surfaces with minute colorless glands; twigs purple to black, yellow glandular; fruit drupaceous, flattened, diameter 1 to 3 mm, wax absent; Region 7, California Sierra Nevada; stream areas.

Comptonia peregrina (Linnaeus) Coulter (*Myrica asplenifolia* Linnaeus), sweetfern(Figure 18.39); small woody shrub, height to 1 m; deciduous; leaves simple, somewhat resemble a fern frond, length 5 to 12 cm, width 0.5 to 2.0 cm, pinnatified (not completely divided into leaflets), lobe number five to fifteen, upper surface glandular (dotted), extremely aromatic when crushed; stipules cordate, acuminate; branchlets reddish brown to gray, inflorescence of male flowers catkin-like; fruit a nutlet, smooth, shiny, hard, surrounded by eight persistent bracts, burlike; Regions 2, New Brunswick to Saskatchewan, and 3b, the Appalachian provinces; xeric and sandy soil. The alternative name refers to *Asplenium*, the genus name for ebony spleenwort, a fern, although *Comptonia* is not a true fern. This species has been strongly associated with and is used in identifying the site (habitat) types for *Pinus banksiana* (jack pine) in the Lake States (Region 2a). Cover type 26.26.

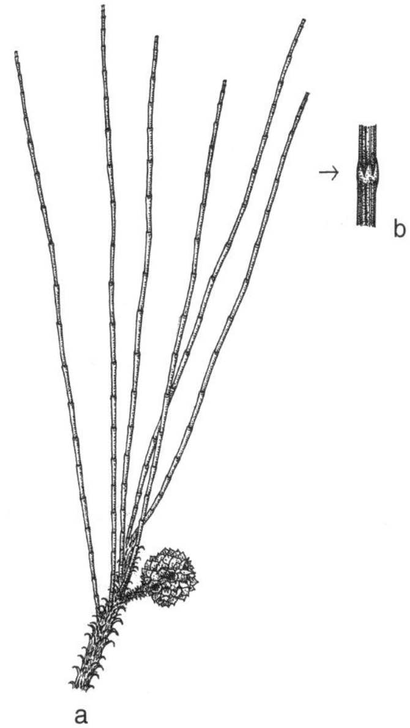

FIGURE 18.40 *Casuarina equisetifolia* (a) twig with nunerous evergreen branchlets ×1/3, (b) minute triangular leaves at an internode ×1.5.

ORDER: CASUARINALES

Casuarinaceae (Beefwood, She-Oak Family)

Casuarina, the only genus in this family (monotypic), is native to the Southern Hemisphere. There are about 65 species distributed in the area from Australia to southeast Asia. However, several species have been introduced into the United States and used as ornamentals, street trees, and windbreaks. The most common species are *C. equisetifolia* Linnaeus (horsetail casuarina; Figure 18.40) and *C. cristata* Miquel (Brazilian beefwood, longleaf casuariana; synonym *C. glauca* Siebold), which have become naturalized and are invading vegetation often to the detriment of native plants. *C. cunninghamiana* Miquel (river-oak casuariana) has also been planted.

The flowers are small, apetalous, and unisexual, with male and female flowers developed on different parts of the tree; plants are monoecious. The male flowers are in aments at the end of a branch, and female flowers in dense heads. The fruit is a samara enclosed in a pair of woody bracts and borne in a conelike multiple fruit structure developed from densely packed flowers.

Casuariana equisetifolia; large tree, height to 35 m; evergreen branchlets cylindrical, length 15 to 20 cm, striated (or grooved), jointed, internode length 3 to 4 cm; leaves scale-like, length 1 to 3 mm, attached in a whorl of six to eight leaves at each node, leaves joined (connate) at the base to form a collar around the branchlet. At a distance, the branchlet appears similar to acicular *Pinus* (pine) needles on a branch. Close examination shows that the branchlets also resemble those of *Ephedra* (ephedra, Mormon-tea) which is a small desert shrub common to the southwestern United States. The branchlets also appear to be a smaller-diameter version of *Equisetum* (horsetail), a plant common in the east. The branchlets and leaves are deciduous as a whole and form a litter mat similar to that of pine needles. *Casuarina cunninghamiana* has eight to ten teeth at an internode and *C. glauca* has twelve to seventeen teeth.

CHAPTER 19

Class: Magnoliopsida; Subclass: Hamamelidae II

ORDER: FAGALES

Fagaceae (Beech and Oak Family)

WORLDWIDE, there are seven to eight genera and 800 to 900 species in this family of trees and shrubs. Species of *Quercus* (oak), *Castanea* (chestnut), and *Fagus* (beech) are common and tend to dominate temperate forests in the Northern Hemisphere, whereas *Nothofagus* (southern beech) is widely distributed in the Southern Hemisphere. A variety of *Quercus* are also extensively distributed in xeric or desert-shrub regions.

The leaves are alternately arranged, simple, pinnately veined, and, depending on the species, vary from lobed to entire. Some species are fully deciduous, others are marcescent, or evergreen. The flowers are small, actinomorphic, imperfect (unisexual, but plant monoecious), apetalous, and anemophilous (except entomophilous in *Castanea*). Male flowers are usually arranged in catkins or small heads. At maturity, the involucre forms a cap or "cupule" at the base of the acorn or completely around the nut. The most common view of the cupule is that it was formed from a cluster of branches and their reduced leaves (Cronquist 1981).

Genus: *Fagus*, Beech

Fagus is represented by a single species in North America, but there are an additional eight to nine species in the Northern Hemisphere, specifically in Europe, Asia, and southwestern Russia (Caucasia). *Fagus sylvatica* (European beech) is an important forest tree in Europe and has been widely planted as a lawn and street tree in the United States and Canada. Dirr (1998) reports as many as 57 cultivars of this species.

Fagus grandifolia Ehrhart, American beech (Figure 19.1); large tree, height to 35 m; marcescent; leaf arrangement alternate; leaf simple, entire, thin, and parchment-like, length 7 to 12 cm, width 3 to 7 cm, elliptical to slightly ovate, apex obtuse to acute; margin remotely serrate, a 1 mm tooth at the end of each lateral vein, surfaces glabrous except for axial tufts on the abaxial surface; twig gray, slender, leaf scar small, circular, stipule scars overlapping; terminal bud linear, length 1.5 to 2.0 cm, width ~3 mm, scales imbricate, glabrous; fruit a nut, length 1.5 to 2.2 cm, cupule valvate, opening along four sutures, outer surface supporting short spines, spine length 1 to 2 mm, seeds two to three, sharply triangular; bark medium gray, smooth; Regions 2, 3, and 4, eastern half of Michigan Upper Peninsula and eastern Wisconsin to Nova Scotia south to Texas and Florida; mesic sites, deep soils, coves, stream terraces, north slopes. Cover type **26.23, 27.8, 27.9, 27.29, 28.11,** 26.1, 26.21, 26.24, 26.28, 26.29, 27.15, 27.22, 27.23, 27.30, 27.31, 28.6, 28.12. Three races have been reported (Camp 1950; Harlow et al. 1996). There are distinct differences in bark color, stem form, and branching characteristics in trees ("red beech") in the Central States Region (southern Illinois, Indiana, and Ohio), as compared with those of the Michigan Upper Peninsula ("gray beech"), which have a dark bark, a straight bole, and few low

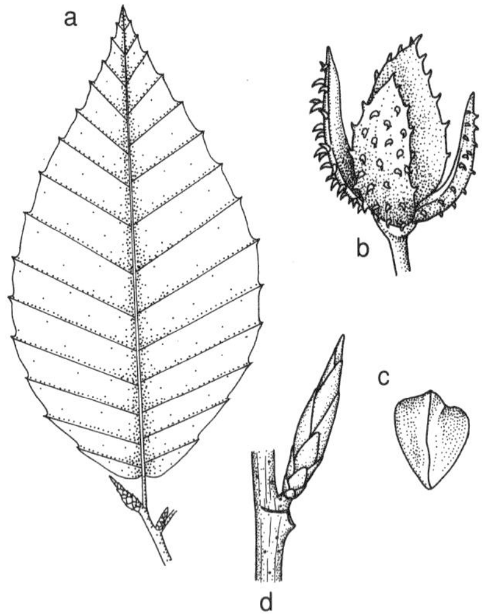

FIGURE 19.1 *Fagus grandifolia* (a) simple leaf ×1/2, (b) cupule ×1, (c) seed ×1, (d) twig with linear bud ×1.

branches. "White beech" is found on poorly drained soil in the Coastal Plain Region. Because *F. grandifolia* is a moisture-demanding mesophyte (and extremely shade tolerant), it has a wide distribution on rolling upland mesic soils in the cool, humid north climate of Region 2, a more restricted distribution primarily in coves and north slopes in the warm, moist climate of Region 3, and a relatively narrow distribution in the drainages and low, wet areas in the hot climate of Region 4. It is apparent that *F. grandifolia* shifts sites, depending on evapotranspiration stress and soil water availability.

Fagus sylvatica Linnaeus, European beech (Figure 19.2); large tree, height to 30 m; marcescent; leaf arrangement alternate; leaf simple, entire, thin and parchment-like, length 7 to 12 cm, width 3 to 7 cm, broadly elliptical, margin remotely serrate to nearly smooth, often ciliated with long hair, blade surfaces often hairy, particularly along the veins; twig gray, slender, leaf scar small, circular, stipule scars overlapping; terminal bud linear, length 1.5 to 2.0 cm, width ~3 mm, scales imbricate, pubescent at the tips; fruit a nut, length 2.5 to 2.2 cm, cupule valvate, opening along four sutures, outer surface supporting long spines, spine length 2 to 5 mm, seeds two to three, sharply triangular; bark medium gray, smooth; introduced.

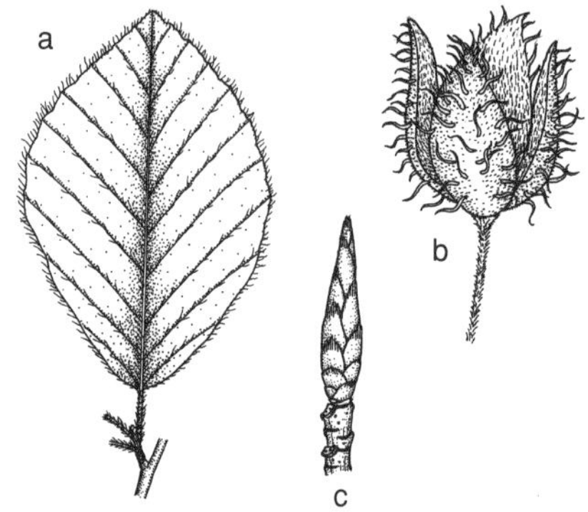

FIGURE 19.2 *Fagus sylvatica* (a) simple leaf ×1/2, (b) cupule ×1, (c) twig with linear bud ×1.

Genus: *Castanea*, Chestnut

This genus of eight to ten species is represented by several species restricted to the eastern United States. There is some debate over the number of species. It is universally agreed that *Castanea dentata* (American chestnut) and *C. pumila* (dwarf chestnut) are distinct species, with the status of *C. ozarkensis* (Ozark chinkapin) and *C. alnifolia* (Florida chinkapin) in dispute. Nixon (1997) lists *C. ozarkensis* as a third species (as does Fernald 1950) and considers *C. alnifolia* as a form of *C. pumila*. Conversely, Foote and Jones (1989) have *C. alnifolia* as a separate species, with *C. ozarkensis* as a variety [var. *ozarkensis* (Ashe) Tucker].

Castanea dentata (Marshall) Borkhausen, American chestnut (Figure 19.3); large tree, height to 35 m; deciduous; leaf arrangement alternate; leaf simple, entire, length 10 to 25 cm, width 4 to 10 cm, elliptical, oblong to slightly oblanceolate, apex acuminate, margin sharply serrate, a 2 to 3 mm tooth tapering to a prickle at the end of each lateral vein, number of lateral veins 12 to 22 per side, proximal veins not reaching the margin, abaxial surface glabrous or with scattered hair along veins; twig reddish brown, terminal bud absent; fruit globose, diameter 4.5 to 7.5 cm, cupule (bur) supporting long spines, spine length ~1 to 2 cm, nuts two to three, often flattened; originally Regions 2a, 3, and 4, Indiana and Michigan to Maine, south to Georgia, Alabama, and Mississippi. Cover type 26.26, 26.28, 27.22, 27.34.

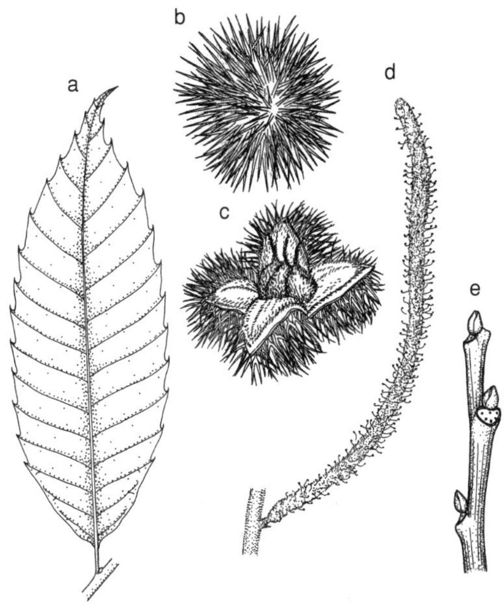

FIGURE 19.3 *Castanea dentata* (a) simple leaf ×1/3, (b) closed cupule (bur) ×1/2, (c) open cupule showing nuts ×1/2, (d) erect catkin of staminate flowers ×1/2, and (e) twig with terminal and lateral buds ×3/4.

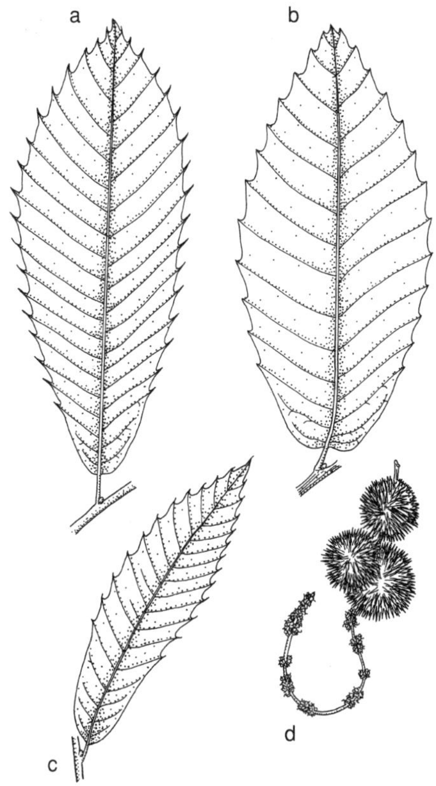

FIGURE 19.4 Simple leaf of (a) *Castanea ozarkensis* ×1/2; (b) *Castanea alnifolia* leaf ×1/2; *C. pumila* (c) simple leaf ×1/2, (d) cupules (burs) enclosing nuts on a stalk ×1/3.

Castanea ozarkensis Ashe, Ozark chinkapin (Figure 19.4); shrub to medium-sized tree, height to 17 m; deciduous; leaf arrangement alternate; leaf simple, entire, length 8 to 20 cm, width 4 to 10 cm, narrowly elliptical to oblanceolate, apex acute, margin sharply serrate, a 2 to 3 mm tooth tapering to a prickle at the end of each lateral vein, number of lateral veins 12 to 22 per side, proximal veins not reaching the margin, abaxial surface lightly to densely white pubescent; twig reddish brown, terminal bud absent; fruit globose or oval, diameter 2 to 5 cm, cupule (bur) supporting long spines, spine length 1 to 2 cm, nuts two to three, often flattened; Region 3, northeast Texas, Oklahoma, Arkansas, southwest Missouri; disjunct in northern Alabama, Ozark and Ouachita Mountains; dry slopes.

Castanea pumila Miller, dwarf chestnut (Figure 19.4); medium-sized tree, height to 17 m; deciduous; leaf arrangement alternate; leaf simple, entire, length 10 to 20 cm, width 4 to 10 cm, narrowly elliptical to oblanceolate, apex acute, margin sharply serrate, a 1 to 2 mm tooth tapering to a prickle at the end of each lateral vein, number of lateral veins 12 to 18 per side, proximal veins not reaching the margin, abaxial surface densely white pubescent; twig reddish brown, terminal bud absent; fruit globose or oval, diameter 2 to 4 cm, cupule (bur) supporting long spines, spine length 5 to 10 mm, nuts two to three, often flattened; Regions 3 and 4, northeastern Texas and Oklahoma, Arkansas, southern Missouri, to southern Indiana, Ohio, and Pennsylvania, south to Florida; dry slopes, rocky soils.

Castanea alnifolia Nuttall, running chinkapin (Figure 19.4); low multi-stemmed shrub; leaves and fruit are similar to those of *C. pumila*, except that the leaf abaxial surface is glabrous and greenish. Region 4, Florida panhandle to Atlantic Coast, disjunct populations to Louisiana and to North Carolina.

Among these species, *Castanea dentata* was once the most important and widespread. The trees grew fast, and the reddish brown wood was light but strong and had a desirable grain and texture. For the present, the species has been eliminated as a forest tree. The chestnut blight, *Cryphonectria parasitica* (*Endothia parasitica*), was introduced by trees imported into the United States in 1904. The fungus spread rapidly through the eastern forests, killing nearly all American chestnut trees. The species may still be occasionally found in the woods as sprouts, inasmuch the fungus does not attack until stems reach tree size. Researchers have attempted to develop a disease-resistant strain through introgression. The Chinese chestnut (*C. mollisima* Blume) is relatively resistant but is limby and has poor form for a forest tree, and *C. dentata* grows to be tall and straight with few low branches. One approach has been to cross these two species, then backcross the hybrid with the American parent, and again backcross the offspring with the America parent. This approach has been relatively successful. Third generation backcrossed individuals have been planted, and it is anticipated that resistant individuals will be reintroduced into the forest by 2010. It should be noted that the fungus also attacks *Quercus coccinea* (scarlet oak) and causes the decay and butt swell seen in nearly all trees.

Genus: *Chrysolepis*, Chinkapin

Chrysolepis is a small genus of the Fagaceae that at one time was included with *Castanopsis* that is native to Asia (Nixon 1997). There are two species native to the United States, both found along the west coast. The nuts are edible.

Chrysolepis chrysophylla (Douglas ex. Hooker) Hjelmquist var. *chrysophylla* [*Castanopsis chrysophylla* (Douglas ex. Hooker) A. de Candolle], giant golden chinkapin (Figure 19.5); large tree, height to 35 m; evergreen; leaf arrangement alternate; leaf simple, entire, thick, coriaceous, length 6 to 14 cm, width 2.0 to 4.5 cm, narrowly elliptical, apex and base acuminate, margin smooth, abaxial surface with gold to brown glandular tomentose hair; twig yellowish, glandular hairy, terminal bud ovoid to nearly globose; fruit similar to that of *C. dentata*, globose, cupule densely spiny, diameter 2 to 6 cm, seeds two to three, rounded or angular; bark thick, extremely rough; Region 7, Washington to central California, coastal to inland mountains,

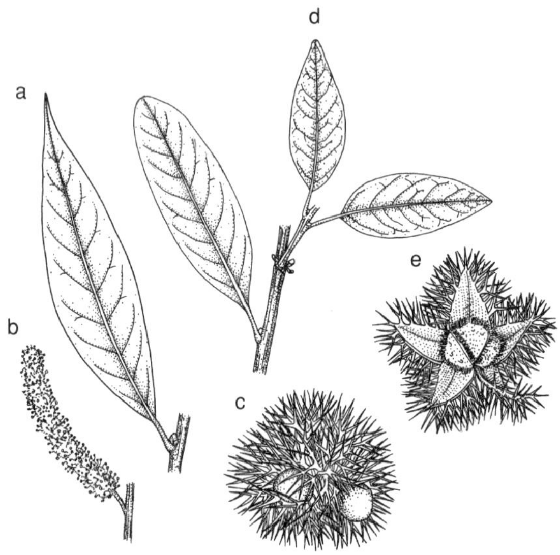

FIGURE 19.5 *Chrysolepis chrysophylla* (a) simple leaf ×1/2, (b) erect inflorescence of staminate flowers ×1/2, (c) cupule (bur) with seeds ×1/2; *Chrysolepsis sempervirens* (d) simple leaf ×1/2 (e) open cupule showing seed cavity ×1/2.

elevation to 2000 m; component of conifer and redwood forests.

Chrysolepis chrysophylla Hjelmquist var. *minor* (Bentham) Munz, golden chinkapin; shrub or arborescent height to 8 m; similar to *C. chrysophylla*, except leaf is moderately folded along midrib, the adaxial surfaces facing each other, abaxial surface golden; Region 7b, southern Oregon to central California; dry slopes and ridges, elevation 300 to 1700 m, open or savanna-like conifer woodland.

Chrysolepis sempervirens (Kellogg) Hjelmquist, bush or Sierra chinkapin (Figure 19.5); shrub, height to 2 m; evergreen; leaf arrangement alternate; leaf simple, entire, thick, coriaceous, length 3 to 7 cm, width 1.5 to 2.0 cm, narrowly oblong or lanceolate, apex and base obtuse, margin smooth, abaxial surface with gold glandular tomentose hair; twig yellowish, glandular hairy; terminal bud ovoid to nearly globose; fruit similar to that of *C. chrysophylla*, globose, cupule densely spiny, yellowish, diameter 2 to 6 cm, seeds two to three, rounded or angular; Region 7, California Sierra Nevada; xeric mountain slopes and rocky ridges, pine forest and chaparral.

Genus: *Lithocarpus*, Tanoak

There are about 300 species worldwide but only one species occurs in western United States. The acorn of *Lithocarpus* is similar to that of *Quercus*, but other characteristics (inflorescence, flowers, and pollen) indicate that is more closely related to *Castanea* and *Chrysolepis* (Nixon 1997).

Lithocarpus densiflorus (Hooker & Arnott) Rehder var. *densiflorus*, tanoak, tanbark-oak (Figure 19.6); large tree, height 30 to 40 m; evergreen; leaf arrangement alternate; leaf simple, entire, length 6 to 11 cm, width 2.5 to 4.5 cm, blade thick, coriaceous, oblong, apex acute, margin nearly smooth (occasionally revolute) to sharply serrate, a 1 to 2 mm tooth tapering at the end of each lateral vein, number of lateral veins 9 to 15 per side, veins sunken, proximal veins not reaching the margin, abaxial surface densely tomentose, hair deciduous at maturity; twig gray brown, densely tomentose, terminal bud present; catkin erect; fruit an acorn, cupule (cap) scales spreading and more or less reflexed; Region 7b, Coastal Ranges of California and Oregon; mesic, fertile mountain slopes and ridges, redwood and mixed pine forest. Cover type 30.37, 30.4, 30.7, 30.18, 30.20, 30.25, 30.30, 30.34, 30.41, 30.42, 30.44, 30.45.

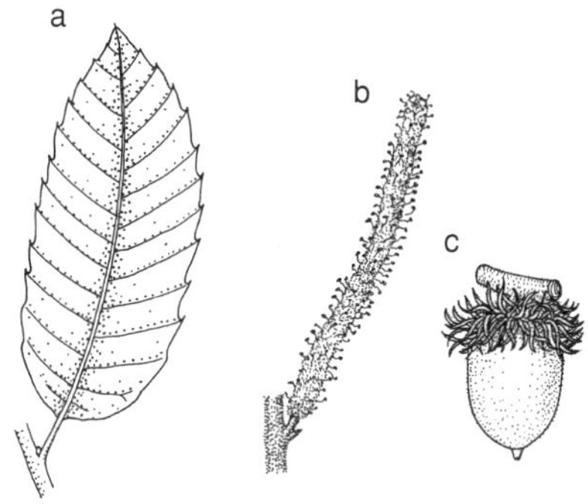

FIGURE 19.6 *Lithocarpus densiflorus* (a) simple leaf × 1/2, (b) erect catkin of staminate flowers × 1/2, (c) cupule and acorn × 1/2.

Lithocarpus densiflorus (Hooker & Arnott) Rehder var. *echinoides*, dwarf tanbark; shrub, height to 2 to 3 m; similar to var. *densiflorus* except: leaf length 3 to 7 cm, secondary veins not sunken, obscure; Region 7b, northern coastal California and Oregon into the Sierra Nevada; open forest or margins, xeric slopes and ridges.

Genus: *Quercus*, Oak

There are approximately 400 to 450 *Quercus* species in North America, Mexico, Central America, the West Indies, South America, Eurasia, and Africa. This genus is the largest woody plant taxon in North America, with a representation of about 85 species of trees and arborescents. Representatives are found in a wide range of soil moisture and nutrient conditions, climatic regimes, and community types.

Characteristics of the family include simple lobed or unlobed leaves, lobes or teeth rounded or bristle tipped, pinnate venation, buds clustered at the twig tip, long male catkins, and an acorn as the fruit type. The large acorn provides substantial food for the developing seedling (Figure 19.7a). *Quercus* species have ring-porous wood; that is, numerous large vessels develop at the start of growth in the spring (spring- or earlywood) when the soil has high water levels, but by early to midsummer, when evapotranspiration rates are high and soil water is limited, the vessels are small and fewer in number (summer- or latewood).

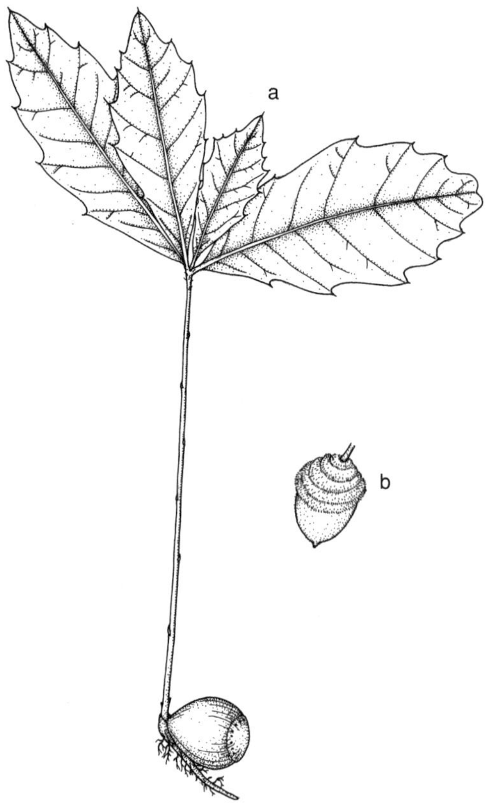

FIGURE 19.7 (a) Red oak seedling ×4/5; (b) *Cyclobalanus* cupule and acorn ×1/2.

Characteristics of the wood are important, as they determine the type of use. Most North American *Quercus* species are subdivided into "white oak" (section *Quercus*) and "red oak" (section *Lobatae*) groups (Nixon 1997) based on the tyloses in the wood and acorn maturation. Tyloses are bladder-like outgrowths from cell membranes of primarily ray parenchyma cells before they die, along with other cell types, to become "heartwood." The outgrowths expand through the pits into the large-diameter vessel cells, effectively blocking water conduction. The result is that only white oak wood can be used for cooperage—the production of barrels used to age whiskey and bourbon.

The acorns of all white oak group species mature annually; most red oak group species have biennial acorn maturation. The red oak group species have awns on the tips of lobes, on the margin, or at the apex (bristle tips), whereas, white oaks lack awns. The red oak species have a dark nearly black bark that is smooth when young and extremely rough, blocky or deeply furrowed, when mature. They have a "tight" bark in that it is strongly attached to the trunk. White oaks have loose, or platy light gray bark regardless of age. Both white and red oak groups contain deciduous and evergreen species.

The third group of North American *Quercus* species are in the section *Protobalanus*, referred to as the golden cup oaks. All are evergreen and located in California, Nevada, Arizona, and Mexico. Although the acorn caps are similar to those of red and white oak groups, in Europe they are more distinctive (Figure 19.7b)

Quercus wood is one of the easiest woods to identify because of the large pores (vessel cells) and long, wide rays (lines that run across the rings from the edge to the center of the stem) visible from the radial (cross-section) surface. Rays can be seen in the top of a stump even if it has substantially decayed. In addition, the height of the rays in the wood (as viewed from the tangential or side of the stem) of the white and red oak groups can be used to separate them. The red oak group has rays that are relatively short, perhaps about 4 to 6 mm high, whereas the white oak has rays that usually are more than a centimeter high.

In the eastern United States, the number of species rapidly increases from southern Ontario, Canada, and Maine, where as many as 8 species occur, to the Gulf Coast where 25 to 30 are found. About 12 to 15 *Quercus* species, along with 6 to 8 *Carya* species, dominate both upland and bottomland forests of the Central States and Appalachian Mountains (Region 3). *Quercus* species also have a strong tendency to replace *Pinus* in the southeastern pine-hardwood forest (Region 4).

In the western United States, *Quercus* species are concentrated in the region extending from Oregon and California (Region 7) into Colorado and western Texas (Region 5); only one or two species extend into Washington. In California, *Quercus* is an important component of tree and shrub plant community types at mid-to-low elevations. Savanna and woodland communities are found in the Sierra Nevada foothills and Coastal Ranges surrounding the Sacramento and San Joaquin river valleys of central California. The shrub-dominated chaparral community is found in the southern part of the state, for example, in the San Bernadino Mountains.

The greatest western concentration of oak species occurs from California, Nevada, and Arizona into New Mexico and western Texas. Here they are mixed with ponderosa pine and juniper-pinyon savanna at low elevations in mountains and are a component of the desert shrub communities of extensive nonmountain areas.

To say that the species within the genus *Quercus* are stable would be a misstatement of substantial proportion. Within each of the oak groups, species freely interbreed, resulting in a multitude of hybrids now naturally found in plant communities and disturbed sites. For example, the state of Missouri recognizes 15 hybrids, which is more than the number of species in the state (Settergren and McDermott 1969). Therefore, the problem related to identification becomes extremely difficult, particularly if backcrossing with one or both parents (introgression) occurs. When there are individuals in a population with characteristics that vary from one parent to the other parent, the problem goes far beyond simple identification and into the realm of taxonomic decisions regarding the range of characteristics to accept for a species and, possibly, the consideration of a new species.

The sections *Lobatae* and *Quercus* have been substantially reviewed and reworked by Jensen (1997) and Nixon and Muller (1997), respectively. A number of names have been discarded and some early species populations have been split into new species or moved into an existing species.

For convenience, and in keeping with the regional/ecological approach, the species are grouped by region (east, southeast, western, California) and then separated by section and several natural subdivisions: white, chestnut, red, willow groups. A line from central Texas to Minnesota generally separates eastern species from southwestern as well as from western species.

Eastern North America (East of the 100th Meridian) Regions 2, 3, and 4

1. Section *Quercus* (Subgenus *Leucobalanus*, *Lepidobalanus*, White Oak Group)—White Oak Species, large trees; large leaves (length >14 cm), sinuses deep.

Quercus alba Linnaeus, white oak (Figure 19.8); large tree, height to 40 m; deciduous; leaves and buds clustered at the twig apex; leaf arrangement alternate; leaf simple, outline elliptical, total length 10 to 20 cm, width 5 to 10 cm, lobe number seven to eleven, apices rounded, middle lobes usually expanded with a tendency to bifurcate, sinuses shallow to deep, extending one quarter to three quarters the distance from the lobe apex to the midrib, abaxial surface generally glabrous except for scattered hair on or near the veins; bud length 2 to 3 mm, essentially glabrous; acorn length 2 to 3 cm, cap rough, warty, pubescent, dull, enclosing about one third of the nut, cap margin thin, fringed with hair; twig surface glabrous, reddish brown, shiny; bark ashy gray, usually developing flat, loose plates above 3 m; Regions 3 and 4, eastern United States north into Michigan and southern Ontario, absent in the Florida peninsula; xeric to mesic upland sites and moist, high stream terraces. *Quercus alba* hybridizes with most of the other "white" and "chestnut" oak species. Trees whose leaves have shallow sinuses appear not to grow as fast nor to be as vigorous and robust as those whose leaves have deep sinuses. This difference may be related to site conditions, with trees with shallow sinuses growing on relatively drier sites and those with deep sinuses growing on north slopes and terraces. Recognized hybrids occur with *Q. bicolor*, *Q. stellata*, *Q. macrocarpa*, *Q. stellata*, and *Q. michauxii*, *Q. prinus*, and *Q. robur* (Miller and Lamb 1985). Cover type **26.28, 27.7, 27.9, 27.16, 27.17, 27.22, 27.26, 28.6, 28.16**, 26.18, 26.29, 27.5, 27.21, 27.23, 27.24, 27.25, 27.32, 27.37, 28.3, 28.4, 28.11, 28.18.

Quercus macrocarpa Michaux, bur oak (Figure 19.9); large tree, height to 40 m; deciduous; leaves and buds clustered at the twig apex; leaf arrangement alternate; leaf simple, outline broadly elliptical to obovate, total length 12 to 25 cm, width 5 to 15 cm, lobe number five to nine, apices rounded, middle lobes usually expanded with a tendency to bifurcate, two paired middle sinuses of the blade deep, extending three quarters or more the distance from the lobe apex to the midrib, C- or U-shaped, proximal one third of leaf narrow with large rounded teeth, distal portion of leaf broad, bullet-like with large rounded teeth, abaxial surface

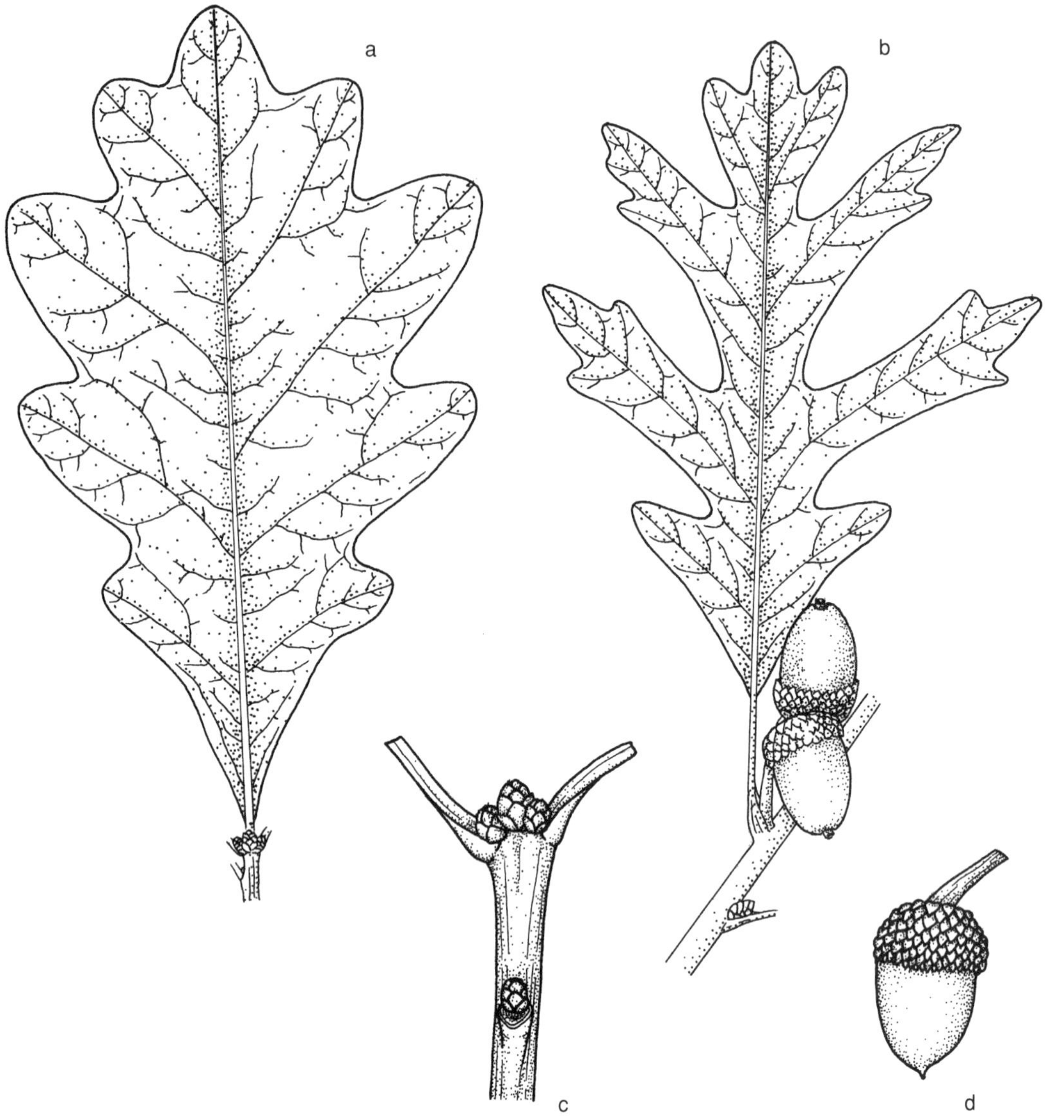

FIGURE 19.8 *Quercus alba* (a), (b) simple leaf ×2/3, (c) twig with terminal and accessory buds ×3, (d) cupule and acorn ×1.

sparingly to densely pubescent; bud length 2 to 4 mm, pubescent, terminal bud usually accompanied by needle-like bracts; acorn length 2 to 5 cm, cap pubescent, dull, enclosing about one half of the nut, cap margin long, fringed; twig surface glabrous, reddish brown, shiny; second-year and older branches often with a thick cork layer or wing; bark ashy gray, often thick; Regions 1, 2 and 3, eastern United States into southern Manitoba, Ontario, and Quebec, Region 4, Texas and Louisiana; xeric to mesic upland sites and moist high stream terraces. Recognized hybrids occur with *Q. bicolor*, *Q. alba*,

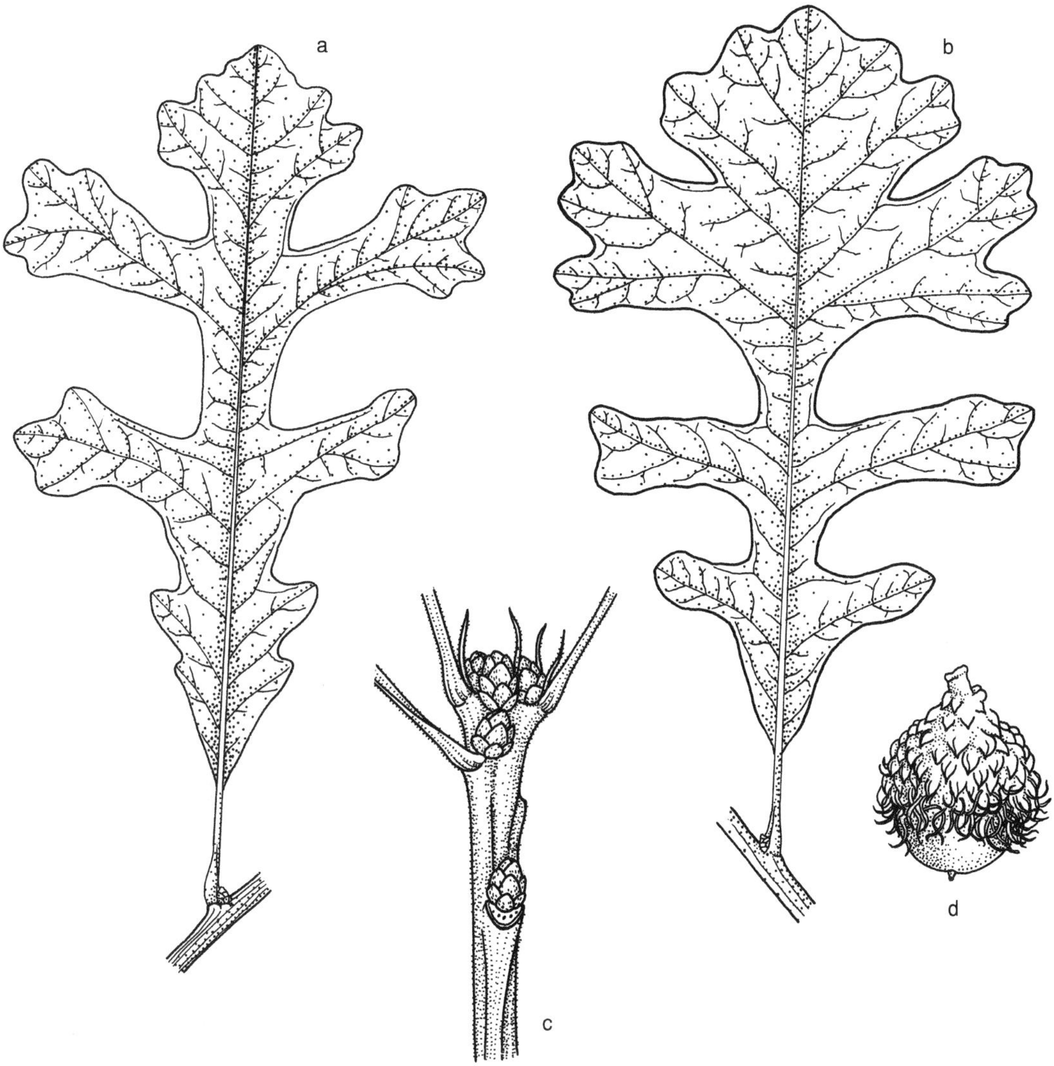

FIGURE 19.9 *Quercus macrocarpa* (a), (b) simple leaves ×1/2, (c) twig with terminal and accessory buds ×2, (d) fringed cupule and acorn ×3/4.

Q. muehlenbergii, *Q. lyrata*, *Q. stellata*, and *Q. michauxii* (Miller and Lamb 1985). Cover type **27.19, 27.20,** 28.9.

Quercus stellata Wangenheim, post oak (Figure 19.10); large tree, height to 28 m; deciduous; leaves and buds clustered at the twig apex; leaf arrangement alternate; leaf simple, shape cruciform, total length 10 to 18 cm, width 7 to 10 cm, lobe number five to seven, apices rounded, middle pair of lobes at right angles to the midrib, squarish and/or bifurcated, creating a cruciform shape, sinuses variable, usually extending one half to two thirds the distance from the lobe apex to the midrib, abaxial surface generally densely

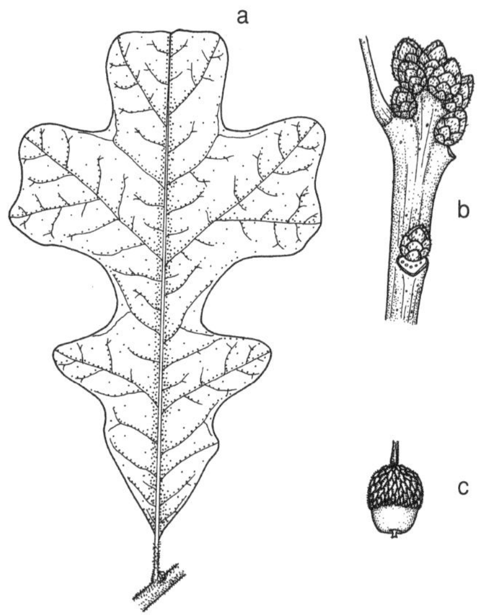

FIGURE 19.10 *Quercus stellata* (a) simple leaf, (b) twig with terminal and accessory buds, (c) cupule and acorn.

pubescent with stellate hair; bud length 3 to 4 mm, yellowish, densely pubescent; acorn length 1.5 to 2.5 cm, cap pubescent, dull, enclosing one half of the nut, cap margin thin; twig surface yellowish, densely pubescent; bark ashy gray; Regions 3 and 4, southern Illinois and Missouri and eastern Texas to Florida and New Jersey, not in the Appalachian Mountains; xeric upland sites, rocky slopes and ridgetops (Fralish et al. 1999). Recognized hybrids occur with *Q. bicolor*, *Q. virginiana*, *Q. macrocarpa*, and *Q. michauxii* (Miller and Lamb 1985). Cover type **27.11, 27.24, 27.25, 28.3, 28.13,** 27.9, 27.18, 27.26, 28.1, 28.4, 28.13, 28.16,

Quercus lyrata Walter, overcup oak (Figure 19.11); large tree, height to 28 m; deciduous; leaves and buds clustered at the twig apex; leaf arrangement alternate, leaf simple, outline elliptical, total length 12 to 25 cm, width 7 to 10 cm, lobe number five to nine, apices rounded, middle lobes often alternate, tending to bifurcate, middle sinuses extending one half to two thirds the distance from the lobe apex to the midrib, abaxial surface generally densely pubescent or glabrous, except for scattered hair on or near the veins; bud length 2 to 3 mm, essentially glabrous; acorn length 1.5 to 2.5 cm, cap pubescent, dull, enclosing more than three quarters of the nut, cap margin thin, fringed with short hair; twig surface reddish to yellowish, glabrous, shiny, terminal buds usually accompanied by narrow bracts; bark ashy gray; Regions 3 and 4, southern Illinois and Missouri and eastern Texas to Florida and New Jersey, not in the Appalachian Mountains or peninsula Florida; mesic-hydric and periodically flooded hydric forest, bottomlands. Recognized hybrids occur with *Q. bicolor*, *Q. virginiana*, *Q. macrocarpa*, *Q. stellata*, and *Q. michauxii* (Miller and Lamb 1985). Cover type **28.17, 28.50,** 27.2, 27.10, 28.8, 28.32, 28.49, 28.55.

2. Section *Quercus* (Subgenus *Leucobalanus*, *Lepidobalanus*, White Oak Group)—White Oak Species, shrubs or small to medium trees, small leaves (length < 14 cm), blade margin entire or lobes poorly developed.

Quercus similis Ashe (*Q. stellata* Wangenheim var. *paludosa* Sargent), delta post oak, bottomland post oak, swamp post oak (Figure 19.12); medium to large tree, height to 25 m; deciduous; leaves and buds clustered at the twig apex; leaf arrangement alternate; leaf simple, total length 7 to 14 cm, width 3 to 8 cm, outline obovate, lobe number seven to nine, all lobe apices rounded or middle pair of lobes squarish and/or bifurcated, creating a weak cruciform shape, sinuses variable, extending one third to two thirds the distance from the lobe apex to the midrib, abaxial surface generally densely stellate pubescent; twig and bud densely pubescent, twig surface yellowish to reddish brown, bud length 3 to 4 mm; acorn length 1.5 to 2.5 cm, cap pubescent, dull, enclosing one half of the nut, cap margin thin, bark ashy gray; Region 4a, Texas to Georgia and South Carolina, excluding Florida, generally absent in 4b and c; mesic-hydric sites, first bottom ridges and terraces.

Quercus austrina Small [*Quercus durandii* Buckley var. *austrina* (Small) E. J. Palmer], bluff white oak, Durand oak (Figure 19.13); medium-sized tree, height to 22 m; deciduous; leaf arrangement alternate; leaf simple, shape elliptical to obovate, total length 4 to 14 cm, width 4 to 8 cm, entire or lobed, apex rounded or obtuse, sinuses shallow, extending one third the distance from the lobe apex to the midrib, abaxial surface essentially glabrous,

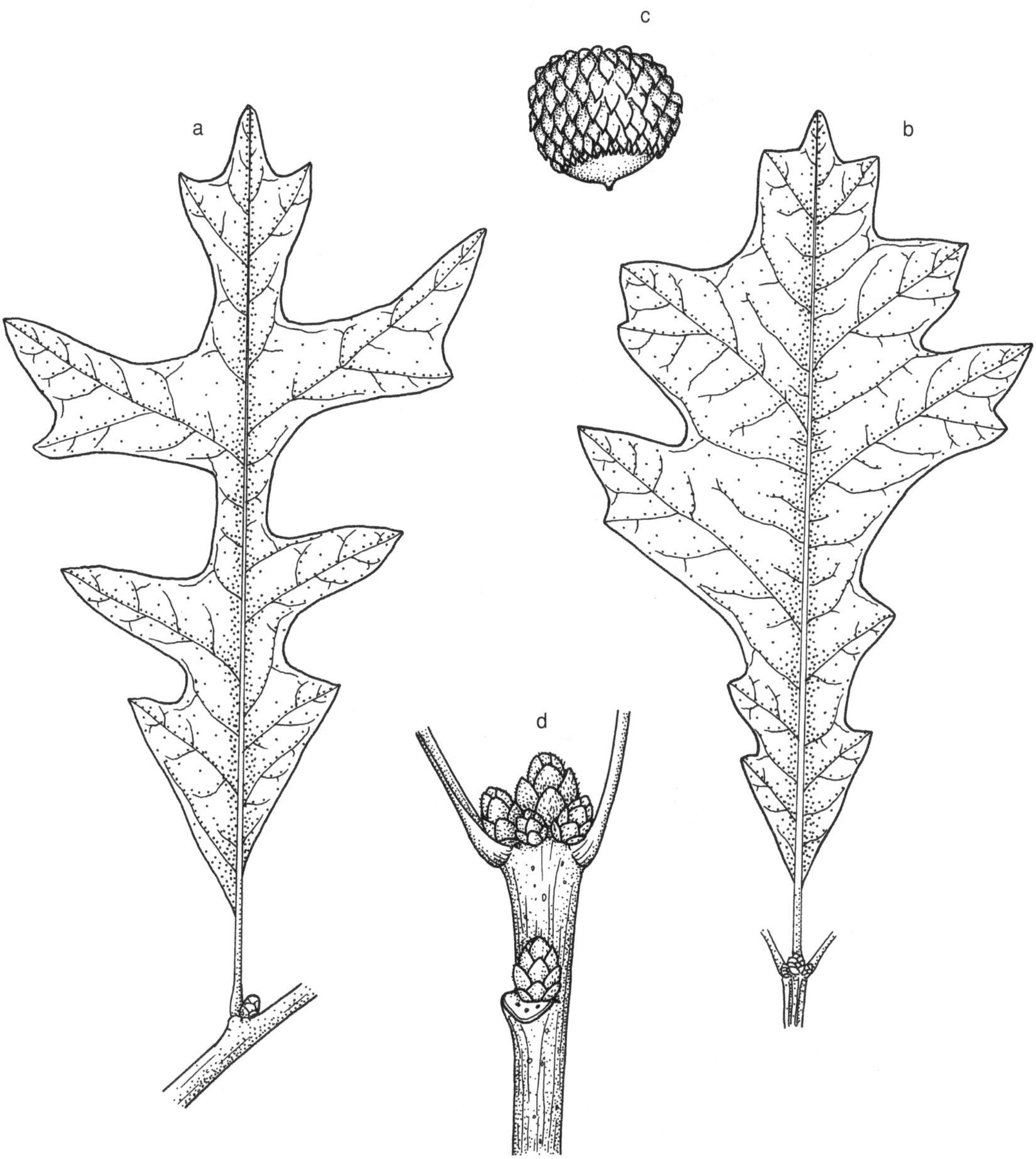

FIGURE 19.11 *Quercus lyrata* (a), (b) simple leaf ×2/3, (c) cupule and acorn ×1, (d) twig with terminal and accessory buds ×2.

except for an occasional scattered hair near the veins; leaves and buds clustered at the twig apex; terminal bud length 2 to 4 mm, lightly pubescent, scale margins ciliated; acorn length 2.0 to 2.5 cm, cap pubescent, dull, enclosing about one third to one half of the nut; twig surface glabrous, reddish brown, shiny; Regions 4a and b, North Carolina to Mississippi except Florida peninsula; mesic-hydric bottomlands and flatwoods. Nixon (1997) has split the exceptionally variable species

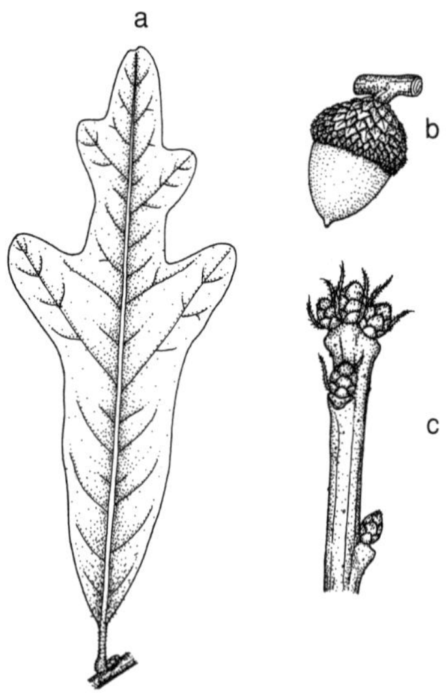

FIGURE 19.12 *Quercus similis* (a) simple leaf ×1/2, (b) cupule and acorn ×3/4, (c) twig with terminal and accessory buds ×1.

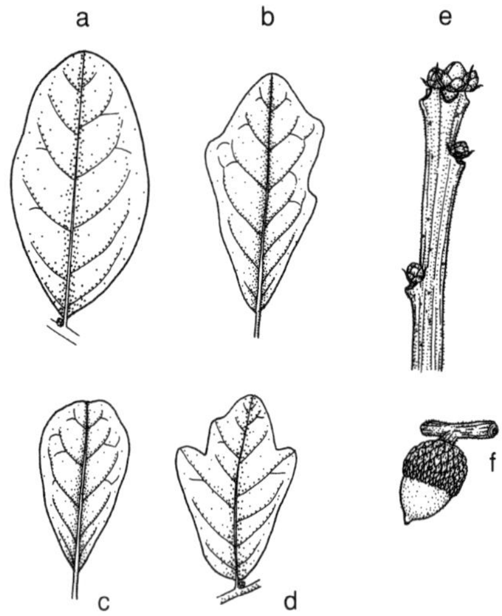

FIGURE 19.14 *Quercus chapmanii* (a), (b), (c), (d) simple leaves ×1/2, (e) twig with terminal and accessory buds ×1, (f) cupule and acorn ×1/2.

population originally known as *Quercus durandii* into *Q. austrina* and *Q. sinuata.*

Quercus chapmanii Sargent, Chapman oak (Figure 19.14); small tree, height to 5 m; evergreen; leaf arrangement alternate; leaf simple, highly variable, total length 3 to 8 cm, width 1.5 to 4.0 cm, blade shape elliptical to narrowly oblong or obovate, entire, margin smooth but with an occasional vestigial lobe, minutely revolute, adaxial surface glabrous, shiny, abaxial surface pubescent, hair appressed, yellowish gray; leaves and buds often clustered at the twig apex; twig and bud surfaces

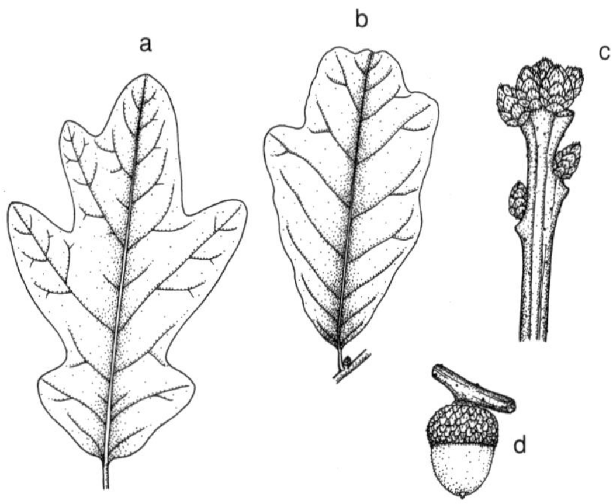

FIGURE 19.13 *Quercus austrina* (a), (b) simple leaf ×3/4, (c) twig with terminal and accessory buds ×1.5, (d) cupule and acorn ×1/2.

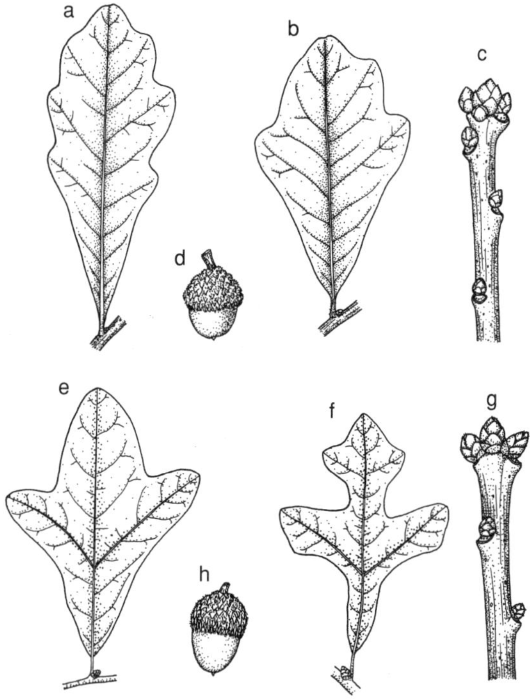

FIGURE 19.15 Simple leaf of *Quercus sinuata* (a), (b) simple leaves ×1/2, (c) twig with terminal and lateral buds ×1, (d) cupule and acorn ×2/3; *Q. margaretta* (e), (f) simple leaves ×1/2, (g) twig with terminal and lateral buds ×1, (h) cuple and acorn ×2/3.

stellate pubescent, hair more or less appressed, light green, terminal bud length 3 to 4 mm; acorn length 1.5 to 2.0 cm, cap enclosing about one half of acorn; Region 4b, Alabama, Georgia, and Florida including the peninsula; xeric sites with deep sandy soils, usually in pine savanna and oak woodland. Cover type 28.34, 28.35.

Quercus sinuata Walter (*Quercus durandii* Buckley), bastard white oak, Durand oak (Figure 19.15); shrub with a height of 20 m (var. *sinuata*) or large tree to a height of 4 m (var. *brevifolia*) (Nixon and Muller 1997); deciduous; leaf arrangement alternate; leaf simple, shape elliptical to obovate, total length 4 to 13 cm, width 4 to 8 cm, entire or lobed, apex rounded or obtuse, sinuses shallow, extending less than one third the distance from the lobe apex to the midrib, abaxial surface densely stellate pubescent; bud length 2 to 4 mm, scale margins ciliated; acorn length 1 to 2 cm, cap pubescent, dull, enclosing about one third to one half of the nut; leaves and buds clustered at the twig apex; twig surface glabrous, tan to reddish brown, shiny. According to Nixon and Muller (1997), the tree var. *sinuata* is found in Regions 4a and b, ranging from North Carolina to central Texas; it is found on mesic-hydric sites and stream terraces. Cover Type 28.19. The shrubby var. *brevifolia* is restricted to central Texas and Oklahoma; it grows on dry sites as well as along streams. Cover Type 29.9. This species was once a population of the species originally known as *Quercus durandii*; also see Q. *austrina*.

Quercus margaretta (Ashe) Ashe in J. K. Small, dwarf or sand post oak (Figure 19.15); shrub or small tree, height to 12 m; deciduous; leaves and buds clustered at the twig apex; leaf arrangement alternate; leaf simple, outline obovate or oblanceolate,

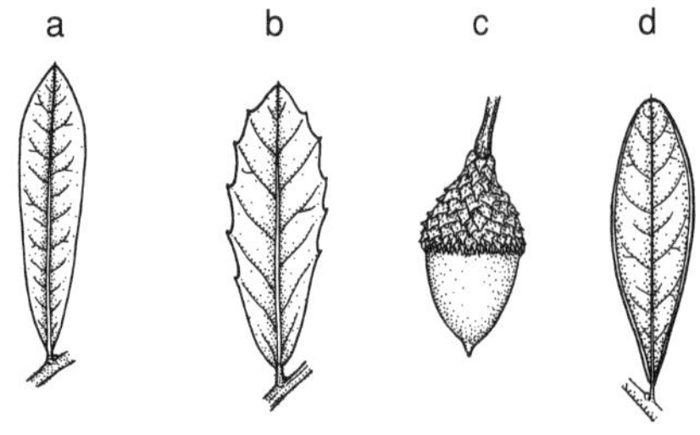

FIGURE 19.16 *Quercus minima* (a), (b) simple leaves ×1/2; (c) cupule and acorn ×3/4; (d) *Q. geminata* leaf ×1/2, note minutely revolute margin.

total length 7 to 11 cm, width 4.0 to 6.5 cm, lobe number three to five, apices usually rounded, medial pair of lobes largest and, if at a right angle to the midrib, creating a cruciform shape, sinuses rounded, usually extending one third to one half the distance from the lobe apex to the midrib, petiole pubescent, abaxial surface stellate pubescent or glabrous with stellate hair along the midrib; terminal bud length 3 to 4 mm, appearing glabrous or with tightly appressed hair; scale margins ciliate; acorn length 1.5 to 2.5 cm, cap pubescent, dull, enclosing one half of the nut, cap margin thin; twig surface yellowish, densely pubescent; bark ashy gray; Region 4a and b but extending northward into Region 3. Cover type **28.39**, northward 28.34, 28.46.

Quercus minima (Sargent) Small (*Quercus virginiana* Miller var. *minima* Sargent), minimal oak (Figure 19.16); multistemmed shrub, height to 1 m; evergreen; leaves and buds often clustered at the twig apex; leaf arrangement alternate; leaf simple, highly variable, total length 2 to 9 cm, width 0.5 to 3.0 cm, blade shape linear, elliptical to narrowly oblanceolate, entire, margin smooth to coarsely toothed to spinose, up to five spines per side, occasionally minutely revolute; adaxial surface glabrous, shiny, abaxial surface densely pubescent, hair appressed, yellowish gray; twig and bud surfaces pubescent, hair appressed, yellowish gray to reddish brown, terminal bud length 1 mm; acorn length 1.5 to 2.0 cm, cap rough, warty, pubescent, dull, enclosing about one half of the cup; Region 4a, Lower Coastal Plain, North Carolina, to Florida and Louisiana, including the Florida peninsula; xeric sites with deep sandy soils; usually in pine savanna and woodland. Cover type 28.34.

Quercus geminata Small [*Quercus virginiana* var. *geminata* (Small) Sargent], sand live oak (Figure 19.16); multistemmed shrub or medium-sized tree, height to 20 m; semi-evergreen, leaves retained through the dormant season; leaves and buds clustered at the twig apex; leaf arrangement alternate; leaf simple, entire, shape elliptical to ovate, total length 3.5 to 7.0 cm, width 1 to 3 cm, margin smooth, minutely to moderately revolute, apex obtuse to rounded, petiole pubescent, abaxial surface pubescent, velvety; twig and bud surfaces pubescent, dark gray to yellowish brown, terminal bud length 1 to 2 mm; acorns usually paired on twig, acorn length 1.5 to 2.0 cm, cap surface pubescent, dull, gray, cap enclosing one third to one half of the nut; Region 4a, Lower Coastal Plain, Louisiana to North Carolina, all of Florida; xeric sands to mesic sites, resistant to salt spray. This species was separated from *Q. virginiana* var. *virginiana* by Nixon and Muller (1997). Cover type **28.44**.

Quercus virginiana Miller, live oak, Virginia live oak, Spanish oak (Figure 19.17); large tree, height to 30 m; semi-evergreen, leaves retained through the dormant season; leaves and buds clustered at the twig apex; leaf arrangement alternate; leaf simple, essentially entire or with one to two vestigial lobes, shape elliptical to ovate, total length 3 to 10 cm, width 1.0 to 4.5 cm, margin smooth or with two to four irregularly spaced teeth, often slightly revolute, apex obtuse to rounded, petiole pubescent or glabrous, abaxial surface pubescent; twig and bud surfaces pubescent, gray to yellowish brown, terminal bud length 3 to 4 mm; acorns borne individually

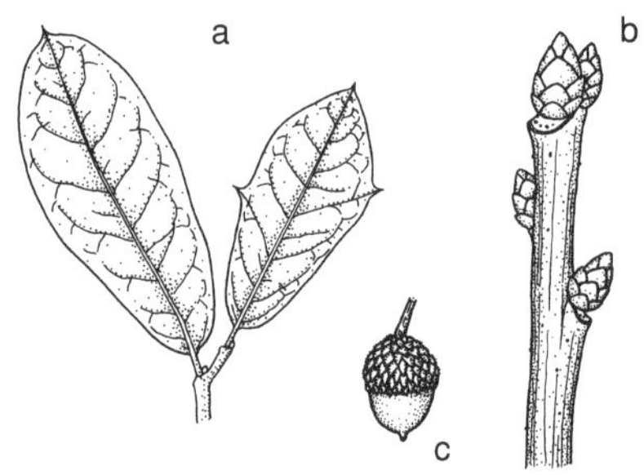

FIGURE 19.17 *Quercus virginiana* (a) simple leaves ×1/2, (b) twig with terminal and lateral buds ×2, (c) cupule and acorn ×2/3.

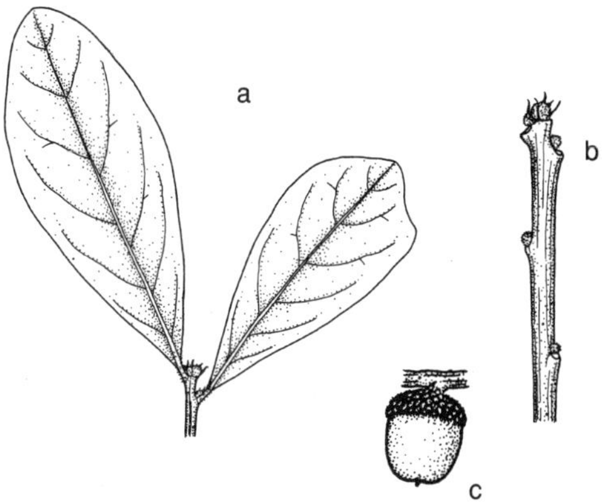

FIGURE 19.18 *Quercus ogelthorprensis* (a) simple leaves × 1/2, (b) twig with terminal and lateral buds × 1, (c) cupule and acorn × 1/2.

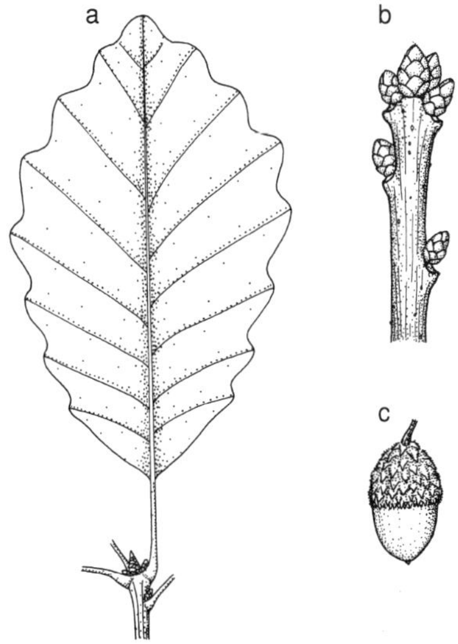

FIGURE 19.19 *Quercus prinus* (a) simple leaf × 1/2, (b) twig with terminal and lateral buds × 1, (c) cupule and acorn × 1/2.

along twig, acorn length 1.5 to 2.0 cm, cap surface pubescent, dull, gray, cap enclosing about one half of the nut; Region 4a, Lower Coastal Plain, eastern Texas to Virginia, all of Florida; xeric sands to mesic sites, resistant to salt spray, forms stable coastal communities. Spanish moss is commonly found on the trees. Cover type **29.51, 29.52,** 28.45.

Quercus oglethorprensis Duncan, Oglethorpe oak (Figure 19.18); medium-sized tree, height to 25 m; deciduous; leaves and buds clustered at the twig apex; leaf arrangement alternate; leaf simple, total length 5 to 13 cm, entire, apex rounded, abaxial surface more or less pubescent; petiole glabrous; bud length 2 to 3 mm, scale surface glabrous to lightly pubescent, margin ciliated, weakly angled, terminal buds often accompanied by narrow bracts that are longer than the buds; acorn length 2 to 3 cm, cap surface pubescent, dull, shallow, enclosing about one third of the nut, bract tips slightly lifted; twig surface glabrous, shiny; Region 4a, c, restricted locations in South Carolina, Georgia and Louisiana; moist slopes and stream terraces. Considered rare and threatened. Named for Oglethorpe County, Georgia.

3. Section *Quercus* (Subgenus *Leucobalanus, Lepidobolanus* White Oak Group)—Chestnut Oak Species, sinuses shallow or margin toothed).

Quercus prinus Linnaeus (*Quercus montana* Willdenow), chestnut oak, rock chestnut oak (Figure 19.19); large tree, height to 30 m; deciduous; leaf arrangement alternate; leaf simple, shape narrowly elliptical to obovate, total length 10 to 20 cm, teeth large, 8 to 15 per side, tooth height 0.5 to 1.0 cm, apex rounded, abaxial surface generally glabrous except for stellate pubescence on the veins; leaves and buds clustered at the twig apex; bud length 3 to 5 mm, essentially glabrous, angled; acorn length 2.5 to 4.0 cm, cap rough, warty, pubescent, dull, enclosing about one third of the nut, cap margin thin, fringed with hair; twig surface glabrous, shiny; Regions 3 and 4c, southern Illinois and eastern Michigan east to Maine, western North and South Carolina, Georgia, to Mississippi; xeric rocky and mesic upland slopes and ridge sites of the interior low plateaus, Appalachian Mountains, and the Piedmont. Nixon and Muller (1997) citing Hardin (1979) suggests that because of past difficulty in separating this species from *Quercus michauxii*, and because the name *Q. prinus* has been applied to *Q. michauxii* material, the name *Q. prinus* should be rejected in favor of *Q. montana.* Hardin (2001), citing Nixon (1997), concurs. However, the name *Q. prinus* is well entrenched in the literature, and in forest communities where these species naturally

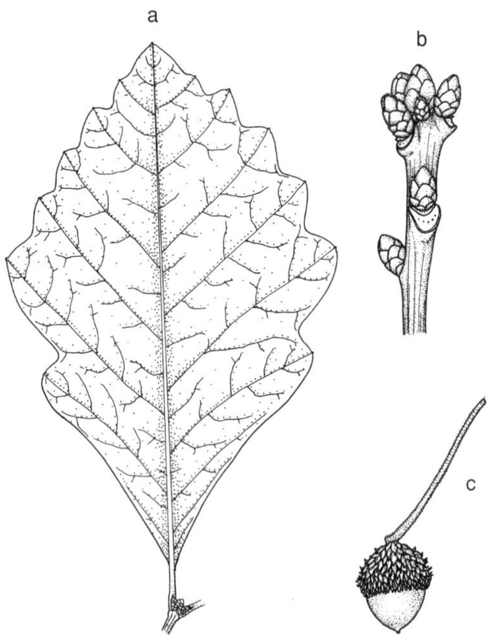

FIGURE 19.20 *Quercus bicolor* (a) simple leaf ×1/2, (b) twig with terminal and lateral buds ×2, (c) cupule and acorn, note the long stalk ×1/2.

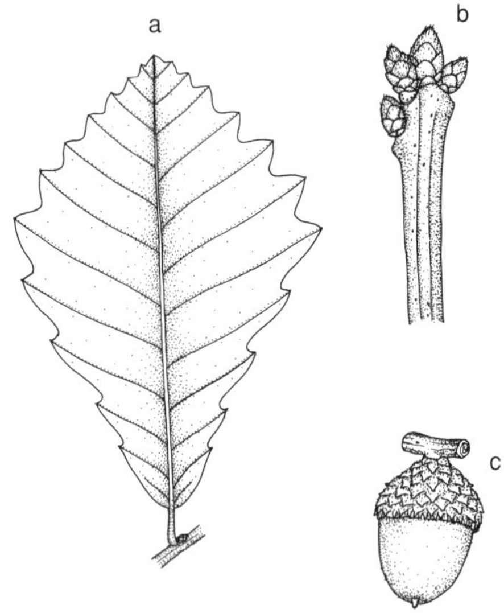

FIGURE 19.21 *Quercus michauxii* (a) simple leaf ×1/3, (b) twig with terminal and accessory buds ×1.2, (c) cupule and acorn ×1/2.

occur, there is little difficulty because ecologically the two species are widely separated. *Quercus michauxii* is a bottomland species that seldom if ever appears on mesic or drier upland sites, and *Q. prinus* is a xeric to xeric-mesic upland species that does not tolerate or appear on bottomland or mesic-hydric sites. Cover type **27.13, 27.22, 27.32, 27.33, 27.34, 28.5, 28.6**, 26.29, 27.23, 27.26, 27.38.

Quercus bicolor Willdenow, swamp white oak (Figure 19.20); large tree, height to 28 m; deciduous; leaves and buds clustered at the twig apex; leaf arrangement alternate; leaf simple, outline obovate, total length 12 to 20 cm, width 7 to 15 cm, blade margin toothed to shallow lobed, tooth height 1.0 to 1.5 cm, rounded, lobes with shallow sinuses extending less than one quarter the distance to the midrib, middle pair of teeth extending wider than others, proximal one third of blade smooth and acute, abaxial surface varying from densely pubescent to glabrous; bud length 2 to 4 mm, glabrous, scales ciliated; acorn length 1.5 to 2.5 cm, cap pubescent, dull, enclosing one third to one half of the nut, cap margin thin, stalk length 4 to 8 cm; twig surface reddish- to gray-brown, glabrous; older branches with exfoliating bark; bark ashy gray; Region 3, southern Minnesota east to Maine and Missouri east to Tennessee and South Carolina; mesic-hydric and periodically flooded hydric forest, bottomlands. Cover type **27.20, 27.21**, 27.2, 27.25.

Quercus michauxii Nuttall, swamp chestnut oak, basket oak, cow oak (Figure 19.21); large tree, height to 30 m; deciduous; leaves and buds clustered at the twig apex; leaf arrangement alternate; leaf simple, total length 10 to 22 cm, width 7 to 11 cm, obovate to oblanceolate; blade margin toothed, 7 to 15 teeth per side, tooth height 0.5 to 1.5 cm, rounded, glandular tipped, abaxial surface with indistinct scattered appressed hair; buds prominent, length 3 to 6 mm, width 3 to 4 mm, scale surface with scattered hair, margin ciliated; acorn stalk length 1 to 2 cm, acorns often paired, acorn length 2.5 to 3.5 cm, cap enclosing one third of the nut, scales pubescent, dull, bract tips lifted, cap margin thin, weakly fringed; twig surface reddish- to gray-brown, lightly pubescent to glabrous; bark ashy gray; Regions 3 and 4, southern Illinois and

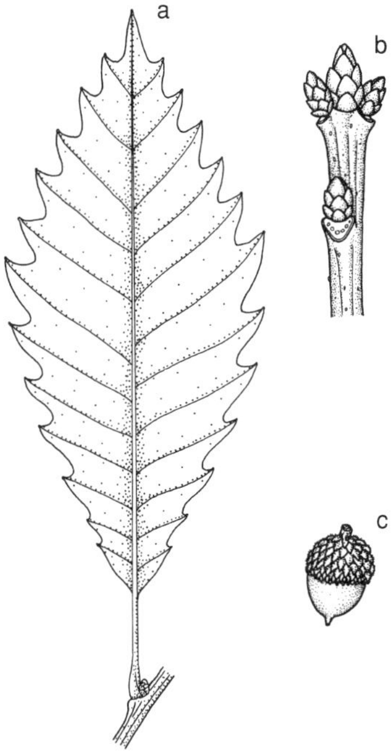

FIGURE 19.22 *Quercus muhlenbergii* (a) simple leaf × 1/2, (b) twig with terminal and accessory buds × 2, (c) cupule and acorn × 1/2.

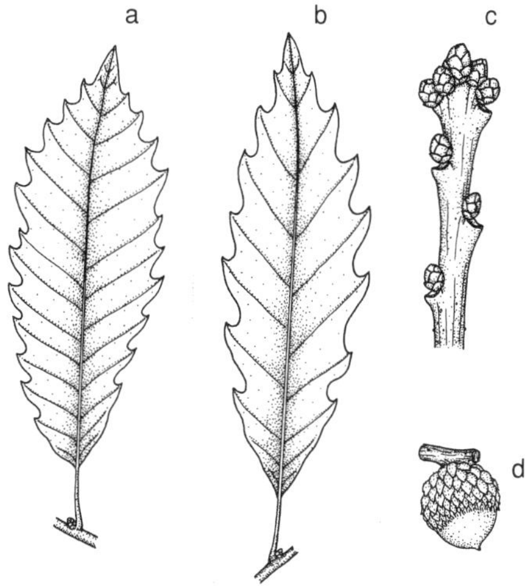

FIGURE 19.23 *Quercus prinoides* (a), (b) simple leaf × 1/2, (c) twig with terminal and lateral buds × 2, (d) cupule and acorn × 1/2.

Indiana, south to Texas, along the coastal states to Florida and New Jersey, absent in the high Appalachian Mountains; mesic-hydric soils of bottomlands. Cover type **28.18**, **28.37**, **28.41**, 27.3, 27.9, 27.14, 28.11, 28.14, 28.20, 28.33, 28.45, 28.49.

Quercus muhlenbergii Engelmann, chinkapin or chinquapin oak, yellow chestnut oak (Figure 19.22); large tree, height to 28 m; deciduous; leaves and buds clustered at the twig apex; leaf arrangement alternate; leaf simple, total length 10 to 19 cm, width 3 to 10 cm, outline elliptical, widest above the middle; blade margin toothed, seven to ten teeth per side, tooth height 0.5 to 1.5 cm, rounded to sharply sawtoothed, abaxial surface with indistinct appressed hair; bud length 2 to 4 mm, scale surface with appressed silvery hair, ciliated; acorn length 1.5 to 2.5 cm, cap pubescent, dull, enclosing one third to one half of the nut, cap margin thin, stalk length < 1 cm, acorns usually paired; twig surface reddish- to gray-brown, glabrous; older branches with exfoliating bark; bark ashy gray; Regions 3 and 4, Iowa to southern Ontario, and Vermont south to Texas and Florida, generally not appearing along the Atlantic Coastal Plain and northern Piedmont; xeric to mesic sites, frequently where limestone is near the soil surface. Cover type **27.5, 27.11, 28.19**, 27.10, 28.22, 29.27, 29.32, 29.35.

Quercus prinoides Willdenow, dwarf chestnut oak (Figure 19.23); small tree, height 5 to 8 m, often stoloniferous; deciduous; leaves and buds clustered at the twig apex; leaf arrangement alternate; leaf simple, total length 6 to 15 cm, width 2 to 5 cm, elliptical to oblanceolate or obovate, blade margin toothed, 7 to 12 teeth per side, tooth height 3 to 5 mm, rounded to saw-toothed, glandular tipped, abaxial surface with indistinct stellate hair; bud length 2 to 3 mm, scale surface thinly pubescent; acorn stalk length 1 to 2 cm, acorns often paired, acorn length 1.5 to 2.5 cm, cap enclosing one third of the nut, scales pubescent, dull, warty, cap margin thin, weakly fringed; twig surface reddish- to gray-brown, lightly pubescent to glabrous; bark ashy gray; Region 3, Arkansas and Iowa to northern Georgia, South Carolina, Massachusetts; xeric sites, sands and shales. This species is a common but minor component of the *Q. stellata-Q. marilandica* type but also is found with other *Quercus* and *Pinus* communities of the Central States. Hybridizes

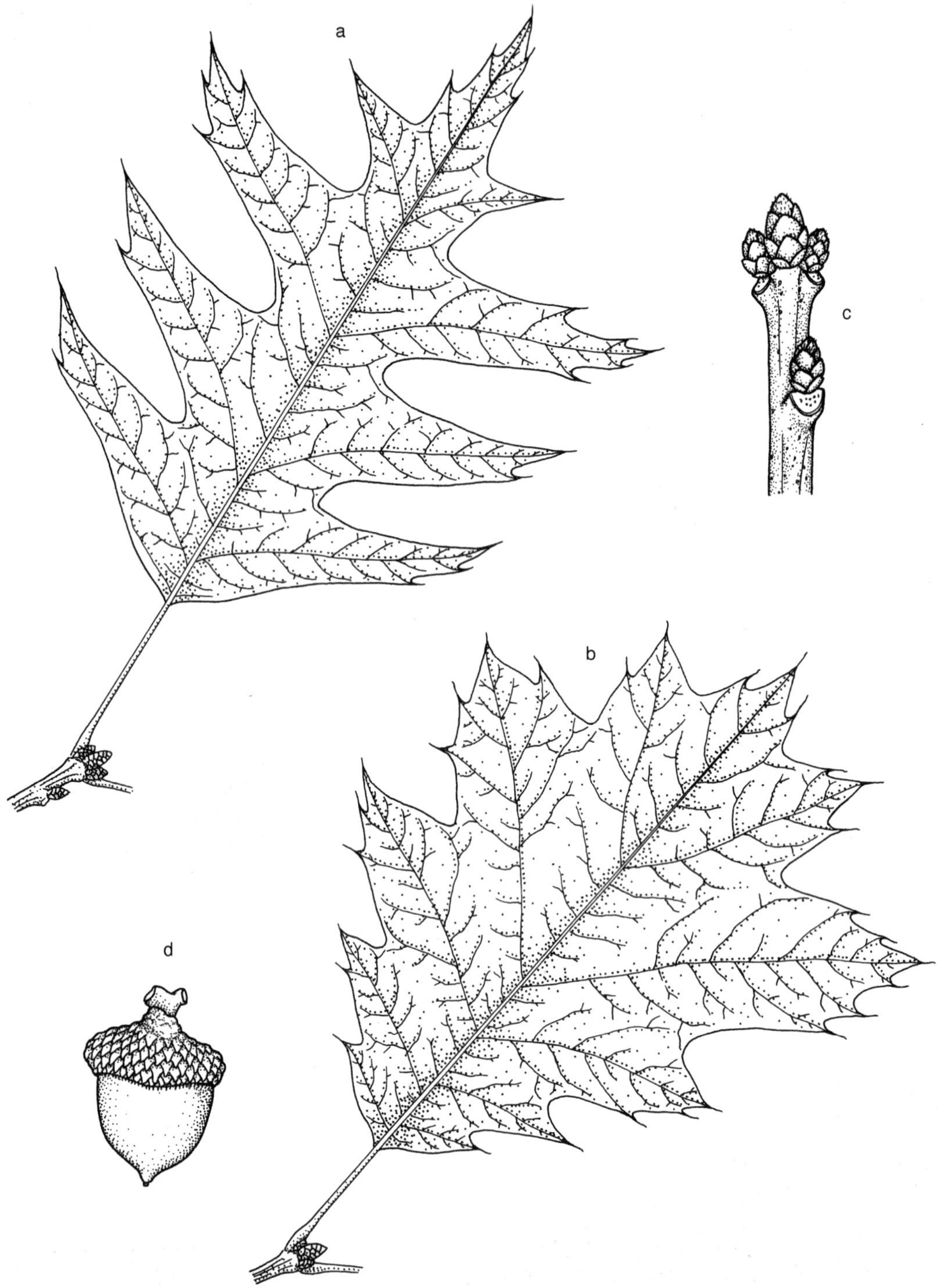

FIGURE 19.24 *Quercus rubra* (a) simple sun leaf ×3/4, (b) simple shade leaf ×3/4, (c) twig with terminal and lateral buds ×2, (d) cupule and acorn ×1.

with *Q. muhlenbergii*, *Q. bicolor*, *Q. stellata*, and *Q. alba* (Miller and Lamb 1985). Visually, this small tree appears similar to *Q. muhlenbergii*, and the two are considered to be closely related (Nixon and Muller 1997). The two species have a similar range in the north and east, but the range of *Q. muhlenbergii* extends some distance into Region 4.

4. Section Lobatae (Subgenus Erythobalanus; Red or Black Oak Group)—Red Oak Species, Regions 2, 3, and 4.

Quercus rubra Linnaeus *Q. borealis* Michaux, northern red oak (Figure 19.24); large tree, height to 33 m; deciduous; leaf arrangement alternate; leaf simple, total length 12 to 25 cm, width 9 to 15 cm, lobes 5 to 11, sun leaf sinuses extending about one half to two thirds the distance from the lobe apex to the midrib, shade leaf sinuses shallow, less than one half the distance to the midrib, several sets of lobes forming a distinct "V" base, each lobe bristle tipped with one to five awns, the middle lobes spreading and tending to bifurcate, lobes at the proximal and distal ends with a single mucro, abaxial surface glabrous except for axillary tufts along primary veins; leaves and buds clustered at the twig apex; bud length 3 to 6 mm, glabrous, weakly angled; acorn length 2 to 3 cm, cap surface glabrous, shiny, shallow, enclosing about one quarter of the nut; twig surface glabrous, shiny, reddish brown; mature bark with tight with alternating rough and smooth vertical reaches; Regions 2, 3 and 4, Minnesota, Wisconsin, Michigan, southern Ontario, and Maritime Provinces south into Oklahoma and Alabama, generally not found on the Coastal Plain; mesic sites, coves, north slopes, stream terraces. This species is of major economic and wildlife importance in the Central States and Lake States forests. Red oak grows fast, with a straight bole, and produces considerable mast (wildlife food; acorns but also buds and new twigs). Resource managers work hard to regenerate this species because of the competition from *Acer saccharum* and other shade-tolerant hardwoods on moist sites in Region 2. These northern populations might be better identified as *Q. borealis* because of differencies in leaf structure and stem form and appearance. Cover type **26.17, 26.28, 27.7, 27.16, 27.22, 27.23, 27.34, 27.35, 28.6**, 26.18, 26.19, 26.23, 26.24, 26.26, 27.8, 27.26, 27.29, 27.30, 27.37, 27.38, 28.3, 28.11.

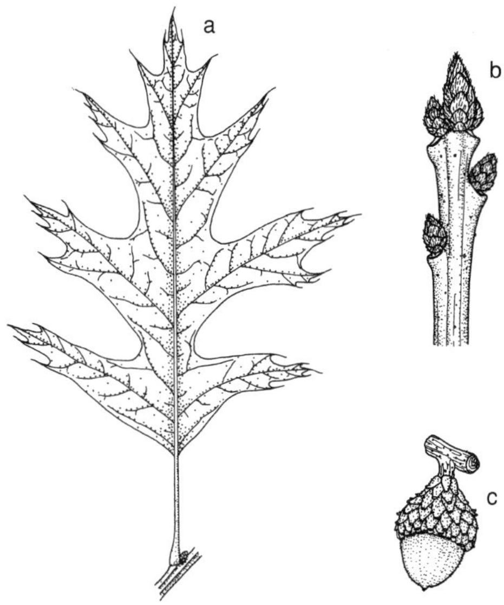

FIGURE 19.25 *Quercus velutina* (a) simple sun leaf ×2/5, (b) twig with terminal and lateral buds ×1, (c) cupule and acorn ×3/4.

Quercus velutina Lamarck, black oak (Figures 19.25 and 14.19); large tree, height to 30 m; deciduous; leaf arrangement alternate; leaf simple, total length 10 to 25 cm, width 9 to 15 cm, sun leaf sinuses extending about one half to three quarters the distance from the lobe apex to the midrib, lobes five to nine, each lobe bristle tipped with three to five awns, lobe apex spreading, abaxial surface pubescent until midsummer or later, hair deciduous and easily brushed off; shade leaf sinuses shallow, less than one half the distance to the midrib, shape often irregular; buds clustered at the twig apex, bud length 6 to 11 mm, pubescent toward the apex, angled; acorn length 1.5 to 2.5 cm, cap surface pubescent, dull, shallow, enclosing about one quarter to one third of the nut; bract tips lifted; twig surface glabrous, shiny; mature bark blocky, tight, nearly black; Regions 3 and 4, eastern United States from Minnesota to southern Ontario and Maine south, not found in the Florida peninsula; xeric, rocky soils, ridgetops and south-facing slopes. Cover type **26.18, 26.28, 27.11, 27.17, 27.22, 27.26, 27.33, 28.6**, 27.18, 27.23, 27.24, 27.37. 28.4. 28.16. *Quercus velutina* readily

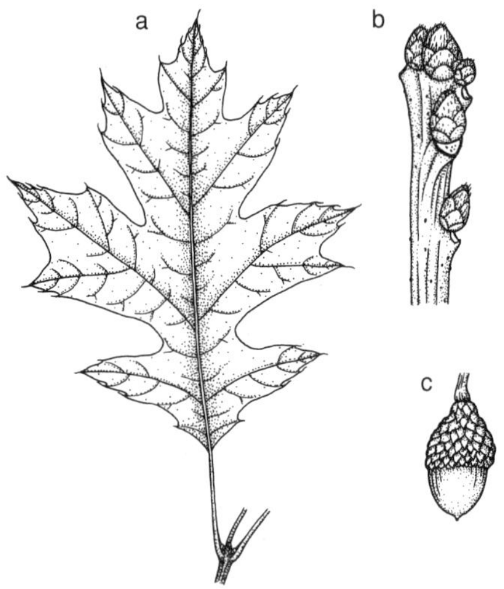

FIGURE 19.26 *Quercus ellipsoidalis* (a) simple leaf ×1/2, (b) twig with terminal and lateral buds ×1, (c) cupule and acorn ×1.

hybridizes with most of the other red oak species, including *Q. rubra* (*Q.* x *hawkinsii* Sudworth).

Quercus ellipsoidalis E. J. Hill, northern pin oak, Hill's oak (Figure 19.26); medium-sized tree, height to 20 m; deciduous; leaf arrangement alternate; leaf simple, total length 8 to 15 cm, width 7 to 10 cm, outline nearly orbicular, sinuses extending about one half to three quarters the distance from the lobe apex to the midrib, lobes seven to nine, each lobe bristle tipped with three to seven awns, middle lobe apices spreading and tending to bifurcate, abaxial surface lacking axillary tufts along the midrib; leaves and buds clustered at the twig apex; terminal bud length 3 to 5 mm, pubescent, grayish toward the tip, margin ciliated; acorn length 1 to 2 cm, bract surface pubescent and dull, cap enclosing about one third to one half of the nut; twig surface glabrous, shiny; Regions 2 and 3a, North Dakota, Minnesota, Iowa, and Wisconsin into northern Illinois, Indiana, and Michigan; xeric outwash sand; oak savanna and barrens. Cover type **26.18**, 26.26. *Q. ellipsoidalis* freely hybridizes with *Q. velutina* (*Q.* x *palaeolithicola* Trelease) and *Q. rubra* (Figure 19.27).

Quercus marilandica Muenchhausen, blackjack oak, jack oak (Figure 19.28); small to medium-sized tree, height to 15 m; leaves and buds clustered at the twig apex; deciduous; leaf arrangement alternate; leaf simple, total length 9 to 17 cm, width 8 to 14 cm, shape broadly spatulate, occasionally three-lobed, occasionally two proximal vestigial lobes, each lobe bristle tipped with zero to two awns, petiole and abaxial surface densely pubescent; terminal bud length 5 to 10 mm, densely pubescent, angled; acorn length 2 to 3 cm, cap densely pubescent, cap enclosing about one third to one half of the nut, bract tips slightly raised; twig surface pubescent, dull; mature bark rough; Regions 3 and 4, Iowa to New Jersey south, except the Florida peninsula; xeric rocky or sandy soil of slopes and ridges. This species hybridizes with a wide variety of species within the red oak group, including *Q. illicifolia* (*Q.* x. *brittonii* W. T. Davis), *Q. imbricaria* [*Q.* x *tridentata* (A. de Candolle) Engelmann], and *Q. phellos* (*Q.* x *rudkinii* Britton). A common hybrid found in the Central States (Region 3) is *Q.* x *bushii* Sargent, a cross and backcross between *Q. velutina* and *Q. marilandica* (Figure 19.29). Cover type **27.11, 27.24, 28.3**, 27.32, 28.1, 28.13, 28.38, 28.46.

Quercus ilicifolia Wangenheim, bear oak, scrub oak (Figure 19.30); arborescent, height to 6 m; deciduous; leaves and buds clustered at the twig apex; leaf arrangement alternate; leaf simple, total length 6 to 13 cm, width 4 to 8 cm, lobes usually five, occasionally three, each lobe bristle tipped with one to two awns, the distal set of lobes larger, petiole and abaxial surface densely pubescent, felty; bud length 2 to 3 mm, pubescent, scale margins ciliated, weakly angled; acorn length 2 to 3 cm, cap surface glabrous, shiny, shallow, enclosing about one quarter to one third of the nut, bract tips appressed; twig surface pubescent, dull; mature bark rough; Region 3, southwestern Maine, New Hampshire, New York, Pennsylvania, West Virginia, and Virginia; acid, sandy soil and xeric sites, cutover, burned, and disturbed areas. According to Jensen (1997), *Q. ilicifolia* hybridizes with most of the other red oak group species in the region, including *Q. velutina*, *Q. coccinea*, *Q. rubra*, and *Q. marilandica* (*Q.* x *brittonii* W. T. Davis) (Figure 19.30).

Quercus coccinea Muenchhausen, scarlet oak (Figure 19.31); large tree, height to 28 m;

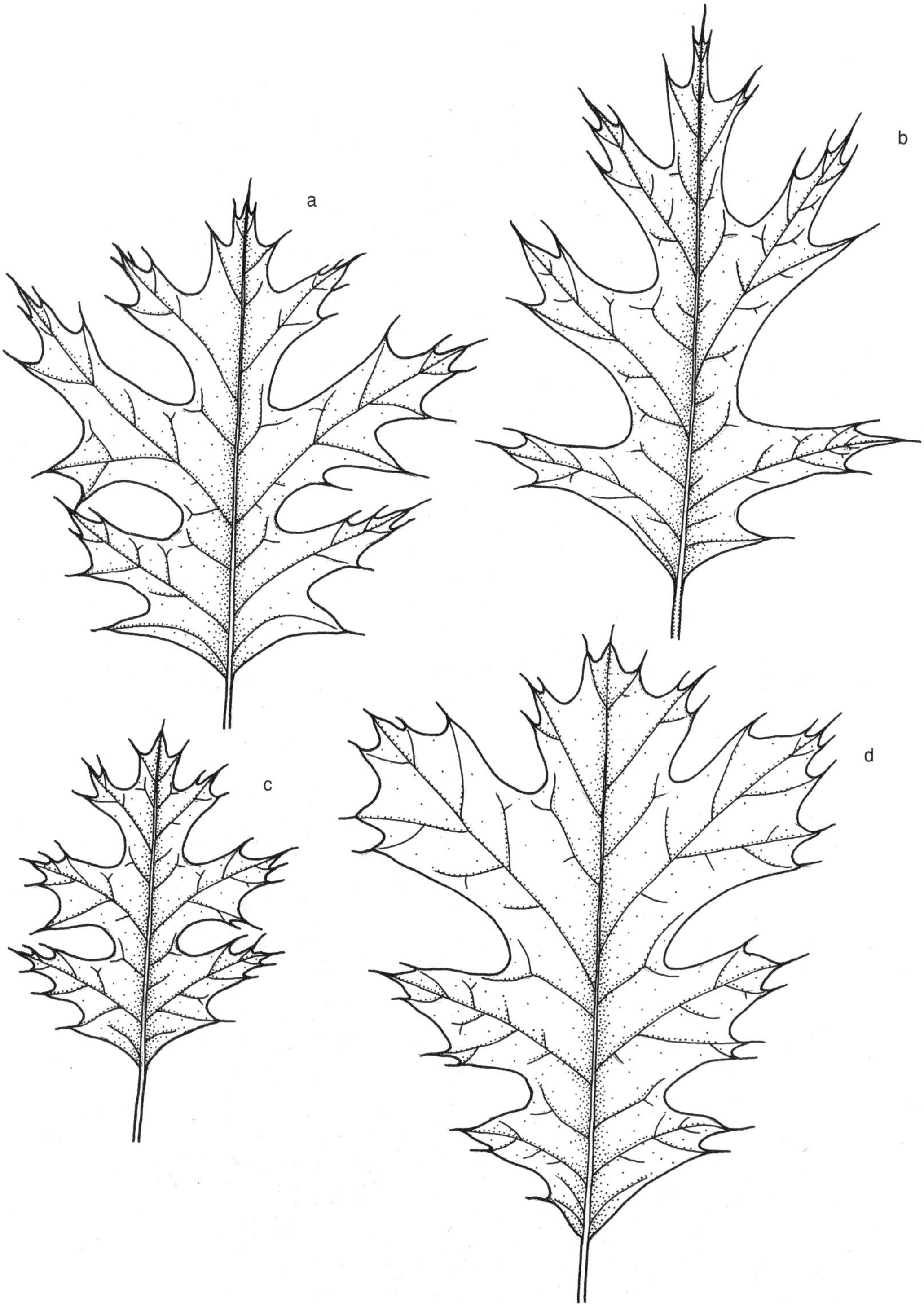

FIGURE 19.27 Hybrid *Quercus* leaves: (a), (b), (c) *Q.* x *paleolithicola* (*Q. velutina* x *Q. ellipsoidalis*) ×2/3; (d) *Q. rubra* x *Q. ellipsoidalis* ×2/3.

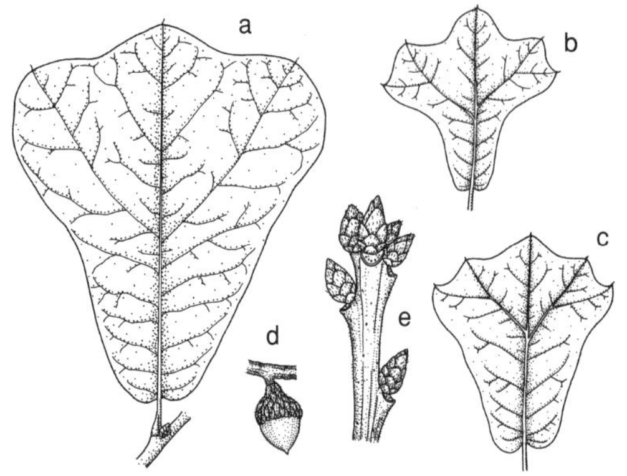

FIGURE 19.28 *Quercus marilandica* (a), (b), (c) leaves ×1/3, (d) twig with terminal and lateral buds ×1, (e) cupule with acorn ×1/3.

deciduous; leaf arrangement alternate; leaf simple, total length 9 to 21 cm, width 7 to 15 cm, lobes five to nine, sun leaf sinuses extending about three quarters to nine tenths the distance from the lobe apex to the midrib, base essentially truncate, each lobe bristle tipped with three to eight awns, the middle lobes spreading and slightly bifurcated, proximal end lobes reduced and bearing two to three awns, abaxial surface glabrous except for tufts along primary veins; leaves and buds clustered at the twig apex; bud length 3 to 7 mm, glabrous, margin ciliated, weakly angled; acorn length 2 to 3 cm, cap surface glabrous, shiny, shallow, enclosing about one half the nut, bract tip appressed; twig surface glabrous, shiny, reddish brown; bole base usually swollen and defective, mature bark below 2 m rough and similar to that of *Q. velutina*, above 3 m similar to that of *Q. rubra*, with alternating rough and smooth vertical reaches; Regions 3 and 4, Missouri, Michigan, and Maine south to central Alabama, Georgia, and the Piedmont, several small outliers found in the Coastal Plain; xeric, warm south- to west-facing slopes and ridgetops, infrequent on mesic north slopes. A common tree species found with a variety of other *Quercus*, *Carya*, and *Pinus* species in the Central States and the Appalachian Mountains. The tree base is usually swollen and defective as a result of infection by chestnut blight. Cover type **27.13, 27.22, 27.32, 28.5**, 27.21, 28.3.

Quercus palustris Muenchhausen, pin oak (Figure 19.32); medium-sized tree, height to 24 m; deciduous; leaf arrangement alternate; leaf simple, total length 8 to 14 cm, sinuses extending about three quarters the distance from the lobe apex to the midrib, lobes five to seven, each lobe bristle tipped with three to five awns, lobe essentially C-shaped or occasionally spreading, abaxial surface glabrous except for small axillary tufts along the midrib; leaves and buds clustered at the twig apex; bud length 2 to 3 mm, glabrous, slightly angled; acorn length 1 to 2 cm, bract surface glabrous, bract tips appressed, cap enclosing about one quarter of the nut; twig surface glabrous, shiny, reddish brown; Region 3, northern boundary from Kansas to southern Michigan and Massachusetts, southern boundary from Oklahoma through Arkansas, Tennessee to North Carolina, generally not in the Appalachian Mountains. Cover type **27.21**, 27.3, 27.25.

Quercus falcata Michaux, southern red oak, Spanish oak (Figure 19.33); medium-sized tree, height to 17 m; deciduous; leaf arrangement alternate; leaf simple, total length 12 to 23 cm, width 5 to 13 cm, lobes three to seven, sides of lobes evenly tapering or bifurcating, apices acute, each lobe bristle tipped with one to three awns, distal sinuses broad, open, and squarish, often extending about three quarters to four fifths the distance from the lobe apex to the midrib, abaxial surface densely pubescent and velvety; leaves and buds clustered at the twig apex; bud length 4 to 7 mm, scale surfaces lightly hairy, margin ciliated; acorn length 1 to 2 cm, bract surface gray pubescent except for reddish margin, bract tips appressed, cap enclosing about one third to one half of the nut; twig surface pubescent, dull, reddish brown; bark dark, extremely rough; Regions 3 and 4, Missouri, Illinois, Ohio, and New Jersey south, except central and southern Florida; thin, rocky soils of xeric slopes and ridges. The three-lobe leaf resembles a bell with an extended ringer. Hybridizes with essentially every Red Oak Group species in its range. The leaves are somewhat similar to those of *Q. laevis*, except that the lower surface of *Q. laevis* leaves are glabrous. Cover type **27.13, 27.18, 28.3, 28.6, 28.13, 28.39**, 27.25, 28.4, 28.13, 28.16, 28.20, 28.38.

Quercus pagoda Rafinesque [*Quercus falcata* var. *pagodaefolia* Elliott; *Quercus pagodaefolia*

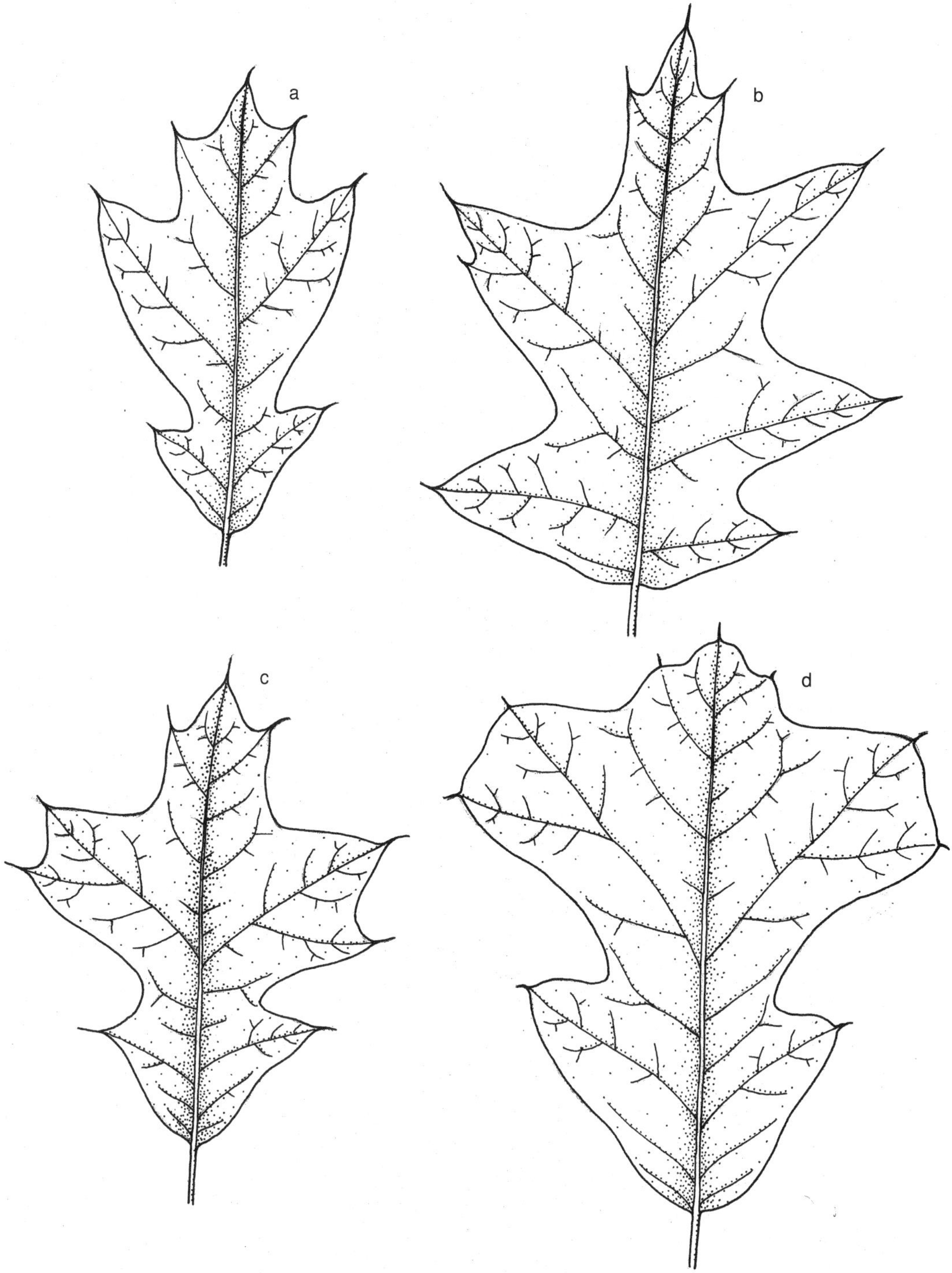

FIGURE 19.29 Leaves of the common *Quercus* x *bushii* (*Q. marilandica* x *Q. velutina*) ×2/3.

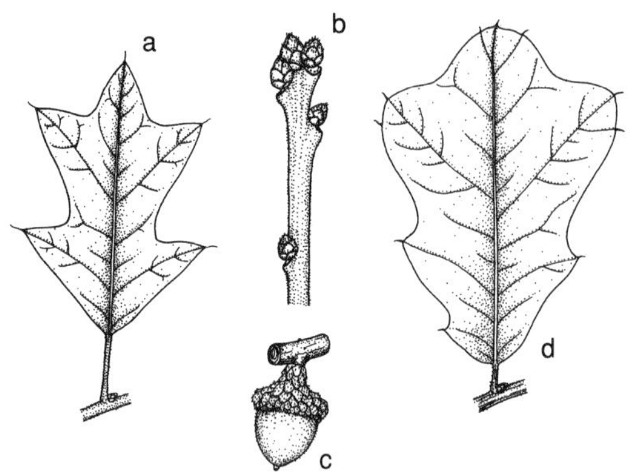

FIGURE 19.30 *Quercus illicifolia* (a) leaf ×1/3, (b) twig with terminal and lateral buds ×1 (c) cupule and acorn ×1/2; (d) *Q.* x *brittonii* (*Q. illicifolia* x *Q. marilandica*) ×1/3.

(Elliott) Ashe], cherrybark oak, swamp red oak (Figure 19.34); large tree, height to 35 m; deciduous; leaf arrangement alternate; leaf simple, total length 12 to 23 cm, width 5 to 14 cm, lobes five to nine, sides of lobes evenly tapering, margin smooth,

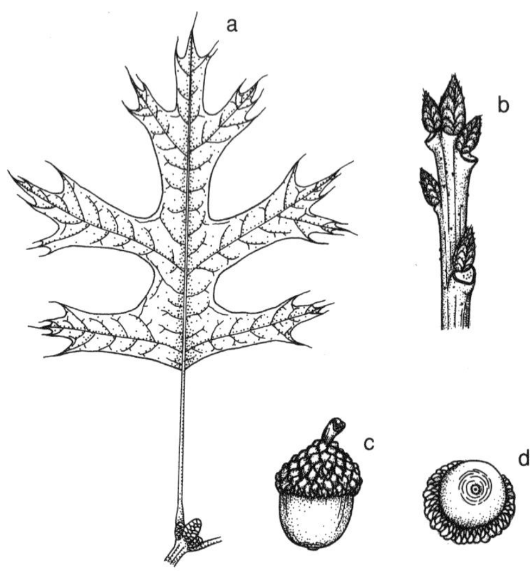

FIGURE 19.31 *Quercus coccinea* (a) simple leaf ×1/3, (b) twig with terminal and lateral buds ×1, (c) cupule and acorn ×2/3, (d) end view of acorn showing concentric rings ×2/3.

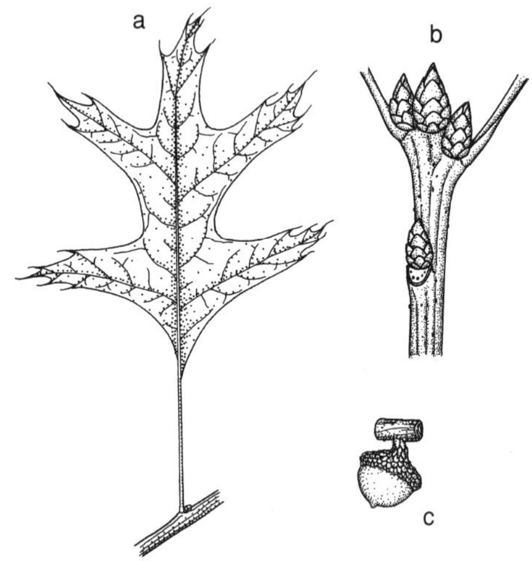

FIGURE 19.32 *Quercus palustris* (a) simple leaf ×1/2, (b) twig with terminal and lateral buds ×1.5, (c) cupule and acorn ×3/4.

apices acute, each lobe bristle tipped with one to three awns, medial sinuses C-shaped, distal sinuses broad, open, and squarish, often extending about three quarters the distance from the lobe apex to the midrib, abaxial surface densely pubescent and velvety; twig and bud surface pubescent, dull, reddish brown; leaves and buds clustered at the twig apex; terminal bud length 4 to 8 mm, scale surfaces densely pubescent, acorn length 1 to 2 cm, bract surface gray pubescent, except for reddish margin, bract tips appressed, cap enclosing about one third to one half of the nut; bark dark, rough, breaking into small plates; Regions 3, Mississippi River embayment, and 4, Gulf and Atlantic Coastal Plains, except Florida peninsula; mesic-hydric sites of bottomlands and terraces. Cover type **28.20, 28.41**, 27.14, 28.14, 28.16, 28.18, 28.49.

Quercus shumardii Buckley, Shumard oak (Figure 19.35); large tree, height to 37 m; deciduous; leaf arrangement alternate; leaf simple, total length 10 to 21 cm, sinuses extending about one half to three quarters the distance or deeper from the lobe apex to the midrib, lobes five to nine, each lobe bristle tipped with five to ten awns, lobe apex spreading and subdivided, on some species adjacent lobes overlapping, abaxial surface glabrous except for axillary tufts along primary veins; leaves and buds

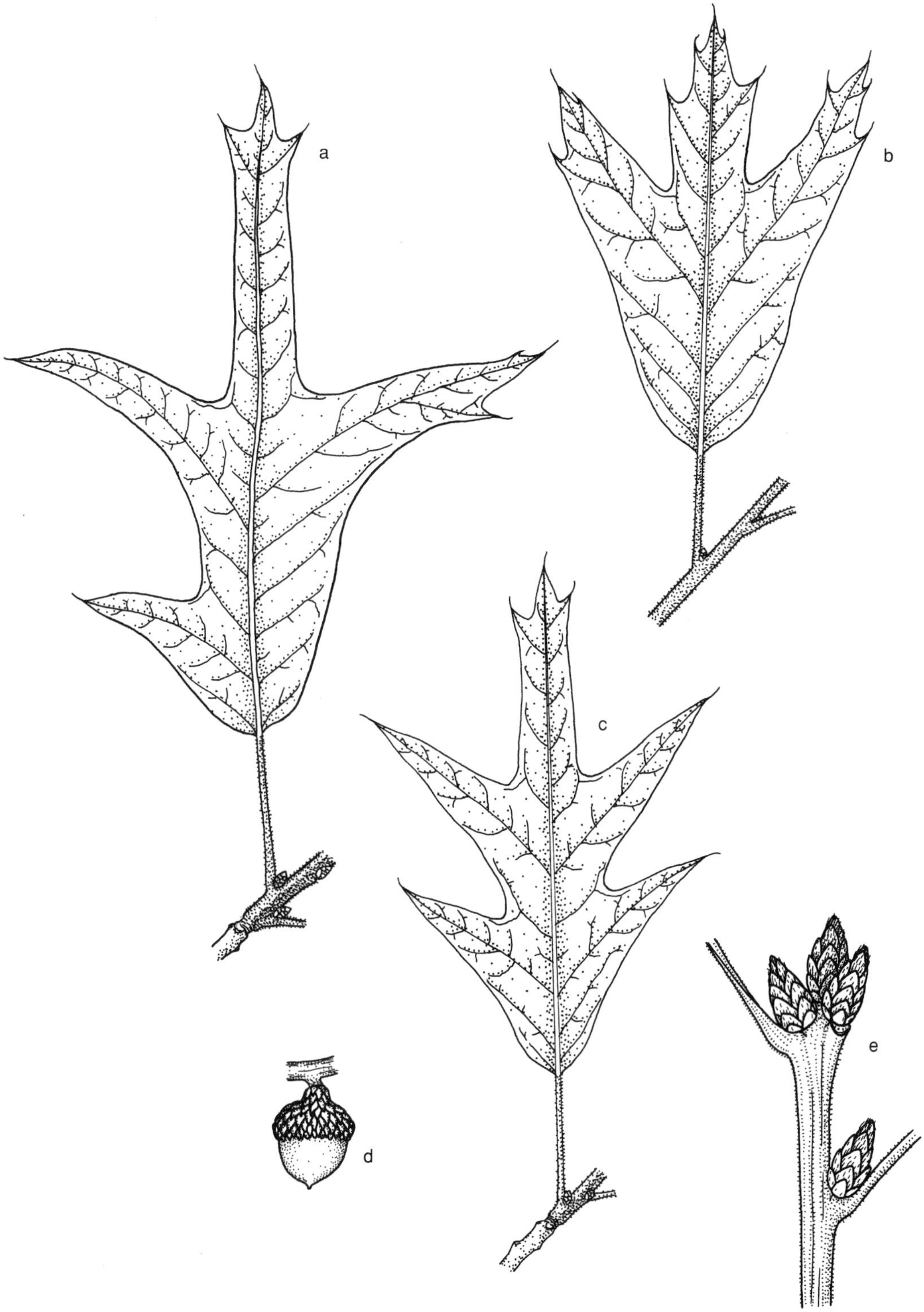

FIGURE 19.33 *Quercus falcata* (a), (b), (c) simple leaves ×3/4, (d) twig with terminal and lateral buds ×2, (e) cupule and acorn ×1.

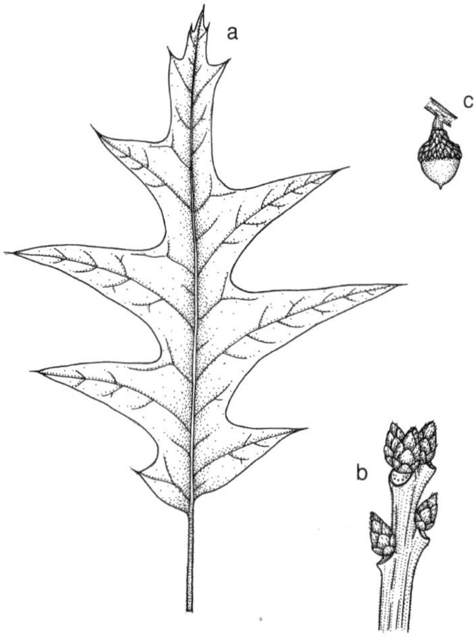

FIGURE 19.34 *Quercus pagoda* (a) simple leaf ×1/2, (b) twig with terminal and lateral buds ×1, (c) cupule and acorn ×1/2.

clustered at the twig apex, bud length 3 to 8 mm, glabrous, weakly angled; acorn length 2 to 3 cm, cap surface pubescent, dull, shallow, enclosing about one quarter of the nut; twig surface glabrous, shiny; mature bark tight with alternating rough and smooth vertical reaches; Regions 3a and 4, Missouri to northern Indiana and Ohio, Texas to northern Florida to North Carolina and Virginia; mesic coves and northern slopes as well as mesic-hydric bottomland sites. Mature bark similar to, and in winter may be confused with, that of *Q. rubra*. Cover type **28.20, 28.41**, 27.20, 28.11, 28.18, 28.19, 28.22.

Quercus laevis Walter, turkey oak (Figure 19.36); medium-sized tree, height to 18 m; deciduous; leaf arrangement alternate; leaf simple, total length 12 to 20 cm, width 8 to 18 cm, lobes three to five, occasionally seven, sides of lobes parallel or evenly converging, apices acute, each lobe bristle tipped with one mucro but occasionally two, distal sinuses broad, open, and squarish, often extending about three quarters to four fifths the distance from the lobe apex to the midrib, abaxial

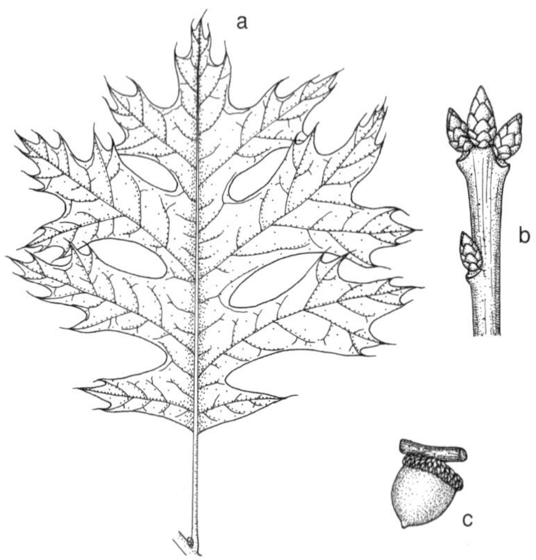

FIGURE 19.35 *Quercus shumardii* (a) simple leaf ×2/5, (b) twig with terminal and lateral buds ×1, (c) cupule and acorn ×2/5.

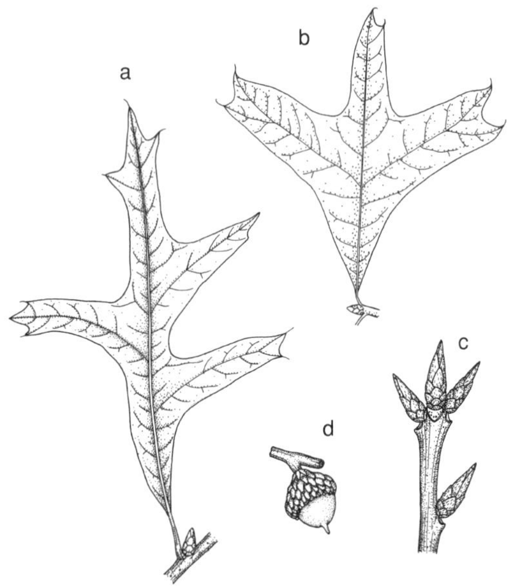

FIGURE 19.36 *Quercus laevis* (a), (b) simple leaves ×2/5, (c) twig with terminal and lateral buds ×1, (d) cupule and acorn ×1/2.

surface glabrous except for axillary tufts; twig surface with scattered hair, dull, dark reddish brown; leaves and buds clustered at the twig apex; terminal bud length 0.8 to 1.2 cm, scale surfaces pubescent, silky; acorn length 1.8 to 2.3 cm, width 1.8 to 2.3; bract surfaces pubescent, reddish brown, bract tips appressed, cap enclosing about one half of the nut; bark dark, extremely rough; Regions 4, Gulf and Atlantic Coastal Plains and the Piedmont; xeric sandy soils of slopes and ridges. The leaves may be similar to those of *Q. falcata*, except that the lower surface of *Q. falcata* leaves are densely pubescent and velvety. Cover type **28.39, 28.46**, 28.34.

Quercus texana Buckley (*Quercus nuttallii* E. J. Palmer), Nuttall oak, Texas red oak, pin oak (Figure 19.37); large tree, height to 28 m; deciduous; leaf arrangement alternate; leaf simple, total length 8 to 19 cm, sinuses extending about one half to three quarters the distance from the lobe apex to the midrib, lobes five to nine, each lobe bristle tipped with three to six awns, lobe sides essentially parallel, apex spreading, abaxial surface glabrous except for small axillary tufts along the midrib; leaves and buds clustered at the twig apex; bud length 3 to 6 mm, pubescent toward the tip, margin ciliated; acorn length 2 to 3 cm, bract surface pubescent and dull except for the reddish brown margin, cap enclosing about one third to one half of the nut; twig surface glabrous, shiny; Regions 3a and 4a, b Texas to Alabama, Arkansas, Tennessee, Missouri bootheel; mesic-hydric sites; poorly drained and periodically flooded bottomland clay soils. Cover type 29.30.

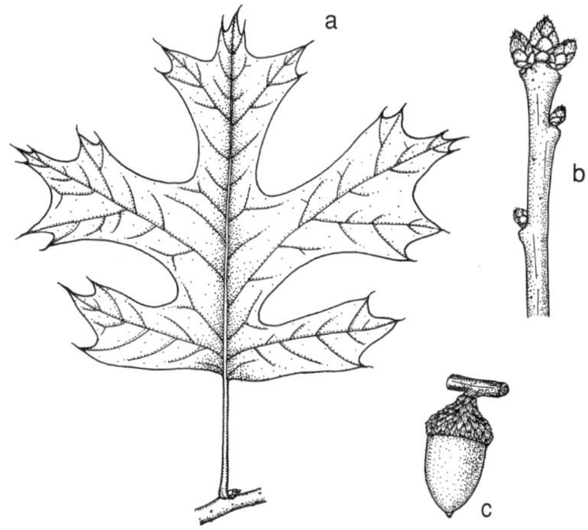

FIGURE 19.38 *Quercus buckleyi* (a) simple leaf ×1/2, (b) twig with terminal and lateral buds ×1, (c) cupule and acorn ×3/5.

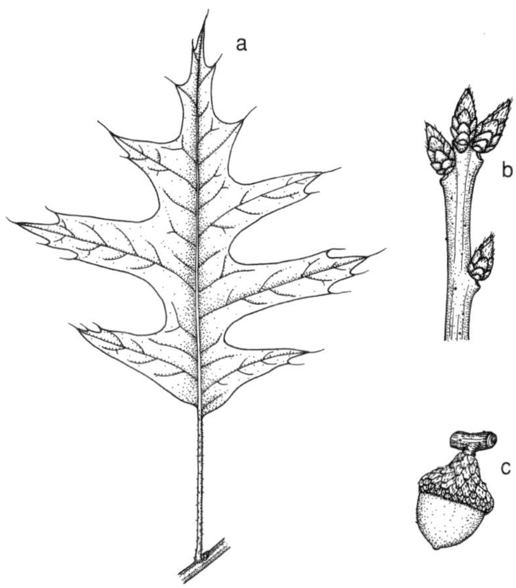

FIGURE 19.37 *Quercus texana (Q. nutallii)* (a) simple leaf ×1/2, (b) twig with terminal and lateral buds ×1.5, (c) cupule and acorn ×2/3.

Quercus buckleyi Nixon & Dorr, Buckley oak (Figure 19.38); medium-sized tree, height to 20 m; deciduous; leaf arrangement alternate; leaf simple, total length 8 to 15 cm, width 7 to 10 cm; outline nearly orbicular, sinuses extending about one half to three quarters the distance from the lobe apex to the midrib, lobes seven to nine, each lobe bristle tipped with three to seven awns, middle lobe apices spreading and tending to bifurcate, abaxial surface lacking axillary tufts along the midrib; leaves and buds clustered at the twig apex; terminal bud length 4 to 6 mm, pubescent, grayish toward the tip, margin ciliated; acorn length 2 to 3 cm, bract surface pubescent and dull, cap enclosing about one third to one half of the nut; twig surface glabrous, shiny; Regions 3, Texas and Oklahoma; xeric calcarious soils, slopes and high terraces. The abaxial leaf surface of *Quercus shumardii*, a closely related

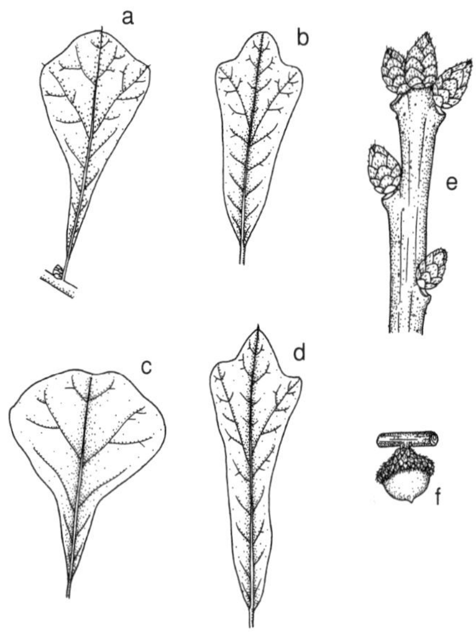

FIGURE 19.39 *Quercus nigra* (a), (b), (c), (d) simple leaves ×1/3, (e) twig with terminal and lateral buds ×1.5, (f) cupule and acorn ×1/2.

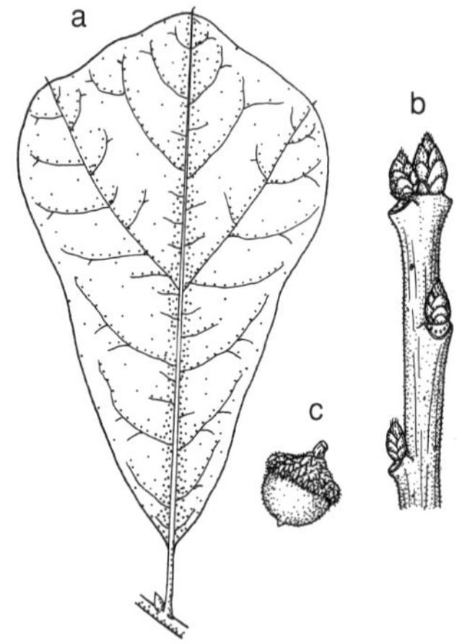

FIGURE 19.40 *Quercus arkansana* (a) simple leaf ×1/2, (b) twig with terminal and lateral buds ×1.5, (c) cupule and acorn ×1/2.

species, has distinct axillary tufts along the midrib. Cover type **29.35**, 28.9.

Quercus nigra Linnaeus, water oak (Figure 19.39); medium to tall tree, height to 30 m; deciduous; leaf arrangement alternate; leaf simple, total length 4 to 13 cm, width 2 to 5 cm, shape broadly spatulate to infrequently oblanceolate, obscurely three-lobed, infrequently distinctly three-lobed, each lobe bristle tipped with zero to two awns, apex usually rounded, petiole densely pubescent, abaxial surface with axillary tufts and scattered hair; leaves and buds clustered at the twig apex; terminal bud length 3 to 6 mm, densely pubescent, angled; acorn length 1.0 to 1.5 cm, cap densely pubescent, gray with glabrous reddish brown areas, shallow, enclosing about one third of the nut, bract tips appressed; twig surface glabrous; mature bark rough; Region 4, Gulf and Atlantic Coastal Plains, southern half of Region 3, Oklahoma and Missouri bootheel east to New Jersey, south to Texas and northern one half of Florida; mesic and mesic-hydric sites, bottomland forests. Cover type **28.16, 28.37, 28.41, 28.49**, 27.10, 28.12, 28.13, 28.14, 28.16, 28.38, 28.42, 28.45, 28.47.

Quercus arkansana Sargent, Arkansas oak, Arkansas water oak (Figure 19.40); medium-sized to tall tree, height to 30 m; deciduous; leaf arrangement alternate; leaf simple, total length 4 to 15 cm, width 2 to 10 cm, shape broadly spatulate to obovate, obscurely lobed, infrequently three-lobed, margin smooth or bristle tipped with zero to two awns per side, apex broadly rounded, petiole densely pubescent, abaxial surface with axillary tufts and scattered stellate hair; leaves and buds clustered at the twig apex; terminal bud length 3 to 6 mm, densely pubescent, angled; acorn length 1.0 to 1.7 cm, cap densely pubescent, gray, shallow, enclosing about one quarter of the nut, bract tips appressed; twig surface pubescent; mature bark rough; Region 4, southern Arkansas, disjunct locations in Alabama, Georgia, and Florida panhandle; xeric soils of slopes and ridges, open forest and barrens.

5. Section Lobatae (Subgenus Erythrobalanus, Red or Black Oak Group)—Willow or Laurel Oak Species; Regions 3 and 4

Quercus imbricaria Michaux, shingle oak (Figure 19.41); large tree, height to 20 m; deciduous; leaf arrangement alternate; leaf simple, shape elliptical or slightly oblanceolate, total length 9 to 21 cm, width 5 to 7 cm, entire, apex aristate, awn

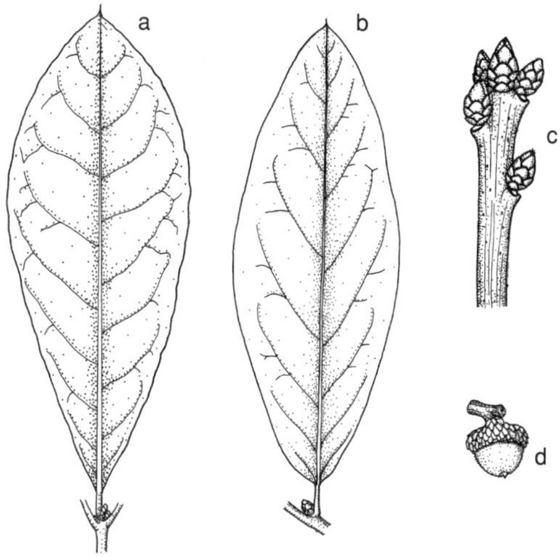

FIGURE 19.41 *Quercus imbricaria* (a), (b) simple leaves ×2/5, (c) twig with terminal and lateral buds ×1.5, (d) cupule and acorn ×1/3.

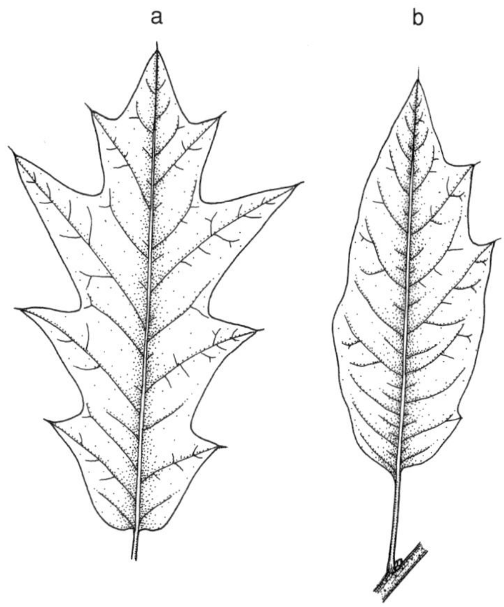

FIGURE 19.42 Leaves of *Quercus* × *leana* ×2/5, a hybrid of *Q. imbricaria* x *Q. velutina*.

length 1 to 2 mm, abaxial surface pubescent; leaves and buds clustered at the twig apex, bud length 2 to 5 mm, scale surface glabrous to lightly pubescent, margin ciliated, weakly angled; acorn length 2 to 3 cm, cap surface pubescent, dull, shallow, enclosing about one quarter to one third of the nut, bract tips slightly lifted; twig surface glabrous to lightly pubescent, shiny; mature bark dark, tight; Region 3a, b, Iowa to Pennsylvania, southwest to Tennessee and Arkansas; xeric-mesic and mesic slopes, mesic-hydric bottomland sites; usually found as individual trees mixed in with a variety of forest community types. *Q. imbricaria* commonly crosses with other red oak species, including *Q. velutina* (*Q.* x *leana* Nuttall) (Figure 19.42).

Quercus phellos Linnaeus, willow oak (Figure 19.43); large tree, height to 30 m; deciduous; leaf arrangement alternate; leaf simple, entire, shape narrowly elliptical, total length 4 to 13 cm, width 0.7 to 2.0 cm, margin smooth, apex acute, mucronate, petiole pubescent or glabrous, abaxial surface glabrous except for continuous bands or scattered tufts along the midrib or blade; leaves and buds clustered at the twig apex; twig surface essentially glabrous, dark brown, terminal bud length 2 to 5 mm, scale surfaces glabrous; acorn length 1.0 to 1.5 cm, cap surface pubescent, dull, gray except for marginal reddish brown glabrous areas, cap enclosing about one third of the nut, bract tips slightly raised; Region 4, Coastal Plain, east Texas and

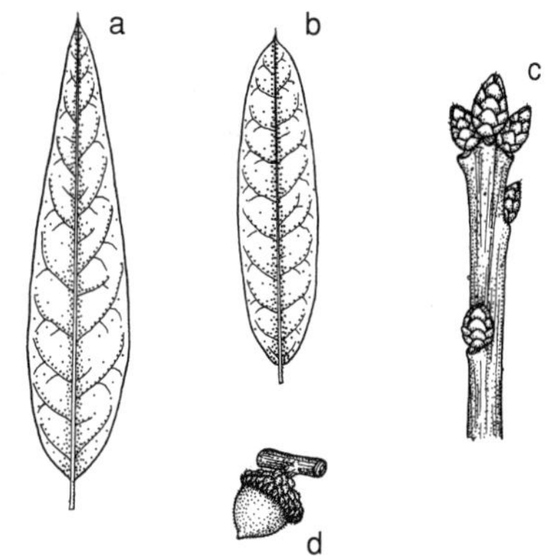

FIGURE 19.43 *Quercus phellos* (a), (b) simple leaves ×1/2, (c) twig with terminal and lateral buds ×1.5, (d) cupule and acorn ×1/2.

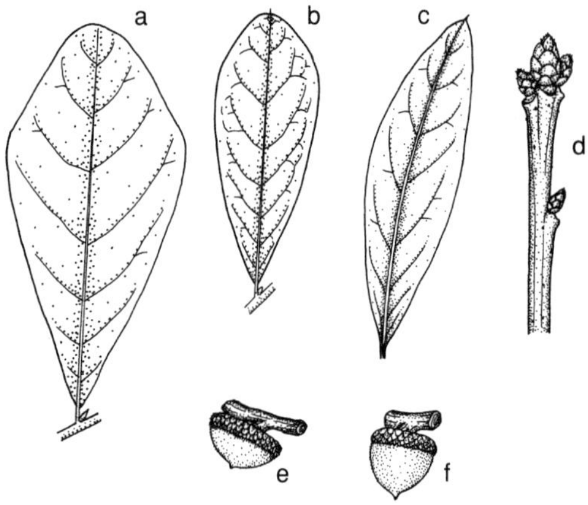

FIGURE 19.44 *Quercus laurifolia* (a), (b), (c) simple leaves ×1/2, (d) twig with terminal and lateral buds ×1, (e), (f) cupule and acorn ×2/3.

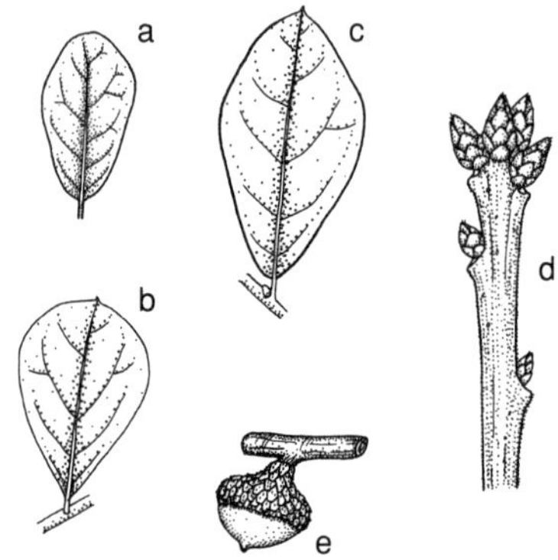

FIGURE 19.45 *Quercus myrtifolia* (a), (b), (c) simple leaves ×1/2, (d) twig with terminal and lateral buds ×2, (e) cupule and acorn ×1.

Oklahoma to New Jersey, except south-eastern Georgia and the Florida peninsula, Region 3a, southern half; mesic-hydric bottomland soils, dense forest. Cover type **28.41, 28.49**, 27.10, 27.14, 28.12, 28.20, 28.26, 28.45, 28.48.

Quercus laurifolia Michaux, swamp laurel oak (Figure 19.44); large tree, height to 35 m; deciduous; leaf arrangement alternate; leaf simple, entire, shape elliptical to oblanceolate, total length 4 to 12 cm, width 2.0 to 4.5 cm, apex rounded or aristate, awn length 1 to 2 cm, abaxial surface glabrous; leaves and buds clustered at the twig apex, bud length 3 to 5 mm, scale surface glabrous, margin ciliated, weakly angled; acorn length 1.0 to 1.5 cm, cap surface pubescent, dull, shallow, enclosing about one quarter of the nut, bract tips appressed; twig surface glabrous, shiny; Region 4, Virginia to Texas; mesic-hydric bottomland sites. Cover **type 28.37, 28.41, 28.47, 28.48, 28.49**, 27.6, 27.10, 28.11, 28.16, 28.32, 28.33, 28.36, 28.38, 28.55.

Quercus myrtifolia Willdenow, myrtle oak (Figure 19.45); small tree, height to 11 m; evergreen; leaf arrangement alternate; leaf simple, blade entire, slightly revolute, shape obovate, oblong, or elliptical, total length 2 to 6 cm, width 1 to 3 cm, apex obtuse, rounded or minutely mucronate, abaxial surface glabrous except for distinct axillary tufts; leaves and buds clustered at the twig apex; bud length 3 to 4 mm, weakly angled, scale surface glabrous, scale margins ciliated; acorn length 0.7 to 1.3 cm, cap surface glabrous, slightly shiny, shallow, enclosing about one half of the nut, bract tips appressed; twig surface dark reddish brown, becoming gray, glabrous; Region 4b, South Carolina to Mississippi; xeric sites, sand hills and ridges. Cover type 28.34, 28.35, 28.44, 28.46.

Quercus hemisphaerica W. Bartram ex. Willdenow, laurel oak (Figure 19.46); large tree, height to 30 m; evergreen or semi-evergreen; leaf arrangement alternate; leaf simple, shape a narrow ellipse, total length 5 to 10 cm, width 1 to 3 cm, entire or infrequently with a vestigial lobe, apex aristate, awn length 1 to 2 mm, abaxial surface glabrous; leaves and buds clustered at the twig apex; bud length 2 to 4 mm, scale surface glabrous, margin ciliated, weakly angled; acorn length 0.7 to 1.5 cm, cap surface glabrous, slightly shiny, shallow, enclosing about one third of the nut, bract tips appressed; twig surface reddish brown, glabrous, shiny; Regions 4, North Carolina to Florida, Peninsula and southern Texas; xeric sites, mesic forest, and hammocks. Hybridizes with many of the southeastern *Quercus*. Cover type **28.45**, 28.51.

Quercus incana, W. Bartram, bluejack oak (Figure 19.46); small tree, height to 12 m; deciduous; leaf arrangement alternate; leaf simple, entire, shape elliptical, total length 4 to 13 cm,

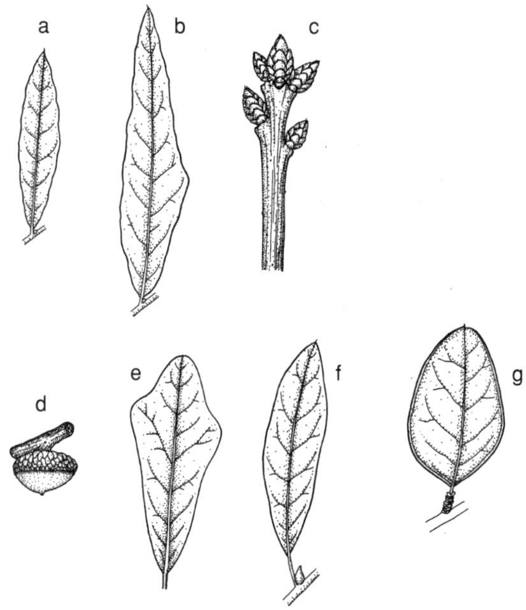

FIGURE 19.46 *Quercus hemisphaerica* (a), (b) simple leaves ×1/2, (c) twig with terminal and lateral buds ×2, (d) cupule and acorn ×1/2; (e) *Q. hemisphaerica* x *Q. nigra* leaf ×1/2; (f) *Q. incana* leaf ×2/3; (g) *Q. inopira* leaf.

width 1 to 3 cm, margin smooth, apex obtuse to acute, awn length 1 to 2 mm, petiole pubescent, abaxial surface pubescent, velvety, grayish; leaves and buds clustered at the twig apex; twig surface pubescent, gray to yellowish, terminal bud length 2 to 5 mm, angular, scale surfaces pubescent; acorn length 1.2 to 1.7 cm, cap surface pubescent, dull, gray except for marginal reddish brown glabrous areas, cap enclosing about one quarter to one third of the nut, bract tips appressed; Regions 4a, b, Coastal Plain, eastern Texas and Oklahoma to Virginia, northern one half of Florida; xeric sandy soil but also flatwoods, disturbance forest where pine has been harvested. Cover type **28.39**, 28.46.

Quercus inopira Ashe, sand scrub oak (Figure 19.46); evergreen; leaf arrangement alternate; leaf simple, blade entire, slightly revolute, shape obovate, oblong, or elliptical; total length 2 to 5 cm, width 1.3 to 2.5 cm, apex rounded, awn length 1 to 2 mm, abaxial surface stellate pubescent; leaves and buds clustered at the twig apex; twig surface dark reddish brown, pubescent, terminal bud length 3 to 4 mm, weakly angled, scale surface pubescent; acorn length 0.7 to 1.3 cm, cap surface pubescent, enclosing about one third to one half of the nut, bract tips appressed; Region 4b, Sand Hills of central Florida; xeric white sands.

Quercus pumila Walter, runner oak; clumped shrub, height to 1 m, stoloniferous; deciduous; leaf arrangement alternate; leaf simple, entire, shape narrowly elliptical, lanceolate, or slightly oblanceolate; total length 4 to 10 cm, width 0.7 to 2 cm, margin entire, minutely revolute, apex acute and mucronate, petiole pubescent, abaxial surface densely tomentose, velvety; leaves and buds clustered at the twig apex; twig surface pubescent, dull, gray, terminal bud length 2 to 5 mm, scale surfaces pubescent; acorn length 1 to 2 cm, cap surface pubescent, dull, enclosing about one third to one half of the nut, bract tips appressed; Region 4, Coastal Plain, eastern Texas to Florida and North Carolina, xeric to mesic soils, associated with upland pine-oak scrub forests but also pine on low

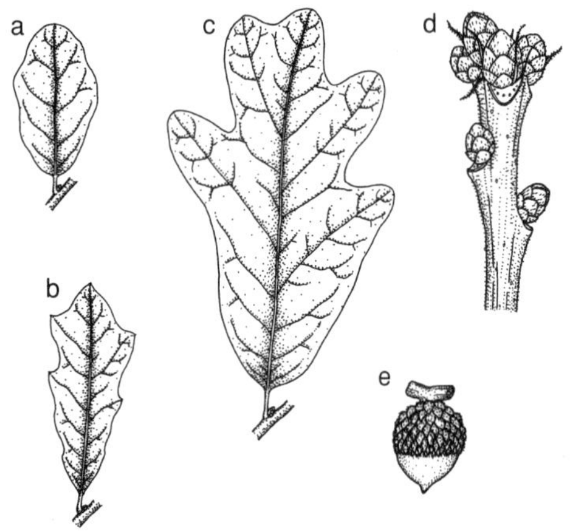

FIGURE 19.47 *Quercus harvardii* (a), (b), (c) simple leaves ×2/5, (d) twig with terminal and lateral buds ×2, (e) cupule and acorn ×2/3.

slope positions. An important wildlife species that responds well to fire management.

Southwestern North America (West of 100th Meridian)—Region 5

1. Section *Quercus* (Subgenus *Leucobalanus*, *Lepidobalanus*, White Oak Group).

Quercus harvardii Rydberg, shinnery or Harvard oak (Figure 19.47); sprawling shrub or arborescent height to 8 m; deciduous; leaf arrangement alternate; leaf simple, outline elliptical to obovate, blade usually five- to seven-lobed or infrequently entire with two to four scattered teeth, sinus depth variable, total length 5 to 12 cm, width 2 to 6 cm, apices rounded, abaxial surface stellate pubescent, somewhat velvety; leaves and buds clustered at the twig apex; twig surface pubescent, dull, brown to reddish brown, terminal bud length 2 to 4 mm, scale surfaces pubescent, acorn length 1 to 2 cm, cap surface pubescent, dull, enclosing about one half of the nut, bract tips slightly raised; Region 5, western Texas and Oklahoma, disjunct in four-corners region of Utah and Arizona; xeric calcarious (limestone-derived) soil. A component of shinnery savanna of low elevations (Dhillion and Mills 1999). Hybridizes with *Q. gambelii* and *Q. turbinella*.

Quercus mohriana, Mohr's oak (Figure 19.48); small tree, height to 7 m; evergreen; leaf arrangement alternate; leaf simple, entire, shape elliptical to oblong, total length 4 to 8 cm, width 1.0 to 3.5 cm, margin entire or infrequently with one or two small lobes or two to six scattered teeth, apices rounded, abaxial surface tomentose and velvety gray; leaves and buds clustered at the twig apex; twig surface pubescent, dull, gray, terminal bud length 2 to 3 mm, scale surface more or less pubescent, margin ciliated with long hair; acorn length 1 to 2 cm, cap surface pubescent, dull, enclosing about one half of the nut, bract tips appressed; Region 5, western Texas, Oklahoma, and Arizona; xeric calcarious (limestone-derived) soil. A component of shinnery savanna of low elevations (Dhillion and Mills 1999). Hybridizes with *Quercus grisea* and *Q. macrocarpa* (Figure 19.48).

Quercus gambelii Nuttall, [*Quercus undulata* Torrey var. *gambelii* (Nuttall) Engelmann], Gambel oak (Figure 19.49); highly variable in growth form, multistemmed and shrubby to single-stem small to medium-sized tree, height to 16 m; deciduous; leaf arrangement alternate; leaf simple, total length 5 to 10 cm, pinnately lobed, sinuses deep and often extending nearly to the midrib, apices rounded, abaxial surface stellate pubescent; leaves and buds clustered at the twig apex; bud length 3 to 5 mm, densely pubescent, angled; acorn length 1.5 to 2.0 cm, cap rough, warty, pubescent, dull, enclosing about one half of the nut; twig surface pubescent, dull; Region 5, northern Utah and Colorado into Mexico; elevation 1000 to 2900 m; Upper Sonoran juniper-pinyon woodland to interior ponderosa pine forest; shade intolerant. Cover type **29.3**, 29.5, 29.6, 29.13, 29.15.

Quercus laceyi Small, Lacey oak (Figure 19.50),); small tree height to 8 m; deciduous; leaf arrangment alternate, leaf simple, outline elliptical to ovate; blade entire of 5 to 7-lobed, sinus shallow; total length 5–9 cm, width 2–5 cm, apex rounded, abaxial surface pubescent, whitish; leaves and buds clustered at twig apex; twig surface pubescent, dull, brown to reddish-brown, terminal bud length 3–5 mm, scale surfaces pubescent, acorn length 1–2 cm, cap surface pubescent, dull, enclosing

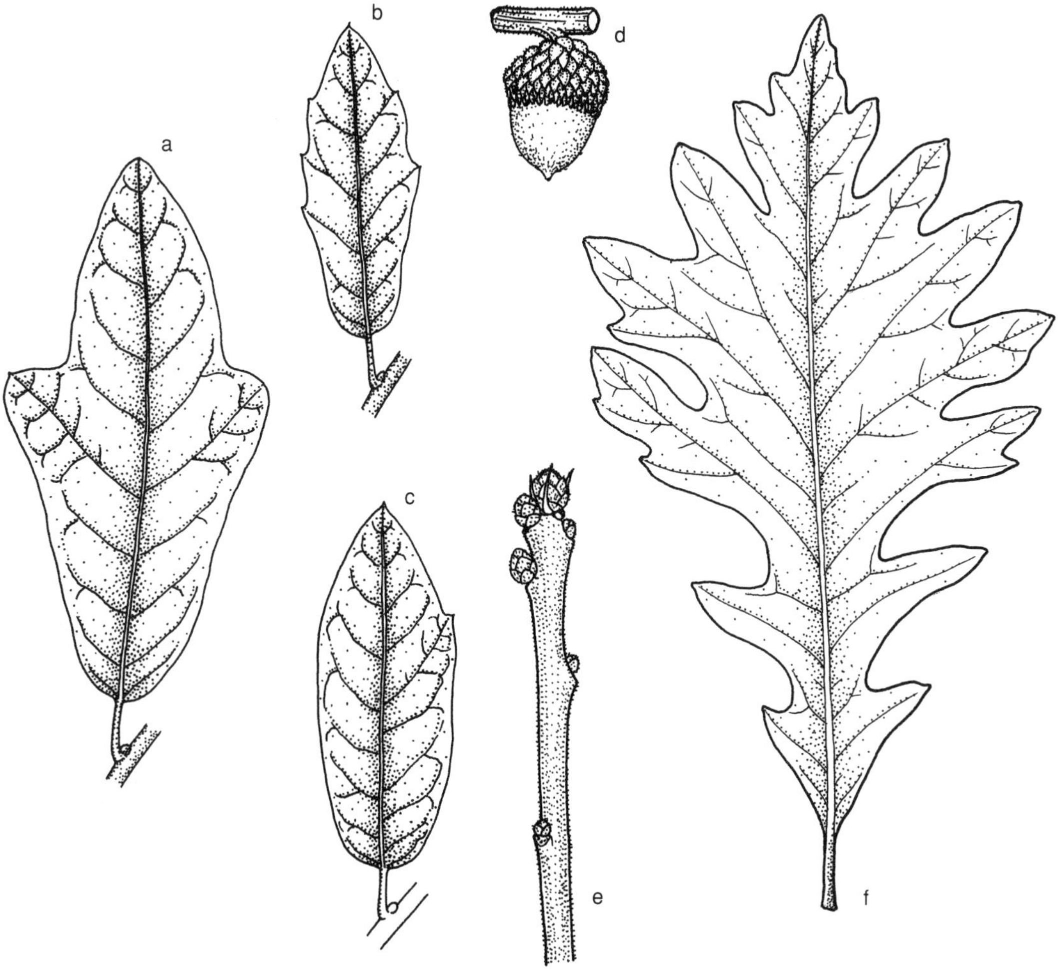

FIGURE 19.48 *Quercus mohriana* (a), (b), (c) simple leaves ×1, (e) twig with terminal and lateral buds ×2, (d), cupule and acorn ×1; (f) hybrid of *Q. mohriana* x *Q. macrocarpa* ×1.

about 1/3 of the nut, bract tips slightly raised; Region 5, southwest Texas; xeric calcarious soil, stream terraces, open woodland.

Quercus fusiformis Small, Texas live oak; medium-sized tree, height to 17 m but often a shrub forming thickets; semi-evergreen, leaves retained through the dormant season; leaf arrangement alternate; leaf simple, essentially entire, shape elliptical, oval, or obovate, total length 3 to 8 cm, width 0.7 to 1.5 cm, margin smooth or infrequently with two to four irregularly spaced teeth, apex obtuse to rounded, petiole pubescent, abaxial surface pubescent, velvety; leaves and buds clustered at the twig apex; twig and bud surfaces pubescent, gray to yellowish brown, terminal bud length 1 to 2 mm; acorns borne individually along twig, acorn length 1.5 to 2.0 cm, cap surface pubescent, dull, gray, cap enclosing about one half of the nut; Region 5, central Texas and Oklahoma; xeric sites, open woodlands. This species is similar to and difficult

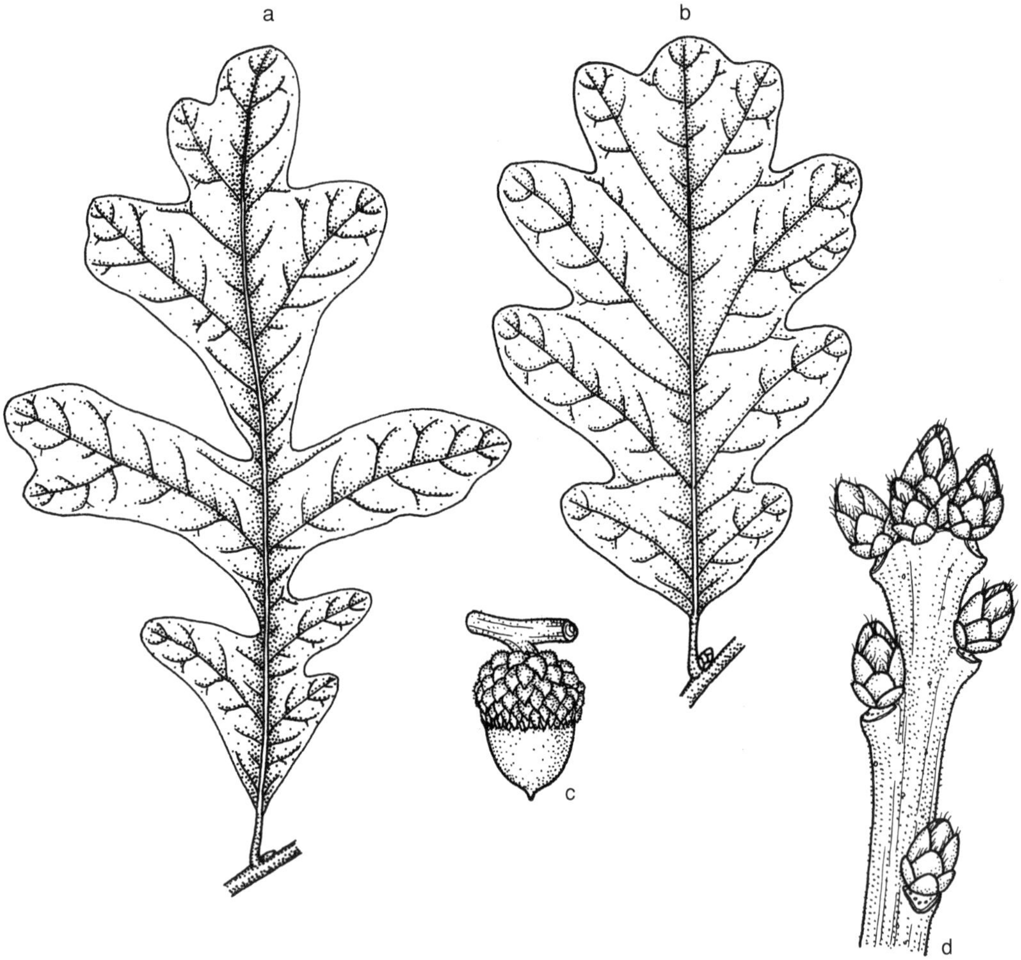

FIGURE 19.49 *Quercus gambelii* (a), (b) simple leaves ×1.2, (c) cupule and acorn ×1, (d) twig with terminal and lateral buds ×4.

to separate from *Q. virginiana*. Cover type 28.9, 29.9.

Quercus grisea Liebmann [*Quercus undulata* Torrey var. *grisea* (Liebmann) Engelmann], gray oak (Figure 19.51); small tree, height to 12 m; evergreen; leaf arrangement alternate; leaf simple, total length 2 to 5 cm, width 1.0 to 2.5 cm, blade entire, margin nearly smooth or undulating, infrequently coarsely toothed with two to four indistinct teeth, adaxial surface glabrous, shiny, petiole pubescent, abaxial surface densely pubescent, gray; leaves and buds clustered at the twig apex; new twig surface densely pubescent, dull, yellowish gray, terminal bud length 1 to 2 mm, surface with scattered hair; acorn length 1.5 to 2.0 cm, cap surface pubescent/scaly, warty, dull gray, enclosing one half of the nut; Region 5, western Texas, New Mexico, and Arizona; xeric slopes and ridges, oak-juniper woodlands and desert shrub communities. Cover type 29.1, 29.4, 29.6.

Quercus oblongifolia Torrey in Sitgreaves, Mexican or Sonoran blue oak (Figure 19.51); small

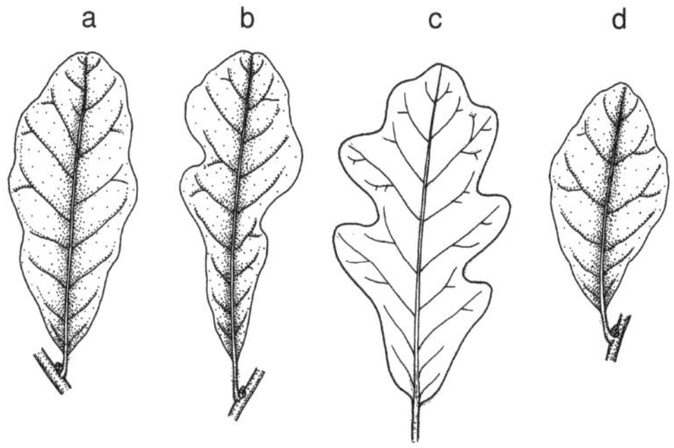

FIGURE 19.50 *Quercus laceyi* leaves ×1/2.

tree, height to 20 m; evergreen; leaf arrangement alternate; leaf simple, total length 2 to 5 cm, width 1 to 2 cm, entire, margin smooth, occasionally with small teeth near the apex, shape oblong with nearly parallel margins, abaxial surface light gray, glabrous to densely pubescent; buds clustered at the twig apex; twig glabrous to pubescent, light brown to gray; acorn length 1 to 2 cm; Region 5, Arizona and New Mexico into Mexico; xeric rocky slopes, elevation 1300 to 1600 m, open savanna.

Quercus arizonica Sargent, Arizona oak, Arizona white oak (Figure 19.52); small to medium-sized tree, height to 16 m; evergreen; leaf arrangement alternate; leaf simple, highly variable, total length 4 to 8 cm, width 1 to 3 cm, blade shape elliptical to obovate, entire, margin smooth to sharply toothed, adaxial surface metallic white, secondary and tertiary veins yellowish and distinct, abaxial surface densely pubescent; leaves and buds clustered at the twig apex; twig surface densely

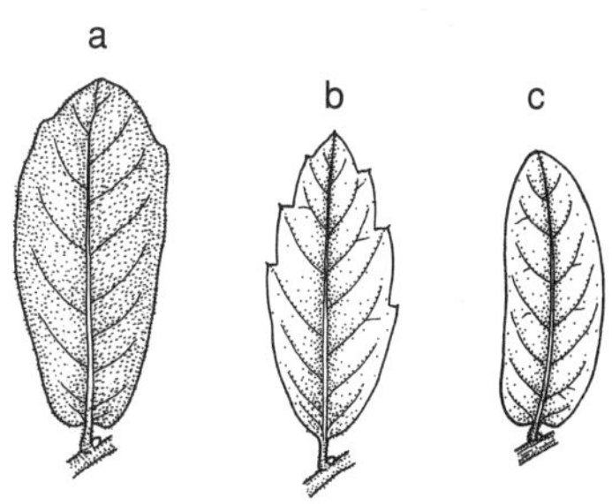

FIGURE 19.51 *Quercus grisea* (a), (b) leaves ×2/3; (c) *Quercus oblongifolia* leaf ×1/2.

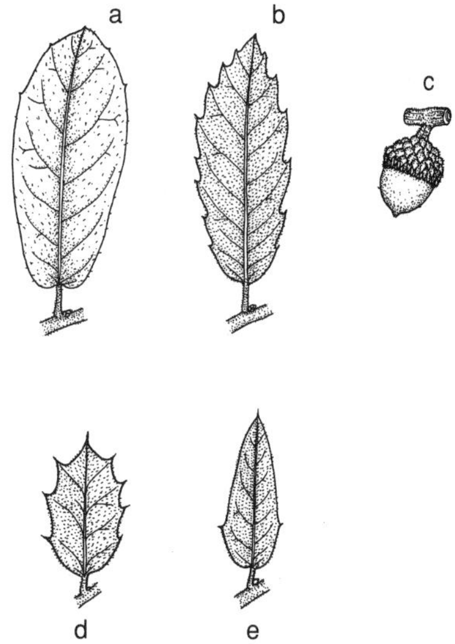

FIGURE 19.52 *Quercus arizonica* (a), (b) simple leaves ×1/2, (c) cupule and acorn ×2/3; (d), (e) *Q. turbinella* leaves ×2/3.

pubescent, dull, yellowish, terminal bud length 1 to 2 mm, reddish brown, pubescence strongly appressed, scale margins ciliated; acorn length 1 to 2 cm, cap rough, densely pubescent, dull, enclosing about one half of the nut; Region 5, central Arizona and New Mexico into Mexico; xeric slopes and moist canyons, found in foothills, elevation 1400 to 2600 m. Grows with a variety of other oak, pinyon, and ponderosa pine. Individual stems in moist canyons are reported to grow to be the largest of the southwestern *Quercus* trees. Cover type 29.5, 29.19.

Quercus turbinella Greene, Sonoran scrub oak, scrub live oak (Figure 19.52); shrub, height to 4 m; evergreen; leaf arrangement alternate; leaf simple, highly variable, total length 1.0 to 3.5 cm, width 0.5 to 1.5 cm, blade shape oval to slightly lanceolate, margin coarsely toothed to spinose, three to five spines per side, apex bristle tipped, adaxial surface stellate pubescent, abaxial surface densely stellate pubescent, velvety, yellowish gray; leaves and buds clustered at the twig apex; twig surface densely pubescent, yellowish gray to light brown, terminal bud length 1 mm, pubescent, scale margins ciliated; acorn length 1.5 to 2.0 cm,

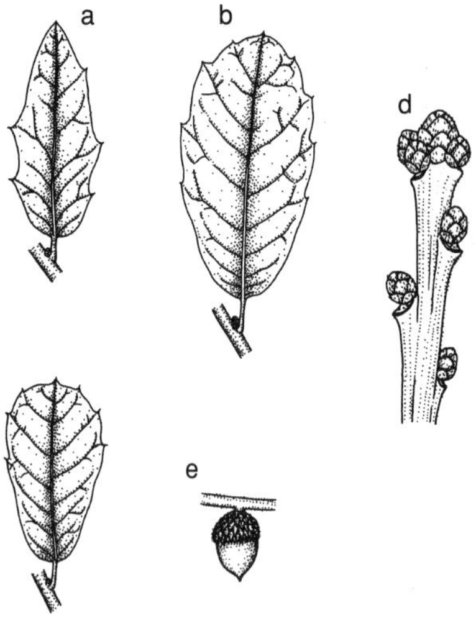

FIGURE 19.53 *Quercus emoryi* (a), (b), (c) simple leaves ×1/2, (d) twig with terminal and lateral buds ×2, (e) cupule and acorn ×1/2.

cap rough, warty, pubescent, dull, enclosing about one third to one half of the nut; Region 5, southern California into Baja California (Mexico), Utah, Arizona, New Mexico, and Colorado into Mexico; xeric slopes and ridges, elevation 850 to 2000 m; pinyon-juniper woodland, savanna, chaparral, and desert-shrub. Cover type 29.5.

2. Section *Lobatae* (Subgenus Erythrobalanus, Red or Black Oak Group).

Quercus emoryi Torrey, Emory oak (Figure 19.53); medium-sized tree, height to 20 m; evergreen; leaf arrangement alternate; leaf simple, total length 2 to 9 cm, width 1.0 to 2.5 cm, outline elliptical, lanceolate, or oval, blade entire, margins vary from nearly smooth to undulate, and weakly spinose, occasionally lobed, when present lobe apex bristle-tipped, adaxial surface lustrous, abaxial surface glabrous and shiny, petiole and basal veins stellate pubescent; leaves and buds clustered at the twig apex, bud length 2 to 4 mm, surface pubescent, scale surface lightly pubescent, margin ciliated; acorn length 1.5 to 2.0 cm, enclosing one third of the nut, bract surface silvery pubescent; twig surface densely pubescent, dull, dark brown to reddish brown; Region 5, primary distribution from central Arizona into northern Mexico, scattered in New Mexico and western Texas; xeric sites at low elevations, 1100 to 2100 m. The acorns are sweet and eaten by wildlife. Cover type 29.19.

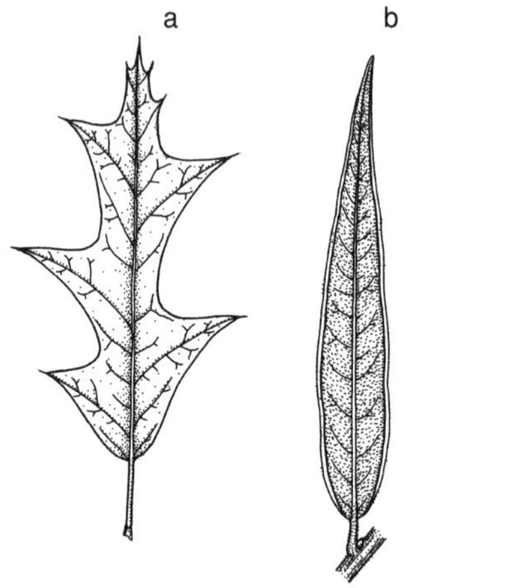

FIGURE 19.54 Leaf of (a) *Quercus gravesii* ×1/2; (b) *Q. hypoleucoides* ×2/3; note minutely revolute margin.

Quercus gravesii, Graves oak, Chisos red oak sudworth (Figure 19.54); small tree, height to 12 m; deciduous; leaf arrangement alternate; leaf simple, total length 4 to 14 cm, width 2 to 10 cm, ovate to elliptical, blade entire, lobes 3–5, falcate, awns 9–19, margins smooth, adaxial surface lustrous, abaxial surface glabrous except of axillary tufts, petiole glabrous; leaves and buds clustered at the twig apex; bud length 2 to 5 mm, glabrous or pubescent at the apex; acorn length 1.5–2.7 cm, enclosing 1/3–1/2 of the nut, bract surface glabrous; twig glabrous, dull, dark brown to reddish brown; Region 5, primary distribution from western Texas into Mexico; xeric sites at low elevations, 1100 to 2100 m Cover type **29.1**, 28.4.

Quercus hypoleucoides A. Camus, silverleaf oak (Figure 19.54); clumped shrub to medium-sized tree, height to 18 m; evergreen; leaf arrangement alternate; leaf simple, entire, shape elliptical to lanceolate, total length 4 to 10 cm, width 1.0 to 2.5 cm, margin entire, minutely revolute, apex acute, abaxial surface densely tomentose and velvety gray; leaves and buds clustered at the twig apex; twig surface pubescent, dull, gray, terminal bud length 1 to

2 mm, scale surface more or less pubescent, margin ciliated with long hair; acorn length 1 to 2 cm, cap surface pubescent, dull, enclosing about one half of the nut, bract tips appressed; Region 5, Mexican border region of western Texas, New Mexico, and Arizona, into Mexico; xeric mountain slopes and canyons, elevation 1500 to 2100 m. A distinctive oak because of its leaves.

West Coast Quercus—Region 7

1. Section Quercus (Subgenus *Leucobalanus*, *Lepidobalanus* White Oak Group).

Quercus garryana Douglas ex. Hooker var. *garryana*, Oregon white oak (Figure 19.55); large tree, height to 30 m, may also appear as a shrub; deciduous; leaf arrangement alternate; leaf simple, total length 11 to 14 cm, pinnately lobed, sinuses extending about one third the distance or deeper from the lobe apex to the midrib, apices rounded, abaxial surface stellate pubescent; leaves and buds clustered at the twig apex; bud length 5 to 8 mm, densely pubescent, angled; acorn length 2 to 3 cm, cap rough, warty, pubescent, dull, enclosing about one third of the nut; twig surface pubescent, dull; Region 7, central California to Washington and Vancouver Island (British Columbia); fog belt to dry slopes and exposed ridges, elevation 0 to 800 m, gravel to clay soils, best development is on bottomlands along the Columbia River. The range of this species extends farther north than any other western *Quercus*. The shrubby populations at higher elevations are considered varieties: *Q. garryana* var. *breweri* (height to 3 m; elevation 1400 to 1850 m) and *Q. garryana* var. *semota* (height to 5 m; elevation 1200 to 1850 m). An important wildlife and grazing food, as leaf protein level nearly equals that of alfalfa (Miller and Lamb 1985). Cover type **30.20**, 30.4, 30.9, 30.17, 30.18, 30.28, 30.37, 30.39.

Quercus douglasii Hooker & Arnott, Douglas or blue oak (Figure 19.56); medium-sized tree, height to 20 m; deciduous; leaf arrangement alternate; leaf simple, total length 3 to 10 cm, width 1 to 5 cm, outline oval, ovate, or obovate, blade entire or lobed, margins vary from nearly smooth to undulate, lobed, and weakly spinose, when present lobes rounded and shallow to deep sinuses, petiole and abaxial surfaces stellate pubescent; leaves and buds clustered at the twig apex, bud length 3 to 6 mm, surface pubescent; acorn length 2 to 3 cm,

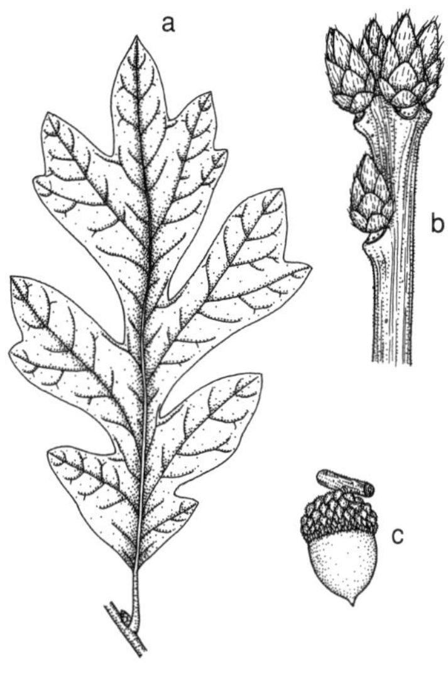

FIGURE 19.55 *Quercus garryana* (a) simple leaf ×1/2, (b) twig with terminal and lateral buds ×1.3, (c) cupule and acorn ×1/2.

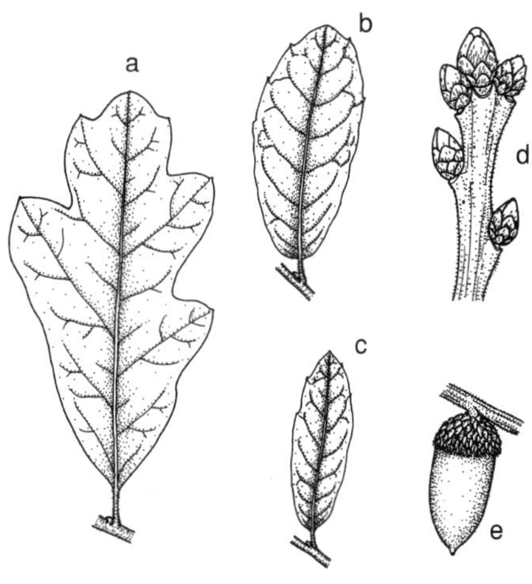

FIGURE 19.56 *Quercus douglasii* (a), (b), (c) simple leaf ×1/3, (d) twig with terminal and lateral buds ×1.3, (e) cupule and acorn ×3/5.

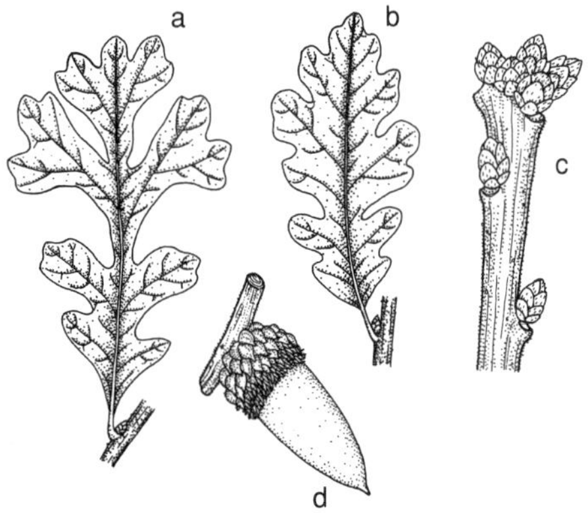

FIGURE 19.57 *Quercus lobata* (a), (b) simple leaf ×1/2, (c) twig with terminal and lateral buds ×1.5, (d) cupule and acorn ×1/2.

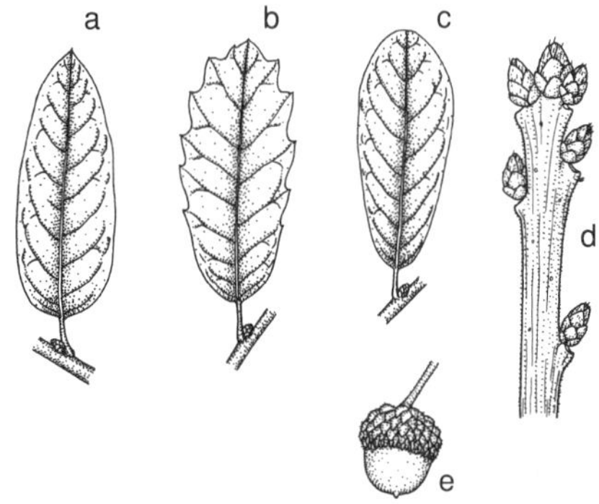

FIGURE 19.58 *Quercus engelmannii* (a), (b), (c) simple leaf ×1/2, (d) twig with terminal and lateral buds ×1.3, (e) cupule and acorn ×1/3.

bract surface silvery pubescent except light brown at apex, apices raised, enclosing one quarter of the nut; twig surface densely pubescent to dull, dark brown to reddish brown; Region 7, Coast Ranges and Sierra Nevada foothills surrounding the central valley, elevation 0 to 1200 m. This is the dominant species of blue oak savanna and woodlands; usually associated with other *Quercus* species and communities. Large, sweet acorns are produced in abundance and are widely used by both wildlife and domestic animals. Cover type **30.38**, 30.29.

Quercus lobata Nee, valley oak (Figure 19.57); medium to large tree, height to 30 m; deciduous; leaf arrangement alternate; leaf simple, total length 4 to 13 cm, width to 5 cm, pinnately lobed, sinuses extending about two thirds the distance or deeper from the lobe apex to the midrib, lobes expanded with one to two indentations at the apex, apices rounded, abaxial surface stellate pubescent; leaves and buds clustered at the twig apex; bud length 3 to 4 mm, lightly pubescent, angled, scale margins ciliated; acorn length 3 to 5 cm, cap rough, warty, pubescent, dull, enclosing about one quarter to one third of the nut; twig surface slightly pubescent, dull; Region 7, California; slopes of open valleys, elevation 0 to 1700 m. George Vancouver first described the large crowns and thick spreading branches of these trees in the oak savanna of the Santa Clara Valley. This species may live to an age of 600 years (Pavlik et al. 1991). Cover type **30.38**.

Quercus engelmannii, Engelmann oak (Figure 19.58); small to medium-sized tree, height to 15 m; evergreen; leaf arrangement alternate; leaf simple, total length 2.5 to 7.0 cm, width 1 to 3 cm, blade entire, margin nearly smooth to coarsely serrated with six to ten scattered, often indistinct, teeth, small leaves infrequently spinose, adaxial surface glabrous, shiny, petiole pubescent, abaxial surface glabrous; leaves and buds clustered at the twig apex; twig surface densely pubescent to scattered tomentose, dull, reddish brown, terminal bud length 2 to 4 mm, surface with scattered hair; acorn length 2 to 3 cm, cap surface pubescent/scaly, warty, dull, shallow, enclosing one half of the nut; Region 7, extreme southwestern California, 200 km north of Baja California south into Baja; xeric sites, elevations up to 1250 m, found in association with a variety of *Quercus* species.

Quercus dumosa Nuttall, coastal sage scrub oak, California scrub oak (Figure 19.59); shrub or small tree, height to 4 m; semi-evergreen; leaf arrangement alternate; leaf simple, highly variable, total length 1.5 to 3.0 cm, width 0.5 to 4.0 cm, blade entire, shape oval to elliptical, margin serrate and/or spinose, slightly revolute, three to ten spines per side, apex rounded, bristle-tipped, abaxial glabrous to pubescent, abaxial surface densely stellate, pubescent to tomentose, velvety, yellowish gray, petiole glabrous to pubescent; leaves and buds clustered at the twig apex; twig surface glabrous to pubescent, reddish brown, terminal bud length 2 to

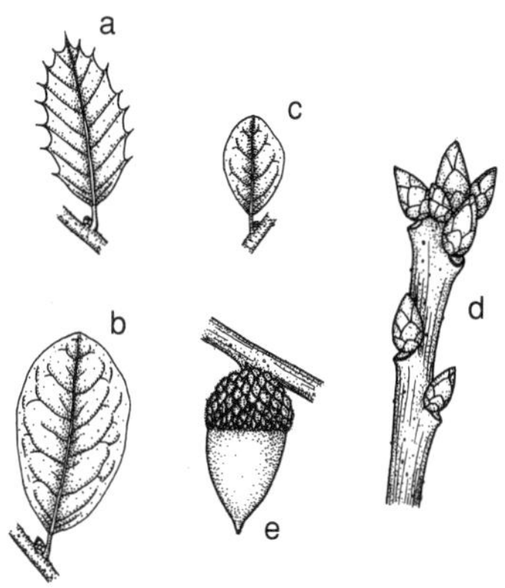

FIGURE 19.59 *Quercus dumosa* (a), (b), (c) simple leaf ×1/2, (d) twig with terminal and lateral buds ×1.5, (e) cupule and acorn ×3/4.

4 mm, pubescent, reddish brown; acorn length 1.5 to 2.5 cm, cap rough, warty, pubescent, dull, enclosing about one third to one half of the nut; Region 7, central California into Baja California (Mexico); xeric sites near the coast, elevation up to 300 m; component of coastal scrub and chaparral. Development along the coast is impacting this species.

Quercus berberidifolia Leibmann, California scrub oak (Figure 19.60); shrub, height to 5 m; semi-evergreen; leaf arrangement alternate; leaf simple, highly variable, total length 1 to 4 cm, width 1 to 2 cm, blade shape oval to elliptical, margin coarsely toothed to spinose, abaxial surface densely pubescent; bud length 1 to 3 mm, densely pubescent; leaves and buds clustered at the twig apex; twig surface pubescent, dull; acorn length 2.0 to 3.5 cm, cap rough, warty, pubescent, dull, enclosing about one fifth to one third of the nut, inner cap surface densely pubescent; Region 7, California; found in the foothills of the Sierra Nevada, in the San Bernardino Mountains, and in the Coast Ranges west of the Sacramento and San Joaquin river valleys, elevation 100 to 1700 m. This species is one of the most important oak shrubs of California and an important component of chaparral and coastal sage shrub.

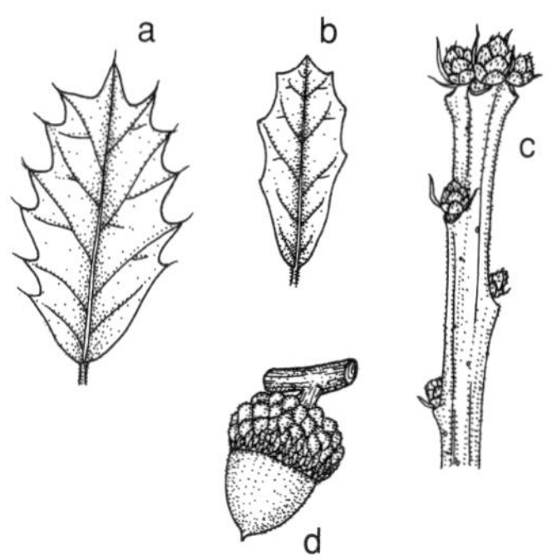

FIGURE 19.60 *Quercus berberidifolia* (a), (b) simple leaf ×1, (c) twig with terminal and lateral buds ×1.5, (d) cupule and acorn ×1/2.

Quercus cornelius-mulleri Nixon & K. P. Steele, Muller oak (Figure 19.61); branched shrub, height to 2.5 m; evergreen; leaf arrangement alternate; leaf simple, highly variable, total length 1.5 to 3.5 cm, width 0.5 to 1.3 cm, blade shape oval to elliptical, margin coarsely toothed to spinose, two to four spines per side, adaxial surface stellate pubescent,

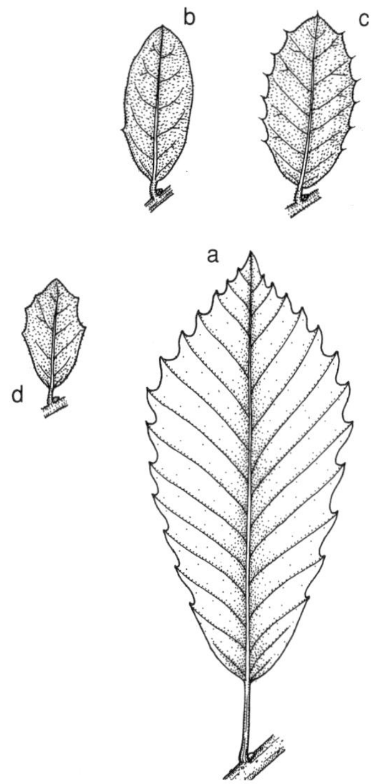

FIGURE 19.61 Leaf of (a) *Quercus sadleriana* ×1; (b), (c) *Q. durata* ×1; (d) *Q. cornelius-mulleri* ×3/4.

abaxial surface densely pubescent, velvety, yellowish gray; leaves and buds clustered at the twig apex; twig surface pubescent, yellowish gray to reddish brown, terminal bud length 1 to 2 mm, densely pubescent, scale margins ciliated; acorn length 1.5 to 2.0 cm, cap rough, warty, pubescent, dull, enclosing about one third to one half of the nut; Region 7, southern California into Baja California (Mexico); xeric sites, elevation 900 to 1850 m, savanna, chaparral, and desert-shrub.

Quercus durata Jepson, leather oak (Figure 19.61); shrub, height to 4 m; evergreen; leaf arrangement alternate; leaf simple, highly variable, total length 1.0 to 3.5 cm, width 0.5 to 2.0 cm, blade entire, shape oval, margin smooth to spinose, slightly revolute, up to seven spines per side, apex bristle-tipped or rounded, adaxial surface stellate pubescent, abaxial surface densely stellate pubescent, velvety, yellowish gray; leaves and buds clustered at the twig apex; twig surface densely pubescent, yellowish gray to light brown, terminal bud length 1 to 2 mm, pubescent, scale margins ciliated; acorn length 1.5 to 2.5 cm, cap rough, pubescent, dull, scale margins ciliated, enclosing about one third to one half of the nut; Region 7, California; serpentine (high-magnesium, low-nutrient) soils on slopes and ridges surrounding the central valley, elevation up to 1700 m, woodland, savanna, and chaparral.

Quercus sadleriana R. Brown, deer oak (Figure 19.61); shrub, height to 3 m, stems clustered, arising from a central root collar to form a spindle-shaped crown; evergreen; leaf arrangement alternate; leaf simple, total length 7 to 15 cm, width 3 to 7 mm, blade entire, shape elliptical to obovate, margin serrated, teeth apices acuminate, abaxial surface with appressed pubescence; leaves and buds clustered at the twig to apex; twig surface glabrous, gray to reddish brown; bud length 3 to 8 mm, tomentose, accessory bracts often present; acorn length 1 to 2 cm, cap densely pubescent, dull, enclosing one third of the nut; Region 7, northern California, southern Oregon; xeric hillsides and ridges, elevations 800 and 2200 m, coniferous forest. The shape and margin of the leaves are reminiscent of the leaf of *Castanea* or a *Viburnum*. Apparently, this species is unrelated to any other western *Quercus*; its leaves are similer to several of the eastern chestnut oak species. Cover type 30.18, 30.24, 30.26, 30.27, 30.30, 30.31.

2. Section *Lobatae* (Subgenus *Erythrobalanus* Red or Black Oak Group)

Quercus kelloggii Newberry; California black oak (Figure 19.62); medium to large tree, height to 30 m; deciduous; leaf arrangement alternate; leaf simple, total length 10 to 22 cm, width 4 to 12 cm, lobes seven to nine, each lobe bristle tipped with three to seven awns, middle lobe apices spreading and bifurcating, on some leaves adjacent lobes overlapping, sinuses extending about one quarter to three quarters the distance from the lobe apex to the midrib, abaxial surface with axillary tufts along primary veins, intervenal regions with widely spaced stellate hairs; twig surface glabrous, shiny; leaves and buds clustered at the twig apex, terminal bud length 3 to 6 mm, glabrous, reddish brown, scale margins ciliated; acorn length 2 to 3 cm, cap surface lightly pubescent, enclosing about one quarter to one third of the nut, bracts loose; Region 7, central California to central Oregon, Coastal Ranges and western slope of the Sierra Nevada, elevation 300 to 2200 m; xeric to mesic sites. Cover type **30.38, 30.44**, 30.20, 30.26, 30.28, 30.30, 30.34, 30.36, 30.39, 30.43.

Quercus wislizensi A. de Candolle, Sierra or interior live oak (Figure 19.63); medium-sized tree, height to 23 m; evergreen; leaf arrangement alternate; leaf simple, total length 3 to 7 cm, width 1 to 3 cm, blade entire, margin smooth to coarsely toothed with up to ten teeth per side, teeth spinose, adaxial surface glabrous, shiny, petiole densely pubescent, abaxial surface essentially glabrous; twig surface pubescent to glabrous, reddish brown; leaves and buds clustered at the twig apex; terminal bud length 5 to 10 mm, surface glabrous, reddish brown, scale margins ciliated; acorn length 2.0 to 3.3 cm, cap surface pubescent, dull, shallow, enclosing one half to two thirds of the nut; Region 7, Coast Ranges and the Sierra Nevada, surrounding the central valley; xeric soil, lower slopes, elevation 500 to 1500 m, open woodland and chaparral. Cover type **30.38,** 30.29, 30.45.

Quercus agrifolia Nee, coast or California live oak (Figure 19.64); medium to large tree, height to 26 m; evergreen; leaf arrangement alternate; leaf simple, total length 2.5 to 7.0 cm, width 1 to 4 cm, blade entire, shape ovate to oval, generally revolute,

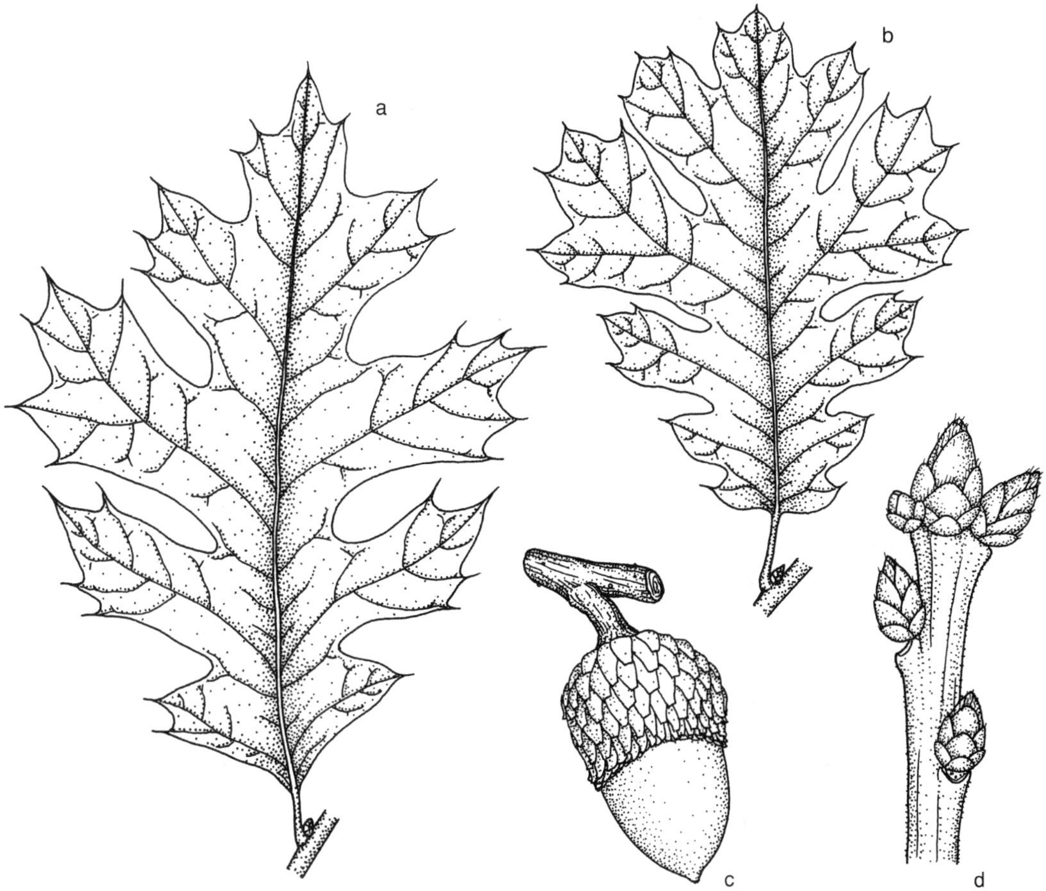

FIGURE 19.62 *Quercus kelloggii* (a), (b) simple leaf ×2/3, (c) cupule and acorn ×1.5, (c) twig with terminal and lateral buds ×2.

margin spinose, adaxial and abaxial surfaces glabrous, waxy, shiny, petiole and basal midrib stellate pubescent; leaves and buds clustered at the twig apex; twig surface densely pubescent to dull, reddish brown; bud length 3 to 6 mm, glabrous; acorn length 2.0 to 3.5 cm, cap surface silvery pubescent except at the light brown margins, dull, enclosing one third to one half of the nut; Region 7, central California (Sonoma County) into Baja California (Mexico); xeric rocky slopes of the Coast Ranges, elevation 0 to 1400 m. This is a common tree between the Coast Ranges and the Pacific Ocean, apparently tolerant of salt spray. A component of mixed evergreen forest and oak woodland communities. In presettlement time, a variety of First Nation peoples used the acorn as a staple food. Early settlers cleared much of the live oak woodlands and savannas. Cover type **30.38**, 30.35, 30.45.

3. Section *Protobalanus* (intermediate or golden-cup oaks).

Quercus tomentella Engelmann, island live or Channel Island oak (Figure 19.65); small to medium-sized tree, height to 15 m; evergreen; leaf arrangement alternate; leaf simple, total length 3 to 7 cm, width 1 to 3 cm, blade entire, margin nearly smooth to coarsely serrated with six to eight

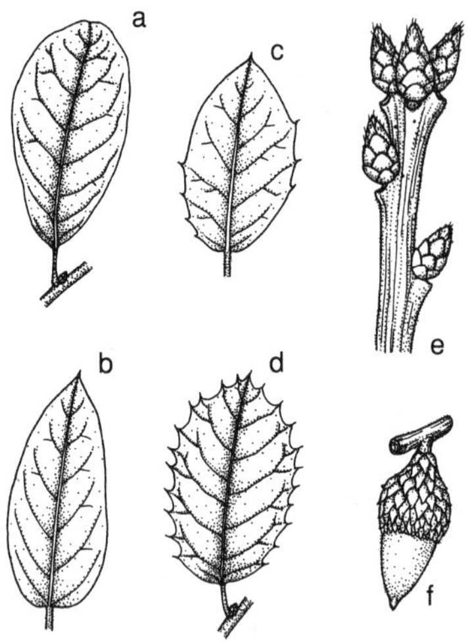

FIGURE 19.63 *Quercus wislizensi* (a), (b), (c), (d) simple leaf ×1/2, (e) twig with terminal and lateral buds ×1, (f) cupule and acorn ×1/2.

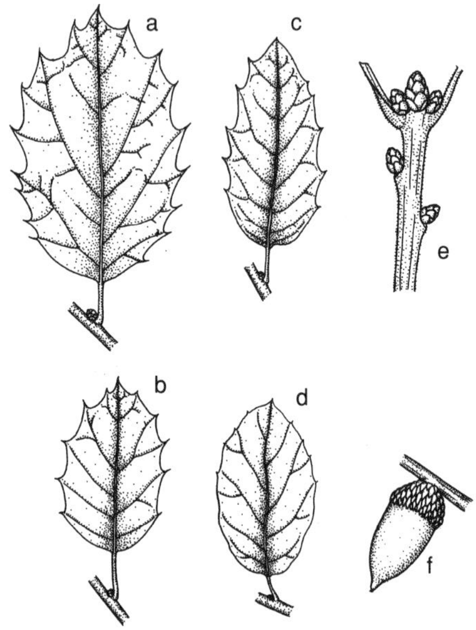

FIGURE 19.64 *Quercus agrifolia* (a), (b), (c), (d) simple leaf ×3/4, (e) twig with terminal and lateral buds ×1, (f) cupule and acorn ×1/2.

scattered teeth, infrequently spinose, adaxial surface glabrous, shiny, petiole and abaxial surface densely pubescent, yellowish; twig surface densely pubescent to scattered tomentose, dull, reddish brown; leaves and buds clustered at the twig apex; bud length 6 to 10 mm, surface pubescent, scale margins ciliated; acorn length 2.5 to 5.0 cm, cap surface pubescent, dull, shallow, enclosing one third of the nut; Region 7. This is a rare species restricted to several islands off the coast of California; considered a relic species (Pavlik et al. 1991). Because of the sea wind and spray, the trees are often deformed or wind-trained (branches swept in the direction of the wind).

Quercus chrysolepis Liebmann, canyon oak, canyon live oak, maul oak (Figure 19.66); shrub to large tree, height to 23 m; evergreen; leaf arrangement alternate; leaf simple, total length 2.5 to 7.0 cm, width 1 to 3 cm, blade entire, margin often revolute, smooth, with three to four scattered teeth or completely spinose, abaxial surface gray, densely pubescent; leaves and buds clustered at the twig apex or a solitary terminal; bud length 3 to 6 mm, glabrous; twig surface densely pubescent to scattered tomentose, dull, reddish brown; bark gray, developing into narrow strips; acorn length 2 to 3 cm, cap surface pubescent, dull, shallow, enclosing less than one quarter of the nut; Region 7, primary concentration from southwestern Oregon to southern California, outliers found in Baja California (Mexico) and Arizona; xeric rocky slopes in the Coastal Ranges and Sierra Nevada foothills, elevation 0 to 2600 m. Not commercially important, but an important component of the savanna in the central valley; also found in conifer forest at higher elevation. The species remains highly variable, although one previously recognized variety (*Q. chrysolepis* var. *vaccinifolia* (Kellogg) is now reported as a species (Manos 1997). Cover type **30.39**, 30.4, 30.7, 30.20, 30.25, 30.26, 30.28, 30.31, 30.34, 30.36, 30.37, 30.38, 30.45.

Quercus vaccinifolia Kellogg ex. Curran, huckleberry oak (Figure 19.67); multistemmed spreading shrub, height < 2 m; evergreen; leaf arrangement alternate; leaf simple, total length 2 to 4 cm, width 7 to 10 mm, blade entire, margin usually smooth, infrequently with several small teeth or points, abaxial surface glabrous or with scattered hair, usually whitish but also greenish or brownish, veins appearing distinctly whitish against the green or brownish intervenal areas; twig surface lightly pubescent to scattered tomentose, reddish brown; leaves and buds clustered at the twig apex or a solitary terminal; bud length 2 to 3 mm, glabrous;

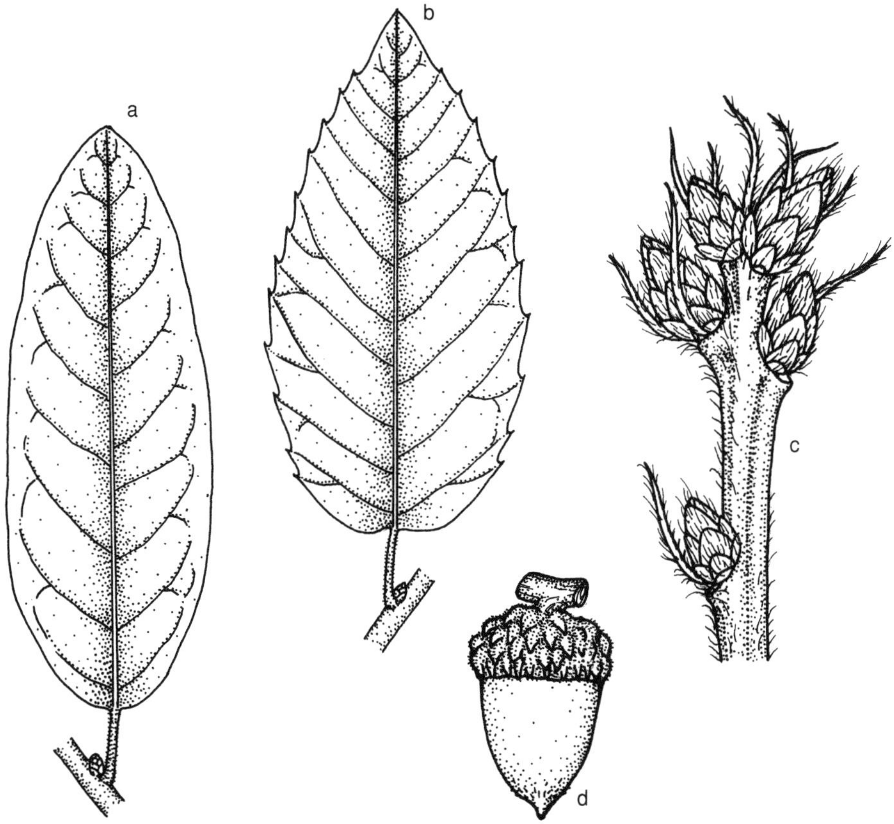

FIGURE 19.65 *Quercus tomentella* (a), (b) simple leaf ×2, (c) twig with terminal and lateral buds ×2, (d) cupule and acorn ×2.

acorn length 1.0 to 1.5 cm, cap surface pubescent, dull, enclosing one half of the nut; Region 7, primary concentration from southwest Oregon to central California; xeric sites, rocky outcrops, elevation 920 to 3000 m. Cover type 30.27, 30.31.

Quercus palmeri Engelmann in S. Watson (*Q. dunnii* Kellogg ex. Curran), Palmer oak (Figure 19.67); shrub or small tree, height to 3 m; evergreen; leaf arrangement alternate; leaf simple, total length 1.4 to 4.5 cm, width 7.0 to 1.5 mm, blade entire, shape broadly oval, margin spinose, infrequently smooth, abaxial surface whitish, glabrous, waxy or with scattered brownish hair; twig surface pubescent, gray to reddish brown; leaves and buds clustered at the twig apex or a solitary terminal; bud length 1 to 3 mm, glabrous; acorn length 1.5 to 3.0 cm, cap surface pubescent, dull, enclosing one sixth of the nut; Region 7, primary concentration in the Coastal Range of central California south in Baja California (Mexico) and to central Arizona; xeric hillsides and desert margins, elevations 900 and 1500 m. This species is a component of chaparral and juniper-pinyon woodland communities.

Introduced Quercus Species

Quercus acutissima Carruthers, saw-tooth oak (Figure 19.68); medium-sized tree, height to 18 m; deciduous; leaf arrangement alternate; leaf simple, blade entire, total length 8 to 20 cm, width 2 to 5 mm, shape narrowly elliptical, margin spinose, secondary veins parallel, each terminating in a 2 mm spine, apex acuminate, abaxial surface glabrous, waxy or with

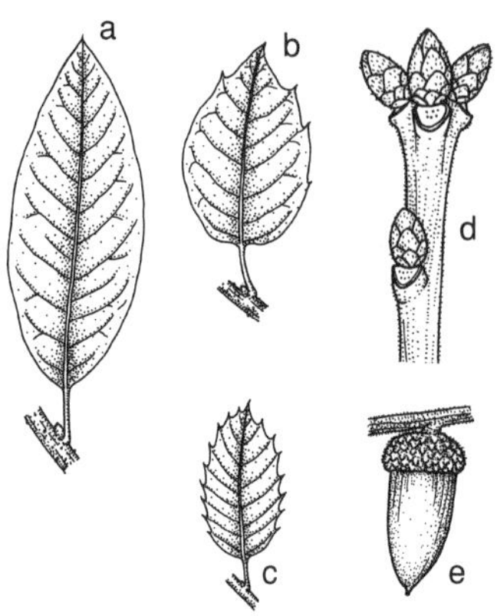

FIGURE 19.66 *Quercus chrysolepis* (a), (b), (c) simple leaves ×2, (d) twig with terminal and lateral buds ×1.5, (e) cupule and acorn ×3/5.

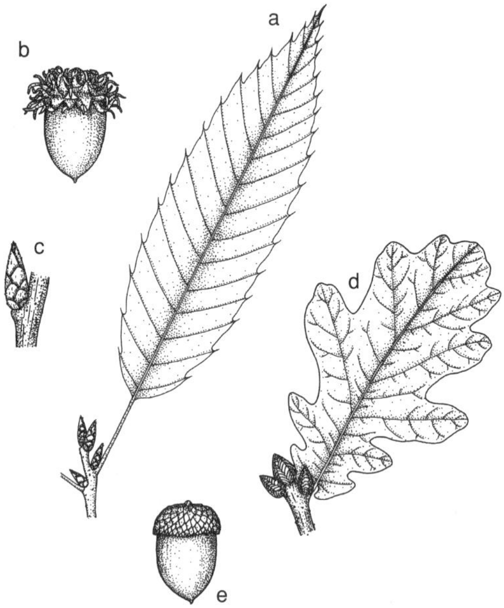

FIGURE 19.68 *Quercus acutissima* (a) simple leaf ×1/2, (b) cupule and acorn ×2/3, (c) twig with terminal and lateral buds ×1; *Q. robur* (d) leaf (e) cupule and acorn ×2/5.

scattered brownish hair; twig surface glabrous, dull, appearing scaly, dark brown, lenticels round and cream-colored; leaves and buds clustered at the twig apex; bud length 3 to 9 mm, scale surface with scattered silky silvery hairy; margin densely ciliated with silvery hair; acorn length 1.5 to 3.0 cm, cap surface pubescent, dull, enclosing about one half of the nut; bracts stout, long, length 3 to 6 mm, strongly reflexed. Asian, introduced as a ornamental tree.

Quercus robur Linnaeus, English oak (Figure 19.68); medium-sized tree, height to 25 m; deciduous; leaf arrangement alternate; leaf simple, total length 6 to 13 cm, width 3 to 7 mm, blade entire, outline obovate, margin lobed, lobe number 8 to 13, abaxial surface glabrous; twig surface glabrous, dull, appearing scaly; leaves and buds clustered at the twig apex; bud length 3 to 4 mm, scale surface glabrous, reddish brown, scale margin ciliated; acorn length 1.5 to 3.0 cm, cap surface pubescent, dull, enclosing about one half of the nut, bracts ciliated. Introduced as a lawn and street tree from Europe. The most well known and widely planted of the European oak species. Hybridizes with North American "white oak" oak species.

There are a large number of rare *Quercus* species that are fully described by Miller and Lamb (1985), Pavlik et al. (1991), Jensen (1997), and Nixon and Muller (1997); these are not considered here beyond noting their general location. Only two species are found in southeastern United States: *Q. georgiana*, Georgia oak (Arkansas, Alabama, Georgia), and

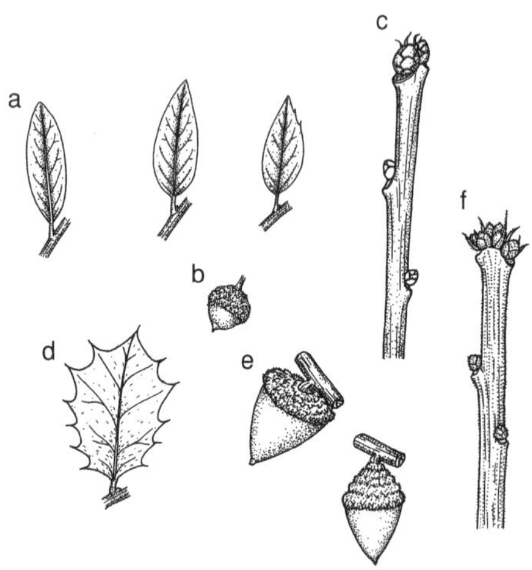

FIGURE 19.67 *Quercus vaccinifolia* (a) simple leaves ×1/2, (b) cupule and acorn ×1/2, (c) twig with terminal and lateral buds ×1; *Q. palmeri* (d) leaf ×1/2, (e) cupules and acorns ×2/3, and (f) twig with terminal and lateral buds ×1.

Q. acerifolia [(E. J. Palmer) Stoynoff & W. J. Hess], maple-leaf oak (Arkansas). The following species are isolated along coastal or in southern California: *Q. pacifica* Nixon & C. H. Muller; *Q. john-tuckeri* Nixon & C. H. Muller, Tucker or desert scrub oak. By far, most of North American rare species appear in only one or two locations in the southwestern United States (Arizona, New Mexico, and western Texas), although some species have larger ranges in Mexico. Rare species include *Q. ajoensis* C. H. Muller, Ajo Mountain scrub oak; *Q. boyntonii* Beadle, Boynton oak; *Q. carmenensis* C. H. Muller; *Q. chihuahuensis* Trelease, Chihuahua oak; *Q. depressipes* Trelease, depressed oak; *Q. graciliformis* C. H. Muller; *Q. hinckleyi* C. H. Muller, Hinckley oak; *Q. intricata* Trelease, intricate oak; *Q. polymorpha* Schlechtendal & Chamisso, netleaf white oak; *Q. pungens* Liebman, pungent oak; *Q. robusta* C. H. Muller; *Q. rugosa* Nee, netleaf oak; *Q. tardifolia*, lateleaf oak; *Q. toumeyi* Sargent, Toumey oak; *Q. vaseyana* Buckley, Vasey oak; *Q. viminea* Trelease.

Betulaceae (Birch Family)

The Betulaceae is one of only a few families distributed primarily in the temperate, cool temperate, and boreal regions of the Northern Hemisphere. However, some species are found as far south as the Andes Mountains in South America, where they occur at high (cool) altitudes. The family contains six genera (*Betula*, birch; *Alnus*, alder; *Carpinus*, hornbeam; *Ostrya*, hophornbeam; *Corylus*, hazel; *Ostryopsis*) and approximately 150 tree and shrub species (Zomlefer 1994). *Ostryopsis* includes two shrubs found in China; the genus is not represented in North America (Furlow 1997).

The differences among genera are sufficiently distinct that two subfamilies are recognized: the subfamily Betuloideae, which include two tribes (Betuleae and Carpineae), and the subfamily Coryloideae (Coryleae). The Carpineae and Coryleae are sometimes elevated to family status (Carpinaceae, Corylaceae), but according to Woodland (1991), recent research indicates that all tribes and genera should continue to be grouped within the Betulaceae. Fernald (1950) includes all five genera in one family but uses the name Corylaceae rather than Betulaceae. There are about 23 to 30 species native to North America.

Members of the family have alternate, simple, serrated leaves and are deciduous. They produce small, imperfect (unisexual) flowers clustered into catkins or aments (spikes of unisexual flowers); the plants are monoecious. For a period of time, the Betulaceae was included, along with the Fagaceae, Juglandaceae, and Salicaceae, in a group called the Amentiferae. The Betulaceae is considered most closely related to Fagaceae (Zomlefer 1994). Staminate flowers have a perianth and occur three in a cluster along the spike, and pistilate flowers have a perianth and occur singly. The pollen-producing catkins often develop in the fall and over winter on the twig, whereas the ovule-producing catkins appear concurrent with leaf expansion. The flowers are wind pollinated (anemophilous). Once pollen has reached the style, it remains dormant until the ovary and ovules mature. The fruit is either a nutlet attached to small wings or bracts, which assist in wind dispersal, or, in *Corylus*, a large sweet nut, readily dispersed by squirrels and other nut-eating animals.

Subfamily: Betuloideae

Genus: *Betula*

The largest genus of the Betuloideae is *Betula* (birch) with about 60 species, of which 12 to 14 occur in North America. Based on leaf, twig, and bark characteristics and structures, species of the genus can be separated into three series (*sensu* Fernald 1950). The Costatae tree species (e.g., *B. alleghaniensis*, *B. lenta*, and *B. nigra*) produce broad pistilate aments (diameter > 1 cm) and reddish to salmon-colored bark, exfoliating in some species. In contrast, the Albae tree species (e.g., *B. papyrifera*, *B. cordifolia*, and *B. populifolia*) have slender aments and white, often exfoliating, bark, and members of the Humiles are shrubs (*B. glandulosa*, *B. nana*, *B. occidentalis*, and *B. pumila*). Most species have spur shoots and resin glands on the twigs and leaves. Economically, *B. alleghaniensis*, *B. papyrifera* and its varieties, and *B. populifolia* are of major importance for veneer and pulpwood (paper) production. Farrar (1995) and Furlow (1997) provide a good overview and description of the birches.

Betula papyrifera Marshall, paper birch, white birch (Figure 19.69); medium-sized to large tree, height to 25 m; deciduous, leaf arrangement alternate; leaves borne on primary twig axis or clustered at the end of a spur shoot on second- and third-year twigs; leaves simple, length 5 to 12 cm, width 4 to 6 cm,

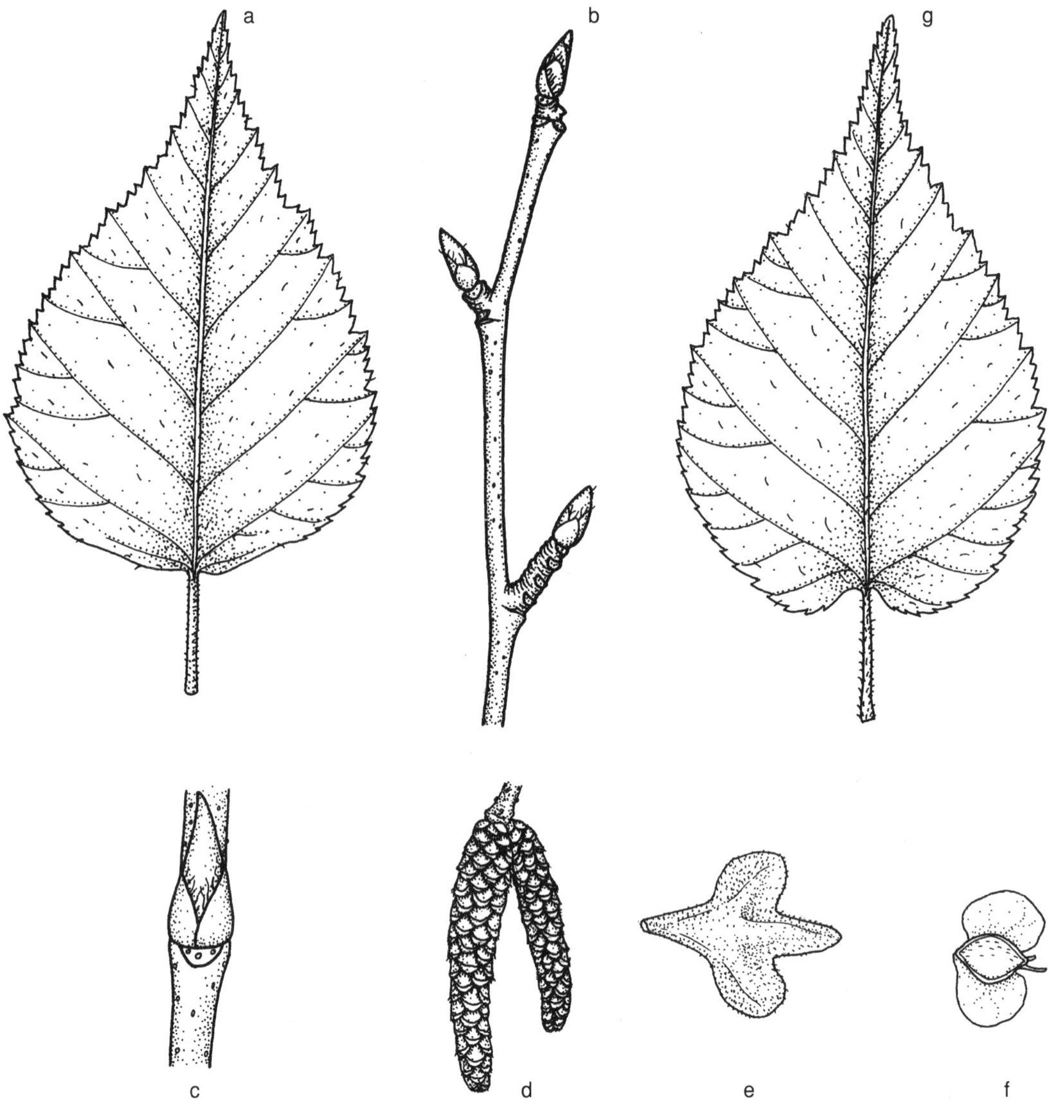

FIGURE 19.69 *Betula papyrifera* (a) simple leaf ×1, (b) twig with pseudoterminal bud and lateral buds on spur shoots ×1, (c) lateral bud ×2, (d) catkin of staminate flowers ×2, (e) bract ×5, (f) winged seed ×5; (g) leaf of *B. cordifolia* ×1.

entire, ovate to somewhat lanceolate, margin serrate to doubly serrate, lateral veins opposite, seven to nine pairs, forked once or twice within 2 cm of the margin, base round to cuneate, apex acuminate; seed small, length 2 mm, lateral wings two, opposite; second- and third-year twigs bearing 0.5 to 1.5 long spur shoots; bud length ~1 cm, greenish, scales ciliated; bark of seedlings and saplings dark reddish brown to nearly black, bark of trees chalky white, often exfoliating on large trees, lenticels black, horizontal and prominent;

Regions 1 and 2, transcontinental in Canada; sandy, usually rocky soil but xeric to mesic and infrequently mesic-hydric sites. Several varieties once included with *B. papyrifera* are now recognized as species. They have been separated from the typical form by differences in twig orientation (horizontal vs. pendulous), leaf thickness (membranous vs. leathery), bark (white vs. brownish), length of bracts and peduncles, and leaf shape. *Betula* x *sandbergii* is a recorded hybrid between *B. papyrifera* and *B. pumila*. Cover type **26.2**, 26.1, 26.4, 26.5, 26.9, 26.10, 26.11, 26.13, 26.14, 26.15, 26.18, 36.29, 26.25, 26.26, 26.29, 27.19, 30.5.

Betula cordifolia Regel [*B. papyrifera* var. *cordifolia* (Regel) Fernald], Mountain or eastern paper birch (Figure 19.69); medium-sized tree, height to 20 m; deciduous; leaf arrangement alternate; leaves borne on primary twig axis or clustered at the end of a spur shoot of second- and third-year twigs; leaves simple, length 5 to 10 cm, width 4 to 6 cm, blade broadly ovate to broadly lanceolate, entire, margin finely serrate, occasionally doubly serrate, lateral veins opposite, 9 to 12 pairs, forked once or twice within 2 cm of the margin, proximal veins forking three to four times, base distinctly cordate, apex acuminate; seed small, length 2 mm, lateral wings two, opposite; second- and third-year twigs bearing 0.5 to 1.5 long spur shoots; bud length ~1 cm, greenish, scales ciliated; bark of seedlings and saplings dark reddish brown to nearly black, bark of trees varying from white to bronze, often exfoliating on larger trees, lenticels black, horizontal and prominent; Regions 1 and 2, Central Ontario, Minnesota, and Wisconsin to New England, Newfoundland, and Prince Edward Island; sandy, usually rocky soil of xeric to mesic sites. Once considered the most common variety of *Populus papyrifera*, this taxon has been upgraded to species status (Furlow 1997). The primary characteristics separating this taxon from *B. papyrifera* are the distinct cordate leaf bases and reddish tint in the bark. Cover type 26.4, 26.15.

Betula alleghaniensis Britton (*Betula lutea* F. Michaux), yellow birch (Figure 19.70); large tree, height to 35 m; leaf arrangement alternate or clustered at the tip of a spur shoot on second- and third-year twigs; leaf simple, length 8 to 12 cm, width 4 to 7 cm, entire, ovate to oblong, margin broadly doubly serrate, apex acute, base slightly cordate, lateral veins subopposite; twigs producing a wintergreen odor

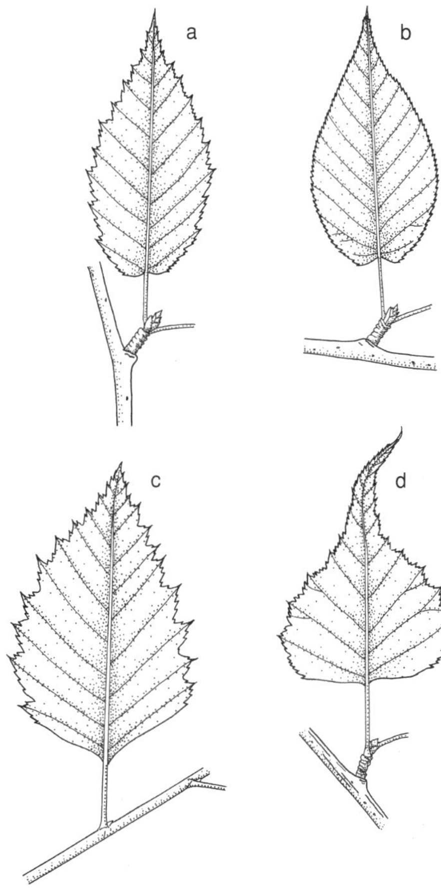

FIGURE 19.70 Leaf and spur shoot of (a) *Betula alleghaniensis* ×1/2; (b) *B. lenta* ×1/2; (c) leaf of *B. nigra* ×1/2; (d) leaf and spur shoot of *B. populifolia* ×1/2.

when crushed; bud scales ciliated; bark yellowish and exfoliating in thin layers, on large (50 cm diameter) trees becoming silvery/gray and peeling in loose, thick plates, lenticels horizontal, prominent; Regions 1 (far eastern), 2, and 3b, southeast Manitoba, Minnesota, and Iowa east to Newfoundland and south, following the Appalachian Mountains into Georgia and South Carolina; deep soil, mesic, fertile upland sites. This tree is a component of the Northern Hardwood-Conifer Forest but is also scattered in mesic-hydric swamps of *Acer rubrum, Abies balsamea,* and *Thuja occidentalis.* Trees with diameters of 1 m or more are not uncommon. The bark has been used as a survival tool, as it will ignite and burn even in cold, wet weather. Stilt-rooted trees are common because the seeds tend to germinate well on decaying logs, tree stumps, and exposed tree-tip mounds. Cover type **26.10, 26.22, 26.23, 27.29**, 26.1, 26.5, 26.8. 26.9, 26.15, 26.24, 26.25, 26.29, 27.25, 26.30, 27.31 27.35, 27.36, 27.37.

Betula lenta Linnaeus, sweet birch, black birch (Figure 19.70); medium-sized tree, height to 22 m; leaf arrangement alternate; leaves borne on primary twig axis or clustered at the tip of a spur shoot on second- and third-year twigs; leaf simple, length 8 to 12 cm, width 4 to 7 cm, entire, ovate to oblong, margin singly serrate, apex acute, base slightly cordate, lateral veins subopposite; twigs producing a wintergreen odor when crushed; bud scales aciliate; bark dark reddish brown to nearly black on large trees, becoming platy, lenticels horizontal, prominent; Regions 2 and 3b, Maine and Quebec south, following the Appalachian Mountains into Georgia and Alabama; deep soil, mesic, fertile sites as a component of the northern hardwood-conifer forest but also growing on dry, rocky soil. Cover type 26.23, 26.25, 26.29, 27.12, 27.15, 27.23, 27.29, 27.35, 27.37.

Betula nigra Linnaeus, river birch (Figure 19.70); medium-sized tree, height to 22 m; leaf arrangement alternate; leaves borne on primary twig axis or clustered at the tip of a spur shoot on second- and third-year twigs; leaf simple, length 6 to 10 cm, width 3 to 6 cm, ovate to rhombic, entire, margin doubly serrate to near dentate, apex acute, base cuneate and smooth, lateral veins alternate; twigs dark brown; bark salmon-colored and exfoliating on large trees, lenticels horizontal, prominent; Regions 3a and 4 excluding central Florida; found as a scattered tree or in small clusters along stream margins, but often in pure stands after disturbance of bottomland sites. This tree grows individually or in clumps of two or three; probably superior to *Betula papyrifera* as a lawn tree in certain regions. Cover type **27.6**, 27.3, 28.12.

Betula populifolia Marshall, gray birch (Figure 19.70); small tree, height to 10 m often multistemmed; deciduous, leaf arrangement alternate; leaves borne on primary twig axis or clustered at the end of a spur shoot; leaves simple, length 5 to 10 cm, width 4 to 6 cm, blade broadly ovate to broadly lanceolate, entire, margin finely serrate, occasionally doubly serrate, lateral veins opposite, 9 to 12 pairs, forked once within 1 cm of the margin, base truncate, apex long acuminate, often angled to one side; seed small, length 2 mm, lateral wings two, opposite; second- and third-year twigs bearing 0.5 to 1.5 cm long spur shoots; bud length ~1 cm, greenish, scales ciliated; bark of seedlings and saplings dark reddish brown to nearly black, bark of trees grayish white, not exfoliating, lenticels black, horizontal and prominent; Region 2b, eastern Ontario to Nova Scotia and Prince Edward Island, south to Pennsylvania and New Jersey; sandy, usually rocky soil of xeric sites. An important pioneer of cutover and burned sites. Cover type 26.25.

Betula neoalaskana Sargent (*B. papyrifera* var. *neoalaska*); resin birch, Alaska paper birch; deciduous, leaf arrangement alternate; leaves borne on primary twig axis or clustered at the end of a spur shoot; leaves simple, length 4 to 7 cm, width 2 to 6 cm, entire, ovate to somewhat deltoid, margin doubly serrate, lateral veins opposite, 7 to 16 pairs, forked at least twice within 2 cm of the margin, base essentially truncate, round, or cuneate, smooth, apex long acuminate; seed small, length 2 mm, lateral wings two, opposite; second- and third-year twigs bearing 0.5 to 1.5 cm long spur shoots; bud length ~1 cm, greenish, scales ciliated; bark of seedlings and saplings dark reddish brown to nearly black, bark of trees white, pinkish, or nearly red, often exfoliating, lenticels black, horizontal and prominent; Region 1, interior Alaska to western Ontario; mesic-hydric bogs and poorly drained soil, occasionally in pure stands or mixed with black spruce, open woodlands.

Betula occidentalis Hooker (*Betula fontinalis* Sargent), water birch (Figure 19.71), red birch, black birch; shrub or small crooked tree, height to 12 m; deciduous; leaf arrangement alternate; leaves borne on primary twig axis or clustered at the end of a spur shoot; leaves simple, length 3 to 6 cm, width 2 to 4 cm, blade broadly ovate to rhombic, entire, margin singly to doubly serrate, lateral veins subopposite, three to six pairs, forked twice within 1 cm of the margin, base cuneate, smooth, apex obtuse; seed small, length 2 mm, lateral wings two, opposite, seed accompanied by three-lobed bract; second- and third-year twigs bearing 0.5 to 1.5 cm long spur shoots, dense resinous warts on surface; bud length ~5 mm, greenish; bark of seedlings and saplings dark brown to nearly black, bark of trees reddish brown, not exfoliating, lenticels gray, horizontal and prominent; Regions 1, 5, and 6, Alaska to central Ontario, south to central California, Arizona and New Mexico; margins of mountain streams and

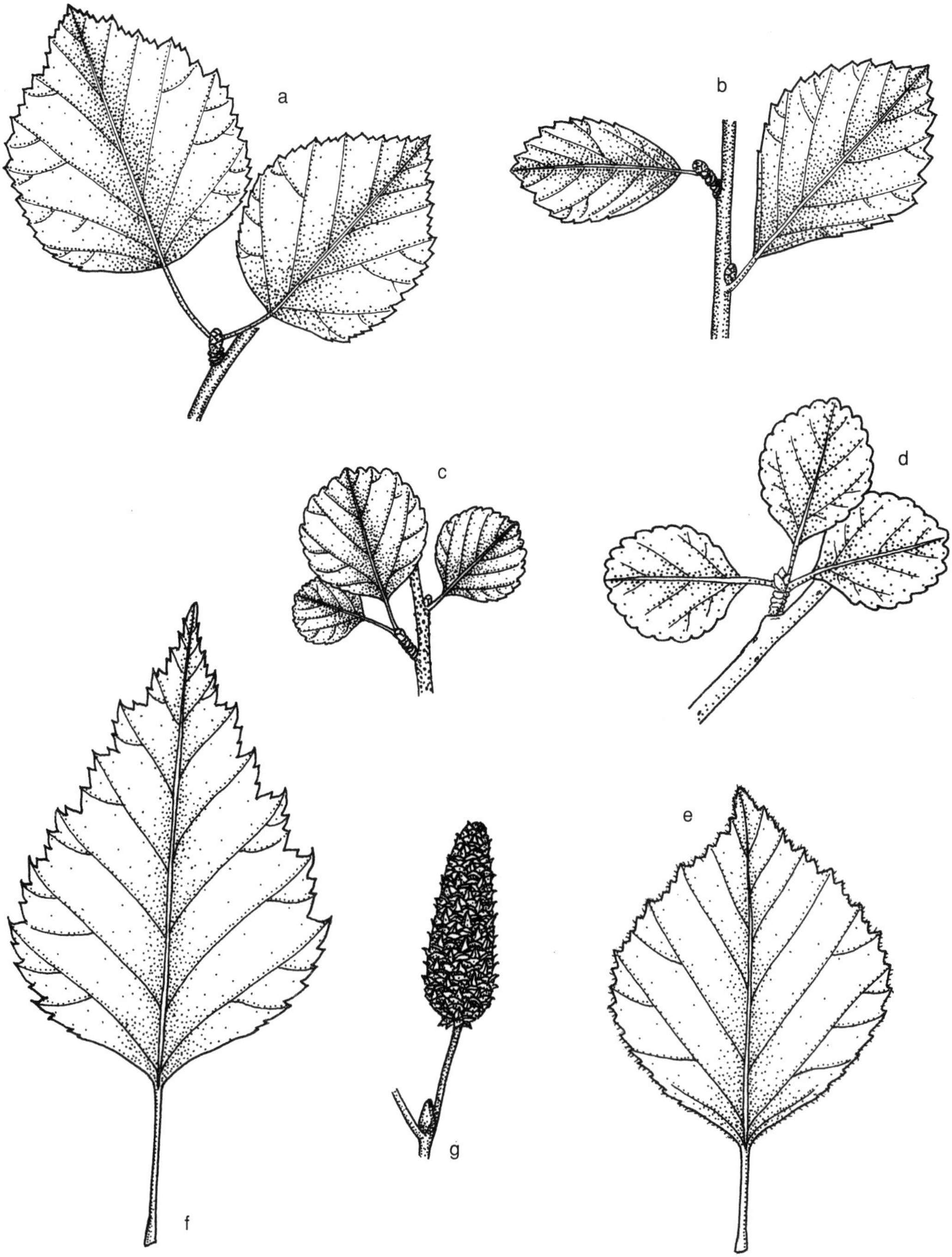

FIGURE 19.71 Leaves and spur shoot of (a) *Betula occidentalis*, (b) *B. pumila* ×1; (c) *B. glandulosa* ×1; (d) *B. nana* ×1; (e) *B. kenaica leaf* ×1, (f) *B. pendula* leaf, (g) catkin of pistillate flowers ×1.

lakes, marshes, elevation 100 to 3000 m. Often found in mesic-hydric woods and in thickets with *Alnus*. Cover type 29.15, 29.28, 30.5.

Betula pumila Linnaeus, bog birch, dwarf birch (Figure 19.71); multistemmed shrub; erect or spreading height < 4 m; deciduous; leaf arrangement alternate; leaves borne on primary twig axis or clustered at the end of a spur shoot; leaves simple, length 1 to 7 cm, width 2 to 4 cm, blade nearly orbicular to broadly ovate, glandular, entire, margin finely serrate or dentate, lateral veins subopposite, three to six pairs, base cuneate to nearly truncate, smooth, apex obtuse, resin glands scattered; seed small, length 2 mm, lateral wings two, opposite, seed accompanied by three-lobed bract, length 3 mm; second- and third-year twigs bearing 0.5 to 1.0 cm long spur shoots; bud length ~0.5 cm, greenish; bark brown, resin glands scattered; Regions 1 and 2, Yukon, Northwest Territories to Newfoundland, south to Michigan and New Jersey, disjunct in the four corner area of Colorado, Kansas, Nebraska, and Wyoming; bogs, muskeg, swamps.

Betula nana Linnaeus, dwarf arctic birch (Figure 19.71); multistemmed shrub; spreading or upright, height < 1 m; deciduous; leaf arrangement alternate; leaves borne on primary twig axis or clustered at the end of a spur shoot; leaves simple, length 1 to 3 cm, width 1 to 2 cm, blade orbicular, often glandular, entire, margin crenate, lateral veins opposite, two to four pairs, base cuneate, smooth, apex rounded to nearly truncate; seed small, length 2 mm, lateral wings two, opposite, seed accompanied by three-lobed bract, length 3 mm; second- and third-year twigs bearing 0.5 to 1.0 cm long spur shoots; bud length ~0.5 cm, greenish; bark brown, resin glands often dense; Region 1, Alaska, Yukon, Northwest Territories to Hudson Bay to Greenland coast; arctic bogs, muskeg, tundra, moist slopes. A close relative, *Betula michauxii* Spach, Newfoundland dwarf birch, is difficult to distinguish from *B. nana*; it is not common within its range in eastern Quebec, Nova Scotia, and Newfoundland.

Betula glandulosa Michaux, dwarf birch (Figure 19.71), resin birch; multistemmed shrub, spreading, length to 3 m, height < 0.2 m; deciduous; leaf arrangement alternate; leaves borne on primary twig axis or clustered at the end of a spur shoot; leaves simple, length 1 to 3 cm, width 1 to 2 cm, blade broadly ovate to orbicular, densely glandular, entire, margin crenate, lateral veins subopposite, two to six pairs, base cuneate, smooth, apex obtuse to rounded; seed small, length 2 mm, lateral wings two, opposite, seed accompanied by three-lobed bract; second- and third-year twigs bearing 0.5 to 1.0 cm long spur shoots; bud length ~0.5 cm, greenish; bark brown, resin glands dense; Regions 1, 5, 6, transcontinental across Canadian subarctic, Alaska south along the Rocky Mountains to Utah and Colorado, and to northern California; marshes, stream banks, muskeg, bogs. Cover type 26.5.

Betula kenaica W. H. Evans [(*B. papyrifera* var. *kenaica* (W. H. Evans) A. Henry], Kenai birch (Figure 19.71); small tree, height to 10 m; deciduous; leaf arrangement alternate; leaves borne on primary twig axis or clustered at the end of a spur shoot; leaves simple, length 5 to 6 cm, width 3 to 5 cm, blade broadly elliptical to nearly deltoid, entire, margin finely serrate, ciliated, lateral veins opposite, three to six pairs, a few forked once near the margin, base truncate, apex short acuminate; seed small, length 2 mm, lateral wings two, opposite; second- and third-year twigs bearing 0.5 to 1.5 cm long spur shoots; bud length ~1 cm, greenish, scales ciliated; bark of seedlings and saplings dark reddish brown to nearly black, bark of trees grayish white, exfoliating, lenticels black, horizontal and prominent; Region 1, Yukon, Alaska, and the Aleutian chain.

Betula pendula Roth, European white birch, weeping birch (Figure 19.71); small tree, height to 10 m; deciduous; leaf arrangement alternate; leaves borne on primary twig axis or clustered at the end of a spur shoot; leaves simple, length 4 to 8 cm, width 3 to 5 cm, blade broadly ovate to rhombic, entire, margin strongly doubly serrate, lateral veins opposite, 6 to 15 pairs, forked once within 1 cm of the margin, base cuneate, apex acuminate; twigs pendulous; seed small, length 2 mm, lateral wings two, opposite; second- and third-year twigs bearing 0.5 to 1.5 cm long spur shoots, bud length ~1 cm, greenish, scales ciliated; branche's drooping; bark of seedlings and saplings dark reddish brown to nearly black, bark of trees grayish white, often exfoliating, lenticels black, horizontal and prominent; resembles *Betula populifolia*. Introduced; naturalized locally in Region 2a, Northeast. Widely planted as a lawn and street tree.

Gleason and Cronquist (1991) outline some of the taxonomic difficulty with the genus *Betula*, which, like *Quercus*, is characterized by hybridization and introgression. They report hybrids for *B. alleghaniensis* and *B. pumila* (*B.* x *purpusii* Schneider), *B. papyrifera* and *B. pumila* (*B.* x *borealis* Spach), *B. cordifolia* and *B. populifolia* (*B.* x *caerulea* Blanchard and *B.* x *caerulea-grandis* Blanchard), and *B. glandulosa* and an alpine form of *B. papyrifera* [*B.* x *minor* (Tuckerman) Fernald]. *Betula* x *murrayana* Barnes & Dancik appears to be an octoploid backcross of *B. purpusii* with *B alleghaniensis.* Gleason and Cronquist report *B.* x *sandbergi* Brit. as a cross between *B. papyrifera* and *B. glandulosa* var. *glandulifera,* which Fernald (1950, corrected 1987 printing) reports as a cross between *B. papyrifera* and *B. pumila* var. *glandulifera.*

Genus: *Alnus*, Alder

Members of the genus *Alnus* (10 North American species; 30 species worldwide) are large shrubs and small trees. Most *Alnus* species are generally found in temperate to northern climates (Regions 1 and 2) and are associated with wet sites, swamps, and stream margins. Ecologically, *Alnus* species are important group because they have a symbiotic relationship with bacteria in the form of root nodules that fix nitrogen. These nodules are important because they increase soil nitrogen so that the land may be revegetated by tree species that require moderately high nitrogen levels. In western clearcut areas, vegetation may be slow and nitrogen levels drop after the fine-grained organic matter has decayed in 2 or 3 years. Therefore, it is not surprising that *Alnus rubra* and other alders are important early successional species that precede forest reestablishment. *Alnus* has simple leaves, stalked buds, and cone-like female reproductive structures.

Alnus incana (Linnaeus) Moench ssp. *rugosa* (DuRoi) Clausen [*A. rugosa* (Du Roi) sprengel]; tag or speckled alder (Figure 19.72); multistemmed shrub or arborescent, height to 7 m; deciduous; leaf arrangement alternate; leaf simple, length 4 to 11 cm, width 3 to 8 cm, blade entire, broadly ovate to elliptical, margin finely serrate to doubly serrate/dentate; twigs and branches weak, bud stalked, scales valvate, visible scales two to three; length 0.5 to 1.0 cm; male catkins cylindrical, often overwintering in clusters on twigs, fruit a woody cone-like structure, oval, length 1 to 2 cm; bark dark brown to nearly black, lenticels prominent and expanded; Regions 1, 2, and 3, Saskatchewan to Newfoundland and Labrador (Region 1) extending southward to northern Illinois, northern Indiana, and Pennsylvania (northern parts of Region 3); a common wetland shrub. The most important eastern/northern alder in northern Wisconsin, Michigan, Minnesota, and southern Ontario. This species invaded hundreds of thousands of hectares of lowland (swamp) forest when *Thuja occidentalis* (northern white cedar), *Abies balsamea* (balsam fir), *Acer rubrum* (red maple), and *Fraxinus nigra* (black ash) were cut and the organic soil frequently burned during dry years of the early 1900s. Because of its root nodules that fix nitrogen, subsp. *rugosa* tolerated these low-nitrogen soils and therefore revegetated extensive areas, where it dominates today. Cover type 26.3, 26.5, 26.6, 26.8, 26.16, 26.8, 26.16, 26.28, 29.6, 29.12, 29.15, 29.19.

Alnus incana (Linnaeus) Moench ssp. *tenuifolia* (Nuttall) Breitung, [*A. tenuifolia* Nuttall]; thinleaf alder; (Figure 19.72); multi-stemmed shrub or arborescent, height to 12 m; deciduous; leaf arrangement alternate; leaf simple, length 4 to 10 cm, width 3 to 8 cm, blade entire, broadly ovate to elliptical, margin with finely serrated large dentate teeth, apex obtuse, base truncate, rounded or cuneate; twig branches weak; bud stalked, scales valvate, visible scales two to three; length 0.5 to 1.0 cm; male catkins cylindrical, often overwintering in clusters on twigs; fruit a woody conelike structure, oval, length 1 to 2 cm; bark dark brown to nearly black, lenticels prominent and expanded; Regions 5, 6, and 7, Alaska and Northwest Territories through the Rocky Mountains to Nevada and New Mexico, and down the Coastal Mountains and Ranges to central California; wet areas, seeps, and stream banks. Cover type 26.13, 29.12.

Alnus rubra Bongard, red alder (Figure 19.73); large tree, height to 27 m; deciduous; leaf arrangement alternate; leaf simple, length 6 to 16 cm, width 3 to 10 cm, blade entire, broadly ovate to elliptical, margin double serrate or serrate/dentate, apex obtuse, base obtuse and smooth; twig branches weak; bud stalked, scales valvate, visible scales two to three; length 0.5 to 1.0 cm; male catkins cylindrical, often overwintering in clusters on twigs; fruit a woody conelike structure, oval, length 1 to 3.5 cm; bark light gray; Region 7, Yukon south to Washington, Oregon, Idaho, and central California; mesic sites; stream banks, floodplains, and open areas, elevation 0 to 800 m. The most important alder in the Pacific Northwest. It often appears in profusion after cutting or fire to initiate community development. Also commercially important. Cover type **30.6**, 30.4, 30.7, 30.9, 30.10, 30.11, 30.15, 30.16, 30.18, 30.19, 30.40.

Alnus oblongifolia Torrey, Arizona alder (Figure 19.73); large tree, height to 28 m; deciduous; leaf arrangement alternate; leaf simple, coriaceous, length

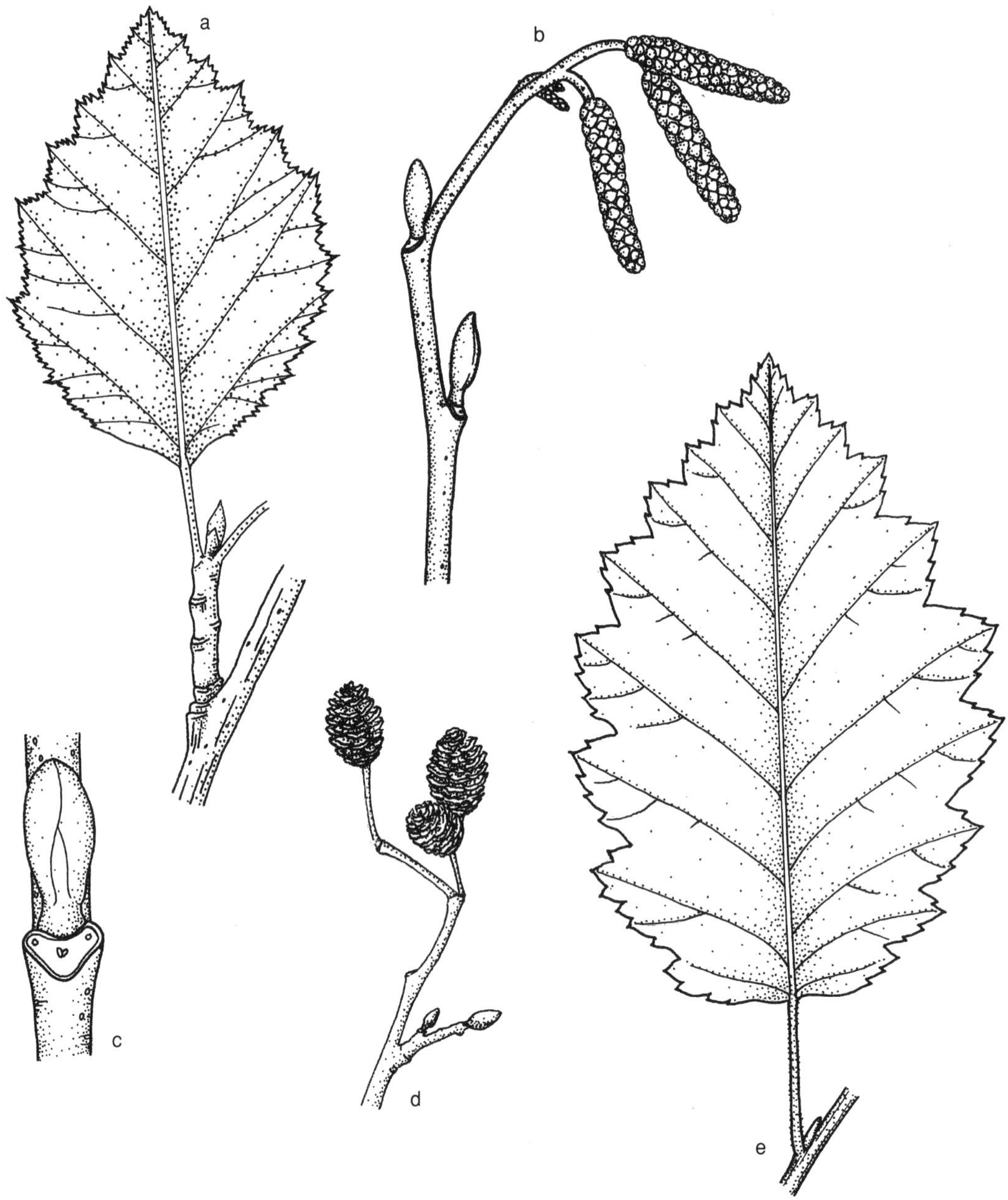

FIGURE 19.72 *Alnus incana* ssp. *rugosa* (a) leaf ×1, (b) twig with stalked buds and staminate catkins ×2, (c) lateral stalked bud and leaf scar ×2.5, (d) twig with seed cones ×1; (e) *A. incana* ssp. *tenuifolia* leaf ×1.

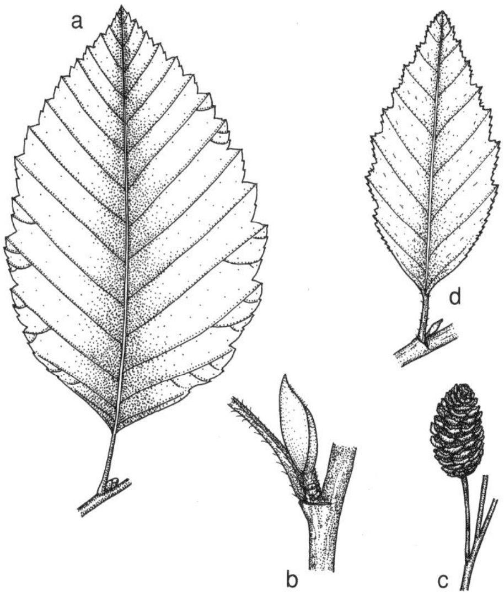

FIGURE 19.73 *Alnus rubra* (a) leaf ×1/2, (b) stalked bud ×2, (c) seed cone ×1; (d) *A. oblongifolia* leaf ×1/2.

5 to 9 cm, width 3 to 6 cm, blade entire, narrowly elliptical to lanceolate, margin doubly serrulate, apex obtuse, base cuneate, resin coated; bud stalked, scales valvate, visible scales two, length 4 to 7 mm; male catkins often overwintering in clusters on twigs, length 3 to 6 cm; fruit a woody cone-like structure, elliptical, length 1.0 to 2.5 cm; bark dark gray to black; Region 5, restricted to a small area of Arizona, New Mexico, and northern Mexico; mesic sites, stream banks and nearby mesic slopes.

Alnus rhombifolia Nuttall; white alder (Figure 19.74); medium-sized tree, height to 24 m; deciduous; leaf arrangement alternate; leaf simple, length 4 to 10 cm, width 2 to 5 cm, blade entire, broadly elliptical to rhomboid, margin finely serrulate or denticulate, base cuneate and often smooth, apex obtuse or rounded; bud stalked, scales valvate, visible scales two, length 4 to 8 mm, resin coated; male catkins often overwintering in clusters on twigs, length 3 to 9 cm; fruit a woody cone-like structure, elliptical, length 1.0 to 2.2 cm; Regions 6 and 7, mountain ranges of California, also in scattered areas where the Oregon, Washington, and Idaho borders meet; rocky stream banks, adjacent slopes. Cover type **29.24, 30.5,** 30.40.

Alnus viridis (Villars) de Candolle ssp. *sinuata* (Regal) Love & Love; sitka or mountain alder (Figure 19.74); multistemmed shrub, height to 5 m; deciduous; leaf arrangement alternate; leaf simple, thin, length 4 to 10 cm, width 3 to 8 cm, blade entire, broadly elliptical to lanceolate, margin serrulate, slightly undulate, apex obtuse, base truncate rounded or slightly cordate, lightly resin coated; bud stalked, scales imbricate, visible scales four to six, length 4 to 7 mm; male catkins often overwintering in clusters on twigs, length 3 to 8 cm; fruit a woody cone-like structure, elliptical; Regions 6 and 7, Aleutian Islands and central Alaska to northern California, Idaho, and Wyoming; mesic to hydric sites, stream banks, drainages, open conifer forest, elevation 0 to 2500 m into high montane. Cover type 29.11, 29.27, 30.10.

Alnus viridis ssp. *crispa* (Aiton) Pursh, green or mountain alder (Figure 19.74); multistemmed shrub, height to 3 m; deciduous; leaf arrangement alternate; leaf simple, thin, length 3 to 6 cm, width 3 to 5 cm, blade entire, narrowly elliptical to ovate, margin serrulate, apex acute, base obtuse, resin coated; bud stalked, scales imbricate, visible scales four to six, length 4 to 7 mm; male catkins often overwintering in clusters on twigs, length 3 to 8 cm; fruit a woody cone-like structure, elliptical; Regions 1 and 2, Alberta and Northwest Territories to Newfoundland, northern Wisconsin and Michigan to New England, disjunct at higher elevations in the Allegheny Mountains, Pennsylvania and North Carolina; bogs, swamps, drainages, stream banks, lake shores. Cover type 26.18.

Alnus viridis (Villars) de Candolle ssp. *fruiticosa* (Ruprecht) Nyman, Siberian alder; multistemmed shrub, height to 3 to 4 m; deciduous; leaf arrangement alternate; leaf simple, thin, length 4 to 8 cm, width 3 to 7 cm, blade entire, broadly ovate, margin finely doubly serrate, apex acute, base truncate to cordate, heavily resin coated; bud stalked, scales imbricate, visible scales four to six, length 4 to 7 mm; male catkins often overwintering in clusters on twigs, length 8 cm; fruit a woody cone-like structure, elliptical; Regions 6 and 7, Saskatchewan, Alberta, British Columbia, Northwest Territories, Yukon, and Alaska; mesic to wet sites, stream banks, shorelines, elevation 0 to 500 m.

Alnus serrulata (Aiton) Willdenow, hazel alder or smooth alder (Figure 19.74); multistemmed shrub or arborescent, height to 7 m; deciduous; leaf arrangement alternate; leaf simple, length 6 to 14 cm,

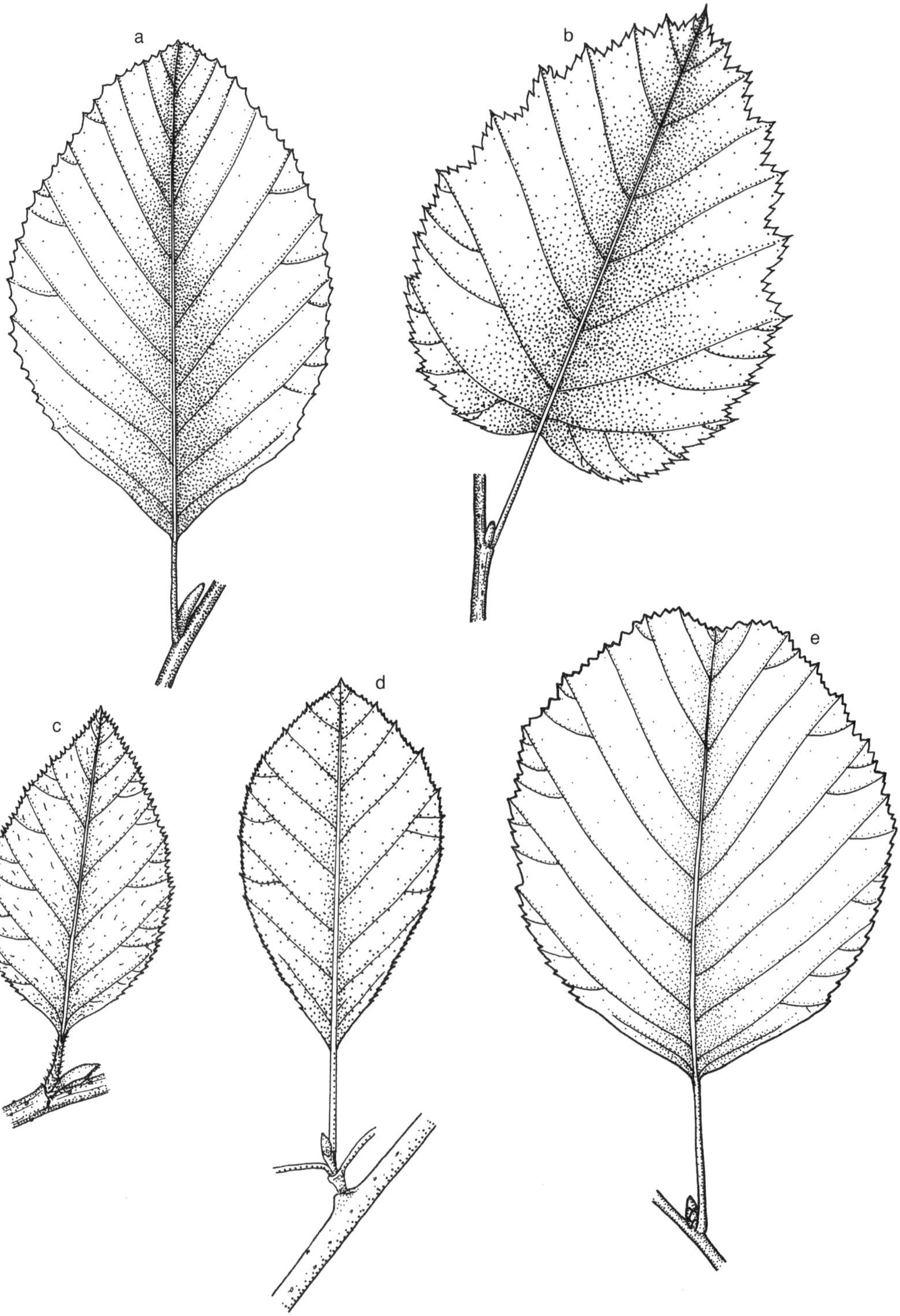

FIGURE 19.74 Leaves and buds of (a) *Alnus rhombifolia* ×1; (b) *A. viridis* ssp. *sinuata* ×1; (c) *A. viridis* ssp. *crispa* ×1; (d) *A. serrulata* ×1; (e) *A. glutinosa* ×1.

width 4 to 8 cm, blade entire, broadly elliptical to obovate, margin finely serrate, slightly undulate; fruit a woody cone-like catkin, elliptical, male catkins often overwintering in clusters on twigs, length 2.5 to 3.5 cm; bud stalked, scales valvate, visible scales two to three; length 0.5 to 1.0 cm; Regions 2, 3 and 4, Missouri and Oklahoma to Quebec, Maine, and Nova Scotia, south to the Gulf and Atlantic coasts, seldom found in the Mississippi River floodplain; stream and lake edges, sloughs, and swamps. Cover types 27.15, 28.11.

Alnus glutinosa (Linnaeus) Gaertner, European alder (Figure 19.74); medium to large tree, height to 23 m; deciduous; leaf arrangement alternate; leaf simple, length 6 to 11 cm, width 3 to 8 cm, blade entire, broadly elliptical to essentially orbicular, apex margin truncate or slightly emarginate, margin finely serrate and undulate or weakly doubly serrate; fruit a woody cone-like structure, elliptical; male catkins often overwintering on twigs, length 5 to 12 cm; bud stalked, scales valvate, visible scales two to three; length 0.5 to 1.0 cm; introduced, escaped cultivation in Regions 2 and 3, Michigan to New York and Massachusetts.

Subfamily: Coryloideae

Genus: *Carpinus*, Hornbeam, Bluebeech, Musclewood

Carpinus caroliniana Walter, American hornbeam (Figure 19.75); small tree, height to 10 m; deciduous; leaf arrangement alternate; leaf simple, entire, length 5 to 11 cm, width 3 to 6 cm, margin doubly serrate, a large tooth appearing where each vein intersects the margin, teeth sharp, adaxial surface shiny, glabrous, veins 11 to 14 pairs, nearly opposite, lateral veins unbranched; fruit a small nutlet (diameter 2 to 4 mm) subtended by a toothed, leafy, three-lobed bract, 5 to 11 clustered on an open strobile; bark smooth, gray, fluted, appearing like exposed human muscles; Regions 2, 3 and 4, several disjunct populations occur in southern Mexico and Central America. The species can be easily recognized in winter by its unusual bark. It is a common component of forest on mesic north slopes, stream terraces, and bottomland sites. Furlow (1997) described two subspecies. *Carpinus carolinana* is the only North American species; worldwide there are about 26 species. Cover type 27.2, 27.8, 27.9, 27.10, 27.12, 28.11, 28.18, 28.37, 28.42, 28.47, 28.48.

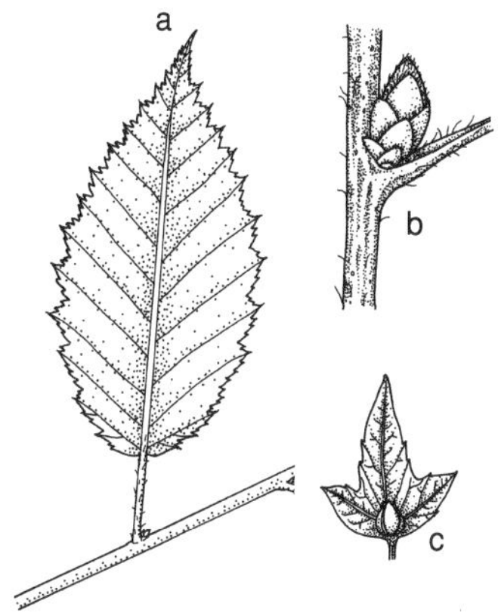

FIGURE 19.75 *Carpinus caroliniana* (a) leaf ×1/2, (b) bud ×2.5, (c) nutlet subtended by bract ×3/5.

Genus: *Ostrya*, Ironwood, Hophornbeam

Three species of *Ostrya* are recognized in the United States, about seven species worldwide. The primary eastern species has a distribution nearly identical to that of *C. caroliniana.*

Ostrya virginiana (Miller) K. Koch, eastern hophornbeam, ironwood (Figure 19.76); small to medium-sized tree, height to 16 m; deciduous; leaf

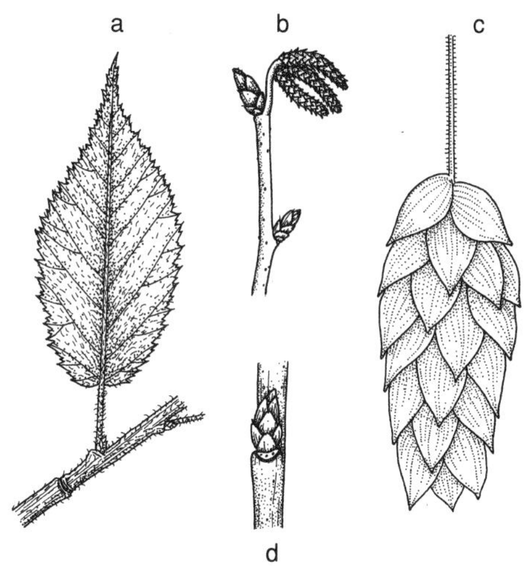

FIGURE 19.76 *Ostrya virginiana* (a) leaf ×1/2, (b) twig with pseudoterminal bud and staminate catkins ×1/3, (c) lateral bud ×2, (d) spike of fruits; each fruit is a nutlet enclosed by a tubular bract ×1.

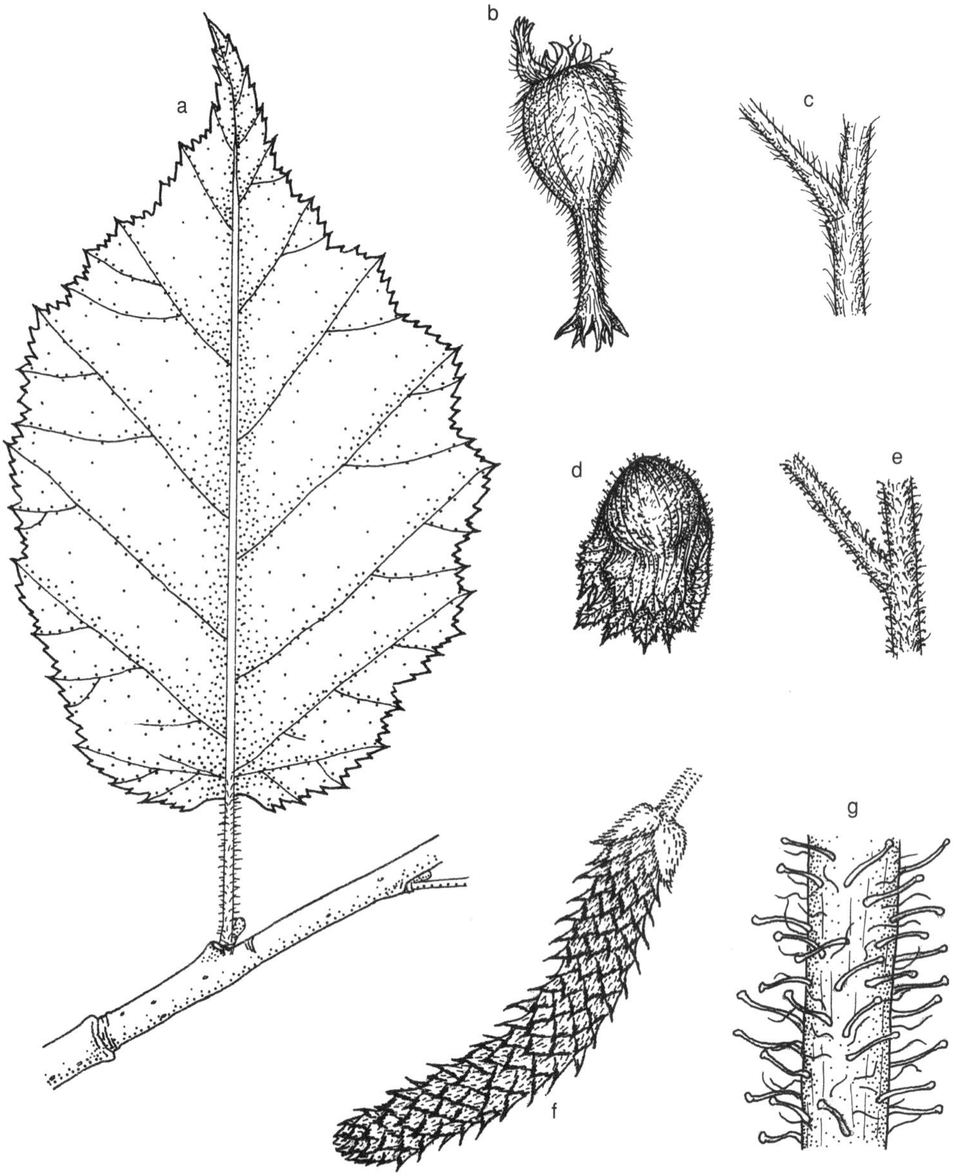

FIGURE 19.77 *Corylus cornuta* (a) leaf ×1, (b) nut with beaked husk ×1, (c) twig with pilose hair ×2; *C. americana* (d) nut with broad-tipped husk ×1, (e) twig with pilose hair and glandular tricomes ×2, (f) twig with pilose hair and glandular tricomes ×5, (g) catkin of staminate flowers ×2.

arrangement alternate; leaf simple, entire, length 5 to 11 cm, width 3 to 6, margin usually singly serrate, occasionally doubly serrate, teeth sharp, adaxial surface dull, softly pubescent, veins 11 to 14 pairs, nearly opposite, lateral veins branching once or sometimes twice near the margin, each fork proceeding to the tip of a tooth; twigs extremely slender; bark breaking into narrow (1.0 to 1.5 cm wide) vertical strips, somewhat shreddy; fruit a small nutlet (diameter 2 to 3 mm) enclosed in a tubular bract, which becomes a closed bladder at maturity, 15 to 30 bladders clustered on an open strobile; Regions 2, 3 and 4, Manitoba to Nova Scotia south to Texas and Florida; xeric-mesic south slopes, mesic north slopes and stream terraces. Found as a common shrub or small tree in most oak-hickory or beech-maple forest communities. Often found in high densities in previously grazed woodlots. Cover type 26.17, 26.23, 27.5, 27.8, 27.9, 27.16, 27.19, 27.23, 27.26, 28.27.

Ostrya chisosensis Correll (Chisos hophornbeam) is similar in size and appearance to *O. virginiana;* it is found only in the Chisos Mountains of southwest Texas near the Mexican border. *Ostrya knowltonii* Coville (Knowlton hophornbeam; cover type 29.27) is a small tree (max. height = 9 m) found in approximately eight locations in Texas, New Mexico, Arizona, and Utah. It is found in canyons with various species of oak, pinyon pine, ponderosa pine, and juniper.

Genus: *Corylus*, Hazel

There are about 15 *Corylus* species worldwide, but only 2 occur in North America. These species are shrubs that are often an important midcanopy component of Lake States and Central States forest communities, where these shrubs may create strong competition for tree seedlings. Stem density may be high because the plant spreads vegetatively by sprouting from the root system. The buds, catkins, and nuts are extensively used by small mammals and birds. Commercially, the nuts are harvested and sold in stores as filberts, usually found in mixed nut combinations.

Corylus americana Walter, American hazel, hazelnut (Figure 19.77); multistemmed shrub, height to 2.5 m; deciduous; leaf arrangement alternate; leaf simple, entire, length 5 to 14 cm, width 4 to 10 cm, margin usually singly serrate, uneven, occasionally doubly serrate or serrate/dentate, teeth sharp, adaxial surface dull, softly pubescent, veins 11 to 14 pairs, alternate, lateral veins branching once or sometimes twice near the margin; twigs and petioles usually pubescent to tomentose, usually with a combination of soft, slender silvery hair and stiff, glandular-tipped, black, hair; male catkins slender, 7 to 12 cm long, overwintering on twigs; fruit an oval nut (diameter 1 to 2 cm) in clusters of one to three, at maturity each nut enclosed in two large leafy bracts (husks), bracts about twice as long as nuts (2 to 3 cm), margin apex broad, ruffled, toothed, and reflexed; Regions 2, 3, and 4, eastern North America, southeastern Saskatchewan and Manitoba to Maine south to Louisiana and Georgia, not along the Gulf Coast or in Florida; in Region 2 often forming a dense midcanopy in forest communities on xeric outwash sands, glacial till loams, and loessal silts; also found in open fields and along fencerows. Cover type 26.18, 26.26, 27.1, 27.4, 27.17, 27.26, 28.12.

Corylus cornuta Marshall, beaked hazel, beaked hazelnut (Figure 19.77); multistemmed shrub, height to 2.5 m; deciduous; leaf arrangement alternate; leaf simple, entire, length 5 to 14 cm, width 4 to 10, margin usually singly serrate, uneven, occasionally doubly serrate or serrate/dentate, teeth sharp, adaxial surface dull, softly pubescent, veins 11 to 14 pairs, alternate, lateral veins branching once or sometimes twice near the margin; twigs and petioles glabrous or pubescent to tomentose usually with only soft, slender silvery hair; male catkins slender, 7 to 12 cm long, overwintering on twigs; fruit an oval nut (diameter 1 to 2 cm) in clusters of one to three, at maturity each nut enclosed in two large leafy bracts (husks), bracts nearly three times as long as nuts (3 to 5 cm), the distal one half of the bracts forming a tubular extension, the apex flaring and cleft; Regions 1, 2, and 3b, British Columbia and Washington across Canada and the Lake States to Newfoundland, south along the Appalachian Mountains to Georgia and Alabama; in Regions 1 and 2 often forming a dense midcanopy in forest communities on xeric outwash sands, glacial till loams, and loessal silts; also found in open fields and along fencerows. The primary identifying characteristics that separate this species from *C. americana* are the absence of the stiff, glandular tipped black hair and the tubular husk. A subspecies [*C. cornuta* var. *californica* (A. Decandolle.) A. J. Sharp] grows in the western mountain ranges from the Sierra Nevada into the Rocky Mountains of southern British Columbia. Cover type 26.2, 26.4, 26.10, 26.19, 26.26, 26.28, 27.17, 27.23.

CHAPTER 20

Class: Magnoliopsida; Subclass: Dilleniidae

ORDER: MALVALES

Tiliaceae (Linden or Basswood Family)

THE Tiliaceae family of 45 to 50 genera and about 450 species has many representatives in the tropics and subtropics but only a few in temperate North America. *Tilia* is the only North American genus, and *Tilia americana* may be the only native species. This species is important in timber production, but its wood is also fine grained and has ideal texture for wood carving. Two conspicuous features of *Tilia* are the numerous root collar sprouts that are ubiquitous on trees and the palmately five-lobed cotyledons of the seedling.

The *Tilia* leaf is deciduous, simple, has an ovate to orbicular shape and an acuminate apex, is palmately veined and sharply serrated, and has a distinct asymmetrical (cordate/truncate) base. Leaf arrangement is alternate. The flowers are borne in cymes attached to a 2 to 4 cm long peduncle subtended by a 7 to 14 cm long leafy, spatulate bract that may be aid in seed distribution. The flower is small, regular, and perfect (plants synecious) with five yellowish petals. The flowers have nectaries and often a strong scent that attracts insect pollinators; the nectar produces a distinctly flavored honey. The fruit is a small (diameter 6 to 9 mm) globose nutlet. The fibers from another genus of the family (*Conchorus*, an herb) are used to make jute.

The problem of determining whether a genus contains several species or a single species with several varieties is nowhere better illustrated than in *Tilia*. In 1922, Sargent (1965 Dover reprint) listed 15 Tilia species. More recently, *Tilia* has been reported as having 4 species (Brown 1945; Fernald 1950; Little 1979). Since then, several populations have been moved from species to variety status or combined. At present, 1 species and two or three varieties are recognized (Geason and Cronquist 1991; Harlow et al. 1996). With its large range, *Tilia americana* is considered the primary species. The present consensus seems to be that all other populations are varieties of *Tilia americana*. Further analysis is necessary to improve the definition of species and varieties.

Tilia americana Linnaeus var. *americana*, American basswood (Figure 20.1); large tree, height to 20 to 25 m; deciduous; leaf arrangement alternate; leaf simple, length 10 to 20 cm, blade slightly more narrow than wide, shape broadly ovate, lower surface of expanding and mature leaves glabrous except for axillary tufts, length 5 to 7 mm; base asymmetrical; terminal bud absent, lateral buds asymmetrical, somewhat divergent, outer scales brownish, inner scales greenish; fruit a globose nutlet, diameter 0.8 to 1.2 cm. Regions 2 and 3, Manitoba to New Brunswick south to Oklahoma, Tennessee, and North Carolina; mesic sites, moist fertile soil, cool slopes, and dense woods. Cover type 26.16, 26.20, 26.22, 26.23, 26.28, 27.5, 27.8. 27.16, 27.19, 27.23.

Tilia americana var. *heterophylla* (Ventenat) Loudon, white basswood (Figure 20.2); large tree, height to 20 to 25 m; deciduous; leaf arrangement alternate; leaf simple, length 10 to 20 cm, blade slightly more narrow than wide, shape broadly ovate, lower surface

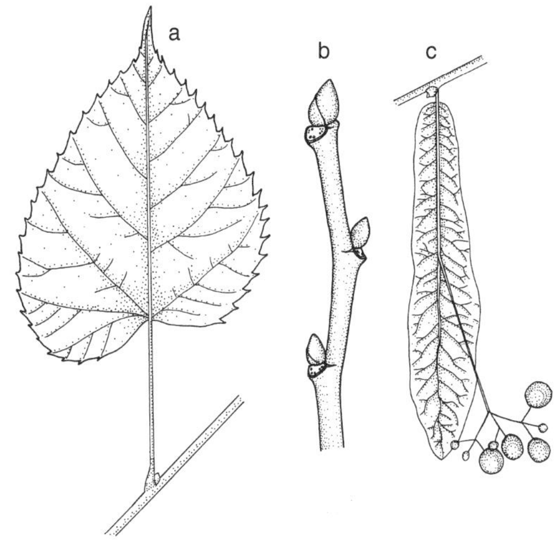

FIGURE 20.1 *Tilia americana* (a) simple leaf ×1/3, (b) twig showing leaf scars and asymmetrical buds ×1, (c) inflorescence subtended by a leafy bract ×1/3.

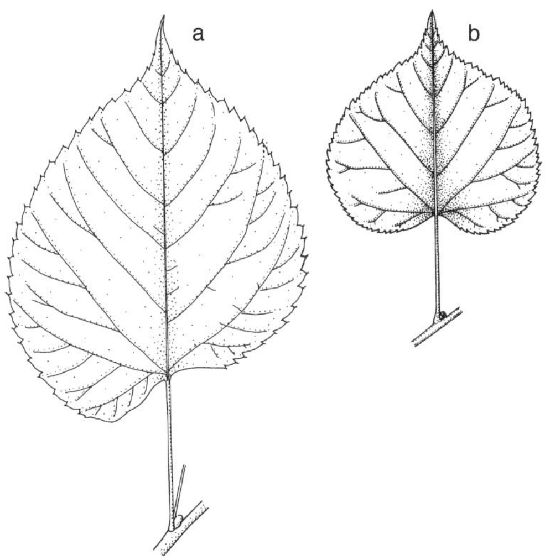

FIGURE 20.2 (a) *Tilia americana* var. *heterophylla* simple leaf ×1/3; (b) *T. cordata* simple leaf ×1/2.

of expanding and mature leaves are covered with dense grayish white stellate hair, base asymmetrical; axillary tufts present; terminal bud absent, lateral buds asymmetrical, somewhat divergent, outer scales brownish, inner scales greenish; Region 3b, Appalachian Mountains from Pennsylvania to northern Georgia and Alabama with outliers are in Florida, New York, and as far west as Missouri and Arkansas. This species differs from *T. americana*, as the lower leaf surface is covered with dense white stellate hair. Once considered a separate species (*Tilia heterophylla* Ventenat), it is now merged with *T. americana* (Gleason and Cronquist 1991; Dirr 1998). Cover type 27.30, 27.12, 27.29, 27.37.

Tilia americana var. *caroliniana* (Miller) Castigl. (*Tilia floridana* Small; *Tilia caroliniana* Miller), Carolina or Florida basswood; small tree, height to 15 m; deciduous; lower leaf surface covered with loose tomentose hair; Regions 3 and 4, southern Missouri to Virginia, Texas, and Florida; mesic sites, high fertility soil. Cover type 28.27, 28.45.

Tilia neglecta (Spach) has loose, sparse stellate hair and appears to be a variety of *T. americana* (Fernald 1950) or an introgressed hybrid of var. *americana* and var. *heterophylla* (Gleason and Cronquist 1991). A highly variable population. Its range is sympatric with that of *T. americana.*

Tilia cordata Miller, small-leaf or littleleaf linden (Figure 20.2); medium-sized tree, height to 22 m; deciduous; leaf arrangement alternate; leaf simple, length 5 to 10 cm, width 3 to 6 cm, blade slightly more narrow than wide, shape broadly ovate, lower surface of expanding leaves glabrous except for axillary tufts, base symmetrical or asymmetrical; terminal bud absent, later buds asymmetrical, somewhat divergent, scales greenish; native of Europe; widely planted as a shade tree in the United States; nearly 25 cultivars have been developed (Dirr 1998).

ORDER: SALICALES

Salicaceae (Willow or Poplar Family)

Members of the Salicaceae are trees, shrubs, and subshrubs. There are three genera, but all but one species belong to either *Salix* (willow, 300 to 500 species) or *Populus* (cottonwood, aspen, poplar, 35 to 40 species).

Unlike that of a number of other families, their center of diversity is in the north temperate, boreal, and subarctic regions, and they are widely distributed other than in parts of the Southern Hemisphere. In North America, there are about 110 *Salix* species and 8 to 11 *Populus* species.

The flowers are unisexual, lack petals, and are borne in catkins. Until recently this family was clustered with other families, such as the Fagaceae and Betulaceae, into a group called the "Amentiferae" (ament- or catkin-bearing plants), although these families apparently developed from different evolutionary lines. *Populus* has a pendulous catkin, lacks nectar and scent, and the flowers are wind pollinated (anemophilous). In contrast, the *Salix* catkin is more rigid, similar to a twig, and the flowers produce large amounts of strongly scented nectar. This nectar attracts bees and moths as well as other insects that pollinate the flowers (entomophilous), although some pollination also occurs by wind. The plants of both genera are dioecious. The tiny seeds (diameter ~1 mm) are wind dispersed.

Because of its small size, the seed is viable for only about 2 weeks and thus must land on open sites. These sites usually have been recently disturbed by logging, fire, wind, or other severe disturbance that has removed the forest and exposed mineral soil. Once on mineral soil, the seed rapidly germinates and initiates the next forest. After the glacier melted about 9000 years ago in the Lake States (Region 2a), the bare landscape was probably revegetated by *Populus tremuloides* Michaux (trembling aspen) seed transported from the Rocky Mountains.

The simple, usually unlobed, leaves have an alternate arrangement. Small stipules accompany the leaves. Species of both genera have fast growth, which results in low density, relatively brittle wood, and easily broken branches. Ecologically, the species are pioneers and extremely shade intolerant.

Genus: *Populus*

Group I: Species have a deltoid, elliptical to nearly ovate leaf shape; the leaf petiole is flattened at the distal end (near the blade base). The petiole is nearly as long as the blade.

Populus deltoides Marshall, eastern cottonwood (Figure 20.3); large tree, height to 30 m; deciduous; leaf arrangement alternate; leaf simple, total length 12 to 20 cm; blade length one half total length, width 6 to 10 cm, length = width, shape deltoid, entire, margin coarsely serrate or crenate, teeth 12 to 20 per side, apex distinctly acuminate or long acuminate, base truncate and lacking teeth, abaxial surface glabrous; petiole length 4.0 to 7.5 cm, vertically flattened at the distal end, glands two to five at distal end; terminal bud length ~1.5 cm, all buds resinous, scales waxy and spotted with yellowish resin; inflorescence a catkin, length 8 to 25 cm; fruit a capsule, length 5 to 8 mm; seed minute, surrounded by long white silky hair; bark in the crown area cream to greenish, smooth; Regions 3 and 4 except the Appalachian Mountains; wet sites, bottomland, and swamps. Cover type **28.15**, 27.3, 27.6, 27.14, 27.20, 27.28, 28.7, 28.21, 29.20.

Populus deltoides var. *occidentalis* Rydberg (*P. sargentii* Sargent), plains cottonwood; medium to large tree, height to 26 to 30 m; deciduous; leaf arrangement alternate, many characteristics similar to *P. deltoides;* leaf simple, total length 12 to 18 cm entire, blade length one half total length, width 7 to 10 cm, width > length, shape deltoid, margin coarsely serrate or crenate, teeth blunt, 12 to 20 per side, apex distinctly acuminate, base truncate or slightly rounded, lacking teeth, abaxial surface glabrous; petiole length 6 to 9 cm, vertically flattened at the distal end; terminal bud length ~1.5 cm, resinous, bud scales pubescent; inflorescence a catkin, length 8 to 20 cm; fruit a capsule, length 5 to 8 mm; seed minute, surrounded by long white silky hair; bark pale gray; Great Plains from north Texas to Alberta and Saskatchewan; water courses, drainages, mesic-hydric sites.

Populus fremontii, S. Watson, Fremont cottonwood (Figure 20.4); large tree, height to 30 m; deciduous; leaf arrangement alternate; leaf similar to *P. deltoides* var. *occidentalis*; leaf simple, total length 10 to 13 cm, blade length one half total length, width 6.0 to 7.5 cm, width > length, shape broadly deltoid, margin coarsely serrate or crenate, teeth blunt, 12 to 20 per side, apex distinctly acuminate, base truncate or slightly rounded, lacking teeth, abaxial surface glabrous; petiole length 4.0 to 5.5 cm, vertically flattened at the distal end; terminal bud length 0.8 to 1.3 cm; inflorescence a catkin, length 8 to 13 cm; fruit a capsule, length 5 to 8 mm, broadly egg shaped; seed minute, surrounded by long white silky hair; bark pale gray, creamy, or brownish; Regions 5 and 7, western Texas and western Colorado to the California

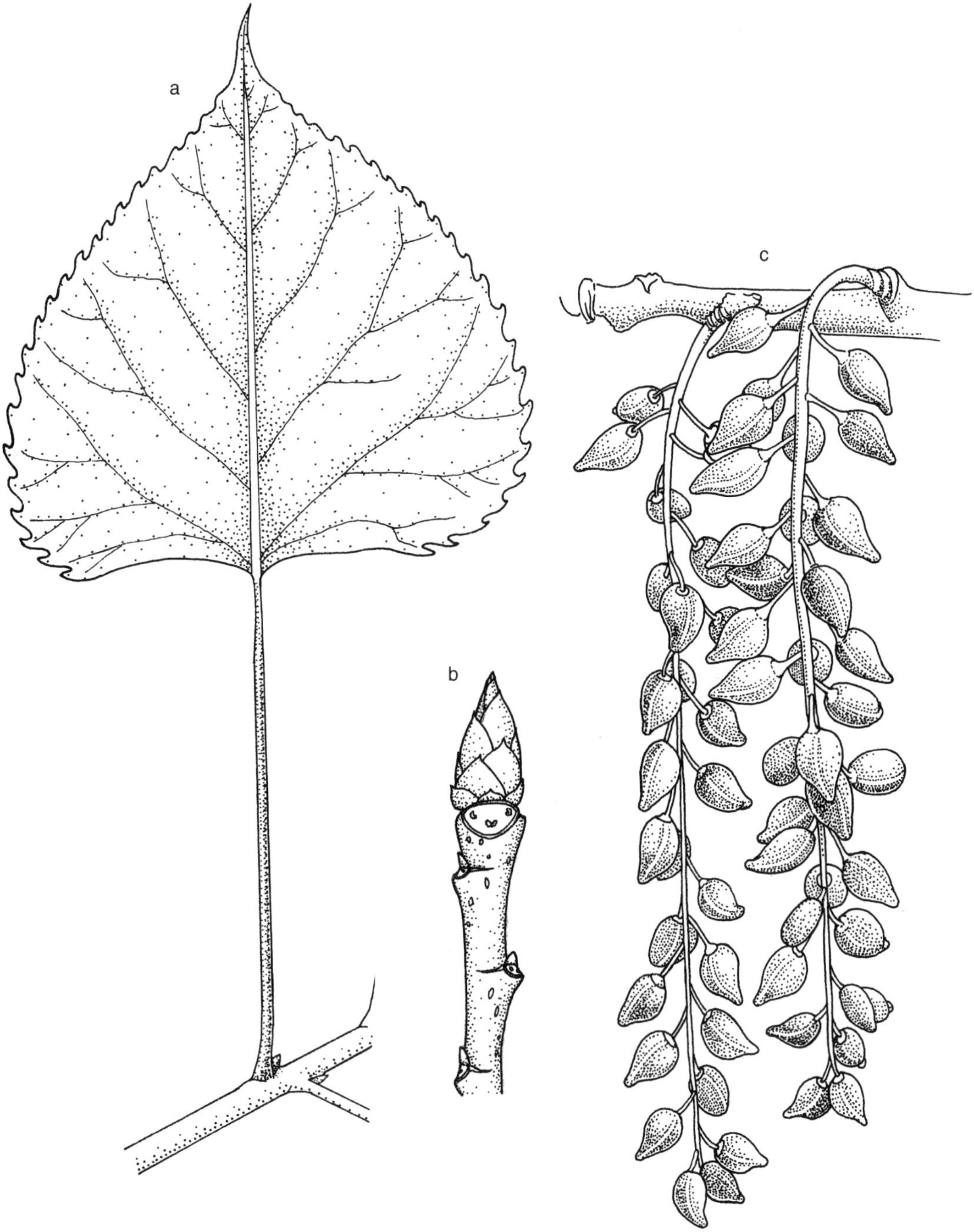

FIGURE 20.3 *Populus deltoides* (a) deltoid leaf ×1, (b) twig with terminal bud and leaf scars ×1.3, (c) female catkins with capsules ×1.3.

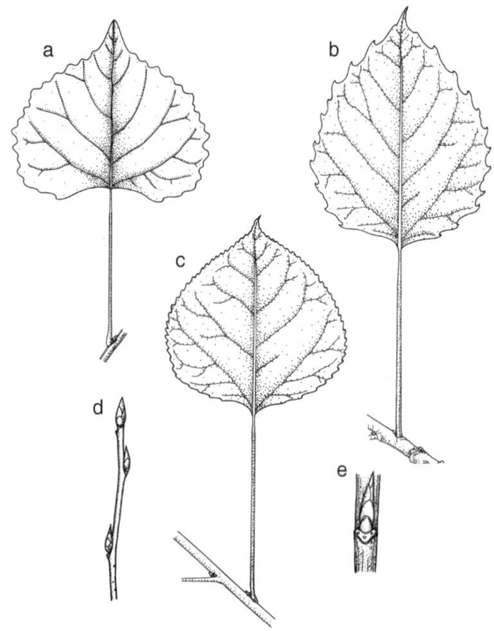

FIGURE 20.4 Simple leaves of (a) *Populus fremontii* × 1/3; (b) *P. grandidentata* ×2/5; *P. tremuloides* (c) leaf ×1/2, (d) twig with terminal and lateral buds ×1/2, (e) lateral bud and leaf scar ×1.

Sierra Nevada and the Coastal Ranges; water courses, drainages, mesic to mesic-hydric soils near streams or seepages, sea level to 2100 m. Cover type **29.8**, 29.2, 29.19, 30.40.

Populus fremontii var. *wislizeni* Watson, Rio Grande cottonwood; one of about six to seven varieties of *P. fremontii*, this one differing primarily in a narrower egg-shaped capsule; Region 5, Rio Grande in Texas. Cover type **29.7**.

Populus tremuloides Michaux, trembling or quaking aspen, popple (Figure 20.4); medium-sized tree, height to 22 m; deciduous; leaf arrangement alternate; leaf simple, total length 5 to 15 cm, blade entire, width 2 to 7 cm, length = width, shape oval to nearly orbicular, margin finely crenate or serrate, teeth blunt, apex acute, base essentially truncate, surfaces glabrous; petiole length 2.5 to 7.0 cm, vertically flat at the distal end; twig color creamy, glabrous; terminal bud reddish, shiny, length 6 to 9 mm, lanceolate, sharp-pointed, lateral buds essentially appressed to the twig, scales five to seven, imbricate; inflorescence a catkin, length 7 to 10 cm; fruit a capsule; seeds minute, surrounded by dense long white silky hair; bark cream colored or nearly white; Regions 1, 2, 3(northern one-half) 5, 6, and 7a, transcontinental in Canada, from Newfoundland to Alaska south into northern California and along the Rocky Mountain chain into Arizona, New Mexico, and Mexico, also Lake States and New England, to the northern limit of trees; xeric and xeric-mesic sands to mesic loam soils, generally in pure stands composed of several clones or mixed with *Betula papyrifera*. The most widely distributed *Populus* and most widely distributed of any North American tree species. A fast-growing but relatively short-lived tree, with ages reaching only 35 to 50 years in warmer parts of the Lake States but nearly 100 years in central Canada. Most new stems come from root suckers (sprouts) after the parent tree has been cut, thus the commonness of clones. Cover type **26.4, 26.7, 26.14, 26.19, 29.12, 29.16, 29.18, 29.26, 29.31, 29.33, 29.34, 29.35, 30.14**, 26.1, 26.2, 26.3, 26.5, 26.10, 26.15, 26.18, 26.26, 29.30, 29.11, 29.13, 29.15.

Populus grandidentata Michaux, bigtooth or largetooth aspen (Figure 20.4), popple; medium to large tree, height to 27 m; deciduous; leaf arrangement alternate; leaf simple, entire, total length 11 to 18 cm, blade width 3 to 8 cm, shape broadly ovate to nearly orbicular, margin coarsely crenate, teeth rounded, apex obtuse, base essentially truncate, surfaces glabrous; petiole length 3.5 to 6.0 cm, vertically flattened at the distal end; twig color greenish gray, glabrous, terminal bud reddish, shiny, length 6 to 9 mm, lanceolate, sharp pointed, lateral buds essentially appressed to the twig, scales 5 to 7, imbricate; inflorescence a catkin, length 7 to 10 cm; fruit a capsule; seeds minute, surrounded by dense, long white silky hair; bark yellowish green, smooth, becoming dark and rough on large trees; Regions 1 and 2, eastern and southern Quebec, Ontario, and Minnesota, but occasionally extending southward as scattered trees into Illinois, Indiana, and Ohio; disjunct populations occur in eastern Tennessee and North Carolina; xeric sandy to mesic loamy well-drained soils. This species is a more robust relative of *P. tremuloides*. Although individual stems of one may be found in stands of the other, these two aspens do not freely mix. *Populus grandidentata* grows to be the larger tree and lives to a greater age, probably

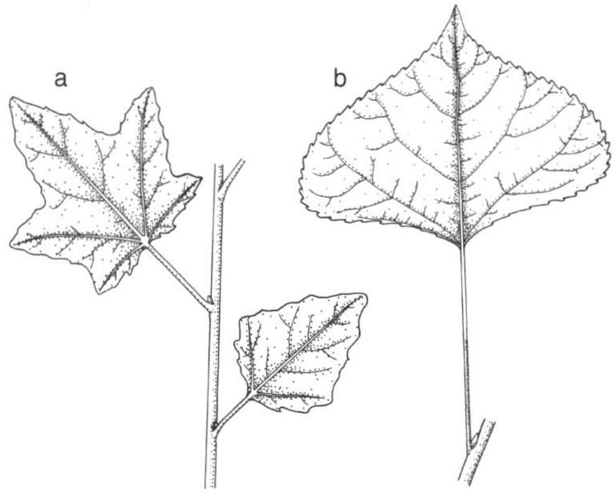

FIGURE 20.5 Leaves of (a) *Populus alba* ×1/2; (b) *P. nigra* ×1/2.

somewhat more than 100 years. It also reproduces by root suckers after the parent tree has been cut. Cover type 26.1, 26.2, 26.15, 26.19, 26.19, 26.29.

Populus alba Linnaeus, white poplar (Figure 20.5); medium-sized tree, height to 16 m; deciduous; leaf arrangement alternate; leaf simple, blade three- to five-lobed, length 3 to 8 cm; abaxial surface, petiole, and new twigs white tomentose, woolly-velvety; bark cream colored becoming dark; introduced as an ornamental, escaped cultivation, often found along fences and disturbed areas in eastern United States and Canada.

Group II: Species have a more or less cordate-shaped leaf and a round leaf petiole at the distal end; petiole length is one third the length of the blade. The buds are resinous and fragrant. The ranges of native species generally are allopatric (do not overlap).

Populus nigra var. *italica* Muenchhausen, Lombardy poplar (Figure 20.5), small to medium-sized tree, height to 12 m; deciduous; leaf arrangement alternate; leaf simple, length 4.5 to 10.0 cm, width 6 to 9 cm, width > length, entire, shape rhomboid to deltoid, margin serrate or crenate, petiole length 4 to 6 cm; introduced from Europe; a common ornamental appreciated for its distinctly columnar (fastigiate) form, resulting from the ascending branch pattern. Fast growing but extremely short-lived.

Populus balsamifera Linnaeus, balsam poplar, balm-of-Gilead (Figure 20.6); medium to large tree, height to 25 m; deciduous; leaf arrangement alternate; leaf simple, total length 8 to 17 cm, blade length two thirds to three quarters total length, width 2.5 to 7.0 cm, shape entire, margin serrulate, crenulate, or nearly smooth, teeth tipped with a brown gland, apex obtuse, base obtuse or rounded, adaxial and abaxial surfaces visibly rusty or orange stained; petiole terete at the distal end and; terminal bud length 1 to 2 cm, buds resinous, aromatic, scales shiny and spotted with yellowish resin; inflorescence a catkin, length 8 to 18 cm; fruit a capsule, length 3 to 9 mm; seed minute, surrounded by cottony hairs; bark in the crown area cream to greenish, smooth; Regions 1 and 2, transcontinental in Canada, Alaska, disjunct in the central Rockies; well-drained stream terraces and margins. Cover type **26.13**, 26.25.

Populus trichocarpa Torrey & Gray, black cottonwood (Figure 20.6); large tree, height to 45 m; deciduous; leaf arrangement alternate; leaf simple, total length 8 to 15 cm, width 6 to 10 cm, entire, shape lanceolate or ovate, margin somewhat crenate or serrate, teeth blunt, 25 to 75 per side, apex long acuminate, base rounded, abaxial surface silvery, often rusty or orange stained, petiole terete at the distal end; terminal bud length 1.5 to 2.0 cm, buds resinous, aromatic, scales shiny and spotted with yellowish resin; inflorescence a catkin, length 8 to 18 cm; fruit a capsule, length 5 to 9 mm; seed minute, surrounded by long white silky hair; bark in the crown area cream to greenish, smooth; Regions 6 and 7, Alaskan coast through British Columbia to Montana, Idaho, and California, disjunct populations in Nevada and Baja California (Mexico), generally coastal; moist slopes, stream terraces and floodplains, from sea level to 2600 m. There is some hybridization with *P. balsamifera* where ranges overlap in northern Rocky Mountains. Cover type **30.11, 30.15**, 29.24, 30.4, 30.6, 30.98, 30.10, 30.19, 30.40.

Populus heterophylla Linnaeus, swamp cottonwood (Figure 20.7); medium to large tree, height to 27 m; deciduous; leaf arrangement alternate; leaf simple, total length 10 to 30 cm, blade length two thirds total length, width 6 to 18 cm, shape extended deltoid to somewhat cordate, entire, margin serrate or somewhat crenate, teeth 25 to 75 per side, apex obtuse, somewhat rounded, base truncate, rounded, or cordate, abaxial veins lined with tan pubescence; petiole terete at the distal end; terminal bud length ~5 mm, buds resinous, scales shiny and spotted with yellowish resin; inflorescence a catkin, length 8 to 18 cm; fruit a capsule, length

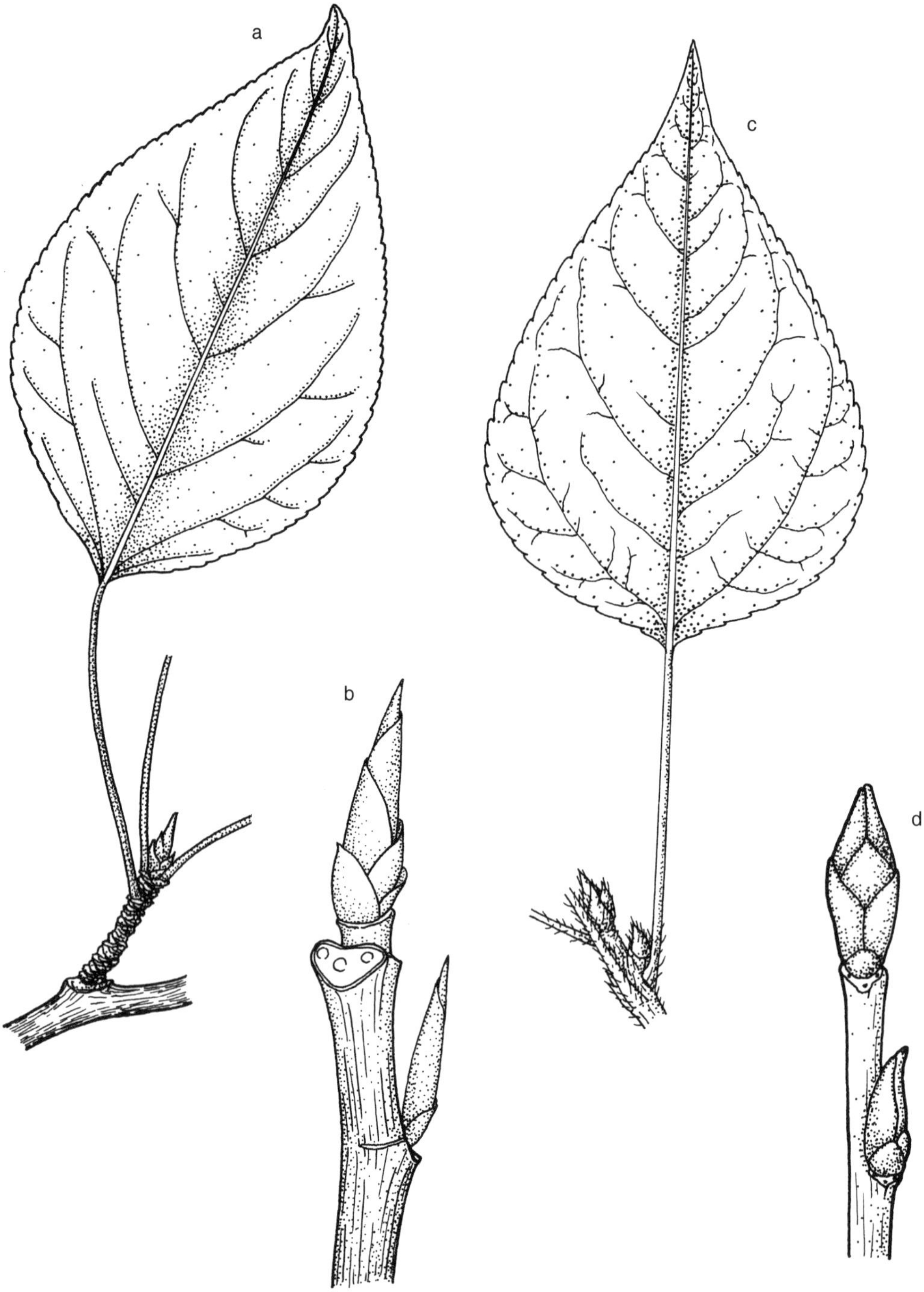

FIGURE 20.6 *Populus balsamifera* (a) simple leaf ×1, (b) twig with terminal and lateral buds and leaf scar ×3; *P. trichocarpa* (c) simple leaf ×1, (d) twig with terminal and lateral buds and leaf scar ×2.

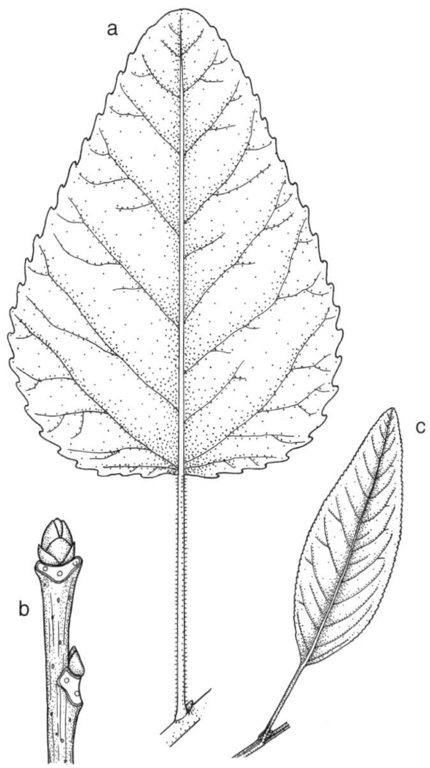

FIGURE 20.7 *Populus heterophylla* (a) simple leaf ×1/2, (b) twig with terminal and lateral buds and leaf scar ×1.5; (c) *P. angustifolia* leaf ×3/4.

5 to 9 mm; seed minute, surrounded by cottony hairs; bark in the crown area cream to greenish, smooth; scattered throughout Regions 3 and 4a; swamps, sloughs and wet sites. Cover type 28.31, 28.43, 28.55.
Group III: Includes only *P. angustifolia*.

Populus angustifolia James; narrowleaf cottonwood (Figure 20.7); medium-sized tree, height to 20 m; deciduous; leaf arrangement alternate; leaf total length 5 to 10 cm, width 1.5 to 3.0 cm, shape lanceolate, entire, margin finely serrate or serrulate, apex distinctly acute to long acuminate, base obtuse and lacking teeth, abaxial surface glabrous; petiole short, length 1.5 to 3.0 cm; terminal bud length 0.6 to 1.2 cm, resinous, scales shiny; inflorescence a catkin, length 3 to 7 cm; fruit a capsule, length 5 to 8 mm; seed minute, surrounded by long white silky hair; bark yellowish green, smooth; Regions 5 and 6, Alberta and Saskatchewan along the Rocky Mountains to Arizona, New Mexico, and Mexico; mountain streams, drainages, and moist terraces. Cover type **29.20**, 29.7, 29.8, 29.15.

Populus tremuloides, P. grandidentata, P. deltoides, and *P. trichocarpa* are economically and ecologically important. Trees of these species are used largely as pulp for making paper and lightweight products such as crates and boxes. The small seed is viable for only about 2 weeks, but because it may be transported for long distances, there is a high probability that the seed will be deposited where a mineral soil has been exposed as a result of cutting, fire, or other disturbance. Because of its seed characteristics and a high light requirement for adequate growth and development, these species are classic (shade-intolerant) pioneers, the first to arrive in a disturbed area to begin the process of revegetation. Extensive stands of *P. tremuloides* in Canada (Region 1) and *P. tremuloides* and *P. grandidentata* in the Lake States (Region 2a) are evidence of the heavy cutting and fire that removed the original forest. *Populus deltoides* is also an early pioneer on disturbed bottomlands or new land (batture) developed along levees in the central and southern states (Regions 3 and 4). In addition, *Populus trichocarpa* forms nearly pure stands on new alluvium and in some cutover areas in the west.

Genus: *Salix*

Salix (willow) species are found in all parts of the world, including Mexico, Guatemala, Puerto Rico, Africa, Australia, New Zealand, Iceland, the Philippines, Madagascar, Malaysia, Europe, Asia, Japan, and China, as well as Canada and the United States. Species occur in a wide variety of climates and habitats, ranging from extremely wet to extremely cold conditions. As a group, these pioneer species are extremely shade intolerant, but a few are able to live in boreal forest communities. As in *Populus*, the light, fluffy seed produced in high amounts may be wind transported long distances, with the probability that a few seeds will land on open or disturbed habitats to begin the process of revegetation.

Although this is a genus of small to large multistemmed shrubs and small trees, the ecological importance of *Salix* should not be underestimated. The species found mostly as scattered individuals or in small thickets along streams and lakes across the United States and into central Canada; however, *Salix* become progressively more important, both on upland and lowland sites and northward from the boreal forest (Region 1) into the subarctic, where they are the major woody species of uplands. Estimates on the number of species in the far north vary from about 50 in Alaska (Viereck and Little 1972) to about 70 in the Yukon (Cody (1996). There stems of the species form extensive, dense thickets across the landscape. These 3 to 4 m high thickets provide excellent cover and food for various birds and animals, including grizzly and black bear, protect the soil surface from erosion, and store nutrients. In the arctic and high alpine zones, individual plants become miniature shrubs. The root often remains woody, whereas the shoot produced during the growing season is more herbaceous. In North Dakota and the southern one third of Manitoba, Saskatchewan, and Alberta, *Salix* shrubs provide nesting and resting habitat for a variety of birds around prairie potholes. In general, buds, twigs, and bark are food for rabbits, mice, deer, porcupines, and other animals. One or two species, called "diamond willow," have a sunken zone (a diamond) around small branches or branch stubs; 2 to 6 cm diameter branches of these species are used to make staffs, walking sticks, and canes. When the bark is peeled, white wood is exposed except in the sunken area which, when cleaned out, exposes a rich, dark reddish brown diamond.

The flattened (duckbill-like) bud of *Salix* is enclosed by a single visible scale, which distinguishes this genus from nearly all other genera. However, often it is difficult to separate species as leaf shape may vary, depending on where it is on the stem (e.g., fertile vs. sterile twigs). In addition, on the same plant stipules will vary in size from large and foliaceous to small and insignificant. Catkin characteristics are relatively consistent, but hybridization creates individuals with intermediate characteristics. Even with these problems accounted for, many species are not distinct and their ranges may have to be reconsidered. At times it may be necessary to examine a cluster of individuals within a larger population.

In general, leaf shape varies from elliptical and broadly lanceolate to narrowly lanceolate. The only species to reach tall tree size is *Salix nigra*. About 25 additional species reach small tree size. Many of the species have a relatively large range. *Salix exigua* Nuttall (coyote willow) has the largest range, extending from Alaska across Canada to the Maritime Provinces, to all regions of the United States except the southeast (Region 4), and into Mexico. *Salix interior* (Rowlee) also grows from Alaska to New England to Texas. Other willow species of Canada, the Lake States, and New England include *S. bebbiana* Sargent (Bebb willow), *S. discolor* Muhlenberg (pussy willow), *S. lucida* Muhlenberg (shining willow), *S. pellita* Andersson (satiny willow), *S. petiolaris* J. E. Smith (meadow willow), *S. pyrifolia* Andersson (balsam willow), and *S. monticola* Bebb ex. Coulter (serviceberry willow). The range of *Salix arbusculoides* Andersson (littletree willow) and *S. alaxensis* (Andersson) Coville (feltleaf willow) extend from James Bay across western Canada into Alaska. Western United States species include *S. hindsiana* Bentham (Hinds willow), *S. hookerana* Barratt (Hooker willow), *S. laevigata* Bebb (red willow), *S. lasiandra* Bentham (Pacific willow), *S. lasiolepis* Bentham (arroyo willow), *S. sessilifolia* Nuttall (northwest willow), *S. sitchensis* Sanson ex. Bongard (sitka willow), and *S. tracyi* Ball (Tracy willow). *Salix amygdaloides* Andersson (peachleaf willow) is a common species ranging from Washington across the Great Plains into New England. A willow restricted to the central and southern states (Regions 3 and 4) is *S. caroliniana* Michaux (Coastal Plain willow).

Argus (1973) and Newsholme (1992) provide a relatively good review of the genus *Salix*. Only a few select species are described here.

Salix nigra Marshall, black willow (Figure 20.8); medium-sized to large tree; height to 28 m; deciduous; leaf arrangement alternate; leaf simple, entire, total length 7 to 12 cm, width 0.5 to 1.8 cm, shape lanceolate or narrow/long elliptical, margin finely serrated, apex long acuminate, surfaces glabrous, abaxial veins lightly hairy; petiole length 3 to 8 mm, glabrous; branchlets tan, deciduous in the fall; stipules somewhat cordate; buds valvate, length 2 to 5 mm, flattened, appressed or nearly appressed to branchlet; catkin length 3 to 5 cm; seed minute, surrounded by dense, long white silky hair; Regions 2, 3, and 4, central Wisconsin to Maine and south, all states; concentrated near streams, terraces, and bottomlands, scattered on open sites. This is a fast-growing but short-lived tree. The wood is soft

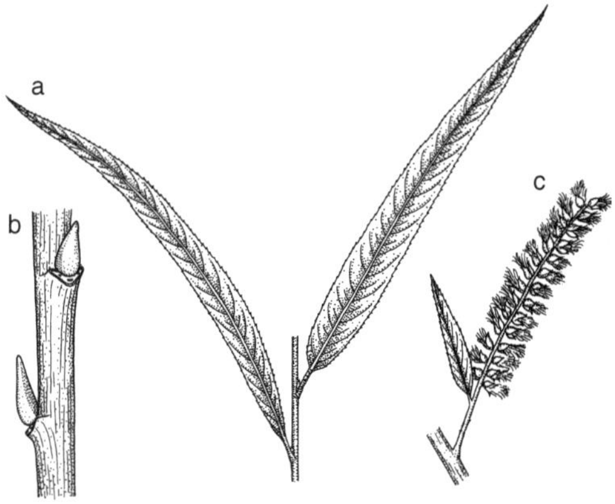

FIGURE 20.8 *Salix nigra* (a) simple leaf ×2/3, (b) twig with lateral buds ×2, (c) twig with female catkin ×3/4.

and usable for light-use furniture, crates, and baskets. Long, thin branchlets are produced each year but, like the leaves, are mostly deciduous. Cover type **27.28, 28.21**, 27.6, 27.10, 27.14, 28.7, 28.15, 28.31, 28.54, 28.55.

Salix exigua Nuttall (*Salix longifolia* Muhlenberg; *Salix interior*, Rowlee), sandbar willow, coyote willow (Figure 20.9); shrub, height to 4 m; deciduous; leaf arrangement alternate; leaf simple, entire, total length 4 to 10 cm, width 0.2 to 1.0 cm, shape lanceolate or narrow/long elliptical, margin smooth, finely serrated, or with small irregularly scattered teeth, apex long acuminate, abaxial surfaces lightly pubescent, abaxial veins lightly hairy; petiole length 3 to 7 mm, glabrous; branchlets reddish brown, eventually glabrous; bud valvate, length 2 to 4 mm, flattened, appressed or nearly appressed to branchlet; catkin length 5 to 8 cm; seed minute, surrounded by dense, long white silky hair; all regions, Alaska and Yukon across Canada, south into California, New Mexico, and Louisiana; colonizes newly exposed sand bars and young islands, not competitive in existing forest. Development of new shoots is primarily from the spreading root system. Extensively browsed by moose. Cover type 28.15, 28.21, 29.8, 30.15.

Salix bebbiana Sargent, Bebb willow (Figure 20.10); shrub or small tree, height to 6 to 7 m;

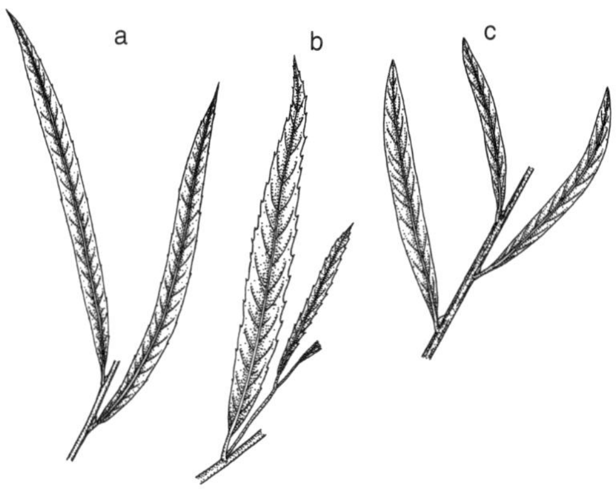

FIGURE 20.9 (a) *Salix exigua* leaves ×1/2; (b) *S. interior* leaves ×1/2; (c) *S. longifolia* leaves ×1/2.

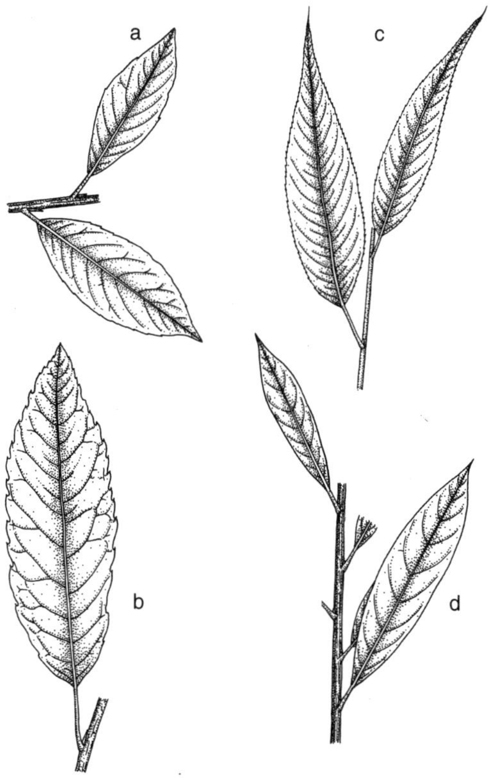

FIGURE 20.10 (a) *Salix bebbiana* leaves ×3/4; (b) *S. discolor* leaves ×3/4; (c) *S. amygdaloides* leaf ×1/2; (d) *S. lucida* leaves ×1/3.

deciduous; leaf arrangement alternate; leaf simple, entire, total length 2 to 6 cm, width 1 to 3 cm, shape elliptical or slightly obovate, margin smooth, uneven, or slightly crenate, apex acuminate, adaxial and abaxial surfaces initially pubescent, becoming glabrous at maturity; petiole length 3 to 9 mm, pubescent; branchlets, reddish brown, densely pubescent; bud valvate, length 2 to 4 mm, flattened, appressed or nearly appressed to branchlet; catkin length 3 to 5 cm; seed minute, surrounded by dense, long white silky hair; Regions 1 and 2, Alaska and Yukon across Canada, south into Region 3, Arizona and New Mexico; well-drained but mesic river margins; uplands as well as mesic-hydric lowlands and prairie. A common component of open forest of spruce and aspen (*Picea glauca, Betula papyrifera*, and *Populus tremuloides*). Cover type 29.34.

Salix discolor Muhlenberg, pussy willow (Figure 20.10), shrub or small tree, height to 6 to 7 m; deciduous; leaf arrangement alternate; leaf simple, entire, total length 5 to 12 cm; width 2 to 4 cm, shape elliptical or slightly obovate, margin undulate, serrate or slightly crenate, apex acute, adaxial surface shiny green, abaxial surface glaucous; petiole length 0.5 to 1.8 cm; branchlets stout, reddish brown, pubescent, becoming glabrous with age; bud valvate, length 2 to 4 mm, flattened, appressed or nearly appressed to branchlet; catkin length 3 to 10 cm, diameter 2.0 to 2.5 cm; seed minute, surrounded by dense, long white silky, hair; Regions 1, 2, and 3, eastern United States and Canada, British Columbia to Labrador.

Salix amygdaloides Andersson, peachleaf willow (Figure 20.10); shrub or small tree, height to 15 m; deciduous; leaf arrangement alternate; leaf simple, entire, total length 5 to 15 cm; width 1.0 to 3.5 cm, shape lanceolate, margin smooth, finely serrated, apex long acuminate, adaxial surface glabrous or with a few long hairs, abaxial surface villous; petiole length 0.6 to 1.9 cm; stipules rapidly deciduous; dark branchlets yellowish brown, villous; bud valvate, length 2 to 4 mm, flattened, appressed or nearly appressed to branchlet; catkin length 4 to 9 cm, diameter 2.0 to 2.5 cm; seed minute, surrounded by dense, long white silky hair; Regions 1, 2, and 3, Washington to Saskatchewan, Wisconsin, New York, and southern Ontario south to across central United States into Mexico; mesic-hydric soil, stream terraces, drainages, and swamps. Cover type 29.20.

Salix lucida Muhlenberg subsp. *lasiandra* (Bentham) E. Murray (*Salix lasiandra* Bentham), yellow willow, shining willow (Figure 20.10); shrub or small tree, height to 8 m; deciduous; leaf arrangement alternate; leaf simple, entire, total length 6 to 13 cm, width 1 to 2 cm, shape lanceolate, margin serrate or crenulate, apex long acuminate, adaxial surface glabrous and shiny, abaxial surface glabrous; petiole length 0.6 to 1.3 cm; branchlets yellowish or tan; bud valvate, length 2 to 4 mm, flattened, appressed or nearly appressed to branchlet; stipules persistent; catkin length 5 to 10 cm; seed minute, surrounded by dense, long white silky hair; Regions 1, 2, 6, and 7, Alaska and the Yukon south to California, east across Saskatchewan to Nova Scotia; lakeshores, stream banks, terraces, drainages, and wet flats. Cover type **30.40**.

Salix sessilifolia Nuttall, northwest willow (Figure 20.11); shrub or small tree, height to 7 m; deciduous; leaf arrangement alternate; leaf simple, entire, total length 4 to 10 cm, width 1.0 to 3.5 cm, shape elliptical, oblong, or slightly lanceolate, margin smooth or serrulate, apex acuminate, adaxial surface glabrous or with a few long hairs, abaxial surface villous; petiole length 1 to 7 mm; branchlets dark reddish brown, villous; bud valvate, length 2 to 4 mm, flattened, appressed or nearly appressed to branchlet; catkin length 3 to 10 cm, diameter 2.0 to 2.5 cm; seed minute, surrounded by dense, long white silky hair; Regions 6 and 7, British Columbia, Washington, Oregon; streams and drainages. Forms thickets by sprouting from the root system.

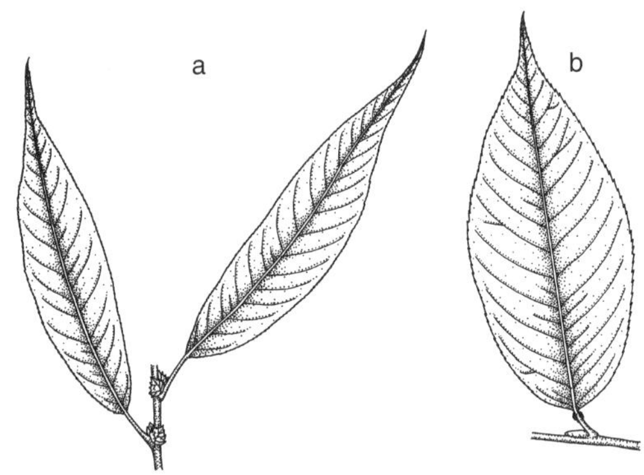

FIGURE 20.11 (a) *Salix sessilifolia* leaves ×4/5; (b) *S. pentandra* leaf ×3/4.

Salix pentandra Linnaeus, bay-leaf willow (Figure 20.11); shrub or small tree, height to 7 m; deciduous; leaf arrangement alternate; leaf simple, entire, total length 4 to 10 cm, width 1.5 to 3.0 cm, broadly ovate to lanceolate, margin serrate, apex long acuminate, adaxial surface glabrous and shiny, abaxial surface glabrous; petiole length 0.6 to 1.0 cm, glandular; branchlets yellowish or tan; bud valvate, length 2 to 4 mm, flattened, appressed or nearly appressed to branchlet, stipules persistent; catkin length 2 to 6 cm; seed minute, surrounded by dense, long white silky hair; Regions 1, 2, and 3, Nova Scotia to Ontario, south to Maryland, Ohio, and Iowa; wet ground, stream banks. Introduced from Europe, escaped from cultivation.

Salix arctica Pallas, arctic willow (Figure 20.12); dwarf shrub, decumbent or matted, height to 40 cm; deciduous; leaf arrangement alternate; leaf simple, entire, total length 2 to 8 cm, width 1.5 to 3.5 cm, shape ovate, slightly obovate, or broadly elliptical, margin smooth, apex obtuse, adaxial and abaxial surfaces glabrous; petiole length 3 to 6 mm; branchlets brown, sometimes rooting at the tip; bud valvate, length 2 to 4 mm, flattened, appressed or nearly appressed to branchlet; stipules absent; catkin length 5 to 10 cm, erect; seed minute, surrounded by dense, long white silky hair; Regions 1, 5, and 6, Alaska to Greenland, south along the Rocky Mountains to New Mexico, circumpolar; tundra.

FIGURE 20.12 *Salix arctica* plant and inflorescence.

ORDER: ERICALES

Ericaceae (Heath Family)

The Ericaceae, as defined here, is a large family of small trees, shrubs, and woody vines widely distributed throughout the world. Depending on whether several subfamilies are included or separated as families, estimates of the number of genera and species range from 99 to 125 and 2245 to 3500, respectively. Shrub species far outnumber tree or vine species. The flowers are perfect (plants synecious) and have fused sepals (connate) and four to five fused petals. The number of stamens usually is ten, twice the number of petals. The stamens and style are exserted and about twice the length of the petals. The leaves are simple, entire, and evergreen or deciduous; the twigs lack stipules.

As with several other large families, there is considerable debate regarding the structure of the Ericaceae. The subfamilies Ericoideae (heather; 17 genera, 850 to 900 species) and Rhododendroideae (rhododendron, azalea; 15 genera, 850 species) have consistently been included within the Ericaceae; only the Rhododendroideae is strongly represented in North America. The Vaccinioideae (blueberry, huckleberry, about 54 genera and 660 species) is usually included as a subfamily of the Ericaceae but at times has been reported as the family Vacciniaceae. Most taxonomists agree that Vaccinioideae includes *Vaccinium* (blueberry) and *Gaylussacia* (huckleberry) and that Rhododendroideae includes *Rhododendron, Ledum, Kalmia, Andromeda,* and *Chamaedaphne.* However, *Oxydendrum, Leucothoe, Lyonia, Epigaea,* and *Gaultheria* may appear in one subfamily or the other, depending on the approach taken (Fernald 1950; Smith 1977; Zomlefer 1994). The Pyrolaceae (*Pyrola*; prince's-pine, shinleaf) and Monotropaceae (*Monotropa*; Indian pipe, an achlorophyllous plant) are closely related families of forest herbs, but at times have been included as subfamilies Pyroloideae and Monotropoideae.

Members of the Ericaceae usually are well adapted to grow in extreme environments, specifically in cold climates, on acid, low-nutrient xeric soil or hydric swamps and bogs. The 25 to 30 species of *Rhododendron* (including these formally named *Azalea*) are concentrated in and reach maximum development in the Appalachian Mountains, but a few range into New England or Florida and as far west as western Kentucky, southern Illinois, and the southwest. Several are endemics with a range of only a few counties.

Genus: *Rhododendron*

Three species have traditionally been referred to as "rhododendron." They are evergreen, and the thick, coriaceous, slightly revolute leaves have a smooth, aciliate margin.

Rhododendron catawbiense Michaux, mountain rosebay (Figure 20.13); shrub, height to 3 to 4 m; evergreen; leaf arrangement alternate, leaves often clustered at twig tip; leaf simple, coriaceous, entire, blade 7 to 14 cm, width 3 to 6 cm, apex rounded or obtuse, slightly revolute, elliptical, or slightly oblanceolate, glabrous; petiole length 1.5 to 3.0 cm, usually pubescent; twig stout, reddish brown or gray, terminal bud length ~1.5 cm; pubescent; flower large, diameter 4 to 6 cm, showy, petals lilac, purple, pink, spotted green; fruit a capsule, carpels five, length ~1.5 cm; Region 3b, Appalachian Mountains, Tennessee, Kentucky, and North Carolina south to Georgia and Alabama; rocky ridge tops, slopes and stream banks.

Rhododendron maximum Linnaeus, rosebay rhododendron (Figure 20.13); shrub or small tree, height to 9 m; evergreen; leaf arrangement alternate; leaves usually clustered at twig tip, blade simple, entire, coriaceous, thick, oblanceolate or oblong, length 8 to 22 cm, width 2 to 6 cm, apex acute or slightly mucronate, slightly revolute, abaxial surface tomentose, velvety; petiole length 1 to 3 cm, pubescent; twig stout, reddish-brown to gray brown, pubescent; flowers large, diameter 3 to 4 cm, showy, petals white to pink, spotted green or orange; fruit a capsule, carpels five, length ~1.5 cm; Region 3b, Appalachian Mountains, southwestern Maine, and southern Ontario south to Tennessee, Kentucky, North Carolina, Georgia, and Alabama; forest on mesic ridgetops, slopes, and stream banks. May form dense thickets.

Rhododendron minus Michaux (*R. carolinianum* Rehder), Carolina, Piedmont, or dwarf rhododendron (Figure 20.13); shrub, height to 3 m; evergreen; leaf arrangement alternate; leaves strongly clustered at twig tips; blade entire, coriaceous, length 4 to 9 cm, width 1 to 5 cm, elliptical to broadly elliptical, apex slightly mucronate or acuminate, slightly revolute, abaxial surface dotted with minute brown scales; petiole length 0.5 to 1.0 cm, dotted with minute brown scales; twig stout, reddish-brown; flowers large, diameter 2 to 3 cm, showy, petals rose, spotted green; fruit a capsule, carpels five, length ~1.5 cm; Region 3b southern Appalachian Mountains, Tennessee and North Carolina south to Georgia and Alabama; wooded slopes and drainages.

Deciduous *Rhododendron* species are called "azalea" and generally have leaves of standard thickness and texture. Leaf margins are ciliate or serrate. The flowers tend to be smaller than rhododendron flowers but form large clusters. Foote and Jones (1989) and Grimm (1993) provide information on most azalea species. The following representative species are fully described and illustrated here.

Rhododendron alabamense Rehder, Alabama azalea; stoloniferous shrub, forming colonies, height to 2 m; deciduous; leaf arrangement alternate; leaves strongly clustered at twig tips; leaf entire, blade length 2.5 to 6.0 cm, width 1 to 2 cm, oblanceolate or narrowly elliptical, margin strigose, adaxial and abaxial veins strigose, distinctly aromatic when crushed; petiole 3 to 5 mm, tomentose; twigs gray, often with stout hair; flowers white, diameter 3 to 6 cm, lemon-scented; fruit a capsule; Region 3b, South Carolina, western Georgia and northern Alabama to Mississippi; dry open forest. Frequently hybridizes with *R. canescens.*

Rhodendron arborscens (Pursh) Torrey, smooth or sweet azalea (Figure 20.14); stoloniferous shrub; height to 6 m; deciduous; leaf arrangement alternate, leaves strongly clustered at twig tips; leaf entire, blade length 2 to 8 cm, width 1 to 3 cm; oblanceolate, margin essentially strigose, glandular, adaxial veins strigose, abaxial midrib with a few coarse hairs, surfaces glabrous; petiole length 4 to 8 mm, essentially glabrous; twigs tan to gray, often pubescent; flowers white with an overtone of pink, spotted yellow, diameter 3 to 5 cm; fruit a capsule, length ~1.5 cm; Region 3b, New York, Tennessee, Georgia, and Alabama; stream banks, woods, terraces, and swamps.

Rhodoendron canescens (Michaux) Sweet, hoary or sweet azalea (Figure 20.14); stoloniferous shrub, forming colonies, height to 5 m; deciduous; leaf arrangement alternate; leaves strongly clustered at twig tips; leaf entire, blade length 2.5 to 7.0 cm, width 1 to 2 cm, narrowly elliptical or oblanceolate, margin strigose, adaxial and abaxial surfaces and veins strigose, velvety; petiole length 4 to 8 mm, tomentose; twigs

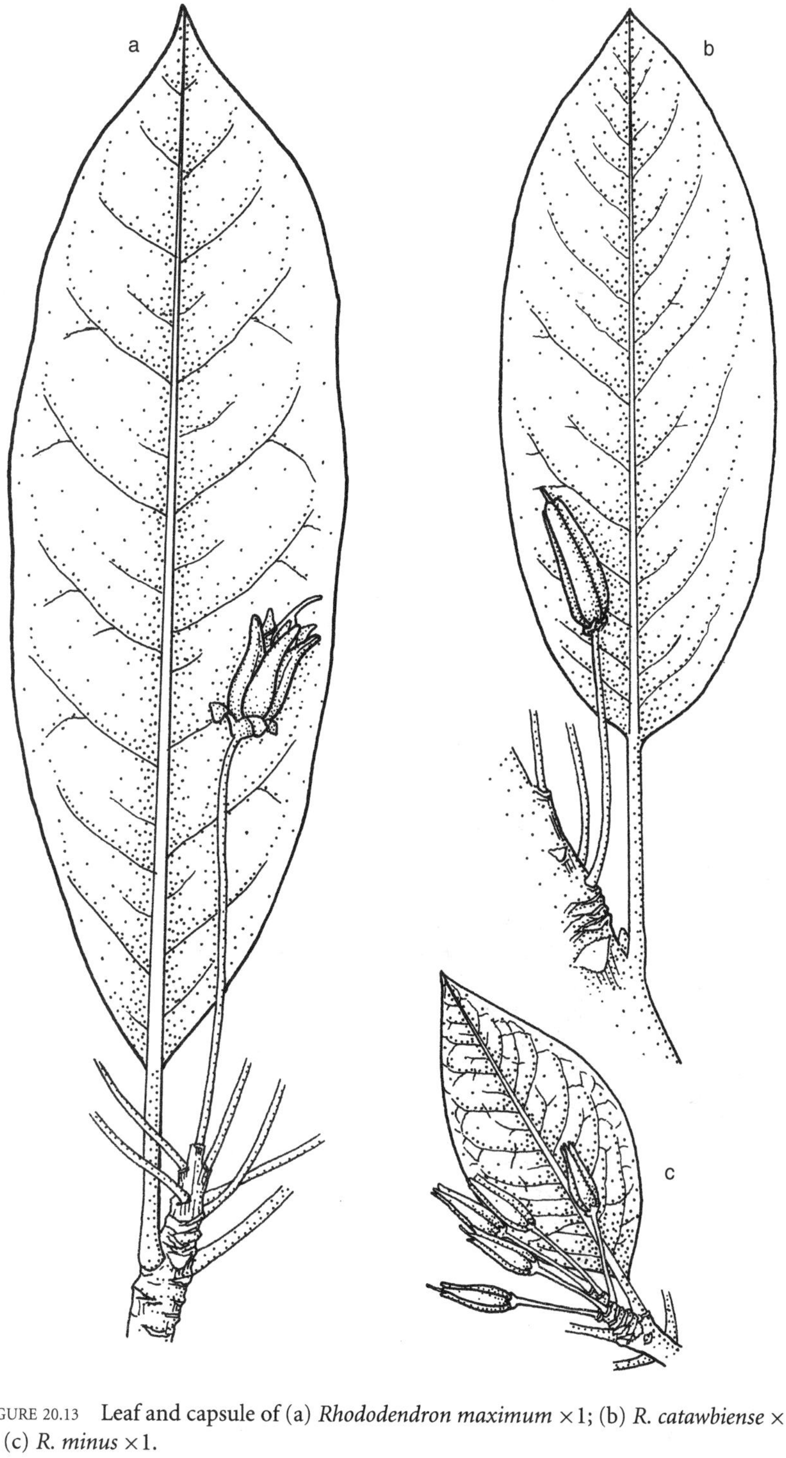

FIGURE 20.13 Leaf and capsule of (a) *Rhododendron maximum* ×1; (b) *R. catawbiense* × 1; (c) *R. minus* ×1.

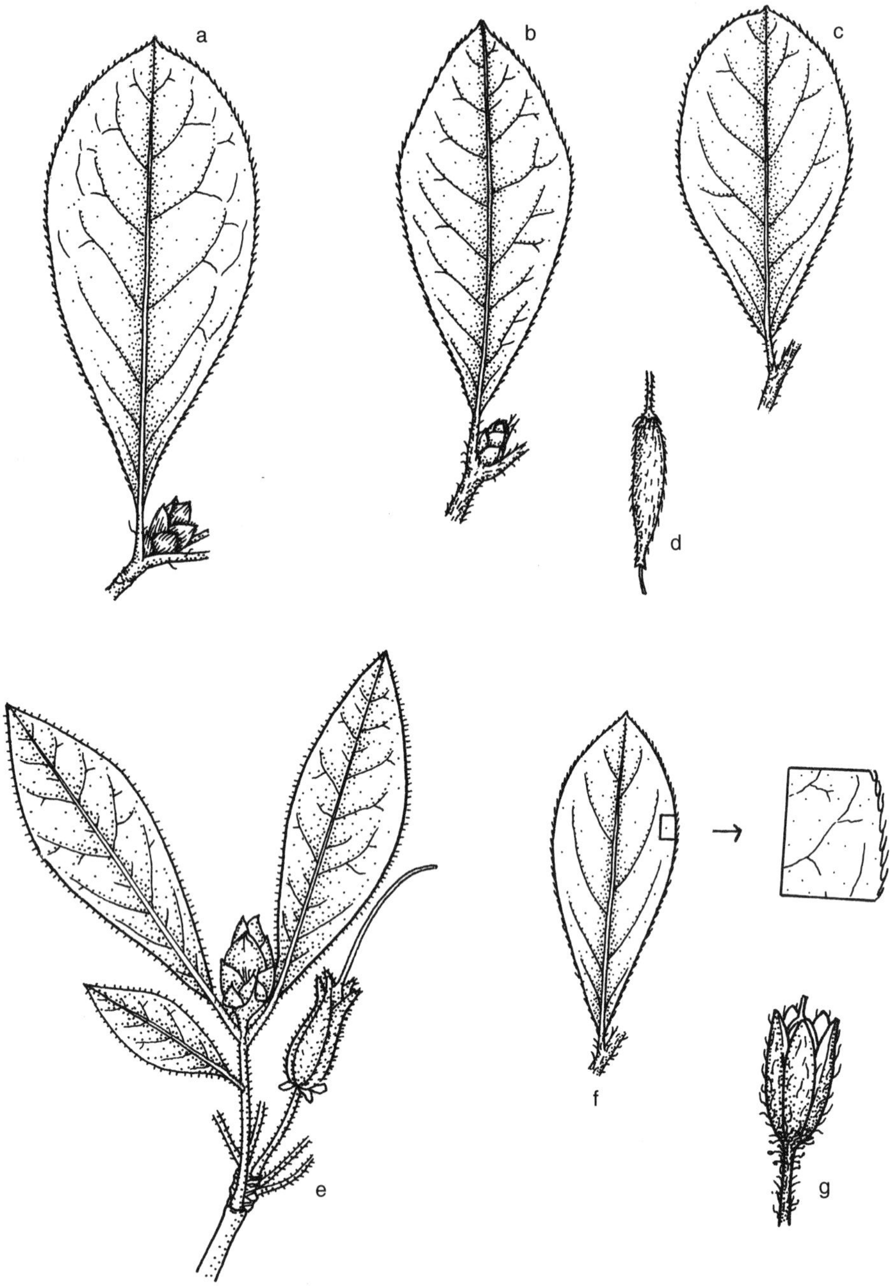

FIGURE 20.14 (a) *Rhododendron arborscens* leaf ×1; *R. canescens* (b), (c) leaves ×1, (d) capsule; (e) *R. calendulaceum* twig with leaves and capsule ×1; *R. viscosum* (f) leaf ×1, (g) capsule ×1.

gray, hairy; flowers white to pink, diameter 3 to 6 cm; fruit a capsule, length ~1 cm; Regions 3b, 4a, and 4c, Virginia and Tennessee south to Florida and Texas, also the Piedmont; mesic and mesic-hydric sites, found in pocosins and savanna. Most abundant of the azaleas.

Rhododendron calendulaceum (Michaux) Torrey, flame azalea (Figure 20.14); nonstoloniferous shrub, height up to 4 m; deciduous; leaf arrangement alternate, leaves strongly clustered at twig tips; leaf entire, blade length 2 to 7 cm, width 1 to 3 cm, elliptical or slightly obovate, margin strigose, adaxial and abaxial veins strigose; petiole 3 to 7 mm, tomentose; twigs gray, often pubescent; flowers yellow, orange, or red, diameter 3 to 6 cm; fruit a capsule, length ~1.5 cm; Region 3b, Appalachian Mountains, Pennsylvania, Tennessee, Kentucky, and North Carolina south to Georgia and Alabama. Common in *Quercus* and *Pinus* communities on mountain heath balds. Cultivated as an ornamental.

Rhododendron viscosum Torrey, swamp azalea, clammy honeysuckle (Figure 20.14); shrub, height to 4 m; deciduous; leaf arrangement alternate, leaves strongly clustered at twig tips; leaf entire, blade length 2 to 6 cm, width 0.7 to 2.0 cm, elliptical, slightly obovate or lanceolate, margin strigose, abaxial surface occasionally whitish, adaxial and abaxial veins strigose; petiole 3 to 5 mm; twigs reddish brown, bearing stiff hair; flowers rose, pink, purple, or red; fruit a capsule, length ~1.5 cm; Region 3b, Maine to Tennessee, Coastal Plain and Piedmont; swamps, stream margins, and wet sandy woods.

The following azaleas generally represent species restricted to the lower elevations of the Coastal Plain province (Regions 4a, b).

Rhododendron atlanticum (Ashe) Rehder, dwarf azalea (Figure 20.15); stoloniferous shrub, forming colonies, height to 7 dm; deciduous; leaf arrangement alternate, leaves strongly clustered at twig tips; leaf entire, blade length 2 to 7 cm, width 1 to 3 cm, elliptical or oblanceolate, strigose, adaxial and abaxial veins strigose; petiole 3 to 5 mm, tomentose; twigs gray, often stiff, hairy; flowers white to purple, diameter 3 to 5 cm; fruit a capsule, length ~1.5 cm; Regions 4a, b, Atlantic Coastal Plain; pine barrens and xeric-mesic woods.

Rhododendron prunifolium (Small) Millais, red or plumleaf azalea; shrub, height to 3 m; deciduous; leaf arrangement alternate, leaves strongly clustered at twig tips; leaf entire, blade length 1.5 to 6.0 cm, width 1 to 3 cm, elliptical or obovate, margin strigose, adaxial and abaxial veins strigose; petiole 3 to 5 mm, tomentose; twigs gray, glabrous; flowers red, diameter 3 to 5 cm; fruit a capsule, length ~1.5 cm; Regions 4a, b, c, Piedmont and Coastal Plain, Georgia and Alabama; drainages and stream banks.

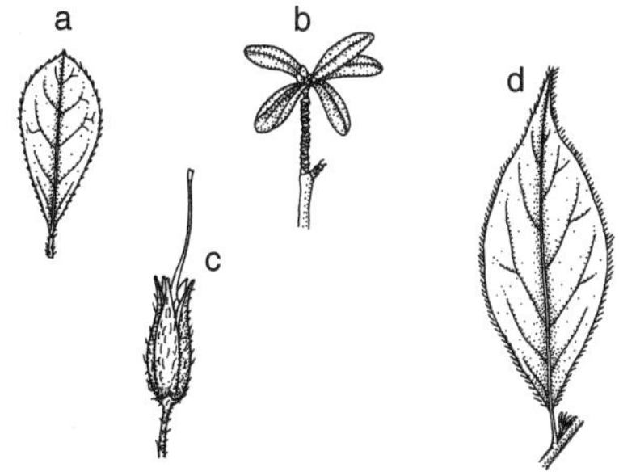

FIGURE 20.15 (a) *Rhododendron atlanticum* leaf ×1/3; (b) *R. lapponicum* leaves ×1/3; (c) *R. periclymenoides* capsule ×1/2; (d) *R. roseum* leaf ×1/2.

Rhododendron serrulatum (Small) Millais, hammock-sweet azalea; shrub, height to 7 m; deciduous; leaf arrangement alternate, leaves strongly clustered at twig tips; leaf entire, blade length 3 to 8 cm, width 1.5 to 4.0 cm, elliptical or slightly obovate, margin strigose, adaxial and abaxial midribs bordered by hair or entirely densely pubescent; petiole 3 to 5 mm, tomentose; twigs reddish brown, bearing stiff hair; flowers red, diameter 3 to 5 cm; fruit a capsule, length ~1.5 cm; Regions 4a, b, Gulf and Atlantic Coastal Plains; swamps, bogs, and ponds.

Other common ericaceous shrubs or small trees that range outside the Appalachian Mountains include the following:

Rhododendron lapponicum (Linnaeus) Wahlenberg; Lapland rose-bay, Lapland rhododendron (Figure 20.15); prostrate, sprawling shrub, infrequently attaining a height of 20 to 30 cm, may form loose mats; evergreen; aromatic; leaf arrangement alternate; leaf coriaceous, blade length 1 to 2 cm, adaxial and abaxial blade surfaces spotted with rust scales; flowers deep purple, diameter 1 to 2 cm; Regions 1 and 2; arctic zone and south, Alaska east to Newfoundland, Quebec, Maine, New Hampshire, and New York into south-central Wisconsin; high elevations in New England and New York.

Rhododendron periclymenoides (Michaux) Shinners (*R. nudiflorum*, pink azalea, pinxter flower (Figure 20.15); stoloniferous shrub, height to 3 m; deciduous; leaf arrangement alternate, leaves clustered near the tip of twigs; leaf entire, blade length 2.0 to 5.5 cm, width 1 to 3 cm, margin strigose, adaxial and abaxial surface veins strigose, adaxial surface with scattered hair, abaxial surface pubescent to nearly glabrous, at times velvety; Regions 3b and 4c, Massachusetts to Tennessee and South Carolina, west to Ohio; mesic woods and swamps.

Rhododendron roseum (Loiseleur) Rehder, rose-shell azalea, early azalea (Figure 20.15); shrub, height to 3 m; deciduous; leaf arrangement alternate; blade length 2.5 to 7.0 cm, width 1 to 3 cm, margin smooth, ciliated to serrulate, adaxial and abaxial surface veins stiff pubescent, surfaces with scattered hair; twigs gray, pubescent; flowers pink to white; Regions 2b and 3a, b, New England south to Virginia and Tennessee, southwest to southern Illinois and Missouri; xeric-mesic and mesic upland forest.

Kalmia latifolia Linnaeus, mountain laurel (Figure 20.16); shrub or small tree, height to 8 m; leaf arrangement alternate, opposite, or three at a node, leaves clustered near the tip of the twig; leaf simple, blade entire, coriaceous, length 4 to 8 cm, width 1.5 to 2.5, elliptical or oblanceolate, apex obtuse or mucronate, on some individuals rounded, abaxial surface yellowish green, glabrous; twigs terete; flowers pink to nearly white; fruit a compressed capsule; Regions 2, 3 and 4, New Brunswick to southern Ontario south to southern Indiana, western Kentucky and west Tennessee, Florida; xeric, often rocky soil, infrequently swamps, upland *Quercus* and *Pinus*-dominated woods. Poisonous to domestic livestock. Planted as an ornamental. Cover type 26.28, 27.13, 27.22, 27.26, 27.32, 27.34, 27.36, 27.38, 28.5, 28.6, 28.11, 28.27.

Kalmia angustifolia Linnaeus, lambkill, sheep laurel, or dwarf laurel (Figure 20.16); shrub, height to 1.5 m; evergreen; leaf arrangement opposite or three at a node; leaf simple, blade entire, coriaceous, length 2.5 to 5.0 cm, width 0.5 to 1.7 cm, glabrous; twigs terete; flowers pink, rose, or purple; fruit a compressed capsule; Regions 1, 2, and 3b, Labrador, Newfoundland, and Nova Scotia to Quebec and Manitoba, south into New England, Pennsylvania, and Michigan, Appalachian Mountains to Georgia; xeric and hydric, acid, low-nutrient soils, drainages, pocosins. In eastern Canada, this species is relatively aggressive and forms

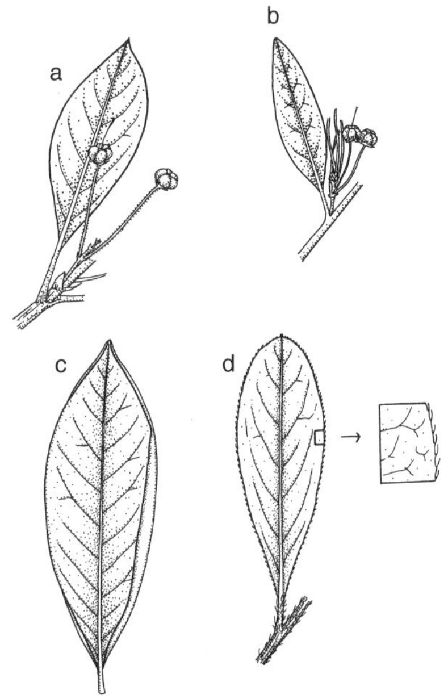

FIGURE 20.16 (a) *Kalmia latifolia* twig with leaf and compressed capsules ×1/2; (b) *K. angustifolia* twig with leaf and compressed capsule ×1/2; (c) *R. macrophyllum* leaf ×1/3; (d) *R. occidentale* leaf ×1/2.

a heath where the forest has been recently cut on moist sand and organic soil. The slow regrowth of forest has been related to the development of this heath community. Poisonous to domestic livestock. Cover type 26.8.

Several species of *Rhododendron* (including azalea) are found in California and the Pacific Northwest.

Rhododendron macrophyllum D. Don, California rosebay; coast rhododendron (Figure 20.16); shrub, height to 8 m; evergreen; leaf arrangement alternate, leaves clustered at the tip of twigs; leaf coriaceous, simple, blade length 7 to 15 cm, width 2.5 to 5.0 cm, slightly revolute, margin smooth, lower surface rusty; petiole length 1.0 to 1.5 cm; twig stout, reddish brown turning dark; flowers white, diameter 3 to 4 cm; Region 7b, west coast forest; lower elevations, to 1100 m.

Rhododendron occidentale (Torrey & Gray) Gray, western azalea (Figure 20.16); shrub, height up to 3 m; deciduous; leaf arrangement alternate, leaves clustered at the tip of twigs; leaf simple, blade entire, length 4.5 to 8.0 cm, width 1 to 3 cm, elliptical, ovate, or slightly oblanceolate, margin strigose, margin bearing short glandular-tipped hair, adaxial surface and abaxial surfaces glabrous or with scattered short whitish hair, abaxial veins strigose; petiole rusty pubescent, length ~5 mm; twig reddish brown turning dark gray, bearing long rust-colored hair; flowers white, diameter 6 to 8 cm; Region 7b, the Sierra Nevada; stream banks and canyons, elevation 1200 to 2500 m.

Rhododendron albiflorum Hooker, Cascades azalea; shrub, height to 1 m; deciduous; leaf arrangement alternate, leaves clustered at the tip of twigs; leaf simple, blade entire, length 2.5 to 6.0 cm, width 1.5 to 3.0 cm, elliptical, margin smooth to irregularly minutely wavy, bearing short glandular-tipped hair, adaxial and abaxial surfaces with scattered long rust-colored hair, veins rust colored; nearly sessile; twig grayish, bearing long rust-colored hair; flowers white, diameter 2 to 3 cm; Region 7a, Pacific Northwest; elevation 1100 to 1800 m.

Ledum glandulosum, Nuttall, Rocky Mountain Labrador tea, trapper's tea (Figure 20.17); shrub, height to 1 m; evergreen; leaf arrangement alternate; blade entire, coriaceous, oval to oblong, length 3 to 6 cm, width 0.3 to 1.0 cm, revolute, abaxial surface whitish but with black-dotted, minute resin droplets appearing as yellow or orange dots; inflorescence a short-terminal raceme; Regions 6 and 7, Wyoming, Montana, Idaho, Washington, Oregon, California; high altitude near timberline. Cover type 29.11, 30.21. Toxic to livestock.

The Bog Community

A number of Ericaceae species are found in the lowland bogs and swamps of Canada (Region 1), the northern and eastern United States (Region 2), and a few scattered bogs in the Central States (Region 3).

Andromeda glaucophylla Link, bog rosemary (Figure 20.17); shrub, height to 70 cm; evergreen; leaf arrangement alternate; blade simple, entire, linear to narrow oblong or elliptical, length 2 to 5 cm, width 0.3 to 1.0 cm, margin revolute, lower surface densely pubescent, whitish; twigs reddish brown, glabrous; flowers pink or white; Regions 1 and 2, Manitoba to Labrador and Greenland, south into Minnesota, Wisconsin, and Michigan; hydric bogs, floating mats, and open swamps. Cover type 26.3.

Andromeda polifolia Linnaeus, wild rosemary shrub, height to 1 m; evergreen; leaf arrangement alternate; blade simple, entire, linear to narrow oblong, length 1 to 2 cm, width 0.2 to 4.0 mm, margin revolute, lower surface glabrous, whitish; twigs reddish brown, glabrous; flowers pink; Region 1 and 2, arctic and boreal forest, south into Minnesota, Wisconsin, and Michigan, New England. Toxic to livestock.

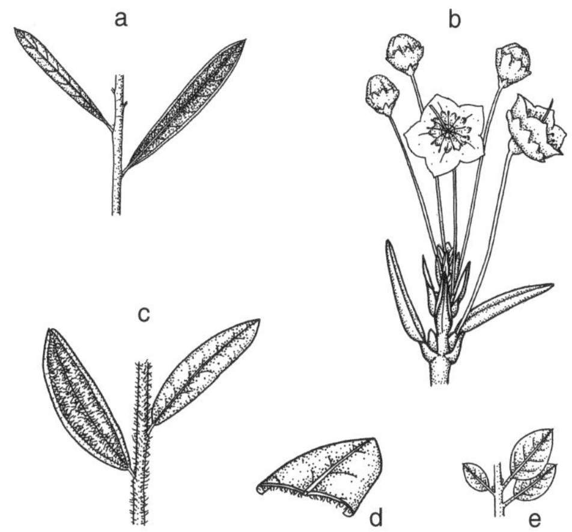

FIGURE 20.17 *Andromeda glaucophylla* (a) twig and leaves ×1/2, (b) capsules and flowers of fused petals ×1/2; *Ledum groenlandicum* (c) twig with leaves ×1/2, (d) leaf cross section showing revolute margin ×3/4; (e) *L. glandulosum* leaves ×1/4.

Ledum groenlandicum Oeder, Labrador tea (Figure 20.17); shrub, height to 1 m; evergreen; leaf arrangement alternate; blade entire, coriaceous, linear to oblong, length 2.0 to 5 cm, width 0.5 to 1.7 cm, revolute, abaxial surface densely rusty tomentose, white tomentose on new leaves; twig tomentose; flowers white; Regions 1 and 2, Greenland and Labrador to Washington, British Columbia, and Alaska, south into Minnesota, Wisconsin, and Michigan, Pennsylvania and New England; hydric bogs, floating mats, and open swamps. Cover type 26.3, 26.16.

Kalmia polifolia Wangenheim, swamp laurel (Figure 20.18); spreading shrub, height up to 60 cm; evergreen; leaf arrangement opposite or three at a node, nearly sessile; blade entire, length 1.0 to 3.5 cm, width 2 to 4 mm, margin revolute, abaxial surface white, glabrous; twigs ridged or two-edged; flowers rose or purple; Regions 1, 2, and 6, Alaska east to Quebec, Newfoundland, and Nova Scotia, south to

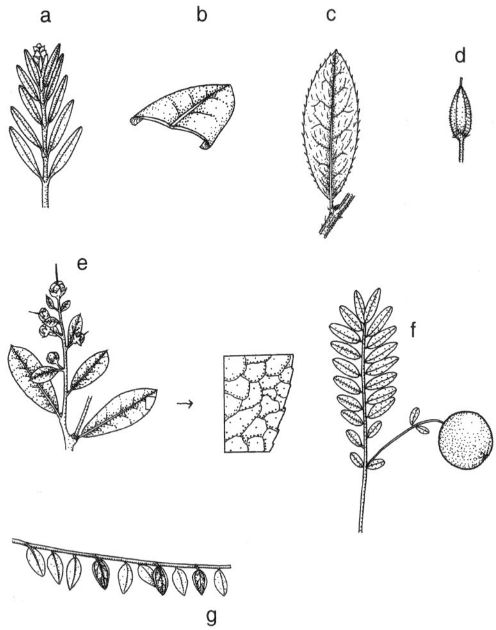

FIGURE 20.18 *Kalmia polifolia* (a) twig and leaves × 1/2, (b) leaf cross section showing revolute margin ×1; *Rhododendron canadense* (c) leaf ×1/2, (d) capsule ×1/2; (e) *Chamaedaphne calyculata* twig with leaves ×1/2; (f) *Vaccinium macrocarpon* twig with leaves and berry (cranberry) ×1/2; (g) *V. oxycoccos* twig with leaves ×1/2.

Maine, New Hampshire, New York, and Pennsylvania, Wisconsin, Minnesota, and Idaho; hydric acid bogs and swamps, sphagnum substrates.

Rhododendron canadense (Linnaeus) Torrey, rhodora (Figure 20.18); shrub, height up to 1 m; deciduous; leaf arrangement alternate; blade entire, coriaceous, oblong, leaf length 2 to 6 cm, width 0.7 to 1.7 cm, margin revolute and ciliated, apex obtuse, adaxial surface sparingly rusty or gray tomentose, abaxial surface sparingly rusty tomentose, strigose along the veins; twig glaucous, sparingly rusty pubescent; flowers pale to deep rose or white; Regions 1 and 2, Labrador and Newfoundland to southern Quebec south to New York, Pennsylvania, and New Jersey; rocky barrens and cold wet bogs. Apparently circumpolar, found in Europe and Asia.

Chamaedaphne calyculata (Linnaeus) Muenchhausen, leather-leaf (Figure 20.18); shrub, height to 1 m; evergreen; leaf arrangement alternate; blade entire, coriaceous, elliptical to oblong, length 1.5 to 4.0 cm, width 0.5 to 1.0 cm, slightly revolute, adaxial and abaxial surfaces dotted with white or rusty scales; flowers white; Regions 1 and 2, Labrador to British Columbia, south into Minnesota, Wisconsin, and Michigan, and New England south to South Carolina; hydric bogs, floating mats, and open swamps. Cover type 26.3.

Vaccinium macrocarpon Aiton, large cranberry (Figure 20.18); creeping shrub, stem length to 1 m; evergreen; leaf arrangement alternate, appearing on both lateral sides of branchlet; leaf coriaceous, blade entire, length 3 to 8 mm, width 1 to 3 mm, ovate or oblong, revolute, abaxial surface strongly whitened or glaucous; petiole 3 to 5 mm; twigs gray, glabrous; flowers pink, four-lobed, diameter 3 to 5 cm; fruit a berry, diameter 0.8 to 1.3 cm; Regions 1, 2, 3c, and 6, Labrador and Nova Scotia to Minnesota, British Columbia, and Oregon, south to Michigan and North Carolina; hydric acid bogs and swamps, sphagnum substrate.

Vaccinium oxycoccos Linnaeus, small cranberry (Figure 20.18); creeping shrub, stem length to 45 cm; evergreen; leaf arrangement alternate, but frequently restricted to one side of branchlet; leaf coriaceous, blade entire, length 0.6 to 1.8 cm, width 2 to 7 mm, elliptical or slightly oblong, slightly revolute, abaxial surface slightly whitened; petiole 3 to 5 mm; twigs gray, glabrous; flowers pink, four-lobed, diameter 3 to 5 cm; fruit a berry, diameter 1 to 2 cm; Regions 1, 2, and 3c, Labrador and Nova Scotia to Minnesota, south to Illinois and North Carolina; hydric acid bogs and swamps, sphagnum substrate.

Bogs (open water) and black spruce/sphagnum swamps have an extremely stressful environment. They occupy the lowest position in the landscape and thus have a high water level, usually within a few decimeters of the bog surface throughout the year. Because of their low position in the landscape, cold air drainage not only shortens their growing season but may produce freezing temperatures during the summer months, when shoot tissue is easily damaged. Because of the high water level and insulating effect of the sphagnum moss above the water, the water may stay frozen into June.

The raw organic "soil" near the center of a bog is developed from various species of sphagnum moss that have accumulated over centuries, usually since glacier melt. Usually, the moss has grown into "cushions", which rise above the water level. Sphagnum moss creates its own environment by excreting hydrogen ions that reduce pH to the range of about 2.8 to 4.0, which is strongly acidic by any standard. In addition, organic (fulvic, humic) acids formed during the decay process and tannin from bark contribute to lowering the pH as well as to the "root beer" color of the water. The low pH, oxygen level, and temperatures reduce the decay process to a minimum. The result is a filling in of the bog and a raising of the forest floor because of the continued buildup of organic matter.

Bogs are low in available nutrients for several reasons. Much of the nutrient capital is tied up in organic matter and living plants, and sphagnum absorbs cations such as calcium directly from the water in exchange for hydrogen ions. Moreover, sphagnum bogs are characteristic of the boreal forest (Regions 1 and 2) and the zone of glaciation. The initial material left by glaciers was highly sandy and low in nutrients, so bogs typically began with a low-nutrient status.

It is in this environment created by the glacier, landscape, and sphagnum moss that ericaceous plants grow. It is of interest that the ericaceous shrubs have thick, coriaceous leaves, some with dense hair on the lower surface where the stomata are located and some with revolute edges that may suggest a xerophytic environment. These adaptations have been linked to the high summer daytime temperatures near the sphagnum surface.

In some bogs and swamps where conditions have been less severe, forest and woody plants have grown and the peat that has developed is a partially decomposed woody peat. It is this material that is packaged and sent to plant nurseries for landscaping purposes.

Genus: *Leucothoe*

The genus *Leucothoe* contains about eight species, four of which are found in the Appalachian Mountains and the Coastal Plain. The other species are found in California, Central America, China, and Japan (Stuart and Sawyer 2001). This genus differs from *Rhododendron* by having small flowers on an upright raceme inflorescence. Although the leaves appear similar in size and shape to those of azalea, the *Leucothoe* lack the stout marginal curved hairs or cilia. There is some debate regarding scientific names.

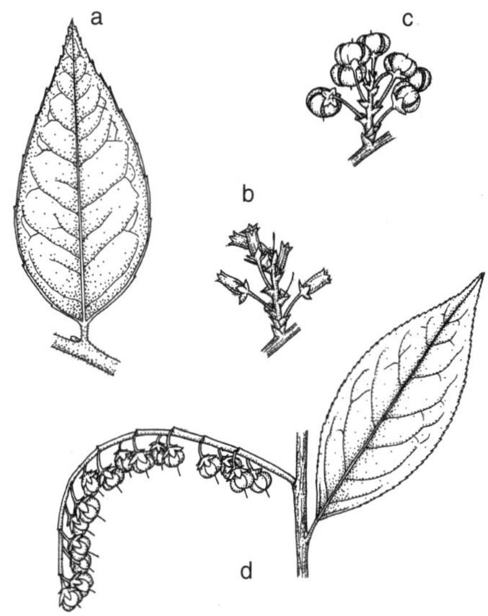

FIGURE 20.19 *Leucothoe axillaris* (a) simple leaf ×1/3, (b) inflorescence showing flowers with fused petals × 1, (c) inflorescence with compressed capsules ×1; (d) *L. recurva* twig with leaf and recurved inflorescence of capsules ×1/3.

Leucothoe axillaris (Lamark) D. Don, swamp dog-laurel, swamp leucothoe, coastal leucothoe (Figure 20.19); shrub, height to 1.5 m; evergreen; leaf arrangement alternate; leaf simple, total leaf length 6 to 13 cm, blade entire, elliptical or slightly ovate, coriaceous, margin serrulate, apex acute, base obtuse, surfaces glabrous; petiole short, length 1 cm, pubescent; twig light reddish brown, pubescent, dull; inflorescence a raceme, axillary, length 1 to 2 cm; flowers 1 to 10 on a stalk, white or slightly pink, petals fused; fruit a small capsule, length 3 to 4 mm, opening along five sutures; Region 4, Atlantic and Gulf Coastal Plain; mesic-hydric woods, swamps and floodplains. Cover type 28.33.

Leucothoe recurva (Buckley) Gray, recurved leucothoe (Figure 20.19); shrub, height to 3 m; deciduous; leaf arrangement alternate; total leaf length 5 to 13 cm, blade simple, entire, elliptical or slightly ovate, margin serrulate, apex acute, base obtuse, adaxial surface hairy along the midrib, abaxial surface hairy along midrib and lateral veins; petiole short, length 5 mm, pubescent; twig light reddish brown, with scattered

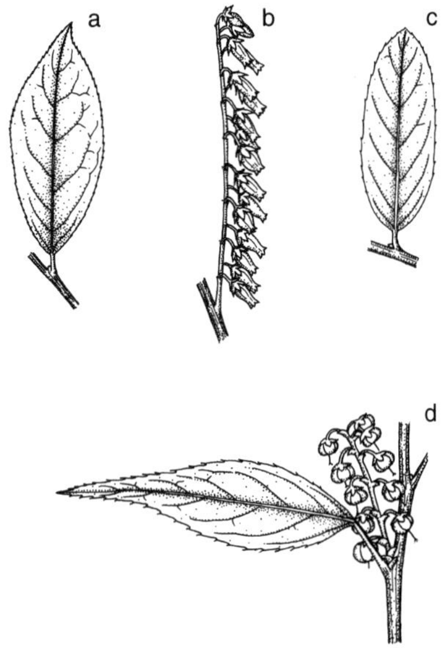

FIGURE 20.20 *Leucothoe racemosa* (a) simple leaf ×1/2, (b) inflorescence of flowers with fused petals ×1/2; (c) *L. davisiae* leaf ×1/2; (d) *L. editorium* twig with leaf and inflorescence with capsules ×2/5.

hair, dull; inflorescence a terminal raceme, basally upright but eventually curved downward, length 4 to 14 cm; flowers 10 to 25, unilateral on a stalk, petals white, fused; fruit a small capsule, length 1 to 3 mm, opening along five sutures; Regions 3b and 4, Appalachian Mountains, West Virginia and Virginia south to Georgia and Alabama; xeric rocky slopes. Cover type 27.22.

Leucothoe racemosa (Linnaeus) Gray, fetter-bush (Figure 20.20); shrub, height to 4 m; deciduous; leaf arrangement alternate; leaf simple, total length 2 to 8 cm, blade entire, elliptical, apex and base acute, margin crenulate to serrulate, upper and lower surfaces lightly pubescent but concentrated along the veins; petiole short, length 1 to 2 mm; twig reddish brown to gray, glabrous; inflorescence a terminal upright raceme, length 4 to 14 cm; flowers 10 to 35 on a stalk, unilateral, petals white, fused, drooping; fruit a small capsule, length 4 to 5 mm, opening along five sutures; Region 4, Atlantic and Gulf Coastal Plains; swamps, bogs, pine flatwoods. Cover type 28.37.

Leucothoe editorium Fernald & Schubert [*L. catesbaei* (Walter) Gray; *L. fontanesiana* (Steudel) Sleumer], dog-hobble (Figure 20.20); height to 2 m; evergreen; leaf arrangement alternate; leaf length 6 to 13 cm, width 1.5 to 3.5 cm, blade entire, margin serrulated, teeth spinulose approaching cilia, apex acuminate, base obtuse to rounded, adaxial and abaxial surfaces lightly hairy, scattered tricomes black tipped (use 10x hand lens); petiole length about 1 cm, angled downward from the plane of the blade; twig reddish brown, pubescent; inflorescence a raceme, axillary or terminal, length 5 to 7 cm; flowers 15 to 40 on a stalk, petals white; fruit a capsule, length 2 to 4 mm, opening along five sutures; Region 3b, Appalachian Mountains, Virginia and Tennessee to Georgia; common in cool mesic wooded mountain coves, forming dense thickets. This species is a southern Appalachian Mountain endemic.

Leucothoe davisiae Torrey, black laurel leucothoe (Figure 20.20); shrub, height to 2 m; evergreen; leaf arrangement alternate; leaf simple, total length 3.5 to 7.0 cm, width 1.8 to 3.0 cm, blade entire, oblong or elliptical, margin serrulate to nearly smooth, apex obtuse or rounded, base obtuse to rounded, adaxial and abaxial surfaces glabrous; petiole length about 2 to 4 mm; twig gray brown, glabrous; inflorescence a terminal raceme, length 6 to 12 cm, peduncle frequently branched; petals white; fruit a capsule, length 2 to 3 cm, opening along five sutures; Region 7, California, the Sierra Nevada; cool mesic soils, riparian zones or lake edges, often rocky sites, elevation 1000 to 2200 m, forms thickets. Cover type 30.27.

Genus: *Lyonia*

Five shrub or small tree species of *Lyonia* are restricted to southeastern United States.

Lyonia ferruginea (Walter) Nuttall, stagger-bush, rusty lyonia (Figure 20.21); shrub or small tree, height 3 to 4 m; evergreen; leaf arrangement alternate; leaf simple, blade coriaceous, entire, oblanceolate or elliptical, length 3 to 7 cm, width 1.0 to 2.5 cm, apex obtuse, base acute; abaxial leaf and twig surfaces rusty scaly, margin minutely revolute; flowers small, each borne on a peduncle from the leaf axil, clustered; fruit a capsule, length 3 to 4 mm, carpels five; Region 4, Coastal Plain,

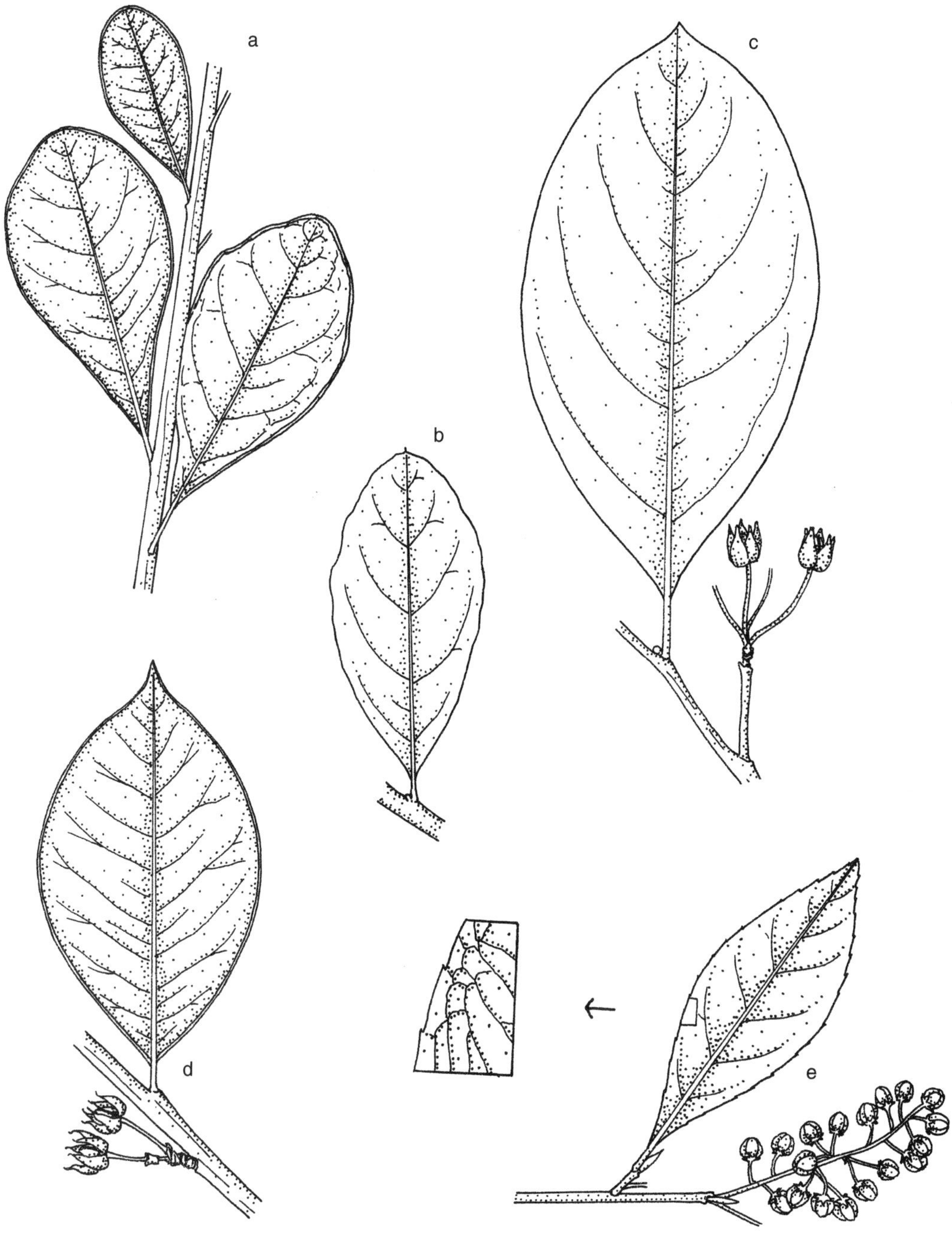

FIGURE 20.21 (a) *Lyonia fruiticosa* twig with leaves ×1; (b) *L. ferruginea* leaf ×1; (c) *L. mariana* twig with leaf and capsules; (d) *L. lucida* twig with leaf and capsules on a spur shoot ×1; (e) *L. ligustrina* twig with leaf and immature capsules ×1.

South Carolina to Florida; xeric sand, *Pinus-Quercus* forest, *Pinus* flatwoods. Cover type 28.44.

Lyonia fruiticosa (Michaux) G. S. Torrey, poorgrub (Figure 20.21); shrub, height to 2 m; evergreen; leaf arrangement alternate, leaf simple, length 1.5 to 5.0 cm, width 0.5 to 2.0 cm, blade entire, elliptical, oblanceolate or obovate, margin minutely revolute, abaxial surface black- and/or rusty-dotted; twig reddish brown to gray, pubescent with grayish white hair, upright, black-dotted; flowers small, borne individually, peduncles clustered on a short shoot; fruit a globose capsule, length 2 to 3 mm; Region 4, Coastal Plain, South Carolina to Florida; bogs. Cover type 28.44.

Lyonia ligustrina (Linnaeus) De Candolle, maleberry (Figure 20.21); shrub, height to 4 m; deciduous; leaf arrangement alternate; leaf simple, length 2 to 7 cm, width 1 to 4 cm, blade entire, oblanceolate or elliptical, apex obtuse, acute, or mucronate, base acute, adaxial surface glabrous or with stout hairs often concentrated along the midrib, abaxial surface bearing stout clear or reddish hairs, margin serrulated; twig reddish brown, glabrous; inflorescence an axillary raceme, length 2 to 4 cm; flowers small; fruit a globose capsule, length 2 to 3 mm, carpels five; Regions 2b, 3, and 4; Maine south to Kentucky, Florida, Louisiana, and Oklahoma to Arkansas, swamps, wet soil, bogs, savanna. Several varieties are recognized.

Lyonia lucida (Lamarck) K. Koch; swamp fetterbush (Figure 20.21); shrub, height to 2 m; evergreen; leaf arrangement alternate; leaf simple, length 3 to 8 cm, width 1.5 to 4.0 cm, blade entire, elliptical, margin minutely revolute, abaxial surface black- and/or red-dotted; twig reddish brown, upright, distinctly three-angled or ridged, black-dotted; flowers small, borne individually, peduncles clustered on a short shoot; fruit a globose capsule, length 2 to 3 mm; Region 4, Coastal Plain, Virginia to Florida and Louisiana; bays and swamps. Cover type 28.024, 28.25, 28.28, 28.30, 28.31, 28.32, 28.33, 28.34, 28.35, 38.36, 38.55.

Lyonia mariana (Linnaeus) D. Don, stagger-bush (Figure 20.21); height to 1.5 m; deciduous; leaf arrangement alternate; leaf simple, length 4 to 8 cm, width 2 to 4 cm, blade entire, broadly elliptical, margin minutely revolute, abaxial surface black- and/or red-dotted; twig tan to gray, stiff, terete, black-dotted; flowers small, borne individually, peduncles clustered on a short shoot; fruit a globose capsule, length 2 to 3 mm; Region 4, Piedmont and Coastal Plain, New England to Florida and Texas; bogs, pond edges, savanna, *Pinus-Quercus*-dominated woods.

Genera: *Gaylussacia* and *Vaccinium*

Gaylussacia (7 North American species) and *Vaccinium* (about 45 species) are small shrubs. Except for *V. arboreum* Marshall (sparkleberry), which reaches a height of 5 to 6 m, most species seldom reach 1 m in height; many are only a few decimeters tall. The leaves are relatively small, usually <5 cm in length. Ecologically, species of these genera provide excellent food (blueberries, huckleberries) for wildlife and humans. They also provide good ground cover and nesting habitat in some areas of the northern Lake States (Region 2), as well as protect the soil and store nutrients on the site. *Vaccinium macrocarpon* and *V. oxycoccos* were described earlier and included with the bog community. In general, the identification of many species is difficult. Only selected species are described here.

Gaylussacia baccata (Wangenheim) K. Koch, black huckleberry (Figure 20.22); multibranched shrub, height to 1 m; deciduous; leaf arrangement alternate; leaf simple, blade entire, oval, elliptical, or slightly lanceolate, length 2.0 to 4.3 cm, width 0.8 to 2.0 cm, apex obtuse, blunt, base acute or obtuse, adaxial and abaxial surfaces yellow-dotted and sticky by minute droplets; branchlets gray brown to nearly black, pubescent; inflorescence a short raceme, length 2 cm; fruit a berry, two to six per raceme, dark blue to black, diameter 0.5 to 1.0 cm; Regions 1, 2, 3, and 4, Newfoundland to Saskatchewan, south to Georgia and Louisiana; xeric upland woods. Cover type 27.17, 27.21, 27.22, 27.32.

Vaccinium myrtilloides Michaux, velvetleaf bilberry (Figure 20.22); dwarf creeping or upright branching shrub, height to 75 cm; deciduous; leaf arrangement alternate; leaf simple, blade entire, elliptical, length 1.5 to 2.5 cm, width 0.5 to 1.0 cm, margin densely ciliated, adaxial and abaxial surfaces pubescent to more or less strigose; branchlet brown, hairy, turning glabrous; fruit a berry, dark blue, diameter 4 to 9 mm; Regions 1, 2, and 3, Newfoundland and Nova Scotia to Saskatchewan and Montana, south into

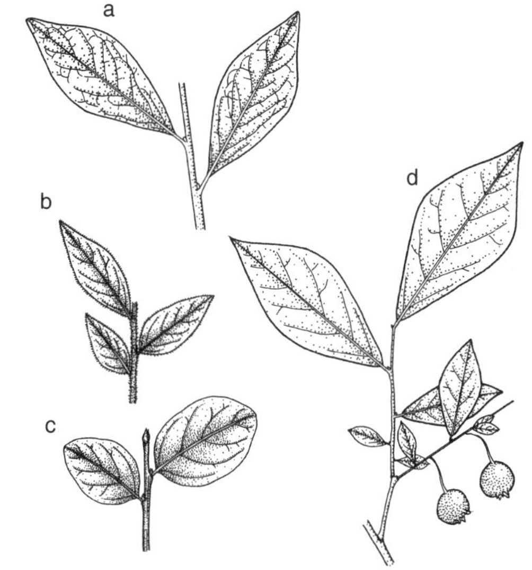

FIGURE 20.22 (a) *Gaylussacia baccata* twig with leaves ×4/5; (b) *Vaccinium myrtilloides* twig with leaves ×2/3; (c) *V. uliginosum* twig with leaves ×1; (d) *V. stamineum* twig with leaves and blueberries ×1/2.

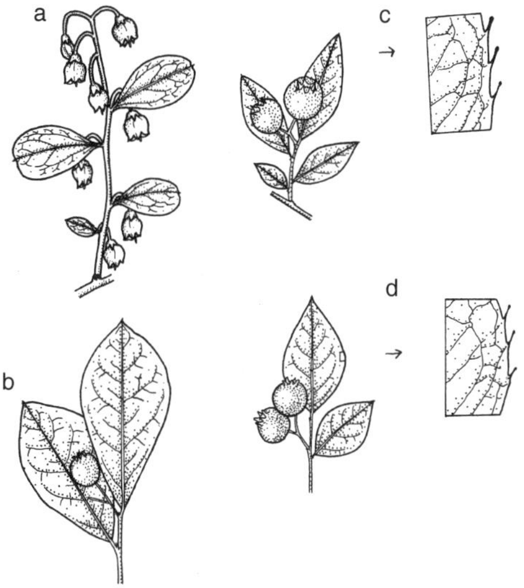

FIGURE 20.23 *Vaccinium arboreum* twigs with (a) leaves and flowers ×1/4, (b) leaves and blueberries ×1/2; (c) *V. angustifolium* leaves and blueberries ×1/2; (d) *V. vacillans* leaves and blueberries ×2/5.

Illinois, Indiana, and Virginia; mesic woods and hydric swamps. Cover type 26.18.

Vaccinium uliginosum Linnaeus, alpine bilberry (Figure 20.22); stoloniferous dwarf shrub, height to 60 cm; deciduous; leaf arrangement alternate; leaf simple, blade entire, ovate or oblanceolate approaching orbicular, length 0.5 to 1.5 cm, width 4 to 9 mm, margin smooth, adaxial and abaxial surfaces and veins pubescent; branchlet green, turning darker, minutely pubescent; fruit a berry, dark blue, diameter 3 to 7 mm; Regions 1 and 2, Labrador and New England to Alaska, into Wisconsin and Minnesota; arctic region, circumpolar, rocky sites and barrens, mountain summits.

Vaccinium stamineum Linnaeus, deerberry (Figure 20.22); multibranched shrub, height to 2 m; deciduous; leaf arrangement alternate; leaf simple, blade entire, long elliptical or slightly lanceolate, length 2.5 to 6.5 cm, width 1.0 to 2.3 cm, glabrous; twig brown, turning darker, generally glabrous; fruit a berry, dark blue, diameter 5 to 8 mm; Regions 2b, 3a, and 4, Massachusetts and southern Ontario to Indiana and Missouri, south to the Gulf Coast. Several varieties are recognized (Fernald 1950). Cover type 27.17, 27.22, 28.6.

Vaccinium arboreum Marshall; sparkleberry, tree blueberry, farkleberry (Figure 20.23); multibranched shrub, height to 7 m; deciduous; leaf arrangement alternate; leaf simple, blade entire, long elliptical or slightly lanceolate, length 2.5 to 6.5 cm, width 1.0 to 2.3 cm, glabrous, slightly coriaceous, margin smooth but with an occasional gland, infrequently strigose, adaxial surface dark green and shiny but midrib often pubescent/strigose, abaxial surface glabrous pubescent/strigose; twig gray, generally glabrous; fruit a berry, dark blue, diameter 3 to 5 mm; Regions 3 and 4; Missouri and southern Illinois to Virginia, south to the Gulf Coast, Florida, Texas, and Oklahoma; extremely xeric and xeric woodlands. This species is the primary midcanopy component of *Quercus stellata-Q. marilandica*-dominated forest in the western portion of the Central States. Cover type 27.17, 27.24, 28.2.

Vaccinium angustifolium Aiton, late sweet blueberry (Figure 20.23); stoloniferous shrub, height to 40 cm; deciduous; leaf arrangement alternate; leaf simple, blade entire, elliptical, length 1 to 3 cm, width

0.4 to 1.0 cm, glabrous, margin serrulated, minute teeth glandular tipped, adaxial and abaxial surfaces glabrous, veins pubescent/strigose; branchlet green, turning darker, generally glabrous; fruit a berry, dark blue, diameter 4 to 6 mm; Regions 1 and 2, Labrador, Newfoundland, New England, New York, and Quebec to Minnesota, Illinois, and Saskatchewan, south to Virginia; xeric sands and rock sites, mountains and hydric bogs, common component of open pine sand barrens in the Lake States. Cover type 26.9, 26.18.

Vaccinium vacillans Torrey, early sweet blueberry (Figure 20.23); stoloniferous shrub, height to 90 cm; deciduous; leaf arrangement alternate; leaf simple, blade entire, oval, elliptical, or obovate, length 2 to 5 cm, width 1.3 to 3.5 cm, glabrous, margin serrulated, minute teeth glandular tipped, adaxial and abaxial surfaces glabrous, veins pubescent/strigose; branchlet green, turning darker, generally glabrous; fruit a berry, dark blue, diameter 5 to 8 mm; Region 3, found primarily on dry and/or sandy soil of upland *Quercus*-dominated forest and savanna of the Central States. This species also ranges into Region 2, Lake States, southern Canada, and New England. Cover type 27.33, 28.2.

Vaccinium crassifolium Andrews, creeping blueberry (Figure 20.24); dwarf creeping or prostrate branching shrub, length to 2 m; evergreen; leaf arrangement alternate; leaf simple, blade entire, coriaceous, elliptical, length 0.5 to 1.5 cm, width 4 to 8 mm, margin serrulate, minutely revolute, adaxial and abaxial surfaces and veins glabrous or minutely pubescent; branchlets green, turning darker, minutely pubescent; fruit a berry, dark blue or black, diameter 3 to 7 mm; Region 4, Virginia to South Carolina; xeric sand to hydric sites, pine barrens and pocosins of the southeast.

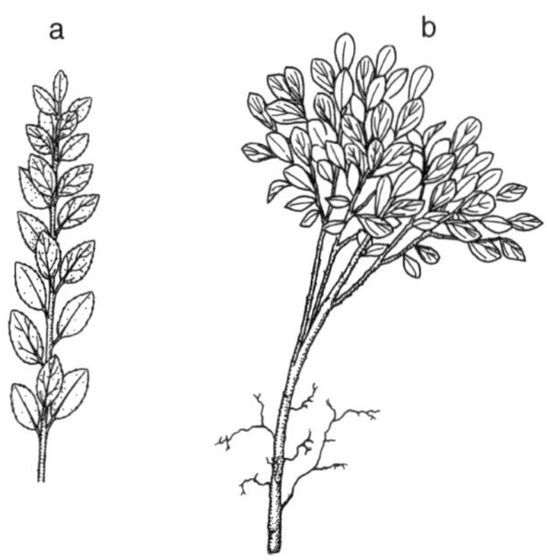

FIGURE 20.24 (a) *Vaccinium crassifolium* twig with leaves ×1/2; (b) *V. cespitosum* plant ×1/4.

Vaccinium cespitosum Michaux, dwarf bilberry (Figure 20.24); stoloniferous dwarf shrub, height to 30 cm; deciduous; leaf arrangement alternate; leaf simple, blade entire, obovate or oblanceolate, length 0.5 to 2.3 cm, width 3 to 8 mm, margin distinctly serrulated, minute teeth glandular tipped, adaxial and abaxial surfaces and veins glabrous; branchlet green, turning darker, minutely pubescent; fruit a berry, dark blue, diameter 3 to 6 mm; Regions 1, 2, 5, 6, and 7, Labrador and New England to Alaska, south to Colorado and California, into Wisconsin and Minnesota; mountain alpine to low-elevation open woods and lakeshores. Circumpolar. Cover type 29.28, 29.11.

Additional notes on species characteristics and distributions of *Vaccinium* in eastern Unites States are found in Fernald (1950), United States Department of Agriculture (1982), Foote and Jones (1989), Gleason and Cronquist (1991), and Nelson (1994). Approximately 13 species occur in the western mountain regions (Bailey and Bailey 1949).

Genus: *Oxydendrum*

Only one *Oxydendrum* species is known in the United States (monotypic). The leaf tissues of this species contain oxalic acid, which can be determined by chewing on a leaf and noting a momentary bitter taste.

Oxydendrum arboreum (Linnaeus) de Candolle, sourwood (Figure 20.25); medium-sized tree, height to 17 m; deciduous; leaf arrangement alternate; leaf simple, blade entire, length 8 to 19 cm, width 3 to 6 cm, oblong or infrequently slightly lanceolate, margin serrate to serrulate, apex acuminate, base obtuse, surfaces glabrous; petiole with widely scattered hair, length 0.5 to 2.0 cm; twigs reddish-brown; inflorescence a branched raceme; flowers small; fruit a capsule, length 3 to 6 mm, reflexed at maturity; Regions 3 and 4, West Virginia and Pennsylvania to Indiana and western Kentucky, south to Louisiana, southwest to Florida; xeric, rocky soils on slopes and ridges. A common component of *Quercus prinus*- or *Pinus*

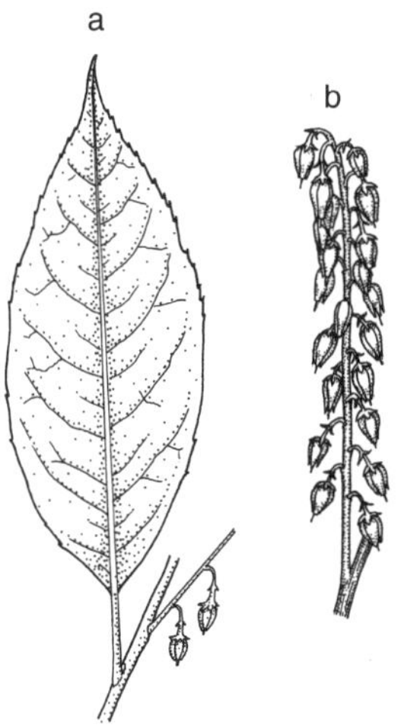

FIGURE 20.25 *Oxydendrum arboreum* (a) leaf ×1/3, (b) inflorescence with capsules ×1/4.

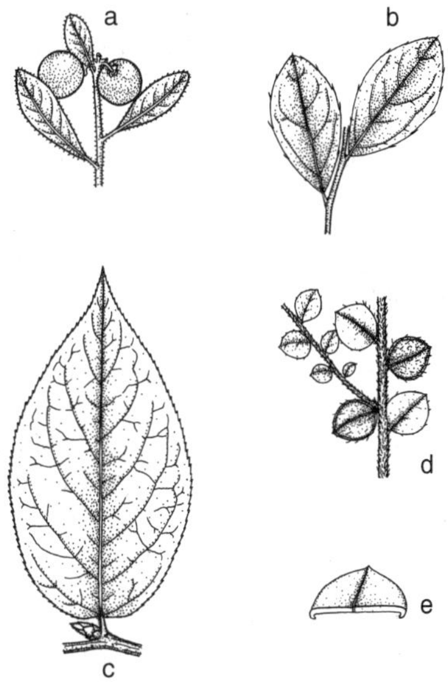

FIGURE 20.26 (a) *Arctostaphylos uva-ursi* twig with leaves and berries ×3/4; (b) *Gaultheria procumbens* leaves ×1/2; (c) *G. shallon* leaf ×2/3; *G. hispida* (d) twig with leaves ×1, (e) leaf cross section showing revolute margin ×2.

rigida-dominated forest. Cover type 27.13, 27.17, 27.22, 27.24, 27.26, 27.34, 28.6, 28.14, 28.16, 28.33.

Genus: *Arctostaphylos*

About 52 *Arctostaphylos* species have been described (United States Department of Agriculture 1982); most have a western distribution, primarily in California. Many are probably varieties of the most widespread and best-known species, which is described here.

Arctostaphylos uva-ursi (Linnaeus) Sprengel, bearberry, kinnikinnick (Figure 20.26); dwarf creeping or prostrate branching shrub, ascending to 15 cm; length to 50 cm; evergreen; leaf arrangement alternate; leaf simple, blade entire, coriaceous, obovate to oblanceolate, length 1.0 to 2.5 cm, width 0.4 to 1.2 cm, margin smooth, ciliated, apex obtuse or rounded, base acute, adaxial and abaxial surfaces glabrous, shiny; branchlet reddish brown, minutely pubescent, branch dark reddish brown to black, exfoliating; fruit a berry, red, diameter 0.5 to 1.0 cm; Regions 1, 3, 5, and 6, distributed from Labrador and Michigan to Alaska to Washington, California, and New Mexico. This widespread prostrate shrub forms dense mats on extremely dry sandy soils and rock outcrops. It is circumpolar. Cover type 26.2, 29.27, 29.30, 29.15, 29.6, 30.14, 30.26.

Genus: *Gaultheria*

The genus *Gaulteria* contains about 200 species in Asia, Australia, and South America. Five species are found in North America.

Gaultheria procumbens, wintergreen, teaberry (Figure 20.26); dwarf creeping or prostrate branching shrub, length to 50 cm, branchlets erect up to 10 cm; evergreen; leaf arrangement alternate; leaf simple, blade entire, coriaceous, oval or broadly elliptical, length 1.5 to 4.0 cm, width 1.0 to 2.5 cm, margin serrulated, apex rounded, base rounded or cordate, adaxial surface glabrous, dark green, abaxial surface glabrous, strong wintergreen odor/taste; petiole length 1 to 2 cm, glabrous; flowers small, two to four appearing from the leaf axil, flower length 1 cm; fruit a berry, red, diameter ~1 cm; Regions 1, 2, 3, and 4, Newfoundland to Minnesota and Manitoba, Lake States, New England south to Georgia and Alabama; xeric sandy soil, open woods, often under conifers. Pleasant chewing. Cover type 26.2, 26.16, 26.27.

Gaultheria hispidula (Linnaeus) Bigelow, creeping snowberry (Figure 20.26); dwarf creeping or prostrate multibranching shrub, length to 50 cm; evergreen; leaf arrangement alternate; leaf simple, blade entire, coriaceous, oval or broadly elliptical, length 4 to 8 mm, width 2 to 5 mm, margin revolute and sparingly ciliated, three to five tricomes per side, apex obtuse, base obtuse, adaxial surface glabrous, dark green, abaxial surface with scattered brown, scale-like oppressed hair, wintergreen odor/taste; petiole length 1 to 2 mm, strigose; flowers small, appearing singly from the leaf axil; fruit a berry, bright white, diameter ~6 mm; Regions 1, 2, 3, 4, and 6, Newfoundland to Minnesota, Manitoba, and Idaho, Lake States, New England south to North Carolina; bogs and moss-covered soil, often under conifers.

Gaultheria shallon Pursh, salal (Figure 20.26); shrub, height to 2 m; evergreen; leaf arrangement alternate; leaf simple, blade entire, coriaceous, broadly ovate, slightly oblong, or nearly orbicular, length 3 to 8 cm, width 2.0 to 5.5 cm, margin finely serrated, apex mucronate or acuminate, at times round, base rounded, adaxial and abaxial surfaces glabrous; petiole length 3 to 4 mm; twig glabrous or sparingly pilose; inflorescence a raceme, length 3 to 8 cm; fruit a berry, purple, diameter 6 to 9 mm, bristly; Region 7, coastal, from Alaska to southern California. Cover type 30.1, 30.2, 30.17, 30.18, 30.19, 30.21, 30.22, 30.26, 30.28, 30.41, 30.42.

Genus: *Epigaea*

Epigaea repens Linnaeus; trailing arbutus (Figure 20.27); dwarf creeping or prostrate branching shrub, length to 50 cm; evergreen; leaf arrangement alternate; leaf simple, entire, coriaceous, blade oblong to oval or nearly orbicular, length 2 to 5 cm, width 1.5 to 4.0 cm, dark green, margin densely ciliated, apex rounded, base rounded or cordate, adaxial surface glabrous or sparingly pubescent, shiny, abaxial surface glabrous except along the veins; petiole length 1 to 2 cm, distinctly pilose; branchlets reddish brown, distinctly pilose, branch dark reddish brown to black, exfoliating; flowers appear in small axillary clusters, flower length 1.0 to 1.5 cm; fruit a berry, red, diameter 0.5 to 1.0 cm; Regions 1, 2, 3, and 4, Newfoundland to Minnesota, south to Florida and Mississippi, west to Saskatchewan; xeric sandy soil, open woods, often under conifers. Protected in some states because of overcollecting for gardens. Cover type 26.26.

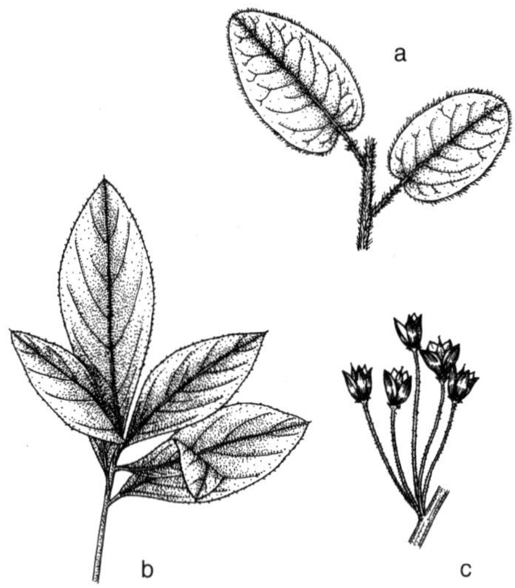

FIGURE 20.27 (a) *Epigaea repens* twig with leaves ×1/2; *Menziesia ferruginea* (b) leaves ×2/3, (c) flowers ×2/3.

Genus: *Menziesia*

Menziesia ferruginea Smith, rusty leaf (Figure 20.27); shrub, height to 5 m; deciduous; leaf arrangement alternate, leaves clustered near the tip of the twig; leaf simple, blade entire, elliptical to slightly oblanceolate, length 2.5 to 6.5 cm, width 1 to 2 cm, margin ciliated, adaxial surface pilose or pubescent, abaxial surface glabrous, occasional hair on the midrib; branchlet tan turning reddish-brown, scattered hair; flowers greenish white or purplish, fruit a capsule, length 5 to 8 mm four to six per axil; Region 7, Washington to Alaska; conifer woods.

Clethraceae (Clethra Family)

Clethraceae is a small family of one genus and about 65 species distributed in tropical and subtropical regions of Asia, the East Indies, and North and South America. These shrubs and small trees have alternate, simple

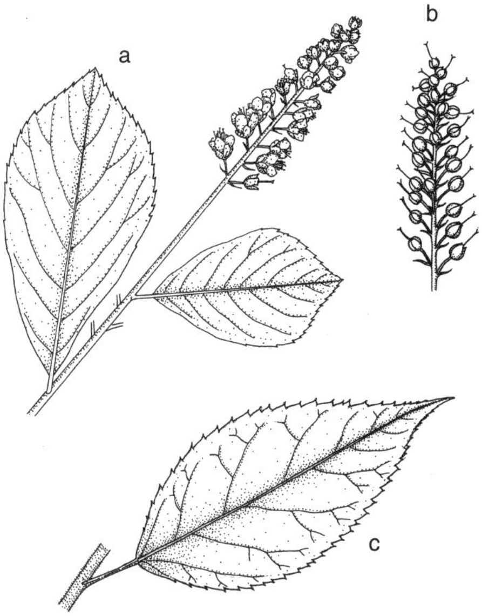

FIGURE 20.28 *Clethra alnifolia* (a) twig with leaves and inflorescence with flowers ×3/5, (b) inflorescence with capsules ×3/5; (c) *C. acuminata leaf* ×1/2.

leaves; the veins are distinctly sunken, giving the leaf a wrinkled appearance. Stipules are absent. The flowers are regular and perfect; plants are synecious. The fruit is a capsule with a pubescent surface. Only the following two species occur in the southeastern United States.

Clethra alnifolia Linnaeus, sweet pepperbush (Figure 20.28); shrub, height to 3 m; deciduous; leaf arrangement alternate; leaf simple, blade entire obovate to oblanceolate, length 4 to 9 cm, width 1.3 to 4.7 cm, margin serrate along the distal two thirds, smooth along the proximal one third, apex obtuse, base cuneate, adaxial surface bearing scattered hair, abaxial surface pubescent or glabrous except for veins; twigs grayish brown, pubescent, dull; inflorescence a raceme, upright, length 5 to 15 cm; flowers white, 15 to 50, exceptionally pleasant odor; fruit a capsule, pubescent, length ~2 mm; Region 4, primarily the Coastal Plain from southern Maine and Massachusetts to Florida and Texas; wet pine savanna, bays, bogs, pocosins, swamp forest. Cover type 28.24, 28.27, 28.28, 28.32, 28.38.

Clethra acuminata Michaux, mountain pepperbush (Figure 20.28); shrub or small tree, height to 5 m; deciduous; leaf arrangement alternate; leaf simple, blade entire, elliptical, oblong or slightly obovate, length 5 to 16 cm, width 2 to 7 cm, margin sharply serrate, apex acuminate, base obtuse or rounded, adaxial surface essentially glabrous, abaxial surface pubescent, densely so along the veins; petiole length 1 to 2 cm; twigs grayish brown, densely pubescent, dull; inflorescence a raceme, upright or bent, pubescent, length 8 to 15 cm; flowers white, 15 to 50, exceptionally pleasant odor; fruit a capsule, pubescent, length ~2 mm; Region 3b, southern Appalachian Mountains, Virginia and Tennessee to Georgia; xeric-mesic rocky forest.

Cyrillaceae (Titi Family)

The Cyrillaceae is an exceptionally small tropical and subtropical family of three genera and 14 species. The two genera of southeastern United States (*Cyrilla, Cliftonia*) are monotypic. The leaves are alternate, simple, and entire.

Cyrilla racemiflora Linnaeus, leatherwood, titi (Figure 20.29); small tree, height to 8 m; semi-evergreen, the leaves remaining green into spring or summer of the following year; leaf arrangement

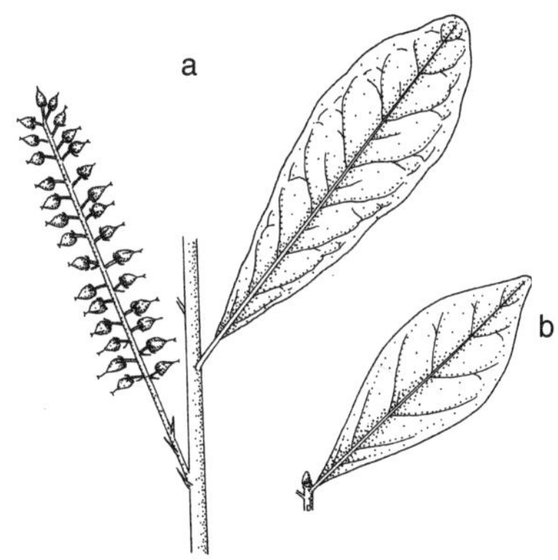

FIGURE 20.29 (a) *Cyrilla racemiflora* twig with leaf and inflorescence of dry drupes ×1/2; (b) *Cliftonia monophylla* leaf ×2/3.

alternate, leaves clustered near the tip of twig; leaf simple, blade entire, oblanceolate, length 3 to 10 cm, width 1.0 to 2.5 cm, margin smooth, minutely revolute, apex obtuse, base acute, adaxial and abaxial surfaces glabrous; inflorescence a long cylindrical raceme, length 5 to 12 cm, attached below the new leaves; fruit a dry drupe, length 2 to 3 mm; Region 4, Coastal Plain, Virginia to Florida to eastern Texas; swamps and wet woods. Cover type 28.11, 28.28, 28.31, 28.36, 28.37, 28.38, 28.53, 28.55.

Cliftonia monophylla (Lamarck) Britton ex. Sargent, titi tree, black titi (Figure 20.29); small tree, height to 8 m; evergreen; leaf arrangement alternate; leaf simple, blade entire, oblanceolate or elliptical, length 3.0 to 5.5 cm, width 1 to 2 cm, sessile, margin minutely revolute, adaxial and abaxial surfaces glabrous, white waxy; inflorescence a terminal raceme, length 2 to 5 cm; fruit dry, length 4 to 5 mm with four longitudinal wings; Region 4, Coastal Plain, Georgia and Florida to eastern Texas; bays, swamps, and pine flatwoods. Cover type 28.23, 28.32, 28.38, 28.55.

FIGURE 20.30 (a) *Bumilia lanuginosa* twig with leaves on a spur shoot and thorn ×3/5; (b) *B. tenax* twig with leaf and thorn ×1/2; (c) *B. reclinata* twig with leaf and berry ×3/4; (d) *B. lycioides* leaf on a thorny twig ×2/3.

ORDER: EBENALES

Sapotaceae (Sapodilla Family)

The Sapodilla family is a pantropical family of as many as 75 genera and 800 species of shrubs and small trees. Twigs are often thorny with zigzag twigs. The leaves are alternate, deciduous, simple, and entire. The sap is milky, the fruit is a berry, and the flowers are sympetalous. Only 7 genera are found in North America, and only *Bumelia* has representatives in temperate and warm temperate forest. The genus name may soon be changed to *Sideroxylon* (Nelson 1996).

Bumelia lanuginosa (Michaux) Pers, gum bumelia, false buckthorn (Figure 20.30); shrub or small tree, height to 10 m; deciduous; leaf arrangment alternate, leaves clustered on spur shoots of older twigs; leaf simple, length 3 to 8 cm, width 1 to 4 cm, blade entire, obovate or oblanceolate, margin smooth, abaxial surface velvety, pubescence dense rusty brown; twigs rusty brown, pubescent, crooked, thorny, thorn length 2 to 8 cm, thorns at the leaf nodes or leaf-bearing twigs becoming pointed; flowers in clusters at the leaf nodes; fruit a berry, fleshy, black, diameter 0.6 to 1.3 cm; Regions 3 and 4; Gulf Coastal Plain, Florida to Texas and north to Missouri, Illinois, and western Kentucky; xeric sandy upland woods.

Bumelia lycioides (Linnaeus) Gaertner f., southern buckthorn (Figure 20.30); medium-sized tree, height 10 to 15 m; semi-evergreen; leaf arrangment alternate; leaf simple, length 8 to 15 cm, width 1 to 5 cm, blade entire, elliptical or lanceolate, margin dark green, glabrous, abaxial surface velvety; twigs glabrous, crooked, thorny, thorn length 2 to 8 cm, thorns at the nodes or leaf-bearing twigs becoming pointed; flowers in clusters at the leaf nodes; fruit a berry, fleshy, black, ovate, length 0.8 to 1.8 cm; Regions 3 and 4, Gulf Coastal Plain north to Illinois and Virginia; moist woods and stream margins.

Bumelia tenax (Linnaeus) Willdenow, toughbumelia (Figure 20.30); shrub or small tree, height to 8 m; deciduous; leaf arrangment alternate; leaf simple, length 2 to 8 cm, width 1 to 3 cm, blade entire, oblanceolate, margin smooth, apex rounded, base acute, adaxial surface dark green, abaxial surface pubsescent, velvety, yellowish to coppery brown; twigs rusty brown, pubescent, crooked, thorny, thorn length 2 to 8 cm, thorns at the leaf nodes or leaf-bearing twigs becoming pointed; flowers white, in clusters at the leaf nodes; fruit a berry, fleshy, black, obovate, diameter 0.9 to 1.3 cm; Region 4, Atlantic Coastal Plain, South Carolina to Florida; coastal dune and interior *Pinus-Quercus* scrub/savanna burned areas.

Bumelia reclinata (Michaux) Ventenat, smooth bumelia (Figure 20.30); height to 5 m; deciduous; leaf arrangment alternate, leaves clustered on spur shoots of older twigs; leaf simple, length 2 to 6 cm, width 0.5 to 2.0 cm, blade entire oblanceolate to spatulare, apex rounded, margin smooth, adaxial surface dark green, glabrous, abaxial surface glabrous; twigs crooked, thorny, thorn length 2 to 8 cm, thorns at the leaf nodes or leaf-bearing twigs becoming pointed; flowers white, in clusters at the leaf nodes; fruit a berry, fleshy, black, oval, diameter 4 to 8 mm; Region 4, Georgia and Florida; upland forest, ravines and stream edges; xeric pine forest.

Ebenaceae (Ebony Family)

The Ebony family has as many as five genera and about 450 species distributed in tropical and subtropical zones. More than 400 species belong to the genus *Diospyros*. Members are shrubs and small trees with alternate, simple, and entire leaves. Stipules are absent. The flowers are regular and usually unisexual; the flower is gamopetalous. Individuals are dioecious. The wood is generally dark, hard, and dense. The family is well known for the beautiful wood of two tropical species, *D. eveneum* and *D. reticulata*, both of which are called ebony. These species have strong wood, denser than that of species found in North America. The primary temperate representative is *Diospyros virginina*.

Diospyros virginiana Linnaeus, persimmon (Figure 20.31); medium-sized tree height to 20 m, deciduous; leaf simple, blade entire, elliptical, length 7 to 15 cm, somewhat thick and leathery, dark green. Upper surface with yellowish midrib, margin smooth; twigs gray, buds short, length ~3 mm, dark chocolate, two visible imbricate scales; fruit a berry, globose, diameter 2 to 5 cm, green when immature, the juice producing a strong astringent sensation on oral tissue,

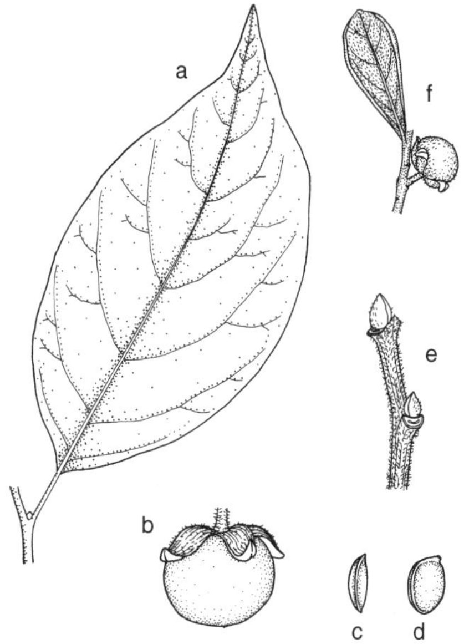

FIGURE 20.31 *Diospyros virginiana* (a) leaf ×2/3, (b) mature berry ×1/2, (c) side view of seed ×1/2, (d) front view of seed ×1/2, (e) twig with lateral and false terminal bud; (f) *D. texana* twig with leaf and berry.

orange and soft when ripe, the pulp used to flavor cakes and cookies, two to four flat seeds, seed length ~1.0 to 1.5 cm; bark of larger stems distinctly blocky; Regions 3 and 4, distributed over most of the Central Hardwood and the Southern Pine-Hardwood Forest regions; xeric to mesic sites, appears most frequently in recently disturbed areas, probably because birds transport the seed to these areas. The hard, dark wood of this species was once important for making golf club heads and other specialty items. A closely related species, *D. texana* Scheele, Texas persimmon, grows from central Texas south into northern Mexico (Figure 20.21).

Styracaceae (Storax Family)

The Storax family of trees and shrubs consists of about ten genera and 160 species distributed in North and South America, southeast Asia, Malaysia, and around the Mediterranean. The flowers are regular, perfect,

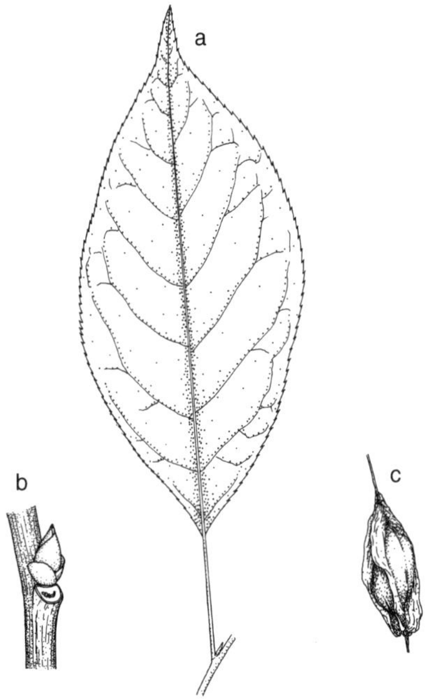

FIGURE 20.32 *Halesia carolina* (a) leaf ×3/4, (b) bud and leaf scar ×2, (c) dry three-winged fruit ×1/2.

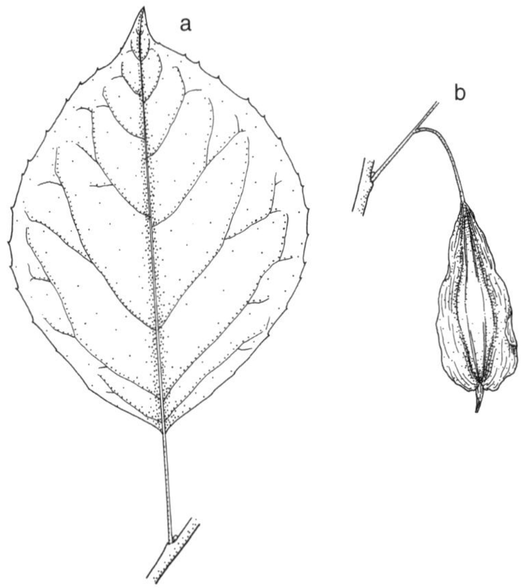

FIGURE 20.33 *Halesia diptera* (a) leaf ×3/4, (b) dry two-winged fruit ×3/4.

with either separate of fused petals. The leaves are alternate, simple, and entire; stipules are absent. In North America the family is represented by *Halesia* (silver bell) with 3 species and *Styrax* (snowbell) with 6 species. The common names refer to the white petals. *Halesia* produces one to three flowers in the leaf axil; *Styrax* has a terminal raceme.

Halesia carolina Linnaeus [Syn. *H. tetraptera* Ellis (Gleason and Cronquist 1991)], silverbell, Carolina silverbell (Figure 20.32); medium-sized tree, height up to 15 m; deciduous; leaf arrangement alternate, leaf simple, length 10 to 15 cm, width 5 to 10 cm, blade entire, broadly elliptical, margin smooth to finely serrate, lower surface stellate pubescent; fruit dry, consisting of a seed cavity and four longitudinally attached wings, total length 3 to 5 cm; wing width 1 cm; Regions 3 and 4, widely scattered areas in its range, southeast from Oklahoma, Illinois, and Ohio; minor component of forest on drier sites. *H. carolina* var. *monticola* grows to a height of 10 m in the Appalachian Mountains. Cover type 28.27.

Halesia parviflora Michaux, little silverbell; small tree, height up to 10 m; deciduous; leaf arrangement alternate; leaf simple, length 5 to 10 cm, width 2 to 5 cm, blade entire, narrowly elliptical, finely serrate; fruit dry, consisting of a seed cavity and four longitudinally attached wings, total length 2 to 3 cm, wing width <5 mm; Region 4, Coastal Plain, Louisiana to South Carolina; dry sandy soil to floodplains.

Halesia diptera Ellis, two-wing silverbell (Figure 20.33); small tree, height to 10 m; leaf arrangement alternate; leaf simple, length 6 to 12 cm, width 5 to 10 cm, blade entire, broadly elliptical, irregularly serrate; fruit dry, consisting of a seed cavity and two longitudinally attached wings, length 3 to 5 cm; Region 4, widely scattered in its range, Coastal Plain, South Carolina to Texas; dry soil and floodplain. Cover type 28.27.

The two *Styrax* species described in the following paragraphs are the major species of the genus. Several other members are found in relatively small regions of the southeast.

Styrax americana Lambert, American snowbell (Figure 20.34); large shrub, height up to 5 m; deciduous; leaf arrangement alternate; leaf simple, length 5 to 8 cm, width 2 to 3 cm, blade entire, broadly elliptical to

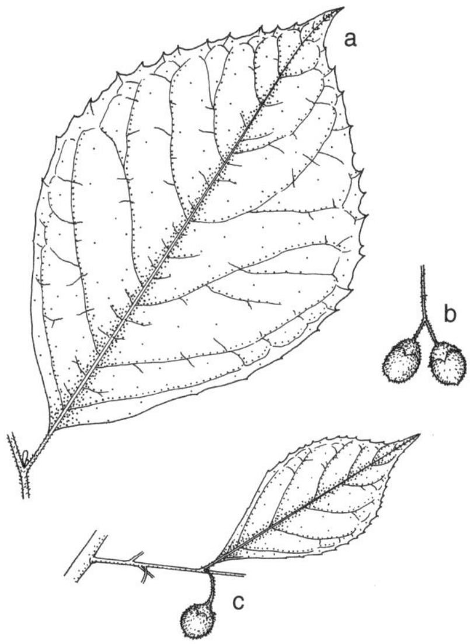

FIGURE 20.34 *Styrax grandifolia* (a) leaf ×1/2, (b) pubescent capsules ×3/4; (c) *S. americana* twig with leaf and pubescent capsule ×1/2.

obovate, serrate toward apex, leaf size decreases from twig apex to older growth; flowers borne in leaf axils; fruit a pubescent capsule, nearly globose, diameter 6 to 8 mm; Regions 3 and 4, primarily in the Coastal Plain but scattered through the Central Hardwood Forest from Indiana and Ohio south; swamps and wet woods. Cover type 28.18, 28.37.

Styrax grandifolia Aiton, bigleaf snowbell (Figure 20.34); large shrub, height up to 6 m; deciduous; leaf arrangement alternate; leaf length 6 to 18 cm, width 5 to 9 cm, blade entire, broadly oval to obovate, serrate toward apex, leaf size decreases from twig apex to older growth; flowers borne in a terminal raceme; fruit a pubescent capsule, nearly globose, diameter 7 to 9 mm, somewhat egg-shaped; Regions 3 and 4, primarily in the Coastal Plain but scattered through the Central Hardwood Forest from Indiana and Ohio south; drainages and upland woods. Cover type 27.9, 28.27.

Symplocaceae (Sweetleaf Family)

Symplocaceae is a monogeneric family of about 300 to 400 species distributed in the tropics and subtropics.

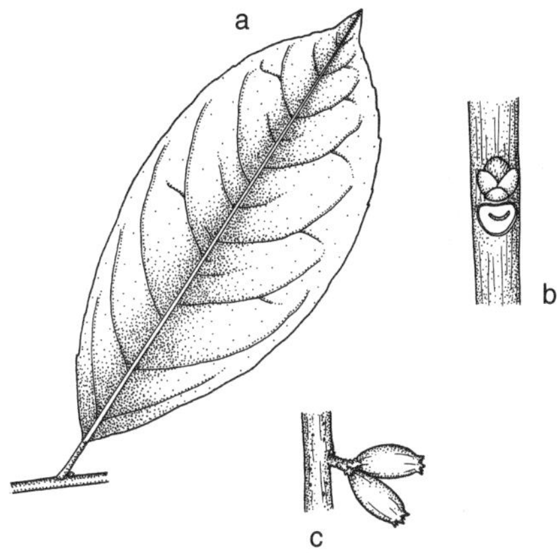

FIGURE 20.35 *Symplocos tinctoria* (a) leaf ×1/2, (b) bud and leaf scar ×2, (c) drupes ×1.

There is only one representative in temperate North America.

Symplocos tinctoria (Linnaeus) L'Her, sweetleaf or horse-sugar (Figure 20.35); shrub, height to 6 m; semi-evergreen, leaves remaining green into the winter months; leaf arrangement alternate; leaf simple, length 5 to 17 cm, width 3.5 to 6.0 cm, blade entire, minutely serrated, somewhat thick or coriaceous; short racemes of yellow flowers borne in leaf axils; fruit a slender, dry, dark orange ellipsoid drupe, 1.0 cm long; Region 4, Coastal Plain, Delaware to Florida to Texas and Oklahoma; upland woods sandhills, swamps and bottomlands. Cover type 27.22, 28.6, 28.14, 28.49.

ORDER: THEALES

Family: Theaceae (Tea Family)

Theaceae is a pantropical family of up to 40 genera and about 500 to 1000 species. Only 3 to 4 species are found in temperate North America, with several more in the subtropical zone at the southern tip of Florida. The leaves are alternate, simple, entire, and have a serrated margin. Stipules are absent. The perfect flowers (plants synoecious) are large (diameter 7 to 10 cm), fragrant, and showy, with five white petals. The flowers are individually attached to the leaf axil. Because of

FIGURE 20.36 *Gordonia lasianthus* twig with leaves and capsules ×1/2.

the large flowers, several species have become popular ornamentals.

Gordonia lasianthus (Linnaeus) Ellis, loblolly bay (Figure 20.36); medium-sized tree, height to 20 m; evergreen; leaf arrangement alternate; leaf simple, length 7 to 15 cm, width 3 to 5 cm, blade entire, shape elliptical to ovate, margin crenulate or finely serrate with blunt teeth, dark green, coriaceous, glabrous; fruit a capsule, nearly globose, length 1.4 to 1.8 cm; Region 4, Coastal Plain, North Carolina to Florida to Louisiana; wet bays and hammocks with *Pinus elliottii, Taxodium, Nyssa aquatica,* and *Acer rubrum.*

Stewartia malacodendron Linnaeus, Virginia stewartia, silky camellia (Figure 20.37); shrub or small tree, height to 6 m; deciduous; leaf arrangement alternate; leaf simple, length 5 to 10 cm, width 3 to 5 cm, blade entire, elliptical to obovate, margin nearly smooth to finely serrate; fruit a capsule, nearly globose, length 1.5 cm; southern Appalachian Mountains and Piedmont; Alabama and Georgia to Virginia and North Carolina, mesic woods.

Stewartia ovata (Cav.) Weatherby, mountain stewartia, mountain camellia (Figure 20.37); shrub or small tree, height to 5 m; deciduous; leaf arrangement alternate; leaf simple, length 6 to 15 cm, width 3 to 7 cm, blade entire, elliptical to oblong, margin nearly smooth to finely serrate, petiole winged; fruit a capsule, ovoid, length 2.0 cm; Region 4, Coastal Plain and Piedmont Virginia to Florida to eastern Texas; mesic woods.

Clusiaceae (Mangosteen or Saint-John's-Wort Family)

The Clusiaceae (Guttiferae) family consists of about 50 genera and 1200 species and is most widespread in tropical regions. Of the two—and more recently three—subfamilies (Judd 1999) that are recognized, the Kielmeyeroideae, Clusioideae, and Hypericoideae, only the Hypericoideae has members well distributed in southern Canada and the United States.

The genus *Hypericum* includes primarily annual and perennial herbs, but a few species are woody. The flowers typically are yellow, with four to five sepals, four to five petals, and numerous stamens. The small, simple leaves have an opposite arrangement. The most distinguishing feature of the genus are the punctate dots that are distributed throughout the leaf blade. These can be easily observed by holding a leaf up to the light and observing through a 10x hand lens.

Although often found in relatively open (nonforest) areas, a few species are associated with the forest. One herb species, or semiwoody species, *Ascyrum hypericoides* Linnaeus [*Hypericum hypericoides* (Linnaeus) Crantz], Saint Andrew's cross, has high fidelity as a forest understory herb, probably more so than woody *Hypericum* species that are found along edges, meadows, and other more open areas. The range of *Ascyrum* and its varieties (Fernald 1950) is from eastern Kansas, Missouri, and southern Illinois to Massachusetts, south to Georgia and Texas. In Region 3 (Central Hardwood Forest), it is common and restricted to *Quercus-Carya* (oak-hickory) communities found only on south slopes where the forest is not so dense and substantial light penetrates to ground level. The fidelity of *Ascyrum* is sufficiently high that it may be used to classify xeric and xeric-mesic sites. The name Saint Andrew's cross comes from the four yellow petals that form a cross and the developing yellow leaves at the shoot apex that also form a cross. Another herb species, *H. gentianoides* (Linnaeus) Britton, Sterns and Poggenberg, pineweed, is a

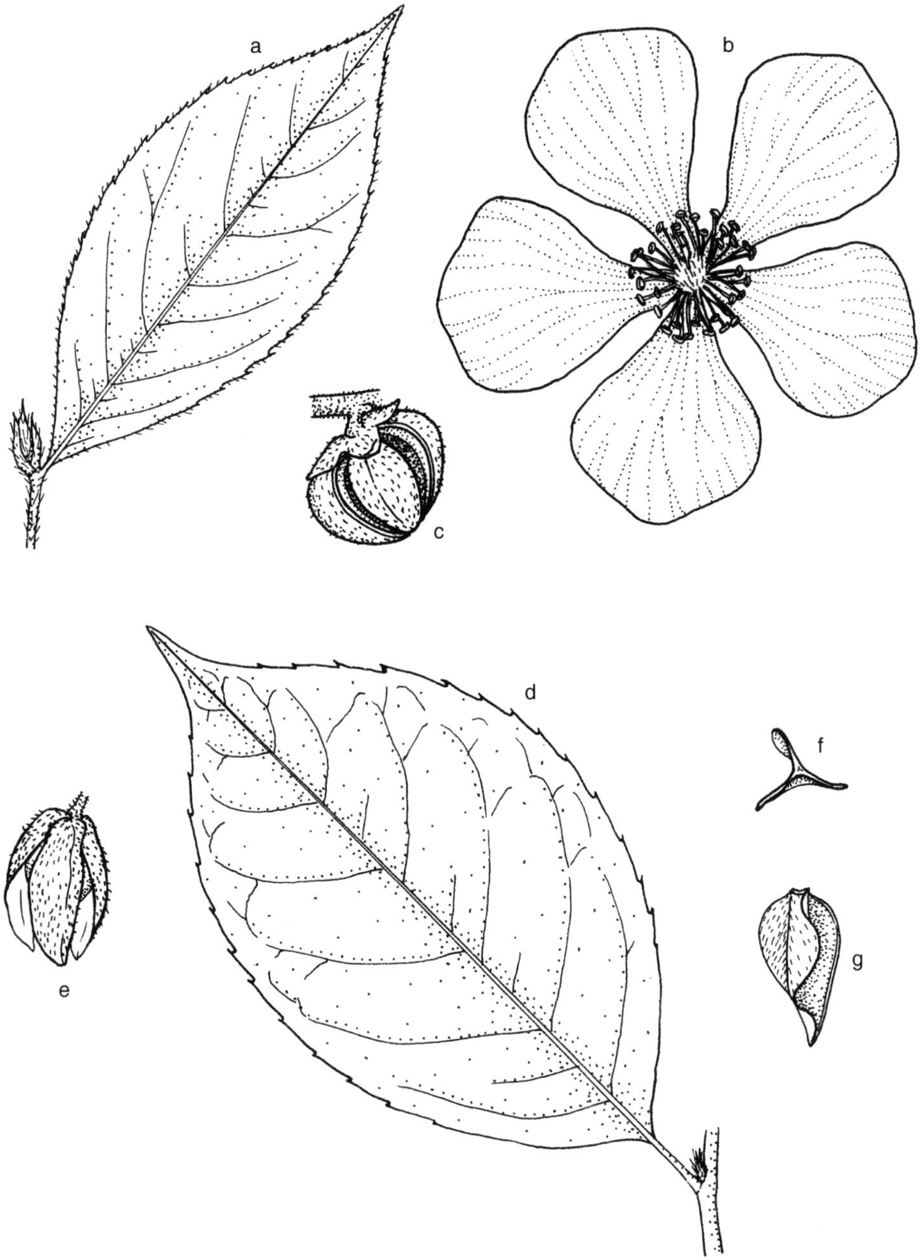

FIGURE 20.37 *Stewartia malacodendron* (a) leaf ×1, (b) flower ×1, (c) capsule ×1; *S. ovata* (d) leaf ×1, (e) capsule ×1, (f) top view of seed ×1, (g) side view of seed ×1.

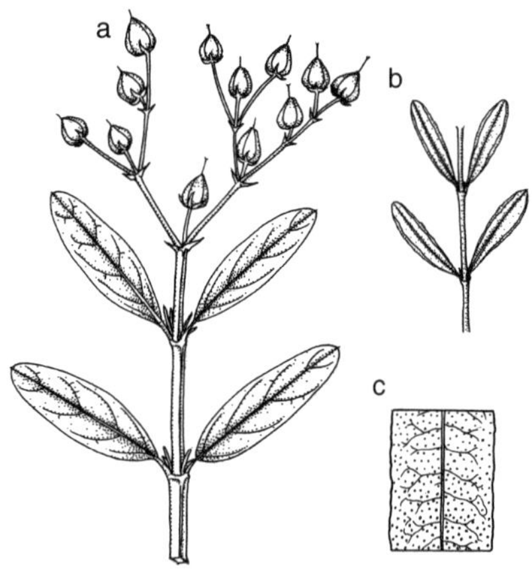

FIGURE 20.38 (a) *Hypericum nudiflorum* twig with leaves and inflorescence ×2/5; (b) *H. densiflorum* leaves showing minutely revolute margin ×1/3; (c) section of leaf showing distribution of punctate dots ×4.

high-fidelity species of sandstone cedar glades in southern Illinois.

Two representative woody plants of the various woodland species are considered here. Other widespread shrub species described by Fernald (1950) and Gleason and Cronquist (1991) include *H. stragulum* P. adams and Robson, *H. prolificum* Linnaeus, *H. stans* (Michaux) P. Adams and Robson, and *H. frondosum* Michaux.

Hypericum nudiflorum Michaux, naked Saint-John's-wort (Figure 20.38); multistemmed shrub, height to 2 m; deciduous; leaf arrangement opposite; leaf simple, length 4 to 7 cm, width 1 to 7 cm, blade entire, oblong or oval to slightly lanceolate, abaxial surface with punctate dots; twigs slender; inflorescence an open cyme, terminal, bearing a few small bracts; flower petals five, yellow; fruit a capsule, oval, length 3 to 7 mm; Regions 3 and 4, Virginia and Tennessee south to Florida and Mississippi; mesic sandy woods and swamps.

Hypericum densiflorum Michaux, naked Saint-John's-wort (Figure 20.38); multistemmed shrub, height to 2.5 m; deciduous; leaf arrangement opposite; leaf simple, length 2 to 5 cm, width 0.5 to 1.0 cm, blade entire, narrowly elliptical to somewhat linear, abaxial surface with punctate dots; twigs slender; inflorescence a cyme, terminal, large, dense, bearing a few small leaves; flowers 7 to 20, flower petals five, yellow; fruit a capsule, lanceolate, length 3 to 7 mm; Regions 3 and 4, Missouri and southern Illinois to Tennessee and New Jersey, south to Florida and Mississippi; mesic to mesic-hydric sandy woods and swamps.

CHAPTER 21

Class: Magnoliopsida; Subclass: Rosidae I

ORDER: ROSALES

Rosaceae (Rose Family)

THE rosaceae is a family of small trees, shrubs, and herbs. Members are found throughout North America, Europe, and Asia. Worldwide, many are associated with forest conditions. There is considerable difficulty in obtaining a close estimate on the number of species because of the problems in identification and because of apomixis (the process of reproduction without the union of pollen with the ovule), which permits the continued existence of new characteristics (e.g., leaf structure) in a population. Species estimates usually report about 85 genera and range from as low as 1500 to about 4700 (Zomlefer 1994; Judd et al. 1999). Much of the problem centers on the major genera, specifically *Rubus* (blackberry, raspberry), *Prunus* (plum, cherry), *Crataegus* (hawthorn), and *Rosa* (rose). Their economic importance is based on their use as fruit crops (e.g., cherry, peach, plum, almonds) and as ornamentals, and for production of medicinal or cosmetic products (e.g., rose hips, perfumes). In the United States and Canada, there are about 62 genera and 850 to 900 species (Zomlefer 1994).

In species considered here, the leaves have an alternate arrangement, are simple or compound, and usually are accompanied by stipules. The flowers are regular (actinomorphic) and perfect, with five sepals, five brightly colored petals, and numerous stamens. The fruit is a drupe, pome, or aggregate of achenes, or a follicle. Based on flower and fruit differences, three or four subfamilies are recognized. However, Judd et al. (1999) indicate that genetic evidence does not clearly distinguish these subfamilies.

Subfamily 1: Rosoideae

The fruit is an aggregate of achenes or miniature drupes.

Genus: *Rubus*

Fernald (1950) describes 205 *Rubus* species (separated into five subgenera), whereas Gleason and Cronquist (1991) describe only 25 and list up to 23 synonyms for some species. Only a few of the more common species are described here. The illustrations, as well as the descriptions, indicate that *Rubus allegheniensis, R. occidentalis,* and *R. idaeus* are related and that *R. odoratus* and *R. arcticus* have a similar affinity.

Rubus allegheniensis Porter, blackberry; stoloniferous shrub (Figure 21.1), multistemmed, height to 2 m, arching; deciduous; leaf arrangement alternate; leaf palmately compound, length 10 to 20 cm; leaflets three or five, infrequently seven, length 6 to 12 cm, width 2 to 4 cm, margin sharply serrated, ovate or lanceolate, terminal leaflet slightly longer than laterals, lateral leaflets somewhat asymmetrical with narrower one half nearer the terminal; petiole and rachis frequently armed with small retrorse prickles, length

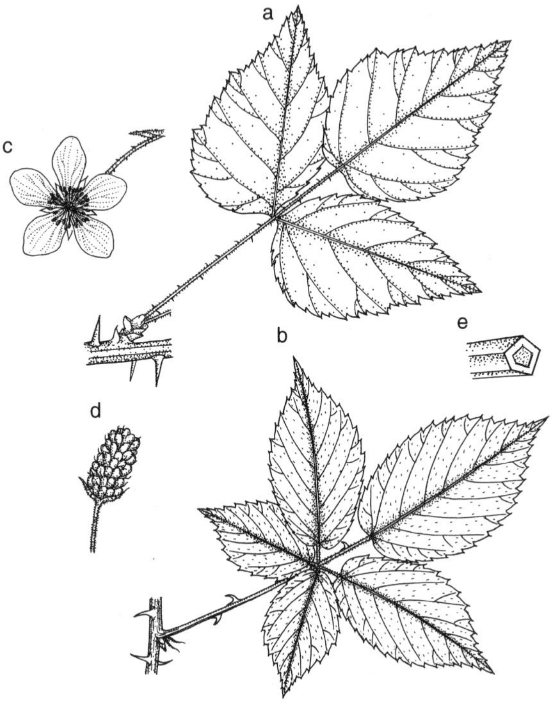

FIGURE 21.1 *Rubus allegheniensis* (a) armed cane and trifoliate leaf ×1/3, (b) palmately compound leaf ×1/3, (c) flower ×1, (d) fruit, an aggregate of drupes ×3/4, (e) cross section of stem ×1.

1 to 2 mm; inflorescence a raceme, flowers white; fruit an aggregate of drupes, red, turning black when ripe, length 1 to 2.5 cm; cane cross section distinctly pentangle, armed with stout prickles, slightly retrorse, deciduous (into clothing and skin), last year canes alive or dead, older canes dead; Regions 2 and 3, New Brunswick and Quebec to Minnesota, south to Tennessee, Georgia, Alabama, and Arkansas; xeric-mesic and mesic soil conditions, fields, roadsides, forest openings, harvested forest. Cover type 26.2, 26.15.

Rubus occidentalis Linnaeus; black raspberry (Figure 21.2); stoloniferous shrub, multistemmed, height to 1.5 m, arching; deciduous; leaf arrangement alternate; leaf trifoliate or pinnately compound, length 13 to 18 cm; leaflets three, infrequently five, length 6 to 12 cm, width 6.5 to 9.0 cm, ovate or lanceolate, margin sharply serrated, occasionally weakly cleft; terminal leaflet slightly longer than laterals, lateral leaflets somewhat asymmetrical with narrower one half nearer the terminal; petiole and rachis frequently armed with small prickles, length 1 to 3 mm; inflorescence a raceme, flowers white; fruit an aggregate of drupes, oval, length 1.5 to 2.0 cm, red, turning black when ripe; cane cross section terete, surface distinctly glaucous, armed with slender prickles, last year canes alive or dead, older canes dead; Regions 2 and 3, New Brunswick and Quebec to Minnesota, and Colorado, south to Tennessee, Georgia, Alabama, and Arkansas; xeric-mesic and mesic soil conditions, fields, roadsides, forest openings, harvest forest. Cover type 26.2, 26.15.

Rubus idaeus Linnaeus; red raspberry (Figure 21.2); stoloniferous shrub, multistemmed, height to 1.5 m, straight; deciduous; leaf arrangement alternate; leaf trifoliate or pinnately compound, length 13 to 18 cm; leaflets three or five, length 6 to 12 cm, width

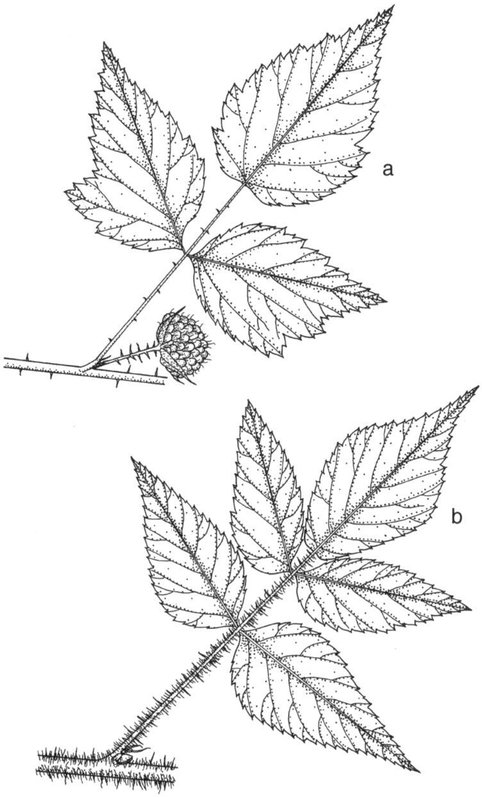

FIGURE 21.2 (a) *Rubus occidentalis* cane, leaf, and aggregate fruit ×2/5; (b) *R. idaeus* cane and leaf ×2/5.

6.5 to 9.0 cm, ovate or lanceolate, margin sharply serrated; terminal leaflet slightly longer than laterals, lateral leaflets somewhat asymmetrical with narrower one half nearer the terminal; petiole and rachis densely armed with slender prickles, length 1 to 3 mm; inflorescence a raceme, flowers white; fruit an aggregate of drupes, nearly globose, diameter 1.0 to 1.5 cm, red when ripe; cane cross section terete, surface densely armed with slender prickles oriented 90° to the stem, last year canes alive or dead, older canes dead; Regions 1, 2, and 3, Newfoundland and Quebec to Minnesota, South Dakota, British Columbia and Alaska, south to Tennessee, Georgia, and Alabama, North Carolina, New Mexico; xeric-mesic and mesic soil conditions, fields, roadsides, forest openings, harvested forest, disturbed places.

Rubus odoratus Linnaeus, thimbleberry (Figure 21.3); stoloniferous, straggling and branching shrub, height 1.0 to 1.5 m; deciduous; leaf arrangement alternate; leaf simple, blade length 8 to 12 cm, width 8 to 12 cm, primary veins and lobes five, margin sharply serrated; petiole length 6 to 10 cm, densely armed with slender prickles, prickle length 1 to 3 mm; inflorescence a panicle, flowers rose or purple, regular, diameter 2.5 to 5.0 cm; fruit an aggregate of drupes, nearly hemispherical, diameter 1.0 to 1.5 cm, red when ripe; cane cross section terete, surface densely armed with slender prickles oriented 90° to the stem; Regions 2 and 3, Nova Scotia, New Brunswick, Quebec, Ontario, and Minnesota, south to Tennessee, Georgia, and Alabama; xeric-mesic and mesic soil conditions, fields, roadsides, forest openings, harvested forest. *Rubus parviflorus* (Figure 21.3) is a species with similar characteristics except it has a white flower and a more western range; Regions 2, 5, and 7 Ontario and Michigan to Alaska, south to California and Arizona.

Rubus arcticus Linnaeus; arctic bramble (Figure 21.3); herb, semiwoody; height to 30 cm; leaf arrangement alternate; leaf simple, length 5 to 12 cm, blade length 3 to 6 cm, width 3 to 6 cm, primary veins and lobes three to five, sinuses shallow, margin serrulated; petiole length 1 to 5 cm; flowers solitary, pink, infrequently white, regular, diameter 2.0 to 2.5 cm; fruit an aggregate of drupes, nearly hemispherical, diameter 1.0 to 1.5 cm, dark purple when ripe; Region 1, Labrador to British Columbia, throughout the arctic, circumpolar.

Genus: *Rosa*, Rose

Erect or climbing shrubs usually having thorns or prickly stems. The flowers are showy, have five sepals, five petals, numerous stamens, and several pistils within the calyx cup. The fruit is called a "hip," which consists of the cup-like receptacle within which are several hard, dry achenes. Rose hips are not only used by wildlife, they also are ingredients in a wide variety of herbal teas and drinks and vitamin tablets.

Rosa multiflora Thunberg; multiflora rose (Figure 21.4); shrub, height to 3 m, stoloniferous and

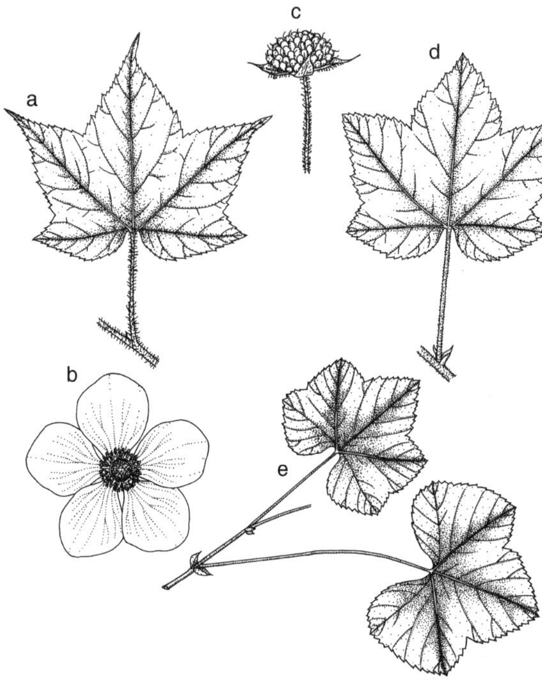

FIGURE 21.3 (a) *Rubus odoratus* cane and attached leaf ×1/4; (b) flower ×3/4, (c) fruit, an aggregate of drupes ×1; (d) *Rubus parviflorus* leaf ×1/4; (e) *R. arcticus* leaves ×1/2.

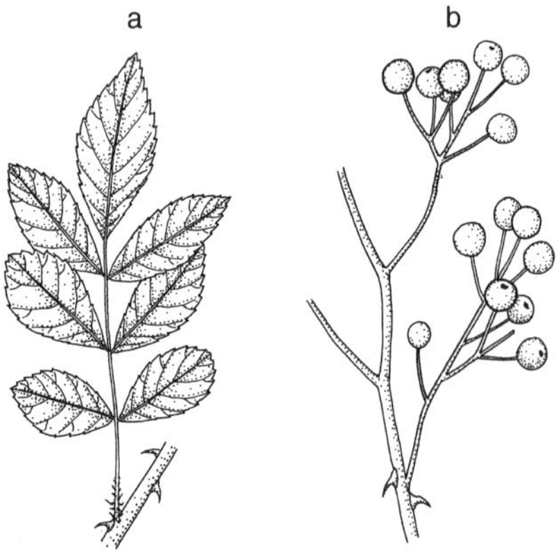

FIGURE 21.4 *Rosa multiflora* (a) cane with retrorse thorns and pinnately compound leaf ×2/5, (b) inflorescence ×2/5 with attached "hips."

forming dense colonies; late deciduous; leaf arrangement alternate; leaf pinnately compound, length 8 to 12 cm, petiole and rachis glabrous, leaflets 5 to 19, length 2 to 4 cm, margin serrated, elliptical or slightly obovate, apex obtuse or acute; stipules often ciliated, length 1 to 2 cm, canes stout, arching, green, terete; thorns stout, not deciduous, retrorse; introduced but escaped cultivation and now naturalized in Regions 2, 3, and 4. Although it creates excellent wildlife habitat (food and cover) and at one time was distributed by state nurseries, this shrub has become a serious invasive pest. The combination of dense contiguous colonies and stout thorns prevents humans from penetrating or passing through an area. It has become dominant along fencerows and in old fields. It has been less successful at invading intact forest communities, usually being limited to a few low stems inside the forest, but forming a dense row along the edge. Sale of this species has been banned in most states.

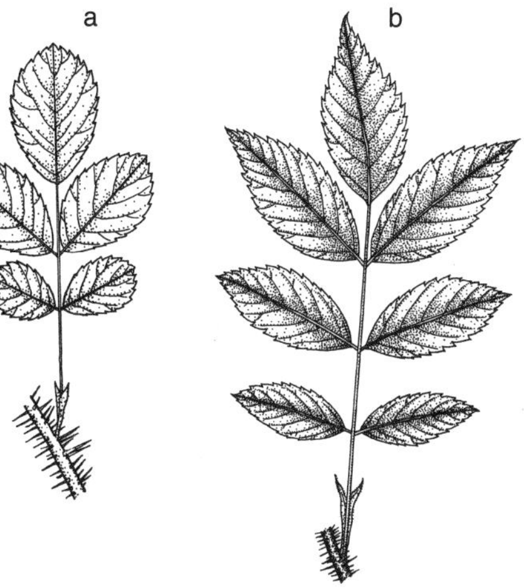

FIGURE 21.5 Prickly cane and pinnately compound leaf of (a) *Rosa carolina* ×1/3 and (b) *R. acicularis* ×1/3.

Rosa acicularis Lindley, prickly rose (Figure 21.5); multistemmed shrub, height to 1.1 m; deciduous; leaf arrangement alternate; leaf pinnately compound, length 8 to 15 cm, leaflets three to seven, length 2 to 5 cm, margin serrated, elliptical, oval or slightly obovate, apex obtuse but pointed; stipules broad, length 1 to 2 cm; stem terete, prickles dense, straight, at right angles to stem, stem reddish or greenish; Regions 2 and 3, New Brunswick and New England to Minnesota, South Dakota, and Colorado, south into West Virginia, also reported as distributed from Ontario to Alaska, may be circumpolar; xeric-mesic upland woods and rocky sites.

Rosa carolina Linnaeus; swamp rose, wild rose, pasture rose (Figure 21.5); bushy shrub, height to 1 m; deciduous; leaf arrangement alternate; leaf pinnately compound, length 8 to 15 cm, petiole and rachis glabrous, leaflets five to seven, length 1 to 3 cm, margin serrated, elliptical, oval or slightly obovate, apex obtuse, approaching rounded; stipules linear, length 1 to 2 cm; stem green, terete, prickles dense, straight, at right angles to stem; Regions 2, 3, and 4, Nova Scotia to Ontario and south to Florida, Mississippi, and Texas; xeric sandy or rocky soil, open upland woods, dunes, prairies.

Rosa blanda Aiton; smooth rose, meadow rose; bushy shrub, height to 15 cm; deciduous; leaf arrangement alternate; leaf pinnately compound, length 8 to 15 cm, petiole and rachis glabrous, leaflets five to seven, length 2 to 5 cm, elliptical, oval or slightly obovate, margin coarsely toothed on the distal one half, apex obtuse, approaching rounded; stipules linear, length 1 to 2 cm, stem green, terete; prickles absent; Regions 2 and 3, Nova Scotia to Ontario and south to Indiana and Missouri; xeric to mesic sandy or rocky soil, open upland woods, dunes, prairies.

Genus: *Potentilla*, Cinquefoil

Potentilla fruiticosa Linnaeus; shrubby cinquefoil (five fingers) (Figure 21.6); multistemmed shrub, height 1 m; deciduous; leaf arrangement alternate; leaf pinnately compound, length ~1 to 2 cm, leaflets usually five, occasionally seven, leaflet length 0.5 to 1.0 cm, pilose; twigs reddish to gray, buds small, length 1 mm, reddish; bark reddish, exfoliating; flowers terminal, regular, diameter ~2 cm, petals yellow; Regions 1, 2, 5, 6, and 7, Greenland to Newfoundland and Alaska, Pennsylvania to Minnesota, south along the Rocky Mountains, California; swamps and mesic forest. Planted as an ornamental because of its unusual foliage and bright yellow flowers.

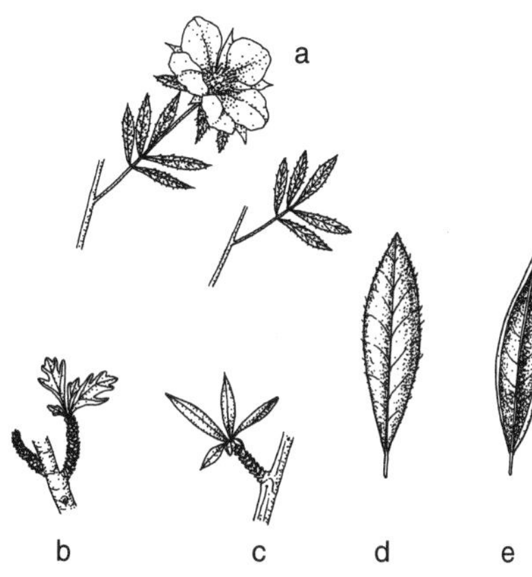

FIGURE 21.6 (a) *Potentilla fruiticosa* compound leaves and flower ×3/4; (b) *Purshia tridentata* spur shoot and leaves ×1/2, *Cercocarpus ledifolius* (c) spur shoot and clustered leaves ×1/3, (d) adaxial surface of the simple leaf ×1, (e) abaxial surface showing revolute margins ×1.

Genus: *Purshia*, Buckbrush

Purshia tridentata de Candolle, antelope-brush, bitterbrush, buckbrush, greasewood (Figure 21.6); shrub, height to 2 m, multibranched; leaves borne at the tips of spur shoots, continuously produced, total length 0.5 to 3.0 cm, cuneate, apex three-lobed, adaxial surface lightly pubescent, abaxial surface white tomentose; flowers regular, sepals five, petals five, yellow, diameter 2 to 3 cm; fruit a coriaceous oblong achene; Regions 5, 6, and 7, British Columbia and Montana to New Mexico and California; xeric slopes, elevation 1300 to 2200 m. *Purshia glandulosa* Curran is a closely related species, but with three- and five-lobed leaves.

Genus: *Cercocarpus*, Mountain Mahogany

There are probably about six or more *Cercocarpus* species in the Rocky Mountain region, and in California, about ten species. Most grow at low elevations in extremely dry soils and climates. They are associated with evergreen sclerophyll vegetation, mixed evergreen forest, oak-laurel forest, oak woodland, pinyon-juniper savanna, and chaparral communities (Barbour and Major 1977).

Cercocarpus ledifolius Nuttall, mountain-mahogany (Figure 21.6); shrub or small tree, height to 12 m, multibranched; evergreen; leaves borne at the tips of spur shoots, leaf length 1.0 to 2.5 cm, narrowly elliptical or lanceolate, apex acute, margin smooth, coriaceous, revolute, adaxial surface lightly pubescent or glabrous, somewhat resinous, abaxial surface rusty tomentose; flowers regular, sepals five, petals five, yellow, diameter 2 to 3 cm; fruit a long villous achene; Regions 5, 6, and 7, Washington and Montana to Colorado and California; xeric slopes, elevation 1300 to 2700 m. Cover type 29.5.

Subfamily 2: Amygdaloideae

Genus: *Prunus*

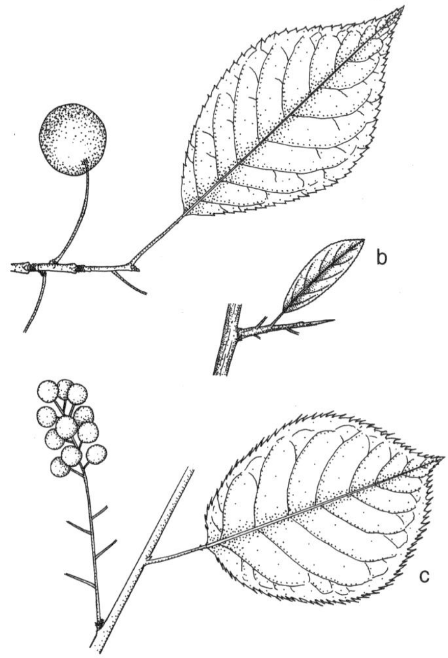

FIGURE 21.7 (a) *Prunus americana* twig with simple leaf and drupe ×3/4; (b) *P. angustifolia* twig with leaf and thorny twig ×1/3; (c) *P. virginiana* twig with simple leaf and inflorescence of attached drupes ×3/5.

Prunus americana Marshall, wild plum (Figure 21.7); shrub or small tree, height to 8 m; deciduous; leaf arrangement alternate; leaf simple, entire, blade length 4 to 8 cm, lanceolate, ovate, or obovate, lateral veins distinct, seven to nine, extending nearly to margin, margin serrate or double serrate, teeth sharp, apex acuminate, base obtuse; petiole length 1.0 to 2.5 cm, glabrous, seldom glandular; flowers white, regular, diameter 2.0 to 2.5 cm, on individual peduncles but clustered three to five at a node of second-year branch; twigs grayish to reddish-brown, glabrous, lacking strong pungent odor, short shoots occasionally thorn-like; fruit a drupe, slightly oval, length 1.7 to 2.5 cm, dark red to yellow; bark of small trees dark; lenticels horizontal, distinct, light brown; Regions 2, 3 and 4; Massachusetts to Ontario to North Dakota, south to Florida and Texas; mesic woods, stream margins, fencerows.

Prunus angustifolia Marshall, narrowleaf crabapple (Figure 21.7); height to 6 m; deciduous; leaf

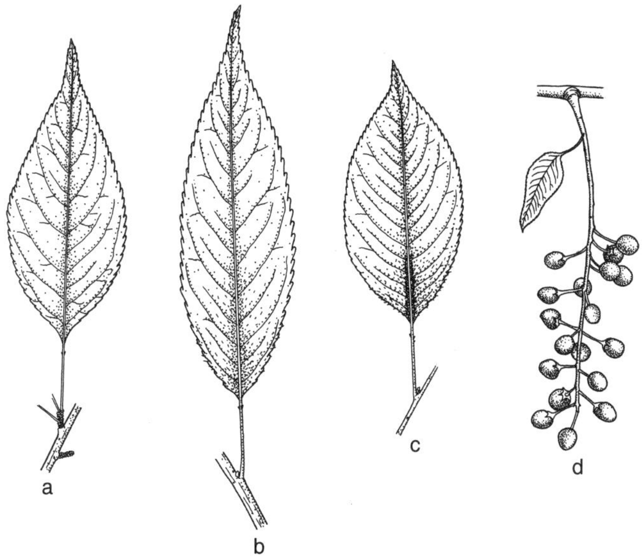

FIGURE 21.8 (a) *Prunus hortulana* leaf attached to a spur shoot ×1/2; *P. serotina* (c) adaxial surface of elliptical leaf ×3/4, (b) abaxial surface of ovate leaf with rows of hair along the midrib near the base ×3/4, (d) inflorescence with attached drupes ×1/2. See Figure 16.4d.

arrangement alternate; leaf simple, entire, blade length 2.5 to 8.0 cm, width 0.5 to 2.0 cm, elliptical-oblong to lanceolate, lateral veins long, extending nearly to margin, margin serrulate, apex obtuse, base obtuse; petiole length 0.5 to 1.0 cm, glabrous, glandular; twigs grayish to reddish brown, glabrous, pungent odor lacking, often thorn-like; inflorescence a cylindrical raceme, length 8 to 18 cm; flowers white, regular, diameter 1 to 2 cm; fruit a drupe, globose, diameter 1.0 to 1.5 cm, red; bark of small trees dark; Regions 3 and 4, Kansas and Missouri to Maryland, south to Florida and Texas; xeric-mesic sites, forest edges.

Prunus virginiana Linnaeus, chokecherry (Figure 21.7); shrub or small tree, height to 4 m; deciduous; leaf arrangement alternate; leaf simple, entire, blade length 4 to 8 cm, obovate, lateral veins distinct, six to eight, extending nearly to margin, margin serrate, teeth sharp, apex abruptly acuminate, base obtuse; petiole length 1.0 to 2.5 cm, glabrous, glandular; twigs grayish to reddish brown, glabrous, pungent odor strong; inflorescence a cylindrical raceme, length 6 to 13 cm; flowers white, regular, diameter 2.0 to 2.5 cm; fruit a drupe, diameter 0.8 to 1.0 cm, red to black; bark of small trees dark; lenticels not distinct; Regions 1, 2 and 3, British Columbia to Newfoundland, south to Tennessee, Missouri, and Oklahoma; common forest shrub, fencerows, edges. The fruits are used by wildlife. The bitter odor and taste of the twigs comes from prussic acid, a strong arsenic compound. Cover type 26.18, 27.19, 29.26, 29.35, 29.13, 30.5.

Prunus hortulana Bailey, wild goose plum, hortulan plum (Figure 21.8); shrub or small tree, height to 9 m; deciduous; leaf arrangement alternate; leaf simple, entire, blade length 4 to 10 cm, lanceolate, ovate, or obovate, lateral veins generally long, approaching the margin, margin serrate or crenate, teeth blunt and glandular tipped, apex acuminate, base obtuse; petiole length 1.0 to 2.5 cm, glabrous, glandular; twigs grayish to reddish brown, glabrous, lacking strong pungent odor; bark of small trees dark, lenticels horizontal, distinct, black; flowers white, regular, diameter 1.7 to 2.5 cm; on individual peduncles but clustered two to four at a node of second-year branch; fruit a drupe, globose, length 2.0 to 2.5 cm, light red; Region 3, Indiana, Ohio, Virginia, and Tennessee, south to Alabama and Texas, east to Iowa, Missouri, and Kansas.

Prunus serotina Ehrhart; black cherry (Figure 21.8); large tree, height to 30 m; deciduous; leaf arrangement alternate; leaf simple, entire, blade length 6 to 11 cm, elliptical, ovate, or lanceolate, lateral veins alternating long and short, margin serrate, teeth blunt, apex acuminate or long acuminate, base obtuse, abaxial midrib near the base bordered on each side by a band of pubescence, pubescence whitish in spring, becoming rusty by midsummer; petiole length 1 to 2 cm, glabrous, glandular; twigs grayish to reddish brown, glabrous, pungent odor strong; inflorescence a cylindrical raceme, length 6 to 15 cm; flowers white, regular, diameter 0.7 to 1.0 cm; fruit a drupe, globose, diameter 0.8 to 1.0 cm, reddish black; bark of small trees dark; lenticels horizontal, distinct, light brown, on larger trees breaking to plates of 2 to 8 cm diameter; Regions 2, 3, and 4, Nova Scotia, New Brunswick, and southern Ontario to Minnesota, Florida, Texas, and Mexico, also New Mexico and Arizona; xeric-mesic and mesic wooded slopes and stream terraces. Large trees of this species are economically important and reach their greatest development in southern Indiana, Ohio, the Appalachian Mountains, and New England. Elsewhere in Region 3 seedlings and saplings are common in the forest understory but seldom attain tree size because they are relatively shade intolerant and small trees are subject to "black knot" disease. The wood is hard, a rich reddish brown with a beautiful grain; thus, it is highly desirable for use as furniture. The fruits are readily eaten by wildlife. The bitter odor and taste of the twigs comes from prussic acid, a strong arsenic compound. Cover type 26.18, 26.23, 26.24, 26.28, 26.29, 27.4, 27.9, 27.23, 27.29, 27.30, 28.14, 29.22.

Prunus nigra Aiton, Canadian plum (Figure 21.9); shrub or small tree, height to 9 m; deciduous; leaf arrangement alternate; leaf simple, entire, blade length 4 to 10 cm, lanceolate, ovate, or obovate, lateral veins alternating long and short, margin serrate or double serrate, teeth blunt and glandular tipped, apex acuminate, base obtuse; petiole length 1.0 to 2.5 cm, glabrous, glandular; twigs grayish to reddish brown, glabrous, lacking strong pungent odor, short shoots occasionally thornlike; flowers white, regular, diameter 1.7 to 2.5 cm, on individual peduncles but clustered two to four at a node of second-year branch; fruit a drupe, slightly oval, length 2.0 to 2.5 cm, light red, orange, or yellow; bark of small trees dark, lenticels horizontal, distinct, light brown; Region 2, New Brunswick through New England to Wisconsin, Minnesota, and Manitoba; occasional in Region 3, Wisconsin, Illinois, and Indiana; pastures, fencerows, borders of woods, riverbanks.

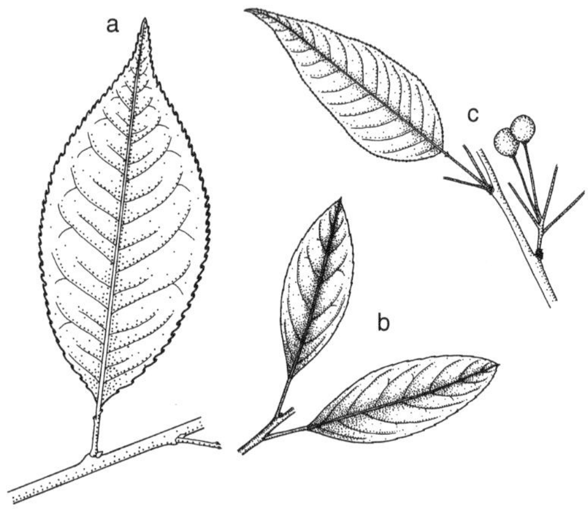

FIGURE 21.9 (a) *Prunus nigra* leaf ×3/5; (b) *P. pumila* leaves ×2/5; (c) *P. pensylvanica* twig with leaf and inflorescence with attached drupes ×1/3.

Prunus pumila Linnaeus, sand cherry; shrub (Figure 21.9), height to 2 m, erect or decumbent; deciduous; leaf arrangement alternate; leaf simple, entire, blade length 4 to 8 cm, width 0.6 to 1.5 cm, somewhat coriaceous, elliptical, narrowly lanceolate or slightly oblanceolate, lateral veins long, six to eight per side, extending nearly to margin, margin finely serrate or serrulate, apex obtuse, base obtuse; petiole length 1.0 to 1.5 cm, glabrous, glands absent; twig reddish brown to grayish, glabrous, pungent odor lacking; inflorescence an umbel, length 8 to 18 cm; flowers white, regular, diameter 1.2 to 1.6 cm; fruit a drupe, globose, diameter 0.8 to 1.0 cm, purplish to black; Regions 1 and 2, New Brunswick to Manitoba, Wisconsin, Michigan, Minnesota, North Dakota, Kansas, Wyoming; xeric sands, sand dunes, component of *P. banksiana* forest. Several varieties are recognized, generally differing in location (Gleason and Cronquist 1991). Cover type 26.18, 26.26.

Prunus pensylvanica Linnaeus f., pin cherry, fire cherry (Figure 21.9); small tree, height to 12 m; deciduous; leaf arrangement alternate; leaf simple, entire, blade length 6 to 12 cm, ovate or lanceolate, lateral veins long, extending nearly to margin, margin finely serrate, apex long acuminate, base obtuse; petiole length 1 to 2 cm, glabrous, glandular; twigs grayish to reddish brown, glabrous, pungent odor lacking; inflorescence a cylindrical raceme, length 8 to 18 cm; flowers white, regular, diameter 1.2 to 1.6 cm; fruit a drupe, globose, diameter 5 to 7 mm, reddish black; bark of small trees dark, lenticels horizontal, distinct, rusty becoming black; Regions 1 and 2, Newfoundland, Labrador, Nova Scotia, and New Brunswick, south to Pennsylvania and Virginia, west to Minnesota, North Dakota, British Columbia, and the Northwest Territories. Occasional tree in open fields and fencerows; forms dense stands after fire in the northeast because of the existence of a seed bank. Cover type 26.15, 26.2, 26.9, 26.14, 26.28, 27.31.

Subfamily 3: Maloideae

Genus: *Crataegus*, Hawthorn

Crataegus appears to be a complicated and unstable genus of small trees, and reasonable estimates of the number of species vary from as few as 35 to more than 100, the largest number of named species in North America. However, Sargent at the Arnold Arboretum (Harvard University) described more than 700 species (Little 1979). Little provides an interesting review. The problem within the genus is due to a variety of active processes, including hybridization (crossing of species), apomixis (producing viable seed without fertilization), polyploidy (doubling of the chromosome number), and aneuploidy (number of chromosomes not a multiple of the haploid number, not 2n, 3n, etc.), all of which are part of speciation (creation of new species). For example, Gleason and Cronquist (1991) reported 65 recognized hybrid species for the northeastern United States. An additional problem is that relatively minute characteristics (e.g., color of anthers) have been used to identify new species. Little (1979) reported that materials from different branches of the same tree were identified as separate species, and thus he accepts only 35 species and 46 hybrids but partitions these into 34 series. Fernald (1950) recognized 103 species and 19 series.

Individuals of most species are small trees, although some may be shrubs. The length of a single bole on individual trees is usually only 1 to 2 m, above which there are multiple branches. Total height seldom exceeds 7 to 8 m. The flower has typical Rosaceae characteristics; that is, it is regular, with five sepals, five petals, and 5 to 25 stamens aligned in one to five rows. The close relationship to *Malus* (apple) is noted by the specialized fruit, a pome. In this genus the fruit is a miniature (diameter 0.6 to 2.0 cm) of the commercial apples found in grocery stores. Only the core is derived from the ovary, and the outer edible material comes from accessory tissues surrounding the inferior ovary. Like that of its larger relative, the core is divided into two to five carpels, with cartilaginous material surrounding the seed and residual floral parts attached at the distal end (the "bottom"). The number of seeds range from one to five. Fruit color varies from reddish to yellow, green, or, infrequently, dark blue.

Ecologically, there is only limited variation in habitat location. Individuals of *Crataegus* are typically found in previously or presently disturbed situations, such as along fencerows, in pastures, and in grazed woodlots frequented by birds. Beyond providing food for wildlife, the next most important use of these trees may be as ornamentals.

For general identification proposes, the most distinguishing characteristic of Crataegus is the 3 to 7 cm long thorn located next to the leaf. Common names vary from hawthorn, haw, redhaw, and thornapple. Because of the genetic and taxonomic problems within the taxon, only four species are illustrated (Figure 21.10) to provide a range in leaf characteristics: *Crataegus punctata* Jacquin, dotted hawthorn, *Crataegus mollis* Scheele, downy hawthorn; *Crataegus spathulata* Michaux, littlehip hawthorn; *Crataegus crus-galli* Linnaeus, cockspur hawthorn.

Genus: *Malus*, *Pyrus*

There is a long-standing debate over whether apple (*Malus*) and pear (*Pyrus*) should be included in one genus or maintained as separate taxa. Depending on the taxonomist, all pears and apples may be included in *Pyrus* (Fernald 1950; Gleason and Cronquist 1991) or *Malus* (Little 1979). Elias (1980) and Judd et al. (1999) maintain separate genera. One consideration for separation has centered on the gritty stone cells in pear fruit, which apple lacks, but the leaves of some species

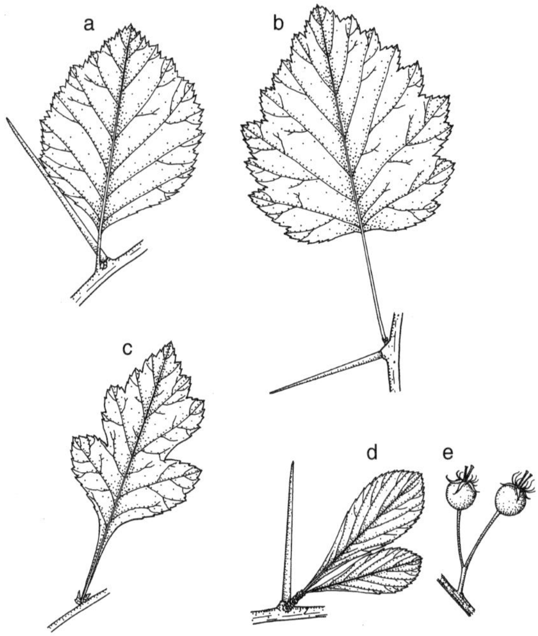

FIGURE 21.10 Leaf and thorn of (a) *Crataegus punctata* ×1/2; (b) *C. mollis* ×1/2; (c) *C. spathulata* ×1/2; *C. crus-galli* (d) twig with thorn and leaves on spur shoot ×1/2, (e) inflorescence with apple-like fruit ×1/2.

from the two taxa are essentially identical. As further justification for his position, Fernald (1950) indicates that members of these two genera, including *Aronia* (chokeberry) and *Sorbus* (mountain-ash), have other overlapping flower and internal fruit characteristics, and there essentially is no difference in wood anatomy. Fernald (1950) clusters these four genera in one genus, *Pyrus*, and recognizes minor differences by dividing them into subgenera.

In addition to their value as producers of wildlife food, crab apple trees are the alternate host of cedar-apple rust, a disease the requires both *Juniperus virginiana* and *Malus* leaves to complete its life cycle. Spores are produced by fruiting bodies on the leaf of *Malus*. When released into the atmosphere, they travel to *Juniperus*, where they germinate to form another fruiting body that produces spores that reinfect *Malus*. The disease in easily identified on *Juniperus* twigs, where the fruiting body is a soft gelatinous mass, having short, protruding tentacles.

Malus pumila Miller [*Malus sylvestris* (Linnaeus) Miller; *Pyrus malus* Linnaeus], apple (Figure 21.11); small tree, height to 10 m, multibranched above 1 m; deciduous; leaf arrangement alternate, but with leaves frequently clustered on spur shoots; leaf simple, blade entire, length 3 to 8 cm, width 2 to 4 cm, elliptical, lanceolate, or ovate, margin nearly smooth to serrate, teeth blunt, apex obtuse or short acuminate, abaxial surface lightly to moderately pubescent; petiole length 1.5 to 4.0 cm; twigs stout, greenish, purplish, or reddish brown; buds usually pubescent; flowers regular, white or pink, petals five, sepals five, diameter about 2.5 cm, each flower on a peduncle usually attached to the end of a spur shoot; fruit a large pome, nearly globose, diameter 6 to 10 cm, reddish or greenish. Introduced from Europe. Many high-yield cultivars have been developed in response to public demand.

Malus ioensis (Wood) Britton [*Pyrus ioensis* (Wood) Bailey], prairie or Iowa crab apple (Figure 21.11); small tree, height to 5 m; multibranched

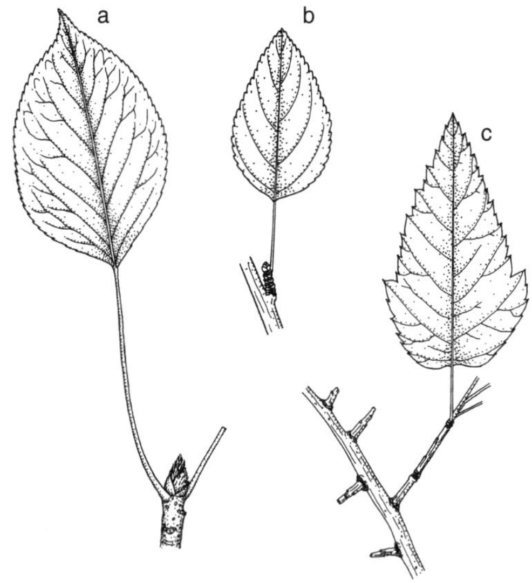

FIGURE 21.11 *Malus pumila* twig with (a) leaf and terminal bud ×1/2, (b) leaf and spur shoot ×1/3; (c) *M. ioensis* leaf and semithorny twigs ×1/2.

above 1 m; deciduous; leaf arrangement alternate, but with leaves frequently clustered on spur shoots; leaf simple, blade entire or with one to three small clefts and large teeth on each side, length 4 to 8 cm, width 2 to 4 cm, generally lanceolate, margin coarsely serrate, abaxial surface lightly to moderately pubescent; petiole length 1.5 to 4.0 cm; twigs stout, greenish, purplish, or reddish brown, often becoming thorns; buds reddish brown, usually pubescent; flowers regular, white or pink, petals five, sepals five, diameter about 2.5 cm, each flower on a peduncle usually attached to the end of a spur shoot; fruit a small pome, globose, diameter 2.5 to 4.0 cm, yellowish-greenish; Region 3, concentrated in Missouri, Illinois, Wisconsin, and Iowa with disjunct populations in Minnesota and south to Louisiana; pastures, abandoned fields, prairies, and bottomland.

Pyrus communis Linnaeus, pear (Figure 21.12); small tree, height to 12 m, multibranched above 1 m; deciduous; leaf arrangement alternate, but with leaves frequently clustered on stout spur shoots; leaf simple, blade entire, length 3 to 8 cm, width 2 to 4 cm, elliptical, ovate, or suborbicular, margin serrate, teeth blunt, apex long acuminate, adaxial and abaxial surfaces glabrous; petiole length 1.5 to 4.0 cm; twigs stout, greenish, purplish, or reddish brown; buds glabrous, conical; flowers regular, white or pink, petals five, sepals five, diameter 2 to 3 cm; fruit a large pome, obovoid or teardrop, length 6 to 12 cm, greenish. Introduced from Europe or Asia.

Pyrus coronaria Linnaeus [*Malus coronaria* (Linnaeus) Miller]; wild or sweet crab apple (Figure 21.12); small tree, height to 9 m; multibranched above 1 m; deciduous; leaf arrangement alternate, but with leaves frequently clustered on spur shoots; leaf simple, blade entire, length 3 to 8 cm, width 2 to 6 cm, lanceolate or elliptic, margin coarsely serrate, occasionally with one to three small clefts and large teeth on each side, adaxial and abaxial surfaces essentially glabrous; petiole length 1.5 to 4.0 cm; twigs reddish brown, slender, often becoming thorns; buds reddish brown, usually pubescent; flowers in clusters of two to three, regular, white or pink, petals five, sepals five, diameter about 2 to 3 cm; fruit a small pome, globose, diameter 2.5 to 4.0 cm, green, waxy; Region 3, concentrated in Indiana, Ohio, Michigan, and Pennsylvania, disjunct populations in adjacent states; fences, roadsides, and other disturbed sites.

Malus angustifolia (Aiton) Michaux (*Pyrus angustifolia* Aiton), southern crab apple, narrowleaf crab apple; small tree; height to 9 m; multibranched above 1 m; deciduous; leaf arrangement alternate, but with leaves frequently clustered on spur shoots; leaf simple, blade entire, length 2 to 4 cm, width 1.0 to 1.5 cm, lanceolate or elliptic, somewhat coriaceous margin serrate, often coarsely serrate or toothed, adaxial and abaxial surfaces essentially glabrous; petiole length 2.0 to 2.5 cm; twigs brownish, slender, often becoming thorns; buds reddish brown; flowers in clusters of two to three, regular, white or pink, petals five, sepals five, diameter about 2 to 3 cm; fruit a small pome, globose, diameter 2.4 to 3.0 cm, yellowish green; Region 4, Coastal Plain species concentrated from North Carolina to Georgia and Mississippi; disjunct populations occur in Region 3.

Pyrus calleryana Decaisne, Bradford pear cultivar (Figure 21.12); small to medium-sized tree, height to 15 m, branched above 1.5 m; deciduous leaf arrangement alternate, leaves often clustered on stout spur shoots; leaf simple, entire, blade length 4 to 6 cm, width 4 to 6 cm, somewhat coriaceous, broadly ovate

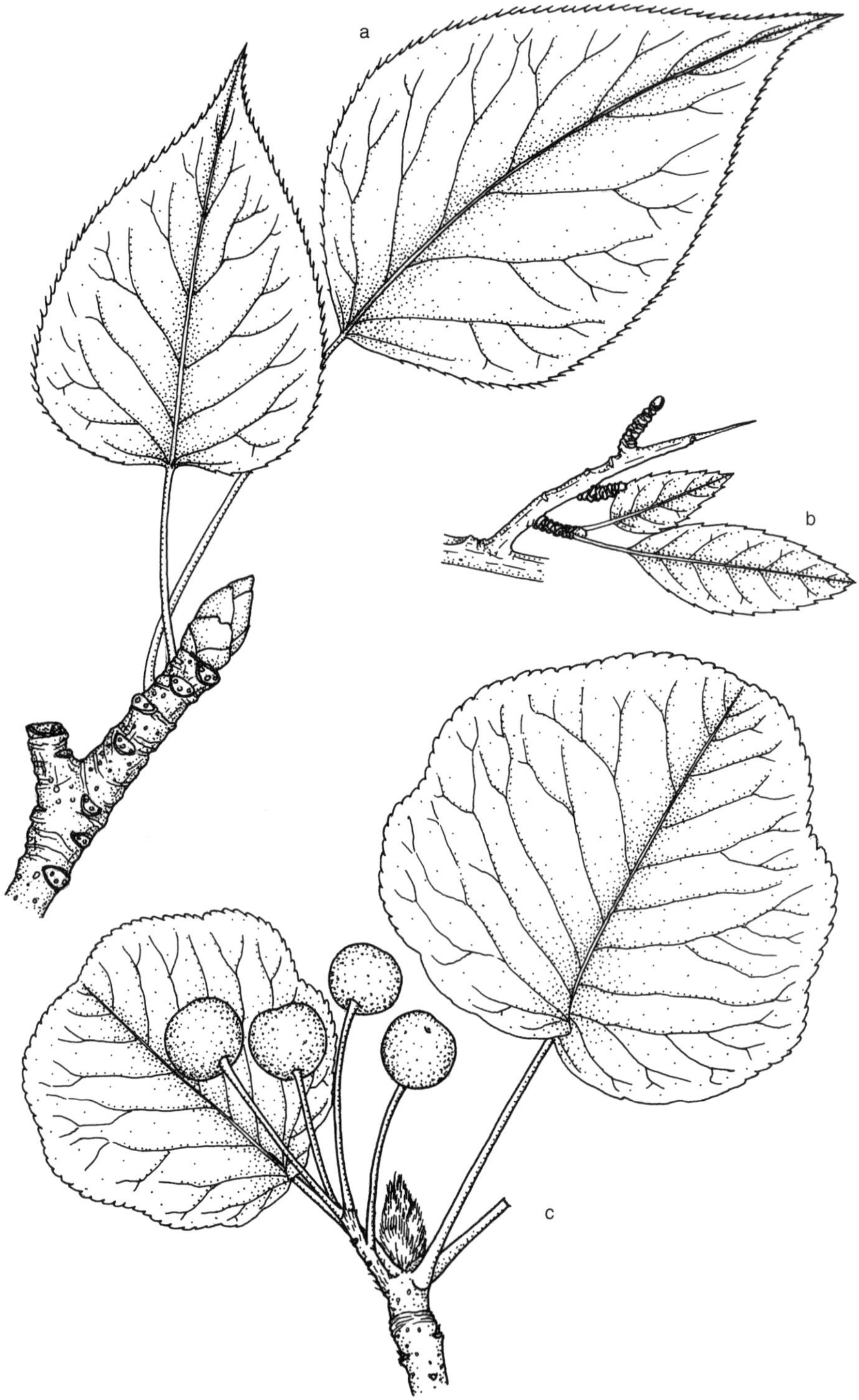

FIGURE 21.12 (a) *Pyrus communis* twig with leaves and terminal bud ×1.2; (b) *P. coronaria* twig with leaves, spur shoots, and thorny twigs ×4/5; (c) *P. calleryana* leaves, terminal bud, and inflorescence with attached pomes ×1.2.

to nearly orbicular, margin crenate, adaxial surface glabrous, shiny, dark green, abaxial surface glabrous; petiole length 2.0 to 4.5 cm; twigs usually stout, gnarled, rough from bud scale scars; buds large, length about 1.5 cm, glabrous; introduced as an ornamental, now found as a lawn tree nationwide. Dirr (1998) has developed some reservations because of over-planting in favor of other native or introduced species. Apparently larger trees tend to split and lose shape because of the many branches that develop within short sections of the main stem.

Genus: *Aronia*

This small genus is often included with *Pyrus* (Fernald 1950; Gleason and Cronquist 1991).

Aronia arbutifolia (L.) Elliott (*Pyrus arbutifolia* Linnaeus), red chokeberry (Figure 21.13); rhizomatous colonial shrub, height to 3 m; deciduous; leaf arrangement alternate; leaf simple, entire, length 4 to 10 cm, width 1.5 to 4.5 cm, broadly elliptical, oblong, or lanceolate, margin serrulate, ciliate, apex obtuse or mucronate, adaxial surface midrib glandular, abaxial surface tomentose, velvety; twigs tomentose; inflorescence a terminal cyme of seven to ten flowers; flower regular, diameter 0.8 to 1.4 cm, sepals five, petals five, white to pink; fruit a pome, bright red, diameter 4 to 7 mm; Regions 2, 3, and 4, Nova Scotia to Michigan, south to Kentucky, West Virginia, Florida, and Texas; swamps, wet woods, bogs and savanna. Cover type 28.36.

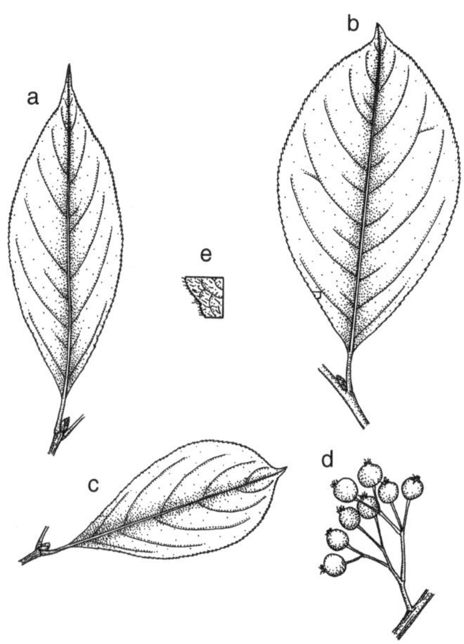

FIGURE 21.13 Leaves of (a) *Aronia melanocarpa* and (b) *A.* x *prunifolia* ×3/4; *A. arbutifolia* (c) leaf ×1/2 and (d) inflorescence with apple-like fruit ×1/4.

Aronia melanocarpa (Michaux) Britton (*Pyrus melanocarpa* Willdenow), black chokeberry (Figure 21.13); rhizomatous colonial shrub, height to 2 m; deciduous; leaf arrangement alternate; leaf simple, entire, length 3 to 8 cm, width 1.5 to 3.0 cm, elliptical, oblong, or lanceolate, margin serrulate, ciliate, apex acuminate or mucronate, adaxial surface midrib glandular and glabrous, abaxial surface glabrous; twigs glabrous; inflorescence a terminal cyme; flower regular, diameter 0.8 to 1.2 cm, sepals five, petals five, white to pink; fruit a pome, purple-black or black, diameter 6 to 8 mm; Regions 2, 3, and 4, Ontario to Minnesota to Missouri, Tennessee, and Georgia, absent on the Coastal Plain and Piedmont; tolerant of soil conditions ranging from bogs (hydric) to bluffs and balds.

Aronia x *prunifolia* (Marshall) Rehder; purple chokecherry (Figure 21.13); rhizomatous colonial shrub, height to 3 m; deciduous; leaf arrangement alternate; leaf simple, entire, length 4 to 8 cm, width 1.5 to 4.0 cm, elliptical, oblong, or oblanceolate, margin serrulate or crenulate, ciliate, apex obtuse or mucronate, adaxial surface midrib glandular, abaxial surface tomentose; twigs lightly hairy; inflorescence a terminal cyme of seven to ten flowers; flower regular, diameter 0.8 to 1.2 cm, sepals five, petals five, white to pink; fruit a pome, red or purplish, diameter 1 to 9 mm; Regions 2, 3b, and 4, New York to Ontario to Florida and Arkansas; swamps and hydric woods. The status of this species is uncertain. Generally, it has been considered a hybrid (*P.* x *prunifolia*) but has also been reported as a species (*A. atrpurpurea* Britton; *Prunus floribunda* Lindley) and a variety (*A. arbutifolia* var. *atropurpurea* Robins).

Genus: *Sorbus*, Mountain-Ash

The genus *Sorbus* contains six small trees and shrubs native to the United States, Canada, and Greenland. Worldwide, there are a total of 80 in the temperate

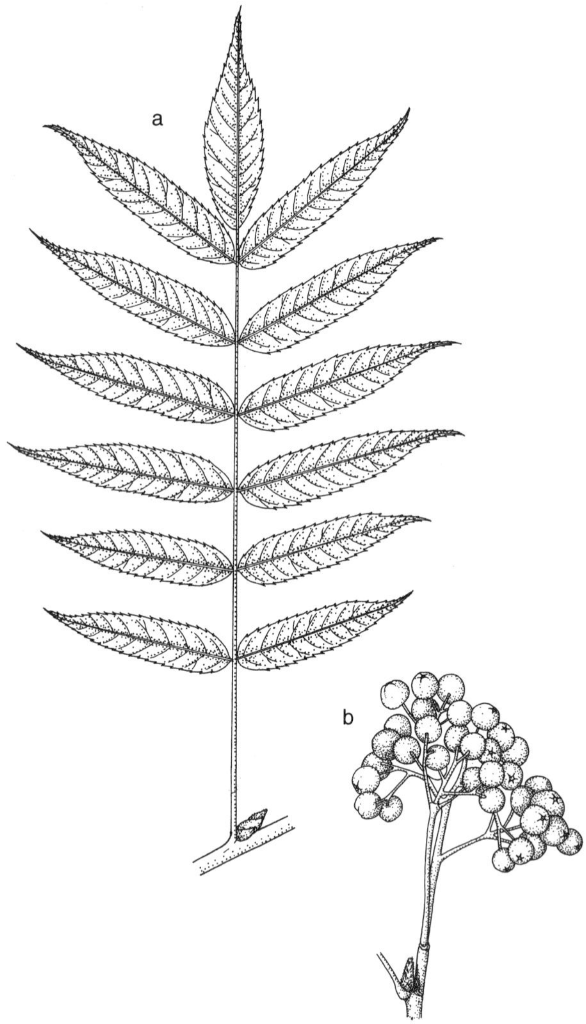

FIGURE 21.14 *Sorbus americana* (a) pinnately compound leaf ×3/4, (b) cyme ×1/2.

region, but in North America, most are found in north temperate climates or in mountains.

Sorbus americana Marshall [*Pyrus americana* (Marshall) de Candolle], American mountain-ash (Figure 21.14); small tree, height to 12 m; deciduous; leaf arrangement alternate; leaf length 15 to 30 cm, pinnately compound; leaflet number 11 to 17, length 3 to 9 cm, width 1.0 to 1.7 cm, oblong to lanceolate, margin serrated from the base to apex, apex long acuminate; twig glabrous; bud reddish brown to gray, cone-shaped, curving at the pointed tip, resinous; inflorescence a cyme or corymbform, number of flowers 15 to 35, flower small, diameter 1 cm, petals five, white; fruit a pome, orange red, globose, diameter 4 to 8 mm; seed flattened, one to two per fruit; crown spreading; Regions 1, 2, and 3b, Ontario and Minnesota to Newfoundland and Labrador, north to the tip of James Bay, south through New England along the Appalachian Mountains to Georgia and South Carolina; cold, wet climate and mesic sites, swamps and uplands. Commonly planted as an ornamental. Cover type 26.1, 26.9, 27.31.

Sorbus decora (Sargent) Schneider [*Pyrus decora* (Sargent) Hyland], showy mountain-ash; small tree, height to 10 m; deciduous; leaf arrangement alternate; leaf length 12 to 25 cm, pinnately compound; leaflet number 11 to 15, length 4 to 8 cm, width 1.0 to 1.7 cm, oblong to lanceolate, margin serrated from base to apex, apex acuminate; twigs glabrous; bud reddish brown to gray, cone-shaped, curving at the pointed tip, scales resinous on the outer surface, pubescent on the inner surface; inflorescence cyme-like number of flowers 15 to 35; flower small, diameter 1.5 cm, petals five, white; fruit a pome, red, globose, diameter 0.6 to 1.0 cm; seed flattened, one to two per fruit; crown spreading; Regions 1 and 2, with disjunct locations in Region 3, Ontario and Minnesota to New England, and Newfoundland to northern Quebec, Labrador, and Greenland. A species similar to and with a range largely overlapping *S. americana*, except that the leaves are not long acuminate and the flowers and fruit are larger and appear later; this phenological difference between species is probably due to the more northern range of *S. decora*. This species has also been considered a variety of *S. americana*.

Sorbus aucuparia Linnaeus, European mountain-ash, Rowan-tree; small tree, height to 10 m; deciduous; leaf arrangement alternate; leaf length 13 to 22 cm, pinnately compound; leaflet number 9 to 15, blade length 2.5 to 7.5 cm, width 1.0 to 1.7 cm, oblong to lanceolate, margin serrated except for the basal one third, apex obtuse, abaxial surface tomentose; twigs lightly pubescent, gray; bud reddish, cone-shaped, curving at the pointed tip, scales white villose, not resinous; inflorescence a cyme; flower small, diameter 1.5 cm, petals five, white; fruit a pome, red, globose, diameter 0.8 to 1.0 cm; seed flattened, one to two per fruit; crown spreading. Introduced.

Sorbus scopulina Greene [*Pyrus scopulina* (Greene) Goodyear], western or green mountain-ash (Figure 21.15); shrub, height to 3 m forming dense clumbs; deciduous; leaf arrangement alternate; leaf length 10 to 15 cm, pinnately compound; leaflet number 9 to 15, length 7 to 15 cm, width 1.5 to 2.5 cm, elliptic to lanceolate, margin doubly serrate from base to apex, apex acute or acuminate; twigs with scattered hair; bud reddish brown to gray, cone-shaped, curving at the pointed tip, resinous; inflorescence a cyme, flowers 15 to 30; flower small, diameter 1 cm, petals five, white; fruit a pome, orange red, globose, diameter 6 to 8 mm; seed flattened, one to two per fruit; crown spreading; Regions 5, 6, and 7; Alaska and the Yukon south to Utah, Arizona, and New Mexico, California; montane woods and open slopes. Reported to hybridize with *Amelanchier alnifolia* (Little 1979). *Sorbus californica* (California mountain-ash) has fewer leaflets than *S. scopulina* (Stuart and Sawyer 2001).

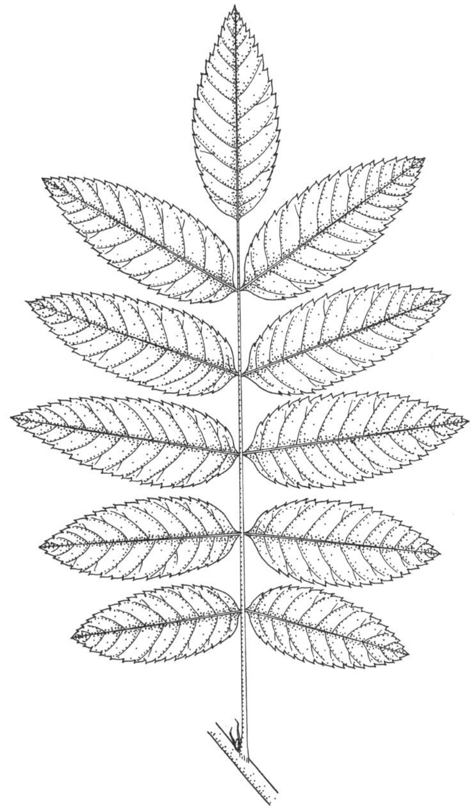

FIGURE 21.15 *Sorbus scopulina* leaf ×3/4.

Sorbus sitchensis Roemer [*S. occidentalis* (S. Watson) Greene], Sitka mountain-ash; shrub, height to 3 m; deciduous; leaf arrangement alternate; leaf length 10 to 15 cm, pinnately compound; leaflet number 9 to 11, length 3 to 5 cm, width 1.0 to 1.7 cm, oblong to lanceolate, margin serrated along the distal one half, apex obtuse to rounded, surfaces glabrous; twigs glabrous; bud reddish brown to gray, cone-shaped, curving at the pointed tip, resinous; inflorescence a cyme, flowers 15 to 30; flower small, diameter 1 cm, petals five, white; fruit a pome, red, diameter 6 to 8 mm; seed flattened, one to two per fruit; crown spreading; Regions 6 and 7; southeast Alaska south to Alberta, Montana, and California; montane woods.

Genus: *Amelanchier*, Serviceberry

Between ten and twenty species have been named in North America. Only the more widely distributed are described here.

Amelanchier arborea (Michaux f.) Fernald; downy serviceberry, Juneberry, shadbush, sarvis (Figure 21.16); small tree, height to 12 m; often two to three stems in a group, leaning; deciduous; leaf arrangement alternate; leaf simple, length 5 to 9 cm, width 2.0 to 3.5 cm, blade entire, ovate, margin sharply serrate, apex short acuminate, base slightly rounded, abaxial surface pubescent; twigs slender, reddish to gray brown, nearly glabrous; bud outer scales reddish brown, inner visible scales greenish, length 0.9 to 1.2 cm, sublinear; inflorescence a short raceme; flowers regular, diameter 3 to 4 cm, sepals five, petals five, white, stamens about 20, in several rows, fully expanded before the leaves; fruit a pome, globose, diameter 0.6 to 1.0 cm, reddish purple, sweet; Regions 2, 3, and 4, Minnesota and Ontario east to Quebec and Nova Scotia and south to the Gulf Coast excluding the Florida Peninsula xeric-mesic upland woods. The fruit is a preferred wildlife food. Cover type 27.22, 27.29, 27.35.

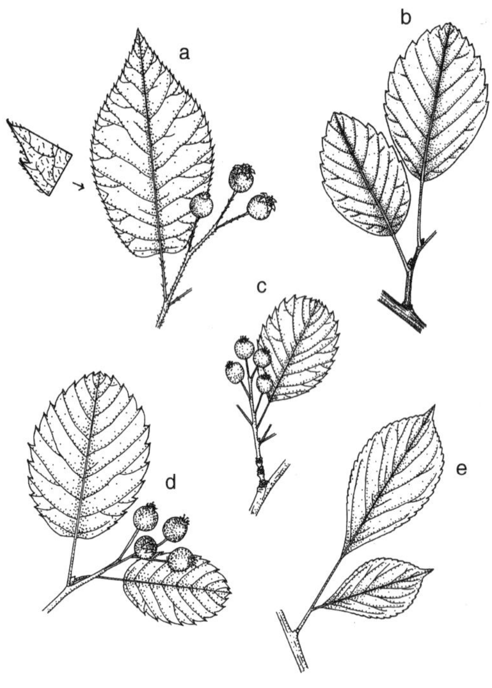

FIGURE 21.16 (a) *Amelanchier arborea* leaf and inflorescence with attached pomes ×1/2; (b) *A. utahensis* leaves ×1; (c) *A. alnifolia* leaf and inflorescence with attached pomes ×1/2; (d) *A. sanguinarea* leaf and inflorescence with attached pomes ×3/5; (e) *A. laevis* leaves ×1/2.

Amelanchier utahensis Koehne (*A. pallada* Greene), Utah serviceberry (Figure 21.16), small tree, height to 8 m; stoloniferous and forming colonies; deciduous; leaf arrangement alternate; leaf simple, length 0.6 to 3.0 cm, width 0.5 to 2.5 cm, blade entire, oval, margin coarsely serrated on the distal one half, slightly coriaceous, apex obtuse to rounded, base rounded, smooth, adaxial and abaxial surfaces pubescent; twigs bright reddish brown to gray; bud reddish brown, length 3 to 8 mm, sublinear; inflorescence a short raceme; flowers regular, diameter 2 to 3 cm, sepals five, petals five, white, stamens numerous, in several rows, fully expanded before the leaves; fruit a pome, ovate, diameter 0.5 to 1.0 cm, light brown becoming purplish black; Regions 5, 6, and 7, Montana, Idaho, and Oregon south into Baja California, Arizona, New Mexico, and Texas; xeric sites, mountain slopes and canyons, elevation 1100 to 2400 m.

Amelanchier alnifolia (Nuttall) Nuttall, Saskatoon-berry, western serviceberry (Figure 21.16); shrub or small tree, height to 4 m; stoloniferous and forming colonies; deciduous; leaf arrangement alternate; leaf simple, length 2 to 5 cm, width 1.5 to 2.5 cm; blade entire, oval to suborbicular, margin sharply serrated distally with teeth decreasing in size at the apex, margin becoming smooth near the base, apex obtuse to rounded, abaxial surface midrib lightly pubescent; twigs reddish to gray brown, nearly glabrous; bud reddish brown, length 2 to 7 mm, sublinear; inflorescence a short raceme; flowers regular, diameter 3 to 4 cm, sepals five, petals five, white, stamens numerous, in several rows, fully expanded before the leaves; fruit a pome, globose, diameter 0.6 to 1.0 cm, red; Regions 1, 5, 6, and 7, also grasslands (G), Iowa, Minnesota, and Manitoba west to Alaska, south into California, Utah, Colorado, extending into Nebraska. This is a common *Amelanchier* species in the western United States and Canada; xeric-mesic upland woods. Cover type 27.19, 29.24, 29.26, 29.27, 29.35, 29.36, 29.13, 29.15, 29.20, 29.6, 30.5, 30.20, 30.16, 30.41.

Amelanchier sanguinarea (Pursh) de Candolle, roundleaf serviceberry; New England serviceberry (Figure 21.16); small tree or shrub, height to 6 m; stoloniferous and forming colonies; deciduous; leaf arrangement alternate; leaf simple, length 3 to 7 cm, width 2.0 to 3.5 cm, blade entire, oval to suborbicular, margin coarsely serrated with teeth slightly decreasing in size at the apex, apex obtuse to rounded, base rounded, smooth, abaxial surface midrib lightly pubescent; twigs bright red; bud reddish brown, length 3 to 9 mm, sublinear; inflorescence a short raceme; flowers regular, diameter 2 to 3 cm, sepals five, petals five, white, stamens numerous, in several rows, fully expanded before the leaves; fruit a pome, globose, diameter 0.5 to 1.0 cm, dark purple or nearly black; Regions 2 and 3, Minnesota and Ontario to Maine, disjunct populations south along the Appalachian Mountains to North Carolina; xeric-mesic sites, upland slopes. The species *A. huronensis* (Wiegand) Rehder probably should be included with this species.

Amelanchier laevis Weigand, smooth serviceberry, Allegheny serviceberry (Figure 21.16); small tree, height to 12 m. Little (1979), with general agreement from Farrar (1995), includes this species with *A. arborea.* The primary difference between these two populations is that the apex of *A. laevis* is abruptly acuminate or mucronate and the base is acute. The ranges are similar in the north, but the range of *A. laevis* extends south into Region 3, Iowa, Indiana, and Maryland, along the Appalachian Mountains to Georgia; xeric-mesic upland woods.

Amelanchier bartramiana (Tausch) Roemer, mountain serviceberry; shrub, height to 2 m; stoloniferous and forming colonies; deciduous; leaf arrangement alternate; leaf simple, length 2.5 to 5.0 cm, width 1.5 to 2.5 cm, blade entire, elliptical, oval, or slightly obovate, margin serrate, apex often mucronate, base acute, adaxial and abaxial surfaces glabrous; twigs bright red; bud reddish brown, length 3 to 9 mm, sublinear; inflorescence a short raceme; flowers regular, diameter 2 to 3 cm, sepals five, petals five, white, stamens numerous, in several rows, fully expanded before the leaves; fruit a pome, oval, length 0.7 to 1.1 cm, dark purple or nearly black; Regions 1 and 2, Manitoba, Ontario from James Bay south to Wisconsin, Michigan, east to Labrador, south to West Virginia; mesic forest and swamps.

Subfamily: Spiraeoideae

Genus: *Spiraea*, *Physocarpus*

Worldwide there are about 75 species of these genera, mostly in the Northern Hemisphere, and these are generally located in the more northern sections of the continents. In North America, there are about 12 to 14 species of *Spiraea* and only 2 *Physocarpus*, with numbers split about equally between western and eastern regions. There are several other genera of only 1 to 2 species in this subfamily.

Spiraea tomentosa Linnaeus; hardhack, steeplebush (Figure 21.17); shrub, height to 1.2 m; deciduous; leaf arrangement alternate; leaf simple, total length 3

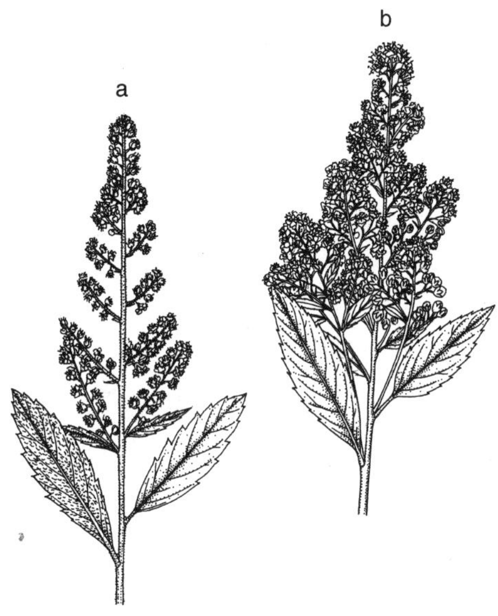

FIGURE 21.17 (a) *Spiraea tomentosa* twig with leaves and inflorescence in flower ×1/2; (b) *S. alba* twig with leaves and inflorescence with attached follicles ×1/2. Note that the abaxial surface of *S. tomentosa* is tomentose.

to 5 cm, width 1.0 to 1.5 cm; blade elliptical, ovate or oblong, serrated, three to six teeth per side, abaxial surface tomentose, dirty white to light tan; twigs bright brownish to dark purple, tomentose, hair deciduous; inflorescence a terminal raceme, length up to 16 cm; flowers regular, small, diameter 4 to 5 mm, sepals five, petals five, white, rose, or purple; fruit a pale brown follicle, tomentose; Regions 1, 2, and 3, Nova Scotia and Quebec to Manitoba south into Arkansas, Tennessee, and Georgia; swamps, mesic-hydric soil, xeric-mesic old fields. Relatively inconspicuous.

Spiraea alba Du Roi, meadow-sweet (Figure 21.17); shrub, height to 1.2 m; deciduous; leaf arrangement alternate; leaf simple, total length 4 to 7 cm, width 1.0 to 1.7 cm, blade elliptical, obovate, or oblanceolate, serrated, 8 to 16 teeth per side, adaxial and abaxial surfaces glabrous or lightly pubescent; twigs bright brownish to dark purple, tomentose, hair deciduous; inflorescence a terminal raceme, length up to 16 cm; flowers regular, small, diameter 6 to 8 mm, sepals five, petals five, white; fruit a pale brown follicle, usually five on a pedicel; Regions 1, 2, and 3, Nova Scotia and Quebec to Alberta south into North Carolina, Ohio, Indiana, Missouri, and North Dakota; swamps, mesic-hydric soil, xeric-mesic old fields. Relatively inconspicuous.

Physocarpus opulifolius (Linnaeus) Maximowicz, ninebark (Figure 21.18); multistemmed shrub, height to 3 m; deciduous; leaf arrangement alternate; leaf simple, total length 3 to 7 cm, width 3 to 6 cm, margin irregularly serrate to crenate, occasionally doubly serrate to three-lobed, apex obtuse, base cuneate, abaxial surface lightly hairy; petiole length 1.0 to 2.5 cm, glabrous;

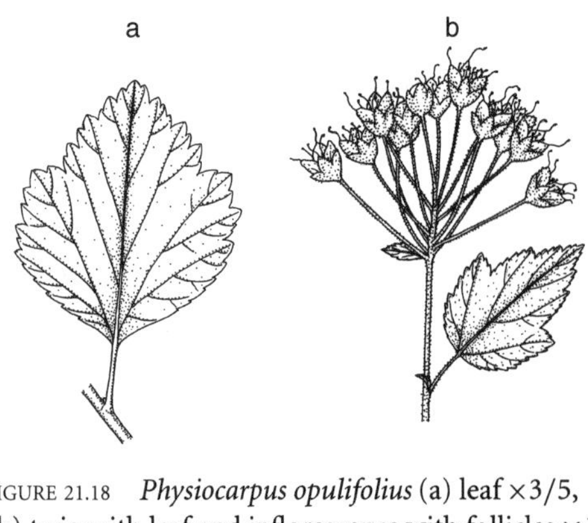

FIGURE 21.18 *Physiocarpus opulifolius* (a) leaf ×3/5, (b) twig with leaf and inflorescence with follicles ×1.

twigs yellowish brown, three-ridged below leaf or leaf scar; inflorescence a round-top corymb; flowers regular, diameter 0.7 to 1.1 cm, white or pinkish, fruit a bladderlike inflated follicle, length 2 to 4 mm, rose or purple, two to five on a pedicel; bark peeling and shredding in thin papery layers; Regions 1, 2, and 3, Quebec and Hudson Bay to Manitoba, North Dakota and Colorado, south to Georgia and Tennessee; stream banks, shores, also xeric-mesic rocky soil. *Physocarpus malvaceus* (Greene) Kuntze; also called ninebark, is a western species with similar leaf, fruit, and bark characteristics; Regions 5 and 6, British Columbia and Alberta south into Oregon, Utah, and Wyoming; montane slopes and woods.

ORDER: FABALES

Fabaceae (Legume, Pea, or Bean Family)

The Fabaceae, renamed from the nonconforming Leguminosae several years ago, is the third largest family of flowering plants after the Orchidaceae (orchids) and Asteraceae (composites, sunflowers, asters, goldenrods, etc.). The Fabaceae contains 700 to 800 genera and 14,000 to as many as 20,000 species, mostly pantropical and subtropical. Three subfamilies are recognized by most taxonomists; Mimosoideae, Caesalpinioideae, and Papilionoideae (=Faboideae or Lotoidae). More recently, Woodland (1991) and Gleason and Cronquist (1991) raised these subfamilies to family status (Mimosaceae, Caesalpiniaceae, and Fabaceae), and some workers use both subfamily and family names simultaneously (Walters and Keil 1977; Zomlefer 1994). However, Judd et al. (1999) used DNA structure and chloroplast gene *rbcL* sequencing to support the monophylitic condition of the Fabaceae. In as much as all genera have trifoliate or pinnately compound leaves (with an infrequent exception), have swollen nodes, and produce legume fruit, the conservative approach of recognizing one family and three subfamilies is used here.

Subfamily: Papilionoideae (Faboideae, Fabaceae)

The Papilionoideae is the largest of the three subfamilies, with about 420 to 440 genera and 10,000 to 14,000 species of trees, shrubs, and herbs. Economically, this subfamily ranks second in importance to the Poaceae (grass) family. The strongly zygomorphic (irregular, bilateral) flower is the typical legume flower consisting of five unequal petals (a standard, two wings, and two basal petals often united into a boat-shaped keel structure, Figure 16.1). Flowers usually have ten stamens, either distinct or nine of which are fused. The stamens are not showy (shorter than petals). North American tree species produce a dry or somewhat fleshy legume, and in some herbaceous species, a loment. A loment (a legume composed of triangular one-seeded sections that separate at maturity) is characteristic of a *Desmodium* species commonly known as tick treefoil or beggar's-ticks, whose achenes stick to clothes. Trees and shrubs are found in the genera *Cladrastis, Robinia*, and *Puereria.*

Cladrastis lutea (Michaux f.) K. Koch [*C. kentukea* (Dumont de Courset) Rudd], yellowwood (Figure 21.19); medium-sized tree, height to 18 m; deciduous; leaf arrangement alternate; leaf pinnately compound, length 13 to 21 cm, petiole base swollen; leaflet number 7 to 11, alternately arranged, petiolated, shape ovate, length 5 to 10 cm, size decreasing from apex to base, apex often mucronate; twigs stout, brownish, terminal bud false, leaf scar U-shaped, bundle scars five to seven; inflorescence a panicle, length 20 to 50 cm; flowers, irregular, petals white, showy; legume length 6 to 10 cm, width 5 to 10 mm, flat, number of seeds four to six, constricted where seeds are absent; bark smooth, gray, wood yellow; Region 3, restricted to widely scattered areas of the central hardwood states from North and South Carolina to southern Indiana, Missouri and Oklahoma, to

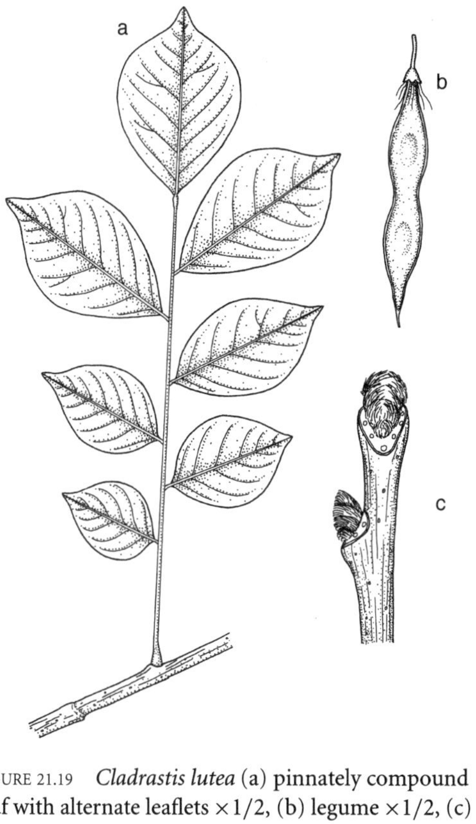

FIGURE 21.19 *Cladrastis lutea* (a) pinnately compound leaf with alternate leaflets ×1/2, (b) legume ×1/2, (c) twig with tomentose lateral and false terminal buds ×2.

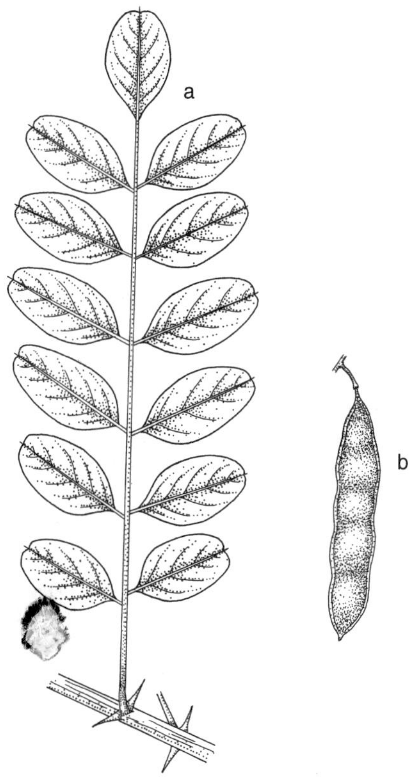

FIGURE 21.20 *Robinia pseudoacacia* (a) twig with pinnately compound leaf and stipular thorns ×1/3, (b) legume ×1/3.

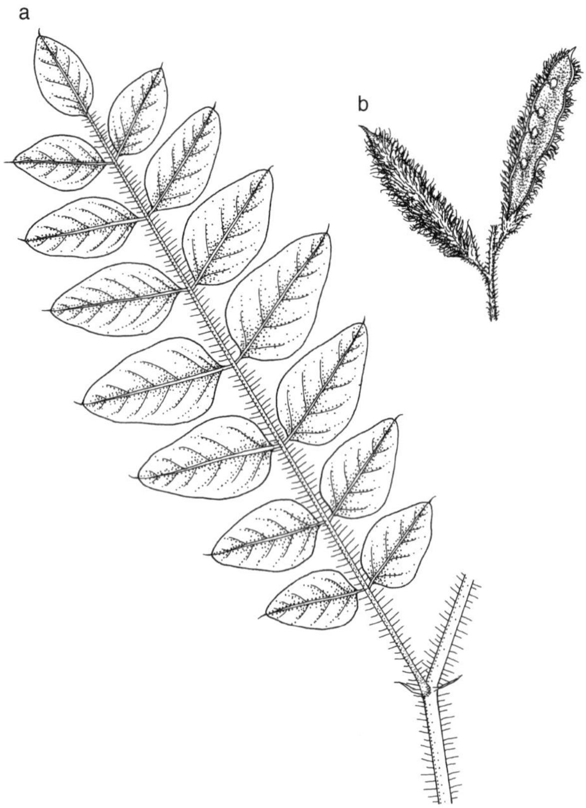

FIGURE 21.21 *Robinia hispida* (a) twig with bristly leaf petiole rachis ×1/2, (b) legumes showing bristly surface and location of seeds ×1/3.

southern Alabama and Mississippi. Not common; listed as rare and endangered in some states.

Robinia pseudoacacia Linnaeus, black locust (Figure 21.20); medium-sized to large tree, height to 25 m; deciduous; leaf arrangement alternate; leaf pinnately compound, length 18 to 35 cm, petiole base swollen; leaflet number 7 to 19, oppositely arranged, petiolated, shape oval, length 3.5 to 5.0 cm, width 1.5 to [illegible].5, apex bristle-tipped, bristle length 1 to 3 mm; twigs stout, dark brown, brittle, glabrous, terminal bud false, leaf scar U-shaped, bundle scars three, buds hidden by leaf scars, buds white tomentose, stipular thorns two, one each side of leaf scar, length 5 to 6 mm, occasionally absent; inflorescence a raceme, length 10 to 20 cm; flowers irregular, petals white, showy, fragrant; legume length 5 to 12 cm, width 1 to 2 cm, thin, flat, number of seeds five to ten; bark platy, gray; Region 3, concentrated in the Appalachian Mountains from Pennsylvania to Georgia and Alabama, in the Ozark mountains of southern Missouri, Arkansas, and eastern Oklahoma, disjunct populations in southern Illinois and Indiana. This is the largest *Robinia*. It has been widely planted in the eastern United States for erosion control and on strip mine spoils. Mid to late summer each year, the locust leafminer (*Odontota dorsalis*) begins to attack the leaf. The entire crown turns yellowish and the tree appears to be dying, but a new set of healthy leaves is produced each spring. Cover type 27.12.

Robinia hispida Linnaeus, bristly locust, rose-acacia (Figure 21.21); shrub, height to 3 m, spreading by stolons; deciduous; leaf arrangement alternate; leaf pinnately compound, length 15 to 30 cm; petiole and rachises bristly-sticky, petiole base swollen; leaflet number 6 to 15, oppositely arranged, petiolated, shape ovate-lanceolate, length 3.5 to 5.0 cm, width 1.5 to 4.5 cm, apex bristle-tipped, bristle length 1 to 3 mm; twigs stout, dark brown, brittle, bristle covered, bristle length 2 to 4 mm, sticky-glandular, terminal bud false, leaf scar U-shaped, bundle scars three, stipular thorns two, one each side of leaf scar, length 1 to 2 cm, occasionally absent; inflorescence a raceme, length 10 to 20 cm; flowers three to eight, irregular, petals white or rose, showy, fragrant, peduncles bristly; legume length 5 to 12 cm, width 1 to 2 cm, thin, flat, number of seeds five to ten; Region 3, Appalachian Mountains, Virginia, and Tennessee south to Georgia and Alabama, cultivated and escaped northward; xeric and xeric-mesic slopes and woods.

Robinia viscosa Ventenat, clammy locust; shrub or small tree, height to 8 m, rhizomatous; deciduous; leaf arrangement alternate; leaf pinnately compound, length 15 to 30 cm, petiole and rachises sticky-glandular, petiole base swollen; leaflet number 6 to 15, oppositely arranged, petiolated, shape ovate-lanceolate, length 3.5 to 5.0 cm, width 1.5 to 4.5 cm, apex bristle-tipped, bristle length 1 to 3 mm; twigs stout, dark brown, brittle, sticky-glandular, terminal bud false, leaf scar U-shaped, bundle scars three, stipular thorns two, one each side of leaf scar, length 1 to 2 cm, occasionally absent; inflorescence a raceme, length 10 to 20 cm; flowers three to eight, irregular, petals pink or rose, showy, not fragrant, peduncles sticky; legume length 5 to 8 cm, width 1 to 2 cm, thin, flat, sticky, number of seeds five to ten; Region 3, Appalachian Mountains, Pennsylvania to Alabama and Georgia.

Robinia elliottii (Chapmann) Ashe, downy locust; shrub, height to 2 m. The leaves and twigs are similar to those of *R. pseudoacacia*, except: leaflets five to seven; leaflet abaxial surfaces lightly pubescent; twigs gray pubescent or short spiny; stipular thorns generally shorter, flowers rose purple; Region 3, North Carolina and Georgia, spreading northward through cultivation; forest edges and thickets.

A fifth *Robinia* (R. neomexicana A. Gray, New Mexican locust) found with pinyon pine, oak, and *Pinus ponderosa* in the southwest (Region 5) is described by Elias (1980).

Pueraria montana (Loureiro) Merrill [*Pueraria lobata* (Willdenow) Ohwi], kudzu, kudzu-vine

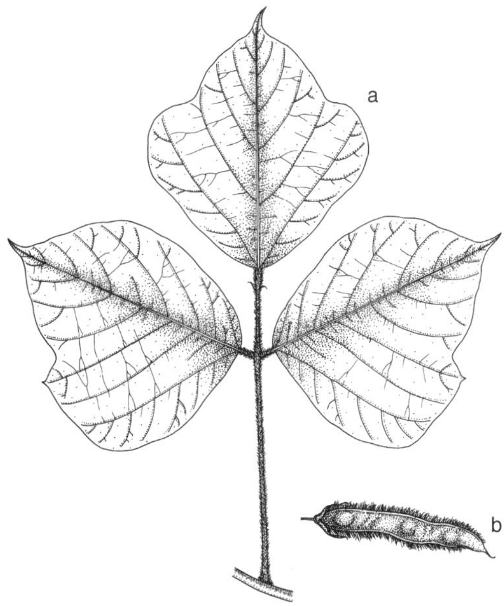

FIGURE 21.22 *Pueraria montana* (a) trifoliate leaf $\times 1/2$, (b) legume $\times 1/2$.

(Figure 21.22); sprawling/climbing vine; deciduous; leaf arrangement alternate; leaf pinnately compound, length 12 to 22 cm, leaflet number three, margin smooth; petiole and petiolules short, bristly; leaflet length 8 to 14 cm, more or less orbicular, apices acuminate; terminal leaflet symmetrical, weakly three-lobed; side leaflets asymmetrical, blade broader to the outside because of a single lobe/large tooth; inflorescence a raceme; flowers purple; legume bristly, length 5 to 7 cm; escaped from cultivation; a serious problem in Region 4, becoming a problem in Region 3. This vine develops a high density over the ground and can climb up and over even medium-sized trees, ultimately killing them. Kills shrubs, small and large trees. Kudzu is not easily controlled, as most herbicides are ineffective.

In addition to *Desmodium*, this subfamily also includes a variety of common herbaceous species associated with forest communities (*Tephrosia*, goat's rue; *Lespedeza*, bush clover; *Apios*, groundnut), prairie or range land (*Lupinus*, lupine; *Amorpha*, lead plant; *Pelalostemon*, prairie clover; *Astragalus*, locoweed), and agriculture (*Medicago*, alfalfa; *Melitotus*, sweet

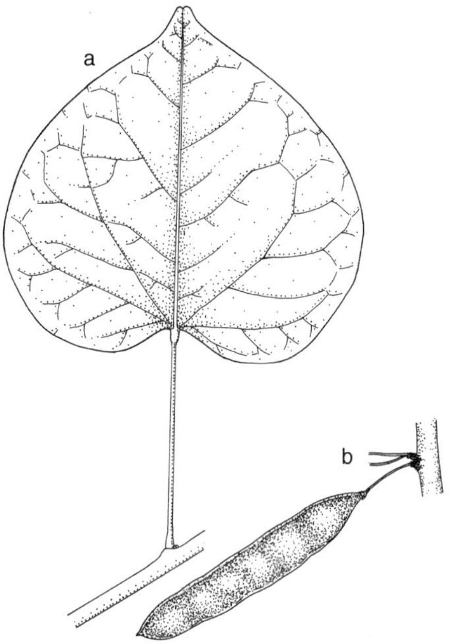

FIGURE 21.23 *Cercis canadensis* (a) simple leaf, (b) legume.

clover; *Trifolium*, clover; *Vicia*, vetch; *Pisum*, pea; *Phaseolus*, bean; *Glycine*, soybean).

Subfamily: Caesalpinioideae

The subfamily caesalpinioideae contains 150 to 180 genera and 2200 to 3000 species, mostly trees and shrubs. The leaves may be simple or pinnately compound. The flower of five unequal petals (standard, two wings, and two basal petals form the keel, Figure 16.1) is similar to the Papilionoideae flower but does not have as strong a zygomorphic character in some species. The flower has ten or fewer, not showy, stamens (shorter than petals). North American tree species produce a dry or fleshy legume. Trees are found in several native and introduced genera. Of the five species considered here, two have a simple leaf, the others pinnately compound leaves.

Cercis canadensis Linnaeus, eastern redbud (Figure 21.23); large multistemmed shrub or small tree with a single stem, height to 12 m; deciduous; leaf arrangement alternate; leaf simple; petiole length 4 to 10 cm, swollen at both ends, blade broadly cordate or reniform, often wider than long, apex obtuse, base cordate, venation mixed with a midrib and usually four accessory veins radiating from the base; twigs dark brown or black, brittle, terminal bud false, buds minute; flowers borne individually on a peduncle, Figure 16.1, irregular, length 2 to 3 cm, color magenta red; legume length 6 to 10 cm, width about 1 cm, long pointed at both ends, glabrous, number of seeds four to six; bark drab brown, smooth; Regions 3 and 4, Nebraska and Iowa to Pennsylvania and New York, south along the coast to west Texas, Oklahoma, and central Mexico; cool mesic sites, deciduous forest along drainages, coves, and hollows. A common arborescent forest tree of the eastern United States but commonly planted as a small lawn or ornamental tree. The numerous bright magenta pink flowers appear in early spring (April) before the leaves expand. Cover type 27.5, 27.9, 27.11, 27.13, 27.22, 27.24, 28.4, 28.18.

Cercis occidentalis Torrey & Gray; California redbud; shrub or small tree, height to 8 m. The leaves, twigs, and legumes are similar to those of *C. canadensis*, except that the petiole length is only 1 to 3 cm and the leaf is somewhat leathery. Regions 7 and 5, California, scattered through Nevada, Utah, and Arizona; xeric and xeric-mesic sites, stream banks, canyons, elevation to 1000 m; also common in chaparral.

Gymnocladus dioica (Linnaeus) K. Koch, Kentucky coffee-tree (Figure 21.24); large tree, height to 25 m; deciduous; leaf arrangement alternate; total leaf length 80 to 100 cm; bipinnately compound with three to six pairs of pinnae, each pinna with 8 to 23 leaflets, pinna nearest the petiole base may have an undivided blade; new twigs pubescent, bud length 6 to 9 mm; inflorescence a raceme, length 7 to 10 cm, flowers unisexual (trees dioecious), male flowers with 10 stamens, female with a single pistil; legume length 15 to 25 cm, width 3 to 4 cm, thick, heavy, hard; seeds two to four; Region 3, southeast South Dakota and Minnesota to western New York and Pennsylvania, along the western boundary of the Appalachian Mountains to Alabama, Mississippi, Louisiana, northeastern Texas, Oklahoma, Missouri, and Iowa. This species is largely restricted to moist, cool sites in the Central States forest, with its greatest geographic concentration between central Ohio and Oklahoma.

Gleditsia triacanthos Linnaeus, honeylocust (Figure 21.25); medium to large tree, height to 30 m;

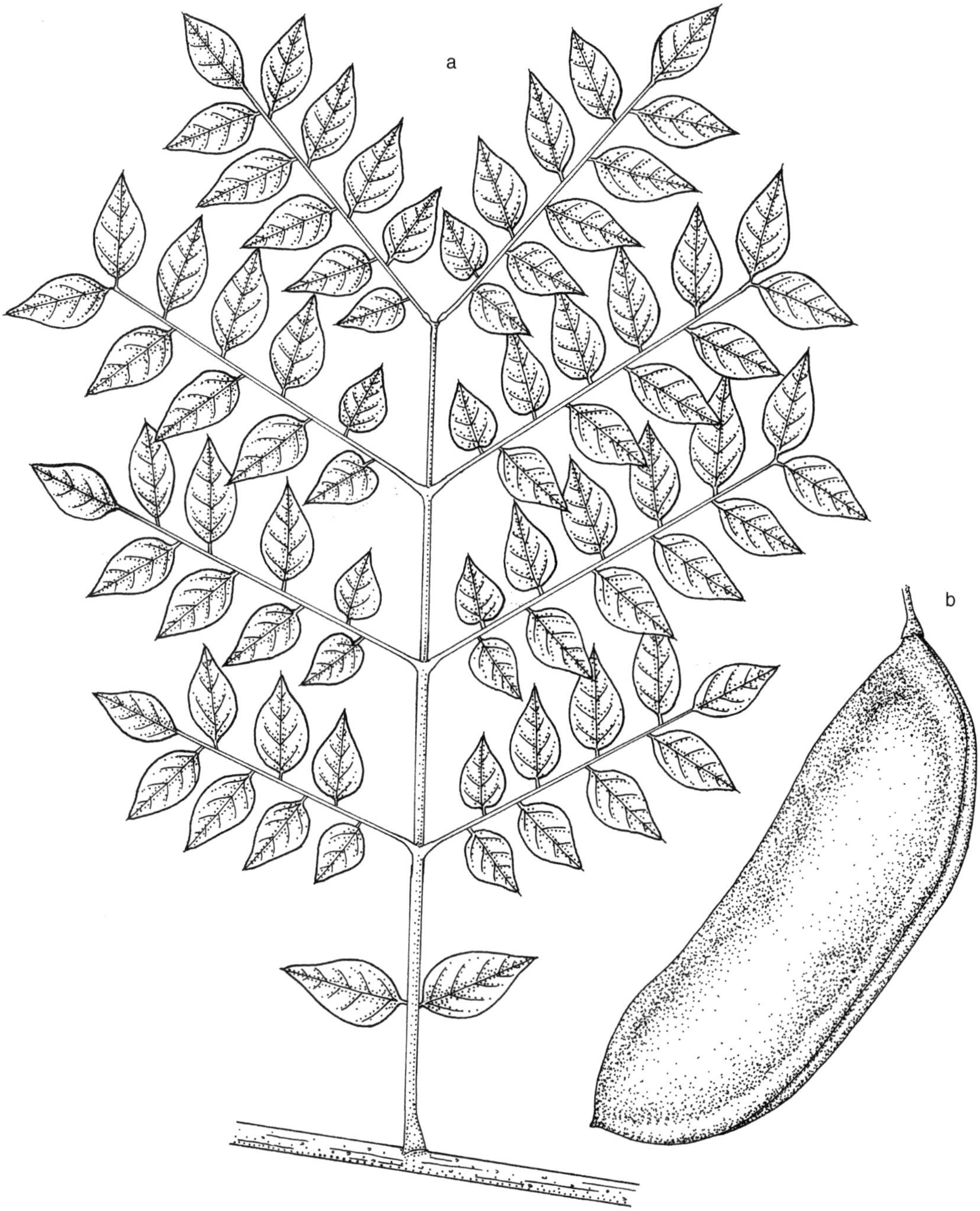

FIGURE 21.24 *Gymnocladus dioica* (a) bipinnately compound leaf ×1/6, (b) legume ×1/2.

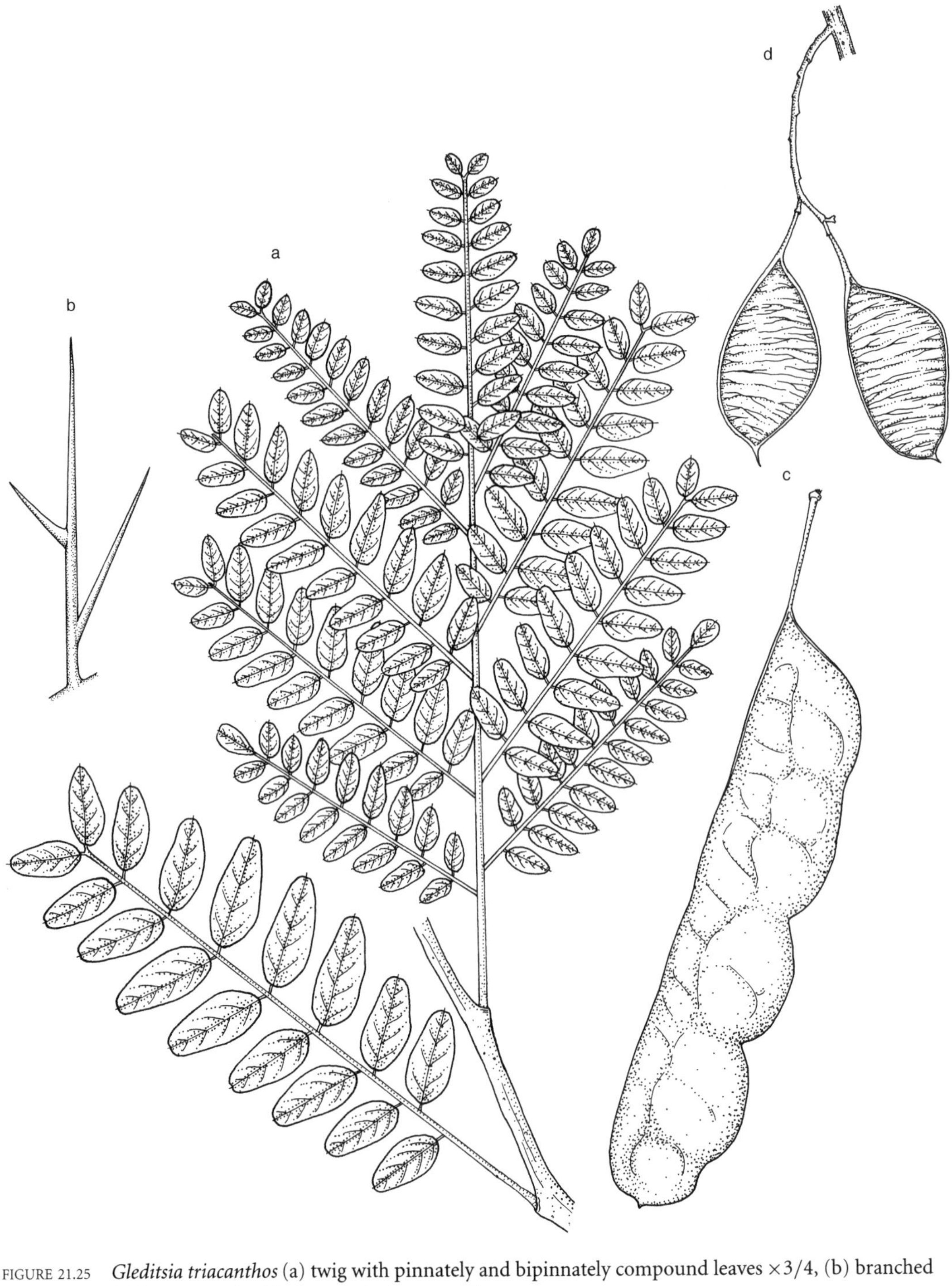

FIGURE 21.25 *Gleditsia triacanthos* (a) twig with pinnately and bipinnately compound leaves ×3/4, (b) branched thorn ×1/2, (c) legume ×1/3; (d) *G. aquatica* legume ×1.

deciduous; leaf arrangement alternate; leaf complexity variable, pinnately or bipinnately compound or both on the same tree or leaf, some leaflets subdivided, others not subdivided, terminal leaflet, pinna, or pinnule often missing; total leaf length 15 to 20 cm; pinna length 7 to 10 cm; pinnule length 1 to 2 cm, width 0.5 to 1.0 cm, apex bristle-tipped; inflorescence a spike or raceme; flowers unisexual, small, male and female flowers on same tree (trees polygamous); legume length 12 to 40 cm, width 2.5 to 4.0 cm, often twisted; seeds 9 to 15; bark gray, smooth or platy, thorny, thorn length up to 15 cm, usually with two to four side points; Regions 3 and 4, Minnesota to western Pennsylvania, along the western boundary of the Appalachian Mountains to Alabama, Mississippi, Louisiana, central Texas, Oklahoma, Missouri, and Iowa; mesic and mesic-hydric sites, swamps and adjacent uplands. Usually found as scattered trees within the forest community but also occurs on disturbed sites and roadsides. The thornless honeylocust *G. triacanthos* var. *inermis* (Linnaeus) Zab. is widely planted as a lawn and street tree. Cover type 27.9, 28.22.

Gleditsia aquatica Marshall, waterlocust; medium-sized tree, height to 18 m; deciduous; leaf arrangement alternate; leaf complexity variable, pinnately or bipinnately compound or both on the same tree or leaf, some leaflets subdivided, others not subdivided, terminal leaflet, pinna, or pinnule often missing; total leaf length 10 to 20 cm; pinna length 7 to 12 cm, pinnule length 1 to 2 cm, width 0.5 to 1.0 cm, apex bristle-tipped; inflorescence a spike or raceme; flowers unisexual, small, male and female flowers on same tree (trees polygamous); legume length 2.5 to 5.0 cm, width 1 to 2 cm, often twisted; seeds one to three; bark gray, smooth or platy, thorny, thorn length up to 15 cm, usually without points; Regions 3 and 4, Mississippi River floodplain from southern Illinois and central Missouri, south to Louisiana and Mississippi to eastern Texas; this species infrequently appears along the Coastal Plain, but has a major concentration from northern Florida to South Carolina; bottomlands and swamps. The range is not continuous but has various gaps and outliers. *Gleditsia triacanthos* and *G. aquatica* are known to hybridize (*Gleditsia* × *texana* Sargent), the hybrid having a pod length of 10 to 15 cm. Cover type **28.8,** 27.28, 28.21, 28.31, 28.50, 28.55.

Other North American genera of this subfamily include *Parkinsonia* and *Cerdidium* (paloverde) of the southwestern deserts and the beautiful introduced *Delonix regia* (Bojer ex. Hooker) Rafinesque (royal poinciana or flamboyant tree) of southern Florida.

Subfamily: Mimosoideae

The Mimosoideae subfamily consists of 50 to 60 genera and more than 3000 species of tropical and subtropical regions. The flower is five-merous, has distinct petals, and is regular (actinomorphic). The numerous small flowers are in tight heads; each flower has numerous (10+) long, brightly colored stamens (Figure 21.26). The leaves are bipinnately compound.

A number of genera are native to North America, but most are restricted to the subtropical or desert regions. Several shrub or small tree *Acacia* species are native to Mexico but extend into parts of Texas [*Acacia berlandierii* Bentham, Berlandier acacia; *A. farnesiana* (Linnaeus) Willdenow, sweet acacia; *A. greggii* A. Gray, catclaw acacia; *A. roemeriana* Scheele. Roemer acacia; *A. tortuosa* (Linnaeus) Willdenow, twisted

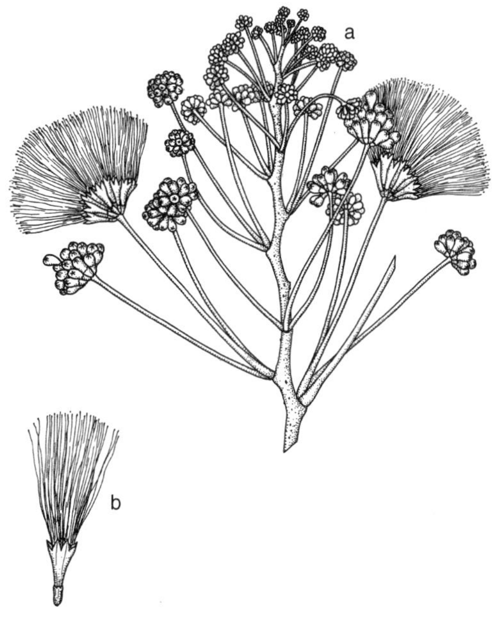

FIGURE 21.26 *Albizzia julibrissin* (a) inflorescence with fruit and flowers ×1/3, (b) flower with long stamens ×1/2.

acacia]. *Acacia greggii* is the most widespread species, extending from Mexico and the Baja into southern California, Arizona, Nevada, New Mexico, and Texas. *Acacia farnesiana* and *A. tortuosa* also occur primarily in Florida. *Acacia macracantha* Humb. & Bonpl. (long spine acacia), *A. choriophylla* Bentham (tamarindillo), *A. pinetorum* Hermann (pine acacia), and *A. smallii* Isely (Small's acacia) are located primarily in southern Florida and the keys. *Acacia* is characterized by strong sharp spines, often paired at the leaf node on the twigs.

Albizzia julibrissin Durazz., mimosa, silk or powderpuff tree (Figure 21.27); small tree, height to 12 m; deciduous; leaf arrangement alternate; leaf even bipinnately compound, lacking a terminal pinna or pinnule, total length 15 to 35 cm; pinnae 6 to 14, paired on the rachis, length 8 to 24 cm; pinnules 30 to 60 per pinna, length 0.5 to 2.5 cm, width 3 to 5 mm, sessile, blade asymmetrical, the primary vein located distally and closely paralleling the margin for its full length, pinnules folding at night; inflorescence cymose (Figure 21.26), flowers numerous, tightly clustered in heads, head spreading to a diameter of about 4 cm, fragrant, stamens showy, pinkish, longer than petals, style singular; legume greenish, becoming whitish and somewhat papery, length 8 to 18 cm; branches and wood weak; introduced from Asia. This species is widely planted as an ornamental as far north as Illinois and Indiana but it is not winter hardy. It has escaped cultivation. In Region 3 it can be found as an occasional stem along roads, fencerows, lawn margins, and woodlots, and in Region 4 it also appears in pinelands and hammocks. A species favored by hummingbirds. Often spelled *Albizia* but corrected by Bentham (Fernald 1950)

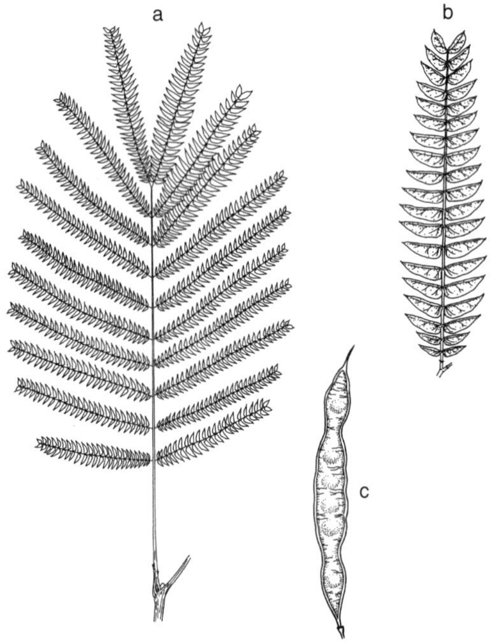

FIGURE 21.27 *Albizzia julibrissin* (a) bipinnately compound leaf ×1/4, (b) pinna with asymmetrical pinnules ×2/3, (c) legume ×1/4.

ORDER: PROTEALES

Elaeagnaceae (Oleaster Family)

This small family of three genera and about 50 species is distributed in the subtropical and temperate regions of the Northern Hemisphere and in Australia. The members are shrubs or small trees; the plants are synecious, dioecious, or polygamous. The leaves of these species are alternate simple, entire, have a smooth margin, and are covered with silvery scales, particularly on the lower surface.

Genus: *Shepherdia*

Three species of the genus *Shepherdia* are native to North American forests and are found in both the west and the east.

Shepherdia canadensis (Linnaeus) Nuttall; soapberry, rabbitberry, buffaloberry (Figure 21.28); multistemmed shrub, height to 2 m; deciduous; leaf

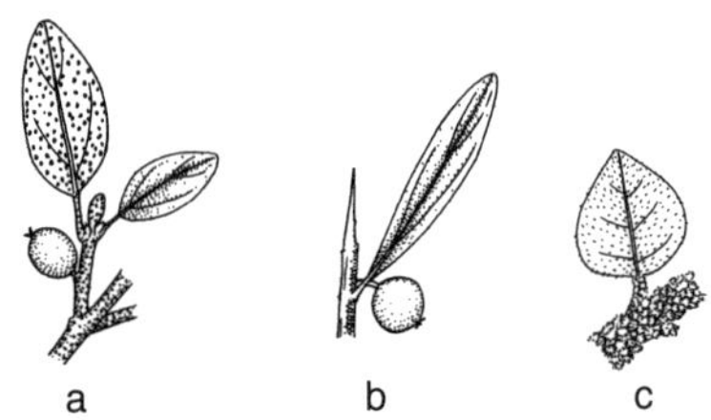

FIGURE 21.28 (a) *Shepherdia canadensis* scaly twigs, leaves, and drupe ×1/4; (b) *S. argentea* leaf and drupe on a thorny twig ×1/3; (c) *S. rotundifolia* leaf and densely scaly twig ×1/2.

arrangement opposite; leaf simple, total length 4 to 9 cm, width 2.0 to 4.5 cm, blade entire, margin smooth, shape elliptical, apex and base obtuse, adaxial surface green and nearly glabrous, abaxial surface densely covered with silvery and rusty stellate scales appearing as minute raised dots; twigs gray, rusty or silvery dotted, scaly; buds valvate, length 4 to 5 mm, rusty dotted; flowers small, yellowish green, bell-shaped, clustered; fruit a drupe, elliptical or oval, reddish or yellowish, one seeded; Regions 1, 2 and 5, Alaska to Newfoundland, New England and Lake States west and south into New Mexico and Arizona. Cover type 26.4, 29.26, 29.28, 29.29, 29.11.

Shepherdia argentea (Pursh) Nuttall; buffaloberry (Figure 21.28); multistemmed shrub, height to 2 m; deciduous; leaf arrangement opposite; leaf simple, total length 1.5 to 7.0 cm, width 0.6 to 1.0 cm, blade entire, margin smooth, shape nearly linear or elliptical, apex rounded, base acute, adaxial and abaxial surfaces densely covered with silvery stellate scales appearing as minute raised dots; twigs silvery scaly, often becoming pointed and thornlike; flowers small, yellowish green, bell-shaped, clustered; fruit a drupe, elliptical or oval, reddish to dark red, one seeded; Regions 5 and 6, Minnesota and Iowa to British Columbia; south into Nevada, Utah, xeric and disturbed sites, slopes and canyons.

Shepherdia rotundifolia Parry, roundleaf buffaloberry (Figure 21.28); multistemmed shrub, height to 2 m; deciduous; leaf arrangement opposite; leaf simple, total length 1.5 to 2.0 cm, width 1.0 to 1.5 cm, blade entire, margin smooth, slightly revolute, shape broadly elliptical to ovate, apex rounded, base acute, adaxial and abaxial surfaces densely covered with silvery stellate scales; twigs noticeably thickened with a layer of silvery scales; flowers small, yellowish tan, bell-shaped, clustered; fruit a drupe, length 1 cm, oval, reddish to dark red, one seeded, Regions 5 and 6, Arizona, Nevada, Utah; xeric and disturbed sites, slopes and canyons.

Genus: *Elaeagnus*

Only one species of *Elaeagnus* is native to the United States (*E. communtata*), but four species have been introduced from Asia and used as ornamentals. *Elaeagnus multiflora* Thunberg is not commonly known and is not illustrated here. The other species have become naturalized and have spread to native forests and lands

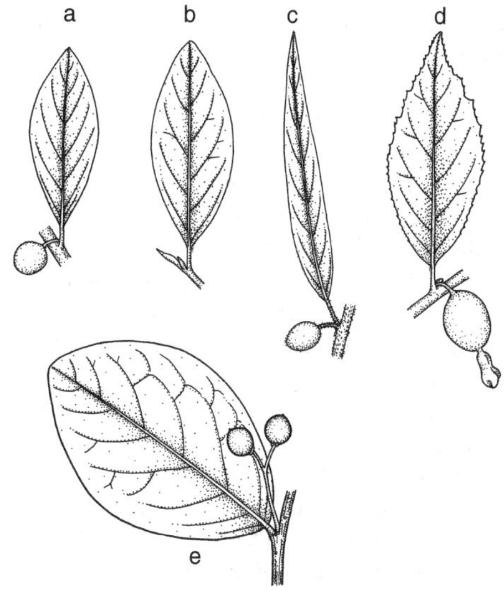

FIGURE 21.29 Leaf and drupe of (a) *Elaeagnus commutata* × 1/2; (b) *E. umbrellata* × 1/2; (c) *E. angustifolia* × 1/2; (d) *E. pungens* × 1/2; (e) *Dirca palustris* oval leaf and drupes × 1/2.

where they often create heavy cover and competition for native species. Their aggressiveness may be related to their symbiotic relationship with bacteria in nitrogen-fixing root nodules. The red drupes are reported to be higher in vitamin A than tomato.

Elaeagnus commutata Bernhardi, silverberry (Figure 21.29); shrub or small tree, height to 7 m, stoloniferous; deciduous; leaf arrangement alternate; leaf simple, total length 2 to 6 cm, width 0.8 to 2.5 cm, blade entire, margin smooth, shape elliptical to slightly lanceolate or obovate, apex and base acute to obtuse, adaxial and abaxial surfaces densely covered with silvery scales, thus appearing minutely pebble-grained (use 10× hand lens), also supporting scattered rusty scales appearing as dots; twigs covered with rusty and silvery scales; fruits a drupe, silvery, oval or nearly globose, length 5 to 9 mm; larger stems becoming less silvery, nearly black; northeastern Quebec to Alaska, southwest Quebec south to Minnesota, North Dakota, and Utah; xeric to xeric-mesic slopes.

Elaeagnus umbellata Thunberg, autumn-olive (Figure 21.29); multistemmed shrub, height to 3 m, stoloniferous; deciduous; leaf arrangement alternate; leaf simple, total length 4 to 9 cm, width 1 to 3 cm, blade entire, margin smooth or minutely uneven, shape elliptical or slightly oblanceolate, apex and base acute to obtuse, adaxial and abaxial surfaces densely covered with silvery scales, thus appearing as minutely pebble-grained (use 10× hand lens), infrequently supporting rusty scales appearing as dots; twigs covered with rusty and silvery scales; fruits a drupe, red but silvery dotted, nearly globose, diameter 5 to 9 mm; larger stems becoming less silvery, nearly black; introduced, escaped cultivation in eastern United States, invading old fields and naturalizing in eastern forests, extremely aggressive.

Elaeagnus angustifolia Linnaeus, Russian-olive (Figure 21.29); similar to *E. umbellata*, except: blade shape nearly linear to narrowly elliptical to narrowly lanceolate, length 2.5 to 8.0 cm; width 0.5 to 1.0 cm; fruit skin red but covered by silvery scales, length 0.8 to 1.2 cm; escaped cultivation in eastern United States but not aggressive.

Elaeagnus pungens Thunberg, silverthorn (Figure 21.29); shrub, similar to *E. umbellata*, except: plant is evergreen; leaf margin uneven, minutely revolute and/or serrulate; fruit skin covered with brown scales, length 1.2 to 1.5 cm; escaped cultivation in eastern United States but not aggressive.

ORDER: MYRTALES

Thymelaeaceae (Mezereum Family)

The approximately 50 genera and 500 species of trees and shrubs (rarely herbs) of Thymelaeaceae are distributed worldwide, including Australia, Africa, Southeast Asia, the Mediterranean region, and Eurasia. The shrubs and trees have tough fibrous bark. Flowers are borne in racemes and lack petals. Only one genus and two species are found in North America.

Dirca palustris Linnaeus, leatherwood, rope-bark (Figure 21.29); single-stem branching shrub, height to 2.5 m; deciduous; leaf arrangement alternate; leaf simple, total length 5 to 9 cm, width 3 to 5 cm, blade entire, papery thin, shape oval to obovate, margin smooth but slightly uneven, apex obtuse to nearly rounded, base obtuse or rounded, abaxial veins often cream colored or tan; twigs reddish brown, becoming dark, enlarged at the nodes, soft and slightly spongy when squeezed; fruit a red drupe, oblong, length 5 to 9 mm; bark tough, peeling in long strips; Regions 2, 3, and 4, Minnesota and southern Ontario to Nova Scotia, south to Florida and Mississippi; mesic woods. The bark is poisonous and reported to produce violent vomiting. Because of its tensile strength, the bark was extensively used by early Americans.

Dirca occidentalis Gray, western leatherwood; essentially identical to *D. palustris*; Region 7, coastal California; low elevations, oak woodland.

ORDER: SAXIFRAGALES

Considerable changes have occurred with this order. Some of the changes have come about through the use of research tools, such as cladograms (similar to dendrograms, or dichotomous figures), and chemical and genetic information. For example, based on genetic evidence, Judd et al. (1999) place the Hamamelidaceae and Grossulariaceae in this order, whereas an earlier classification system by Conquist (1981) positions them in the Rosales. Species of the three families considered here (Hydrangeaceae, Grossulariaceae, and Iteaceae) were once part of the Saxifragaceae, a relatively large family about 40 genera and 700 species, mostly herbs with a few woody plants.

Hydrangeaceae (Hydrangea or Mock-Orange Family)

The Hydrangeaceae is a small family of 17 genera and about 170 species, widely distributed in temperate and subtropical areas of the Northern Hemisphere. *Hydrangea, Philadelphus, Decumaria,* and *Fendlera* are native to North America. This group is now considered a separate family (Zomlefer 1994; Judd et al. 1999) and relatively distinct from the Saxifragaceae in which it was a subfamily (Hydrangeoideae). At times, *Philadelphus* (mock orange or syringa) has been placed in its own family (Philadelphaceae).

The species are large multistemmed, spreading shrubs with simple leaves. Stipules are absent. The flowers are unusual and arranged in a flat or convex cyme or appearing to be a corymb. Most of the fertile flowers are small and have no corolla. The small number of scattered marginal flowers that do have a corolla (diameter ~2 cm) are sterile, fruit an

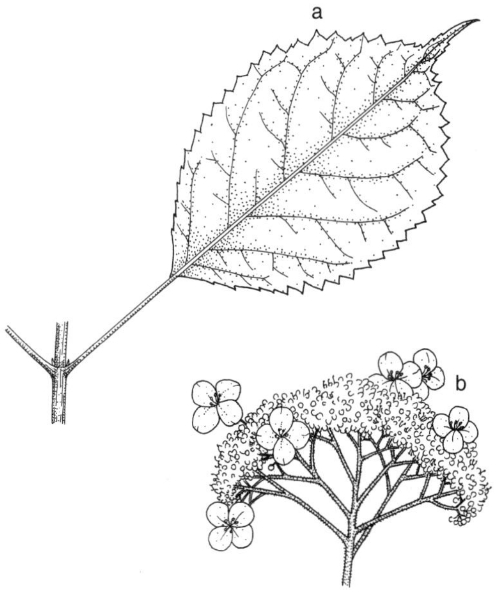

FIGURE 21.30 *Hydrangea arborescens* (a) simple leaf ×3/5, (b) inflorescence with centrally located fertile flowers and peripherally located sterile flowers with petals ×2/5.

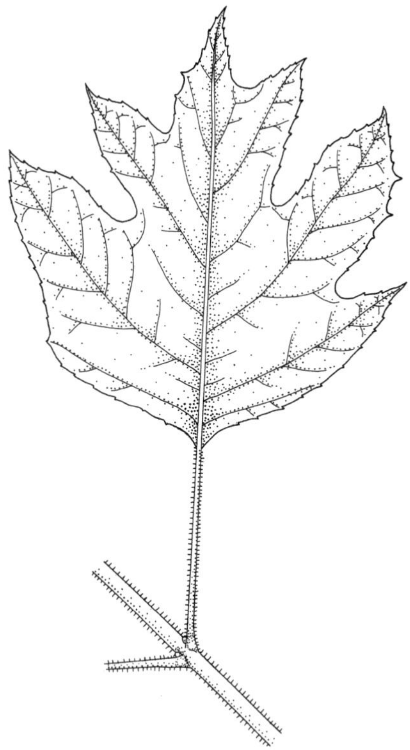

FIGURE 21.31 *Hydrangea quercifolia* leaf ×1/2.

urn-shaped capsule (*hydro* = water; *aggeion* = vessel). The stems have a large pith and are weak.

Hydrangea arborescens Linnaeus, wild hydrangea; mountain hydrangea (Figure 21.30); multistemmed shrub, height to 3 m; deciduous; leaf arrangement opposite; leaf simple, length 8 to 15 cm, width 6 to 10 cm, shape elliptical, ovate, or broadly lanceolate, thick, entire, margin coarsely serrate or nearly dentate, apex acuminate, abaxial surface tomentose, whitish; inflorescence a cyme (5 to 15 cm broad) with numerous small fertile green flowers filling the center, accompanied by a few (two to six) peripheral sterile flowers identified by their large, showy cream white sepals; fruit a small ribbed capsule, two locular (compartments), resembling a double-spout water vessel; stem gray, weak, pith large, bark peeling; Regions 3 and 4, Oklahoma, Iowa, Missouri, and Illinois to North Carolina and West Virginia and south to Georgia; cool, moist sites, coves, drainages, stream terraces, beneath rocky cliffs, elevation to 1300 m in the southern Appalachian Mountains. Occasionally an important forest shrub. A number of varieties and forms are recognized (Fernald 1950; Dirr 1998). The cultivated "Hills of Snow" form bears only sterile flowers. Cover type 27.5, 27.29, 27.30, 27.37.

Hydrangea quercifolia Bartram, oak-leaf hydrangea (Figure 21.31); multistemmed shrub, height to 2.5 m; deciduous; leaf arrangement opposite; leaf simple, total length 12 to 25 cm, width 8 to 14 cm, petiole length 10 to 15 cm, blade shape broadly elliptical or ovate to orbicular, thick, irregularly lobed, lobes three to seven, margin serrulated, apex acuminate, abaxial surface tomentose, whitish; inflorescence a cyme (5 to 15 cm broad) with numerous small fertile green flowers filling the center, accompanied by a few (two to six) peripheral sterile flowers identified by their large showy sepals (see Figure 21.30); fruit a small ribbed capsule, two locular (compartments), resembling a double-spout water vessel; stem tan to orange colored, bark peeling, weak, pith large; Introduced about 1800, escaped cultivation, Region 4, Piedmont and Coastal Plain, Oklahoma, Georgia, and Florida west to Louisiana; cool, moist sites, coves, drainages, stream terraces, base of rock outcrops. Widely used as an ornamental shrub. An occasional forest understory shrub. The leaf somewhat resembles that of *Quercus* (oak). Cover type 28.27.

Decumaria barbara Linnaeus, climbing hydrangea, cow-itch vine; vine, climbing by means of aerial roots developing from the main stem; deciduous; leaf arrangement opposite; leaf simple, length 10 to 12 cm, width 4 to 7 cm, shape broadly ovate or lanceolate, entire, margin smooth or wavy, occasionally serrated distally, apex obtuse; inflorescence a cyme (5 to 12 cm broad); flowers small, sepals white, lacy; fruit a single-ribbed capsule resembling a water vessel, stem gray; Region 4, Virginia to Florida and Louisiana; mesic-hydric and hydric sites, swamps, springs,

A large number of *Philadelphus* (about a dozen plus hybrids, varieties, and four dozen cultivars) are distributed across both the eastern and western United States. *Carpenteria californica* Torrey (tree anemone) occurs only in a limited area of the Sierra Nevada foothills, and *Whipplea modesta* Torrey (yerba de selva) of the Pacific Coast is distributed from the Olympic Peninsula into southern California (Region 7).

Grossulariaceae (Currant Family)

Until a few years ago the Grossulariaceae, a family of about 25 genera and 350 species, was included in the Saxifragaceae (Saxifrage family), in which *Ribes* (gooseberry, currant) was identified as a subfamily (Ribesioideae). *Ribes* is the largest genus in the Grossulariaceae, with about 150 species primarily in the Northern Hemisphere, but members also grow along the Andes Mountains into South America. In North America there are about 12 to 15 northern/eastern species, some with a range that includes large transcontinental areas. Three or four of these species are found in the Rocky Mountains along with six additional (Rydberg 1922), and possibly as many as another ten to fifteen in California (Jepson 1925; Barbour and Major 1977). Several *Ribes* species have been introduced.

From the 1950s into the 1970s, student foresters and biologists found work on *Ribes* eradication crews in the western United States. The idea behind eliminating these species from the forest was to prevent the spread of white pine blister rust, a disease which, like cedar-apple rust, alternates between two hosts to complete its life cycle. The disease completes one half of its life cycle on the leaves of *Ribes*, which are not damaged by the organism. Spores from the fruiting bodies on *Ribes* leaves travel by air to the bark of two of the most commercially important white pines in the western United States: *Pinus monticola* (western white pine) and *Pinus lambertiana* (sugar pine). In the eastern United States, *Pinus strobus* (eastern white pine) is also severely attacked. The disease infects the tree trunk around dead branch stubs and grows under the cambium, creating a dead zone or lesion. Because these pine species have multiple branches at a node (a whorl), the lesions eventually girdle and kill the tree. Infected trees can be identified by a white resinous flow down the trunk from the branch stubs. On the beneficial side, the fruit is used to make currant preserves and pie filling, which may be purchased in almost every grocery store.

All native species are small shrubs with prickles (gooseberry) or without prickles (currant) on the stems. The flowers are regular, perfect, and borne solitary or in a raceme. The fruit is a berry, which may have a smooth surface or may be prickly; the remains of the calyx are attached. The leaves are simple, more or less cleft, and have a coarsely serrate, dentate or toothed margin. The leaves are alternately arranged. Spur shoots are common. Some species may be difficult to separate because they are distinguished on the basis of flower characteristics (e.g., length of calyx tubes vs. lobes; length of stamens vs. length of calyx lobes).

Stems Armed at the Nodes

Ribes oxyacanthoides Linnaeus, northern gooseberry (Figure 21.32); shrub, height to 1 m, erect or prostrate; leaf length 5 to 10 cm, lobes three to five or seven, sinuses a narrow V, one half the distance to midrib; marginal teeth somewhat rounded, base cordate, abaxial surface glandular; stem nodes thorny, internodes prickly; flowers greenish white to purplish; berry reddish purple, lacking spines, diameter 0.6 to 1.0 cm; Regions 1, 2, 6, 7, and G; Alaska to James Bay (Ontario), south to Pennsylvania, Minnesota, Nebraska, Wyoming, British Columbia, Yukon, and Alaska; mesic upland woods into prairie.

Ribes setosum Lindley, bristly northern gooseberry (Figure 21.32); shrub, height to 60 cm, erect

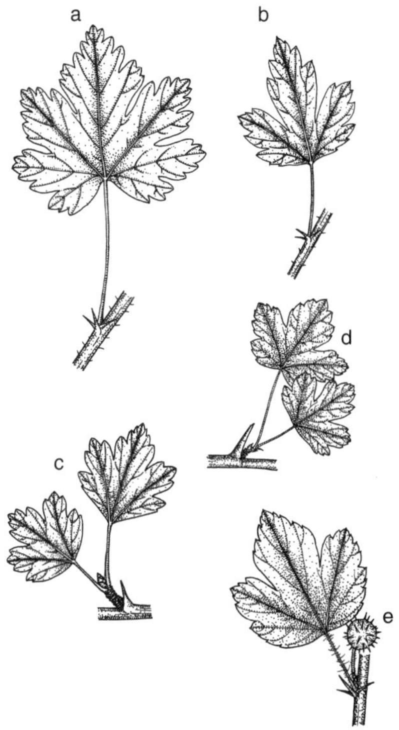

FIGURE 21.32 Leaves of (a) *Ribes oxyacanthoides* × 1/2, (b) *R. setosum* × 1/2, (c) *R. rotundifolium* × 1/2, (d) *R. missouriense* × 1, (e) *R. cynosbati* leaf and spine-covered berry × 1/2. These species have thorns at the nodes (gooseberry).

or prostrate; leaf length 5 to 10 cm lobes three, sinuses cleft, greater than one half the distance to midrib; marginal teeth sharply pointed, base rounded to cuneate, nodes thorny, length to 2 cm, internodes prickly; flowers greenish white; berry red or black, sometimes lacking spines, diameter 0.6 to 1.0 cm; Regions 1, 2, 5, and G; Manitoba to James Bay (Ontario), south to Wisconsin, Nebraska, and Wyoming; xeric-mesic upland woods, rocky sites into prairie. Once considered a phase or subspecies of *R. oxyacanthoides.*

Ribes rotundifolium Michaux, roundleaf gooseberry (Figure 21.32), shrub, height to 1 m, erect or prostrate; leaf length 2 to 5 cm lobes three to five, sinuses a narrow V, one third the distance to midrib; marginal teeth rounded, base rounded truncate, nodes thorny, internodes prickly; flowers greenish purple; berry bluish purple, lacking spines, diameter 0.6 to 1.0 cm; Region 3b, Massachusetts to Kentucky, West Virginia, and North Carolina; mountains, rocky upland woods.

Ribes missouriense Nuttall, Missouri gooseberry (Figure 21.32); shrub, height to 60 cm, erect; leaf length 1.5 to 3.5 cm, lobes three to five, sinuses a narrow V, less than one half the distance to midrib; marginal teeth somewhat rounded, base cordate, node thorns stout, reddish, internode prickles scattered; flowers white to greenish white; berry bluish purple, lacking spines, diameter 0.0 to 1.3 cm; Regions 2 and 3, Connecticut to Minnesota and North Dakota, south to Kansas, Arkansas, Tennessee, and Wyoming; open upland woods.

Ribes cynosbati Linnaeus, prickly gooseberry, dogberry (Figure 21.32); shrub, height to 1.3 m; leaf length 2 to 6 cm, lobes three, sinuses cleft, less than one half the distance to midrib; marginal teeth somewhat rounded, base cordate or truncate, nodes thorny, internode prickles scattered; flowers greenish; berry reddish purple, prickly, diameter 0.6 to 1.0 cm; Regions 1, 2, and 3; New Brunswick to Manitoba, south to Georgia and Alabama, west to Missouri; upland woods and openings.

Ribes lacustre Persoon (Poiret), bristly black gooseberry, spiny swamp gooseberry; erect shrub, height to 1 m; leaf length 3 to 6 cm, blade wider than long, distinct lobes three to five, sinuses a narrow V, less than one half the distance to midrib; margin serrate, teeth sharp, base truncate, abaxial surface glandular; stem nodes thorny, internodes prickly; flowers greenish to purplish; berry black, lacking spines, diameter 0.8 to 1.0 cm, not sweet; Regions 1, 2, 5, 6, and 7, Alaska to Newfoundland and New England, Wisconsin, Colorado, Utah, and California; mesic-hydric woods and swamps.

Ribes menziesii Pursh, canon gooseberry (Figure 21.33); erect shrub, height to 2.5 m; leaf length 1 to 3 cm, blade wider than long, distinct lobes three to five, sinuses a narrow V, cleft, less than one half the distance to midrib; margin serrate, teeth sharp, base truncate, abaxial surface glandular; stem nodes thorny, internodes prickly; flower sepals purple-tipped, petals white; berry black, spiny, diameter 0.8 to 1.0 cm, not sweet; berry red, spiny, diameter 0.8 to 1.0 cm; Region 7, central California to southern Oregon; mixed evergreen forest of the Coastal Ranges.

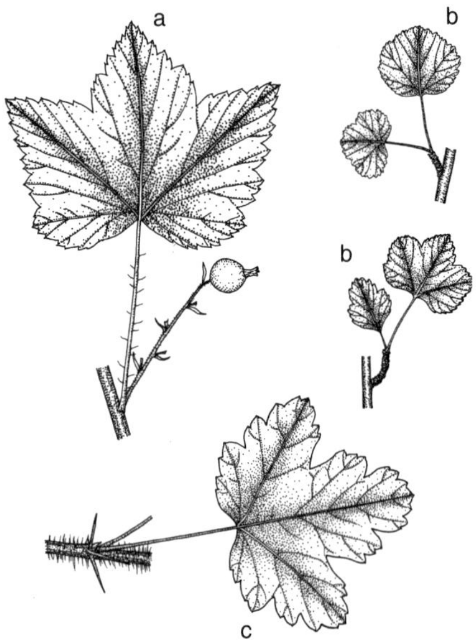

FIGURE 21.33 (a) *Ribes americanum* leaf and berry ×3/4; (b) *R. cereum* leaves on a spur shoot ×2/3; (c) *R. menziesii* leaf and spiny twig ×1.

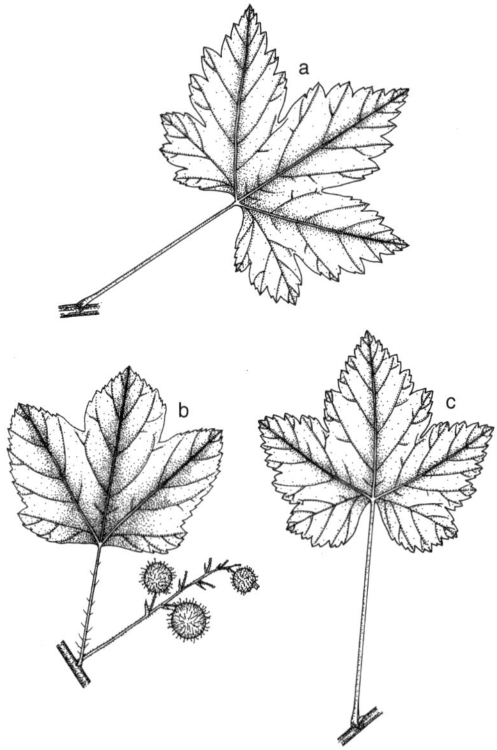

FIGURE 21.34 (a) *Ribes glandulosum* leaf ×1; (b) *R. triste* leaf and hairy fruit ×3/5; (c) *R. hudsonianum* leaf ×3/5.

Stems Unarmed

Ribes americanum Miller, wild black currant (Figure 21.33); erect shrub, height to 1.3 m; leaf length 4 to 7 cm, blade wider than long, lobes three, infrequently five, sinuses a shallow V, less than one fifth the distance to midrib; margin finely serrate, teeth sharp, base rounded or truncate, abaxial surface bearing yellow resin/glands; stem nodes lacking thorns, internodes lacking prickles; flowers greenish white to yellowish; berry black, spiny, diameter 0.8 to 1.0 cm; Regions 1, 2, 3b, and 6, Alberta to Labrador, south to New Jersey, West Virginia, Michigan, Ohio, South Dakota, Nebraska, and New Mexico.

Ribes cereum Douglas, wax currant (Figure 21.33), shrub, height to 1.5 m; leaves fragrant, length 1.5 to 2.5 cm, blade small, length 1.0 to 2.5 cm, slightly wider than long, entire and orbicular or with three lobes, sinuses a broad V, less than one fourth the distance to midrib; margin serrate, teeth blunt, base truncate, abaxial surface glandular; stem nodes lacking thorns, internodes lacking prickles; flowers greenish to purplish; berry red, lacking spines, diameter 0.8 to 1.0 cm; Region 7, California; high montane, 2000 to 5000 m.

Ribes glandulosum Grauer, skunk currant (Figure 21.34); prostrate shrub, all parts producing a skunk-like odor when crushed; leaf length 3 to 7 cm, blade wider than long, lobes five to seven, sinuses a narrow V, one half the distance to midrib; margin serrate or doubly serrate, teeth pointed, base deeply cordate, abaxial surface glandular; stem nodes lacking thorns, internodes lacking prickles; flowers yellowish greenish to purplish; berry red, spiny, diameter 0.6 to 1.0 cm; Regions 1, 2, and 3, Newfoundland, to British Columbia, Maine to North Carolina; rocky upland and wet woods, swamps, cool sites. Makes excellent preserves.

Ribes hudsonianum Richards, northern black currant (Figure 21.34); shrub, height to 1 m; leaf length 8 to 10 cm, blade wider than long, lobes three to five, sinuses a narrow V, less than one quarter the distance to midrib; margin serrate, doubly serrate, or dentate, teeth blunt, base deeply cordate, abaxial surface glandular; stem nodes lacking thorns, internodes lacking prickles; flowers white; berry black, bitter, lacking spines, diameter 0.8 to 1.2 cm; Regions 1, 2, 6, and 7, Alaska to James Bay (Ontario), south to Minnesota, Montana, and Washington; mesic-hydric woods, wet woods, swamps, and rocky upland sites.

Ribes triste Pallas, swamp red currant (Figure 21.34); reclining or sprawling shrub; leaf length 5 to 9 cm, blade wider than long, lobes three, infrequently five, sinuses a narrow V, less than one quarter the distance to midrib; margin finely serrate, teeth sharp, base truncate, abaxial surface glandular; stem nodes lacking thorns, internodes lacking prickles; flowers white; berry red, spiny, diameter 0.8 to 1.4 cm; Regions 1, 2, and 3b, Alaska and Alberta to Quebec, south to South Dakota, Wisconsin, Michigan, and New Jersey; swamps, bogs mesic woods, ravines.

Other northern/eastern *Ribes* species not described here include *R. inerme* Rydberg, *R. viscosissimum* Pursh (sticky currant), *R. curvatum* (granite gooseberry), *R. odoratum* Wendland (buffalo-currant), and *R. hirtellum* Michaux (wild gooseberry). Some western species not described include *R. aureum* Pursh (golden currant), *R. viscosissimum* Pursh, *R. nevadense* Kell, *R. sanguineum* Pursh, *R. laxiflorum* Pursh, and *R. californicum* Hooker & Arnott. Several additional species have been introduced from Europe References that give descriptions include Jepson (1925), Fernald (1950), Rydberg (1922), Gleason and Cronquist (1991), Grimm (1993), Cody (1996), and Kurz (1997).

Iteaceae (Itea Family)

In addition to *Ribes* and other genera being moved into the Grossulariaceae, the Hydrangiaceae has been segregated and the genera *Itea* and *Escallonia* moved into the Escalloniaceae (Zomlefer 1994), or *Itea* into its own family (Judd et al. 1999). The 15 genera and 200 species in the Escalloniaceae are shrubs or trees, generally found in the montane and temperate climates of the Southern Hemisphere; only a few, including *Itea*, occur in North America (Zomlefer 1994). Fernald (1950) considered *Itea* in the subfamily Escallonoideae, whereas Gleason and Conquist (1991) position it in the Grossulariaceae (gooseberry) family.

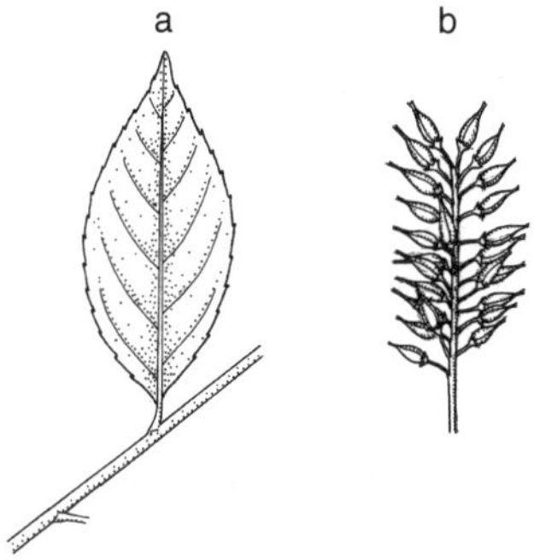

FIGURE 21.35 *Itea virginica* (a) simple leaf ×2/5, (b) inflorescence with attached capsules ×1/3.

Itea virginica Linnaeus, Virginia willow (Figure 21.35); shrub, height to 2.5 m; deciduous; leaf arrangement alternate; leaf simple, total leaf length 5 to 10 cm, width 2 to 4 cm; blade shape elliptical, entire, margin serrulated, apex and base acute, venation essentially arcuate, the tips of the lateral veins curving to parallel the margin; twigs green to slightly reddish; inflorescence a raceme, length 0.5 to 15 cm; flowers small, petals five, white; fruit a capsule, shape lanceolate, length 1 cm; branches slender, bark gray; Regions 3 and 4, Missouri, southern Illinois, and Kentucky to Pennsylvania and the New Jersey pine barrens, south to Louisiana and Florida, primarily on the Coastal Plain; swamps and bottomland forests.

ORDER: CORNALES

Cornaceae (Dogwood Family)

The Cornaceae is one of few families whose members are found primarily in temperate and north temperate regions. The 75 to 80 species in about five genera (excluding the Nyssaceae) are trees, shrubs, and, infrequently, herbs. The leaves are opposite, simple, and entire and have a smooth margin. Stipules are absent. The flowers are regular, perfect or imperfect, and borne in cymes, which appear in spring or early summer. The fruit is a drupe containing one to two seeds. The fruits are often clustered into a tight group. A comparison of published material indicates that there is some debate over the status and naming of species.

The North American *Cornus* species have a shrub or small tree form, except for *Cornus canadensis* L. (Canada cornel, bunchberry), which is an herb but still recognizable by its leaves as a member of the Cornaceae.

Members of the genus are found in both eastern coniferous and hardwood forests and western coniferous forests. Depending on the species and its range, the fruits may be eaten by numerous songbirds and larger birds such as ruffed and sharptail grouse, bobwhite quail, turkeys, pheasants, and pigeons. Deer, rabbits, elk, and moose browse on the twigs, and beavers eat the inner bark. Many of the dogwood species are used in landscaping.

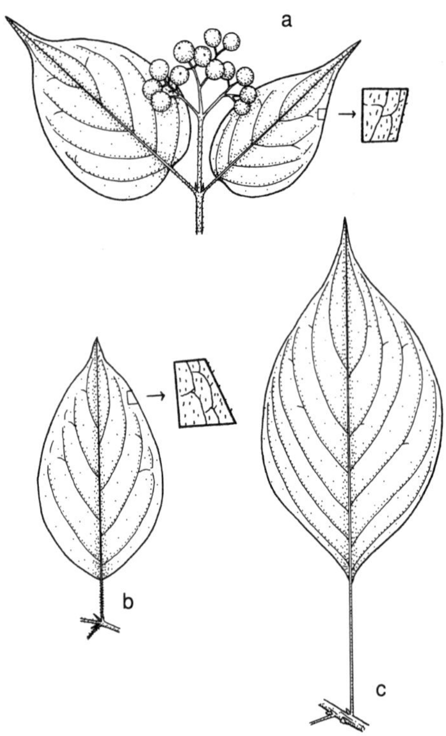

FIGURE 21.36 (a) *Cornus drummondii* simple leaves and inflorescence with attached drupes × 1/2; (b) *C. obliqua* simple leaf × 1/2; (c) *C. alternifolia* simple leaf × 3/4.

Eastern/Northern Species

Cornus alternifolia Linné f., alternative-leaf dogwood (Figure 21.36); shrub or small tree, height to 7 m; deciduous; leaves opposite, except usually clustered and alternate at tip of twig; leaf simple, total length 6 to 14 cm, width 3 to 7 cm, blade entire, shape broadly elliptical to lanceolate, margin smooth, apex mucronate to acute, adaxial surface glabrous, abaxial surface whitish with scattered appressed white hair, venation arcuate, ends of veins paralleling blade margin; twigs greenish; buds valvate, visible scales two; inflorescence a terminal cyme; flowers small, white, large showy bracts absent; fruit a drupe, globose, diameter 5 to 8 mm, bluish; stem gray brown, pith white; Regions 2 and 3, scattered in Region 4, Manitoba, Ontario, and Minnesota to Nova Scotia and Newfoundland, south to Missouri, Georgia, and Alabama, scattered along the Gulf and Atlantic Coasts; mesic upland interior forest and margins. Cover type 26.24, 26.28, 27.5, 27.29, 27.30.

Cornus drummondii C. A. Meyer (*C. asperifolia* Michaux), roughleaf dogwood (Figure 21.36); shrub, height to 4 m; deciduous; leaf arrangement opposite; leaf simple, total length 5 to 11 cm, width 2 to 5 cm, blade shape lanceolate, ovate or elliptical, entire, margin smooth, adaxial surface softly scabrous with stiff white hair, abaxial surface whitish with scattered appressed white hair, venation arcuate, ends of veins paralleling blade margin, vein pairs three to four; new twigs reddish, turning gray, pubescent; buds valvate, visible scales two; inflorescence a terminal cyme; flowers small, white, large showy bracts absent; fruit a drupe, white, globose, diameter 5 to 8 mm; stem gray- to reddish-brown, pith usually brown, infrequently white; Regions 3a and 4, South Dakota and Iowa to Michigan and southern Ontario, south into Louisiana, Mississippi, and central Texas, Oklahoma, and Kansas; mesic upland forest and high stream terraces. Cover type 27.10, 28.10, 28.15, 28.17, 28.21, 28.22, 28.51.

Cornus obliqua Rafinesque (*C. amomum* Miller), silky dogwood, pale dogwood, narrowleaf dogwood (Figure 21.36); shrub, height to 3 m; deciduous; leaf arrangement opposite; total leaf length 5 to 12 cm, width 2.0 to 4.5 cm, blade entire, shape lanceolate, ovate or elliptical, entire, margin smooth, apex acuminate, base acute, adaxial surface glabrous, abaxial surface whitish with scattered appressed white hair, venation arcuate, ends of veins paralleling blade margin, vein pairs; new and older twigs reddish purple, glabrous or with scattered hair, buds valvate, visible scales two; inflorescence

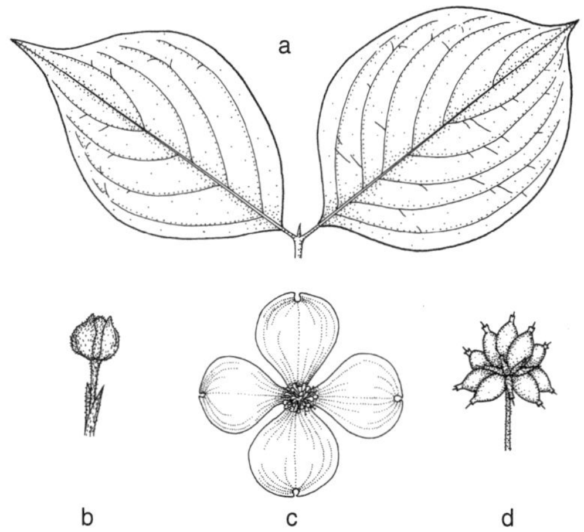

FIGURE 21.37 *Cornus florida* (a) simple leaves ×2/5, (b) flower bud ×1, (c) cluster of flowers subtended by four showy bracts ×1/3, (d) cluster of drupes ×1/2.

a terminal cyme; flowers small, white, large showy bracts absent; fruit a drupe, bluish, globose, diameter 3 to 6 mm; bark greenish gray, pith brown; Regions 2, 3, and 4, Michigan, southern Ontario, and New Brunswick, south to Florida and Texas; mesic and mesic-hydric sites. On the basis of leaves, *C. amomum* is separated from *C. obliqua* in some earlier works. *Cornus amomum* has a wider leaf with a rounded base. These species probably cannot be distinguished in winter.

Cornus florida Linnaeus, flowering dogwood (Figure 21.37, 15.4); small tree, height to 13 m, diameter to 25 cm; deciduous; leaf arrangement opposite; leaf simple, length 7 to 14 cm, width 3 to 6 cm, blade shape elliptical or broadly elliptical, entire, margin smooth or slightly wavy, adaxial and abaxial surfaces smooth to the touch but supporting white appressed hair, venation arcuate, ends of veins paralleling blade margin; new twigs green or reddish, turning gray, terminal leaf bud lance-shaped, visible scales two, length 5 to 8 mm, terminal flower bud globose, stalked, length 0.5 to 1.5 cm; inflorescence a terminal cyme, flowers small, green, surrounded by four large white (sometimes pink) showy bracts, bracts notched at the tips, length 3 to 6 cm; fruit a drupe, bright red, oblong, length 0.7 to 1.0 cm; stem gray, pith white; bark of older trees broken into distinct polygonal pattern; Regions 2, 3, and 4, Missouri, Michigan, southern Ontario to Maine

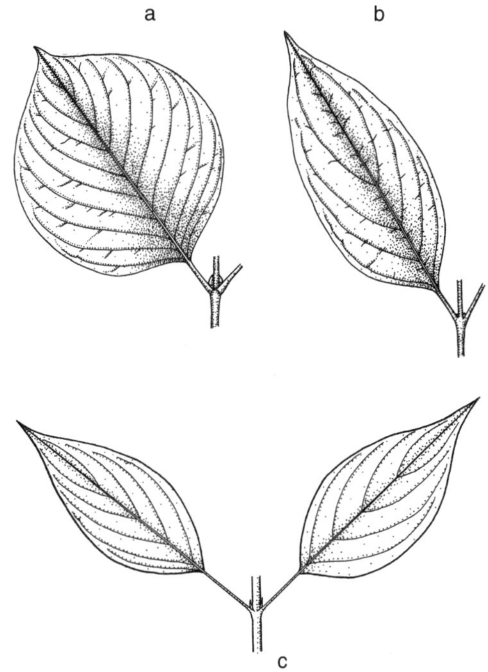

FIGURE 21.38 Simples leaves of (a) *Cornus rugosa* ×1/3; (b) *C. foemina* ×1/2; (c) *C. stolonifera* ×1/2.

and south to the Gulf and Atlantic Coasts and Mexico; stream terraces, xeric-mesic and mesic upland slopes onto stream terraces. An anthracnose disease has killed the species over substantial portions of its range, particular in the far east. Cover type 27.1, 27.4, 27.5, 27.9, 27.13, 27.15, 17.16, 27.17, 27.22, 27.23, 27.26, 28.4, 28.6, 28.11, 28.14, 28.16, 28.18, 28.27, 28.38.

Cornus rugosa Lamarck, roundleaf dogwood (Figure 21.38); shrub, height to 3 m; deciduous; leaf arrangement opposite; leaf simple, total length 6 to 14 cm; width 5 to 12 cm, petiole pubescent, blade shape oval or nearly round, entire, margin smooth, apex and base obtuse, lower surface often white pubescent, slightly velvety, venation arcuate, ends of veins paralleling blade margin, vein pairs five to nine; twig greenish or reddish brown, spotted with purple, pubescent; bud valvate, visible scales two; inflorescence a cyme; flowers small, white, large showy bracts absent; fruit light blue, globose, diameter 5 to 6 mm; stem gray, pith white; Regions 2 and 3, Manitoba and Ontario to eastern Quebec and Nova Scotia, south to Virginia, Illinois, Iowa, and North Dakota; moist and sandy soil.

Cornus racemosa Lamarck, panicled dogwood; multi-stemmed shrub, height to 2.5 m; deciduous; leaf arrangement opposite; leaf simple, total length 4 to 8 cm, width 2 to 4 cm, blade lanceolate to ovate, entire, apex acuminate, base obtuse, margin minutely rough and irregular, adaxial and abaxial surfaces smooth to the touch but supporting scattered appressed white hair, venation arcuate, ends of veins paralleling blade margin, vein pairs three to five; buds valvate, visible scales two; inflorescence a cyme; flowers small, white, large showy bracts absent; fruit a drupe, globose, diameter 5 to 6 mm, bright red, on bright red pedicels; stems gray, pith light brown; Regions 2 and 3, Minnesota and southern Ontario to Maine, south into West Virginia, Kentucky, Missouri, and Oklahoma; xeric to mesic sites, woods, stream terraces, and disturbed areas, roadsides and fields. A common shrub.

Cornus foemina Miller (*Cornus stricta* Lamarck), stiffcornel dogwood, panicled dogwood (Figure 21.38); shrub, height to 5 m; deciduous; leaf arrangement opposite, blade simple, total length 5 to 11 cm, width 2.0 to 4.5 cm, blade shape lanceolate, ovate, or elliptical, entire, margin smooth, apex long acuminate, base obtuse, adaxial and abaxial surfaces smooth to the touch but supporting scattered appressed white hair; venation arcuate, ends of veins paralleling blade margin, vein pairs four to five; new and older twigs reddish, becoming gray; buds valvate, visible scales two; inflorescence a terminal cyme; flowers small, white, large showy bracts absent; fruit a drupe, bluish white, globose, diameter 4 to 6 mm; bark gray, pith white; Regions 2, 3, and 4, southeast Missouri and southern Indiana to Virginia, south to Florida and Texas; mesic-hydric and hydric sites, swamps, riverbanks, and bottomland forest. May be confused with *C. racemosa*. The names stiffcornel and *stricta* are references to the stiff upright branches. Cover type 27.10, 28.11.

Cornus stolonifera Michaux (*Cornus sericea* Linnaeus), red-osier dogwood (Figure 21.38); multi-stemmed shrub, height to 3 m; deciduous; leaf arrangement opposite; leaf simple, total length 4 to 10 cm, width 2 to 4 cm, blade lanceolate to ovate, entire, margin smooth, apex acuminate, base round, adaxial and abaxial surfaces smooth to the touch but supporting scattered appressed white hair; venation arcuate, ends of veins paralleling blade margin, vein pairs four to seven; twigs bright red, buds valvate, visible scales two; inflorescence a cyme; flowers small, white, large showy bracts absent; fruit a drupe, white or gray white, globose, diameter 6 to 7 mm; stem color distinctly red to purplish red, pith white; Region 1, Newfoundland to the Yukon and Alaska; Region 2, Lake States and New England; Regions 5 and 6, Arizona, New Mexico, and Mexico, Region 7, British Columbia south to northern California; hydric sites. One of the most common shrubs of swamps, marshes, and stream margins. The var. *baileyi* (Coulter & Evans) Drescher, which occurs from Pennsylvania to Manitoba, has a white tomentose abaxial surface of the leaf. The western populations have been at times called *C. interior* Rydberg. Cover type 26.3, 26.5, 26.13.

Western Species

Cornus nuttallii Audubon, Pacific dogwood, western flowering dogwood, mountain dogwood (Figure 21.39); small tree, height to 20 m; deciduous; leaf arrangement opposite; leaf simple, total leaf length 7 to 14 cm, width 4 to 9 cm, blade simple, entire,

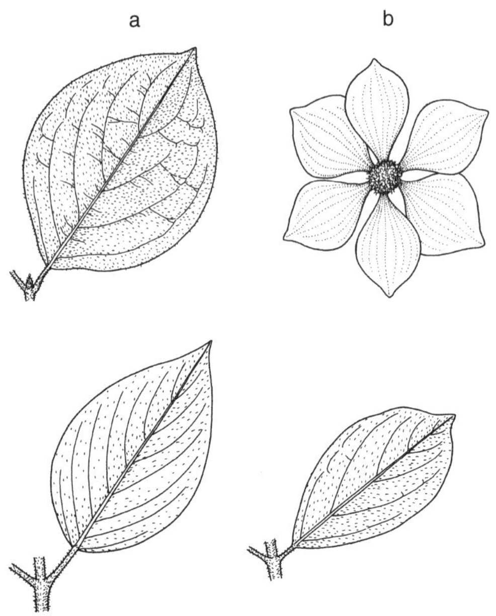

FIGURE 21.39 *Cornus nuttallii* (a) leaf ×2/5, (b) cluster of flowers subtended by six showy bracts ×1/4; leaves of (c) *C. occidentalis* ×2/5, (d) *C. glabrata* ×1/2.

shape elliptical or ovate, margin smooth, ciliated, apex acuminate, base acute, lower surface white pubescent, slightly velvety; venation arcuate, ends of veins paralleling blade margin, vein pairs four to six; twigs greenish or reddish, bud valvate, visible scales two; inflorescence a cyme; flowers white, the central cluster surrounded by four to six petal-like bracts, bracts pointed; fruit a drupe, bright orange to red, oval, length 1.0 to 1.3 cm, in tight clusters of about 30; bark gray brown, pith brown; Region 7, southern California into southern British Columbia, occasionally found in Idaho (Region 6); mesic soil, bottomland, drainages, and stream edges, elevation to 1200 m. One of the most colorful ornamentals in the far western coniferous forest. Cover type 30.16, 30.28, 30.34, 30.41, 30.44.

Cornus occidentalis Torrey & Gray, western dogwood (Figure 21.39); shrub or small tree, height up to 5 m; deciduous; leaf arrangement opposite; leaf simple, total leaf length 6 to 12 cm, width 2 to 4 cm, blade shape broadly or narrowly elliptical, infrequently lanceolate, entire, margin smooth, apex acute, base rounded, lower surface white pubescent, slightly velvety; venation arcuate, ends of veins paralleling blade margin, vein pairs four to seven; twigs reddish purple, buds valvate, visible scales two; inflorescence a cyme; flowers lacking large bracts; fruit white, globose, diameter 0.6 to 1.0 cm; bark brown to reddish, pith white to light brown; Region 7, southern California counties into Oregon and Washington to the Olympic Peninsula; coastal, mesic sites, drainages, valleys on lower mountain slopes, elevation 2400 m. This species is not found in the Sierra Nevada. (Hitchcock and Cronquist 1973) considered this species as *C. stolonifera* var. *occidentalis* (Torrey & Gray) Hitchcock because of its abaxial pubescence (spreading or curled hair). Leaf shape similar to that of *C. alternifolia.* Stuart and Sawyer (2001) list this species as *C. Sericea* (*C. stolonifera*) ssp. *occidentalis.*

Cornus glabrata Bentham, brown or smooth dogwood (Figure 21.39); shrub or small tree, height up to 6 m; deciduous; leaf arrangement opposite; leaf simple, total leaf length 3 to 8 cm, width 1 to 3 cm, blade shape elliptical to slightly obovate, entire, margin smooth, apex usually acute, sometimes minutely acuminate, base acute, adaxial and abaxial surfaces smooth to the touch but supporting scattered appressed white hair; venation arcuate, ends of veins paralleling blade margin, vein pairs three to four; twigs reddish purple, often drooping and often rooting where tips touch the ground, buds valvate, visible scales two; inflorescence a cyme; flowers lacking large bracts; fruit bluish to nearly white, globose, diameter 6 to 8 mm; bark brown to reddish, pith brown; Region 7, southern California counties into southern Oregon; coastal, mesic sites, drainages, valleys on lower mountain slopes, elevation 1700 m.

Cornus sessilis Torrey & Durand, blackfruit dogwood; small tree, height to 5 m; deciduous; leaf arrangement opposite; leaf simple, total leaf length 5 to 10 cm, width 4 to 9 cm, blade simple, entire, shape elliptical or ovate, margin smooth, apex acuminate, base acute, lower surface white pubescent; venation arcuate, ends of veins paralleling blade margin, vein pairs four to five; twigs green accented with red blotches, bud valvate, visible scales two; inflorescence a cyme; flowers greenish white, the central cluster surrounded by four short (1 cm long) bracts; fruit a drupe, black at maturity, oval, length 1.0 to 1.3 cm, in tight clusters of about 30; bark gray brown, pith brown; Region 7, stream banks, drainages, elevation to 1500 m. Leaf shape is similar to those of *C. stolonifera* and *C. foemina.*

Garryaceae (Silktassel Family)

A taxon closely related to *Cornus* but now separated into its own family is the single North American genus, *Garrya*, the silktassel genus of about 15 species.

Garrya fremontii Torrey, Fremont silktassel (Figure 21.40); shrub, height to 3 m; evergreen; leaf arrangement opposite; leaf simple, total length 3 to 6 cm, width 1 to 3 cm, blade elliptical to slightly obovate, entire, margin smooth but minutely revolute, abaxial surface glabrous or with a few scattered hairs; twigs reddish to brownish, glabrous; flowers borne in a catkinlike spike or tassel, drooping, male and female flowers in separate spikes (plants dioecious); fruit a drupe, globose, blue at maturity, surface glaucous; bark reddish; Region 7, California; xeric uplands and lower montane zone of the Sierra Nevada.

Garrya elliptica Douglas, wavyleaf silktassel (Figure 21.40); shrub, height to 3 m; evergreen; leaf arrangement opposite; leaf simple, total length 3.5 to 8.0 cm, width 2 to 3 cm, blade elliptical to oval, entire, margin revolute or intermittently revolute or smooth, abaxial surface densely pubescent; twigs reddish to brownish, glabrous or pubescent; flowers borne in a catkin-like spike or tassel, drooping, male and

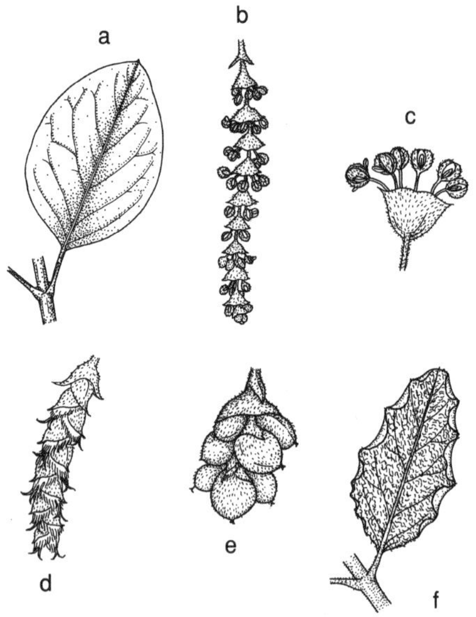

FIGURE 21.40 *Garrya fremontii* (a) leaf ×1/2, (b) ament of male flowers, (c) male flower with exserted stamens ×1/3, (d) ament of female flowers ×1/2, (e) cluster of drupes ×1/3; (f) *G. elliptica* leaf with partially revolute margin and pubescent abaxial surface ×1/2.

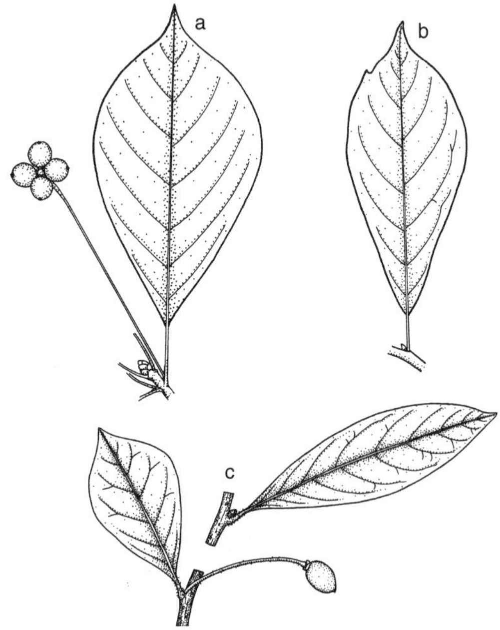

FIGURE 21.41 *Nyssa sylvatica* (a) leaf and inflorescence with attached drupes ×2/5, (b) common leaf shape of understory stems ×2/5; (c) *N. biflora* leaf shapes and drupe ×1/3.

female flowers in separate spikes (plants dioecious); fruit a drupe, globose, yellowish gray at maturity, surface densely pubescent; bark reddish; Region 7, California Coastal Ranges into Oregon; xeric uplands. Cover type 30.45.

Nyssaceae (Tupelo Family)

The Nyssaceae is a family of three genera and about eight species of trees and shrubs found in eastern North America and China. Only the genus *Nyssa*, with its three tree species, is native to the United States. Until recently this genus was placed in the Cornaceae.

The staminate flowers appear in dense heads, and the pistilate flowers are solitary or several in a cluster. The fruit is an oblong drupe with a thin ectoderm and fleshy mesoderm enclosing a hard ridged endoderm (pit). Each fruit is borne on a long peduncle. *Nyssa* has a diaphragmed pith that may be used to confirm identification.

Nyssa sylvatica Marshall; blackgum (Figure 21.41); medium-sized to large tree, height to 25 m; deciduous; leaf arrangement alternate; leaf simple, total length 6 to 18 cm, width 3 to 7 cm; petiole usually pubescent or nearly tomentose, length ~1 cm; blade broadly obovate to oblanceolate or elliptical, entire, margin smooth, infrequently one to two large (1 cm long) marginal teeth appear along the distal one half, apex mucronate, base obtuse, abaxial surface lightly hairy, except midrib pubescent notably the proximal section; twigs glabrous or with scattered hair, terminal bud length 3 to 5 mm, buds imbricated, visible scales six to eight, lightly pubescent and ciliated, scales bicolored with reddish margin forming a border around the green center; pith diaphragmed (green cross-walls in cream-colored matrix); flowers clustered 2 to 14; fruit a drupe, oval, dark blue to black, length ~1 cm, one two three on a peduncle, peduncle length 2 to 3 cm, endocarp (stone) flattened and ridged; mature bark having a distinct polygonal (pentangle, hexangle) pattern; Regions 3 and 4, eastern United States and Canada, Maine to southern Ontario, southwestern Michigan, Illinois, Missouri, Arkansas, and eastern Oklahoma and Texas; upland xeric-mesic and mesic sites and stream terraces. Cover type 27.9, 27.16, 27.21, 27.22, 27.24, 27.33, 27.36, 28.2, 28.3, 28.4, 28.13, 28.16, 28.18, 28.24, 28.29, 28.38. Although *N. sylvatica* may grow to a large size (diameter 50 to 70 cm), most stems appear as seedlings and saplings in the forest understory, particularly on more xeric (south) slopes and soils. Many of these stems appear to be sprouts, not seedlings, growing slowly until tree death in the overstory and the increased sunlight permit several to rapidly grow into the gap. Only on moist sites may stems eventually grow into the tree overstory.

Nyssa bifora Walter [*Nyssa sylvatica* var. *biflora* (Walter) Sargent], swamp tupelo (Figure 21.41); medium-sized tree, height to 20 m; deciduous; leaf arrangement alternate; leaf simple, total length 6 to 13 cm, width 1.5 to 4.0 cm; petiole pubescent, petiole length ~1 cm; entire; blade elliptical, narrowly elliptical, oblong, or narrowly oblanceolate, margin smooth, apex and base obtuse, abaxial surface midrib pubescent; new twigs pubescent, terminal bud length 2 to 4 mm, buds imbricated, visible scales six to eight, densely pubescent and ciliated, uniformly gray or gray brown; twig pith diaphragmed (green cross-walls in cream-colored matrix); flowers clustered one two three; fruit a drupe, oval, dark blue to black, length ~1 cm, one to two on a peduncle, peduncle length 1.0 to 3.5 cm, endocarp (stone) flattened and distinctly ridged; base of mature trees buttressed (swollen); Region 4, Atlantic and Gulf Coastal Plains, New Jersey and Delaware to Florida, Louisiana, and east Texas; swamps and bottomland sites. Cover type **28.29, 28.32, 28.33, 28.36, 28.42, 28.48, 28.55,** 28.23, 28.24, 28.25, 28.36, 28.43, 28.47, 28.49, 28.54. This lowland species, usually considered a variety of *N. sylvatica*, appears sufficiently different morphologically to be distinguished as a species. Based on the specimens examined at the Missouri Botanical Garden, St. Louis, the buds and leaves are relatively distinct and the number of flowers on a cluster is different, but not unlike *N. sylvatica* var. *sylvatica*, this species may have one or two fruits on a peduncle. *Nyssa biflora* appears on wet sites, in contrast to *N. sylvatica*, which is a terrace or upland species.

Nyssa aquatica Linnaeus, water tupelo (Figure 21.42); medium-sized to large tree, height to 30 m;

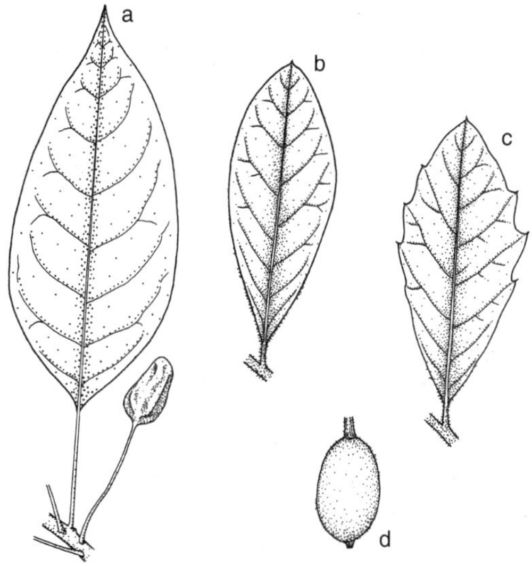

FIGURE 21.42 (a) *Nyssa aquatica* leaf and drupe ×1/3; *N. ogeche* (b), (c) leaves ×1/3, (d) drupe ×1/2.

deciduous; leaf arrangement alternate; leaf simple, total length 10 to 25 cm, width 3 to 8 cm; petiole lightly hairy to densely pubescent or nearly tomentose, length 3.0 to 4.5 cm; blade entire, broadly elliptical or ovate, margin smooth, frequently one two three large (1 cm long) marginal teeth appear near midleaf, teeth oriented perpendicular to midrib, apex acuminate, apex and teeth bristle-tipped, base obtuse or rounded, abaxial surface midrib and veins pubescent, adaxial surface midrib pubescent at the proximal end; twigs glabrous or with scattered hair, terminal bud length 3 to 5 mm, buds imbricated, visible scales five to six, lightly pubescent or glabrous, ciliated with tan hair, scales dark blue or nearly black, pith diaphragmed (green cross-walls in cream-colored matrix); flowers clustered 2 to 14; fruit a drupe, oblong, dark blue to black, length 1.5 to 3.5 cm, one on a peduncle, peduncle length 3 to 5 cm, endocarp (stone) flattened and ridged; base of mature trees buttressed (swollen); Region 4, Coastal Plain, Virginia to Florida and east Texas, Mississippi floodplain to southern Illinois and Indiana; flooded conditions, cypress swamps and bottomland, in pure stands or mixed with cypress or other bottomland species. Cover type **28.31, 28.55,** 28.14, 28.30, 28.18, 28.41, 28.50, 28.53, 28.54.

Nyssa ogeche Bartram ex. Marshall, Ogeche tupelo (Figure 21.42); shrub tree, height to 13 m; deciduous; leaf arrangement alternate; leaf simple, total length 8 to 17 cm, width 3 to 6 cm; petiole yellow pubescent or nearly tomentose, length 1 to 2 cm; blade entire, elliptical, ovate, or oblanceolate, margin typically smooth but occasionally irregularly toothed, number of teeth two to five, apex obtuse and bristle-tipped, base obtuse or rounded, abaxial surface lightly pubescent, abaxial midrib and veins densely yellowish, pubescent to tomentose, adaxial surface midrib pubescent at the proximal end, cream colored; new twigs pubescent, terminal bud length 2 to 3 mm, buds imbricated, visible scales three to four, densely pubescent and ciliated with tan hair, scales dark blue or nearly black, pith diaphragmed (green cross-walls in cream-colored matrix); flowers clustered 2 to 14; fruit a drupe, oblong, dark blue to black, length 2.5 to 3.5 cm, one on a peduncle, peduncle length ~1 cm, endocarp (stone) flattened and deeply ridged; base of mature trees buttressed (swollen); Region 4, tristate, southeastern South Carolina, southern Georgia, and Northern Florida; wet or flooded soil.

ORDER: CELASTRALES

Aquifoliaceae (Holly Family)

The Aquifoliaceae is a small family of three genera and about 400 species of trees and shrubs distributed worldwide, but with few species in Africa and Australia. The largest genus is *Ilex* (holly), with nearly 400 species, about 20 of which are native to eastern North America (Regions 2, 3, and 4). *Nemopanthus mucronatus* (Linnaeus) Trelease (mountain-holly), once thought to be the only representative of the genus, is distributed around the Great Lakes and eastward to New England and Newfoundland, to West Virginia (primarily Region 2). Foote and Jones (1989) include a second closely related species, *N. collinus* (Alexander) Clark, in the genus, although it had been placed in *Ilex* (Edwin 1957). This species is found in the mountains of West Virginia, Virginia, and North Carolina. *Bryonia* is not represented in North America. Zomlefer (1994) and Judd et al. (1999) consider the family to be monotypic with *Ilex* the only genus.

The small white or greenish flowers are regular, 4-9-merous, and clustered in axils of leaves or at the base of young branchlets. The fruit is a red, orange yellow, or black berry like drupe enclosing four nutlets; fruit diameter is about 5 to 7 mm. Differences in fruit color may be helpful in identifying several species. The leaves are simple, entire, serrated or spiny, relatively small (length <10 cm), and deciduous or evergreen, depending on the species.

Individuals of *Ilex* are common in the forest communities of the southeastern United States and may be so dense that they are strong competition for tree seedlings and saplings. Prescribed burning is often used to reduce forest understory density. On the positive side, the drupes appear late in the growing season and provide food for various birds and animals, and the dense low canopy may provide nesting habitat and cover. *Ilex glabra* is reported to produce sufficient nectar to be a good honey plant. *Ilex opaca* is the only species whose stems grow large enough to be used for cabinet wood. Because it is the familiar "Christmas holly," some states have strict laws that prohibit cutting this species along roadsides. Although a strong mixture of *I. vomitoria* will cause vomiting, weak solutions have been used as a tea substitute (Brown 1945).

Evergreen Species

Ilex opaca Aiton, American holly (Figure 21.43); shrub or small tree, height to 2 m; evergreen; leaf arrangement alternate; leaf simple, entire, blade length 5 to 9 cm, width 2.5 to 4.5 cm, coriaceous, elliptical or oval, margin revolute, remotely spiny, spine length 2 to 3 mm, apex spine tipped, adaxial surface dark green, waxy, shiny, abaxial surface glabrous; petiole length 2 to 3 mm, puberulent; twigs light brown, turning gray, puberulent; flowers individual, cauline, peduncle length 4 to 7 mm; fruit a drupe, bright red, globose, diameter 4 to 8 mm, nutlet ribbed; Regions 3 and 4, southern Illinois and Missouri to Pennsylvania and Virginia, south to Florida and Louisiana, west to Oklahoma; mesic and mesic-hydric woods, drainages and stream margins. This species produces the Christmas holly and red fruits that are so familiar. However, the species is dioecious and only the female trees produce fruit. Cover type 27.9, 28.11, 28.14, 28.18, 28.27, 28.29, 28.36, 28.42, 28.45.

Ilex coriacea (Pursh) Chapman, large or sweet gallberry (Figure 21.43); shrub, height to 4 m;

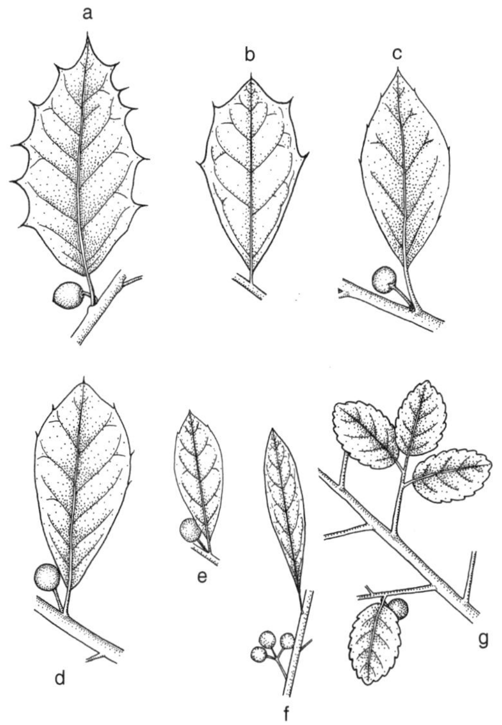

FIGURE 21.43 Leaves and drupes of (a), (b) *Ilex opaca* ×3/5; (c) *I. coriacea* ×3/5; (d) *I. cassine* ×3/5; (e) *I. glabra* ×3/5; (f) *I. myrtifolia* ×3/4; (g) *I. vomitoria* ×2/5.

evergreen; leaf arrangement alternate; leaf simple, entire, blade length 2.5 to 7.5 cm, width 1 to 3 cm, elliptical, oval or ovate, apex mucronate, margin remotely serrulate on the distal one half, abaxial surface glabrous or infrequently finely pubescent, red or black dotted (use 10× hand lens); petiole length 2 to 3 mm, pubescent; twigs gray, puberulent; flowers axillary, clustered one two six, peduncle length 5 to 7 mm; fruit a drupe, nearly black, globose, diameter 3 to 6 mm, nutlet ribbed; Region 4, Virginia to Florida and Louisiana; wet forest, bays, and swamps. Cover type 28.25, 28.30, 28.33, 28.36, 28.38.

Ilex cassine Linnaeus, dahoon (Figure 21.43); shrub, height to 8 m; evergreen; leaf arrangement alternate; leaf simple, entire, blade length 2.5 to 7.5 cm, width 1.0 to 3.5 cm, elliptical, oblong, oval or obovate, apex obtuse, margin smooth or remotely serrulate on the distal one half, minutely revolute, adaxial veins bearing scattered hair, abaxial surface veins finely pubescent; petiole length 3 to 9 mm, pubescent; twigs gray green; flowers axillary, clustered two to three at a node, peduncle length 3 to 5 mm; fruit a drupe, red, globose, diameter 4 to 6 mm; Region 4a, b, North Carolina to Florida and Louisiana, Coastal Plain; bogs, bays, cypress swamps, and pocosins.

Ilex glabra (Linnaeus) Gray, inkberry (Figure 21.43); shrub, height to 2 m; evergreen; leaf arrangement alternate; leaf simple, entire, blade length 1.5 to 5.0 cm, width 0.6 to 2.0 cm, coriaceous, elliptical or oblanceolate, apex obtuse and tipped with a 0.5 mm gland, margin showing two to three teeth near the apex, abaxial surface glabrous, red or black dotted (use 10× hand lens); petiole length 3 to 7 mm, puberulent; twigs gray, puberulent; inflorescence a cyme; flowers two to six, peduncle length 3 to 4 mm; fruit a drupe, nearly black, globose, diameter 4 to 7 mm, nutlet ribbed; Regions 2, 3, and 4, Nova Scotia to Virginia, Florida, and Louisiana, Coastal Plain; wet forest swamps, flatwoods, bogs, savanna. Cover type 28.2, 28.13, 28.28, 28.30, 28.35, 28.36, 28.38.

Ilex vomitoria Aiton, yaupon (Figure 21.43; shrub or small tree, height to 7 m; evergreen; leaf arrangement alternate; leaf simple, entire, blade length 1 to 4 cm, width 0.5 to 2.0 cm, coriaceous, elliptical, oval or ovate, apex obtuse, often round, margin crenate, teeth strongly rounded, abaxial surface veins glabrous; petiole length 2 to 3 mm, puberulent; twigs gray, puberulent, lateral twigs numerous, stiff, and short, length 2 to 15 cm; flowers axillary, numerous, peduncle length 2 to 3 mm; fruit a drupe, red or infrequently yellow, globose, diameter 4 to 8 mm, nutlet ribbed; Region 4, Coastal Plain and Mississippi River valley, Virginia to Florida and Louisiana, north to Arkansas and Oklahoma; hydric swamps, wet soil. Cover type 28.2, 28.13, 28.27, 28.32, 28.36, 28.38, 28.44, 28.53, 28.55.

Ilex myrtifolia Walter, myrtle-leaf holly (Figure 21.43); shrub, height to 3 m; evergreen; leaf arrangement alternate; leaf simple, entire, blade length 1.0 to 4.5 cm, coriaceous, width 0.5 to 0.8 cm, narrowly elliptical to slightly lanceolate, apex acute, margin smooth or bearing two to three minute teeth near the apex, minutely revolute, surfaces glabrous; petiole length 2 to 3 cm, puberulent; twigs reddish brown to gray, puberulent; flowers axillary red to yellow, one

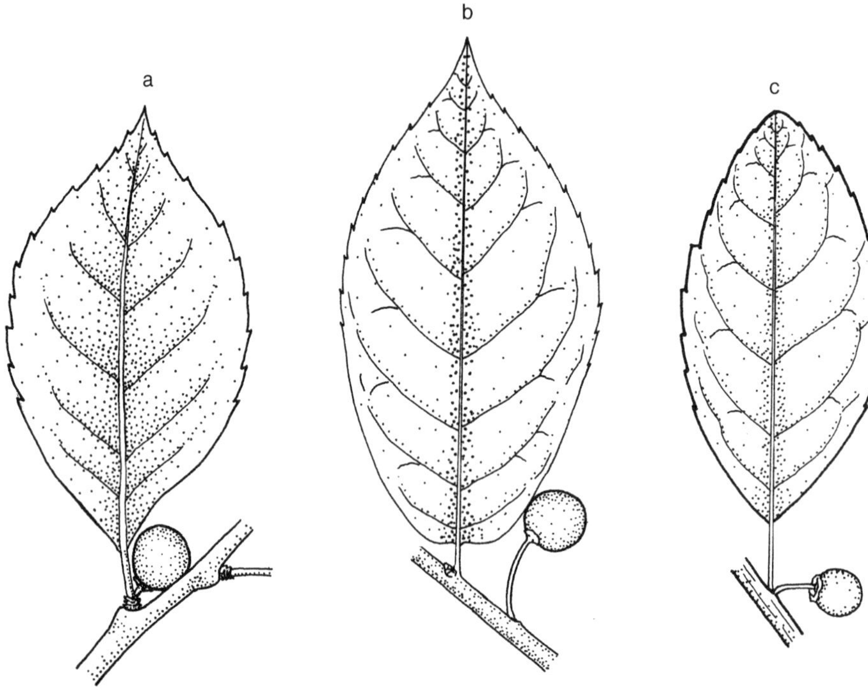

FIGURE 21.44 Leaves and drupes of (a) *Ilex ambigua* ×1.2; (b) *I. amelanchier* ×1.2; (c) *I. laevigata* ×1.

to three per peduncle, peduncle length 0.5 to 0.8 cm; fruit a drupe, globose, diameter 0.4 to 1.0 cm, nutlet ribbed; Regions 4a, b, North Carolina to Florida and Louisiana; swamps, hydric woods, pine flatwoods.

Deciduous Species

Ilex ambigua (Michaux) Torrey, Carolina holly (Figure 21.44); tall shrub, height to 5 m; deciduous; leaf arrangement alternate; leaf simple, entire; size highly variable, blade length 1 to 8 cm, width 1 to 5 cm, oval or elliptical, apex usually mucronate, margin serrated, ciliated or aciliate, smooth at the base, adaxial surface glabrous or bearing scattered whitish hair, abaxial surface or veins glabrous, sparingly pubescent or pubescent, hair whitish and stiff; petiole length 0.5 to 1.0 cm, pubescent; twigs reddish brown or gray, becoming, glabrous, lenticels raised; flowers axillary, clustered in groups of two to ten, peduncle length 2 to 3 mm; fruit a drupe, red to orange, globose, diameter 5 to 7 mm, nutlet ribbed; Region 4, Piedmont and Coastal Plain, West Virginia to Georgia and Alabama; xeric upland woods.

Ilex amelanchier Curtis, sarvis holly (Figure 21.44); shrub, height to 2 m; deciduous; leaf arrangement alternate, leaves often clustered at nodes or tips of spur shoots; leaf simple, entire, blade length 3.0 to 7.5 cm, width 1.0 to 3.5 cm, oblong or obovate, slightly coriaceous, often rugose, margin uneven to remotely serrulated but smooth at the base, apex obtuse and sharp-pointed to rounded, abaxial surface and veins pubescent; petiole length 4 to 6 mm, pubescent; twigs light reddish brown, becoming gray, puberulent, lenticels raised, whitish; flowers axillary, clustered in groups of two to seven, one two per axil, peduncle length 6 to 8 mm; fruit a drupe, red, globose, diameter 0.5 to 1.0 cm, Regions 3 and 4, Virginia to Georgia and Louisiana; swamps, floodplains; rare.

Ilex laevigata (Pursh) Gray, smooth winterberry (Figure 21.44); shrub, height to 4 m; deciduous; leaf arrangement alternate, leaves often at the tip of spur shoots or at a node; leaf simple, entire, blade length 4 to 8 cm, width 1.5 to 3.0 cm, elliptical or lanceolate, membranous, occasionally rugose, apex obtuse to acute, margin uneven to remotely serrated but smooth

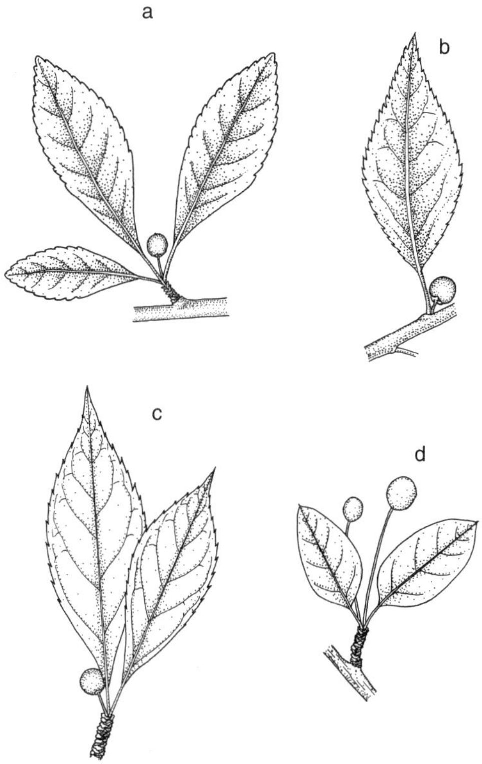

FIGURE 21.45 Leaves and drupes of (a) *Ilex decidua* ×3/5; (b) *I. verticullata* ×3/5; (c) *I. montana* ×1/2; (d) *Nemopanthus mucronata* ×2/5.

at the base, teeth sharp or slightly rounded, abaxial surface veins pubescent; petiole length 0.5 to 1.0 cm, glabrous; twigs light brown, becoming gray, glabrous, lenticels raised; flowers axillary, clustered in groups of two to five, orange red, one to two per axil, peduncle length 0.5 cm; fruit a drupe, red or infrequently yellow, globose, diameter 0.5 to 1.0 cm, nutlet ribbed; Regions 2, 3, and 4a, New England south to south Carolina, many disjunct populations; swamps and mesic-hydric woods.

Ilex decidua Walter, deciduous holly (Figure 21.45); tall shrub or small tree, height to 9 m; deciduous; leaf arrangement alternate, leaves often clustered at nodes or tips of spur shoots; leaf simple, entire, blade length 2.5 to 8.0 cm, width 1 to 3 cm, elliptical, oblanceolate, or obovate, base apex acuminate, margin crenulate but smooth of the base, adaxial surface glabrous, abaxial surface veins usually densely pubescent, hair whitish; petiole length 0.5 to 1.0 cm, pubescent; twigs gray or light brown, becoming dark gray, glabrous or at times puberulent, lenticels raised; flowers axillary or at the tip of spur shoots, peduncle length 2 to 3 mm; fruit a drupe, red, globose, diameter 5 to 7 mm; Regions 3a and 4, southern Illinois to Virginia, south to Florida and Texas; hydric woods, water margins, stream terraces. Cover type 27.2, 27.3, 27.10, 27.28, 28.10, 28.12, 28.18, 28.31, 28.42, 28.49, 28.54, 28.55.

Ilex verticillata (Linnaeus) Gray, common winterberry (Figure 21.45); tall shrub or small tree, height to 8 m; deciduous; leaf arrangement alternate, leaves often at the tip of spur shoots or at a node; leaf simple, entire, blade length 2.5 to 8.0 cm, width 1 to 4 cm, elliptical, ovate, or obovate, apex acuminate, margin serrated apex to base, teeth sharp or somewhat blunt and incurved, abaxial surface veins sparingly pubescent; petiole length 0.5 to 1.0 cm, pubescent or nearly glabrous; twigs gray or reddish brown, becoming gray, glabrous, lenticels raised, cream colored; flowers axillary, clustered in groups of two to four, peduncle length 2 to 3 mm; fruit a drupe, red, infrequently yellow or white; globose, diameter 5 to 7 mm, nutlet smooth; Regions 2, 3, and 4, Minnesota to Nova Scotia and south, scattered in Florida, Alabama, and Louisiana; swamps, hydric woods, water margins. Several varieties are recognized. Cover type 26.20.

Ilex montana Torrey & Gray, mountain winterberry (Figure 21.45); tall shrub or small tree, height to 10 m; deciduous; leaf arrangement alternate, leaves often clustered at the tip of spur shoots; leaf simple, entire, blade length 4 to 14 cm, width 1.5 to 5.0 cm, elliptical, nearly orbicular, ovate, or oblanceolate, apex acuminate, margin sharply serrated but smooth at the base, abaxial surface veins sparingly pubescent, abaxial surface usually pubescent; petiole length 0.5 to 1.5 cm, essentially glabrous; twigs light brown, becoming gray; flowers axillary or at the tip of spur shoots, peduncle length 5 mm; fruit a drupe, red, globose, diameter 0.5 to 1.0 cm, nutlet ribbed; Regions 2, 3b 4c, Massachusetts and New York, south to Georgia and Alabama, Many disjunct populations; mesic sites, mountain forests. Cover type 26.21, 26.27, 27.29, 27.35.

Ilex longipes Chapman, Georgia holly; tall shrub or small tree, height to 8 m; deciduous; leaf arrangement alternate, leaves often at the tip of spur shoots; leaf simple, entire, blade length 2.5 to 6.0 cm, width 1 to 3 cm, elliptical, ovate, or obovate, apex acuminate, infrequently rounded, margin serrated, teeth sharp or somewhat blunt and incurved, abaxial surface veins pubescent or sparingly pubescent; petiole length 0.5 to 1.0 cm, pubescent or nearly glabrous; twigs gray, glabrous, lenticels raised, tan; flowers axillary, peduncle length 1.0 to 1.8 cm; fruit a drupe, globose, red or yellow; diameter 5 to 9 mm, nutlet smooth; Regions 3 and 4, Tennessee, West Virginia, and North Carolina, south to Florida and Louisiana; xeric-mesic slopes to hydric bogs.

Nemopanthus mucronata Rafinesque, mountain holly (Figure 21.45); tall shrub or small tree, height to 2 m; deciduous; leaf arrangement alternate, leaves often clustered at nodes or tips of spur shoots; leaf simple, entire, blade length 2 to 5 cm, width 0.8 to 2.0 cm, elliptical to slightly obovate, apex obtuse, tipped with a 1 mm long point, margin minutely wavy and irregular, adaxial surface glabrous, abaxial surface glabrous, venal pattern finely reticulate; petiole length 6 to 8 mm, glabrous; twigs reddish brown, becoming gray, glabrous; flowers axillary or at the tip of spur shoots, peduncle length 2.0 to 2.7 cm; fruit a drupe, red, globose, diameter 5 to 7 mm; Regions 2 and 3, Nova Scotia to Minnesota, south to Virginia, Ohio, and northern Illinois; mesic upland forest.

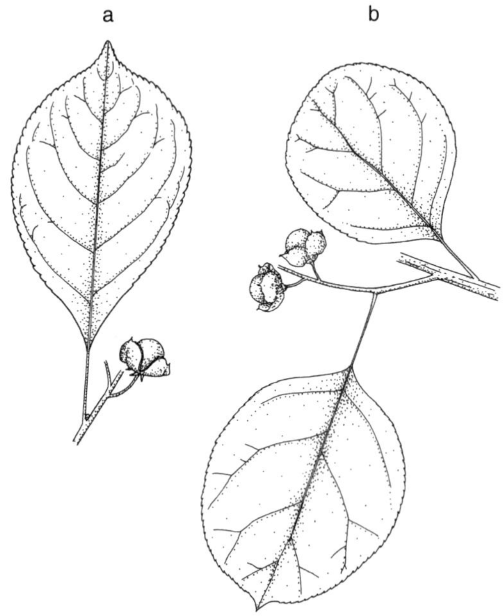

FIGURE 21.46 Leaves and capsules of (a) *Celastris scandens* ×3/5; (b) *C. orbiculatus* ×2/5.

Celastraceae (Bittersweet Family)

The Celastraceae family of about 55 genera and 850 species is found primarily in tropical and subtropical regions with a few representatives in the temperate zone.

Genus: *Celastris*

Celastris scandens Linnaeus, American bittersweet (Figure 21.46); vine, twining and climbing to 5 m; deciduous; leaf arrangement alternate; leaf simple, entire, blade length 5 to 9 cm, width 2.5 to 6.0, broadly elliptical to obovate, margin serrate, apex mucronate, base obtuse, adaxial and abaxial surfaces glabrous; petiole length 0.5 to 2.5 cm, glabrous; twigs gray; inflorescence a terminal raceme; flowers greenish; fruit a capsule, orange, splitting to expose the scarlet seeds; Regions 2, 3, 4, and 5, New England to Quebec, south to Georgia, Alabama, and Oklahoma, Kansas, and New Mexico; mesic upland soil. Over-collecting for use in indoor bouquets has increased concern for this once common species. It is now protected in some states, even where it is abundant; picking it should be avoided. The clearing and widening of roads has eliminated fencerows where this species once grew unrestricted.

Celastris orbiculatus Thunberg, Oriental bittersweet (Figure 21.46); vine, climbing and twining and forming thickets; height to 7+ m; deciduous; leaf arrangement alternate; leaf simple, entire, blade length 5 to 10 cm, width 2.5 to 6.0, orbicular to obovate, margin serrated, apex usually abruptly mucronate, base obtuse to truncate, adaxial and abaxial surfaces glabrous; petiole length 1.4 to 3.0 cm, glabrous; twigs reddish brown to gray; flowers axillary, individual or in a cyme of two to three, flowers greenish; fruit a capsule, green or orange, splitting to expose the yellow green or scarlet seeds. Introduced, native of Asia. This vine has escaped cultivation and is essentially naturalized in the eastern United States. It is a strong competitor and aggressive, so that it forms dense thickets by growing over the tops of seedlings, saplings, shrubs, and small trees.

Genus: *Euonymous*

Euonymous americanus Linnaeus, strawberry-bush, (Figure 21.47); multibranching shrub, straggling, height to 2 m; deciduous; leaf arrangement opposite; leaf simple, entire, blade length 3 to 9 cm, width 1.3 to 3.0, elliptical, ovate, or lanceolate, margin serrulated, apex mucronate to acuminate, base obtuse, adaxial and abaxial surfaces glabrous except for scattered veinal hair; petiole length 1 to 2 mm, glabrous; new twigs green, usually four-ridged, turning gray; inflorescence an axillary cyme with two to five flowers, regular, petals five, diameter 8 to 9 mm; fruit a capsule, bright red, nearly globose, four-lobed, diameter 1.0 to 1.5 cm, surface rough with minute projections; Regions 3 and 4, New York and West Virginia to southern Indiana, Missouri, and Oklahoma, south to Florida and Texas; mesic woods, canyons, drainages. Cover type 27.15, 28.12, 28.16.

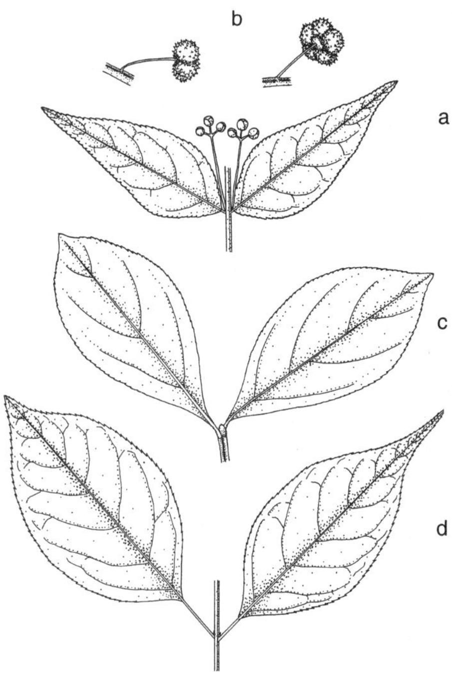

FIGURE 21.47 *Euonymous americanus* (a) twig with leaf and immature fruit ×1/2, (b) mature fruit ×1; (c) *E. obovatus* leaf ×1/2; (d) *Euonymous atropurpureus* leaf ×2/5.

Euonymous obovatus Nuttall, running strawberry-bush, (Figure 21.47); low, trailing, vine, but climbing to 3 m; deciduous; leaf arrangement opposite; leaf simple, entire, blade length 3 to 9 cm, width 1.3 to 3.0, elliptical, lanceolate, obovate, or oblanceolate, margin serrulated, apex mucronate, base obtuse, adaxial and abaxial surfaces glabrous, except for scattered veinal hair; petiole length 3 to 4 mm, glabrous; new twigs green, usually four-ridged, turning gray, bud green; inflorescence an axillary cyme with two to five flowers, regular, petals five, diameter 8 to 9 mm; fruit a capsule, bright red, nearly globose, four-lobed, diameter 1.0 to 1.5 cm, surface rough with minute projections; Region 3, New York to Michigan, south to West Virginia, Tennessee, Illinois, and Missouri mesic woods, canyons, drainages.

Euonymous atropurpureus Jacquin, eastern wahoo, burning-bush (Figure 21.47); multibranching shrub, height to 5 m; deciduous; leaf arrangement opposite; leaf simple, entire, blade length 3.5 to 13.0 cm, width 2.5 to 6.0 cm, elliptical, oblong, lanceolate obovate, margin serrulated, apex abruptly mucronate to slightly acuminate, base obtuse, adaxial and abaxial surfaces glabrous; petiole length 1.0 to 1.7 cm, glabrous; new twigs green, often ridged, becoming round, turning gray; inflorescence an axillary cyme with 3 to 12 flowers; fruit purple, oval, length ~1 cm; seed enclosed by an orange aril; introduced as an ornamental, escaped cultivation in the eastern United States and southern Canada.

Euonymous atatus (Thunberg) Siebold, winged euonymous, is an introduced shrub species widely planted as an ornamental. The stems have two to four prominent corky wings (0.5 to 1.0 cm wide).

ORDER: RHAMNALES

Rhamnaceae (Buckthorn Family)

Representatives of the Rhamnaceae family of 55 to 58 genera and 900 species are generally trees, shrubs, and vines, distributed worldwide but with some concentration in tropical regions. The flowers are small, inconspicuous, regular, and perfect. The leaves are simple, and twigs often are modified into stiff spines or thorns. In North America, the major genera are *Rhamnus* and *Ceanothus*, both of which have a simple, alternate, pinnately veined leaf with the veins tending to have an arcuate pattern. The leaf of *Rhamnus* has one primary vein, and the leaf of *Ceanothus* has three primary veins.

Genus: *Rhamnus*

The largest genus is *Rhamnus*, with more than 150 species. The 10 or so species native to North America, as well as several introduced species, are shrubs, but some approach small trees, and they occur in a wide range of plant community types. Three to seven flowers appear in axillary cymes and produce three to four purple to black drupes (diameter 6 to 10 mm). Yellow and green dyes are obtained from the fruit and bark of some species. *Rhamnus alnifolia, R. betulifolia, R. californica, R. caroliniana,* and *R. crocea* may reach small tree size (7 to 9 m).

Eastern Species

Rhamnus alnifolia L'Heritier De Brutelle, alderleaf buckthorn (Figure 21.48); low shrub, height to 5 m; deciduous; leaf arrangement alternate; leaf simple, blade length 4 to 9 cm, width 1.5 to 3.5 cm, entire, elliptical, margin serrated, adaxial surface midrib with scattered hair and sunken at the proximal end, apex obtuse or mucronate, lateral veins distinct and tan on the abaxial surface, venation essentially arcuate, five to eight veins per side; petiole length 0.4 to 1.0 cm, bearing scattered hair; twigs deep reddish brown, lightly pubescent, not thorny, buds appressed to twig; fruit a drupe, dark blue black, diameter 3 to 5 mm, three-seeded; Regions 1, 2, 6, and 7, Newfoundland to British Columbia, south to New Jersey and northern Illinois, Nebraska, Wyoming, and California; swamps, bogs, wet meadows, stream banks. Cover type 26.20.

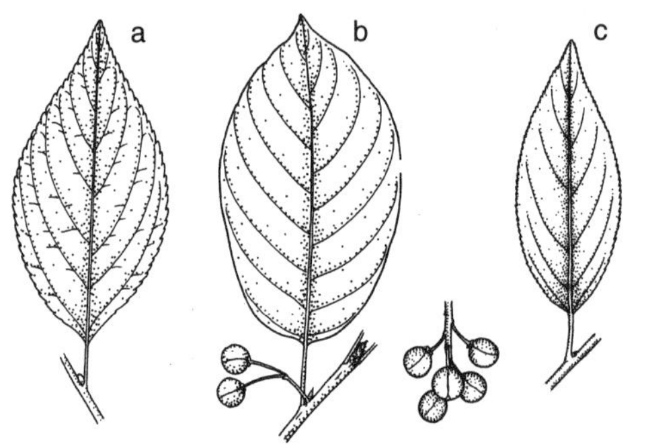

FIGURE 21.48 (a) *Rhamnus alnifolia* leaf ×3/5; (b) *R. caroliniana* leaf and drupes ×1/2; *R. lanceolata* (c) leaf ×1/2, (d) inflorescence with attached drupes ×1/2.

Rhamnus caroliniana Walter, Carolina buckthorn (Figure 21.48); usually a shrub, infrequently a small tree, height to 5 m; deciduous; leaf arrangement alternate; leaf simple, blade length 5 to 12 cm, width 2.0 to 4.5 cm, entire, oblong or long obovate, margin serrulated, apex obtuse, rounded or nearly mucronate, adaxial surface midrib with scattered hair and sunken at the proximal end, abaxial surface veins bearing scattered hair, lateral veins distinct and tan on the abaxial surface, nearly straight but arcuate at the margin, six to ten veins per side; petiole length 0.5 to 1.4 cm, sparingly pubescent; twigs deep reddish brown, lightly pubescent, not thorny, buds appressed to twig; fruit a drupe, red, diameter 4 to 6 mm, three-seeded; Regions 3 and 4; Virginia to southern Illinois, southeast Missouri and Nebraska, south and east to Florida and Texas; xeric-mesic to mesic upland forest.

Rhamnus lanceolata Pursh, lanceleaf buckthorn (Figure 21.48); low shrub, height to 5 m; deciduous; leaf arrangement alternate; leaf simple, blade length 4 to 7 cm, width 1.8 to 3.0 cm, entire, oblong, elliptical, or slightly lanceolate, margin finely serrated, adaxial surface midrib with scattered hair and sunken at the proximal end, apex usually narrowly mucronate or acuminate with a rounded tip, lateral veins distinct and tan on the abaxial surface, venation essentially arcuate, six to nine veins per side; petiole length 4 to 6 mm, pubescent; twigs grayish, greenish, or slightly reddish, lightly pubescent or puberulent, not thorny, buds appressed to twig; fruit a drupe, dark blue black, diameter 4 to 6 mm, three-seeded; Regions 3 and 4, central United States, Pennsylvania, and West Virginia to Iowa and Nebraska, south to Alabama and Texas; wooded slopes.

Western Species

Rhamnus betulifolia Greene, birchleaf buckthorn (Figure 21.49), height to 5 m; deciduous; leaf arrangement alternate or subopposite (Figure 15.3d); leaf simple, blade length 4 to 10 cm, width 1.5 to 5.0 cm, entire, broadly elliptical, oblong, or obovate, margin

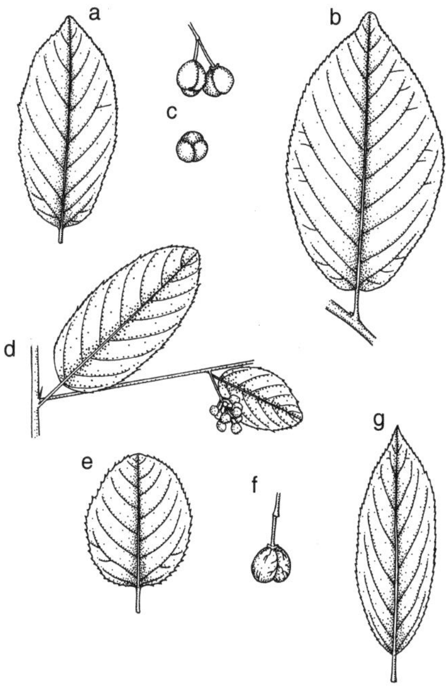

FIGURE 21.49 *Rhamnus betulifolia* (a), (b) leaves × 1/2, (c) three-seeded drupe ×4/5; (d) *R. californica* leaves and inflorescence with immature drupes × 3/5; *R. crocea* (e) leaf ×4/5, (f) three-seeded drupe ×1; (g) *R. rubra* leaf ×1.

irregularly serrate or nearly smooth, apex obtuse to weakly mucronate, adaxial surface glabrous to sparingly pubescent, midrib sunken at the proximal end, abaxial surface densely pubescent, yellowish green, velvety, lateral veins distinct, 6 to 11 veins per side; petiole length 4 to 8 mm, densely pubescent; twigs yellowish gray or green becoming reddish brown, pubescent, buds appressed to twig, densely pubescent; fruit a drupe, dark reddish brown, diameter 4 to 7 mm, three-seeded; Region 5, Mexico to Utah; xeric oak woodland and ponderosa pine savanna.

Rhamnus californica Eschscholtz, California buckthorn or coffeeberry (Figure 21.49); shrub, height to 5 m; deciduous; leaf arrangement alternate or subopposite; leaf simple, blade length 2 to 6 cm, width 1.5 to 3.0 cm, entire, broadly elliptical or oval, infrequently approaching linear, margin smooth or spinulate, apex obtuse to rounded, adaxial surface yellowish green, glabrous, sunken at the proximal end, abaxial surface densely pubescent, yellowish green, velvety; venation essentially arcuate, lateral veins distinct, 4 to 11 veins per side; petiole length 2 to 6 mm, densely pubescent or puberulent; twigs yellow gray or green becoming reddish brown, puberulent, buds appressed to twig, pubescent; fruit a drupe, black, diameter 4 to 7 mm, three-seeded; Regions 5 and 7, California, Arizona, and New Mexico; chaparral, scrub woodland, redwood forest; serpentine soil. Cover type 30.38.

Rhamnus crocea Nuttall, hollyleaf buckthorn, spiny redberry (Figure 21.49); shrub, height to 5 m; deciduous; leaf arrangement alternate or subopposite; leaf simple, blade length 1.0 to 3.5 cm, width 1 to 3 cm, entire, broadly elliptical, oval or orbicular, margin finely spiny, spine length 1 to 2 mm, apex obtuse, rounded, or emarginate, adaxial and abaxial surfaces shiny, adaxial surface midrib pubescent; venation essentially arcuate, lateral veins distinct, 4 to 11 veins per side; petiole length 2 to 6 mm, densely pubescent; twigs reddish brown, pubescent or puberulent, buds appressed to twig, pubescent; fruit a drupe, dark blue black, diameter 3 to 5 mm, three-seeded; Regions 5 and 7, California, Arizona, and Mexico; xeric scrub woodland, chaparral, coniferous forest, elevation to 700 m.

Rhamnus rubra Greene, Sierra buckthorn, redstem buckthorn, Sierra coffeeberry (Figure 21.49); low shrub, height to 5 m; deciduous; leaf arrangement alternate; leaf simple, blade length 1.0 to 4.5 cm, width 0.8 to 1.5 cm, entire, oblong, elliptical, or slightly oblanceolate, margin serrulated, adaxial surface glabrous, apex obtuse to nearly rounded; venation essentially arcuate, lateral veins straight, distinct, arcuate near the margin, six to nine veins per side; petiole length 4 to 7 mm, pubescent or glabrous; twigs reddish brown becoming gray, lightly puberulent, not thorny, buds pubescent; fruit a drupe, dark blue black, diameter 0.7 to 1.1 cm, three-seeded; Region 6 and 7, California Sierra Nevada and several states to the east, 1200 to 1800 m.

Rhamnus purshiana de Candolle, cascara buckthorn (Figure 21.50); usually a shrub, infrequently a small tree, height to 5 m; deciduous; leaf arrangement opposite, subopposite, or alternate, leaves often somewhat clustered near the twig apex; leaf simple, blade length 5 to 10 cm, width 2 to 6 cm, entire, broadly elliptical, oblong to slightly obovate, margin finely serrated, apex obtuse, rounded or weakly mucronate, adaxial surface midrib pubescent and sunken at the proximal end, abaxial surface and veins pubescent, usually velvety, venation essentially arcuate, 10 to 14 veins per side; petiole length 1.0 to 1.8 cm, densely pubescent; twigs gray, glabrous or pubescent, tips of twigs becoming sharp thorns; bud length 4 to 5 mm, appressed to twig, densely pubescent; fruit a drupe, dark blue black, diameter 5 to 8 mm, three-seeded; Regions 6 and 7, northern California to British Columbia; montane coniferous forest; source of a strong laxative.

Two other California species, *R. illicifolia* and *R. tomentella* are described by Stuart and Sawyer (2001).

Introduced Species
Rhamnus cathartica Linnaeus, European buckthorn (Figure 21.50); usually a shrub, infrequently a small tree, height to 5 m; deciduous; leaf arrangement opposite or subopposite (Figure 15.3d), leaves frequently clustered at nodes or on spur shoots; leaf simple, blade length 2 to 6 cm, width 2.0 to 4.5 cm, entire, broadly elliptical or oval to nearly orbicular or obovate, margin finely serrulate or crenate, apex obtuse, rounded or nearly mucronate, adaxial surface midrib with scattered hair and sunken at the proximal end, abaxial surface veins lightly pubescent, venation essentially arcuate, four to five veins per side; petiole length 1 to 2 cm, sparingly pubescent; twigs gray, glabrous, tips of twigs becoming sharp thorns; bud length 4 to 5 mm, appressed to twig; fruit a drupe, black, diameter 4 to 6 mm, three-seeded; Regions 2, 3 and 4, Nova Scotia to Minnesota and south; introduced from Europe; open woods, fencerows, and pastures.

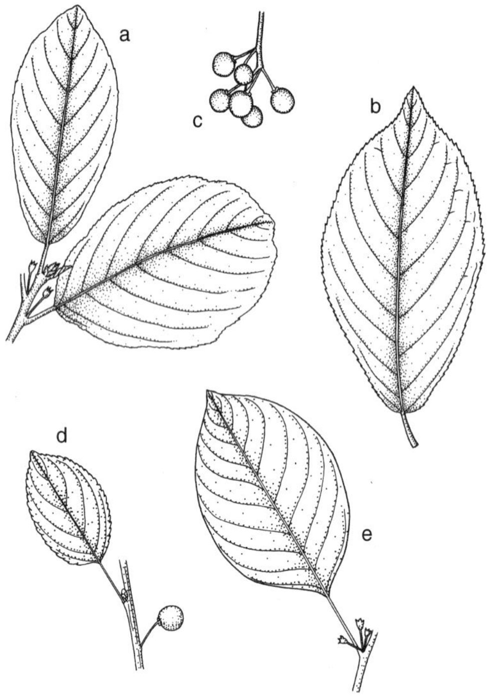

FIGURE 21.50 *Rhamnus purshiana* (a) twig with leaves and developing flowers ×1, (b) leaf ×1, (c) inflorescence with attached three-seeded drupes ×1/2; (d) *R. cathartica* and drupe ×3/5; (e) *R. frangula* twig with leaf and developing flowers ×1/2.

Rhamnus frangula Linnaeus, glossy buckthorn (Figure 21.50); a shrub or small tree, height to 5 m; deciduous; leaf arrangement alternate; leaf simple, blade length 3 to 10 cm, width 2.5 to 6.5 cm, entire, elliptical to broadly obovate, margin smooth, apex obtuse, rounded or weakly mucronate, adaxial surface midrib with scattered hair and sunken at the proximal end, abaxial surface veins bearing scattered hair, lateral veins distinct and tan on the abaxial surface, nearly straight but arcuate at the margin, 7 to 11 veins per side; petiole length 1 to 2 cm, puberulent; twigs deep reddish brown, lightly puberulent, not thorny, buds appressed to twig; fruit a drupe, black, diameter 4 to 6 mm, three-seeded; introduced from Europe; rapidly expanding in the eastern forest, a serious competitor and noxious weed in the forests of southern Wisconsin and northern Illinois.

Genus: *Ceanothus*

Ceanothus species are an important component of California chaparral. Of some 54 *Ceanothus* species and recognized hybrids, 48 are restricted to California (Region 7), 3 are only in the eastern United States, and 3 are found in both regions. The leaves are characterized by three primary veins, the two large lateral veins extending to three quarters the leaf length; the leaf venation is mixed palmate and pinnate. The inflorescence is a compact panicle borne in the axils of leaves. The seeds are often propelled from the splitting pit (endocarp) of the drupe. In western forests, *Ceanothus* may quickly dominate a site after cutting and a wildfire or a hot prescribed fire, either of which will crack the hard seed coat and permit the seed to germinate; relatively cool fires will not scarify the seed coat. Seed coat scarification after a hot fire, combined with nitrogen-fixing actinomycetes in the roots, appears to produce the large abundance of these species on some sites, particularly in chaparral, which burns frequently. The widely distributed species in the east are *C. americanus* L. (New Jersey tea) and *C. ovatus*. These species are associated with dry open woodland and prairie.

Ceanothus americanus Linnaeus, New Jersey tea, redroot (Figure 21.51); shrub, height to 1 m; deciduous; leaf arrangement alternate; leaf simple, blade entire, length 3.0 to 9.5, width 1.5 to 5.5 cm, lanceolate or ovate, margin finely serrated, each tooth tipped with a black gland, apex slightly acuminate, venation mixed, adaxial surface glabrous to lightly pubescent, abaxial surface lightly to densely pubescent and velvety; petiole length 0.4 to 4.0 cm, pubescent; inflorescence a panicle on a long, leafless peduncle borne in the axils of upper leaves; twigs reddish brown, usually pubescent; fruit a three-lobed drupe, length/diameter ~4 mm, a cluster of dry pedicels terminated by cuplike receptacles remaining on the plant after seed dispersion; Regions 2, 3, and 4, southern Manitoba to Ontario, south to Texas, Alabama, and Florida. Cover type 26.18.

Ceanothus herbaceus Refinesque (*C. ovatus* Desfontaines), small redroot, inland New Jersey tea (Figure 21.51); shrub, height to 1 m; deciduous; leaf arrangement alternate; leaf simple, blade entire, length 1.5 to 5.5, width 0.5 to 1.8 cm, elliptical or oblong, margin serrulated, each tooth tipped with a black gland, apex slightly acuminate, venation mixed, adaxial surface glabrous to lightly pubescent, abaxial surface usually densely pubescent and velvety; petiole length 2 to 5 mm, pubescent; twigs reddish brown to gray, usually pubescent; inflorescence a panicle on a long, leafless peduncle borne in the axils of upper leaves or may be terminal; fruit a three-lobed drupe, length/diameter ~2 to 3 mm, a cluster of dry pedicels terminated by cuplike receptacles remaining on the plant after seed dispersion; Regions 2, and 3, Vermont, New York, southrern Ontario, southwest to Texas and Arkansas, west to Manitoba and Montana; xeric soil, open sites, prairie.

Ceanothus arboreus Greene, California ceanothus (Figure 21.52); shrub or small tree, height to 3 m; deciduous; leaf arrangement alternate; leaf simple, blade entire, length 1.5 to 6.5, width 1.5 to 5.5 cm, lanceolate or ovate, margin serrulated, each tooth tipped with a black gland, apex obtuse, venation mixed, adaxial surface glabrous to lightly pubescent, abaxial surface lightly to densely pubescent and velvety, light yellowish brown; petiole length 0.4 to 1.0 cm, pubescent; twigs light yellowish brown turning dark gray, pubescent; inflorescence a panicle on a long, leafless peduncle borne in the axils of upper leaves or may be terminal; fruit a three-lobed drupe, length/diameter 3 to 6 mm, a cluster of dry pedicels terminated by cuplike receptacles remaining on the plant after seed dispersion; Region 7, California; low elevations, xeric chaparral, savanna, open land.

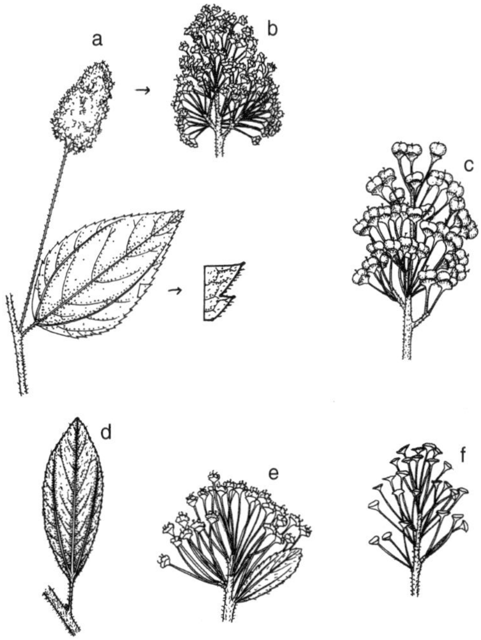

FIGURE 21.51 *Ceanothus americanus* (a) twig with leaf and inflorescence ×1/2, (b) inflorescence with flowers ×1/2, (c) inflorescence with attached drupes ×1/2; *C. herbaceus* (d) twig and leaf ×3/4, (e) inflorescence with flowers ×1/2, (f) inflorescence with drupes dispersed ×1/2.

Ceanothus cuneatus (Hooker) Nuttall, buckbrush (Figure 21.52); shrub or small tree, height to

FIGURE 21.52 (a) *Ceanothus arboreus* leaf ×1; *C. cuneatus* (b) leaves ×1, (c) inflorescence with attached three-seeded drupes ×1; (d) *C. cordulatus* thorny twig and leaf ×1; (e) *Berchemia scandens* twig with leaves and inflorescence with attached drupes ×1.

4 m; evergreen; leaf arrangement alternate; leaf simple, blade entire, length 0.5 to 1.5, width 4 to 9 mm, elliptical or obovate, margin smooth, apex obtuse, rounded or slightly emarginate, venation indistinct, adaxial surface glabrous to lightly pubescent, abaxial surface lightly to densely pubescent and velvety, light yellowish brown; petiole length 1 mm, pubescent; twigs dark gray, lateral twigs stout, extremely stiff, and rough; inflorescence a panicle, short, length 3 cm; fruit a three-lobed drupe, length/diameter 4 to 6 mm, a

cluster of dry pedicels terminated by cuplike receptacles remaining on the plant after seed dispersion; Region 7, California; chaparral, xeric open slopes, forming dense thickets. Cover type 30.29.

Ceanothus cordulatus Keller, snowbrush (Figure 21.52); thorny shrub, height to 1.5 m, spreading horizontally; evergreen; leaf arrangement alternate; leaf length 1.0 to 2.5 cm, width 0.5 to 1.5 cm, entire, oval or elliptical, primary veins three, glabrous; twigs tapered to a thorny projection; flowers white; fruit a three-lobed drupe; Region 7, California; open slopes and ridges.

Genus: *Berchemia*

Berchemia scandens (Hill) K. Hoch, rattan vine, supplejack (Figure 21.52); high-climbing vine; deciduous; leaf arrangement alternate; leaf simple, blade length 2 to 6 cm, width 0.6 to 4.0 cm, entire, elliptical, or ovate, margin minutely wavy and revolute, adaxial and abaxial surfaces glabrous, apex obtuse, lateral veins distinct, essentially arcuate, 8 to 12 veins per side; petiole length 4 to 8 mm, glabrous; inflorescence a terminal panicle; fruit a drupe, dark blue black, oval or oblong, length 4 to 7 mm; Regions 3 and 4, primarily the Coastal Plain, Virginia and Tennessee to Florida and Texas, following the Mississippi River valley to Missouri, Illinois, and Kentucky; swamps, mesic-hydric forest. Cover type 27.10, 28.7, 28.14, 28.22, 28.31, 28.49, 28.50.

Vitaceae (Grape Family)

This family of 11 or 12 genera and about 700 species is primarily tropical and subtropical in distribution, but several genera extend into the temperate zone of North America. Members of the family typically are climbing vines. Species of *Vitis, Ampelopsis,* and *Parthenocissus* are common in eastern forests (Regions 2, 3 and 4). The largest genus, *Cissus,* with about 300 species, is restricted to the tropics and is usually found only in the temperate zone as cultivated plants, although the United States Department of Agriculture (1982) lists several species as native to the southeastern United States.

Vitis and *Ampelopsis* have alternate, simple, coarsely serrated, palmately veined leaves generally with a long (8 to 10 cm) petiole. Some species have lobed leaves. The stipules are deciduous. The branched tendrils used for climbing are opposite the leaves. The main stem is a trailing vine, and in *Vitis* species may reach 30 m in length and 15 cm in diameter. The pith of *Vitis* is brown; *Vitis* may be dioecious or polygamo-dioecious and *Ampelopsis* polygamo-dioecious or polygamo-monoecious. *Vitis* has a fleshy, usually dark blue berry with three to six seeds.

Genus: *Vitis*

Of 23 *Vitis* species, only 2 or 3 are native to the far western United States, with 4 other species extending into the southwest (Arizona and New Mexico). The common eastern *Vitis* species collectively occupy an area from eastern Texas to Minnesota, east to the coast. The problem with the use of common names for *Vitis* is that one name may apply to several species.

Eastern Species

Vitis aestivalis Michaux, summer grape (Figure 21.53); vine; deciduous; leaf arrangement alternate or leaves opposite the inflorescence or tendril; leaf simple, blade

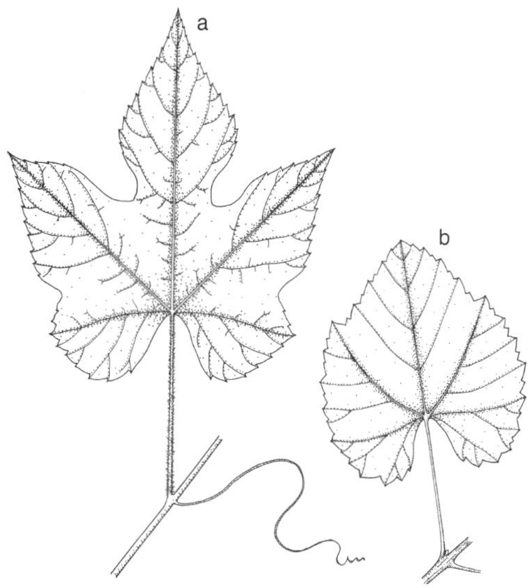

FIGURE 21.53 *Vitis aestivalis* leaf shapes ×1/2.

often strongly three-lobed, infrequently entire, length 7 to 16 cm, width 7 to 16 cm, palmately veined, margin coarsely dentate, apex obtuse to long acuminate, base shallow to deeply cordate, adaxial surface glabrous to pubescent, abaxial surface densely rusty pubescent; petiole length 4 to 11 cm, densely rusty pubescent; twigs reddish brown, new leaf-bearing twigs glabrous, glaucous; inflorescence a panicle; fruit a berry, blue black, diameter 0.6 to 1.3 cm; tendrils usually forked; Regions 2, 3, and 4, Wisconsin and Michigan to Massachusetts, south to Georgia and Texas; xeric woods, slopes, roadsides, open sites. Cover type 28.22.

Vitis cinerea Engelmann, winter or graybeard grape (Figure 21.54); vine; deciduous; leaf arrangement alternate or opposite the inflorescence or tendril; leaf simple, blade entire or weakly three-lobed, length 7 to 18 cm, width 7 to 12 cm, palmately veined, margin coarsely dentate, apex obtuse to long acuminate, base shallow to deeply cordate, adaxial surface glabrous to pubescent, abaxial surface densely grayish tomentose; petiole length 4 to 11 cm, densely pubescent; twigs light brown to reddish brown, leaf-bearing twigs densely gray pubescent or puberulent; inflorescence a panicle; fruit a berry, blue black, diameter 0.6 to 1.8 cm; tendrils usually forked; Regions 3 and 4, Virginia to Illinois and Mississippi, south to Florida and Texas; mesic woods, north slopes, terraces.

Vitis labrusca Linnaeus, fox grape; vine; deciduous; leaf arrangement alternate or opposite the inflorescence or tendril; leaf simple, blade weakly to strongly three-lobed, length 5 to 12 cm, width 6 to 13 cm, sinuses shallow, or if deep, then U-shaped, palmately veined, margin dentate, apex obtuse to acuminate, base shallow to deeply cordate, adaxial surface glabrous, abaxial surface densely rusty tomentose, felty; petiole length 4 to 8 cm, tomentose; twigs reddish brown, rusty tomentose, new leaf-bearing twigs usually distinctly ridged; inflorescence a panicle; fruit a berry, blue black, diameter 0.8 to 2.0 cm; tendrils usually forked; Regions 3 and 4, New England to southern Michigan and south, uncommon in Region 4; upland woods.

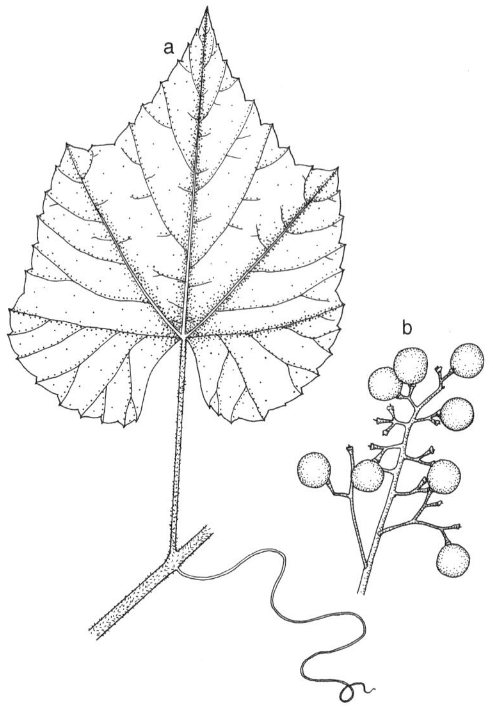

FIGURE 21.54 *Vitis cinerea* (a) twig and leaf ×1/2, (b) inflorescence with attached grapes (berries) ×1/3.

FIGURE 21.55 *Vitis palmata* twig showing opposite arrangement of leaf and tendril ×3/5.

Vitis palmata Vahl, Missouri or red grape (Figure 21.55); vine, leaf arrangement alternate or opposite the inflorescence or tendril; leaf simple, blade generally three-lobed, length 5 to 8 cm, width 5 to 8 cm, palmately veined, margin coarsely and deeply toothed, teeth acuminate, apices long to exceptionally long acuminate, base shallow cordate, adaxial surface glabrous or with some venal hair, abaxial surface usually sparingly pubescent but with dense axillary tufts; petiole length 3 to 5 cm, sparingly pubescent; twig reddish brown, new leaf-bearing twigs usually ridged but not twisted; inflorescence a panicle; fruit a berry, blue black, diameter 5 to 8 mm; tendrils usually forked; Regions 3 and 4; southern Indiana, central Illinois and Missouri south to Texas and Lousiana; swamps.

Vitis riparia Michaux, riverside or frost grape (Figure 21.56); vine; deciduous; leaf arrangement alternate or opposite the inflorescence or tendril; leaf simple, blade often strongly three-lobed or infrequently entire, length 5 to 12 cm, width 4 to 7 cm, sinuses U- or C-shaped, palmately veined, margin coarsely and deeply dentate, teeth acuminate, apex obtuse to long acuminate, base shallow to deeply cordate, adaxial surface glabrous or with venal pubescent, abaxial surface usually sparingly pubescent but with dense rusty axillary tufts; petiole length 4 to 11 cm, sparingly pubescent; twigs reddish brown, new leaf-bearing twigs usually distinctly ridged to nearly flattened and twisted; inflorescence a panicle; fruit a berry, blue black, diameter 0.6 to 1.2 cm; tendrils usually forked; Regions 2, 3, and 4, New Brunswick, New England to Wisconsin and Manitoba, southwest to Virginia, Tennessee, Missouri, Texas, and New Mexico; riverbanks, terraces mesic woods.

Vitis rotundifolia Michaux, muscadine grape (Figure 21.57); vine, seldom climbing but often forming dense ground cover; deciduous; leaf arrangement alternate or opposite the inflorescence or tendril; leaf

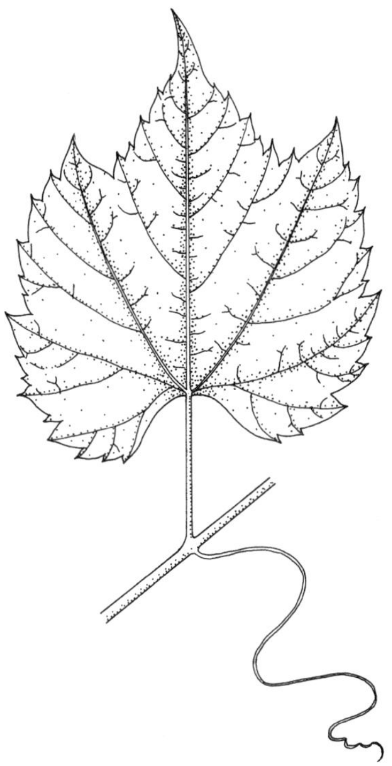

FIGURE 21.56 *Vitis riparia* twig showing opposite arrangement of leaf and tendril ×1/2.

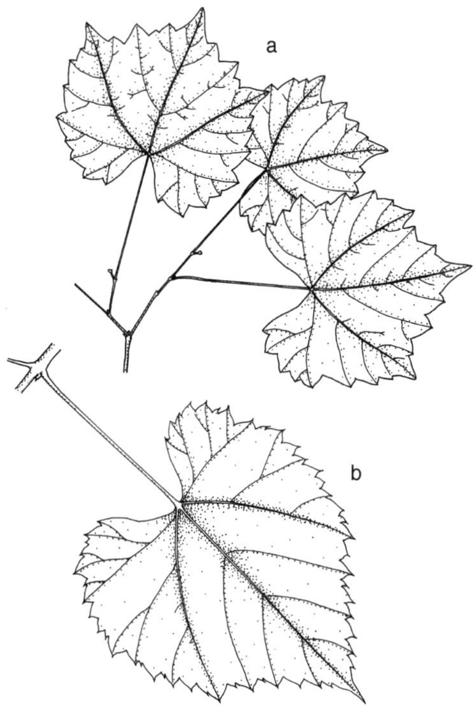

FIGURE 21.57 (a) *Vitis rotundifolia* leaves ×1/2; (b) *V. vulpina* leaf ×2/5.

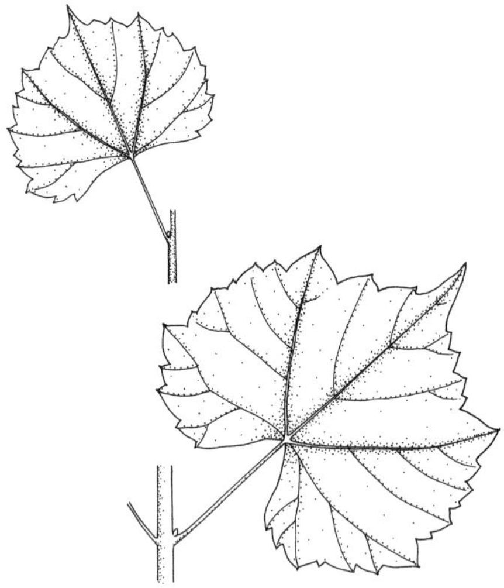

FIGURE 21.58 *Vitis rupestris* leaf shapes ×2/5.

simple, blade usually weakly three-lobed, infrequently entire, length 2.5 to 7.0 cm, width 3.5 to 7.0 cm, deltoid to orbicular, often wider than long, palmately veined, margin coarsely dentate, teeth sharp-pointed or blunt, apices acuminate, base shallow to deeply cordate, adaxial surface glabrous to sparingly pubescent, abaxial surface sparingly pubescent with dense axillary tufts; petiole length 1 to 5 cm, scattered hair; twigs reddish brown or gray; inflorescence a panicle; fruit a berry, blue black, diameter 1.2 to 2.5 cm; tendrils usually forked; Regions 3 and 4, Delaware and Virginia to southern Indiana and Missouri, south to Florida, Oklahoma, and Texas; xeric-mesic upland woods. Cover type 27.10, 28.49.

Vitis vulpina Linnaeus (*Vitis cordata* Michaux), chicken or frost grape (Figure 21.58); vine; deciduous; leaf arrangement alternate or opposite the inflorescence or tendril; leaf simple, blade usually weakly three-lobed to entire, length 7 to 11 cm, width 5 to 11 cm, generally deltoid, often as wide as long, palmately veined, margin coarsely dentate, teeth sharp-pointed, apices acuminate, base shallow cordate, adaxial surface glabrous, veins strigose, abaxial surface veins densely strigose with dense axillary tufts; petiole length 4 to 10 cm, pubescent; twigs light reddish brown; inflorescence a panicle; fruit a berry, blue black, diameter 3 to 9 mm; tendrils usually forked; Regions 3 and 4, New York and West Virginia to Illinois and Missouri, south to Florida and Texas; bottomlands and stream banks.

Vitis rupestris Scheele, sand, bush, or sugar grape (Figure 21.58); deciduous; leaf arrangement alternate or opposite the inflorescence; leaf simple, blade usually weakly three-lobed, infrequently entire, length 3.5 to 8.0 cm, width 4 to 13 cm, wider than long, palmately veined, margin coarsely dentate, teeth sharp-pointed, apices acuminate but often obscure, base shallowly cordate to truncate, adaxial surface essentially glabrous, abaxial surface with scattered strigose hair on the veins and dense axillary tufts; petiole length 2.5 to 7.0 cm, glabrous or with scattered hair; twigs yellowish- to reddish-brown; inflorescence a panicle; fruit a berry, blue black, diameter 0.8 to 1.0 cm; tendrils absent; Regions 3 and 4, Delaware and Virginia to southern Indiana and Missouri, south to Florida, Oklahoma, and Texas; stream banks to upland woods.

Western Species

Vitis californica Bentham, California wild grape (Figure 21.59); trailing vine; deciduous; leaf arrangement alternate or opposite the inflorescence or tendril; leaf simple, blade weakly to distinctly three-lobed, length 5 to 10 cm, width 6 to 11 cm, sinuses shallow, palmately veined, margin dentate,

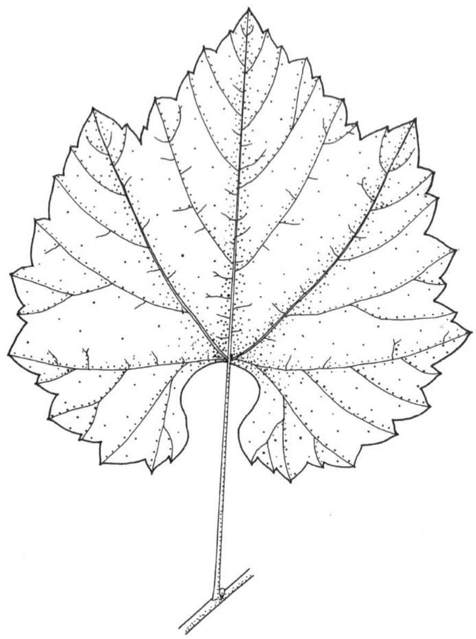

FIGURE 21.59 *Vitis californica* leaf ×3/4.

teeth apices sharp to somewhat rounded, base cordate, adaxial surface glabrous, or lightly tomentose, abaxial surface white tomentose, generally velvety; petiole length 4 to 8 cm, white tomentose; twigs gray brown, white tomentose; inflorescence a panicle; fruit a berry, blue black; tendrils usually forked; Region 7, California.

Vitis girdiana Munson; trailing vine; similar to *V. californica* but abaxial surface is densely white tomentose and felty; Region 7, California.

Genus: *Ampelopsis*

There are two species of *Ampelopsis* in North America. Their fruits are similar to those of *Vitis* but are not edible. The vines have a white pith.

Ampelopsis cordata Michaux, simple-leaf ampelopsis (Figure 21.60); trailing or climbing vine; deciduous; leaf arrangement alternate or opposite the inflorescence or tendril at each node; leaf simple, entire or weakly three-lobed, blade length 4 to 9 cm, width 3.5 to 8.0 cm, somewhat deltoid, margin coarsely

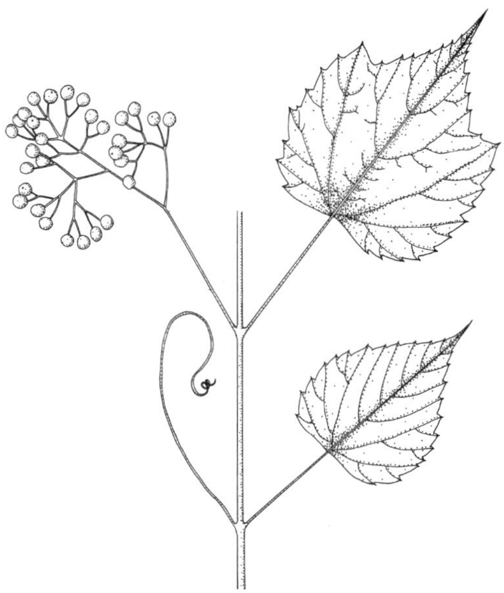

FIGURE 21.60 *Ampelopsis cordata* twig with leaves, tendril, and attached inflorescence with attached berries.

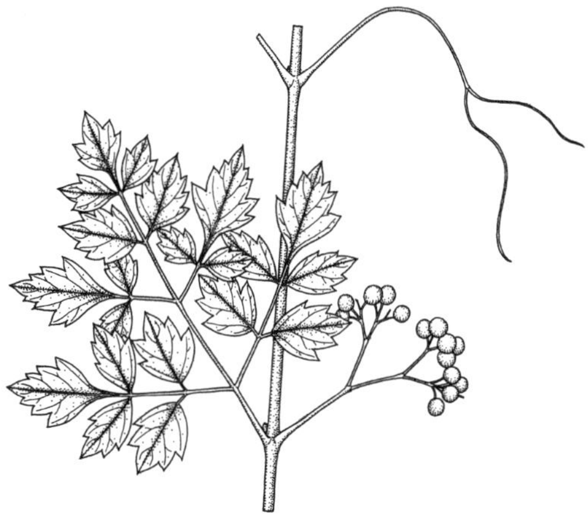

FIGURE 21.61 *Ampelopsis arborea* twig with leaves, tendril, and attached inflorescence with attached berries $\times 1/3$.

dentate to serrate, apex acute or acuminate, base truncate or cordate; petiole length 2 to 4 cm, all surfaces glabrous; inflorescence a panicle, open, length 1.5 to 4.0 cm; flowers 5 to 20 in an inflorescence; fruit a berry, bluish, diameter 3 to 7 mm, seeds one to two, not edible; vine nodes swollen, tendrils common, pith white; Regions 3 and 4, South Carolina and Virginia to southern Illinois, south to Texas and Florida, west to Iowa and Nebraska; mesic and mesic-hydric woods, forming a dense blanket-type thicket along forest margins and open areas. Cover type 28.50.

Ampelopsis arborea (Linnaeus) Koehne, peppervine (Figure 21.61); trailing or climbing vine; deciduous; leaf arrangement alternate or opposite the inflorescence; leaf bipinnately compound, some leaves partially tripinnately compound, total length 10 to 20 cm; leaflet blade length 3 to 7 cm, width 1.0 to 2.5 cm, margin with two to three sharp-pointed lobes, sinuses cleft, deep clefts creating a trifoliate condition, apices acuminate, veins usually bearing hair; petiole length 2 to 4 cm, sparingly pubescent; inflorescence a panicle, open, length 1.5 to 4.0 cm; flowers 5 to 20 in an inflorescence; fruit a berry, bluish, diameter 3 to 7 mm, seeds one to two, not edible; vine nodes swollen, tendrils infrequent; Regions 3 and 4, Maryland to southern Illinois and Missouri, south to Florida, west to Texas and Oklahoma; mesic-hydric sites, swamps. Cover type 28.10, 28.12, 28.15, 28.21, 28.31, 28.49, 28.50.

Genus: *Parthenocissis*

Parthenocissis is a small genus of only three to four species. Virginia creeper and woodbine are common names applied to both *Parthenocissus quinquefolia* and *P. inserta*. A close relative, *Parthenocissus tricuspidata*, is introduced from Asia. The flowers are small, regular, and have a minute calyx and five greenish petals. *Parthenocissus* may have perfect flowers or may be polygamo-monoecious. The fruit of *Parthenocissus* is not edible.

Parthenocissus quinquefolia and *P. inserta* have palmately compound leaves and a long (15 to 20 cm) petiole. The stem usually grows along the ground, or occasionally up the bark of a tree, and remains slender, in contrast to *Vitis* vines. Both *P. quinquefolia* and *P. tricuspidata* are noted for their climbing ability on buildings through the use of tendrils that terminate in enlarged adhesive disks. *Parthenocissus tricuspidata* can be distinguished from *Hedera helix* Linnaeus (English ivy), which has a somewhat similar leaf but climbs by use of small but numerous aerial roots that appear as soon as there is contact with a wall. Clearing these vines after they have largely covered a wall can leave considerable unsightly residue on brick, and damaged paint and siding on wood walls.

Parthenocissus quinquefolia (Linnaeus) Planch; Virginia creeper, woodbine (Figure 21.62); vine, prostrate or climbing on trees or buildings by means of a slender, many-branched (four to eight) tendril, an adhesive disk terminates the end of each branch of the tendril; deciduous; leaf arrangement alternate; leaf palmately compound, leaflets five, essentially elliptical, total leaf length 5 to 15 cm, blade coarsely toothed, elliptical; inflorescence a panicle, fruit a berry, dark blue to black, diameter 4 to 6 mm; Regions 2, 3, and 4, New England, Great Lakes region south into Mexico, Cuba, and the Caribbean; upland forest, open sites, fencerows. Cover type 28.4, 28.7, 28.10, 28.49.

Parthenocissus inserta (Kerner) K. Fritscher [*P. vitacea* (Knerr) A. S. Hitchcock], woodbine (Figure 21.63); differs from *P. quinquefolia* by having tendrils with reduced forking (three to five) and more slender tips without disks, leaflet margin more nearly lobed, and with acuminate teeth; Regions 2, 3, 4, and 5, New England to Utah and Arizona.

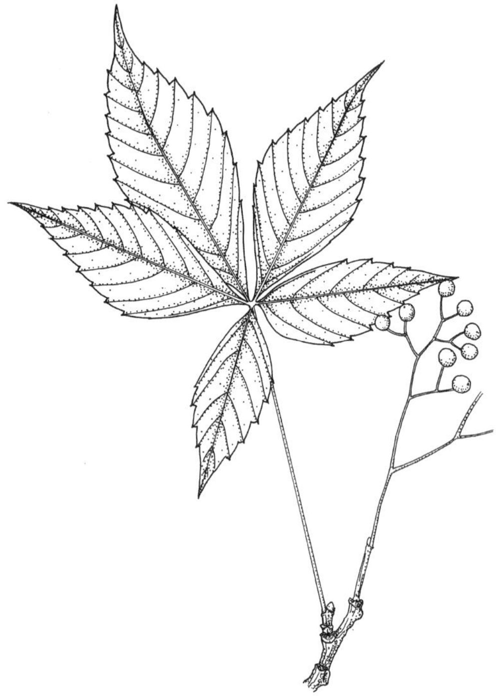

FIGURE 21.62 *Parthenocissus quinquefolia* twig with palmately compound leaf and inflorescence with attached berries ×1/3.

Parthenocissus tricuspidata (Siebold & Zuccarini) Planchon, Boston ivy (Figure 21.63); climbing vine, simple leaf with three lobes, but basal sprouts may trifoliate, 5 to 9 cm long, 5 to 8 cm wide, apices acute, base slightly cordate; introduced from China, escaped cultivation.

ORDER: APIALES

Araliaceae (Ginseng Family)

The Araliaceae includes about 65 genera and 750 species of trees, shrubs, and herbs worldwide, mostly in the tropics. This family is closely aligned with the Apiaceae (formerly Umbelliferae), the parsnip or carrot family, and has frequently been included with the Apiaceae (Benson 1957; Zomlefer 1994; Judd et al.

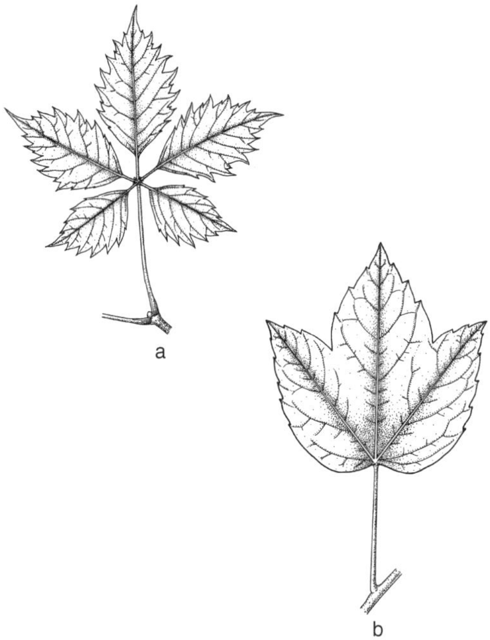

FIGURE 21.63 (a) *Parthenocissus inserta* leaf ×1/5; (b) *P. tricuspidata* leaf ×1/2.

FIGURE 21.64 A portion (about 1/2) of an *Aralia spinosa* leaf showing simple and pinnately compound leaflets at the nodes of the rachis ×1/4.

1999). The Apiaceae, including the Araliaceae, contains about 460 genera and 4250 species.

Both woody and herbaceous species have alternately arranged pinnately or palmately compound leaves, a large pith, and hollow internodes. The inflorescence may be an umbel, raceme, or panicle. *Daucus carrota* Linnaeus (Queen Anne's lace or wild carrot), the most common roadside plant in summer, is easily recognized by the flat-topped umbel of white flowers. Although it has a small unrecognizable root (which smells like a carrot), this species is the wild parent of the large carrots found in grocery stores. Other commercially important plants include *Petroselinum* (parsley), *Apium* (celery), *Anethum* (dill), *Pastinaca* (parsnip), and *Carum* (caraway), to name a few. Some species, such as water hemlock (*Cicuta*) and poison hemlock (*Conium*), contain resins or alkaloids and are poisonous. *Panax quinquefolia* (American ginseng) is a forest herb that is dug for its roots, which are purported to have a variety of beneficial effects in humans. The roots bring a high price, and there is concern in the midwest for the conservation of this species (Anderson et al. 1993). Laboratory research has shown ginseng to be effective in killing cancer cells (Southern Illinois University Medical School, Carbondale).

Aralia spinosa Linnaeus; devil's walking stick, Hercules'-club (Figure 21.64; 21.65); small tree, height to 8 m; deciduous; leaf arrangement alternate, bipinnately compound, total leaf length 40 to 90 cm; pinnule blade length 3 to 9 cm, width 1.5 to 7.0 cm, margin sharply serrated, petiole, rachis, and petiolules usually armed with spines, spine length 1 to 3 mm, apex narrowly long acuminate; inflorescence a raceme, length 10 to 20 cm; fruit a drupe, red, diameter 3 to 4 mm; main stem densely armed with stiff, brittle spines, spine length 1.0 to 1.5 cm; Regions 1, and 2, New England, New York south to Florida, Texas, Missouri, and Illinois; xeric-mesic woods, cool slopes, a frequent component of cut, burned, or otherwise disturbed forest

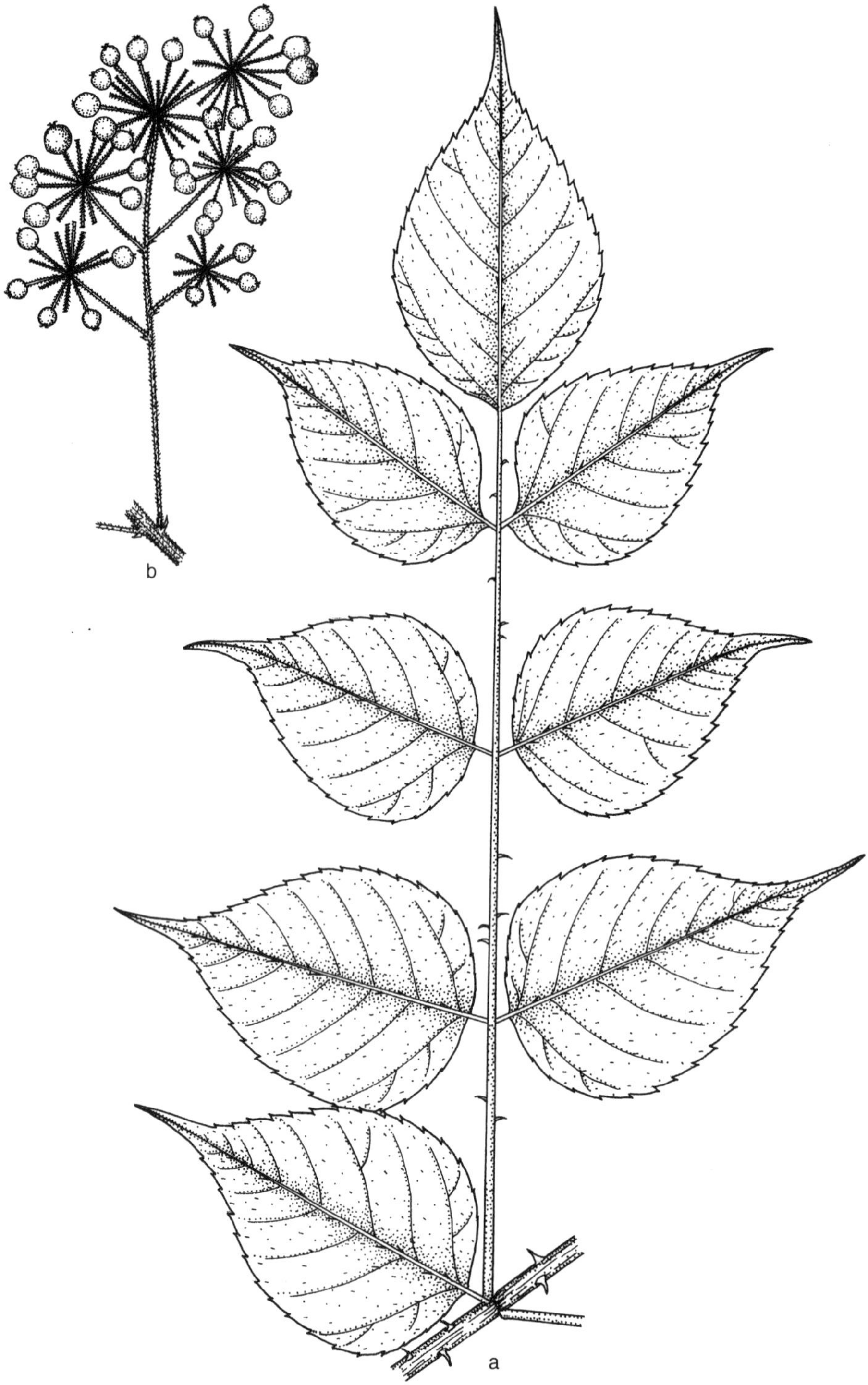

FIGURE 21.65 *Aralia spinosa* (a) leaflet and compound leaflet (pinnules) ×1, (b) inflorescence with attached drupes ×1/2, (c) main stem with thorns ×1.3.

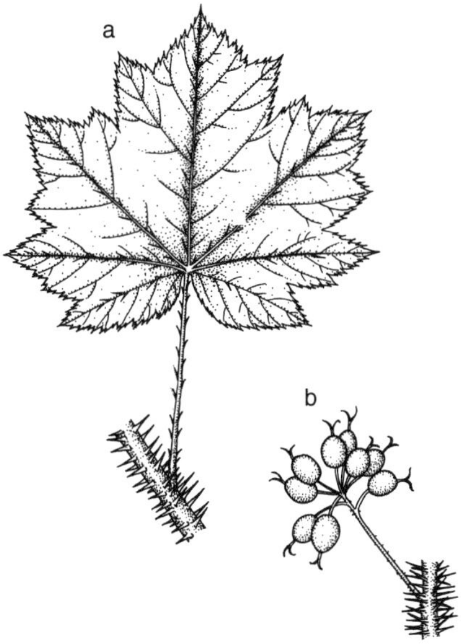

FIGURE 21.66 *Oplopanax horridus* (a) twig with thorns and spiny leaf ×4/5, (b) inflorescence with attached drupes ×1.

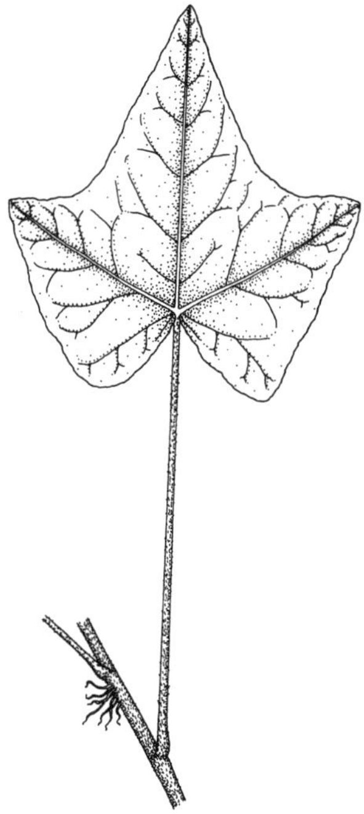

FIGURE 21.67 *Hedera helix* twig with leaf and aerial roots ×3/4.

communities. The common name, devil's walking stick, refers to a situation in which a biologist in the woods needs help walking up a hill or to be stopped from falling after tripping on a downed branch. The response is to reach out and clutch a sapling-size stem, which so often turns out to be *A. spinosa* or *Oplopanax horridus.* The result is a palm with embedded thorns or holes left by the thorns typically found on such stems. Cover type 27.1, 27.4, 28.18.

Oplopanax horridus (J. E. Smith) Miquel, devil's club (Figure 21.66); shrub or small tree, height to 4 m; deciduous; leaf arrangement alternate; leaf simple, blade length 9 to 30 cm, width 10 to 25 cm; blade palmately lobed, lobes and primary veins five to seven, sinuses shallow, broadly V-shaped, margin sharply serrated, adaxial and abaxial surface veins usually armed with spines, spine length 3 to 7 mm; petiole length 6 to 17 cm; inflorescence a raceme, length 10 to 20 cm; fruit a drupe, red, diameter 4 mm; main stem upright, densely armed with stiff, brittle spines, spine length 1.0 to 1.5; Regions 1, 2, 6, and 7; southern Ontario, Michigan, Wisconsin, and Minnesota, west to Montana, Alaska, and California; most rocky sites, coves, drainages, and woods. Cover type 30.1, 30.4, 30.10, 30.19, 30.22.

Hedera helix Linnaeus, English ivy (Figure 21.67); climbing vine, to 30 m; evergreen; leaf arrangement alternate; leaf simple, dimorphic, three to five lobes, outer lobes combining to become squarish or star-shaped; blade length 2 to 5 cm, margin uneven, somewhat serrulate; inflorescence an umbel. Introduced from Europe, escaped cultivation.

CHAPTER 22

Class: Magnoliopsida; Subclass: Rosidae II

ORDER: SAPINDALES

Aceraceae (Maple Family)
Alt: Sapindaceae (Soapberry Family)

Genus: *Acer*

THE natural grouping of plants continues to be examined as new taxonomic information becomes available. One reported change is the transfer of the Aceraceae (maple) into the Sapindaceae (Judd et al. 1999). Originally, the Aceraceae was a family of two genera, *Acer* (maple) and *Dipteronia*, the latter genus composed of two species native to central China. The number of *Acer* species has been estimated at about from 102 (Woodland 1991) and 112 (Harlow et al. 1996) to approximately 150 species (Lawrence 1951; Woodland 1991). A recent definitive work (DeJong et al. 1994) places the number at 124 but adds 95 subspecies and eight varieties, many of which are cultivars.

Acer flowers are borne in racemes, corymbs, or panicles (Figure 22.1). The flowers are unisexual, or, if bisexual, they function as unisexual; in either case the plants are polygamodioecious. The flowers have four to five sepals, four to five petals or none, and four to ten stamens. There are two styles and two carpels per ovary. The fruit is a double samara or a samaroid schizocarp, implying that it splits into two samaras or one-winged mericarps at maturity.

Leaves and twigs of all species have an opposite arrangement. The leaf is deciduous, simple, and palmately lobed, except for *A. negundo*, which has a trifoliate leaf or a pinnately compound leaf of five leaflets. The simple leaves of most species have three to five lobes; exceptions include *A. circinatum*, *A. japonicum*, and *A. palmatum*, which may have as many as nine lobes.

In the eastern United States, *Acer* tree species are informally split into "hard" and "soft" maple. These groupings derive in part from the early pioneer days, when one- or two-person crosscut saws were used to fell trees in the mixed northern hardwood-conifer forest of new England, the Lake States (Minnesota, Wisconsin, and Michigan), and several surrounding states and provinces (Pennsylvania, Ohio, Ontario, Quebec). Loggers found the "hard wood" of *A. saccharum* and *A. nigrum* trees difficult to saw, but the "soft" wood of *A. rubrum* and *A. saccharinum* trees made them relatively easy to cut down. There are additional characteristics associated with these two groups. Soft maple grows fast, but the weak branches and main stem are subject to considerable breakage; thus, as lawn trees, they rapidly provide shade, but their life span is often limited because of damage from wind, ice, or wet snow. Hard maple produces a dense shade, can withstand climatic elements, and lives for several hundred years, but it grows more slowly. The hard maple leaf has large teeth, a smooth margin, and U-shaped sinuses, whereas the soft maple leaf has small teeth, a finely serrated margin, and V-shaped (cleft) sinuses.

Acer macrophyllum and *A. platanoides* leaves have petioles with milky sap; the other species have clear sap. In New England and the Lake States, trees of

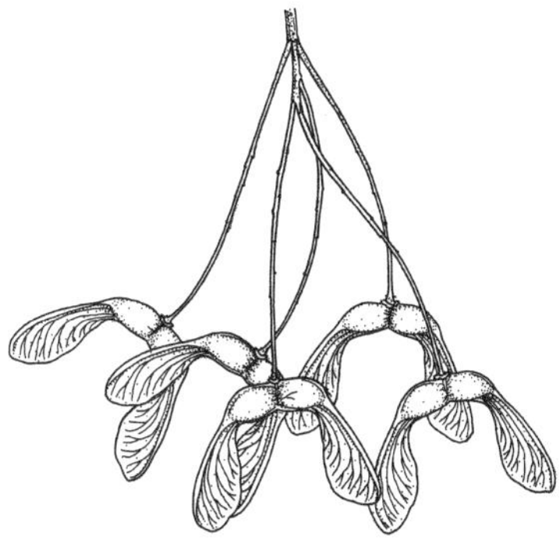

FIGURE 22.1 Panicle of *Acer* with attached mericarps (double samara) ×1/2.

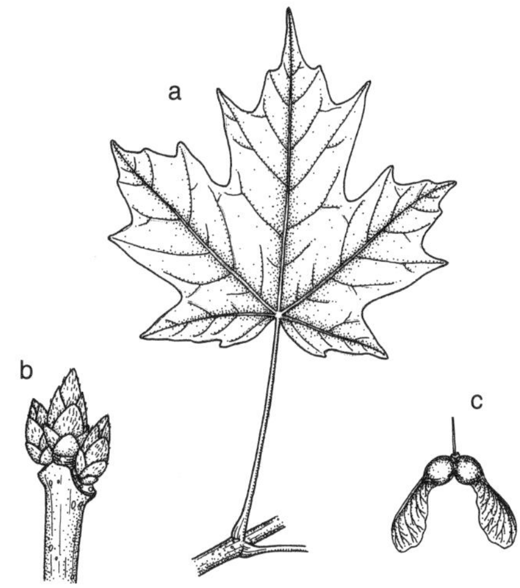

FIGURE 22.2 *Acer saccharum* var. *saccharum* (a) leaf × 1/2, (b) terminal and associated lateral buds ×2, (c) double samara ×1/2.

A. saccharum and *A. nigrum* are tapped for sap during the spring months for maple syrup. Some "sugar bushes" are fitted with an elaborate system of plastic tubing to drain the sap to a central collecting point so that it can be easily collected and boiled down to maple syrup or sugar. Researchers at Southern Illinois University are testing the use of *A. saccharinum* for maple syrup production.

In the United States, *Acer saccharum*, *A. rubrum*, *A. spicatum*, and *A. pensylvanicum* are major components of upland forest in the Lake States and New England, a region of moderate to high rainfall (summer precipitation about 30 to 60 cm). In southern Canada (Ontario, Quebec), where precipitation may be somewhat lower (20 to 30 cm), cooler temperatures reduce evapotranspiration and increase the effectiveness of the soil moisture reserve. These *Acer* species are an important component of the forest as well in the moist Canadian Maritime Provinces of New Brunswick and Nova Scotia. Both *A. saccharinum* and *A. rubrum* are found throughout the bottomland forests of the eastern United States. Several species of *Acer* are components of the montane forest regions of the western United States and southern British Columbia and Alberta. Representatives of *Acer* are also found in Mexico, Japan, southeastern China, Europe, and northern Africa.

Eastern Species

Acer saccharum Marshall, sugar maple (Figure 22.2); large tree, height to 35 m; leaf arrangement opposite; leaf simple, total length 10 to 20 cm, petiole length 7 to 12 cm, blade palmately lobed, lobes five, lobes with large paired teeth, primary veins five, apices acute, sides of middle lobe parallel or divergent sinuses U-shaped, margin smooth, petiole and abaxial blade surface glabrous or with a few scattered hairs; twigs gray- to reddish-brown, lenticels prominent, light gray, terminal buds sharp-pointed, gray brown, visible scale pairs five to seven, scales decussate, leaf scar V-shaped with three distinct bundle scars; fruit a double samara, wings angled at 70°, samaras green, length 2.5 to 4.0 cm; Regions 2 and 3, also Gaspé Peninsula but not on the Piedmont; moist sites; *A. saccharum* Marsh. is one of the most important timber species of the northern hardwood-conifer forest (Region 2) and is quickly becoming more important in the central hardwood forest (Region 3a) as it replaces stands of pioneer *Quercus* (oak) and *Carya* (hickory) species. Several forms of *A. saccharum* have also been reported (Steyermark 1963), and the species frequently crosses with *A. nigrum*; see *A. nigrum* for more detail on this hybridization problem. In the central and southeastern forests (Regions 3 and 4) there are either two varieties or subspecies of *A. saccharum* that had previously been considered separate species (Nelson 1994). Dirr (1998) retains them as species. Cover type **26.17, 26.23, 26.24, 27.1, 27.4, 27.5, 27.8,**

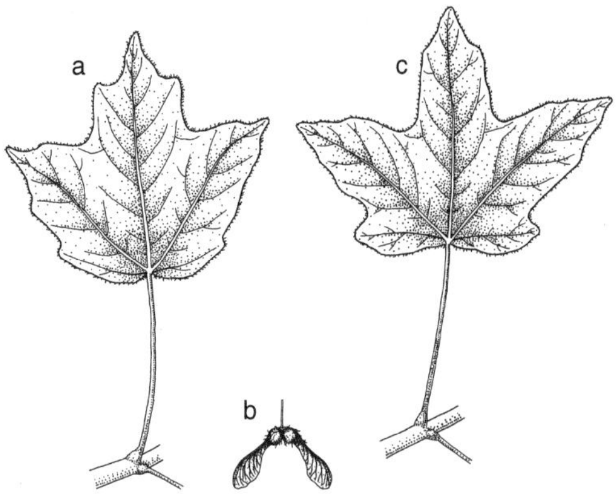

FIGURE 22.3 Acer *saccharum* var. *floridanum* (a) leaf × 1/2, (b) double samara × 1/2; (c) *Acer saccharum* var. *leucoderme* leaf × 1/2.

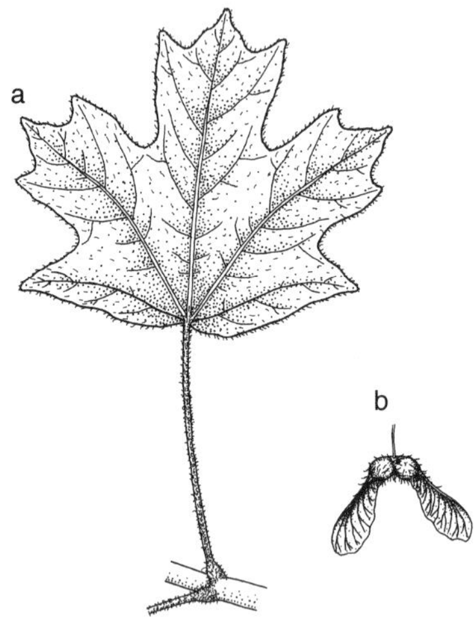

FIGURE 22.4 *Acer nigrum* (a) leaf × 1/2, (b) double samara × 1/2.

27.14, 27.23, 27.30, 26.21, 26.22, 26.27, 26.28, 26.29, 27.12, 27.16, 27.29, 27.31, 27.37, 28.29, 28.22.

Acer saccharum var. *floridanum* (Chapman) Small and Heller (*Acer barbatum* Michaux), southern sugar maple, Florida maple (Figure 22.3); medium-sized tree, height to 20 m; leaf arrangement opposite; leaf simple, total length 8 to 12 cm, petiole length 3 to 5 cm, blade palmately lobed, primary veins five, lobes 3 to 5, middle lobe narrow at the base with parallel sides, lobes lacking teeth, margin smooth to undulate, ciliated, apices obtuse to rounded, abaxial surface whitish, abaxial blade surface glabrous or with a few scattered hairs; twigs light gray brown, terminal buds sharp-pointed, light gray brown scale, pairs five to seven, scales decussate, leaf scar V-shaped with three distinct bundle scars; inflorescence a terminal raceme; fruit a double samara, wings angled at 60 to 70°, samaras green, length 2.5 to 4.0 cm; bark light gray to whitish; Regions 3 and 4, Coastal Plain and Piedmont from Virginia to Florida and east to Texas, northward into Missouri; bottomland and mesic slopes.

Acer saccharum var. *leucoderme* (*Acer leucoderme* Small), chalk maple (Figure 22.3); small tree, height to 13 m; leaf and twig characteristics are similar to those of *A. saccharum* var. *floridanum*, except abaxial surface green and sides of middle lobe convergent; Region 4, Upper Coastal Plain of Alabama and Georgia and the Piedmont, generally not on Lower Coastal Plain, disjunct populations in Louisiana and Arkansas; hydric-mesic bottomland. Less common than *A. saccharum* var. *floridanum.*

Acer nigrum Michaux f., black maple (Figure 22.4); large tree, height to 35 m; leaf arrangement opposite; leaf simple, total length 8 to 16 cm, petiole length 4 to 7 cm, blade palmately lobed, lobes three to five, primary veins five, apices and teeth of proximal pair of lobes rounded, distal lobes with paired rounded teeth, apices acute to rounded, sinuses U-shaped, margin smooth, petiole and abaxial blade surface with scattered hair or densely pubescent; blade margins and lobes often drooping or slightly revolute; twigs gray brown, lenticels prominent, light gray, terminal buds sharp-pointed, visible scale pairs five to seven, scales decussate, leaf scar V-shaped with three distinct bundle scars; bark often charcoal black; inflorescence a terminal raceme; fruit a double samara, wings angled at 70°, samaras green, length 2.5 to 4.0 cm; Region 3a, extending into western Ontario and Quebec, New York, and Pennsylvania; moist sites. This species is less common and less important than *A. saccharum.* Its range generally coincides with the Central Hardwood Forest (Region 3a). Steyermark (1963) lists several forms of *A. nigrum*, as well as varieties and forms of *A. saccharum.* At least part of this subspecies problem may come from introgression between the two species; that is, parents of the two pure species may cross with hybrids backcrossed several times with either parent. Desmarais (1952) found that pure *A. nigrum* was located in the Ozark region of southern Missouri (western portion of Region 3) and that pure *A. saccharum* was found in the northern and eastern Great Lakes region (Region 2). In the area between Missouri and Wisconsin, the two species were found to hybridize and backcross so that species characteristics were mixed (cited from Curtis 1959). In southern Illinois, the full mixture of characteristics can be found, with some trees with short pubescent petioles and a glabrous abaxial blade surfaces, and some with long glabrous or pubescent petioles and pubescent abaxial surfaces. Some trees have drooping leaves and the leaves of others are flat. Desmarais (1952) also reported that the frequency of *A. nigrum* characteristics decreased from the Ozarks into central Wisconsin (an introgressive hybridization gradient). Cover type **27.1**.

Acer rubrum Linnaeus, red maple (Figure 22.5); medium-sized to large tree, height to 30 m; leaf arrangement opposite; leaf simple, total length 9 to 19 cm, petiole length 4 to 9 cm, blade palmately lobed, lobes three to five, primary veins five, apices obtuse, sinuses V-shaped or cleft, margin serrated, abaxial blade surface glabrous; twigs and buds reddish, terminal bud rounded, visible scale pairs three to four, decussate, scales ciliated, leaf scar V-shaped with three distinct bundle scars; fruit borne individually on a 4 to 8 cm long peduncle; fruit a double samara angled at 75 to 100°, immature samaras pink, turning green, length 2 to 4 cm; immature bark smooth, becoming platy with age; Regions 2, 3, and 4, entire eastern United States, including the Canadian Gaspé Peninsula; xeric and mesic to mesic-hydric sites. Drummond's red maple [*A. rubrum* var. *drummondii* (Hooker and Arnold) sargent; *A. drummondi* Hooker and Arnold] has a white pubescence on the abaxial surface and is the variety frequently found on mesic-hydric and hydric sites near bottomlands and swamp margins in Region 3a. *Acer rubrum* var. *trilobum* Torrey & Gray ex. K. Koch (*A. rubrum* var. *tridens* A. Wood) is a three-lobed variety found in southeastern swamps and pocosins (Figure 22.5) (cover type 28.23, 28.31, 28.48). In the northeastern United States, *A. rubrum* has become an aggressive invader of *Quercus* communities and is threatening to replace them by forming a dense understory, reducing the light level below that required for the survival of oak seedlings. Cover type **26.20, 26.25, 27.2, 27.36, 28.12, 28.33, 28.42,** 26.8, 26.9, 26.10, 26.14, 26.17, 26.18, 26.19, 26.22, 26.23, 26.24, 26.26, 26.28, 26.29, 27.9, 27.10, 27.12, 27.16, 27.21, 27.22, 27.23, 27.28, 27.30, 27.33, 27.34, 28.4, 28.6, 28.7, 28.10, 28.11, 28.14, 28.21, 28.23, 28.27, 28.28, 28.29, 28.30, 28.36, 28.38, 28.40, 28.49, 28.53, 28.55.

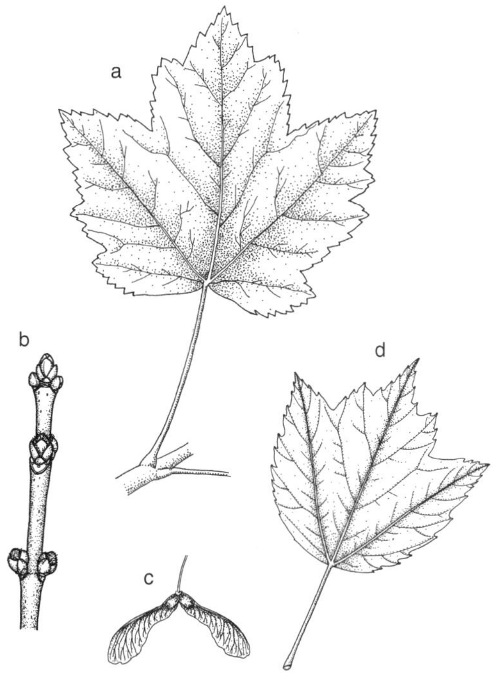

FIGURE 22.5 *Acer rubrum* var. *rubrum* (a) leaf ×2/5, (b) twig with terminal and lateral buds ×1, (c) double samara × 1/2, (d) *Acer rubrum* var. *trilobum* leaf × 1/2.

Acer saccharinum Linnaeus, silver maple (Figure 22.6); medium-sized to large tree, height to 30 m; leaf arrangement opposite; leaf simple, total length 10 to 20 cm, petiole length 4 to 9 cm, blade palmately lobed, lobes five, primary veins five, apices acute to acuminate, sinuses V-shaped or cleft, margin serrated, abaxial blade surface glabrous; twigs and buds reddish to reddish brown, terminal bud rounded, visible scale pairs three to four, decussate, scales ciliated, leaf scar V-shaped with three distinct bundle scars; fruit borne singly on a 4 to 7 cm long peduncle; fruit a double samara, wings angled at 80 to 100°, samaras green, turning brown, wing terminal, length 3.5 to

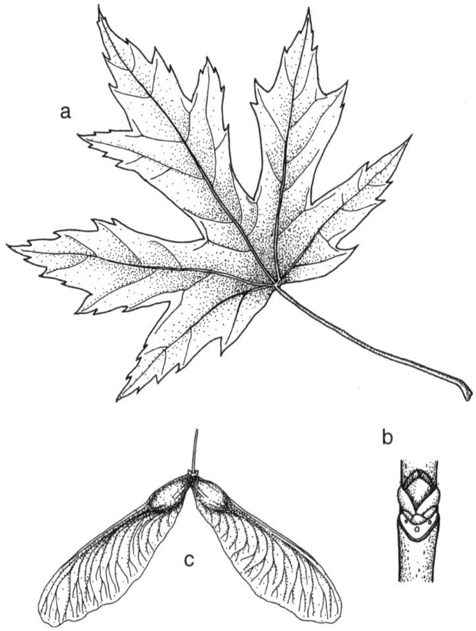

FIGURE 22.6 *Acer saccharinum* (a) leaf ×1/2, (b) lateral bud leaf scar ×2, (c) double samara ×1/2.

6.0 cm; bark smooth, becoming platy with age; Regions 2, 3, and 4, eastern United States, including the Gaspé Peninsula, absent along most of the Atlantic and Gulf Coasts; bottomland and poorly drained sites. Trees of this species are common along the margins of streams and lakes and may dominate communities on stream terraces and bottomland sites. Because of its fast growth, *A. saccharinum* is often planted as a lawn or street shade tree. However, this tree has extremely brittle wood and is subject to severe damage from wind, ice, and snow. Cover type **27.3**, 26.25, 27.6, 27.10, 27.14, 28.7, 28.15, 28.21.

Acer negundo Linnaeus, boxelder, Manitoba maple (Figure 22.7); medium-sized tree, height to 18 m; leaf arrangement opposite; total leaf length 12 to 19 cm, petiole length 4 to 7 cm, blade compound, leaflets three to five, length 5 to 9 cm, apices acute to acuminate, margin toothed to nearly lobed along the distal one half, smooth at the base; twigs medium green to purple, glaucous, bud scales white tomentose, lateral buds short stalked, leaf scar V-shaped with three distinct bundle scars; inflorescence a terminal raceme; fruit a double samara angled at 40 to 60°, samaras

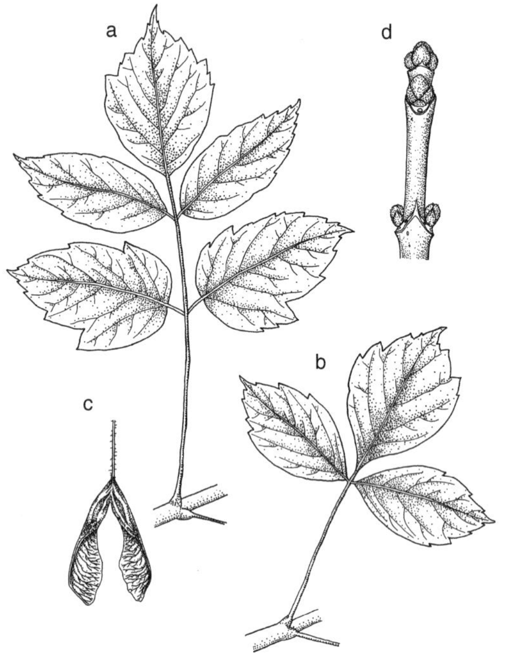

FIGURE 22.7 *Acer negundo* (a) pinnately compound five-leaflet leaf ×4/5, (b) trifoliate leaf ×4/5, (c) double samara ×1/2, (d) twig with terminal and lateral buds ×1.

green, length 2.5 to 4.0 cm, wing terminal; Regions 2, 3, 4, 5, and 7, throughout the United States except Washington, Oregon, and Idaho, extending north into southern Manitoba, Saskatchewan, and Alberta; the leaves and twigs of the California subspecies are heavily pubescent; mesic-hydric sites particularly disturbed sites; a poorly formed tree found along streams and in bottomlands; also grows well on uplands, where it has been used as a lawn tree and in windbreaks. Not recommended as a shade tree because of its short life span and potential for breakage. Cover type **28.7, 29.18,** 27.6, 27.12, 27.14, 27.28, 28.9, 28.15, 28.21, 29.20.

Acer spicatum Lambert, mountain maple (Figure 22.8); shrub or small tree, height to 7 m; leaf arrangement opposite; leaf simple, total length 9 to 20 cm, blade palmately lobed, lobes three to five, primary veins five; lobes tapered to a long acuminate apex, base rounded, margin coarsely serrated, densely ciliated; inflorescence a terminal raceme; fruit a double samara angled at about 90° samaras reddish, length

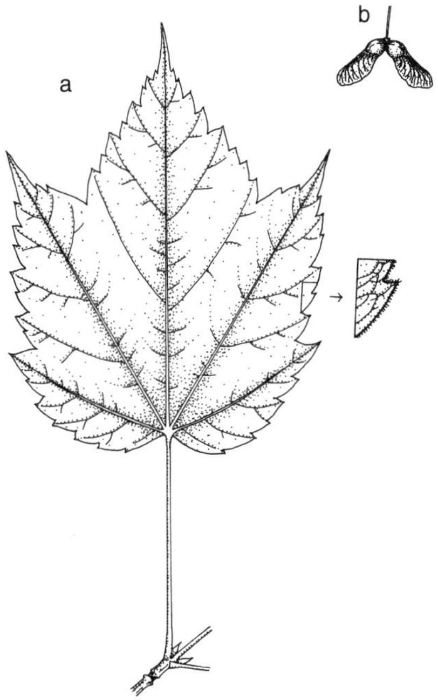

FIGURE 22.8 *Acer spicatum* (a) leaf ×1/2, (b) double samara ×1/2.

1.5 to 2.5 cm; bark gray brown; Regions 1, 2, and 3b, Newfoundland west to Saskatchewan, Minnisota, and Wisconsin, south along the Appalachian Mountains; moist sites; a common understory component of the Northern Hardwood-Conifer forest. Difficult to distinguish from *A. rubrum*, except for the slightly narrower blade and the margin cilia. Cover type 26.9, 26.13, 26.19, 26.21, 26.22, 27.29, 27.31.

Acer pensylvanicum Linnaeus, striped maple; moosewood (Figure 22.9); shrub or small tree, height to 9 m; leaf arrangement opposite; leaf simple, total leaf length 10 to 19 cm, palmately lobed, lobes three, primary veins three, lobe apices tapered to a long acuminate apex, base rounded to truncate, margin finely to doubly serrated; inflorescence a terminal raceme; fruit a double samara angled at 90 to 130°, samaras green, length 2 to 3 cm, wing terminal; bark greenish with pale yellow or cream-colored vertical stripes; Regions 2 and 3b, also Gaspé Peninsula; moist sites; a common understory component of the Northern Hardwood-Conifer Forest. Cover type 26.9, 26.21, 26.22, 26.23, 26.29, 27.29, 27.34, 27.35.

FIGURE 22.9 *Acer pensylvanicum* twig with leaf, raceme with attached double samaras ×1/2.

Western Species

Acer macrophyllum Pursh, bigleaf maple (Figure 22.10); medium-sized to large tree, height to 35 m; leaf arrangement opposite; leaf simple, total length 10 to 30 cm, petiole length 7 to 10 cm, blade palmately lobed, primary veins five, primary lobes five, primary lobes with large paired or nonpaired blunt teeth and secondary lobes, apices acute, sinuses weakly U-shaped, sinus margin smooth, sap milky, petiole and abaxial blade surface glabrous; twig moderately stout to stout, twig and buds reddish to greenish brown, terminal bud blunt, often absent, being replaced by the inflorescence, visible bud scale pairs three to four, scales decussate, leaf scar V-shaped with seven distinct bundle scars; inflorescence a terminal raceme of 15 to 30 flowers; fruit a double samara, length 2.5 to 4.0 cm, green, wings terminal, angled at 30 to 40° and often overlapping; Region 7, west coast from southern California, Coastal Ranges, and the Sierra Nevada through British Columbia to the Alaskan border; moist slopes and terraces. One of a few commercially important hardwood timber trees in the western United States, used for veneer and furniture. Cover type **30.4, 30.16,** 30.6, 20.7, 30.10, 30.18, 30.19, 30.20, 30.34, 30.39, 30.40, 30.41.

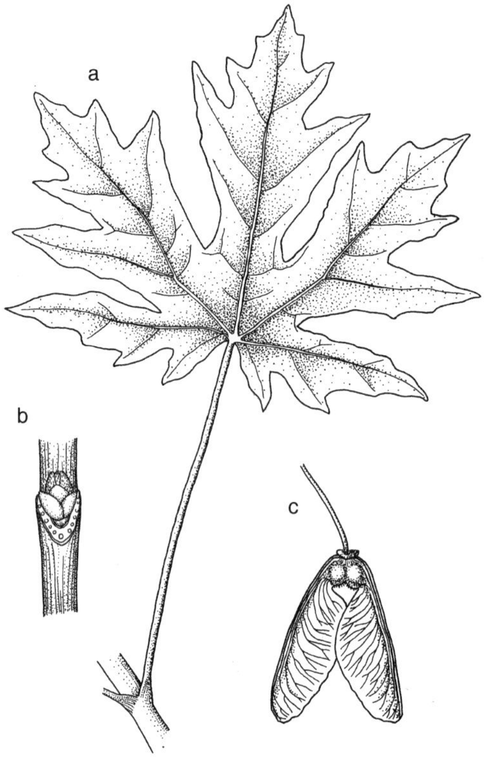

FIGURE 22.10 *Acer macrophyllum* (a) leaf ×1/2, (b) lateral bud and leaf scar ×2, (c) double samara ×3/4.

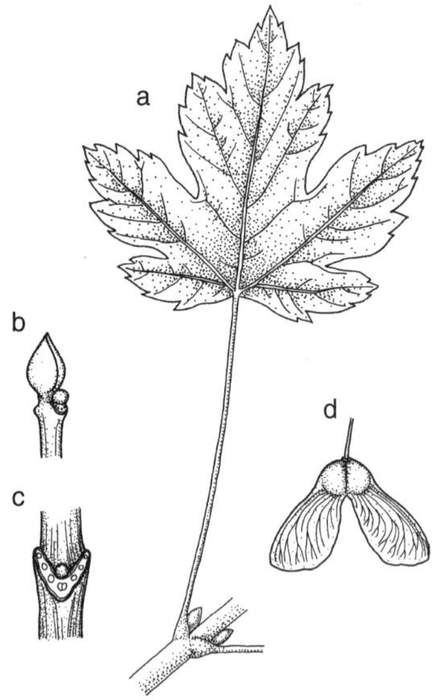

FIGURE 22.11 *Acer glabrum* (a) leaf ×1/2, (b) terminal bud ×1.5, (c) lateral bud and leaf scar ×2, (d) double samara ×1.

Acer glabrum Torrey, Rocky Mountain maple, Sierra maple, Douglas maple (Figure 22.11); shrub or small tree, height to 10 m; leaf arrangement opposite; leaf simple, infrequently divided into leaflets, total length 6 to 17 cm, palmately lobed, primary veins three with two additional veins splitting from the outside primary veins to form the proximal lobes, lobes three to five; lobes widest at the middle, then tapered to an obtuse or acute apex, margin coarsely to finely serrated, base truncated to rounded; twigs reddish brown, glabrous, bud scales two; leaf scar V-shaped with seven distinct bundle scars; inflorescence a raceme of five to ten flowers; fruit a double samara, wings angled at about 70 to 80°, samara reddish, length 1.5 to 2.5 cm; Regions 5, 6, and 7, southern Rocky Mountains of Arizona and New Mexico and the Sierra Nevada of California into southeast Alaska; moist sites. Largest size reached on bottomland and mesic sites. Most widespread of the western *Acer* species. Cover type 29.11, 29.12, 29.23, 29.27, 29.28, 29.36, 30.3.

Acer grandidentatum Nuttall, bigtooth maple, canyon maple, western sugar maple (Figure 22.12); medium-sized tree, height to 13 m; leaf arrangement opposite; leaf simple, total length 5 to 12 cm, petiole length 2 to 6 cm, blade palmately lobed, primary veins five, primary lobes three to five, primary lobes with large paired blunt teeth, apices acute to obtuse, sinuses U-shaped, sinus margin smooth, sap clear, abaxial blade surface glabrous or slightly pubescent; new twigs and buds reddish brown, somewhat stout, older twigs becoming light gray; terminal bud sharp, visible scale pairs two to three, scales decussate, leaf scar V- or U-shaped with three groups of one to two bundle scars; inflorescence a terminal raceme of 5 to 15 flowers; fruit a double samara, wings angled from 90 to 125°, samaras green, length 2 to 4 cm, wings terminal; Region 5, scattered from southern Idaho to Utah, Arizona, New Mexico, and Western Texas, disjunct populations in Mexico; mountains and sides of canyons. A close relative of *A. saccharum* and occasionally has been listed as *A. saccharum* var. *grandidentatum* (Nuttall) Sudworth. Cover type **29.1, 29.13.**

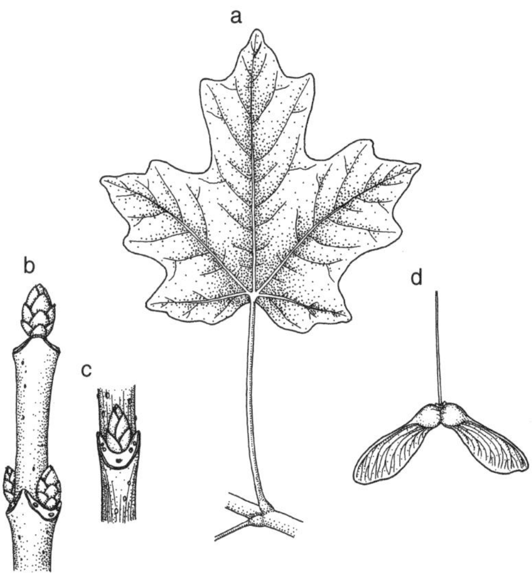

FIGURE 22.12 *Acer grandidentatum* (a) leaf ×3/4, (b) twig with terminal bud and lateral buds ×2, (c) lateral bud and leaf scar ×2, (d) double samara ×1/2.

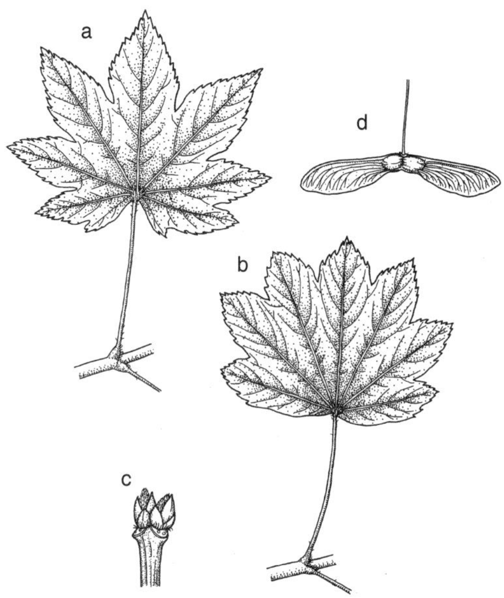

FIGURE 22.13 *Acer circinatum* (a), (b) leaves ×2/5, (c) terminal bud ×1.5, (d) double samara ×1/2.

Acer circinatum Pursh, vine maple (Figure 22.13); small tree, height to 13 m but often reclining and vine-like; leaf arrangement opposite; leaf simple, total length 8 to 17 cm, petiole length 2 to 7 cm, blade palmately lobed, primary veins seven, primary lobes seven to nine, margin serrated, apices acute, sinuses cleft, sap clear, petiole and blade surfaces glabrous except for tufts of white hair at the bases; twigs and buds greenish to purplish, surface with scattered patches of long white hair; buds accompanied by long white hair; terminal buds blunt, often two at the twig tip, visible scale pairs two to three, scales decussate, leaf scar V-shaped with three distinct bundle scars; fruit a double samara, angled at 170 to 180°, samaras green, length 1.0 to 3.5 cm, wings terminal; Region 7, west coast mountains from southern British Columbia to northern California elevations up to 1200 m. This species is abundant where the forest has been disturbed by cutting or wind. The name "vine maple" comes from its liana-like stems that twist around downed trees and logs. Cover type 30.1, 30.2, 30.3, 30.6, 30.16, 30.18, 30.19, 30.21, 30.22, 30.28.

Introduced Species

Several *Acer* species have been introduced into North America and are commonly used as street trees and ornamentals. These include species from Japan, Korea, east Asia, and Europe.

Acer platanoides Linnaeus, Norway maple (Figure 22.14); large tree, height to 25 m; leaf arrangement opposite; leaf simple, total length 15 to 30 cm, petiole length 12 to 18 cm, blade palmately lobed, green to dark red, primary veins five, primary lobes five, primary lobes with large paired or non-paired blunt teeth, apices acute to acuminate, sinuses weakly U-shaped, sinus margin smooth, sap milky, petiole and abaxial blade surface glabrous; twigs and buds dark reddish maroon, terminal buds sharp blunt, visible scale pairs two to three, scales decussate, leaf scar V-shaped with three distinct bundle scars; fruit a double samara, angled at 160 to 180°, samaras green, length 3 to 5 cm, wings terminal. Introduced from Europe and widely planted as a street tree because it is relatively resistant to pollution and extreme soil conditions. The dark red variety is *A. platanoides* var. *schwedleri* (K. Koch) Schwedler, Schwedleri Norway maple.

Acer pseudoplatanus Linnaeus, planetree maple, sycamore maple (Figure 22.15); large-tree in North America, height to 25 m; leaf arrangement opposite;

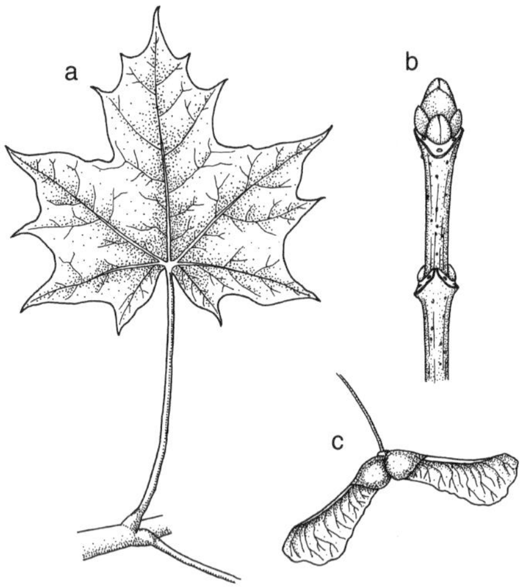

FIGURE 22.14 *Acer platanoides* (a) leaf ×4/5, (b) twig with terminal and lateral buds ×2, and (c) double samara ×4/5.

leaf simple, total length 7 to 16 cm, blade width great than length, petiole length 4 to 8 cm, blade palmately lobed, primary veins five, lobes 3 to 5, apices obtuse to acute, sinuses V-shaped, margin irregularly serrated to crenate, abaxial blade surface glabrous or slightly hairy;

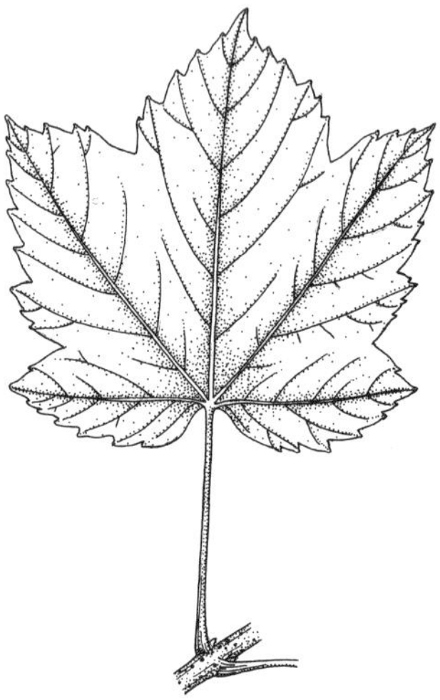

FIGURE 22.15 *Acer pseudoplatanus* leaf ×1/2.

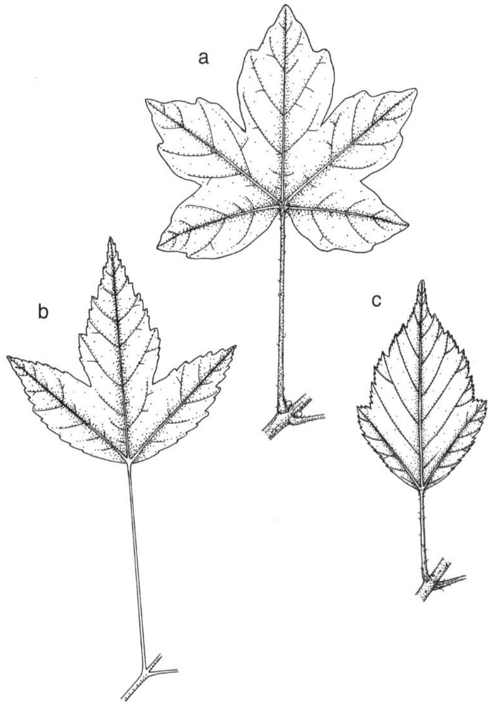

FIGURE 22.16 Leaf of (a) *Acer compestre* ×3/4; (b) *A. ginnala* ×4/5; (c) *A. ginnala* var *maximum* ×4/5.

twigs and buds greenish, terminal bud large, length 1.0 to 1.5 cm, visible scale pairs three to four, decussate, leaf scar V-shaped with three distinct bundle scars; fruit a double samara, angled at 100°, samaras green, length 3 to 5 cm. Introduced from Europe as a lawn and street tree.

Acer compestre Linnaeus, hedge maple (Figure 22.16); small to medium-sized tree, height to 20 m; leaf arrangement opposite; leaf simple, total length 7 to 12 cm, petiole length 3 to 6 cm, blade palmately lobed, lobes five, primary veins five, lobes with paired rounded teeth, apices obtuse, sinuses essentially V-shaped, margin smooth, petiole and abaxial blade surface with scattered hair or pubescent; twig gray brown, often with corky fissures, terminal bud short, length 3 to 6 mm, gray brown, visible scale pairs three to four, scales decussate, leaf scar V-shaped with three distinct bundle scars; fruit a double samara, length 3 to 4 cm, wings terminal, angled at essentially 180°; introduced from Europe and the Middle East.

Acer ginnala Maximowicz, Amur maple (Figure 22.16); small tree, height to 8 m; leaf arrangement opposite; leaf simple, total length 3 to 8 cm, petiole length 2 to 4 cm, blade palmately lobed, lobes three or blade entire, primary veins 1 to 3, apices acuminate, sinuses V-shaped, margin finely serrated, abaxial blade surface glabrous; twigs and buds reddish to gray brown, terminal bud rounded, visible scale pairs one to two, decussate, leaf scar V-shaped with three distinct bundle scars; inflorescence a terminal raceme; fruit a double samara with wings nearly parallel to a 30° angle, samaras green, length 2 cm. Introduced in the mid 1800s from China as a lawn and street tree. Although the leaf is generally three-lobed, the side lobes may be reduced to small teeth on some leaves (var. *maximum*; Figure 22.16), and thus this species may be confused with *A. rataricum* Linnaeus, tartarian maple, a closely related species of southwestern Asia and southeastern Europe (Hillier Nurseries 1998).

Acer palmatum Thunberg, Japanese maple (Figure 22.17); small tree, height to 8 m, multiple branching above 1 m; leaf arrangement opposite; leaf simple, total length 5 to 13 cm, petiole length 3 to 6 cm, blade palmately lobed, primary lobes seven to nine, lobes widest in the middle, apices acuminate, margin finely serrated, sinuses broadly cleft, extending more than one half the distance from the apex to the midrib, sap clear, petiole and abaxial blade surface glabrous; twigs and buds greenish to reddish or purplish, terminal buds blunt, visible scale pairs one (nearly valvate) or two (decussate), leaf scar V-shaped with three distinct bundle scars; fruit a double samara, angled at 75 to 100°, samaras green, length 1 to 2 cm, wings terminal. Introduced as a lawn tree; the var. *atropurpureum* turns a bright crimson or purple in the fall.

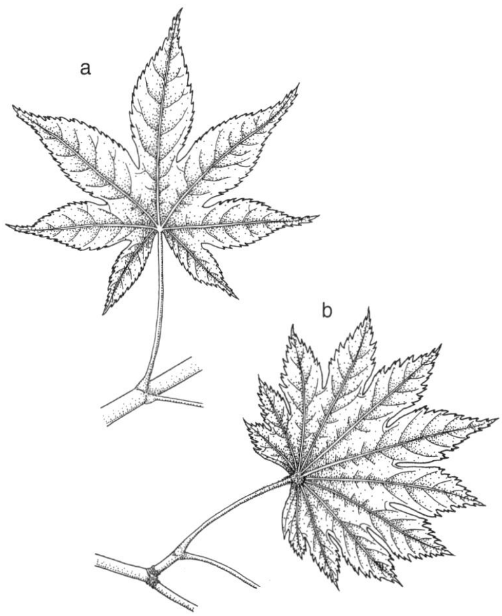

FIGURE 22.17 Leaf of (a) *Acer palmatum* ×1/2; (b) *Acer japonicum* ×1/3.

Acer japonicum Thunberg, fullmoon maple (Figure 22.17); medium-sized tree, height to 15 m, multiple branching above 1 m; leaf arrangement opposite; leaf simple, total length 8 to 17 cm, petiole length 4 to 7 cm, blade palmately lobed, primary lobes 7 to 11, lobes widest in the middle, apices acuminate, margin finely serrated, sinuses essentially cleft, extending about one third the distance from the apex to the midrib, sap clear, petiole and abaxial blade surface glabrous; twig and buds greenish to reddish or purplish, terminal buds blunt, visible scale pairs one (nearly valvate) or two (decussate), leaf scar V-shaped with three distinct bundle scars; fruit a double samara, angled at 75 to 100°, samaras green, length 1 to 2 cm, wings terminal. Introduced from Asia as a lawn tree; some varieties turn a deep crimson or purple in the fall.

Sapindaceae (Soapberry Family)

The Sapindaceae is a family of about 140 genera and 1500 species concentrated in the tropical and subtropical regions. The common name, soapberry, derives from a species of tropical America, *Sapindus saponaria* Linnaeus, whose berries form a lather and are used as soap. Members of the family are trees, shrubs, and, occasionally, non-woody vines.

Genus: *Koelreuteria*

Koelreuteria paniculata Laxmann, Golden raintree, Pride of India (Figure 22.18); small to medium-sized tree, height to 13 m, deciduous; leaf arrangement alternate; leaf total length 15 to 40 cm, pinnately compound, leaflets 5 to 15, leaflet outline generally ovate to oblong, often strongly pinnatified, sinuses often deep,

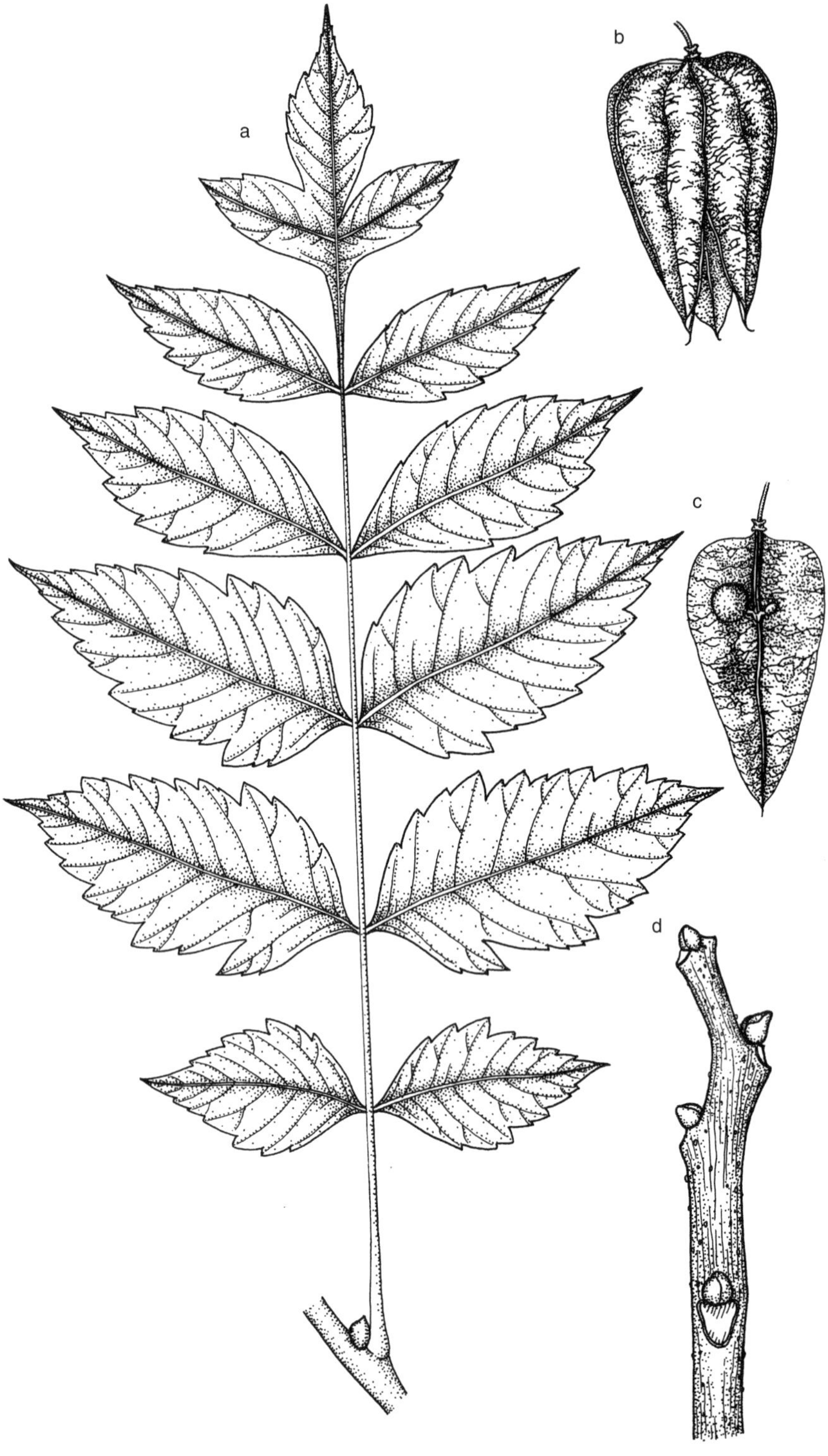

FIGURE 22.18 *Koelreuteria paniculata* (a) leaf ×3/4, (b) capsule showing partial dehiscence ×1, (c) capsule cross section showing seed position ×1, (d) twig with terminal and lateral buds ×3/4.

cleft; twigs greenish to light brown, twig scar present, lateral buds short, length 6 to 8 mm, visible scales two; fruit a large capsule, length 3.5 to 5.0 cm, walls paper-thin cellulose, three locular; seeds hard, black, diameter 3 to 4 mm. Introduced from Asia, planted as an ornamental.

Anacardiaceae (Cashew or Sumac Family)

The Anacardiaceae family contains 60 to 80 genera and about 600 tree species, most of which are found in the tropics. Three genera are found in the temperate zone, and two as far north as southern Ontario and Quebec. The most well known genera include *Rhus*, with 100 to 125 pantropical species but only about 15 in North America, and *Toxicodendron*, with about 17 species in Asia and 4 or 5 in North America. The third genus is represented only by *Cotinus obovatus* (smoke-tree), a small tree of the central hardwood forest (Region 3). *Cotinus coggygria* Scopoli, the Old World smoke-tree introduced from Eurasia, is also commonly planted. One other species, *Metopium toxiferum* (Linnaeus) Krug & Urban (Florida poisonwood) is restricted to southern (subtropical) Florida.

The flowers are typically small, perfect or imperfect (plants polygamodioecious), and usually regular. There are three to seven sepals and petals, and five to ten stamens. In North American, nonpoisonous *Rhus* species have numerous (hundreds of) flowers clustered in a single dense panicle located at the end of the branch (terminal) (Figure 22.19); however, the inflorescence of *Rhus aromatica* is an ament (spike). At maturity the fruit is a small (4 to 5 mm diameter), dry red pubescent drupe. In *Toxicodendron*, the 10 to 30 small (5 to 7 mm diameter) gray or cream-colored glabrous drupes are borne on a loose axillary panicle.

Individuals of most *Rhus*, *Cotinus*, and *T. vernix* become small trees (arborescent) that grow to a maximum height of 5 m, although *R. copallina* and *R. lanceolata* may reach 8 to 10 m. However, *R. aromatica* and *R. trilobata* remain as erect low (<1 m) shrubs. Over much of its range, the form of *T. radicans* varies from a woody shrub (height 2 to 3 m) to a vine climbing a tree by means of dense root clusters (hence the name *radicans*, referring to roots). In the more northern sections of its range (Region 2), it generally appears as an herb, with a single upright trifoliate leaf extending

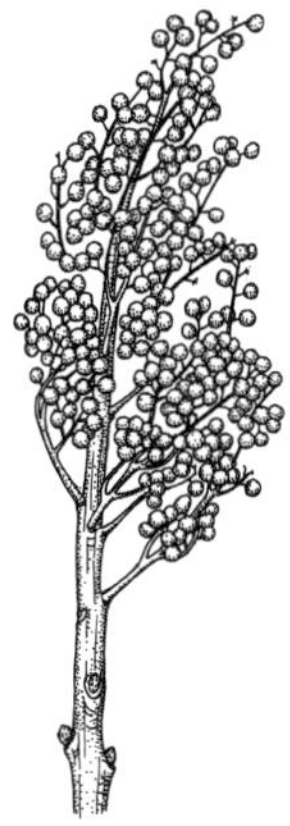

FIGURE 22.19 Terminal inflorescence (panicle) and drupes of *Rhus* ×1/3.

10 to 20 cm from the underground root system, or as a weak trailing vine. *Rhus typhina* and *Rhus glabra* have milky sap.

Although the North American species are shrubs or arborescent and have limited commercial value, several tropical members are important and well known for their nuts, fruits, and chemicals. The most important are *Anacardium occidentale* (cashew nut), *Pistacia vera* (pistachio nut), and *Mangifera indica* (mango), which are widely sold in North America. The gums and resins extracted from members of this family have industrial applications. *Rhus*, *Schinus* (pepper-tree), *Cotinus*, and *Pistacia* are also used as ornamentals. A number of North American species are important for erosion control and deer browse.

The genus name *Toxicodendron* (toxic = poison; dendron = tree) is reserved for species that contain a resinous sap or oil that causes a skin rash or dermatitis. The sap of species in the genus *Rhus* do not cause a skin rash. However, some older book editions (*e.g.*, Fernald 1950) still list all species as *Rhus*. Both *Rhus* and *Toxicodendron* are represented by close relatives in both eastern and western United States.

Genus: *Rhus*

Several groups of *Rhus* can be recognized on the basis of their leaf characteristics. Deciduous species that have a pinnately compound leaf of 5 to 31 leaflets include *R. glabra, Rhus copallina, Rhus typhina,* and *Rhus lanceolata.* Species that have a trifoliate leaf are *R. aromatica* and *R. trilobata. Rhus ovata, R. integrifolia, R. laurina,* and *R. kearneyi* have simple, entire, evergreen leaves.

Rhus copallina Linnaeus, winged or shining sumac (Figure 22.20); shrub or small tree, height to 3 m; deciduous; leaf arrangement alternate; leaf pinnately compound, length 10 to 25 cm, leaflet number 5 to 13, leaflet length 5 to 8 cm, sessile, shape ovate, elliptical, or lanceolate, abaxial surface light green, apex acuminate, base round, margin smooth; rachises winged; twig surface dull, pubescent with prominent lenticels; pith spongy, large; bud short, 1.0 to 1.5 mm, densely pubescent, nearly surrounded by leaf scar; inflorescence a dense panicle, length 10 to 20 cm; fruit drupaceous, slightly hairy, red at maturity; Regions 3 (Central Hardwood Forest) and 4 (Southeastern Pine-Hardwood Forest), but found only as widely scattered disjunct populations in Region 2. United States Department of Agriculture (1982) incorrectly lists the species name as *copallinum.* A close relative with similar leaf structure, *R. lanceolata* (Gray) Britton (prairie sumac), is distributed from Oklahoma and eastern Texas to Arizona and Mexico. Cover type 28.13.

Rhus glabra Linnaeus, smooth sumac (Figure 22.21); shrub or small tree, height to 3 m; deciduous; leaf arrangement alternate; leaf pinnately compound, length 25 to 40 cm, leaflet number 15 to 22,

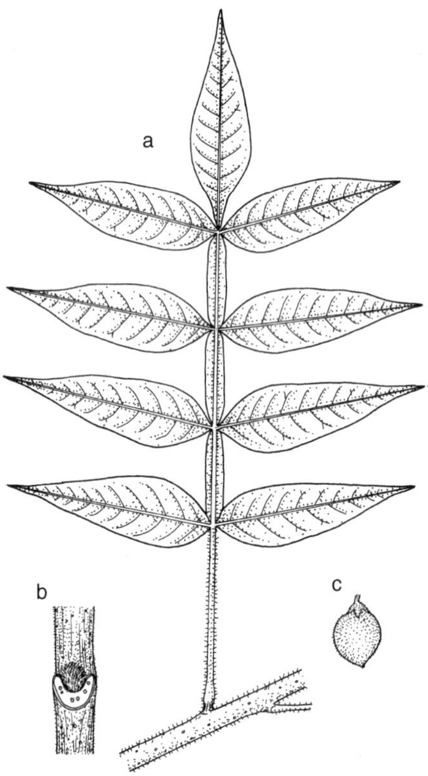

FIGURE 22.20 *Rhus copallina* (a) leaf ×1/2, (b) lateral bud and leaf scar ×1, (c) drupe ×2.

FIGURE 22.21 *Rhus glabra* (a) leaf ×1/3, (b) lateral bud and leaf scar ×1, (c) drupe ×2.

leaflet length 4 to 11 cm, width 1 to 3 cm, essentially sessile, shape narrowly lanceolate, abaxial surface whitish, apex acuminate, base round, margin serrated; petiole and rachis glabrous, glaucous; twig surface glabrous, glaucous, but becoming shiny/waxy when rubbed, lenticels prominent; sap milky; pith large, spongy; terminal bud absent, lateral bud length 2 to 4 mm, densely tomentose, white, nearly surrounded by leaf scar; inflorescence a dense panicle, length 10 to 25 cm; fruit drupaceous, scaly/sticky, deep red at maturity; Regions 2 and 3, the Great Lakes and New England south to, but extending into the Upper Coastal Plain (Region 4a) and Piedmont (Region 4c) (Little 1997). The most widely distributed *Rhus* species. Petrides (1992) also shows widely scattered disjunct populations occurring in the Rocky Mountains from Arizona to British Columbia.

Rhus typhina Linnaeus, staghorn sumac (Figure 22.22); shrub or small tree, height to 3 m; deciduous; leaf arrangement alternate; leaf pinnately compound, length 20 to 48 cm, leaflet number 13 to 29, leaflet length 4 to 12 cm, width 1.5 to 3.0 cm, essentially sessile, shape narrowly lanceolate, often slightly falcate, abaxial surface white, apex acuminate, base round, margin serrated; petiole and rachis densely tomentose; twig surface densely tomentose, lenticels obscure; sap milky; pith large, spongy, less than 1 mm of woody material surrounding the pith, terminal bud absent, lateral bud length 3 to 5 mm, densely tomentose, white, nearly surrounded by leaf scar; inflorescence a dense panicle, length 10 to 21 cm; fruit drupaceous, densely tomentose, deep red at maturity; Region 2 and the Appalachian Mountains (Region 3b) with some disjunct populations in the central states (Region 3a).

Rhus aromatica Aiton, aromatic or fragrant sumac (Figure 22.23); multistemmed shrub, height to 1.5 m, forming thickets; deciduous; leaf arrangement alternate; leaf pinnately trifoliate, length 3 to 8 cm, terminal leaflet length 2.5 to 5.0 cm, side leaflet length 1.5 to 3.0 cm, adaxial and abaxial surfaces velvety pubescent, margin ciliated, with five to seven lobes/large teeth on the distal one half, entire/smooth on the proximal one half, base acute, concave, apex obtuse, aromatic when crushed; twigs pubescent, gray brown, medium gray, lightly pubescent, buds naked, yellowish, tomentose; inflorescence a panicle, small catkin like, reddish tan, length 5 mm, width 1.5 mm; fruit drupaceous, globose, pubescent, reddish at maturity, diameter 4 to 6 mm; Regions 2b, 3 and 4,

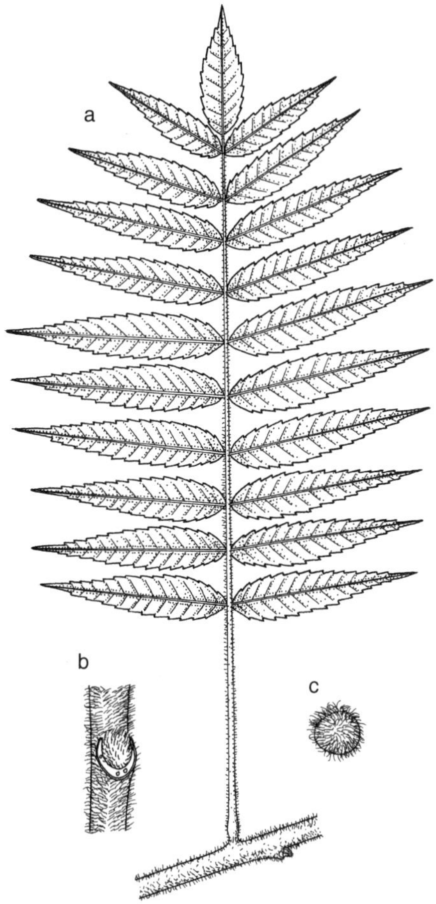

FIGURE 22.22 *Rhus typhina* (a) leaf × 1/3, (b) lateral bud and leaf scar × 1, (c) drupe × 2.

Nebraska, and Indiana, to Vermont and Quebec, south into Georgia and Texas. Fruit and buds are eaten by turkeys and ruffed grouse; deer eat the twigs. Cover type 27.11.

R. trilobata Nuttall, squawbush (Figure 22.23); multistemmed shrub, height to 1 m, forming thickets; deciduous; leaf arrangement alternate; leaf pinnately trifoliate, length 2 to 6 cm, terminal leaflet length 2.5 to 6.0 cm, side leaflet length 1.5 to 3.0 cm, adaxial and abaxial surfaces velvety pubescent, margin ciliated, with five to seven lobes/large teeth on the distal one half, entire/smooth on the proximal one half; base acute, concave, apex obtuse, aromatic when

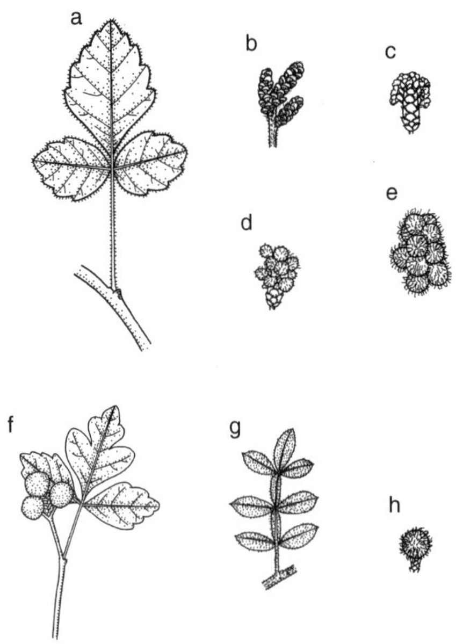

FIGURE 22.23 *Rhus aromatica* (a) leaf ×1/2, (b) staminate ament ×1, (c) pistilate ament with distal flowers ×1, (d) pistilate ament with developing fruit ×1, (e) inflorescence of mature drupes ×1; (f) *R. trilobata* leaf and fruit ×1/2; *R. microphylla* (g) leaf ×1; (h) drupe ×1.

crushed; twigs pubescent, gray brown, medium gray, lightly pubescent, buds naked, yellowish, tomentose; fruit drupaceous, globose, pubescent, reddish at maturity, diameter about 5 mm; Region 5, Colorado, Utah, California, Nevada, Arizona, New Mexico, and Mexico. This species is the western counterpart of *R. aromatica.*

Rhus microphylla Engelmann, littleleaf sumac (Figure 22.23); shrub, height to 5 m; deciduous; leaf arrangement alternate, leaf pinnately compound, length 1.0 to 2.5 cm, leaflets seven to nine, leaflet length 4 to 8 mm, densely tomentose, rachises winged; short twigs numerous, stiff, length 5 to 20 cm, often tapered at the apex, medium gray, densely pubescent; fruit drupaceous, globose, pubescent, red at maturity, diameter about 5 mm; Region 5b, southwestern Oklahoma, western Texas, Arizona, and New Mexico.

Other *Rhus* include *R. choriophylla* Wooten & Standley (evergreen or Mearns sumac) of New Mexico, Arizona, and Mexico, and *R. ovata S. Watson, R. integrifolia* (Nuttall) Bentham & Hooker (lemonade sumac), and *R. laurina* Nuttall (laurel sumac) of Arizona, California, and lower (Baja) California. *Rhus kearneyi* Barkley (kearney sumac) is a rare species of extreme southwestern Arizona and lower California.

Genus: *Toxicodendron*

The genus *Toxicodendron* includes a sumac as well as poison-ivy and poison-oak. The resinous sap or oil (called urushiol) of *Toxicodendron* causes a dermatitis characterized by a sequence of itching, swelling, and a rash with oozing blisters, beginning one hour to a day after contact. Washing with soap and warm water immediately after contact has usually been the recommended preventative procedure, although a recent recommendation calls for washing with cold water to keep the skin pores closed so that the urushiol cannot enter. Over-the-counter steroid creams are effective in many cases, but the most sensitive person may require medication prescribed by a physician. Some people are nearly immune to the oil. Occasionally, a rash may unexpectedly appear in a person who has not been near a living plant that contains urushiol. In this case, contact with residual oil on boots, clothing, or equipment, such as a shovel, is usually the cause. The chemical is extremely stable, and even if outside the plant for a year or more, it may still produce dermatitis. The oil is also carried in smoke from a fire, and if inhaled, can cause serious problems with the eyes, nose, throat, and lungs. Professionals working around prescribed fires should use extreme caution. People who are exceptionally sensitive may have a reaction after eating cashew nuts or mangos.

Toxicodendron vernix Linnaeus, poison-sumac (Figure 22.24); shrub or small tree, height to 5 m; deciduous; leaf arrangement alternate, leaf pinnately compound, length 8 to 35 cm, leaflet number 7 to 13, leaflet length 5 to 15 cm, sessile, shape elliptical to slightly oblanceolate, abaxial surface light green to whitish, apex acuminate, base obtuse, margin smooth; twig smooth, gray to reddish brown, lenticels on new twigs distinct, raised; pith spongy, large; bud short, 1.0 to 1.5 mm, densely pubescent, nearly surrounded by leaf scar; inflorescence an open axillary panicle, length 10 to 20 cm; fruit drupaceous, smooth, cream to light yellowish green at maturity; Regions 2, 3b, and 4b, Minnesota to Maine and down the Appalachian

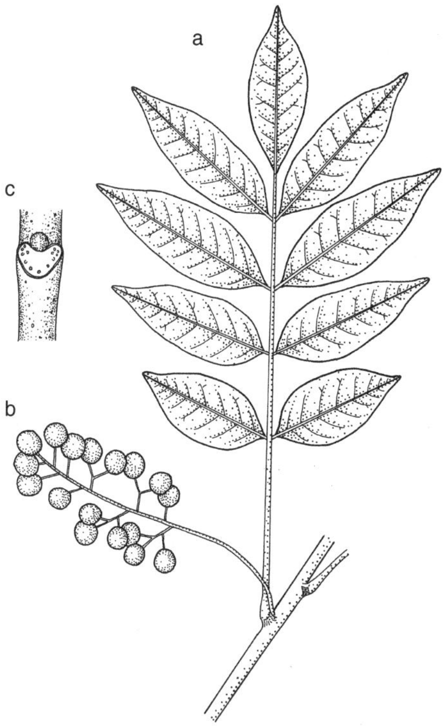

FIGURE 22.24 *Toxicodendron vernix* twig with (a) leaf ×1/3, (b) inflorescence of drupes ×1/3, (c) bud and leaf scar ×2.

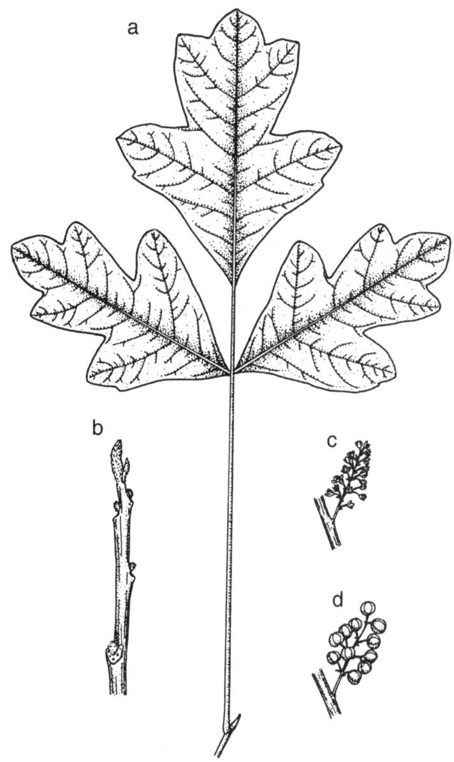

FIGURE 22.25 *Toxicodendron quercifolia* (a) leaf ×2/3, (b) twig with terminal and lateral buds ×1, (c) inflorescence of flowers ×1/2, (d) drupes ×1/2.

mountains to Florida and eastern Texas; wetlands and bogs. Cover type 28.31, 28.55.

Toxicodendron quercifolia (Michaux) Greene (*Toxicodendron toxicarium* Salisbury, I. diversilobum (Torrey & Gray) Green), poison-oak (Figure 22.25); vine or shrub, height to 2 m; deciduous; leaf arrangement alternate; leaf pinnately trifoliate, length 7 to 20 cm, terminal leaflet length 5 to 9 cm, margin with large teeth or lobed on the distal one half, occasionally deeply cleft, side leaflet length slightly smaller, inside margin often smooth or entire; twig reddish brown, bud naked, yellowish tan to light yellow, densely pubescent, leaf scar V-shaped, bundle scars five; inflorescence an axillary panicle; fruit a drupe, globose, diameter 3 to 5 mm, cream to light yellowish green, surface smooth with a few scattered hairs; Regions 5, 6, and 7. As with *T. radicans*, all parts of the plant are poisonous.

Toxicodendron radicans (Linnaeus) Kunse, poison-ivy (Figure 22.26); vine, shrub, or arborescent, height to 3 m; deciduous; leaf arrangement alternate; leaf pinnately trifoliate, length 7 to 20 cm, terminal leaflet length 5 to 9 cm, margin with large teeth or lobed on the distal one half, seldom cleft, side leaflet length slightly smaller, inside margin often smooth or entire; twig reddish brown, bud naked, yellowish tan to light yellow, densely pubescent, leaf scar V-shaped, bundle scars five; inflorescence an axillary panicle; fruit a drupe, globose, diameter 3 to 5 mm, cream to light yellowish green, essentially glabrous; Regions 2, 3, and 4, eastern United States; xeric-mesic, mesic, mesic-hydric sites, ridge tops, slopes, terraces, bottomland forest, also fencerows, old fields, and disturbed sites. There is sufficient variability in the leaf that several subspecies have been recognized at one time or another: *T. radicans* subsp. *negundo* (Greene) Gillis (lobes/teeth rounded), *T. radicans* subsp. *verrucosum*

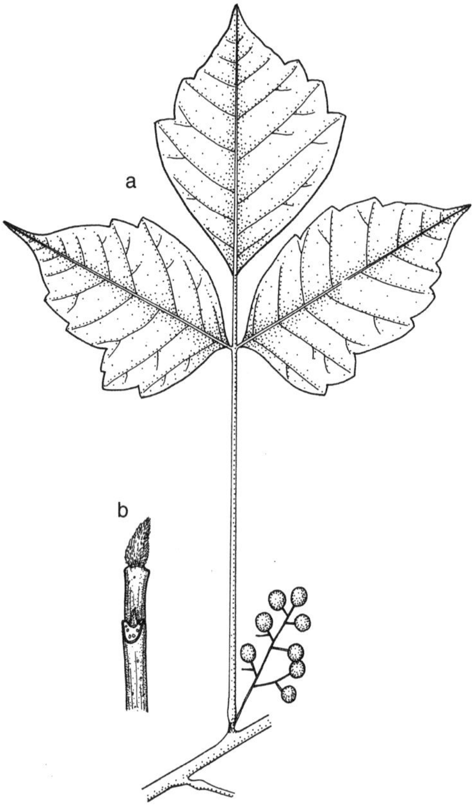

FIGURE 22.26 *Toxicodendron radicans* (a) twig with leaf and inflorescence of drupes ×2/3, (b) twig showing terminal bud and lateral bud with leaf scar ×1.5.

(Schelle) Gillis (teeth/lobes pointed), and *T. radicans* (Linnaeus) Kuntze subsp. *radicans* (margin smooth, without teeth or lobes). *Toxicodendron radicans* is a woodland species and frequently may become the major herb/shrub and a strong competitor in the understory. In this circumstance, tree seedling density and herbaceous richness are extremely low. The density of *T. radicans* tends to decrease rapidly with periodic fire. All parts of the plant are poisonous. Cover type 27.1, 27.10, 28.4, 28.10, 28.12, 28.15, 28.21, 28.31, 28.36, 28.49, 28.50, 29.19.

There has been substantial confusion not only regarding the distribution of "poison ivy" (not a true ivy) and "poison-oak," (not a true oak) but also in the nomenclature. In older books, poison-ivy may be listed as *Rhus radicans* Linnaeus or *Rhus toxicodendron* (Fernald 1950). According to Whitney (1985) *T. diversilobum* (Torrey & Gray) Green (poison-oak) is found only along the west coast from Baja, California, to Washington, and is so common in California that it could qualify as a state shrub. East of the Pacific Coast region it is replaced by *T. radicans*, which appears to be listed by Bailey and Bailey (1949) as *T. rydbergii* (Small) Green [syn. *T. radicans* Linnaeus var. *rydbergii* (Small) Rehd]. Gleason and Cronqist (1991) refer to *T. rydbergii* as "western poison ivy" because it is widespread in the western United States. Bailey and Bailey (1949) and the United States Department of Agriculture (1982) also restrict *T. diversilobum* to the west coast, but also indicate that poison-oak, *T. quercifolia* (Michaux) Green, occurs throughout Regions 3 and 4 of the central and southeastern United States. Hunter (1989) lists poison-oak as *T. toxicarium* (Salisbury) Gillis, and to futher confuse the issue, Fernald (1950) uses the name *Rhus toxicodenron* Linnaeus, the older name sometimes given to poison-ivy.

Genus: *Cotinus*

Cotinus obovatus Rafinesque, smoke-tree (Figure 22.27); small tree, height to 10 m; deciduous; leaf

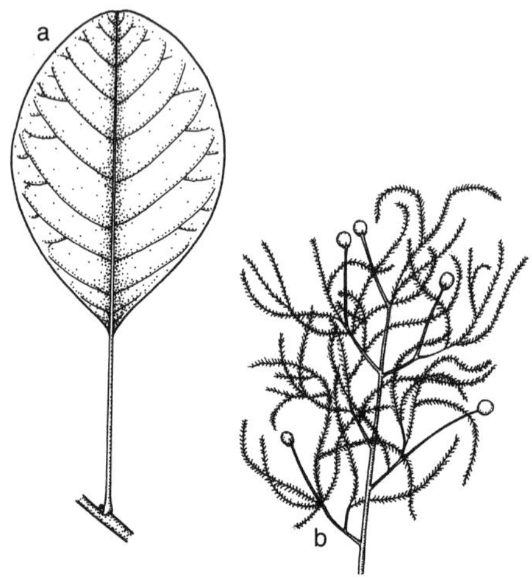

FIGURE 22.27 *Cotinus obovatus* (a) leaf ×1/2, (b) inflorescence (panicle) with drupes ×1/5.

arrangement alternate; leaf simple, entire, length 10 to 15 cm, oval or slightly elliptical, margin smooth, apex rounded, surfaces glabrous or abaxial surface lightly hairy; petiole length 3 to 5 cm, glabrous; inflorescence a panicle, length 0.8 to 25.0 cm, sterile pedicels plumose, numerous; fruit drupaceous, dry, length 5 to 8 mm; Region 3, widely scattered locations of the central United States; woods of xeric and xeric-mesic sites, calcarious soil.

Hippocastanaceae (Buckeye Family)

The Hippocastanaceae, an extremely small family of trees and shrubs, includes two genera largely restricted to the north temperate zone of the Western Hemisphere. The genus *Aesculus* (buckeye) has about 13 species. Three species of *Billia* range from southern Mexico across Central America into South America.

Aesculus leaves are opposite and palmately compound with five to eleven leaflets. Stipules are absent. The twigs are relatively stout (diameter 4 to 10 mm on new twigs), accompanied by a large terminal bud (length 1 to 2 cm). The terminal bud is nonresinous, except in *A. hippocastanum*. Collectively, the twigs of the past few years generally present a crooked or gnarled appearance, resulting from the death of the terminal bud, as evidenced by a large terminal twig scar.

Some of the species are difficult to separate when not in flower. However, the problem is somewhat reduced because of stem size at maturity, and a number of species have a limited range and do not naturally overlap. *Aesculus flava* (previously *A. octandra*) and *A. glabra* are the only species that grow to be medium to large trees. *Aesculus flava* is generally restricted to the moist slopes and deep coves of the Appalachian Mountains, and the primary range of *A. glabra* is west of the Appalachians through the central portion of the central hardwood region. The range of *A. glabra* overlaps the range of *A. flava* only along the edge of the western slope of the Appalachian Mountains (Pennsylvania, Ohio, western Kentucky, and middle Tennessee).

Shrub and small tree species include *Aesculus pavia* L. (red buckeye) of the southeastern Coastal Plain Region, *A. sylvatica* Bartram (painted buckeye) of the outer Piedmont Plateau and Coastal Plain, and *A. parviflora* (bottlebrush buckeye) of Alabama. *Aesculus hippocastanum* L. (horsechestnut) was introduced from eastern Europe. It is widely planted as a street and lawn tree and is noted for its white flowers. A cross between *A. pavia* and *A. hippocastanum* is marketed as *A.* × *carnea* Hayne (red horsechestnut), a popular ornamental with pink to reddish flowers. (Dirr 1998). *Aesculus californica* is the only western species.

The irregular (bilaterally symmetrical) perfect flowers of four to five petals are borne in a terminal panicle or raceme (Figure 22.28) and appear in late spring or early summer. Flower color by species is as follows: dark yellow, *A. flava*; light yellow, *A. glabra*; light yellow to creamy, *A. sylvatica*; dark red, *A. pavia*; and white, *A. hippocastanum*. The sepals are often connate. The fruit is a globose leathery capsule (diameter 3 to 4 cm) containing two or three large, smooth wingless seeds. The capsule of *A. glabra* is prickly, and that

FIGURE 22.28 Terminal inflorescence (panicle) of *Aesculus* ×1/2.

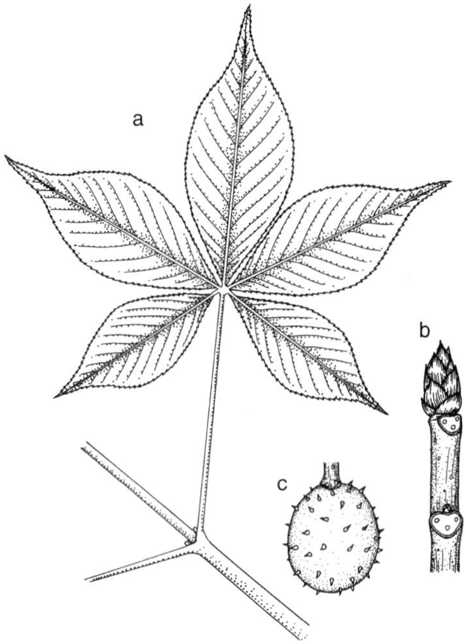

FIGURE 22.29 *Aesculus glabra* (a) leaf ×1/3, (b) twig showing terminal bud and leaf scars ×1, (c) capsule ×2/5.

of *A. hipposcastanum* is spiny; the capsules of other *Aesculus* species have a smooth surface. The fruits are poisonous.

Aesculus glabra Willdenow, Ohio buckeye (Figure 22.29); medium-sized tree, height to 22 m; deciduous; leaf arrangement opposite; leaf palmately compound, petiole length 6 to 13 cm, leaflets five, leaflet length 5 to 14 cm, shape elliptical to slightly oblanceolate, apex long acuminate, base acute, margin finely serrated, number of veins per side 14 to 22, veins unbranched, varying from alternate to opposite, crushed leaves and twigs pungent; twig and buds stout, new twigs often 5 to 10 mm in diameter, glabrous, leaf scars shield-shaped, not notched at the top, width greater than or equal to length, bundle scars in three groups of one or two; terminal bud length 1.0 to 1.5 cm, nonresinous, scales decussate and distinctly striated, the middle lines raised, forming a weak ridge; inflorescence a terminal raceme, length 8 to 16 cm, peduncle densely pubescent, flowers light yellow or greenish yellow; fruit a leathery capsule, globose, diameter 3 to 4 cm, surface characterized by scattered short spines, seeds two to three, brown; Region 3a, Kansas and Iowa to western Pennsylvania, south to northern Alabama and Texas, scattered locations in the Mississippi Embayment; mesic woods, stream terraces and low slopes of drainages. Cover type 28.22.

Aesculus flava Soland (*A. octandra* Marshall), yellow buckeye (Figure 22.30); large tree, height to 30 m; deciduous; leaf arrangement opposite; leaf palmately compound, petiole length 8 to 16 cm, leaflets five, leaflet length 8 to 23 cm, shape elliptical, apex long acuminate, base acute, margin finely serrated, number of veins per side 20 to 30, veins unbranched, varying from opposite to alternate, crushed leaves and twigs lacking pungent odor; twigs and buds stout, new twigs often 5 to 10 mm in diameter, glabrous, leaf scars shield-shaped, usually not notched at the top, length greater than width, bundle scars in three groups of one or two; terminal bud length 1.0 to 1.7 cm, nonresinous, scales decussate and indistinctly striated;

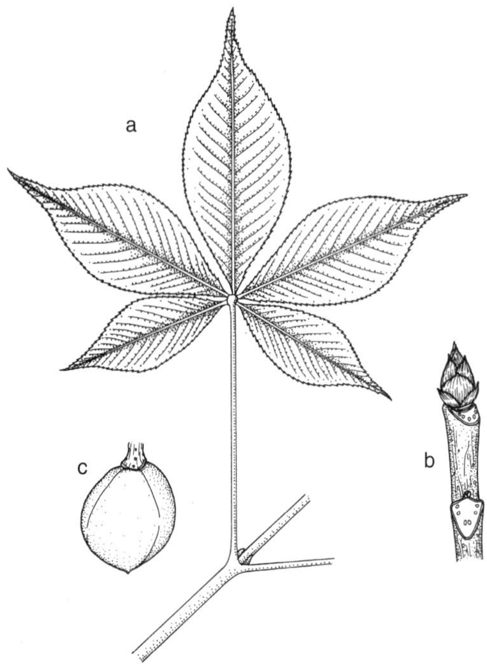

FIGURE 22.30 *Aesculus flava* (a) leaf ×1/3, (b) twig showing terminal bud and leaf scars ×1, (c) capsule ×2/5.

FIGURE 22.31 *Aesculus pavia* leaf × 1/3.

inflorescence a terminal raceme, length 8 to 19 cm, peduncle densely pubescent, flowers yellow; fruit a leathery capsule, globose, diameter 3 to 4 cm, surface smooth, seeds two to three, brown; Regions 3b, Appalachian Mountains, southwestern Pennsylvania, West Virginia, Virginia south to northern Georgia and Alabama, Region 3b, extending along the Ohio River to Illinois; mesic woods, cool slopes, coves and drainages. Cover type **27.29, 27.30**.

Aesculus pavia Linnaeus, red buckeye (Figure 22.31); shrub or small tree, height to 5 m; deciduous; leaf arrangement opposite; leaf palmately compound, petiole length 8 to 14 cm, leaflets five, leaflet length 7 to 18 cm, width 2 to 6 cm, narrowly elliptical to oblanceolate, apex long acuminate, base acute, margin finely serrated, number of veins per side 14 to 24, veins unbranched, varying from opposite to alternate; twigs and buds stout, new twigs often 5 to 10 mm in diameter, glabrous, leaf scars shield-shaped, usually notched at the top, length greater than width, bundle scars three to six, in three groups of one or two; bud length 1.0 to 1.5 cm, nonresinous, scales decussate, several central striations developed into a keel; inflorescence a terminal panicle, length 12 to 18 cm, peduncle densely pubescent, flowers red but some may be yellow; fruit a leathery capsule, globose, diameter 1.5 to 3.0 cm, surface smooth, seeds two to three, brown; spreading by underground runners; Region 4, Coastal Plain, North Carolina to Florida and Texas, north in to Region 3a, Kentucky, southern Illinois, Missouri, and Tennessee; mesic sites, Coastal Plain woods. *Aesculus discolor* Purshiana (particolored buckeye) is now included with this taxon. Cover type 27.9, 28.22. *Aesculus pavia* is one of the first plants to bloom in the spring, sometimes as early as February. The panicle of bright red flowers is distinctive (Nelson 1994).

Aesculus sylvatica Bartram, painted or dwarf buckeye; multistemmed shrub or small tree, height to 5 m; deciduous; leaf arrangement opposite; leaf palmately compound, petiole length 7 to 15 cm, leaflets five, leaflet length 7 to 23 cm, width 2.5 to 8.5 cm, oblanceolate to elliptical, apex long acuminate, base acute, margin finely serrated, number of veins per side 20 to 32, veins unbranched, varying from opposite to alternate; twigs and buds stout, new twigs often 0.4 cm in diameter, glabrous, leaf scars shield-shaped, notched at the top, width greater than or equal to length, bundle scars three to six, in three groups of one or two; terminal bud length 1.0 to 1.5 cm, nonresinous, scales decussate, indistinctly striated; inflorescence a terminal raceme, length 7 to 17 cm, peduncle lightly pubescent, occasionally densely pubescent, flowers yellowish, cream, or pink; fruit a leathery capsule, more or less globose to oblong, diameter 1.5 to 3.0 cm, surface smooth, seeds two to three, brown; Region 4c, Piedmont Plateau, Virginia to Alabama and Region 3a, central Tennessee; mesic and hydric-mesic sites, bottomland, stream margins. Cover type 27.15, 28.12, 28.18.

Aesculus parviflora Walter, bottlebrush buckeye; shrub, height to 3 m; deciduous; leaf arrangement opposite; leaf palmately compound, petiole length 7 to 18 cm, leaflets five, leaflet length 8 to 21 cm, width 3 to 9 cm, oblanceolate to obovate, apex long acuminate, base acute, margin finely serrated, number of veins per side 14 to 22, veins unbranched, varying from opposite to alternate; twigs and buds stout, new twigs often 0.4 cm in diameter, glabrous, leaf scars shield-shaped, moderately to deeply notched, width greater than or equal to length, bundle scars three to six, in three groups of one or two; terminal bud length 1.0 to 1.5 cm, nonresinous, scales decussate, indistinctly striated; inflorescence a terminal panicle, length 25 to 37 cm, peduncle lightly pubescent, flowers have white petals with long exserted red stamens, petals occasionally tinged with red; fruit a leathery capsule, more or less globose, diameter 1.5 to 3.0 cm, surface smooth, seeds two to three, brown; Region 4, central Alabama and western Georgia; mesic sites.

Aesculus californica (Spach) Nuttall, California buckeye (Figure 22.32); small tree, height to 10 m; deciduous; leaf arrangement opposite; leaf palmately compound, petiole length 4 to 12 cm, leaflets five, leaflet length 5 to 14 cm, width 2 to 4 cm, narrowly elliptical or oblanceolate, apex long acuminate, base acute, margin finely serrated to crenate, number of veins per side 13 to 16, veins occasionally branched near the margin, varying from opposite to alternate; twigs and buds stout, new twigs often 5 mm in diameter, glabrous, leaf scars shield-shaped, slightly notched at the top, width greater than or equal to length, bundle scars five, evenly distributed to form a U; bud length 1.0 to 1.5 cm, nonresinous, scales decussate, several central striations developed into a keel; inflorescence a terminal panicle, length 13 to 25 cm, peduncle densely pubescent, flowers creamy white or tinted with pink; fruit a leathery capsule, more or less globose, diameter 1.5 to 3.0 cm, surface smooth, seeds two to three, brown; Region 7, Coastal Ranges and Sierra Nevada surrounding the central valley; moderately xeric sites, oak woodland and chaparral.

Aesculus hippocastanum Linnaeus, horsechestnut (Figure 22.33); medium sized tree, height to 20 m; deciduous; leaf arrangement opposite; leaf palmately compound, petiole length 8 to 17 cm, leaflets 7 to 11,

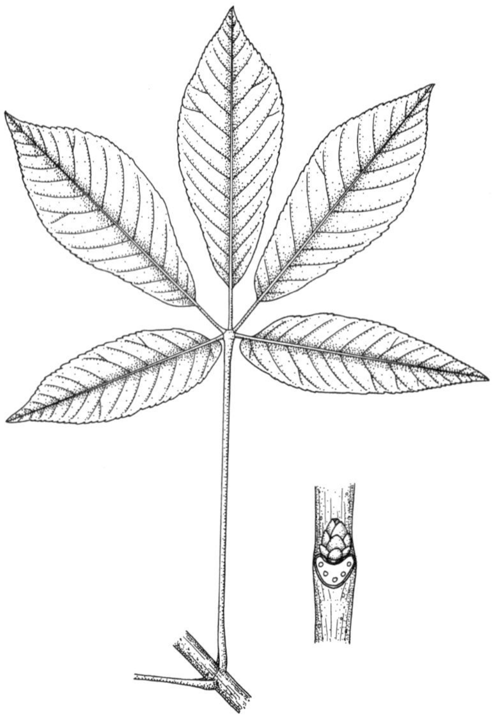

FIGURE 22.32 *Aesculus californica* (a) leaf ×1/2, (b) twig showing lateral bud and leaf scar ×1.

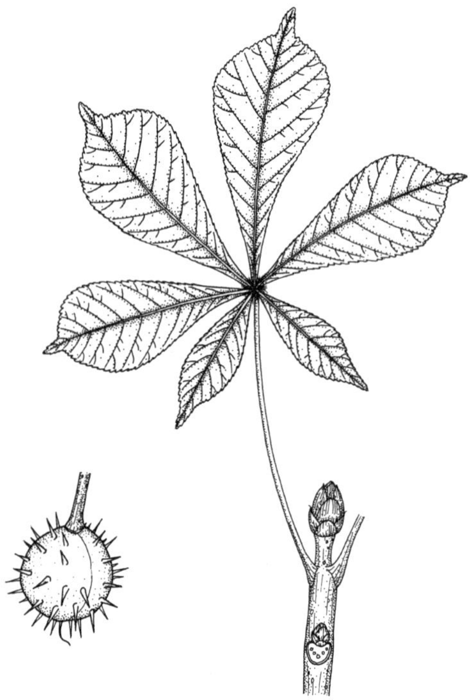

FIGURE 22.33 *Aesculus hippocastanum* (a) leaf and twig showing buds and leaf scar ×1/3, (c) capsule.

leaflet length 8 to 28 cm, width 4 to 10 cm, distinctly oblanceolate, apex long acuminate, base acute, margin finely serrated to crenate, number of veins per side 13 to 16, veins often branched near the margin, varying from opposite to alternate; twigs and buds stout, new twigs often 7 to 8 mm in diameter, glabrous, leaf scars shield-shaped, slightly notched at the top, width about equal to length, bundle scars five to seven, evenly distributed to form a U; terminal bud length 1.5 to 2.0 cm, resinous, shiny, scales decussate; inflorescence a terminal panicle, length 13 to 25 cm, peduncle densely pubescent, flowers creamy white and may be blotched with yellow, then red; fruit a leathery capsule, more or less globose, diameter 1.5 to 3.0 cm, surface with sharp spines, spine length to 8 mm; seeds one to two, brown; introduced as a lawn and street tree, hardy to cold and pollution. This species is native to the border region of Greece and Albania. It was introduced in Europe in 1576, in England in the 1600s (Hillier Nurseries 1998), and probably into the United States in the 1800s.

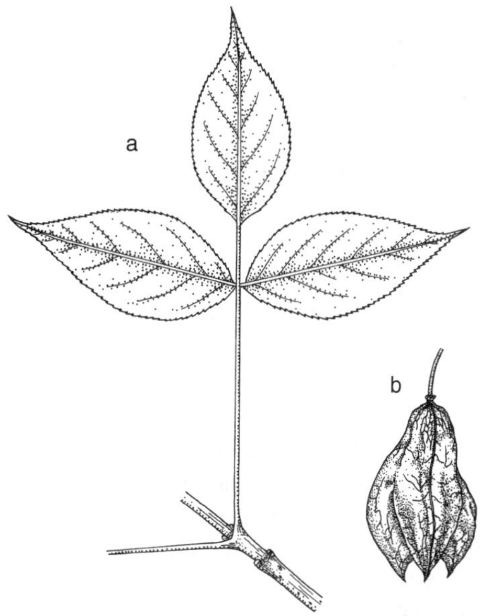

FIGURE 22.34 *Staphylea trifolia* (a) leaf ×2/5, (b) bladder-like capsule ×1/2.

Staphyleaceae (Bladdernut Family)

Staphyleaceae is a family composed of five genera and 50 to 60 species of trees and shrubs, distributed mostly in Europe and across Asia to China and Japan; only one genus (*Staphylea*) and two species are found in North America. The flowers are perfect and regular, with five sepals and five stamens, the latter alternating with the five petals. The panicles of white flowers appear in April or May. The fruit is a capsule. The species have limited wildlife and economic value. Occasionally they are planted as ornamentals.

Staphylea trifolia Linnaeus, bladdernut (Figure 22.34); small tree, height to 8 m; leaf arrangement opposite; deciduous; leaf pinnately compound, leaflets three to five, stipulate; leaflet length 5 to 10 cm, ovate, elliptical, or lanceolate, margin serrulated, side leafets nearly sessile; twigs slender, green; fruit an inflated bladder-like capsule with thin cellulose walls, length 4 to 6 cm, three-pointed; Region 3, range closely follows the northern and western regional boundary, extends into Region 4; forest communities on moist sites, north slopes and stream terraces. Cover type 26.24, 27.12.

Staphylea bolanderii A. Gray, Sierra bladdernut; shrub or small tree, height to 6 m; differs from *A. trifoliata* in that leaflet length for *S. bolanderii* is 2.5 to 6.0 cm, the shape is ovate, and capsule length is 2.5 to 5.0 cm; Region 7, range restricted to California Sierra Nevada foothills; chaparral and woodlands, elevation 700 to 1400 m.

Rutaceae (Citrus or Rue Family)

The Rutaceae, a large family of 150 genera and 1500 to 1600 species, is centered in the tropical and warm temperate zones. The greatest number of species are in Australia and South Africa. Members are trees and shrubs and, infrequently, herbs. The largest genus is *Zanthoxylum*, with more than 200 species.

The leaves are alternate, simple (as in the genus *Citrus*), or compound (in *Ptelea, Zanthoxylum*), have ethereal oil glands (appearing translucent as pallucid dots when held up to light), and emit a strong, often pungent, odor reminiscent of citrus in many genera. The presence of glands separates this family from other closely related families. The flowers are yellowish green in color, perfect, regular, with three to five petals, and borne singly (*Citrus*) or in cymes (*Ptelea, Zanthoxylum*). Plants are dioecious.

The economic importance of the family is related to the genus *Citrus*, which includes *C. sinensis* (Linnaeus) Osbeck (sweet orange), *C. aurantium* Linnaeus (sour orange), *C. aurantifolia* (Christman) Swingle (lime), *C. limon* (Linnaeus) Burmann f. (lemon), *C. medica* Linnaeus Seville (orange), *C. paradisi* MacFadyen (grapefruit), and *C. reticulata* Blanco (mandarin orange and tangerine). Other genera occur in subtropical southern Florida and California.

Representatives of only two genera (*Ptelea* and *Zanthoxylum*) are found in North American forests.

Genus: *Ptelea*

In addition to the eastern *Ptelea trifoliata*, there are two other *Ptelea* species in the United States. *Ptelea tomentosa* var. *mollis* Torrey & Gray is a strongly pubescent variety of the southeast (Region 4). *Ptelea crenulata* Greene is a shrub restricted to California's San Joaquin and Sacrameno river valleys.

Ptelea trifoliata Linnaeus; hoptree, water-ash (Figure 22.35); shrub or small tree, height to 7 m; deciduous; leaf arrangement alternate; leaf pinnately compound, total length 10 to 18 cm leaflets three or five, broadly elliptical, length 5 to 10 cm margin smooth, apex acuminate, base obtuse or acute; twigs reddish brown; fruit an oval samara, seed cavity surrounded by a lateral wing, diameter 2 cm; Regions 3 and 4, Quebec, New York, and southern to Nebraska, south to Florida and central Mexico. Cover type 28.22.

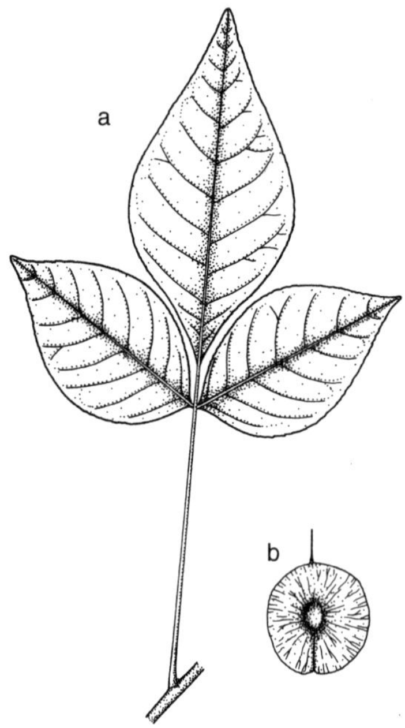

FIGURE 22.35 *Ptelea trifoliata* (a) leaf ×3/4, (b) samara ×3/4.

Genus: *Zanthoxylum*

Zanthoxylum (or Xanthoxylum) is a genus of six species of shrubs or small trees (height 8 m). *Zanthoxylum* species not described here include *Z. hirsutum* Buckley (Texas Hercules-club; central Texas), *Z. fagara* (Linnaeus) Sargent (lime prickly-ash; southern Texas and Florida), *Z. flavum* Vahl (yellowheart, southern Florida), and *Z. coriaceum* A. Rich (Biscayne prickly-ash, southern Florida).

Zanthoxylum americanum Miller, common or northern prickly-ash, toothache-tree (Figure 22.36); shrub or small tree, height to 8 m; deciduous; leaf arrangement alternate; leaf pinnately compound, length 15 to 30 cm; leaflets 5 to 11, length 3 to 7 cm, ovate, oval, or lanceolate, margin smooth, serrulate or crenulate;

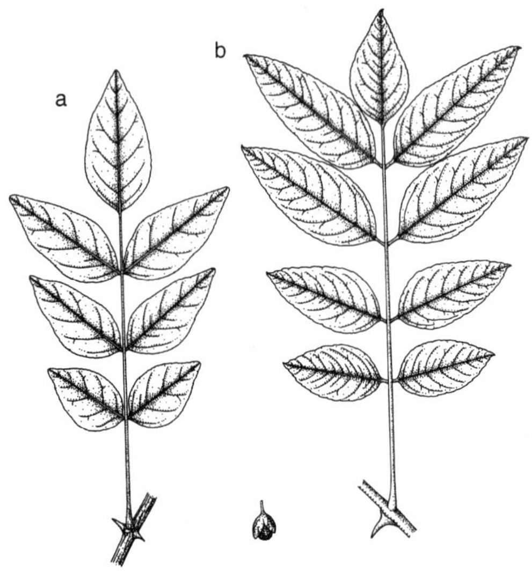

FIGURE 22.36 Leaf of (a) *Zanthoxylum americanum* × 1/4, (b) Follicle ×1, (c) *Z. clava-herculis* ×1/4.

petiole often prickly; crushed foliage producing an orangelike odor; twigs with stipular thorns, one on each side of the petiole or leaf scar; flower petals yellowish green; fruit a follicle, length 5 mm; Regions 2 and 3, southern Quebec to North Dakota, south to Oklahoma and Georgia; xeric-mesic and mesic sites; forms dense thickets in open areas, along fencerows, and in woods. Usually indicative of disturbance, particularly grazing. Apparently, leaves, fruit, or stems were chewed by First Nation people for toothache, the sour juice having a numbing effect.

Zanthoxylum clava-herculis Linnaeus (Figure 22.36); Hercules-club; southern prickly ash; shrub or small tree, height to 9 m; deciduous; leaf arrangement alternate; leaf pinnately compound, length 15 to 30 cm; leaflets 7 to 17, length 3 to 9 cm, ovate, oval, or lanceolate, margin smooth or crenulate; petiole often prickly; crushed foliage producing an orange-like odor; twig with stipular thorns, one on each side of the petiole or leaf scar; flower petals yellowish green; fruit a follicle, length 5 mm; Region 4, Atlantic and Gulf Coastal Plain from Virginia to Texas and Oklahoma; xeric or mesic woods, thicket forming.

Simaroubaceae (Quassia Family)

The Simaroubaceae, a small pantropical family of 25 genera and 150 tree and shrub species, has few subtropical and fewer temperate zone representatives. Several genera occur in the subtropical zone of southern Florida and the keys, and 3 species of *Holacantha* (crucifixion-thorn) are native to the deserts of the southwest. The leaves, bark, wood, seeds, and flowers of the Simaroubaceae have a very bitter (vile and extremely offensive) taste and odor, due to characteristic chemicals called simaroubalides, which include quassin and other antimalaria drugs. The flowers of male trees have a particularly strong odor. The family is similar to the Rutaceae but lacks resin glands in the leaves.

Ailanthus altissima (Miller) Swingle, tree-of-heaven (Figure 22.37); small to medium-sized tree, height to 15 m; deciduous; leaf arrangement alternate; leaf pinnately compound, length 45 to 60 cm; leaflets 11 to 41, lanceolate, length 7 to 13 cm, margin smooth except for basal tooth with a gland on the abaxial surface; twigs stout, leaf scars shield-shaped, length 0.5 to 1.0 cm, lenticels prominent; inflorescence a panicle, terminal, length 20 to 40 cm; flower petals five, yellowish green, usually unisexual or perfect (plants are polygamo-dioecious); fruit a samara, length 3 to 5 cm, width 0.8 to 1.4 cm, wings at each end of the seed cavity, pointed, twisted; introduced, escaped cultivation in Regions 2, 3, and 4. This species spreads aggressively by both seed and root suckers and grows rapidly. Fortunately, its distribution is generally restricted to cities, roads, and along fences.

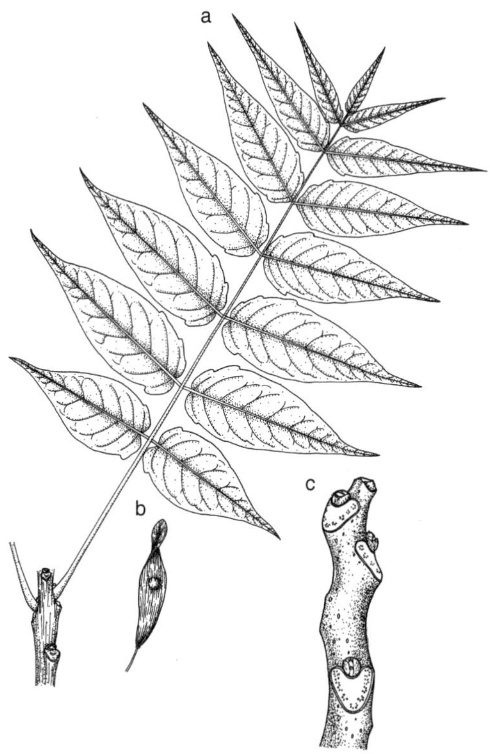

FIGURE 22.37 *Ailanthus altissima* (a) leaf and twig ×1/6, (b) samara ×1/2, (c) twig with twig scar, pseudoterminal bud, and lateral buds ×1.

CHAPTER 23

Class: Magnoliopsida; Subclass: Asteridae

ORDER: SCROPHULARIALES

Oleaceae (Olive or Ash Family)

THE olive family is well distributed in both the temperate and tropical regions, with strong representation ranging from southeast Asia to Australia. The family includes 30 genera and about 600 species. Reproductive and vegetative characters widely differ between genera. Some authors have considered this family to be an unnatural assemblage, with *Fraxinus* (ash) more closely related to the Sapindaceae, and *Ligustrum* (privet) to the Loganiaceae (Lawrence 1951). The major genera include *Jasminum* (jasmine; 200 species), *Chionanthus* (fringe tree; 125 species), and *Fraxinus* (60 species). Other genera include *Syringa* (lilac; 30 species), *Olea* (olive; 20 species), and *Ligustrum* (privet; 35 species native to the Old World).

The largest genus in North America is *Fraxinus*, represented by 15 species about equally divided among eastern and western forests and the extreme southwest. The flower is imperfect, with staminate and pistilate flowers on separate trees (trees dioecious). The fruit is a samara with a terminal wing; fruits are usually borne in a loose cluster (panicle). The leaves are pinnately compound, except in *F. anomala* Torrey (single-leaf ash). All the species grow to tree size, most into large trees (height 25 to 35 m). *Fraxinus* is an important forest tree on mesic sites in the Lake States and New England (Region 2) and in the Central States (Region 3). The following is a summary of *Fraxinus* in North America.

Fraxinus americana Linnaeus, white ash (Figure 23.1); large tree, height 35 m; deciduous; leaf pinnately compound, length 20 to 30 cm; leaflets usually seven (five to nine), shape elliptical to lanceolate, petioluled, margin smooth, undulate, or serrate, length 5 to 13 cm, width 3 to 6 cm; twigs usually stout, dark gray to green; bud scaly and inserted in V notch of shield-shaped leaf scar, terminal bud broad, short (width ~8 m), brown; fruit a samara (length 3 to 5 cm) with a terminal wing (width 5 to 9 mm), seed cavity short (1.0 to 1.7 cm) but broad (3 to 4 mm); Regions 2, 3, and 4, all states and provinces except in the Coastal Plain along the Gulf and Atlantic coasts; deep soil and moist north slopes and stream terraces. Cover type **26.24, 27.7**, 26.17, 26.23, 26.28, 26.29, 27.5, 27.8, 27.9, 27.11, 27.16, 27.23, 27.24,27.30, 27.37, 28.7, 28.14, 28.18, 28.19, 28.21, 28.27.

Fraxinus pennsylvanica Marshall, green ash, red ash (Figure 23.2); large tree, height to 40 m; deciduous; leaf pinnately compound, length 20 to 30 cm; leaflets usually seven (five to nine), shape elliptical to lanceolate, petioluled, margin serrate, smooth, or undulate, length 5 to 13 cm, width 3 to 5 cm; twigs usually stout, gray to greenish; lateral bud positioned on top of shield-shaped leaf scar, top margin of scar straight or slightly crescent-shaped, terminal bud broad, short (length <1 cm), pubescent; fruit a samara (length 3 to 5 cm) with a terminal wing (width 5 to 9 mm), seed cavity narrow (2 mm) and long (length ~2 cm); Regions 2, 3, and 4, all states and provinces west to Montana, Wyoming, Nebraska, Kansas, and Manitoba, best development in Region 4;

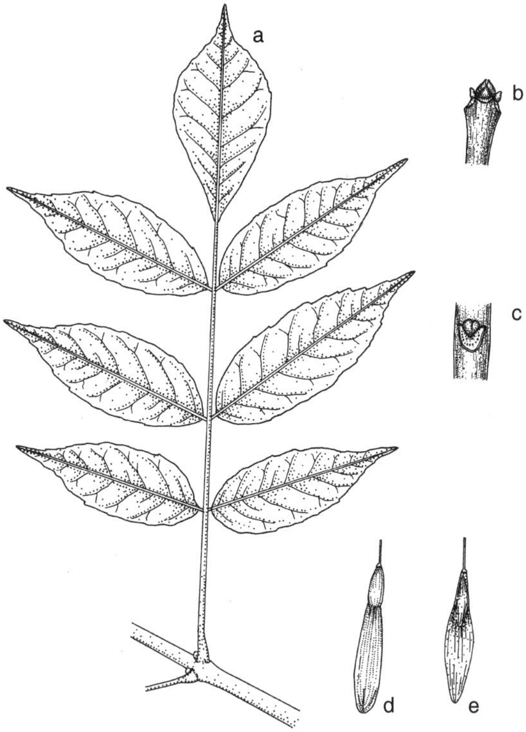

FIGURE 23.1 *Fraximus americana* (a) pinnately compound leaf with elliptical leaflets ×1/3, (b) terminal bud ×1, (c) lateral bud with V-notch in the shield-shaped leaf scar ×1, (d), (e) samaras ×1/2.

deep moist soil of terraces and bottomland swamps. There is variability within this taxon. The typical *F. pennsylvanica*, which has velvety tomentose twigs, petioles, and lower leaf surfaces, is called red ash. The variety *F. pennsylvanica* var. *subintegerrima* (Wahl) Fernald (synonym: *F. lanceolata*) is glabrous and is called green ash (Figure 23.3). Cover type **27.2, 27.10, 27.14**, 27.3, 27.14, 27.19, 27.20, 27.28, 28.10, 28.11, 28.12, 28.15, 28.17, 28.18, 28.32, 28.33, 28.49, 28.50, 28.55.

Fraxinus quadrangulata Michaux, blue ash (Figure 23.4); large tree, height to 25 m; deciduous; leaf pinnately compound, length 20 to 30 cm; leaflets 7 to 11, ovate to lanceolate, short petioluled, margin serrate; twigs four-angled, occasionally nearly winged; fruit a samara (length 2.5 to 4.5 cm), terminal wing short (length 1 to 2 cm), shorter than seed cavity (length 1.5 to 2.5 cm), somewhat spatulate; Region 3, range narrow, from northern Arkansas and kansas through central Missouri, central Illinois, Indiana, western Ohio, and south to northern Alabama; deep mesic soil. Cover type 27.11, 28.19.

Fraxinus nigra Marshall, black ash (Figure 23.5); medium-sized tree, height to 20 m; deciduous; leaf pinnately compound, length 20 to 30 cm; leaflets three to seven, elliptical-lanceolate, sessile, finely serrated; twigs usually stout, grayish, buds dark brown; fruit a samara (length 2 to 4 cm), terminal wing (width 1.0 to 1.3 cm) extends around seed cavity, seed cavity length ~1.5 cm, width ~0.5 cm; Region 2, but extending into southern one half of Region 1 and northern section; of Region 3; Illinois, Indiana, Ohio, and west Virginia north; wet swamps and bogs. Cover type **26.25**, 26.1, 26.20.

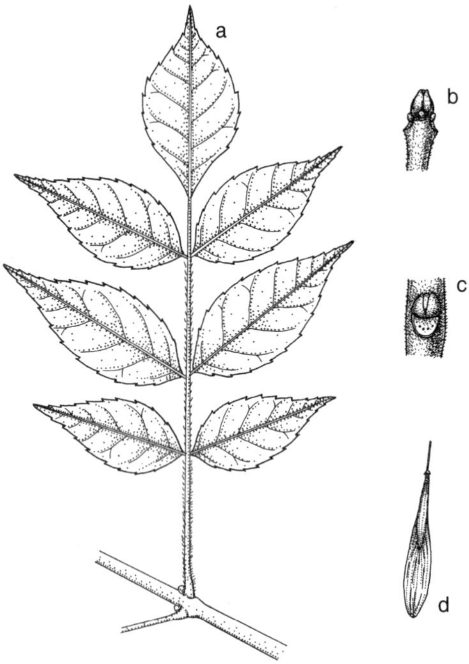

FIGURE 23.2 *Fraxinus pennsylvanica* (a) pinnately compound leaf with elliptical leaflets ×1/3, (b) terminal bud ×1, (c) lateral bud with shield-shaped leaf scar ×1, (d) samara ×1/2.

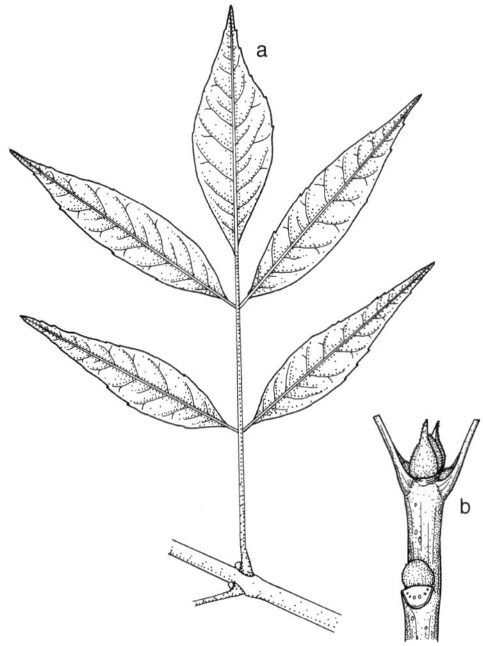

FIGURE 23.3 *Fraxinus pennsylvanica* var. *subintegerrima* (a) compound leaf with lanceolate leaflets ×1/3, (b) twig with terminal and lateral buds and shield-shaped leaf scar ×1.5.

Fraxinus caroliniana Miller, Carolina ash, pop ash (Figure 23.6); small tree, height to 10 m; deciduous; leaf pinnately compound, length 16 to 30 cm; leaflets five to seven, length 8 to 12 cm, width 2 to 5 cm, shape oval, elliptical, or lanceolate, base asymmetrical, margin smooth or serrated, petioluled; twigs slender, pubescent, gray to green, buds dark brown; fruit a samara, terminal wing (width ~1.4 to 1.7 cm) extends around the seed cavity, seed cavity length 2.0 to 2.5 cm, width ~0.5 cm; Region 4, but generally not on the Piedmont (Region 4c), Texas to Virginia; wet swamps and bottomland; trunk base may be swollen or buttressed. Cover type 28.22, 28.31, 28.43, 28.47, 28.53, 28.55.

Fraxinus tomentosa Michaux [*Fraxinus profunda* (Bush) Bush], pumpkin ash (Figure 23.7); large tree, height to 30 m; deciduous; leaf pinnately compound, length 15 to 35 cm; leaflets seven to nine, shape ovate to lanceolate, length 7 to 15 cm, width 4 to 7 cm, petioluled, margin smooth to serrated; petiole, rachises, and petiolules tomentose; twigs stout, gray, new twigs pubescent; fruit a samara, length 4 to 7 cm, terminal wing partially surrounding the elongated seed cavity; scattered in Regions 3 and 4, greatest concentration in upper Mississippi and Ohio River valleys, northern Florida, and along the coast from Georgia to Virginia; wet swamps and bottomland. Cover type 27.2, 28.31, 28.50, 28.55.

Fraxinus latifolia Bentham, Oregon ash (Figure 23.8); large tree, height to 25 m; deciduous; leaf pinnately compound, length 15 to 35 cm; leaflets five to seven, sometimes three, length 4 to 9 cm, width 2 to 4 cm, shape ovate or elliptical, margin serrulate to smooth, lower surface tomentose; twigs brown, bud pubescent; fruit a samara, length 3 to 5 cm, terminal wing partially enclosing seed cavity; Region 7b, Washington and Oregon to southern part of California; moist soil along streams, rivers, and canyons, elevation to 500 m. Cover type **30.9**, 30.20.

Fraxinus dipetala Hooker & Arnott, two-petal ash; small tree, height to 8 m; deciduous; leaf pinnately

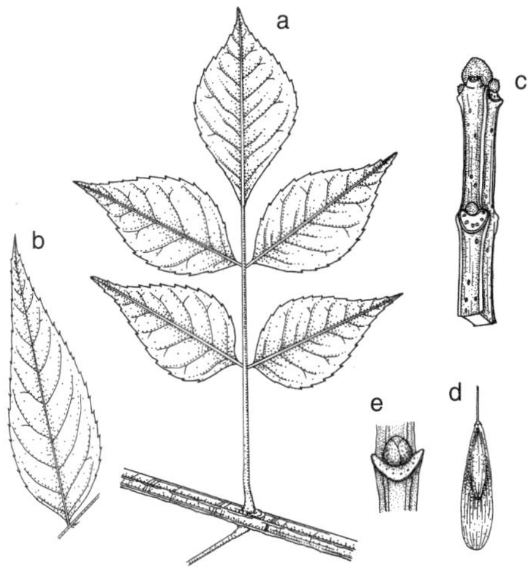

FIGURE 23.4 *Fraxinus quadrangulata* (a) pinnately compound leaf with ovate leaflets ×1/4, (b) lanceolate leaflet found on some individuals × 1/3, (c) four-angled twig with terminal and lateral buds and shield-shaped leaf scar ×1, (d) lateral bud and leaf scar ×1.5, (e) samara ×1/2.

compound, length 8 to 12 cm; leaflets three to seven (infrequently nine); shape ovate to lanceolate; length 2 to 5 cm, width 1.0 to 2.5 cm; Region 7b into lower California; dry slopes of coastal ranges and Sierra Nevada foothills, elevations to 1000 m. Unlike apetalous flowers of other *Fraxinus* species, the *F. dipetala* flower has two white petals, hence its common and species names.

There are a number of *Fraxinus* species, each with a different range, scattered througout the area from southern and lower California and Nevada through Utah, Arizona, New Mexico, Texas, and Mexico. Most are not associated with dense forest communities. *Fraxinus velutina* Torrey (velvet ash; cover type 29.2, 29.19), *F. anomala* Torrey (single-leaf ash), and *F. berlandierana* A. De Candolle (Berlandier ash) grow on riverbanks and along streams and moist washes, whereas *F. cuspidata* Torrey (fragrant ash), *F. greggii* A. Gray (Gregg or little-leaf ash), *F. papillosa* (Chihuahua ash), *F. texensis* (Gray) Sargent (Texas ash; cover type 29.9) are found on dry, rocky slopes, bluffs, and canyons. *Fraxinus gooddingii* Little (Goodding ash) is found in dry desert scrub or grassland. Most are small trees (height up to 9 m), except *F. velutina* (13 m) and *F. texensis* (height <17 m).

Of these ash species, *Fraxinus velutina* has the largest range, from California, Utah, and Nevada to Texas and Mexico, and is the most common ash in the southwest desert country. This species occurs in moist washes with cottonwood and in canyons up to an elevation of 2100 m (Peattie 1953), and in oak woodland and ponderosa pine forest, occasionally forming nearly pure communities (Elias 1980). The pinnately compound leaf has three to five elliptical (shape) leaflets; the lower surfaces are tomentose. The range of *Fraxinus anomala* extends from southern California to Colorado and Arizona. Its leaf may be simple and resemble the leaflet of other ashes, or the leaf may be compound with three leaflets. This species also grows along stream banks and washes. Additional information and range maps can be found in Elias (1980) and Peattie (1953).

Other North American species in this family have a simple, entire leaf, and the flowers of most have petals.

Chionanthus virginicus Linnaeus, fringe-tree (Figure 23.9); multistemmed shrub or small tree, height to 10 m; deciduous; leaf arrangement opposite; leaf simple, margin smooth, shape ovate or oblong, length 10 to 15 cm; twigs resemble those of *Fraxinus*; fruit is a drupe, dark blue; Regions 3b and 4, low elevation Appalachians, Piedmont and Coastal Plain, and southern one half of Region 3a, disjunct areas in Ohio, Kentucky, Missouri, and Arkansas; deep, moist soils of slopes and terraces. The narrow, 2 cm long white petals give the tree its common name and make it a favored ornamental. Cover type 27.24, 28.14, 28.27.

Two related deciduous shrubs (height <8 m) of the genus *Foresteria* are found from the southern one half of Region 3 through Region 4.

Foresteria acuminata (Michaux) Poir, swamp privet (Figure 23.10); shrub or small tree, height to 8 m; deciduous; leaf arrangement opposite; leaf simple, length 4 to 10 cm, width 1 to 5 cm, elliptical, blade base and apex acuminate, base smooth, becoming minutely serrate toward the apex; some twigs tapered to appear as small thorns; flowers borne in an axillary panicle; fruit a black drupe with a thin mesocarp, narrowly ellipsoid, length 1.5 cm; Regions 3 and 4, Missouri, Illinois, and Indiana south to the Gulf Coast; disjunct populations from south Carolina to Texas; bottomland swamps and stream terraces. Cover type 27.2, 27.3, 27.28, 28.8, 28.15, 28.17, 28.21, 28.31, 28.50, 28.53, 28.55.

Foresteria ligustrina (Michaux) Poir, upland privet (Figure 23.10); shrub, height to 3 m; deciduous; leaf

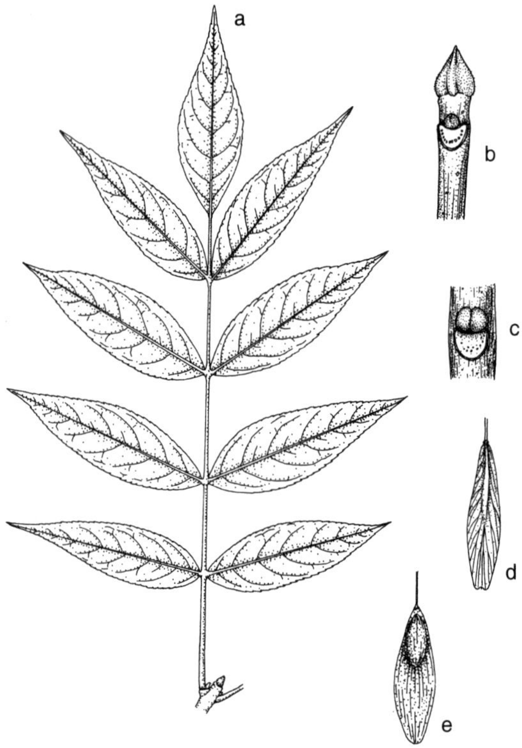

FIGURE 23.5 *Fraxinus nigra* (a) pinnately compound leaf with lanceolate leaflets ×1/3, (b) twig with terminal and lateral buds and leaf scar ×1, (c) lateral bud with shield-shaped leaf scar ×1.5, (d) (e) samaras ×1/2.

arrangement opposite; leaf simple, length 1 to 4 cm, shape elliptical to oblong, serrulated; fruit a drupe, oval, length ~0.5 cm, dark purple, sessile; Regions 3 and 4, Kentucky and Tennessee south to Alabama and Florida; mesic and drier sites, from stream banks to rocky upland slopes and barrens. Cover type 27.11.

Osmanthus americanus (Linnaeus) Gray, devilwood, wild olive (Figure 23.11); shrub to medium-sized tree, height to 15 m; evergreen; arrangement opposite; leaf simple, length 10 to 15 cm, width 2 to 6 cm, shape elliptical to oblanceolate, coriaceous, glabrous, margin smooth; twigs slender, buds and twigs reddish brown to gray; fruit a drupe, oval, olivelike, diameter 1.0 to 1.5 cm, dark blue; Region 4, following the coast from North Carolina to Florida and Louisiana; mesic woods, swamps, and hammocks. Cover type 28.44.

Various Oleaceae species have been introduced. Four or five *Ligustrum* species have escaped cultivation to become naturalized from New England southward. These include *L. vulgare* Linnaeus (privet, Europe, open woods), *L. ovaliforium* Hasskarl (oval-leaf privet, Japan, roadsides), *L. amurense* Carriere (amur privet, Asia, open woods), and *L. obtusifolium* Siebold & Succarini (obtuse-leaf privet, Japan, roadsides and woods).

Ligustrum vulgare Linnaeus, European or common privet (Figure 23.11); shrub, height to 4 m; deciduous; leaf arrangement simple; leaf simple, length 2 to 5 cm, margins smooth, secondary veins obscure; panicles terminal, fruit a one- to three-seeded berry, black.

Syringa vulgare Linnaeus, common lilac (Figure 23.12); multistemmed shrub, height to 5 m;

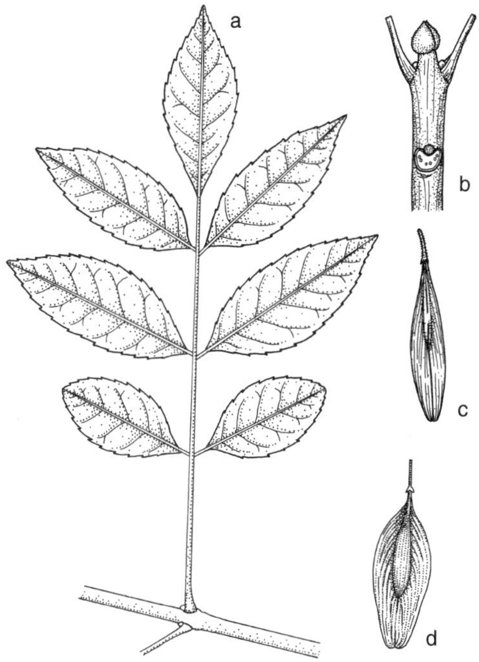

FIGURE 23.6 *Fraxinus caroliniana* (a) pinnately compound leaf with elliptical leaflets ×1/3, (b) twig with terminal and lateral buds and leaf scar ×1, (c), (d) samaras ×1/2.

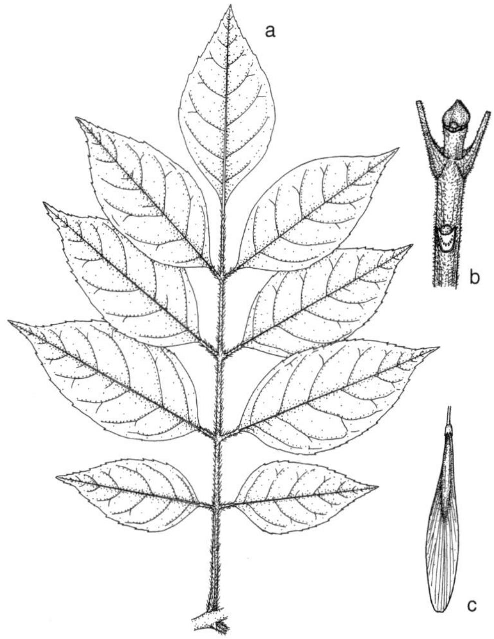

FIGURE 23.7 *Fraxinus tomentosa* (a) pinnately compound leaf with elliptical leaflets ×1/3, (b) twig with terminal and lateral buds and shield-shaped leaf scar ×1, (c) samara ×1/2.

deciduous; leaf arrangement opposite; leaf length 5 to 12 cm, shape broadly lanceolate to ovate, apex acuminate, base truncate to subcordate, margin smooth; flower corolla tubular, four-lobed, purple to white. Many different varieties have been developed for their floral characteristics and are widely planted as ornamentals.

Several species of *Forsythia* are ornamental shrubs well known for their flowers of narrow yellow petals that appear before the leaves in early spring. The leaves tend to remain green into the late fall.

Forsythia viridissima Lindley, greenstem forsythia (Figure 23.12); height 2 to 3 m; deciduous; leaf arrangement opposite; leaf simple, length 4 to 9 cm, shape ovate, elliptical, or lanceolate; twigs green, pith chambered at nodes and through internodes in young twigs.

Forsythia suspensa (Thunberg) Vahl, weeping forsythia (Figure 23.12); multistemmed shrub; height 2 to 3 m; deciduous; leaf arrangement opposite; leaf simple or blade pinnately cleft to produce two basal lobes, length 5 to 10 cm; simple leaf shape ovate, elliptical, or lanceolate; twigs yellowish brown, pith hollow in the internodes and solid at the nodes.

Forsythia × *intermedia* Zab., border forsythia; multistemmed shrub; height 2 to 3 m; deciduous; leaf arrangement opposite; leaf simple, 7 to 13 cm, shape ovate, elliptical, or lanceolate; twig yellow brown to light green, pith chambered through the internodes and solid at the nodes.

Bignoniaceae (Bignonia or Catalpa Family)

The Bignoniaceae family of trees, shrubs, and, occasionally, vines is relatively large, with 110 to 120 genera and at least 650, perhaps as many as 800, species distributed pantropically. However, only four genera and

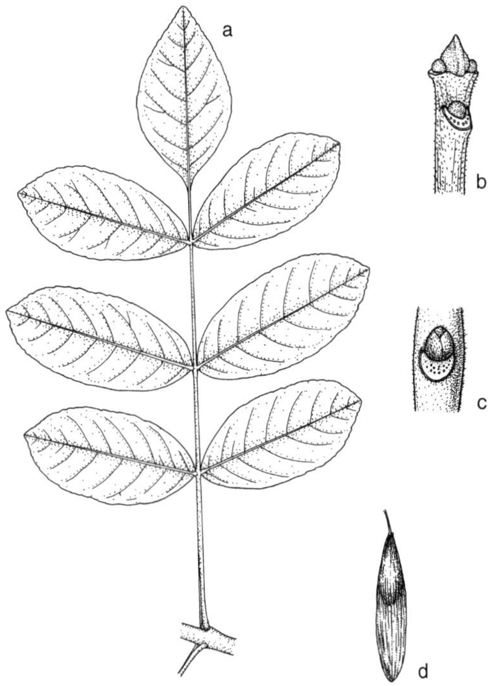

FIGURE 23.8 *Fraxinus latifolia* (a) pinnately compound leaf with oblong leaflets ×1/3, (b) twig with terminal and lateral buds and leaf scar ×1, (c) lateral bud with shield-shaped leaf scar ×1.5, (d) samara ×1/2.

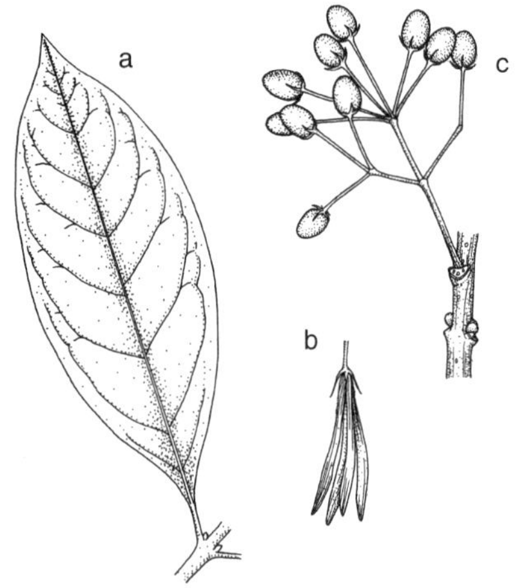

FIGURE 23.9 *Chionanthus virginicus* (a) elliptical-oblong leaf, opposite leaf arrangement ×1/2, (b) flower with 2 to 3 cm long (yellow) petals ×1. (c) inflorescence with drupes ×1.

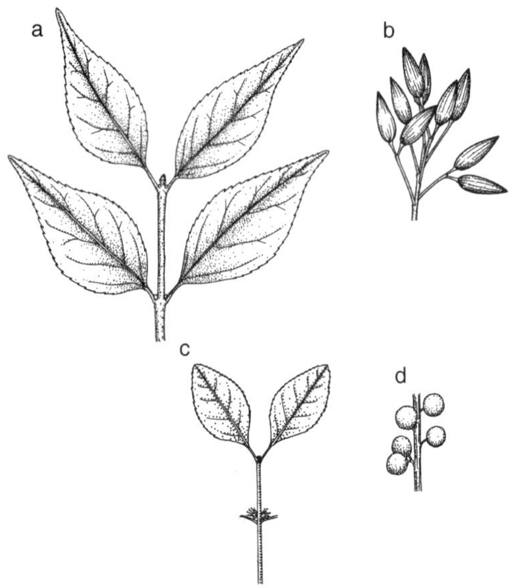

FIGURE 23.10 *Foresteria acuminata* (a) twig bearing opposite leaves ×1/3, (b) inflorescence with ellipsoid drupes ×1/2; *F. ligustrina* (c) leaf ×1/2, (d) globose drupes.

only a few more species are found in the temperate zone. *Catalpa*, *Bignonia*, and *Campsis* are common to eastern United States. The flower has a brightly colored floral tube of up to nearly 10 cm in length in some species (*Campsis*). The fruit is a capsule of varying size and shape.

Catalpa speciosa Warder, northern catalpa (Figures 23.13, 16.2a); large tree, height to 30 m; deciduous; leaf arrangement whorled and opposite; leaf simple, length 15 to 30 cm, width to 20 cm, shape cordate, apex long acuminate, margin smooth, lower surface softly pubescent, petiole long but slightly shorter than blade; twig brittle, pith large, white; flower zygomorphic, sepals fused into a tube, two-lipped; petals fused into a floral tube, length 4 to 5 cm, five-lobed, lobes spreading, white with yellowish lines and purple spots, lower lobe emarginate; fruit a terete capsule (pod), length to 20 to 35 cm, 2 locular, septum

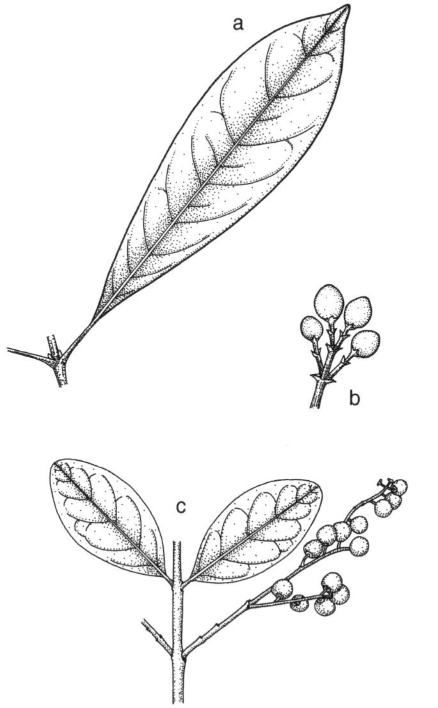

FIGURE 23.11 *Osmanthus americanus* (a) oblanceolate leaf × 1/2, (b) inflorescence with drupes × 1/2; (c) *Ligustrum vulgare* twig with oval leaves and inflorescence with drupes × 2/3.

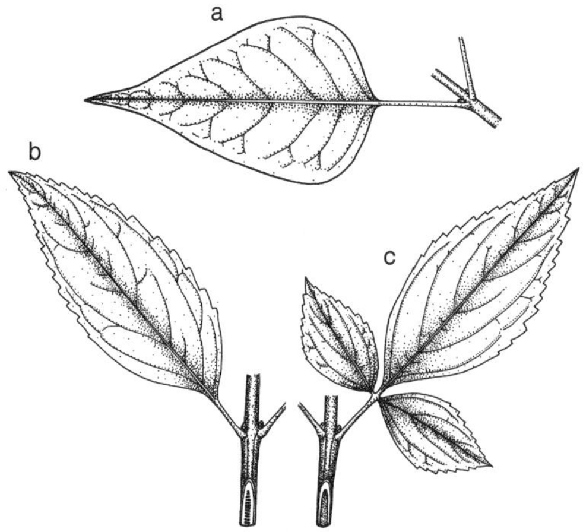

FIGURE 23.12 (a) *Syringa vulgare* lanceolate leaf × 1/2; (b) *Forsythia viridissima* leaf and twig with chambered pith × 1/2; (c) F. *suspensa* leaf and twig with hollow pith × 1/2.

central; seeds bearded at each end; primarily Region 3, range follows the Mississippi River and Ohio River floodplains from Arkansas into Missouri, southern Illinois, southern Indiana, Kentucky, and Tennessee, naturalized eastward to Ohio and West Virginia, widely planted outside its range from Wisconsin to New York and the southeastern United States.

Catalpa bignonioides Walter, southern catalpa (Figure 23.14); medium-sized tree, height to 15 m; deciduous; leaf arrangement whorled and opposite; leaf simple, length to 26 cm, shape cordate, apex short acuminate, margin smooth, blade malodorous when crushed, petiole slightly shorter than blade; twig brittle, pith large, white; flower zygomorphic; sepals fused into a tube, petals fused into a floral tube, length 2.5 to 4.0 cm, five-lobed, lobes spreading, white with yellowish lines and some purple spots, lower lobe entire; fruit a flattened capsule (pod), length to about 30 cm, 2 locular, septum central; seeds bearded at each end; Region 4, central Mississippi, central and southern Alabama, southwestern Georgia, and the Florida panhandle; floodplains and bottomlands.

Bignonia capreolata Linnaeus, cross-vine (Figure 23.14); slender vine climbing to 10 m; evergreen; leaf arrangement opposite; leaf compound, leaflets two, or one leaflet and a slender branched tendril with a terminal adhesive disk, leaflet shape elliptical to oblong, dark green, length 5 to 18 cm, width 2 to 6 cm, petioluled length 1 to 2 cm, apex acute, base cordate to slightly auriculate; pith cross section cross-shaped; flower calyx fused into a tube, bright reddish orange; fruit a capsule, length up to 20 cm, width to 2.5 cm, 2 locular, septum central; Regions 3a and 4, moist woods, stream terraces and swamps of central and southern United States. Cover type 27.10, 28.22, 28.31, 28.49. The vascular bundles in the stem form a cross, hence the name cross-vine (Gleason and Cronguist (1991).

Campsis radicans (Linnaeus) Seem, trumpet-creeper (Figure 23.15); climbing vine, potential diameter to 8 cm, may have aerial rootlets; deciduous; leaf arrangement opposite; leaf pinnately compound, length

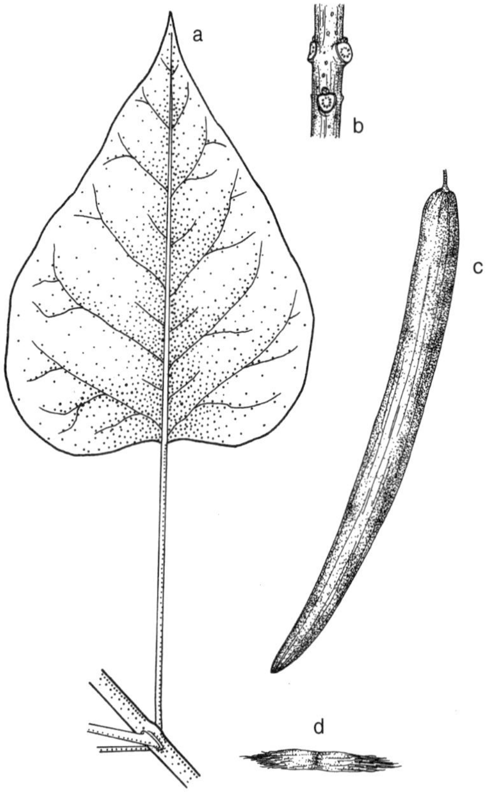

FIGURE 23.13 *Catalpa speciosa* (a) twig with cordate leaf, note long acuminate apex ×1/3, (b) twig with whorled leaf scars ×1, (c) elongated capsule ×1/4, (d) bearded seed ×1/2.

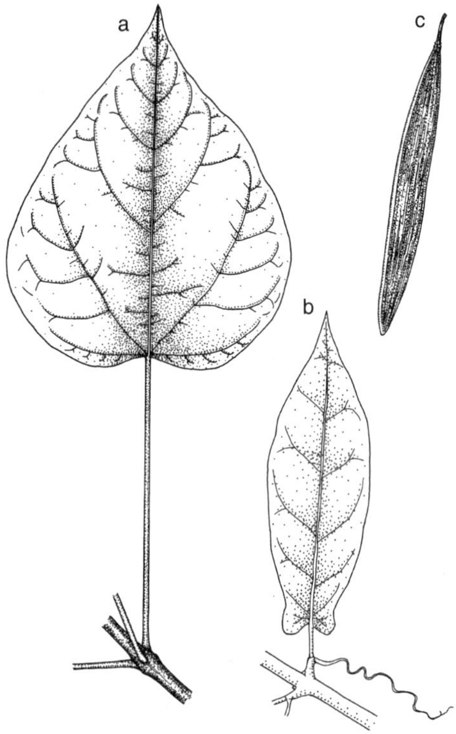

FIGURE 23.14 (a) *Catalpa bignonioides* cordate leaf ×1/3; *Bignonia capreolata* (b) compound leaf with one leaflet and a tendril ×1/2, (c) capsule ×1/3.

to 35 cm; leaflets 7 to 15, length up to 8 cm, width 4 cm, broadly ovate, margin coarsely toothed; flower petals calyx, trumpet-like, bright reddish orange, length to 10 cm, fruit a capsule, length 10 to 20 cm; Regions 3 and 4, Illinois to West Virginia and south to the Texas and Florida; upland and bottomland forest, but this species thrives and may form dense thickets along fencerows where there is a high light environment. Cover type 27.10, 28.10, 28.12, 28.15, 28.21, 28.50.

Temperate members of this family have limited economic and ecological importance except as ornamentals. Because of their showy flowers and large leaves that provide shade and fast growth, both *Catalpa* species are widely planted as ornamentals and may now be naturalized from Wisconsin and New York southward (Regions 2, 3, and 4). However, the branches are extremely brittle and trees may be damaged by wind. *Campsis radicans* is a good ornamental or garden species. The tropical species *Amiphitecna latifolia* (Miller) Gentry (calabash tree) has been reported as native to southern Florida, where it is widely cultivated. The fruit of this tree is large (diameter 10 to 15 cm) and has a hard shell that can be used as a drinking cup. *Chilopsis linearis* (Cavanilles) Sweet (desert-willow; not a true willow) is a nonforest species found in desert washes from west Texas to southern California.

Scrophulariaceae (Figwort Family)

Paulownia tomentosa is an Asian species that is rapidly becoming naturalized. In can be found along roads and

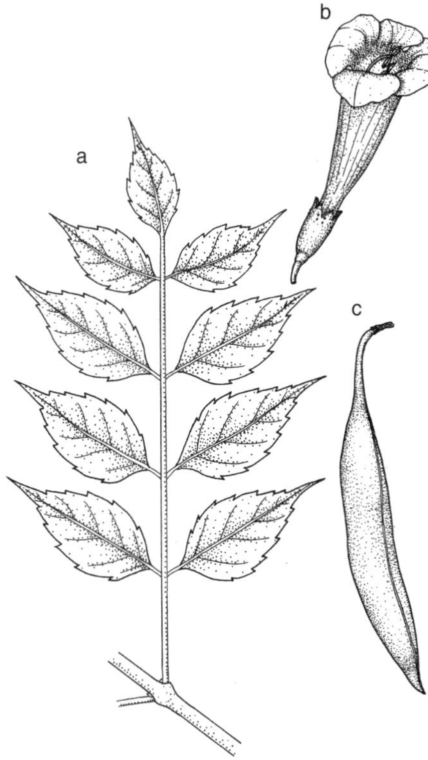

FIGURE 23.15 *Campsis radicans* (a) compound leaf ×1/3, (b) flower with a tubular corolla ×1/2, (c) capsule ×1/3.

fencerows in the Central States (Region 3). *Paulownia* has been shifted between the Bignoniaceae and Scrophulariaceae (figwort) families and presently is placed in the latter. It is considered here because it closely resembles, and in the absence of fruit, may be confused with *Catalpa*. In eastern Asia the wood is used to make quality chairs, tables, wardrobes, and tea chests, the latter used to protect dishes and place settings periodically used in honoring the anniversaries of ancestors. The overseas demand for the cabinet wood of *Paulownia tomentosa* has increased its importance in the past few years; this species is extremely fast growing, possibly up to 4 to 5 m per year.

Paulownia tomentosa (Thunberg) Siebold & Steud, princess-tree, royal paulownia (Figure 23.16); medium-sized tree, height to about 15 m; deciduous; leaf arrangement opposite; leaf simple, total length to 40 cm, blade width to 30, width nearly equal to length, margin smooth, shape cordate to reniform, base cordate, apex obtuse, pubescent; twigs and branches brittle, pith chambered or hollow; flowers showy, calyx tube five-cleft, floral tube with five unequal lobes, violet with yellow stripes, spreading; fruit a capsule, oval, length 3 to 4 cm, 2 locular, septum central; planted, escaped from cultivation.

ORDER: RUBIALES

Rubiaceae (Madder or Coffee Family)

The Rubiaceae is a large family (about 450 to 600 genera, 6500 to 9000 species), almost exclusively confined to tropical climates. Most are trees and shrubs; however, most members in the temperate zone are herbaceous. In North America, several genera and several species are associated with natural ecosystems. Probably the most well known member of this family is *Cephalanthus*.

Cephalanthus occidentalis Linnaeus, buttonbush (Figure 23.17); shrub, height to 3 m; deciduous; leaf arrangement opposite, occasionally in whorls of 3 or 4; leaf simple, length 3 to 15 cm, elliptical to ovate, entire, margin smooth, apex acuminate, stipules often well developed; flower petals fused into a tube, actinomorphic; flowers densely packed into a globose head (a button, hence the common name), diameter 3 cm, peduncle length 2 to 5 cm; fruit a nutlet; Regions 3 and 4, most of eastern North America east of a line from Texas to southern Ontario, also in California; Region 2b but absent in the mountains; swamps and stream and lake margins, may be an aggressive pioneer in shallow water (<30 cm) conditions where silt accumulates. Cover type 27.2, 27.28, 28.28, 28.53, 28.54, 28.55.

The most common species of this family are the woodland plants known as "bedstraws," or cleavers, belonging to the genus *Galium* (~300 species). At least 30 species occur in eastern U.S. forests, with many representatives in the Central States (Region 3). Plants are usually trailing and can be identified by the herbaceous, distinctly square green stem and what appears to be four, six, or eight leaves at a node. Actually, the stipules are so well developed that they are nearly indistinguishable from leaves, and thus *Galium* appears to have whorls of leaves. Some of

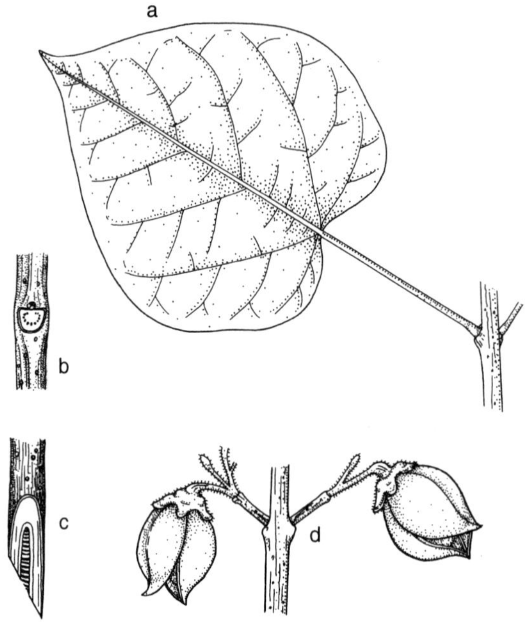

FIGURE 23.16 *Paulownia tomentosa* (a) cordate leaf ×1/4, (b) twig with leaf scar ×1, (c) twig showing chambered pith ×1, (d) capsules within an inflorescence ×1/2.

the common *Galium* species include *G. circaezans* Michaux (wild licorice; four ovate leaves, one primary vein, two shorter basal veins), *G. pilosum* Aiton (hairy bedstraw; four ovate leaves, one vein), *G. boreale* Linnaeus (northern bedstraw; four linear-lanceolate leaves, three primary veins. Cover type 29.20, 30.14), *G. concinnum* Torrey & Gray (six linear leaves, bristle-tipped, veinless), *G. triflorum* Michaux (sweet-scented bedstraw; six elliptical-lanceolate leaves, one primary vein), *G. aparine* Linnaeus (goosegrass; eight oblanceolate leaves, one primary vein, hispid).

Other taxa of the family include *Michella repens* Linnaeus (partridge-berry), a somewhat woody, prostrate plant most common in the northern forest. There is only one representative of *Mitchella* in the United States and Canada. One other species exists in Japan and northeast Asia.

Mitchella repens Linnaeus, partridge-berry (Figure 23.17); creeping vine, length to 30 cm; evergreen; leaf arrangement opposite; leaf simple, total length 2 cm; blade round, ovate, or slightly deltoid, dark green, shiny; petiole length ~5 mm; fruit a berry, red, diameter 5 to 8 mm, often two on a peduncle, stem rooting at the node; Regions 1, 2, 3, and 4, Nova Scotia to Ontario, south to Florida and Texas; upland forest, often pine woods, xeric-mesic sites.

In the tropics are found *Coffea* (coffee), the most important economic tropical family member, and *Cinchona*, which produces quinine. *Gardenia* is a common cultivated flower. *Hedyotis* (*Houstonia*), the common bluets of woodlands, are also members of the Rubiaceae.

ORDER: DIPSACALES

Caprifoliaceae (Honeysuckle Family)

Members of the Caprifoliaceae are primarily shrubs and vines having a northern distribution, particularly in eastern North America and eastern Asia; a few representatives are found in South America

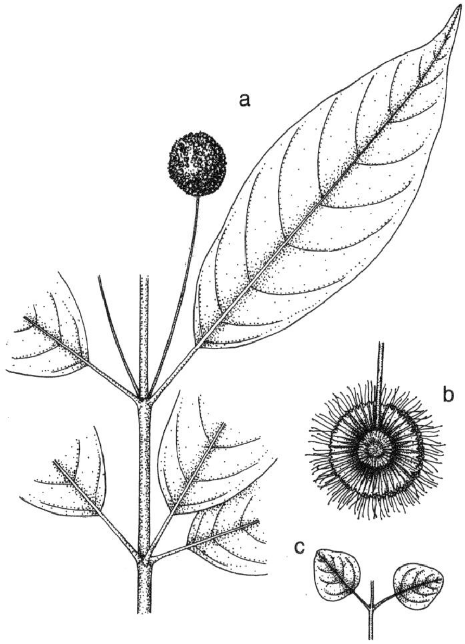

FIGURE 23.17 *Cephalanthus occidentalis* (a) twig with elliptical leaf and fruit head ×1/2, (b) cross section of flower head showing long stamens and pistils ×1/2; (c) *Mitchella repens* opposite leaves ×1/2.

and New Zealand. The family has been closely allied, and sometimes placed in the same order (Rubiales), with the Rubiaceae, which has similar flower characteristics but differences in stipules and endosperm; there has been considerable debate on the status of the family and the genera within the family. For example, Walters and Keil (1977) and Gleason and Cronquist (1991) placed them in different orders. Until recently, *Sambucus* (elderberry) and *Viburnum* (viburnum) were included in the Caprifoliaceae, but recent research indicates that because of differences in flower and vegetative characteristics, they belong in the Adoxaceae family (Zomlefer 1994; Judd et al. 1999).

Narrowly defined (excluding *Sambucus* and *Viburnum*), the Caprifoliaceae includes 12 genera and about 450 species. In North America, two species of *Diervilla* (bush-honeysuckle), several native and introduced *Lonicera* (honeysuckle), several *Symphoricarpos* (snowberry), and one *Linnaea* (twinflower) species are found in boreal and northern forest communities (Regions 1 and 2). Characteristics of the family include opposite simple leaves, a five-lobed or toothed calyx, and petals fused into a narrow five-lobed corolla tube (funnel form). The stamens are as many as the lobes (five) and attached to the corolla within the tube. The flowers are zygomorphic.

Diervilla lonicera Miller, bush-honeysuckle (Figure 23.18); woodland shrub, height to 50 cm; deciduous; leaf arrangement opposite; leaves simple borne on slightly four-angled branchlets, branchlets terminated by two to three small yellow flowers; leaves simple, length 5 to 14 cm, oblong-lanceolate, margin serrated and ciliated, petioled; Regions 1 and 2, Newfoundland and Nova Scotia to Saskatchewan, south to Iowa Illinois, Ohio, Tennessee and North Carolina; xeric and xeric-mesic pine and aspen communities and roadsides. The species name indicates its resemblance to honeysuckle. Cover type 26.2, 26.18, 26.19, 26.26, 27.8.

Diervilla sessifolia Buckley, sessile-leaf bush-honeysuckle; height to 2 m; similar to *D. lonicera*, except that the branchlets are more distinctly four-angled and the leaves are sessile (Fernald 1950); Region 3b, northern Georgia and Alabama to Virginia; moist rocky sites and bogs.

Lonicera is represented by at least 27 species in North America, about one third of which have

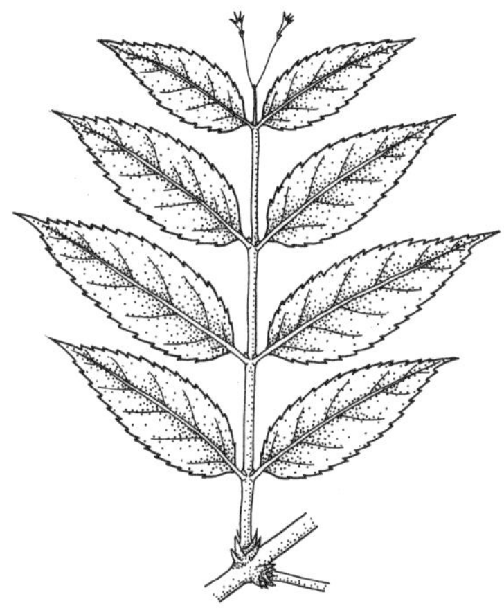

FIGURE 23.18 *Diervilla lonicera* leaves on a branchlet terminated by small (yellow) flowers ×1/3.

escaped from cultivation after being imported from Asia. About 20 species occur in the eastern United States and about an equal number in the west. The leaves are small (length 3 to 8 cm), smooth margined, ciliated, and usually ovate to somewhat lanceolate in shape, except for *L. sempervirens* (trumpet honeysuckle; native vine; Regions 3 and 4), which has connate (perfoliate) leaves at the stem apex. *Lonicera canadensis* is a common woodland shrub that has been used for habitat typing forest sites in Region 2. Descriptions of the eastern species can be found in Grimm (1957, 1993), Dirr (1977), Fernald (1950), and Gleason and Cronquist (1991).

Lonicera canadensis Marshall, American fly honeysuckle (Figure 23.19); straggling shrub, height to 2 m; deciduous leaf arrangement opposite; leaf simple, length 3 to 8 cm, ovate, margin smooth but ciliated; twigs extremely slender; fruit bright a berry, red, length ~1 cm, borne two to three on an axillary peduncle; pith solid; Regions 1 and 2, Nova Scotia west to Saskatchewan, New England and the northern Lake States; Region 3b, Appalachian Mountains. Cover type 26.19, 26.24, 16.29.

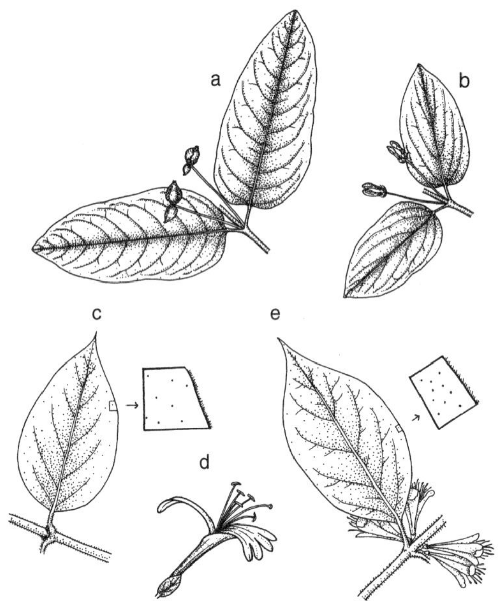

FIGURE 23.19 (a) *Lonicera canadensis* twig with ovate leaves and (red) berries ×1/2; (b) *Lonicera tartarica* twig with ovate leaves and flowers ×1/2; *L. japonica* (c) leaf ×1/2, (d) zygomorphic flower with tubular corolla ×1/2; (e) *L. maackii* twig with leaf and flowers ×1/2.

The introduced species (usually escaped from gardens) have created a serious ecological problem in eastern forest communities. Like other *Lonicera*, these species are shrubs or vines with oppositely arranged, simple, ciliated leaves. *Lonicera japonica* is an aggressive woody vine particularly in pine plantations and disturbed areas, but is becoming increasingly common in deciduous forest communities in the Central States (Region 3). Its rapid, dense growth over tree and shrub seedlings and saplings usually kills them and prevents establishment of new individuals. Its leaves remain green and supposedly functional over winter; therefore it may build a large food reserve that is used the following spring and summer. *Lonicera maackii*, a much branched Asian shrub, has appeared in the deciduous forest understory of Regions 3 and 4. During the fall months, this upright shrub (height 3 m) produces red berries that are widely distributed by birds, so the problem is expected to increase across the region. *Lonicera tartarica* has also escaped and may dominate the forest understory in the east. The introduced species described here have hollow pith whereas *L. canadensis* has a solid pith.

Lonicera tartarica Linnaeus, tartarian honeysuckle (Figure 23.19); shrub to 4 m; deciduous; leaf arrangement opposite, leaf simple, ovate-elliptical to broadly lanceolate, length 2 to 7 cm, margin ciliated, surfaces glabrous; twigs brown, glabrous, buds closely appressed; pith hollow, cylinder brown; noxious weed escaped from cultivation and becoming naturalized in the eastern United States.

Lonicera japonica Thunberg, Japanese honeysuckle (Figure 23.19); twining vine extending to 10 m, but branchlets may root; leaf arrangement opposite; semi-deciduous; climbing vine, semievergreen, leaves green into winter; leaf simple, length to 7 cm, elliptical to lanceolate, early spring leaves often lobed, margin ciliated; flowers fragrant, zygomorphic; fruit a berry, red; pith hollow, cylinder white; Regions 3 and 4; noxious weed escaped from cultivation, found in disturbed areas and plantations.

Lonicera maackii (Ruprecht) Maxim, Amur honeysuckle (Figure 23.19); multibranched shrub, height to 4 m; leaf simple, ovate-elliptical to broadly lanceolate, length 5 to 8 cm, margin ciliated, veins pubescent; fruit a berry, red; twigs gray, new growth pubescent; pith hollow, cylinder brown; noxious weed escaped from cultivation and becoming naturalized in the eastern United States.

Symphoricarpos is a small genus (about 15 species worldwide) of low shrubs, mostly distributed in the southwest or along the west coast. Elmore (1976) describes several southwestern species. The three to four eastern species are similar and separated mostly by flower characteristics. *Symphoricarpos orbiculatus* and *S. albus* are nearly allopatric species. *Symphoricarpos occidentalis* is primarily a species of the Great Plains and the prairie.

Symphoricarpos orbiculatus Moench, coralberry (Figure 23.20); low shrub, height to 1 m; deciduous; leaf arrangement opposite; leaf broadly ovate, nearly as wide as long, length 2 to 4 cm, margin slightly wavy, lower surface somewhat pubescent; twigs light brown; fruit a drupe, red, diameter ~0.3 cm; Regions 3 and 4, west of the Appalachians into Colorado; disturbed areas and open woods. This species responds to and is a good indicator of past human disturbance. Old clearings and, often, wooded homesites in the Central States are overgrown by this species.

Symphoricarpos albus (Linnaeus) Blake, snowberry; low shrub, height to 1.5 m; similar to *S. orbiculatus*; Regions 1, 2, 5, 6 and 7, British Columbia to Quebec, Massachusetts, and Virginia, west to Michigan, Wisconsin, Colorado and California; xeric wooded slopes; at least one recognized variety introduced from

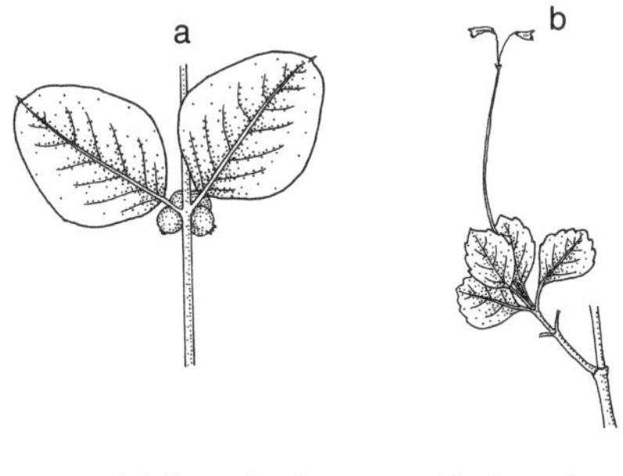

FIGURE 23.20 (a) *Symphoricarpos orbiculatus* leaves and fruit ×1/2, (b) *Linnaea borealis* clustered leaves and minute flowers ×1/2.

the Pacific slope (Fernald 1950). Cover type 29.23, 29.27, 29.32, 29.34, 30.4, 30.9, 30.15, 30.17, 30.21.

The genus *Linnaea* is monotypic; that is, it contains only one species that is found in North America. It is described here because it is common in many forest communities and has been used in habitat typing or ecological land classification (Kotar et al. 1988).

Linnaea borealis Linnaeus, twinflower (Figure 23.20); small trailing semiwoody plant, height to 25 cm; evergreen; leaf arrangement opposite; leaf simple, length ~1.5 cm, shape oval, margin irregularly serrated; the upright main stalk supports two to three sets of leaves and terminates with two flowers with nearly white petals; Regions 1 and 2, Canada, New England, and Lake States, Asia. Cover type 29.23, 29.27, 29.15, 30.2,30.3, 30.19, 30.22.

Adoxaceae (Elderberry Family)

The Adoxaceae, a small family of five genera, now includes *Viburnum* and *Sambucus*, two genera that have been considered closely related to each other but not to the other members of the Caprifoliaceae. Zomlefer (1994) and Judd et al. (1999) cites supporting evidence for the transfer of *Viburnum* and *Sambucus* to this family. It seems likely that this transfer will not solve the taxonomic problem of linking these genera to other closely related taxa, and in the final analysis, the solution may be to separate them into their own family (Sambucaceae) as a few taxonomists have suggested. Including these two genera, the Adoxaceae contains more than 245 species. Their distribution is primarily in the Northern Hemisphere.

The species of this family differ from those of the Caprifoliaceae by having small, sympetalous, actinomorphic flowers; small, simple or large, compound leaves, stipules, and a fleshy drupe for fruit. Zomlefer (1994) lists additional flower differences. *Sambucus* and *Viburnum* species are common forest shrubs, the former with 25 to 30 species worldwide and the latter with 150 species.

In North America, *Sambucus* is represented by a genus of eight species, two of which are found in eastern North America and six species in the west and southwest. Varieties of several species are recognized, and based on the various names and varieties, the status of some western species appears to be uncertain.

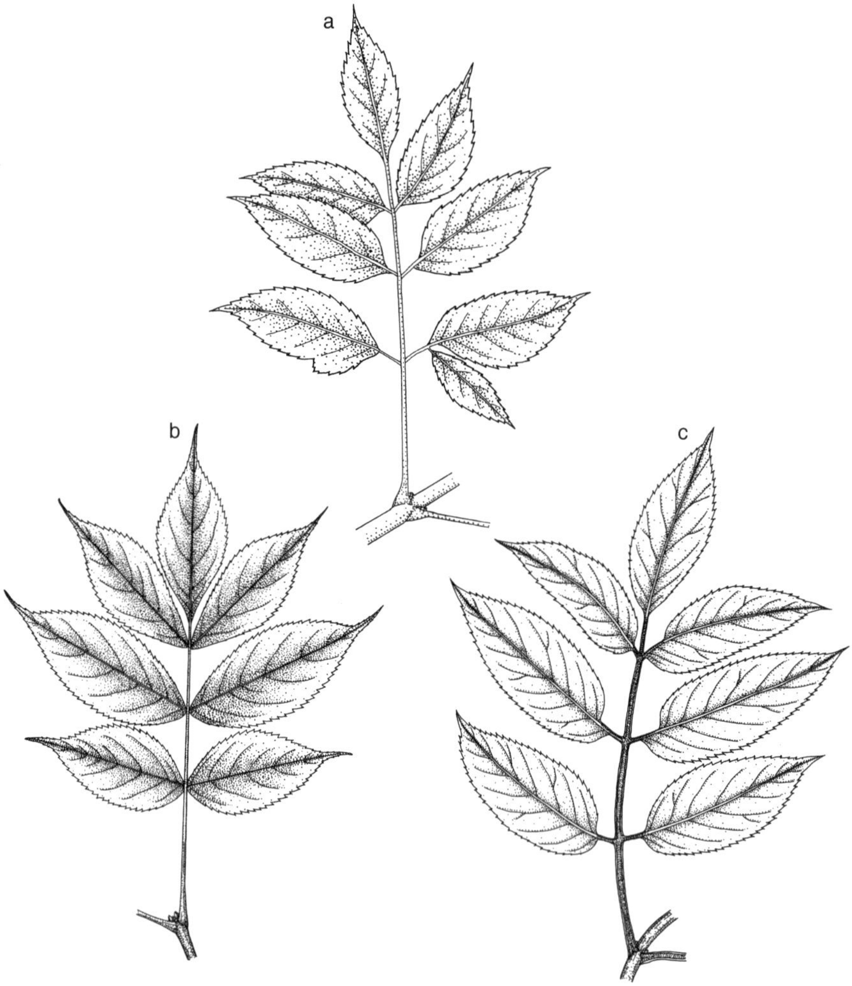

FIGURE 23.21 (a) *Sambucus canadensis* pinnately compound leaf × 1/3; (b) S. *pubens* leaf × 1/3; (c) S. *cerulea* leaf × 1/3.

These multistemmed shrubs or small trees have a large spongy pith that results in weak, easily broken stems, and the stems often die back to ground level. The gray stem has distinct, raised, dark lenticels. The leaf is pinnately compound, with lanceolate to ovate leaflets; the leaflet margin is serrated. The petioles are shaped like celery, with a groove on the upper surface. There is no terminal bud. The inflorescence is a compound cyme. The fruit is a small drupe that is an important wildlife food, and in a few species, also makes excellent elderberry preserves and wine.

Sambucus canadensis Linnaeus, American elderberry (Figure 23.21); shrub, height to 3 m; deciduous; leaf arrangement opposite; leaf pinnately compound, leaflet number 5 to 11, leaflet shape elliptical to lanceolate, length 5 to 12 cm, serrated, lower leaflets often cleft or divided; drupe color purple to nearly black, white bloom, diameter 3 to 6 mm; pith cylinder large, spongy, white; bud shape elliptical, several pair of scales visible; Regions 1, 2, 3 and 4, eastern North America from Manitoba east to Nova Scotia and south to Texas and Florida, also Colorado, Montana, Saskatchewan; moist woods. Two varieties have been distinguished (Fernald 1950); fruit edible.

Sambucus pubens Michaux [synonyms *S. racemosa* Linnaeus or *S. racemosa* var. *pubens* (Michaux) Koehne], red elderberry (Figure 23.21); shrub, height to 4 m; deciduous; leaf arrangement opposite; leaf pinnately compound, leaflet number five to seven, leaflet shape elliptical to lanceolate, length 5 to 12 cm, serrated; pith cylinder large, spongy, salmon-colored; drupe red, yellow, or whitish, diameter 4 to 6 mm; Regions 1, 2, and 3b, Alaska to British Columbia and Newfoundland, south to Colorado and North Carolina; rocky sites, moist woods. Considered not edible. Cover type 26.28.

Sambucus cerulea Rafinesque (synonyms *S. glauca* Nuttall; *S. mexicana* C. Presl.), blue elderberry (Figure 23.21); shrub or small tree, height to 10 m; leaf pinnately compound, leaflet number five to nine, leaflet shape elliptical to lanceolate, length 5 to 14 cm, serrated, malodorous when crushed; pith cylinder large, spongy, white, bud shape elliptical; drupe bluish black, glaucous, diameter 3 to 5 mm; Regions 5, 6 and 7, Idaho, Utah, southern British Columbia, along the coast to northern Mexico; dry stony soil and open woods. Largest of the *Sambucus* species. Cover type 30.5, 30.6. Fruit edible.

Sambucus racemosa var. *arborescens* (Torrey & Gray) Gray [synonym *Sambucus callicarpa* Greene], red elderberry; leaf arrangement opposite leaf pinnately compound; leaflet number five to seven, leaflet shape elliptical to lanceolate, length 6 to 15 cm, serrated; bud shape globose; pith cylinder large, spongy, white; drupe red, occasionally yellow to white, diameter 4 to 6 mm; Region 7, coastal area from central California to southwestern Alaska, elevation up to 600 m; moist sites. Not edible, poisonous. Cover type 26.15.

Sambucus racemosa var. *melanocarpa* (Gray) McMinn (synonym *Sambucus melanocarpa* A. Gray), blackberry elder; leaf arrangement opposite; leaf pinnately compound; leaflet number five to seven, leaflet shape elliptical to lanceolate, length 5 to 12 cm, serrated; pith cylinder large, spongy, white; drupe shiny black; Regions 5, 6 and 7, British Columbia, the Sierra Nevada, New Mexico; along streams. Cover type 30.6, 30.16. The cluster of drupes resembles shiny black beads. Not edible.

Sambucus racemosa var. *microbotrys* (Rydberg) Kearn & Peeb (synonym *Sambucus microbotrys* Rydberg), dwarf elderberry; small shrub, height to 1 m; leaf arrangement opposite; leaf pinnately compound, leaflet number five to seven, leaflet shape oval to elliptical, length 3–10 cm, serrated; drupe red; Regions 5a and 6a, Sierra Nevada and Rocky Mountain montane forest.

Sambucus velutina Durand & Hilgard, blueberry elder; shrub or small tree, height to 6 m; leaf arrangement opposite leaf pinnately compound, length 13 to 17 cm; leaflet number three to five, leaflet shape elliptical to lanceolate, length 3 to 7 cm, serrated, base asymmetrical, apex mucronate; leaves and twig pubescent; drupe shiny black, glaucous; Region 7, primarily the Sierra Nevada, 1000 to 2400 m, scattered in Nevada and Arizona.

Sambucus mexicana K. Presley, Mexican elderberry; leaf arrangement opposite; leaf pinnately compound, length 13 to 17 cm, leaflet number three to five, leaflet shape elliptical to lanceolate, length 3 to 7 cm, serrated, base asymmetrical, apex acuminate; leaves and twig glabrous; drupe shiny black glaucous; Regions 5 and 7 and surrounding areas, southern California, Arizona, Nevada, New Mexico, and Mexico; montane forests of mountain ranges located in the Mojave Desert.

Viburnum is a genus of 24 to 25 shrub (height 3 to 4 m) species native to North America, but an additional 6 to 7 species have been introduced as ornamentals. The leaves are simple, usually small (length 5 to 15 cm), and toothed or serrated to various degrees. The stems have a small pith. Leaf scars are more or less V-shaped and have three vascular bundle scars, much like those of *Acer*. Buds are thin and usually long (length 1.0 to 1.5 cm); scales may be valvate or naked. White flowers are in compound cymes. The 1 to 2 cm long drupe is oval (flattened) in cross section and may be blue, black, or red. The species can be grouped by leaf and bud characteristics.

Group I. Leaf palmately three-lobed, 3 primary veins; margin medium to coarsely serrate or toothed, buds covered by two visible scales.

Viburnum acerifolium Linnaeus, maple-leaf viburnum (Figure 23.22); shrub, height 1 to 2 m;

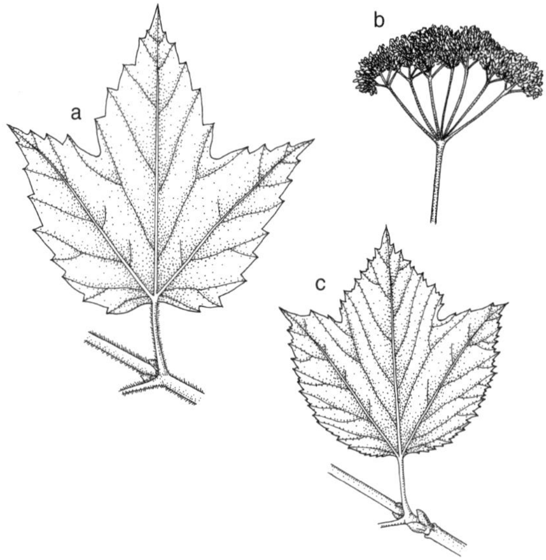

FIGURE 23.22 (a) *Viburnum acerifolium* (a) leaf × 1/2, (b) inflorescence × 1/2; (c) *V. edule* leaf × 1/2.

deciduous; leaf arrangement opposite, leaf palmately three-lobed, three-veined, length 6 to 12 cm, coarsely toothed (2 teeth per 2 cm margin length), lobe apex acuminate, lower surface lightly pubescent and dotted; petiole and twig finely pubescent; buds valvate; fruit a drupe, slightly flattened, purple black; Regions 2, 3, and 4; Minnesota to Quebec and New England, south to Georgia and Mississippi, xeric and mesic forest. Cover type 26.21, 26.24, 26.28, 27.26, 27.37, 28.2, 28.6, 28.12.

Viburnum edule (Michaux) Rafinesque, squashberry, mooseberry (Figure 23.22); shrub, height 1 to 2 m, deciduous; leaf arrangement opposite; leaf palmately three-lobed, three-veined, length 5 to 10 cm, margin medium serrated (six to seven teeth per 2 cm margin length), lobe apex acuminate; petiole and twig glabrous; buds valvate; fruit a drupe, slightly flattened, red or orange; Regions 1 (transcontinental), 2, 5b (Colorado, central Rocky Mountains north), 7 (Oregon to Alaska); xeric and moist forest.

Viburnum trilobum Marshall (Synonym *V. opulus* var. *americanum* Aiton), highbush cranberry (Figure 23.23); shrub or small tree, height to 4 m; deciduous; leaf arrangement opposite; leaf palmately three-lobed, three-veined, length 5 to 9 cm, margin coarsely serrated (one to three teeth per 2 cm margin length) or dentate but smooth at the base and in the sinuses, teeth usually blunt; petiole glabrous with one to several glands; twig glabrous; buds valvate; fruit a drupe, slightly flattened, red; Regions 1, 2, and 7; mesic woods.

Viburnum opulus Linnaeus (var. *opulus*), Guelder-rose (Figure 23.23); leaf palmately three-lobed, three-veined, margin coarsely serrated (three to four teeth per 2 cm margin length) or including at the base and in the sinuses, teeth sharp pointed; petiole pubescent, fruit red; introduced from Europe, escaped from cultivation in the eastern United States.

Group II. Leaf entire, one-veined, margin dentate; buds with two visible scales.

Viburnum dentatum Linnaeus, southern arrowwood (Figure 23.24); shrub, height to 3 m; deciduous; leaf arrangement opposite; leaf simple, shape broadly ovate, length 5 to 10 cm, width 4 to 8 cm, petiole and blade glabrous or often heavily pubescent, veins somewhat paired, number of pairs 5 to 11, margin dentate (two to six teeth per 2 cm margin length), 11 to 20 teeth per side; twig pubescent, buds valvate; fruit a drupe, slightly flattened, blue black; Regions 2b, 3, and 4, Texas to Illinois, Massachusetts, and Florida;

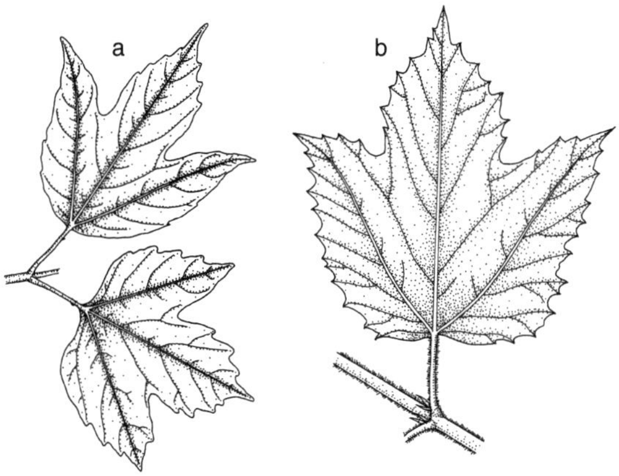

FIGURE 23.23 (a) *V. trilobum* leaves ×1/2; (b) *Viburnum opulus* leaf ×1/2.

xeric and mesic woods. A variable species with several recognized varieties (Cronquist 1981). Cover type 28.2, 28.18.

Viburnum recognitum Fernald, northern arrowwood (Figure 23.24); essentially similar to *V. dentatum*,

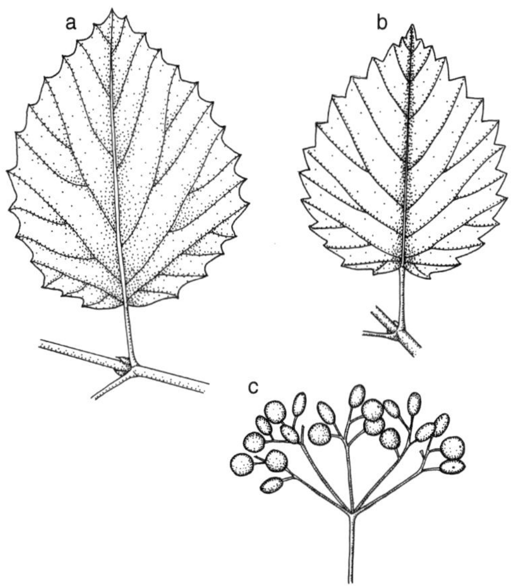

FIGURE 23.24 (a) *Viburnum recognitum* leaf ×1/2; (b) *V. dentatum* leaf ×1/2.

except that the petiole, leaf and twig are glabrous, lower veins may be pilose. Regions 2 and 3, New Brunswick to Michigan, south to South Carolina; mesic woods. Considered a separate species by Fernald (1950) but listed as a variety of *V. dentatum* (var. *lucidum* Aiton) by Gleason and Cronquist (1991).

Viburnum molle Michaux, Missouri or soft arrowwood (Figure 23.25); shrub or small tree, height to 4 m; deciduous; leaf arrangement opposite; leaf simple, length 7 to 13 cm, width 5 to 9 cm, shape broadly ovate

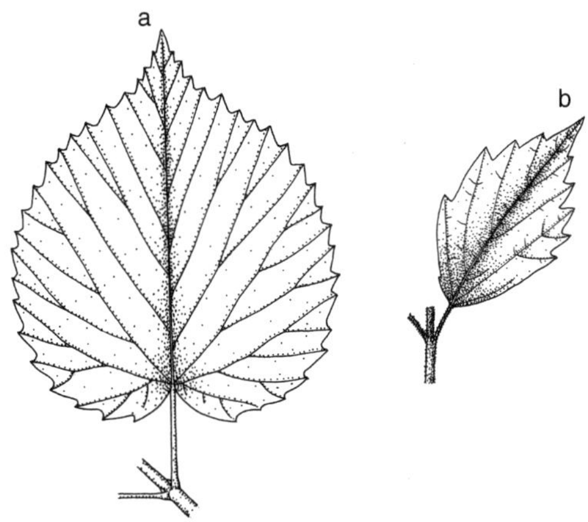

FIGURE 23.25 (a) *Viburnum molle* leaf ×1/2; (b) *V. rafinesquianum* leaf ×1/2.

to cordate, margin coarsely dentate (two to five teeth per 2 cm margin length), 20–30 teeth per side; buds valvate; fruit a drupe, slightly flattened, blue black; Region 3a, Missouri to Kentucky and Arkansas; xeric woods and bluffs.

Viburnum rafinesquianum Schultes [*V. affine* (Bush) House], downy arrowwood (Figure 23.25); shrub, height to 2 m; deciduous; leaf arrangement opposite; leaf ovate-lanceolate, length 5 to 10 cm, width 3 to 6 cm, coarsely dentate, four to ten teeth per side, abaxial surface pubescent; twig and petiole pubescent, buds valvate; fruit a drupe, slightly flattened, dark purple; Regions 2 and 3, Quebec and Manitoba to Missouri and Georgia. A variety, *V. rafinesquianum* var. *affine*, is sometimes elevated to species level. Cover type 27.7.

Group III. Leaf entire, one primary vein, margin finely serrated or crenulate; buds with two visible scales.

Viburnum lentago Linnaeus, nannyberry, wild raisin (Figure 23.26); shrub or small tree, height to 10 m; deciduous; leaf arrangement opposite; leaf simple, shape elliptical to lanceolate, length 5 to 9 cm, width 1 to 6 cm, margin finely toothed (teeth per 2 cm length), teeth sharp; twig dull, terminal bud scaly, buds valvate; fruit a drupe, flattened, blue black Regions 2, 3a (northern one half), 3b, and 5b, Montana; Colorado; music woods and terraces.

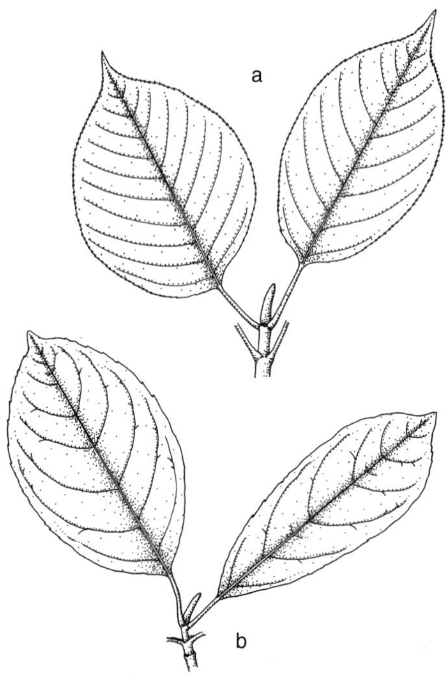

FIGURE 23.26 (a) *Viburnum lentago* leaves ×1/2; (b) *V. nudum* leaves ×1/2.

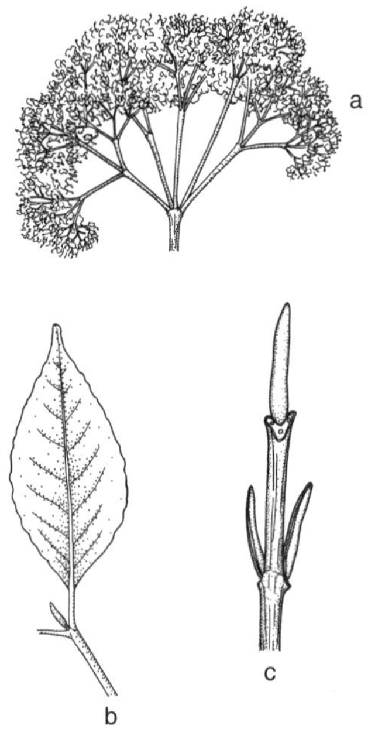

FIGURE 23.27 *Viburnum cassinoides* (a) inflorescence × 1/2, (b) lanceolate leaf ×1/2, (c) twig with long terminal and lateral buds ×1.

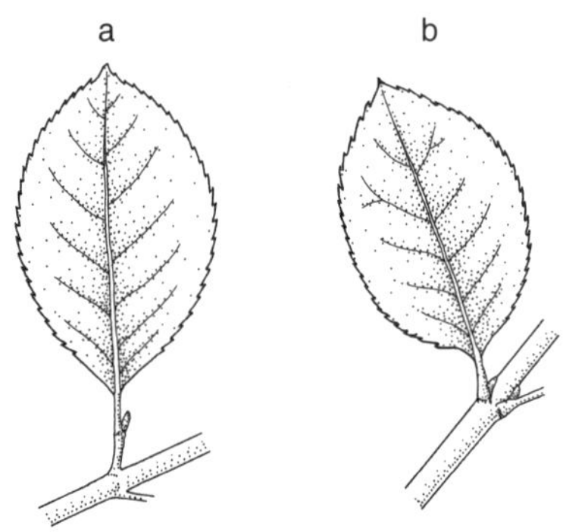

FIGURE 23.28 (a) *Viburnum prunifolium* leaf ×1/2, (b) *V. rufidulum* leaf ×1/2.

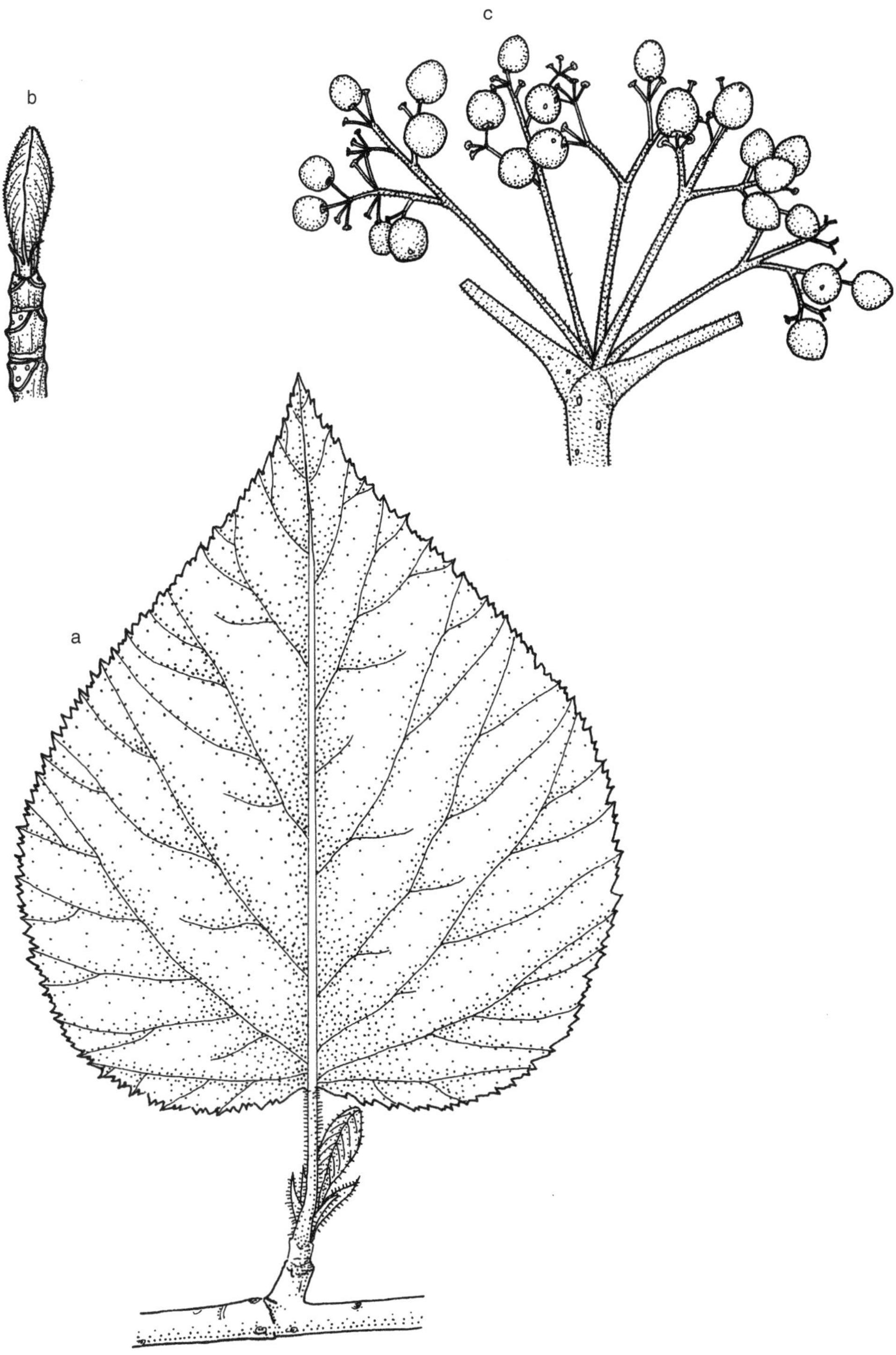

FIGURE 23.29 *Viburnum alnifolium* (a) leaf ×1, (b) naked terminal bud ×1.3, (c) inflorescence with drupes ×1.

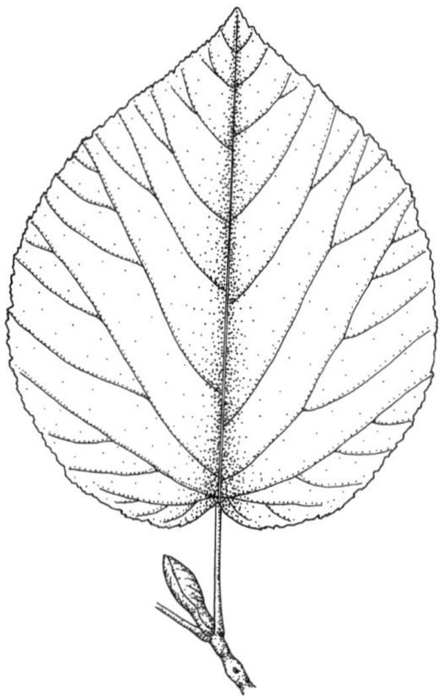

FIGURE 23.30 *Viburnum lantana* leaf and naked terminal bud ×3/4.

Viburnum nudum Linnaeus, possum-haw, swamp-haw, southern witherod (Figure 23.26); shrub or small tree, height to 5 m; deciduous; leaf arrangement opposite; leaf simple, shape oval or oblong, length 6 to 13 cm, width 3 to 7 cm, upper surface shiny, coriaceous, margin entire to crenate or slightly serrulate, apex mucronate but rounded; twig shiny, buds valvate, brownish to reddish; fruit blue black; Regions 3, southern one half, and 4, Arkansas to Connecticut south; swamps and wet pinelands, bogs.

Viburnum cassinoides Linnaeus, northern witherod or wild raisin (Figure 23.27); extremely similar to *V. nudum*, shrub or small tree height to 3.5 m; deciduous; leaf simple, narrowly elliptical to slightly lanceolate, length 5 to 12 cm, width 2 to 6 cm, margin crenulate, apex acuminate but rounded; twig dull, buds valvate with one pair of yellow or golden scurfy scales; Regions 2 and 3b; swamps, upland woods and clearings. Sometimes listed as *V. nudum* var. *cassinoides* (Linnaeus) Torrey & Gray (Gleason and Cronquist 1991). Cover type 26.5, 26.8.

Viburnum prunifolium Linnaeus (*V. bushii* Ashe), black-haw (Figure 23.28); shrub or small tree, height to 8 m; deciduous; leaf arrangement opposite; leaf simple, ovate to oval, length 3 to 9 cm, width 1 to 3 cm, margin finely serrate (six to eight teeth per 2 cm margin length), petiole narrow; buds valvate; fruit a drupe, slightly flattened, blue black; Regions 3a and 4; xeric-mesic upland woods Cover type 28.12.

Viburnum rufidulum Rafinesque, rusty black-haw (Figure 23.28); shrub or small tree, height to 9 m; deciduous; leaf arrangement opposite; leaf simple, length 5 to 8 cm, width 2.5 to 4.0 cm, shape elliptical, oval to obovate, dark green, upper surface shiny; serrulate (eight to ten teeth per 2 cm margin length), petiole broad and winged; buds valvate with two pairs of visible bud scales; fruit blue black; Regions 3 and 4, Kansas to Missouri and southern Ohio and south; xeric-mesic upland woods. Cover type 27.11, 28.12, 28.22, 18.49.

Group IV. Leaf one-veined, entire; buds naked, minature leaf visible.

Viburnum alnifolium Marshall (synonym *V. lantanoides* Michaux), hobblebush, witchhobble (Figure 23.29); shrub, height to 1 to 2 m; deciduous; leaf arrangement opposite; leaf broadly ovate, length 8 to 18 cm, width nearly equal to length, lower surface pubescent, margin serrated (8 to 15 teeth per 2 cm margin length), petiole and abaxial veins rusty stellate; twig rusty-hairy, buds naked; fruit a drupe, slightly flattened, red becoming nearly black; Regions 2 and 3b, Ontario to New Brunswick, south to Ohio and northern Georgia; upland woods. Cover type 26.10, 26.21, 26.23, 26.27, 26.29, 27.31, 27.37.

Viburnum lantana Linnaeus, wayfaring tree (Figure 23.30); shrub or small tree, height to 4 m; deciduous; leaf arrangement opposite; leaf oblong to ovate, length 5 to 12 cm, margin serrulated (8 to 15 teeth per 2 cm margin length), petiole pubescent, abaxial leaf surface pubescent or tomentose; twig gray-hairy, buds naked; fruit a drupe, slightly flattened, red becoming nearly black; Region 2. Introduced from Europe and Asia, occasionally escaping cultivation.

CHAPTER 24

Class: Liliopsida (Monocotyledoneae)

THE taxon referred to as "monocots" (Liliopsida; Cronquist 1981) contains 19 orders, 65 families, and about 50,000 species. A large proportion of species are grasses or herbs; only a few are woody or semiwoody and shrubby, or arborescent (bamboo, palm). The term "monocotyledon" refers to the single cotyledon present in the seed. The leaves of woody monocots may have parallel venation or one to several large veins extending from the base to the leaf tip. The basal sheath of the leaf may encircle the stem (e.g., *Arundinaria gigantea* (Walter) Chapman, giant cane), or the leaf may have a standard petiole or leaf stalk (e.g., *Smilax*, greenbrier). The flowers are generally three-merous (three sepals, three sepals, and six stamens).

SUBCLASS: ARECIDAE

ORDER: ARECALES

Arecaceae (Palm Family)

The 200 to 215 genera and about 3000 (estimates range from 2700 to 4000) species of this family are primarily pantropical. Plants of this family are woody shrubs, trees, and, occasionally, vines. The trees may be 20 m tall, but the rattan palm vine may be 60 m long. Palms (including palm trees) do not have a vascular cambium (lateral meristem), thus there is no secondary growth and the main stem or trunk does not increase in diameter after primary growth has ceased. The stems are unbranched, but part or all of the trunk may be covered with persistent leaf bases (sheaths). In the savannas of Belize, Central America, the leaf bases protect palmetto (*Paurotis wrightii*) stems from fire damage through a number of fires. With each fire, the stem becomes smaller in diameter as more of the sheath is removed. The result is that the tallest (oldest) trees have the smallest diameter (Anderson and Fralish 1975).

Palms are evergreen, with leaves clustered at the terminal end of the stem. The leaves are large (length up to 10 m in some species), including the long petiole, and have either a simple pinnately veined leaf, a simple fanlike (flabellate) leaf, or a pinnately compound leaf with many pinna (leaflets; the feather palms). The feather palm leaf develops differently from pinnately compound dicot leaves. The feather palm blade is simple before it expands, and as it expands, the blade tears along zones of weak cells, ripping the blade into leaflets (Walters and Keil 1977).

A few species extend into the subtropical forests of southeastern North America and into California and Arizona (southern edge of Region 5). *Serenoa repens* (Bartram) Small (saw palmetto; cover type 28.13, 28.34, 28.35, 28.38, 28.39, 28.44) is found in the Lower Coastal Plain of Florida and in southern South Carolina, Georgia, Alabama, and Louisiana (Region 4), where it forms a dense undergrowth that creates major competition problems for pine seedlings. *Serenoa repens* is commonly associated with *Pinus elliottii* (slash pine), *Pinus palustris* (longleaf pine) and wire grass (*Aristida stricta* Michaux.) in flatwoods communities. This palm species is recognized by its

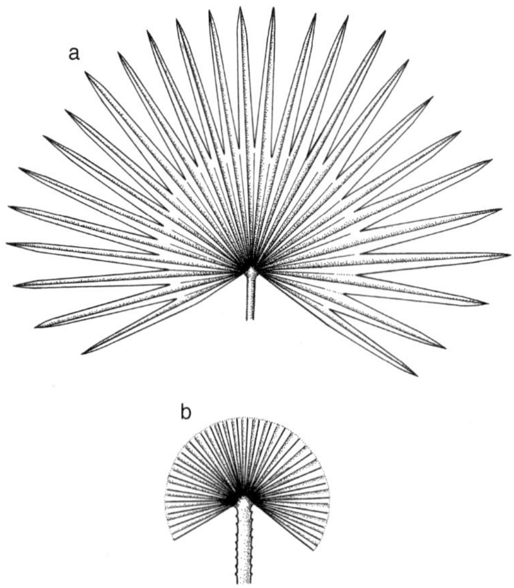

FIGURE 24.1 (a) *Sabal palmetto* leaf $\times 1/30$; (b) *Serenoa repens* leaf base and petiole.

numerous fan-shaped leaves (Figure 24.1) attached to a low, nearly buried, trunk (height <0.5 m). The 50 to 100 cm long leaf petioles are lined with sharp recurved spines similar to a saw blade. Research from the Archbold Research Station reported by Nelson (1996) indicates that the plants may grow to be older than 700 years. The yellow drupes are being tested for use in the treatment of prostate cancer. This potential use and other reported medical benefits have made this plant commercially important.

Sabal palmetto (Walter) Loddiges (sabal palm; cabbage palmetto; cover type **28.52**) is the only native palm tree found throughout nearly all of Florida (except in the panhandle), and is the state tree of both Florida and South Carolina. It has approximately the same distribution as *Serenoa repens.* Trees may grow to a height of 18 m but often remain in shrub form where fire is common. This species is distinguished from *S. repens* by the lack of recurved spines on the petiole (Figure 24.1) and is typically associated with pine woodlands.

Sabal minor (Jacquin) Pers (dwarf palmetto; cover type 28.18, 28.47) is more widely distributed than *S. repens*, extending north into North Carolina and west to eastern Texas and Arkansas. This species is a component of forest on bottomland and low slopes, its 60 to 70 cm long leaves appearing directly out of the ground where they are attached to a subterranean stem. The absence of spines on the petiole of *S. palmetto* separates small individuals of this species from the low-growing *S. repèns.*

SUBCLASS: COMMELINIDAE

ORDER: CYPERALES

Poaceae (Grass Family)

The Poaceae family, formerly named Gramineae, is the largest and most economically important food-plant family. Worldwide, it consists of 600 to 650 genera and about 7500 to 10,000 grass and, occasionally, woody species. Species occur in all climates, on all soil types, and in all plant community types. Grass is the major, but not the only, component of prairie and steppe communities, which also have a mixture of herbs and forbs. Forbs are semiwoody species such as *Solidago* spp. (goldenrod), *Aster* spp., *Ceanothus americanus* (New Jersey tea), and *Rhus copallina* (winged sumac). A savanna community is composed of scattered trees with the ground stratum primarily dominated by grasses similar in mixture to those of a prairie (Anderson et al. 2000). The grasses have a fibrous (finely divided) root system that permeates the soil, and as the roots continuously die and regrow, organic matter is incorporated into the upper layer of the soil, thus forming the characteristic dark, deep (~0.5 m) surface horizon (A_1). However, other than in savannas, so few grasses and other monocots appear in forest-dominated communities that they have little effect on soil formation.

It has been estimated that worldwide, grass-dominated communities are 20 percent of total vegetation cover (Woodland 1991). Economically important species include *Triticum* spp. (wheat), *Zea mays* (corn), *Oryza sativa* (rice), *Avena sativa* (oats), *Hordeum vulgare* (barley), *Seceale ceareale* (rye), *Saccharum officinarum* (sugar cane), *Sorghum bicolor* (sorghum), and *Pennisetum glaucum* (millet). Bamboo, also a member of the Poaceae, is important as a building material in Asia.

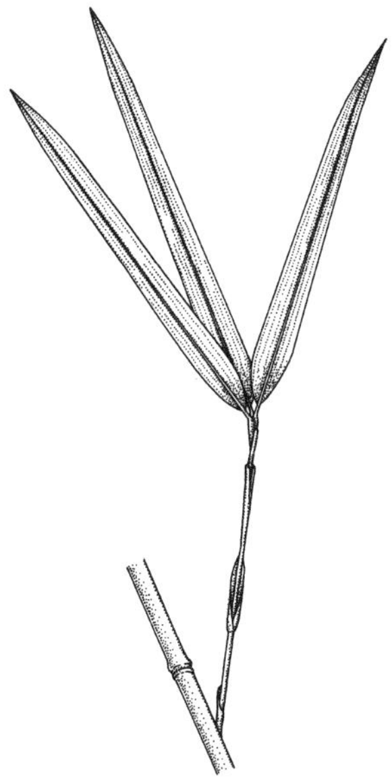

FIGURE 24.2 *Arundinaria gigantea* culm with grasslike leaves, hollow internodes, swollen nodes.

In eastern North America, *Arundinaria gigantea* (giant cane; subfamily Bambusoideae; cover type 27.14, 28.22, 28.36), a more or less woody member of the Poaceae, is found south of a line extending from southern Missouri to southern Ohio to Delaware (Regions 3 and 4). The arborescent bamboo-like culm (stem) of *A. gigantea* has grasslike leaves, hollow internodes, and swollen nodes (Figure 24.2). It is easily recognized because it resembles a small bamboo or an exceptionally large grass. Maximum height is about 8 m, but average height is closer to 2 to 3 m. Like most grasses, the leaf (length 5 to 15 cm; width 1.0 to 3.0 cm) is alternate, simple, and linear and has sheathing around the stem. The flowers are small, irregular, imperfect, and in spikelets arranged in a panicle composed of a few simple racemes; the spikelets are 2.5 to 6.0 cm long.

Arundinaria gigantea is a common component of bottomland forest, but it is occasionally found on low, moist upland slopes. Although this species is now generally scattered in small patches, descriptions from early explorers and settlers indicate that it once occurred in large, dense groves. The endangered Swainson's warbler (*Limnothlypis swainsonii*) and canebrake rattlesnake (*Crotalus horridus*) are associated with sites where cane is presently found.

SUBCLASS: LILIIDAE

ORDER: LILIALES

Smilacaceae (Greenbrier or Catbrier Family)

During the 1980s taxonomists split 10 to 12 genera from the Liliaceae into a new family called the Smilacaceae. Most of the 300 to 375 species are widely distributed in tropical and subtropical regions, particularly in the Southern Hemisphere. By far, the largest genus is *Smilax*, with about 325 species, of which about 12 woody and herbaceous species are found in the temperate forests of North America (primarily in Regions 3 and 4).

The flowers are regular, unisexual, and occur in umbels. The fruit is a berry. Unlike the Liliaceae, members of the Similacaceae are slender (diameter <10 mm), green, woody, perennial vines that often have numerous prickles and tendrils. The vines, which may be several meters long, arise from tuber-bearing rhizomes. The name "brier" refers to a stem with prickles or thorns, thus "greenbrier" refers to the armed green stem and "catbrier" refers to the cat-claw-like sharp prickles. These species have stems armed with sharp prickles to varying degrees.

This group of species is identified by their green, woody vines and simple, alternately arranged leaves that have smooth margins. The leaves have three to seven veins radiating (palmate venation) from the point of petiole attachment at the base of the blade, then curving and converging simultaneously at the blade apex. The primary veins are connected by a network of small veins. The leaves are set opposite the tendrils that encircle other stems or leaf petioles and permit the vine to climb. *Smilax* has little economic importance. These species often form dense thickets that make for extremely difficult and unpleasant walking in disturbed or cutover areas.

There are three additional *Smilax* species that have herbaceous stems that lack prickles. Not considered here are *S. pseudochina* Linnaeus and

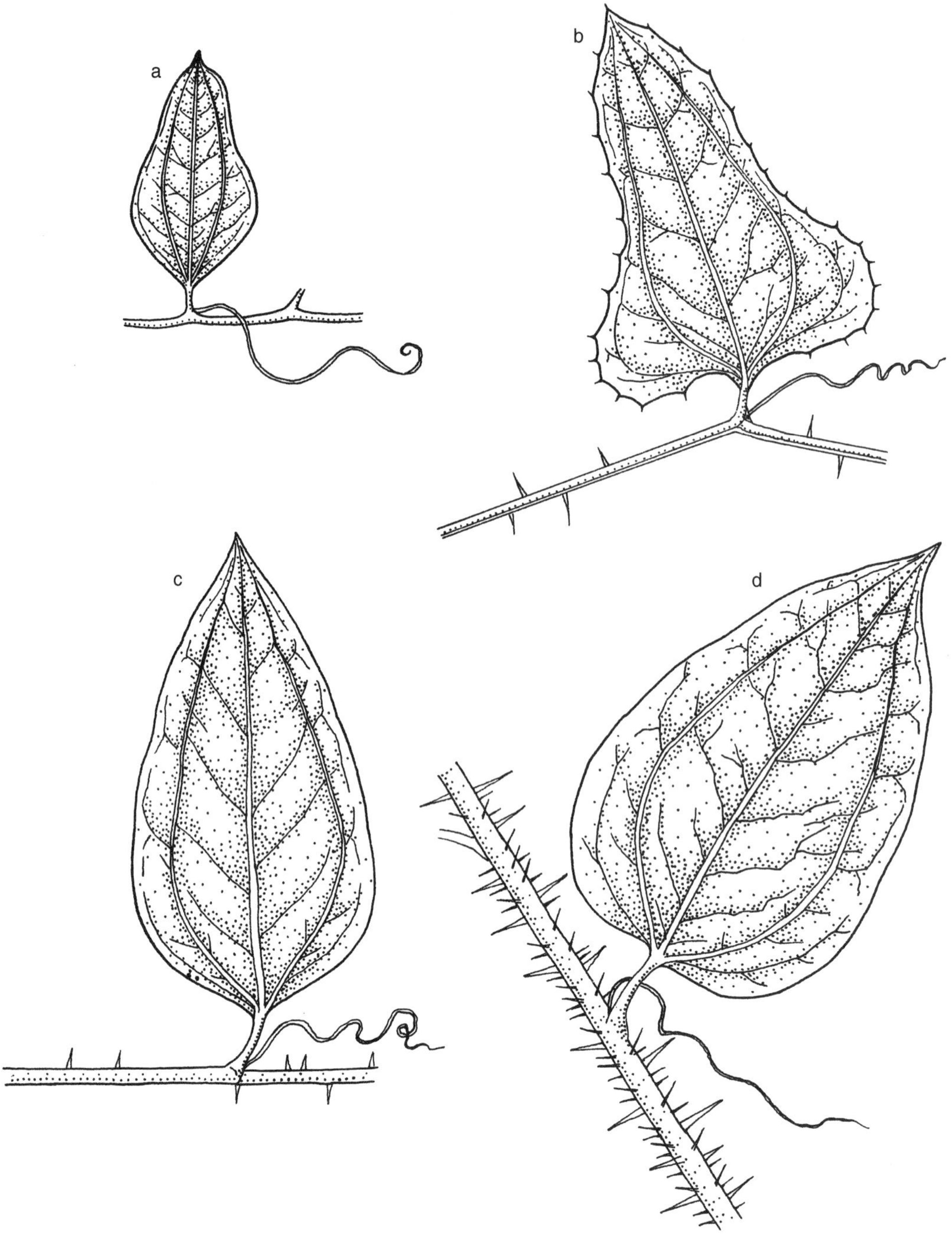

FIGURE 24.3 Twig, tendril, and leaf of (a) *Smilax auriculata* ×3/4; (b) *Smilax bona-nox* ×3/4; (c) *Smilax glauca* ×1; (d) *Smilax hispida* ×1.

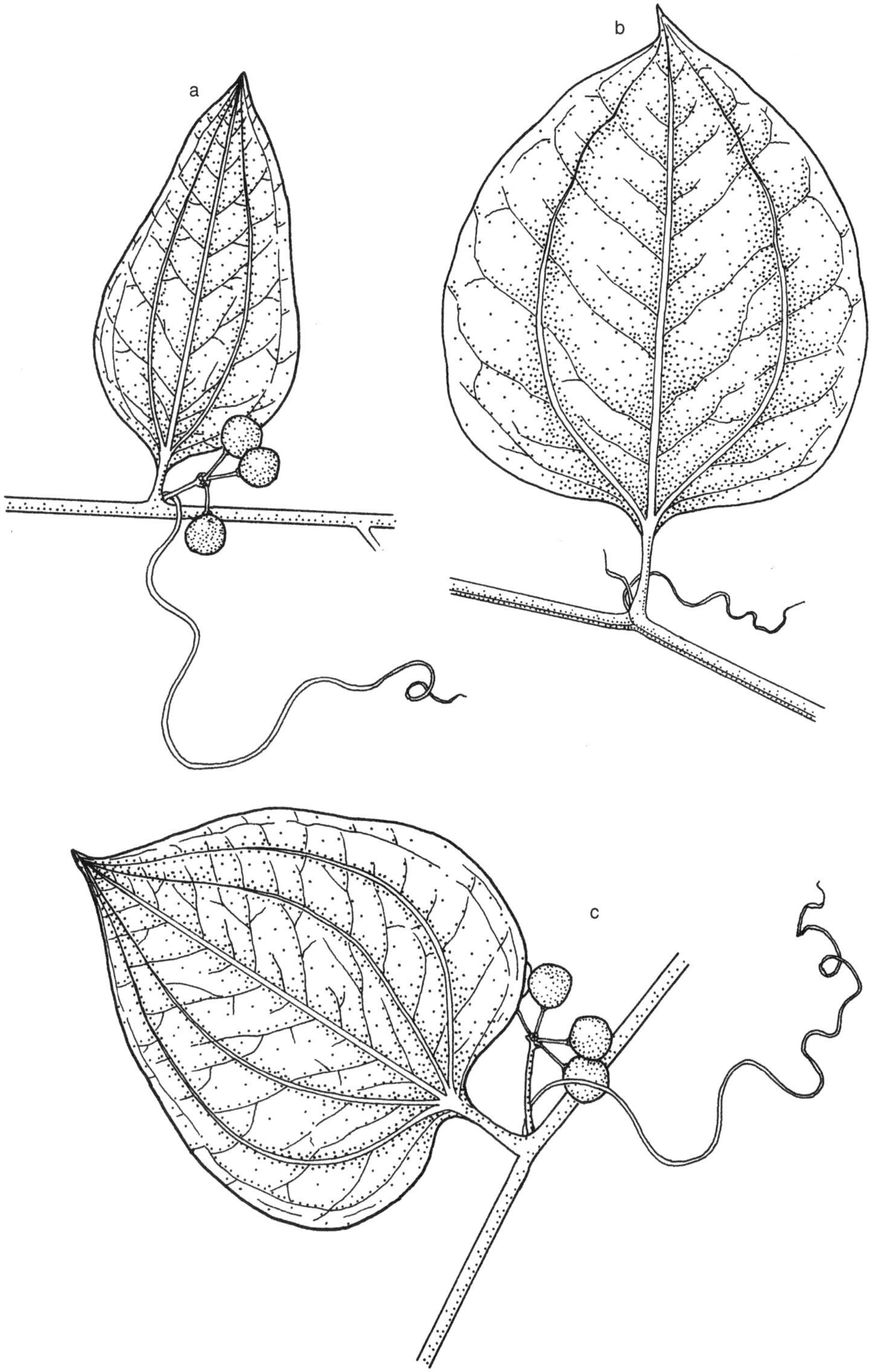

FIGURE 24.4 Twig, tendril, and leaf of (a) *Smilax laurifolia* ×1/2; (b) *Smilax rotundifolia* ×3/4; (c) *Smilax hispida* ×3/4.

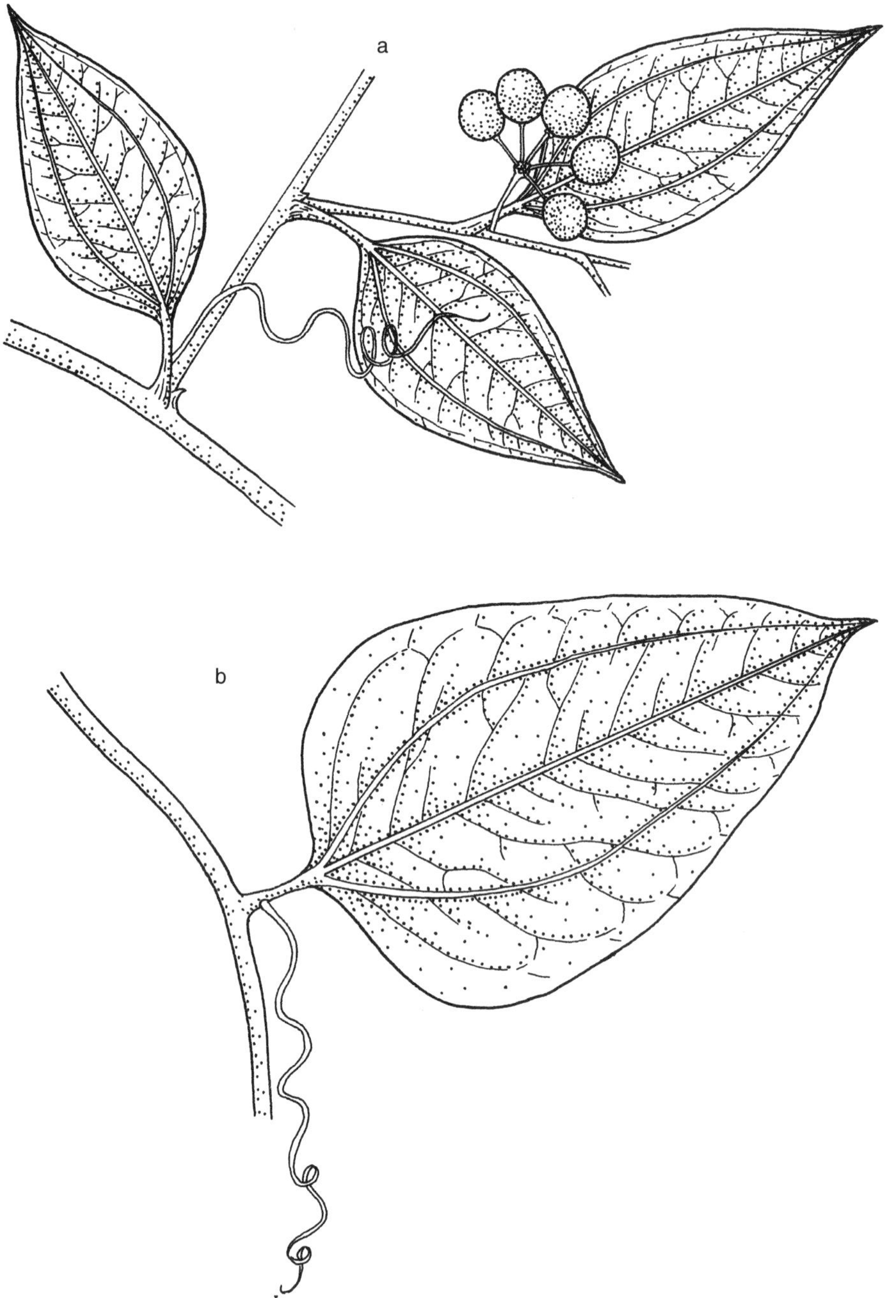

FIGURE 24.5 (a) *Smilax smallii* leaf, twig, and berries ×1; (b) *Smilax walteri* leaf and twig ×1.

S. pulverulenta (synonym: *S. herbacea*, carrion flower), which have tendrils, and *S. ecirrata* (an annual), which lacks tendrils.

The following are the most commonly encountered species in eastern forest, particularly in disturbed forest.

Smilax auriculata Walter, catbrier, wild bamboo (Figure 24.3); green woody vine with scattered prickles and tendrils; leaf arrangement alternate; evergreen; leaf simple, blade entire, coriaceous, lanceolate to panduriform, length 3 to 10 cm, width 1 to 4 cm; margin smooth, abaxial surface green, veins three to five, marginal veins forming an intermarginal groove; stem more or less four-angled; fruit a glaucous berry, purple at maturity; Region 4, Texas to Florida, New Jersey, and Tennessee; swamps and wet sites.

Smilax bona-nox Linnaeus, catbrier (Figure 24.3); green woody vine, prickles and tendrils frequent; semi-evergreen leaves, remaining green and on the stem into winter; leaf arrangement alternate; leaf simple, blade shape panduriform, length 6 to 13 cm, width 3 to 10 cm abaxial surface green, veins three to five, surface often appearing water-spotted, tiny prickles on margin and on lower surface of midrib; fruit a glabrous berry, black at maturity; Regions 3 and 4. Cover type 27.10, 28.22.

Smilax glauca Walter, common greenbrier; wild sarsaparilla (Figure 24.3); green woody vine with scattered prickles and tendrils; leaves semi-evergreen, remaining green and on the stem into winter; leaf arrangement alternate; leaf simple, blade entire, shape ovate to lanceolate, length 5 to 12 cm, width 3 to 5 cm, veins three to five, upper surface often appearing water-spotted, abaxial surface white glaucous, margin smooth; fruit a glaucous (immature) or shiny (mature) berry; dark blue to black at maturity; Regions 3 and 4; xeric to mesic sites. Cover type 28.49.

Smilax hispida Muhlenberg (*Smilax tamnoides* Linnaeus), bristly greenbrier (Figure 24.3, 24.4); green woody vine, often with 1.0 cm long dark brown to black prickles, densely arranged on stem, tendrils present; leaf arrangement alternate; leaves semi-evergreen, remaining green and on the stem into winter; leaf simple, blade simple, shape ovate to slightly panduriform, length 5 to 12 cm, width 3 to 6 cm, abaxial surface green with minute prickles on the midrib, veins three to five, margin rough and serrulated; fruit a berry, black at maturity; Regions 2, 3, and 4, north into southern Ontario and New Brunswick; mesic to mesic-hydric sites.

Smilax laurifolia Linnaeus, laural-leaf greenbrier, bamboo-vine (Figure 24.4); green woody vine with numerous prickles; tendrils frequent; leaf arrangement alternate; leaves evergreen; leaf simple, blade entire, coriaceous, shape oblong to linear-oblong, length 6 to 17 cm, width 3 to 7 cm abaxial surface green; fruit a black berry; Regions 3b and 4; swamps and bottomland.

Smilax rotundifolia Linnaeus, round-leaf greenbrier (Figure 24.4); green woody vine with scattered stout, flattened green prickles, tendrils frequent; leaf arrangement alternate; leaves semi-evergreen, remaining green and on the stem into winter; leaf simple, blade entire, shape ovate or nearly orbicular, often wider than long, margin smooth or slightly rough, abaxial surface green, length 8 to 14 cm, width 7 to 10 cm, veins five to seven; petioles red; fruit a berry, reddish to black at maturity; Regions 2, 3, and 4, north into southern Ontario and New Brunswick; xeric to mesic-hydric sites. Cover type 28.49.

Smilax smallii Morong, lanceleaf greenbrier (Figure 24.5), green woody vine with scattered prickles, tendrils, and evergreen leaves; leaf simple, entire shape lanceolate, length 4 to 10 cm, width 1 to 3 cm, veins three to five; Region 4, restricted to the Coastal Plain; mesic woods.

Smilax walteri Pursh, Walter's or red-berried greenbrier (Figure 24.5); vine slender, flexible, terete; prickles essentially absent, tendrils numerous; leaves deciduous; leaf simple, entire shape ovate to broadly lanceolate; length 5 to 12 cm, width 2 to 6 cm, margin smooth veins three to five; petioles red; fruit a berry, bright red at maturity; Region 4, restricted to the Coastal Plain; swamps and marshes.

PART IV

Forest Community Ecology: Combining Species into Communities

CHAPTER 25

Introduction to Forest Ecology and Classification

Every natural aggregation of plants is the product of definite conditions, present and past, and can exist only when these conditions are filled.

J. Braun-Blanquet, 1928
(in his preface to Pflanzensoziologie, translated into Plant Sociology: The Study of Plant Communities by Fuller and Conard, 1932)

PRINCIPLES OF FOREST ECOLOGY THAT RELATE TO CLASSIFICATION

ALL plants are adapted to specific environmental conditions, based on their physiological capabilities. Certain amounts of particular resources are required (water, light, nutrients), and plants possess specific ways of obtaining those resources (plant adaptations niche). For example, *Taxodium distichum* is capable of withstanding the low oxygen levels found in waterlogged soils and several *Juniper* species are able to withstand long droughts during the growing season. Thus, the distribution and abundance of a given species is determined by its response to various environmental factors in the landscape (Cain 1944). There are, of course, several intermediate adaptations to a gradient of resources.

When moderate moisture conditions persist and many taxa have the ability to grow and thrive under these conditions, competition for light becomes an important consideration. Those species able to assimilate the most carbon sugars with the least light will ultimately replace species less tolerant of shade. As early as 1950, Baker placed tree species into shade tolerance classes. There is often a trade-off between the physiological adaptations necessary for survival in stressful (high and low) moisture conditions and the ability to compete for light and nutrients. Finally, species may also be adapted to certain types of disturbance and will maintain themselves as long as the disturbance reoccurs. *Pinus contorta* and *Pinus banksiana*, because of their serotinous cones, reproduce prolifically following fire.

Plant species distributions are differentiated according to environmental influences because species have adapted to a given range of environmental conditions. If two or more species have adapted to an environment in similar ways, then they are often found together within a community. A community is defined as an assemblage of species (populations) that occurs repeatedly under similar environmental conditions across the landscape. The purpose of this section of the book is to describe the framework for community classification and to explore the environmental conditions under which these communities occur. This chapter discusses the major factors controlling vegetation structure and composition and introduces the classification of vegetation. Chapters 26 to 30 describe some of the major forest communities of the United States and Canada.

The controlling factors of vegetation (Figure 25.1) are discussed within a hierarchical framework.

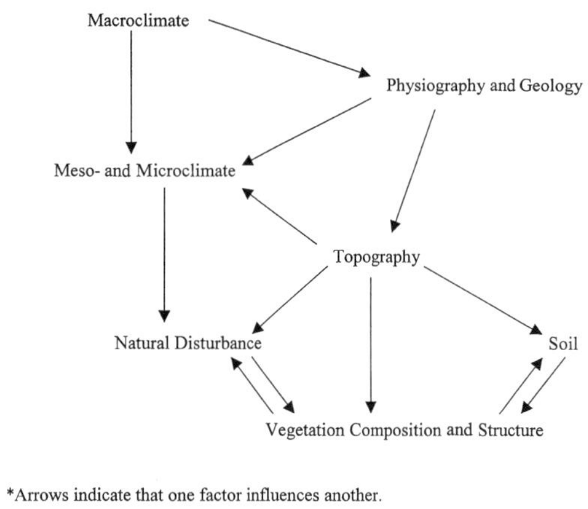

FIGURE 25.1 Relationships of factors controlling the distribution of vegetation.

Hierarchy theory suggests that certain environmental forces have greater influence than others on the distribution of forest vegetation. For example, the great biomes of the world are controlled by macroclimates (Scale 1). One of those biomes is the temperate deciduous forest. If the entire biome had the same climate, the same soils, no disturbance, and no influence from topography, then the vegetation would be similar throughout. However, topography modifies the macroclimate regime, yielding a different mesoclimate (regional climates within the macroclimate; Scale 2, nested in Scale 1). Within the temperate deciduous forest, the Appalachian Mountains have a spruce-fir forest at their peaks and oak-pine in their foothills. The mesoclimate, amount of soil moisture, geological parent material, and current vegetation affect the soil-forming processes. Within the temperate deciduous forest at lower elevations (foothills) in the Appalachian Mountains, the vegetation is strikingly different on Coastal Plain geologic formations (conifer-dominated) than on limestone or sandstone formations (oak-hickory dominated) (soil characteristics or types within regional climate; Scale 3, nested in Scale 2). Disturbance may influence the current structure and composition of vegetation by altering microclimate and microenvironmental conditions (disturbance of forest on different soils in different microclimates; Scale 4, nested in Scale 3). Within the temperate deciduous forest at lower elevations (foothills) of the Appalachian Mountains in areas generally dominated by oak-hickory forest that are often of second growth recovering from disturbance. These sites may go through a grass-and-shrub stage before reverting to forest (Oliver 1980). In addition, the younger forests often have different species than the older forests (e.g., the common pioneer species tulip poplar (*Liriodendron tulipifera*) evenually disappears from older forest stands.

The hierarchy concept suggests that the different layers (scales) of influence are nested, such that the major factor constrains the possible responses of the nested factor (Allen and Starr 1982). For example, fire plays a role in almost every terrestrial system, but the responding taxa change based on the biome and topography. Lodgepole pine (*Pinus contorta*) regenerates following fire in the west, jack pine (*Pinus banksiana*) in the northeast, and various yellow pines in the south. Fire maintains hill prairies on southwest-facing upper slopes, but every species of an oak-hickory forest on lower slopes and along northern faces in the Central Hardwood Region of the Eastern United States resprouts, and the burned site quickly reverts back to forest.

Mountainous regions, windthrow disturbance, and serpentine soils are found in various biomes, but each respective controlling factor has a different assemblage of species in different biomes. The following is a short introduction to the major factors controlling the distribution of forests in North America. The goal of this introduction is to outline the characteristics used to describe the forest communities in the following chapters and the logic behind community classification.

Climate: Controller of Temperature and Moisture

Climate is defined as the long-term atmospheric conditions of a region. Globally, climate is the strongest determining factor of vegetation structure and soils in natural systems. The production of new biomass and the decomposition of dead organic matter are processes constrained by moisture and temperature fluctuations. The diurnal and seasonal fluxes of energy and moisture (climatic regime) are important factors in the development of the major forest biomes. For example, some tropical forests lack a seasonal pattern of temperature or moisture, whereas other regions have pronounced temperature fluctuations and specific periods of abundant rainfall.

Three important factors determining the climate regime are latitude, altitude, and proximity to large bodies of water. Both latitude and altitude influence the radiation budget, the movement of light energy into and out of a particular system. Direct radiation along the equator yields warmer temperatures, and temperatures decrease as one moves toward either pole. Three major temperature zones can be distinguished by latitude. The humid tropic zone has mean monthly temperatures greater than 18° C, and the northern summerless zone has mean monthly temperatures of less than 10° C. Between these two extremes lies the temperate climate, which has summer and winter seasons. High-elevation mountain areas have lower average temperatures than low-elevation areas, with air temperatures decreasing at a rate of ~4° C per 100 m of increasing mountain elevation. The air at lower elevations is denser and holds more latent energy than the air at higher elevations.

Large bodies of water moderate temperature extremes because water has greater thermal capacity than land and thus heats and cools more slowly. The coast offers an escape from the midcountry winter weather, reflecting the difference between maritime and continental climates. Average mean maximum and minimum January temperatures for San Francisco (37°62′ north latitude and 122°38′ west longitude) from 1990 to 1996 were 13.83° C and 6.69° C, respectively; in St. Louis (38°75′ north latitude and 90°37′ west longitude) the averages were 4.99° C and −3.79° C, respectively. Large lakes may also moderate climate. The climate of the continental polar air mass in the northeastern United States is altered by the Great Lakes. Average mean maximum and minimum January temperatures for Toronto (43°67′ north latitude and 79°40′ west longitude) from 1980 to 1990 were −0.95° C and −6.97° C, respectively; in Rochester, Minnesota, (43°92′ north latitude and 92°50′ west longitude), the averages were −4.65° C and −13.99° C, respectively (data from National Oceanic and Atmospheric Administration). Large bodies of water also increase precipitation on their leeward sides (prevailing winds).

Temperature can limit the distribution of forests in four ways: (1) Low winter temperatures may cause mechanical injury directly, by freezing tissues, and indirectly, through the freeze-thaw fluctuations in soil. (2) Short and cool summers result in an insufficient amount of gross photosynthesis to overcome the amount of respiration and litterfall of plants, resulting in no net growth. (3) High summer temperatures increase water loss and reduce available soil water. (4) The lack of a winter chilling period to break dormancy or stimulate flowering restricts many temperate species from migrating south (Kimmins 1987).

Plants not only respond to maximum and minimum temperatures, but also to daily and seasonal temperature fluctuations. These major influencing factors can be plotted along a latitudinal and elevational gradient, thus providing fairly predictable changes in the physiognomy of the vegetation.

It is impossible to separate the effects of moisture from those of temperature when examining vegetation-climate relationships. Indeed, moisture also decreases with latitude. One ratio used for classification of general patterns of vegetation is the precipitation-potential evaporotranspiration ratio (P/E ratio). The P/E ratio indicates the drougtiness of a region, balancing the amount of precipitation against the amount of evaporation from land and water surfaces and transpiration from plants (temperature is also a major factor). Where potential evapotranspiration is higher than precipitation, droughts are usually annual occurrences and affect vegetation structure and function (e.g., grasslands; $P/E < 1$). The lower the P/E ratio, the more severe the drought conditions (deserts P/E ratios range from 0.1 to 0.3) (Figure 25.2). Deciduous forests have P/E ratios > 1, and west coast

FIGURE 25.2 Dry shrubland in an intermountain basin, southern California.

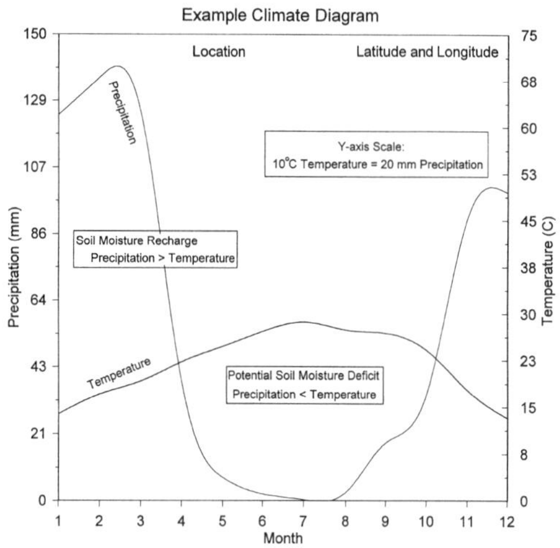

FIGURE 25.3 Climate diagram illustrating average precipitation and temperature throughout the year.

coniferous forests have P/E ratios >2. Climate diagrams are often used to express the relationship of precipitation and evaporation when potential evapotranspiration is unknown, indirectly depicting the soil moisture deficit or excess typical of growing season conditions (Figure 25.3). The figures in this book give the mean monthly maximum values of temperature against the mean monthly precipitation for the decade 1980 to 1989; the y-axis scale of 10°C = 20 mm rain holds for all diagrams. Periods of high temperature and low precipitation potentially produce a soil moisture deficit. These figures are given at the beginning of each of the following chapters on the various regions. All figures give the name, latitude, and longitude for the location of the data. All climate data were acquired from National Oceanic and Atmospheric Administration.

The availability of moisture is closely related to plant nutrition. Soil water stress can result in reduced photosynthesis, because stomates are closed and cannot uptake atmospheric gases. Without sufficient energy and an active transpiration stream and energy transport system, roots cannot absorb nutrients, the internal movement of nutrients and metabolic products is reduced; thus metabolism and growth suffer. Within the soil column, low soil water limits plant water uptake and thus, nutrient uptake. In addition, decomposition and mineralization processes are slowed, reducing the levels of available plant nutrients. The results of the link between water and nutrients on forests are evidenced through site index analyses, an assessment of site productivity.

Physiography and Geology: Precursor to Soils and Topographic Patterns

Fenneman (1939) classified landscapes on the basis of their geologic and glacial history. Different rocks vary in their resistance to erosion, and in their porosity, permeability, and chemistry. Thus, soils will have properties obtained from their parent material. Because each physiographic region has a relatively specific geology, bedrock, and glacial history, soil types tend to be related and are generally classified in broad soil groups (orders). Soils derived from weathered coastal plain gravels (the Coastal Plain Region of the southeastern United States) are typically of poor nutrient and textural quality. Conversely, soils derived from limestone tend to be loamy and high in nutrients, both excellent soil properties for forest growth.

In addition, the physiographic provinces have different dominant patterns of topographic relief. Broad-scale landform patterns can be classified as plains, plains with mountains, hills, and tablelands. Past geologic activity, such as uplifting, created various mesoclimates. For example, mountain chains strongly affect the temperature, precipitation, and growing season as the result of an orographic effect (see "Topography" later in this chapter). In addition, carving by glaciers and erosion by glacial waters or general precipitation have created strong topographic relief in many states. Because of these various mechanisms, the land of North America ranges in elevation from 6194 m on Mount McKinley, Alaska, to −86 m in Death Valley, California. A variety of taxa seem specific to certain physiographic provinces, and the greatest diversity of vegetation consistently occurs where several physiographic provinces merge (e.g., the Oregon-California border).

Soil: The Growth Medium

Soil is the relatively thin layer of mineral and organic materials that cover much of the earth's crust and can support living matter and vegetation. Soil conditions that affect plant growth and community associations include moisture and nutrients. Soil moisture and nutrients are determined by climate, as noted earlier, soil texture and structure, and vegetation. While the parent material provides the base for soils, the development of the soil profile, which is a reflection of texture and nutrient percolation, is controlled by topographic position (a stream terrace versus a ridgetop), time, vegetation, and climate. Thus, soils have characteristics that indicate site productivity and potential vegetation. For example, mottled soil horizons characteristic of wetlands, indicate alternating wet and dry periods due to a fluctuating water table.

The three components of mineral soil are sand (0.05 mm to 2.0 mm diameter particle size), silt (0.002 to 0.05 mm diameter particle size), and clay (particle size diameter <0.002 mm). Relative amounts of sand, silt, clay, and organic matter determine the soil water available for plant growth. Generally, the smaller the particle size, the smaller the pore spaces and the greater the ability of a soil to hold water because of capillary action and the adhesive properties of water. However, very small pores hold water so tightly that plants are unable to extract it. Thus, pure clay soils may hold considerable water, but only a portion of it is available to plants. Likewise, water easily percolates through sandy soils, and they become dry in a very short time. Clays and organic matter are required for soil nutrient retention. Both of these materials have negative charges and so attract essential cations. Thus, soils with moderate amounts of silt, sand, and clay (loams) and adequate amounts of organic matter have the greatest growth potential, as they hold the greatest amount of plant-available water and nutrients.

The quality of soil affects forest productivity, and in turn the forest (or vegetation) on a site affects the soil. This feedback, along with the driving force of climate, regulates the development of a soil profile in any given area. A soil profile is developed by the vertical separation of texture, organic matter, iron and aluminum oxides, carbonates, and nutrients into soil horizons according to their differential leaching properties (Figure 25.4). A soil profile includes the *O*, *A*, *E*, *B*, *C*, and *R* horizons. The *O* horizon consists of the organic matter undecomposed and partially decomposed on the soil surface. The *A* horizon is the portion of the soil profile in which most of the decomposed

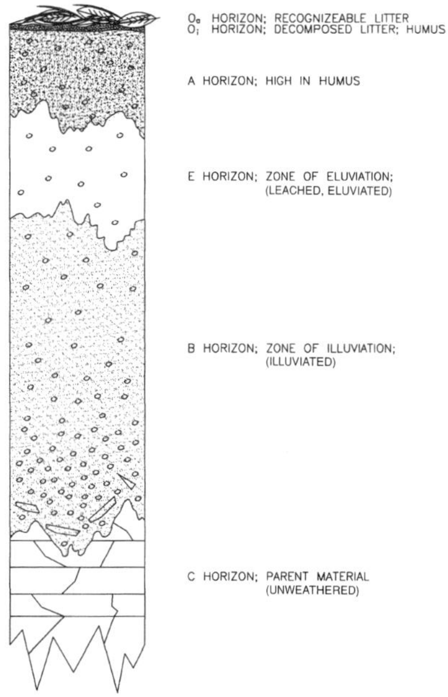

FIGURE 25.4 Soil profile showing typical horizons and their characteristics.

organic matter (OM) is incorporated and thus has a distinctive dark color. The *E* horizon is a zone of maximum eluviation (leaching), typical of forest soils (but not under those of grasslands) and is characteristically bleached or gray in color. The *B* horizon is the zone of maximum illuviation (accumulation) of iron, aluminum, clay, and OM. Rooting occurs in these master horizons, but primarily within the *A* and *B* horizons. The *C* horizon consists of unconsolidated parent material and mineral soil below the root zone. The *R* horizon represents consolidated bedrock (the geology). The process of soil development (Figure 25.5) and resulting profiles have been classified into soil orders (Figure 25.6); the characteristics of some of the major orders are discussed in the following paragraphs.

Oxisols are typical of tropical forests. Horizons are strongly developed as a result of intense leaching through the soil profile (Figure 25.5). Laterization is the dominant soil-forming process. This process occurs under hot, moist conditions in which silicon is weathered and leached more than iron and aluminum. Thus, iron and aluminum are concentrated in the

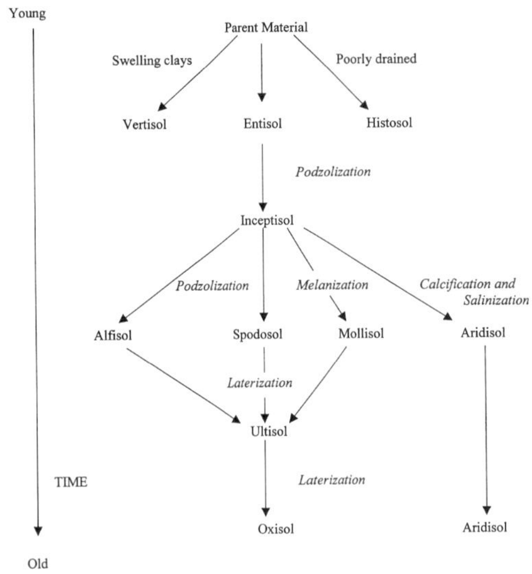

FIGURE 25.5 Soil development showing processes of development (italics) and soil type. Adapted from Birkeland (1999) with permission from Oxford University Press.

E horizon and a *B* horizon may not develop. The soil profile is characterized by a deep *E* horizon, high in iron, aluminum, and clays, large stores of soil organic matter and nitrogen, and little surface organic matter (due to fast decomposition and nutrient cycling). Most nutrients are tied up in the biomass, and the soil is very poor. A similar soil order, Ultisols, is typical of subtropical forests where a dry season occurs, such as in the southeastern United States. Laterization also dominates the formation of this soil, yielding a red clay *E* horizon rich in iron and aluminum.

Spodosol and Alfisol soils developed under the forests of central and northern North America. Spodosols are typically found with coniferous vegetation, and Alfisols are found with temperate deciduous forest. The predominant soil-forming process of Spodosols is podzolization. Podzolization occurs because the coniferous litter has low concentrations of nitrogen, is high in lignin, and tends to be highly acidic, and thus decomposition occurs slowly. In addition, a greater relative amount of nutrient leaching occurs in these soils than in soils under deciduous litter. A Spodosol soil profile is characterized by large accumulations of organic matter, a thin *A* horizon, a distinct *E* horizon, and a high sand content. Accumulations of iron, aluminum, and organic matter occur in the *B* horizon (Figure 25.7). Decomposition occurs more quickly in deciduous forests, and there are higher levels of animal activity particularly earthworm, building a more well developed *A* horizon in these Alfisols. In addition, Alfisols are characterized by the leaching of primarily clays, as well as aluminum and iron, to the *B* horizon, an enriched zone that often has fine roots. The *E* horizon in Alfisols is not as well developed as in Spodosols (Figure 25.7).

Aridisols and Mollisols are typical of shrubland/desert and grassland systems, respectively. Here, greater potential for evaporation and a decreased amount of precipitation reduce leaching, thus slowing soil-forming factors. The *A* horizon is typically quite deep but much darker in the Mollisols as the result of the increased turnover of fine roots of grasses. Aridisols develop in arid regions because of the soil-forming processes of calcification and salinization, whereby percolation is minimal and easily soluble compounds remain within the soil profile or concentrated at the surface. Mollisols develop in temperate regions because of the soil-forming process of melanization, whereby clays leach to the *B* horizon and the *A* horizon is enriched and darkened by the addition of organic matter. Mollisols may lack an *E* horizon.

Inceptisols (weak profile development) and Entisols (essentially no profile development) have large *O* horizons and minor *E* horizons. Although podzolization is the dominant soil-forming process, the soil profiles are not developed because they have been recently deposited. Also, in the north, temperatures can be extremely cold. Thus, decomposition and leaching to form soil profiles are greatly slowed (Figure 25.8). These soils are typical in areas of recent glacial retreat, and recently deposited alluvial (deposition from flood events) and colluvial (deposition from adjacent uplands) material. Entisols have a predominance of mineral soil materials, and Inceptisols are often shallow.

Histosols develop when large amounts of organic matter, such as *Sphagnum* moss, occur over rock or frozen ice (permafrost). Decomposition is extremely slow because of waterlogged soils throughout the year,

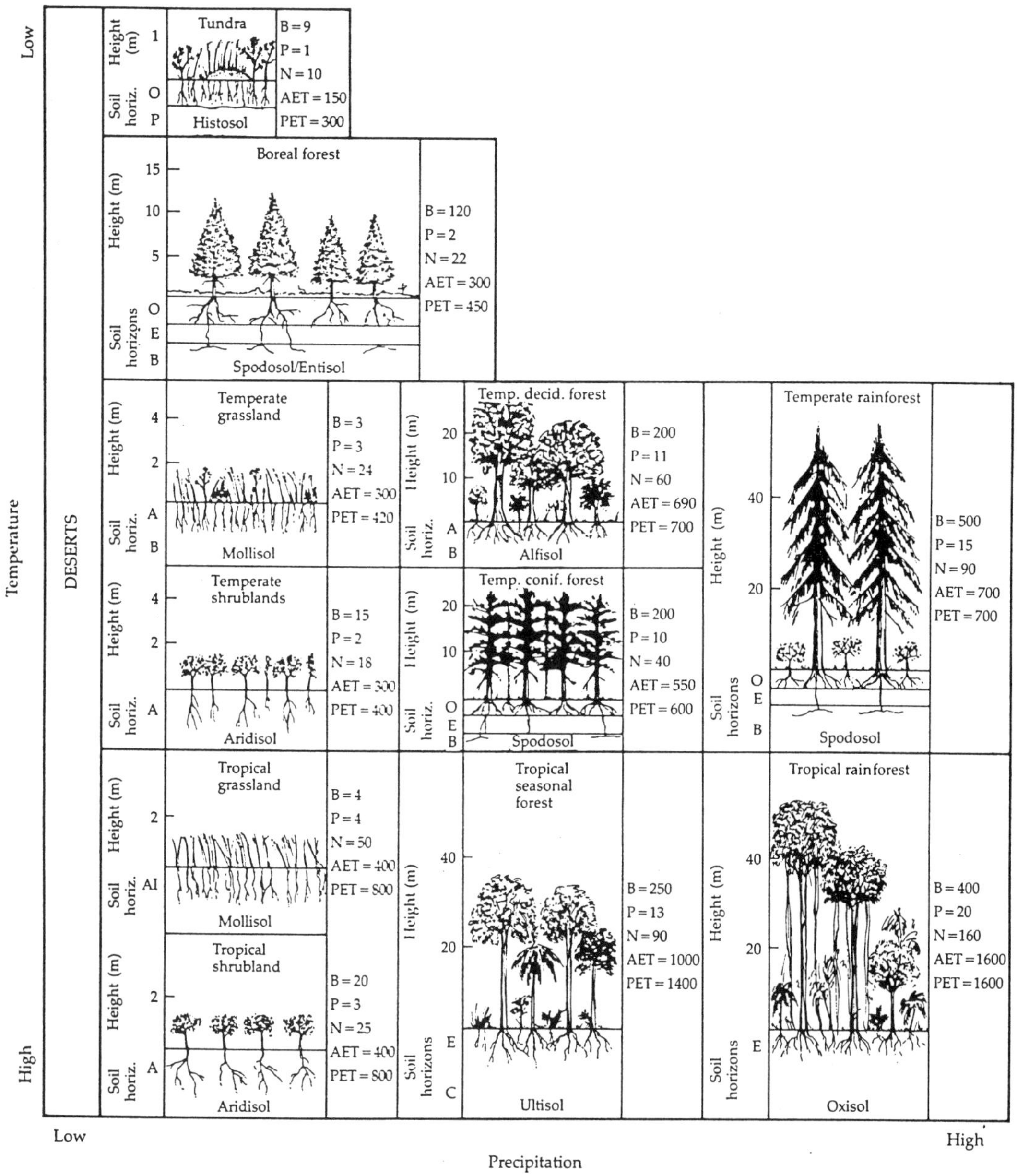

FIGURE 25.6 Soil orders in relation to climate and the dominant physiognomic vegetation for each soil order. In the legend, B is total plant biomass (Mgms/ha), P is total above ground production (Mgms/ha/yr), N is nitrogen uptake by plants (Kgms/ha/yr), AET is actual evapotranspiration (mm H_2O/yr), and PET is potential evapotranspiration (mm H_2O/yr). Reproduced from Aber and Mellilo (1991) with permission from Saunders College Publishing.

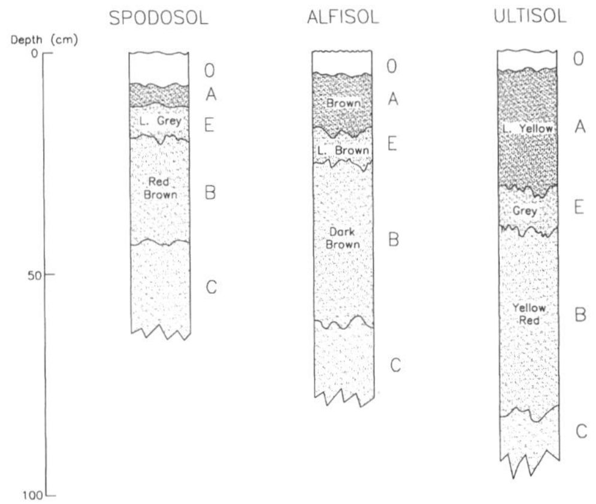

FIGURE 25.7 Horizons of Spodosol, Alfisol, and Ultisol soil profiles (adapted from P.E. James, *A Geography of Man*, 1959, with permission from John Wiley & Sons).

presence of toxic substances (methane, hydrogen sulfite), and/or very cold temperatures with frost occurring even during the growing season. Rooting is concentrated toward the top of the organic matter where oxygen may be available. Nutrient levels are typically very low, supporting only the heartiest trees, insect-eating plants, and shrubs of the Ericaceae and Salicaceae families.

Another special category of soil, although not a soil order, includes the serpentines. Serpentine soils

FIGURE 25.8 An Entisol soil in the Rocky Mountains along Bear Tooth Pass, Montana.

form from metamorphic magnesium silicate rock. The growth of plants is generally limited because of low amounts of nitrogen, calcium, potassium, and phosphorus and the extremely high level of magnesium. In addition, these soils are high in toxic elements such as nickel and chromium. Although serpentine soils can be found throughout the world, they are most common in Pacific Coast states (Barbour et al. 1998) with additional areas along the Atlantic Coast (Tyndall and Hull 1999). These soils provide a unique habitat to which many species have adapted, including several endemics (i.e., found in only a specific locale) and insectivorous plants. Scattered Jeffrey pine (*Pinus jeffreyi*) and knobcone pine (*Pinus attenuata*) are common on serpentines of northwest United States.

Topography: Mesoclimate and Microclimate

An increase in elevation produces a cooler atmosphere, which holds less moisture and an increase in precipitation. In addition, growing seasons are shorter. Increased latitude causes a similar change, referred to as a climatic gradient. The changing climate yields vegetation zones that are often conspicuous, although they seldom follow perfect elevational contours. Instead, vegetation zones intermingle as a result of local climate and soils differences due to topography and geology. The zone boundary is thus blurred, with plants that require a cooler, more moist environment occurring in protected drainages and north slopes within a warmer, drier zone. The reverse situation also occurs.

As a rule, vegetation zones in the Northern Hemisphere extend to higher elevations on south-facing slopes than on north-facing slopes, Southern exposures receive more hours of sunlight and higher intensities of radiation than northern exposures. Consequently, southern slopes are warmer, free of snow earlier, and have longer growing seasons. Within a vegetation zone, ridges are generally sunnier and free of snow earlier, whereas valleys or coves are cooler and keep snow longer. On lower slopes, the same phenomenon tends to restrict mesophytic species to northern exposures, with the driest and hottest conditions found on upper southwest-facing sites (Figure 25.9). Southwestern slopes are hottest because they receive the most direct solar radiation during the hottest portion of the

FIGURE 25.9 Aspect can have a direct impact on vegetation. South and west aspects are the driest (seen as grassland in picture) in North America, whereas north and east slopes have cooler and more moist conditions that permit trees to survive. Photo taken in low hills north of the Grand Canyon, Arizona.

day. Correspondingly, southwestern aspects have the greatest evaporative potential. This phenomenon was quantified for use in modelling by Beers et al. (1966).

Large mountains generally alter regional precipitation patterns as well. In North America the general wind direction is to the east, because of the influence of the westerlies. When an air mass presses against a mountain, it is forced up and over the obstruction, creating an orographic effect (Figure 25.10). As the air mass rises, it cools, increases in density, and has less ability to hold water; thus, precipitation increases at high elevations of the western slope of mountain ranges. The air mass passes over the mountain peak and continually warms and expands as it travels down the other side. This warmer air holds more water and has a greater potential for evaporation, thus creating what is referred to as a rainshadow. Here, little precipitation falls and the potential for evaporation is quite high. Within the rainshadow occur some of the driest climates in North America: the cold deserts of the Intermountain Region and the short grass prairie on the eastern side of the Rocky Mountains (Figure 25.10) The orographic effect controls the zonal communities in mountainous regions, with similar communities found at higher elevations on the drier rainshadow slopes (Peet 1988; Barbour et al. 1998).

Vegetation Dynamics: Response To Disturbance

The study of community dynamics developed around successional theories in the late nineteenth century. Although change is one of the central tenets of community ecology, no general model of community dynamics has been universally accepted. This is largely due to the variety of influences affecting vegetation dynamics, starting with disturbance. Disturbances can vary in frequency (how often they occur), magnitude (the intensity of the disturbance), and severity (the direct effects on the flora and fauna). Differences in magnitude and severity can occur within the same disturbance event. For example, the fires that burned through Yellowstone National Park in 1988 ranged from extremely hot canopy fires that blackened entire areas (Figure 25.11) to ground fires that swept through the understory, leaving the overstory essentially unscathed. Disturbance frequency tends to be inversely related to magnitude and severity. The Corps of Engineers has designated annual floods, 10-year floods, and 100-year floods for the Mississippi River, based on the average occurrence of these events, in the order of increasing magnitude and severity.

Plant responses to disturbance are based on their life history characteristics, their adaptations for germinating, growing, reproducing, and disseminating seed. Thus, disturbed populations respond in different ways to the same disturbance because they have adapted unique life history strategies. Although it is generally known that different communities hold different populations (by definition), it should be noted that within communities, differences are also the norm. Rarely do two communities have the same densities and biomass of the same species. Thus, the responses of specific stands classified as belonging to the same general community may be quite different because of inherent differences in numbers and densities of populations. This inherent variability makes it difficult to develop general statements about the effects of disturbance. However, researchers have developed classifications for understanding or describing population responses to disturbance.

Pianka (1970) split species into two categories, r-selected and K-selected. Essentially, the r-selected species are true pioneer species that have evolved to take advantage of disturbance. These species have fast

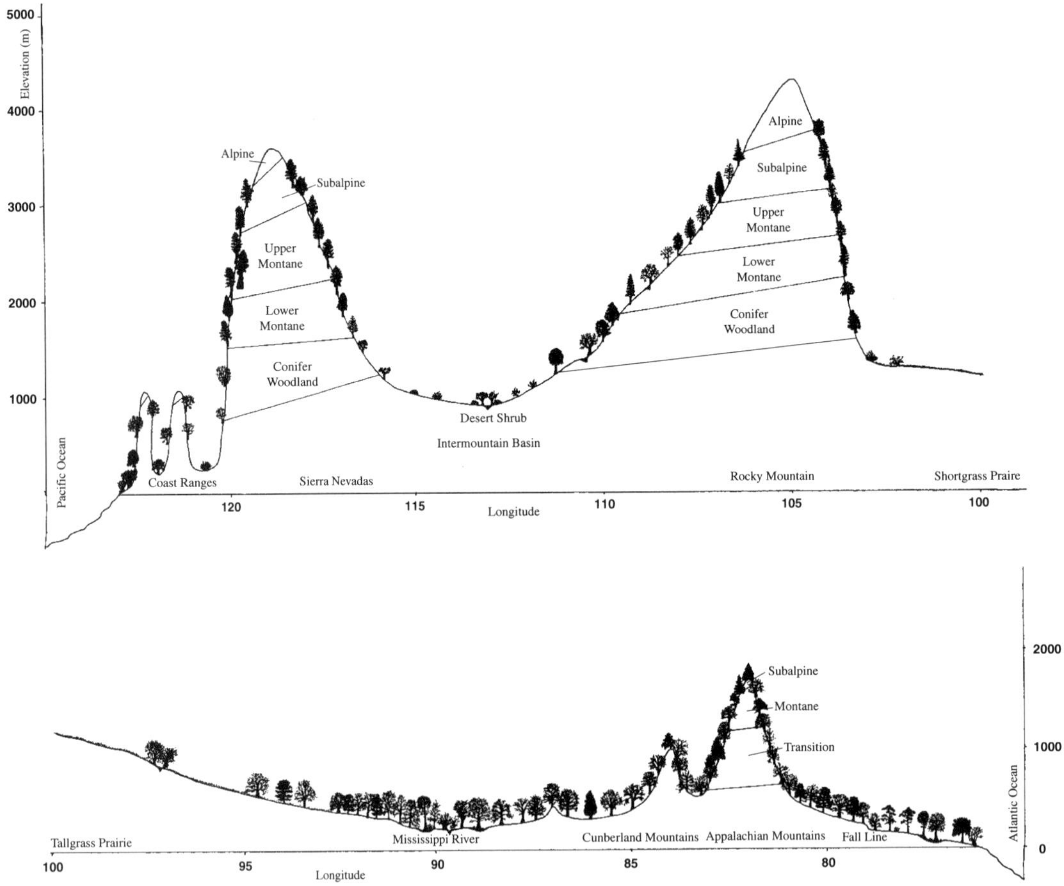

FIGURE 25.10 Schematic diagram showing the longitudinal distribution of vegetation across the conterminous United States, approximately 36° north latitude. Drawing by Deanna Boswell Franklin.

growth rates and short life spans. They tend to produce abundant crops of light seed designed for long distance dispersal. Finally, they are generally not competitive. That is, they are intolerant of shade. The r-selected species have adapted to colonize open sites and maximize their rate of growth (the r in population growth models). K-selected species, on the other hand, maximize their ability to compete and maintain their populations at the carrying capacity (the K in population growth models) of the system. They have the opposite characteristics of r-selected, pioneer species. The truth is that these descriptions are the extremes, and many species have characteristics that lie somewhere in the middle. Grime et al. (1990) categorized plants in a similar way, adding one other group. He developed the CRS classification, referring to plant strategies as competitive (C), ruderal (R), and stress-tolerating (S). The competitive and ruderal strategies match the K-selected and r-selected strategies, respectively. The stress-tolerator strategy refers to individuals that have adapted an avoidance to competition by growing where few other organisms grow in very stressful environments. The carnivorous bog plants are good examples. These classifications of life history strategies

FIGURE 25.11 Remnants of a lodgepole pine (*Pinus contorta*) forest following the 1988 wildfire season in Yellowstone National Park. Photo courtesy of Larry Franklin.

FIGURE 25.12 Beaver dam in Canada. The hydrologic changes created by a beaver dam produce changes in vegetation as well. Spruce and fir forests may be deadened, whereas willow species are able to take advantage of the new conditions. Beavers move around the landscape, thus creating new patches. The forest recovers as beaver move to another location.

allow resource managers to predict the response of populations to disturbance or lack of disturbance (i.e., succession).

Disturbance is an important functional part of every ecosystem. Major natural disturbances include windthrow (including events ranging from a mild storm blowing one old tree down to microbursts, tornadoes, and hurricanes), disease, pest outbreaks, fire, and flooding. Natural disturbances are periodic, and most species have adapted to them in some way. Fires are prevalent in coniferous forests of the north, west, western Coastal Range, and southeast. Windthrow is a common disturbance component of deciduous forests. Examples range from single tree gaps in maple-beech forests to community destruction by tornadoes and microbursts in oak forests. Hurricanes are common along the Atlantic coast, especially in the southern portion. Flooding is typical along all river courses and their connected floodplains. It is also a natural result of beaver activity (Figure 25.12), which drastically modifies local hydrology and selectively alters vegetation composition through herbivory.

In addition to the temporal scale of disturbances, whose magnitude and severity are inversely related to frequency, there is a hierarchy of disturbances, ranging from major climatic alterations producing long-term vegetation change to annual fire cycles resulting in cyclic succession. Long-term vegetation change has been studied with the use of pollen preserved in sediments. These records document the continuing migration of vegetation as it receded from glacial expansion southward and as it advanced northward following glacial retreat. The pollen record shows the cycle occurring several times throughout the Pleistocene. During the height of the Wisconsinan glaciation, approximately 18,000 years before present (B.P.), *Picea* spp. and *Pinus banksiana* dominated most of the central eastern United States (Figure 25.13). Southern hardwoods and pines were restricted to the Coastal Plain. Northern hardwoods were restricted to a rather narrow band in between boreal and southern associations.

The changes in vegetation that have occurred during the last 18,000 years can be illustrated with an example from Land Between The Lakes in western Kentucky and Tennessee. (Figure 25.14). At the height of glaciation (Figure 25.14a), *Picea* species dominated the floodplains and a mix of northern and central hardwood species (as they occur today) dominated the uplands. The component species then migrated north, following glacial retreat. During an extremely warm and dry period (Hypsithermal, Figure 25.14b), the ridges and upper slopes were open savanna and prairie. These open forests were kept open, despite a continuously cooling climate, by periodic disturbances such as large ungulate grazing and fires

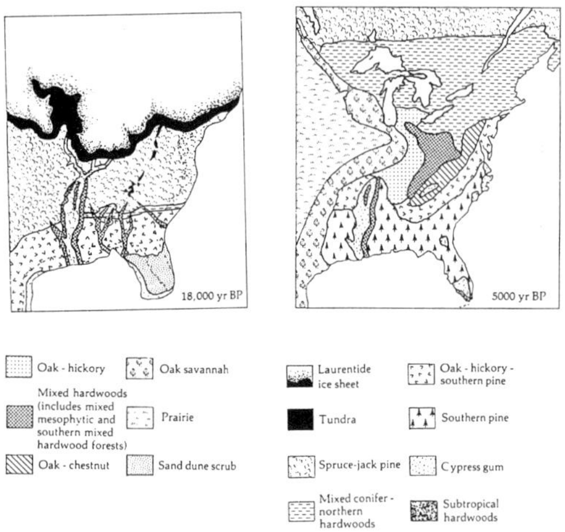

FIGURE 25.13 Reconstruction of the vegetation of the eastern United States from pollen records (reprinted from Delcourt and Delcourt (1981) with permission from Plenum Publishers). Vegetation at 18,000 years before present (B.P.) occured at the last major glacial peak, the Wisconsinan. Notice the constriction of current vegetation zones below the glacial extensions. Vegetation at 5000 years B.P. represents the last major warming trend (Hysithermal of the Holocene), when the prairie expanded eastward. Vegetation has continuously migrated north and south of advancing and retreating glacial extensions.

(natural and anthropogenic). The current vegetation (Figure 25.14c) is a closed forest, resulting from the suppression of fire. Because of large-scale logging of forests in the 1800s, which must have been followed by excellent oak regeneration at the time, uplands are dominated by *Quercus* species. However, the closed forests of the Central Hardwoods impede oak regeneration and the succession to more shade-tolerant mesophytic species (e.g., *Acer saccharum*) seems a certainty. This example illustrates three main points: (1) climate is continually changing, (2) disturbances can be significantly altered by human activity, and (3) vegetation is in constant flux.

Some of the most influential disturbances in forests have been pest and disease outbreaks, whether natural (e.g., spruce budworm caterpillar, *Choristoneura fumiferana*) or exotic (e.g., gypsy moth (*Lymantria dispar*), and chestnut blight (*Cryphonectria parasitica*)) (Figure 25.15). In addition, these disturbances may be related and synergistic. For instance, the demise of *Abies fraserii*, due to the woolly adelgid in the Great Smoky Mountains National Park, may have been stimulated by the exposure of older trees to pollution, which caused their stressed condition. Moreover, droughts within older-aged stands are often related to insect outbreaks.

Knowledge of the successional status of a community is important in determining proper forest management. Succession is defined as the natural change in populations, resulting in a different community. Such change often occurs because the adaptations necessary for pioneering a site are not the same as those necessary to compete for light. Pioneer species (those species that colonize a site quickly after disturbance) tend to be wind dispersed, which helps to distribute them to open disturbed areas, fast growing to compete with herbaceous plants, and light demanding as they usually need exposure to full sun to reach reproductive age quickly and avoid the developing competitors; also they are short-lived. Mid-seral species tend to be more shade tolerant, as they usually grow under pioneer species. Mid-seral species grow more slowly and reach reproductive age later, because they are longer-lived and dominate the site for some time.

Species that can reproduce beneath themselves and maintain themselves on a site for long periods of time are referred to as "climax" species. The term simply means that without climate change or disturbance, the climax species will dominate the area. A forest community of *climax species* implies that it will eventually be *compositionally stable.* There are three main factors that impinge upon the development of the climax. First, climate is always fluctuating. Second, there are disturbances in every ecosystem, and rarely is a climax reached. Third, even sites within the same climate may not necessarily have the same dominants. Two sites under the same environmental conditions may have two different sets of dominants, both able to maintain their dominance over long periods of time (McCune and Allen 1985).

Various communities that may be recognizable during succession occur in stages, referred to as seral stages. For example, if fire is a regular and severe disturbance in a northern forest, jack pine may likely dominate. This is due to its various adaptations to fire,

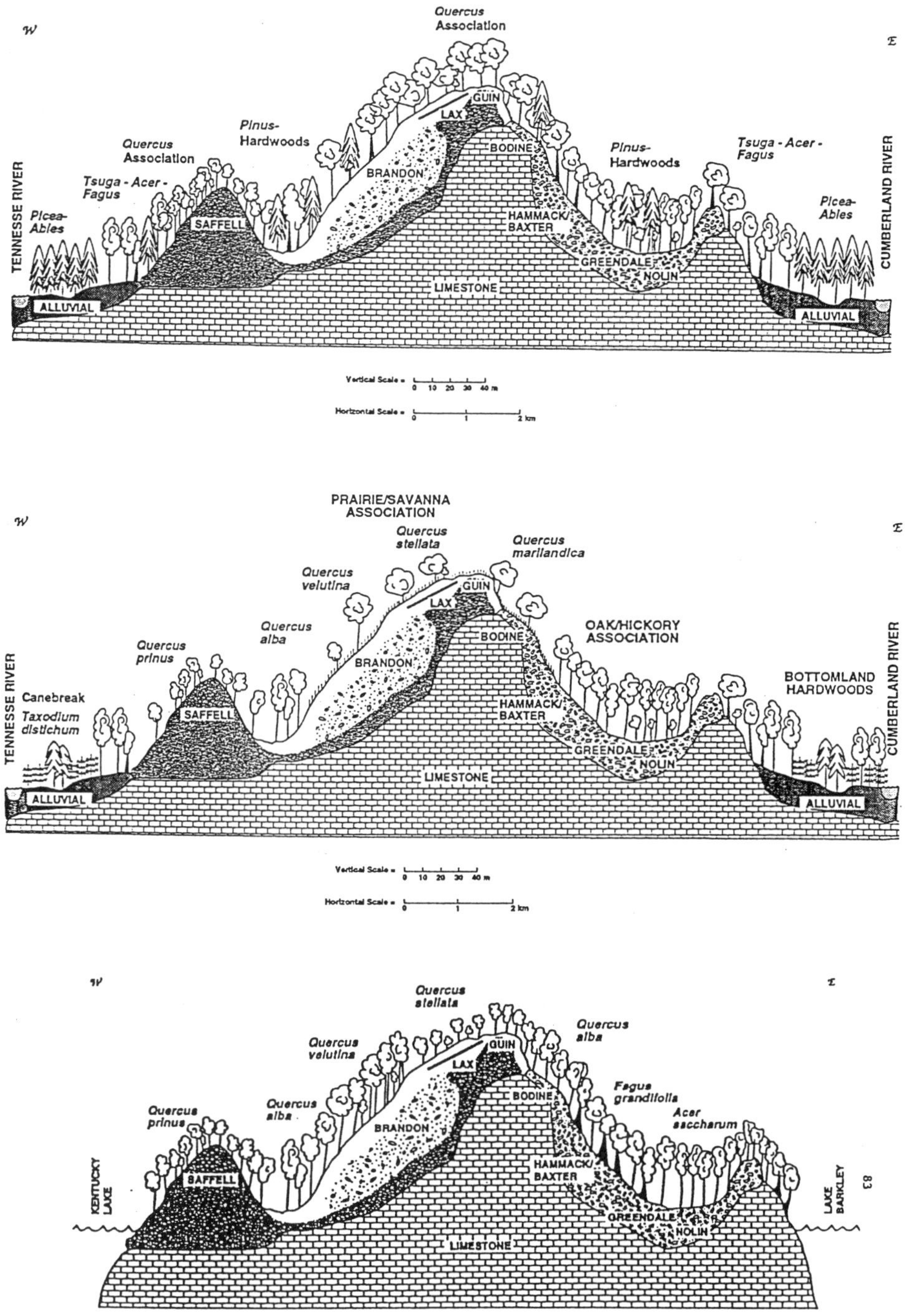

FIGURE 25.14 Schematic reconstruction of the vegetation for Land Between the Lakes, Kentucky and Tennessee, based on pollen evidence, historical records (Franklin 1994), and current vegetation analyses (Franklin et al. 1993). Diagram (a) is based on known pollen records and the autecology of species. Diagram (b) is based on historical records. Diagram (c) is based on data from 133 forest stands.

FIGURE 25.15 Standing dead American chestnut (*Castanea dentata*). The chestnut blight (*Cryphonectria parasitica*) eliminated this important tree from the eastern forests by 1930; now only sprouts remain. Photo courtesy of Jack Grubaugh.

including serotinous cones that require fire for them to open and disseminate seed. However, if longer intervals occur between fire events or if a fire event is less severe, succession to white pine and red maple can occur. Thus, the magnitude and frequency of a disturbance is an essential element in determining the influence of that disturbance on vegetation.

If vegetation is used as an indicator of site quality (see the discussion of diagnostic species in "Types of classification"), it must be understood that the seral species are generalists. That is, they tolerate a wide range of site conditions and are found along an extensive portion of the gradient. Disturbance is important because it can "set back" the successional sequence to an earlier stage and maintain pioneer populations that may otherwise be lost during succession.

It should be apparent that the major controlling factors of vegetation distributions are not independent of one another (Figure 25.1). There are feedback mechanisms between these major influences that result in a continuum of growing conditions, that is, gradients of environmental parameters. Gradients can be defined as the entire range of an influencing factor—such as moisture, ranging from very wet to very dry. These gradients are then used to classify the forest vegetation for management.

CLASSIFICATION OF FOREST SYSTEMS

Forest systems have been classified for the purpose of management for the past 75 or so years. Specific forest types are managed with definite goals in mind, whether those goals are for production of paper, building materials, or veneer, for the maintenance of biotic-abiotic relationships, or for conservation of biodiversity. Until recently, classifications were specific to each piece of public land. Currently, national vegetation classifications in the United States and Canada are under way in an effort to merge classifications into a uniform format, set of terms, and set of communities. In the United States the collaborative efforts between state and federal agencies and The Nature Conservancy may promote better communication between management specialists throughout the country.

Why Classify Forests?

In the eighteenth century, Baron von Humboldt was one of the first individuals to classify vegetation. He organized plants and animals according to latitude and elevation zones (i.e., climate). Around the turn of the nineteenth century, several other European ecologists discovered that plants are normally found in recognizable associations. Based on the efforts of these researchers, Rexford Daubenmire pioneered the use of plant associations to classify sites in North America. Since then, classification has been used as a tool to identify the extraordinary diversity of North American forests. Classification also has been used to relate vegetation to abiotic components. The response of vegetation to abiotic conditions can be thought of as a continuum. Classification groups the vegetation/environment continuum in the same way that elements of the light continuum are grouped into specific colors (Gaugh 1982). A primary reason for classification is to provide managers and researchers with a common language for describing and managing forests communities.

Another reason to classify forests is to examine the vegetation-environment relationships. Classification provides a framework for systematic gathering and interpretation of research data and empirical knowledge. Often, the specific vegetation can provide information on site characteristics that are difficult to sample (see "Techniques" later in this chapter).

A third reason to classify forests is to create a resource inventory. The Resources Planning Act of 1974 (Public Law 93-378) requires federal land management agencies to inventory the renewable resources of the nation. The information gained from inventories includes timber volume, pounds of forage, plant composition, wildlife habitat characteristics, slope, aspect, and topography, as well as current conditions and productive potential for each resource. An inventory is essential for scrutinizing various management strategies for cost-effectiveness, returns, and sustainability (Bailey 1996). Because sustainability is important, land stewards must examine options on an individual basis. Inventorying endangered or threatened species and communities has been a task of most public land stewards for decades.

Finally, classifying the structure and function of forests is important for examining mesoscale and macroscale carbon and climate models (Nemani and Running 1997). Without a solid knowledge of current flora and their ecological relationships, it is impossible to predict the effects of global climatic change, insect outbreaks, catastrophic floods, or any other major event.

How Are Forests Classified?

Types of Classification

In classifying forests, it is necessary to determine the objectives of the classification and how the classification will be used. There are several types of classification, and classification results will vary greatly depending on the user's viewpoint and the basis of the classification scheme. It is important for the user to understand the basic premises of the classification. One question is whether to examine natural, or potential, vegetation (Küchler 1964; Vankat 1979) or current vegetation (Braun 1950; Weakley et al. 1998). The difference between these two approaches to classification is that Küchler telescoped succession for his classification and Vankat ignored any impact of European settlement, whereas Braun considered disturbance (e.g., chestnut blight) and Weakley incorporated post-European developments (e.g., pine plantations).

In addition, there are several other known approaches to developing a classification. Recognized approaches include habitat typing, forest ecosystem classification, forest cover typing, and ecological land classification. Forest cover types (Eyre 1980), developed by the Society of American Foresters, were the standard for planning forest management for decades (see Burns 1983). There are limitations to the use of forest cover types for this purpose, however, because they are based on current, often successional, vegetation (Kotar et al. 1988). This method generally renders cover types that encompass a wide range of environmental conditions that respond differently to management techniques. The cover types were named according to the dominant species of the canopy strata, that is, the species with the greatest basal area or wood volume.

The habitat type classification system uses potential natural vegetation in an effort to recognize ecologically equivalent landscape units (see Daubenmire and Daubenmire 1968). A habitat type includes all sites capable of producing similar plant communities at "climax." Habitat types are generally developed for specific areas and not mapped (e.g., Pfister and Arno 1980; Kotar et al. 1988). Habitat typing assumes that understory vegetation progresses more rapidly toward climax than successional woody species, and that understories become relatively stable soon after the canopy closes. The habitat type name generally uses one or two dominant climax species with one or two diagnostic understory species. A diagnostic species is a species that appears to be relatively abundant and consistent on one group of sites, but not on others. This does not necessarily mean that it is dominant. The species typically has a narrow tolerance range along the complex continuum of environmental factors, and thus is found in association with specific species in specific areas. It is generally recommended that more than one diagnostic species be used for classifying communities. The use of diagnostic species developed from the Braun-Blanquet and Zurich-Montpellier phytosociological schools, and they are currently used for classification in many areas around the world (e.g. Becking 1957; Mueller-Dombois and Ellenberg 1974; Cajander 1926).

Ecosystem classification and ecological land classification are similar in their use of both biotic and abiotic factors to delineate groupings. Generally, the classification is hierarchical. Ecological land classification uses specific landforms and soils first (e.g., climate and soils), followed by vegetation. The vegetation of the ecological land classification units is represented by the potential climax association (i.e., habitat type). Ecological land classifications have been developed for specific regions in the United States and most of Canada. Ecosystem classification has broadened its view in an attempt to classify every ecosystem on earth. The hierarchy of this classification system includes climate, physiography, physiognomy, and floristics. Ecological land classifications and ecosystem classifications are often mapped, because they are based on ecosystem elements that have previously been mapped (e.g., climate, physiographic units, soils). A general increase may be noticed in the scale to which each of these systems pertain, as well as an inevitable increase in the hierarchical structure of the classification.

Classification Techniques

At coarser scales, the classification of the landscape generally includes information derived from remote sensing. Current continental and global scale classifications are based on the magnitude and temporal evolution of spectral vegetation indices (SVI). These indices are combinations of red and near-infrared reflectances. A common example is the Normalized Difference Vegetation Index (NDVI).

At finer levels, there are a variety of ordination and classification techniques that are available to resource managers. Ordination and classification approaches can be used together or separately, but they often reveal similar groupings. Ordination techniques simply "order" the samples along an axis based on their similarity. The ordination may be derived strictly from the vegetation data (indirect ordination; e.g., detrended correspondence analysis and nonmetric multidimensional scaling) or from the combination of vegetation and environmental data (direct ordination; e.g., canonical correspondence analysis). Although direct analyses have tended to dominate in the literature for the past decade or so, indirect analyses are currently being supported as more appropriate techniques. Researchers should first target the composition of the associations, then examine the relationships of the associations and abiotic factors. Much information may be lost if investigation is limited to a one-step approach.

Classification techniques group samples on the basis of their similarity. The approach may be dichotomous or clustering. The dichotomous techniques (e.g., TWINSPAN, Two-way indicator species analysis) split the samples at the most dissimilar point, then split these subgroups in a like manner, and continue until dissimilarities are not strong. Indicator species are used for these analyses, which differentiate the two split groups either by presence/absence or amount, which may be based on cover, density, or biomass. Clustering techniques put the most similar samples together and then build a tree of similarity from this starting point.

There is an enormous amount of literature on ordination and classification techniques. These techniques have their share of problems, and several articles have seriously criticized their effectiveness. Ordination and classification are exploratory analyses that look for relationships and should be thought of as a starting point for classifying vegetation. There are so many choices of how to perform the analyses, one can almost think of these exploratory methods as yielding an artistic expression of the data. The interpretation of the results of such analyses must incorporate ecological sense. For further exploration of this subject, we recommend the books by Ludwig and Reynolds (1988) and Jongman, et al. (1987).

Drawbacks of Forest Classification

There are two main drawbacks to classification: (1) Classifications are abstract, in the sense that a continuum is being split (discussed earlier), and (2) classifications are temporal "snapshots" of a dynamic system. Even when classifications extrapolate succession to the climax community in an attempt to avoid the confounding effects of disturbance, the classification ignores climate change, a point made by Küchler (1964).

National Vegetation Classification

The classification followed in this text is the National Vegetation Classification Standard (NVCS). The NVCS has been adopted by all federal agencies, many state agencies, and some private organizations. Thus, a great majority of land stewards will be working under this classification umbrella. Although it is still being drafted, we can describe the current classification and how major forestry species fit into the system (in the following chapters). This broad overview is meant to provide a base for exploration of greater detail in specific regions. Because the NVCS is still being developed and implementation is expected to take some time, we have developed our own naming strategy for the various communities described, which follows more closely with the current Society of American Forester's (SAF) cover type system (Eyre 1980) followed by the specific NVCS alliance name (e.g., Jack Pine Cover Type (NVCS *Pinus banksiana* Forest Alliance)). We hope that this helps the transition for foresters.

Ecoregion Classification

The National Vegetation Classification Standard is one of many classification standards being developed by the Federal Geographic Data Committee (FGDC). It is important to link the NVCS with other classifications to yield insight on geographic location and controlling factors of vegetation distribution. One such classification is Bailey's Ecoregion Classification. The Ecoregion Classification stems from the concept that all components in an ecosystem are linked. In this classification, interrelationships between components are stressed rather than the individual components. In addition, ecosystems are recognized at various scales, so they may be nested within one another (Klijn and Udo de Haes 1994). Ecoregion classification is based on two principles: (1) Ecosystems are defined by multiple factors (termed *natural classification*), and (2) classification is better done according to the causes (e.g., minimum winter temperatures) of the class rather than according to the effects that such differences produce (units derived from such a classification are termed *genetic;* Bailey 1996). The regions were classified by the factors described in the preceding section, "Principles of Forest Ecology That Relate to Classification." The Forest Service is currently developing software (NRIS Terra) incorporating these various factors in a standard fashion for all regions.

Within this text, the divisions of Bailey's regions are modified for the purposes of developing separate, easy-to-use keys for specific regions (Table 25.1). We follow these divisions in describing communities in the following chapters. Although some communities can be found in several regions, we describe them in their region of greatest concentration.

National Vegetation Classification Standard

The objective of the National Vegetation Classification Standard (NVCS; http://biology.usgs.gov/fgdc.veg/standards/vegstd.htm; last updated August 15, 2000) is to support the use of a consistent national vegetation classification to produce uniform statistics for vegetation resources from vegetation cover data. Thus, an accurate assessment (inventory) of the nation's forests can be developed. The "standard" represents the minimum required during inventory studies, rather than the ideal or maximum. The NVCS recognizes the substantial efforts of The Nature Conservancy (TNC) and its classification system and regards that system as the base for the NVCS. The Ecological Society of America has a panel in place to further develop the classification standards (esa.sdsc.edu/orgwebpg.htm).

There are nine levels in the NVCS system: (1) Division, (2) Order, (3) Class, (4) Subclass, (5) Group, (6) Subgroup, (7) Formation, (8) Alliance, and (9) Association (Figure 25.16). Levels 3 to 7 are physiognomic, Physiognomic Class (3) being defined by life form and percent cover. Physiognomic Subclass is determined by the predominant leaf phenology of woody plants and the leaf type and periodicity of herbaceous plants. Climate, leaf morphology, and leaf phenology define the Physiognomic Group, and Physiognomic Subgroups simply separate natural/seminatural types from planted/cultivated types. Physiognomic Formation identifies ecological groupings of vegetation units with broadly defined environmental factors and additional physiognomic factors.

The Alliance and Association levels are floristically defined. The Association, the finest floristic level classified in this hierarchy, is characterized by diagnostic species that occur in all strata. The diagnostic species are primarily the dominant species. The NVCS system most closely resembles the Braun-Blanquet method in the use of diagnostic species and is essentially the same at the Association level. Alliances represent an aggregation of Associations and are characterized by one or a group of diagnostic species that occur in the dominant or uppermost stratum. The Alliance level is thought to be similar to SAF cover types, except that they are based on dominant species, and therefore the NVCS is more detailed. The NVCS classification scheme represents current vegetation (unlike the habitat typing methods), including all combinations occurring as a result of past disturbances, stochastic natural successional patterns, and cultivation of

Table 25.1. Hierarchical Classification of the Major Forests of North America as Separated by This Volume in Comparison with the Ecoregion Classification of Bailey (1996)

	Ecoregion Classification		Separations by Fralish and Franklin	
Forest Type	**Bailey's Domains**	**Bailey's Divisions**	**Dominant Taxa**	**Subregions**
Boreal	Polar	Subarctic	*Picea-Abies-Populus*	
Northern Hardwood-Conifer	Humid temperate	Warm continental	*Acer-Betula-Pinus*	A. Lake States B. New England
Central Hardwood	Humid temperate	Hot continental	*Quercus-Carya*	A. Interior lowlands B. Appalachian Mountains
Southern Pine-Hardwood	Humid temperate	Subtropical	*Pinus-Quercus*	A. Piedmont B. Upper Coastal Plain C. Lower Coastal Plain
Coastal Conifer				
Northern	Humid temperate	Maritime	*Picea-Tsuga-Abies*	
Southern	Humid temperate	Medditerranean	*Abies-Pinus-Quercus*	
Rocky Mountain Conifer	Dry	Temperate steppe	*Pinus-Pseudotsuga-Picea*	A. Northern Rocky Mountain conifer B. Southern Rocky Mountain conifer
Pine-Juniper	Dry	Tropical/subtropical steppe	*Pinus ponderosa-Pinus edulis-Juniperus*	

Physiognomic Categories

Category	Example
Class	Woodland
Subclass	Mainly Evergreen Woodlands
Group	Evergreen Needle-leaved Woodlands
Subgroup	Natural/Seminatural
Formation	Evergreen Coniferous Woodlands with Rounded Crowns

Floristic Categories

Alliance	*Juniperus occidentalis*
Association	*Juniperus occidentalis/Artemesia tridentata*

FIGURE 25.16 Structure of the National Vegetation Classification System. The first five levels of this hierarcical classification are physiognomic, and the last two are floristic (From *An Initiative for a Standardized Classification of the Vegetation of the United States*, Review Draft v.5.1, 1999).

specific species. For this reason, the classification yields a rather daunting number of communities (as of August 1999, there were >4700 Associations and ~2000 Alliances in the TNC system; personal communication, Dennis Grossman, senior ecologist, TNC). The Nature Conservancy's working draft recognizes 306 forest and woodland Alliances and 883 forest and woodland Associations for the southeastern United States alone. The NVCS should not be confused with either a potential vegetation classification or a site classification, but is instead a floristic classification of the current vegetation based on diagnostic taxa.

CHAPTER 26

Forest Environment and Cover Types—Regions 1 and 2

NORTHERN CONIFER-HARDWOOD (BOREAL) FOREST AND NORTHERN HARDWOOD-CONIFER FOREST (BAILEY'S POLAR DOMAIN/SUBARCTIC DIVISION, *PICEA-ABIES-POPULUS*, AND HUMID TEMPERATE DOMAIN/WARM CONTINENTAL DIVISIONS, *ACER-BETULA-PINUS*)

INTRODUCTION

THE boreal forest (Region 1, Figure 2.1; Bailey's Polar Domain/Subarctic Division) climates are controlled chiefly by polar and arctic air masses. These climates have low temperatures, severe winters, and small amounts of precipitation. Most of the precipitation falls in the summer. The growing season ranges from 90 to 100 days. There are two divisions in this domain, tundra and subarctic. The subarctic division is the only division with considerable forest coverage. This domain makes up most of Canada and Alaska.

The Subarctic Division is south of the tundra zone between 50° and 70° north latitude, usually referred to as the boreal forest region. The climate shows great seasonal range in temperature, with the northern boundary of the boreal forest falling along the 13° C July isotherm. Net radiation may also be a factor governing this border. The southern boundary corresponds roughly to the 18° C July isotherm. In addition, the climate is moist all year, with cool, short summers (Figure 26.1). Average monthly temperatures are subfreezing for 6 to 7 months. When permafrost is present, seasonal thaw penetrates the soil between 0.5 m and 4 m of depth, depending on latitude, aspect, and soil type. The growing season ranges from 90 to 100 days. Annual precipitation ranges from 40 to 90 cm yr^{-1} (Bailey 1998).

Bedrock varies from Precambrian granitic and gneissic rocks of the Canadian Shield in eastern Canada to Paleozoic and Cretaceous sedimentary rocks in western Canada and Alaska. However, the bedrock often has less effect on forest composition than surficial deposits (e.g., alluvium, till, and outwash). The eastern Shield and the interior plain have periodically been glaciated, the last major extension occurring 18,000 years ago. Soils are Spodosols with pockets of wet organic Histosols. Soil development via podzolization is slow, and the soil profile is shallow. Spodosols have a black humus layer on top of the soil profile; it is not incorporated deeper because of the lack of biotic activity. This humus layer is highly acidic. Below the humus, the mineral soil is strongly leached, creating an

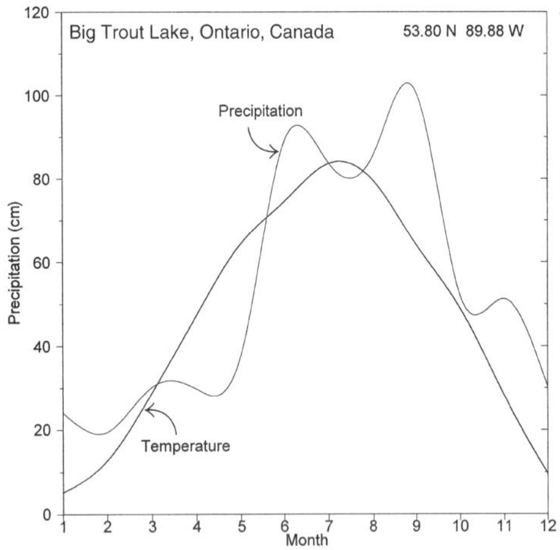

FIGURE 26.1 Climate diagram for Big Trout Lake, Ontario, Canada, representing the Subarctic Division of the Polar Domain (*sensu* Bailey 1998).

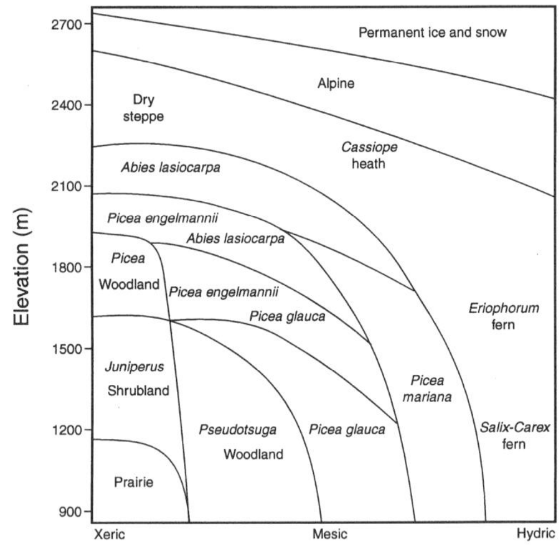

FIGURE 26.2 Gradient mosaic diagram illustrating variations in forest vegetation composition with elevation and topographic positions for Jasper National Park, Alberta, Canada. Diagram represents the boreal forest type (see Table 25.1). Redrawn from Peet (1988) with permission from Cambridge University Press and the Ecological Society of America (original illustration was from La Roi and Hnatiuk (1980)).

E horizon, and the *B* horizon is reddish because of the accumulation of iron and aluminum oxides. The soils are naturally infertile. Histosols can form in any region in wetlands where decomposition is restricted and large amounts of organic matter accumulate. There is little to no profile development in Histosol soils, but they may exceed 6 m depths (Walker 1999). These soils are mainly peat accumulations over rock or frozen water (permafrost).

Boreal forest or taiga stands (closed forests dominated by coniferous species) dominate the vegetation, following a gradient ranging from closed forest at the southern border to open lichen woodlands to the north (Figure 26.2). The physiognomy of the boreal forest is rather uniform throughout its range, due to the strong dominance of conifers (Elliot-Fisk 1988). Forests that are tightly packed, consisting only of *Picea* and *Abies*, are often referred to as taiga, a Siberian Russian term meaning "land of little trees." *Picea mariana* and *P. glauca* occur throughout the boreal forest, *Abies balsamea, Pinus banksiana* are found only in the central and eastern boreal zones, and *Picea rubens* only in the far eastern boreal zone. *Larix laricina* is also less important in the far west. To the north, the proportion of *Picea mariana* and *Larix laricina* increases because of the more rigorous climate. *Picea mariana* dominates the timberline between forest and tundra.

The Northern Hardwood-Conifer Forest (Region 2, Figure 2.1; NVCS Warm Continental Division; *Acer-Betula-Pinus*) is characterized by a humid, warm climate where average temperatures exceed 10° C for 4 to 7 months (Bailey 1998). Average temperature for the warmest month never exceeds 22° C, and this isotherm separates the Warm Continental Division from the Hot Continental Division to the south. Precipitation occurs year-round but is more substantial during the summer. There is no dry season. We have divided this region into two subregions: the Lake States and New England.

The Lake States' climate is continental and is determined by air masses from the north and west. Winters are cold and summers pleasant but short. Mean annual precipitation is 61 to 81 cm, and mean temperatures vary between 0° and 28° C (Figure 26.3). The New England climate near the coast is influenced by moist, cool air masses of the north Atlantic Ocean, and the inland climate is controlled by continental air masses moving with the westerlies (Figure 26.4). The result is a precipitation gradient ranging from west,

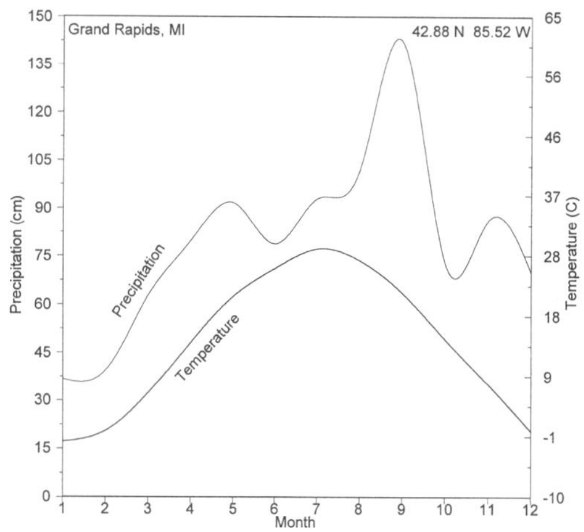

FIGURE 26.3 Climate diagram for Grand Rapids, Michigan, representing the Lake States climate within the Warm Continental Division of the Humid Temperate Domain (*sensu* Bailey 1998).

about 76 cm yr^{-1}, to east, about 127 cm yr^{-1}. Mean monthly temperature ranges from 34° to 58° C. The growing season of the Lake States and New England ranges from 100 to 160 days, north to south (Barrett 1980).

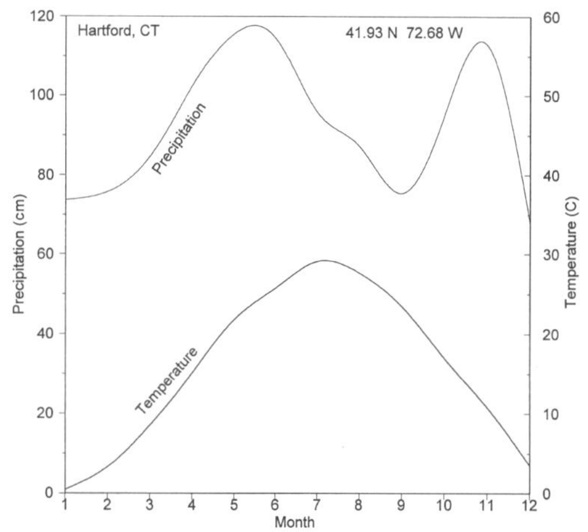

FIGURE 26.4 Climate diagram for Hartford, Connecticut, representing the New England climate within the Warm Continental Division of the Humid Temperate Domain (*sensu* Bailey 1998).

Glaciation, which impacted nearly all of the Lake States and New England subregions, left a gently to steeply rolling topography and a mosaic of deep outwash sands, morainic tills, and lake-bed clays. Loess deposits (up to 1 m thick) may overlay the glacial sediments directly east of the Mississippi River. Although there is little physiographic relief in the Lake States, several ranges occur in the New England. The air masses are influenced by various mountain ranges (Catskill, Adirondack, White, Green) that are extensions of the Appalachian Mountains. At elevations between 300 and 1000 m, mean annual temperature is 5° C and annual precipitation 90 cm. Above 1400 to 1700 m, weather can be severe, and krummholtz (Figure 26.5) and alpine vegetation dominate. The tree line is at about 1212 m (Barrett 1980). Glacial deposits are common throughout. Topography ranges from rolling to steep and precipitous.

Needleleaf and mixed needleleaf-deciduous forests grow on spodosol soils. These soils are generally acidic with a distinct A horizon, a strongly leached *E* horizon, and accumulations of iron and aluminum in the *B* horizon. They are often referred to as the gray-brown podzolics, *sensu* Great Soil Groups. Glacial deposits are common, creating a mosaic of deep outwash sands and morainic tills.

The predominant species for both the Lake States and New England subregions are *Acer saccharum, Tsuga canadensis, Betula alleghaniensis, Acer rubrum,*

FIGURE 26.5 High elevation/high latitude krummholz forest. The trees are dwarfed and flagged by continuous erosive winds. Survival is often dependent on being below the snowpack during the winter months. James S. Fralish is pictured.

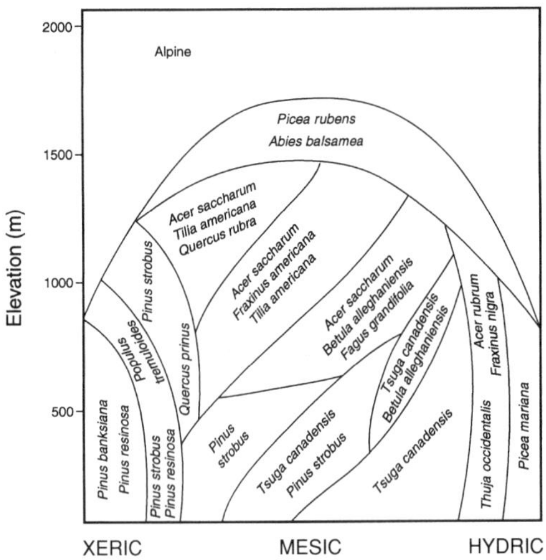

FIGURE 26.6 Gradient mosaic diagram illustrating variations in forest vegetation composition with elevation and moisture for the New England forest subregion, United States.

Quercus rubra, Pinus resinosa, and *Pinus strobus.* In the Lake States, *Pinus banksiana, Quercus ellipsoidalis, Betula alleghaniensis, Populus grandidentata,* and *Populus tremuloides* are common on dry sites and as pioneer species on moist sites following disturbance. They are of lesser importance in New England (Figure 26.6). *Pinus* species are generally restricted to sandy soils and disturbed sites (including abandoned agriculture land), whereas *Acer saccharum, Fagus grandifolia, Tsuga canadensis,* and *Betula alleghaniensis* dominate on more mesic soils (Barrett 1980; Walker 1999). The Northern Hardwood-Conifer Forest forms an ecocline between the boreal forest of the north and the deciduous hardwoods of the south.

The use of forest products throughout history has greatly altered the Northern Hardwood-Conifer Forest. Intensive cutting began in the east at about 1800 and moved steadily west. By 1850, the New England subregion was importing lumber from the Lake States subregion. By 1900, northern states were importing lumber from southern and western states. The sandy soils were often depleted of potassium by the cutting and burning. *Populus* spp., *Pinus resinosa,* and *P. banksiana* reproduced on these vast areas. *Pinus strobus* followed after potassium levels increased (Barrett 1980). This forest type extends down along the Appalachian Mountains because of the orographic ameolioration of the Hot Continental climate.

Why Conifers Instead of Hardwoods?

The most likely reason for the dominance of conifers over hardwoods in northern climates is their ability to survive on sterile soils (Pielou 1988), rather than their tolerance for the severe cold conditions. The Spodosols of the north are often quite young and have not fully developed (see the discussion in the preceding section). The answer to the dominance of conifers in nutrient-poor conditions seems to be that the leaves of the conifers last for several years. A thick cuticle, a strong epidermis, and sunken stomates are adaptations to summer and winter droughts. Having adapted to such stress, the leaves survive for more than 1 year (Vankat 1979). Thus, conifer trees do not have to grow a whole new set of leaves each spring. Instead, they only supplement their current leaves with some new leaves, and therefore their requirements for nutrient resources are much lower than those of deciduous species. Conifers seem to be more "frugal" (*sensu* Pielou 1988). Conifers are able to retain a higher proportion of their minerals (i.e., they cycle materials slowly), whereas hardwoods loose higher proportions of these essential elements when they drop their leaves in the fall. Thus, hardwoods require a more nutrient-rich soil for growth and cycle materials more quickly. The greater thickness of the organic horizons (*O*) in conifer forests as compared with hardwood forests further exemplifies the slower materials cycling under conifers, due to the combination of nutrient-poor litter and shorter growing seasons.

Remaining Heathy Within the Boreal Forest

Cold hardiness is an important adaptation for northern tree species. Whereas herbaceous species simply die back and let the blanket of snow protect them, trees must endure the lowest of temperatures. Although all of the species found in the north are frost tolerant, some are more so than others. Mild (−15° to −40° C) and severe (below −40° C) frosts separate the species into hardy and very hardy (Pielou 1988; Woodward 1987). Hardy species escape frost injury by supercooling, extremely hardy trees by extracellular freezing. Supercooling is possible because there are no nuclei (no minute particles or roughness to the inside of the cells) around which ice crystals can begin to form. Crystallizing water, snow or ice, needs a nucleus to attach to and form around. Thus, cells stay liquid well below the normal freezing point. Extracellular freezing involves the draining of cellular liquids into the many empty spaces within a living tree that are outside the cells. The liquids may freeze, and do, without injuring the tree, because the liquids are not within the cells. However, according to Pallardy et al. (1995) the tracheid cells of conifer wood are more resistance to cavitation (development of air bubbles and loss of transpirational stream) than are the vessels of angiosperms; thus conifers have greater protection from freeze-thaw cycles and greater ability to withstand cold climates. The extremely hardy tree species include *Pinus banksiana, Larix laricina, Picea mariana, Picea glauca, Abies balsamea, Betula papyrifera, Populus tremuloides,* and *Populus balsamifera.* There are several shrubs that are also extremely hardy (e.g., *Alnus* spp. and *Salix* spp.).

COVER TYPES

The cover types are in alphabetical order and numbered. The numbers are used to link the two main sections of the book, species and communities. All cover types in which a species occurs are listed at the end of the species description.

Boreal Forest (NVCS Polar Domain/Subarctic Division; *Picea-Abies-Populus*)

26.1 Balsam Fir Cover Type (NVCS *Abies balsamea* Forest Alliance): The Balsam Fir Cover Type is found on moist and wet sites and is most common in the Maritime Provinces and northern New England. *Abies balsamea* occupies optimum sites in the boreal region, generally on Spodosols, Histosols, or Inceptisols of varying acidity (Eyre 1980). Although it may form pure stands in swamps, it grows best on adjacent flats with *Picea* spp. It occurs on a wide range of organic and glaciated soil, from heavy clays to sandy glacial till, but is most common on shallow, glacial tills. *Abies balsamea* is considered a strong competitor; it is not as tolerant of shade as *Picea rubens,* but more tolerant than *Picea glauca* or *Picea mariana. Abies balsamea* is a prolific seeder with larger seeds (viability is 1 to 2 years) than *Picea* spp., often developing a monoculture regeneration understory. It can produce seed at 15 years of age (spread by wind, mainly, and some rodents) and have good seed crops every 2 years thereafter (Burns and Honkala 1990a). However, the spruce budworm generally prefers *Abies* spp. and often seriously damages old stands. *Abies balsamea* can also vegetatively propagate by layering. Layering occurs when lower limbs become buried in organic matter or sphagnum and develop a root system. *Abies balsamea* is intolerant of fire and has a moderate age span (90 to 200 years). The balsam woolly adelgid is another chief insect pest. Associates in the boreal forest region of Canada include *Picea mariana, Picea glauca, Betula papyrifera,* and *Populus tremluloides.* On upland sites of the northern forest region, *Populus tremuloides, Populus grandidentatum, Sorbus americana, Betula alleghaniensis,* and *Fagus grandifolia* are common associates. In flats and swamps, *Picea mariana, Larix laricina, Fraxinus nigra, Acer rubrum,* and *Thuja occidentalis* co-occur. *Picea rubens* may be an important associate in the eastern part of the range (e.g., Maine and New Brunswick).

26.2 Paper Birch Cover Type (NVCS *Betula papyrifera* Forest Alliance): *Betula papyrifera* is a pioneer type that revegetates following disturbance. It is strongly adapted to fire disturbance, having a thin, sloughing bark that burns readily even when wet. It does not sucker like poplar, but does produce multiple stems from burned stumps (important browse for animals). Its characteristics for pioneering include stump sprouting, seed production at about age 15, wind-dispersed seeds, quick growth requiring about 50 percent full sun, and a short life span, generally <100 years (Eyre 1980; Burns and Honkala 1990b). In addition, it has thin, sloughing bark that may fuel periodic fires. It grows best on deep, well-drained soils with good fertility, but can occur anywhere. Common associates are *Populus tremuloides* and *P. grandidentatum,* and *Prunus pensylvanica. Abies balsamea, Picea rubens, Picea glauca, Acer saccharum,* and *Tsuga canadensis* are common understory associates within their respective ranges, indicating the successional status of this type. *Rubus occidentalis* and *Rubus alleghaniensis* make up a large component of the understory. Common woody shrubs include *Corylus cornuta, Arctostaphylos uva-ursi, Diervilla lonicera, Gaultheria procumbens, Aralia nudicaulis*, and *Vaccinium* spp. *Dryopteris* spp., *Oxalis stricta*, *Cornus canadensis*, *Maianthemum dilatatum*, and *Pteridium aquilinum* are herbs common in the understory (NVCS 1999).

26.3 Tamarack Cover Type (NVCS *Larix laricina* Saturated Forest Alliance): This wetland type occurs in even-aged stands because of its shade intolerance. In addition, this shallow-rooted species is tolerant of low soil temperatures and high soil moisture. It is a pioneer burned organic matter and unburned open bogs. *Larix laricina* can occur on sites ranging from well-drained to very wet (sphagnum bogs, cool swamps, and lakeshores), but is especially dominant on poorly drained sites with organic Histosol soil (usually wetter sites than favored by *Picea mariana*) (Eyre 1980). However, it grows best in fertile loam soils along water courses. Like other bog species, *Larix laricina* can vegetatively propagate by layering (Burns and Honkala 1990a). It may bear seed at age 12 to 15 years, and the wind-dispersed seeds remain viable for 2 to 4 years; there is a good seed crop every 2 to 6 years. *Larix laricina* can live for more than 300 years, but has two serious insect pests: larch sawfly (*Pristiphora erichsonii*) and European casebearer (*Coleophora laricella*) (Harlow et al. 1996). Associates include *Picea mariana, Picea glauca, Abies balsamea*, and *Populus tremuloides*. Both tall (e.g., *Alnus rugosa, Salix* spp., and *Cornus stolonifera*) and low shrubs (e.g., *Ledum groenlandicum, Andromeda glaucophylla*, and *Chamaedaphne calyculata* var. *angustifolia*) are common in the understory. Ground cover is usually composed of *Sphagnum* spp.

Note on spruce and fir forests: The three *Picea* spp. (*P. rubens* (red), *P. glauca* (white), and *P. mariana* (black) and *Abies balsamea* are often found together, have similar silvicultural characteristics, and *Picea* species often hybridize. *Picea glauca* and *Picea mariana* are ubiquitous throughout the Subarctic Division. The distribution of *Picea mariana* is found even farther south. *Picea rubens* is an eastern species and does not reach the northern tree line. Shade tolerance decreases in the following order: *Picea rubens* > *Abies balsamea* > *Picea mariana* and *Picea glauca* > *Larix laricina* and deciduous taxa. Although there is no one true successional pattern evident in boreal forests, the tendency is the reverse of the shade tolerance order. Succession often proceeds through an open meadow community, to a broadleaf tree seral stage, and finally to an evergreen spruce-fir forest within about 300 years.

26.4 White spruce-Balsam Fir-Aspen Cover Type (NVCS *Picea glauca-Abies balsamea-Populus* spp. Forest Alliance): Inland, this cover type occurs along riverbanks (alluvial soils) and eutrophic sites, which can temporarily be flooded. The soils, often calcareous, are typically well drained, deep, and of various textures. However, this cover type is also extensive on acidic rocky and sandy sites. On the coast, *Picea glauca* can form pure stands within the fog belt (Figure 26.7) (NVCS 1999). *Picea glauca* trees can grow under extreme conditions and are able to vegetate sites

FIGURE 26.7 A white spruce (*Picea glauca*) forest near the coast of Labrador, Canada.

following glacial retreat. *Populus* occurs in younger stands and can be a very important component to this system (see the Note on hardwood taxa later in this section). *Picea glauca* is a common component near the tree line and is able to maintain itself in many areas because of its competitive ability. Thus, it is a common component of older forests, attaining ages to 350 years. However, it is extremely intolerant of fire, the primary cause of its loss (Harlow et al. 1996). Associated species include *Picea mariana, Picea rubens, Betula papyrifera, Shepherdia canadensis, Salix* spp., and *Empetrum* spp. *Corylus cornuta* is a common understory shrub. Feather mosses are common to abundant on the forest floor. The herbaceous layer is extensive in sand dune areas but limited elsewhere.

26.5 Black Spruce Cover Type (NVCS *Picea mariana* Saturated Forest Alliance): *Picea mariana* often occurs in pure stands, occupying sites ranging from moderately dry to very wet. This particular alliance is found on organic soils (Histosols) and is generally stable (Figure 26.8). Its stability is likely due to the fact that *Picea mariana* tolerates moister conditions than *Picea glauca.* Semi-serotinous cones also make *Picea mariana* more fire tolerant. Common associates include *Picea glauca, Populus tremuloides, Abies balsamea, Betula papyrifera,* and *Larix laricina. Alnus rugosa* is the most common shrub, but *Cornus stolonifera, Rubus strigosus,* and *Betula glandulosa* may also be present. On nutrient-poor mucky peat sites, *Viburnum cassinoides* and *Nemopanthus mucronatus* replace *Alnus rugosa* (NVCS 1999). Layering is an important regeneration strategy on the wetter sites (Burns and Honkala 1990a).

FIGURE 26.8 Black spruce (*Picea mariana*) forest surrounding a lake in Canada.

26.6 Black Spruce-Tamarack Cover Type (NVCS *Picea mariana-Larix laricina* Saturated Forest Alliance): This cover type occurs on low, wet, minerotrophic sites along small streams and peat bogs. Soils are generally acidic and poorly or very poorly drained. *Abies balsamea* and *Thuja occidentalis* are associates in the southern part of the range; *Alnus rugosa* and *Salix* spp. are common in the understory. Ground cover is usually composed of *Sphagnum* moss (NVCS 1999).

26.7 Black Spruce-Trembling Aspen Cover Type (NVCS *Picea mariana-Populus tremuloides* Forest Alliance): This cover type is found on mineral soils, often originating after a wildfire (NVCS 1999). *Picea mariana* can form stable communities in a variety of habitats. *Pinus banksiana* can be a common associate on very dry sites. The understory is typically dominated by mosses (e.g., feather mosses and *Pleurozium schreberi*) and lichens (e.g., *Cladina* spp.).

26.8 Red Spruce Cover Type (NVCS *Picea rubens* Forest Alliance): Red spruce is an eastern species that extends down the southern Appalachians (see the NVCS *Picea rubens-Abies fraserii* Forest Alliance no. 26.9). This is the most common type (because of its need for more abundant moisture; $\sim$100 cm yr^{-1}) in eastern Canada and Maine, occurring on moderately well-drained to poorly drained sites and thin-soiled upper slopes. These soils are mainly Spodosols and Inceptisols (Eyre 1980). Soils are developed from acidic till, derived from granite or sandstone. *Picea rubens* is long-lived and shade tolerant, forming stable stands. It has a shallow, fibrous root system that is weak and slow growing during establishment. The thin bark and flammable leaves of *Picea rubens* make it especially vulnerable to fire. Common associates are *Acer rubrum, Betula alleghaniensis, Pinus strobus, Tsuga canadensis, Picea glauca,* and *Thuja occidentalis* (*Picea mariana* on wet sites). Common shrubs include *Vaccinium* spp., *Viburnum cassinoides, Rhododendron canadense, Kalmia angustifolia, Nemopanthus mucronatus,* and *Alnus rugosa.* Bryophytes and lichens form a typical ground cover. On higher slopes, *Oxalis stricta* and feather mosses are also common, and on flats, feather mosses, *Maianthemum canadense,* and *Cornus canadensis* are common understory components. The most important insect enemy is the spruce budworm (*Choristoneura fumiferana*) (Burns and Honkala 1990a; Harlow et al. 1996).

26.9 Red Spruce-Balsam Fir Cover Type (NVCS *Picea rubens-Abies balsamea* Forest Alliance): This cover type is most abundant in the Maritime Provinces and Maine, occurring on strongly to weakly podzolized soils derived from glacial debris. Sites of occurrence can be described as well-drained upper mountain slopes that are steep and rocky with shallow soils. Principal associates are *Picea glauca, Pinus strobus, Tsuga canadensis, Thuja occidentalis, Betula papyrifera, Betula alleghaniensis, Acer rubrum, Acer spicatum, Acer pensylvanicum, Sorbus americana*, and *Prunus pensylvanica.* Shrubs associated with forest openings include *Vaccinium angustifolium, Symphoricarpos mollis, Lonicera involucrata, Ribes hirtellum*, and *Nemopanthus mucronatus.* The ground vegetation is generally mosses (*Dicranum* spp., *Hylocomium splendens, Hypnum crista-catrensis*, and *Pleurozium shreberi*) and liverworts, and sparse unless there are canopy gaps. Within gaps, common herbaceous species are *Cornus canadensis, Oxalis stricta, Maianthemum dilatatum, Clintonia borealis, Dryopteris austiaca* var. *spinulosa*, and *Trientalis borealis* (NVCS 1999). This type may also be found as a saturated forest, with or without *Abies balsamea* as a strong codominant (NVCS *Picea rubens-Abies balsamea* Saturated Forest Alliance and *Picea rubens* Saturated Forest). Sites of occurrence can be described as moderately (or less so) well-drained soils on flats, low ridges, or knolls surrounding some form of water (e.g., lakes and streams). *Sphagnum* moss may occur in the understory.

26.10 Red Spruce-Yellow Birch Cover Type (NVCS *Picea rubens-Betula alleghaniensis* Forest Alliance): This cover type occurs in the Maritime Provinces of Canada, growing primarily on acidic till soils and moist alluvial soils on lower valley slopes (NVCS 1999). Principal associates include *Tsuga canadensis, Abies balsamea, Acer rubrum, Betula papyrifera*, and *Thuja occidentalis. Corylus cornuta* and *Viburnum alnifolium* are common shrubs, and *Cornus canadensis, Oxalis stricta, Coptis groenlandica, Trientalis borealis, Aralia nudicaulis*, and *Medeola virginiana* are common herbs. Feather mosses and club mosses make up the typical ground cover.

26.11 Jack Pine Cover Type (NVCS *Pinus banksiana* Forest Alliance): This type of forest generally originates after forest fires because of *Pinus banksiana*'s serotinous cones (see the following Note on serotiny) and silvical characteristics (e.g., very shade intolerant) (Burns and Honkala 1990a). For this reason, it is usually found in even-aged stands (Figure 26.9). The environment ranges from well-drained coarse sands, including dunes, to poorly drained sphagnum-dominated areas (Eyre 1980). The tree thrives on Spodosol and dry, well-drained muskeg soils. Associates include *Populus tremuloides, Betula papyrifera, Abies balsamea, Pinus strobus*, and *Picea mariana;* these species may encroach onto and succeed *Pinus banksiana* stands. Only on very dry outwash sands can *Pinus banksiana* maintain some temporal stability. Young trees may be attacked by the jack pine budworm (*Choristoneura pinus*), the jack pine shoot moth (*Eucosma sonomana*), and the pine tip moth (*Rhyaciona adana*).

FIGURE 26.9 Jack pine (*Pinus banksiana*) stand in northern Wisconsin.

Note on serotiny: The species typically thought of as serotinous (e.g., *Pinus banksiana, Pinus rigida, Pinus elliotii*, and *Pinus contorta*) are not serotinous throughout all populations (Figure 26.10). Indeed, only a percentage of the species are truly serotinous and that percentage affects whether species recover on a site following a fire (Whitney 1997). Further, there is evidence in *Pinus banksiana* communities that they have greater portions of serotinous individuals under more mesic site conditions, where succession to other species is inevitable, competition is greater, and the need to specifically time seed dispersal with a disturbance is more important. Where sandy soils and drier climate prevail, there is apparently less need for serotiny and populations have few serotinous individuals

FIGURE 26.10 Jack pine (*Pinus banksiana*) seedlings germinating and growing following a wildfire in northern Michigan.

(Pielou 1988). Likewise, nonserotinous individuals of *Pinus contorta* are much more common in stands on bare, rocky ridges with rugged topography. These landscapes do not carry fire well, and the serotinous adaptation would certainly be inhibitive to the long-term sustainability of *Pinus contorta* on these sites (Peet 1988). Further, serotiny in *Pinus contorta* may be age dependent. Studies show that the species does not acquire serotiny until about 30 years of age, perhaps when competition becomes more prevalent in the understory.

26.12 Jack Pine-Trembling Aspen Cover Type (NVCS *Pinus banksiana-Populus tremuloides* Forest Alliance): This is a common alliance found following fires on droughty sites than those of pure *Pinus banksiana* stands. Both of these pioneers are well adapted to periodic fire disturbance, and the dominant species following a fire may simply be a matter of which one gets there first (Pielou 1988). Without fire, this type would succeed to dominance by *Picea glauca* and *Abies balsamea.*

26.13 Balsam Poplar Cover Type (NVCS *Populus balsamifera* Temporarily Flooded Forest Cover Type): *Populus balsamea* is a typical pioneer species associated with floodplain forests of the boreal zone, growing primarily on Inceptisols. This species dominates shortly (within 25 years) following the formation of a sand or gravel bar, but usually succeeds to spruce within 100 years (Burns and Honkala 1990b). Like other pioneer species, *Populus balsamea* is very intolerant of shade, grows quite quickly, has a moderately short life span (<200 years), and has prolific seed production (viability of seeds is only 4 to 6 weeks). Thus, these communities are generally even-aged stands with several associates, including *Salix* spp., *Alnus tenuifolia, Betula papyrifera, Picea mariana, Picea glauca,* and *Abies balsamea.* Shrub associates include *Cornus stolonifera, Cornus canadensis,* and *Acer spicatum.* Herbaceous associates include *Equisetum* spp., *Calamagrostis canadensis, Galium* spp., *Epilobium angustifolium, Mertensia paniculata,* and *Actaea rubra.*

26.14 Quaking Aspen Cover Type (NVCS *Populus tremuloides* Forest Alliance): *Populus tremuloides* is found on most soils, including droughty and some relatively wet sites. This pioneer revegetates following fire and clear-cutting (Figure 26.11) and is the most widely distributed tree species in North America (Eyre 1980). *Populus tremuloides* produces shallow, wide-spreading root systems. Stands continuously reproduce root suckers that await a disturbance, but *Populus tremuloides* is also able to produce seed as soon as age 10 years (younger if a clone) (Burns and Honkala 1990b). Although the viability of the seeds is high, they are extremely short-lived (2 to 4 weeks), and the environmental requirements for germination and growth must be exact at the time of dispersal. Such an occurrence is rare, of course, which necessitates the use of a vegetative dispersal strategy, as noted earlier. For this reason, *Populus tremuloides* is often found in small to large clonal clumps (Figure 26.12). *Populus tremuloides* often invades grasslands along northern margin of the Prairie Division. *Betula papyrifera, Prunus pensylvanica,* and *Pinus banksiana* are common associates. *Betula papyrifera* is often a strong codominant

FIGURE 26.11 A clear-cut in Newfoundland, Canada.

(NVCS *Populus tremuloides-Betula papyrifera* Forest Alliance). This transition type also is replaced by conifers or hardwoods.

Populus tremuloides is short-lived (living for as few as 55 to 60 years in the east and to possibly more than 100 years in the west) and very intolerant of shade, succeeding to *Pinus resinosa, Acer rubrum,* or *Quercus* spp. on dry sites, to *Pinus strobus* on intermediate sites, to northern hardwoods, *Picea glauca,* to *Abies balsamea* on fertile sites, and to *Abies balsamea, Picea mariana,* and *Thuja occidentalis* on relatively wet upland sites.

FIGURE 26.12 Quaking aspen (*Populus tremuloides*) stand in southern Nova Scotia, Canada.

Thus, the regenerating understory may be more indicative of the site environment than the overstory. Pioneer species are generally not good indicators of site conditions, but merely of the time elapsed since the last disturbance (i.e., less than 100 years). Typical understory shrub and herbaceous species are those adapted to fire in some way. *Populus tremuloides* stands have the greatest herbaceous diversity of boreal forests because of the deep shade in coniferous-dominated communities. In addition, during the growing season this type may be inundated for brief periods by overflow from adjacent waterways (NVCS *Populus tremuloides* Temporally Flooded Forest Alliance).

Note on hardwood taxa: All hardwood taxa (*Populus* spp., *Betula* spp., *Alnus* spp., and *Salix* spp.) that occur in northern climates have similar reproductive structures, catkins. Wind pollination is likely a better strategy than insect pollination in this climate. Pollination must occur early in the short growing season for the species to have time to develop fruits; this strategy is remarkably similar to that of the conifers (Pielou 1988). Whereas all other taxa depend on wind, *Salix* species do attract, and can use, insect pollinators during periods of mild temperatures. In addition, the seral hardwood taxa provide a cover under which the shade tolerant *Picea* spp. and *Abies* spp. are protected during the initial development of a conifer seral stage. The poplars provide a damp, shady environment during warm months, a leafy humus that conserves soil moisture in the fall, and a snow fence effect (greater amounts of snow) in the winter that buries young conifers and protects them from the cold (Pielou 1988). The poplars are referred to as "nurse trees" for this reason and, in essence, are eliminated from the stand by creating an environment favorable to competitors.

26.15 Pin Cherry Cover Type (NVCS *Prunus pensylvanica* Forest Alliance): *Prunus pensylvanica* is a short-lived (~30 years), early successional species (Burns and Honkala 1990b). Populations develop quickly following fire, windthrow, and clear-cutting (Eyre 1980). The species invades from viable (75 to 150 years) seed found in the seed pool (up to 1,000,000 seeds ha^{-1} in the soil). The *Prunus pensylvanica* component dies out and is succeeded by other species within 40 years. *Prunus pensylvanica* grows on infertile rocky ledges, sand plains, moist loamy soils, and rich loams; generally on Spodosols and Alfisols, but also on Inceptisols in the Appalachian Mountains. Frequent associates include *Populus grandidentatum* and *Populus tremuloides, Betula papyrifera* and *B. alleghaniensis, Abies balsamea* and *Picea rubens. Rubus occidentalis, Rubus alleghaniensis,* and *Sambucus racemosa* var. *arborescens* are common initial associates prior to crown closure. Fruits of the species are important for birds and small mammals (seed dispersers), and its leaves and twigs are important browse.

26.16 Northern White-Cedar Cover Type (NVCS *Thuja occidentalis* Saturated Forest Alliance): *Thuja occidentalis* grows on limestone-derived soils that are neutral or slightly alkaline, and moist to somewhat poorly-drained. However, it occurs on a wide variety of organic or mineral soils (Inceptisols and Entisols) of swamps, likely because of less competition (Figure 26.13). The best swamp areas have actively moving soil water. *Thuja occidentalis* is more shade tolerant and succeeds *Larix laricina* and *Picea mariana. Thuja occidentalis* is a long-lived species (>400 years), but is susceptible to fires. Its common name is arborvitae, meaning "tree of life," because of the durability of its wood when in contact with soil (although one source suggests that this name derives from the fact that Native Americans used the foliage to treat scurvy) (Eyre 1980; Harlow et al. 1996). *Abies balsamea* and *Larix laricina* are common associates. *Alnus rugosa* is commonly the most important shrub on better sites (mainly swamp). *Ledum groenlandicum, Vaccinium* spp. and *Gaultheria procumbens* are typical on poorer sites. Ground cover is usually a mosaic of *Sphagnum* and other mosses and leafy liverworts (NVCS 1999). There are no serious pests.

FIGURE 26.13 A northern white cedar (*Thuja occidentalis*) bog in northern Michigan. Note the large tip mound due to the shallow structure of the roots.

Northern Hardwood-Conifer Forest (NVCS Humid Temperate Domain/Warm Continental Division; *Acer-Betula-Pinus*)

Lake States

26.17 Sugar Maple-Basswood-Northern Red Oak Cover Type (NVCS *Acer saccharum-Tilia americana-(Quercus rubra)* Forest Alliance): This cover type is most frequently found on well-drained loams and silt loams, mostly Alfisols. Topography includes lower slopes and ravines in the upper midwest (typically northern or eastern facing) and coves of the Appalachian Mountains. Soils are generally acidic and have high nutrient concentrations (Eyre 1980; NVCS 1999); *Tilia americana* is an nitrogen-demanding species (Burns and Honkala 1990b). Although both *Acer saccharum* and *Tilia americana* are prolific seeders, *T. americana* rarely reproduces from seed. Instead, stump sprouts (a helpful field identification character) are produced, seemingly without defect or loss of growth vigor. Slow succession toward *Acer saccharum* may be occurring. Common associates include *Fraxinus americana, Ostrya virginiana, Ulmus americana, Pinus strobus, Tsuga canadensis,* and *Acer rubrum.*

26.18 Jack Pine-Northern Pin Oak-Black Oak Cover Type (NVCS *Pinus banksiana-Quercus (ellipsoidalis, velutina)* Forest Alliance): This association occurs chiefly in central Michigan and central

Wisconsin on dry, acid (infertile), sandy soils with a very thin organic layer formed on outwash plains and on gravelly morainal slopes (NVCS 1999). Harvesting of pines has allowed the oak to become more dominant in this cover type, but oak wilt has caused heavy mortality in some stands (Harlow et al. 1996). *Quercus ellisoidalis* is extremely shade intolerant and scrubby in appearance. Current stands are generally of sprout origin (Burns and Honkala 1990b), because seed crops are seldom extensive enough to produce seedlings. Herbivory and weevil damage are quite high for these seed crops. This seral community may succeed to more tolerant *Quercus* spp. (*Q. velutina*, and *Q. rubra*) or *Pinus strobus*. Other associates include *Pinus resinosa. Populus tremuloides, Populus grandidentatum, Acer rubrum, Prunus serotina*, and *Betula papyrifera*. Common undergrowth includes *Alnus viridis* ssp. *crispa, Vaccinium myrtilloides, Vaccinium angustifolium* (very common), *Diervilla lonicera, Ceonothus americanus, Prunus pensylvanica, Prunus pumila, Prunus virginiana, Crateagus* spp., *Corylus americana, Juniperus* spp., *Rosa* spp., *Amelanchier* spp., *Chimaphila* spp., *Pyrola* spp., and *Salix humilus*.

26.19 White Pine-Red Pine-Quaking Aspen Cover Type (NVCS *Pinus strobus-(Pinus resinosa)-Populus tremuloides* Forest Alliance): This alliance is found on flat to rolling topography of glacial outwash plains and moraines. Soils are generally well-drained sands or loamy sands, ranging from moderately infertile to moderately fertile. The cover type is composed of a mix of coniferous and deciduous species developing after a disturbance. Large acreages were cleared and burned in the late 1800s, and the sandy soils were depleted of many nutrients, especially the easily leached potassium. *Pinus resinosa* tolerates lower K^+ levels, so they typically regenerated on cut and burned sites, followed later by *Pinus strobus* when K^+ levels increased (Barrett 1980). Associates include *Acer rubrum, Betula papyrifera, Populus grandidentata*, and *Quercus rubra*. The conifers are generally taller than the deciduous trees (NVCS 1999). The tall shrub layer is moderately dense, including *Acer spicatum, Corylus cornuta, Diervilla lonicera*, and *Lonicera canadensis*.

26.20 Northern White-Cedar-Red Maple Cover Type (NVCS *Thuja occidentalis-Acer rubrum* Saturated Forest Alliance): This type is found on somewhat poorly and poorly drained mineral and half-bog soils. Other associates include *Betula alleghaniensis*,

FIGURE 26.14 Old-growth hemlock (*Tsuga canadensis*)-hardwood forest on a moist terrace in northern Wisconsin.

Fraxinus nigra, and *Abies balsamea*. Shrub associates include *Alnus* spp., *Cornus sericea, Ilex verticillata, Lonicera* spp., *Nemopanthus mucronatus*, and *Rhamnus alnifolia* (NVCS 1999).

26.21 Eastern Hemlock Cover Type (NVCS *Tsuga canadensis* Forest Alliance and Saturated Forest Alliance): This type occurs at elevations ranging from sea level to 750 m in the north, 300 to 900 m on the Allegheny Plateau, and 600 to 1500 m in the southern Appalachians (Eyre 1980). It is generally found with moist soil conditions (its shallow roots are susceptible to drought), on thin, rocky till, or on colluvial or shale deposits (Figure 26.14). It regenerates on moist, shaded humus or partially decomposed logs and stumps. It may also colonize sandy, marine, or fluvial banks. Soils are generally acidic, well-drained or imperfectly drained lithic regosols, lithic and gleyed podzols, or orthic gleysols. *Tsuga canadensis* is very tolerant of low light intensities, its seedlings surviving with as little as 5 percent sunlight, and is long-lived. The average age longevity is >600 years, and the record age is near 1000 years (Burns and Honkala 1990a). This species has a remarkable ability to withstand some 200 years of suppression and grow very well following a canopy opening (Harlow et al. 1996). Thus, *Tsuga canadensis* replaces most cohorts over long periods of time. However, detrimental fires (because of its thin bark) and logging have greatly reduced the dominance of *Tsuga canadensis*. Currently, hemlock stands are being infested by the hemlock woolly aldegid (*Adelges*

FIGURE 26.15 Yellow birch (*Betula alleghaniensis*) growing on an old stump. The seed germinated on the stump, and the roots grew to the soil, avoiding the dense litter layer. The stump is decaying and leaving the birch with a "prop root" structure.

tsugae), which kills as much as 80 percent of the stems in forest stands. Common associates include *Pinus strobus, Abies balsamea, Picea rubens, Acer saccharum, Fagus grandifolia,* and *Liriodendron tulipifera.* A sparse understory may include *Acer pensylvanicum, A. spicatum, Viburnum alnifolium, Viburnum acerifolium, Ilex montana,* and *Hamamelis virginiana.* Two hemlock loopers consume foliage: *Lambdina fiscellaria* and *L. athasaria* (Walker 1999).

26.22 Eastern Hemlock-Yellow Birch Cover Type (NVCS *Tsuga canadensis-Betula alleghaniensis* Forest Alliance): This type occurs on moist flats along the Canadian-United States border and at elevations between 600 and 1500 m in eastern mountain ranges. More southern occurrences are restricted to cool, moist sites (Eyre 1980). High water tables in various soil textures with a thick, mor humus layer typify older stands. A dense overstory impedes sunlight, and the lack of sunlight and dense humus layer prevent reproduction from becoming established by seed. Regeneration often takes place on partially decayed stumps and logs. In fact, a good field indicator of yellow birch is a "prop root" structure following reproduction on a log, with the log decaying and disappearing beneath (Figure 26.15). *Betula alleghaniensis* is intolerant of shade and short-lived, so this species is replaced by *Tsuga canadensis* and other longer-lived associates (Burns and Honkala 1990a). Regeneration requires a canopy opening. Common associates include *Acer rubrum, Acer saccharum,* and *Tilia americana. Acer saccharum* dominates in drier locales, and *Pinus strobus* in moister locales. Undergrowth associates includes *A. pensylvanicum* and *A. spicatum.*

New England

26.23 Sugar Maple-Yellow Birch-(American Beech) Cover Type (NVCS *Acer saccharum-Betula alleghaniensis-*(*Fagus grandifolia*) Forest Alliance): This cover type occurs at elevations from almost sea level to 800 m, where it abuts the boreal forest in southern Canada, and at elevations between 800 and 1400 m in the southern Appalachians (Eyre 1980). The type grows on moderately deep to deep sandy loam, clay loam, or loamy sand soils with a thick *O* horizon over glacial till. *Acer saccharum* tolerates a wide variety of soils but grows best on moist, well-drained loams

FIGURE 26.16 Old-growth sugar maple (*Acer saccharum*) forest in northern Nova Scotia, Canada. James S. Fralish is pictured.

(Figure 26.16) (Eyre 1980). Topography is generally flat to moderate slopes. *Fagus grandifolia* is gradually being reduced as a primary component because of the beech-Nectria complex, i.e., beech bark disease (Harlow et al. 1996). Root suckers of beech are common in these stands, a sign of the complex, which attacks older stems. Selective deer herbivory has favored *Fagus grandifolia* and *Picea* spp. over other species in many stands. The type is considered able to maintain itself through time, although *Tsuga canadensis* may outlive *Fagus grandifolia* and *Acer saccharum*. Associates include *Acer rubrum, Tsuga canadensis, Fraxinus americana, Prunus serotina, Tilia americana, Betula lenta, Quercus rubra, Pinus strobus, Abies balsamea, Ulmus americana, Ulmus thomasii, Picea glauca*, and *Picea rubens*. Understory associates include *Acer pensylvanicum, Viburnum alnifolium, Ostrya virginiana, Hamamelis virginiana, Viburnum* spp., and *Amelanchier* spp.

26.24 Sugar Maple-White Ash-Basswood Cover Type (NVCS *Acer saccharum-Fraxinus americana-Tilia americana* Forest Alliance): This type occurs on flat to rolling topography. The fertile soils are generally deep, moderately to well drained, and derived generally from calcareous parent materials (Alfisols or Spodosols). Although *Acer saccharum* is found in varying conditions, both *Fraxinus americana* and *Tilia americana* require high soil fertility and moisture. *Fraxinus americana* sprouts vigorously after cutting and fire (Figure 26.17). Soils range from sand to silt loam, have high soil moisture capacity, and have high cation exchange capacity. Associates include *Quercus rubra, Ulmus rubra, Acer rubrum, Betula alleghaniensis, Fagus grandifolia, Juglans nigra, Liriodendron tulipifera*, and *Prunus serotina*. Shrubs may include *Cornus alternifolia, Hamamelis virginiana, Lonicera canadensis, Rhododendron nudiflorum, Staphylea trifolia*, and *Viburnum acerifolium*.

FIGURE 26.17 A white ash (*Fraxinus americana*) sprout following a burn. All of the hardwoods in the Eastern Forest have the ability to sprout following a top-kill, as long as their stems are below a certain diameter threshold.

26.25 Black Ash-Red Maple Cover Type (NVCS *Fraxinus nigra-Acer rubrum* Saturated Forest Alliance): This cover type occurs in a zone extending 300 to 500 km around the Great Lakes and the St. Lawrence River. Soils are poorly drained, and either moist to wet muck or shallow peat found in swamps, peat bogs, gullies, riverbanks, or small depressions (Histosols) (Eyre 1980). Soils are frequently acidic and of medium nutrient value. *Fraxinus nigra* has a very shallow, fibrous root system. It is shade tolerant, and has a slow growth rate resulting from edaphic conditions (Burns and Honkala 1990b). Slow succession is generally to *Thuja occidentalis, Larix laricina*, or *Picea mariana*, unless soil drainage improves; succession is then toward northern hardwood species. Associates vary regionally. Lake States associates include *Ulmus americana, Abies balsamea, Populus balsamea*, and *Betula alleghaniensis*. The dominance of *Ulmus americana* has declined because of Dutch elm disease. In New England and eastern Canada, associates include *Betula lenta, B. papyrifera, B. populifolia, Acer saccharinum*, and *Picea mariana*.

26.26 Red Pine Cover Type (NVCS *Pinus resinosa* Forest Alliance): This type occurs throughout the Lake States on gently rolling sand plains, most commonly at elevations between 200 and 430 m (Figure 26.18) and up to 1280 m in West Virginia; it grows well along

FIGURE 26.18 View of the red pine (*Pinus resinosa*)-dominated sand plains of Wisconsin.

the southern fringe of the boreal forest. *Pinus resinosa* is also found on well-drained sandy soils of flatwoods, ridges, and rock outcrops. *Pinus resinosa* is intolerant of shade, growing primarily in even-aged stands following disturbance. It has fast, vigorous growth when young, and apparently succeeds *Pinus banksiana* on cutover sites. *Pinus resinosa* efficiently recycles nutrients and can tolerate lower levels of K^{+} and Mg^{++}, a common result of past cutting and burning practices, than *Pinus strobus* and most hardwoods.

Shade tolerance for northern pine decrease as follows: *Pinus strobus* > *Pinus resinosa* > *Pinus banksiana* [Burns and Honkala 1990a, Harlow et al. 1996]. As the stand ages, shorter-lived (maximum age for *Pinus resinosa* is 350 years) associates such as *Pinus banksiana, Betula papyrifera,* and *Populus* spp. die out (Figure 26.19). Eventually, more shade-tolerant species (*Pinus strobus, Picea glauca, Acer rubrum* and *Abies balsamea*) succeed the *Pinus resinosa.* Wildfires often disrupt this ecological succession (see the following Note on succession in Lake States). In addition, *Pinus resinosa* has also invaded many sites formerly dominated by the American chestnut (*Castanea dentata*), which was eliminated by the fungus *Cryphonectria parasitica.* Associates include *Pinus strobus, Acer rubrum, Quercus rubra, Picea glauca,* and *Abies balsamea*; on drier stands, *Pinus banksiana, Populus tremuloides, P. grandidentatum, Betula papyrifera,* and *Quercus ellipsoidalis.* Undergrowth shrubs include *Corylus americana, C. cornuta, Vaccinium myrtilloides, Amelanchier* spp., *Rubus* spp., *Comptonia peregrina, Salix humilus, Prunus pumila, Diervilla lonicera, Epigaea repens,* and *Spirea* spp. About ten insect species feed on *Pinus resinosa,* but none causes serious injury (Walker 1999).

FIGURE 26.19 A red pine (*Pinus resinosa*) forest in Maine.

Note on succession in the Lake States: Fire and large-scale windthrows have been common disturbances in the Lake States for centuries. Rarely does succession continue to an end point (if one exists) with these ongoing disturbances. Indeed, this is one reason supporting the argument against classifying communities as climax. Succession, in the sense of a floristic relay (Egler 1954), in the Lake States in the absence of fire is as follows: *Pinus banksiana or Pinus resinosa* → *Pinus strobus, Acer rubrum,* and *Quercus rubra; Pinus strobus* or *Populus* spp. → *Acer saccharum.*

26.27 White Pine Cover Type (NVCS *Pinus strobus* Forest Alliance): This type is found on a wide variety of soils, mostly derived from granites, gneisses, schists, or sandstones. Elevations range from sea level to ~450 m in New England and from 365 to 1065 in the southern Appalachians (Eyre 1980; NVCS 1999). The type is common on moderately moist soils. *Pinus strobus* is an ecological pioneer, especially in old fields (abandonment began in the 1860s), but has a intermediate tolerance to shade. It is less tolerant than *Acer saccharum, Fagus grandifolia,* and *Tsuga canadensis,* but more tolerant than other *Pinus* spp., *Betula* spp., and *Populus* spp. *Pinus strobus* is fairly long-lived, at least 450 years (Burns and Honkala 1990a). Seed production begins between ages 30 and 40 years, with 3- to 10-year intervals between good seed crops. Seed germination is best on moist mineral soil, and growth is fast. Establishment is easier on poorer sites because of less hardwood competition. On moist sites *Pinus strobus* can maintain itself if the soil is repeatedly scarified, but generally succeeds to *Acer saccharum* and *Tsuga canadensis.* Understories are generally sparse and change according to latitude, but may include *Vaccinium* spp., *Gaylussacia* spp., *Rhododendron* spp., *Gaultheria procumbens, Kalmia latifolia, Rubus* spp., *Viburnum alnifolium,* and *Ilex montana. Pinus strobus* has two serious enemies: white pine blister rust (a fungus that needs *Ribes* spp. as an alternate host, *Cronartium ribicola*) and white pine weevil (*Pissodes strobi*), which attacks and girdles young stems. In addition, high concentrations of ozone and sulfur dioxide are known to damage the trees.

26.28 White Pine-Oak Cover Type (NVCS *Pinus strobus-Quercus (alba, rubra, velutina)* Forest Alliance): The type generally occupies moist, well-drained soils up to elevations of 450 m in the north and up to elevations of 750 m in the Appalachian Mountains. Soil texture ranges from sandy loam to sandy, but sites may include gravel knolls within large swamps. Soils are typically acidic and relatively nutrient poor (NVCS 1999). The type often develops from second-growth *Pinus strobus* forests on old fields where hardwood seedlings and saplings have invaded, or on dry rocky soils never cleared for agriculture. Hardwoods compete better on heavier soils. The type is a seral community succeeding to more shade-tolerant hardwoods. Cutting and windthrow often hasten this succession (Harlow et al. 1996). Strong associates include *Acer rubrum* and *Fraxinus americana.* Other common trees include *Tsuga canadensis, Betula* spp., *Prunus serotina, Tilia americana, Acer saccharum,* and *Fagus grandifolia.* A once prominent component, *Castanea dentata,* has been eliminated by chestnut blight. The well-developed understory shrub layer associates include *Hamamelis virginiana, Cornus alternifolia, Viburnum acerifolium, Corylus cornuta, Alnus rugosa, Kalmia latifolia, Gaylussacia* spp., *Rubus* spp., *Vaccinium* spp., and *Sambucus pubens.*

26.29 White Pine-Hemlock Cover Type (NVCS *Pinus strobus-Tsuga canadensis* Forest Alliance): This type is generally found at elevations between sea level and 610 m in the northern part of its range, and between 610 and 1070 m in the southern Appalachians (NVCS 1999). The type is common on moderately well drained moist sites on midslopes, but site conditions are variable. In addition, the type is common on old fields with a dense pan in the soil. Toward the south, the type is restricted to narrow bottoms and cove sites. In the southern Appalachians, *Pinus strobus, Tsuga canadensis,* and hemlock form a transition between the white pine type of lower elevations and the *Picea rubens-Abies fraserii* of higher elevations. Both *Pinus strobus* and *Tsuga canadensis* regenerate following disturbances such as drought, fire, and windthrow. Both species are long-lived, but *Pinus strobus* will eventually drop out, leaving *Tsuga canadensis* to dominate (Eyre 1980). *Acer rubrum* is a common associate throughout the range of this cover type. Other associates common to the northern portion of the range include *Betula papyrifera, Quercus rubra, Fagus grandifolia, Acer saccharum, Betula alleghaniensis, Betula lenta, Picea rubens, Fraxinus americana,* and *Abies balsamea.* In the south, associates include *Quercus alba, Betula lenta, Prunus serotina,* and *Quercus prinus.* Seedlings of *Acer rubrum* and *A. saccharum* are common in the understory. *Acer pensylvanicum, Viburnum alnifolium, Lonicera canadensis,* and *Vaccinium corymbosum* may also be found in the understory.

CHAPTER 27

Forest Environment and Cover Types—Region 3

CENTRAL HARDWOOD FOREST (NVCS HUMID TEMPERATE DOMAIN/HOT CONTINENTAL DIVISION, *QUERCUS-CARYA*)

INTRODUCTION

THE Central Hardwood Forest (Region 3, Figure 2.1; NVCS Hot Continental Division; *Quercus-Carya*) has pronounced seasons of hot summers and cool winters. The Hot Continental Division is separated from the Warm Continental Division by the isotherm of 22° C for the warmest month. The Hot Continental Division is distinguished by both maritime tropical and continental polar air masses. Cool, dry air masses from the north meet warm, moist air masses from the south, resulting in rapid and frequent changes in weather and severe thunderstorms in both spring and fall. There is a gradient from north to south in growing season length (3 to 5 months in the north to 6 months in the south) and snow cover. The forests consist of broadleaf deciduous trees, needleleaf evergreen trees, and various mixtures of the two. We have split this forest type into two subregions, Interior Lowlands and Appalachian Mountains.

The regional climate is quite variable. Mean annual temperature ranges between 8° and 13° C; annual precipitation is between 75 and 110 cm (Figure 27.1). The north-south movement of air masses produces frequent changes in weather and severe thunderstorms. Soil parent material is glacial till in the Central Lowlands, from loess and coastal plain deposits near the Mississippi and Missouri Rivers, and from residuum deposits in the Interior Low Plateau. Loess deposits are on the eastern side of the rivers, starting sometimes 10 m thick and gradually decreasing in depth toward the east. The soils are chiefly Inceptisols in the bottomlands, and Alfisols (the majority of soils) and Ultisols on the uplands. Ultisols have been weathered some 100 to 110 cm. Alfisols are rich in humus with a distinct, light-colored leached zone (*E* horizon) under the dark *A* horizon. Base saturation is moderate to high. Deciduous forests dominate this region, but extensive clearing for agriculture, grazing, and logging have reduced, modified, and fragmented the original vegetation (Figure 27.2).

Near the western prairie-forest border, the forests are composed of few tree species—generally, *Quercus stellata* and *Q. marilandica* (Figure 27.3). Eastward, near the center of the region, the forest has high species richness and is composed of *Quercus alba, Q. stellata, Q. velutina, Q. rubra, Q. falcata,* and *Q. coccinea* on middle slopes to ridgetops, along with several *Carya* spp. (e.g., *C. glabra* and *C. tomentosa*). On lower slopes and in cool ravines, *Acer saccharum, Fagus grandifolia, Quercus rubra, Carya ovata, C. cordiformis, Nyssa sylvatica, Q. shumardii,* and *Liriodendron tulipifera* are common. The proportion of *Fagus grandifolia* increases to the east and north. *Pinus rigida* in the east

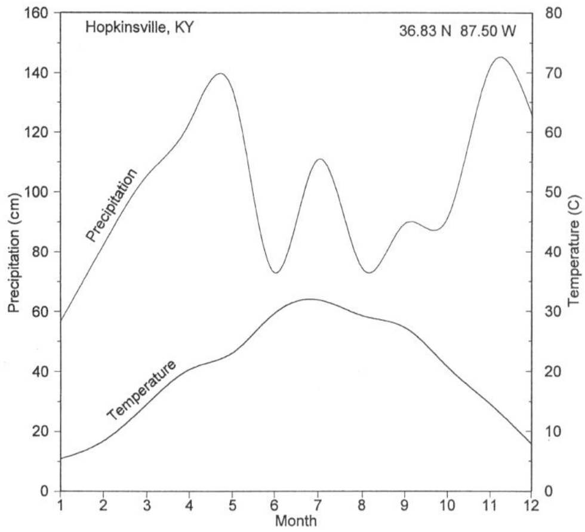

FIGURE 27.1 Climate diagram for Hopkinsville, Kentucky, representing the Interior Low Plateau climate within the Hot Continental Division of the Humid Temperate Domain (*sensu* Bailey 1998).

and *Pinus echinata* in the south may occur on the driest sites. The bottomlands, especially developed within the Coastal Plain physiographic areas of this region, are also dominated by a variety of *Quercus* spp. (e.g., *Q. michauxii, Q. palustris, Q. lyrata, Q. pagoda, Q. phellos*), as well as *Liquidambar styraciflua, Acer rubrum, Celtis* spp. and *Platanus occidentalis.* Pioneer species include *Populus deltoides, Salix nigra,* and *Acer negundo. Acer sacchrinum, Ulmus americana,* and *Acer negundo* increase in dominance from south to north on the floodplains, and overall forest diversity decreases. *Taxodium distichum-Nyssa aquatica* forests are found in the wettest areas in the southern portion of this region, but lack many of the associates of the more southern depressional forests.

The Appalachian Mountains subregion includes the Blue Ridge, Ridge and Valley, and the Appalachian Plateau. Elevation ranges from 200 to 2050 m. Annual precipitation is largely dependent on altitude; the area averages between 90 and 140 cm (Figure 27.4), but the highest peaks may receive 200 cm. Mean annual temperatures are between 10° and 15° C at lower elevations. The Appalachian uplift consists of very old igneous and metamorphic rocks, generally resistant to weathering. The Blue Ridge is dominated by crystalline rocks such as granite, gneiss, schist, mica schist and greenstone.

FIGURE 27.2 Aerial view of fragmented landscape typical throughout the eastern United States.

In constrast, the Appalachian Plateau and Ridge and Valley, areas to the northwest, are dominated by sedimentary rocks: sandstone, shale, limestone, and dolomite. With the variety of parent materials and soil-forming processes at work, the soil series are extremely diverse, even within small areas.

The vegetation of the mountains includes mixed hardwood and *Tsuga canadensis* stands in the coolest coves, *Acer saccharum* and *Quercus rubra* on cool northern-facing slopes, *Quercus alba* and *Q. velutina* on intermediate sites, *Q. prinus* and *Q. coccinea* on dry sites, and *Pinus echinata* and *Pinus virginiana* on the driest sites (Figure 27.5). At the highest elevations, *Abies fraserii* and *Picea rubens* dominate (Figure 27.6). Several of the *Quercus* spp. have developed stands following the demise of the American chestnut (*Castanea dentata*). These forests were severely cut, and some severly burned, during the 1800s, and most forests

FIGURE 27.3 Dry, open oak (*Quercus stellata*) forest.

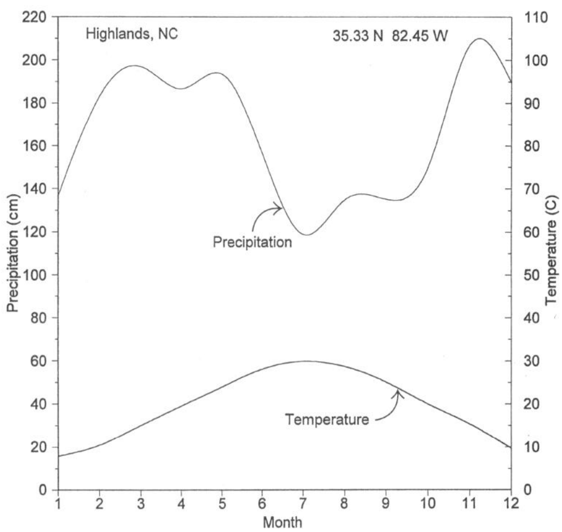

FIGURE 27.4 Climate diagram for Highlands, North Carolina, representing the Appalachian Mountains climate within the Hot Continental Division of the Humid Temperate Domain (*sensu* Bailey 1998).

FIGURE 27.5 View of three types of vegetation in Fall Creek Falls State Park, Blue Ridge Province, Tennessee. The dark trees near the valley stream are hemlock (*Tsuga canadensis*), and the dark-colored trees on the ridge are shortleaf pine (*Pinus echinata*) and virginia pine (*Pinus virginiana*). The gray slopes in between these coniferous communities are dominated by maple (*Acer saccharum*), beech (*Fagus gandifolia*), tulip polpar (*Liriodendron tulipifera*), and a variety of oak species.

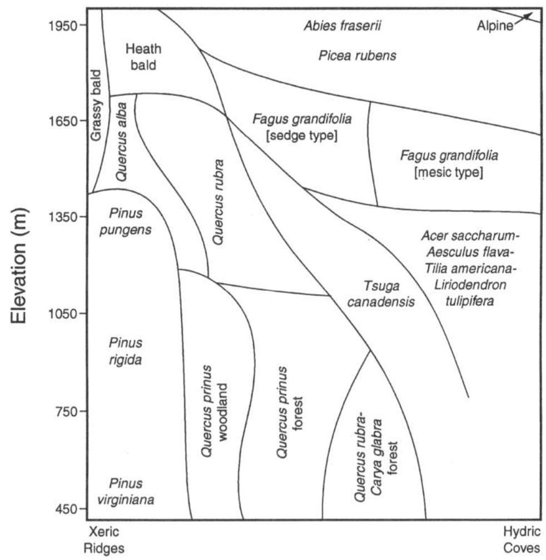

FIGURE 27.6 Gradient mosaic diagram illustrating variations in forest composition with elevation and topographic positions for Great Smoky Mountains National Park, North Carolina and Tennessee, representing the Central Hardwoods forest type (see Table 25.1). Modified from Whittaker 1956, copyright Ecological Society of America, reprinted by permission.

today are a result of stump sprouting; 80 percent of the trees of the Great Smoky Mountains National Park are estimated to be stump sprouts.

COVER TYPES

Central Hardwood Forest (NVCS Humid Temperate Domain/Hot Continental Division; *Quercus-Carya*)

Interior Lowlands

27.1 Maple-Bitternut Hickory Cover Type (NCS *Acer* (*nigrum, saccharum*)-*Carya cordiformis* Forest Alliance): Stands occur bordering floodplains, on northern aspects, on moist, well drained to moderately well drained soils. Soils are medium-textured silt (Spodosols) and colluvial, deep clay loams (Inceptisols and Entisols). Stands are moist year-round, with occasional flooding. The stress incurred during periodic flooding can create canopy openings, allowing regeneration of shade-intolerant species (Eyre 1980). *Acer nigrum* withstands warmer and drier climates than *Acer saccharum*, as well as moister soils. Other than that, *Acer saccharum* and *A. nigrum* are quite similar in ecology and response. Both species are very shade tolerant and can grow and develop after long periods of suppression (Harlow et al. 1996). The seeds and buds of *Acer* spp. are eaten by birds; *Acer* spp. are also browsed by deer (Burns and Honkala 1990b). Common associates include *Ulmus americana, U. rubra,* and *Morus rubra.* Typical shrubs include *Asimina triloba, Lindera benzoin, Aralia spinosa, Corylus americana,* and *Cornus florida.* Lianas include *Toxicodendron radicans* and *Vitis* spp.

27.2 Red Maple-Green Ash Cover Type (NVCS *Acer rubrum-Fraxinus pennsylvanica* Seasonally Flooded Forest Alliance): This type is found on level to nearly level alluvial soils of clayey or loamy sediments on floodplains of the Mississippi Valley and Coastal Plain. Flooding occurs during the winter and spring (up to 1 m in depth), and soils are flooded or saturated for a significant portion of the growing season; water may be ponded for most of the year in shallow depressions (Weakley et al. 1998). The community is considered diverse, and other common associates include *Fraxinus profunda, Liquidambar styraciflua, Quercus lyrata, Quercus bicolor,* and *Ulmus americana.* The shrub layer may include *Carpinus caroliniana, Cephalanthus occidentalis, Forestiera acuminata, Ilex decidua,* and *Itea virginica.*

27.3 Silver Maple Cover Type (NVCS *Acer saccharinum* Temporarily Flooded Forest Alliance): This type occurs primarily on well-drained moist sites along the margins of river bottoms and floodplains. It is more common near larger rivers. Sites are flooded at times (infrequently), but seedlings are adapted to survive long periods of inundation. Soils are generally organic but may also be medium- to fine-textured deep silts (Inceptisols; Eyre 1980; Weakley et al. 1998; NVCS1999). The type is rare on clays and gravels but can occur on levees. *Acer saccharinum* is a particularly fast growing tree that ranges from very intolerant to moderately intolerant of shade (Burns and Honkala 1990b). It bears seed relatively quickly (by age 11 years) and is fairly short-lived; maturity occurs at about 125 years. Succession for this pioneer community is generally to stronger dominance of bottomland *Quercus* spp. Major associates include *Liquidambar styraciflua, Quercus palustris, Q. michauxii, Populus deltoides, Betula nigra, Platanus occidentalis,* and *Fraxinus pennsylvanica.* Common shrubs include *Lindera benzoin, Ilex decidua, Asimina triloba,* and *Forestiera acuminata.* The damaging agents are many, partially due to the brittle wood of the species. Wind and ice often cause broken branches.

27.4 Sugar maple-Bitternut Hickory Cover Type (NVCS *Acer saccharum-Carya cordiformis* Temporarily Flooded Forest Alliance): Stands of this alliance are found on moist, well drained to moderately well drained soils, primarily on northern aspects of gentle slopes. Soils are medium-textured silt and colluvial, deep clay loams, derived mainly from sandstone and shales (Weakley et al. 1998). Sites are moist throughout the growing season and floods of short duration may occur. Associates include *Ulmus americana, Ulmus rubra, Morus rubra,* and *Prunus serotina.* Shrub associates include *Asimina triloba, Toxicodendron radicans, Vitis* spp., *Lindera benzoin, Aralia spinosa, Corylus americana,* and *Cornus florida.*

27.5 Sugar Maple-Chinkapin Oak Cover Type (NVCS *Acer saccharum-Quercus muhlenbergii* Forest Alliance): The community occurs in dry to mesic, well-drained soils derived from limestone or dolomite. *Quercus muhlenbergii* is known for its affinity to limestone- and dolomite-derived soils. These calcareous soils have a high pH (low acidity). Limestone outcrops may be present, as well as karst features (Weakley et al.

1998). *Quercus muhlenbergii* is also found on moist soils of higher bottomlands and low to mid slopes. Although never very abundant (perhaps because of its very sweet acorns), *Q. muhlenbergii* is found sporadically in these regional forests (Burns and Honkala 1990b). The presence of *Acer saccharum* may be due to the suppression of fire during the twentieth century. Associates include *Tilia americana, Fraxinus americana, Carya* spp., *Quercus alba,* and *Liriodendron tulipifera.* Understory shrubs may include *Cornus florida, Cercis canadensis, Calycanthus floridus, Cornus alternifolia, Ostrya virginiana,* and *Hydrangea arborescens.*

27.6 River Birch-Sycamore Cover Type (NVCS *Betula nigra-(Platanus occidentalis)* Temporarily Flooded Forest Alliance): This type is found next to watercourses on natural levees or other sites slightly elevated from the floodplain flats. Site elevations are generally less than 500 m. The community develops on riverfronts with seasonal, short-duration flooding. *Betula nigra* seedlings can withstand no more than 3 months of inundation. Soils tend to be deep and well drained to moderately well drained, with a layer of sand periodically deposited during flood events (Eyre 1980). Soils are low in organic matter and have little or no clay, but moderate to high available water capacity. *Betula nigra* is a shade-intolerant pioneer species that stump sprouts vigorously. It is also the only *Betula* species with fruit maturing in late spring, an adaptation timing seed dispersal to the exposure of alluvial shorelines as spring floods recede (Burns and Honkala 1990b). Codominants may include *Acer negundo, Populus deltoides, Acer saccharinum, Salix nigra, Celtis laevigata, Quercus laurifolia, Liriodendron tulipifera,* and *Liquidambar styraciflua.* The sparse shrub layer may contain *Asimina triloba, Lindera benzoin, Crataegus marchallii,* and *Crataegus viridis.* No serious diseases or pests are associated with this cover type.

27.7 Hickory-Ash-Oak Cover Type (NVCS *Carya (glabra, ovata)-Fraxinus americana-Quercus (alba, rubra)* Forest Alliance): This type occurs on dry upper slopes or ridge tops with relatively nutrient-rich soils. *Fraxinus americana* is considered a nitrogen-demanding species. Soils are generally well-drained loams or sandy loams. The hickories are susceptible to damage by fire. *Carya* spp. are common, but not abundant throughout the east, perhaps because of their nuts are a staple food source for many birds and mammals. However, *Carya* spp. are not considered a good browse food. Hickories are also considered to be "soil improvers" because of the high calcium content in their leaves (Burns and Honkala 1990). This facilitation, along with their intermediate shade tolerance, generally renders a succession to more shade-tolerant hardwoods. The canopy includes a variety of oak (e.g., *Quercus alba*) and hickory (e.g., *Carya ovalis*) species. The shrub layer may be absent or may include *Viburnum rafinesquianum* or an occasional *Vaccinium* spp. (Weakley et al. 1998). The most serious disease for hickories is canker rot (*Poria spiculosa*), and the hickory bark beetle (*Scolytus quadrispinosus*) is the most important insect enemy.

Note on exotic invasions: The flora of the United States is now currently composed of more than 25 percent exotic species (Figure 27.7). Exotics are species found beyond their natural ranges and are foreign in nature, establishing in new habitats (Barbour et al. 1998). Exotic species often outcompete native species because they are aggressive in their growth habits, have no natural controls on population growth such as disease or predators, and put out large amounts of seed with long viability (many are annuals) (Devine 1999). Not all exotic species are invasive, but the few that are can cause much trouble. Several agencies (e.g., the Forest Service) and organizations (e.g., the Tennessee Exotic Pest Plant Council) are currently developing lists of exotic species and their potential of invasiveness—that is, the extent to which specific exotic species may threaten our natural systems. Invasive exotics may harm the environment, the economy, and, in some cases, human health (e.g., West Nile virus). Former Interior Secretary Bruce Babbit has stated, "Invasive plant and animal species pose very real threats to America's natural heritage and our economic future." Commerce Secretary Norma Mineta estimated that invasive species cost the economy more than $100 billion each year. Exotic invasions are considered the number one threat to our National Parks. A recent Executive Order (#13112) signed by President Bill Clinton established an Invasive Species Council (ISC), which will oversee the development and implementation of a national management plan (submitted for public input in October 2000; www.invasivespecies.gov/council/nisc/nmp.html).

FIGURE 27.7 The invasive exotic western yellow star (*Centaurea solstitialis*). This species can dominate thousands of hectares following disturance.

FIGURE 27.8 Beech (*Fagus grandifolia*) forest in western Tennessee.

The invasion of an alien species into natural ecosystems has been shown to negatively affect species diversity and ecosystem function. Some of the tree species considered threatening to our natural systems include the princess tree (*Paulownia tomentosa,* although there is a current market for the wood), tree of heaven (*Ailanthus altissima*), and melaleuca (*Melaleuca quinquenervia*).

27.8 Beech-Maple Cover Type (NVCS *Fagus grandifolia-Acer saccharum-*(*Liriodendron tulipifera*) Forest Alliance): This type is found on flat to rolling topography in the north and on northern-facing sites of dissected topography in the south (Figure 27.8). Soils are fertile and well drained, generally loamy silt or sand (Weakley et al. 1998). Stands in the north are on glacial till; in the south, parent materials include alluvium, sandstone, and shale (Eyre 1980). Other overstory trees include *Fraxinus americana, Quercus rubra, Tilia americana,* and *Ulmus americana.* Common midcanopy trees include *Ostrya virginica* and *Carpinus caroliniana.* Shrub associates differ from north (e.g., *Diervilla lonicera* and *Sambucus* spp.) to south (e.g., *Asimina triloba* and *Lindera benzoin*).

27.9 Beech-White Oak Cover Type (NVCS *Fagus grandifolia-Quercus alba* Forest Alliance): This type occupies xeric-mesic to mesic sites of lower slopes or small stream bottoms (Figure 27.9). Soils are often calcareous and range from silty clays to loamy fine sands (Weakley et al. 1998). *Fagus grandifolia* is more sensitive to reduced soil moisture than *Quercus alba.* Both sprout well from stumps, and *Fagus grandifolia* may also develop a large number of root sprouts and suckers (in the "beech gaps" of the mountains, regeneration

FIGURE 27.9 A beech (*Fagus grandifolia*) and white oak (*Quercus alba*) cove in western Kentucky.

is almost entirely by suckering; Burns and Honkala 1990b). *Fagus grandifolia* is considered very shade tolerant, because of its low respiration and quick stomatal response, but is slow growing. Both species are fairly long-lived (up to 400 years). The presence of *Quercus alba* is likely due to past disturbance, as well as the selective logging practices of the early 1900s that left *Fagus grandifolia* standing alone on cut sites. Overstory cohorts include *Acer rubrum, Carya tomentosa, Carya myristiciformis, Carya ovata, Carya texana, Celtis laevigata, Diospyros virginiana, Fraxinus americana, Gleditsia triacanthos, Liquidambar styraciflua, Morus rubra, Nyssa sylvatica, Prunus serotina, Quercus michauxii, Quercus stellata, Ulmus americana,* and *Ulmus rubra. Magnolia acuminata* and *M. grandiflora* may occur on lower, moister sites. Shrubs include *Crateagus marshallii, Crateagus spathulata, Aesculus pavia, Asimina triloba, Ostrya virginiana, Carpinus caroliniana, Cornus florida, Cercis canadensis, Ilex opaca,* and *Styrax grandifolia.*

27.10 Green ash-American Elm-Hackberry Cover Type (NVCS *Fraxinus pennsylvanica-Ulmus americana-Celtis (occidentalis, laevigata)* Temporarily Flooded Forest Alliance): This type occurs along major river courses within the first bottoms and sloughs. Sites are generally flat. Soils range from organic Entisols to clays and silt loams, alluvial in origin (Inceptisols). *Celtis occidentalis* dominates in the central portion of the region, and *Celtis laevigata* dominates southern portions. Both species are found on rich alluvial soils and limestone outcrops. They are tolerant of shade, and fire is very damaging to these species (Burns and Honkala 1990b). This type is considered a midsuccessional community, often following some of the earliest pioneer communities. For example, the type succeeds *Salix nigra* riverfront communities along the Mississippi Valley. The fruits of *Celtis* spp. are important wildlife food, and birds are mainly responsible for their distribution (Harlow et al. 1996). Associates are extensive and include *Carya aquatica, Quercus texana, Quercus phellos, Quercus nigra, Quercus lyrata, Quercus laurifolia, Quercus muhlenbergii, Taxodium distichum, Siderxylon lanuginosum, Liquidambar styraciflua, Acer rubrum, Acer saccharinum, Platanus occidentalis,* and *Salix nigra.* Midcanopy trees include *Carpinus caroliniana, Asimina triloba, Planera aquatica,* and *Cornus stricta.* Common shrubs include *Ilex decidua* and *Cornus drummondii.* Lianas are important and common in these forests and include such species as *Berchemia scandens, Campsis radicans, Smilax bona-nox, Bignonia capreolata, Vitis rotundifolia, Brunnichia ovata, Cocculus carolinianus,* and *Toxicodendron radicans.*

27.11 Redcedar-Oak Cover Type (NVCS *Juniperus virginiana-Quercus (muehlenbergii, stellata)* and *Juniperus virginiana-Quercus (stellata, velutina, marilandica)* Forest Alliances): These types are generally found on calcareous soils, with *Quercus muhlenbergii* indicating the most calcareous soils. *Juniperus virginiana* and the other *Quercus* species may also be found on sandstone, on sandstone chert, or as a seral community during old-field succession (birds are very important as seed dispersers of *Juniperus virginiana;* Burns and Honkala 1990a). After about 60 years of old-field succession, *Quercus* species begin to dominate. Soils may become very thin where this type grades into a woodland alliance (open canopy structure). The presence of *Juniperus virginiana* is one of the major reasons for all the Egyptian names found throughout the region (e.g., Memphis) and in Kentucky. Early Explorers suggested the landscape and physiognomy were similar to the region around Egypt. Cedar Barrens of Lebanon State Park lies in middle Tennessee on exposed Ordovician limestone. *Juniperus virginiana* and *dry Quercus* species dominate this area and are also quick to invade native prairies and barrens. *Juniperus virginiana* is very drought tolerant with low water loss. The presence of *Juniperus virginiana* generally represents the suppression of fire for long periods of time or over-grazing by livestock. The species is slow growing but may live as long as 300 years (Harlow et al. 1996). Over time, *Quercus* species will dominate these sites because of their greater shade tolerance. *Juniperus virginiana* is most permanent on poor sites with thin, rocky soils. On calcareous soils, associates include *Carya carolinae-septentronalis, Fraxinus americana, Fraxinus quadrangulata,* and *Celtis laevigata.* Shrubs and midcanopy species include *Cercis canadensis, Bumelia lyciides, Forestiera ligustrina, Rhus aromatica, Hypericum frondosum, Cheilanthes lanosa,* and *Viburnum rufidulum.* On noncalcareous soils and old fields, associates include *Carya texana* and *Celtis tenuifolia,* as well as those already mentioned. *Juniperus virginiana* is an important browse food.

27.12 Tulip Poplar Cover Type (NVCS *Liriodendron tulipifera* Forest Alliance): This type is found on middle to lower slopes, sheltered coves or gentle concave slopes, and river terraces. *Liriodendron tulipifera* is typical of high-quality sites below 900 m elevation (Figure 27.10), dominating following disturbance. The type is common following clear-cuts and fire, as well as in old fields. *Liriodendron tulipifera* is one of the principal pioneer species in the eastern United States. It has very fast growth and excellent wind-dispersed seeds (bumper crops produced annually starting at age 15) with high germination capabilities (seeds may remain viable for many years; Harlow et al. 1996). These pioneer characteristics include extreme shade intolerance, so this species must have some canopy opening in order to reproduce (Eyre 1980). Its shade intolerance leads to a self-pruning that renders a limbless (upside-down smiles where limbs once occurred), usually straight, bole. This is an excellent character for field identification. *Liriodendron tulipifera* can have a long life span. Several stems at Joyce Kilmer National Forest are more than 300 years old (supposedly its maximum age) and more than 6 m in diameter. All are windtopped because of the brittleness of the fast-growing wood. Succession generally favors more shade-tolerant associates. Associates include *Acer rubrum, Magnolia acuminata, Carya tomentosa, Robinia pseudoacacia, Betula lenta, Acer saccharum,* and *Acer negundo;* in the Appalachians, also *Halesia tetraptera, Tsuga canadensis, Tilia heterophylla,* and *Magnolia fraserii.* The midcanopy may include *Asimina triloba, Carpinus caroliniana,* and *Staphylea trifolia.*

FIGURE 27.10 A tulip poplar (*Liriodendron tulipifera*) and mixed mesophytic hardwoods forest in western Tennessee.

27.13 Shortleaf Pine-Oak Cover Type (NVCS *Pinus echinata-Quercus* (*coccinea, falcata, alba, velutina*): This type is essentially a transition from *Pinus echinata* to *Quercus* spp., as a result of the pine's lesser tolerance of shading. Soils are generally coarse textured, well drained, and often shallow and droughty. Soils are derived from sandstones, shales, granites, and coastal plain gravels. Soils are near neutral in pH and are mainly Ultisols. The type is more common on steep, south-facing slopes, at elevations between 200 m and 900 m. On higher, drier sites, *Quercus coccinea* and *Q. prinus* associates are more common. On moister, more fertile sites, *Quercus falcata* and *Q. alba* is more common. *Pinus echinata* is less shade tolerant than *Pinus taeda*, and *P. taeda* is found on wetter soils. As compared with *P. taeda*, *P. echinata* has a larger root system, lower tolerance to poor soil aeration, and lower demand for soil nutrients. *Pinus echinata* is thus a pioneer, growing moderately fast and short-lived (rarely >150 years). *Pinus echinata* produces good seed crops every 3 to 10 years, but also can resprout well because of a peculiar juvenile growth habit (Burns and Honkala 1990a). Stems of seedlings may grow prostrate, then turn upward from just above the cotyledon, thus keeping the cotyledonary portion of the stem near the ground. Dormant buds in the axils of the leaves above the cotyledons start to grow and develop new sprouts when the bole is cut or injured by fire. Small-stature tree and shrub associates include *Oxydendrum arboreum, Kalmia latifolia, Cornus florida, Cercis canadensis, Vaccinium* spp., *Viburnum* spp., *Rhododendron nudiflorum*, and *Gaylussacia* spp. *Pinus echinata* has high resistance to fusiform rust. Its greatest threat comes from littleleaf disease, but the southern pine beetle is also a serious enemy.

27.14 Sycamore-Mixed Bottomland Hardwoods Cover Type (NVCS *Platanus occidentalis-(Fraxinus pennsylvanica, Celtis laevigata, Acer saccharum)* Temporarily Flooded Forest Alliance): This type occupies riverfronts of the first bottoms of major rivers and banks of small rivers and large creeks. Riverfronts are slightly elevated sites with poorly drained to well-drained silty soil of alluvial origin. In small creek bottoms, the type may occur on coarser-textured nonalluvial soils. Both soils are typically rich, moderately well drained, and have adequate moisture throughout the growing season. *Platanus occidentalis* must have good aeration during the growing season. Inundation for more than 2 weeks during the growing season usually kills stems. This type succeeds *Populus deltoides* stands on riverfronts, succeeds *Alnus* in the Piedmont, and colonizes old fields and cut sites on bottoms. Stands eventually succeed to *Quercus michauxii, Q. pagoda*, and/or *Q. phellos*. *Platanus occidentalis* has a strongly branched root system and is root-firm. It is one of the largest eastern hardwoods and has an open, spreading crown. Historical records show that *Platanus occidentalis* reached diameters of 4 m. It is an intermediate to intolerant pioneer that grows fast, but is long-lived (up to 500 to 600 years) (Harlow et al. 1996). Common associates include *Fraxinus pennsylvanica, Celtis laevigata, Acer negundo, Acer saccharinum, Populus deltoides, Salix nigra, Ulmus americana*, and *Carya illinoensis*. *Asimina triloba* is a shrub commonly found with this cover type. *Arundinaria gigantea* (Figure 27.11) in also a common understory associate (Eyre 1980). Defoliation by anthractnose is common, but serious disease and insect problems seem restricted to plantations (Burns and Honkala 1990b).

FIGURE 27.11 A "cane break" (*Arundinaria gigantea*) understory along a small drainage in southern Illinois.

27.15 Sycamore Cover Type (NVCS *Platanus occidentalis-(Liquidambar styraciflua-Liriodendron tulipifera)* Temporarily Flooded Forest Alliance): This type occurs on the alluvial deposits of relatively high-gradient rivers. Associated species include *Aesculus sylvatica, Asimina triloba, Cornus florida, Alnus serrulata, Ulmus americana*, and *Fagus grandifolia*. In more montane zones, *Pinus strobus, Tsuga canadensis, Betula alleghaniensis*, and *Betula lenta* may be present. At lower elevations, *Euonymous americanus* is often present, and at higher elevations *Rhododendron maximum* and *Leucothoe fontanesiana* are common.

27.16 White Oak-Red Oak-Hickory Cover Type (NVCS *Quercus alba-(Quercus rubra, Carya* spp.) Forest Alliance): This type occurs on mesic upland slopes, often with a northern and/or eastern exposure. Elevation ranges from 150 to 1220 m (Figure 27.12).

FIGURE 27.12 A white oak (*Quercus alba*) forest.

Conditions for this type are more mesic than for the White Oak-Black Oak Cover Type. Soils are generally deep and vary from silts to clays to loams. Parent materials include glacial till, limestone, shale, and sandstone. *Quercus alba* is one of the most importantlumber trees in the east. However, reproduction of *Q. alba* is often limited by the absence of good seed years and the high herbivory of the seeds by many forest animals. Mast crops, produced in these species between the ages of 50 and 200 years, occur only every 4 to 10 years. In extensive areas of the Central Hardwood region, this type is succeeding to *Acer saccharum, Fagus grandifolia,* and other mesic species (Shotola et al. 1992). Associated species include *Carya glabra, Carya ovata, Carya tomentosa, Fraxinus americana, Acer saccharum* var. *leucoderme, Cornus florida, Nyssa sylvatica, Ostrya virginiana, Tilia americana, Acer saccharum,* and *Acer rubrum.* Squirrels consume a large quantity of acorns from the white oak group (Leucobalanus). Acorns of the black oak group (Erythrobalanus) require 2 years to mature and germinate in the spring, as opposed to the white oak group acorns that mature in their first year and germinate in the fall. Acorns of the black oak group are more often avoided by rodents, perhaps because of the chemicals in the seed (Walker 1999).

27.17 White Oak-Black Oak Cover Type (NVCS *Quercus alba-Quercus* (*coccinea, rubra, velutina, falcata*) Forest Alliance): These forests are commonly successional following logging or agriculture. The type is most prevalent on middle to upper south slopes. Soils are typically well-drained silt (loess) sandy loams, but clay loams are also common (Alfisols). Parent materials include acid sandstones and shales, but the type may also be found in karst topography (i.e., limestone). Although oak complexes may maintain themselves on drier sites, there has been a succession toward domination by the white oak group simply because of their longer life spans (300 to 500 years for white oaks as compared with 100 to 200 years for black oaks) (Harlow et al. 1996). Most of the mortality of overstory oaks is found in the black oak group (Figure 27.13; also see Wellbaum 1989), a signature of the amount of

FIGURE 27.13 An old-growth forest in southern Illinois Weaver's Woods, Shotola et al. 1993. The three individuals are standing by dead black oak (*Quercus velutina*) and northern red oak (*Quercus rubra*) trees. Most of the black oak group have shorter life spans than the white oak group. Most of the canopy deaths in the Central Hardwood forests are in the black oak group. All of the small stems in the understory are sugar maple (*Acer saccharum*).

disturbance in the late nineteenth century that established the current older forests. Associates include several other *Quercus* and *Carya* species. The subcanopy and shrub layer includes *Oxydendrum arboreum, Ulmus alata, Vaccinium arboreum, Juniperus virginiana, Cornus florida, Sassafras albidum, Gaylussacia frondosa* var. *frondosa, Gaylussacia baccata, Vaccinium vascillans,* and *Vaccinium stamineum* (Eyre 1980). Both *Quercus falcata* and *Quercus stellata* are indicators of the driest conditions (Weakley et al. 1998; *Quercus alba-Quercus* (*falcata, stellata*) Forest Alliance) for this type, but with more mesic site conditions, this type is a seral community following disturbance. *Quercus* species are an important browse source, which hinders their regeneration. In addition, several pests may inhibit their success. The most serious insect pests include the gypsy moth (*Lymantria dispar*), the orange-striped oakworm (*Anisota senatoria*), and the oak-leaf caterpillar (*Heterocampa mantso*). Oak wilt (*Ceratocystis fagacearum*) is potentially the most destructive disease (Burns and Honkala 1990b).

27.18 Southern Red Oak Cover Type (NVCS *Quercus falcata* Forest Alliance): This type is found on upper slopes under xeric topographic conditions (south- and west-facing slopes and ridge tops) (Figure 27.14). Soils are well-drained sandy loam or clay-loam and often shallow (Ultisols and Alfisols). Sites may also be found on upland flats under xerohydric conditions because of a hardpan that creates a perched water table during the winter. These sites are generally very dry in late summer (i.e., flatwoods) (Weakley et al. 1998). *Quercus falcata* is highly susceptible to fire, but sprouts vigorously and has rapid growth. Seeds are produced from ages 25 to 75 years, and maximum age is 150 years. Common associates include *Quercus stellata, Q. velutina,* and *Q. coccinea. Vaccinium* species are common in the understory. *Quercus falcata* is highly susceptible to oak wilt (*Ceratocystus fagacearum*), but oak wilt is not found below the 35th parallel. Below the 35th parallel, *Hypoxylon* spp. colonize weakened trees (Burns and Honkala 1990b).

FIGURE 27.14 A southern red oak (*Quercus falcata*) forest.

27.19 Bur Oak Cover Type (NVCS *Quercus macrocarpa* Forest Alliance): This alliance occurs on the northern Great Plains and Black Hills on rolling hills, swales, breaks, and mountain slopes and along watercourses. These sites have more mesic microclimatic conditions than surrounding areas of prairie and are protected from frequent fires. Soils are moderately deep sandy loams or loams (moderately well drained to well drained), derived mainly from aeolian or alluvial sand and silt, but also from siltstone and glacial till (Weakley et al. 1998). *Quercus macrocarpa* is resistant to drought, diseases, and fire, often invading grasslands of the Great Plains. It is intermediate in shade tolerance (Burns and Honkala 1990b). Associates include *Tilia americana, Ulmus americana, Fraxinus pennsylvanica, Betula papyrifera,* and *Pinus ponderosa* (only in the Black Hills). A subcanopy may include *Ostrya virginiana* and *Juniperus virginiana.* Shrubs include *Symphocarpos occidentalis, Prunus virginiana, Ribes* spp., *Amelanchier alnifolia, Corylus cornuta,* and *Corylus americana.*

27.20 Bur Oak-Swamp White Oak Cover Type (NVCS *Quercus macrocarpa-Quercus bicolor-*(*Carya laciniosa*) Temporarily Flooded Forest Alliance): This floodplain type occurs on slack water deposits with moderately well drained to poorly drained, neutral to basic (often of calcareous origin) alluvial soils (Wekley et al. 1998). *Quercus macrocarpa* is relatively tolerant of flooding, but is not resistant to prolonged floods during the growing season. Moist soil is required for germination. Good seed crops occur every 2 to 3 years, and fruit are borne up to age 400 years, longer than in any other oak species (Burns and Honkala 1990b). Seeds are dispersed mainly by squirrels and water. Overstory associates include *Fraxinus pennsylvanica, Quercus shumardii, Carya illinoinensis, Populus deltoides,* and *Ulmus americana.*

27.21 Pin Oak-Swamp White Oak Cover Type (NVCS *Quercus palustris-(Quercus bicolor)* Seasonally Flooded Forest Alliance): This cover type occurs on floodplains with poorly drained soils, including wet clay flats and clay ridges of first bottoms (Entisols). These sites are referred to as "flatwoods." A perched water table is created by either massive brittle fragipans or impermeable clays in depressions. The soils are typically saturated during winter and spring and dry during late summer and early fall. *Quercus palustris* is not tolerant of growing season flooding (Harlow et al. 1996). *Quercus bicolor* is highly tolerant of poorly aerated soils, including glei subsurface soils (Burns and Honkala 1990b), but is generally not very abundant. The soils are also typically acidic (pH < 5.5). Species able to tolerate these extreme conditions dominate the sites. An interesting mixture of swamp-tolerant and drought-tolerant species often prevails because of the similarities in the physiological adaptations necessary to tolerate the two extremes of the moisture gradient. Reproduction for *Q. palustris,* as with many oaks, is difficult to establish. Poor seed crops occur every 3 to 4 years, and seeds are under heavy herbivore pressure, especially from waterfowl (Harlow et al. 1996). However, seeds tolerate long periods under water and are dispersed well by birds and rodents. In addition, *Quercus palustris* sprouts vigorously, although this declines with increasing stem diameter. *Quercus palustris* is intolerant of shade. Succession is generally toward longer-lived bottomland hardwoods (Eyre 1980). Other overstory species include *Liquidambar styraciflua, Quercus alba, Quercus coccinea, Acer rubrum, Carya ovata,* and *Nyssa sylvatica.* Shrubs include *Gaylussacia baccata* and *Vaccinium corymbosum. Quercus palustris* is considered a "most preferred" host for the gypsy moth (*Lymantria dispar*) and is subject to most oak diseases. Diseases in *Q. bicolor* are essentially the same as those in *Q. alba* (Harlow et al. 1996).

27.22 Chestnut Oak-Mixed Oak Cover Type (NVCS *Quercus prinus-Quercus (alba, falcata, rubra, velutina)* and *Quercus prinus-(Quercus coccinea, Quercus velutina)* Forest Alliances): These xeric to dry-mesic forests often occur on upper slopes with southern and western exposures (with stronger dominance of *Quercus coccinea* and *Q. velutina*) to low ridges, flats, and valleys (with stronger dominance of *Q. alba* and *Q. rubra*). Elevation ranges from 450 to 1,400 m. Soils are typically rocky, well drained, and acidic under the driest conditions, but may be developed over limestone as well. Sites may have a deep oak litter due to slow decomposition) or have relatively open canopies (Figure 27.15) with no litter because of the steep slopes. Many of these forests are of sprout origin. *Quercus*

FIGURE 27.15 A dry chestnut oak (*Quercus prinus*) forest. Note the open forest canopy and the common shrub, sparkleberry (*Vaccinium arboreum*).

prinus sprouts are persistent and prolific, with rapid initial growth (Eyre 1980). *Quercus prinus* is intermediate in shade tolerance and generally has slow growth because of site conditions. This type has replaced many forests dominated by *Castanea dentata* prior to the blight. Other overstory species include *Acer rubrum, Carya tomentosa, Carya glabra, Carya ovata, Carya pallida, Fagus grandifolia, Liriodendron tulipifera, Pinus strobus, Nyssa sylvatica, Pinus rigida,* and *Oxydendrum arboreum.* The midcanopy may include *Cornus florida, Hamamelis virginiana, Amelanchier arboreum,* and *Cercis canadensis.* Shrub species include *Kalmia latifolia, Symplocos tinctoria, Gaylussacia baccata, Leucothoe recurva, Vaccinium stamineum, Vaccinium vascillans,* and *Rhododendron maximum.* In addition, *Castanea dentata* sprouts may be common.

27.23 Northern Red Oak-Sugar Maple Cover Type (NVCS *Quercus rubra-(Acer saccharum)* Forest Alliance): This type is typical of mesic-xeric conditions on middle slopes and coves with northerly/easterly aspects. Soils are generally deep, moist to well-drained loams and silt-loams. Soils are often base-rich, being derived from limestone or dolomite. This type grades into a *Quercus rubra*-Mixed Oak Type [Weakley et al. 1998; Type *Quercus rubra-(Quercus alba, Quercus prinus)*] under slightly drier conditions or disturbance. Many of the more mesophytic stands are succeeding to shade-tolerant species, especially *Acer saccharum.* The current dominance of *Quercus* species is likely due to past disturbance that favored oak regeneration (Figure 27.16). *Quercus rubra* has intermediate shade tolerance, good seed crops occur every 2 to 5 years, and seedlings have fast growth (Harlow et al. 1996). The species may live 200 to 300 years. Overstory associates include *Quercus alba, Quercus velutina, Quercus prinus, Betula lenta, Fraxinus americana, Tilia americana, Ulmus americana, Acer rubrum,* and *Fagus grandifolia.* The midcanopy may include *Cornus florida, Ostrya virginiana,* and *Prunus serotina.* Shrub associates include *Hamamelis virginiana, Lindera benzoin, Amelanchier* spp., and *Corylus cornuta.* The gypsy moth (*Lymantria dispar*) is the most destructive insect pest to these species. Oak wilt and shoestring root rot (*Armillaria mellea*) are potentially destructive, and wildfires seriously damage stems (Burns and Honkala 1990b).

FIGURE 27.16 A prescribed fire in an upland oak forest of western Kentucky. This fire was prescribed to suppress succession of oak forest to sugar maple (*Acer saccharum*).

27.24 Post Oak-Blackjack Oak Cover Type (NVCS *Quercus stellata-Quercus marilandica* Forest Alliance): This xeric community is typical of upper slopes and ridges and exposed topographic positions (i.e., westerly and southerly) (Figure 27.17). The type can grow on a variety of soils, but the soils are generally shallow because of some impermeable substratum (such as a fragipan in loess) or are dry and gravelly (Ultisols and Alfisols). *Quercus marilandica* is typical of sterile soils (Harlow et al. 1996). *Quercus stellata* is very drought resistant (Harlow et al. 1996), partly because of the drought tolerance of the leaves and very thick taproot. Both *Q. stellata* and *Q. marilandica* can take on a scrubby (even pygmy) growth form under the harshest site conditions. In fact, *Q. marilandica* is typically scrubby in appearance. In addition, the aboveground stem may die back because of drought and resprout several times (Burns and Honkala 1990b). As a trade-off for its drought tolerance, *Quercus stellata* is shade intolerant and slow growing. *Quercus stellata* can attain ages of >400 years (Figure 27.18). The maximum age for *Q. marilandica* is much less. Other overstory species include *Carya glabra, Carya texana, Quercus alba, Quercus velutina, Pinus virginiana, Pinus echinata, Fraxinus americana, Ulmus alata, Nyssa sylvatica,* and *Oxydendrum arboreum.* Subcanopy species may include *Cercis canadensis, Juniperus virginiana, Diospyros virginiana, Vaccinium arboreum,* and *Chionanthus virginicus. Quercus stellata* is susceptible to the chestnut blight fungus (*Cryphonectria parasitica*) and most other insects and diseases common on oaks.

FIGURE 27.17 A post oak (*Quercus stellata*) forest on a ridge.

FIGURE 27.18 A survivor from cutting, the post oak (*Quercus stellata*) is more than 400 years old; Panther Hollow, southern Illinois.

27.25 Post Oak Flatwoods (NVCS *Quercus stellata* Flatwoods Forest Alliance): This type occurs on nearly level soil with a well-developed hardpan. The hardpan creates a shallow soil profile, trapping water during the wet winter months (perched water table) and drying during the warm, dry summers. This wide fluctuation in moisture is a very stressful condition, and often only *Quercus stellata* can tolerate the extremes. Trees are often stunted as a result of the poor growing conditions. Associates may include *Quercus alba, Q. bicolor, Q. falcata, Q. imbricaria,* and *Q. palustris.* The shrub layer is generally very sparse.

27.26 Black Oak-White Oak Cover Type (NVCS *Quercus velutina-Quercus alba-*(*Quercus coccinea*) Forest Alliance): This type occurs on upper slopes and sheltered ridge tops under xeric-mesic conditions (Eyre 1980). Soils are often sandy or rocky, and thus well drained. Soils are mainly Alfisols and Ultisols (Burns and Honkala 1990b). Common parent materials include loess, sandstone, siltstone, chert, and shale. Disturbance is likely the driving force behind the establishment this community. Windthrows and logging favor the shade-intolerant *Q. velutina* in these forests. In fact, Eyre (1980) suggests that the occurrence of this type is likely due to a fortuitous combination of circumstances, as opposed to inherent site characteristics. *Quercus velutina,* one of the most common of the eastern upland oaks, is intermediate in shade tolerance and sensitive to competition. Advanced regeneration is important, as the number of seeds even in good years (mast crops every 2 to 3 years) may be low. As in many oaks, the top may die back because of shade intolerance and resprout several times. Meanwhile, the root system continues to grow. Thus, when released, the expansive root system stimulates faster, stronger growth of the seedlings that successfully compete with other vegetation. The technique of clear-cutting works only when this advanced regeneration is present. Other overstory species include *Quercus rubra, Q. stellata, Q. prinus, Carya tomentosa, Carya glabra, Carya ovata, Pinus virginiana,* and *Pinus echinata. Cornus florida, Corylus americana, Ostrya virginiana, Oxydendrum arboreum, Sassafras albidum, Kalmia latifolia, Rhododendron calendulaceum, Gaylussacia ursina, Viburnum acerifolium, Vaccinium* spp., and *Hamamelis virginiana* are typical of the subcanopy. Fire seriously damages these trees. Oak wilt (*Ceratocystis fagacearum*), shoestring root rot (*Armillaria mellea*), and the gypsy moth (*Lymantria dispar*) are the most serious pests. The maximum age of *Quercus velutina* is <200 years; most die before they are 150 years old.

27.27 Carolina Willow Cover Type (NVCS *Salix caroliniana* Temporarily Flooded Forest Alliance): Also found in the Atlantic Coastal Plain, this type occurs on newly exposed soils (e.g., sandbars) along rivers and lakes. For this reason, the community is often found in linear or crescent-shaped patches. Flooding disturbance is required to maintain this pioneer community. *Salix caroliniana* is short-lived, and the community generally succeeds to more shade-tolerant, longer-lived bottomland hardwood species. This type provides an important wildlife cover and a food source for beaver (Leopold et al. 1998).

27.28 Black Willow Cover Type (NVCS *Salix nigra* Temporarily Flooded Forest Alliance): *Salix nigra* is a pioneer species on newly formed sandbars and river margins (along with *Populus deltoides*), often growing so densely that it excludes any other species. Propagation by natural roots and root suckers is excellent (Harlow et al. 1996), making *Salix nigra* a favorite for bank stabilization projects. *Salix nigra* may also be found throughout the first bottoms on poorly drained soils. *Populus deltoides* will outgrow *Salix nigra* unless flooding is frequent and extended within the growing season. *Salix nigra* has the ability to tolerate large amounts of sediment deposition, up to 1.5 m (much greater than most species can tolerate). Adventitious roots are generated at the surface of the latest deposition and grow parallel to the soil surface. In fact, it is typical to find that each of these adventitious root collars represents a different depositional event. *Salix nigra* also propagates through root sprouting, but not to the extent of *Salix exigua. Salix nigra* is a very shade-intolerant pioneer that produces good seed crops every year between the ages of 25 and 75 years; it is short-lived. The stringent requirements for seed germination (best on nearly flooded, exposed mineral soil) and establishment limit *Salix nigra* to wet, alluvial soils (Entisols) near watercourses. *Populus deltoides* is the chief associate, especially in the early stages, with lesser associates including *Fraxinus pennsylvanica, Platanus occidentalis, Carya illinoensis, Diospyros virginiana, Gleditsia aquatica, Ulmus americana, Taxodium distichum, Acer rubrum, Celtis laevigata,* and *Acer negundo.* Shrubs include *Planera aquatica, Forestiera acuminata, Cephalanthus occidentalis,* and *Ilex decidua.* There are no serious disease or insect pests of *Salix nigra.*

Appalachian Mountains

27.29 Yellow Birch-American Beech-Yellow Buckeye Cover Type (NVCS *Betula alleghaniensis-Fagus grandifolia-Aesculus flava* Forest Alliance): This type occurs in cool, humid climates at high elevations (1066 to 1822 m). Sites are variable and include upper concave slopes, steep boulder fields, talus slopes, flat ridge tops, and saddles between ridges. This type generally occurs as a patch in a matrix of other forest or grassland types because of microtopographic variations and disturbance. Other canopy species include *Acer saccharum, Betula lenta, Halesia tetraptera* var. *monticola, Picea rubens, Prunus serotina, Quercus rubra,* and *Tilia heterophylla.* The subcanopy may include *Acer pensylvanicum, Acer spicatum, Amelanchier arborea, Aristolochia macrophylla, Cornus alternifolia, Crataegus punctata, Hydrangea arborescens, Ilex montana, Ribes* spp., *Rubus* spp., *Vaccinium erythrocarpon,* and *Viburnum alnifolium.*

27.30 Tulip Poplar-Mesophytic Hardwoods Cover Type (NVCS *Liriodendron tulipifera-Tilia heterophylla-Aesculus flava-Acer saccharum* Forest Alliance): This alliance occurs in broad coves and on lower slopes (protected mesic sites) at elevations between 610 and 1370 m. Soils are deep, moist, well-drained, medium-to-fine-textured soils, derived from sandstone, shale, and limestone. *Liriodendron tulipifera* is extremely shade intolerant, so an opening in the canopy is necessary for reproduction (a gap may be enough). In addition, the seeds have the unusual ability to lie dormant in humus for many years (Eyre 1980). The canopy has a mixture of mesophytic species, often without a clear dominant. Locally, these are referred to as "cove forests" (Figure 27.19). Other species present may include *Prunus serotina, Fraxinus americana, Acer*

FIGURE 27.19 A cove forest in the Great Smoky Mountains National Park. Such forests have a high diversity of mesophytic species.

rubrum, Quercus rubra, Fagus grandifolia, Carya cordiformis, Beluta alleghaniensis, Halesia tetraptera, and *Magnolia fraserii.* Shrubs may include *Hydrangea arborescens, Lindera benzoin, Calycanthus floridus,* and *Cornus alternifolia.*

27.31 Red Spruce-Fraser Fir Cover Type (NVCS *Picea rubens-Abies fraseri* Forest Alliance): This type occurs at elevations between 1400 and 1700 m in the Appalachian Mountains of North Carolina and Tennessee. Annual rainfall at these elevations is between 1900 and 2540 mm, and average summer temperatures are typically below 16° C (Burns and Honkala 1990a). Soils are generally shallow, rocky, and intensely weathered Inceptisols. The soils are highly acidic (pH about 4.0), low in exchangeable bases, and high in exchangeable aluminum. *Abies fraseri* is closely related to *Abies balsamea,* found to the north, and is probably its ecological equivalent. Windthrows are quite common in the high-elevation forests, where the trees are shallow rooted. Stands may also be very dense, with a basal area up to 60 m^2/ha. In addition, fire and cutting may produce a younger forest dominated by *Prunus pensylvanica* and *Betula alleghaniensis.* These young stands are invaded by *Abies* and *Picea,* the hardwoods acting as nurse trees, but the replacement process is generally slow. In general, *Picea rubens* dominates at lower elevations and *Abies fraseri* dominates at higher elevations. However, *Abies fraseri* has been impacted by the balsam woolly adelgid (*Adelges piceae*). The adelgid was first discovered in North Carolina on Mount Mitchell (Great Smoky Mountains National Park) in 1957 and has since spread to and decimated all areas of the endemic *Abies fraseri* (Figure 27.20). Research suggests that the fir were weakened by pollution, allowing a quicker and more intense invasion by the exotic adelgid. *Picea-Abies* forests throughout the Appalachian Mountains have been negatively affected by ozone, sulfur dioxide, and nitrous oxides. In addition, the woolly adelgid infection weakens the tree, rendering it susceptible to many other pests and shortening *Abies* natural age from ~150; however, seedlings are becoming established. *Tsuga canadensis* and *Betula alleghaniensis* are the principal associates, and less often *Fagus grandifolia, Acer saccharum, Aesculus octandra, Sorbus americana,* and *Acer spicatum* cooccur. At higher elevations, only *Betula alleghaniensis,*

FIGURE 27.20 Fraser fir (*Abies fraserii*) on Clingman's Dome, Great Smoky Mountains National Park. The balsam woolly adelgid (*Adelges piceae*) has decimated this forest.

Acer spicatum, and *Prunus pensylvanica* co-occur. Shrubs include *Vaccinium* spp., *Rhododendron* spp., and *Viburnum alnifolium*. Bryophytes are abundant. Red squirrels are strong herbivores of seeds and terminal buds of *Abies fraseri*.

27.32 Xeric Pine-Oak Cover Type (NVCS *Pinus* (*rigida, echinata*)-*Quercus* (*montana, coccinea*) Forest Alliance): This type occurs on southerly, upper, convex slopes and ridges. Elevation ranges from 515 to 1151 m. The *Quercus* development in this community is due to fire suppression. Thus, a natural succession from pine to oak is now occurring. Under a natural fire regime, *Pinus* species would be favored on these poor sites. *Pinus virginiana*, *Quercus alba*, and *Quercus marilandica* may also be present. The generally dense shrub layer includes *Vaccinium vascillans*, *Kalmia latifolia*, *Gaylussacia ursina*, *Gaylussacia baccata*, and *Rhododendron maximum*.

27.33 Pitch Pine-Oak Cover Type (NVCS *Pinus rigida*-*Quercus* (*velutina, prinus*) Forest Alliance): This type has a spotty distribution, restricted to sandy soils or infertile ridges and slopes in the northeast and throughout the Appalachian Mountains (Burns and Honkala 1990a; Weakley et al. 1998; Harlow et al. 1996). It occurs at elevations below 600 m in the north and between 430 and 1400 m in the southern Appalachians. *Pinus rigida* is intolerant of shade and one of the first species to invade following fire and abandonment of agricultural lands. The deep root system helps the species tolerate drought (Harlow et al. 1996). In addition, *Pinus rigida* has an outstanding ability to tolerate fire and browsing. Even if the foliage is killed by fire, the crown may "green up" from dormant buds capable of active growth when stimulated. Even the lowermost buds on the bole are protected because of the characteristic basal crook. This growth is caused by the stem growing prostrate for the first part of its life prior to turning upward (Burns and Honkala 1990a). Thus, stumps may sprout following fire, a rare trait in conifers. Because good seed crops occur only every 4 to 9 years, this species relies heavily on vegetative propagation. However, individuals may bear seeds at 3 years of age and some individuals have serotinous cones, thus colonizing well following fire. *Quercus* spp. and *Carya* spp. succeed *Pinus rigida*, except on very poorly drained sites, where *Acer rubrum*, *Nyssa sylvatica*, and *Magnolia virginica* succeed *Pinus rigida*. *Vaccinium vascillans* and *Gaylussacia* spp. are common understory associates. The large amount of pitch inhibits serious and extensive damage by pests. The maximum age for *Pinus rigida* is 200 years. Stems may be heavily browsed.

27.34 Chestnut Oak-Red Oak Cover Type (NVCS *Quercus prinus*-*Quercus rubra* Forest Alliance): This type occurs at moderate elevations between 900 and 1200 m, generally on protected slopes. Soils are often rocky. It also occurs at intermediate aspects and slope positions that are not too hot and not too cool. *Castanea dentata* sprouts are common in the understory (Figure 27.21), an indication of the species historical dominance of these forests. *Liriodendron*

FIGURE 27.21 Chestnut (*Castanea dentata*) sprouts under the canopy of an oak and beech forest. Sprouts are found throughout the former range of chestnut, one of the dominant canopy trees of the eastern deciduous forest before its demise due to chestnut blight (*Endothia parasitica*).

tulipifera and *Acer rubrum* are common overstory species, and the midcanopy may include *Hamamelis virginiana, Oxydendrum arboreum,* and *Acer pensylvanicum.* Shrubs include *Rhododendron maximum, Rhododendron minus, Kalmia latifolia, Gaylussacia* spp., and *Vaccinium* spp.

27.35 Northern Red Oak Cover Type (NVCS *Quercus rubra* Montane Forest Alliance): This type is found at higher elevations (1070 to 1525 m) than the Chestnut Oak-Red Oak cover type, above the elevational limit of *Quercus prinus.* Sites include broad ridges, middle to upper slopes, and steep rocky slopes. Soils are well drained, loamy textured, and underlain by Precambrian schists, gneisses, and granites. Soils generally have a high base status (Weakley et al. 1998). Associates include *Acer rubrum, Crateagus punctata, Crataegus flabellata, Betula alleghaniensis,* and *Betula lenta.* The tree form may be stunted because of wind exposure, ice damage, and poor growing conditions. Subcanopy species include *Hamamelis virginiana, Amelanchier arborea, Acer pensylvanicum, Halesia tetraptera,* and *Ilex montana.*

27.36 Eastern Hemlock-Red Maple Cover Type (NVCS *Tsuga canadensis-Acer rubrum* Saturated Forest Alliance): These palustrine forests occur interspersed with *Sphagnum*-dominated depressions in poorly drained bottomlands at elevations <1200 m. Other canopy species include *Liriodendron tulipifera, Nyssa sylvatica,* and *Betula alleghaniensis.* Dominant shrubs include *Rhododendron maximum, Kalmia latifolia,* and *Leucothoe fontanesiana* (Weakley et al. 1998).

27.37 Eastern Hemlock-Tulip Polar Cover Type (NVCS *Tsuga canadensis-Liriodendron tulipifera* Forest Alliance): This type occurs in deep coves, moist flats, and ravines (and thus with infrequent fires) at elevations between 400 and 1200 m in the mountainous areas of the central eastern United States (Figure 27.22) (Eyre 1980; Weakley et al. 1998). Soils are Spodosols. The type originates from some form of disturbance, inasmuch as *Liriodendron tulipifera* requires mineral soil and sufficient light (as well as adequate moisture) for germination and establishment. *Tsuga canadensis,* however, tends to regenerate better on a litter layer or rotten log (not mineral soil) under some shade. The combination thus arises from ground fires or cutting. Associates include *Fraxinus americana, Tilia heterophylla, Betula alleghaniensis, Beluta lenta, Magnolia acuminata, Acer saccharum, Quercus rubra, Q. velutina,*

FIGURE 27.22 A hemlock (*Tsuga canadensis*) forest in the Great Smoky Mountains National Park.

and *Q. alba.* Undergrowth may include *Rhododendron* spp., *Viburnum acerifolium, Viburnum alnifolium,* and *Hydrangea arborescens* (Weakley et al. 1998).

27.38 Carolina Hemlock Cover Type (NVCS *Tsuga caroliniana* Forest Alliance): This type occurs mostly on cliffs, rocky slopes, and ridges at elevations between 700 and 1200 m (Harlow et al. 1996). Soils are generally nutrient poor and rocky and have exposed rock at the surface. Tree form in this type may be wind shorn, gnarled, and twisted because of exposure. *Tsuga caroliniana* is an endemic of the southern Appalachians that is more suited to drier sites than *Tsuga canadensis.* Associates may include *Quercus prinus, Quercus rubra, Pinus rigida,* and *Pinus pungens.* An ericaceous shrub layer may include *Rhododendron maximum, Rhododendron catawbiense, Rhododendron minus, Kalmia latifolia, Gaylussacia* spp., and *Vaccinium* spp. (Weakley et al. 1998).

CHAPTER 28

Forest Environment and Cover Types—Region 4

SOUTHERN PINE-HARDWOOD FOREST (BAILEY'S HUMID TEMPERATE DOMAIN/SUBTROPICAL DIVISION, *PINUS-QUERCUS*)

INTRODUCTION

HIGH humidity and the absence of cold winters characterize the Southern Pine-Hardwood Forest (Region 4, Figure 2.1; Bailey's Subtropical Division, *Pinus-Quercus*). These conditions occur throughout the southern Atlantic and Gulf Coastal states. The Subtropical Division generally has no dry season, with at least 3 cm of rain in the driest month. Rainfall is ample and markedly greater during the middle one-half of the year because of frequent thunderstorms, tropical cyclones, and hurricanes. Average temperatures of the warmest month are >22° C. Soils become more strongly leached toward the south, generally changing from Spodosols to Ultisols. Ultisols are often referred to as red-yellow podzolics, *sensu* Great Soil Groups. They are weathered to more than 150 cm and rich in oxides of both iron and aluminum. Coastal Plain soils are formed from deep deposits of sand, clay, and gravel, and decayed bedrock or saprolite underlies Piedmont and Ozark Plateau soils. These soils are quite poor of most essential nutrients. There are seven major *Pinus* spp. (*P. echinata, P. taeda, P. virginiana, P. rigida, P. palustris, P. elliottii, P. clausa*). Whereas inland areas have deciduous forest species (especially *Quercus* spp.), *Pinus palustris, Pinus taeda,* and *Pinus elliottii* typically dominate coastal areas. Prescribed fires are necessary to suppress the succession from pine to oak, as *Quercus* species are generally more shade tolerant than the *Pinus* spp. We have split this division into three subregions: Piedmont, Upper Coastal Plain, and Lower Coastal Plain (see Table 25.1).

The climate of the three subregions is some what similar (Figures 28.1, 28.2, and 28.3). A pronounced dry season occurs in the fall, from September (in the Piedmont) and November (in the Lower Coastal Plain). The bulk of the precipitation (almost all rain) falls in the summer in both the Piedmont and Upper Coastal Plain subregions, and the Lower Coastal Plain receives significant moisture in the winter as well.

The Piedmont subregion is a narrow eroded zone in the easternmost section of the Appalachian highlands. Elevations range between 183 and 457 m. Parent materials are very old and mostly metamorphic (gneisses, schists, marble, quartzite, and granite). Soils, mainly Alfisols and Ultisols, are loam or clay loam and slightly acidic. Mineral nutrient content is fairly high except for nitrogen. Settlement began in the mid-1700s and by the mid-1800s, 87 percent of the land was under cultivation. Continuous cultivation has left the Piedmont seriously eroded, some portions having lost up to 75 percent of the loamy topsoil. The eroded reddish subsoils are poorly drained and low in organic

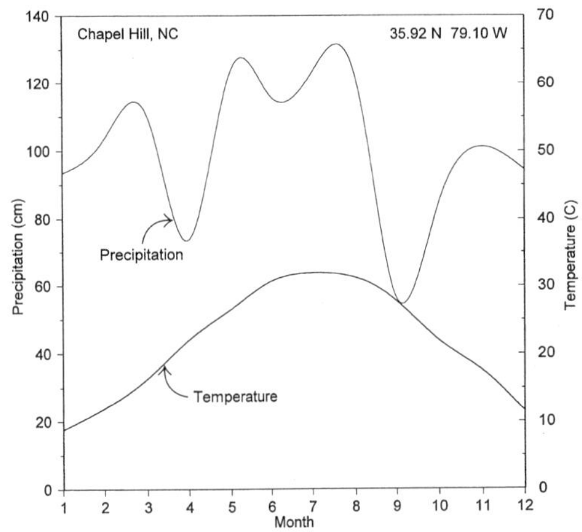

FIGURE 28.1 Climate diagram for Chapel Hill, North Carolina, representing the Piedmont climate within the Subtropical Division of the Humid Temperate Domain (*sensu* Bailey 1998).

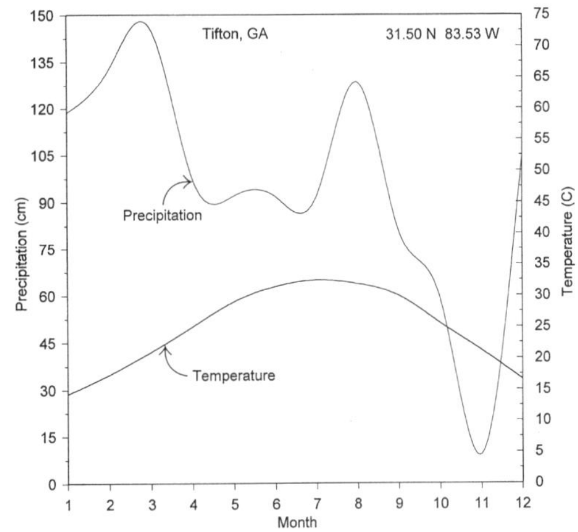

FIGURE 28.3 Climate diagram for Tifton, Georgia, representing the Lower Coastal Plain climate within the Subtropical Division of the Humid Temperate Domain (*sensu* Bailey 1998).

matter (Barrett 1980). In addition, there was extensive logging for the turpentine and pulpwood industries. Fire is thought to be the key factor in maintaining pine forests and suppressing the succession to oak and hickory. The main pine species are *Pinus taeda, P. echinata, P. rigida,* and *P. virginiana.*

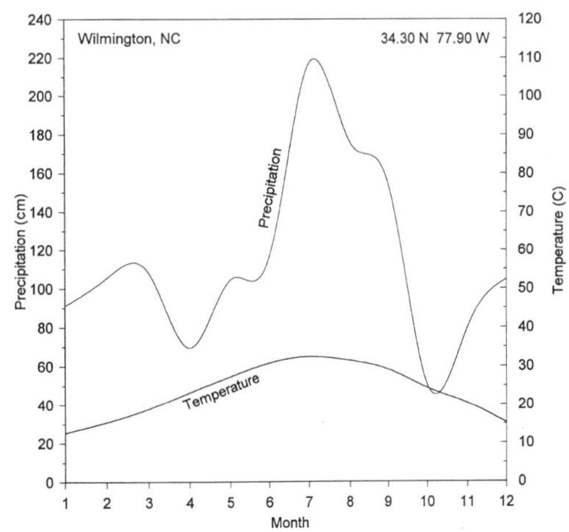

FIGURE 28.2 Climate diagram for Wilmington, North Carolina, representing the Upper Coastal Plain climate within the Subtropical Division of the Humid Temperate Domain (*sensu* Bailey 1998).

The Piedmont interfaces with the Coastal Plain at the fall line (Figure 25.10). Topography changes rather abruptly from rolling hills to flat plain. Parent materials also abruptly change (Christensen 1988). The Upper Coastal Plain consists of a series of terraces oriented almost parallel to the coastline, the Upper Coastal Plain being higher in elevation (about 30 to 183 m) than the Lower Coastal Plain. Soils are mainly Ultisols, varying from sand to clay, all waterlaid deposits. They are generally acidic and poor in organic matter and nutrients. Soils horizons appear mainly yellowish or reddish, hence the older name of this soil was red and yellow podzolics. Some soils contain hardpans. Unlike in the Piedmont, floodplains tend to be broad and well developed within the coastal plain subregion. Both *Pinus taeda* and *P. echinata* are common on old fields. *Pinus taeda* was named by pioneers based on its typical location in "loblollies," or muddy holes. Early explorers described extensive, majestic forests and abundant fires.

The Lower Coastal Plain tends to be flatter and more poorly drained than the Upper Coastal Plain. Elevation ranges from 0 to about 30 m. Bottomland hardwoods are diverse and generally have several canopy dominants. In areas that are continuously wet or nearly so, *Taxodium distichum* dominates.

FIGURE 28.4 A bottomland forest in the Mississippi alluvial valley.

Pinus palustris is the most widespread pine, followed by *P. taeda* and *P. elliottii*. *Pinus palustris* and *P. elliottii* dominate sandy flatwoods where hardpans underlie the soil and impede drainage. Several unique wetland conditions are found near the coast, including bayheads and pocosins. Pocosins are swamps of organic matter on ridges between streams and where shoreline dunes impede drainage. The term comes from First Nation people, meaning "swamp-on-a-hill."

Note on the bottomland hardwoods of the southeast: Although not a specific physiographic region, the bottomland forests (Figure 28.4) of the eastern United States, generally restricted to the Coastal Plain province, are both economically and ecologically important. The bottomland hardwood forests are primarily located on floodplains adjacent to rivers and streams, but also on poorly drained areas. There are a large number of species found in this system. The soils vary from well drained to poorly drained and usually have some horizon development (Inceptisols). The major site types are the first bottoms, which are periodically flooded, and the second bottoms, which no longer flood periodically (Figure 28.5). Within the bottoms, because of the past meandering of the main channel and its tributaries, occur secondary site types due to micro- and mesotopography: ridges, flats, sloughs, and swamps (Table 28.1). A typical profile of a southeastern bottomland ecosystem shows the importance of microtopographic variation, resulting in dramatic changes in vegetation with submeter changes in elevation (Figure 28.5). The importance of the topography is directly related to the mean depth of the water table, and thus to the growing conditions for various tree species (Robertson et al. 1978). The wetlands have long acted like a buffer, mitigating the effects of upland activities (e.g., fire and cutting) on adjacent river and stream systems by catching nutrients and eroding sediments (Sharitz and Mitsch 1993). However, many of these forests have been converted to farmland.

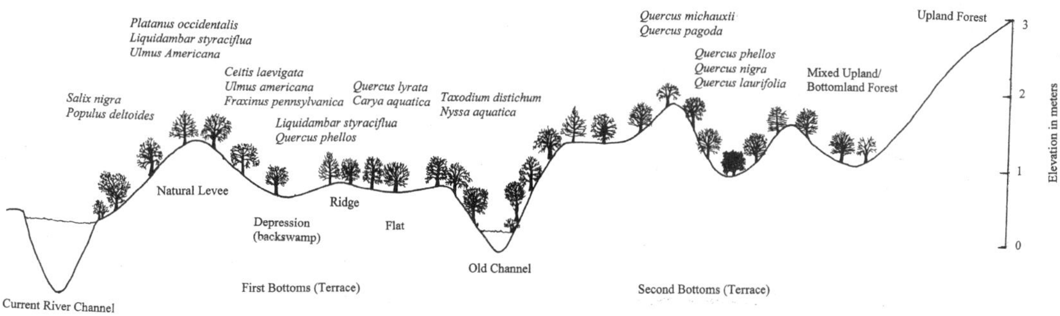

FIGURE 28.5 Schematic diagram of a typical bottomland topographic gradient and the species associated with the various microtopographic components (adapted from Wharton et al. 1982). Differences of a meter or less may produce an entirely different community composition in these bottoms. Drawing by Deanna Boswell Franklin.

Table 28.1. Characteristics of floodplain zones

	Zones				
	Depressions		Flats		Ridges
Characteristics	II	III	IV	V	VI
Degree of Inundation and Saturation	Nearly permanent inundation	Semipermanently inundated	Seasonally inundated	Temporarily inundated	Intermittently inundated
Timing of Flooding	Year-round	Spring and summer; most of growing season	Spring; 1 to 2 months of growing season	Periodically; up to 1 month of growing season	During exceptionally high floods
Probability of Annual Flooding	100%	51%–100%	51%–100%	10%–50%	1%–20%
Duration of Flooding; Percentage of Growing Season	100%	>25%	12.5%–25%	2%–12.5%	<2%

Adapted from Wharton et al. (1982).

COVER TYPES

Southern Pine-Hardwood Forest (NVCS Humid Temperate Domain/Subtropical Division; *Pinus-Quercus*)

Piedmont

28.1 Shortleaf Pine Cover Type (NVCS *Pinus echinata* Forest Alliance): This type occurs on dry, coarse-textured, nutrient-poor sites. Soils are either excessively drained or very shallow, thus soil moisture is quite low. *Pinus echinata* is intolerant of poorly aerated sites. Soils are derived from sandstones, shales, granites, and coastal plain gravels. Soils are near neutral and mainly Ultisols. Topography includes upland ridges and south-facing slopes between 200 and 900 m in elevation. This type is almost always successional to more shade-tolerant hardwoods, as a result of *Pinus echinata*'s intolerance of shade, moderate growth rate, and short life span (rarely >150 years). The community succeeds to *Quercus* species (NVCS *Pinus echinata-Quercus stellata-Quercus marilandica* Forest Alliance), but succession is slower on the most xeric, nutrient-poor sites. *Pinus echinata* is less shade tolerant than *Pinus taeda*, and *P. taeda* is found on wetter soils. As compared with *P. taeda*, *P. echinata* has a larger root system, lower tolerance to poor soil aeration, and lower demand for soil nutrients. *Pinus echinata* produces good seed crops every 3 to 10 years, but also can resprout well because of a peculiar juvenile growth habit (Burns and Honkala 1990a). Stems of seedlings may grow prostrate, then turn upward from just above the cotyledon, thus keeping the cotyledonary portion of the stem near the ground. Dormant buds in the axils of the leaves above the cotyledons start to grow and develop new sprouts when the bole is cut or injured by fire.

28.2 Loblolly Pine-Shortleaf Pine-(Redcedar) Cover Type [NVCS *Pinus taeda-Pinus echinata-(Juniperus virginiana)* Forest Alliance]: *Pinus taeda* generally dominates, except on the driest sites. This type occurs on moist sites, including poorly drained soils of the Coastal Plain, and dry, shallow, infertile soils of higher elevations (the community is typical on abandoned farmland) that were historically dominated by *Pinus palustris. Pinus taeda* does not tolerate dry, thin soils, and *Pinus echinata* does not tolerate very wet soils. This seral community succeeds to hardwoods without disturbance because the hardwoods have greater shade tolerance. The regenerating hardwoods (e.g., *Quercus* spp., *Nyssa sylvatica, Carya tomentosa,* and *Liriodendron tulipifera*) impede almost all pine regeneration. The understory of moist sites may contain a dense shrub stratum, dominated by *Myrica cerifera, Ilex glabra, Ilex vomitoria, Viburnum acerifolium, Viburnum dentatum, Vaccinium vascillans,* and *Vaccinium arboreum*. Herbaceous taxa are sparse but may include *Uniola sessilifolia, Panicum* spp, *Andropogon* spp., and *Carex* spp. (NVCS 1999).

28.3 Loblolly Pine-Oak Cover Type [NVCS *Pinus taeda-Quercus (marilandica, falcata, stellata)* Forest Alliance]: This community occurs on upland sites of xeric-mesic to xeric conditions. Other pine associates may include *Pinus echinata* and *Pinus virginiana*. Succession is strongly toward *Quercus* species, including *Q. falcata, Q. marilandica, Q. alba, Q. rubra,* and *Q. coccinea*), although *Pinus taeda* can live up to 300 years. *Carya ovata, C. glabra, C. tomentosa,* and *Nyssa sylvatica* are also common associates.

28.4 Virginia Pine Cover Type (NVCS *Pinus virginiana* Forest Alliance): This type occurs under a variety of circumstances, but generally on moderately to well drained Spodosols and Inceptisols. *Pinus virginiana* can tolerate acidic soils that are moderately high in organic matter. In the Appalachian region, the type occurs on upper slopes and ridges, on dry shale barrens, and on shallow rocky soils where drought-resistant pines compete successfully with the hardwoods (Figure 28.6). Stands occur at elevations between 15 and 760 m (Eyre 1980). The community is well developed on abandoned and cut-over land with exposed mineral soil, a necessity for germination (Burns and Honkala 1990a). In the Piedmont, it occupies old fields, many of which are missing a soil horizon because of extensive erosion problems (see the Note on abandoned farmland). As compared with associated *Pinus* spp., it is generally more successful on poorer sites. *Pinus virginiana* is shallow rooted and shade intolerant, eventually replaced by hardwoods (NVCS 1999; *Pinus virginiana-Quercus (alba, stellata, falcata, velutina)* Forest Alliance). The more shade-tolerant, longer-lived *Quercus* species are usually present in the understory of the forest as advanced regeneration. Any cutting activity simply releases them, but also allows *Pinus virginiana* to revegetate from a seed pool. Tree associates include a variety of dry *Quercus* species, as

FIGURE 28.6 A virginia pine (*Pinus virginiana*) forest on a ridge.

well as *Pinus rigida, P. pungens, Liriodendron tulipifera, Nyssa sylvatica, Acer rubrum, Diospyros virginiana, Ulmus alata,* and *Carya* spp. Shrub associates include *Rhus* spp., *Cornus florida, Cercis canadensis, Rubus* spp., *Toxicodendron radicans, Parthenocissus quinquefolia, Chimaphila* spp., and *Smilax* spp. Heart rot (*Phellinus pini*) is often present in stands more than 60 years old. Another serious disease is pitch canker (*Fusarium monoliforme* var. *subglutinans*). A variety of insects are also pests, mainly the southern pine beetle (*Dendroctonus frontalis*), *Ips* spp., the Virginia pine sawfly (*Neodiprion pratti pratti*), and the redheaded pine sawfly (*Neodiprion lecontei*).

28.5 Virginia Pine-Xeric Oak Cover Type [NVCS *Pinus virginiana-Quercus* (*coccinea, prinus*) Forest Alliance]: This type may be maintained under harsh environmental conditions created by topography (xeric upper slopes and ridges) and soils (shallow rocky soils and dry shale substrates) (Weakley et al. 1998). The type also occurs following disturbances at elevations below 900 m in the Piedmont and Appalachian Mountains. The type generally has an ericaceous shrub layer, including *Kalmia latifolia* and *Vaccinium* spp.

Note on abandoned farmland: Agricultural and logging practices in the eastern United States, especially prior to the dust bowl of the 1930s, often led to serious erosional problems. Large areas were stripped of their protective vegetation layer, exposing soils for long periods of time to both precipitation and wind erosion. The problems were exacerbated by the compaction of soil with large machinery, which decreased the infiltration of water into the soil and the percolation of water through the soil. Thus, water was forced to flow over the soil surface (overland flow), causing greater erosion. The result was that much of the topsoil was washed away. In southern Illinois, stumps from a cut early in the twentieth century are exposed some 2 m down from the current stream terrace. The uplands have lost entire slopes and ridges (Figure 28.7), and the bottomlands have been choked with excess soil. Pine species (*Pinus taeda* and *Pinus echinata*) were often planted to help stabilize the soil because they are fast-growing and tolerate lower soil nutrient levels. Many of the (now public) lands of the eastern United States have a history of mismanagement and erosion problems.

28.6 Chestnut Oak-White Oak Cover Type [NVCS *Quercus prinus-Quercus* (*alba, falcata, rubra, velutina*) Forest Alliance]: This type is found on slightly more mesic sites than the Chestnut Oak-Mixed Oak Cover Type (see Chapter 27) and is common on upper mesic, dry slopes of the Piedmont. It is also found on the Cumberland Plateau and Blue Ridge at elevations between 762 and 1036 m. Canopy species may also include *Acer rubrum, Carya tomentosa, Carya glabra, Carya ovalis, Carya ovata, Carya pallida, Fagus grandifolia, Liriodendron tulipifera,* and *Nyssa sylvatica. Cornus florida* and *Oxydendron arboreum* are common understory species. *Juniperus virginiana* may be common on drier sites and disturbed sites. Other possible

FIGURE 28.7 A deep gully resulting from mismanaged agricultural land. Prior to the 1930s little attention was given to the problems of erosion. Poor agricultural practices have greatly altered the topography and soil fertility. Kudzu (*Pueraria montana*) was planted with loblolly pine (*Pinus taeda*) in Natchez Trace State Park, Tennessee, to help stop erosion.

understory species include *Kalmia latifolia, Hamamelis virginiana, Rhododendron calendulaceum, Rhododendron maximum, Stewartia ovata, Symplocos tinctoria, Vaccinium stamineum,* and *Viburnum acerifolium.*

Upper Coastal Plain

28.7 Boxelder Cover Type (NVCS *Acer negundo* Temporarily Flooded Forest Alliance): This type is common near large rivers in the active floodplain or on sandbars. The distribution of *Acer negundo* is wide, due in part to its drought tolerance and cold tolerance and in part to its flood tolerance. Flooding frequency is high, and *Acer negundo* can tolerate prolonged flood. Water tables remain near the soil surface in the spring, but drop in the summer. Soils are clay loams to sandy loams. *Acer negundo* is a pioneer on disturbed sites (old fields) or newly formed river bars. It generally remains subdominant to *Populus deltoides* when that species is present, and follows *P. deltoides* and *Salix nigra* in succession (Burns and Honkala 1990b). *Acer negundo* is a prolific reproducer (reproduction begins at about 8 years of age). Seeds are distributed by wind from October through the spring, and germination is generally good. It is tolerant of shade, but succession is generally toward other bottomland hardwoods, including *Fraxinus americana, Ulmus americana, Celtis* spp., and *Carya illinoensis.* Associates include *Platanus occidentalis, Acer rubrum, Liquidambar styraciflua, Acer saccharinum, Ulmus alata, Celtis laevigata,* and *Populus deltoides.* Shrubs may be sparse or heavy and vary with location. Lianas include *Berchemia scandens, Toxicodendron radicans,* and *Parthenocissus quinquefolia.*

28.8 Water Hickory-Water-Locust Cover Type [NVCS *Carya aquatica-Gleditsia aquatica* Seasonally Flooded Forest Alliance]: This type occurs along the Mississippi River alluvial plain, on sites slightly drier than those of the Bald Cypress-Gum Cover Type. *Carya aquatica* is typical of clay flats and backwater on Inceptisol soils (Burns and Honkala 1990b). Flooding is important for its seed dispersal, and reproduction is aggressive. This species is considered to be intermediately shade tolerant. Logging has increased its abundance. Associates include *Quercus lyrata* and *Celtis laevigata. Planera aquatica* and *Forestiera acuminata* are common understory shrubs. Lianas are common and important.

28.9 Pecan-Sugarberry Cover Type [NVCS *Carya illinoinensis-(Celtis laevigata)* Temporarily Flooded Forest Alliance]: This alliance is typical near large rivers that frequent flood. The stands develop best where soils are heavily textured and calcareous. Stands of *Carya illinoensis* are best developed on sites that are well drained (Burns and Honkala 1990b). Both *Carya illinoensis* and *Celtis laevigata* have been affected by the altered flooding regimes of the lower Mississippi River. Other canopy species include *Quercus macrocarpa, Ulmus americana, U. crassifolia, Juglans nigra, Fraxinus* spp., *Quercus buckleyi, Q. fusiformis,* and *Acer negundo.*

28.10 Sugarberry Cover Type (NVCS *Celtis laevigata* Temporarily Flooded Forest Alliance): This community occurs along the floodplains of major rivers throughout the southeastern United States. The community is common especially on low ridges, broad flats, and shallow sloughs in first bottoms (Burns and Honkala 1990b). Soils are generally Inceptisols or Entisols. *Celtis laevigata* produces abundant fruit crops starting at 15 years of age and continuing until age 70. Fruits are dispersed by birds, mainly, but are also dispersed by water. *Celtis laevigata* is short-lived (about 150 years), but shade tolerant and reproduces readily, so the community may be relatively stable. Common associates include *Ulmus americana* and *Fraxinus pennsylvanica,* and these species are also shade tolerant when small and reproduce well. In addition, all three species sprout prolifically. *Celtis laevigata* often dominates on recently disturbed sites. Other tree associates include *Quercus texana, Quercus lyallii, Acer rubrum,* and *Carya aquatica.* Shrub species include *Cornus drummondii, Crategus* spp., *Ilex decidua,* and *Morus rubra.* Several vines are important: *Campsis radicans, Ampelopsis arborea, Brunnichia cirrhosa, Berchemia scandens, Cocculus carolinus, Parthenocissus quinquefolia, Vitis* spp., and *Toxicodendron radicans.* Herbaceous associates include *Galium* spp., *Viola* spp., *Daucus* spp., *Lactuca* spp., *Amsonia* spp., *Polygonum* spp., and *Amorpha fruticosa.* The range of *Celtis laevigata* overlaps with that of *Celtis occidentalis* between 35° and 37° north latitude. *Celtis occidentalis* replaces *Celtis leavigata* north of their overlap.

28.11 American Beech-Bottomland Hardwoods Cover Type (NVCS *Fagus grandifolia* Temporarily Flooded Forest Alliance): This type occurs along small streams that are intermittently flooded, but not for long durations. The 1993 flood along the Mississippi killed many large *Fagus grandifolia* due to its duration and depth. Soils are well-drained Inceptisols. Associates include *Quercus alba, Q. rubra, Q. laurifolia, Q. shumardii, Q. michauxii, Acer saccharum* var. *floridanum, A. rubrum, Liquidambar styraciflua, Fraxinus pennsylvanica,* and *Magnolia grandiflora.* Shrubs include *Asimina triloba, Ilex opaca, Cornus florida, Cornus stricta, Magnolia acuminata, Carpinus caroliniana, Hamamelis virginiana, Alnus serrulata,* and *Vaccinium elliottii. Cyrilla racemiflora* and *Kalmia latifolia* occur in the Piedmont.

28.12 Sweetgum Cover Type [NVCS *Liquidambar styraciflua-*(*Acer rubrum*) Seasonally Flooded Forest Alliance, and *Liquidambar styraciflua-*(*Liriodendron tulipifera, Acer rubrum*) Temporarily Flooded Forest Alliance]: This community is widespread throughout the alluvial floodplains of major rivers in the southeast (Figure 28.8) and is especially extensive in the Mississippi River delta. *Liquidambar styraciflua* is a pioneer species on these floodplains (e.g., old fields and logged sites), with a fruit well adapted to dispersal by wind or water (multiple of achenes). The species is also adapted to a wide variety of site conditions, growing fast to moderately fast (Burns and Honkala 1990b). The community perpetuates itself on first-bottom ridges and terrace flats where inundation occurs through much of the growing season (seasonally flooded), and on disturbed sites with frequent, but short-duration, flooding (temporarily flooded).

FIGURE 28.8 A sweetgum (*Liquidambar styraciflua*) forest on a floodplain terrace.

This community generally succeeds to *Celtis laevigata, Ulmus americana,* and/or *Fraxinus pennsylvanica,* typical associates in the seasonally flooded sites, and to a variety of bottomland hardwoods (including *Quercus*) on temporarily flooded sites. *Liquidambar styraciflua* is shade intolerant. Other canopy associates include *Acer saccharum* var. *floridanum, Aesculus sylvatica, Quercus nigra, Quercus phellos, Carya cordiformis, Platanus occidentalis, Betula nigra, Juglans nigra,* and *Fagus grandifolia.* The shrub stratum, which may be well developed, includes *Euonymous americanus, Lindera benzoin, Corylus americana, Viburnum acerifolium, V. prunifolium, V. rufidulum, Hamamelis virginiana, Asimina triloba,* and *Ilex decidua.* The understory includes several vines: *Smilax* spp., *Brunnichia cirrhosa, Ampelopsis arborea, Campsis radicans,* and *Toxicodendron radicans. Liquidambar styraciflua* has few insect or disease problems, although sweet gum blight may become a problem during droughts (Harlow et al. 1996). It is an important commercial hardwood species that may invade moist upland sites.

28.13 Longleaf Pine-Shortleaf Pine-Southern Oak Cover Type [NVCS *Pinus palustris-Pinus echinata-*(*Pinus taeda*)*-Quercus* (*stellata, falcata*) Forest Alliance]: This is the Upper Coastal Plain variant of the mixed *Pinus palustris* communities (Refer to type 28.38). The community is found north of the range of *Pinus elliottii* (Lower Coastal Plain) on exposed slopes and ridges of the clay hills and mountains. These communities are generally mixed because of fire suppression over the last 200 years. Prior to settlement, extensive stands of pure *Pinus palustris* existed throughout the Coastal Plain, estimated at between 12 and 24 million hectares. The community is limited to areas that burn, which may be explained by the life history of *Pinus palustris.* This species is adapted to fire at all ages. Seeds are wind dispersed and require mineral soil for germination and high amounts of light for growth. Once they have germinated, a grass stage is formed, whereby the long needles surround and protect the apical meristem (Figure 28.9) from annual surface fires of low to moderate intensity, (even during the growing season). During a short period of growth, seedlings of 0.6 to 0.9 m in height are susceptible to fire, and subsequently become fire resistant again (Burns and Honkala 1990a). Another positive result of surface fires is that they decrease the abundance of brown spot disease, a disease to which *Pinus palustris* is especially susceptible. Because of the community's dependence on fire, stands typically occur on drier sites conducive to fire, such as ridge tops and southwest-facing slopes. The density of associates increases with decreases in the frequency of fires. Principal associates include *Quercus marilandica, Q. falcata, Q. stellata,* and *Carya tomentosa.* In addition, *Cornus* spp., *Quercus nigra, Nyssa sylvatica, Sassafras albidum, Diospyros virginiana,* and *Liquidambar styraciflua* are common (Eyre 1980). Shrub associates include *Ilex glabra, Vaccinium* spp., *Rhus copallina, Ilex vomitoria, Serenoa repens, Myrica cerifera, Gaylussacia* spp., and *Chrysobalanus oblingifolius. Aristida stricta* (wiregrass) dominates the understory, especially in the eastern portion of the community's range. *Andropogon* spp. and *Panicum* spp. can predominate in the western portion of the community's range.

FIGURE 28.9 The apical meristem of the longleaf pine (*Pinus palustris*) is protected from fire by the surrounding cluster of needles, which are pulled away for the photo.

28.14 Loblolly Pine Cover Type (NVCS *Pinus taeda* Forest Alliance): *Pinus taeda* is a widespread species of the southeast and a favorite for pine plantations. It is the leading commercial timber species of the southeastern United States. Its fast growth and tolerance to a wide range of conditions make it especially suitable to the paper industry (Eyre 1980). Rotation age for a particular variant (there are many hybrids) may be less than 12 years. The type is found mainly on Ultisols that are moderately acidic (Burns and Honkala 1990a). Soils are generally deep, but have fine-texture subsoils that impede surface drainage. *Pinus taeda* communities are most common where soil moisture is favorable. *Pinus taeda* is more shade tolerant than *P. palustris* or *P. elliottii*, but less shade tolerant than its hardwood associates. This community may be stable where fires occur frequently or where soil water tables exclude hardwood competition (e.g., *Pinus taeda* Temporarily and Seasonally Flooded Forest Alliances). Otherwise, the community is seral to more shade-tolerant hardwood species. Typical hardwood associates on wet sites include broadleaf evergreens (*Magnolia virginiana* and *Persea borbonia*), *Nyssa aquatica*, and *Acer rubrum*, and on moist sites include *Liquidambar styraciflua*, *Quercus nigra*, *Quercus pagoda*, *Quercus michauxii*, *Fraxinus americana*, *Liriodendron tulipifera*, and *Ulmus americana*. The undergrowth of these moist sites is rich in species and has a high density of regenerating trees and shrubs, but few in herbaceous taxa. Typical small tree and shrub associates include *Prunus serotina*, *Cornus florida*, *Ilex opaca*, *Sassafras albidum*, *Crataegus* spp., *Oxydendrum arboreum*, *Chionanthus virginicus*, and *Symplocos tinctoria*. Woody vines, such as *Berchemia scandens* and *Gelsemium sempervirens*, may also be common. Pests include fusiform rust (*Heterobasidion annosum*), black root rot (*Phaeolus schweinitzii*), and pinebark beetles (e.g., *Dendroctonus frontalis*).

28.15 Cottonwood Cover Type (NVCS *Populus deltoides* Temporarily Flooded Forest Alliance): This cover type is found on the frontland and banks of major streams in central and southern forests (Figure 28.10). The community is especially common along the Missouri, Ohio, and Mississippi Rivers (Eyre 1980). *Populus deltoides* is a pioneer species capable of fast growth. It germinates wherever bare, moist soils occur, but does not tolerate long floods during the growing season (Burns and Honkala 1990b). These conditions are typical of new sandbars (Figure 28.11) and well-drained flats after disturbance (also of well-drained old fields). Soils are generally deep, infertile sands and clays (Entisols and Inceptisols). Some of the best sites are now in the batture (front) lands, the areas between levees and rivers. Sites are generally within 6 m elevation from the average water level of the stream (Burns and Honkala 1990b). *Populus deltoides* shows several typical pioneer characteristics: light, wind-dispersed seeds that travel great distances (or seeds may be dispersed by water for miles), fast growth (sometimes phenomenal growth), very low shade tolerance, and a short life span. Trees >70 years old rapidly deteriorate, and an age of 200 years seems to be the maximum (Harlow et al. 1996). *Salix nigra* and *Salix exigua* occur with *Populus deltoides* during the early stages of succession, but *Populus* will generally outgrow the *Salix* species and dominate. If frequent or extended flooding occurs during the growing season and covers the trees, however, *Populus deltoides* does not survive (only *Salix* spp.). Associated trees include *Acer saccharinum*, *Ulmus americana*, *Celtis laevigata*, *Acer negundo*, *Platanus occidentalis*, *Liquidambar styraciflua*, and *Fraxinus pennsylvanica*. Shrubs include *Cornus drummondii* and *Forestiera acuminata*. Vines are important components and may include *Toxicodendron radicans*, *Smilax* spp., *Campsis radicans*, *Ampelopsis arborea*, and *Vitis* species. *Rubus* spp., *Phytolacca americana*, and *Urtica dioica* are herbaceous species commonly associated (NVCS 1999). Levee systems and stream channelization have restricted the number of new

FIGURE 28.10 A cottonwood (*Populus deltoides*) grove forest along the Ohio River, southern Illinois.

FIGURE 28.11 Young cottonwood (*Populus deltoides*) on a sandbar of the Mississippi River.

areas for possible colonization by *Populus deltoides* (see Note on altered watercourses at the end of this chapter). *Populus deltoides* is susceptible to many diseases and insects and is very vulnerable to fire.

28.16 White Oak-Water Oak Cover Type [NVCS *Quercus alba*-(*Quercus nigra*) Forest Alliance]: This type occurs on a variety of sites, including sandy swamp islands, coastal plain slopes, ridge tops, and drier areas of acidic soils. *Quercus nigra* is found on most bottomland types except swamps, and most commonly on Inceptisols (Burns and Honkala 1990b). Its growth is fast, but seedlings require abundant moisture and light (it is shade intolerant). Seedlings do not tolerate prolonged inundation. *Quercus nigra* produces good seed crops every other year. Animals and floodwaters disseminate seeds. Associates include *Carya tomentosa, Carya glabra, Carya ovalis, Carya pallida, Liquidambar styraciflua, Pinus taeda, Pinus echinata, Quercus velutina, Nyssa sylvatica, Quercus stellata, Quercus falcata* and, on more mesic sites, *Quercus laurifolia, Quercus pagoda, Magnolia grandiflora, Magnolia macrophylla, Hamamelis virginiana, Asimina triloba, Callicarpa americana, Gelsemium sempervirens,* and *Euonymous americanus. Oxydendrum arboreum* and *Cornus florida* are common midcanopy species.

28.17 Overcup Oak-Water Hickory Cover Type [NVCS *Quercus lyrata*-(*Carya aquatica*) Seasonally Flooded Forest Alliance]: This type occurs on floodplains, especially in the backwater basins of principal rivers on first bottoms and swamp edges at elevations of <150 m. Soils are generally impermeable clays and silty clays (Inceptisols). Late spring flooding is common, so water may stand into the growing season. *Quercus lyrata* and *Carya aquatica* are the last to leaf out in the spring, thus avoiding the problems associated with the typical late spring flooding. *Quercus lyrata* is a slow-growing species. *Quercus lyrata* can reproduce well and consistently, but good seed years may occur only every 4 to 5 years. The seeds also seem less desirable to wildlife, and dissemination is by floodwaters. Seeds remain dormant until the water recedes, and they can remain viable up to 14 months (Harlow et al. 1996). *Carya aquatica* is a prolific sprouter (Eyre 1980). Overstory associates include *Quercus texana, Fraxinus pennsylvanica,* and *Celtis laevigata.* The subcanopy may include *Planera aquatica, Cornus drummundii,* and *Forestieria acuminata.* Wood borers are a common problem for *Quercus lyrata,* but there are no serious diseases (Burns and Honkala 1990b). The borers include carpenter worm (*Prionoxystus robiniae*) and red oak borer (*Enaphalodes rufulus*).

28.18 Swamp Chestnut Oak Cover Type (NVCS *Quercus michauxii* Temporarily Flooded Forest Alliance): This community is scattered in occurrence throughout the southeast on the floodplains of the major rivers, except for the Mississippi River, where it is scarce (NVCS 1999). Soils are either Inceptisols or Entisols (Burns and Honkala 1990b). The sites are relatively well drained, loamy ridges, and silty clay and loamy terraces in first bottomlands (Harlow et al. 1996). The sites are rarely covered with standing water during the growing season. On older alluvium, the community is stable. *Quercus michauxii* seeds are sweet and heavily utilized by wildlife. Many other hardwood species are associated with this community, including

Fraxinus americana, F. pennsylvanica, Carya ovata, C. laciniosa, C. tomentosa, C. cordiformis, Quercus alba, Q. pagoda, Q. shumardii, Q. stellata var. *paludosa*, and *Nyssa sylvatica*. The understory may include *Aesculus sylvatica, Asimina triloba, Carpinus caroliniana, Cornus florida, Sabal minor, Salix caroliniana, Styrax americana, Viburnum dentatum, Ilex decidua, Aralia spinosa, Cercis canadensis*, and *Ilex opaca*. There are many diseases and pests of *Quercus michauxii*.

28.19 Chinkapin Oak Cover Type (NVCS *Quercus muhlenbergii* Forest Alliance): This type occurs in mesic and dry-mesic site conditions on slopes generally not of northern exposure. The type can be found near prairie openings on blackland soil and along the loess bluffs of the Mississippi River. Soils are generally neutral to basic, often associated with a limestone parent material (Burns and Honkala 1990b; Weakley et al. 1998), and classed as Alfisols, Ultisols, and Inceptisols. *Quercus muhlenbergii* is intolerant of shade. Other canopy species include those typical of high base substrates, such as *Fraxinus americana, Fraxinus quadrangulata*, and *Quercus shumardii*, throughout the range of this type, and *Quercus sinuata* var. *sinuata* and *Carya myristiciformis* in the southwestern portion of the its range. *Acer saccharum* and *Acer saccharum* var. *floridanum* may also be found in the overstory.

28.20 Shumard Oak-Cherrybark Oak Cover Type (NVCS *Quercus shumardii-Quercus pagoda* Forest Alliance): These forests occur on loamy soils derived from limestone (calcareous parent material) and on loess bluffs. The type is found on lower slopes above the bottomland hardwood forests (also around prairies in Louisiana), on first bottom ridges, and on well-drained terraces and colluvial sites (Figure 28.12) (Burns and Honkala 1990b). Flooding occurs only during the dormant season, but groundwater may remain high throughout the year. Thus, conditions are generally mesic. *Quercus shumardii* is intolerant of shade, and *Quercus pagoda* has intermediate tolerance to shade (Table 28.2). Hence, the community develops following disturbance. Both species are fairly fast growing and may colonize old fields. The canopy can be a mixture of swamp species and slope species. Common associates include *Quercus phellos, Quercus michauxii, Quercus falcata*, and *Quercus stellata*.

28.21 Black Willow Cover Type (NVCS *Salix nigra* Seasonally Flooded and Semipermanently Flooded Forest Alliances): This community occurs along first bottom frontlands, sloughs, flats, and banks of most

FIGURE 28.12 Bottomland forest dominated by cherrybark oak (*Quercus pagoda*).

major streams throughout the central and southern forests (and extends into northern forests as well). It is less common along the east coast. *Salix nigra* is a pioneer species capable of very fast growth. It is the first to appear on newly formed sandbars and river margins along with *Populus deltoides*. Like *Populus, Salix nigra* has several pioneeristic adaptations, including light, wind-dispersed seeds, fast growth, shade intolerance, and a short life span. An additional characteristic is its ability to withstand burial during flood events. Sites may silt to a depth of 6 m during the life of an individual. Individual floods have increased elevations as much as 1.5 m. *Salix nigra* and *S. exigua* have extensive adventitious buds in their stems, capable of forming roots spreading perpendicular to the stem upon burial. The resulting below-ground physiognomy is a taproot-like system with specific "nodal" secondary root growth. This plastic response to the dynamic flooding regime of large rivers enables pure stands of *Salix* to withstand the most severe flooding conditions. As soils build up, especially those incorporating silts and clays rather than only sand, the community will succeed to a variety of lowland hardwoods (discussed in the next section). *Populus deltoides* is the major associate of this community because of its similar pioneer life-history characteristics, but loses dominance under more severe flooding conditions. Other associates may include *Fraxinus pennsylvanica, Platanus occidentalis, Carya illinoensis, Diospyros virginiana, Gleditsia aquatica, Ulmus americana, Taxodium distichum, Acer rubrum, Celtis laevigata, Acer negundo*, and *Acer saccharinum*. Small-stature

Table 28.2. Selected southeastern bottomland species and their tolerances to competition and flooding

Species	Tolerance to Competition	Tolerance to Flooding
Cherrybark Oak	Intolerant	Very Intolerant
Shumard Oak	Intolerant	
Swamp Chestnut Oak	Moderately Intolerant	Intolerant
Hawthorn	Very Tolerant	
Sweet Gum	Intolerant	Intermediately Tolerant
Sugarberry	Very Tolerant	
Nuttall Oak	Intolerant	
Water Oak	Intolerant	
Pecan	Moderately Intolerant	
River Birch	Intolerant	
American Elm	Tolerant	
Pin Oak	Intolerant	
Roughleaf Dogwood	Tolerant	
Green Ash	Intolerant	Tolerant
Pumpkin Ash	Intolerant	
Eastern Cottonwood	Very Intolerant	
American Sycamore	Very Intolerant	
Silver Maple	Intolerant	
Willow Oak	Intolerant	
Black Willow	Very Intolerant	
Red Maple	Tolerant	
Laurel Oak	Intolerant	
Overcup Oak	Moderately Intolerant	
Water Hickory	Moderately Tolerant	
Swamp-privet	Tolerant	
Possumhaw	Tolerant	
Boxelder	Moderately Tolerant	Moderately Tolerant
Baldcypress	Moderately Tolerant	Very Tolerant
Swamp Tupelo	Moderately Intolerant	
Water Tupelo	Intolerant	
Buttonbush	Moderately Intolerant	

Adapted from Barrett (1980).

trees include *Forestiera acuminata, Cornus drummondii,* and *Planera aquatica.* Vines may be important, including *Brunnichia cirrhosa, Ampelopsis arborea, Campsis radicans,* and *Toxicodendron radicans.* The herbaceous stratum is relatively sparse as a result of the frequency and duration of flooding. However, following cutting and partial opening of the stands, a heavy undergrowth of annual graminoids and *Xanthium pungens* (can be almost 3 m high and loaded with seed on Mississippi River islands) may occur.

Lower Coastal Plain

28.22 Southern Sugar Maple Cover Type (NVCS *Acer saccharum* var. *floridanum-Fraxinus americana-Juglans nigra* Forest Alliance): This type occurs on moist flats and lower slopes of bluffs over calcareous substrates. Soils are fertile and moist, but well drained (Inceptisols, Entisols, and Ultisols) (Burns and Honkala 1990b). *Acer saccharum* var. *floridanum* is tolerant of shade and has no serious diseases or pests. Where *Acer saccharum* var. *floridanum* and *Acer saccharum* var. *saccharum* are found together, *Acer saccharum* var. *floridanum* distributes its seeds 2 weeks earlier than *A. saccharum* (Harlow et al. 1996), an adaptation to longer growing seasons in the south. Overstory species include *Aesculus glabra, Carya carolinae-septentrionalis, Carya laciniosa, Carya myristiciformis, Celtis laevigata, Gleditsia triacanthos, Quercus shumardii,* and *Quercus muhlenbergii.* Shrub species include *Aesculus pavia, Cornus drummundii, Crataegus calpodendron, Crataegus marshallii, Frangula caroliniana, Ptelea trifoliata,* and *Viburnum rufidulum. Berchemia scandens, Bignonia capreolata, Smilax bona-nox,* and *Vitis aestivalis* are common lianas (NVCS 1999). *Arundinaria gigantea* may also be present (Figure 27.11).

28.23 Atlantic White-Cedar Cover Type (NVCS *Chamaecyparis thyoides* Saturated Forest Alliance): This type occurs on saturated, peaty substrates (Histosols). The pH is neutral to slightly acidic. *Chamaecyparis thyoides* tolerates only brief inundation. Thus, sites are typical of small black-water or spring-fed streams, seepages, or constantly saturated basins with few flooding events. Fires are uncommon in these stands, but gap succession is necessary for regeneration of this seral type. Regeneration will follow fires when they occur, as *Chamaecyparis thyoides* is a prolific seeder. Good seed crops are produced every year, and seeds are wind dispersed. Suitable seedbeds include moist, rotting logs, sphagnum moss, and muck soil. The forests were heavily cut in the nineteenth century, decreasing abundance of this type (Harlow et al. 1996). Although the species is shade tolerant, succession is to a bay forest in the absence of disturbance. This may take some time, as *Chamaecyparis thyoides* is long-lived (250+ years). Associate species include *Pinus elliottii* var. *elliottii, Pinus serotina, Acer rubrum, Nyssa biflora, Persea palustris,* and *Liriodendron tulipifera.* The subcanopy may include *Acer rubrum* var. *trilobum, Nyssa biflora, Magnolia virginiana,* and *Cliftonia monophyla* (NVCS 1999).

28.24 Atlantic White-Cedar-Red Maple Cover Type (NVCS *Chamaecyparis thyoides-Acer rubrum* Saturated Forest Alliance): *Chamaecyparis thyoides* often occurs in pure, dense stands, but *Acer rubrum* is also commonly present. This community is found intermittently along the Atlantic coast from southern Maine to northern Florida, as well as along the Florida panhandle. The type is found principally on peat deposits overlying sandy soils that are a part of stream swamps (Entisols). The peat ranges from a few centimeters to 12 m deep and is generally infertile. Soils are continuously wet. The community is seral, developing following disturbance and succeeding to associated hardwoods when favorable conditions occur. *Chamaecyparis thyoides* is a prolific seed producer. Peat, rotten logs, and moist mineral soil are suitable sites for the germination of the wind-dispersed seeds. Although slow growing, this species has intermediate shade tolerance and can grow up through a moderate shrub canopy. Other associates include *Nyssa sylvatica, Betula lutea, Pinus strobus,* and *Tsuga canadensis* in the northern part of its range, and *Pinus serotina, Pinus elliottii, Magnolia virginiana, Nyssa biflora, Persea borbonia,* and *Gordonia lasianthus* in the southern part of its range. Shrubs commonly form dense stands along the edges of the community. *Clethra alnifolia, Rhododendron viscosum, Vaccinium corymbosum, Gaylussacia frondosa,* and *Lyonia lucida* are common. Various liverworts, mosses (especially *Sphagnum*), *Sarracenia* spp., and *Drosera* spp. are common.

28.25 Black Titi Cover Type (NVCS *Cliftonia monophylla* Saturated Forest Alliance): This evergreen type occurs over peat at least 1 m thick and on nutrient-poor, sandy, acidic, saturated mineral soils. The hydroperiod of this type may be longer than other baygall types. The forest has a low, thicket-type structure (Little 1998). *Pinus serotina* or *Pinus elliottii* var. *elliottii* may co-occur. The substrata may be dense or sparse, consisting of *Myrica inodora, Gaylussacia mosieri, Lyonia lucida, Ilex coriacea,* and *Nyssa biflora.*

28.26 Hawthorn Cover Type [NVCS *Crataegus* (*aestivalis, opaca, rufula*) Seasonally Flooded Forest Alliance]: This type occurs in seasonally flooded depressions of uplands or floodplains. All are small trees, up to only 12 m in height. Other associate species may include *Quercus phellos, Pinus taeda,* and *Gleditsia* spp. This type provides important deer browse (Harlow et al. 1996).

28.27 American Beech-Southern Magnolia Cover Type (NVCS *Fagus grandifolia-Magnolia grandiflora* Forest Alliance): This type occurs on mesic to dry-mesic forests of lower slopes and flats that receive no flooding (not on first bottoms) or receive occasional temporary flooding. Sites occur on the oldest alluvium and outwash areas of floodplains (Spodosols, Alfisols, Vertisols, and Ultisols) (Burns and Honkala 1990b). *Magnolia grandiflora* cannot tolerate long inundations (Harlow et al. 1996). Generally, the type is protected from fire because of its topographic position (elevations <60 m) or mesic litter and soils. Winter droughts can have serious effects on individuals and can cause mortality (Burns and Honkala 1990). Birds and small mammals eat *Magnolia grandiflora* seeds. Overstory associates include *Acer saccharum* var. *floridanum, Carya tomentosa, Carya glabra, Carya pallida, Celtis laevigata, Fraxinus americana, Ilex opaca, Juglans nigra, Liquidambar styraciflua, Magnolia macrophylla, Persea borbonia, Pinus taeda, Pinus glabra,* several *Quercus* species, *Sabal palmetto, Sapindus marginatus, Tilia caroliniana, Ulmus alata,* and *Ulmus americana.* The understory may include *Cornus florida, Halesia tetraptera, Halesia diptera, Magnolia ashei, Magnolia pyramidata, Ostrya virginiana, Acer rubrum, Prunus caroliniana,* and, rarely, *Taxus floridana* (Figure 28.13) and *Torreya taxifolia.* Shrubs include *Chionanthus virginicus, Hamamelis virginiana, Ilex vomitoria, Hydrangea quercifolia, Kalmia latifolia, Sebastiana fruticosa, Stewartia malacodendron, Styrax grandifolia, Clethra alnifolia, Asimina parviflora,* and *Vaccinium elliottii.* Epiphytes such as *Tillandsia usneoides* and *Pleopeltis polypodioides* ssp. *michauxiana* often occur.

28.28 Loblolly-Bay Cover Type (NVCS *Gordonia lasianthus* Saturated Forest Alliance): This type occurs in stream-head pocosins, other peaty stream-head situations, peat dome pocosins, peat-filled bays, and seeps at slope bases. Soils are acidic and generally organic (Histosols), but may be mineral. *Gordonia lasianthus* is shade tolerant and slow growing. It regenerates well by sprouting. *Gordonia lasianthus* is one of the four species associated with "bay forests"; the others are *Persea palustris, Persea borbonia,* and *Magnolia virginica* (Harlow et al. 1996). Additional associates include *Acer rubrum, Pinus serotina,* and *Pinus elliottii* var. *elliottii.* Shrubs include *Cyrilla racemiflora, Clethra alnifolia, Itea virginica, Lyonia lucida, Ilex glabra,* and *Cephalanthus occidentalis.*

28.29 Sweetbay-Swamp Tupelo Cover Type (NVCS *Magnolia virginica-Nyssa biflora* Seasonally Flooded Forest Alliance): This community occurs throughout the southeastern Coastal Plain, where soils are saturated throughout the growing season. The persistent soil saturation produces a stable community. Sites include branch heads and narrow bottoms of streams, pocosins, and poorly drained upland depressions (e.g., peat bogs and borders of swamps). Soils are sandy and colluvial. The wettest sites are also more acidic, favoring species such as *Persea borbonia, Pinus serotina, Gordonia lasianthus,* and *Chamaecyparis thyoides.* Deeper flooding favors *Nyssa biflora, Taxodium distichum* var. *imbricarium* and var. *distichum,* and *Acer rubrum.* Sites with greater drainage characteristically contain *Nyssa sylvatica, Liriodendron tulipifera, Liquidambar styraciflua, Ilex opaca,* and *Magnolia grandiflora.*

FIGURE 28.13 Florida yew (*Taxus floridana*) lying dead along the bluffs near the Gulf of Mexico in western Florida. This endemic species is in danger of extinction.

28.30 Sweetbay-Red Bay Cover Type (NVCS *Magnolia virginiana-Persea palustris* Saturated Forest Alliance): This type occurs where soils are saturated throughout the growing season, such as at branch heads, narrow bottoms of small perennial or intermittent streams, pocosins, and poorly drained upland depressions (peat bogs, edges of small ponds and swamps). Soils are generally sandy, very acidic (pH = 4.0 to 4.5), and sterile (Eyre 1980). Most stands of this type are found at elevations of <61 m (Burns and Honkala 1990b). *Magnolia virginica* is slow growing, but flowering begins very early. Seed crops are small and dispersed by wind and birds. It is intermediate in shade tolerance and very resistant to fire. Fires are common and may completely change the community composition, depending on the amount of peat consumed; shallow burns favor *Pinus serotina,* and deep burns favor *Taxodium distichum* var. *imbricarium* and *Nyssa aquatica.* Other associated species include *Taxodium distichum* var. *distichum* and *Acer rubrum.* Shrubs include *Ilex coriacea, Ilex glabra, Illicium floridanum,* and *Lyonia lucida* (NVCS 1999). The bay forests are important deer browse.

28.31 Water Tupelo Cover Type [NVCS *Nyssa aquatica-*(*Taxodium distichum*) Semipermanently Flooded Forest Alliance]: This type is always found on wet sites where surface water stands for a good portion (or all) of the growing season. The community is restricted to deep swamps, sloughs, and low flats of the southeastern alluvial floodplains. Water depth may be 6 m. Soils are Alfisols, Entisols, Inceptisols, and Histosols, moderately to strongly acidic (Burns and Honkala 1990b). *Nyssa aquatica* has several adaptations to this environment. It has a swollen buttress for support and roots that can oxidize the rhizosphere, thereby decreasing the quantity of compounds in their reduced, toxic form. Roots can also control anaerobic respiration. Prolific seed crops are produced every year. The seed germinates when the water recedes, staying viable up to 14 months. The seedlings can survive as long as their tops are above the water. *Nyssa aquatica* is intolerant of shade, and fire can cause serious damage. The *Nyssa aquatica* community may include *Acer rubrum, Salix nigra, Planera aquatica, Fraxinus profunda, Fraxinus caroliniana, Populus heterophylla,* and *Gleditsia aquatica.* Shrub associates include *Ceanothus occidentalis, Forestiera acuminata, Itea virginica, Cornus stricta, Cyrilla racemiflora, Ilex decidua, Leucothoe* spp., *Rosa palustris,* and *Toxicodendron vernix.* Common woody vines include *Smilax* spp., *Berchemia scandens, Bignonia capreolata, Decumaria barbara, Ampelopsis arborea,* and *Toxicodendron radicans.*

28.32 Swamp Tupelo Cover Type (NVCS *Nyssa biflora* Seasonally Flooded and Semipermanently Flooded Forest Alliances): The *Nyssa biflora* community occurs on slightly drier sites (i.e., shallower and of shorter flood durations) than the *Nyssa aquatica* community. In addition, *Nyssa biflora* sites tend to be more acidic, have less silt and clay content, and have higher amounts of organic matter at the soil surface. *Nyssa biflora* thrives under flooded conditions with shallow, moving water throughout the growing season. The "water roots" tolerate high CO_2 concentrations, oxidize the rhizosphere, and carry on anaerobic respiration (Burns and Honkala 1990b). *Quercus lyrata, Carya aquatica, Ulmus americana, Fraxinus pennsylvanica, Quercus laurifolia, Quercus texana, Liquidambar styraciflua, Diospyros virginiana,* and *Magnolia virginica* are common overstory associates. Understory associates include *Cliftonia monophylla, Ilex vomitoria,* and other species, *Myrica cerifera, Lyonia lucida, Clethra alnifolia,* and *Crataegus* spp.

28.33 Swamp Tupelo-Red Maple Cover Type [NVCS *Nyssa biflora-Acer rubrum-*(*Liriodendron tulipifera*) Saturated Forest Alliance]: The hydrology characteristic of this type is constant flooding. Sites include extensive peat flats, seepage-fed floodplain edges, and seepage slopes. Soils are generally acidic and organic (Histosols). Some of the peat flat forests may be recovering from logging of *Chamaecyparis thyoides* forests that previously dominated these sites. Associated species include *Liquidambar styraciflua, Magnolia virginiana, Oxydendrum arboreum, Persea palustris, Fraxinus pennsylvanica, Quercus laurifolia, Ulmus rubra, Quercus michauxii,* and *Pinus taeda.* Dominant shrubs include *Leucothoe axillaris, Ilex coriacea, Myrica heterophylla,* and *Lyonia lucida.*

28.34 Sand Pine Cover Type (NVCS *Pinus clausa* Forest Alliance): This community represents a pioneer stage on droughty, infertile, strongly acidic sands of marine origin (Entisols), generally in central Florida (although it has been planted elsewhere). There are two types recognized: *P. clausa* var. *clausa* has serotinous cones and relies on fire for regeneration, and *P. clausa* var. *immuginata* has cones that open after maturing. Fires must be periodic for the serotinous cone variety, occurring frequently enough for cones to be opened but infrequently enough for seedlings to mature to a fire-resistant stage. However, older stems of this species are much less fire tolerant than those of *Pinus palustris* or *Pinus elliotii*. *Pinus clausa* is intermediate in shade tolerance, but can be quite tolerant at an early age (Burns and Honkala 1990a). It can bear fruit by 5 years of age. Scrub oaks (*Quercus minima, Q. myrtifolia, Q. margaretta, Q. chapmanii, Q. laevis*) and *Diospyros virginiana* are the main associates. Shrub associates include *Chrysobalanus oblingifolius, Sabal etonia, Serenoa repens,* and *Lyonia lucida*. *Opuntia* spp., *Andropogon virginicus,* and *Aristida stricta* are common herbaceous components.

28.35 Slash Pine Cover Type (NVCS *Pinus elliottii* Saturated Forest Alliance): *Pinus elliotii* communities are restricted to moist sites, mainly because of the susceptibility of young stems to fire (older stems are quite tolerant). Soils must be well aerated (Burns and Honkala 1990a). Following the suppression of fire, the community can extend onto drier sites. Good seed crops occur every 3 years. A second variety grows in south Florida, *P. elliottii* var. *densa*. The community is confined to three main habitats when fires occur at regular intervals: bay swamps, boggy flatwoods, and coastal pine flatwoods (Erye 1980; Harlow et al. 1996). The Slash Pine-Hardwood Cover Type, which dominates the bay swamps, is discussed next. Boggy flatwoods border creek swamps and acid depressions, and common associates here include *Pinus palustris* and *Sarracenia* spp. (Figure 28.14). *Pinus elliottii* alternates with *Pinus palustris* across the landscape, *Pinus palustris* dominating the drier knolls. The coastal pine flatwoods are interdunal swales and flats with a common undergrowth of *Aristida* spp., *Andropogon* spp., *Serenoa repens, Ilex glabra,* and *Lyonia lucida*. The sites associated with dunes may also have *Quercus myrtifolia, Q. chapmanii,* and *Ceratiola ericoides*. The most damaging agent is fusifom rust (*Cronartium quercuum* f. *fusiforme*), but annosus root rot (*Heterobasidion annosum*) may also be detrimental to stands.

28.36 Slash Pine-Hardwood Cover Type (NVCS *Pinus elliottii-Magnolia virginica-Taxodium distichum* var. *imbricarium-Nyssa biflora* Saturated Forest Alliance): Communities of this type occur in fire-protected situations, such as in depressions in a

FIGURE 28.14 An open pine flatwood. These wet, poor-nutrient sites often have insectivorous plants such as the pitcher plant shown (*Sarracenia* spp.).

pine flatwoods landscape. The community is typically found in bayhead swamps and adjacent saturated small stream forest. These sites are highly acidic and poorly drained. Associated species include *Magnolia virginiana, Nyssa biflora, Acer rubrum, Liriodendron tulipifera, Liquidambar styraciflua, Quercus laurifolia,* and *Quercus nigra*. Understory woody species include *Sorbus arbutifolia* var. *arbutifolia, Arundinaria gigantea, Cyrilla racemiflora, Gaylussacia mosieri, Gelsemium rankinii, Hypericum galioides, Ilex coriacea, Ilex glabra, Ilex opaca, Itea virginica, Ilex vomitoria, Lyonia lucida, Myrica cerifera, Myrica heterophylla, Persea palustris, Toxicodendron radicans,* and *Viburnum nudum* var. *nudum*. Typical herbaceous species include *Polygala cymosa, Carex glaucescens, Eriocaulon compressum, Viola primulifolia, Sarracenia* spp., *Woodwardia areolata,* and *Xyris fimbriata*. *Sphagnum* spp. are abundant and, on some sites, may be the only herbaceous taxa present. This alliance grades to vegetation of slightly less wet sites with greater fire frequency.

28.37 Spruce Pine-Mixed Oak Cover Type [NVCS *Pinus glabra-Quercus (laurifolia, michauxii, nigra)* Temporarily Flooded Forest Alliance]: This type occurs along streams on natural levees or floodplain flats that are infrequently flooded. Soils are acidic, sandy loams high in organic matter (Burns and Honkala 1990a). These sites are in the intermediate range between dry, sandy ridge soils and alluvial bottoms. *Pinus glabra* is very shade tolerant and establishes in the shade of hardwoods, growing slowly into the canopy. *Liquidambar styraciflua* may also be found in the overstory. The subcanopy is typically dominated by *Carpinus caroliniana.* Shrubs include *Hypericum hypericoides, Sebastiana fruticosa, Leucothoe racemosa, Cyrilla racemosa, Styrax americana, Crataegus marshallii,* and *Rhododendron canescens.* In saturated conditions, *Pinus glabra* and *Quercus laurifolia* dominate (NVCS *Pinus glabra-Quercus laurifolia* Saturated Forest Alliance) (NVCS 1999). *Pinus glabra* is less susceptible to insects and disease than many of the other southern pine, partially because of its naturally sparse distribution.

28.38 Longleaf Pine-Slash Pine Cover Type (NVCS *Pinus palustris-Pinus elliottii* Forest Alliance): This is the southern variant of the mixed *Pinus palustris* community (Refer to type 28.13). Both species tolerate a wide range of conditions, from dry sandy ridges to poorly drained flatwoods (Figure 28.14). The two occur together only in sites where there has been a break in the fire cycle. Although *Pinus palustris* stems are fire tolerant within 2 years (Figure 28.9), *Pinus elliottii* stems must reach a height of 3 m (usually about 5 years' growth) before they can tolerate even light burns. For this reason, the community is more common in wet areas harboring *Pinus elliotii* from fire and maintaining a seed source. Associated species depend on topography (Eyre 1980). *Cornus florida, Quercus stellata, Q. marilandica, Q. falcata, Carya* spp., *Ilex vomitoria, Diospyros virginiana,* and *Crataegus* spp. are typical associates on well-drained upland soils. On the Coastal Plain flatwoods, the community includes *Acer rubrum, Liquidambar styraciflua, Nyssa sylvatica, Quercus nigra,* and *Q. laurifolia.* Periodically flooded sites include *Taxodium distichum, Taxodium distichum* var. *imbricarium,* and *Nyssa aquatica.* The most common understory species are *Serenoa repens* and *Ilex glabra.* Other species common only on wetter sites include *Magnolia virginiana, Cliftonia monophylla, Lyonia* spp., *Smilax* spp., *Cyrilla racemiflora, Clethra alnifolia, Hypericum hypericoides,* and *Ilex coriacea.* Herbaceous cover is dominated by *Aristida stricta* (wiregrass), *Andropogon* spp., *Panicum* spp., *Sporobolus curtissii, Paspalum* spp., *Trilisa odoratissima, Desmodium* spp., and *Cassia fasciculata.* Well-drained sites may include *Heterotheca graminifolia* and *Aster* species.

28.39 Longleaf Pine-Scrub Oak Cover Type [NVCS *Pinus palustris-Pinus (echinata, taeda)-Quercus (incana, margaretta, falcata, laevis*]: This community is especially common on the sandhills (acid sands of marine origin) of the Carolinas, Georgia, and northern Florida. Low moisture, low soil fertility, and periodic fires (Figure 28.15) characterize the community. Increased logging and fire suppression have led to a decline in pine and an increase in scrub oak. The scrub oaks inhibit pine regeneration. All of these scrub oak species have stress-adapted leaves to tolerate the dry conditions (Harlow et al. 1996). The small leathery leaves decrease surface area exposure, thus decreasing transpiration. Leaf thickness also decreases desiccation. In addition, leaves are generally hairy, creating a boundary zone where humidity is higher and wind speed is slower, resulting in reduced transpiration. *Quercus incana* is characteristic of dry, sandy uplands, and *Quercus margaretta* is characteristic of the driest,

FIGURE 28.15 A longleaf pine (*Pinus palustris*) forest under prescribed burning management at the Zuni Pine Barrens, North Carolina. The understory of such forests can be very diverse in herbaceous species.

deepest sands (Harlow et al. 1996). *Diospyros virginiana, Rhus* spp., and *Crataegus* spp. may occur, along with a scattering of shrubs, including *Chrysobalanus oblongifolius, Myrica cerifera, Ceratiola ericoides, Vaccinium* spp., and *Serenoa repens.* The dominant ground cover is *Aristida stricta* (wiregrass), and *Andropogon* spp., *Panicum* spp., and *Galactia* spp. are common (Weakley et al. 1998).

28.40 Pond Pine Cover Type (NVCS *Pinus serotina* Forest Alliance): This community is common on the broad, poorly drained interstream areas of peaty soils (pocosins), bays, pond margins, and poorly drained depressions. *Pinus serotina* exists under conditions that effectively exclude other pine species: a combination of poor aeration, high soil acidity, slow nitrogen fixation and nitrification, and wildfires. The species is unusual, with both serotinous cones and resprouting ability. *Pinus serotina* is a prolific sprouter, an unusual trait for conifers. It can recover; that is, adventitious buds along stems sprout new leaves following complete defoliation by fire (Harlow et al. 1996). *Pinus serotina* is fairly intolerant of shade, with about the same tolerance as *Pinus taeda,* less tolerance than *Pinus elliottii,* and greater tolerance than *Pinus palustris.* Periodic years of drought produce intense fires as a result of the accumulation of fuels between drought events (Burns and Honkala 1990a). Depending, generally, on the characteristics and length of the recovery period since the last fire, community associates may include *Pinus elliottii, Pinus taeda, Taxodium distichum* var. *imbricarium* and var. *distichum, Nyssa aquatica, Acer rubrum, Liquidambar styraciflua, Magnolia virginica, Gordonia lasianthus,* and *Persea borbonia.*

28.41 Loblolly Pine-Lowland Oak Cover Type [(several NVCS types: *Pinus taeda-Quercus nigra; Pinus taeda-Quercus* (*pagoda, michauxii, shumardii*) Temporarily Flooded; *Pinus taeda-Quercus* (*phellos, nigra, laurifolia*) Temporarily Flooded; *Pinus taeda-Quercus laurifolia-Chamaecyparis thyoides* Temporarily Flooded Forest Alliances)]: *Pinus taeda* has a wide tolerance of moisture conditions and is thus found in a variety of site types. In addition, the species is only moderately tolerant of shade and thus is seral in most cases. The associates of this community are then dependent on sometimes slight topographic variations and the age and successional status of the stand. *Quercus* species are typically associated with *Pinus taeda* on moist inland floodplain sites (black-water rivers). Succession is toward oak dominance because of *Quercus* species being generally more shade tolerant than *Pinus* species.

28.42 Loblolly Pine-Sweetgum Cover Type (NVCS *Pinus taeda-Liquidambar styraciflua-Nyssa biflora* Temporarily Flooded and *Pinus taeda-Liquidambar styraciflua-Acer rubrum* Saturated Forest Alliances): *Pinus taeda* is a widespread species that often associates with *Liquidambar styraciflua* on moist inland sites following disturbance. Both species are adapted to colonize following disturbance, but this seral community will succeed to other bottomland hardwood species over time. This community is typical of slightly elevated ridges on floodplains. Canopy associates include *Diospyros virginiana, Quercus nigra, Carya cordiformis, Ulmus americana, Celtis laevigata,* and *Carya tomentosa,* and understory associates include *Asimina triloba, Ilex decidua, Carpinus caroliniana, Crataegus viridis, Crataegus spathulata,* and *Ilex opaca.*

28.43 Planer-tree Cover Type (NVCS *Planera aquatica* Seasonally Flooded Forest Alliance): This type occurs in the backswamps along rivers that are flooded annually. Emergents from this small statured forest include *Taxodium distichum* var. *imbricarium* and var. *distichum, Nyssa biflora, Populus heterophylla, Fraxinus caroliniana, Liquidambar styraciflua,* and *Platanus occidentalis.* Rapid siltation on the floodplains attributable to eroding agricultural lands has deadened the overstory of many floodplain forests by suffocating them. The occurrence of this type may be partially due to the hydrologic modifications that follow the silt aggradation (NVCS 1999). The herbaceous layer is sparse.

28.44 Sand Live Oak Cover Type (NVCS *Quercus geminata* Forest Alliance): This type occurs along the Atlantic coast on xeric hammocks (coastal dunes). Soils are deep (thus well drained), infertile sands. Other overstory species include *Quercus myrtifolia, Lyonia ferruginea, Osmanthus americanus* var. *americanus, Persea borbonia,* and *Myrica cerifera.* Shrub species may include *Ilex vomitoria, Lyonia fruticosa, Serenoa repens,* and *Bumelia lanuginosa* (NVCS 1999).

28.45 Sand Laurel Oak Cover Type (NVCS *Quercus hemisphaerica* Forest Alliance): This evergreen forest is typical of dry, sandy areas, sometimes called high hammocks (often surrounded by swamps). The type occurs in fire-protected topographic locations. The type may also occur where human disturbance has suppressed the frequency of fire. *Quercus hemisphaerica* is tolerant of shade, but not abundant except in Florida (Harlow et al. 1996). Associated species of *Quercus hemisphaerica* vary according to latitude (tropical to temperate maritime hammocks) and site type (e.g., protection from salt spray). Overstory species include *Pinus taeda, Quercus virginiana, Q. phellos, Quercus nigra, Magnolia grandiflora, Ilex opaca, Carya glabra, Carya pallida,* and *Carya tomentosa.* The type is successional to the aforementioned species and to *Tilia caroliniana* and *Quercus michauxii.*

28.46 Turkey Oak Cover Type (NVCS *Quercus laevis* Forest Alliance): This type occurs in xeric sandhill communities. The sands are deep and may lack clay particles to a depth of 3 m (Entisols) (Burns and Honkala 1990b). Soils are generally strongly acidic, are low in organic matter, and have a water table at least 1.5 m deep. The plane surface of *Quercus laevis* leaves is oriented relatively vertically, an adaptation to the hot sun and highly reflective white sandy soils (Harlow et al. 1996). This type may occur as a scrub community (along some of the driest sand ridges of central Florida) or forest succeeding *Pinus palustris* forest when fire is suppressed (NVCS 1999). The type also dominates if *Pinus palustris* forests are cut for logging or turpentining. *Quercus laevis* has moderately fast to fast growth. Associated species include *Quercus incana, Q. margaretta, Q. myrtifolia, Carya floridana,* and *Quercus marilandica. Gaylussacia dumosa* may also be present.

28.47 Diamondleaf Oak Cover Type (NVCS *Quercus laurifolia* Seasonally Flooded Forest Alliance): This type occurs in seasonally flooded blackwater areas. Soils are alluvial Inceptisols or Ultisols. *Quercus laurifolia* is fast growing and short-lived. It can produce large acorn crops that are disseminated by squirrels. *Quercus laurifolia* is shade tolerant, often growing up through the dense canopy of swamp borders (Burns and Honkala 1990b). Prolonged flooding during the growing season and fire have negative effects on stands, causing high mortality. *Quercus laurifolia* can sprout from its stump. The type may succeed to a *Taxodium-Nyssa* forest. Overstory associates include *Taxodium distichum, Nyssa biflora, Quercus nigra,* and *Fraxinus caroliniana. Carpinus caroliniana* is common in the subcanopy, and *Itea virginica* and *Sabal minor* are common shrubs.

28.48 Diamondleaf Oak-Swamp Gum Cover Type (NVCS *Quercus laurifolia-Nyssa biflora* Saturated Forest Alliance): This type occurs on nonriverine flats with poor drainage resulting from high water tables (e.g., at the edge of peatlands) or sheetflow from adjacent pocosins (NVCS 1999). Soils are generally mineral soils and fine textured. Other overstory species include *Quercus phellos* and *Acer rubrum* var. *trilobum. Carpinus caroliniana* is common in the subcanopy.

28.49 Willow Oak-Water Oak-Diamondleaf Oak Cover Type [NVCS *Quercus (phellos, nigra, laurifolia)* Temporarily Flooded Forest Alliance]: This type occurs mainly along flats of blackwater or low-sediment, low-nutrient rivers. Soils are typically fine-textured (silt loams and clay loams) alluvial Entisols, varying from very acidic to weakly basic. The soils are relatively poorly drained. *Quercus phellos* is intolerant of shade, regenerating under small to large openings in the canopy. Although this species produces a heavy annual acorn crop, advanced regeneration is important in maintaining its presence in these forests (Burns and Honkala 1990). This type is commonly found on drier sites than the *Quercus lyrata-Carya aquatica* stands, and on wetter sites than *Quercus pagoda* and *Quercus michauxii.* Associates vary with geographic locations, but some of the main associates include *Fraxinus pennsylvanica, Quercus michauxii, Liquidambar styraciflua, Acer rubrum, Nyssa biflora, Carya tomentosa, Quercus pagoda, Taxodium distichum,* and *Celtis laevigata.* Shrubs include *Ilex decidua, Rhododendron canescens, Symplocos tinctoria, Viburnum rufidulum,* and *Illicium floridanum.* Woody vines are important in this forest, including *Toxicodendron radicans, Bignonia capreolata, Smilax rotundifolia, Vitis rotundifolia, Parthenocissus quinquefolia, Trachelospermum difforme, Berchemia scandens, Smilax glauca,* and *Ampelopsis arborea.*

28.50 Nuttall Oak-Overcup Oak Cover Type [NVCS *Quercus texana-(Quercus lyrata)* Seasonally Flooded Forest Alliance]: This type occurs in the floodplains of rivers, under the most hydric conditions of all oaks. Depressions are common within the community, where *Taxodium distichum* and *Nyssa aquatica* may cooccur in the overstory. Soils are generally organic dark brown silt loams and clays of alluvial origin (Inceptisols and Entisols). The soils are poorly drained and acidic (Burns and Honkala 1990b). *Quercus texana* is intolerant of shade and must regenerate in canopy openings. Acorn crops rarely fail, and they are an important food source for waterfowl and rodents. Seeds are also dispersed by water. Seedlings tolerate high water during the growing season (Burns and Honkala 1990b). Other overstory species include *Liquidambar styraciflua, Fraxinus profunda, Fraxinus pennsylvanica, Celtis laevigata, Carya aquatica,* and *Gleditsia aquatica. Forestiera acuminata* may occur as a shrub. Woody vines are important, such as, *Campsis radicans, Toxicodendron radicans, Ampelopsis arborea, Ampelopsis cordata,* and *Berchemia scandens.*

28.51 Live Oak Cover Type (NVCS *Quercus virginiana* Forest Alliance and *Quercus virginiana-Sabal palmetto* Forest Alliance): This evergreen forest dominates barrier islands and coastal hammocks in xeric to moist soil conditions. Some of these forests are sculpted by wind and salt spray from the oceans, giving them the appearance of leaning inland (Figure 28.16). Communities are found on sand flats, lower slopes, and stabilized sand dunes protected from saltwater flooding. Overstory species include *Quercus hemisphaerica, Pinus taeda,* and *Juniperus virginiana* var. *silicicola. Juniperus virginiana* var. *silicicola* is typical of moist to wet soils near limestone outcroppings or Native American shell middens bordering tidal marshes. This species was harvested heavily to make pencils in the nineteenth century, so it is less common than it was in historical forests. Young stems of *Quercus virginiana* are susceptible to fire but sprout abundantly. This species is intermediate in shade tolerance, and thus, is not a good competitor, but extremely tolerant of salt spray and high levels of soil salinity. For this reason, *Quercus virginiana* dominates forests near the coast (Burns and Honkala 1990b). Individuals more than 300 years old have been found. Abandoned old fields succeed through a weed stage to shrubs in about 5 years. The shrubs (*Myrica cerifera* and *Cornus drummondii*) appear as a young forest by age 25. *Quercus virginiana* dominates an old field in about 50 years. Several other species may co-occur in more protected areas. *Sabal palmetto* codominates on sites directly behind frontal dunes, often as a strand between the shrub communities and inland forests. Soils are sandy, poorly developed, and low in natural fertility and organic matter. Fire is a common component of the type, which thus favors the fire-resistant *Sabal palmetto* over the fire-sensitive *Quercus* species.

FIGURE 28.16 Typical structure of a coastal forest. The smallest individuals take the full impact of the wind and salt spray (ocean is to the right of the photograph), thus protecting the more inland individuals.

28.52 Cabbage Palmetto Cover Type (NVCS *Sabal palmetto-Quercus virginiana* Saturated Forest Alliance): This type occurs along the Atlantic coast in small patches (<10 ha) at elevations of <30 m (the approximate shoreline during the Pleistocene; Eyre 1980). Soils are poorly to very poorly drained with a high calcium content. The presence of calcareous deposits frequently characterizes this community. *Sabal palmetto* can tolerate a wide range of soil salinity, pH, and drainage, but does not tolerate lengthy tidal inundations. However, it can withstand 2 m fluctuations in water table, as it develops abundant adventitious rootlets from its trunk up to the high-water mark. The community can be found on coastal dunes, along floodplains of major rivers, and on coastal marsh islands, marsh shorelines, bay heads, swales, and hammocks. *Sabal palmetto* is shade tolerant and produces abundant fruit crops that are dispersed by wind. It is fire tolerant unless the fire consumes the deep organic soils. *Sabal palmetto* is the most wind-resistant tree on the Atlantic coast (Burns and Honkala 1990a), an adaptation to periodic hurricanes. *Myrica cerifera* is a typical small shrub or tree associate. *Quercus virginiana* and *Juniperus virginiana* var. *silicicola* may also be present.

28.53 Pond Cypress Cover Type (NVCS *Taxodium distichum* var. *imbricarium* Seasonally Flooded Forest

Alliance): This type occurs (as domes) in shallow depressions and poorly drained flatwoods near the Gulf Coast at elevations of <30 m. Sites are not generally along river courses, as with *Taxodium distichum* var. *distichum* (Burns and Honkala 1990a). The soils are fine sands, sometimes marls underlain by limestone (Eyre 1980)—Spodosols and Ultisols. *Nyssa aquatica* and *Taxodium distichum* var. *distichum* are the principal associates. Along their margins or within ponds on slightly elevated sites, *Pinus elliottii, P. serotina, P. taeda, Acer rubrum, Persea borbonia, Gordonia lasianthus,* and *Magnolia virginica* are common associates. Shrubs include *Cephalanthus occidentalis, Viburnum* spp., *Cyrilla racemiflora, Ilex vomitoria, Fraxinus caroliniana, Forestiera acuminata, Myrica cerifera,* and *Cornus stricta.* The wood of *Taxodium* species is highly resistant to decay, often referred to as "wood everlasting."

28.54 Bald Cypress Cover Type (NVCS *Taxodium distichum* var. *distichum* Semipermanently Flooded Forest Alliance): This type dominates sites with frequent, prolonged flooding. Floodwater may be 1 m deep or more, with stagnant to slow-moving water. Sites are generally in depressions that prolong the flood beyond that of the adjacent flats (Figure 28.17). However, a period of dryness is required for regeneration, and the permanency of some ponded areas, due to road construction and man-made levees, has eliminated reproduction in many of these communities; they are essentially turning to open marsh. *Taxodium* spp. and *Nyssa* spp. will regenerate deeper into swamps during unusually dry periods. *Taxodium distichum* has several adaptations for periods of inundation, similar to those of its main associates. Their roots have been shown to oxidate the rhizosphere (soil area immediately adjacent to roots) to depths of 1.2 m (Eyre 1980). This facilitates oxidation in a highly reductive soil, thus minimizing compounds toxic in their reduced form. In addition, the tree has several descending roots (Harlow et al. 1996) and a swollen base to help with stability. Where water levels fluctuate, roots produce "knees" that extend upward out of the soil (Figure 28.18). Although these knees are full of aerenchyma tissue, there is no evidence that they are used to help with gas exchange. Seedlings grow quickly, up to 30 cm in the first season and up to 5 m by age 10. Once leaves are above the water, stems can tolerate lengthy inundation. Older stems grow slowly, with maximum ages reaching possibly 1000 years (Figure 28.19). The main associates include *Nyssa biflora* in the alluvial floodplains and *Nyssa aquatica* in the swamps and estuaries. Other associates include *Taxodium distichum* var. *imbricarium* and *Salix nigra.* Common shrubs include *Cephalanthus occidentalis, Cornus stricta, Ilex decidua, Viburnum nudum,* and *Viburnum obovatum.* The fruit is eaten by many animals (Burns and Honkala 1990a).

FIGURE 28.17 Pondcypress (*Taxodium distichum* var. *imbricarium*)-dominated forest of the Okefenokee Swamp, Georgia.

28.55 Bald Cypress-Gum Cover Type [NVCS *Taxodium distichum* var. *distichum Nyssa biflora*-(*Nyssa aquatica*) Saturated Forest Alliance]: This type occurs in sites with water tables above the soil surface either well into or throughout the growing season. The sites are poorly drained. The mineral soils of the alluvial bottoms range from silt loam to almost pure clay. Soils are generally acidic, but *Nyssa biflora* cannot survive where soil acidity is too high and waters are too brackish. The tupelos (*Nyssa* spp.) are better regenerators than *Taxodium distichum* because of their

FIGURE 28.18 Bald Cypress (*Taxodium distichum* var. *distichum*) and red maple (*Acer rubrum*) forest. Note the "knees" (aerial extensions of the roots) at the bottom of the photo.

FIGURE 28.19 Large bald cypress (*Taxodium distichum* var. *distichum*) that survived cutting in southern Illinois, estimated to be more than 800 years old.

greater seed production and faster growth when young. In addition, stump sprouting by *Nyssa* spp. is an important regeneration strategy in these sites, but *Taxodium* does not sprout. In deep alluvial swamps, common associates include *Acer rubrum, Salix nigra, Fraxinus caroliniana, F. profunda, Populus heterophylla, Planera aquatica,* and *Gleditsia aquatica;* in shallower margins, *Quercus lyrata, Carya aquatica, Ulmus alata, Fraxinus pennsylvanica, Quercus texana, Q. laurifolia, Liquidambar styraciflua, Diospyros virginiana,* and *Magnolia virginiana.* A wide variety of wet shrubs may be found in the understory, including *Cephalanthus occidentalis, Itea virginica, Forestiera acuminata, Cyrilla racemiflora, Cliftonia monophylla, Cornus stricta, Lyonia lucida, Leucothoe* spp., *Ilex vomitoria, Ilex decidua, Rosa palustris, Myrica cerifera,* and *Toxicodendron vernix.*

Note on altered watercourses. Since the turn of the nineteenth century, county, state, and federal agencies have straightened, deepened, widened, and leveed (Figure 28.20) portions of nearly all of the large rivers in western Tennessee, along with many of their tributaries (Simon and Hupp 1992; Shankman 1996). Among other things, these channel modifications were made to accelerate storm water drainage, to increase the overbank flood stage, to protect agricultural land from flooding, and to lower the water table in bottomland areas, thereby increasing the amount of acreage suitable for cultivation. The effects of channelization on ecosystem conditions and processes within a river and its associated riparian zone are diverse and vary geographically according to variability in factors such as climate, soil, channel morphology, and drainage basin characteristics (Malanson 1993). Given a low base level, the increase in stream gradient and power brought about by channelization will often cause channel incision (Daniels 1960). Riparian vegetation may be left stranded above the active floodplain, promoting a floristic shift away from flood-tolerant bottomland species (e.g., Fredrickson 1979). In addition, the channel no longer meanders, and the typical successional patterns of bottomland forests are strongly modified (Kupfer and Malanson 1993; Figure 28.21a). The creation of

FIGURE 28.20 Aerial view of a channelized river in western Tennessee. The old meander channel can be seen in the middle-right of the photo. Channelization was used to drain wetlands for agricultural purposes.

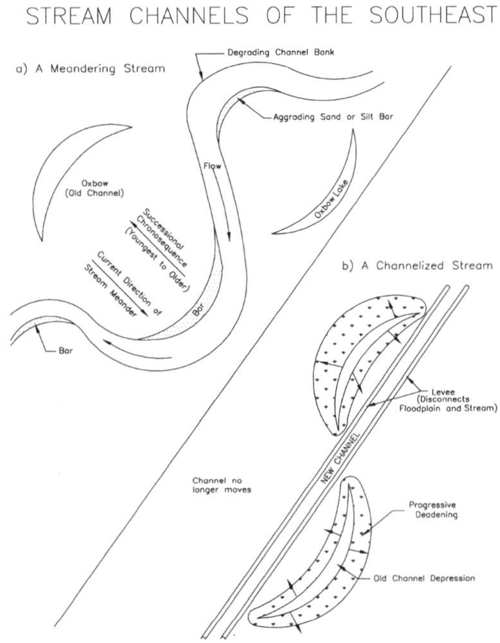

FIGURE 28.21 The differences between a naturally flowing and meandering stream (a) and a channelized and leveed stream (b). There is a loss of habitat following channelization, because point bars are no longer created and stream flow is not as variable. In addition, levees can act as dams, prohibiting water from precipitation or overbank flow to return to the channel, thus creating sloughs in the lowest areas in the floodplain (typically old channels). These low, wet areas may progressively grow over time, deadening large areas of bottomland hardwood forest.

levees, which further isolate the floodplain from the river, can exacerbate the changes in ecosystem structure and function resulting from channelization. Two inputs can occur: precipitation and flooding over the levee. Neither of the inputs can return to the channel, because the levee acts as a dam, ponding water on the floodplain. A zone of deadened timber soon develops, generally from the lowest spot on the floodplain—that is, the site of the channel prior to modification (Figure 28.21b). Where levees do not occur, or have been breached, sediment inputs via overland transport from agricultural fields, as well as headcutting and lateral bank failures of channelized tributaries, have led to channel aggradation in some main stem rivers (Nabb and Shankman 1997). Channel blockage due to bed aggradation in turn has led to increased ponding of water, resulted in elevated water tables and extended periods of flooding, and forced federal and state authorities to rechannelize or redredge stream reaches as a means of hydrologic management (Hupp 1992).

CHAPTER 29

Forest Environment and Cover Types—Regions 5 and 6

PINE-JUNIPER FOREST AND ROCKY MOUNTAIN CONIFER FOREST (BAILEY'S DRY DOMAIN/TROPICAL/SUBTROPICAL STEPPE DIVISION, *PINUS PONDEROSA-PINUS EDULIS-JUNIPERUS,* AND DRY DOMAIN/TEMPERATE STEPPE DIVISION, *PINUS-PSEUDOTSUGA-PICEA*)

INTRODUCTION

THE essential feature for the Dry Domain climate is that annual losses of water through evaporation at the earth's surface exceed annual input through precipitation (P/E < 1). The Rocky Mountain Conifer Forest is split into two forest regions or divisions: (1) Montane Conifer Forest (Bailey's Dry Domain/Temperate Steppe Division, *Pinus-Pseudotsuga-Picea)* and (2) Pine-Juniper Forest (Bailey's Dry Domain/Tropical/Subtropical Steppe Division, *Pinus ponderosa-Pinus edulis-Juniperus).* The Dry Domain forests consist of all areas of the western United States except along the Pacific coast and especially where mountains occur. Average temperatures can drop 1.7° C for every 300 m decent in the Rocky Mountains, resulting in an additional 2 to 3 weeks in the growing season at low elevation (Vankat 1979).

The Pine-Juniper Forest [(Region 5a) Figure 2.1; Bailey's Tropical/Subtropical Steppe Division, *Pinus ponderosa-Pinus edulis-Juniperus*] is found in hot, semiarid climates where average temperatures above 0° C occur throughout the year and precipitation is concentrated during the summer months (Figure 29.1). The P/E ratio is <1, and soils are primarily Aridisols. The Tropical/Subtropical Steppe vegetation is composed of shortgrasses and other herbaceous taxa intermixed with shrub and woodland components (Figure 29.2). The Pinyon-Juniper (*Pinus edulis-Juniperus*) Forest is the primary type. This region, sometimes referred to as the Madrean region, harbors unusual species (Peet 1988).

The Montane Conifer Forest (Regions 5b and 6; NVCS Temperate Steppe Division) has a cooler climate, with at least 1 month of average temperature below 0° C. The Rocky Mountains have a strong continental climate. Winters are cool and dry, summers are warm and hot, and droughts are common. The Rocky Mountains support a wide variety of forest communities, with an elevational gradient (1500 to 4400 m) ranging from ponderosa pine dominated woodlands to spruce-fir dominated forests. High-elevation forests have a krummholz structure (Figure 26.5) near the treeline. Emergents at these higher elevations are often flagged (Figure 29.3). The abrupt topography of the Rocky Mountains produces a dramatic zonation of

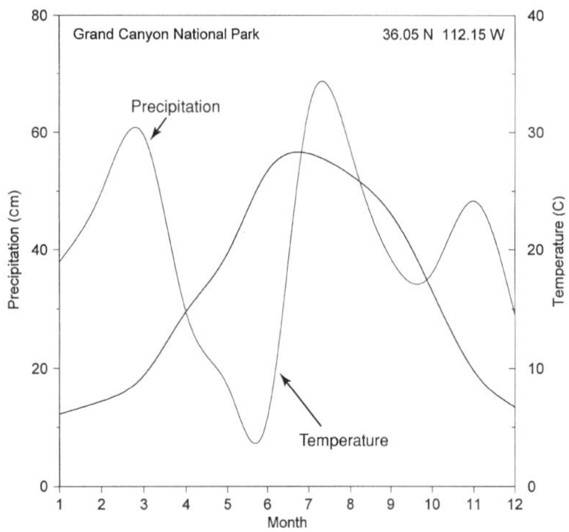

FIGURE 29.1 Climate diagram for Grand Canyon National Park, Arizona, representing the Tropical/Subtropical Steppe Division of the Dry Domain (sensu Bailey 1998).

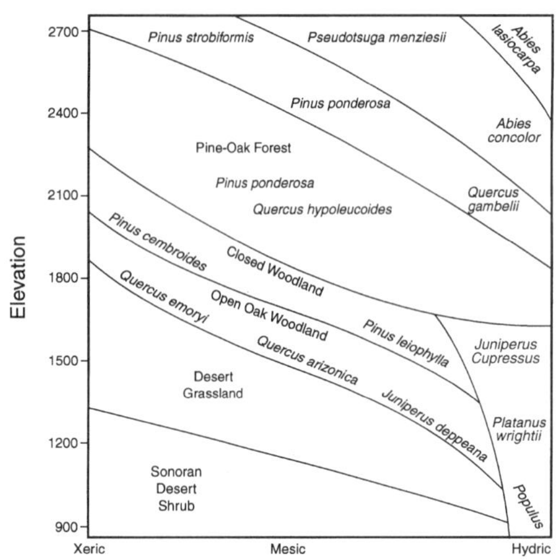

FIGURE 29.2 Gradient mosaic diagram illustrating variations in forest composition with elevation and topographic positions for the Santa Catalina Mountains, representing the *Pinus Ponderosa-Pinus edulis-Juniperus* forest subregion (see Table 28.1). Redrawn from Whittaker and Niering (1965) with permission from Cambridge University Press.

FIGURE 29.3 Flagging, a process by which high-elevation trees loose their limbs on the windward side because of erosion by wind with ice crystals.

grasslands to the west to forest to grasslands on the east because of the vertical stratification of climate and soils (Vankat 1979; Peet 1988). In addition, a given vegetation type tends to be found on drier sites at higher elevations (Peet 1988; Barbour et al. 1998). Soils are generally young, derived from precambrian granites or limestone. The limestone-derived soils tend to have a more xeric vegetation assemblage than granitic soils, likely due to soil texture differences (Wentworth 1981). Thus, Peet (1988) suggests three main gradients for the Rocky Mountains: (1) elevation, (2) topographic moisture, and (3) soils (Figure 29.4).

The Montane Conifer Forest has been split into two subregions (Northern and Southern) based on climatic patterns and forest composition. Additional

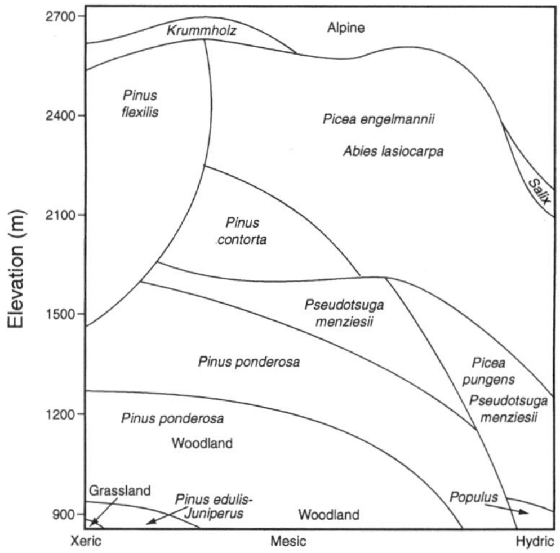

FIGURE 29.4 Gradient mosaic diagram illustrating variations in forest vegetation composition with elevation and topographic positions for the Colorado Front Range, representing the Southern Rocky Mountains subregion (see Table 28.1). Redrawn from Peet (1988) with permission from the Ecological Society of America.

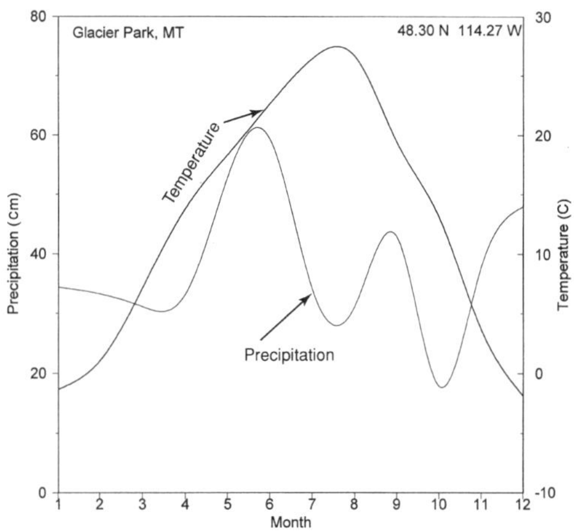

FIGURE 29.5 Climate diagram for Glacier Park, Montane, representing the northern Rocky Mountain portion, Temperate Steppe Division of the Dry Domain (*sensu* Bailey 1998).

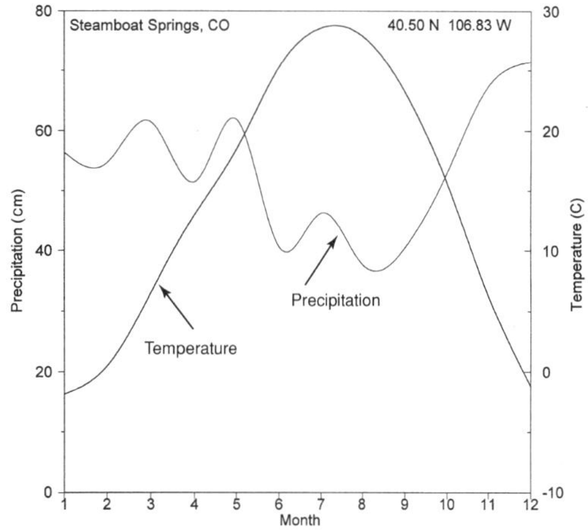

FIGURE 29.6 Climate diagram for Steamboat Springs, Colorado, representing the southern Rocky Mountain portion, Temperate Steppe Division of the Dry Domain (*sensu* Bailey 1998).

storm tracks penetrate east, dropping greater amounts of precipitation on the northern Rockies (Peet 1988; Bailey 1998) (Figures 29.5 and 29.6). The southern Rockies (Region 5b) receive little influence from the Pacific airmass, and a warmer continental climate dominates. There is a northward decrease in temperature, resulting in a decrease in the elevation for zones of specific communities. The general elevational sequence is from Pinyon-Juniper at low elevation to *Pinus ponderosa,* then to *Pseudotsuga menziesii,* and finally to *Picea engelmanii* at high elevations. At least seven to ten species that occur in the northern Rockies (Region 6, Figure 2.1) are not found in the southern Rockies (Region 5b). Included here are various *Pinus, Larix, Tsuga, Thuja,* and *Abies.*

COVER TYPES

Pine-Juniper Forests (NVCS Dry Domain/Tropical/Subtropical Steppe Division; *Pinus ponderosa-Pinus edulis-Juniperus*) (Region 5a)

29.1 Xalapa Madrone-Bigtooth Maple-Oak Species Cover Type (NVCS *Arbutus xalapensis-Acer grandidentatum-Quercus* spp. Forest Alliance): This community occurs in mesic canyons of southern Texas mountain ranges. The sites occur in protected habitats with lower fire frequencies than the surrounding hillsides. The *Quercus* associates include *Q. muhlenbergii, gravesii, grisea,* and *rugosa* (netleaf oak). Other associates include various juniper (including *Juniperus deppeana*) and pine (including *Pinus edulis* and *Pinus ponderosa* var. *scopulorum*) species, *Cupressus arizonica* in some locales, *Prunus serotina* and *Ostrya knowltonii* in some locales. These communities grade into the more mesophytic forests strongly dominated by *Acer grandidentatum* (see the next section on Southern Rocky Mountain Conifer Forests).

29.2 Arizona Cypress Cover Type (NVCS *Cupressus arizonica* Forest Alliance): This community is found on mountain slopes and coves along intermittent streams, with coarse, rocky soils at elevations between 910 and 1800 m (Harlow et al. 1996; Little 1998). Reproduction is usually scanty because of the generally inhospitable conditions for germination and growth. Fire is necessary, as it opens serotinous cones, kills competitors (mainly *Pinus edulis* and *Juniperus* spp.), and prepares regeneration sites (Eyre 1980). This community may include *Pinus arizonica, Pinus engelmanii, Pinus leiophylla* var. *chihuahuana, Platanus wrightii, Juglans major, Arbutus arizonica, Celtis reticulata, Fraxinus velutina, Robinia neomexicana, Populus fremontii,* and *Morus microphylla.* On the drier, rocky sites *Juniperus deppeana* may be found. Several live oaks may also be present. Shrubs include *Rhus trilobta, Cercocarpus betuloides, Fendlera rupicola, Dalea formosa,* and *Eriodictyon angustifolium.* Forbs may include *Erigeron* spp., *Oenothera* spp., *Verbena* spp., *Sphaeralcea* spp., and *Senecio* spp. (Eyre 1980; Walker 1999).

29.3 Rocky Mountain Juniper Cover Type (NVCS *Juniperus scopulorum* Forest Alliance): *Juniperus scopulorum* is an important component of the pinyon-juniper woodlands and forests (Harlow et al. 1996) and has the largest range of any western juniper except *Juniperus communis.* These communities occur on rocky soils, especially on exposed limestone and lava outcrops of southern exposure, within the foothills at altitudes between 1520 and 2700 m. The rocky sites burn less often than the surrounding areas. Soils tend to be poorly developed mollisols that are slightly alkaline. *Juniperus scopulorum* is drought enduring but less drought resistant than other junipers. As with other species of the pinyon-juniper forest, seed dissemination by birds is extremely important (West 1999). *Juniperus scopularum* and associates will invade sagebrush dominated areas where fire has been excluded. Co-occurring species include *Pinus edulis, Pinus ponderosa,* and *Pinus flexilis.* This type mixes with *Pinus edulis-Juniperus* or *Populus-Salix* communities (on moist sites) in the south, and *Pseudotsuga menziesii* or *Pinus ponderosa* communities in the north (Eyre 1980). *Quercus gambelii* communities dominate upper elevations (Walker 1999). Shrub associates include *Cercocarpus* spp. and *Artemesia tridentata.* Graminoids may include *Agropyron* spp., *Stipa occidentalis, Bouteloua gracilis, Bromus tectorum,* and *Elymus* spp.

29.4 Mexican Pinyon-Chisos Red Oak Cover Type (NVCS *Pinus cembroides-Quercus gravesii* Forest Alliance): This community occurs on canyon slopes at high elevations in the extreme southwest and, likely, into Mexico. The community sits between oak-juniper woodlands, at lower elevations, and coniferous forests at higher elevations. Several *Pinus* species (e.g., *cembroides* and *arizonica*), *Juniperus* species (e.g., *deppeana, flaccida,* and *erythrocarpa*), evergreen *Quercus* species (e.g., *grisea* and *tardifolia*), and deciduous *Quercus* species (e.g., *graciliformis*) may be present (NVCS 1999).

29.5 Pinyon Pine Cover Type (NVCS *Pinus edulis* Forest Alliance): The *Pinus edulis* community is common throughout the southwest; locations include canyons within the desert and breaks and escarpments within the Great Basin. Elevations range from 1500 to 2440 m. Droughts are common in this semiarid climate. Soils are typically shallow, rocky, calcareous, and alkaline, derived from limestone, sandstone, or shale. Soils are Aridisols (with pedogenic horizons and moderate to low organic matter) or Entisols (in the Basin and Range provinces). This community is frequently found adjacent to sagebrush and readily invades overgrazed areas. Several species of *Juniperus* (e.g., *osteosperma, scopulorum*) or *Quercus* (e.g., *arizonica, gambelii, turbinella*) may be present, depending on the geographic location. *Pinus edulis* is extremely drought tolerant and slow growing. Although young stems are susceptible to fire, older trees are generally not killed. Woodland communities are apparently adapted to surface fires burning fine fuels. However, the lack of fuel and decreased wind under closed-canopy forests inhibit fire movement, and thus closed stands do not readily burn. However, closed stands will revert back to grass or brush following major disturbance. The understory is shaded by the canopy and thus sparse, but characteristic species include *Artemesia tridentata, Cercocarpus montanus, Cercocarpus ledifolius, Coleogyne ramosissima, Ephedra viridis,* and *Gutierrezia sarothrae. Agave* spp. and *Yucca baccata* may associate in southern latitudes. Herbaceous grasses, although also sparse, include *Bouteloua curtipendula, Bouteloua gracilis, Aristida* spp., *Erioneuron pilosa, Festuca arizonica, Koelaria macrantha, Oryzopsis hymenoides,* and *Poa fendleriana* (NVCS 1999). The *Pinus edulis-Juniperus* forests are apparently extending their range into the woodlands and grasslands of the west as a result of fire suppression (West 1999).

29.6 Ponderosa Pine Cover Type [NVCS *Pinus ponderosa* var. *scopulorum* (interior species) Forest Alliance]: The *Pinus ponderosa* community is widespread throughout the midwestern and southwestern United States at the lowest elevations of the coniferous zones. These are among the driest forests of the western United States, always experiencing some seasonal drought (Whitney 1997). The community occurs on a variety of aspects, but especially northeast- to northwest-facing slopes of gentle grades between the plains and the *Picea-Abies* forests. The species can be found at elevations between 1000 and 2000 m in the Black Hills (the Black Hills were named for the dark canopy of *Pinus ponderosa* forests), between 1800 and 2600 m in the central Rockies, and >1800 m in the southwestern states. Characteristic soil features include good aeration and drainage, neutral to only slightly acidic conditions, and an abundance of mineral material. Parent materials are variable. Fire is a key factor in maintaining the typical openness of these communities. Many of the closed forest stands were likely woodlands historically, but fire suppression has led to a closed canopy (see Note on the 10:00 A.M. policy, Smokey Bear, and Bambi). Fire suppression has led to increases in more shade-tolerant species in the understory, more brush, and less graminoids, which will ultimately affect fire behavior and its effects (Barbour et al. 1998). However, soil drought and infertility control the occurrence of this community as well, and these stands may be sustained for long periods of time (Robertson 1992). *Pinus ponderosa* is drought tolerant and shade tolerant. However, it seems to require certain conditions for regeneration, namely, adequate soil moisture, good seed crops, and bare mineral soil. Thus, its apparent episodic regeneration requires good climate and good seed years to coincide, perhaps occurring only every 20 years (Whitney 1997; NVCS 1999). These conditions are most typical in the Black Hills. Competition from grasses and forbs during early growth is a concern (Burns and Honkala 1990). Associated trees include *Pseudotsuga menziesii, Abies concolor, Picea pungens,* and *Pinus strobiformis* in the Rockies. Associated species in the south vary with elevation, but include *Quercus gambelii, Pinus strobiformis, Juniperus deppeana, Quercus grisea, Quercus muehlenbergii, Pinus arizonica, Cupressus arizonica, Juniperus flaccida, Arctostaphylos uva-ursi,* and *Cercocarpus montanus.* Shrubs may include *Physocarpus capitatus, Spirea betulifolia, Holodiscus discolor, Crateagus douglasii,* and *Amelanchier alnifolia* (Whitney 1997). No understory plants are unique to the *Pinus ponderosa* forest, but species may include *Festuca arizonica, Schizachyrium scoparium* var. *neomexicanum, Panicum bulbosum,* and *Piptochaetium pringlei* (NVCS 1999). The understory may be depauperate because of *Pinus ponderosa* utilizing all the soil resources (Whitney 1997).

> *Note on the 10:00 A.M. policy, Smokey Bear, and Bambi:* The Forest Service began fire suppression early in the twentieth century. One of the policies included completely controlling wildfires by 10:00 A.M. the following day. If this goal was not met, the goal was set for 10:00 A.M. the next day. In other words, complete suppression was the rule for several decades. The mascot Smokey Bear, a baby bear found stranded following a fire, and movies such as *Bambi* embedded the evils of wildfire in the minds of the American public. However, the changes in forests resulting from this policy were rather quickly realized. Increased ladder fuels from more shade-tolerant taxa encroaching on sites, loss of fine fuels, and increases in dead and downed fuels produced more severe fires that were very difficult to contain and extremely dangerous for firefighters. The Forest Service began to rethink its policies and, within the last two decades, implemented a "let burn" policy where it was appropriate. In addition, prescribed fires are currently utilized to reduce fuels during less dry conditions, thus allowing the fires to be controlled. A continual problem faced by forest stewards is the urban-wildland interface.

Managers must make decisions about letting areas burn, but cannot risk the loss of private residences that are often interspersed with, or are found on the edges of, public lands. The problems pursuant to making wrong decisions were brought to the forefront with the uncontrollable prescribed burn set in Bandelier National Park in May 2000. The Cerro Grande fire escaped, destroyed 300 homes, and threatened New Mexico's Los Alamos National Laboratory. In addition, 18,000 people were evacuated and 48,000 acres were burned.

29.7 Rio Grande Cottonwood Cover Type (NVCS *Populus Fremontii* spp. *wislizeni* Temporarily Flooded Forest Alliance): This community occurs along moderately large rivers below 1800 m in elevation. Typically, the community dominates only new point bars of alluvial sands and gravels, but mature stands may be found several meters from the channel on alluvial soils that are deep, cobbly, stratified sandy loams. *Populus angustifolia* may also be present. Typical shrubs include *Juniperus monosperma, Forestiera neomexicana, Fallugia paradoxa, Baccharis glutinosa, Baccharis sarothroides,* and *Gutierrezia sarothrae. Clematis ligusticifolia* may be present, along with graminoids such as *Muhlenbergia asperifolia, Muhlenbergia enersleyi, Eleocharis palustris,* and *Distichilis spicata.*

Note on cottonwoods and regulated rivers: The seral forests (the various cottonwood forests) of floodplains typically require a meandering river that exposes new areas for colonization and a periodic flooding regime. Water regulation (controlling the flow via dams and reservoirs) may be the greatest threat to these communities. The regulation of water alters the timing of flooding, the duration of flooding, and the severity of flooding, thus affecting river movement and the riparian community structure. The frequency and duration of any disturbance are key components to assessing disturbance effects. The alterations of the flooding regime produces fewer new point bars and increased encroachment by later successional species. In addition, the cottonwood and willow species are important ecosystem components, as they effectively hold soil in place (Walker 1999). Loss of this streamside community often increases bank sloughing and erosion.

29.8 Fremont Cottonwood Cover Type (NVCS *Populus fremontii* Temporarily Flooded Forest Alliance): Stands of this community occur along the valley floors of the large rivers in New Mexico at elevations between 1500 and 2100 m (lower elevations for than *Populus angustifolius;* Peet 1988) and along low-gradient rivers in Arizona. Periodic droughts may result in dry streams. As with other cottonwood communities, *Populus fremontii* requires specific flood regime characteristics for germination and establishment (Stromberg 1993). Spring floods often deposit large amounts of silt and sand. The community is typically found on the most mesic or hydric sites of the floodplain where water tables remain high throughout the year and spring floods occur annually. *Salix gooddingii* is usually present, and shrub species include *Amorpha fruticosa, Baccharis salicifolia,* and *Salix exigua.* The sparse herbaceous layer may contain *Anemopsis californica, Distichilis spicata,* and *Juncus balicus.*

29.9 Buckley Oak Cover Type (NVCS *Quercus buckleyi* Forest Alliance): This community is found on mesic, calcareous soils in Oklahoma and Texas upslope from floodplains. Associated species include *Ulmus crassifolia, Celtis laevigata, Quercus fusiformis, Quercus muhlenbergii, Quercus sinuata* var. *breviloba, Juniperus ashei, and Fraxinus texensis.*

29.10 Gambel Oak Cover Type (NVCS *Quercus gambelii* Forest Alliance): This community occurs on slopes and valleys of mountains and foothills at elevations between 1500 and 2400 m; it is often found on disintegrated granite. *Quercus gambelii* is a frequent associate of *Pinus ponderosa* forests (Harlow et al. 1996) and is clearly successional to *Pinus edulis* and *Pinus Ponderosa* (Peet 1988). *Quercus gambelii* is a valuable mast-producing species for deer and other animals.

Rocky Mountain Conifer Forests (NVCS Dry Domain/Temperate Steppe Division; *Pinus-Pseudotsuga-Picea*)

Southern Rocky Mountain Conifer Forests (Region 5a)

29.11 Subalpine Fir Cover Type (NVCS *Abies lasiocarpa* Forest Alliance): *Abies lasiocarpa* forests are common communities of the upper montane and subalpine zones of the western mountains. Summers are cool and frosts are common. The location of this community in the ranges nearer the Pacific coast is typically in a rainshadow, with substantially less precipitation than surrounding areas. Elevation ranges between 1300 and 1950 m in the north (eastern Cascades) and between 2000 and 3500 m in the central and southern Rockies, where snowpack can be deep. Soils are derived from variable parent materials, are typically shallow and poorly developed, and are relatively stony. The soils are Inceptisols, but may show evidence of podzolization (NVCS 1999). *Picea engelmannii* is often codominant, dominating especially the moister sites and more mature stands. However, *Abies lasiocarpa* seems to establish under shade and on organic substrates better than *Picea engelmannii* (Peet 1988). *Pinus contorta* and *Populus tremuloides* are also associates, due to their earlier dominance of these sites following disturbance. Other associates include *Pinus albicaulis, Larix lyallii, Abies concolor, Abies procera, Larix occidentalis,* and *Pseudotsuga menziesii.* The understory is dominated by grasses or dry sedges on drier sites and ericaceous shrubs or moist forbs under closed canopies. Typical shrubs include *Vaccinium myrtillus, Vaccinium scoparium, Vaccinium cespitosum, Mahonia repens, Shepherdia canadensis, Physocarpus monogyra, Jamesia americana, Juniperus communis,* and *Acer glabrum.* Under semiriparian and seep slope conditions (NVCS *Abies lasiocarpa* Temporarily Flooded and Seasonally Flooded Forest Alliances), associations with *Picea pungens* and *Pseudotsuga menziesii* are common and *Alnus viridis* spp. *sinuata* is the dominant shrub (develops on the sites following fire). On the wettest soils, *Ledum glandulosum, Vaccinium membranaceum, Alnus incana,* and *Salix* species dominate the shrub layer. The diverse herbaceous layer includes *Cardamine cordifolia, Caltha leptosepala, Ligusticum porteri, Mertensia ciliata,* and *Mitella pentandra.* Windthrow is a more common disturbance in the wetter sites because of shallow root systems.

29.12 White Fir-Quaking Aspen Cover Type (NVCS *Abies concolor-Populus tremuloides* Forest Alliance): This community type is seral, establishing after fire or windthrow and typically succeeding to mixed conifers already regenerating in the understory. Gap-forming disturbances favor *Populus tremuloides* over *Abies concolor,* as *Abies concolor* is much more shade tolerant. The forests occur at montane elevations of 2000 to 2900 m (Figure 29.7). Soils are slightly acidic, well-drained loams or sandy loams. Typical shrubs include *Acer glabrum, Arctostaphylos patula, Rosa woodsii, Symphoricarpos oreophilus, Mahonia repens,* and

FIGURE 29.7 A white fir (*Abies concolor*) and Engelmann's spruce (*Picea engelmannii*) forest in the high-elevation mountains near Logan, Utah.

Juniperus communis. The herbaceous stratum, which is much richer than those of adjacent conifer forests, includes *Festuca arizonica, Poa fendleriana, Carex foenea, Carex rossii, Thalictrum fendleri, Achillea millefolium, Rudbeckia occidentalis,* and *Pteridium aquilinum.*

29.13 Bigtooth Maple Cover Type (*Acer grandidentatum* Forest Alliance): This community consists mainly of deciduous species. Typical sites are lower montane (elevations of 1300 to 1800 m) north-facing valleys or canyons that have moderate insolation and favorable soil moisture. These sites appear to be maintained by periodic fire that prevents conifer encroachment. Associates include *Quercus gambelii, Prunus virginiana, Amelanchier alnifolia, Physocarpus malvaceus, Artemesia tridentata, Pachistima myrsinites, Symphoricarpos* spp., and *Populus tremuloides.* The understory, which is relatively sparse, contains short evergreen shrubs, both annual and perennial grasses, and forbs.

29.14 Boxelder Cover Type (NVCS *Acer negundo* Seasonally Flooded Forest Alliance): *Acer negundo* dominates many floodplain systems of the southeastern United States, but is also found sporadically in the west. These western riparian forests occur along nearly level floodplains below elevations of 1800 m. Coarse loamy and fine loamy soils typify this community. *Equisetum arvense* is a strong component of the herbaceous stratum. Other herb species include *Maianthemum stellatum, Mentha arvensis,* and *Elymus glaucus.*

29.15 Blue Spruce Cover Type (NVCS *Picea pungens* Forest Alliance, Temporarily Flooded Forest

Alliance, and Seasonally Flooded Forest Alliance): This community occupies protected coves of mainly north-facing slopes at elevations between 1800 and 3000 m. *Picea pungens* and associates can create a well-developed stream bank community (Eyre 1980). Soils are generally young and derived from glacial outwash or alluvial deposits. Soil pH tends to be slightly alkaline. *Picea pungens* is slow growing and long-lived, so the community is considered to be a late seral stage of succession. Shade tolerance of the species is about that of *Pseudotsuga menziesii*. *Picea* and *Abies* are more shade tolerant and dominate the stand later in succession. Common associates include *Pseudotsuga menziesii, Abies lasiocarpa, Abies concolor, Picea engelmannii, Pinus ponderosa, Pinus contorta,* and *Pinus strobiformis*. Next to streams, *Alnus incana, Betula occidentalis, Populus angustifolia,* and *Populus tremuloides* are common. Some shrubs occur, but do not form a great deal of cover. Cold-deciduous and shrub species include *Quercus gambelii, Amelanchier alnifolia, Acer glabrum, Arctostaphylos uva-ursi, Juniperus communis, Rubus parviflorus,* and *Mahonia repens*. Adequate soil moisture permits a well-developed herbaceous layer consisting of *Carex foenea, Festuca arizonica, Arnica cordifolia, Senecio cardamine, Fragaria virginiana, Linnaea borealis,* and *Pseudoroegneria spicata* (NVCS 1999).

29.16 Blue Spruce-Quaking Aspen Cover Type (NVCS *Picea pungens-Populus tremuloides* Forest Type): This community type is common on moderately steep middle to lower mountain slopes and plateaus at elevations between 2300 and 2800 m. Soils are either alluvium or colluvium. The community is seral, with sustainability dependent on gap-forming fires. The dominance of *Populus tremuloides* lessens with the increasing age of the stand. Other canopy species include *Pseudotsuga menziesii, Abies concolor, Pinus flexilis,* and *Pinus contorta*. The shrub layer is often dominated by *Juniperus communis*. The herbaceous layer is often sparse, but may contain *Bromus anomalus, Carex geyeri, Carex rossii, Elymus trachycaulus, Achillea millefolium, Antennaria microphylla, Fragaria vesca,* and *Galium boreale*.

29.17 Bristlecone Pine Cover Type (NVCS *Pinus aristata* Forest Alliance): *Pinus flexilis, Pinus albicaulis,* and *Pinus aristata* typify the dry ridges and exposed southern slopes of the southern Rocky Mountains (Peet 1988). These sites are characterized by high light levels, thin soils with little available soil moisture (Inceptisols), and rapid runoff. *Pinus aristata* occurs at elevations of 2300 to 3500 m, subalpine to treeline, dominating these areas in the southern portions of the Rockies. Stands are most frequently on igneous or volcanic substrates. *Pinus aristata* var. *longaeva*, well known for its longevity, is among the oldest living species on earth (>4600 years). The species requires open habitat because it is shade intolerant and, because of site conditions, slow growing (Little 1998). The understory is sparse, but may include *Jamesia americana, Ribes* spp., *Cercocarpus* spp., *Shepherdia* spp., and *Juniperus communis* (Eyre 1980).

29.18 Lodgepole Pine-Quaking Aspen Cover Type (NVCS *Pinus contorta-Populus tremuloides* Forest Alliance): This community is found on mountain slopes and plateaus at elevations between 1700 and 3000 m. Soils are alluvial or colluvial, of various parent materials, mainly sandstone and quartzite, and not very moist. This is a seral community that establishes after periodic fires. Both codominants are thin barked and readily killed by fire. However, their strategies for survival are quite different. *Pinus contorta* has serotinous cones and establishes from seed following burns, and the clonal *Populus tremuloides* readily sprouts after being top-killed. *Pinus contorta* litter builds up as a result of slow decomposition, yielding stand-replacement fires. *Populus tremuloides* stands develop rapid decomposition and nutrient cycling, which encourages growth of a mesophytic understory and a forest of lower flammability. Codominant canopy species include *Pseudotsuga menziesii, Abies lasiocarpa, Picea engelmannii, Picea pungens,* and *Pinus flexilis*. The succession pattern is from mixed *Populus deltoides/Pinus contorta* to *Pinus contorta* to either *Picea-Abies* (subalpine) or *Pseudotsuga menziesii* (montane). The landscape cover of *Populus tremuloides* communities has decreased as a result of fire suppression (Walker 1999). *Juniperus communis* and *Symphoricarpos oerophilus* typically dominate the shrub layer. Common herbs include *Calamagrostis rubescens, Carex geyeri, Lathyrus lanszwertii,* and *Thalictrum fendleri*.

29.19 Arizona Sycamore Cover Type (NVCS *Platanus wrightii* Temporarily Flooded Forest Alliance): This type occurs along perennial and seasonally intermittent streams in the southwestern states. Sites include narrow stream valleys and sloping alluvial terraces at elevations between 600 and 1800 m. Soils are generally a gravelly alluvium, must contain adequate moisture through a reliable stream flow during the winter and spring, and are periodically inundated. The major codominants are *Fraxinus velutina, Juglans major,* and *Populus fremontii.* The subcanopy may include *Celtis reticulata, Prosopis velutina, Quercus arizonica, Quercus emoryi, Robinia neomexicana,* and *Salix gooddingii. Rhus trilobus, Toxicodendron radicans, Amorpha fruticosa,* and *Baccharis salicifolia* are common shrubs. The herbaceous layer includes *Aristida* spp., *Bouteloua curtipendula, Carex* spp., *Erigonum* spp., *Gutierrezia sarothrae, Elymus elymoides, Sporobolus cryptandrus,* and *Vitis arizonica.*

29.20 Narrowleaf Cottonwood Cover Type (NVCS *Populus angustifolia* Temporarily Flooded Forest Alliance): This community occurs along terraces and floodplains (at elevations between 1350 and 2400 m) immediately adjacent to streams of the western United States. Soils are typically well-drained clayey loams with high amounts of coarse fragments. The soils are frequently reworked by flooding and beaver. This community is sustained by the typical dynamics of a floodplain. Newly exposed sandbars resulting from flooding events are required for this shade-intolerant, fast-growing species to reproduce. Following colonization, further accumulation of sand may lift this site above the annual flood zone. The stand will then mature and, without disturbance succeed to other species. The maintenance of the community thus requires an intact floodplain that is meandering and creating newly exposed bars. Associated trees include *Populus deltoides, Acer grandidentatum, Salix amygdaloides,* and *Acer negundo. Amelanchier alnifolia, Lonicera involucrata,* and *Rosa woodsii* dominate the shrub layer. Hay grasses dominate the herbaceous layer with fewer forbs, including *Maianthemum stellatum* and *Thalictrum fendleri.*

29.21 Red Willow Cover Type (NVCS *Salix bonplandiana* Temporarily Flooded Forest Alliance): *Salix bonplandiana* stands occur on rocky soils and disturbed areas (either severely scoured or heavily grazed) where the water table is well below the soil surface. Elevations for this community range from 850 to 1200 m. Flooding episodes occur during the growing season and are brief. *Baccharis salicifolia* is typical in the shrub layer.

29.22 Wing-Leaf Soapberry Cover Type (NVCS *Sapindus saponaria* Temporarily Flooded Forest Alliance): This community occurs within riparian zones at elevations between 1100 and 1500 m. *Juglans major* may codominate, and *Arbutus texana* and *Juniperus deppeana* may be present in the subcanopy. Shrub associates include *Berberis fremontii, Celtis reticulata, Morus microphylla, Prunus serotina, Rhus trilobata,* and *Ptelea angustifolia.* There is a significant vine layer dominated by *Vitis arizonica.* Common herbs include *Bouteloua curtipendula, Brickellia brachyphylla, Chenopodium album, Croton pottsii, Phaseolus wrightii,* and *Sitanion hystrix.*

Northern Rocky Mountain Conifer Forests (Region 6)

29.23 Grand Fir Cover Type (NVCS *Abies grandis* Temporarily Flooded Forest Alliance): This alliance is typical of benches, toe slopes, and valley bottoms along mountain streams. Elevation ranges from 790 to 1410 m. Soils range from very cobbly to sandy loam and silt loam. Soils are well-drained, but with a shallow water table, generally within 1 m. *Abies grandis* is a long-lived, shade-tolerant species able to regenerate under dense shade, and thus maintains itself without disturbance. Canopy cover ranges from 50 to 80 percent, and downed logs are abundant, acting as 'nurse logs' for regeneration (NVCS 1999). Periodic fires and other disturbances allow early seral species to co-occur. *Picea engelmannii* often codominants; other codominants include *Pseudotsuga menziesii, Larix occidentalis,* and *Abies lasiocarpa.* Shrub associates include *Acer glabrum, Rubus parviflorus, Symphoricarpos albus, Linnaea borealis,* and *Rosa gymnocarpa.*

29.24 Grand Fir-White Alder Cover Type (NVCS *Abies grandis-Alnus rhombifolia* Forest Alliance): This community is found in north central Idaho along streams at low elevations (600 to 700 m). *Alnus rhombifolia* is an obligate wetland species found in riparian areas of foothill streams. *Abies grandis* is typically at higher elevations, but cold air drainage and moist alluvial soils allow the two species to coexist. Such communities are regularly disturbed by flooding and are generally within 5 m of a stream. Associated trees include *Pseudotsuga menziesii, Pinus ponderosa,* and *Populus trichocarpa.* Tall shrubs include *Taxus brevifolia* and *Philadelphus lewisii;* other shrubs include *Rosa woodsii, Crateagus douglasii, Amelanchier alnifolia, Salix scouleriana, Spiraea betulifolia,* and *Cornus sericea.* The herbaceous layer commonly includes *Maianthemum stellatum, Trillium ovatum, Adenocaulon bicolor, Polystichum munitum,* and *Rubus parviflorus* (NVCS 1999).

29.25 Subalpine Fir-Alpine Larch Cover Type (NVCS *Abies lasiocarpa-Larix lyalii* Forest Alliance): This high-elevation (2000 to 2800 m) subalpine community occurs at or near the treeline on north-facing sites, where snowfields remain until June or July. Snow accumulation ranges from 2 to 3 m, and abrasion from wind-driven snow results in stunted or flagged trees. Soils are poorly developed and acidic Inceptisols, resulting from fractured granitic or quartzite rock. These sites were subjected to heavy alpine glaciation within the last 12,000 years. The community grades into *Larix lyalii* woodlands at higher elevations, and *Abies lasiocarpa-Picea engelmanii* at lower elevations. The *Pinus albicaulis* cover type may occur on adjacent south-facing slopes (NVCS 1999). *Larix lyalii* retains the tree form higher into the timberline zone than any other species in North America (Barbour et al. 1998). *Larix lyalli* is a very slow growing, long-lived tree, outlasting *Abies lasiocarpa* by several generations. Fire is limited in upper elevation stands because of sparse canopies, cool, moist habitats, and rocky terrain, but windthrow and snow avalanches can be typical disturbances in this community (NVCS 1999). Shrubs include *Vaccinium scoparium* and *Phyllodoce empetriformis. Luzula hitchcockii* and *Arnica latifolia* may occur in the herbaceous layer.

29.26 Subalpine Fir-Quaking Aspen Cover Type (NVCS *Abies lasiocarpa-Populus tremuloides* Forest Alliance): This seral community is found on mountain slopes and plateaus at elevations between 1700 and 3000 m. These riparian forests are found on alluvial or colluvial soils of various parent materials. *Populus tremuloides* looses dominance over time as the community succeeds to *Abies lasiocarpa* and *Picea engelmanii.* Periodic stand-replacing disturbance, mainly fire, is required for sustaining the community. Other tree taxa include *Picea engelmannii, Pseudotsuga menziesii, Abies concolor, Picea pungens, Pinus contorta,* and *Pinus flexilis.* Tall shrubs (*Amelachier alnifolia, Acer grandidentatum,* and *Prunus virginiana*) may be present. Dominant shrubs include *Juniperus communis, Shepherdia canadensis,* and *Symphoricarpos oreophilus.* The comparatively sparse herbaceous layer includes *Bromus carinatus, Calamagrostis rubescens, Carex geyeri, Carex rossi, Pedicularis racemosa,* and *Thalictrum fendleri.*

29.27 Western Larch Cover Type (NVCS *Larix occidentalis* Forest Alliance): The western larch forests are found on mesic sites at elevations between 500 and 1600 m and where there is an influence from the Pacific maritime climate. There is a cold, short growing season in May and June, followed by 2 dry months. Annual precipitation is about 75 cm, but only 20 percent falls during the growing season. This community is located on low mountain slopes, valley bottoms, and benches. Soils are often glacial till or colluvium and loamy. *Larix occidentalis* is shade intolerant (in fact, one of the least shade-tolerant conifers) and develops even-aged stands following crown fires and other major disturbances. It is the most important seral species at high elevations (Peet 1988). In the absence of disturbance, shade-tolerant codominants (e.g., *Pseudotsuga menziesii, Abies lasiocarpa,* and *Picea engelmannii*) will quickly overtop western larch. *Larix occidentalis* can maintain itself for some period of time, however, because of its longevity (>700 years) and various attributes that inhibit crown fires (pruning of lower branches, low-flammability foliage, and thick bark with low levels of resins) (Burns and Honkala 1990a). Any survivors of fire quickly supply seed, but good seed crops occur sporadically (every 4 to 5 years) (Eyre 1980). Shrub associates include *Acer glabrum, Alnus viridis* ssp. *sinuata, Amelanchier alnifolia, Pachistima myrsinites, Physocarpus malvaceous, Rubus parviflorus, Symphoricarpos albus, Vaccinium* spp., and *Salix scouleriana.* The herbaceous layer includes *Aralia nudicaulis, Artostaphylos uva-ursi, Arnica latifolia, Calamagrostis rubescens, Clintonia uniflora, Epilobium angustifolium, Linnaea borealis,* and *Xerophyllum tenax* (NVCS 1999).

29.28 Engelmann Spruce Cover Type (NVCS *Picea engelmanii* Forest Alliance, Temporarily Flooded Forest Alliance, Seasonally Flooded Forest Alliance, and Saturated Forest Alliance): This community dominates upper montane and subalpine forest, the coolest and wettest climates of the interior west. The elevation of this community ranges from 1000 to 3500 m, depending on latitude. Precipitation is usually greater than 75 cm, and snowpacks are deep and late melting. Soils are generally in the process of pozolization with an organic layer present (Inceptisol). The community grades into *Pinus albicaulis* communities at upper elevations, and *Abies grandis/Tsuga heterophylla* at lower elevations. There is no seasonal drought. Canopy associates include *Pinus contorta, Abies lasiocarpa, Pinus flexilis, Larix occidentalis,* and *Pseudotsuga mensiezii. Abies lasiocarpa* is often associated, but *Picea* seems to be more tolerant of extreme conditions (Figure 29.8). *Picea engelmannii* also appears more successful than *Abies lasiocarpa* in establishing on mineral soil after fire (Peet 1988). *Picea engelmannii* is very shade tolerant and long-lived. Following disturbance, however, faster-growing shade-intolerant species may invade and, because of their longevity, remain on the site for some time. A moderately dense ericaceous shrub layer may be present. Shrub associates include *Vaccinium cespitosum, Vaccinium scoparium, Physocarpus malvaceous, Vaccinium myrtillus, Shepherdia canadensis, Jamesia americana, Acer glabrum,* and *Juniperus communis.* A lush herbaceous community includes *Calamagrostis rubescens, Calamagrostis canadensis, Clintonia uniflora, Actaea rubra, Arnica cordifolia, Senecio cardamine, Senecio triangularis, Mertensia ciliata,* and *Carex foenea.* The community is also found on level sites with water tables < 50 cm from the surface— that is, in low-elevation (about 900 m) riparian areas (NVCS *Picea engelmannii* Saturated Forest Alliance). Here, *Alnus incana, Betula occidentalis,* and *Cornus sericea* are the common shrubs, and the herbaceous layer is dominated by *Lysichiton americanum.*

FIGURE 29.8 Subalpine fir (*Abies lansiocarpa*) and Engelmann's spruce (*Picea engelmanii*) forest in Washington.

29.29 Whitebark Pine Cover Type (NVCS *Pinus albicaulis* Forest Alliance): This community occurs in the subalpine zone at elevations between 2000 and 3200 m on dry ridges of exposed southern slopes (Peet 1988). Snowpack is often deep and may persist into the summer. Stands typically occur where disturbances such as avalanches or fire have created gaps in the forest, thus reducing the more shade-tolerant tree species. Stands also occur in protected, moist pockets at the treeline (tree islands), but may still experience summer drought conditions. Substrates are generally well-drained, coarse-textured lithic, soils derived from colluvium, glacial till, or residuum. *Pinus albicaulis* is a slow-growing, long-lived conifer. Its colonization of high subalpine tree islands often moderates site conditions (successional facilitation) for less hardy tree species. The community is seral at lower subalpine areas, regenerating after large disturbances, including stand-replacing fire and avalanches. Without disturbance, stands-succeed to *Abies-Picea, Tsuga mertensiana,* or *Pseudotsuga menziesii. Pinus flexilis* may be present in the southern communities, *Pinus contorta* and *Pseudotsuga menziesii* in lower subalpine communities, and *Abies lasiocarpa* and *Picea engelmannii* in Canadian communities. Both the shrub layer and the herbaceous layer are often sparse. Shrubs include *Juniperus communis, Ribes montigenum, Shepherdia canadensis, Symphoricarpos oreophilus,* and *Vaccinium scoparium.* Herbaceous species include *Achillea millefolium, Antennaria microphylla, Arnica* spp., *Carex geyeri, Carex rossi, Festuca idahoensis, Lupinus laetus, Poa nervosa, Potentilla diversifolia,* and *Solidago multiradiata.* The mountain pine beetle (*Dendroctonus ponderosae*) and an exotic pathogen, white pine blister rust (*Conartium ribicola*), can be very detrimental to this community.

29.30 Lodgepole Pine Cover Type (NVCS *Pinus contorta* Forest Alliance): This community is

FIGURE 29.9 Lodgepole pine (*Pinus contorta*) forest in Washington; Photo taken during a field trip of the National Vegetation Classification Workshop, Ecological Society of America annual meeting, Spokane, Washington.

widespread and occupies a wide array of habitats, because of the broad range of temperature and moisture tolerances of *Pinus contorta* (Figure 29.9). There are at least three varieties recognized: (1) Rocky Mountain lodgepole pine, *Pinus contorta* var. *latifolia,* (2) Sierra lodgepole pine, *Pinus contorta* var. *murryana,* and (3) shore lodgepole pine, *Pinus contorta* var. *contorta.* The community occupies large areas in the lower subalpine zone of the Rockies. *Pinus contorta* is a shade-intolerant conifer with serotinous cones (the Sierra variety does not have the closed cone habit, but fire is still necessary to prepare a seed bed; Eyre 1980) and thin, resinous, extremely flammable bark (Mutch 1970; Whitney 1997). These forests are adapted to periodic stand-replacing fires and generally develop as even-aged forests (Figure 29.10). Succession (if there is no disturbance) generally leads to dominance by *Pseudotsuga menziesii, Picea engelmanii,* and *Abies lasiocarpa,* depending on site conditions, within 200 years. Extensive attacks of bark beetles (*Dendroctonus ponderosae*) often accelerate the process (Eyre 1980). As with *Pinus banksiana* and *Pinus pungens,* not all individuals have serotinous cones (see Note on serotiny in "Cover Types," Chapter 26). Scattered open-coned individuals are common and allow the population to disperse into openings in the forest canopy not created by fire (Whitney 1997). On stressful sites that exclude serious competitors, *Pinus contorta* may

FIGURE 29.10 A stand replacement fire in a lodgepole pine (*Pinus contorta*) forest. Note the standing dead lodgepole pine and the strong regenerating cohort.

maintain itself for 400+ years. Associated trees include *Pseudotsuga menziesii, Abies concolor, Abies lasiocarpa, Picea engelmannii, Picea pungens, Pinus flexilis, Larix occidentalis,* and *Populus tremuloides.* Common shrub species include *Arctostaphylos uva-ursi, Vaccinium* spp., *Menziesii ferruginea, Symphoricarpos* spp., and *Mahonia repens.* Graminoids and shade-tolerant forbs dominate the understory; typical species include *Carex* spp., *Deschampsia cespitosa, Elymus glaucus, Calamagrostic rubescens, Arnica cordifolia,* and *Thalictrum* spp.

29.31 Limber Pine-Quaking Aspen Cover Type (NVCS *Pinus flexilis-Populus tremuloides* Forest Alliance): This mixed community occurs on montane and subalpine slopes at elevations between 2650 and 3000 m. The community is typically found on steep slopes of windswept southern aspects—under very stressful growing conditions. The more widely spaced *Pinus flexilis* stems are generally of small to medium size with crooked, multistemmed trunks. Soils are rocky colluvium and residuum from a variety of parent materials, but *Pinus flexilis* has an affinity for calcareous soils. *Pinus flexilis* exists on these dry, low-quality soils because it can not compete with other species. Reproduction is sporadic, as birds and mammals consume nearly all of the seeds (Eyre 1980). Development of new individuals can occur from forgotten seed caches. *Abies concolor, Picea engelmanii,* and *Pseudotsuga menziesii* may codominate. Short shrubs include *Arctostaphylos patula, Symphoricarpos oreophila, Juniperus communis, Rhus trilobata, Rosa woodsii,* and *Mahonia repens.* The herbaceous stratum is sparse but may include *Bromus anamolous, Carex rossii, Elymus glaucus, Elymus trachycaulus, Melica spectabilis, Achillea millefolium, Arnica cordifolia, Astragalus miser, Geranium viscosissimum,* and *Ligusticum filicinum* (NVCS 1999). Specific species depend on elevation.

29.32 Western White Pine Cover Type (NVCS *Pinus monticola* Forest Alliance): This community represents a midseral assemblage maintained by high-intensity, stand-replacement fires at long intervals. The community is found throughout the northern Rocky Mountains at elevations between 450 and 1200 m. Stands are found in valley bottoms and midslope positions with favorable soil moisture. Soils are generally well drained, but arise from a variety of parent materials. Fire suppression has also lead to a decrease in the dominance of this community on the landscape. Succession is toward other canopy species including *Larix occidentalis, Abies grandis, Thuja plicata,* and *Tsuga heterophylla.* Successional conversion has been hastened by continued attacks of the mountain pine beetle and the introduced white pine blister rust. Dense, young stands may develop after these attacks (Eyre 1980). Ericaceous shrubs dominate the herbaceous layer; species include *Vaccinium* spp. *Chimaphila umbellata, Symphoricarpos albus,* and *Menziesia ferruginea.* Herbaceous species include shade tolerants such as *Asarum caudatum, Streptopus amplexicaulis, Cornus canadensis, Clintonia uniflora,* and *Trillium ovatum.* The youngest regeneration stands develop with *Epilobium angustifolium, Carex* spp., and many forbs, as well as shrubs (e.g., *Salix* spp., *Ceanothus* spp., and *Sorbus* spp.).

29.33 Ponderosa pine-Quaking Aspen Cover Type (NVCS *Pinus ponderosa-Populus tremuloides* Forest Alliance): This type develops following stand-replacement fire on gentle to steep slopes or in riparian areas at elevations between 2400 and 2700 m. Soils are often coarse textured but may range from sandy loams to clay loams derived from colluvium or alluvium. As the stand ages, *Pinus ponderosa* increases in dominance as denity of the shorter-lived *Populus tremuloides* decreases.

29.34 Quaking Aspen Cover Type (NVCS *Populus tremuloides* Forest Alliance): These seral communities are widespread in the western United States and in the northern Great Plains and Lake States regions at elevations between 900 and 3350 m. Associated vegetation differs between these regions. Topography is important, the species occurring on upland sites and stream terraces above the floodplain in the west, on swales and canyons below seeps in the Great Basin, and on rolling topography such as glacial moraines between grasslands and other forest communities (prairie-forest border) in the northern Great Plains. The soils are deep, loamy, often rocky, and must contain adequate moisture. Parent materials may vary, but limestone, basalt, and calcareous or neutral shales are common. Water tables are at or near the surface in spring for some communities (NVCS *Populus tremuloides* Seasonally Flooded Forest Alliance), but may drop to 1 m below the surface in midsummer. As a rule, aspens indicate the presence of moisture within a meter of the soil surface (Whitney 1997). These communities develop following stand-replacing disturbances caused by wind, fire, beaver, or disease. Quaking aspen is short-lived (100 to 150 years, and has major heart rot within 100 years), but has a strong resprouting ability when top-killed. It is clonal and can reproduce up to 20,000 stems ha^{-1} (Vankat 1979), and thus, with adequate disturbance, clones may live for many centuries. Aspen has the ability to root sucker, its main form of propagation, inasmuch as seeds rarely survive droughts. The regenerating sprouts are an important browse. Grazing has a heavy impact within many of the *Populus tremuloides* forests. The negative effects of grazing include invasion by exotics, and many of these aspen stands now have the exotics *Taraxacum officinale* and *Poa pratensis* as dominant floral components. These aggressive exotics often lead to decreased diversity of the communities and may affect the function of the community. In the absence of fire, stands generally succeed to longer-lived conifer forest because aspen is extremely intolerance to shade. Conifers such as *Pseudotsuga menziesi*, *Pinus contorta*, *Picea engelmannii*, *Abies concolor*, and *Abies lasiocarpa* are typical associates. *Populus tremuloides* and *Pinus contorta* are found under similar site conditions and disturbance regimes. Although *Pinus contorta* apparently can out-compete aspen within its range, the range of aspen is much wider and thrives without competition from lodgepole pine (Peet 1988). Penetration of light through the canopy and ample moisture permit a lush understory. Shrubs include *Alnus incana*, *Alnus tenuifolia*, *Lonicera involucrata*, *Ribes montigenum*, *Rosa woodsii*, *Salix bebbiana*, *Salix geyeriana*, *Salix monticola*, and *Symphocarpos albus*. *Calamogrostis canadensis*, *Equisetum arvense*, and *Ranunculus alismifolius* often dominate the herbaceous stratum.

29.35 Quaking Aspen-Douglas-fir Cover Type (NVCS *Populus tremuloides-Pseudotsuga menziesii* Forest Alliance): This community is found at montane elevations of 1700 to 2800 m. Sites occur on gentle to steep slopes with soils derived from alluvium, colluvium, and residuum (from various parent materials). The community depends on periodic disturbance, without which it succeeds to *Pseudotsuga menziesii*. *Pseudotsuga menziesii* is relatively tolerant of fire after a certain age, because of self-pruning and thick bark. Other canopy species include *Abies concolor*, *Abies lasiocarpa*, *Picea engelmannii*, *Picea pungens*, *Pinus contorta*, *Pinus flexilis*, and *Pinus ponderosa*. The understory of this community is best developed in younger stands that have more aspen, allowing more light to penetrate to the forest floor. Common shrubs include *Amelanchier* alnifolia, *Prunus virginiana*, *Acer grandidentatum*, *Symphoricarpos oreophila*, *Juniperus communis*, and *Mahonia repens*. Graminoids typically dominate the understory, with species such as *Bromus carinatus*, *Calamagrostis rubescens*, *Carex geyeri*, and *Elymus glaucus*. Typical forbs include *Achillea millefolium*, *Arnica cordifolia*, *Aster engelmanii*, and *Erigeron speciosus*.

29.36 Douglas-fir Cover Type (NVCS *Pseudotsuga menziesii* Forest Alliance see cover types 30.16, 30.17, 30.18, and 30.19 in the Northern Coastal Conifer Forest, Chapter 30): The Douglas-fir communities of the Rocky Mountains differ from those of the northwest, occurring on drier sites (50 to 100 cm precipitation) under a continental climate. The community is found at elevations between 100 and 2900 m. As a result of grazing and fire suppression, *Pseudotsuga menziesii* has been aggressively invading lower-elevation forests for the past several decades (Whitney 1997). Soils are derived from sedimentary rock, are well drained and slightly acidic, and are usually gravelly or rocky. *Pseudotsuga menziesii* communities dominate midslope forests above *Pinus ponderosa* forests and below *Picea-Abies* forests, but widespread disturbances have generated mixed stands throughout the *Pseudotsuga mensiezii* range (Whitney 1997). Typical associates include *Abies lasiocarpa*, *Picea engelmannii*, *Pinus contorta*, *Pinus ponderosa*, *Pinus monticola*, and *Pinus flexilis*. The shrub layer is well developed, including species such as *Physocarpus malvaceous*, *Prunus emarginata*, *Prunus virginiana*, *Amelanchier alnifolia*, *Artemesia tridentata*, and *Acer glabrum*. Common herbs include *Arnica cordifolia*, *Lathyrus nevadensis*, *Calamagrostis rubescens*, *Pseudoroegneria spicata*, *Carex gereyi*, and *Poa nervosa* (NVCS 1999).

CHAPTER 30

Forest Environment and Cover Types—Region 7

NORTHERN AND SOUTHERN COASTAL CONIFER FOREST (BAILEY'S HUMID TEMPERATE DOMAIN/MARINE DIVISION, *PICEA-TSUGA-ABIES*, AND MEDITERRANEAN DIVISION, *ABIES-PINUS-QUERCUS*)

INTRODUCTION

THE Coastal Conifer Forest (Region 7, Figure 2.1), located at middle latitudes (30° to 60° N), has distinct seasons with strong cycles of temperature and moisture. A distinctive winter season separates this climate from tropical climates. The Northern Coastal Conifer Forest (Bailey's Marine Division; *Picea-Tsuga-Abies*, Region 7a) and the Southern Coastal Conifer Forest (Bailey's Mediterranean Division; *Abies-Pinus-Quercus*, Region 7b) have average temperatures above freezing during the coldest month of the year and are situated within 200 km of the Pacific Ocean. Coastal fogs extend many kilometers inland, resulting in decreased radiation, which lowers temperatures and evapotranspiration, and increases precipitation due to fog drip (moisture in the air condensing on leaves and stems and falling to the forest floor) (Vankat 1979). This process ameliorates the typical summer droughts. The central and southern Sierra Nevada Mountain forests are included in the Southern Coastal Conifer Forest.

The Marine Division receives abundant rainfall (80 to 300 cm annually, ~80 percent occurring between October 1 and March 31) from maritime polar air masses and has a narrow temperature range because of its close proximity to the ocean. Precipitation is abundant throughout the most of year, but markedly lower during the summers (in fact, they are relatively dry) (Figure 30.1). Mild winters and cool summers are typical, yielding a damp, humid climate with much cloud cover. Soils are generally strongly leached (especially in the temperate rain forest), acidic Inceptisols and Ultisols. Needleleaf forest is the natural vegetation. These forests exhibit the greatest biomass accumulations and some of the highest productivity levels in the world (Franklin 1998). Trees are often >75 m tall and 2 m in diameter (Vankat 1979). Disturbance regimes are typically low-frequency catastrophic events, unlike conditions in much of the other western forests, where frequent, low-intensity disturbances prevail (or, at least, have historically prevailed).

The junction between Marine and Mediterranean Divisions is ~43° north latitude. The division between the Marine and Mediterranean climates, and the juncture of a high number of physiographic provinces (Franklin and Dyrness 1988), produces some of the most diverse forest assemblages of the world (Figure 30.2). The coastal fog belt, Siskiyou Mountains, Klamath Mountains, Cascade Mountains, and basin and range areas have a wide range of topographic conditions (steeply dissected to flat) and parent

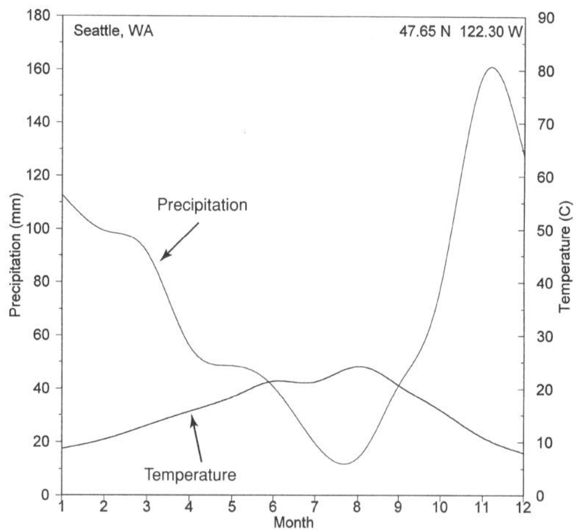

FIGURE 30.1 Climate diagram for Seattle, Washington, representing the Maritime Division of the Humid Temperate Domain (*sensu* Bailey 1998).

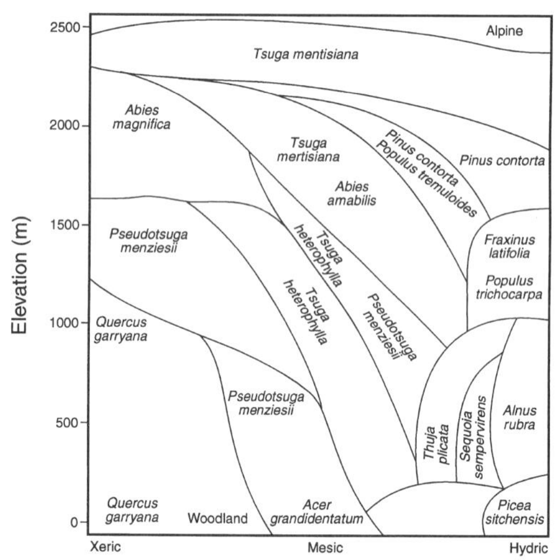

FIGURE 30.2 Gradient mosaic diagram illustrating variations in forest vegetation composition with elevation and topographic position for Siskiyou Mountains, Oregon and California, representing the Northern Coastal Forest subregion (see Table 25.1). Redrawn from Whittaker (1960) with the permission from the Ecological Society of America.

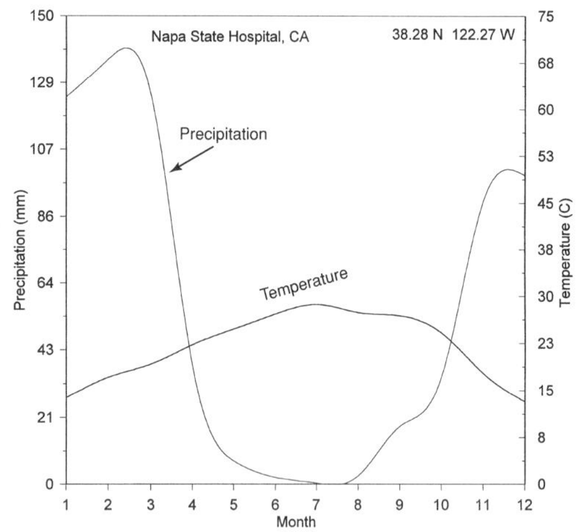

FIGURE 30.3 Climate diagram for Napa State Hospital, California, representing the Mediterranean Division of the Humid Temperate Domain (*sensu* Bailey 1998).

materials (sedimentary to lava flows) from which soils are derived. The result is a spectacular diversity of conifer species, some of which are endemic to this area (e.g., Port-Orford-cedar, *Chamaecyparis lawsoniana*, and Brewer spruce, *Picea breweriana*).

The Mediterranean Division is subject to alternating dry and wet seasons. Wet winters and dry, hot summers result in a unique set of schlerophyllous vegetation: hard-leaved evergreen trees and shrubs. The trees and shrubs must withstand 2 to 4 months of severe summer drought and severe evaporation (Figure 30.3). The coastal and mountainous portions of this zone receive greater amounts of precipitation and are generally more humid (Figure 30.4). About 80 to 85 percent of the precipitation falls in winter. The Sierra Nevada host some lush conifer forest, including stands of giant sequoia (*Sequoiadendron giganteum*), and another giant, the Coast Redwood (*Sequoia sempervirens*), occurs along the northern California coast. The communities grade from oak woodlands to mixed evergreen forest (*Pseudotsuga menziesii* and mixed *Quercus* spp.) to midmontane (*Pinus ponderosa* and *Abies lowiana*) to upper montane conifer (*Pinus contorta, Populus tremuloides*, and *Abies magnifica*) forest to a woodland subalpine forest (*Pinus balfouriana, Pinus albicaulis, Pinus flexilis*, and *Tsuga mertensiana*) (Figure 30.5) (Barbour 1988).

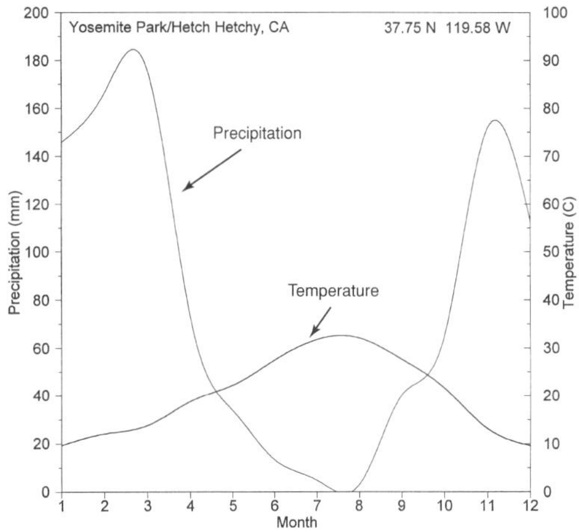

FIGURE 30.4 Climate diagram for Yosemite Park, California, representing mountainous areas within the Mediterranean Division of the Humid Temperate Domain (*sensu* Bailey 1998).

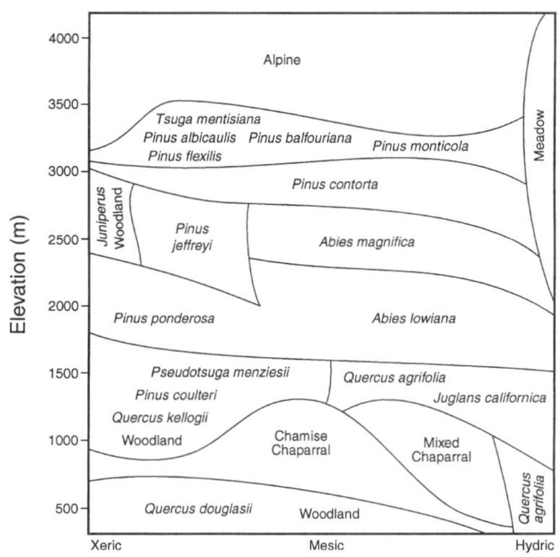

FIGURE 30.5 Gradient mosaic diagram illustrating variations in vegetation composition with elevation and topographic positions for the Sierra Nevada California, representing the Southern Coastal Forest subregion (see Table 25.1). Redrawn from Barbour (1988) with the permission of Cambridge University Press.

Why Conifers Instead of Hardwoods?

The mild winters and abundant precipitation of the West Coast provide optimum growing conditions for both deciduous and coniferous species. Hardwoods usually dominate the mesic, temperate forests of the world, so why do conifers dominate the forests here? Whereas most deciduous species grow to greater sizes than their inland counterparts, it is the massive conifer species that seem to have evolved to acquire resources best, and this is seemingly due to their evergreen habit. In fact, conifers outnumber hardwoods at an estimated 1000:1 (Küchler 1964 in Franklin 1988). Because they have green leaves year-round, they may also synthesize carbon year-round when the weather is warm enough. Thus, they can take advantage of warm periods in late fall, early spring, and, sometimes, midwinter (Pielou 1988). In fact, as much as half of the net carbon assimilation by lowland Douglas-fir (*Pseudotsuga menziesii*) may occur between October and May (Edmonds 1982). Another feature is their long, conical crowns, which are better at intercepting radiation of low winter sun angles than deciduous crowns. In addition, the most stressful time within this climate is during the summer drought. Hardwoods are less capable of withstanding the drought because of their thin, wide leaves, which loose moisture more readily, than the small needle leaves of conifers. Moreover, large conifers (their size is due to their longevity) store large amounts of water in their sapwood and have large leaf areas that condense fog and dew, both of which help to offset the summer drought (Edmonds 1982).

COVER TYPES

Northern Coastal Conifer Forest (NVCS Humid Temperate Domain/Marine Division; *Picea-Tsuga-Abies*) (Region 7a)

30.1 Pacific Silver Fir Cover Type (NVCS *Abies amabilis* Forest Alliance): This community occupies cool, moist habitats with a significant winter snowpack of 1 to 3 m (although less snowpack than experienced by *Tsuga mertensiana* communities) at elevations between 600 and 1300 m. Soils are generally deep and moist and may be seasonally flooded. *Abies amabilis* may occur with *Abies procera, Tsuga heterophylla*, and *Tsuga mertensiana*. Both *Abies amabilis* and *Tsuga heterophylla* are very shade tolerant, forming persistent communities. *Abies procera* is not shade tolerant and is usually replaced by *Abies amabilis* in older stands. This is generally a montane forest type, intermediate between temperate mesophytic (e.g., *Pseudotsuga menziesii* and *Tsuga heterophylla*) and subalpine (e.g., *Tsuga mertensiana*) types. It occurs on the western slopes of the Cascade range. Other common associates are *Pseudotsuga menziesii, Thuja plicata, Chamaecyparis nootkatensis*, and *Pinus monticola*. Northern sites in wetter habitats have a well-developed evergreen shrub understory. Typical understory plants include *Vaccinium alaskaense, Vaccinium ovalifolium, Vaccinium membranaceum, Vaccinium parvifolia, Oplopanax horridus, Cornus canadensis, Clintonia uniflora, Acer circinatum*, and *Berberis nervosa*. Herbs on rich sites include *Streptopus roseus, Smilacina stellata, Vancouveria haxandra, Tiarella trifoliata*, and various ferns; on poorer sites, *Xerophyllum tenax, Vaccinium membranaceum*, and *Gaultheria shallon*. This species can be found in stands attaining heights greater than 125 m (200 ft).

30.2 Pacific Silver Fir-White Fir Cover Type (NVCS *Abies amabilis-Abies lowiana* Forest Alliance): This community exists where the ranges of the two species overlap in the southern Oregon Cascades. The community occurs in a narrow zone of elevation between 1400 and 1600 m, with precipitation exceeding 150 cm (1 to 3 m of snow accumulation). This is typically a mixed community with associates including *Pseudotsuga menziesii, Abies procera, Tsuga heterophylla, Calocedrus decurrens*, and *Abies magnifica*. The shrub understory may include *Rhododendron macrophyllum, Acer circinatum, Menziezia ferruginea, Gaultheria shallon, Mahonia nervosa*, and *Vaccinium* spp. The herbaceous layer includes shade-tolerant species such as *Maianthemum stellatum, Achlys triphylla, Linnea borealis, Polystichum munitum, Cornus canadensis*, and *Clintonia uniflora*.

30.3 Grand Fir Cover Type (NVCS *Abies grandis* Forest Alliance): This community occurs in montane areas (750 to 2000 m elevation) with 50 to 100 cm precipitation year^{-1} and is found in a variety of topographic positions. The community is generally on moister sites than forests dominated by *Pseudotsuga menziezii* and *Pinus ponderosa*, and less less moist sites than forests dominated by *Tsuga heterophylla* or *Thuja plicata*. Parent materials are variable, but volcanic ash is commonly present in the soil profile (NVCS 1999). *Abies grandis* is typically found in lowland valleys. The species has deep, spreading roots that protect it from windthrow, but its thin bark is susceptible to fire. The community was likely restricted to moist coves and stream terraces during pre-European-settlement fire regimes. During succession, *Abies grandis* often forms ladder fuels, leading to the occurrence of canopy fires (NVCS 1999). *Abies grandis* is shade tolerant but looses its tolerance with age (Walker 1999). This long-lived species (>600 years) attains maturity at about age 350. Trees greater than 125 m in height are possible. Where found along the northern California coast, *Sequoia sempervirens* is a common associate. Other associates include *Pinus contorta, P. monticola* (perhaps more important prior to white pine blister rust), and *Larix occidentalis*. Shrub associates include *Pachistima myrsinites, Linnaea borealis, Menziezia ferruginea, Physocarpus malvaceus, Vaccinium membranaceum, Acer glabrum, Spiraea betulifolia*, and *Acer circinatum*. Common herbs include *Achlys triphylla, Asarum caudatum, Clintonia uniflora, Coptis occidentalis, Maianthemum stellatum, Orthilia secunda, Polemonium pulcherrimum, Trautvetteria caroliniensis, Trientalis borealis*, and *Xerophyllum tenax*.

30.4 Bigleaf Maple Cover Type (NVCS *Acer macrophyllum* Forest Alliance): *Acer macrophyllum* is found along major rivers and streams and in moist canyon soils, sometimes in pure stands, within 100 km of the coast. The community occurs to elevations of 300 m in the north and between 900 and 1700 m in the south. Flooding and scouring may be necessary for the establishment of these communities. They colonize following periodic flooding disturbances. Sites flood from winter into late spring or early summer (NVCS *Acer macrophyllum* Temporarily Flooded and Seasonally Flooded Forest Alliances). Some sites may include backwaters (e.g., oxbow lakes) and often have heavy, poorly drained soils. Companion trees include *Lithocarpus densiflorus, Arbutus menziesii, Quercus chrysolepsis*, and *Quercus garryana*. Other associates include *Alnus rubra, Populus trichocarpa*, and *Betula papyrifera*. A typically dense shrub layer includes *Corylus cornuta, Rubus parviflorus, Rubus spectabilis, Oplopanax horridus,* and *Symphoricarpos albus.* The herbaceous stratum includes *Oxalis oregana, Polystichum munitum, Anthyrium filix-femina, Trillium ovatum, Maianthemum dilatatum,* and *Achlys triphylla.*

30.5 White Alder Cover Type (NVCS *Alnus rhombifolia* Temporarily Flooded Forest Alliance): This community occurs in mountainous regions at elevations between 600 and 2700 m, and in riparian areas between 400 and 700 m. *Alnus rhombifolia* communities are found along canyon walls, steep slopes, cliffs, rocky slopes, and areas of slumping or mud-rock flows. Thus, the stands characterize areas prone to slippage and erosion where periodic severe rains saturate the soil. This frequent disturbance, along with a solid canopy cover, limits the herbaceous community. The climate inside the canyons may be quite different from that outside the canyons. Summers are hot and dry, and winters are mild with limited snow accumulation. Annual precipitation ranges from 33 to 46 cm. Associated canopy species include *Amelanchier alnifolia, Betula occidentalis,* and *Celtis reticulata. Sambucus cerulea, Prunus virginiana, Philadelphus lewisii,* and *Rosa woodsii* var. *ultramontana* dominate the understory.

30.6 Red Alder Cover Type (NVCS *Alnus rubra* Forest Alliance): *Alnus rubra* can form pure stands near water (NVCS *Alnus rubra* Seasonally Flooded Alliance) or in early stages of succession. *Alnus rubra* is well adapted to wet soil conditions. It is a shade-intolerant pioneer species, and thus this type is always seral. This cover type occurs on bottomlands of the Pacific Coast at elevations below 1000 m. The community is the initial successor on finely ground granite and other igneous rocks following the retreat of glaciers. *Alnus rubra* modifies the environment for the succession to more shade-tolerant species. *Alnus rubra* not only acts as a nurse tree, but also improves the soil by providing nitrogen fixation (through a symbiotic relationship with the bacterium genus *Frankia*) and large amounts of litter (Walker 1999). *Tsuga heterophylla* and *Picea sitchensis* often succeed *Alnus rubra,* just as *Alnus rubra* often invades these sites following cutting, fire, and landslides. Common associates are *Acer macrophyllum, Populus trichocarpa, Salix lasiandra, Prunus emarginata, Pseudotsuga menziesii, Picea sitchensis, Tsuga heterophylla, Thuja plicata,* and *Abies grandis.* The forest undergrowth is generally species-rich and well developed. Common shrub associates include *Sambucus callicarpa, Sambucus cerulea, Rubus spectabilis, Rubus parviflorus, Oplopanax horridus, Osmaronia cerasiformis, Sambucus racemosa* var. *melanocarpa, Acer circinatum,* and *Menziesia ferruginea.* Herbaceous species include *Oxalis oregana, Polystichum munitum,* and *Stachys mexicana.*

30.7 Port-Orford-Cedar Cover Type (NVCS *Chamaecyparis lawsoniana* Forest Alliance): *Chamaecyparis lawsoniana* occurs primarily in a varied mixture of conifers on a wide range of sites. This species is conspicuously dominant on moist serpentine-derived soils (Eyre 1980). This cover type is typical of northerly aspects and on midslopes with dependable summer moisture. It is also interspersed with meadow communities on ultramafic soils, and occurs along watercourses and in cool, moist locations. The thick-barked *Chamaecyparis lawsoniana* appears more fire resistant than many of its associates, is a fire-resistant seed source, and is very shade tolerant. In addition, it pioneers on open, moist sites. Typical coastal associates include *Tsuga heterophylla, Picea sitchensis, Abies grandis, Pinus contorta, Thuja plicata, Sequoia sempervirens, Lithocarpus densiflorus, Alnus rubra, Acer macrophyllum,* and *Umbellularia californica.* Further inland and at higher elevations, associates include *Pinus lambertiana, Pinus monticola, Pinus jeffreyi, Pinus ponderosa, Calocedrus decurrens, Abies lowiana, Arbutus menziesii, Quercus chrysolepsis,* and *Castanopsis chrysophylla. Taxus brevifolia* is a common understory evergreen tree. In the northern part of its distribution, herbaceous associates include *Polystichum munitum, Rhododendron macrophyllum,* and *Xerophyllum tenax.* Within the inland mixed evergreen zone, *Lithocarpus densiflorus* and mixed pine are common associates. At upper elevations in the *Abies lowiana* zone, it is found in mixed *Abies* and *Tsuga* communities. A Temporarily Flooded Alliance (NVCS 1999) occurs in the Klamath/Siskiyou Mountain region (NVCS 1999).

30.8 Alaska-Cedar Cover Type (NVCS *Chamaecyparis nootkatensis* Forest Alliance and Temporarily Flooded Forest Alliance): *Chamaecyparis nootkatensis* occurs on Histosol soils (e.g., muskeglike soils) where ample moisture is available. The soils are shallow, acidic, and nutrient poor. The community is found at sea level in the far north, but occurs at elevations between 600 and 2000 m for much of its range.

Precipitation ranges from 150 to 400 cm annually and is mainly in the form of snow. Stands are often even-aged, developing from stand-replacement disturbances (Franklin and Dyeness 1988). Associates include other shade-tolerant species: *Picea sitchensis, Tsuga heterophylla, Abies amabilis, Abies lasiocarpa,* and *Thuja plicata*. Shrub cover associates include *Vaccinium* spp., *Cornus canadensis,* and *Arctostaphylos.* The generally sparse herbaceous layer consists of *Blechnum spicant, Cornus canadensis, Clintonia uniflora, Tiarella trifoliata,* or *Polystichum munitum.* A Temporarily Flooded Forest Alliance (NVCS 1999) occurs in Oregon.

30.9 Oregon Ash Cover Type (NVCS *Fraxinus latifolia* Seasonally Flooded and Temporarily Flooded Forest Alliances): *Fraxinus latifolia* occurs in wet soils along streams and in canyons (at elevations up to 1700 m) with *Alnus rubra, Populus trichocarpa, Salix* spp., and *Quercus garryana.* The soils are typically poorly drained silts and clays or sandy/gravelly soils with a high water table. Scour associated with spring floods from melting snow generates soil fines and fine-scale disturbance patterns. In fact, the community depends on the flooding disturbance to open sites for establishment. On drier sites, associates include *Acer macrophyllum* and *Quercus garryana. Symphoricarpos albus* is the most abundant understory shrub. The lush herbaceous layer is composed of perennial grasses, forbs, and ferns, including *Urtica dioica, Athyrium felix-femina, Polystichum munitum, Ranunculus uncinatus,* and *Geum macrophyllum.*

30.10 Sitka Spruce Cover Type (NVCS *Picea sitchensis* Forest Alliance): The most common associate of *Picea sitchensis* is *Tsuga heterophylla*; others include *Pseudotsuga menziesii, Abies amabilis, Thuja plicata, Chamaecyparis lawsoniana, Tsuga mertensiana, Chamaecyparis nootkatensis, Abies lasiocarpa, Alnus rubra, Acer macrophyllum, Alnus uiridis* ssp. *sinuata,* and *Populus trichocarpa.* The broadleaf species are associates in early seral stages. This type is limited to a narrow oceanside strip (generally only a few kilometers in width) characterized by mild winters, cool summers, and abundant moisture during the growing season (due to the influence of the Japanese ocean current). Rain is supplemented by summer fog along the coast, and overall precipitation usually exceeds 250 cm. The precipitation is distributed throughout the year except for dry midsummer periods. Two of every three days are overcast. Generally, this community is restricted to elevations below 150 m unless mountain masses are immediately adjacent to the ocean. The best growth is seen on well-drained soils derived from colluvium, alluvium, or glacial till. These soils are some of the finest in the region: deep, relatively rich, fine textured, and with a rich humus layer. Surface soils are typically acidic, high in organic matter (soils are about 20 percent nonmineral matter) and nitrogen, and low in base saturation. *Picea sitchensis* will invade recently exposed land following beach uplift, glacial recession, stream channel change, or landslide. It requires mineral soil or organic seedbeds (e.g., moss and logs) in the open or in light shade for seed germination. *Picea sitchensis* is more tolerant of salt spray than many species and can form pure stands along beaches and headlands. This type will usually succeed to *Tsuga heterophylla,* but may maintain itself for long periods. Understory plant associates include *Oplopanax horridus, Lysichituon americanus, Rubus spectabilis, Rubus parviflorus, Sambucus callicarpa, Oxalis oregana, Polystichum munitum, Vaccinium* spp., ferns, mosses, and lichens. This community can achieve very large sizes (>125 m heights). This type can also dominate floodplains and perennially waterlogged, mucky-peat soils on beach plains and marine terraces (NVCS *Picea sitchensis* Temporarily Flooded/Seasonally Flooded Forest Alliances). *Picea sitchensis* tolerates higher water tables than *Pseudotsuga menziesii.* Although small stems can grow and maintain themselves in bogs (Walker 1999), regeneration in the permanently wet sites often requires "nurse logs." These wetter sites may include *Cornus sitchensis, Vaccinium ovalifolium, Salix sitchensis,* and *Salix hookeriana* as understory associates (NVCS 1999).

30.11 Sitka Spruce-Black Cottonwood Cover Type (NVCS *Picea sitchensis-Populus trichocarpa* Forest Alliance): *Picea sitchensis* and *Populus trichocarpa* are often associates in the early stages of succession, both pioneering recently exposed lands, or in a seral stage following *Alnus rubra.* This alliance is typically found along river terraces of the western Olympic Peninsula.

30.12 Whitebark Pine Cover Type (NVCS *Pinus albicaulis* Forest Alliance): This type is the highest-elevation arborescent cover type in western North America, found at elevations between 1800 and 3000 m (Eyre 1980). Soils are typically strongly acidic Inceptisols. However, soils lacking fine material (coarse talus, exposed bedrock, and lava flows) may develop Entisols. Stands are typically open with clustered tree stems. This species often forms krummholtz with *Abies lasiocarpa* on highly windy ridges. The pine seeds are an excellent food source for high-elevation creatures; squirrel caches are often raided by bears, and Clark's nutcracker can disseminate seeds over great distances. Wildfire is important at long intervals of 50 to 300 years. *Pinus albicaulis* can reach an age of 500 years or more, but is otherwise successional to the associate species of this community. Common associates include *Larix lyallii, Abies lasiocarpa,* and *Picea engelmannii. Larix lyallii* tends to dominate north aspects, and *Pinus albicaulis* dominates adjacent south-facing slopes. In some areas, *Pinus albicaulis* has been severely damaged by white pine blister rust (*Cronartium ribicola*). Several larger stems in the northern Rocky Mountains have been killed by mountain pine beetle (*Dendroctonus ponderosae*) epidemics (Eyre 1980).

30.13 Lodgepole Pine Flooded Cover Type (NVCS *Pinus contorta* Temporarily Flooded Forest Alliance): This community is found at elevations between 1220 and 2280 m, where *Pinus contorta* (the only tree) is invading *Deschampsia cespitosa*-dominated meadows (Kovalchik 1987; NVCS 1999). Soils are derived from pumice alluvium at lakesides, streamsides, edges of meadows, and concave drainages and basins (NVCS 1999). Forbs include *Thalictrum occidentale, Fragaria* spp., *Trifolium longipes, Aster occidentalis, Mimulus primuloides, Achillea millifolia,* and *Antennaria* spp. Graminoids, although dominated by *Deschampsia cespitosa,* may include *Juncus balticus, Koeleria macrantha, Calamagrostis canadensis,* and *Danthonia californica.*

30.14 Lodgepole Pine-Quaking Aspen Cover Type (NVCS *Pinus contorta* var. *murryana-Populus tremuloides* Seasonally Flooded Forest Alliance): Communities of this type are riparian forests at elevations ranging from 1100 to 2135 m. Soils are generally deep and alluvial, derived from pumice parent materials, with textures ranging from loamy sand to silty clay loam. These communities are flooded in the spring and dry during the summer. The water table can fluctuate 1 m (NVCS 1999). During the dry season, fire may be an important determinant of stand structure and composition. *Pinus contorta, Populus tremuloides,* and *Spirea douglasii* are all adapted to regenerating following fire. The shrub layer is typically dominated by *Spiraea douglasii,* but also may contain *Arctostaphylos uva-ursi, Rosa woodsii,* and several *Salix* species. *Carex angustata* often dominates the herbaceous stratum; other species include *Achillea millifolium, Elymus glaucus, Fragaria virginiana, Galium boreale,* and *Maianthemum stellatum.*

30.15 Black Cottonwood Cover Type (NVCS *Populus trichocarpa* Temporarily Flooded Forest Alliance): *Populus trichocarpa,* also referred to as *Populus balsamifera* var. *trichocarpa,* grows along streams, forming bottomland floodplain communities. *Populus trichocarpa* often grows in association with *Alnus rubra,* both being pioneer species on floodplains and young glacial moraines. Soils are generally Entisols with up to 1 m of mineral soil overlying gravel. Vegetative propagation can occur when small branches fall to the ground and develop roots (Walker 1999). This seral community is succeeded by conifers that are also associates, including *Pinus ponderosa* and *Picea* spp. Shrub associates include *Crataegus douglasii, Salix exigua,* and *Symphoricarpos albus.*

30.16 Douglas-Fir-Bigleaf Maple Cover Type (NVCS *Pseudotsuga menziesii-Acer macrophyllum* Forest Alliance): This type is found at the extremes of the *Pseudotsuga menziesii* range of tolerance. It occurs on damp terraces and dry talus and scree of Pacific Coast foothills at elevations <1000 m. The community is seral, developing from past logging or wildfires. On all but the wettest sites, dominance shifts toward *Pseudotsuga menziesii* with age, and then to *Tsuga heterophylla.* Other broad-leaved associates include *Alnus rubra, Umbellularia californica, Cornus nuttallii,* and *Betula papyrifera.* Common shrubs include *Rubus spectabilis, Sambucus racemosa* var. *melanocarpa, Corylus cornuta* var. *californica, Acer circinatum,* and *Rhododendron macrophyllum.* The generally well-developed herbaceous layer may be quite diverse. *Oxalis oregana, Polystichum munitum, Trillium ovatum, Hydrophyllum tenuipes,* and *Asarum caudatum* are common.

30.17 Douglas-Fir-Pacific Madrone Cover Type (NVCS *Pseudotsuga menziesii-Arbutus menziesii* Forest Alliance): This community occurs in fire-prone, relatively dry areas west of the Cascades, in pronounced rainshadows of the Olympics or Coastal Ranges. The sites are generally southern exposures of low elevation (0 to 1000 m). Soils usually drain quickly, as they occur on skeletal bedrock material or have an extremely sandy texture. Both species are shade intolerant and require periodic disturbance. *Arbutus menziesii* stump sprouts following fire, and these communities likely had regularly occurring fires prior to European settlement (NVCS 1999). Associates in the canopy include *Taxus brevifolia, Juniperus scopulorum, Pinus contorta,* and *Quercus garryana*. Typical shrub species include *Gaultheria shallon, Holodiscus discolor, Symphoricarpos albus, Lonicera hispidula, Vaccinium ovatum*, and *Corylus cornuta* var. *californica*. The herbaceous layer is generally well developed and dominated by xerophytic grasses and forbs, including *Vicia americana, Lathyrus nevadensis, Sanicula crassicaulis, Festuca occidentalis, Elymus glaucus*, and *Bromus vulgaris* (NVCS 1999).

30.18 Douglas-Fir Cover Type (NVCS *Pseudotsuga menziesii* Forest Alliance): This type is widespread on elevations from sea level to 500 m in the northern part of its range and to 2400 m in the southern part of its range, occurring on moderately deep and well-drained soils (Figure 30.6). It is generally a xeric-mesic-to mesic-site species in locations where precipitation may be as low as 58 cm annually (Walker 1999). This community occupies sites where soil features or local rainshadows induce a drought. The community falls between the drier *Quercus*-dominated woodlands and the moister *Sequoia sempervirens* and *Tsuga heterophylla* forests (NVCS 1999). Communities of this type show no real affinities to soil parent material and occur on various geologic substrates. Soils are typically slightly acidic (pH 5.0 to 6.0), well drained and well aerated, and tend to be gravelly or rocky (NVCS 1999). *Pseudotsuga menziesii* naturally regenerates following fire in the shade cast by standing dead stems. Although *Pseudotsuga menziesii* may maintain itself in drier conditions (where it is mixed with hardwoods), it is seral to more shade-tolerant species in moist conditions. Stands were likely subjected to recurring fires every 30 to 100 years (NVCS 1999). *Pseudotsuga menziesii* may occur with *Tsuga heterophylla, Abies grandis, Abies amabilis, Abies procera, Abies magnifica, Thuja plicata, Acer macrophyllum, Arbutus menziesii, Alnus rubra, Lithocarpus densiflorus, Quercus garryana, Castanopsis chrysophylla, Acer circinatum, Sequoia sempervirens, Chamaecyparis lawsoniana, Pinus ponderosa, Pinus monticola, Calocedrus decurrens, Pinus lambertiana*, and *Picea sitchensis*. On seral sites, common associates are *Abies grandis, Tsuga heterophylla, Acer circinatum, Berberis nervosa*, and *Polystichum munitum*. The understory is usually species-rich and well developed. Common shrub taxa include *Gaultheria shallon, Toxicodendron diversilobum, Rhododendron macrophyllum, Quercus sadleriana, Holodiscus discolor, Mahonia nervosa*, and *Acer circinatum*. In moist areas, species of this community may exceed 50 m in height.

FIGURE 30.6 Douglas fir (*Pseudotsuga menziesii*) forest in the Klamath Mountains, Oregon. Note the regeneration utilizing the fallen log as a substrate.

Note on alterations to the disturbance regime: Many of the western forest communities evolved after a fire or some other disturbance. Adaptations of species to burn (*Pinus contorta*) or to withstand burns (*Quercus agrifolia*) confirm this evolution and long-term relationship. Although anthropogenic influences (especially burning by Native Americans) were certainly strong before European settlement, the extensive alterations in land stewardship practices since the arrival of Europeans have had a major impact. Suppression of fire (apparently the opposite of Native American stewardship practice), high rates of logging, alterations to flooding regimes (see Note on cottonwoods and regulated rivers in chapter 29), and alterations to grazing patterns are the main changes. The modification of disturbances has modified the forests. The remaining area of old-growth forest in California, Oregon, and Washington is only 20 percent of the current forest area, a decrease from the 50 percent of the total forest area only 70+ years ago (Bolsinger and Waddell 1993). More than 80 percent of the old-growth forest is on federal land. This was what Gifford Pinchot had in mind when he established the Forest Service at the turn of the twentieth century (Franklin 1998). However, this may not be the most significant change that has occurred. Throughout the western forests, fire suppression has led to remarkable changes in forest composition and structure. The changes to the disturbance regimes have not only altered community composition and structure, but have also changed landscape scale mosaic patterns (Arno et al. 1993). The small-scale frequent disturbances of the past seem to be changing to large-scale, stand-replacing disturbances (Barrett et al. 1991). In addition, the alteration of one disturbance may affect another. There is good evidence that fire suppression has led to an increase in insect outbreaks. An example is the extensive damage to *Pinus contorta* and *Pinus ponderosa* in the Rockies by the mountain pine beetle (Peet 1988).

30.19 Douglas-Fir-Western Hemlock Cover Type (NVCS *Pseudotsuga menziesii-Tsuga heterophylla* Forest Alliance): This type occurs west of the Cascade range, between 41° and 52° north latitude, from sea level to 1100 m. The type thrives in mild (with neither winter nor summer temperature extremes), humid climates with a mean annual precipitation between 1200 and 4000 mm. Soils are mostly moderately deep and of medium acidity. Surface horizons are well aggregated and porous, with moderate to high organic content. *Pseudotsuga menziesii* is usually predominant, but *Tsuga heterophylla* may dominate on moister or less fertile sites. *Pseudotsuga menziesii* occupies south-facing slopes in the northern part of its range, and north-facing slopes in the southern part of its range (Walker 1999). The type is often seral to more shade-tolerant species. However, *Pseudotsuga menziesii* can live to 500 years or more and is thus a common feature of older forests. The most common associate is *Thuja plicata*, but *Abies grandis, Pinus monticola*, and *Abies amabilis* are also common associates. Several hardwood species may occur, although in small quantities: *Alnus rubra, Acer macrophyllum, Populus trichocarpa, Castanopsis chrysophylla, Arbutus menziesii* (at elevations below 450 m), *Acer circinatum*, and *Rhamnus purshiana.* Common shrub associates include *Gaultheria shallon, Berberis nervosa, Vaccinium parvifolia, Vaccinium ovalifolium, Rubus spectabilis, Rubus parviflorus, Oplopanax horridus*, and *Rhododendron* spp. In the Pacific part of its range, *Ceonothus* spp. and *Arctostaphylos* spp. shrubs are common (Walker 1999). Herbaceous associates include *Polystichum munitum, Blechnum spicant, Linnaea borealis*, and *Oxalis oregana.*

30.20 Oregon White Oak Cover Type (NVCS *Quercus garryana* Forest Alliance): *Quercus garryana* ranges from closed stands (pure or mixed with conifers) to an open savanna-type community structure (considered a Woodland Alliance). This type is found primarily in inland valleys and on lower slopes between the coast and the Cascade or Sierra Nevada ranges, ranging in elevation from near sea level to 1200 m. Although this type may be stable in drier locales, it is typically succeeded by *Pseudotsuga menziesii* in most areas. Suppression of fire has resulted in two-stage stands: large, spreading scattered trees approximately 300 years old, interspersed with smaller, narrow-crowned trees (60 to 120 years old) (Eyre 1980). In the moister northern portions of its range, common associates include *Pseudotsuga menziesii, Acer macrophyllum*, and *Fraxinus latifolia.* In southern drier locales, major associates are *Pseudotsuga menziesii, Pinus ponderosa, Calocedrus decurrens, Quercus kellogii, Arbutus menziesii, Quercus chrysolepsis*, and *Lithocarpus densiflorus.* On moister sites, understory associates include *Corylus cornuta var. californica* and *Polystichum munitum.* On moderate sites, associates include *Prunus avium, Symphoricarpos albus*, and *Amelanchier alnifolia.* On drier sites, *Toxicodendron diversiloba* dominates the understory (although *Toxicodendron diversiloba* is ubiquitous throughout this community type) (NVCS 1999).

30.21 Western Redcedar Cover Type (NVCS *Thuja plicata* Forest Alliance): *Thuja plicata*, also called the western arborvitae, generally occupies moist and wet sites on flats, mountain slopes, and valley bottoms. Elevations range from sea level to 1200 m near the coast, and to 2100 m inland. Precipitation ranges from 75 to 150 cm year^{-1}. Most of the precipitation falls in the winter, with little in the form of snow. This cover type occurs on organic soils along the coast and is more prevalent on mineral soils inland, but 12 percent soil moisture is necessary and controls the species distribution (Walker 1999). *Thuja plicata* can tolerate high water tables and thus dominates wet to very wet sites, while yielding to other species on drier sites because of its shade intolerance. *Thuja plicata* also self-prunes. It can be a prolific seed producer, but tends to be sporadic in this regard (Walker 1999). Major associates include *Picea sitchensis, Tsuga heterophylla, Pseudotsuga menziesii, Abies grandis, Taxus brevifolia,* and *Larix occidentalis.* The shrub layer consists of *Acer circinatum, Mahonia nervosa, Rubus spectabilis,* and *Whipplea modesta.* A dense layer of shade-tolerant herbaceous species includes *Oxalis oregana, Blechnum spicant, Viola sempervirens, Adiantum pedatum, Cornus canadensis, Trillium ovatum,* and *Athyrium filix-femina.* Trees of this type may reach a height of 50 m. This cover type will dominate in wetland forests and where soils remain saturated, such as in old mudflows and basins (NVCS *Thuja plicata* Seasonally Flooded/Saturated Forest Alliances). Associated shrubs of these wetter sites include *Vaccinium membranaceum, Menziesia ferruginea, Gaultheria shallon, Vaccinium ovalifolium, Ledum glandulosum,* and *Berberis nervosa.*

30.22 Western Hemlock Cover Type (NVCS *Tsuga heterophylla* Forest Alliance): The most common tree associates are *Picea sitchensis, Thuja plicata,* and *Pseudotsuga menziesii. Tsuga heterophylla* forms stable communities on moist sites (Figure 30.7), often replacing *Pseudotsuga menziesii.* The reverse occurs on drier sites. This species grows best in the Olympic Peninsula temperate rain forest, receiving up to 380 cm year^{-1}, at elevations from sea level to 600 m. The highest inland elevation is 1500 m. Although these communities typically occur in high montane or subalpine zones, they can also occur on slope benches or canyon bottoms where cold-air drainage is important (NVCS 1999). Soils derived from sandstone, shale, and igneous rock are the parent materials, forming soils of good growing conditions for *Tsuga heterophylla.* The soils are Alfisols. In addition, they are strongly acidic with a substantial organic component as a result of high moistures and low temperatures, which suppress decomposition. The community often attains very old ages, due to low disturbance (e.g., fire) frequency. Associated plants include *Gaultheria shallon, Pachistima myrsinites, Berberis nervosa, Rhododendron*

FIGURE 30.7 Western hemlock (*Tsuga heterophylla*) forest in the eastern mountains of Washington.

macrophyllum, Oxalis oregana, Menziesia ferruginea, Rubus spectabilis, Vaccinium spp., ferns, mosses, and lichens. Shrubs include *Acer circinatum, Mahonia nervosa, Arctostaphylos nevadensis, Holodiscus discolor,* and *Oplopanax horridus. Polystichum munitum, Trientalis latifolia, Achlys triphylla,* and *Linnaea borealis* are common forbs (NVCS 1999). *Tsuga heterophylla* dominates wet to very wet sites, including stream terraces, depressions, impounded waters, and areas with impermeable layers in the soil (NVCS *Tsuga heterophylla* Temporarily Flooded/Seasonally Flooded/Saturated Forest Alliances). Following disturbance, *Alnus rubra* vigorously colonizes. On the wettest sites, large logs are common and serve as nurse logs for regeneration. Sites that are perennially saturated may be transitional to nonforested wetlands (NVCS 1999).

30.23 Mountain Hemlock-Pacific Silver Fir Cover Type (NVCS *Tsuga mertensiana-Abies amabilis* Forest

Alliance): This community occurs in a transition zone separating the subalpine forest of *Tsuga mertensiana* from montane forests. Species from both zones occur in the ecotone. The two species are often the only tree species, although *Chamaecyparis nootkatensis* may codominate in Washington (NVCS 1999). Trees in this community can reach >50 m at lower elevations with *Tsuga heterophylla* and *Pseudotsuga menziesii*. Ericaceous shrubs dominate the shrub layer. Undergrowth is often depauperate as a result of low light levels.

30.24 Mountain Hemlock Cover Type (NVCS *Tsuga mertensiana* Forest Alliance): *Tsuga mertensiana* is a slow-growing, long-lived species that can endure snowpacks up to 6 m deep and is commonly found in pure stands (Walker 1999). Thus, it is a characteristic species of high mountain subalpine sites in the northwest where growing conditions are poor. The community is found at elevations between 1200 and 1700 m in the Washington Cascades (Eyre 1980), between sea level and 1000 m in Alaska, and at higher elevations (up to 3500 m) in the northern Rockies. At upper elevations, it becomes a dwarf, creeping shrub. The community grows on moist, coarse or rocky (well drained), acidic, and organic soils, including Inceptisols. This type may also dominate subalpine environments along streams with seasonal inundation. Soils are typically sandy loam, silt loam, or silt. Forests rarely burn and can be very old. Common canopy associates include *Abies amabilis, A. lasiocarpa, Picea engelmannii, Pinus albicaulis*, and *Chamaecyparis nootkatensis*. The understory is dominated by ericaceous shrubs, including *Vaccinium membranaceum, V. ovalifolium, V. scoparium, Quercus sadleriana, Menziesia ferruginea*, and *Rhododendron albiflorum. Xerophyllum tenax, Orthilia secunda, Carex* spp., *Luzula glabrata, Clintonia uniflora*, and *Chimaphila umbellata* dominate the herbaceous stratum.

Southern Coastal Conifer Forest (NVCS Humid Temperate Domain/Mediterranean Division, *Abies-Pinus-Quercus*) (Region 7b)

30.25 Bristlecone Fir Cover Type (NVCS *Abies bracteata* Forest Alliance): This species also called Santa Lucia fir, is known only from the Santa Lucia Mountains of Monterey and San Luis Obispo counties, California. The association is found at elevations between 250 and 1400 m on steep, rough terrain where fires are less common. Site aspect is typically northern or eastern, and soils are well drained and rocky. Parent materials are sandstones and shales. *Quercus chrysolepsis* is a common codominant. Other overstory associates include *Pinus coulterii, P. ponderosa, P. lambertiana, Arbutus menziesii*, and *Lithocarpus* densiflorus. The shrub and herbaceous layers are not well developed.

30.26 White Fir Cover Type (NVCS *Abies lowiana* Forest Alliance): *Abies lowiana* inhabits rich soils that are moist and well drained on cool north-facing slopes (Walker 1999). Forests of this cover type occur at relatively low elevations (370 to 1500 m; >200 cm precipitation year^{-1}) in southwestern Oregon and at medium to high elevations (1200 to 3150 m; <75 cm precipitation year^{-1}) in the southern Sierra Nevada (NVCS 1999). *Abies lowiana* is tolerant of shade and can form stable communities. Its long life span is indicated by its age of maturation, about 300 years (Walker 1999). Trees greater than 125 m in height are possible. Fire suppression over the past century has allowed *Abies lowiana* to succeed many other important conifers at lower elevations and on apparently drier sites (e.g., *Pinus ponderosa* woodlands). *Abies lowiana* also replaces *Pinus lambertiana* and *Sequoiadendron giganteum* as a result of being very shade tolerant. However, it may also recover rapidly after disturbance, and extensive stands can be found in the Sierra Nevada where logging and burning, or mining, have occurred (Eyre 1980). Co-occurring species include *Pinus ponderosa* on drier sites and *Pinus lambertiana* under moister conditions. Other possible codominants include *Calocedrus decurrens* and *Pseudotsuga menziesii*. Cold-deciduous and schlerophyllous broad-leaved trees are sometimes common, including *Arbutus menziezii, Quercus kellogii, Q. chrysolepsis, Q. sadleriana*, and *Castanopsis chrysophylla*. Ericaceous and cold-deciduous shrubs include *Amelanchier alnifolia, Arctostaphylos patula, A. uva-ursi, Arctostaphylos nevadensis, Ceonothus velutinus, Gaultheria shallon*, and *Holodisus discolor*. Shade-tolerant forbs and ferns dominate the understory: *Pyrola picta, Orthilia secunda, Ribes viscosissimum, Ribes roezlii, Symphoricarpos oreophilus, Chimaphila umbellata, Achlys triphylla, Polystichum munitum, Pteridium aquilinum,* and *Maianthemum stellatum.*

30.27 California Red Fir Cover Type (NVCS *Abies magnifica* Forest Alliance): *Abies magnifica* attains the largest size of the North American balsam firs. Its range is the southern portion of the Marine Division and south, where most of the precipitation falls between November and May. Summers are warm and dry. The species grows at elevations between 1500 and 3000 m. *Abies magnifica* often reproduces after disturbance or is preceded by a *Pinus contorta* pioneer stage (Eyre 1980). Older, stable stands of *Abies magnifica* may be very dense. Common associates include *Abies lowiana* and *Pinus lambertiana* at lower elevations, *Pinus monticola, Pinus contorta,* and *Tsuga mertensiana* at upper elevations, and *Pinus jeffreyi* on drier sites. Typical shrub and forb associates include *Leucothoe davisiae, Pteridium aquilinum, Castanopsis sempervirens, Symphoricarpos mollis, Arnica cordifolia, Quercus vaccinifolia, Quercus sadleriana, Orthilia secunda, Arctostaphylos nevadensis, Chimaphila umbellata, Rhododendron macrophyllum, Lupinus albifrons,* and *Penstemon gracilentus* (NVCS 1999). Vegetation is generally more lush in the disturbed stands, with *Pedicularis semibarbata* common and *Ribes viscosissimum, Chrysopsis breweri, Gayophytum ramosissimum, Pyrola picta,* and *Monardella* spp. sometimes present (Eyre 1980).

30.28 Incense-Cedar-Douglas-Fir Cover Type (NVCS *Calocedrus decurrens-Pseudotsuga menziesii* Forest Alliance): *Calocedrus decurrens* grows best on cool, moist sites in the transition zone between *Quercus* woodlands at lower elevations and *Abies lowiana* forests at higher elevations. The association is found at elevations between 600 and 1200 m in the mountains of the Oregon-California border. These were previously fire-maintained woodlands, with fire intervals between 20 and 30 years (NVCS 1999). These are seral communities, possibly succeeding to *Abies lowiana.* However, *Calocedrus decurrens* is shade tolerant, slow growing and long-lived, reaching maturity at age 300. Principal associates include *Pinus lambertiana, Pinus ponderosa, Pinus jeffreyi,* and *Abies lowiana.* In addition, diverse broad-leaved species may associate: *Quercus garryana, Q. kellogii, Q. chrysolepsis,* and *Arbutus menziesii.* The community may have a developed shrub layer on mesic sites, including *Acer circinatum, Mahonia nervosa, Gautheria shallon,* and *Cornus nuttallii.* Herbaceous species range from drought-tolerant species at lower elevations (e.g., *Polystichum munitum* and *Achlys triphylla*) to mesophytic species at higher elevations (e.g., *Hieracium albiflorum* and *Festuca idahoensis*).

30.29 McNab Cypress Cover Type (NVCS *Cupressus macnabiana* Forest Alliance): The cypress community forms an open to continuous canopy <10 m in height. The community occurs on upper slopes at elevations between 300 and 800 m in the northern Sierra Nevada and southern Cascade Mountains. Soils are typically infertile, but may be derived from a variety of substrates, including basalt and serpentine. The soils are generally rocky and range from deep to shallow. Fire is a necessary component for this community, as *Cupressus macnabiana* has serotinous cones. Possible codominants include *Pinus attenuata, Pinus sabiniana, Cupressus sargentii, Quercus douglasii,* and *Quercus wislizenii.* The shrub component includes *Ceonothus cuneatus, Arctostaphylos viscida,* and *Eriodictyon californicum.* The community lacks a typifying herbaceous layer.

30.30 Brewer Spruce Cover Type (NVCS *Picea breweriana* Forest Alliance): Brewer spruce is endemic to southwestern Oregon and northwestern California. The community is found in rugged, dissected terrain at elevations ranging from 1200 to 1560 m. Soils are typically shallow and rocky (derived from metamorphic rocks) or low in essential nutrients (derived from granitic rocks). On sites with abundant soil moisture and high atmospheric moisture due to fog, *Chamaecyparis lawsoniana* is a common associate. The community is structurally diverse, with *Picea breweriana* a subcanopy under *Abies lowiana, Pseudotsuga menziesii,* or *Pinus lambertiana.* In addition, the subcanopy may include *Quercus sadleriana, Q. kelloggii, Castanopsis chrysophylla,* and *Lithocarpus densiflorus.* Shrubs include *Arctostaphylos nevadensis, Chimaphila umbellata, Holodiscus discolor, Mahonia nervosa,* and *Pachistima myrsinites.* The herbaceous associates include *Arnica cordifolia, Galium ambiguum, Goodyera oblongifolia,* and *Pyrola picta* (NVCS 1999).

30.31 Knobcone Pine Cover Type (NVCS *Pinus attenuata* Forest Alliance): *Pinus attenuata* has a wide distribution through various edaphic and climatic zones, from near sea level to an elevation of 2200 m. Thus, the associated flora are diverse. Drier sites at lower elevations often include *Pinus sabiniana* and *Quercus chrysolepsis.* At higher elevations in open woodland, a variety of oaks may codominate, and at higher elevations, *Pinus attenuata* associates with *Pinus ponderosa, Pseudotsuga menziesii,* and

Chamaecyparis lawsoniana. Typical soils are shallow, dry, stony, or high in magnesium and are thus too harsh for effective conifer competition. *Pinus attenuata* is short-lived, rarely reaching 80 years. Regeneration typically occurs following burns, because of the serotinous cones. Thus, a typical stand is even-aged with a shrub or grass understory. Common shrubs in the community's northern range include *Quercus vaccinifolia, Quercus sadleriana, Amelanchier alnifolia, Ceanothus velutinus, Holodiscus discolor*, and *Castanopsis sempervirens*. In the southern portion of the range, shrubs of this community include several *Arctostaphylos* spp., *Ceonothus* spp., and shrub *Quercus* spp. (Eyre 1980).

30.32 Jeffrey Pine Cover Type (NVCS *Pinus jeffreyi* Forest Alliance): The *Pinus jeffreyi* community is found throughout the coastal and Sierra Nevada range of California, typically at elevations between 150 and 2900 m. *Pinus jeffreyi* is very tolerant of drought, is cold hardy, and is adapted to short growing seasons. This community is generally found on harsher sites than *Pinus ponderosa*, including soils that are shallow, rocky, and infertile (ultramafic soils) (Eyre 1980). The slow-growing, shade-intolerant *Pinus jeffreyi* is out-competed by almost all species under less stressful conditions. However, adaptations such as efficient nutrient uptake from infertile soils, completing growth and entering dormancy sooner, and requiring greater cold exposure to break bud dormancy allow *Pinus jeffreyi* to exist where *Pinus ponderosa* cannot. Associated trees include *Pinus ponderosa, Calocedrus decurrens, Abies lowiana, Abies procera, Abies magnifica, Pinus lambertiana, Pinus monticola, Pinus contorta*, and *Juniperus occidentalis*. The undergrowth varies from north to south, but often includes scrub *Quercus* spp., *Vaccinium* spp., and *Arctostaphylos* spp.

30.33 Bishop Pine Cover Type (NVCS *Pinus muricata* Forest Alliance): This community occurs along the coast, within 12 km of the ocean, on low hills and plains at elevations of 0 to 1000 m from northern California south to Baja California. Three quarters of the summer days are foggy, offsetting the dry months and maintaining ample moisture through the growing season. The community seems to thrive on swampy sites and peat bogs (Harlow et al. 1996). Soils are shallow, acidic, and poorly drained, but occur on a variety of substrates (Histosols). Some soils are podzolized (Inceptisols) and flooded with perched water over a hardpan. The community originates following fire, as *Pinus muricata* pine is short-lived and can be destroyed by disease by 80 years of age. Only in the northern portion of its range does this species mix with other conifers. Two common shrubs in this community are *Vaccinium ovatum* and *Toxicodendron diversilobum*, and two common forbs are *Polystichum munitum* and *Pteridium aquilinum*.

30.34 Pacific Ponderosa Pine-Douglas-Fir Cover Type (NVCS *Pinus ponderosa* var. *ponderosa* [coastal species]-*Pseudotsuga menziesii* Forest Alliance): This community extends from the Klamath Mountains of southern Oregon through middle California at elevations between 450 and 1200 m. *Pinus ponderosa* var. *ponderosa* is adapted to long, hot, dry summers and cool, moist winters. Associates include *Pinus lambertiana, Calocedrus decurrens, Arbutus menziesii, Quercus kellogii, Quercus chrysolepsis, Lithocarpus densiflorus, Cornus nuttallii*, and *Acer macrophyllum*. Succession favors *Pseudotsuga menziesii*, whereas disturbances reverse the trend. The pine can self-perpetuate in areas too dry for *Pseudotsuga menziesii*. This community grades into the Sierra Nevada mixed conifer (where *Abies lowiana* becomes common) community at higher elevations (Eyre 1980) and *Quercus* communities at lower elevations.

30.35 Monterey Pine Cover Type (NVCS *Pinus radiata* Forest Alliance): *Pinus radiata* is restricted to marine terraces and headlands of southern California within 8 km of the coast. *Pinus radiata* is adapted to periodic fires. *Quercus agrifolia* may be an important canopy species, and other associated trees include *Pinus muricata, Pseudotsuga menziesii, Pinus ponderosa*, and *Sequoia sempervirens*. Shrub associates include *Arctostaphylos tomentosa, Vaccinium ovatum, Diplacus aurantiacus* spp. *aurantiacus*, and *Symphoricarpos mollis*.

30.36 Bigcone Douglas-Fir Cover Type (NVCS *Pseudotsuga macrocarpa* Forest Alliance): *Pseudotsuga macrocarpa* is the dominant of a montane association in the California central Coast Ranges, found at elevations between 270 and 2400 m. The soils are well drained and rocky, derived from schists and sandstones. Droughts of several years may be common. There is little other vertical structure to this forest. This association is subjected to recurrent fires, likely of low intensity in the past. *Pseudotsuga macrocarpa* has thick bark to withstand such fires and may also trunk sprout. Tree associates include *Quercus kellogii, Q. chrysolepsis, Calocedrus decurrens, Pinus coulterii, P. ponderosa,*

P. monophylla, P. lambertiana, and *Abies lowiana. Quercus chrysolepsis* may in some cases be a strong codominant, creating a two-tiered older forest (NVCS *Pseudotsuga macrocarpa-Quercus chrysolepsis* Forest Alliance).

30.37 Douglas-Fir-Tanoak-Pacific Madrone Cover Type (NVCS *Pseudotsuga menziesii-Lithocarpus densiflorus-Arbutus menziesii* Forest Alliance): This type represents a diverse assemblage of species due to various historical disturbances, consisting mainly of *Pseudotsuga menziesii, Arbutus menziesii*, and *Lithocarpus densiflorus*. The community is found in the mild, moist climate within 2 km of the coast at elevations between 150 and 600 m (Walker 1999). Cool temperatures and high precipitation are necessary. *Pseudotsuga menziesii* is moderately shade tolerant and fast growing, *Lithocarpus densiflorus* is slow growing and very shade tolerant, can sprout well, and is a prolific seed producer, and *Arbutus menziesii* is a vigorous sprouter and strong competitor on drier sites. All three require periodic fire for their regeneration. Recent land stewardship has greatly altered the success of *Pseudotsuga menziesii*. In the past, low to moderately severe surface fires recurred every 15 to 30 years, allowing *Pseudotsuga menziesii* to become established. The practice of salvaging (cutting remaining useable timber following a disturbance) after a wildfire appears to have favored hardwood species. Once the hardwood community is established, *Pseudotsuga menziesii* may have difficulty establishing (Stuart et al. 1993). Associates include *Pinus lambertiana, Pinus ponderosa, Chamaecyparis lawsoniana, Tsuga heterophylla, Taxus brevifolia, Pinus jeffreyi, Quercus garryana, Umbellularia californica*, and *Castonopsis chrysophylla. Quercus chrysolepsis* may also be present (Eyre 1980).

30.38 Mixed Oak Cover Type [NVCS Quercus (*agrifolia, douglasii, kellogii, lobata, wislizenii*) Forest Alliance]: Coast live oak, blue oak, California black oak, valley oak (California white oak), and interior live oak trade dominance on sites with moderately deep soils in gently sloping valleys. They can occur as woodland or forest. The climate is strongly Mediterranean, without fog to alleviate the summer drought. Most of the precipitation falls between November and April and saturates the soils. Most of the oak species grade into woodlands on drier sites in valleys and foothills (Pavlik et al. 1991). Riparian areas are dominated by *Quercus lobata* and *Quercus wislizeni* within the valleys, and by *Quercus agrifolia* and *Q. chrysolepsis* along the coast (Pavlik et al. 1991). *Quercus agrifolia* occurs on the western side of the Coastal Range, where it is exposed to the moist Pacific Ocean breeze, and on loamy to sandy well-drained soils; or on steep north-facing slopes that provide a cool, moist microclimate (Eyre 1980). This species is long-lived (up to 1000 years) and shade tolerant, regenerating under its own canopy (Walker 1999). The tree has the remarkable ability to sprout from its larger stems after being completely charred by fire. The *Quercus lobata* community grades into *Quercus douglasii* stands at elevations of 30 to 90 m, and the *Quercus douglasii* mix with *Pinus sabiniana* up to 1000 m. Both species are deep rooted, an adaptation for tolerating the annual summer droughts of the climate (Eyre 1980). *Quercus douglasii* is the most extensive hardwood type in California, covering nearly 486,000 ha (1,200,000 acres) (Borchert et al. 1993). *Pinus sabiniana* extends farther upslope and occupies dry, south-facing exposures. The *Quercus douglasii-Pinus sabiniana* community grades into dense shrub cover, *Pinus ponderosa*, or *Quercus kellogii. Quercus kellogii* communities occupy sandy, gravelly and rocky soils of foothills and mountains (elevation 300 to 2400 m; Little 1998). Other common canopy species of this mixed type include *Umbellularia californica, Aesculus californica, Pseudotsuga menziesii, Pinus ponderosa, Arbutus menziesii*, and *Heteromeles arbutifolia.* Periodic fires favor oak over pine species. *Rhamnus californica, Baccharis pilularis*, and *Toxicodendron diversilobum* are common shrubs (NVCS 1999). The herbaceous layer is often grass dominated. The more open the stands, the more grasses and forbs (which can be highly diverse) in the understory.

30.39 Canyon Oak Cover Type (NVCS *Quercus chrysolepsis* Forest Alliance): This cover type is a lowland or submontane evergreen schlerophyllous forest. *Quercus chrysolepsis* has the ability to maintain its leaves for 3 to 4 years (Harlow et al. 1996). The community can be found on canyon bottoms and along boulder-strewn streams, on cool, moist canyon walls, or on rocky slopes at elevations between 450 and 2000 m (Little 1998; NVCS 1999; Walker 1999). Soils are shallow and sandy or rocky and well drained. *Quercus chrysolepsis* is the most common component of the mixed evergreen forest (Barbour 1988). Emergent conifers are often present, including *Pseudotsuga macrocarpa, Pseudotsuga menziesii, Calocedrus decurrens, Abies lowiana, Pinus coulteri*, and *Pinus lambertiana.* The main tree canopy includes *Acer macrophyllum, Quercus kellogii, Quercus garryana*, and

Umbellularia californica. An infrequent shrub component may include *Rubus* spp., *Ceonothus integerrimus, Pentagrama triangularis, Arctostaphylos mewukka, Silene lemmonii, Arbutus menziesii*, and *Polystichum imbricans*.

30.40 Pacific Willow Cover Type (NVCS *Salix lasiandra* Seasonally Flooded Forest Alliance): The *Salix lasiandra* community occurs along floodplains and low-gradient streamsides from sea level to an elevation of 2700 m in California. The community requires low-intensity inundation at least every few years. Canopy dominance can be shared by *Acer macrophyllum, Populus trichocarpa, Populus fremontii, Salix gooddingii, Salix hookeriana, Salix laevigata, Salix sitchensis, Alnus rubra*, and *Alnus rhombifolia*.

30.41 Coast Redwood Cover Type (NVCS *Sequoia sempervirens* Forest Alliance): The coastal *Sequoia sempervirens* trees can attain heights of 120 m at elevations between 10 and 600 m and are considered the tallest organisms on earth (Figure 30.8). The tall crowns enhance seed dispersal by wind, measured at distances of 60 m uphill and 120 m downhill (Walker 1999). These giant trees are supported by deep moist soils (on upland or alluvial streamsides), a summer coastal fog that offsets dry months, and very wet winters (November to May) that produce >150 cm precipitation. The population is restricted to locations between 35° and 42° north latitude. *Sequoia sempervirens* strongly dominates terrace sites but often shares dominance on uplands with *Acer macrophyllum, Arbutus menziesii, Umbellularia californica, Pseudotsuga menziesii, Abies grandis, Lithocarpus densiflorus*, and *Tsuga heterophylla*. Forests are subjected to catastrophic fires at long intervals. *Sequoia sempervirens* responds by crown sprouting as well as stump sprouting, and the stump sprouts can develop in any season within 2 to 3 weeks following a fire (Eyre 1980). Sprouts soon develop their own root systems. Thus, the species seems to perpetuate itself when burned or logged. Shrubs and forbs, which may or may not be common, include *Ceanothus parryi, Ceonothus thyrsiflorus, Cornus nuttallii, Corylus cornuta* var. *californica, Gaultheria shallon, Physocarpus capitatus, Rosa gymnocarpa, Hierochloe* spp., shade-loving ferns, and a variety of herbs.

30.42 Coast Redwood-Western Hemlock Cover Type (NVCS *Sequoia sempervirens-Tsuga heterophylla* Forest Alliance): This cover type is restricted to the fog belt in northern California and southern Oregon at elevations below 1000 m, where the two dominants are often >60 m tall. These communities occur inland from pure stands of *Sequoia sempervirens*. Soils are generally shallow and well drained. In addition, both fire and windthrow are important components of this forest. The subcanopy is dominated by *Lithocarpus densiflorus* (also common is *Umbellularia californica*), and the shrub layer is dominated by *Rhododendron macrophyllum* and/or *Vaccinium ovatum*. Canopy cover of the shrub layer can be quite extensive, up to 70 percent. Other common species in the understory include *Gaultheria shallon* and *Berberis nervosa* and, on moister sites, *Polystichum munitum* and *Oxalis oregana*.

30.43 Giant Sequoia Cover Type (NVCS *Sequoiadendron giganteum* Forest Alliance): This type occurs on the western slopes of the Sierra Nevada,

FIGURE 30.8 Coast redwood (*Sequoia sempervirens*) forest along the coast of Oregon. A coast redwood stands behind the author's brother and sister. Note the "propped" Douglas fir (*Pseudotsuga menziesii*) roots resulting from establishment of a seedling on a log that has since decomposed.

FIGURE 30.9 Giant sequoia (*Sequoiadendron giganteum*) forest in Yosemite National Park, California.

at elevations of about 1400 to 2100 m (Figure 30.9). Most of the precipitation falls as snow, with 2 m accumulations common. Soils are generally derived from granite-based residuum, alluvium, or glacial outwash that creates deep, well-drained sandy loams. The soil remains moist year-round despite the Mediterranean climate, and this moisture is essential for the trees to survive. The *Sequoiadendron giganteum* trees are considered to be the largest organisms on earth, but they are very shade intolerant. Fire is necessary to maintain the community by providing openings for regeneration. The suppression of fire over the past century has had a negative impact on the regeneration of this community. The mixed conifer forest of the region typifies the subcanopy. Common species include *Quercus kellogii, Pseudotsuga menziesii, Calocedrus decurrens, Pinus jeffreyi, Pinus contorta* var. *murrayana, Pinus ponderosa, Abies magnifica, Pinus lambertiana,* and *Abies lowiana.* Shrubs may include *Castonopsis sempervirens, Chamaebatia foliolosa, several Ceonothus* species, *Arctostaphylos patula,* and *Rhododendron occidentale.*

30.44 Sierra Nevada Mixed Conifer (NVCS *Abies lowiana-Pinus-Quercus* Forest Alliance): The mixture of *Abies lowiana, Pinus ponderosa, Pinus lambertiana, Calocedrus decurrens, Quercus kellogii,* and *Pseudotsuga menziesii* typify western slopes of the Sierra Nevada between the elevations of 900 and 1800 m. Less common hardwood associates include *Arbutus menziesii, Lithocarpus densiflorus,* and *Cornus nuttallii.* Succession depends on the location of the stand and the type of disturbance. Generally, disturbance favors *Pinus ponderosa, Pinus lambertiana,* and *Pseudotsuga menziesii,* whereas lack of disturbance favors *Abies lowiana* and *Calocedrus decurrens.* Shade tolerance is greatest in *Abies lowiana; Pinus lambertiana* and *Pseudotsuga menziesii* are intermediate; and *Pinus ponderosa* is the most intolerant. *Quercus kellogii* tolerates shade but needs considerable light for best development. Thus, the oak maintains itself on harsh sites (ridge tops and south slopes) and is replaced by conifers on other slopes in the absence of disturbance. Numerous shrub species occur in this community, including *Arctostaphylos patula, Ribes nevadense, Ceanothus integerrimus, Ceonothus cordulatus, Ceanothus velutinus, Ceonothus prostratus, Chamaebatia foliosa, Prunus emarginata,* and *Castanopsis sempervirens.* The herbaceous layer consists of a rich composition of grasses and forbs.

30.45 California Bay Cover Type (NVCS *Umbellularia californica* Forest Alliance): California bay coastal forests (elevation from 0 to 500 m) occur on intermittently flooded stream terraces. The periodic floods are necessary for the community to establish and maintain itself. Soils are primarily derived from sandstone or schist. Codominants include *Quercus chrysolepsis, Quercus agrifolia, Quercus wislizeni, Arbutus menziesii, Sequoia sempervirens,* and *Lithocarpus densiflorus.* The shrub layer is not significant, but *Garrya elliptica* is common. The herbaceous layer is sparse.

Glossary of Technical Terms

A- A prefix indicating not, different from, or absent.

Ab- A prefix meaning a turning away from.

Abaxial The lower surface (e.g., of a leaf); facing away from the axis (stem). Compare with *Adaxial.*

Abortive A term used to describe a structure that is lacking or not fully developed.

Abrupt Suddenly ending; e.g., a truncated apex.

Abscission The loss of a leaf or flower due to the formation of a layer of cells (abscission layer) preventing the movement of water or nutrients into the organ. The loss of a leaf produces a leaf scar on the twig.

Accessory bud(s) One to several additional buds, usually a flower bud above or to the side of the primary (foliage) bud in the axil of the leaf or surrounding the terminal bud.

Achene A dry indehiscent fruit in which a single seed is attached to the ovary wall.

Acicular Extremely long and narrow with essentially parallel sides; i.e., needle-shaped.

Aciliate Lacking hair along the margin. Compare with *Ciliate.*

Acorn The fruit of oak. A dry, hard, oval to nearly globose indehiscent fruit accompanied by a caplike involucre.

Acrid Having a bitter taste.

Actinomorphic Formed so that two equal halves (mirror images) occur along any plane drawn through the center; radically symmetrical; regular. Compare with *Zygomorphic.*

Acuminate Gradually tapering to a pointed apex at a progressively decreasing angle so the sides are slightly concave. Compare with *Acute* and *Obtuse.*

Acute Tapering to a pointed apex at an angle of less than 90°, the sides are usually straight. Compare with *Acuminate* and *Obtuse.*

Ad- A prefix meaning to turn toward.

Adaxial The side of the leaf toward the stem. Compare with *Abaxial.*

Adventitious Developed in an unusual location on a plant, such as roots developed from a buried branch or buds under the bark.

Adventitious bud A bud not located at a twig node. For example, buds found under the bark of trees; these are usually suppressed buds that grow at the same rate as the tree, so they are not covered over by newly formed bark (phloem).

Adventitious roots Roots that develop other than below ground.

Aerial roots Roots that develop above ground, such as those that hold the stem of poison ivy to tree trunks.

Aestival Reference to summer.

Agamospermy Seed development without fertilization; parthenocarpy.

Aggregate A dense clustering; usually referring to aggregate fruit that is formed by the small fleshy fruit of tightly clustered individual pistils of a single flower. Compare with *multiple Fruit.*

Alate With wings; e.g., *Ulmus alata*, the elm with corky wings on the branches.

Allelopathy Interference with plant growth due to production of a chemical by an adjacent plant; e.g., juglone production by black walnut.

Alliance A physiognomically uniform group of plant associations sharing one or more dominant diagnostic species. This level of the classification is similar to the Society of American Foresters' Cover Types (Eyre 1980).

Allopatric Species whose ranges (geographic distributions) do not overlap. Compare with *Sympatric.*

Alluvial Organic or mineral soil material associated with running water (creeks, rivers).

Alpine The area on a mountain above timberline. Compare with *Arctic.*

Alternate Leaves located singly at each twig node such that leaves are separated along the twig and appear to alternate from one side of the twig to the other. Compare with opposite. Alternate leaf arrangement produces an alternate bud and twig arrangement.

Ament A pendent or erect inflorescence composed of a compacted spike of unisexual, apetalous flowers; a catkin; the characteristic inflorescence of the Fagaceae, Betulaceae, and Salicaceae.

Amorphous Without form.

Anastomosing Interlacing; e.g., braided ridges of bark on some tree trunks.

Andro-dioecious A species having stamen-bearing and perfect flowers on separate plants.

Androecious A collective term referring to the stamens of a flower.

Anemophilous A species in which pollen is carried from the anther to the stigma of the pistil by wind; wind pollinated.

Anther The saclike pollen-bearing structure of the stamen.

Anthocyanin The blue, purple, and red water-soluble pigments in plant cells.

Apetalous A flower without petals.

Apex The distal point or point farthest away from the area of attachment. Compare with *Base.*

Apical Located near or at the apex.

Apical meristem The actively growing regions of shoot and root tips; this meristem increases twig and root length.

Apomixis A general term referring to the setting of seeds or seed production without fertilization. Agamospermy.

Apophysis The abaxial area of a cone scale visible when the cone is closed; usually differentiated by color, but in some species may be swollen and support a prickle or claw.

Appressed Pressed against or lying close to, as a bud against the twig or hair on a leaf surface.

Aquatic Associated with or growing in water. Compare with *Terrestrial.*

Arboreous A tree with a single tall, generally straight stem (trunk or bole) growing to a height of 10 to 30 m or more.

Arborescent Treelike, but with a short or poorly formed main stem or stems; usually less than 10 m in height.

Arboretum A parklike garden where trees and shrubs are cultivated for science and education.

Arctic The treeless area north of the boreal (conifer-hardwood or spruce-fir) forest; geographically, the area between the arctic circle and the north pole.

Arcuate Curved or arched; pinnate veins that curve to follow the leaf margin.

Aril A seed surrounded by fleshy, thickened integuments of the seed coat; e.g., fruit of *Taxus.*

Aristate Tipped with a stiff bristle of material different from that of the supporting organ; e.g., an oak leaf blade (lobe) with a bristle tip.

Armed A twig, cane, or cone with one or more thorns, spines, or prickles.

Artificial classification A system based on arbitrary but convenient morphological characteristics rather than on characteristics that group related plants.

Ascending Growing upward, usually toward the tip or apex of the stem.

Asepalous Lacking sepals.

Association A plant community type of definite floristic composition, uniform habitat conditions, and uniform physiognomy.

Asymmetric Unequal halves, as in comparing the two sides of a leaf blade split along the midrib or two sides of a bud split along the midline.

Attenuate Tapering to a long apex or base, as with a leaf, usually where the apex is distinct from the blade

Atypical Not typical.

Auricle An earlobe-like appendage.

Auriculate Bearing an auricle.

Autogamy Self-pollination.

Autumnal Appearing during the autumn.

Awl-shaped A flat, short (1 to 2 cm) triangular-shaped structure; shaped like an awl.

Awn A short needle-like structure; e.g., bristle tips on a red oak leaf.

Axial Referring to or located in the axil or on the axis.

Axil The adaxial area (or angle) formed between the twig and leaf petiole or leaf stalk at the point of attachment; e.g., the bud is located in the axil formed by the twig and petiole.

Axillary Referring to or located in the axil.

Axis The longitudinal supporting structure to which organs are attached; e.g., leaves to a twig.

Balsamiferous Producing a sticky resin; e.g., balsam fir; balsam-poplar.

Barb Short, sharp bristle often pointing backward toward the base.

Bark A general term collectively referring to annual layers of phloem tissue produced outside the vascular cambium and layers of cork produced by the cork cambium.

Base The proximal point or point nearest the point of attachment.

Basal Located at the base.

Basipetal Developing from the apex downward toward the base; flowers in an inflorescence produced and maturing in sequence from the apex to the base.

Beak A long tip found on some flowers, fruits, or seeds.

Bearded Bearing long, stiff hairs; e.g., the bearded seeds of catalpa.

Berry A fleshy fruit containing several seeds and developed from a single pistil; e.g., in wild grape or persimmon.

Bi- A prefix meaning two.

Bifid Split into two parts.

Bicolor A structure having two colors; e.g. abaxial and adaxial leaf surfaces.

Bifurcate A characteristic in which a single structure splits into two; a leaf lobe spliting into two tips.

Biomass The amount of plant material within a unit of area (hectare, acre).

Bipinnate Twice pinnately compound; each leaflet is again subdivided; e.g., honey locust.

Bisexual A perfect flower; both stamens (male reproductive organ) and a pistil (female reproductive organ) are present; hermaphrodite.

Bladder An inflated organ formed by thin cellulose or paperlike walls; e.g., the fruit of bladdernut.

Blade The green, flat expanded tissue of a leaf or petal.

Bloom A whitish or bluish waxy material on the surface of a twig, leaf, flower, or fruit.

Blotches Irregular spots that are distinctly different from the surface backround.

Bog A depressional area accumulating peat, which has no inflows or outflows.

Bole The main stem of a tree.

Bordered The edge of a structure which has a characteristic (e.g., color) that differs from the interior portion.

Boreal Northern; usually referring to far north.

Boreal forest zone The spruce-fir-hardwood forest of central Canada.

Boss A raised but not sharp projection.

Botanical name The Latin binomial name of a plant.

Bottomland Forested lowlands along streams and rivers; alluvia floodplains periodically inundated unless they have been cut off from the river by levees.

Bract A modified (small) leaf or leaflike structure subtending (at the base of) the flower or inflorescence.

Branchlet An extremely small twig, often deciduous with the leaves; e.g., baldcypress.

Bristle A sharp, stiff hair-like structure.

Bristle-tipped A leaf lobe or apex bearing a bristle.

Bud A miniature shoot supporting leaves or flowers, or a combination, enclosed by modified leaves (scales).

Bud scale scar The scar remaining after a bud scale falls as the bud expands in the spring; numberous scars form a ring around the twig, which marks the beginning of yearly growth.

Bud scales The modified leaves covering a bud; after bud break and shoot elongation, the ring of scars on the twig indicates the point at which yearly growth begins.

Bundle scar The scars produced by the vascular bundles within the leaf scar; the scars may form a series of dots or a continuous line, circle, or other pattern. The pattern is usually consistent for a species.

Bur A structure armed with numerous spines or barbs; e.g., the fruit husk of American chestnut.

Buttress The flared base of a tree; e.g., bald cypress, swamp tupelo.

Calcicole A plant growng on a calcareous (e.g., limestone-derived) soil or substrate.

Calcifuge A plant that does not survive or grow well on a calcareous substrate.

Calyx Collectively, the sepals of a flower or the outermost row of flower parts.

Calyx tube The structure formed by united (fused) sepals.

Cambium A zone of actively dividing cells. See *Vascular cambium* and *Cork cambium.*

Canescent Densely covered with colored hair; e.g., rusty red buds of pawpaw and silvery buds of magnolia.

Capsule A dry, many-seeded fruit of more than one carpel dehiscing (spliting open) at maturity; e.g., catalpa and honeylocust pods.

Carotene The yellow and orange pigments found in leaf cells.

Carpel That portion of a simple pistil that contains seeds (the old ovary) or a section of a compound pistil usually, but not always, separated from other sections by walls.

Catkin See *Ament.*

Cauliflorous The characteristic of a tree that produces leaves on the main bole or larger branches; e.g., shortleaf pine.

Chambered pith A pith subdivided into empty sections by radial cross-walls (diaphragms); e.g., found in black walnut and butternut. See *Diaphragmed, Spongy,* and *Excavated Pith.*

Chaparral The vegetation defined by dense thickets of sclerophyllous (hard evergreen leaf) shrubs located in southwestern North America, primarily in California.

Characteristic A morphological structure (e.g., leaf arrangement) or an ecological tolerance/response to environment (e.g., shade tolerance).

Chemotaxonomy An area of science that relies on plant chemical components to improve the classification of plants.

Chlorophyll The green pigment found in plant cells and responsible for photosynthsis.

Cilia The hairs along the margin of a structure.

Ciliate Characteristic of a structure with a marginal fringe of hair.

Circum- (prefix) Around, as in circumpolar; e.g., species of spruce and fir are found around the earth (North America, Scandanavia, Siberia) near the north pole.

Cladistics The process of defining evolutionary relationships between plant groups.

Clasping Term applied to a leaf petiole that largely, partially, or completely surrounds the twig.

Class The name for a formal cluster of plant groups with a rank between Subdivision and Order.

Classification The formal arrangement of plants into groups and groups into various hierachial levels.

Clavate Club or bat-shaped. See *Spatulate.*

Cleft An extremely narrow sinuslike split in a leaf blade that does not extend beyond one half the distance between the margin and the midrib. Compare with *Parted.*

Climax community See *Compositionally stable community.*

Climbing A term describing a vine or plant whose stem or tendrils twine around another structure for support.

Clone A group of genetically identical individuals originating from a single parent by means of root or stump sprouts or by branch cuttings.

Community The collection of trees, shrubs, vines, and herbs, along with microbiotic and zoologic communities, mixed and interacting (primarily competing) on a specific area of land (a site). An assemblage of species (populations) found together repeatedly under similar environmental conditions across the landscape.

Complanate Flattened.

Complete Refers to a flower with all standard parts present, including sepals, petals, stamens, and a pistil.

Compositionally stable community An assemblage of species that have the ability to maintain themselves (regenerate, grow, and survive) on a site over a long period of time. Some people refer to this as the a climax community, because of the influence of climate in determining the community composition.

Compound Composed of two or more similar components.

Compound leaf A leaf blade that has been subdivided into leaflets; a leaf is identified as simple or compound by the axillary bud located at the base of the petiole.

Compressed Flattened.

Concave The characteristic shape of a structure that has been hollowed out or dented inward; e.g., the shape of a sinus between two lobes of a leaf.

Cone A reproductive structure of a dense cluster of sporophylls on a central stalk; the structure may produce pollen or bear seed. Also called a strobilus (pl. strobili).

Cone scale An ovule-bearing scale; e.g., the scale of a pine or spruce cone.

Conical A term describing a structure having a cone shape; e.g., the crown of a spruce tree.

Conifer A cone-bearing tree; a gymnosperm.

Coniferous Producing cones.

Connate A term describing flower parts that are fused or joined from the time of development; e.g., the fusion of petals to form a corolla tube. See *Perfoliate.*

Contiguous Adjoining.

Copious Plentiful in number; e.g., a copious amount of seed.

Coppice A large number of stems (collectively) arising from a tree stump; a vegetative method used to reproduce species such as willow, cottonwood, and silver maple.

Cordate Heart-shaped; e.g, the shape of a leaf or leaf base.

Coriaceous A term describing a tissue that has a leathery texture; e.g., a thick, tough leaf blade that is stiff or has a pebble-grained or wrinkled surface.

Cork cambium A meristem developing in the outer layers of the phloem; it produces the material that develops the bark pattern of mature trees.

Corolla The collective name applied to the petals.

Corolla tube The cylindrical structure formed by united petals.

Corrugated The surface formed by alternating furrows and ridges.

Corymb A simple, indeterminate, nearly flat-topped inflorescence similar to a raceme; the nearer the flowers to the apex, the shorter the pedicels.

Cotyledon The primary leaf or leaves of the embryo; a cotyledon may store food as in the seeds of various species, or it may enlarge, become green, and produce food when the shoot emerges from the seed.

Creeping A term describing a species that produces a horizontal stem that grows along the ground surface.

Crenate A term describing a leaf margin with rounded teeth.

Crenulate A term describing a leaf margin with extremely small rounded teeth.

Crown The area of a tree or shrub outlined by the foliage.

Cruciform Cross-shaped.

Cuneate Wedge- or triangle-shaped; a shape of a leaf base.

Cupule A small cup; specifically, an acorn cap or involucre formed from the coalescence of bracts.

Cusp A short, sharp point formed from leaf blade tissue.

Cuspidate A term describing a leaf whose apex ends in a cusp; compare with *Aristate.*

Cuticle The wax layer on the upper epidermis of a leaf.

Cyme A flat- or round-topped determinate inflorescence in which the oldest flower (first to bloom) is at the center of the cluster.

Cymose Cyme-like.

DBH Diameter breast height; the diameter measurement taken on a tree at 4.5 ft or 1.37 m above ground level.

Deciduous A term describing a woody plant whose leaves senesce and fall to the ground in autumn; the deciduous habit permits the plant to avoid overwinter freeze damage.

Decurrent Extending downward below or basipetally from the point of insertion.

Decussate A term describing leaves or cone scales arranged in an opposite fashion, with adjacent pairs set at 90° from each other; four-ranked.

Dehiscence The splitting open of some fruits at maturity; e.g., the capsule of catalpa.

Deliquescent A term describing the pattern of diffuse branching in the absence of a central bole that disappears in the crown of most angiosperm or deciduous trees.

Deltoid The shape of an equilateral triangle; the shape of the Greek letter *delta.*

Dendritic A term describing the branching pattern of a tree; in geology, describing the pattern created by the streams in a drainage system or watershed.

Dentate A term describing a toothed margin with the teeth pointed at a right angle to the midrib or margin.

Denticulate Dentate with extremely small teeth.

Depauperate Poorly developed; e.g., a forest with few tree or herbaceous species.

Depressed Flattened; e.g., viburnum has a nearly round fruit that is depressed on two opposing sides.

Descending Pointed downward.

Determinate Characterized by the formation of an inflorescence or terminal bud that prevents further elongation of a twig, usually for the rest of the growing season. The terminal flower in the inflorescence blooms first, thereby eliminating the terminal bud.

Di- A prefix meaning two or twice.

Diaphragmed pith A pith composed of spongy material divided into sections by radial crosswalls (diaphragms); common in yellow poplar and black gum. See *Chambered, Spongy,* and *Excavated pith.*

Dichotomous Branching into two equal parts; e.g., the veins in a ginkgo leaf.
Dicotyledonous Bearing two cotyledons.
Dicotyledons The two primary leaves or storage organs found in a seed.
Diffuse Wide-spreading or pointing in various directions; e.g., the leaf arrangement on spruce trees.
Digitate Arising or appearing to arise from a central point and spreading like fingers on a hand; lobe arrangement in maple or leaflet arrangement in buckeye; see *Palmate.*
Dimorphic Having two forms; e.g., sequoia leaves on a cone-bearing twig are different from those on a sterile twig.
Dioecious A term describing stamens and pistil that are borne in flowers (unisexual) on different plants. From the root *ecos* meaning "two houses".
Disjunct A term describing a group of plants that is geographically separated from the primary population range of the same species.
Dissected Divided by deep sinuses into many narrow segments.
Distal Toward the tip or end opposite the point of attachment. Compare with *Proximal.*
Distichous Two-ranked; e.g., the leaf arrangement of baldcypress.
Distinct Separated.
Divergent Angling away from the primary axis.
Division The name for a formal cluster of plant groups with a rank between Kingdom and Subdivision. A subdivision of a Domain determined by isolating areas of definite vegetation affinities that fall within the same regional climate.
Domain Groups of ecoregions with related climates.
Dorsal The back or outward surface in relation to an axis; the lower surface of a cone or leaf. See *Ventral.*
Dorsiferous Located on the back.
Double serrate A term describing the margin of a leaf that has small teeth positioned on large teeth.
Downy A term describing a surface covered with extremely fine, soft hair.
Drupaceous Similar to a drupe or producing a drupe.
Drupe An indehiscent fruit consisting of three layers, collectively called a pericarp: an outer skin (exocarp), a middle fleshy pulp (mesocarp), and an inner hard shell (endocarp, a pit), enclosing one or two seeds, e.g., a peach or cherry.
Drupelet A small drupe; e.g., the individual units of a raspberry or mulberry.
E- A prefix meaning without or lacking; see *Emarginate.*
Ecological land classification The process of evaluating and sub dividing land based on potential community composition and productivity using soil and topography and/or the presence of specific shrubs and herbaceous plants.
Ecology The science dealing with relationships between plants or plant populations, and between plants and their environment.
Ecoregion Major ecosystem, resulting from large-scale predictable patterns of solar radiation and moisture.
Ecotype A subdivision of a species population adapted to a local environment.
Effectively published A term describing a new Latin name published in a technical botanical journal or book.
Elliptical In the shape of an ellipse; an oval.
Emarginate A term describing a leaf or petal that has a notch or indentation in place of a pointed apex; an apex with no margin.
Endangered species A species that may become extinct without proper conservation, protection, or management.
Endemic species A plant confined to a local area and found nowhere else.
Endocarp The innermost layer of the pericarp; see *Drupe.*
Endogenous Developing from inside or within.
Endosperm The nutritive tissue within a seed that is used to support the developing embryo for a short time after germination.
Entire A term describing a leaf blade or petal without lobes or large teeth.
Entomophilous A term describing a species in which pollen is carried from the anther to the stigma of the pistil by an insect; insect pollinated.
Ephemeral Present for only a short time; e.g., spring woodland wildflowers that take advantage of the sunlight at the forest floor to complete their life cycle before the tree leaves are fully expanded and the floor is shaded.
Epi- A prefix meaning upon; e.g., epicormic branching, branches formed on top of the bark of the main stem.

Epidermis The outer layer of cells in a plant organ such as a leaf; the outer skin.
Epigeal A term describing the position of cotyledons somewhat above the ground surface after germination; e.g., basswood, maple, yellow poplar, pine, spruce.
Epiphyte A plant physically growing on another plant but not parasitic.
Equilateral Having equal sides if divided down the middle; e.g., the identical sides of a simple leaf if divided along the midrib.
Erect Oriented vertically.
Erose A term describing a rough, uneven, somewhat serrated margin of a cone scale or leaf.
Estipulate Lacking stipules.
Eu- A prefix meaning true.
Even-pinnate Pinnately compound (leaf) with an even number of leaflets resulting from the loss of the terminal leaflet.
Evergreen A species of plant that retains its leaves over winter.
Ex- A prefix that means without; e.g., exstipulate, without stipules; estipulate.
Excavated pith Empty or hollow pith; e.g., excavated pith in persimmon. See *Spongy, Diaphragmed,* and *Chambered pith.*
Excurrent Extending beyond; e.g., the structure of a conifer tree whose single main stem extends from the ground to the highest apex and supports small lateral branches; a leaf vein extending beyond the blade margin.
Exfoliate To peel in layers; e.g., the bark of sycamore or river birch.
Exocarp The outer layer of the pericarp; see *Drupe.*
Exogenous Developing from the outside.
Exserted Projecting beyond; e.g., a long bract extending beyond the end of the cone scale in some conifers.
Exstipulate See *Estipulate.*
Extant Now living.
Extinct Completely eliminated.
Extirpated species A species that has been eliminated from a portion of its natural range.
Falcate Sickle-shaped; e.g., a curved leaf or leaflet, sharp-pointed at the apex.
Fascicle A tight group or bundle; e.g., white pine has a fascicle of five needles.
Fascicle sheath A thick cellulose tissue that surrounds the base of needles within a fascicle; it is persistent in the hard pines, deciduous in the soft pines.
Fertile Actually or potentially producing seeds.
Fertilization The process of joining the male gamete with an egg to produce a new individual.
Fetid Having a strong, extremely unpleasant odor.
Fiber A long, thick-walled xylem cell, pointed at both ends, that provides support and strength to wood; found primarily in angiosperms but also in gymnosperms. Compare with *Vessel* and *Tracheid.*
Fibrous roots A root system of fine, extensively branched roots of about the same size; usually found in grasses and herbs.
Flabellate Fan-shaped; e.g., palm tree or saw palmetto leaves.
Fleshy A term describing moist and soft tissue; also refers to the thick leaves of succulent plants.
Floral tube The tube formed when petals are fused together during development; e.g., in the catalpa flower.
Flower The reproductive portion of a plant supporting the stamens (male) and pistil (female) organs.
Foliaceous Similar to a leaf either in texture or color; producing leaves.
Foliage A term referring collectively to the leaves of one or more plants.
Foliar Of the foliage; e.g., a foliar insect; one that attacks a leaf.
Follicle A dry, many-seeded fruit dehiscing (opening) along a single suture line at maturity; e.g., the individual units in compound magnolia fruit.
Fringed With a row of hairs along the margin.
Fruit A ripened ovary and associated attached structures.
Fruitescent Like the form of a shrub.
Fruiticose The form of a shrub.
Fusiform Spindle-shaped; cylinder-like but somewhat broader in the middle.
Gamopetalous A term describing a flower with partially or completely united petals; see *Floral tube.*
Glabrate Nearly glabrous; a surface with scattered hair.
Glabrous Lacking hair.
Glandular Bearing glands; e.g., the petiole of *Prunus* (plum species).
Glaucous Covered with a white, waxy coating; a bloom; e.g., the twig surface of smooth sumac.
Globose Round, like a ball or globe.

Gymnosperm A plant that produces a seed not borne in an ovary (no fruit), but naked, as on the surface of a cone scale; a general term collectively refering to conifers.

Gynoecium The female part of a flower; the pistil or, collectively, the pistils of a flower.

Habit The general form of a plant; i.e., tree, shrub, herb; also used in the sense of a deciduous vs. an evergreen habit.

Habitat The type of site conditions in which an organism grows and develops.

Hastate The general shape of an arrowhead, but with diverging lobes.

Head A dense cluster of sessile or nearly sessile flowers.

Heartwood In trees and other woody stems, the inner, darker, central cylinder of wood (xylem) composed of dead cells used for storage of materials. Compare with *Sapwood.*

Hemisphere Half of a sphere; the northern or southern half of the earth.

Herb A plant with no persistent woody stem, usually reproducing by seed each year (an annual) or every other year (a biennial).

Herbaceous A term referring to a plant with the characteristics of an herb.

Herbarium A collection of dried, pressed, and preserved plant specimens; a museum of dead plants.

Hermaphrodite A plant with pistils and stamens in the same flower; a perfect flower; bisexual.

Heterogamous A term describing a plant producing both male and female (unisexual) flowers.

Hirsute A term used to describe a surface covered with coarse, stiff hair.

Hispid A term used to describe a surface covered with bristle-like hair.

Homogamous A term used to describe a plant producing flowers of one sex.

Homologous A term refering to organs that have the same origin but differ in appearance and function; e.g., leaf, bud scales, stipules, and flower petals.

Hydric Wet, generally referring to soil or site conditions.

Hydrophyte A plant that grows in or near water. Compare with *Xerophyte* and *Mesophyte.*

Hypogeal A term describing the position of cotyledons below the soil surface after germination.

Imbricate Overlapping, as with bud scales. Compare with *Valvate.*

Imperfect flower A flower in which either stamens or the pistil is absent. See *Staminate* or *Pistillate flower.*

Incised Deeply and irregularly cut; usually used to describe a leaf blade margin.

Incomplete flower A flower that lacks one or more of the common parts: sepals, petals, stamens, or pistil.

Incurved Curved inward; e.g., on a leaf margin, referring to the curved teeth that point toward the apex.

Indeterminate A term used to describe an inflorescense in which the lower flowers bloom first, thus permitting continued growth and development of flowers by the apical meristem.

Indigenous Native to an area.

Inequilateral Having unequal sides if divided down the middle; e.g., the different sides of a simple leaf if divided along the midrib.

Inflorescence Collectively, the flowers and the structure supporting plant flowers.

Inflorescence type The arrangement of flowers within an inflorescence.

Inodorous Lacking an odor.

Inserted Attached to and/or shorter than.

Integument The layer of the ovary that becomes the seed coat.

Inter- A prefix meaning between or among.

Internode The twig area between leaf nodes; the stem area between branch nodes.

Intervenous The area between veins in a leaf blade.

Intolerant Unable to grow and develop in limiting environmental conditions; e.g., low light, low soil water, low soil nutrients, flooding, etc.

Introduced plant A plant planted in an area or county where it is not native; e.g., Japanese honeysuckle introduced from Japan.

Introgression The introduction of genes from one species into the genome of another species by backcrossing a hybrid with a parent.

Involucre A collection or whorl of bracts subtending a flower.

Irregular See *Zygomorphic.* Compare with *Actinomorphic.*

Keel A central, dorsal, longitudinal ridge; e.g., on the surface of a bud scale; the lowest united petals of a papillionaceous flower (legume).

Krummholz German word meaning dwarf or crooked (*krum*) forest (*holz*); describes trees found at the timberline in subalpine conditions.

Lacustrine Associated with lakes.
Lamina The blade of a leaf.
Lanceolate Lance-shaped; the shape of a leaf that has a broad base tapering to a sharp point at the apex.
Latex A milky sap; a characteristic of the Moraceae family.
Leaf The green, photosynthetic organ of a plant. A leaf is identified as simple or compound by the axillary bud at the base of the petiole.
Leaflet The leaflike unit of a compound leaf; a leaflet does not have a bud at the base of its stalk; also called a pinna.
Leaf scar The visible structure formed on a twig where a leaf petiole was attached.
Legume A dry, dehiscent fruit of a single carpel; at maturity, it splits along two suture lines; e.g., a honeylocust pod.
Lenticel Small (<10 mm long), rough, often lens-shaped area on twig and bark surfaces.
Lepidote Covered with scurfy scales.
Levee A buildup of local soil as a dam against the movement of water from a main channel to a floodplain. Levees are typically associated with barrow pits, from which the soil material for the levees was excavated.
Liana A climbing woody vine.
Linear The shape of a flat leaf or petal that is many times longer than wide, such that the sides are essentially parallel; e.g., the leaf of fir.
Littoral Near the shore line.
Lobe A rounded segment of a leaf or petal separated from an adjacent lobe by a sinus.
Lobed Divided into lobes.
Locule A chamber of an ovary.
Lustrous A term used to describe a highly shiny surface that reflects solar radiation; e.g., the surfaces of post oak leaves.
Lyrate Shaped like a lyre; a term applied to a leaf that has a large terminal lobe and smaller, irregularly shaped, smaller lobes positioned basipetally; e.g., the leaf of overcup oak.
Macrosporophyll A leaf modified into a sporangia-producing organ.
Malodorous Producing a strongly repulsive odor.
Marcescent Refers to an organ (especially leaves) dying but remaining on the tree through winter.
Margin The edge of a leaf, petal, or other organ or tissue.
Marginate Having a distinct margin.
Mast Collectively, hickory nuts, oak acrons, beech nuts, and similar fruits used by wildlife as food.
Megasporangium The structure producing an ovule.
Megasporophyll A modified leaf supporting the structure producing an ovule.
Membranous A term used to describe a tissue that is thin, paperlike, and possibly translucent.
Mericarp One section of a schizocarp, a fruit that splits into two halves at maturity; the term is sometimes applied to maple fruit.
Meristem The actively growing regions of a plant; see *Apical meristem, Vascular cambium,* and *Cork cambium.*
-merous A suffix meaning in sets of; e.g., trimerous, a set of three; pentamerous, a set of five.
Mesic Moist, generally referring to soil or site conditions. Compare with *Xeric* and *Hydric.*
Mesic-hydric An intermediate soil moisture condition between mesic (moist) and hydric (wet)
Mesocarp The middle, fleshy layer of the pericarp; see *Drupe.*
Mesophyll The central tissues located between the epidermal layers of a leaf.
Mesophyte A plant adapted to moist soil or site conditions. Compare with *Xerophyte* and *Hydrophyte.*
Micro- A prefix meaning small.
Microsporangium A structure producing pollen.
Microsporophyll A leaf supporting a structure producing pollen.
Midrib The large central vein extending from the base to the apex in a leaf or leaflet.
Mono- A prefix meaning one.
Monocotyledon A plant with a single cotyledon or seed leaf.
Monoecious A term used to describe a species with imperfect (unisexual) flowers with staminate and pistillate flowers on the same plant; mono- = one, ecos = house.
Monotypic A term used to describe a genus containing one species, or a family with only one genus.
Montane Referring to mountains; growing on a mountain.
Mottled With blotches of different colors superimposed on a single dominant color.
Mucilaginous Slippery, moist.
Mucro A short, small, relatively sharp-pointed tip.

Mucronate A leaf or similar organ tip having a mucro.

Multiple fruit A compound fruit developed from multiple flowers packed on a single receptacle; e.g., mulberry.

Mycorrhiza The symbiotic association between plant roots and a fungus.

Naked A term indicating the absence of a protective structure; e.g., the naked bud (no bud scales) of pawpaw; seed borne naked (no ovary) on a pine cone scale.

Needle The acicular leaf of pine, spruce, and larch.

Net-veined A term indicating a network of winding veins, as in a leaf; reticulate.

Node The point at which a leaf or leaves is attached to a twig, or a branch or branches to the main bole of a tree.

Nodule A swelling; e.g., a root nodule formed when nitrogen-fixing bacteria form a symbiotic relationship with the roots of a legume such as black locust.

Nut A fruit in which the seed is protected by a dry, hard, indehiscent shell.

Ob- A prefix meaning "the reverse of".

Oblanceolate A term applied to a lance-shaped leaf blade with the petiole attached at the narrow end; compare with *Lanceolate.*

Oblique A term used to describe a leaf base with unequal sides; refer to Inequilateral.

Oblong A term used to describe a leaf that is two to four times longer than broad, with nearly parallel sides.

Obovate Egg-shaped, with the point of attachment at the narrow end; compare with *Ovate.*

Obtuse Rounded or blunt at the apex, with the margins meeting at an angle of greater than 90°.

Odd-pinnate Pinnately compound (leaf) with a terminal leaflet.

Odoriferous Having a distinct odor.

Opposite In positions across from each other at a node; e.g., two leaves on a twig or two leaflets on a rachis.

Opulifolius Leaves similar to *Viburnum opulus*

Orbicular Nearly circular in outline.

Organ A plant part composed of several tissues and having a specific function; e.g., leaf, root, twig, bud.

Oval Broady elliptical, the length about twice the width.

Ovary The lower expanded portion of the pistil that contains the seed.

Ovate Egg-shaped, with the point of attachment at the broad end; compare with *Obovate.*

Ovule The structure that becomes the seed after fertilization; the megasporangium and integuments of a seed plant.

Palmate Initiating from a central point and spreading like fingers on a hand; lobe arrangement in maple or leaflet arrangement in buckeye; digitate.

Palmate-pinnate A term indicating a type of venation in which several pirmary veins arise from a central point with secondary veins pinnately distributed from the large midrib.

Paludal Of the marsh.

Paludify To become marshy, wet.

Pan A term meaning entire or whole; pantropical, pantemperate.

Panicle A branched, elongated raceme with flowers developing in sequence from the base to the apex.

Papilionaceous A term used to describe a butterfly-like flower with five petals: a banner, two wings, and a keel fused from two petals.

Papilla (plural *papillae*) A short bump or projection.

Parallel venation An arrangement of multiple primary veins in a leaf, running nearly parallel to each other from the base to the apex; common in monocots. Compare with *Net-veined.*

Parted A term indicating an extremely narrow sinus-like split occurring in a leaf blade and extending to a point more than one half the distance from the margin to the midrib. Compare with *Cleft.*

Parthenocarpy Development of a fruit without fertilization; agamospermy.

Parthenogenesis Development from an egg without fertilization; parthenocarpy.

Pedicel The stalk of a single flower within an inflorescence. Compare with *Peduncle.*

Peduncle The stalk that attaches an individual flower or an inflorescence to a twig.

Peltate Mushroom- or shield-shaped; a flattened cap attached to a stalk; e.g., peltate cone scales of bald cypress.

Pendent Hanging down.

Penniveined A term indicating parallel secondary veins that arise from the leaf midrib and extend to the margin.

Pentagonal Five-angled.

Pentamerous A term indicating a set of five; e.g., stamens in a flower.

Perennial A plant that lives longer than 2 years.

Perfect flower A flower that has both anthers (male reproductive organ) and a pistil (female reproductive organ); a hermaphrodite.

Perfoliate Having two sessile leaves connate at the base and encircling the twig; the twig appears to perforate the middle of a single leaf.

Perianth Collectively, the flower calyx and corolla.

Pericarp The wall of a fruit. In a drupe it is composed of three tissues: exocarp, mesocarp, and endocarp.

Persistent Contining to remain attached; e.g., leaves of evergreens.

Petal The usually colored, showy segment of a flower; a modified leaf.

Petiole The stalk attaching a simple leaf blade to the twig or, in a compound leaf, attaching the rachis to the twig. See *Rachis.*

Petiolate With a petiole. Compare with *Sessile.*

Petiolule A leaflet stalk.

Phloem The tissue through which food flows from the leaves to the roots; it is produced outside the vascular cambium. Refer to *Bark* and *Cork Cambium.*

Phreatophyte A plant closely associated with streams and lakes; a plant rooted in saturated or waterlogged soil.

Phyllotaxy The manner in which leaves are arranged on a twig.

Pilose Covered with soft, long hair.

Pinna (plural pinnae) A leaflike unit of a compound leaf. See *Leaflet.*

Pinnate A term used to describe a compound leaf with leaflets arranged (usually in opposite fashion) along the rachis.

Pinnatilobate A term indicating pinnately arranged lobes, as in a simple leaf.

Pinnule The leaflike unit of a compound leaflet (pinna) in a bipinnately compound leaf; a secondary pinna.

Pistil The female (seed-bearing) organ of a flower; consists of a stigma, a style and an ovary.

Pistillate flower A flower in which only a pistil or pistils are present; a unisexual flower. Compare with *Staminate flower.*

Pith The nonwoody central cyclinder of a twig; it usually is composed of spongy parenchyma cells and continuous in the first- and second-year stems. Also see *Chambered, Diaphragmed,* and *Excavated pith.*

Plane A flat surface.

Pocosin An evergreen shrub-dominated bog found on the Atlantic Coastal Plain. The word *pocosin* comes from an Algonquin phrase, "swamp on a hill."

Pod See *Legume.*

Pollen Mature microspores (male reproductive cells or grains) produced and released from the microsporangium of a gymnosperm or from the anther of a angiosperm.

Pollination The transfer of pollen to the pistil stigma by various vectors, such as wind, water, insects, or animals.

Polygamo-dioecious Polygamous but primarily dioecious. A species in which an individual plant (tree) may have bisexual (perfect) flowers and unisexual (either staminate or pistillate) flower types.

Polygamo-monoecious Polygamous but primarily monoecious. A species in which an individual plant (tree) may have bisexual (perfect) flowers and both unisexual (staminate and pistillate) flower types.

Polygamous A term used to describe a species in which an individual plant (tree) may have both bisexual (perfect) and unisexual (imperfect) flower types.

Polypetalous A term used to describe a flower with individual petals; petals not fused. Compare with *Gamopetalous* or *Sympetalous.*

Polyploid A plant or species with more than two sets of chromosomes in each cell.

Polystichous Having many ranks or rows.

Pome A fleshy fruit in which the modified floral tube surrounds a central core; e.g., apple.

Precocious Developing early in the season (flowers before the leaves; e.g., redbud and flowering dogwood) or at an early age (jack pine producing cones at age 5 to 7).

Prickle A slender, sharp outgrowth of the epidermis or from cone scales. Compare with *Spine* and *Thorn.*

Primary First and largest; e.g., the primary vein of a leaf, the midrib.

Procumbent A term used to describe the stem of a plant lying flat on the ground but without rooting developing along the stem; prostrate. Also called trailing.

Prominent A term used to describe an easily visible projection or protuberance, distinctly rising from a surface.

Prostrate See *Procumbent.*

Protuberance A rounded swelling abruptly rising from a surface.

Proximal Toward the base or the end of an attachment. Compare with *Distal.*

Puberulent A term used to describe a surface with a few scattered hair.

Pubescence A covering of short, soft hair.

Pubescent A term used to describe a surface covered with dense soft, short hair.

Punctate Dotted with minute pits, depressions, or translucent glands.

Pungent Sharp; used in the sense of a sharp thorn or disagreeable odor.

Quadrangular Four-sided or-angled.

Raceme A simple indeterminate inflorescence in which the individual flowers are borne on pedicels of about equal length and are evenly distributed along the central axis; the flowers develop in sequence from the base to the apex.

Racemose Resembling a raceme.

Rachis The primary stalk of a compound leaf from the first basal leaflet to the apex.

Radical Associated with roots.

Radicle The primary root first appearing from the seed.

Range The geographic distribution of a species.

Ranked Positioned in line, as in a row; e.g., two horizontal rows of leaves on a twig.

Receptacle The expanded end of the pedicel that supports the flower parts.

Recurved Bent or curved backward.

Reflexed Abruptly bent back or downward; e.g., reflexed leaves of eastern hemlock are bent backward to expose the abaxial surface.

Regular See *Actinomorphic.*

Remote Spaced at a distance; e.g., remotely serrate leaf of beech in which the teeth are spaced at 1 cm intervals along the margin edge.

Reniform Kidney-shaped; e.g., leaf of eastern redbud.

Repand Having a slightly uneven, wavy, or slightly sinuate margin.

Resinous Producing resin; sticky.

Reticulate venation See *Net-veined venation.* Compare with *Parallel venation.*

Retrorse Directed downward; e.g., spines pointed down the stem toward the base rather than toward the apex.

Revolute Margin rolled under; in a leaf, rolled toward the abaxial side; e.g., southern magnolia.

Rhizome A horizontal subterranean stem.

Rhombic Diamond-shaped figure; a four-sided equilateral figure with oblique lateral angles.

Rhomboid; rhomboidal A four-sided equilateral figure with obtuse lateral angles.

Riparian Associated with or growing along the banks of a river or spring.

Root The below-ground portion of a plant.

Rosette A dense cluster of leaves radiating from a central point on a twig or from the base of a plant.

Ruderal Growing in a disturbed environment; e.g., fields, roadsides.

Rufus A reddish brown color.

Rugose Wrinkled; e.g., the wrinked surface of a leaf due to sunken veins.

Samara A dry, indehiscent, winged fruit; e.g., fruit of ash.

Sap Plant fluid; clear in most plants, milky latex as in the mulberry family.

Sapwood The outer few living growth layers of xylem through which water is transported to the leaves. Compare with *Heartwood.*

Saturated forest A forest in which surface water is seldom present, but substrate is saturated to the surface for extended periods during the growing season.

Scaberulose Slightly rough to the touch because of the presence of short, stiff hair.

Scabrous Strongly rough to the touch because of stiff hair.

Scar See *Leaf scar* and *Bud scale scar.*

Schizocarp A dry, indehiscent fruit that splits at maturity; e.g., the double samara of maple.

Scurfy Covered with small scales.

Seasonally flooded A term used to describe a site in which surface water is present for extended periods during the growing season, but is absent by the end of the growing season in most years. The depth to the water table after flooding ceases is variable.

Seed A ripened ovule.

Semipermanently flooded A term used to describe a site in which surface water persists throughout the growing season except during drought years.

Sepal An element of the outermost floral envelope.

Septate Divided by a partition or septum; e.g., the capsule of catalpa is divided internally by a septum.

Septum A dividing wall or partition as in a multilocule ovary or capsule.

Seral community An assemblage of species, repeated across the landscape, but unable to maintain itself over a long period of time. In the absence of disturbance will change to another assemblage of species, often as a result of competition, e.g., for light.

Serotinous Late in flowering or leafing out; e.g., the flowers of redbud and dogwood appear before the leaves.

Serrate Having medium to large sawlike teeth.

Serrulate Having minute teeth.

Sessile A term used to describe a leaf, flower, or fruit which lacks a petiole, peduncle, or stalk.

Shade tolerance The ability of a plant to grow and develop in low light conditions.

Sheath The portion of the leaf base that surrounds the stem. See *Fascicle sheath.*

Shoot The above ground part of a plant consisting of a stem that bears leaves, buds, thorns, flowers, and fruit.

Shrub A short (height < 8m), multistemmed woody plant.

Silky Having long, soft, straight, silvery, lustrous, appressed hair.

Simple Undivided; a leaf with a single undivided blade.

Sinuate A term used to describe a deeply wavy margin. Compare with *Repand.*

Sinus The space between two lobes of a leaf.

Slough A shallow swamp or marsh where water is essentially stagnant.

Smooth A term used to describe a surface lacking hair or a leaf margin without teeth or spines.

Spermatophyte A seed-producing plant.

Spherical See *Globose.*

Spike An unbranched, elongated inflorescence of sessile unisexual flowers.

Spine A stiff, sharp, more or less woody outgrowth from the stem; usually a modified stipule or leaf.

Spinose Bearing spines.

Spongy pith A continuous pith composed of soft parenchyma cells. See *Diaphragmed, Chambered,* and *Excavated pith.*

Sporophyll A modified leaf that produces spores; i.e., pollen or ovules.

Spray A twig with leaves and flowers or fruit.

Spreading Extending to a nearly horizontal position; e.g., a shrub with widely extended branches.

Stalk See *Petiole* or *Peduncle.*

Stamen One of the male reproductive (pollen-producing) organs of a flower; it consists of an anther and a filament.

Staminate flower A flower in which only stamens are present; a unisexual flower. See *Imperfect flower.* Compare with *Pistillate flower.*

Standard The large, upper petal in a papilionaceous flower (legume).

Stellate A term used to describe a branched hair in which the branches radiate from a central point; star-shaped.

Stem The above ground part of a plant that bears leaves, buds, thorns, flowers, and fruit. Compare with *Shoot.*

Sterile Incapable of producing offspring; infertile.

Stigma The top part of the pistil that receives pollen.

Stipular Pertaining to stipules.

Stipulate Having stipules.

Stipule A leaflike appendage attached to each side of the twig and beside the bud and petiole.

Stipule scar The mark produced when the stipule falls off the twig. There may be two scars (one each side of the bud), one continuous scar encircling the twig, or no scar in the case of estipulate species.

Stolon A basal branch that roots, such as the tip of a raspberry or blackberry cane. Also defined as a horizontal creeping stem.

Stoloniferous Rooting at the tip.

Stomate An opening in the epidermis of the leaf that permits the exchange of oxygen and carbon dioxide for photosynthesis.

Striate Covered with fine parallel longitudinal lines.

Strobilus A clublike cluster of microsporophylls (pollen producing) or a conelike cluster of megasporophylls (ovule producing); found in gymnosperms.

Style The narrow section of the pistil between the stigma and the ovary.

Sub- A prefix meaning below or less than; e.g., subtending, or attached below.

Subalpine In mountains, the zone with an upper boundary marked by the treeline and directly below the alpine zone.

Subapical Directly below the apex.

Subcordate Nearly cordate.

Subshrub See *Suffrutescent.*

Subtropical zone Almost tropical forest zones; e.g., southern tip of Florida, which has numerous tropical plant species and few temperate species. Compare with *Temperate zone* and *Tropical zone.*

Subulate Awl-Shaped.

Succession The natural change in populations resulting in a different community. Such change is generally brought about through competition e.g., for light.

Suffrutescent Almost a shrub; a small shrub; a subshrub.

Suture A seam or line of dehiscence.

Swamp A wetland dominated by trees and shrubs.

Symmetrical Having equal halves, as in comparing two sides of a leaf blade split along the midrib or the shape of a bud split along the midline; mirror image.

Sympatric A term used to indicate species that have the same or nearly the same geographic range. Compare with *Allopatric.*

Sympetalous A term used to describe a flower with united petals; gamopetalous. Compare with *Polypetalous.*

Syn-, sym- A prefix meaning united or together.

Synoecious Having both staminate and pistillate flowers.

Taproot A large descending root that grows vertically to penetrate the soil; common to trees such as pine, oak, hickory, and walnut, soil conditions permitting.

Taxon (plural Taxa) A taxonomic division or unit such as an order, family, genus, or species.

Temperate In terms of climate, generally, neither extremely hot nor cold.

Temperate zone In the Northern Hemisphere, that area between the Tropic of Cancer and the area of the United States extending from central Florida to southern Canada. Compare with *Boreal forest zone, Tropical zone,* and *Subtropical zone.*

Temporarily flooded A term used to describe a site where surface water is present for brief periods during the growing season, but whose water table usually lies well below the soil surface. Often characterizes floodplain wetlands.

Tendril A modified petiole or slender twig providing support by twining or clasping; usually produced by vines.

Tepel A unit of the flower perianth not identifiable as a petal or sepal; e.g., in *Magnolia* flowers.

Terminal At the end or apex.

Ternate Appearing in groups of three; e.g., a trifoliate leaf.

Terrestrial In ecology, a reference to dry, upland areas as opposed to lakes and oceans. Compare with *Aquatic.*

Thorn A rigid, woody, sharp-pointed modified twig. Compare with *Spine* or *Prickle.*

Tissue An organized cluster of cells that performs a specific function or process; e.g., leaf mesophyll or palisade tissue containing chlorophyll for photosynthesis.

Tolerance The ability of a plant to grow and develop under the influence of a given set of limiting conditions. See *Shade tolerance.*

Tomentose Having long, soft, intertwining or matted hair.

Tooth A single projection of a serrated or toothed margin.

Toothed margin A leaf margin bearing large teeth; each tooth about 1 cm across.

Tracheid The type of xylem cell usually composing conifer wood; the cell is long, slender, tapered at both ends, and thick walled and is used for support and strength as well as for water transport. Found primarily in conifers.

Trailing See *Procumbent.*

Transpiration Loss of water from a plant primarily through the stomata of the leaf surface.

Tri- A prefix meaning three

Trichome A hair-or spinelike outgrowth from an epidermal cell.

Trifid Three-cleft.

Trifoliate A term used to describe a leaf of three leaflets.

Trimerous Having a set of three; e.g., stamens in a flower.

Tripetalous A term used to describe a flower with three petals.

Tropical zone The area between the Tropic of Cancer and the Tropic of Capricorn; the area of hot, moist climate supporting tropical forest.

Truncate Ending abruptly; appearing to be cut off.

Trunk The bole or main stem of a tree.

Twig The smallest woody axis of a tree or shrub to which leaves and flowers are attached.

Twining Encircling another stem or object for support and climbing.

Two-ranked In two rows; a term usually applied to two opposite horizontal rows of leaves, as found in hemlock and fir.

Type specimen The herbarium specimen used to describe and name a species.

Umbel A somewhat flat- to round-topped inflorescence in which the pedicels are attached to the tip of a common peduncle.

Umbo The projection from an apophysis of a cone scale; boss.

Unarmed An absence of thorns, spines, or prickles; e.g., unarmed white pine cone scale.

Undulate Having a somewhat wavy, but less than sinuate, margin. See *Repand.*

Uni- A prefix meaning one.

Unisexual flower A flower with either stamens or a pistil. See *Staminate, Pistillate,* or *Imperfect flower.*

Valvate Opening by means of a valve or similar mechanism, as in a capsule or anther; also describes the meeting of the bud scales or flower sepals along the edges before opening; compare with *Imbricate.*

Variegated Characterized by a mixture of different colors; in green plant leaves, a pattern of light yellow areas often along the veins or blade margin.

Vascular bundle A strand consisting of both xylem and phloem conducting tissue visible in the leaf or flower as a vein.

Vascular bundle scar The scar formed within the area of the leaf scar that represents the ends of xylem and phloem conductive tissue.

Vascular cambium The actively growing lateral meristem in a tree that produces xylem (wood for water transport) to the inside and phloem (bark for food transport) to the outside. Compare with *Cork cambium* and *Apical meristem.*

Vegetative Pertaining to the nonreproductive parts of a plant: roots, stem, leaves, branches, twigs.

Vegetative reproduction The development of new individuals from vegetative plant parts.

Vein See *Vascular bundle.*

Velutinous Covered with velvet-like short, soft hair.

Venation The arrangement or pattern of veins within a leaf.

Ventral See *Adaxial.*

Vernal Associated with spring.

Vernation The folding of leaves within the bud.

Verticillate Term used to describe three or more leaves, twigs, or branches attached at a node; whorled. Compare with *Alternate* or *Opposite.*

Vessel A short, broad, thick-walled xylen cell having a perforation plate at both ends. Its primary function is transport of water within the woody plant. A common type of cell found in angiosperm wood. Compare with *Fiber* and *Tracheid.*

Vestigial An organ or tissue that is greatly reduced compared to a normal organ and is more or less nonfunctional.

Vine A plant whose main stem is procumbent, climbs, or twines.

Wart A corky swelling or bulge; e.g., the warty bark of hackberry.

Whorled See *Verticillate.*

Wing A thin, flat structure.

Winged Having wings; e.g., fruit of maple, twigs of winged elm or sweet gum.

Wooly A term used to describe extremely, dense, soft, usually tangled or matted hair.

x A hybrid designation when placed between genus and species names.

Xanthophyll The yellow, orange, and red pigments found in the leaf cells.

Xeric Extremely dry, a reference to site conditions.

Xeric-mesic An intermediate soil moisture conditions between xeric (dry) and mesic (moist).

Xeromorphic The form of a green plant, modified and adapted for surviving in an arid or xeric environment, including succulent stems for water storage and early dormancy with leaf drop to reduce transpiration.

Xerophyte A plant adapted to xeric or dry environments.

Xylem The wood of trees and shrubs; the water-conducting tissue of plants. Refer to *Sapwood* and *Heartwood.* Compare with *Phloem.*

Zygomorphic Formed so that two equal halves (mirror images) occur along only one plane; bilaterally symmetrical; irregular. Compare with actinomorphic.

Glossary of Genus and Species Names

abies (L) Fir-like.
Abies (L) Classical name for fir.
abramsiana Named for Abrams.
Acacia Greek name for an Egyptian species.
Acer (L) Aboriginal name for maple.
Acoelorraphe From the word *acoelo* (without) and *raphe* (a stalk); not having a true free stalk for the seed.
acumin (L) Point.
acuminata (L) Possessing an acumin or sharp point.
acut (L) Sharp, a small angle.
acutissima (L) Sharp-tooth leaves, as in sawtooth oak.
aden (G) Gland.
adenostroma (G) Large glands.
Aesculus Ancient name for a mast-bearing tree.
agri (L) Wild, growing in a field.
agrifolia (L) Wild-looking leaves.
Ailanthus A Moluccan common word meaning tree-of-heaven.
alat, alata (L) Wing.
alatus (L) Winged.
alba (L) White.
albertiana Of Alberta.
albicaulis (L) White bark or stem.
albidum (L) Whitish.
Albizzia Named for Filippo degli Albizzi, who introduced the genus into Europe in the 1700s. Corrected from *Albizia* by Bentham.
-alis, -ale (L) Pertaining to, used to form an adjective.
alleghaniensis Of the *Alleghenies.*
alni (L) Alder-like.
alnifolia (L) Foliage or leaf like an alder.
alnus (L) Alder.
Alnus (L) Ancient name for alder.
altern (L) To follow in succession; root of *alternate.*
alternifolia (L) Leaves arranged in an alternate fashion on a twig.
alti (L) Tall.
altissima (L) Rapidly becoming tall.
amabil (L) Lovely.
amabilis (L) Graceful.
ambar (Arabic) Amber.
ambigu, ambigua (L) Doubtful, uncertain.
Amelanchier Provincial French name for a European species.
americana, americanum, americanus Of America.
amomum (L) Latin name of a shrub.
Amorpha Genus in which four petals are missing; deformed flowers.
amorphos (G) Deformed.
Ampelopsis From the word *ampelos* (vine) and *opsis* (appearance); vine-like.
ampelos (G) Vine.
amurense; amurensis From the Amur River area of northern China and Siberia.
amygdal (G) Almond, peach.
amygdaloides, amygdalas (G) Resembling an almond or peach.
Anacardium Genus name for cashew.
Andromedah (G) Named (by Linnaeus) for Andromeda of mythology.
angust (L) Narrow.
angustifolia (L) Narrow-leaved, long, tapering.
Annona From the Hispaniolan *anon* or *hanon.*

apo (L) Away, different.
Apocarya (L,G) Different from true hickory or Eucarya.
aqua (L) Water.
aquatica (L) Growing in or near water.
aquifolium (L) Water-leaf.
arbor (L) Tree.
arborescens (L) Becoming tree-like.
arboreum (L) Tree-like.
Arbutus (L) Strawberry tree.
Arceuthobium (G) From the words *arceuthos* (juniper) and *bios* (life); living on or a parasite of juniper and other conifers.
Arctostaphylos (G) From *arctos* (bear) and *staphyle* (a bunch of grapes); bear grape.
arguta sharp sharp toothed.
arista (L) A bristle.
aristata (L) Bristly.
arizonica Of Arizona.
arkansana Of Arkansas.
armeniaca Of Armenia.
aroma (G) Odor.
aromatica (G) Fragrant; aromatic.
ashei Named for William Ashe (1872–1932), chair of Forest Service Tree Name Committee.
Asimina From the Indian name *assimin*;
asper (L) Rough.
asperata (L) Rough twigs.
asplenia A fern.
asplenifolia Having fernlike leaves.
atro (L) Terrible; bad.
atropurpureus (L) Dark purple.
attenuata (L) Narrow; thin.
aucaparia (L) Attractive to birds.
aucup (L) Bird-catching.
aurantifolia (L) Golden-leaved.
aurantium (L) Golden.
aurea, aureum (L) Golden.
auri (L) Gold (metal color).
australis (L) Southern, of the south
Avalia (Fr.-Cana.) From *Avalie*, the original name assigned by a Quebec physician.
avi (L) Bird.
avium (L) Of birds.
babylonica Of Babylon.
bacca (L) Berry.
baccata (L) Berrylike.
bakeri Named for Baker.
balanus (G) Acorn.
balforiana Named for John Balfour (1808–1884), a Scottish doctor and botanist, director of the Royal Botanical Gardens, Edinburgh.
balsam (G) Balsam, an aromatic resin of the balsam tree.
balsamea (G) Balsam-producing.
balsamifera (G) Possessing balsam; in balsam poplar and balsam fir the buds are resinous.
banksiana Named for Sir Joseph Banks (1743–1820).
barbat (L) Beard.
barbatum (L) Having a beard (in maple at the summit of the flower).
bartramiana Named for William Bartram (1739–1823), American botanist and ornithologist.
Bebbiana Named for Michael Bebb (1833–1895), an American expert on the genus.
benzoin An old name for a genus of the Lauraceae.
berberis From the Arabic name *berberys*, the name of the fruit.
Berchemia Named for Berthout van Berchem, an eighteen-century Dutch botanist.
berlandierana Named for Berlandier.
besseyi Named for Charles Edwin Bessey (1845–1915).
Betula (L) Ancient name for birch.
betulus, betuloid Birchlike.
betulifolia (L) Having leaves like a birch.
bi (L) Two.
bicolor (L) Two-colored; in swamp white oak, a reference to the upper and lower leaf surfaces.
bidwillii Named for Bidwill.
biflora (L) Two-flowered.
Bignonia Named for the Abb'e Jean Paul Bignon (1662–1743), a friend of Tournefort, an early botanist.
bignoniodes Appearing like Bignonia.
biloba (L) Two-lobed.
bland (L) Mild.
blanda (L) With reference to *Rosa blanda*, the species lacks prickles.
bolanderi Named for Bolander.
bona-nox (S) Good night; from the Spanish *buenas noches*.
borbonia The old name for *Persea*.
bract (L) A modified leaf.
bracteata (L) With bracts (subtending the cone scales).
brevi (L) Short, abreviated.
brevifolia (L) Short-leaved.

breweriana Named for Brewer.
Broussonetia Named for Auguste Broussonet (1761–1807), a Montpellier physician.
Buckleya Named for Samual Buckley (1809–1884), an American botanist.
Bumelia (G) Aboriginal name for ash.
bunda Many.
Caesalpinia After Andreas Cassalpini (1519–1603), an Italian botanist.
californica Of California.
Calocedrus (G) From the words *calo* (beautiful) and *cedrus* (cedar).
calyc (G) Calyx.
calyco (G) From *calyx*, a cup.
Calycocarpum (G) From the words *calyx* (cap) and *carpus* (fruit): a fruit shaped like a cap.
calyculata (G) Possessing an outer calyx.
campestre (L) Found in flat areas or plains.
camphora Camphor.
Campsis (G) A reference to the curved stamens of the trumpet flower.
canadensis (L) Of Canada.
canbyi Named after Wiliam Marriott Canby (1831–1904).
canin (L) Dog.
canina (L) Of a dog; especially one that bites; nasty.
canthos (L) Thorn.
capreol (L) A tendril or support.
caribaea Of the Carribbean.
carica Named for Caria in Asia Minor.
carolinian (L) Of Carolina, Carolinian.
Carpinus (L) Hornbeam, an early name.
carpusa (G) A fruit.
cary (G) Nut.
Carya (G) Hickory; an ancient name of the walnut.
cassine After Count Henri de Cassini (1781–1832), an Italian botanist.
castana (G) Chestnut.
Castance An ancient Latin name taken from *castana.*
Castanea Ancient Latin name taken from the Greek.
Castanopsis (L) Chestnut-like; chinkapin.
Casuarina Drooping branches resemble the feathers of a cassowary bird, *Casuarnius.*
Catalpa (L) Aboriginal name.
catawbiense Named for the Catawba River, North Carolina.
cathart (G) Cleansing.
cathartica (G) Having cleansing properties.
caulis (G) Stem.
Ceanothus (G) Name for a Spiny shrub.
Cedrus (G) Cedar.
celastrina (G) Celastris-like.
Celastrus (G) Aboriginal name for an evergreen tree.
cembra An old name for the arolla pine or Swiss stone pine (*Pinus cembra*).
cembroides Resembling *Pinus cembra.*
cephala (G) A head.
ceras, cerasus (L) Classical name for cherry.
cerc (G) A type of poplar.
Cercidium (G) From *Kerkis*, Judas-tree
Cercis Ancient name for the Judas-tree (oriental).
cercocarpus (G) *Kerkos*, tail, and *karpos*, fruit, refers to the plums-like style.
cereum (L) Waxy, wax-bearing.
ceri (L) Wax.
Chamaecypress (G) From the words *chamae* (on the ground) and *cypress*; dwarf cypress.
Chamaedaphyne (G) From *chamae* (on the ground) and *daphne* (laurel); ground laurel.
chapmanii Named after Chapman.
chihuahuana Named for Chihuahua.
Chionathus (G) From *chion* (snow) and *anthos* (flower); the genus has snow white flowers.
chryso (L) Gold.
Chrysolepis (L, G) *chrysos*, gold, lepis, scale; golden scales; canyon live oak.
chrysophylla (L, G) Golden leaves.
cimer (L) Ashes.
cimerea (L) Ash-colored, the bark of butternut.
cinnamomum (L) Cinnamon-colored.
circi (L) Circle.
circinatum (L) Coiled from the tip downward, as when a leaf unrolls from the bud.
cissos (G) Ivy.
Cissus (G) Name for ivy.
Citrus (L) Name for citran, a *Citrus* used in medicine.
clados,cladus (G) Branch.
Cladrastis (G) From the words *clados* (branch) and *thraustos* (brittle); easily broken.
clausa Closed; closed cones of *Pinus clausa.*
Cliftonia Named for william Clifton, who collected this plant about 1764.
clove-herculis (L) Literally, Hercules'-club.
coccin (L) Scarlet.
coccinea (L) In scarlet oak, a reference to the leaf petioles in mid to late summer.

Cocculus (G) From the world *coccus*, a berry.
columna (L) Pillar or column.
columnaris (L) Pillarlike or cylindrical.
common (L) Common.
communis (L) In clumps.
communitata (L) Changing.
contort (L) To twist.
contorta (L) Twisted, short.
copal (Mex) Blunt.
copallina Reference to blunt, round lenticels.
cordata (L) Heart-shaped.
cordi (L) Heart.
cordiformis (L) Heart-shaped; the nut of bitternut hickory.
coria (L) Leather.
coriacea (L) Leatherlike (leaves).
coron (L) Crown.
coronaria (L) Suitable for a crown or wreath.
Cornus (L) From the word *corn*, which means a horn; probably a reference to the extreme hardiness of the wood; wood from a European species was used as a skewer. (See Fernald 1950, for more detail.)
cornut (L) Horned.
cornuta (L) Beaked; in hazel, a reference to the elongated involucre or husk of the fruit.
Corylus (G) Hazel; a reference to the protective involucre.
corymb (G) A type of inflorescense.
corymbosa (G) Bearing a corymb.
corys (G) A helmet.
Cotinus (G) Wild olive.
Cotoneaster (L) From the name *cotonea*, for quince, and aster, meaning "a type of."
coulteri Named for John Coulter, a western botanist of the late 1800s.
Cowania For James Cowan, an English merchant who brought plants from Central and South America. Died in 1823.
crass (L) Thick.
crassifolia (L) Thick-leaved.
Crataegus (G) From the name *Kratos*, meaning strength.
crenata, crenulata (L) Bearing rounded teeth.
crenate (L) Dentate with blunt teeth.
crocea, croceus (G) (Rhamnus) yellow, saffron
Cryptomeria (G) From the words *crypt* (hidden) and *meri* (part); hidden parts.
cune (L) Wedge.
cuneatus (L) Wedge-shaped.
cunninghamiana, cunninghamii: Named for Cunningham.
Cupressus (L) cypress.
cuspid (L) A point.
cuspidata (L) Pointed.
cyno- (G) Dog.
Cyrilla (G) Named for Pomenico Cirillo (1734–1799), a professor of medicine at Naples.
davurica Of Dahuria (Asia).
de (L) Down, out.
decid (L) Falling off.
decidu (L) Dropping off.
decidua (L) Deciduous leaves.
decurren (L) From the words *de* (down) and *curren* (running): decurrent, branching down and outward.
delt (G) Triangular, like the letter delta.
deltoid (G) Similar to a equilateral triangle.
deltoides (G) Leaves shaped like a delta orisosoles triangle.
dendron (G) Tree.
dens (L) Dense, thick.
densa (L) Dense.
densiflora; densiflorous (L) Thick foliage.
dent (L) Tooth.
dentata (L) Toothed.
depressa Depressed, low to the ground.
derm (G) Skin.
di (G) Two.
dioicus (G) Dioecious; male and female flowers borne on different trees.
Diospyros (G) From the words *dios* (noble and devine) and *pyros* (grain); beautiful wood, similar to ebony.
dipetala (G) Two-petaled.
diptera (G) Two-winged.
dis (G) Double.
discolor (G) Two colors.
disso (G) Double.
dist (L) Distant, separate.
distichophylla (G) Leaves two-ranked on the twig.
distichum Two-ranked (leaves).
diversi (L) Various, many.
diversiflorus (L) Variously flowered.
diversifolia (L) Variously leaved.
diversilobum (L,G,) Variously lobed.
douglasii Named for David Douglas (1798–1934), a Scottish botanical explorer of North America; described many oak and pine species.

drummondii Named for Thomas Drummond (1980–1835).
dum (L) A bramble.
dumosa (L) Bramble-like.
dunnii Named after Dunn.
durandii Named after Durand.
echin (G) Sea urchin which is spiny.
echinata (G) Armed with spines or prickles; the scale of a pine cone.
editorum Of highlands.
eglanteria From the old English name Eglantine.
Elaeagnus (G) From the words *elaeo* (olive) and *agno* (meaning pure).
elastica (L) Elastic.
elata (L) Tall or high.
Elliottia Named after Elliott.
elliottii Named for Stephen Elliott (1771–1830).
ellips (G) Elliptical, not round.
ellipsoidalis In Hill's oak, a reference to ellipsis-shaped acorns.
elliptica (G) Elliptical, not round.
em (L) Into.
emarginata (L) Into the margin.
emoryi Named after Emory.
engelmannii Named after Engelmann.
Ephedra (G) Named for a similar plant
Epigaea (G) From *epi* (upon) and *gaea* (earth); on the ground.
equi (L) Of equal units, pertaining to a horse.
equisetifolia (L) Similar to Equisetum, horsetail.
erica (G) Heath.
erytho (G) Red.
erythobalanus (G) Red oaks with bitter acorns.
Erythrina (G) Red; refering to flower color.
eu (G) Good, true.
Eucarya (G) True hickory.
Euonymus (G) Aboriginal name; from *eu* (good) and *onoma* (name); used as a tongue-in-cheek word as the plant was known to be poisonous. (Fernald 1950).
europaea (L) Of Europe.
excelsior (L) High or lofty.
exigua (L) Small, little.
faba (L) Bean.
fagara (L) Beechlike.
Fagus (L) Beech; from the Greek root *phag*, (eat); a reference to the nuts.
falcata (L) In Southern red oak, a reference to the sickle-like leaf or lobes.
falcate (L) Sickle-like.
fendleri Named for Fendler.
feruginea (L) Rusty; light brown with a tinge of red.
Ficus (L) An ancient name for fig.
fili (L) Thread.
filiformis (L) Threadlike.
firma (L) Strong.
flau (L) Yellow.
flauescens (L) Yellowish.
flex (L) Bend.
flexilis (L) Bending, flexible.
flora (L) Flower.
floribunda (L) Many flowered.
florida, floridana, floridanum profusely flowering
florum pertaining to the flower; flowering.
folia, folium (L) Referring to a leaf.
forbesii Named for Forbes.
Forestiera Named for Forestier, a physican of Saint-Quentin, France, and a teacher of botany; he died in 1820.
forma (L) Shape, outline, form.
Forsythia Named for William Forsyth (1737–1804).
Fothergilla Named for John Fothergill (1712–1780), a well-known London physician and botanist.
fragilis (L) Fragile, brittle, easily broken.
frang (L) Break.
frangula (L) An old generic name.
fraseri Named for John Fraser (1750–1811).
Fraxinus (L) Classical Latin name for ash.
fremontii Named for Fremont.
fusca (L) Brown, light brown.
fusi (L) Poured out, sprawling.
gale An old generic name.
gambelii Named after Gambel.
gardeni Named for Alexander Garden (~1730–1791).
Garrya Named for Nicholas Garry of the Hudson Bay Co.
garryana Named for Garry.
Gaultheria Named for Jean-François Gaulteria (1708–1756), a Quebec naturalist and physician.
Gaylussacia Named for Joseph-Louis Gay-Lussac (1778–1850), a French chemist.
georgiana Of Georgia.
gigan (G) Giant.
giganteum (G) Giant, huge.
Ginkgo An old Asian name.
glabr, glabrous, glabrata, glabra, glabrum (L) Smooth, without hair.
glandulosa (L) With glands.

glans (L) Gland.
glauc (G) Gray, bluish gray, or whitish gray.
glauca (G) With a whitish bloom.
glaucophylla (G) Blue-green-leaved.
Gleditsia Named for Johann Gottlieb Gleditsch, 1714–1786.
glora, glori, gloriasa (L) Glory.
glut (L) Glue.
glutinosa (L) Sticky, gummy.
gooddingii Named for Goodding.
goveniana Cupressus name for Robert Gowen (died 1862), a British horticulturalist.
gracili (L) Slender.
grandi, grandis (L) Large, magnificent.
grandidentata (L) Big-toothed.
grandiflora (L) Large-flowered.
grandifolia (L) Attractive leaves.
greggii Named for Gregg.
grise (L) Gray.
grisea (L) Grayish.
groenlandicum Of Greenland.
gymno (G) Naked.
gymnocarpa Naked fruit.
Gymnocladus Naked branch; being stout and prominent in winter.
haemato (G) Blood.
haemato carpa (G) Crimson-colored fruit.
Halesia Named for Stephen Hales (1677–1761), an exceptional plant physiologist.
Hamamelis An ancient Greek name.
haplo (G) One, single.
Haploxylon (G) From the words *haplo* and *xylum* (wood); one vascular strand.
harvardii Named after Harvard.
Hedera (L) Aboriginal name.
helix (G) Spiral or coil.
herba (L) Grass.
herbacea (L) Resembling an herb.
hetero (G) Different, various.
heterophylla (G) Many different leaf shapes.
hindsii Named after Hinds.
hippo (G) Horse.
hippocastanum (G) Aboriginal name meaning horse chestnut.
hirsute (L) With stiff, coarse hair.
hirsutum (L) Being hirsute.
hispid (L) Bristle.
hispida, hispidus (L) Bristly, hairy, hispid.
homo (G) Alike; resembling one another.
homolepis (G) Identical scales (on a cone).
hookeriana Named for Sir Joseph Hooker (1817–1911), who developed a taxonomic classification system with George Bentham.
horizontalis (G) Horizontal, close to the ground.
horrid (L) Prickly.
horridus (L) Prickly-like.
hort (L) Garden.
hortulana (L) Found in a garden.
hudsonianum Of Hudson Bay.
humi (L) Ground.
humifusa (L) Sprawling on the ground.
hydrangia (G) From the words *hydro* (water) and *aggeian* (a vessel); a capsule shaped like a "water vessel."
hydro (G) Water.
hypo (G) Below, under, beneath.
hypoleucoides (G) Whitish on the lower side, as on the leaf of *Q. hypoleucoides.*
idaeus Named for Mount Ida in Crete or in Turkey.
ifera (G) To possess or be similar to, to be able to.
Ilex (L) Aboriginal name for holly oak, not holly (Fernald 1950).
ilici (L) Entice; holly.
ilicifolia (L) Holly-leaved.
Illicium (L) From *illicio,* meaning to attract, a reference to flower fragrance.
illinoensis Of Illinois.
imbricaria (L) A shingle; shingle oak was used by early settlers for making shingles.
in (L) Without, not.
in (L) In.
indie (L) Indigo; Indian.
indiea (L) Indigo-colored.
inerme ( ) Unarmed.
inodora (L) Without an odor.
institia (L) Used in grafting.
integ (L) Entire, whole, complete.
integerrimus (L) Quite entire or complete.
inter (L) In the middle of.
interior (L) Between the coasts; inland.
ioensis Of Iowa.
irrig (L) Bring water to.
irriguum (L) Watered; growing by a stream.
issima (L) A superlative ending.
Itea (G) Willow.
japonica, japonicum Of Japan.
jeffreyii Named for Jeffrey.
juarezensis Named for Juarez, Mexico.

Juglans (L) Contraction for *Jupiter's glans*, a reference to the seeds on the walnut tree.
julibrissin A Persian name.
Juniperus (L) The ancient name for evergreen conifers.
Kalmia Named after Pehr Kalm (1716–1779), a student of Linnaeus who studied American flora.
kearneyi Named after Kearney.
kelloggii Named after Kellogg.
kentukea Of Kentucky.
koreana Of Korea.
koyami Of Koyama, Japan.
lacini (L) Peeling, a flap.
lacinosa, laciniaus (L) peeling or having flaps; torn, shaggy, slashed.
lacustre (L) Of lakes.
laevi (L) Smooth, light, agile.
laevigat (L) Smooth, slick, slippery.
laevigata (L) Being smooth.
laevis (L) Smooth (as of leaves).
lamarkiana Named for Jean-Baptiste Antoine de Monnet de Lamarck (1744–1829), a French botanist who wrote the flora of France; developed theory of acquired characteristics.
lambertiana Named for Aylmer Lambert (1761–1842), an English botanist who specialized in pines.
lance (L) A spear.
lanceolata (L) Spear-shaped.
lanug (L) Wool or down.
lanuginosa (L) Woolly.
laricina (L) Larch like; originally *Larix laricina* was thought to be a pine.
Larix (L) Classical name for a European species.
lasi (G) Hairy.
lasiandra (G) Hairy stamens.
lasio (G) Shaggy.
lasiocarpa (G) Shaggy cones.
lauri (L) Laurel.
laurifolia Laurel-like leaves.
laurina (L) Similar to laurel.
lawsoniana Named for Lawson.
Ledum (G) From *Ledon*, the name of a species of *Cistus* that produced a resin similar in odor to that of *Ledum*.
legumin (L) Legume, pod.
leio (G) Smooth.
leiophylla (G) Smooth-leaved.
Leitneria Named for E.F. Leitner, a German naturalist who was killed in Florida during the Seminole War.
lent (L) Thick.
lenta (L) Tough, flexible (twigs).
lepis (G) Scale.
lepto (G) Slender.
leptolepis (G) Slender-scaled.
Lespedeza Named for the Spanish governor of East Florida during the exploration of Michaux. According to Fernald (1950), the name Vincente Manuel de Cespedes was misspelled as de Lespedez.
Leucaena A reference to white.
leuco (G) White.
leucobalanus (G) White oak with sweet, edible acorns.
leucoderme (G) Light-colored skin or epidermis.
leucophala (G) White head.
Leucothoe Named after Leucothoe, the daughter of Orcharmus, king of Babylon.
Ligustrum (L) Classical name for privet.
lili (L) A lily.
limon, lemon (L) Meadow.
Lindera Named for John Linder (1676–1723), a Swedish botanist.
lindheimeri Named for Ferdinand Jacob Lindheimer, a botanist (1801–1879).
linea (L) A line.
linearis (L) Narrow, straight, like a line.
liqui, liquidus (L) Liquid.
Liquiambar (L) Amber-colored fluid.
lirio (G) Lily.
liriodendron (G) Tree leaf that resembles a lily.
litho (G) A stone; stone-like.
Lithocarpus (G) Stone-like fruit.
Litsea Of Chinese origin; meaning uncertain.
lob (G) Lobe.
lobata (L) Lobed as with oak leaves.
long (L) Long.
longaeua (L) Elongated, tall.
lucid (L) Clear, shinning.
lucida (L) Shinning, leaves in Salix.
lute (L) Yellow.
lutea (L) Yellowish.
lyalli ( ) Larix.
Lyonia, lyonii Named for John Lyon, an English botanist who explored North America prior to 1818.
Maclura Named after William Maclure (1763–1840), an American geologist.
macnabiana Named for McNab.
macro (G) Large, big.
macrocarpa (G) Large fruit.

macropetalis (G) Large-petaled.
macrophylla, macrophyllum (G) Large-leaved.
magnifera, magnifica (L) Large, tall.
magnolia Named for Pierre Magnol (1638–1715), a botany professor at Montpellier.
mahaleb An Arabic name.
major (L) Greater, larger.
Malus (G) An apple tree.
margin (L) Border, edge.
mari (L) The sea.
mariana (L) Of wet lands. Of Maryland, broadly interpreted to mean North America (according to Fernald 1950).
marilandica Of Maryland.
marsiesii Named for Marie.
maxim (L) Largest, greatest.
maximum (L) Greatest.
meni (G) Moon.
menispermum (G) From the words *meni* (moon) and *sperma* (seed); moon-shaped seed.
menziesii Named for Archibald Menzies (1754–1842), who first colected species from the northwest coast of North America.
mertensiana Named for Franz Karl Mertens (1764–1831), a German botanist.
mexicana Of Mexico.
michauxii Named after François André Michaux (1770–1855), a French botanist and explorer who first described the species.
micrantha Small-flowered.
micro (G) Small.
microcarpa (G) Small-fruited.
microphylla (G) Small-leaved.
Mimosoideae From mimosa, a leguminous tree.
minor (L) Small.
missouriense Of Missouri.
mohriana Named after Mohr.
molli (L) Soft.
mollissima (L) Soft-texured leaves.
mono (G) One.
monophylla (G) One-leaved.
monosperma (G) One-seeded (cone).
mont (L) Mountain.
montanus (L) Of the mountain.
monti, montana, montan (L) Mountain.
monticola, montana (L) Found on the mountain.
Morus (L) Classical name for mulberry.
mucro (L) Sharp point.
mucronata, mucronatus (L) Abrupt, sharp-pointed (leaf).
muhlenbergii Named for Gothhilf Heinrich Ernst Muhlenburg (1753–1815), a German-educated American botanist who cataloged native plants.
multi (L) Many.
multiflora (L) Many-flowered.
muricat (L) Point.
muricata (L) Pointed.
murrayara Named for Murray.
myrist (G) Ointment, anoint.
myristiciformis From the terms *myrio* (many) and *forma* (forms); many forms.
myrt (G) Myrtle.
myrtifolia Myrtle-like, small oval leaves.
nan (G) Dwarf.
nana (G) Dwarfed.
negundo (L) Aboriginal name for box elder.
Nemopanthus (G) A combination of *nema* (thread), *pous* (foot) and *anthos* (flower).
neo (G) New.
neo mexicana Of New Mexico.
nerv (L) Nerve
nervuosa (L) With many veins (nervus).
nevadense, nevadanensis Of Nevada.
nevenii Named for Neven.
nig (L) Dark.
nigra, nigrum (L) Black.
nootkatensis Of Nootka Sound, Vancouver Island, British Columbia.
nordmanniana Named for Nordmann.
nudi (L) Naked.
nudiflorum (L) Nude, flowers borne on a separate stalk, not the primary leaf-bearing stem.
nut (L) Sway, nod.
nutans (L) Nodding branchlets.
nutkana Of Nootka, Alaska.
nuttallii Named for Thomas Nuttall (1786–1857), an English botanist who pioneered American paleontology; curator of Harvard Botanical Gardens.
Nyssa (L) From mythology; the nurse of Baccus, god of wine.
ob (L) The opposite of.
oblongifolia (L) Oblong.
oblongus (L) Oblong, angle, obtuse.
obtusifolium (L) Obtuse elongated.
obovatus (L) Inverse, ovate.
obtus (L) Blunt, e.g., a broad leaf.

occidentale, occidentalis (L) Western.
oct (L) Eight.
octandra (L) In *Aesculus*, eight stamens.
oglethorpe *Quercus oglethropensis* was dicovered in Ogelthorpe County, Georgia, in 1940.
oliganthus (L) From the words *oligo* (small, few) and *authus* (flower); small-flowered.
onoma (G) Name.
opac, opaca (L) Not light-transmitting; thick, opaque leaves.
Oplopanax (G) A combination of *hoplon* (weapon), *pas* (all) and *akos* (cure).
Opuntia (G) A name used by Theophrastus for a plant that grew near Opus (Opuntis).
orbic (L) A circle.
orbiculatus (L) Round.
orientalis Oriental.
orni (G) Bird.
ornus (G) ash.
Osmanthus (G) From the terms *osme* (odor) and *anthos* (flower); fragrant flowers.
osteo (G) Bone.
osteosperma (G) Hard, boney cone (seed).
Ostrya (G) A tree having extremely hard wood.
oval (L) Oval, egg-shaped.
ovalis (L) Oval.
ovalifolium (L) Oval-shaped leaves.
ovat, ovata (L) Oval, egg-shaped.
Oxydendrum (G) From the words *oxys* (sour) and *dendron* (tree).
ozarkensis Of the Ozarks.
padus From Padus, the River Po, used by Theophrastus.
palmeri Named for Palmer.
palmetto Palm like.
panicle (L) A type of inflorescense.
paniculata (L) Possessing panicles.
papil, papilio A butterfly-like corolla.
papill (L) A nipple.
papillosa, papillatus (L) Covered with nipples.
persica Of *Persia.*
petal (G) Leafy appendage of a flower.
Philadelphus Named for Ptolemy Philadelphus, king of Egypt from 283 to 247 B.C. (Fernald 1950).
platy (G) Broad.
platyphyllos (G) Broad-leaved.
poli (G) Gray.
polita, politus elegant, polished.
procer, procera (L) Tall.
Pseudotsuga From the words pseudo (false) and *Tsuga* (hemlock).
puber (L) Mature, downy.
pubera (L) Somewhat downy, hairy.
quad four.
quadalupensis Of Guadalupe, Mexico.
Quercus Classical name for oak.
radic (L) Root.
radicans (L) A reference to having many aerial roots, as in *Rhus radicans*, poison ivy.
Rhus (L) Aboriginal name for sumac.
rydbergii Named For Pehr Axel Rydberg, a botanist who wrote a flora of the Rocky Mountains.
sabal Probably derived from an American vernacular name.
sabiniana Named for Joseph Sabine (1770–1837), a British horticulturalist.
sacchar (G) Sugar.
saccharinum (G) Sugary.
saccharum (G) Sugar.
salicifolia (L) Willow-leafed.
salix (L) Willow, classical name.
sangui (L) Blood.
sanguinea, sangineum, sanguineus (L) Bloodlike, crimson, reddish.
sargentii Named for Charles Sargent (1841–1927), professor of horticulture and director of the Arnold Arboretum, Harvard University.
Sassafras The aborigiral name.
sativa (L) Cultivated, sown.
Saxifragaceae From the Latin words *saxum*, a stone, and *frangere*, to break; a plant whose bulblets were supposed to dissolve kidney stones (Fernald 1950).
scand (L) To climb.
scandens (L) Climbing.
Schisandra (G) From the words *schizo* (to divide) and *andr* (anther); divided anthers.
Schrankia Named for Franz von Paula von Schrank, a German botanist (1747–1835).
scopul (L) Small broom.
scopulina, scopulorum (L) Broom like.
scouleriana Named after Scouler.
segreg (L) Separate, distinct.
segregata (L) Separated or distinct from.
Sequoia Named for Sequoiah (1770–1843), son of a British merchant and a Cherokee Indian woman; transcribed the Cherokee language.

sempervirens (L) From the words *semper* (always) and *vivus* (alive).
Serenoa Named for Sereno Watson (1826–1892), a North American botanist.
serotin (L) Late, delayed.
serotina (L) Late-ripening or opening.
serr (L) Saw.
serrulata (L) Like a saw; finely serrated.
sessil (L) Without a stalk or petiole.
sessilis (L) Attached directly to the twig.
seti (L) Bristle.
setigera (L) Bristly.
Shepherdia Named for John Shepherd (1776–1836), a curater at the Liverpool Botanic Garden.
shumardii Named for Benjamin Franklin Shumard (1820–1869), state geologist of Texas in 1860.
sibirica Of Siberia.
simonii Named after Simon.
sinensis (L) Of China.
sinua (L) Bent.
sinuata (L) To be sinuate.
sinuate (L) Wavy, sinuate, as along the margin of a leaf blade.
sitchensis Named for Sitka, Alaska.
smallii Named for John Kunkel Small, who wrote a flora of the southeastern United States (1903) and the *Manual of the Southeastern Flora* (1935).
Smilax (G) An ancient name for this plant.
smithiana Named for Smith.
sorediatus neglected or dirty-looking.
speci (L) Special, attractive to look at, showy species.
speciosa (L) Showy.
spectabilis (L) An impressive sight.
spectrum (L) Image, a sight.
sperm (G) Seed.
spica (L) A point or spike.
spicata (L) Spiked.
Spiraea (G) From spirea; a plant used by Theophratus for making wreaths.
stolon (L) A runner (basal surface stem) that roots.
stolonifera (L) Possessing stolons.
strict, stricta (L) Erect, upright, stiff.
strob (G) Twist, turn.
strobiformis (L) Twisted form or shape; a reference to deformed shapes at high altitudes.
strobus (G) An ancient term; probably a reference to the rows of scales that twist about the stalk in the cone of white pine.
stroma (G) Spread out, a covering.
styrac (G) A shrub.
styraciflua (G) Old generic name meaning flowering gum.
styrax (G) Aboriginal name for the tree that produces storax.
suber (L) Corky.
suspensa (L) Suspended.
taeda An ancient name for resinous pine.
Tamarindus tamrhindi, a dried date
tataricum () Tartarian.
tax (G) Yew.
taxifolia (G) Leaved like yew.
Taxodium (G) Yew like.
Taxus (G) Yew tree.
tenax (L) Tough.
tens (L) Stretched as to lengthen.
tenuifolia (L) Slender leaves.
texana, texensis Of Texas.
thomasii Named for David Thomas, who discovered *Ulmus thomasii.*
thraustos (Ratinesquian) Brittle, easily broken.
Thuja Ancient name of a conifer.
thunbergii Named for Carl Thunberg (1743–1828), a Swedish botanist and pupil of Lainnaeus.
thyoides Thuja-like.
thyrs (G) A staff.
thyrsiflous Thyrse-flowered.
Tilia (L) Classical name for basswood, linden.
tinto (It.) Tint, color.
tintoria (It.) Used as a dye.
tomentella (L) Dense hair, woolly, usually in a leaf or twig.
tomentosa (L) Dense woolly matted hair.
torreya, torreyana Named for John Torrey (1776–1873), an important early botanist who worked on the plants of central and eastern North America.
toumeyi Named after Toumey.
toxi (L) Poison.
Toxicodendron (L,G) Poison tree.
toxiferum (L) Poisoned.
tremul (L) Tremble, shake, or quiver.
tremula (L) To quiver or shake in the wind.
tremuloides To quiver or shake in the wind.
tri, tria (L) Three.
triacanthos (L) From *tria* (three) and *canthos* (thorn); a thorn with two branch barbs (three points).
tricho (G) Hairy.

tricho carpa (G) Hairy fruit.
tridentata (L) Three-toothed.
trifolia, trifoliata (L) Three-leaved.
triloba Three-lobed.
trilobata (L,G) Three-lobed.
tripetula Three-petaled.
Tsuga Hemlock.
tulipifera (L) An old generic name meaning tulip-bearing, a reference to the flower.
turbin (L) Whirling or spinning as a top.
turbinella Similar to a top.
Typha (G) Cattail, with velvety fruit, "cattails."
typhina (G) Fruit and twigs velvety, similar to stalks of *Typha.*
uber (L) Fruitful.
Ulmus (L) The classical Latin name for elm.
umbel (L) An inflorescense in which the pedicals are of the same length and attached at a central point.
umbellata (L) At a central point with umbells.
Umbellularia A genus whose members produce an unbel-like inflorescense.
ussuriensis Of Ussuria.
utahensis Of Utah.
uva-ursi (L) From *uva* (berry) and *urs* (bear).
vaccin (L) A dun (grayish brown) color. Reference uncertain; a hill (Gaelic).
vaccinifolia Having leaves similar to a blueberry.
Vaccinum (L) Aboriginal name; plants with dun-colored stems or found on a hill; blueberry.
varia (L) Variable, subject to change.
variablilis (L) Considerable leaf variability.
veitchii Named for Veitch.
velutin, velutina, velutinus (L) Velvet like.
vern (L) Spring.
vernix (L) Varnished; perhaps a reference to a shiny leaf.
verticill (L) A whorl.
verticilate (L) Whorled.
villos (L) Hairy.
villosa (L) Hairy.
vimen (L) A flexible twig.
viminalis Bearing flexible twigs.
vire, virens (L) Green, greenish.
virginiana, virginieus Of Virginia.
viridissima (L, G) Very green.
viscos, viscosa (L) Gummy, sticky.
viscosissimum (L) Being sticky.
vitacea (L) Grape like.
Vitus (L) Old classical name for grape.
vomit (L) A discharge from the stomach.
vomitoria (L) Causing vomiting, as when the tea of *Ilex* is drunk to excess.
vulgar, vulgare, vulgaris (L) Common, widespread.
wichuraiana Named for Max Wichura (1817–1866).
wislizenii Named for Wislizen.
woodsii Named for Joseph Woods (1776–1864), an English botanist who studied the genus *Rosa.*
wrightii Named for charles Wright (1811–1886) a botanical collector of southwestern species
Xanthoxylum (G) From *xanthos,* meaning yellow, and *xylem,* meaning wood; yellow wood.
yucca A native Haitian or Carib name.
Zamia (L) From *zamia,* meaning loss (Hyam and Pankhurst 1995). Possible reference to possession of cones.

References

Aber, J. D., and J. M. Mellilo. 1991. *Terrestrial Ecosystems.* Philadelphia: Saunders College Publishing.

Adams, R. P. 1993. "*Juniperus.*" In *Flora of North America North of Mexico.* Vol. 2. *Pteridophytes and Gymnosperms,* edited by Flora of North America Editorial Committee, 412–420. New York: Oxford University Press. 475 pp.

Allen, T. F. H., and T. B. Starr. 1982. *Hierarchy: Perspectives for Ecological Complexity.* Chicago: University of Chicago Press.

Anderson, E. 1949. *Introgressive Hydridization.* New York: John Wiley & Sons, Inc.

Anderson, J. P. 1974. *Flora of Alaska and Adjacent Parts of Canada.* Ames: Iowa State University Press. 543 pp.

Anderson, R. C., and J. S. Fralish. 1975. "An Investigation of Palmetto (*Paurotis wrightii*) Communities in Belize, Central America." *Turrialba* 24: 37–44.

Anderson, R. C., J. S. Fralish, J. E. Armstrong, and P. K. Benjamin. 1993. "The Ecology and Biology of *Panax quinquefolium* L. (Araliaceae) in Illinois." *The American Midland Naturalist* 129:357–372.

Anderson, R. C., J. S. Fralish, and J. M. Baskin, editors. 1999. Savannas, Barrens, and Rock Outcrop Plant Communities of North American. Cambridge, UK: Cambridge University Press. 470 pp.

Argus, G. W. 1973. *The Genus* Salix *in Alaska and the Yukon.* Botanical Publication No. 2. Ottawa: National Museum of Canada. 279 pp.

Arno, S. F., E. D. Reinhardt, and J. H. Scott. 1993. *Forest Structure and Landscape Patterns in the Subalpine Lodgepole Pine Type: A Procedure for Quantifying Past and Present Conditions.* USDA Forest Service General Technical Report INT-294. Ogden, Utah: Intermountain Research Station.

Arno, S. F., D. G. Simmerman, and R. E. Keane. 1984. *Forest Succession on Four Habitat Types in Western Montana.* USDA Forest Service General Technical Report INT-34. Ogden, Utah: Intermountain Forest and Range Experiment Station.

Arnold, C. A. 1948. "Classification of gymnosperms." *Botanical Gazette* 110:2–12.

Bailey, D. K. 1970. "Phytogeography and taxonomy of *Pinus* subsection *Balfourniae.*" *Annals of the Missouri Botanical Garden* 57:210–249.

Bailey, L. H., ed. 1923. *The Cultivated Evergreens.* New York: Macmillan. 434 pp.

Baily, L. H. 1963. *How Plants Get Their Names.* New York: Dover Publications.

Bailey, R. G. 1996. *Ecosystem Geography.* New York: Springer-Verlag.

Bailey, R. G. 1998. *Ecoregions: The Ecosystem Geography of the Oceans and Continents.* New York: Springer-Verlag.

Bailey, V. L., and H. E. Bailey. 1949. *Woody Plants of the Western National Parks.* Notre Dame, IN: Notre Dame University Press. 274 pp.

Barbour, M. G. 1988. "California Upland Forests and Woodlands." In *North American Terrestrial Vegetation,* edited by M. G. Barbour and W. D. Billings, 131–164. New York: Cambridge University Press.

Barbour, M. G., and W. D. Billings, eds. 1988. *North American Terrestrial Vegetation.* New York: Cambridge University Press.

Barbour, M . G., and J. Major. 1977. *Terrestrial Vegetation of California.* New York: John Wiley & Sons, Inc. 1002 pp.

Barbour, M. G., J. H. Burk, W. D. Pitts, M. W. Schwartz, and F. Gilliam. 1998. *Terrestrial Plant Ecology.* 3d ed. Menlo Park, CA: Benjamin Cummings Publishing.

Barker, W. T. 1997. "Planera and Celtis." In *Flora of North America North of Mexico: Magnoliophyta: Magnoliidae and Hamamelidae,* edited by Flora of North America Editorial Committee, 376–379. New York: Oxford University Press. 590 pp.

Barnes, B. V., and W. H. Wagner Jr. 1981. *Michigan Streets.* Ann Arbor: University of Michigan Press. 383 p.

Barrett, J. W. 1980. *Regional Silviculture of the United States.* New York: John Wiley & Sons, Inc.

Barrett, J. W., S. Arno, and C. Key. 1991. "Fire Regimes of Western Larch-Lodgepole Pine Forests in Glacier National Park, Montana." *Canadian Journal of Forest Research* 21:1711–1720.

Barringer, K. 1997. "*Aristolochia.*" In *Flora of North America North of Mexico.* Vol. 3. *Magnoliophyta: Magnoliidae and Hamamelidae,* edited by Flora of North America Editorial Committee, 45–50. New York: Oxford University Press. 590 pp.

Baskin, J. M., and C. C. Baskin. 1999. "Cedar Glades of the Southeastern United States." In *Savannas, Barrens, and Rock Outcrop Plant Communities of North America,* edited by R. C. Anderson, J. S. Fralish, and J. M. Baskin, 206–219. Cambridge, UK: Cambridge University Press. 470 pp.

Baumgardt, J. P. 1982. *How to Identify Flowering Plant Families.* Portland, OR: Timber Press. 269 pp.

Becking, R. W. 1957. "The Zurich-Montpellier School of Phytosociology." *Botanical Review* 23: 411–488.

Beers, T. W., P. E. Dress, and L. C. Wensel. 1966. "Aspen Transformation in Productivity Research." *Journal of Forestry* 64:691–692.

Bell, A. D. 1991. *Plant Form: An Illustrated Guide to Flowering Plant Morphology.* New York: Oxford University Press. 341 pp.

Benson, L. 1957. *Plant Classification.* Boston: D. C. Heath & Co. 688 pp.

Benson, L. D. 1943. "Revision of Status of Some Southwestern Trees and Shrubs." *American Journal of Botany* 30:230–240.

Billington, C. 1977. *Shrubs of Michigan.* 2d ed. Bloomfield Hills, MI: Cranbrook Institute of Science. 339 pp.

Birkeland, P. W. 1999. *Soils and Geomorpology.* 3d ed. New York: Oxford University Press.

Blackburn, B. 1952. *Trees and Shrubs in Eastern North America.* New York: Oxford University Press. 358 pp.

Boivin, B. 1966. "Enumeration des plantes du Canada." *Naturaliste Canada* 93:583–646.

Bold, H. C. 1977. *The Plant Kingdom.* 4th. ed. Englewood Cliffs, NJ: Prentice-Hall. 310 pp.

Braun, E. L. 1950. *Deciduous Forests of the Eastern United States.* New York: Hafner Publishing.

Bolsinger, C. L., and K. L. Waddell. 1993. *Area of Old-Growth Forests in California, Oregon, and Washington.* USDA Forest Service Resource Bulletin PNW-RB-197. Portland, Oregon: Pacific Northwest Forest Experiment Station.

Borchert, M. I., N. D. Cunha, P. C. Krosse, and M. L. Lawrence. 1993. *Blue Oak Plant Communities of Southern San Luis Obispo and Northern Santa Barbara Counties, California.* USDA Forest Service General Technical Report PSW-GTR-139.

Borror, D. J. 1960. *Dictionary of Word Roots and Combining Forms.* Palo Alto, CA: Mayfield Publishing. 134 pp.

Braun-Blanquet, J. 1964. *Pflanzensoziologie. Grundzüge der Vegetationskunde.* Biologishe Studienb, cher 7. 3rd. ed. New York: Springer-Verlag.

Brayshaw, T. C. 1965. *Native Poplars of Southern Alberta and Their Hybrids.* Department of Forestry, Canada. Publication No. 1109. Ottawa: Queen's Printer and Controller of Stationery. 40 pp.

Britton, N. L., and A. Brown. 1970. *An Illustrated Flora of the Northern United States and Canada.* New York: Dover Publications. Vol. 1, 680 pp.; Vol. 2, 735 pp.; Vol. 3. 637 pp.

Brockman, C. F. 1968. *Trees of North America.* Golden Field Guide Series. New York: Golden Press. 280 pp.

Brown, C. A. 1945. *Louisiana Trees and Shrubs.* Louisiana For. Comm. Bull. No. 1. Reprint, Baton Rouge, LA: Claitor's Publication Division, 1972. 262 pp.

Brown, H. P. 1922. *Trees of New York, Native and Naturalized.* Technical Bulletin No. 15. Syracuse, NY: Syracuse University, College of Forestry.

Brunsfield, S. J., P. S. Soltis, D. E. Soltis, P. A. Gadek, C. J. Quinn, D. D. Strenge, and T. A. Ranker. 1994.

"Phylogenetic Relationships Among the Genera of Taxodiaceae and Cupressaceae: Evidence from *rbcL* sequences." *Systematic Botany* 19: 253–262.

Buchholz, J. T. 1939. "The Genetic Segregation of the Sequoias." *American Journal of Botany* 26: 535–538.

Burger, D. 1993. *Revised Site Regions of Ontario: Concepts, Methodology and Utility.* Forest Research Report No. 129. Ontario Forest Research Institute.

Burns, R. M. 1983. *Silvicultural Systems for the Major Forest Types of the United States.* USDA Agriculture Handbook 445. Washington, DC: Government Printing Office.

Burns, R. M., and B. H. Honkala. 1990a. *Silvics of North America: Conifers.* Vol. 1. USDA Forest Service Handbook No. 654. Washington, DC: Government Printing Office. 675 pp.

Burns, R. M. and B. H. Honkala. 1990b. *Silvics of North America: Hardwoods.* Vol. 2. USDA Forest Service Handbook No. 654. Washington, DC: Government Printing Office. 877 pp.

Cain, S. A. 1944. *Foundations of Plant Geography.* New York: Harper & Brothers.

Cajander, A. K. 1926. "The Theory of Forest Types." *ACTA Forest Fenneca* 29:1–108.

Camp, W. H. 1950. "A Biogeographical and Paragenetic Analysis of the American Beech." *Yearbook of the American Philosophical Society* 1950:166–169.

Charlet, D. A. 1996. *Atlas of Nevada Conifers.* Reno: University of Nevada Press. 320 pp.

Christensen, N. L. 1988. "Vegetation of the Southeastern Coastal Plain." In *North American Terrestrial Vegetation*, edited by M. G. Barbour and W. D. Billings, 317–364. New York: Cambridge University Press.

Cliburn, J., and G. Klomps. 1990. *A Key to the Missouri Trees in Winter.* Columbia: Conservation Commission, State of Missouri. 46 pp.

Cody, W. J. 1996. *Flora of the Yukon Territory.* Ottawa: National Research Council of Canada Press. 643 pp.

Cohn, L. 1992. "Searching for Harmony." *Forest Perspectives* 2:6–9.

Coombes, A. J. 1985. *Dictionary of Plant Names.* Portland, OR: Timber Press. 207 pp.

Coombes, A. J. 1992. *Trees.* New York: Dorling Kindersley. 320 pp.

Covington, W. W., and L. F. DeBano. 1994. *Sustainable Ecological Systems: Implementing and Ecological Approach to Land Management.* General Technical Report RM-247. U. S. Forest Service. Fort Collins, CO: Rocky Mountain Forest and Range Experiment Station. 363 pp.

Cox, C. B., and P. D. Moore. 1993. *Biogeography: An Ecological and Evolutionary Approach.* 5th ed. Oxford, UK: Blackwell Science.

Cox, P. W., and P. Leslie. 1988. *Texas Trees.* San Antonio: Corona Publishing.

Critchfield, W. B., and E. L. Little Jr. 1966. *Distribution of the Pines of the World.* Washington, DC: Government Printing Office.

Crittenden, M. 1977. *Trees of the West.* Millbrae, CA: Celestial Arts. 212 pp.

Cronquist, A. 1981. *An Integrated System of Classification of Flowering Plants.* New York: Columbia University Press. 1262 pp.

Curtis, C. C., and S. C. Bausor. 1929. *The Complete Guide to North American Trees.* New York: Collier Books. 342 pp.

Curtis, J. T. 1959. *The Vegetation of Wisconsin.* Madison: University of Wisconsin Press. 657 pp.

Daniels, R. B. 1960. "Entrenchment of the Willow Drainage Ditch, Harrison County, Iowa." *American Journal of Science* 258:167–76.

Daubenmire, R. 1968. "Some Geographic Variation in *Picea sitchensis* and Their Ecological Interpretation." *Canadian Journal of Botany* 46: 787–798.

Daubenmire, R., and J. B. Daubenmire. 1968. *Forest Vegetation of Eastern Washington and Northern Idaho.* Washington Agriculture Experiment Station Technical Bulletin 60. Pullman: Washington State University.

DeJong, P. C., D. M. van Gelderen, H. J. Oterdoom, and J. R. Smith. 1994. *Maples of the World.* Portland, OR: Timber Press. 478 pp.

Delcourt P. A., and H. R. Delcourt. 1981. "Vegetation Maps for Eastern North America: 40,000 Years B.P. to the Present." *Geobotany II*, edited by R. C. Romans, 123–165. New York: Plenum Press.

Desmarais, Y. 1952. "Dynamics of Leaf Variations in the Sugar Maples." *Brittonia* 7:347–388.

Devine, B. 1999. "Clear-Cut Missions." *Nature Conservancy* 49:12–17.

Dhillion, S. S., and M. H. Mills. 1999. "The Sand Shinnery Oak (*Quercus harvardii*) Communities of the Llano Estacado: History, Structure, Ecology, and Restoration." In *Savannas, Barrens, and Rock*

Outcrop Plant Communities of North America, edited by R. C. Anderson, J. S. Fralish, and J. M. Baskin, 262–274. Cambridge, UK: Cambridge University Press. 470 pp.

Dirr, M. A. 1998. *Manual of Woody Landscape Plants.* 5th ed. Champaign, IL: Stipes Publishing. 536 pp.

Eastman, J. 1992. *The Book of Forest and Thicket: Trees, Shrubs, and Wildflowers of Eastern North America.* Harrisburg, PA: Stackpole Books. 212 pp.

Eckenwalder, J. E. 1976. "Re-evaluation of Cupressaceae and Taxociaceae: A Proposed Merger." *Madrono* 23:237–256.

Eckenwalder, J. E. 1993. "*Cupressus.*" In *Flora of North America North of Mexico.* Vol. 2. *Pteridophytes and Gymnosperms*, edited by Flora of North America Editorial Committee, 405–408. New York: Oxford University Press. 475 pp.

Edmonds R. L., ed. 1982. Introduction to *Analysis of Coniferous Forest Ecosystems in the Western United States*, 1–27. Stroudsburg, PA: Hutchinson Ross Publishing.

Edwin, G. 1957. "Notes on Ilex." *Rhodora* 59:20–23.

Egler, F. E. 1954 "Vegetation Science Concepts. I. Initial Floristics Composition: A Factor in Old-Field Vegetation Development." *Vegetation* 4:412–417.

Elias, T. S. 1972. "The Genera of Juglandaceae in the Southeastern United States." *Journal of Arnold Arboretum* 53:26–51.

Elias, T. S. 1980. *The Complete Trees of North America.* New York: Van Nostrand Reinhold. 948 pp.

Elliot-Fisk, D. L. 1988. "The Boreal Forest." In *North American Terrestrial Vegetation*, edited by M. G. Barbour and W. D. Billings, 131–164. New York: Cambridge University Press.

Elmore, J. R. 1976. *Shrubs and Trees of Southwest Uplands.* Tucson, AZ: Southwest Parks and Monuments Association. 214 pp.

Engler, A., and H. Prantl. 1926. *Die Naturlichen Pflanzenfamilien.* Vol. 2. Leipzig.

Esau, K. 1977. *Anatomy of Seed Plants.* New York: John Wiley & Sons, Inc. 550 pp.

Eyre, F. H., ed. 1980. *Forest Cover Types of North America.* Washington, DC: Society of American Foresters. 148 pp.

Farjon, A. 1984. *Pines: Drawings and Descriptions of the Genus* Pinus. Leiden: E. J. Brill. 219 pp.

Farjon, A., J. A. Perez de la Rosa, and B. T. Styles. 1997. *A Field Guide to the Pines of Mexico and Central America.* Kew, UK: The Royal Botanic Gardens. 147 pp. (Range maps).

Farrar, J. L. 1995. *The Trees of Northern United States and Canada.* Ames: Iowa State University Press. 502 pp.

Fenneman, N. M. 1928. "Physiographic Divisions of the United States." *Annals, Association of American Geographers* 18:261–253.

Fernald, M. L. 1950. *Gray's Manual of Botany.* 8th. ed. Corrected printing 1987. Portland, OR: Dioscorides Press. 1632 pp.

Flora of North America Editorial Committee, ed. 1993a. *Flora of North America North of Mexico.* Vol. 1. *Introduction.* New York: Oxford University Press. 372 pp.

Flora of North America Editorial Committee, ed. 1993b. *Flora of North America North of Mexico.* Vol. 2. *Pteridophytes and Gymnosperms.* New York: Oxford University Press. 475 pp.

Flora of North America Editorial Committee, ed. 1997. *Flora of North America North of Mexico.* Vol. 3. *Magnoliophyta: Magnoliidae and Hamamelidae.* New York: Oxford University Press. 590 pp.

Food and Agricultural Organization of the United Nations. 1979. *Poplars and Willows in Woody Production and Land Use.* FAO Forestry and Forest Products Studies No. 12. Rome: FAO. 328 pp.

Foote, L. E., and S. B. Jones. 1989. *Native Shrubs and Woody Vines of the Southeast.* Portland, OR: Timber Press. 199 pp.

Ford-Robertson, F. C., ed. 1983. *Terminology of Forest Science Technology Practice and Products.* Washington, DC: Society of American Foresters.

Forman, R. T. T. 1995. *Land Mosaics: The Ecology of Landscapes and Regions.* Cambridge, UK: Cambridge University Press. 632 pp.

Forman, R. T. T., and M. Godron. 1986. *Landscape Ecology.* New York: John Wiley & Sons, Inc. 619 pp.

Foster, A. S., and E. M. Gifford Jr. 1959. *Comparative Morphology of Vascular Plants.* San Francisco: W. H. Freeman. 555 pp.

Fowell, H. A. 1965. *Silvics of Forest Trees of the United States.* USDA Forest Service Handbook No. 276. Washington, DC: Government Printing Office. 762 pp.

Fralish, J. S., S. B. Franklin, and D. D. Close. 1999. "Open Woodland Communities of Southern

Illinois, Western Kentucky, and Middle Tennessee." In *Savannas, Barrens, and Rock Outcrop Plant Communities of North America*, edited by R. C. Anderson, J. S. Fralish, and J. M. Baskin, 171–189. Cambridge, UK: Cambridge University Press. 470 pp.

Franklin, D. B. 1998. *Historical Overview of United States Forest Practice Law: Specific Emphasis on Public Participation Law.* Working Paper. European Forest Institute.

Franklin, J. F., and C. T. Dyrness. 1988. *Natural Vegetation of Oregon and Washington.* Corvallis, Oregon; Oregon State University Press. 452 pp.

Franklin, S. B. 1994. "Late-Pleistocene and Holocene Vegetation History of Land Between the Lakes, Kentucky and Tennessee." *Transactions of the Kentucky Academy of Sciences* 55:6–19.

Franklin, S. B., P. A. Robertson, J. S. Fralish, and S. M. Kettler. 1993. "Overstory Vegetation and Successional Trends of Land Between the Lakes, USA." *Journal of Vegetation Science* 4:509–520.

Fredrickson, L. H. 1979. "Lowland Hardwood Wetlands: Current Status and Habitat Values for Wildlife." In *Wetland Functions and Values: The State of Our Understanding*, edited by P. E. Greeson, J. R. Clark, and J. E. Clark, 296–306. Minneapolis. American Water Resources Association.

Frodin, D. G., and R. Govaerts. 1996. *World Checklist and Bibliography of Magnoliaceae.* Kew, UK: Royal Botanic Gardens. 72 pp.

Fuller, G. D., and H. S. Conrad. 1932. *Plant Sociology: The Study of Plant Communities.* New York: McGraw-Hill.

Fuller, H. J., and O. Tippo. 1954. *College Botany.* New York: Henry Holt. 993 pp.

Furlow, J. J. 1997. "Betulaceae." In *Flora of North America North of Mexico:* Vol. 3. *Magnoliophyta: Magnoliidae and Hamamelidae*, edited by Flora of North America Editorial Committee, 507–538. New York: Oxford University Press. 590 pp.

Gale, R. P., and S. M. Cordray. 1991. "What Should Forests Sustain: Eight answers." *Journal of Forestry* 89:31–36.

Gauch, H. G. 1982. *Multivariate Analysis in Community Ecology.* Cambridge, UK: Cambridge University Press.

Gerlach, L. P., and D. N. Bengston. 1994. "If Ecosystem Management Is the Solution, What's the Problem?" *Journal of Forestry* 92:18–21.0

Gibson, D. J., R. A. Zampella, and A. G. Windisch. 1999. "New Jersey Pine Plains: The 'True Barren' of the New Jersey Pine Barren." In *Savannas, Barrens, and Rock Outcrop Plant Communities of North America*, edited by R. C. Anderson, J. S. Fralish, and J. M. Baskin, 52–66. Cambridge, UK: Cambridge University Press. 470 pp.

Gleason, H. A. 1926. "The Individualistic Concept of the Plant Association." *Bulletin of the Torrey Botanical Club* 53:7–26.

Gleason, H. A. 1935. "The Individualistic Concept of the Plant Association." *American Midland Naturalist* 21:92–110.

Gleason, H. A., and A. Cronquist. 1991. *Manual of Vascular Plants of Northeastern United States and Adjacent Canada.* Bronx, NY: The New York Botanical Garden. 910 pp.

Glenhill, D. 1989. *The Names of Plants.* New York: Cambridge University Press. 202 pp.

Graves, A. H. 1956. *Illustrated Guide to Trees and Shrubs.* Reprint, New York: Dover Publications, 1992. 271 pp.

Greller, A. M. 1988. "Deciduous Forest." In *North American Terrestrial Vegetation*, edited by M. G. Barbour and W. D. Billings, 287–316. Cambridge, UK: Cambridge University Press.

Grime, J. P. 1977. "Evidence for the Existence of Three Primary Strategies in Plants and Its Relevance to Ecological and Evolutionary Theory." *American Naturalist* 11:1169–1194.

Grime, J. P., J. G. Hodgson, and R. Hunt. 1990. *The Abridged Comparative Plant Ecology.* London: Unwyn Hyman.

Grimm, W. C. 1983a. *The Illustrated Book of Trees.* Harrisburg, Pennsylvania: Stackpole Books. 493 pp.

Grimm, W. C. 1957. *The Book of Shrubs.* New York: Bonanza Books. 522 pp.

Grimm, W. C. 1993. *The Illustrated Book of Wildflowers and Shrubs.* Harrisburg, PA: Stackpole Books. 637 pp.

Habeck, J. R. 1958. "White Cedar Types in Wisconsin." *Ecology* 39:457–463.

Hardin, J. W. 1979. "*Quercus prinus L.—nomen ambiguum.*" *Taxon* 28:355–357.

Hardin, J. W., D. J. Leopold, and F. M. White. 2001. *Harlow and Harrar's Textbook of Dendrology.* 9th. ed. New York: McGraw-Hill. 534 pp.

Harlow, W. M., E. S. Harrar, and F. M. White. 1979. *Textbook of Dendrology.* New York: McGraw-Hill. 510 pp.

Harlow, W. H. 1946. *Fruit Key to Northeastern Trees.* New York: Dover Publications. 56 pp.

Harlow, W. M., E. S. Harrar, J. W. Hardin, and F. M. White. 1996. *Textbook of Dendrology.* 8th ed. New York: McGraw-Hill. 534 pp.

Harrington, H. D. 1957. *How to Identify Plants.* Chicago: Swallow Press. 203 pp.

Harris, J. G., and M. W. Harris. 1994. *Plant Identification Terminology: An Illustrated Glossary.* Payson, UT: Spring Lake Publishing. 197 pp.

Heikens, A. L. 1999. "Savanna, Barrens, and Glade Communities of the Ozark Plateaus Province." In *Savannas, Barrens, and Rock Outcrop Plant Communities of North America,* edited by R. C. Anderson, J. S. Fralish, and J. M. Baskin, 220–230. Cambridge, UK: Cambridge University Press. 470 pp.

Hickey, M., and C. King. 1997. *Common Families of Flowering Plants.* Cambridge, UK: Cambridge University Press. 212 pp.

Hillier Nurseries. 1998. *The Hillier Manual of Trees and Shrubs.* Devon, UK: David and Charles. 928 pp.

Hitchcock, C. L., and A. Cronquist. 1973. *Flora of the Pacific Northwest.* Seattle: University of Washington. 730 pp.

Hosie, R. C. 1979. *Native Trees of Canada.* Don Mills, ON: Fitzhenry & Whiteside. 380 pp.

Hunt, R. S. 1993. "*Abies.*" *In Flora of North America North of Mexico.* Vol. 2. *Pteridophytes and Gymnosperms,* edited by Flora of North America Editorial Committee, 354–362. New York: Oxford University Press. 475 pp.

Hunter, C. G. 1989. *Trees, Shrubs, and Vines of Arkansas.* Little Rock: The Ozark Society Foundation. 207 pp.

Hupp, C. R. 1992. "Riparian Vegetation Recovery Patterns Following Stream Channelization: A Geomorphic Perspective." *Ecology* 73:1209–1226.

Hyam, R., and R. Pankhurst. 1995. *Plants and Their Names: A Concise Dictionary.* New York: Oxford University Press. 545 pp.

Ingram, G., and M. R. Moss, eds. 1992. *Landscape Approaches to Wildlife and Ecosystem Management.* Morin Heights, Canada: Polyscience Publications. 267 pp.

Jacobs, B. F., C. R. Werth, and S. I. Guttman. 1984. "Genetic Relationships in *Abies* (Fir) of Eastern United States: An Electophretic Study." *Canadian Journal of Botany* 62:609–616.

James, P. E. 1952. *A Geography of Man.* 2d ed. Boston: Ginn.

Jeffrey, C. 1982. *An Introduction to Plant Taxonomy.* 2d ed. New York: Cambridge University Press. 154 pp.

Jensen, E. C., D. J. Anderson, J. C. Tappeiner Jr., and J. C. Zasada. 1995a. *The Reproductive Ecology of Broadleaved Trees and Shrubs: Vine Maple,* Acer circinatum *Pursh.* Research Contribution 9b. Forest Corvallis: Research Laboratory, Oregon State University. 6 pp.

Jensen, E. C., D. J. Anderson, J. C. Tappeiner Jr., and J. C. Zasada. 1995b. *The Reproductive Ecology of Broadleaved Trees and Shrubs: Red Alder,* Alnus rubra *Bong.* Research Contribution 9c. Corvallis: Forest Research Laboratory, Oregon State University. 7 pp.

Jensen, E. C., D. J. Anderson, J. C. Tappeiner Jr., and J. C. Zasada. 1995c. *The Reproductive Ecology of Broadleaved Trees and Shrubs: Tanoak,* Lithocarpus densiflorus *(Hook. & Arn.) Rehd.* Research Contribution 9d. Corvallis: Forest Research Laboratory, Oregon State University. 7 pp.

Jensen, E. C., D. J. Anderson, J. C. Tappeiner Jr., and J. C. Zasada. 1995d. *The Reproductive Ecology of Broadleaved Trees and Shrubs: Salmonberry,* Rubus spectabilis *Pursh.* Research Contribution 9e. Corvallis: Forest Research Laboratory, Oregon State University. 7 pp.

Jenson, R. J. 1997. "*Quercus Linnaeus* sect. *Lobatae Louden*: Red or Black Oaks." In *Flora of North America North of Mexico:* Vol. 3. *Magnoliophyta: Magnoliidae and Hamamelidae,* edited by Flora of North America Editorial Committee, 447–471. New York: Oxford University Press. 590 pp.

Jepson, W. L. 1925. A *Manual of Flowering Plants of California.* Berkeley.

Johnson, D., L. Kershaw, A. MacKinnon, and J. Polar. 1995. *Plants of the Western Boreal Forest and Aspen Parkland.* Edmonton, AL: Lone Pine Publishing. 392 p.

Jones, G. N. 1950. *Flora of Illinois.* Notre Dame, IN: University of Notre Dame Press. 368 pp.

Jongman, R. H. G., C. J. F. ter Braak, and O. F. R. van Toneren. 1995. New Edition *Data Analysis in Community and Landscape Ecology.* Cambridge, UK: Cambridge University Press.

Jordan, W. R., III, M. E. Gilpin, and J. D. Aber, eds. 1987. *Restoration Ecology: A Synthetic Approach to Ecological Research.* New York: Cambridge University Press. 342 pp.

Judd, W. S., C. S. Campbell, E. A. Kellogg, and P. F. Stevens. 1999. *Plant Systematics: A Phylogenetic Approach.* Sunderland, ME: Sinauer. 464 pp.

Keeler, H. 1969. *Our Northern Shrubs and How to Identify Them.* New York: Dover Press. 539 pp.

Kessell, S. R., and W. C. Fischer. 1981. *Predicting Post-fire Plant Succession for Fire Management Planning.* USDA Forest Service General Technical Report INT-94. Ogden, Utah: Intermountain Forest and Range Experiment Station.

Kimmins, J. P. 1987. *Forest Ecology.* New York: Macmillan Publishing.

Klijn, J. M., and H. A. Udo de Haes. 1994. "A Hierarchical Approach to Ecosystems and Its Applications for Ecological Land Classification." *Landscape Ecology* 9: 89–104.

Knobel, E. 1972. *Identify Trees and Shrubs by Their Leaves.* New York: Dover Publications. 47 pp.

Kosztarab, M. 1988. "Biological Diversity: National Biological Survey." In *Proceedings of the First Annual Symposium on the Natural History of the Lower Tennessee and Cumberland River Valleys,* edited by D. H. Synder, 1–25. Clarksville, TN: Center for Field Biology, Austin Peay State University.

Kotar, J., and T. L. Burger. 1996. *A Guide to Forest Communities and Habitat Types of Central and Southern Wisconsin.* Madison: Department of Forestry, University of Wisconsin.

Kotar, J., J. A. Kovach, and C. T. Loey. 1988. *Field Guide to Forest Habitat Types of Northern Wisconsin.* Madison: Department of Forestry, University of Wisconsin.

Kovalchik, B. L. 1987. *Riparian Zone Associations of the Deshutes, Ochoco, Fremont, and Winema National Forests.* Ecology Technical Paper 279–287. USDA Forest Service, Pacific Northwest Region. Corvallis, OR.

Kral, R. 1993. "*Pinus.*" In *Flora of North America North of Mexico.* Vol. 2. *Pteridophytes and Gymnosperms,* edited by Flora of North America Editorial Committee, 373–398. New York: Oxford University Press. 475 pp.

Kral, R. 1997. "*Annonaceae.*" In *Flora of North America North of Mexico:* Vol. 3. *Magnoliophyta: Magnoliidae and Hamamelidae,* edited by Flora of North America Editorial Committee, 11–20. New York: Oxford University Press. 590 pp.

Kramer, P. J., and T. T. Kozlowski. 1979. *Physiology of Woody Plants.* New York: Academic Press. 811 pp.

Küchler, A. W. 1964. *Potential Natural Vegetation of the Conterminous United States.* New York: American Geographical Society.

Kupfer, J. A., and G. P. Malanson. 1993. "Observed and Modeled Directional Change in Riparian Forest Composition at a Cutbank Edge." *Landscape Ecology* 8:185–199.

Kurz, D. 1997. *Shrubs and Woody Vines of Missouri.* Jefferson City: Missouri Department of Conservation. 387 pp.

Kurz, H., and R. K. Godfrey. 1982. *Trees of Northern Florida.* Gainesville: University of Florida. 311 pp.

Lanner, R. M. 1981. *The Pinon Pine.* Reno: University of Nevada Press. 208 pp.

LaRoi, G. H., and R. J. Hnatiuk. 1980. "The *Pinus contorta* Forests of Banff and Jasper National Parks: A Study of Comparative Synecology and Syntaxonomy." *Ecological Monographs* 50:1–29.

Lauriault, J. 1989. *Identification Guide to the Trees of Canada.* Richmond Hill, ON: Fitzhenry & Whiteside. 479 pp.

Lawrence, G. H. M. 1951. *Taxonomy of Vascular Plants.* New York: Macmillan. 823 pp.

Lehmkuhl, J. F., P. F. Hessburg, R. L. Everett, M. H. Huff, and R. D. Ottmar. 1994. *Historical and Current Forest Landscapes of Eastern Oregon and Washington.* Part 1. *Vegetation Pattern and Insect and Disease Hazards.* USDA Forest Service General Technical Report PNW-GTR-328. Portland, Oregon: Pacific Northwest Forest Experiment Station.

Le Master, D. C., and R. A. Sedjo. 1992. *Modeling Sustainable Forest Ecosystems.* Washington, DC: American Forests. 181 pp.

Lense, A. N. 1959. *Common Trees of New Jersey.* New Brunswick: College of Agriculture, Rutgers State University.

Leopold, D. J., W. C. McComb, and R. N. Muller. 1998. *Trees of the Central Hardwood Forests of North America.* Portland, OR: Timber Press. 469 pp.

Linnaeus, C. 1730. *Hortus uplandicus.* Stockholm.

Linnaeus, C. 1737. *Genera plantarum.* Lugd. Bat.

Linnaeus, C. 1753. *Species plantarum.* 2 vols. Stockholm.

Lipscomb, B. 1993. "*Pseudotsuga*." 1993. In *Flora of North America North of Mexico*. Vol. 2. *Pteridophytes and Gymnosperms*, edited by Flora of North America Editorial Committee, 365–366. New York: Oxford University Press. 475 pp.

Little, E. L., Jr. 1950. *Southwestern Trees: A Guide to the Native Species of New Mexico and Arizona*. Agriculture Handbook No. 9. USDA Forest Service. Washington, DC: Government Printing Office. 109 pp.

Little, E. L., Jr. 1966. *Geographic Distribution of the Pines of the World*. USDA Forest Service Miscellaneous Publication No. 991. Washington, DC: Government Printing Office. 97 pp.

Little, E. L., Jr. 1971. *Atlas of United States Trees: Conifers and Important Hardwoods*. Vol. 1. USDA Forest Service. Miscellaneous Publication No. 1146. Washington, DC: Government Printing Office. 202 pp.

Little, E. L., Jr. 1976. *Atlas of United States Trees: Minor Western Hardwoods*. Vol. 3. USDA Forest Service. Miscellaneous Publication No. 1314. Washington, DC: Government Printing Office. 214 pp.

Little, E. L., Jr. 1977a. *Atlas of United States Trees: Minor Eastern Hardwoods*. Vol. 4. USDA Forest Service. Miscellaneous Publication No. 1342. Washington, DC: Government Printing Office. 168 pp.

Little, E. L., Jr. 1977b. *Rare and Local Trees in the National Forests*. USDA Forest Service Conservation Report. No. 21. Washington, DC: Government Printing Office. 14 pp.

Little, E. L., Jr. 1978. *Atlas of United States Trees: Florida*. Vol. 5. USDA Forest Service Miscellaneous Publication No. 1361. Washington, DC: Government Printing Office. 260 pp.

Little, E. L., Jr., 1979a. *Checklist of United States Trees*. USDA Forest Service Agriculture Handbook No. 541. Washington, DC: Government Printing Office. 375 pp.

Little, E. L., Jr. 1979b. *Forest Trees of the United States and Canada and How to Identify Them*. New York: Dover Publications. 70 pp.

Little, E. L., Jr. 1980a. *Field Guide to North American Trees: Eastern Region*. New York: Alfred A. Knopf. 714 pp.

Little, E. L., Jr. 1980b. *Field Guide to North American Trees: Western Region*. New York: Alfred A. Knopf. 639 pp.

Little, E. L., Jr. 1981. *Atlas of United States Trees: Supplement*. Vol. 6. USDA Forest Service. Miscellaneous Publication No. 1410. Washington, DC: Government Printing Office. 214 pp.

Little, E. L. 1997. *National Audubon Society Field Guide to North American Trees: Eastern Region*. New York: Adolph A. Knopf.

Little, E. L. 1998. *National Audubon Society Field Guide to North American Trees: Western Region*. New York: Adolph A. Knopf.

Little, E. L., Jr., and W. B. Critchfield. 1969. *Subdivisions of the Genus* Pinus. USDA Forest Service Miscellaneous Publication No. 1144. Washington, DC: Government Printing Office. 51 pp.

Liu, T. 1971. *A Monograph on the Genus* Abies. Taipei: National Taiwan University. 608 pp.

Longyear, B. O. 1927. *Trees and Shrubs of the Rocky Mountain Region*. New York: G. P. Putnam's Sons. 244 pp.

Lubchenco, J. A., M. Olson, L. B. Brubaker, S. R. Carpenter, M. M. Holland, S. P. Hubbell, S. A. Levin, J. A. MacMahon, P. A. Matson, J. M. Melillo, H. A. Mooney, C. H. Peterson, H. R. Pulliam, L. A. Real, P. J. Regal, and P. G. Risser. 1991. "The Sustainable Biosphere Initiative: An Ecological Research Agenda." *Ecology* 72:371–412.

Ludwig, J. A., and J. F. Reynolds. 1988. *Statistical Ecology: A Primer on Methods and Computing*. New York: John Wiley & Sons, Inc.

Lyon, L. J. 1971. *Vegetational Development Following Prescribed Burning of Douglas Fir in South-Central Idaho*. USDA Forest Service Research Paper INT-105. Ogden Utah: Intermountain Forest and Range Experiment Station.

MacKinnon, A., J. Pojar, and R. Coupe, eds. 1992. *Plants of British Columbia*. Vancouver: Lone Pine Publishing. 351 pp.

Maisenhelder, L. C. 1969. *Identifying Juvenile Seedlings in Southern Hardwood Forests*. Research Paper SO-47. USDA Forest Service. New Orleans: Southern Forest Experiment Station. 77 pp.

Malanson, G. P. 1993. *Riparian Landscapes*. Cambridge, UK: Cambridge University Press.

Maser, C. 1994. *Sustainable Forestry: Philosophy, Science and Economics*. Delray Beach, FL: St. Lucie Press. 373 pp.

Mathews, F. S. 1915. *Field Book of American Trees and Shrubs*. New York: G. P. Putnam's Sons. 537 pp.

Mathews, J. 1991. "Sustain [Defined]." *Journal of Forestry* 89:29–30.

Matzenko, A. E. 1968. "Conspectus generis *Abies*." *Mill. Novosti Sist. Vyssh. Rast.* 5:9–12.

McCune, B., and T. F. H. Allen. 1985. "Will Similar Forests Develop into Similar Sites?" *Canadian Journal of Botany* 63:367–376.

Meyer, F. G. 1997. "Hamamelidaceae." In *Flora of North America North of Mexico:* Vol. 3. *Magnoliophyta: Magnoliidae and Hamamelidae,* edited by Flora of North America Editorial Committee, 358–367. New York: Oxford University Press. 590 pp.

McCoyle, D. 1981. *Roadside Trees and Shrubs of Oklahoma.* Norman: University of Oklahoma Press. 116 pp.

Miller, H., and S. Lamb. 1985. *Oaks of North America.* Happy Camp, CA: Naturegraph Publishers. 327 pp.

Miller, H. A. 1944. *How to Know the Trees.* Dubuque, IA: William C. Brown. 263 pp.

Minsch, H. 1994. *A Reference Guide to the Gymnosperms of the World.* Champaign, IL: Keoltz Scientific Books. 99 pp.

Mirov, N. T., and J. Hasbrouck. 1976. *The Story of Pines.* Bloomington: Indiana University Press.

Mohlenbrock, R. H. 1986. *Guide to the Vascular Flora of Illinois.* Revised. Carbondale: Southern Illinois University Press. 507 pp.

Mohlenbrock, R. H. n. d. *Forest Trees of Illinois.* Springfield: Division of Forestry, Illinois Department of Conservation. 328 pp.

Montgomery, F. H. 1970. *Trees of Canada and the Northern United States.* Toronto: McGraw-Hill Ryerson. 144 pp.

Morin, N. R., R. D. Whetstone, D. Wilkem, and K. L. Tomlinson. 1988. *Floristics for the 21st Century.* St. Louis: Missouri Botanical Garden. 163 pp.

Mucina, L. 1997. "Classification of Vegetation: Past, Present and Future." *Journal of Vegetation Science* 8:751–760.

Mueller-Dombois, D., and H. Ellenberg. 1974. *Aims and Methods of Vegetation Ecology.* New York: John Wiley & Sons, Inc.

Muenscher, W. C. 1950. *Keys to Woody Plants.* Ithaca, NY: Comstock Publishing. Associates, Cornell University Press. 108 pp.

Mutch, R. W. 1970. "Wildland Fires and Ecosystems: a Hypothesis." *Ecology* 51:146–1051.

Nabb, E. J., and D. Shankman. 1997. "Hydrogeomorphic Response of a Coastal Plain Stream to Tributary Channelization." *Geographic Bulletin* 39:81–89.

National Research Council. 1990. *Forestry Research: A Mandate for Change.* Washington, DC: National Academy Press. 84 pp.

National Vegetation Classification Standard (NVCS). 1999. Working draft v. 5.1. Washington DC: The Nature Conservancy.

Naveh, Z., and A. S. Lieberman. 1994. *Landscape Ecology: Theory and Application.* New York: Springer-Verlag. 360 pp.

Nelson G. 1994. *The Trees of Florida.* Sarasota: Pineapple Press. 338 pp.

Nelson G. 1996. *The Shrubs and Woody Vines of Florida.* Sarasota: Pineapple Press. 391 pp.

Nelsson, P. W. 1987. *The Terrestrial Natural Communities of Missouri.* Revised. Jefferson City: Missouri Department of Natural Resources. 197 pp.

Nemani, R., and S. Running. 1997. "Land Cover Characterization Using Multitemporal Red, Near-ir, and Thermal-ir Data from NOAA/AVHRR." *Ecological Applications* 7:79–90.

Newsholme, C. 1992. *Willows: The Genus* Salix. London: B. T. Batsford. 224 pp.

Niemiec, S. S., G. R. Ahrens, S. Willits, and D. E. Hibbs. 1995. *Hardwoods of the Pacific Northwest.* Research Contribution 8. Corvallis: Forest Research Laboratory, Oregon State University. 115 pp.

Nixon, E. S. 1985. *Trees, Shrubs, and Woody Vines of East Texas.* Nacogdoches, TX: Bruce Lyndon Cunningham Productions. 240 pp.

Nixon, K. C. 1997. "Fagaceae." In *Flora of North America North of Mexico:* Vol. 3. *Magnoliophyta: Magnoliidae and Hamamelidae,* edited by Flora of North America Editorial Committee, 436–447. New York: Oxford University Press. 590 pp.

Nixon K. C., and C. H. Muller. 1997. "*Quercus: Linnaeus,* sect. *Quercus*: White Oaks." In *Flora of North America North of Mexico:* Vol. 3. *Magnoliophyta: Magnoliidae and Hamamelidae,* edited by Flora of North America Editorial Committee, 471–506. New York: Oxford University Press. 590 pp.

Oliver, C. D. 1980. "Even-Aged Development of Mixed-Species Stands." *Journal of Forestry* 78:201–203.

Oosting, H. J. 1956. *The Study of Plant Communities.* 2d ed. San Francisco: Freeman.

Otis, C. H. 1931. *Michigan Trees*. Ann Arbor: University of Michigan Press. 362 pp.

Pallardy, S. G., J. Cermak, F. W. Ewers, M. R. Kaufmann, W. C. Parker, and J. S. Sperry. 1995. "Water Transport Dynamics in Trees and Stands." In *Resource Physiology of Conifers*, edited by W. K. Smith and T. M. Hinckley, 301–389. San Diego: Academic Press. 396 pp.

Parker, E. L. 1963. "The Geographic Overlap of Noble Fir and Red Fir." *Forest Science* 9:209–216.

Parker, W. H. 1993. "*Larix*." In *Flora of North America North of Mexico*. Vol. 2. *Pteridophytes and Gymnosperms*, edited by Flora of North America Editorial Committee, 366–368. New York: Oxford University Press. 475 pp.

Patterson, P. A., K. E. Neiman Jr., and J. R. Tonn. 1983. *Field Guide to Forest Plants of Northern Idaho*. Missoula, MT: USDA Forest Service. 243 pp.

Pavlik, B. M., P. C. Muiek, S. G. Johnson, and M. Popper. 1991. *Oaks of California*. Los Ovivos, CA: Cachuma Press. 184 pp.

Peattie, D. C. 1953. *A Natural History of Western Trees*. Lincoln: University of Nebraska Press. 751 pp.

Peet, R. K. 1988. "Forests of the Rocky Mountains." In *North American Terrestrial Vegetation*, edited by M. G. Barbour and W. D. Billings, 63–102. Cambridge: Cambridge University Press.

Perry, J. P., Jr. 1991. *The Pines of Mexico and Central America*. Portland, OR: Timber Press. 231 pp.

Peterson, P. V., and P. V. Peterson Jr. 1975. *Native Trees of the Sierra Nevada*. Berkeley: University of California Press.

Petrides, G. A. 1972. *A Field Guide to Trees and Shrubs*. Boston: Houghton Mifflin.

Petrides, G. A. 1988. *A Field Guide to Eastern Trees*. Peterson Field Guide Series. Boston: Houghton Mifflin. 272 pp.

Petrides, G. A. 1992. *A Field Guide to Western Trees*. Peterson Field Guide Series. Boston: Houghton Mifflin. 308 pp.

Pfister, R. D., and S. F. Arno. 1980. "Classifying Forest Habitat Types Based on Potential Climax Vegetation." *Forest Science* 26:52–70.

Pfister, R. D., B. L. Kovalchik, S. F. Arno, and R. C. Presby. 1977. *Forest Habitat Types of Montana*. USDA Forest Service General Technical Report INT-34. Ogden, Utah: Intermountain Forest and Range Experiment Station.

Philips, R. 1978. *Trees of North America and Europe*. New York: Random House. 225 pp.

Pianka, E. R. 1970. "On r- and K-selection." *American Naturalist* 104:592–597.

Pielou, E. C. 1988. *The World of Northern Evergreens*. Ithaca, NY. Comstock Publishing Associates.

Pilarski, M., ed. 1994. *Restoration Forestry: An International Guide to Sustainable Forestry Practices*. Durango, CO: Kivaki Press. 525 pp.

Popovich, S. J., W. D. Shepperd, D. W. Reichert, and M. A. Cone. 1993. *Flora of the Fraser Experimental Forest, Colorado*. USDA Forest Service General Technical Report. RM-233. Fort Collins: Rocky Mountain Forest Experiment Station.

Porter, C. L. 1962a. *A Flora of Wyoming: Introduction, Vegetation Zones, and Gymnospermae*. Part I. Agricultural Experiment Station Bulletin 402. Laramie: University of Wyoming Press. 39 pp.

Porter, C. L. 1962b. *A Flora of Wyoming: Dicotyledoneae*. Part V. Agricultural Experiment Station Research Journal 14. Laramie: University of Wyoming Press. 37 pp.

Porter, C. L. 1964. *Wyoming Trees*. Agricultural Extension Service. Laramie: University of Wyoming Press. WY. 55 pp.

Porter, C. L. 1967. *Taxonomy of Flowering Plants*. San Francisco: W. H. Freeman & Co. 472 pp.

Preston, R. J., Jr. 1989. *North American Trees*. 4th ed. Ames: Iowa State University Press. 407 pp.

Price, R. A., and J. M. Lowenstein. 1989. "An Immunological Comparison of the Sciadopytiaceae, Taxodiaceae, and Cupressaceae." *Systematic Botany* 14: 141–149.

Pringle, J. 1997. "*Clematis*." In *Flora of North America North of Mexico: Magnoliophyta: Magnoliidae and Hamamelidae*, edited by Flora of North America Editorial Committee,164–176. New York: Oxford University Press. 590 pp.

Quicke, D. L. J. 1993. *Principles and Techniques of Contemporary Taxonomy*. London: Blackie Academic & Professional. 311 pp.

Ravin, P. H. 1966. *Native Shrubs of Southern California*. Berkeley: University of California Press. 132 pp.

Rhodes, D. G. 1997. "Menispermaceae." In *Flora of North America North of Mexico: Magnoliophyta: Magnoliidae and Hamamelidae*, edited by Flora of North America Editorial Committee, 295–299. New York: Oxford University Press. 590 pp.

Robertson, P. A. 1992. "Characteristics of Spruce-Fir and Lodgepole Pine Old-Growth Stands in the Arapaho-Roosevelt National Forest, Colorado." In *Old-Growth Forest in the Southwest and Rocky Mountain Regions: Proceedings of a Workshop*, edited by M. R. Kaufmann, W. H. Moyr, and R. L. Bassett, 128–134. USDA Forest Service General Technical Report RM-213. Washington, DC: Government Printing Office.

Robertson, P. A., G. T. Weaver, and J. A. Cavanaugh. 1978. "Vegetation and Tree Species Patterns Near the Northern Terminus of the Southern Floodplain Forest." *Ecological Monographs* 48: 249–267.

Roland, A. E., and E. C. Smith. 1969. *The Flora of Nova Scotia.* Halifax: The Nova Scotia Museum.

Rowe, J. S. 1961. "The Level of Integration Concept and Ecology." *Ecology* 42:420–427.

Rowe, J. S. 1972. *Forest Regions of Canada.* Department of the Environment, Canadian Forestry Service Publication No. 1300.

Rydberg, P. A. 1922. *Flora of the Rocky Mountains and Adjacent Plains.* 2d ed. New York.

Salwasser, H. 1994. "Ecosystem Management: Can It Sustain Diversity Productivity?" *Journal of Forestry* 92:6–10.

Sargent, C. S. 1965. *Manual of the Trees of North America.* Vol. 1. 1922 reprint. New York: Dover Publications. 433 pp.

Schimper, A. F. W. 1903. *Plant Geography upon a Physiological Basis,* Clarendon, Oxford, translated by W. R. Fisher. 839 pp.

Schropmeyer, C. S. 1974. *Seeds of Woody Plants in the United States.* Agricultural Handbook No. 450. U. S. Forest Service. Washington, DC: Government Printing Office, 883 pp.

Settergren, C., and R. E. McDermott. 1969. *Trees of Missouri.* Columbia: University of Missouri Press. 120 pp.

Shankman, D. 1996. "Stream Channelization and Changing Vegetation Patterns in the U. S. Coastal Plain." *Geographical Review* 86:216–232.

Shanks, R. E., and A. J. Sharp. 1963. *Summer Key to Trees of Tennessee and the Great Smokies.* Knoxville: University of Tennessee Press. 24 pp.

Sharitz, R. R., and W. J. Mitsch. 1993. Southern Floodplain Forests. In *Biodiversity of the Southeastern United States: Lowland Terrestrial Communities,* edited by W. H. Martin, S. G. Boyce, and A. C. E. Echternacht, 311–372. New York: John Wiley & Sons Inc.

Shaw, G. R. 1914. *The Genus* Pinus. Arnold Arboretum Publication No. 5. Cambridge, MA: Harvard University Press. 96 pp.

Shenefelt, R. D. 1970. *A Key to Wisconsin Trees.* Madison: Department of Natural Resources. 93 pp.

Sherman-Broyles, S. L. 1997. "Ulmaceae." In *Flora of North America North of Mexico:* Vol. 3. *Magnoliophyta: Magnoliidae and Hamamelidae,* edited by Flora of North America Editorial Committee, 369–375. New York: Oxford University Press. 590 pp.

Shetler, S. G., and L. E. Skog, eds. 1978. *A Provisional Checklist of Species for Flora North America.* St. Louis: Missouri Botanical Garden. 199 pp.

Shotola, S. J., G. T. Weaver, P. A. Robertson, and W. C. Ashby. 1992. "Sugar Maple Invasion of an Old-Growth Oak-Hickory Forest in Southwestern Illinois." *American Midland Naturalist* 127: 125–138.

Simon, A. S., and Hupp, C. R. 1992. *Geomorphic and Vegetative Recovery Processes Along Modified Stream Channels of West Tennessee.* Denver: U. S. Geological Survey.

Sims, R. A., I. G. W. Corns, and K. Klinka. 1996. "Introduction ñ Global to Local: Ecological Land Classification." *Environmental Monitoring and Assessment* 39:1–10.

Sims, R. A., H. M. Kershaw, and G. M. Wickware. 1990. *The Autecology of Major Tree Species in the North Central Regions of Ontario.* NWOFTDU Technical Report 48. Thunder Bay: Ontario Ministry of Natural Resources. 126 pp.

Sivarajan, V. S. 1991. *Introduction to the Principles of Plant Taxonomy.* Edited by N. K. P. Robson. New York: Cambridge University Press. 292 pp.

Small, J. K. 1933. *Manual of Southeastern Flora.* Lancaster, PA: Science Press.

Smith, J. P., Jr. 1977. *Vascular Plant Families.* Eureka, CA: Mad River Press. 320 pp.

Smith, N. F. 1995. *Trees of Michigan and the Upper Great Lakes.* 6th. ed. Lansing, MI: Thunder Bay Press. 178 pp.

Soule, M. E., ed. 1986. *Conservation Biology: The Science of Scarity and Diversity.* Sunderland, ME: Sinauer Associates.

Sporne, K. R. 1974. *The Morphology of Gymnosperms.* 2d ed. London: Hutchinson University Library. 216 pp.

Stace, C. A. 1980. *Plant Taxonomy and Biosystematics.* London: Edward Arnold. 279 pp.

Stearn, W. T. 1992. *Botanical Latin.* 4th. ed. Portland, OR: Timber Press. 546 pp.

Steyermark, S. A. 1963. *Flora of Missouri.* Ames: Iowa State University Press. 1725 pp.

Stone, D. E. 1997. "Juglandaceae: *Carya.*" In *Flora of North America North of Mexico:* Vol. 3. *Magnoliophyta: Magnoliidae and Hamamelidae,* edited by Flora of North America Editorial Committee, 416–425. New York: Oxford University Press. 590 pp.

Stromberg, J. C. 1993. "Fremont Cottonwood-Goodding Willow Riparian Forests: A Review of Their Ecology, Threats, and Recovery Potential." *Journal of Arizona-Nevada Academy of Sciences* 23:97–110.

Stuart, J. D., M. C. Grifantini, and L. Fox III. 1993. "Early Successional Pathways Following Wildfire and Subsequent Silvicultural Treatment in Douglas Fir/Hardwood Forests, NW California." *Forest Science* 39:51–572.

Stuessy, T. F., and S. H. Sohmer, eds. 1966. *Sampling the Green World.* New York: Columbia University Press. 289 pp.

Stupka, A. 1968. *Trees, Shrubs, and Woody Vines of Great Smoky Mountains National Park.* Knoxville: University of Tennessee Press. 186 pp.

Sudworth, G. B. 1908. *Forest Trees of the Pacific Slope.* 1967 Reprint. New York: Dover Publications. 265 pp.

Swanson, F., and D. Berg. 1991. "The Ecological Roots of New Approaches to Forestry." *Forest Perspectives* 1:6–8.

Symonds, G. W. D. 1958. *The Tree Identification Book.* New York: William Morrow. 379 pp.

Symonds, G. W. D. 1963. *The Shrub Identification Book.* New York: William Morrow. 379 pp.

Takhtajan, A. 1968. "Classification and Phylogeny, with Species Reference to the Flowering Plants." London: *Proceedings of the Linnaeian Society* 179:221–227.

Taylor, R. J. "*Tsuga.*" 1993a. In *Flora of North America North of Mexico.* Vol. 2. *Pteridophytes and Gymnosperms,* edited by Flora of North America Editorial Committee, 362–365. New York: Oxford University Press. 475 pp.

Taylor, R. J. "*Picea.*" 1993b. In *Flora of North America North of Mexico:* Vol. 2. *Pteridophytes and Gymnosperms,* edited by Flora of North America Editorial Committee, 369–373. New York: Oxford University Press. 475 pp.

Thor, E., and P. E. Barnett. 1974. "Taxonomy of *Abies* in the Southern Appalachians: Variation in Balsam Monoterpenes and Wood Properties." *Forest Science* 20:32–40.

Tippo, O. 1942. "A Modern Classification of the Plant Kingdom." *Chronicles of Botany* 7: 203–206.

Trelease, W. 1931. *Winter Botany.* New York: Dover Publications. 396 pp.

Tull, D., and G. O. Miller. 1991. *A Field Guide to Wildflowers, Trees, and Shrubs of Texas.* Houston: Gulf Publishing. 344 pp.

Turesson, G. 1922. "The Genotypical Response of the Plant Species to the Habitat." *Hereditas* 3: 211–350.

Turesson, G. 1925. "The Plant Species in Relation to Habitat and Climate." *Hereditas* 6:141–236.

Tyndall, R. W., and J. C. Hull. 1999. "Vegetation, Flora, and Plant Physiological Ecology of Serpentine Barrens of Eastern North America." In *Savannas, Barrens, and Rock Outcrop Plant Communities of North America,* edited by R. C. Anderson, J. S. Fralish, and J. M. Baskin, 67–82. Cambridge, UK: Cambridge University Press.

United Nations Food and Agricultural Organization. 1979. *Poplars and Willows in Wood Production and Land Use.* Rome, Italy: United Nations Food and Agricultural Organization. 328 pp.

United States Department of Agriculture. 1974. *Seeds of Woody Plants in the United States.* U. S. Forest Service. Agriculture Handbook No. 450. Washington, DC: Government Printing Office. 883 pp.

United States Department of Agriculture. 1982. *National List of Scientific Plant Names.* Bulletin No. SCS-TP-159. Soil Conservation Service. Vol. I. Washington, DC: Superintendent of Documents. 416 pp.

Van Dersal, W. R. 1938. *Native Woody Plants of the United States: Their Erosion-Control and Wildlife Values.* USDA Soil Conservation Service. 362 pp.

van der Werff, H. 1997. "*Litsea, Sassafras, Umbellularia, Cinnamomum, Licaria,* and *Nectandra.*" In *Flora of North America North of Mexico: Magnoliophyta:* Vol. 3. *Magnoliidae and Hamamelidae,* edited by Flora of North America Editorial Committee, 29–33. New York: Oxford University Press. 590 pp.

Vankat, J. L. 1979. *The Natural Vegetation of North America: An Introduction.* New York: John Wiley & Sons, Inc.

Vankat, J. L. 1982. "A Gradient Perspective on the Vegetation of Sequoia National Park, California." *MaroÒo* 29:200–214.

Viereck, L. A., and E. L. Little Jr. 1972. *Alaska Trees and Shrubs.* USDA Forest Service Agricultural Handbook No. 410. Washington, DC: Government Printing Office. 265 pp.

Viereck, L. A., and E. L. Little Jr. 1975. *Atlas of United States Trees: Alaska Trees and Common Shrubs.* Vol. 2. USDA Forest Service Miscellaneous Publication No. 1293. Washington, DC: Government Printing Office. 84 pp.

Viereck, L. A., and E. L. Little Jr. 1986. *Alaska Trees and Shrubs.* Fairbanks: University of Alaska Press.

Viertel, A. T. 1959. *Trees, Shrubs, and Vines.* Syracuse, NY: State University College of Forestry. 166 pp.

Vines, R. A. 1977. *Trees of East Texas.* Austin: University of Texas Press. 538 pp.

Walker, L. C. 1999. *The North American Forests: Geography, Ecology, and Silviculture.* Boca Raton, FL: CRC Press.

Walters, D. R., and D. J. Keil. 1977. *Vascular Plant Taxonomy.* 3d ed. Dubuque, IA: Kendall/Hunt Publishing. 488 pp.

Warming, E. 1909. *Oecology of Plants: An Introduction to Plant Communities.* Oxford, UK: Clarendon Press. 422 pp.

Watson, F. D. 1993. "*Sequoia, Sequiodendron, and Taxodium.*" In *Flora of North America North of Mexico. Pteridophytes and Gymnosperms,* edited by Flora of North America Editorial Committee, 401–404. New York: Oxford University Press. 475 pp.

Watson, F. D., and J. E. Eckenwalder. 1993. "Cupressaceae." ln *Flora of North America North of Mexico.* Vol. 2. *Pteridophytes and Gymnosperms,* edited by Flora of North America Editorial Committee, 399–401. New York: Oxford University Press. 475 pp.

Weakley, A., K. D. Patterson, S. Landaal, M. Pyne, et al. (compilers). 1998. (Working draft.) "International Classification of Ecological Communities: Terrestrial Vegetation of the Southeastern United States." The Nature Conservancy, Southeast Regional Office, Southern Conservation Science Department, Community Ecology Group, Chapel Hill, NC.

Weber, W. A. 1976. *Rocky Mountain Flora.* Boulder: Colorado Associated University Press. 479 pp.

Wellbaum, E. M. 1989. "*Site, Stand, and Tree Characteristics Associated with Oak Decline and Mortality.*" Master's thesis. Department of Foresty, Southern Illinois University. 63 pp.

Wentworth, T. R. 1981. "Vegetation on Limestone and Granite in the Mule Mountains, Arizona." *Ecology* 62:469–482.

West, N. E. 1999. "Juniper-Pinyon Savannas and Woodlands of Western North America." In *Savannas, Barrens, and Rock Outcrop Communities of North America,* edited by R. C. Anderson, J. S. Fralish, and J. M. Baskin, 228–308. Cambridge: Cambridge University Press.

Wharton, C. H., W. M. Kitchens, E. C. Pendelton, and T. W. Sipe. 1982. *The Ecology of Bottomland Hardwood Swamps of the Southeast: A Community Profile.* Biological Sciences Program FWS/OBS-81/37. U.S. Fish & Wildlife Service.

Whitman, A. H., ed. 1987a. *Familiar Trees of North America: East.* New York: Chanticleer Press. 191 pp.

Whitman, A. H., ed. 1987b. *Familiar Trees of North America: West.* New York: Chanticleer Press. 191 pp.

Whitney, S. 1997. *Western Forests.* New York: Alfred A. Knopf.

Whittaker, R. H. 1956. "Vegetation of the Great Smoky Mountains." *Ecological Monographs* 26:1–80.

Whittaker, R. H. 1960. "Vegetation of the Siskiyou Mountains, Oregon and California." *Ecological Monographs* 30: 279–338.

Whittaker, R. H., and W. A. Niering. 1965. "Vegetation of the Santa Catalina Mountains, Arizona: A Gradient Analysis of the South Slope." *Ecology* 46:428–452.

Whittemore, A. T. 1997. "*Berberis.*" In *Flora of North America North of Mexico:* Vol. 3. *Magnoliophyta: Magnoliidae and Hamamelidae,* edited by Flora of North America Editorial Committee, 276–286. New York: Oxford University Press. 590 pp.

Wilkenson, K. 1990. *A Habitat Field Guide to Trees and Shrubs of Alberta.* Edmonton: Lone Pine Publishing. 191 pp.

Wilson, E. O, ed. 1988. *Biodiversity.* Washington, DC: National Academy Press. 521 pp.

Wood, C. E., Jr. 1974. *A Student's Atlas of Flowering Plants: Some Dicotyledons of Eastern North America.* New York: Harper & Row. 120 pp.

Woodin, H. E., and A. A. Lindsey. 1954. "Juniper-Pinyon East of the Continental Divide, as Analyzed by the Line-Strip Method." *Ecology* 35:473–489.

Woodland, D. W. 1991. *Contemporary Plant Systematics.* Englewood Cliffs, NJ: Prentice-Hall. 582 pp.

Woods, R. S. 1966. *An English-Classical Dictionary for the Use of Taxonomists.* Claremont, CA: Pomona College Press. 331 pp.

Woodward, F. I. 1987. *Climate and Plant Distribution.* Cambridge, UK: Cambridge University Press.

Woodward, F. I. 1995. "Ecophysiological Controls of Conifer Distributions." In *Ecophysiology of Coniferous Forests*, edited by W. K. Smith and T. M. Hinckley, 79–94. San Diego: Academic Press. 338 pp.

Zomlefer, W. B. 1994. *Guide to Flowering Plant Families.* Chapel Hill: University of North Carolina Press. 430 pp.

Index of Species

Numbers in bold indicate the page for the species description; page numbers in brackets indicate the page(s) for the cover type where the species name appears in the cover type name.

Subject Index